# ATOMIC MASSES OF THE ELEMENTS

| Name | Symbol | Atomic Number | Atomic Mass[a] | Name | Symbol | Atomic Number | Atomic Mass[a] |
|---|---|---|---|---|---|---|---|
| Actinium | Ac | 89 | (227) | Neodymium | Nd | 60 | 144.2 |
| Aluminum | Al | 13 | 26.98 | Neon | Ne | 10 | 20.18 |
| Americium | Am | 95 | (243) | Neptunium | Np | 93 | (237) |
| Antimony | Sb | 51 | 121.8 | Nickel | Ni | 28 | 58.69 |
| Argon | Ar | 18 | 39.95 | Niobium | Nb | 41 | 92.91 |
| Arsenic | As | 33 | 74.92 | Nitrogen | N | 7 | 14.01 |
| Astatine | At | 85 | (210) | Nobelium | No | 102 | (259) |
| Barium | Ba | 56 | 137.3 | Osmium | Os | 76 | 190.2 |
| Berkelium | Bk | 97 | (247) | Oxygen | O | 8 | 16.00 |
| Beryllium | Be | 4 | 9.012 | Palladium | Pd | 46 | 106.4 |
| Bismuth | Bi | 83 | 209.0 | Phosphorus | P | 15 | 30.97 |
| Bohrium | Bh | 107 | (264) | Platinum | Pt | 78 | 195.1 |
| Boron | B | 5 | 10.81 | Plutonium | Pu | 94 | (244) |
| Bromine | Br | 35 | 79.90 | Polonium | Po | 84 | (209) |
| Cadmium | Cd | 48 | 112.4 | Potassium | K | 19 | 39.10 |
| Calcium | Ca | 20 | 40.08 | Praseodymium | Pr | 59 | 140.9 |
| Californium | Cf | 98 | (251) | Promethium | Pm | 61 | (145) |
| Carbon | C | 6 | 12.01 | Protactinium | Pa | 91 | 231.0 |
| Cerium | Ce | 58 | 140.1 | Radium | Ra | 88 | (226) |
| Cesium | Cs | 55 | 132.9 | Radon | Rn | 86 | (222) |
| Chlorine | Cl | 17 | 35.45 | Rhenium | Re | 75 | 186.2 |
| Chromium | Cr | 24 | 52.00 | Rhodium | Rh | 45 | 102.9 |
| Cobalt | Co | 27 | 58.93 | Roentgenium | Rg | 111 | (272) |
| Copper | Cu | 29 | 63.55 | Rubidium | Rb | 37 | 85.47 |
| Curium | Cm | 96 | (247) | Ruthenium | Ru | 44 | 101.1 |
| Darmstadtium | Ds | 110 | (271) | Rutherfordium | Rf | 104 | (261) |
| Dubnium | Db | 105 | (262) | Samarium | Sm | 62 | 150.4 |
| Dysprosium | Dy | 66 | 162.5 | Scandium | Sc | 21 | 44.96 |
| Einsteinium | Es | 99 | (252) | Seaborgium | Sg | 106 | (266) |
| Erbium | Er | 68 | 167.3 | Selenium | Se | 34 | 78.96 |
| Europium | Eu | 63 | 152.0 | Silicon | Si | 14 | 28.09 |
| Fermium | Fm | 100 | (257) | Silver | Ag | 47 | 107.9 |
| Fluorine | F | 9 | 19.00 | Sodium | Na | 11 | 22.99 |
| Francium | Fr | 87 | (223) | Strontium | Sr | 38 | 87.62 |
| Gadolinium | Gd | 64 | 157.3 | Sulfur | S | 16 | 32.07 |
| Gallium | Ga | 31 | 69.72 | Tantalum | Ta | 73 | 180.9 |
| Germanium | Ge | 32 | 72.64 | Technetium | Tc | 43 | (98) |
| Gold | Au | 79 | 197.0 | Tellurium | Te | 52 | 127.6 |
| Hafnium | Hf | 72 | 178.5 | Terbium | Tb | 65 | 158.9 |
| Hassium | Hs | 108 | (269) | Thallium | Tl | 81 | 204.4 |
| Helium | He | 2 | 4.003 | Thorium | Th | 90 | 232.0 |
| Holmium | Ho | 67 | 164.9 | Thulium | Tm | 69 | 168.9 |
| Hydrogen | H | 1 | 1.008 | Tin | Sn | 50 | 118.7 |
| Indium | In | 49 | 114.8 | Titanium | Ti | 22 | 47.87 |
| Iodine | I | 53 | 126.9 | Tungsten | W | 74 | 183.8 |
| Iridium | Ir | 77 | 192.2 | Uranium | U | 92 | 238.0 |
| Iron | Fe | 26 | 55.85 | Vanadium | V | 23 | 50.94 |
| Krypton | Kr | 36 | 83.80 | Xenon | Xe | 54 | 131.3 |
| Lanthanum | La | 57 | 138.9 | Ytterbium | Yb | 70 | 173.0 |
| Lawrencium | Lr | 103 | (260) | Yttrium | Y | 39 | 88.91 |
| Lead | Pb | 82 | 207.2 | Zinc | Zn | 30 | 65.41 |
| Lithium | Li | 3 | 6.941 | Zirconium | Zr | 40 | 91.22 |
| Lutetium | Lu | 71 | 175.0 | — | — | 112 | (285) |
| Magnesium | Mg | 12 | 24.31 | — | — | 113 | (284) |
| Manganese | Mn | 25 | 54.94 | — | — | 114 | (289) |
| Meitnerium | Mt | 109 | (268) | — | — | 115 | (288) |
| Mendelevium | Md | 101 | (258) | — | — | 116 | (292) |
| Mercury | Hg | 80 | 200.6 | — | — | 118 | (293) |
| Molybdenum | Mo | 42 | 95.94 | | | | |

[a] Values in parentheses are the mass number of the most stable isotope.

# Mastering**CHEMISTRY**™

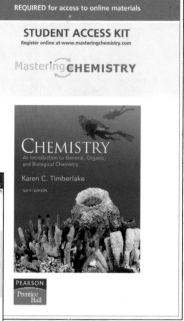

With your purchase of a new copy of Timberlake, *An Introduction to General, Organic, and Biological Chemistry*, 10e, you should have received a Student Access Kit for **MasteringChemistry**™. The kit contains instructions and a code for you to access **MasteringChemistry**. Your Student Access Kit will look something like the image on the right.

## Don't Throw Your Access Kit Away!

If you did not purchase a new textbook or cannot locate the Student Access Kit and would like to access the wealth of self-study resources on **MasteringChemistry**, or if your instructor has required **MasteringChemistry** as part of your course, you can purchase your subscription online with a major credit card.

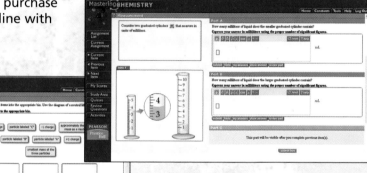

## What Is MasteringChemistry?

**MasteringChemistry** provides you with two learning systems: the most widely used self-study media available, and the most educationally effective chemistry homework and tutorial system (if your instructor chooses to make online assignments part of your course).

For Self-Study Activities: **MasteringChemistry** provides you with learning resources such as PowerPoint™ lecture notes, case studies, review questions and quizzes, and much more.

For Instructor-Assigned Homework: **MasteringChemistry** is able to coach you with feedback specific to your needs, and hints when you get stuck. The result is targeted tutorial help to optimize your study time and maximize your learning.

---

## Using MasteringChemistry

### How to Register for Self-Study Activities

To use **MasteringChemistry**, you first register online at **www.masteringchemistry.com**. For self-study activities, you need only a valid e-mail address and the Student Access Code that is printed beneath the pull-tab in your Student Access Kit (you can also purchase access online). After you register, you can log in right away. Return to **www.masteringchemistry.com** and log in using your personal Login Name and Password that you created during registration. When asked, be sure to identify the correct title and edition of your textbook.

### How to Register for Online Assignments from Your Instructor

If your instructor requires you to complete online homework assignments, your instructor will provide you with his or her **MasteringChemistry** Course ID. Follow the instructions above to register and log in. Your instructor may have guidelines on completing your registration, such as how to input your name, format your Student ID, etc. When prompted, enter your instructor's Course ID.

### Minimum System Requirements

(Subject to change. See Website for up-to-date requirements.)

**Windows:** Microsoft® Windows 2000, XP, Vista

**Macintosh:** Apple® Mac OS® 10.2, 10.3, 10.4, 10.5

**Both:**
- **RAM:** 64 MB
- **Screen resolution:** 1024 x 768
- **Browser (OS dependent. Check Website for more detail):** Firefox 1.5, 2.0; Internet Explorer 6.0, 7.0; Safari 1.3, 2.0

**Browser settings and players/plug-ins:**
- **Javascript** must be enabled in your browser.
- Any **popup blocker** should either be disabled or set to allow popups from: session.masteringchemistry.com
- **Flash™ Player** 9.0 or higher (downloadable from www.adobe.com)

**Technical Support**
www.masteringchemistry.com/support

# CHEMISTRY

## An Introduction
## to General, Organic,
## and Biological Chemistry

# CHEMISTRY

## An Introduction to General, Organic, and Biological Chemistry

TENTH EDITION

Karen C. Timberlake

PEARSON

Prentice Hall

Upper Saddle River, NJ 07458

**Library of Congress Cataloging-in-Publication Data**

Timberlake, Karen.
  Chemistry : an introduction to general, organic and biological chemistry /
Karen C. Timberlake. — 10th ed.
    p. cm.
  ISBN 978-0-13-601970-1 (hardcover)
  1. Chemistry—Textbooks.  I. Title.
  QD31.3.T55 2009
  540—dc22
                                        2007048775

*Editor in Chief:* Nicole Folchetti
*Senior Editor:* Kent Porter Hamann
*Editor in Chief, Development:* Ray Mullaney
*Assistant Editor:* Jessica Neumann
*Director of Team-Based Project Management:* Vince O'Brien
*Assistant Managing Editor, Science:* Gina Cheselka
*Project Manager:* Shari Toron
*Art Director:* Maureen Eide
*Interior and Cover Designer:* Suzanne Behnke
*Cover Photo Credits: Diver photo:* Jeff Hunter/Photographer's Choice/Getty Images, Inc.
                    *Coral:* Wayne Lawler/Ecosene/CORBIS
*Proofreader:* Michael Rossa
*Copy editor:* Fay Ahuja
*Director of Marketing:* Christy Lawrence
*Chemistry Marketing Manager:* Dawn Giovanniello
*Associate Media Producer:* Kristin Mayo
*Compositor:* ICC Macmillan Inc.
*Art Production Manager:* Connie Long
*Illustration Art Director:* Jay McElroy
*Art Studios:* Production Solutions, Precision Graphics
*Production Manager, Production Solutions:* Sean Hogan
*Assistant Production Manager, Production Solutions:* Ronda Whitson
*Illustrators, Production Solutions:* Stacy Smith, Audrey Simonetti, Mark Landis
*Senior Production Manager, Precision Graphics:* J.C. Morgan
*Lead Illustrator, Precision Graphics:* Gary Hunt
*Operations Specialist:* Alan Fischer

Printed in the United States of America
10 9 8 7 6 5 4 3

ISBN-10:      0-13-601970-6
ISBN-13: 978-0-13-601970-1

Pearson Education Ltd., *London*
Pearson Education Australia Pty. Limited, *Sydney*
Pearson Education Singapore Pte. Ltd.
Pearson Education North Asia Ltd., *Hong Kong*
Pearson Education Canada, Ltd., *Toronto*
Pearson Educación de Mexico, S.A. de C.V
Pearson Education—Japan, *Tokyo*
Pearson Education Malaysia, Pte. Ltd.

# BRIEF CONTENTS

# CONTENTS

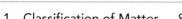

# 3
# Atoms and Elements  82

# 4
# Compounds and Their Bonds    120

# 5
# Chemical Quantities and Reactions    161

# 6
# Gases   209

# 7
# Solutions   241

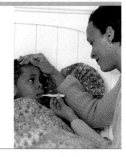

# 11

## Unsaturated Hydrocarbons   382

# 12

## Organic Compounds with Oxygen and Sulfur   405

# 13
## Carboxylic Acids, Esters, Amines, and Amides    447

# 14
## Carbohydrates    483

# 15
## Lipids    513

# 16

## Amino Acids, Proteins, and Enzymes   552

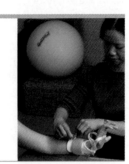

# 17

## Nucleic Acids and Protein Synthesis   591

# 18
## Metabolic Pathways and Energy Production   623

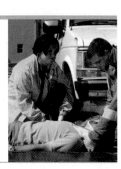

# APPLICATIONS AND ACTIVITIES

## Explore Your World

## Health Note

*Career Focus*

*Environmental Note*

*Green Chemistry Note*

# Guide to Problem Solving

# ABOUT THE AUTHOR

**Karen Timberlake** is professor emeritus of chemistry at Los Angeles Valley College, where she taught chemistry for allied health and preparatory chemistry for 36 years. She received her bachelor's degree in chemistry from the University of Washington and her master's degree in biochemistry from the University of California at Los Angeles.

Professor Timberlake has been writing chemistry textbooks for 30 years. During that time, her name has become associated with the strategic use of pedagogical tools that promote student success in chemistry and the application of chemistry to real-life situations. More than one million students have learned chemistry using texts, laboratory manuals, and study guides written by Karen Timberlake. In addition to *Chemistry: An Introduction to General, Organic, and Biological Chemistry*, tenth edition, she is also the author of *General, Organic, and Biological Chemistry, Structures of Life*, second edition, and *Basic Chemistry*, second edition, along with the accompanying *Study Guide with Solutions for Selected Problems, Laboratory Manual*, and *Essentials Laboratory Manual*.

Professor Timberlake belongs to numerous scientific and educational organizations including the American Chemical Society (ACS) and the National Science Teachers Association (NSTA). She was a Western Regional Winner of Excellence in College Chemistry Teaching Award given by the Chemical Manufacturers Association. In 2004, she received the McGuffey Award in Physical Sciences from the Text and Academic Authors Association for her textbook *Chemistry: An Introduction to General, Organic, and Biological Chemistry*, eighth edition, which has demonstrated excellence over time. In 2006, she received the Textbook Excellence Award for the first edition of *Basic Chemistry*. She has participated in education grants for science teaching including the Los Angeles Collaborative for Teaching Excellence (LACTE) and a Title III grant at her college. She speaks at conferences and educational meetings on the use of student-centered teaching methods in chemistry to promote the learning success of students.

Her husband, Bill, is also a professor emeritus of chemistry and has contributed to writing this text. He taught preparatory and organic chemistry at Los Angeles Harbor College for 36 years. When the Professors Timberlake are not writing textbooks, they relax by hiking, traveling, trying new restaurants, cooking, playing tennis, and taking care of their grandchildren, Daniel and Emily.

## Dedication

*I dedicate this book to*

- *My husband for his patience, loving support, and preparation of late meals*
- *My son, John, daughter-in-law, Cindy, grandson, Daniel, and granddaughter, Emily, for the precious things in life*
- *The wonderful students over many years whose hard work and commitment always motivated me and put purpose in my writing*

## Favorite Quotes

*The whole art of teaching is only the art of awakening the natural curiosity of young minds.*
—Anatole France

*One must learn by doing the thing; though you think you know it, you have no certainty until you try.*
—Sophocles

*Discovery consists of seeing what everybody has seen and thinking what nobody has thought.*
—Albert Szent-Gyorgi

*I never teach my pupils; I only attempt to provide the conditions in which they can learn.*
—Albert Einstein

# PREFACE

## TO THE STUDENT

Welcome to the tenth edition of *Chemistry: An Introduction to General, Organic, and Biological Chemistry*. This is a chemistry text designed to help you prepare for a career in a health-related profession, such as nursing or respiratory therapy, and a text that assumes no prior knowledge of chemistry. This new edition introduces more problem-solving strategies and new conceptual and challenge problems.

This text contains fundamental topics of general, organic, and biological chemistry. My goal is to provide you with a learning environment that makes the study of chemistry an engaging and positive experience. It is also my goal to help you become a critical thinker by understanding scientific concepts that will form a basis for making important decisions about issues concerning health and the environment. Thus, I have utilized materials that

- Motivate you to learn and enjoy chemistry
- Relate chemistry to careers that interest you
- Develop problem-solving skills that lead to your success in chemistry
- Promote your learning and success in chemistry

I hope that this textbook helps you discover exciting new ideas and gives you a rewarding experience as you develop an understanding and appreciation of the role of chemistry in your life.

## FEATURES OF THE TEXT

You may wonder why your career path includes a class in chemistry. A common view is that chemistry is just a lot of facts to be memorized. To change this perception, I have included many features to help you learn about chemistry in your life and chosen career and to give you the skills to learn chemistry successfully. These features include connections to real life, visual guides to problem solving, and in-chapter problem sets to work immediately that reinforce the learning of new concepts. A successful learning program in this text provides you with many learning tools, which are discussed below.

**Chapter Opening Interviews with Scientists and Health Care Professionals** Every chapter begins with an interview with a professional in a career such as nursing, forensics anthropology, physical therapy, or respiratory therapy. These professionals in science and health fields discuss the importance of chemistry in their careers.

**Learning Goals** Learning Goals in every section state learning outcomes by previewing the concepts the student is to learn.

**Connections to Real Life** Health Notes, Green Chemistry Notes, and Environmental Notes throughout the text relate chemistry chapters to real-life topics in science and medicine that are interesting and motivating to students and support the role of chemistry in the real world. Explore Your World Features allow students to explore chemistry topics using everyday materials. The Career Focus interviews provide additional examples of professionals using chemistry.

**Guides to Problem-Solving Visualizations** As part of the comprehensive learning program, many Sample Problems and Study Checks model successful problem-solving techniques for you. Guides to Problem Solving (GPS) illustrate the steps needed to solve problems. In the Sample Problems, color blocks visually guide you to a successful solution. You can find the Answers to all the Study Checks at the end of the chapter.

**Questions and Problems in Every Section** At the end of every section, a set of Questions and Problems allows you to apply problem solving to the concepts in that section. By working the problems after each section, you immediately reinforce newly learned concepts rather than waiting until you get to the end of the chapter. I encourage you to be an active learner by working problems as you progress through each chapter.

**Matched Pairs of Problems** Every question or problem in the text is matched with a similar type of question or problem. For each pair, the answer for the odd-numbered problem is located at the end of each chapter rather than at the end of the textbook. Checking your answers gives you immediate feedback to problem-solving success. The answers to the even-numbered problems are not in the text, but your instructor may use these problems for homework and/or quiz questions.

**End-of-Chapter Questions** At the end of every chapter, more problems, beginning with Understanding the Concepts, encourage you to think about the concepts you have learned. Additional Questions and Problems provide in-depth questions that integrate the topics from the entire chapter to improve your understanding and critical thinking. Challenge Questions provide in-depth study and may be utilized in cooperative learning environments. The *Study Guide* contains additional learning exercises for every section, a practice exam, and worked-out solutions for all the odd-numbered problems. The *Instructor's Manual* has the solutions to all the problems in the text.

**Macro-to-Micro Art illustrations of Atomic Organization** Throughout the text, a robust and vibrant illustration program visually connects the real-life world with atomic-level representations. Macro-to-micro illustrations show you that everyday things have an atomic level of organization and structure that determines their behavior and functions. Every figure in the text also contains a question that asks you to study the figure and relate it to the content in the text. The plentiful use of three-dimensional structures of molecules stimulates the imagination and helps you understand atomic and molecular structures.

**End-of-Chapter Aids**   At the end of each chapter, a Concept Map visually connects the topics and concepts in the chapter. Chapter Reviews give you an overview of the important concepts in each section. Key Terms and their definitions will help you review the new vocabulary presented in each chapter.

# NEW FOR THE TENTH EDITION

New features that have been added to every chapter of this tenth edition include the following:

- More Guides to Problem Solving (GPS) that illustrate problem-solving strategies.
- New Green Chemistry Notes including "Biodiesel: An Alternative Fuel," "Greenhouse Gases," and "Energy-Saving Light Bulbs."
- New Health Notes, including "Brachytherapy" and "Protein Structure and Mad Cow Disease."
- New Career Focus interviews, including "Geologist."
- Molecular models in problem sets that improve visual understanding of chemical reactions.
- Measurement of parts per million (ppm) and parts per billion (ppb).
- New Sample Problems and Study Checks to model problem-solving strategies.
- New Understanding the Concepts add more visual examples to conceptual learning.
- New inter-chapter Combining Ideas problem sets provide problems with greater depth by combining topics from several chapters.

# CHAPTER ORGANIZATION OF THE TENTH EDITION

## Prologue

The **Prologue** introduces students to the concepts of chemicals and chemistry, discusses the scientific method, and asks students to develop a study plan for learning chemistry. New items added to the Prologue are *Learning Goals, Concept Map, Understanding the Concepts, Additional Questions and Problems, Challenge Questions,* and *Answers.* A *Health Note* on "Early Chemists: The Alchemists" was added.

## Chapters 1 and 2

**Chapter 1, Measurements**, looks at measurement and the need to understand numerical structures of the metric system in

the sciences. Section 1.2, *Scientific Notation* is a new section. Items added include time in seconds, tera and pico prefixes to metric and SI prefixes, parts per million and parts per billion, and a new feature *Green Chemistry Note*, "Toxicology and Risk-Benefit Assessment." *Guides to Problem Solving (GPS)* use color blocks as visual guides through the solution pathway.

**Chapter 2, Energy and Matter**, looks at energy and heat, nutritional energy values, temperature conversions, states of matter, and energy involved in changes of state. New energy problems use the SI unit the joule (J). *Green Chemistry Notes* update the content of "Carbon Dioxide and Global Warming" and add "Fuel Cells and Clean Energy for the Future." *Sample Problems* were reworked to utilize the Guide to Problem Solving strategy. New to this edition is an inter-chapter problem set, *Combining Ideas from Chapters 1 and 2.*

## Chapters 3, 4, and 5

**Chapter 3, Atoms and Elements**, covers the classification of matter, elements, atoms, and subatomic particles. Section 3.3, *The Periodic Table*, emphasizes the numbering of groups from 1 to 18. Elements with atomic numbers 116 and 118 have been added. New items include a *Health Note* on toxicity of mercury, the discovery of electrons by J. J. Thomson using the cathode ray tube, and the calculation of the average atomic mass of an element using percent abundance and isotope mass. Section 3.9, *Periodic Trends*, discusses periodic properties of elements, including valence electrons, atomic size, and ionization energy. A new *Green Chemistry Note*, "Energy-Saving Light Bulbs," has been added.

**Chapter 4, Compounds and Their Bonds**, describes how atoms form ionic and covalent bonds and compounds. Students learn to write formulas and to name ionic compounds, including those with polyatomic ions, and covalent compounds. A discussion on sizes of atoms and their corresponding ions was added. Section 4.6, *Electronegativity and Bond Polarity*, leads to a discussion of the polarity of bonds and molecules. Section 4.9, *Attractive Forces in Compounds*, has been added.

**Chapter 5, Chemical Quantities and Reactions**, introduces moles and molar masses of compounds, which are used in calculations to determine the mass or number of particles in a given quantity. In Section 5.5, *Types of Reactions*, we look at collisions and interactions between atoms and molecules. Students learn to balance chemical equations and to recognize the types of chemical reactions. There is a new *Green Chemistry Note*, "Fuel Cells: Clean Energy for the Future." Section 5.7, *Mole Relationships in Chemical Equations*, and Section 5.8, *Mass Calculations for Reactions*, prepare students for the quantitative relationships in reactions. The chapter concludes with Section 5.9, *Energy in Chemical Reactions*, in which activation energy and exothermic and endothermic reactions are discussed. New items include more problems in *Understanding the Concepts* and in *Challenge Questions.* New to this edition is an inter-chapter problem set, *Combining Ideas from Chapters 3 to 5.*

## Chapters 6, 7, and 8

**Chapter 6, Gases**, discusses the properties of gases and calculates changes in gases using the gas laws. Problem-solving strategies enhance the discussion and calculations with gas laws. New items include predicting changes in gas variables, a GPS titled "Using Molar Volume for Reaction" for Sample Problem 6.12, and Table 6.4, Summary of Gas Laws. A new *Green Chemistry Note* on greenhouse gases is also added.

**Chapter 7, Solutions**, describes solutions, saturation, solubility, electrolytes, concentrations, and osmosis and dialysis. New problem-solving strategies clarify the use of concentrations to determine volume or mass of solute. The volumes and molarities of solutions are used in chemical reactions. The discussions of dilution, titration, and osmosis have been updated, and new problems on concentrations and dilution were added.

**Chapter 8, Acids and Bases**, discusses acids and bases and their strengths, conjugate acid–base pairs, ionization of water, pH, and buffers. Section 8.1, *Acid and Bases*, now includes Brønsted–Lowry acids and bases. New items include calculating $[H_3O^+]$ from pH, a *Green Chemistry Note* on acid rain, and new problems related to acid rain. Acid–base titration uses the neutralization reactions to calculate quantities of acid in a sample. New to this edition is an inter-chapter problem set, *Combining Ideas from Chapters 6 to 8*.

## Chapters 9, 10, 11, and 12

**Chapter 9, Nuclear Radiation**, looks at the types of radioactive particles that are emitted from the nuclei of radioactive atoms. Equations are written and balanced for both naturally occurring radioactivity and artificially produced radioactivity. Discussion of the biological effects of radiation were added to the text. A new *Health Note*, "Brachytherapy," was added along with an update of the discussion on radon in the home. The half-lives of radioisotopes are discussed, and the amount of time for a sample to decay is calculated. Radioisotopes, important in the field of nuclear medicine, are described.

**Chapter 10, Introduction to Organic Chemistry: Alkanes**, introduces organic compounds, starting with the alkane family. The visual GPS guides clarify the rules for nomenclature. The chapter concludes with the introduction of *Functional Groups* for each family of organic compounds, which forms a basis for understanding the biomolecules of living systems.

**Chapter 11, Unsaturated Hydrocarbons**, discusses alkenes and alkynes, cis–trans isomers, addition reactions, polymers of alkenes, and aromatic compounds. *Addition Reactions* include hydrogenation and hydration, which are important in biological systems. Synthetic polymers that are used in everyday items are also discussed.

**Chapter 12, Organic Compounds with Oxygen and Sulfur**, discusses alcohols, ethers, thiols, aldehydes, and ketones. The discussion of *Fischer projections* and Section 12.6, *Chiral Molecules*, have been rewritten to better prepare

students for the discussions on carbohydrates in Chapter 14 and amino acids in Chapter 16, along with chiral drugs and their behavior in living systems. New to this edition is an inter-chapter problem set, *Combining Ideas from Chapters 9 to 12*.

## Chapters 13, 14, and 15

**Chapter 13, Carboxylic Acids, Esters, Amines, and Amides**, completes the discussion of organic chemistry. Families and chemical reactions discussed are most applicable to biochemical systems. Sections 13.1, *Carboxylic Acids*, and 13.2, *Properties of Carboxylic Acids*, discuss the carboxyl group and carboxylic acids as weak acids. Sections 13.3, *Amines*, and 13.4, *Amides*, emphasize the nitrogen atom in their functional groups and their names. A *Health Note* on alkaloids discusses the naturally occurring amines in plants.

**Chapter 14, Carbohydrates**, applies organic chemistry to biochemistry, which relates the study of chemistry to health and medicine. Sections 14.2, *Fischer Projections of Monosaccharides*, and 14.3, *Haworth Structures of Monosaccharides*, have new titles and new coverage. Many structures were redrawn as Fischer projections. More examples of sugars were added to the discussion on sweeteners in Section 14.5, *Disaccharides*.

**Chapter 15, Lipids**, continues our study of biomolecules with discussions of fatty acids, triacylglycerols, glycerophospholipids, and steroids. Arachidonic acid was added to Table 15.1, Fatty Acids. The discussion of cis and trans isomers has been rewritten for easier identification of isomers. A new *Green Chemistry Note* discusses biodiesel as an alternative fuel because it is a nonpetroleum-based fuel. *Health Notes* of interest to students include olestra, trans fatty acids, and lipoproteins. The role of lipids in cell membranes is discussed as well as the lipids that function as steroid hormones. In the *Challenge Questions*, students are asked to do calculations for the hydrogenation and saponification of a triacylglycerol. New to this edition is an inter-chapter problem set, *Combining Ideas from Chapters 13 to 15*.

## Chapters 16, 17, and 18

**Chapter 16, Amino Acids, Proteins, and Enzymes**, discusses amino acids, formation of proteins, structural levels of proteins, enzymes, and enzyme action. New items include an updated discussion of zwitterions and isoelectric points and a new *Green Chemistry Note*, "Protein Structure and Mad Cow Disease." Amino acids are written as their ionized forms in physiological solutions. Section 16.5, *Levels of Protein Structure*, describes the importance of protein structure at the primary, secondary, tertiary, and quaternary levels. Enzymes are discussed as biological catalysts, along with the impact of substrate concentration, cofactors, inhibitors, and denaturation on enzyme action.

**Chapter 17, Nucleic Acids and Protein Structure**, describes the components and structures of nucleic acids. The role of complementary base pairing is highlighted in both DNA replication and the formation of mRNA during protein synthesis.

New items include replacing the term "nitrogen base" with "base" and the comparison of percentages of A and T, G and C in DNA. A new *Health Note* describes DNA fingerprinting and its role in medicine and criminology. Section 17.4, *RNA and the Genetic Code*, discusses the relationship of the genetic code to the order of amino acids in a protein. Mutations describe ways in which the nucleotide sequence is altered in genetic diseases. In Section 17.7, *Viruses*, we look at how DNA or RNA in viruses utilizes host cells to produce more viruses.

**Chapter 18, Metabolic Pathways and Energy Production**, describes the metabolic pathways of biomolecules from the digestion of foodstuffs to the synthesis of ATP. Students look at the stages of metabolism and the digestion of carbohydrates along with the coenzymes required in metabolic pathways. The breakdown of glucose to pyruvate is described using the glycolytic pathway, which is followed by the decarboxylation of pyruvate to acetyl CoA. We look at the entry of acetyl CoA into the citric acid cycle and the production of reduced coenzymes. We describe electron transport, oxidative phosphorylation, and the synthesis of ATP. The oxidation of lipids and the degradation of amino acids are also discussed. New to this edition is an inter-chapter problem set, *Combining Ideas from Chapters 16 to 18*.

# INSTRUCTIONAL PACKAGE

*Chemistry: An Introduction to General, Organic, and Biological Chemistry,* tenth edition, is the nucleus of an integrated teaching and learning package of support material for both students and professors.

# NEW ONLINE TUTORIAL AND HOMEWORK SYSTEM

**MasteringChemistry™**
**www.masteringchemistry.com**

## MasteringChemistry

MasteringChemistry is the most advanced tutorial, homework, and self-study system ever built.

For Instructors, MasteringChemistry allows tutoring and assessment of classes like never before. Instructors can

- Provide effective homework and tutorial assignments with automatic grading. Quickly build your own assignments, choosing from an unprecedented variety of ranking and sorting tasks, dynamic visualization activities, interactive tutorials, and end-of-section and end-of-chapter problems.
- Generate ideal exams and tests. Unique tools instantly identify assessment questions of ideal difficulty, duration, and concept coverage for the course.
- View class statistics at a glance, compare results with the national average, identify the most difficult problem for the class, or zero in on time spent by each student. Sophisticated analysis of student performance allows

every item to be refined systematically for educational effectiveness and assessment accuracy.

For Students, MasteringChemistry provides immediate guidance.

- Self-tutoring problems provide specific feedback to common wrong answers and give students hints for partial credit.
- Whether instructors assign homework activities or not, students still have access to a wealth of self-study media. Interactive activities foster learning through manipulating applets and making predictions. Other targeted self-assessment aids are provided, including quizzes, review questions, flashcards, and real-world case studies.
- Throughout *Chemistry: An Introduction to General, Organic, and Biological Chemistry,* tenth edition, icons direct students to ChemPlace, which students find by following MasteringChemistry's Study Area link. The self-study activities, cases, quizzes, and more pertain to concepts the students are currently exploring.
- **SELF-STUDY ACTIVITY**
  Students may explore rich interactive tutorials on key concepts in GOB chemistry. These engaging self-paced activities invite students to make predictions, generate graphs, play animations, and more, as they confront their common misconceptions and build upon their foundation of chemistry knowledge.
- **SELF-PACED CASE STUDIES**
  Modern case studies are posted online to help students see how chemistry is relevant, fascinating, and vital to the world around them. These popular cases motivate students to see how their GOB studies can be put to good use in the real world. These cases cover topics such as kidney stones, hyperventilation, and food irradiation.

## For Students

**Student Study Guide and Selected Solutions** for *Chemistry: An Introduction to General, Organic, and Biological Chemistry,* tenth edition, by Karen Timberlake, is keyed to the learning goals in the text and designed to promote active learning through a variety of exercises with answers as well as mastery exams. The *Study Guide* also contains complete solutions to odd-numbered problems. (ISBN: 0-13601999-4)

**Special Edition of the Chemistry Place™** for *Chemistry: An Introduction to General, Organic, and Biological Chemistry,* tenth edition (Access through http://www.masteringchemistry.com in the Study Area)

 For students who wish to utilize resources for additional practice, the Chemistry Place Study Area contains **tutorials**, **review questions**, quizzes, **career resources**, a **glossary**, **flashcards**, an **interactive periodic table**, and **case studies** that show how chemical concepts apply to familiar situations.

**Laboratory Manual** by Karen Timberlake. This best-selling lab manual coordinates 42 experiments with topics in general, organic, and biological chemistry, uses new terms during the lab, and explores chemical concepts. Laboratory investigations develop skills of manipulating equipment, reporting data, solving problems, making calculations, and drawing conclusions. (ISBN: 0-805349049)

**Essentials Laboratory Manual** by Karen Timberlake. This lab manual contains 25 experiments for the standard course sequence of topics. Features safety procedures, pre-lab questions, and repost pages. (ISBN: 0805330232)

**Walkthrough and Media Icons** The Walkthrough following the Preface helps students and instructors find the book's features easily. Media Icons in the margins throughout the text direct students to tutorials and case studies on The Chemistry-Place™ Web site for *Chemistry*.

# FOR INSTRUCTORS

**Instructor's Solutions Manual** by Karen Timberlake. The solutions manual highlights chapter topics, includes answers and solutions for all problems in the text, and contains suggestions for the laboratory. (ISBN 0-13601997-8)

Also visit the catalog page for Timberlake, *Chemistry: An Introduction to General, Organic, and Biological Chemistry*, tenth edition, to download available instructor supplements.

**Online Instructor's Manual to Laboratory Manual** contains answers to report pages for the *Laboratory Manual* and *Essentials Laboratory Manual*. [www.prenhall.com]

**Printed Test Bank** Over 1500 questions in multiple-choice, matching, true-false, and short-answer format. A computerized version of the test item file is available on the Instructor's Resource Center on CD/DVD-File Folder Format. (ISBN 0-13500907-3)

**The Chemistry Place** for *Chemistry: An Introduction to General, Organic, and Biological Chemistry*, tenth edition (Access at www.masteringchemistry.com through the Study Area)

The Chemistry Place Study Area offers the best teaching and learning resources all in one place. Conveniently organized by type, these compiled resources help you save time and help your students reinforce and apply what they have learned in class.

Resources for Students include:

- Case Studies that show how chemical concepts apply to familiar situations.
- PowerPoints—complete with art, photos, and tables from the text—teach key concepts from every section of the book

- Review questions and Quizzes
- And much more.

**Course Management. Blackboard and WebCT** course management systems provide powerful course management capability. All of the content available in the Chemistry Place Study Area is also available in **WebCT** and **Blackboard**. Prentice Hall offers content cartridges for these text-specific Classroom Management Systems. Visit **http://cms.prenhall.com** or contact your Prentice Hall sales representative for details.

**WebAssign** (Access at: http://www.webassign.com) WebAssign's homework delivery service allows you to create assignments from a database of end-of-chapter Questions and Problems from *Chemistry*, tenth edition, or to write and customize your own.

**Transparency Acetates** A set of 200 color transparencies is available. (ISBN 0-13500905-7)

**Instructor's Resource Center on CD/DVD-File Folder Format** This CD/DVD includes all the art and tables from the book in JPG format for use in classroom projection or when creating study materials and tests. In addition, instructors can access PowerPoint lecture outlines, written by Karen Timberlake, or create their own with the art provided in PPT format. Also available on the discs are downloadable files of the Instructor Solutions Manual and Test Bank.

# ACKNOWLEDGMENTS

The preparation of a new edition is a continuous effort of many people. As in my work on other textbooks, I am thankful for the support, encouragement, and dedication of many people who put in hours of tireless effort to produce a high quality book that provides an outstanding learning package. The editorial team at Pearson has done an exceptional job. I appreciate the work of my senior editor, Kent Porter-Hamann, who supported my vision of this tenth edition text with a new visual learning strategy and art program. Ray Mullaney, editor in chief, Development, continually encouraged me at each step during the development of this edition, while skillfully coordinating reviews, art, Web site materials, and all the things it takes to make a book come together. Jessica Neumann, assistant editor, also worked carefully and swiftly on the supplements. Fay Ahuja, copy editor, precisely edited the manuscript to make sure the words were correct to help students learn chemistry.

I am especially proud of the art program in this text, which lends beauty and understanding to chemistry. I would like to thank Maureen Eide, art director, and Suzanne Behnke, book designer, whose creative ideas provided beautiful art and an outstanding design for the cover and pages of the book. Rachel Lucas, photo researcher, was invaluable in researching and selecting vivid photos for the text so that students can see the beauty of chemistry. The macro-to-micro illustrations designed by *Production Solutions* and *Precision Graphics* give students visual impressions of the atomic and molecular organization of everyday things and are a fantastic learning tool. I want to thank Michael Rossa, proofreader, for the hours of proofreading all the pages. I also appreciate all the hard work in the field put in by the marketing team, Christy Lawrence, director of marketing, and Dawn Giovanniello, chemistry marketing manager. Without them, no one would know about this text.

This text also reflects the contributions of many professors who took the time to review and edit the manuscript and provide outstanding comments, help, and suggestions. A special thanks to Dr. MaryKay Orgill, University of Nevada, Las Vegas, for her outstanding review of the initial manuscript. Her keen eye and thoughtful comments were extremely helpful in the development of GOB 10e. I am extremely grateful to an incredible group of peers for their careful assessment of all the new ideas for the text; for their suggested additions, corrections, changes, and deletions; and for providing an incredible amount of feedback about the best direction for the text. In addition, I am grateful for the time scientists took to let us take photos and discuss their work with them. I admire and appreciate every one of you.

If you would like to share your experience with chemistry or have questions and comments about this text, I would appreciate hearing from you.

*Karen Timberlake*
Email: khemist@aol.com

# REVIEWERS

### Tenth Edition Reviewers

Kristopher M. Baker, *SUNY, Rockland CC*
Daniel Bernier, *Riverside City College*
David Canoy, *Chemeketa Community College*
Ron Choppi, *Chaffey College*
Susan E. Cordova, *Central New Mexico Community College*
Alan D. Earhart, *Southeast Community College*
Coretta Fernandes, *Lansing Community College*
Eric Goll, *Brookdale Community College*
John G. Griggs, *Troy University*
Byron Howell, *Tyler Junior College*
Shelli Hull, *Tarrant County Community College, South*
Timothy Kreider, *University of Medicine & Dentistry of New Jersey*
Nancy Leigh, *Spoon River College*
Chunmei Li, *Stephen F. Austin State University*
Chris Massone *Molloy College*
Susanne McFadden, *Tarrant County Community College, NW*
Felix Ngassa, *Grand Valley State University*
Dr. MaryKay Orgill, *University of Nevada, Las Vegas*
Gordon A. Pomeroy, *York Technical College*
Lesley Putman, *Northern Michigan University*
Mark Quirie, *Algonquin College*

Kathleen Thrush Shaginaw, *Particular Solutions, Inc.*
Mark Thomson, *Ferris State University*
Rod Tracey, *College of the Desert*
Eileen Reilly-Wiedow, *Fairfield University*

### Ninth Edition Reviewers

Stephan Angel, *Washburn University*
Martin Brock, *Eastern Kentucky University*
Salim Diab, *University of St. Francis*
Adeliza Flores, *City College of San Francisco*
Mushtaq Khan, *Union County College*
Da-Hong Lu, *Fitchburg State College*
Janice O'Donnell, *Henderson State University*
Paris Powers, *Volunteer State Community College*
Kellie Summerlin, *Troy State University*

### Eighth Edition Reviewers

Mamta Agarwal, *Chaffey Community College*
David Ball, *Cleveland State University*
Bal Barot, *Lake Michigan University*
Gerald Bergman, *Northwest State College*
Ildy Boer, *County College of Morris*
Ana Ciereszko, *Miami-Dade Community College*
Bob Eierman, *University of Wisconsin-Eau Claire*
Sandra Etheridge, *Gulf Coast Community College*
Jean Gade, *Northern Kentucky University*
Eric Goll, *Brookdale Community College*
Kevin Gratton, *Johnson County Community College*
Denise Guinn, *Regis College*
Michael Hauser, *St. Louis Community College-Meramec*
John Havrilla, *University of Pittsburgh-Johnstown*
Jack Hefley, *Blinn College*
Larry Jackson, *Montana State University*
T. G. Jackson, *University of South Alabama*
Sharon Kapica, *County College of Morris*
Colleen Kelley, *Northern Arizona University*
Peter Krieger, *Palm Beach Community College*
Kathryn MacNeil, *Kent State University-Tuscarawas*
Hank Mancini, *Paradise Valley Community College*
Larry McGahey, *College of St. Scholastica*
R. John Muench, *Heartland Community College*
Thomas Nycz, *Broward Community College*
Jung Oh, *Kansas State University-Salina*
Steve Samuel, *State University of New York-Old Westbury*
Karen Sanchez, *Florida Community College-Jacksonville*
Bahar Sheikh, *Barton County Community College*
Mark Sinton, *Clarke College*
Scott Smith, *Montcalm Community College*
Heinz Stucki, *Kent State University-Tuscarawas*
Eric Taylor, *University of Louisiana at Lafayette*
Beth Wise, *Brookdale Community College*
Cheryl Wistrom, *Saint Joseph's College*

# CAREER FOCUS AND REAL-WORLD APPLICATIONS

**THIS TEXT WAS DESIGNED TO HELP STUDENTS ATTAIN THEIR CAREER GOALS.**

# 5 Chemical Quantities and Reactions

the **Chemistry** place

Visit **www.chemplace.com** for extra quizzes, interactive tutorials, career resources, PowerPoint slides for chapter review, math help, and case studies.

*"In our food science laboratory I develop a variety of food products, from cake doughnuts to energy beverages," says Anne Cristofano, senior food technologist at Mattson & Company. "When I started the doughnut project, I researched the ingredients, then weighed them out in the lab. I added water to make a batter and cooked the doughnuts in a fryer. The batter and the oil temperature make a big difference. If I don't get the right taste or texture, I adjust the ingredients, such as sugar and flour, or adjust the temperature."*

*A food technologist studies the physical and chemical properties of food and develops scientific ways to process and preserve it for extended shelf life. The food products are tested for texture, color, and flavor. The results of these tests help improve the quality and safety of food.*

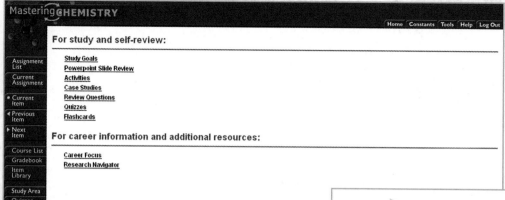

## On the Web
The **MasteringChemistry Study Area** features in-depth resources for each of the health professions featured in the book and takes students through interactive case studies.

## Career Focus
Within the chapters are additional examples of allied health professionals using chemistry.

### Physical Therapist

"Physical therapists need to understand how the body, muscles, and joints function in order to recognize when something is not working and which area needs strengthening," says Vincent Leddy, physical therapist. "Chemistry is important to understanding the body's physiology and how chemical changes in the body affect movement. I went into physical therapy because I enjoy teaching movement to children. I am guiding Maggie into her chair but allowing her to move as much as she can by herself. I help her by giving gentle pressure and reassurance that she'll be safe in the transition. I work on getting Maggie to use her body, and the occupational therapist works on Maggie's fine motor skills and how she uses her hand on the switch or picks up objects. By using both physical and occupational therapy, we enhance the child's performance."

## ◀ Chapter Opener
Each chapter begins with an interview with a professional in allied health or other relevant fields. These interviews illustrate how health professionals interact with chemistry in their careers.

# STUDENTS WILL LEARN CHEMISTRY
# USING REAL-WORLD EXAMPLES.

## *Green Chemistry Note*

### Fuel Cells: Clean Energy for the Future

Fuel cells are of interest to scientists because they provide an alternative source of electrical energy that is more efficient, does not use up oil reserves, and generates products that do not pollute the atmosphere. Fuels cells are considered to be a clean way to produce energy.

Unlike a battery that runs down, fuel cells are provided with new reactants continually to provide an electrical current. One type of hydrogen–oxygen fuel cell has been used in automobile prototypes. In this hydrogen cell, gas enters the fuel cell and comes in contact with a platinum catalyst embedded in a plastic membrane. The catalyst assists in the oxidation of hydrogen atoms to protons and electrons.

$$2H_2(g) \longrightarrow 4H^+(aq) + 4e^- \quad \text{Oxidation}$$

The electrons produce an electric current as they travel through the wire. The protons move through the plastic membrane to react with oxygen molecules. The oxygen is reduced to oxide ions that combine with the protons to form water.

$$O_2(g) + 4H^+(aq) + 4e^- \longrightarrow 2H_2O(l) \quad \text{Reduction}$$

The overall hydrogen–oxygen fuel cell reaction can be written as

$$2H_2(g) + O_2(g) \longrightarrow 2H_2O(l)$$

Fuel cells have already been used to power the space shuttle and may soon be available to produce energy for cars and buses.

A major drawback to the practical use of fuel cells is the economic impact of converting cars to fuel cell operation. The stor-

**Electric current**

H$_2$ gas — H$^+$ — O$_2$ gas

O$^{2-}$

H$^+$ → H$^+$ — O$^{2-}$

H$^+$ — H$^+$ — O$^{2-}$

H$^+$ → H$^+$

**Plastic membrane**

Cathode — Anode

**Catalyst**

Oxidation
$2H_2(g) \longrightarrow 4H^+(aq) + 4e^-$

Reduction
$O_2(g) + 4H^+(aq)$

## NEW Green Chemistry Notes

The **Green Chemistry Notes** highlight the practical applications of chemistry that are beneficial to human health and the environment. The new green chemistry approach that chemists, engineers, scientists, health professionals, and researchers are taking focuses on practices and products that are "benign by design" and provide sustainability.

## *Environmental Note*

### Dating Ancient Objects

A technique known as radiological dating is used by geologists, archaeologists, and historians as a way to determine the age of ancient objects. The age of an object derived from plants or animals (such as wood, fiber, natural pigments, bone, and cotton or woolen clothing) is determined by measuring the amount of carbon-14, a naturally occurring radioactive form of carbon. In 1960, Willard Libby received the Nobel Prize in Chemistry for the work he did developing carbon-14 dating techniques during the 1940s. Carbon-14 is produced in the upper atmosphere by the bombardment of $^{14}_{7}N$ by high-energy neutrons from cosmic rays.

$$^{1}_{0}n + ^{14}_{7}N \longrightarrow ^{14}_{6}C + ^{1}_{1}H$$

Neutron from cosmic rays — Nitrogen in atmosphere — Radioactive carbon-14 — Proton

The carbon-14 reacts with oxygen to form radioactive carbon dioxide: $^{14}CO_2$. Carbon dioxide is continuously absorbed by living plants, incorporating carbon-14 into the plant material. The uptake of carbon-14 stops when the plant dies.

$$^{14}_{6}C \longrightarrow ^{14}_{7}N + ^{0}_{-1}e$$

As the carbon-14 decays, the amount of radioactive carbon-14 in the plant material steadily decreases. In a process called **carbon dating**, scientists use the half-life of carbon-14 (5730 years) to calculate the length of time since the plant died. As the plant material ages, there is less radioactive carbon-14 remaining, and the approximate age of the sample can be determined. For example, a wooden beam found in an ancient Indian dwelling might have one-half of the carbon-14 found in living plants. Because one half-life of carbon-14 is 5730 years, the dwelling was constructed about 5730 years ago. Carbon-14 dating was used to determine that the Dead Sea Scrolls are about 2000 years old.

A radiological dating method used for determining the age of rocks is based on the radioisotope uranium-238, which decays through a series of reactions to lead-206. The uranium-238 isotope has a very long half-life, about $4 \times 10^9$ (4 billion) years. Measurements of the amounts of uranium-238 and lead-206 enable geologists to determine the age of rock samples. The older rocks will have a higher percentage of lead-206 because more of the uranium-238 has decayed. The age of rocks brought back from the moon by the *Apollo* missions was determined using uranium-238. They were found to be about $4 \times 10^9$ years old, approximately the same age calculated for Earth.

## Environmental Notes

**Environmental Notes** throughout the text relate chemistry to real-life topics in science and medicine that are interesting and motivating to students and support the role of chemistry in the real world. They delve into issues such as global warming, radon in our homes, acid rain, and biodiesel fuel.

# Health Note

## Smog and Health Concerns

There are two types of smog. One, photochemical smog, requires sunlight to initiate reactions that produce pollutants such as nitrogen oxides and ozone. The other type of smog, industrial or London smog, occurs in areas where coal containing sulfur is burned.

Photochemical smog is most prevalent in cities where people are dependent on cars for transportation. On a typical day in Los Angeles, for example, nitrogen oxide (NO) emissions from car exhausts increase as traffic increases on the roads. The nitrogen oxide is formed when $N_2$ and $O_2$ react at high temperatures in car and truck engines.

$$N_2(g) + O_2(g) \xrightarrow{\text{Heat}} 2NO(g)$$

Then NO reacts with oxygen in the air to produce $NO_2$, a reddish-brown gas that is irritating to the eyes and damaging to the respiratory tract.

$$2NO(g) + O_2(g) \longrightarrow 2NO_2(g)$$

When $NO_2$ is exposed to sunlight, it is converted into NO and oxygen atoms.

$$NO_2(g) \xrightarrow{\text{Sunlight}} NO(g) + O(g)$$
$$\text{Oxygen atoms}$$

Oxygen atoms are so reactive that they combine with oxygen molecules in the atmosphere, forming ozone.

$$O(g) + O_2(g) \longrightarrow O_3(g)$$
$$\text{Ozone}$$

In the upper atmosphere (the stratosphere), ozone is beneficial because it protects us from harmful ultraviolet radiation that comes from the sun. However, in the lower atmosphere, ozone irritates the eyes and respiratory tract, where it causes coughing, decreased lung function, and fatigue. It also causes deterioration of fabrics, cracks rubber, and damages trees and crops.

Industrial smog is prevalent in areas where coal with a high sulfur content is burned to produce electricity. During combustion, the sulfur is converted to sulfur dioxide:

$$S(s) + O_2(g) \longrightarrow SO_2(g)$$

The $SO_2$ is damaging to plants, suppressing growth, and it is corrosive to metals such as steel. $SO_2$ is also damaging to humans and can cause lung impairment and respiratory difficulties. The $SO_2$ in the air reacts with more oxygen to form $SO_3$. Acidic rain occurs when $SO_3$ combines with water in the air to form sulfuric acid.

$$2SO_2(g) + O_2(g) \longrightarrow 2SO_3(g)$$
$$SO_3(g) + H_2O(l) \longrightarrow H_2SO_4(aq)$$
$$\text{Sulfuric acid}$$

The presence of sulfuric acid in rivers and lakes causes an increase in the acidity of the water, reducing the ability of animals and plants to survive.

**Health Notes** A rich array of **Health Notes** in each chapter apply chemical concepts to relevant topics of health and medicine. These topics include weight loss and weight gain, artificial fats, anabolic steroids, alcohol, genetic diseases, viruses, and cancer.

**Explore Your World** **Explore Your World** includes hands-on activities that use everyday materials to encourage students to actively explore selected chemistry topics, either individually or in group-learning environments. Each activity is followed by questions to encourage critical thinking.

# Explore Your World

## Oxidation of Fruits and Vegetables

Freshly cut surfaces of fruits and vegetables discolor when exposed to oxygen in the air. Cut three slices of a fruit or vegetable such as apple, potato, avocado, or banana. Leave one piece on the kitchen counter (uncovered). Wrap one piece in plastic wrap and leave on the kitchen counter. Dip one piece in lemon juice and leave uncovered.

### QUESTIONS

1. What changes take place in each sample after 1–2 hours?
2. Why would wrapping fruits and vegetables slow the rate of discoloration?
3. If lemon juice contains vitamin C (an antioxidant), why would dipping a fruit or vegetable in lemon juice affect the oxidation reaction on the surface of the fruit or vegetable?
4. Other kinds of antioxidants are vitamin E, citric acid, and BHT. Look for these antioxidants on the labels of cereals, potato chips, and other foods in your kitchen. Why are antioxidants added to food products that will be stored on our kitchen shelves?

# STUDENT-FRIENDLY APPROACH

## KEEPING STUDENTS ENGAGED IS THE ULTIMATE GOAL.

---

**LEARNING GOAL**

Use Avogadro's number to determine the number of particles in a given number of moles.

## 5.1 THE MOLE

At the store, you buy eggs by the dozen. In an office, pencils are ordered by the gross, and paper by the ream. In a restaurant, soda is ordered by the case. The terms *dozen, gross, ream,* and *case* are used to count the number of items present. For example, when you

---

## CHAPTER REVIEW

### 5.1 The Mole

**Learning Goal:** Use Avogadro's number to determine the number of particles in a given number of moles.

One mole of an element contains $6.02 \times 10^{23}$ atoms; a mole of a compound contains $6.02 \times 10^{23}$ molecules or formula units.

### 5.2 Molar Mass

**Learning Goals:** Determine the molar mass of a substance, and use molar mass to convert between grams and moles.

The molar mass (g/mole) of any substance is the mass in grams equal numerically to its atomic mass, or the sum of the atomic masses, which have been multiplied by their subscripts in a formula. It becomes a conversion factor when it is used to change a quantity in grams to moles or to change a given number of moles to grams.

### 5.3 Chemical Changes

**Learning Goal:** Identify a change in a substance as a chemical or a physical change.

A chemical change occurs when the atoms of the initial substances rearrange to form new substances. When new substances form, a chemi-

A chemical equation shows the formulas of the substances that react on the left side of a reaction arrow and the products that form on the right side of the reaction arrow. An equation is balanced by writing the smallest whole numbers (coefficients) in front of formulas to equalize the atoms of each element in the reactants and the products.

### 5.5 Types of Reactions

**Learning Goal:** Identify a reaction as a combination, decomposition, or replacement.

Many chemical reactions can be organized by reaction type: combination, decomposition, single replacement, or double replacement.

### 5.6 Oxidation–Reduction Reactions

**Learning Goal:** Define the terms *oxidation* and *reduction*.

When electrons are transferred in a reaction, it is an oxidation–reduction reaction. One reactant loses electrons, and another reactant gains electrons. Overall, the number of electrons lost and gained is equal.

### 5.7 Mole Relationships in Chemical Equations

**Learning Goal:** Given a quantity in moles of reactant or

---

## Learning Goals

At the beginning of each section, a **Learning Goal** clearly identifies the key concept of the section, providing a roadmap for studying. All information contained in that section relates back to the Learning Goal. The Learning Goals for each section are also repeated in the Chapter Review so students can make sure they have mastered the key concepts.

**Writing Style** Karen Timberlake is known for her accessible writing style, based on a carefully paced and simple development of chemical ideas, suited to the background of allied health students. She precisely defines terms and sets clear goals for each section of the text. Her clear analogies help students to visualize and understand key chemical concepts.

**Concept Maps** Each chapter ends with a **Concept Map** that reviews the key concepts of each chapter and shows how they fit together.

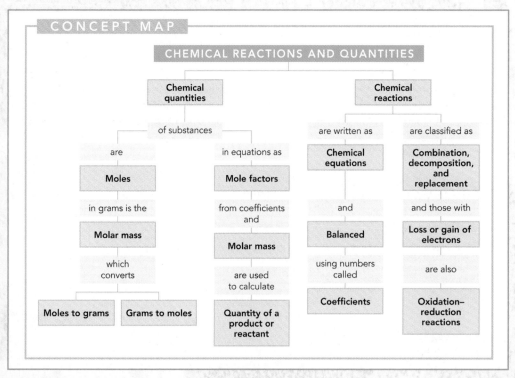

xxviii

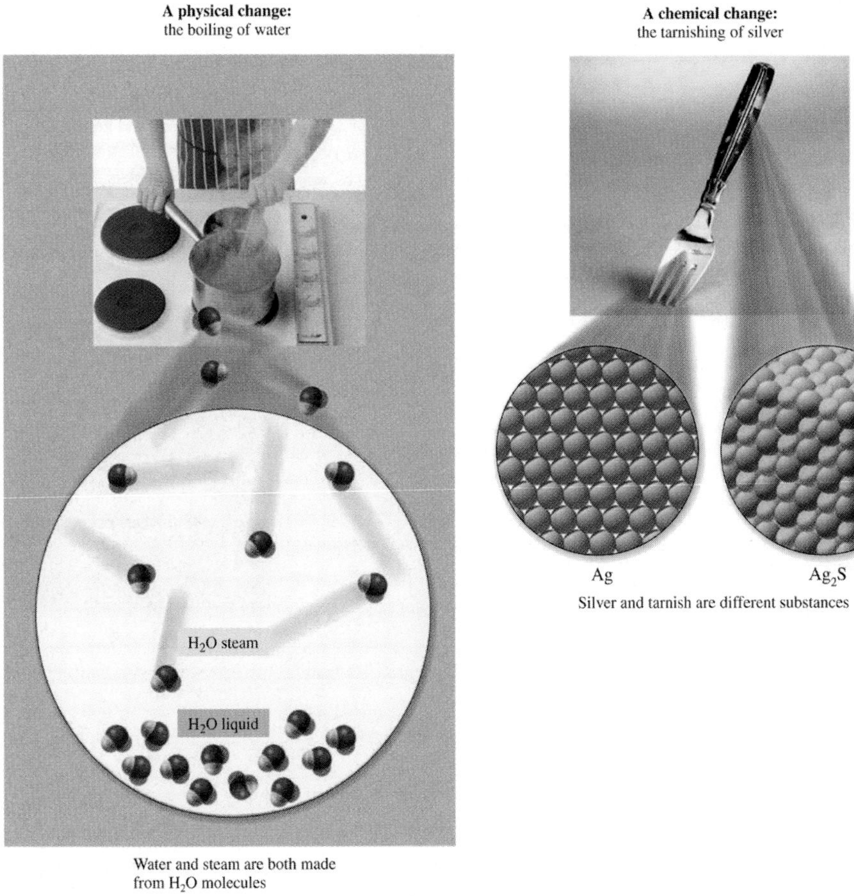

**A physical change:**
the boiling of water

**A chemical change:**
the tarnishing of silver

H₂O steam

H₂O liquid

Water and steam are both made
from H₂O molecules

Ag          Ag₂S

Silver and tarnish are different substances

**FIGURE 5.2** A chemical change produces new substances; a physical change does not.
**Q** Why is the formation of tarnish a chemical change?

## Illustrations that Help Students Visualize Chemistry
The art program is not only beautifully rendered, but pedagogically effective as well.

## Macro-to-Micro Art
Photographs and drawings illustrate the atomic structure of recognizable objects, putting chemistry in context and connecting the atomic world to the macroscopic world. Questions paired with figures challenge students to think critically about photos and illustrations.

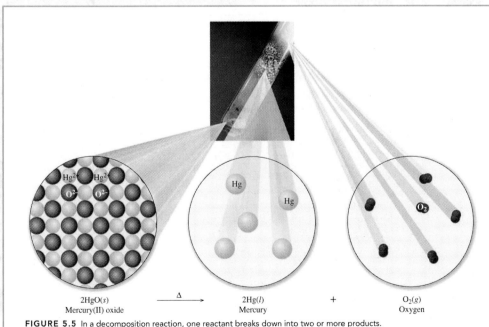

Hg²⁺   Hg²⁺
O²⁻    O²⁻

Hg

Hg

O₂

2HgO(s)                  $\xrightarrow{\Delta}$          2Hg(l)          +          O₂(g)
Mercury(II) oxide                              Mercury                     Oxygen

**FIGURE 5.5** In a decomposition reaction, one reactant breaks down into two or more products.
**Q** How do the differences in the reactant and products classify this as a decomposition reaction?

# PROBLEM SOLVING

## MANY TOOLS SHOW STUDENTS HOW TO SOLVE PROBLEMS.

### A Visual Guide to Problem Solving

The author understands the learning challenges facing students in this course, so she walks students through the problem-solving process **step by step**. For each type of problem, she uses a unique, color-coded flow chart that is coordinated with parallel worked examples to visually guide students through each problem-solving strategy.

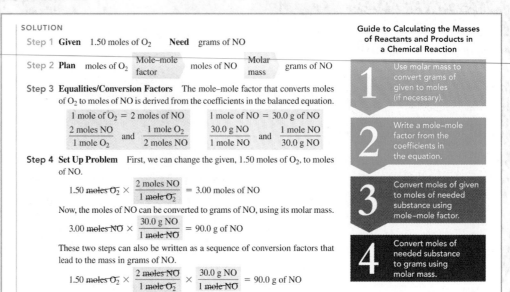

**SOLUTION**

**Step 1 Given** 1.50 moles of $O_2$    **Need** grams of NO

**Step 2 Plan**   moles of $O_2$   $\boxed{\text{Mole–mole factor}}$   moles of NO   $\boxed{\text{Molar mass}}$   grams of NO

**Step 3 Equalities/Conversion Factors**   The mole–mole factor that converts moles of $O_2$ to moles of NO is derived from the coefficients in the balanced equation.

1 mole of $O_2$ = 2 moles of NO

$\dfrac{2 \text{ moles NO}}{1 \text{ mole } O_2}$ and $\dfrac{1 \text{ mole } O_2}{2 \text{ moles NO}}$

1 mole of NO = 30.0 g of NO

$\dfrac{30.0 \text{ g NO}}{1 \text{ mole NO}}$ and $\dfrac{1 \text{ mole NO}}{30.0 \text{ g NO}}$

**Step 4 Set Up Problem**   First, we can change the given, 1.50 moles of $O_2$, to moles of NO.

$$1.50 \text{ moles } O_2 \times \dfrac{2 \text{ moles NO}}{1 \text{ mole } O_2} = 3.00 \text{ moles of NO}$$

Now, the moles of NO can be converted to grams of NO, using its molar mass.

$$3.00 \text{ moles NO} \times \dfrac{30.0 \text{ g NO}}{1 \text{ mole NO}} = 90.0 \text{ g of NO}$$

These two steps can also be written as a sequence of conversion factors that lead to the mass in grams of NO.

$$1.50 \text{ moles } O_2 \times \dfrac{2 \text{ moles NO}}{1 \text{ mole } O_2} \times \dfrac{30.0 \text{ g NO}}{1 \text{ mole NO}} = 90.0 \text{ g of NO}$$

**Guide to Calculating the Masses of Reactants and Products in a Chemical Reaction**

**1** Use molar mass to convert grams of given to moles (if necessary).

**2** Write a mole–mole factor from the coefficients in the equation.

**3** Convert moles of given to moles of needed substance using mole–mole factor.

**4** Convert moles of needed substance to grams using molar mass.

**STUDY CHECK**

Using the equation in Sample Problem 5.13, calculate the grams of $CO_2$ that can be produced when 25.0 g of $O_2$ reacts.

**Sample Problems with Study Checks**   Numerous **Sample Problems** appear throughout the text to immediately demonstrate the application of each new concept. The worked-out solution gives step-by-step explanations, provides a problem-solving model, and illustrates required calculations. Each Sample Problem is followed by a **Study Check** question that allows students to test their understanding of the problem-solving strategy.

---

### QUESTIONS AND PROBLEMS

**Mass Calculations for Reactions**

**5.53** Sodium reacts with oxygen to produce sodium oxide.

$4Na(s) + O_2(g) \longrightarrow 2Na_2O(s)$

**a.** How many grams of $Na_2O$ are produced when 2.50 moles of Na react?

**b.** If you have 18.0 g of Na, how many grams of $O_2$ are required for reaction?

**c.** How many grams of $O_2$ are needed in a reaction that produces 75.0 g of $Na_2O$?

**5.54** Nitrogen gas reacts with hydrogen gas to produce ammonia by the following equation:

$N_2(g) + 3H_2(g) \longrightarrow 2NH_3(g)$

**a.** If you have 1.80 moles of $H_2$, how many grams of $NH_3$ can be produced?

**b.** How many grams of $H_2$ are needed to react with 2.80 g of $N_2$?

**c.** How many grams of $NH_3$ can be produced from 12.0 g of $H_2$?

**5.55** Ammonia and oxygen react to form nitrogen and water.

$4NH_3(g) + 3O_2(g) \longrightarrow 2N_2(g) + 6H_2O(g)$

Ammonia

**a.** How many grams of $O_2$ are needed to react with 8.00 moles of $NH_3$?

**b.** How many grams of $N_2$ can be produced when 6.50 g of $O_2$ reacts?

**c.** How many grams of water are formed from the reaction of 34.0 g of $NH_3$?

**5.56** Iron(III) oxide reacts with carbon to give iron and carbon monoxide.

$Fe_2O_3(s) + 3C(s) \longrightarrow 2Fe(s) + 3CO(g)$

**5.57** Nitrogen dioxide and water react to produce nitric acid, $HNO_3$, and nitrogen oxide.

$3NO_2(g) + H_2O(l) \longrightarrow 2HNO_3(aq) + NO(g)$

**a.** How many grams of $H_2O$ are required to react with 28.0 g of $NO_2$?

**b.** How many grams of NO are obtained from 15.8 g of $NO_2$?

**c.** How many grams of $HNO_3$ are produced from 8.25 g of $NO_2$?

**5.58** Calcium cyanamide reacts with water to form calcium carbonate and ammonia.

$CaCN_2(s) + 3H_2O(l) \longrightarrow CaCO_3(s) + 2NH_3(g)$

**a.** How many grams of water are needed to react with 75.0 g of $CaCN_2$?

**b.** How many grams of $NH_3$ are produced from 5.24 g of $CaCN_2$?

**c.** How many grams of $CaCO_3$ form if 155 g of water react?

**5.59** When the ore lead(II) sulfide burns in oxygen, the products are lead(II) oxide and sulfur dioxide.

**a.** Write the balanced equation for the reaction.

**b.** How many grams of oxygen are required to react with 0.125 mole of lead(II) sulfide?

**c.** How many grams of sulfur dioxide can be produced when 65.0 g of lead(II) sulfide react?

**d.** How many grams of lead(II) sulfide are used to produce 128 g of lead(II) oxide?

**5.60** When the gases dihydrogen sulfide and oxygen react, they form the gases sulfur dioxide and water vapor.

**a.** Write the balanced equation for the reaction.

**b.** How many grams of oxygen are required to react with

### Integrated Questions and Problems

**Questions and Problems** at the end of each section encourage students to apply concepts and begin problem solving after each section.

**Paired Problems** of each even-numbered problem with a matching odd-numbered problem guide students through solving problems.

**Answers** to odd-numbered problems are given at the end of each chapter.

# End-of-Chapter Questions and Problems

**Understanding the Concepts** questions encourage students to think about the concepts they have learned. **Additional Questions and Problems** integrate the topics from the entire chapter to promote further study and critical thinking. **Challenge Questions** are designed for group work in cooperative learning environments.

## UNDERSTANDING THE CONCEPTS

**5.71** Using the following models of the molecules (black = C, light blue = H, yellow = S, green = Cl), determine each of the following.

1.

2.

    **a.** formula
    **b.** molar mass
    **c.** number of moles in 10.0 g

**5.72** Using the following models of the molecules, (black = C, light blue = H, yellow = S, red = O), determine each of the following:

1.      2.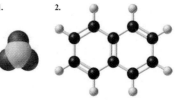

    **a.** formula    **b.** molar mass    **c.** number of moles in 10.0 g

## ADDITIONAL QUESTIONS AND PROBLEMS

**5.87** Calculate the molar mass of each of the following:
    **a.** $FeSO_4$, ferrous sulfate, iron supplement
    **b.** $Ca(IO_3)_2$, calcium iodate, iodine source in table salt
    **c.** $C_5H_8NNaO_4$, monosodium glutamate, flavor enhancer
    **d.** $C_6H_{12}O_2$, isoamyl formate used to make artificial fruit syrups

**5.88** Calculate the molar mass of each of the following:
    **a.** $Mg(HCO_3)_2$, magnesium hydrogen carbonate
    **b.** $Au(OH)_3$, gold(III) hydroxide, used in gold plating
    **c.** $C_{18}H_{34}O_2$, oleic acid from olive oil
    **d.** $C_{21}H_{26}O_5$, prednisone, anti-inflammatory

**5.92** How many moles are in 4.00 g of each of the following compounds?
    **a.** $NH_3$    **b.** $Ca(NO_3)_2$    **c.** $SO_3$

**5.93** During heavy exercise and workouts, lactic acid, $C_3H_6O_3$, accumulates in the muscles where it can cause pain and soreness.
    **a.** How many molecules are in 0.500 mole of lactic acid?
    **b.** How many C atoms are in 1.50 moles of lactic acid?
    **c.** How many moles of lactic acid contain $4.5 \times 10^{24}$ atoms of O?
    **d.** What is the molar mass of lactic acid?

**5.94** Ammonium sulfate, $(NH_4)_2SO_4$, is used in fertilizers to provide

## CHALLENGE QUESTIONS

**5.107** A gold bar is 2.31 cm long, 1.48 cm wide, and 0.0758 cm thick.
    **a.** If gold has a density of 19.3 g/mL, what is the mass of the gold bar?
    **b.** How many atoms of gold are in the bar?
    **c.** When the same mass of gold combines with oxygen, the oxide product has a mass of 5.61 g. How many

moles of oxygen atoms are combined with the gold?
    **d.** What is the formula of the oxide product?

**5.108** A toothpaste contains 0.24% by mass sodium fluoride used to prevent dental caries and 0.30% by mass triclosan $C_{12}H_7Cl_3O_2$,

**Combining Ideas** This new feature appears after every 2–4 chapters as a set of integrated problems designed to test students' understanding of the previous chapters.

# Combining Ideas from Chapters 3 to 5

**CI.7** The following reaction occurs between a metal and a nonmetal.

    X    Y    Y

    **a.** Which spheres represent a metal? A nonmetal?
    **b.** Which reactant has the higher electronegativity?
    **c.** What are the ionic charges of X and Y in the product?
    **d.** If these elements are both in period 3,
        **1.** Write the electron arrangement of the atoms.
        **2.** Write the electron arrangement of their ions.
        **3.** Give the names of the noble gases with the same electron arrangement as these ions.
        **4.** Write the formula and name of the product.
    **e.** Match the spheres below with atoms of Li, Na, K, and Rb.

    A.    B.    C.    D.

    **b.** When 20.0 g of ethanol at $-55\ °C$ is heated to 37 °C, how much energy in calories is required?
    **c.** If the density of ethanol is 0.796 g/mL, how many kilojoules are needed to vaporize 1.00 L of ethanol at 78 °C?
    **d.** If a 15-gallon gas tank is filled with E10, how many liters of ethanol are in the gas tank?
    **e.** Write the balanced chemical equation for the reaction of ethanol with oxygen gas to produce carbon dioxide and

# THE WORLD'S MOST POWERFUL ONLINE HOMEWORK AND TUTORIAL SYSTEM

## MasteringChemistry™

MasteringChemistry is the first adaptive-learning online homework system. It provides selected end-of-chapter problems from the text, as well as over a hundred tutorials with automatic grading, immediate answer-specific feedback, and simpler questions on request. Based on extensive research of concepts students struggle with, MasteringChemistry uniquely responds to your immediate learning needs, thereby optimizing your study time. (See the MasteringChemistry insert at the front of the book.)

## ADDITIONAL SUPPLEMENTS

### Valuable Ancillaries for Both Students and Instructors.

## THE CHEMISTRY PLACE

### Interactivity

The Chemistry Place for *Chemistry: An Introduction to General, Organic, and Biological Chemistry*, Tenth Edition (access at www.masteringchemistry.com through the Study Area).

The Chemistry Place Study Area offers the best teaching and learning resources all in one place. Conveniently organized by type, these compiled resources help students reinforce and apply what they have learned in class.

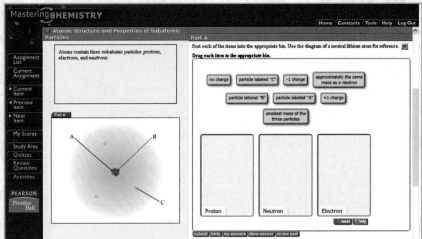

Resources for Students include:

- Case Studies that show how chemical concepts apply to familiar situations.

- PowerPoints™—complete with art, photos, and tables from the text—teach key concepts from every section of the book.

- Review Questions and Quizzes

- And much more.

### Applications

The website, like the text, is committed to students. In-depth looks at Health Careers and Case Studies will provide a relevant context for learning.

# SUPPLEMENTS FOR STUDENTS

## Study Guide and Selected Solutions

**by Karen Timberlake**
The Study Guide is keyed to the learning goals in the text and designed to promote active learning. It includes complete solutions to odd-numbered problems. (ISBN: 0-136019994)

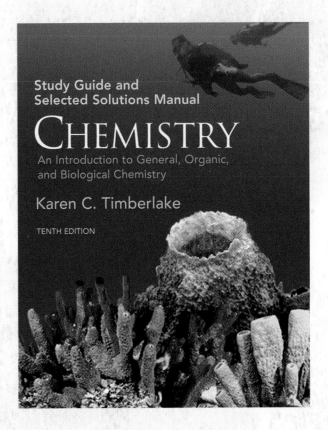

## Essential Laboratory Manual

**by Karen Timberlake**
The Essential Laboratory Manual contains 25 experiments for the standard course sequence of topics. (ISBN: 0-80-533023-2)

## Laboratory Manual for General, Organic, and Biological Chemistry

**by Karen Timberlake**
This best-selling Laboratory Manual contains 42 experiments for the standard course sequence of topics. (ISBN: 0-80-534904-9)

# SUPPLEMENTS FOR INSTRUCTORS

## Image Presentation Tools

## Instructor Resource Center on CD/DVD-File Folder Format

This CD/ DVD includes all the art and tables from the book in JPG format for use in classroom projection or when creating study materials and tests. In addition, the instructor can access PowerPoint lecture outlines, written by Karen Timberlake, or create their own with the art provided in PPT format. (ISBN: 0-13-500906-5)

## Transparency Pack

This package includes 200 full-color transparency acetates. (ISBN: 0-13-500905-7)

## Instructor's Solutions Manual

**By Karen Timberlake**
Highlights chapter topics and includes suggestions for the laboratory. Contains answers and solutions to all problems in the text. (ISBN: 0-13-601997-8)

## Printed Test Bank

Over 1500 questions in multiple-choice, matching, true-false, and short-answer format. Available in print or on the Instructor Resource Center on CD/ DVD-File Folder Format. (ISBN: 0-13-500907-3)

## Online Instructor's Manual to Laboratory Manual

Contains answers to report pages for the Laboratory Manual and Essential Laboratory Manual. Also, visit the Prentice Hall catalog page for Timberlake's Chemistry, Tenth Edition, at www.prenhall.com to download available instructor supplements.

# PROLOGUE
# Chemistry in Our Lives

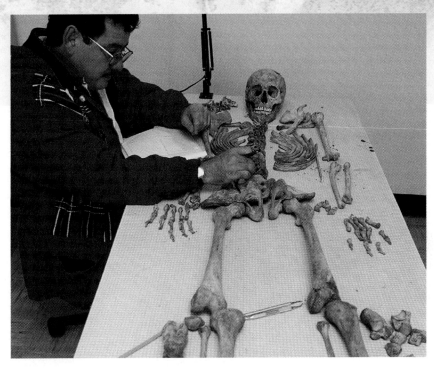

Visit **www.chemplace.com** for extra quizzes, interactive tutorials, career resources, PowerPoint slides for chapter review, math help, and case studies.

*"Chemistry plays an integral part in all aspects of medical legal death investigation," says Charles L. Cecil, forensic anthropologist and medical legal death investigator, San Francisco Medical Examiner's Office. "Crime-scene analysis of blood droplets determines whether they are human or nonhuman; the analyses of toxicological samples of blood and/or other fluids help determine the cause and time of death. Specialists in forensic anthropology can analyze trace-element ratios in bones to identify the number of individuals in mixed human bone situations. These conditions are quite often found during investigations of massive human-rights violations, such as the site of El Mozote in El Salvador."*

*Forensic anthropologists help police identify skeletal remains by determining the gender, approximate age, height, and cause of death. Analysis of bone can also provide information about illnesses or trauma a person may have experienced.*

# W

hat are some questions in science you have been curious about? Perhaps you are interested in how smog is formed, what causes ozone depletion, how nails form rust, or how aspirin relieves a headache. Just like you, scientists are curious about the world we live in.

- How does car exhaust produce the smog that hangs over our cities? One component of car exhaust is nitrogen oxide (NO) formed in car engines, where high temperatures convert nitrogen gas ($N_2$) and oxygen gas ($O_2$) to NO. In chemistry, these reactions are written in the form of equations such as $N_2(g) + O_2(g) \rightarrow 2NO(g)$. The reaction of NO with oxygen in the air produces $NO_2$, which gives the reddish-brown color to smog.

- Why has the ozone layer been depleted in certain parts of the atmosphere? In the 1970s, scientists associated substances called chlorofluorocarbons (CFCs) with the depletion of ozone over Antarctica. As CFCs are broken down by ultraviolet (UV) light, chlorine (Cl) is released that causes the breakdown of ozone ($O_3$) molecules and destroys the ozone layer:

$$Cl + O_3 \rightarrow ClO + O_2.$$

- Why does an iron nail rust when exposed to air and rain? When iron (Fe) in a nail reacts with oxygen ($O_2$) in the air, the oxidation of iron forms rust ($Fe_2O_3$): $4Fe(s) + 3O_2(g) \rightarrow 2Fe_2O_3(s)$.

- Why does aspirin relieve a headache? When a part of the body is injured, substances called prostaglandins are produced, which cause inflammation and pain. Aspirin acts to block the production of prostaglandins, thereby reducing inflammation, pain, and fever.

Some scientists design new fuels and more efficient ways to use them. Researchers in the medical field look for evidence that will help them understand and design new treatments for diabetes, genetic defects, cancer, AIDS, and aging. For the chemist in the laboratory, the physician in the dialysis unit, or the agricultural scientist, chemistry plays a central role in providing understanding, assessing solutions, and making important decisions.

## LEARNING GOAL

Define the term *chemistry*, and identify substances as chemicals.

# P.1 CHEMISTRY AND CHEMICALS

**Chemistry** is the study of the composition, structure, properties, and reactions of matter. *Matter* is another word for all the substances that make up our world. Perhaps you imagine that chemistry is done only in a laboratory by a chemist wearing a lab coat and goggles. Actually, chemistry happens all around you every day and has a big impact on everything you use and do. You are doing chemistry when you cook food, add chlorine to a swimming pool, or start your car. A chemical reaction has taken place when a nail rusts or an antacid tablet fizzes when dropped into water. Plants grow because chemical reactions

**TABLE P.1** Chemicals Commonly Used in Toothpaste

| Chemical | Function |
|---|---|
| Calcium carbonate | Acts as an abrasive to remove plaque |
| Sorbitol | Prevents loss of water and hardening of toothpaste |
| Carrageenan (seaweed extract) | Keeps toothpaste from hardening or separating |
| Glycerin | Makes toothpaste foam in the mouth |
| Sodium lauryl sulfate | Acts as a detergent used to loosen plaque |
| Titanium dioxide | Makes toothpaste base white and opaque |
| Triclosan | Inhibits bacteria that cause plaque and gum disease |
| Sodium monofluorophosphate | Prevents formation of cavities by strengthening tooth enamel with fluoride |
| Methyl salicylate | Gives a pleasant flavor of wintergreen |

convert carbon dioxide, water, and energy to carbohydrates. Chemical reactions take place when you digest food and break it down into substances that you need for energy and health.

All the things you see around you are composed of one or more chemicals. A **chemical** is any material used in or produced by a chemical process. Chemical processes take place in chemistry laboratories, manufacturing plants, and pharmaceutical labs as well as every day in nature and in the body. A **substance** is a chemical that consists of one type of matter and always has the same composition and properties wherever it is found. Often the terms *chemical* and *substance* are used interchangeably to describe a specific type of matter.

Each day you use products containing substances that were developed and prepared by chemists. When you shower in the morning, chemicals in soaps and shampoos combine with oils on your skin and scalp and are removed by rinsing with water. When you brush your teeth, the chemicals in toothpaste clean your teeth and prevent plaque formation and tooth decay. Toothpaste contains chemicals such as abrasives, antibacterial agents, enamel strengtheners, colorings, and flavorings. Some of the substances used to make toothpaste are listed in Table P.1.

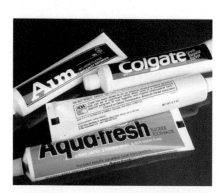

In cosmetics and lotions, chemicals are used to moisturize, prevent deterioration of the product, fight bacteria, and thicken the product. Your clothes may be made of natural polymers such as cotton or synthetic polymers such as nylon or polyester. Perhaps you wear a ring or watch made of gold, silver, or platinum. Your breakfast cereal is probably fortified with iron, calcium, and phosphorus, while the milk you drink is enriched with vitamins A and D. Antioxidants are chemicals added to your cereal to prevent it from spoiling. Some of the chemicals you may encounter when you cook in the kitchen are shown in Figure P.1.

Silicon dioxide (glass)
Caffeine (coffee)
Chemically treated water
Polymer material
Metal alloy
Baking carbohydrates, fats, and proteins
Natural gas
Fruits grown with fertilizers and pesticides

**FIGURE P.1** Many of the items found in a kitchen are obtained using chemical reactions.

**Q** What are some other chemicals found in a kitchen?

SAMPLE PROBLEM  P.1

■ **Everyday Chemicals**

Identify the chemical described by each of the following statements:

**a.** Aluminum used to make cans.
**b.** Salt (sodium chloride) used throughout history as a preservative.
**c.** Sugar (sucrose) used as a sweetener.

SOLUTION

**a.** aluminum      **b.** salt (sodium chloride)      **c.** sugar (sucrose)

STUDY CHECK

Which of the following are chemicals?
**a.** iron      **b.** tin      **c.** a low temperature      **d.** water

## QUESTIONS AND PROBLEMS

### Chemistry and Chemicals

The answers for all the Study Checks and the magenta odd-numbered exercises are given at the end of this Prologue. Checking your answers will let you know if you understand the material.

**P.1**  Obtain a vitamin bottle, and observe the list of ingredients. List four. Which ones are chemicals?

**P.2**  Obtain a box of breakfast cereal, and observe the list of ingredients. List four. Which ones are chemicals?

**P.3**  A "chemical-free" shampoo includes the ingredients: water, cocomide, glycerin, and citric acid. Is the shampoo "chemical-free"?

**P.4**  A "chemical-free" sunscreen includes the ingredients: titanium dioxide, vitamin E, and vitamin C. Is the sunscreen "chemical-free"?

**P.5**  Pesticides are chemicals. Give one advantage and one disadvantage of using pesticides.

**P.6**  Sugar is a chemical. Give one advantage and one disadvantage of eating sugar.

# P.2  SCIENTIFIC METHOD: THINKING LIKE A SCIENTIST

## LEARNING GOAL

Describe the activities that are part of the scientific method.

When you were very young, you explored the things around you by touching and tasting. When you got a little older, you asked questions about the world you live in. What is lightning? Where does a rainbow come from? Why is water blue? As an adult, you may have wondered how antibiotics work. Each day you ask questions and seek answers as you organize and make sense of the world you live in.

When Nobel Laureate Linus Pauling described his student life in Oregon, he recalled that he read many books on chemistry, mineralogy, and physics. "I mulled over the properties of materials: why are some substances colored and others not, why are some minerals or inorganic compounds hard and others soft." He said, "I was building up this tremendous background of empirical knowledge and at the same time asking a great number of questions." Linus Pauling won two Nobel Prizes: the first, in 1954, was in chemistry and for his work on the structure of proteins, and the second, in 1962, was the Peace Prize.

### Scientific Method

Although the process of trying to understand nature is unique to each scientist, there is a set of general principles called the **scientific method** that describes the thinking of a scientist.

**1. Observations**  The first step in the scientific method is to observe, describe, and measure some event in nature. Observations based on measurements are called *data*.

2. **Hypothesis**   After sufficient data are collected, a *hypothesis* is proposed that states a possible interpretation of the observations. The hypothesis must be stated in a way that it can be tested by experiments.

3. **Experiments**   Experiments are tests that determine the validity of the hypothesis. Often many experiments are performed, and a large amount of data collected. If the results of experiments provide different results than predicted by the hypothesis, a new or modified hypothesis is proposed and a new group of experiments are performed.

4. **Theory**   When experiments can be repeated by many scientists with consistent results that confirm the hypothesis, the hypothesis becomes a *theory*. Each theory, however, continues to be tested and, based on new data, sometimes needs to be modified or even replaced. Then a new hypothesis is proposed, and the process of experimentation takes place once again.

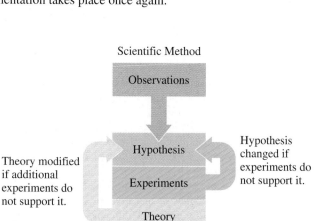

## Using the Scientific Method in Everyday Life

You may be surprised to realize that you use the scientific method in your everyday life. Let's suppose that you visit a friend in her home. Soon after you arrive, your eyes start to itch and you begin to sneeze. Then you observe that your friend has a new cat. Perhaps you ask yourself why you are sneezing and form the hypothesis that you are allergic to cats. To test your hypothesis, you leave your friend's home. If the sneezing stops, perhaps your hypothesis is correct. You test your hypothesis further by visiting another friend who also has a cat. If you start to sneeze again, your experimental results support your hypothesis that you are allergic to cats. However, if you continue sneezing after you leave your friend's home, your hypothesis is not supported. Now you need to form a new hypothesis, which could be that you have a cold.

---

**SAMPLE PROBLEM   P.2**

■ **Scientific Method**

Identify each of the following statements as an observation or a hypothesis.

**a.** A silver tray turns a dull gray color when left uncovered.
**b.** It is warmer in summer than in winter in the Northern Hemisphere.
**c.** Ice cubes float in water because they are less dense.

SOLUTION

**a.** observation          **b.** observation          **c.** hypothesis

**STUDY CHECK**

The following statements are found in a student's notebook. Identify each as an observation (O), a hypothesis (H), or an experiment (E).

a. "Today, I planted two tomato seedlings in the garden. Two more tomato seedlings were placed in a closet. I will give all the plants the same amount of water and fertilizer."

b. "After 50 days, the tomato plants in the garden are 3 feet high with green leaves. The plants in the closet are 8 inches tall and yellow."

c. "Tomato plants need sunlight to grow."

# Health Note

## Early Chemists: The Alchemists

For many centuries, chemistry has been the study of changes in matter. From the time of the Greeks to about the sixteenth century, alchemists described matter in terms of four components of nature: earth, air, fire, and water, with the qualities of hot, cold, damp, or dry. By the eighth century, alchemists believed that they could rearrange these qualities in such a way as to change metals such as copper and lead into gold and silver. They searched for an unknown substance called a philosopher's stone that they thought would turn metals into gold as well as prolong youth and postpone death. Although these efforts failed, the alchemists did provide information on the processes and chemical reactions involved in the extraction of metals from ores. The alchemists also designed some of the first laboratory equipment and developed early laboratory procedures. These early efforts were some of the first observations and experiments using the scientific method.

One alchemist, who used the name Paracelsus (1493–1541), thought that alchemy was not about producing gold, but preparing new medicines. Using observation and experiments, he viewed the body as a series of chemical processes that could be unbalanced by certain chemical compounds and rebalanced by using minerals and medicines. For example, he determined that the lung diseases of miners were caused by inhalation of dust, not by underground spirits. He also thought that goiter was a problem caused by contaminated drinking water, and he treated syphilis with compounds of mercury. His opinion of medicines was that the right dose makes the difference between a poison and a cure. Today, this idea is part of the risk analysis of medicines. Paracelsus changed alchemy in ways that helped to establish modern medicine and chemistry.

FAMOSO DOCTOR    PARESELSVS

# Science and Technology

When scientific information is applied to industrial and commercial uses, it is called technology. Such uses have made the chemical industry one of the largest industries in the United States. Every year, technology provides new materials or procedures that produce more energy, cure diseases, improve crops, and produce new kinds of synthetic materials. Table P.2 lists some of the important scientific discoveries, laws, theories, and technological innovations that were made over the past 300 years.

However, there have also been improper applications of scientific research. The production of some substances has contributed to the development of hazardous conditions in our environment. We have become concerned about the energy requirements of new products and how some materials may cause changes in our oceans and atmosphere. We want to know if the new materials can be recycled, how they are broken down, and whether there are processes that are safer. The ways in which we continue to utilize scientific research will strongly impact our planet and its community in the future. These decisions can best be made if every citizen has an understanding of science.

**TABLE P.2** Some Important Scientific and Technological Discoveries

| Discovery | Date | Name | Country |
|---|---|---|---|
| Law of gravity | 1687 | Isaac Newton | England |
| Oxygen | 1774 | Joseph Priestley | England |
| Electric battery | 1800 | Alessandro Volta | Italy |
| Atomic theory | 1803 | John Dalton | England |
| Anesthesia, ether | 1842 | Crawford Long | U.S. |
| Nitroglycerine | 1847 | Ascanio Sobrero | Italy |
| Germ theory | 1865 | Louis Pasteur | France |
| Antiseptic surgery | 1865 | Joseph Lister | England |
| Discovery of nucleic acids | 1869 | Friedrich Miescher | Switzerland |
| Radioactivity | 1896 | Henri Becquerel | France |
| Discovery of radium | 1898 | Marie and Pierre Curie | France |
| Quantum theory | 1900 | Max Planck | Germany |
| Theory of relativity | 1905 | Albert Einstein | Germany |
| Identification of components of RNA and DNA | 1909 | Phoebus Theodore Levene | U.S. |
| Insulin | 1922 | Frederick Banting, Charles Best, John Macleod | Canada |
| Penicillin | 1928 | Alexander Fleming | England |
| Nylon | 1937 | Wallace Carothers | U.S. |
| Discovery of DNA as genetic material | 1944 | Oswald Avery | U.S. |
| Synthetic production of transuranium elements | 1944 | Glenn Seaborg, Arthur Wahl, Joseph Kennedy, Albert Ghiorso | U.S. |
| Determination of DNA structure | 1953 | Francis Crick, Rosalind Franklin, James Watson | England |
| Polio vaccine | 1954 | Jonas Salk | U.S. |
|  | 1957 | Albert Sabin | U.S. |
| Laser | 1958 | Charles Townes | U.S. |
|  | 1960 | Theodore Maiman | U.S. |
| Cellular phones | 1973 | Martin Cooper | U.S. |
| MRI | 1980 | Paul Lauterbur | U.S. |
| Prozac | 1988 | Ray Fuller | U.S. |
| HIV protease inhibitor | 1995 | Joseph Martin, Sally Redshaw | U.S. |

# *Environmental Note*

## DDT, Good Pesticide, Bad Pesticide

DDT (dichlorodiphenyltrichloroethane) was once one of the most commonly used pesticides. Although it was first made in 1874, it was not used as an insecticide until 1939. Before DDT was widely used, insect-borne diseases such as malaria and typhus were rampant in many parts of the world. Paul Mueller, who discovered that DDT was an effective pesticide, was recognized for saving many lives and received the Nobel Prize for medicine and physiology in 1948. DDT was considered the ideal pesticide because it was toxic to many insects, had a low toxicity to humans and animals, and was inexpensive to prepare.

In the United States, DDT was extensively used on home gardens as well as farm crops, particularly cotton and soybeans. Because of its stable chemical structure, DDT did not break down quickly in the environment, which meant that it did not have to be applied as often. At first, everyone was pleased with DDT as crop yields increased and diseases such as malaria and typhus were under control.

However, in the early 1950s, problems attributed to DDT began to arise. Insects were becoming more resistant to the pesticide. At the same time, there was increasing public concern about the long-term impact of a substance that could remain in the environment for many years. The metabolic systems of humans and animals cannot break down DDT, which is soluble in fats, but not in water, and is stored in the fatty tissues of the body. Although the concentrations of DDT applied to crops were very low, the concentrations of DDT

found in fish and the birds that ate fish were as much as 10 million times greater. Although the birds did not directly perish, DDT was found to reduce the calcium in their eggshells. As a result, the incubating eggs cracked open early. Because of this difficulty with reproduction, the populations of birds such as the bald eagles and brown pelicans dropped significantly.

By 1972, DDT was banned in the United States. The EPA reported that the DDT levels were reduced by 90% in fish in Lake Michigan by 1978. Today new types of pesticides that are more water soluble and do not persist in the environment have replaced the long-lasting pesticides such as DDT. However, these new pesticides are much more toxic to humans.

## QUESTIONS AND PROBLEMS

### Scientific Method: Thinking like a Scientist

**P.7** Identify each as an observation (O), a hypothesis (H), an experiment (E), or a theory (T). At a popular restaurant, where Chang is the head chef, the following occur:
a. Chang determines that sales of the chef's salad have dropped.
b. Chang decides that the chef's salad needs a new dressing.
c. In a taste test, four bowls of lettuce are prepared with four new dressings: sesame seed, oil and vinegar, blue cheese, and anchovies.
d. The tasters rate the dressing with sesame seeds the best.
e. After two weeks, Chang notes that the orders for the chef's salad with the new sesame dressing have doubled.
f. Chang decides that the sesame dressing improved the sales of the chef's salad because the sesame dressing improved the taste of the salad.

**P.8** Identify each as an observation (O), a hypothesis (H), an experiment (E), or a theory (T). Lucia wants to develop a process for dyeing shirts so that the color will not fade when the shirt is washed. She proceeds with the following activities:
a. Lucia notices that the dye in a design on T-shirts fades when the shirt is washed.
b. Lucia decides that the dye needs something to help it set in the T-shirt fabric.
c. She places a spot of dye on each of four T-shirts and then places each one separately in water, salt water, vinegar, and baking soda and water.
d. After 1 hour, all the T-shirts are removed and washed with a detergent.
e. Lucia notices that the dye has faded on the T-shirts in water, salt water, and baking soda, while the dye held up in the T-shirts soaked in vinegar.
f. Lucia thinks that the vinegar binds with the dye so it does not fade when the shirt is washed.

# P.3 A STUDY PLAN FOR LEARNING CHEMISTRY

Here you are taking chemistry, perhaps for the first time. Whatever your reasons are for choosing to study chemistry, you can look forward to learning many new and exciting ideas.

## Features in This Text Help You Study Chemistry

This text has been designed with a variety of study aids to complement different learning styles. On the inside of the front cover is a periodic table of the elements that provides information about the elements. On the inside of the back cover are tables that summarize useful information needed throughout the study of chemistry. Each chapter begins with *Looking Ahead*, which outlines the topics in the chapter. A *Learning Goal* at the beginning of each section previews the concepts you are to learn. At the end of the text, there is a comprehensive glossary/index.

Before you begin to read a chapter, obtain an overview of the topics by reviewing the *Looking Ahead* list of topics. As you prepare to read a section of the chapter, look at the section title and turn it into a question. For example, for Section P.1 "Chemistry and Chemicals," you could write a question that asks "What is chemistry?" or "What are chemicals?" When you are ready to read that section, review the *Learning Goal*, which tells you what you need to accomplish. As you read, try to answer the question you wrote. When you come to a *Sample Problem*, take the time to work it through, try each *Study Check*, and check your answer at the end of the chapter. If your answer does not match, you may need to study the section again. When you finish each section, work through *Questions and Problems* to practice problem solving immediately.

Throughout each chapter, boxes called *Health Notes* and *Green Chemistry Notes* connect the chemical concepts you are studying to real-life situations. Many of the figures and diagrams use macro-to-micro illustrations to depict the atomic level of organization of ordinary objects. These visual models illustrate the concepts described in the text and allow you to "see" the world in a microscopic way.

At the end of each chapter, you will find several study aids that complete the chapter. *Chapter Reviews* and *Concept Maps* at the end of each chapter give a summary and show the connections between important concepts. The *Key Terms* are boldface in the text and listed again at the end of the chapter. *Understanding the Concepts* is a set of questions that utilize art and structures to help you visualize concepts. *Additional Questions and Problems* and *Challenge Problems* at the end of each chapter provide more problems to test your understanding of the topics in the chapter. Answers to all the *Study Checks* and *Answers to Selected Questions and Problems* are provided at the end of the chapter.

## Using Active Learning to Learn Chemistry

A student who is an active learner continually interacts with the chemical ideas while reading the text and attending lecture. Let's see how this is done.

As you read and practice problem solving, you remain actively involved in studying, which enhances the learning process. In this way, you learn small bits of information at a time and establish the necessary foundation for understanding the next section. You may also note questions you have about the reading to discuss with your professor and laboratory instructor. Table P.3 summarizes these steps for active learning. The time you spend in lecture can also be useful as a learning time. By keeping track of the class schedule and reading the assigned material before lecture, you become aware of the new terms and concepts you need to learn. Some questions that occur during your reading may be answered during the lecture. If not, you can ask for further clarification from your professor.

Many studies now indicate that studying with a group can be beneficial to learning for many students. In a group, students motivate each other to study, fill in gaps, and

**TABLE P.3** Steps in Active Learning

1. Read the set of *Looking Ahead* topics for an overview of the material.
2. Form a question from the title of the section you are going to read.
3. Read the section looking for answers to your question.
4. Self-test by working *Sample Problems* and *Study Checks* within each section.
5. Complete the *Questions and Problems* that follow each section, and check the odd-numbered answers.
6. Proceed to the next section, and repeat the above steps.

correct misunderstandings by teaching and learning together. Studying alone does not allow the process of peer correction that takes place when you work with a group of students in your class. In a group, you can cover the ideas more thoroughly as you discuss the reading and problem solving with other students. Waiting to study until the night before an exam does not give you time to understand concepts and practice problem solving. Ideas may be ignored or avoided that turn out to be important on test day.

## Thinking Scientifically About Your Study Plan

As you embark on your journey into the world of chemistry, think about your approach to studying and learning chemistry. You might consider some of the ideas in the following list. Check those ideas that will help you learn chemistry successfully. Commit to them now. Your success depends on you.

**My study of chemistry will include the following:**

———— Reviewing the *Learning Goals*
———— Keeping a problem notebook
———— Reading the text as an active learner
———— Self-testing by working the chapter problems and checking solutions in the text
———— Reading the chapter before lecture
———— Being an active learner in lecture
———— Going to lecture
———— Organizing a study group
———— Seeing the professor during office hours
———— Attending review sessions
———— Organizing my own review sessions
———— Studying a little bit as often as I can

## SAMPLE PROBLEM P.3

■ **A Study Plan for Learning Chemistry**

Which of the following activities would you include in a study plan for learning chemistry successfully?

a. skipping lecture
b. forming a study group
c. keeping a problem notebook
d. waiting to study the night before the exam
e. becoming an active learner

SOLUTION

Your success in chemistry can be helped if you

**b.** form a study group
**c.** keep a problem notebook
**e.** become an active learner

STUDY CHECK

Which of the following would help you learn chemistry?

**a.** skipping review sessions
**b.** working assigned problems
**c.** attending the professor's office hours
**d.** staying up all night before an exam
**e.** reading the assignment before lecture

## QUESTIONS AND PROBLEMS

### Scientific Method: Thinking like a Scientist

**P.9**  A student in your class asks you for advice on learning chemistry. Which of the following might you suggest?
  **a.** Form a study group.
  **b.** Skip lecture.
  **c.** Visit the professor during office hours.
  **d.** Wait until the night before an exam to study.
  **e.** Become an active learner.

**P.10**  A student in your class asks you for advice on learning chemistry. Which of the following might you suggest?
  **a.** Do the assigned problems.
  **b.** Don't read the book; it's never on the test.
  **c.** Attend review sessions.
  **d.** Read the assignment before lecture.
  **e.** Keep a problem notebook.

## CONCEPT MAP

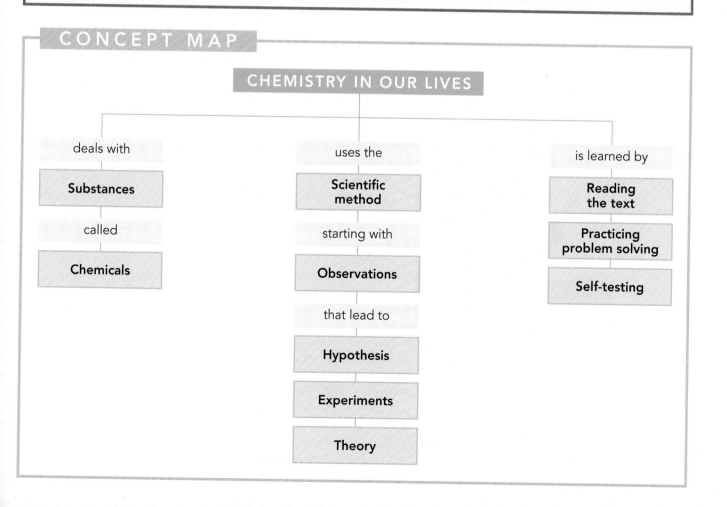

**CHEMISTRY IN OUR LIVES**

deals with
**Substances**
called
**Chemicals**

uses the
**Scientific method**
starting with
**Observations**
that lead to
**Hypothesis**
**Experiments**
**Theory**

is learned by
**Reading the text**
**Practicing problem solving**
**Self-testing**

# CHAPTER REVIEW

### P.1 Chemistry and Chemicals

**Learning Goal:** Define the term *chemistry*, and identify substances as chemicals.

Chemistry is the study of the composition of substances and the way in which they interact with other substances. A chemical is any substance used in or produced by a chemical process.

### P.2 Scientific Method: Thinking like a Scientist

**Learning Goal:** Describe the activities that are part of the scientific method.

The scientific method is a process of explaining natural phenomena, beginning with observations, hypothesis, and experiments, which may lead to a theory when experimental results support the hypothesis. Technology involves the application of scientific information to industrial and commercial uses.

### P.3 A Study Plan for Learning Chemistry

**Learning Goal:** Develop a study plan for learning chemistry.

A study plan for learning chemistry utilizes the visual features in the text and develops an active learning approach to study. By using the *Learning Goals* in the chapter and working the *Sample Problems* and *Study Checks* and the problems at the end of each section, the student can successfully learn the concepts of chemistry.

# KEY TERMS

**chemical** A substance used in or produced by a chemical process.

**chemistry** A science that studies the composition of substances and the way they interact with other substances.

**experiment** A procedure that tests the validity of a hypothesis.

**hypothesis** An unverified explanation of a natural phenomenon.

**observation** Information determined by noting and recording a natural phenomenon.

**scientific method** The process of making observations, proposing a hypothesis, testing the hypothesis, and developing a theory that explains a natural event.

**substance** A particular kind of matter that has the same composition and properties wherever it is found.

**theory** An explanation of an observation that has been validated by experiments that support a hypothesis.

# UNDERSTANDING THE CONCEPTS

**P.11** According to Sherlock Holmes, "one must follow the rules of scientific inquiry, gathering, observing, and testing data, then formulating, modifying and rejecting hypotheses, until only one remains." Did Sherlock use the scientific method? Why or why not?

**P.12** In "A Scandal in Bohemia," Sherlock Holmes receives a mysterious note. He states, "I have no data yet. It is a capital mistake to theorize before one has data. Insensibly one begins to twist facts to suit theories instead of theories to suit facts." What do you think Sherlock meant?

# ADDITIONAL QUESTIONS AND PROBLEMS

**P.13** Classify each of the following statements as an observation or a hypothesis:
   **a.** Aluminum melts at 660 °C.
   **b.** Dinosaurs became extinct when a large meteorite struck Earth and caused a huge dust cloud that severely decreased the amount of light reaching Earth.
   **c.** The 100-yard dash was run in 9.8 seconds.

**P.14** Classify each of the following statements as an observation or a hypothesis:
   **a.** Analysis of ten ceramic dishes showed that four dishes contained lead levels that exceeded federal safety standards.
   **b.** Marble statues undergo corrosion in acid rain.
   **c.** Statues corrode in acid rain because the acidity is sufficient to dissolve calcium carbonate, the major substance of marble.

# CHALLENGE QUESTIONS

**P.15** Classify each of the following as an observation, a hypothesis, or an experiment:
   **a.** The bicycle tire is flat.
   **b.** If I add air to the bicycle tire, it will expand to the proper size.
   **c.** When I added air to the bicycle tire, it was still flat.
   **d.** The bicycle tire must have a leak in it.

**P.16** Classify each of the following as an observation, a hypothesis, or an experiment:
   **a.** A big log in the fire does not burn well.
   **b.** If I chop the log into small wood pieces, it will burn better.
   **c.** The smaller pieces of wood burn brighter and make a hotter fire.
   **d.** The small wood pieces are used up faster than burning the big log.

# ANSWERS

## Answers to Study Checks

**P.1** a, b, and d

**P.2** **a.** E     **b.** O     **c.** H

**P.3** b, c, and e

## Answers to Selected Questions and Problems

**P.1** Many chemicals are listed on a vitamin bottle such as vitamin A, vitamin $B_3$, vitamin $B_{12}$, folic acid, etc.

**P.3** No. All of the ingredients listed are chemicals.

**P.5** One advantage of a pesticide is that it gets rid of insects that bite or damage crops. One disadvantage is that a pesticide can destroy beneficial insects or be retained in a crop that is eventually eaten by animals or humans.

**P.7** **a.** O     **b.** H     **c.** E     **d.** O
     **e.** O     **f.** T

**P.9** a, c, and e

**P.11** Yes. Sherlock's investigation includes observations (gathering data), formulating a hypothesis, testing the hypothesis, and modifying it until one of the hypotheses is validated.

**P.13** **a.** observation     **b.** hypothesis     **c.** observation

**P.15** **a.** observation     **b.** hypothesis     **c.** experiment
     **d.** hypothesis

# 1 Measurements

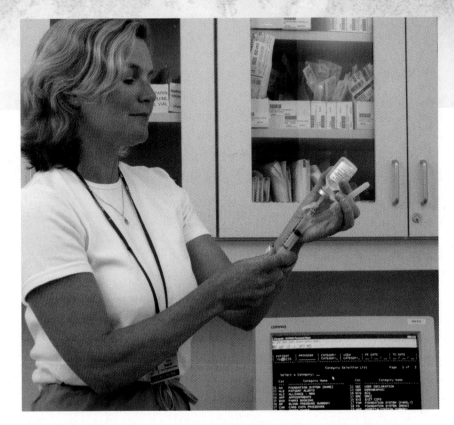

*"I use measurement in just about every part of my nursing practice,"* says registered nurse Vicki Miller. *"When I receive a doctor's order for a medication, I have to verify that order. Then I draw a carefully measured volume from an IV or a vial to create that particular dose. Some dosage orders are specific to the size of the patient. I measure the patient's weight and calculate the dosage required for the weight of that patient."*

*Nurses use measurement each time they take a patient's temperature, height, weight, or blood pressure. Measurement is used to obtain the correct amounts for injections and medications and to determine the volumes of fluid intake and output. For each measurement, the amounts and units are recorded in the patient's records.*

C hemistry and measurement are an important part of our everyday life. Levels of toxic materials in our air, soil, and water are discussed in our newspaper. We read about radon in our homes, holes in the ozone layer, trans fatty acids, global warming, and DNA analysis. Understanding chemistry and measurement helps us make proper choices about our world.

Think about your day; you probably made some measurements. Perhaps you checked your weight by stepping on a scale. If you did not feel well, you may have taken your temperature. To make some soup, you added 2 cups of water to a package mix. If you stopped at the gas station, you watched the gas pump measure the number of gallons of gasoline you put in the car.

Measurement is an essential part of health careers such as nursing, dental hygiene, respiratory therapy, nutrition, and veterinary technology. The temperature, height, and weight of a patient are measured and recorded. Samples of blood and urine are collected and sent to a laboratory where glucose, pH, urea, and protein are measured by the lab technicians.

By learning about measurement, you will develop skills for solving problems and learn how to work with numbers in chemistry. If you intend to go into a health career, measurement and an understanding of measured values will be an important part of your evaluation of the health of a patient.

# 1.1 UNITS OF MEASUREMENT

**LEARNING GOAL**

Write the names and abbreviations for the units used in measurements of length, volume, and mass.

The **metric system** is used by scientists and health professionals throughout the world. It is also the common measuring system in all but a few countries in the world. In 1960, a modification of the metric system called the *International System of Units*, Système International (**SI**), was adopted by scientists to provide additional uniformity for units used in the sciences. In this text, we will use metric units and introduce some of the SI units that are in use today, as listed in Table 1.1.

## Length

The metric and SI unit of length is the **meter (m)**. It is 39.4 inches (in.), which makes a meter slightly longer than a yard (yd). A smaller unit of length, the **centimeter (cm)**, is more commonly used in chemistry and is about as wide as your little finger. For comparison, there are 2.54 cm in 1 in. (See Figure 1.1.)

$$1 \text{ m} = 100 \text{ cm}$$
$$1 \text{ m} = 39.4 \text{ in.}$$
$$2.54 \text{ cm} = 1 \text{ in.}$$

## Volume

**Volume** is the amount of space a substance occupies. A **liter (L)**, which is slightly larger than the quart (qt), is commonly used to measure volume. The **milliliter (mL)** is more

**FIGURE 1.1** Length in the metric (SI) system is based on the meter, which is slightly longer than a yard.
**Q** How many centimeters are in a length of 1 inch?

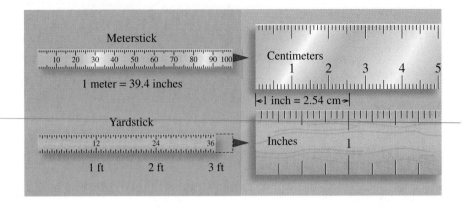

convenient for measuring smaller volumes of fluids in hospitals and laboratories. A comparison of metric and U.S. units for volume appears in Figure 1.2.

$$1 \text{ L} = 1000 \text{ mL}$$
$$1 \text{ L} = 1.06 \text{ qt}$$
$$946 \text{ mL} = 1 \text{ qt}$$

## Mass

The **mass** of an object is a measure of the quantity of material it contains. You may be more familiar with the term *weight* than with mass. Weight is a measure of the gravitational pull on an object. On Earth, an astronaut with a mass of 75.0 kg has a weight of 165 lb. On the moon where the gravitational pull is one-sixth that of Earth, the astronaut has a weight of 27.5 lb. However, the mass of the astronaut is the same as on Earth, 75.0 kg. Scientists measure mass rather than weight because mass does not depend on gravity.

**FIGURE 1.2** Volume is the space occupied by a substance. In the metric system, volume is based on the liter, which is slightly larger than a quart.
**Q** How many milliliters are in 1 quart?

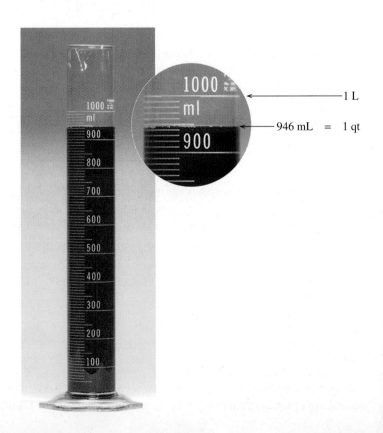

In the metric system, the unit for mass is the **gram (g)**. The SI unit of mass, the **kilogram (kg)**, is used for larger masses such as body weight. It takes 2.20 lb to make 1 kg, and 454 g is needed to equal 1 pound.

$$1 \text{ kg} = 1000 \text{ g}$$
$$1 \text{ kg} = 2.20 \text{ lb}$$
$$454 \text{ g} = 1 \text{ lb}$$

In a chemistry laboratory, a balance is used to measure the mass of a substance as shown in Figure 1.3.

**FIGURE 1.3** On an electronic balance, mass is shown in grams as a digital readout.
**Q** How many grams are in 1 pound of candy?

## Temperature

You probably use a thermometer to see how hot something is, or how cold it is outside, or perhaps to determine if you have a fever. (See Figure 1.4.) The **temperature** of an object tells us how hot or cold that object is. In the metric system, temperature is measured on the Celsius temperature scale. On the **Celsius (°C) scale**, water freezes at 0 °C and boils at 100 °C. On the Fahrenheit (°F) scale, water freezes at 32 °F and boils at 212 °F. In the SI system, temperature is based on the **Kelvin (K) temperature scale**, where the lowest temperature possible is assigned as 0 K. We will discuss the relationships between these temperature scales in Chapter 2.

## Time

We typically measure time in units of years, days, minutes, or seconds. Of these, the SI and metric unit of time is the **second (s)**. The standard now used to determine a second is an atomic clock. A comparison of metric and SI units for measurement is shown in Table 1.1.

**FIGURE 1.4** A thermometer is used to determine the temperature of a substance.
**Q** What kinds of temperature readings have you made today?

**TABLE 1.1** Units of Measurement

| Measurement | Metric | SI |
|---|---|---|
| Length | Meter (m) | Meter (m) |
| Volume | Liter (L) | Cubic meter ($m^3$) |
| Mass | Gram (g) | Kilogram (kg) |
| Time | Second (s) | Second (s) |
| Temperature | Celsius (°C) | Kelvin (K) |

---

SAMPLE PROBLEM   1.1

### ■ Units of Measurement

State the type of measurement (mass, length, volume, temperature, or time) indicated by the unit in each of the following:

**a.** 45.6 kg    **b.** 1.895 L    **c.** 14 s    **d.** 45 m    **e.** 315 K

SOLUTION

**a.** mass    **b.** volume    **c.** time    **d.** length    **e.** temperature

STUDY CHECK

Write the name of the metric unit and symbol you would use to express each of the following:

**a.** length of a football field
**b.** daytime temperature
**c.** mass of salt in a shaker

Answers to all of the Study Checks can be found at the end of this chapter in the Answer section. Checking your answers will let you know if you understand the material in this section.

## Explore Your World

### Units Listed on Labels

Read the labels on a variety of products such as sugar, salt, soft drinks, vitamins, and dental floss.

**QUESTIONS**

1. What metric or SI units of measurement are listed on the labels?
2. What type of measurement (mass, volume, etc.) do they indicate?
3. Write the metric or SI amounts in terms of a number plus a unit.

---

## QUESTIONS AND PROBLEMS

### Units of Measurement

In every chapter, each magenta, odd-numbered exercise in Questions and Problems is paired with the next even-numbered exercise. The answers for the magenta, odd-numbered Questions and Problems are given at the end of this chapter. The complete solutions to the odd-numbered Questions and Problems are in the *Study Guide*.

1.1  Compare the units you would use and the units that a student in Mexico would use to measure the following:
   **a.** mass
   **b.** height
   **c.** amount of gasoline to fill a gas tank
   **d.** temperature

1.2  Suppose that a friend tells you the following information. Why are the statements confusing, and how would you make them clear using metric (SI) units?
   **a.** I rode my bicycle for 15 today.
   **b.** My dog weighs 25.

   **c.** It is hot today. It is 30.
   **d.** I lost 1.5 last week.

1.3  What is the name of the unit and the type of measurement (mass, volume, length, temperature, or time) indicated for each of the following quantities?
   **a.** 4.8 m          **b.** 325 g
   **c.** 1.5 mL        **d.** 480 s
   **e.** 28 °C

1.4  What is the name of the unit and the type of measurement (mass, volume, length, temperature, or time) indicated for each of the following quantities?
   **a.** 0.8 L          **b.** 3.6 m
   **c.** 4 kg           **d.** 35 g
   **e.** 373 K

---

**LEARNING GOAL**

Use scientific notation to express large and small numbers.

# 1.2  SCIENTIFIC NOTATION

## Scientific Notation

In chemistry we use numbers that are very small or very large. We might measure something as tiny as the width of a human hair, which is 0.000 008 m. Or perhaps we want to count the number of hairs on the average scalp, which is about 100 000 hairs. (In this section we have added spaces to help make the places easier to count.) (See Figure 1.5.) It is more convenient to write these small and large numbers in scientific notation.

| Item | Value | Scientific Notation |
|---|---|---|
| Width of a human hair | 0.000 008 m | $8 \times 10^{-6}$ m |
| Hairs on a human scalp | 100 000 hairs | $1 \times 10^5$ hairs |

**FIGURE 1.5** Humans have an average of $1 \times 10^5$ hairs on their scalps. Each hair is about $8 \times 10^{-6}$ m wide.
**Q** Why are large and small numbers written in scientific notation?

$8 \times 10^{-6}$ m

## Writing a Number in Scientific Notation

When a number is written in **scientific notation**, there are two parts: a coefficient and a power of 10. For example, the number 2400 in scientific notation is $2.4 \times 10^3$. The coefficient is 2.4 and $10^3$ shows the power of 10. The coefficient is determined by moving the decimal point three places to the left to give a number from 1 to 9. Because we moved the decimal point three places to the left, the power of 10 is a positive 3, which is written as $10^3$. For a number greater than one, the power of 10 is positive.

$$2\,4\,0\,0 = 2.4 \times 1000 = 2.4 \times 10 \times 10 \times 10 = 2.4 \quad \times \quad 10^3$$
$\longleftarrow$ 3 places                                          Coefficient    Power of 10

When a number less than one is written in scientific notation, the power of 10 is negative. For example, to write the number 0.000 86 in scientific notation, the decimal point is moved to the right four places to give a coefficient of 8.6. By moving the decimal point four places to the right, the power of 10 is a negative 4 or $10^{-4}$.

$$0.000\,86 = \frac{8.6}{10\,000} = \frac{8.6}{10 \times 10 \times 10 \times 10} = 8.6 \quad \times \quad 10^{-4}$$
4 places $\longrightarrow$                                          Coefficient    Power of 10

Table 1.2 gives some examples of numbers written as positive and negative powers of 10. The powers of 10 are really a way of keeping track of the decimal point in the decimal number. Table 1.3 gives examples of writing measurements in scientific notation.

**TABLE 1.2** Some Powers of 10

| Number | Multiples of Ten | Scientific Notation | |
|---|---|---|---|
| 10 000 | $10 \times 10 \times 10 \times 10$ | $1 \times 10^4$ | |
| 1 000 | $10 \times 10 \times 10$ | $1 \times 10^3$ | Some positive powers of 10 |
| 100 | $10 \times 10$ | $1 \times 10^2$ | |
| 10 | 10 | $1 \times 10^1$ | |
| 1 | 0 | $1 \times 10^0$ | |
| 0.1 | $\dfrac{1}{10}$ | $1 \times 10^{-1}$ | |
| 0.01 | $\dfrac{1}{10} \times \dfrac{1}{10} = \dfrac{1}{100}$ | $1 \times 10^{-2}$ | Some negative powers of 10 |
| 0.001 | $\dfrac{1}{10} \times \dfrac{1}{10} \times \dfrac{1}{10} = \dfrac{1}{1\,000}$ | $1 \times 10^{-3}$ | |
| 0.0001 | $\dfrac{1}{10} \times \dfrac{1}{10} \times \dfrac{1}{10} \times \dfrac{1}{10} = \dfrac{1}{10\,000}$ | $1 \times 10^{-4}$ | |

**TABLE 1.3** Some Measurements Written in Scientific Notation

| Measured Quantity | Measurement | Scientific Notation |
|---|---|---|
| Diameter of Earth | 12 800 000 m | $1.28 \times 10^7$ m |
| Volume of gasoline used in U.S./yr | 550 000 000 000 L | $5.5 \times 10^{11}$ L |
| Time for light to travel from the Sun to Earth | 500 s | $5 \times 10^2$ s |
| Mass of a typical human | 68 kg | $6.8 \times 10^1$ kg |
| Mass of a hummingbird | 0.002 kg | $2 \times 10^{-3}$ kg |
| Length of a poxvirus | 0.000 000 3 m | $3 \times 10^{-7}$ m |
| Mass of a bacterium (mycoplasma) | 0.000 000 000 000 000 000 1 kg | $1 \times 10^{-19}$ kg |

## Scientific Notation and Calculators

You can enter numbers in scientific notation on most calculators using the EE or EXP key. After you enter the coefficient, push the EXP (or EE) key and enter only the power of 10 because the EXP function key includes the $\times$ **10** value. To enter a negative power of ten, push the plus/minus ($+/-$) key or the minus ($-$) key (depending on your calculator), but **not** the key that performs the subtraction operation "$-$". Some calculators require entering the sign before the power.

| Number to Enter | Method | Display Reads |
|---|---|---|
| $4 \times 10^6$ | 4 EXP (EE) 6 | 4 06 or $4^{06}$ or 4E06 |
| $2.5 \times 10^{-4}$ | 2.5 EXP (EE) $+/-4$ | $2.5 - 04$ or $2.5^{-04}$ or 2.5E$-$04 |

When a calculator answer appears in scientific notation, it is usually shown in the display as a number from 1 to 9 followed by a space and the power of 10. To express this display in scientific notation, write the number, insert $\times$ 10, and use the power of 10 as an exponent.

| Calculator Display | Expressed in Scientific Notation |
|---|---|
| $7.52^{04}$ or 7.52E04 | $7.52 \times 10^4$ |
| $5.8^{-02}$ or 5.8E$-$02 | $5.8 \times 10^{-2}$ |

On many scientific calculators, a number can be converted into scientific notation using the appropriate keys. For example, the number 0.000 52 can be entered followed by hitting the $2^{nd}$ or $3^{rd}$ function key and the SCI key. The scientific notation appears in the calculator display as a coefficient and the power of 10.

0.000 52     [2nd or 3rd function key]     [SCI]     $=$     $5.2^{-4}$ or $5.2 - 04$     $= 5.2 \times 10^{-4}$
　　　　　　　　　　　 Key 　　　　　　　　　 Key 　　　　　　　 Display

---

**SAMPLE PROBLEM**  **1.2**

■ **Scientific Notation**

Write the following measurements in scientific notation:

**a.** 350 g        **b.** 0.000 16 L         **c.** 5 220 000 m

SOLUTION

**a.** $3.5 \times 10^2$ g        **b.** $1.6 \times 10^{-4}$ L         **c.** $5.22 \times 10^6$ m

STUDY CHECK

Write the following measurements in scientific notation:

**a.** 425 000 m        **b.** 0.000 000 8 g

## QUESTIONS AND PROBLEMS

### Scientific Notation

The answers for all the Study Checks and the magenta, odd-numbered Questions and Problems are given at the end of this chapter. The complete solutions to the odd-numbered Questions and Problems are in the *Study Guide*.

1.5  Write the following measurements in scientific notation:
   **a.** 55 000 m        **b.** 480 g        **c.** 0.000 005 cm
   **d.** 0.000 14 s        **e.** 0.0072 L

1.6  Write the following measurements in scientific notation:
   **a.** 180 000 000 g        **b.** 0.000 06 m        **c.** 750 000 g
   **d.** 0.15 m        **e.** 0.024 kg

1.7  Which number in each pair is larger?
   **a.** $7.2 \times 10^3$ or $8.2 \times 10^2$
   **b.** $4.5 \times 10^{-4}$ or $3.2 \times 10^{-2}$
   **c.** $1 \times 10^4$ or $1 \times 10^{-4}$
   **d.** 0.000 52 or $6.8 \times 10^{-2}$

1.8  Which number in each pair is smaller?
   **a.** $4.9 \times 10^{-3}$ or $5.5 \times 10^{-9}$
   **b.** 1250 or $3.4 \times 10^2$
   **c.** 0.000 000 5 or $5 \times 10^{-8}$
   **d.** $4 \times 10^8$ or $5 \times 10^{-2}$

# 1.3 MEASURED NUMBERS AND SIGNIFICANT FIGURES

### LEARNING GOAL

Determine the number of significant figures in measured numbers.

Whenever you make a measurement, you compare an object to a standard. For example, you may use a meterstick to measure your height, a scale to check your weight, and a thermometer to take your temperature. **Measured numbers** are the numbers you obtain when you use a measuring tool to determine your height, weight, or temperature.

## Measured Numbers

Suppose you are going to measure the lengths of the objects in Figure 1.6. You would select a ruler with a scale marked on it. By observing the lines on the scale, you determine the measurement for each object. Perhaps the divisions on the scale are 1 cm. Another ruler might be marked in divisions of 0.1 cm. To report the length, you would first read the numerical value of the marked line. Finally, you *estimate* between the smallest marked lines. This estimated value is the final digit in a measured number.

For example, in Figure 1.6a, the end of the object falls between the lines marked 4 cm and 5 cm. That means that the length is 4 cm plus an estimated digit.

If you estimate that the end is halfway between 4 cm and 5 cm, you would report its length as 4.5 cm. However, someone else might report the length as 4.4 cm. The last digit in a measured number may differ because people do not always estimate in the same way. The ruler shown in Figure 1.6b is marked with lines at 0.1 cm. With this ruler, you can estimate to the hundredths place (0.01 cm). Perhaps you would report the length of the object as 4.55 cm, while someone else may report its length as 4.56 cm. Both results are acceptable.

There is always some *uncertainty* in every measurement. When a measurement lines up with a marked line, a zero is written for the estimated digit. For example, in Figure 1.6c, the length measurement is written as 3.0 cm, not 3.

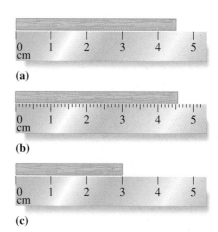

**FIGURE 1.6** The lengths of the rectangular objects are measured as (a) 4.5 cm and (b) 4.55 cm.
**Q** What is the length of the object in (c)?

**TABLE 1.4** Significant Figures in Measured Numbers

| Rule | Measured Number | Number of Significant Figures |
|---|---|---|
| **1. A number is a *significant figure* if it is** | | |
| **a.** not a zero | 4.5 g | 2 |
| | 122.35 m | 5 |
| **b.** a zero between nonzero digits | 205 m | 3 |
| | 5.082 kg | 4 |
| **c.** a zero at the end of a decimal number | 50. L | 2 |
| | 25.0 °C | 3 |
| | 16.00 g | 4 |
| **d.** any digit in the coefficient of a number written in scientific notation | $4.0 \times 10^5$ m | 2 |
| | $5.70 \times 10^{-3}$ g | 3 |
| **2. A zero is *not significant* if it is** | | |
| **a.** at the beginning of a decimal number | 0.0004 lb | 1 |
| | 0.075 m | 2 |
| **b.** used as a placeholder in a large number without a decimal point | 850 000 m | 2 |
| | 1 250 000 g | 3 |

## Significant Figures

In a measured number, the **significant figures** are all the digits including the estimated digit. All nonzero numbers are counted as significant figures. Zeros may or may not be significant depending on their position in a number. Table 1.4 gives the rules and examples of counting significant figures.

When one or more zeros in a large number are significant digits, they are shown more clearly by writing the number using scientific notation. For example, if the first zero in the measurement 500 m is significant, it can be shown by writing the measurement as $5.0 \times 10^2$ m. In this text, we will place a decimal point after a significant zero at the end of a number. For example, a measurement written as 250. g has three significant figures, which includes the zero. It could also be written as $2.50 \times 10^2$ g. We will assume that zeros at the end of large standard numbers are not significant; that is, we would interpret 400 000 g as $4 \times 10^5$ g.

## Exact Numbers

**Exact numbers** are numbers obtained by counting items or from a definition that compares two units in the same measuring system. Suppose a friend asks you to tell her the number of coats in your closet or the number of bicycles in your garage or the number of classes you are taking in school. Your answer would be given by counting the items. It was not necessary for you to use any type of measuring tool. Suppose someone asks you to state the number of seconds in one minute. Without using any measuring device, you would give the definition: 60 seconds in 1 minute. Exact numbers are not measured, do not have a limited number of significant figures, and do not affect the number of significant figures in a calculated answer. For more examples of exact numbers, see Table 1.5.

**TABLE 1.5** Examples of Some Exact Numbers

| Counted Numbers | Defined Equalities | |
| | U.S. System | Metric System |
|---|---|---|
| 8 doughnuts | 1 ft = 12 in. | 1 L = 1000 mL |
| 2 baseballs | 1 qt = 4 cups | 1 m = 100 cm |
| 5 capsules | 1 lb = 16 ounces | 1 kg = 1000 g |

## SAMPLE PROBLEM 1.3

■ **Measured Numbers and Significant Figures**

Identify each of the following numbers as measured or exact; give the number of significant figures in the measured numbers.

**a.** 42.2 g          **b.** 3 eggs          **c.** $5.0 \times 10^{-3}$ cm
**d.** 450 000 km      **e.** 8 pencils

**SOLUTION**

**a.** measured, three      **b.** exact          **c.** measured, two
**d.** measured, two        **e.** exact

**STUDY CHECK**

State the number of significant figures in each of the following measured numbers:

**a.** 0.000 35 g        **b.** 2000 m          **c.** 2.0045 L

## QUESTIONS AND PROBLEMS

### Measured Numbers and Significant Figures

**1.9** Are the numbers in each of the following statements measured or exact?
 **a.** A patient weighs 155 lb.
 **b.** The basket holds 8 apples.
 **c.** In the metric system, 1 kg is equal to 1000 g.
 **d.** The distance from Denver, Colorado, to Houston, Texas, is 1720 km.

**1.10** Are the numbers in each of the following statements measured or exact?
 **a.** There are 31 students in the laboratory.
 **b.** The oldest known flower lived $1.20 \times 10^8$ years ago.
 **c.** The largest gem ever found, an aquamarine, has a mass of 104 kg.
 **d.** A laboratory test shows a blood cholesterol level of 184 mg/dL.

**1.11** In each set of numbers, identify the measured number(s), if any.
 **a.** 3 hamburgers and 6 oz of meat
 **b.** 1 table and 4 chairs
 **c.** 0.75 lb of grapes and 350 g of butter
 **d.** 60 s = 1 min

**1.12** In each set of numbers, identify the exact number(s), if any.
 **a.** 5 pizzas and 50.0 g of cheese
 **b.** 6 nickels and 16 g of nickel
 **c.** 3 onions and 3 lb of potatoes
 **d.** 5 mi and 5 cars

**1.13** For each measurement, indicate if the zeros are significant figures.
 **a.** 0.0038 m      **b.** 5.04 cm      **c.** 800. L
 **d.** $3.0 \times 10^{-3}$ kg      **e.** 85 000 m

**1.14** For each measurement, indicate if the zeros are significant figures.
 **a.** 20.05 g      **b.** 5.00 m      **c.** 0.000 02 L
 **d.** 120 000 years      **e.** $8.05 \times 10^2$ L

**1.15** How many significant figures are in each of the following measured quantities?
 **a.** 11.005 g      **b.** 0.000 32 m      **c.** 36 000 000 km
 **d.** $1.80 \times 10^4$ g      **e.** 0.8250 L      **f.** 30.0 °C

**1.16** How many significant figures are in each of the following measured quantities?
 **a.** 20.60 mL      **b.** 1036.48 g      **c.** 4.00 m
 **d.** 20.8 °C      **e.** 60 800 000 g      **f.** $5.0 \times 10^{-3}$ L

# 1.4 SIGNIFICANT FIGURES IN CALCULATIONS

**LEARNING GOAL**

Adjust calculated answers to the correct number of significant figures.

In the sciences, we measure many things: the length of a bacterium, the volume of a gas sample, the temperature of a reaction mixture, or the mass of iron in a sample. The numbers obtained from these types of measurements are often used in calculations. The number of significant figures in the measured numbers limits the number of significant figures that can be given in the calculated answer.

Using a calculator will usually help do calculations faster. However, calculators cannot think for you. It is up to you to enter the numbers correctly, press the right function keys, and adjust the calculator display to give an answer with the correct number of significant figures.

## Rounding Off

To calculate the area of a carpet that measures 5.5 m by 3.5 m, you multiply 5.5 times 3.5 to obtain the number 19.25 as the area in square meters. Each measurement of length and width has only two significant figures. This means that the calculated result must be *rounded off* to give an answer that also has two significant figures, 19 m². All four digits cannot be given in the answer because they are not all significant. When you obtain a calculator result, determine the number of significant figures for the answer and round off using the following rules.

### Rules for Rounding Off

1. If the first digit to be dropped is *4 or less,* it and all following digits are simply dropped from the number.
2. If the first digit to be dropped is *5 or greater,* the last retained digit of the number is increased by 1.

|  | Three Significant Figures | Two Significant Figures |
|---|---|---|
| Example 1: 8.4234 rounds off to | 8.42 | 8.4 |
| Example 2: 14.780 rounds off to | 14.8 | 15 |
| Example 3: 3256 rounds off to | 3260 | 3300 or $3.3 \times 10^3$ |

---

## SAMPLE PROBLEM 1.4

### ■ Rounding Off

Round off each of the following numbers to three significant figures:

**a.** 35.7823 m       **b.** 0.002 627 L
**c.** $3.8268 \times 10^3$ g       **d.** 1.2836 kg

SOLUTION

**a.** 35.8 m       **b.** 0.002 63 L
**c.** $3.83 \times 10^3$ g       **d.** 1.28 kg

STUDY CHECK

Round off each of the numbers in Sample Problem 1.4 to two significant figures.

---

## Multiplication and Division

In multiplication or division, the final answer is written so it has the same number of significant figures as the measurement with the *fewest significant figures* (SFs).

### Example 1

Multiply the following measured numbers: $24.65 \times 0.67$.

| 24.65 | ☒ | 0.67 | ☰ | 16.5155 | ⟶ | 17 |
|---|---|---|---|---|---|---|
| Four SFs | | Two SFs | | Calculator display | | Final answer, rounded to two SFs |

The answer in the calculator display has more digits than the data allows. The measurement 0.67 has the least number of significant figures, two. Therefore, the calculator answer is rounded off to two significant figures.

## Example 2

Solve the following:

$$\frac{2.85 \times 67.4}{4.39}$$

To do this problem on a calculator, enter the numbers and then press the operation keys. In this case, we might press the keys in the following order:

2.85  ×  67.4  ÷  4.39  =  43.756264  ⟶  43.8

Three SFs    Three SFs    Three SFs    Calculator display    Final answer, rounded to three SFs

All of the measurements in this problem have three significant figures. Therefore, the calculator result is rounded off to give an answer, 43.8, that has three significant figures.

## Adding Significant Zeros

Sometimes, a calculator displays a small whole number. To give an answer with the correct number of significant figures, significant zeros may need to be written after the calculator result. For example, suppose the calculator display is 4, but you used measurements that have three significant numbers. The answer 4.00 is obtained by placing two significant zeros after the 4.

$$\frac{8.00}{2.00} = \quad 4 \quad \longrightarrow \quad 4.00$$

Three SFs    Calculator display    Final answer, two zeros added to give three SFs

---

### SAMPLE PROBLEM  1.5

■ **Significant Figures in Multiplication and Division**

Perform the following calculations of measured numbers. Give the answers with the correct number of significant figures.

**a.** $56.8 \times 0.37$    **b.** $\dfrac{71.4}{11}$    **c.** $\dfrac{(2.075)\,(0.585)}{(8.42)\,(0.0045)}$    **d.** $\dfrac{25.0}{5.00}$

**SOLUTION**

**a.** 21    **b.** 6.5    **c.** 32    **d.** 5.00 (Add significant zeros)

**STUDY CHECK**

Solve:

**a.** $45.26 \times 0.010\ 88$    **b.** $2.6 \div 324$    **c.** $\dfrac{4.0 \times 8.00}{16}$

---

## Addition and Subtraction

In addition or subtraction, the final answer is written so it has the same number of decimal places as the measurement with the *fewest decimal places*.

**Example 3**

Add:

$$2.045 \quad \text{Three decimal places}$$
$$\boxed{+}\ 34.1 \quad \text{One decimal place}$$
$$36.145 \quad \text{Calculator display}$$
$$36.1 \quad \text{Answer, rounded to one decimal place}$$

When numbers are added or subtracted to give answers ending in zero, the zero does not appear after the decimal point in the calculator display. For example, 14.5 g − 2.5 g = 12.0 g. However, if you do the subtraction on your calculator, the display shows 12. To give the correct answer, a significant zero is written after the decimal point.

**Example 4**

Subtract:

$$14.5\ \text{g} \quad \text{One decimal place}$$
$$\boxed{-}\ 2.5\ \text{g} \quad \text{One decimal place}$$
$$12. \quad \text{Calculator display}$$
$$12.0\ \text{g} \quad \text{Answer, zero written after the decimal point}$$

---

**SAMPLE PROBLEM   1.6**

■ **Decimal Places in Addition and Subtraction**

Perform the following calculations and give the answers with the correct number of decimal places:

**a.** 27.8 cm + 0.235 cm        **b.** 153.247 g − 14.82 g

**SOLUTION**

**a.** 28.0 cm        **b.** 138.43 g

**STUDY CHECK**

Solve:

**a.** 82.45 mg + 1.245 mg + 0.000 56 mg        **b.** 4.259 L − 3.8 L

---

## QUESTIONS AND PROBLEMS

### Significant Figures in Calculations

**1.17** Why do we usually need to round off calculations that use measured numbers?

**1.18** Why do we sometimes add a zero to a number in a calculator display?

**1.19** Round off each of the following numbers to three significant figures.
 **a.** 1.854          **b.** 184.2038
 **c.** 0.004 738 265  **d.** 8807
 **e.** $1.832 \times 10^5$

**1.20** Round off each of the numbers in problem 1.19 to two significant figures.

**1.21** Write each of the following in scientific notation with three significant figures:
 **a.** 5080 L        **b.** 37 400 g
 **c.** 104 720 m     **d.** 0.000 250 82 m

**1.22** Write each of the following in scientific notation with two significant figures:
 **a.** 5 100 000 L   **b.** 26 711 s
 **c.** 0.003 378 m   **d.** 56.982 g

**1.23** For the following problems, give answers with the correct number of significant figures:
 **a.** $45.7 \times 0.034$        **b.** $0.00278 \times 5$
 **c.** $\dfrac{34.56}{1.25}$      **d.** $\dfrac{(0.2465)\,(25)}{1.78}$

**1.24** For the following problems, give answers with the correct number of significant figures:
 **a.** $400 \times 185$        **b.** $\dfrac{2.40}{(4)\,(125)}$
 **c.** $0.825 \times 3.6 \times 5.1$   **d.** $\dfrac{3.5 \times 0.261}{8.24 \times 20.0}$

**1.25** For the following problems, give answers with the correct number of decimal places:
   **a.** 45.48 cm + 8.057 cm
   **b.** 23.45 g + 104.1 g + 0.025 g
   **c.** 145.675 mL − 24.2 mL
   **d.** 1.08 L − 0.585 L

**1.26** For the following problems, give answers with the correct number of decimal places:
   **a.** 5.08 g + 25.1 g
   **b.** 85.66 cm + 104.10 cm + 0.025 cm
   **c.** 24.568 mL − 14.25 mL
   **d.** 0.2654 L − 0.2585 L

# 1.5 PREFIXES AND EQUALITIES

The special feature of the metric system of units is that a **prefix** can be attached to any unit to increase or decrease its size by some factor of ten. For example, the prefixes *milli* and *micro* are used to make the smaller units, milligram (mg) and microgram ($\mu$g). Table 1.6 lists some of the metric prefixes, their symbols, and their decimal values.

The prefix *centi* is like cents in a dollar. One cent would be a centidollar or $\frac{1}{100}$ of a dollar. That also means that one dollar is the same as 100 cents. The prefix *deci* is like dimes in a dollar. One dime would be a decidollar or $\frac{1}{10}$ of a dollar. That also means that one dollar is the same as 10 dimes.

The U.S. Food and Drug Administration has determined the daily values (DV) of nutrients for adults and children age four or older. Some of these recommended daily values, which use prefixes, are listed in Table 1.7.

The relationship of a prefix to a unit can be expressed by replacing the prefix with its numerical value. For example, when the prefix *kilo* in *kilometer* is replaced with its value of 1000, we find that a kilometer is equal to 1000 meters. Other examples follow.

1 **kilo**meter (1 km) = **1000** meters (1000 m) = $10^3$ meters ($10^3$ m)

1 **kilo**liter (1 kL)  = **1000** liters (1000 L)  = $10^3$ liters ($10^3$ L)

1 **kilo**gram (1 kg)  = **1000** grams (1000 g)  = $10^3$ grams ($10^3$ g)

## LEARNING GOAL

Use the numerical values of prefixes to write a metric equality.

the
**C**hemistry
place

WEB TUTORIAL
Metric System

**TABLE 1.6** Metric and SI Prefixes

| Prefix | Symbol | Numerical Value | Scientific Notation | Equality |
|--------|--------|-----------------|---------------------|----------|
| **Prefixes That Increase the Size of the Unit** | | | | |
| tera | T | 1 000 000 000 000 | $10^{12}$ | 1 Tg = 1 × $10^{12}$ g |
| giga | G | 1 000 000 000 | $10^9$ | 1 Gm = 1 × $10^9$ m |
| mega | M | 1 000 000 | $10^6$ | 1 Mg = 1 × $10^6$ g |
| kilo | k | 1 000 | $10^3$ | 1 km = 1 × $10^3$ m |
| **Prefixes That Decrease the Size of the Unit** | | | | |
| deci | d | 0.1 | $10^{-1}$ | 1 dL = 1 × $10^{-1}$ L |
|  |  |  |  | 1 L = 10 dL |
| centi | c | 0.01 | $10^{-2}$ | 1 cm = 1 × $10^{-2}$ m |
|  |  |  |  | 1 m = 100 cm |
| milli | m | 0.001 | $10^{-3}$ | 1 ms = 1 × $10^{-3}$ s |
|  |  |  |  | 1 s = 1 × $10^3$ ms |
| micro | $\mu$ | 0.000 001 | $10^{-6}$ | 1 $\mu$g = 1 × $10^{-6}$ g |
|  |  |  |  | 1 g = 1 × $10^6$ $\mu$g |
| nano | n | 0.000 000 001 | $10^{-9}$ | 1 nm = 1 × $10^{-9}$ m |
|  |  |  |  | 1 m = 1 × $10^9$ nm |
| pico | p | 0.000 000 000 001 | $10^{-12}$ | 1 ps = 1 × $10^{-12}$ s |
|  |  |  |  | 1 s = 1 × $10^{12}$ ps |

**TABLE 1.7** Daily Values for Selected Nutrients

| Nutrient | Amount Recommended |
|----------|--------------------|
| Vitamin C | 60 mg |
| Vitamin B$_{12}$ | 6 $\mu$g |
| Calcium | 1000 mg |
| Iron | 18 mg |
| Iodine | 150 $\mu$g |
| Magnesium | 400 mg |
| Potassium | 3500 mg |
| Sodium | 2400 mg |
| Zinc | 15 mg |

## SAMPLE PROBLEM    1.7

### ■ Prefixes

Fill in the blanks with the correct prefix.

**a.** 1000 grams = 1 _____ gram
**b.** 0.01 meter = 1 _____ meter
**c.** $1 \times 10^6$ liter = 1 _____ liter

### SOLUTION

**a.** The prefix for 1000 is *kilo;* 1000 grams = 1 kilogram
**b.** The prefix for 0.01 is *centi;* 0.01 meter = 1 centimeter
**c.** The prefix for $1 \times 10^6$ is *mega;* $1 \times 10^6$ liters = 1 megaliter

### STUDY CHECK

Write the correct prefix in the blanks.

**a.** 1 000 000 000 seconds = 1_____ second
**b.** 0.01 meter = 1_____ meter

---

| First quantity | | Second quantity | |
|---|---|---|---|
| 1 | m | = 100 | cm |
| ↑ | ↑ | ↑ | ↑ |
| Number + unit | | Number + unit | |

**TABLE 1.8** Some Typical Laboratory Test Values

| Substance in Blood | Typical Range |
|--------------------|---------------|
| Albumin | 3.5–5.0 g/dL |
| Ammonia | 20–150 $\mu$g/dL |
| Calcium | 8.5–10.5 mg/dL |
| Cholesterol | 105–250 mg/dL |
| Iron (male) | 80–160 $\mu$g/dL |
| Protein (total) | 6.0–8.0 g/dL |

## Measuring Length

An ophthalmologist may measure the diameter of the retina of an eye in centimeters (cm), whereas a surgeon may need to know the length of a nerve in millimeters (mm). When the prefix *centi* is used with the unit *meter,* it indicates the unit *centimeter,* a length that is one-hundredth of a meter (0.01 m). A *millimeter* measures a length of 0.001 m. There are 1000 mm in a meter.

If we compare the lengths of a millimeter and a centimeter, we find that 1 mm is 0.1 cm; there are 10 mm in 1 cm. These comparisons are examples of **equalities**, which show the relationship between two units that measure the same quantity. For example, in the equality 1 m = 100 cm, each quantity describes the same length but in a different unit. Note that each quantity in the equality expression has both a number and a unit.

### Some Length Equalities

$$1 \text{ m} = 100 \text{ cm} = 1 \times 10^2 \text{ cm}$$
$$1 \text{ m} = 1000 \text{ mm} = 1 \times 10^3 \text{ mm}$$
$$1 \text{ cm} = 10 \text{ mm}$$

Some metric units for length are compared in Figure 1.7.

## Measuring Volume

Volumes of 1 L or smaller are common in the health sciences. When a liter is divided into 10 equal portions, each portion is a deciliter (dL). There are 10 dL in 1 L. Laboratory results for blood work are often reported in mass per deciliter. Table 1.8 lists typical laboratory tests for some substances in the blood.

When a liter is divided into a thousand parts, each of the smaller volumes is called a milliliter. In a 1-L container of physiological saline, there are 1000 mL of solution (see Figure 1.8).

### Some Volume Equalities

$$1 \text{ L} = 10 \text{ dL}$$
$$1 \text{ L} = 1000 \text{ mL} = 1 \times 10^3 \text{ mL}$$
$$1 \text{ dL} = 100 \text{ mL} = 1 \times 10^2 \text{ mL}$$

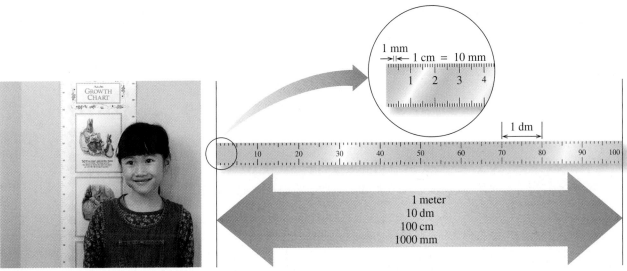

**FIGURE 1.7** The metric length of 1 meter is the same length as 10 dm, 100 cm, or 1000 mm.
**Q** How many millimeters (mm) are in 1 centimeter (cm)?

The **cubic centimeter** (**cm³, cc**) is the volume of a cube whose dimensions are 1 cm on each side. A cubic centimeter has the same volume as a milliliter, and the units are often used interchangeably.

$$1 \text{ cm}^3 = 1 \text{ cc} = 1 \text{ mL}$$

When you see *1 cm*, you are reading about length; when you see *1 cc* or *1 cm³* or *1 mL*, you are reading about volume. A comparison of units of volume is illustrated in Figure 1.9.

## Measuring Mass

When you get a physical examination, your mass is recorded in kilograms, whereas the results of your laboratory tests are reported in grams, milligrams (mg), or micrograms ($\mu$g).

**FIGURE 1.8** A plastic intravenous fluid container contains 1000 mL.
**Q** How many liters of solution are in the intravenous fluid container?

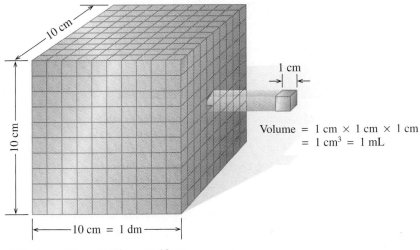

Volume = 1 cm × 1 cm × 1 cm
= 1 cm³ = 1 mL

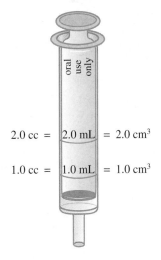

2.0 cc = 2.0 mL = 2.0 cm³

1.0 cc = 1.0 mL = 1.0 cm³

Volume = 10 cm × 10 cm × 10 cm
= 1000 cm³
= 1000 mL
= 1 L

**FIGURE 1.9** A cube measuring 10 cm on each side has a volume of 1000 cm³, or 1 L; a cube measuring 1 cm on each side has a volume of 1 cm³ (cc) or 1 mL.
**Q** What is the relationship between a milliliter (mL) and a cubic centimeter (cm³)?

A kilogram is equal to 1000 g. One gram represents the same mass as 1000 mg, and one mg equals 1000 $\mu$g.

**Some Mass Equalities**

$$1 \text{ kg} = 1000 \text{ g} = 1 \times 10^3 \text{ g}$$
$$1 \text{ g} = 1000 \text{ mg} = 1 \times 10^3 \text{ mg}$$
$$1 \text{ mg} = 1000 \mu\text{g} = 1 \times 10^3 \mu\text{g}$$

---

### SAMPLE PROBLEM 1.8

#### ■ Writing Metric Relationships

Complete the following list of metric equalities:

**a.** 1 L = _____ dL      **b.** 1 km = _____ m

**c.** 1 m = _____ cm      **d.** 1 cm$^3$ = _____ mL

**SOLUTION**

**a.** 10 dL      **b.** $1 \times 10^3$ m

**c.** 100 cm      **d.** 1 mL

**STUDY CHECK**

Complete the following equalities:

**a.** 1 kg = _____ g      **b.** 1 mL = _____ L

---

## QUESTIONS AND PROBLEMS

### Prefixes and Equalities

**1.27** The speedometer at right is marked in both km/h and mi/h, or MPH. What is the meaning of each abbreviation?

**1.28** In Canada, a highway sign gives a speed limit as 80 km/h. If you were going the maximum speed allowed in Canada, would you be exceeding the speed limit in the United States of 55 mph?

**1.29** How does the prefix *kilo* affect the gram unit in *kilogram*?

**1.30** How does the prefix *centi* affect the meter unit in *centimeter*?

**1.31** Write the abbreviation for each of the following units:
a. milligram      b. deciliter
c. kilometer      d. kilogram
e. microliter      f. nanogram

**1.32** Write the complete name for each of the following units:
a. cm      b. kg
c. dL      d. Gm
e. $\mu$g      f. mL

**1.33** Write the numerical values for each of the following prefixes:
a. centi      b. kilo
c. milli      d. deci
e. mega      f. pico

**1.34** Write the complete name (prefix + unit) for each of the following numerical values:
a. 0.10 g      b. $1 \times 10^{-6}$ g
c. 1000 g      d. 0.01 g
e. 0.001 g      f. $1 \times 10^9$ g

**1.35** Complete the following metric relationships:
a. 1 m = _____ cm      b. 1 km = _____ m
c. 1 mm = _____ m      d. 1 L = _____ mL

**1.36** Complete the following metric relationships:
a. 1 kg = _____ g      b. 1 mL = _____ L
c. 1 g = _____ kg      d. 1 g = _____ mg

**1.37** For each of the following pairs, which is the larger unit?
a. milligram or kilogram      b. milliliter or microliter
c. cm or km      d. kL or dL
e. nanometer or picometer

**1.38** For each of the following pairs, which is the smaller unit?
a. mg or g      b. centimeter or millimeter
c. mm or $\mu$m      d. mL or dL
e. mg or Mg

# 1.6 WRITING CONVERSION FACTORS

Many problems in chemistry and the health sciences require a change of units. You make changes in units every day. For example, suppose you spent 2.0 hours (h) on your homework, and someone asked you how many minutes that was. You would answer 120 minutes (min). You must have multiplied 2.0 h × 60 min/h because you knew an equality (1 h = 60 min) that related the two units. When you expressed 2.0 h as 120 min, you did not change the amount of time you spent studying. You changed only the unit of measurement used to express the time. Every equality can be written in the form of a fraction called a **conversion factor**, in which one of the quantities is the numerator and the other is the denominator. Two conversion factors are always possible from an equality because a factor can be inverted. Be sure to include the units when you write the conversion factors.

### Two Conversion Factors for the Equality 1 h = 60 min

$$\frac{\text{Numerator}}{\text{Denominator}} \longrightarrow \frac{60 \text{ min}}{1 \text{ h}} \quad \text{and} \quad \frac{1 \text{ h}}{60 \text{ min}}$$

These factors are read as "60 minutes per 1 hour," and "1 hour per 60 minutes." The term *per* means "divide." Some common relationships are given in Table 1.9. It is important that the equality you select to construct a conversion factor is a true relationship.

When an equality shows the relationship for two units from the same system (metric or U.S.), it is considered a definition and exact. It is not used to determine significant figures. When an equality shows the relationship of units from two different systems, the number is measured and counts toward the significant figures in a calculation. For example, in the equality 1 lb = 454 g, the measured number 454 has three significant figures. The number one in 1 lb is considered as exact. An exception is the relationship of 1 in. = 2.54 cm: the value 2.54 has been defined as exact.

## Metric Conversion Factors

We can write metric conversion factors for the metric relationships we have studied. For example, from the equality for meters and centimeters, we can write the following factors:

| Metric Equality | Conversion Factors |
|---|---|
| 1 m = 100 cm | $\frac{100 \text{ cm}}{1 \text{ m}}$ and $\frac{1 \text{ m}}{100 \text{ cm}}$ |

LEARNING GOAL

Write a conversion factor for two units that describe the same quantity.

**TABLE 1.9** Some Common Equalities

| Quantity | U.S. | Metric (SI) | Metric–U.S. |
|---|---|---|---|
| **Length** | 1 ft = 12 in.<br>1 yard = 3 ft<br>1 mile = 5280 ft | 1 km = 1000 m<br>1 m = 1000 mm<br>1 cm = 10 mm | 2.54 cm = 1 in.<br>1 m = 39.4 in.<br>1 km = 0.621 mi |
| **Volume** | 1 qt = 4 cups<br>1 qt = 2 pt<br>1 gallon = 4 qt | 1 L = 1000 mL<br>1 dL = 100 mL<br>1 mL = 1 cm$^3$ | 946 mL = 1 qt<br>1 L = 1.06 qt |
| **Mass** | 1 lb = 16 oz | 1 kg = 1000 g<br>1 g = 1000 mg | 1 kg = 2.20 lb<br>454 g = 1 lb |
| **Time** | | 1 h = 60 min<br>1 min = 60 s | |

**FIGURE 1.10** In the U.S., the contents of many packaged foods are listed in both U.S. and metric units.

**Q** What are some advantages of using the metric system?

Both are proper conversion factors for the relationship; one is just the inverse of the other. The usefulness of conversion factors is enhanced by the fact that we can turn a conversion factor over and use its inverse.

### Metric–U.S. System Conversion Factors

Suppose you need to convert from pounds, a unit in the U.S. system, to kilograms in the metric (or SI) system. A relationship you could use is

$$1 \text{ kg} = 2.20 \text{ lb}$$

The corresponding conversion factors would be

$$\frac{2.20 \text{ lb}}{1 \text{ kg}} \quad \text{and} \quad \frac{1 \text{ kg}}{2.20 \text{ lb}}$$

Figure 1.10 illustrates the contents of some packaged foods in both metric and U.S. units.

---

**SAMPLE PROBLEM 1.9**

### ■ Writing Conversion Factors from Equalities

Write conversion factors for the relationship for the following pairs of units:

**a.** milligrams and grams
**b.** minutes and hours
**c.** quarts and milliliters

SOLUTION

| Equality | Conversion Factors | | |
|---|---|---|---|
| **a.** $1 \text{ g} = 1000 \text{ mg}$ | $\dfrac{1 \text{ g}}{1000 \text{ mg}}$ | and | $\dfrac{1000 \text{ mg}}{1 \text{ g}}$ |
| **b.** $1 \text{ h} = 60 \text{ min}$ | $\dfrac{1 \text{ h}}{60 \text{ min}}$ | and | $\dfrac{60 \text{ min}}{1 \text{ h}}$ |
| **c.** $1 \text{ qt} = 946 \text{ mL}$ | $\dfrac{1 \text{ qt}}{946 \text{ mL}}$ | and | $\dfrac{946 \text{ mL}}{1 \text{ qt}}$ |

STUDY CHECK

A zeptosecond (zs) is a very small quantity of time. As an equality it is written

$$1 \text{ zs} = 1 \times 10^{-21} \text{ s}$$

Write the conversion factors for this equality.

## Conversion Factors Stated Within a Problem

Many times, an equality is specified within a problem that is true only for that problem. It might be the cost of one kilogram of oranges or the speed of a car in kilometers per hour. Such equalities are easy to miss when you first read a problem. Let's see how conversion factors are written from statements made within a problem.

**1.** The motorcycle was traveling at a speed of 85 kilometers per hour.

   **Equality:**     $1 \text{ h} = 85 \text{ km}$

   **Conversion factors:**     $\dfrac{85 \text{ km}}{1 \text{ h}}$   and   $\dfrac{1 \text{ h}}{85 \text{ km}}$

**2.** One tablet contains 500 mg of vitamin C.

   **Equality:**     $1 \text{ tablet} = 500 \text{ mg vitamin C}$

   **Conversion factors:**     $\dfrac{500 \text{ mg vitamin C}}{1 \text{ tablet}}$   and   $\dfrac{1 \text{ tablet}}{500 \text{ mg vitamin C}}$

## Conversion Factors for a Percentage, ppm, and ppb

Sometimes a percentage is given in a problem. The term percent (%) means parts per 100 parts. To write a percentage as a conversion factor, we choose a unit and express the numerical relationship of the parts of this unit to 100 parts of the whole. For example, an athlete might have 18% (percent) body fat by mass. The percent quantity can be written as 18 mass units of body fat in every 100 mass units of body mass. Different mass units such as grams (g), kilograms (kg), or pounds (lb) can be used, but both units in the factor must be the same.

   **Percent quantity:**     18% body fat by mass

   **Equality:**     $18 \text{ kg body fat} = 100 \text{ kg body mass}$

   **Conversion factors:**     $\dfrac{100 \text{ kg body mass}}{18 \text{ kg body fat}}$   and   $\dfrac{18 \text{ kg body fat}}{100 \text{ kg body mass}}$

or

   **Equality:**     $18 \text{ lb body fat} = 100 \text{ lb body mass}$

   **Conversion factors:**     $\dfrac{100 \text{ lb body mass}}{18 \text{ lb body fat}}$   and   $\dfrac{18 \text{ lb body fat}}{100 \text{ lb body mass}}$

When scientists want to indicate ratios with very small percent values, they use parts per million (ppm) or parts per billion (ppb). The ratio of parts per million indicates the milligrams of a substance per kilogram (mg/kg). The ratio of parts per billion gives the micrograms per kilogram ($\mu$g/kg). For example, the maximum amount of lead allowed by the FDA in pottery glaze on small bowls is 5 ppm, which is 5 mg/kg.

   **ppm quantity:**     5 ppm lead in glaze

   **Equality:**     $5 \text{ mg lead} = 1 \text{ kg glaze}$

   **Conversion factors:**     $\dfrac{5 \text{ mg lead}}{1 \text{ kg glaze}}$   and   $\dfrac{1 \text{ kg glaze}}{5 \text{ mg lead}}$

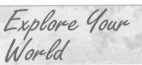

## SI and Metric Equalities on Product Labels

Read the labels on some food products on your kitchen shelves and in your refrigerator, or use the labels in Figure 1.10. List the amount of product given in different units. Write a relationship for two of the amounts for the same product and container. Look for measurements of grams and pounds or quarts and milliliters.

### QUESTIONS

**1.** Use the stated measurement to derive a metric–U.S. conversion factor.

**2.** How do your results compare with the conversion factors we have described in this text?

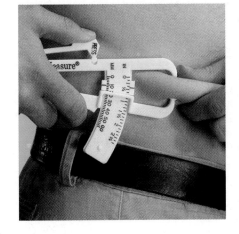

## SAMPLE PROBLEM 1.10

### ■ Conversion Factors Stated in a Problem

Write the possible conversion factors for each of the following statements:

**a.** There are 325 mg of aspirin in 1 tablet.

**b.** One kilogram of bananas costs $1.25.

**c.** The EPA has set the maximum level for mercury in tuna at 0.1 ppm.

### SOLUTION

**a.** $\dfrac{325 \text{ mg aspirin}}{1 \text{ tablet}}$    and    $\dfrac{1 \text{ tablet}}{325 \text{ mg aspirin}}$

**b.** $\dfrac{\$1.25}{1 \text{ kg bananas}}$    and    $\dfrac{1 \text{ kg bananas}}{\$1.25}$

**c.** $\dfrac{0.1 \text{ mg mercury}}{1 \text{ kg tuna}}$    and    $\dfrac{1 \text{ kg tuna}}{0.1 \text{ mg mercury}}$

### STUDY CHECK

What conversion factors can be written for the following statements?

**a.** A cyclist in the Tour de France bicycle race rides at the average speed of 62.2 km/h.

**b.** The permissible level of arsenic in water is 10 ppb.

# Green Chemistry Note

## Toxicology and Risk-Benefit Assessment

Each day we make choices about what we do or what we eat, often without thinking about the risks associated with these choices. We are aware of the risks of cancer from smoking or the risks of lead poisoning, and we know there is a greater risk of having an accident if we cross a street where there is no light or crosswalk.

A basic concept of toxicology is the statement of Paracelsus that the right dose is the difference between a poison and a cure. To evaluate the level of danger from various substances, natural or synthetic, a risk assessment is made by exposing laboratory animals to the substances and monitoring the health effects.

Often, doses much greater than humans might encounter are given to the test animals. Many hazardous chemicals or substances have been identified by these tests. One measure of toxicity is the $LD_{50}$, or "lethal dose," which is the concentration of the substance that causes death in 50% of the test animals. A dosage is typically measured in milligrams per kilogram (mg/kg) of body mass or micrograms per kilogram ($\mu$g/kg).

| Dosage | Units |
|---|---|
| parts per million (ppm) | milligrams per kilogram (mg/kg) |
| parts per billion (ppb) | micrograms per kilogram ($\mu$g/kg). |

There are other evaluations that need to be made, but it is easy to compare $LD_{50}$ values. Parathion, a pesticide, with an $LD_{50}$ of 3 mg/kg would be highly toxic. That means that half the test animals given 3 mg/kg body mass would be expected to die. But salt (sodium chloride) with an $LD_{50}$ of 3 000 mg/kg would have a much lower toxicity. By comparison with parathion, a huge amount of salt would need to be ingested before any toxic effect would be observed. Although risk to animals based on dose can be evaluated in the laboratory, it is more difficult to determine the impact in the environment since there is also a difference between continued exposure and a single, large dose of the substance.

Table 1.10 lists some $LD_{50}$ values comparing pesticides and common substances in our everyday life in order of increasing toxicity.

**TABLE 1.10** Some $LD_{50}$ Values for Pesticides and Common Materials Tested in Rats

| Substance | $LD_{50}$ (mg/kg) |
|---|---|
| Table sugar | 29 700 |
| Baking soda | 4 220 |
| Table salt | 3 000 |
| Ethanol | 2 080 |
| Aspirin | 1 100 |
| Caffeine | 192 |
| Sodium cyanide | 6 |
| Parathion | 3 |

## QUESTIONS AND PROBLEMS

### Writing Conversion Factors

**1.39** Why can two conversion factors be written for an equality such as 1 m = 100 cm?

**1.40** How can you check that you have written the correct conversion factors for an equality?

**1.41** Write the equality and two conversion factors for each of the following:
    **a.** centimeters and meters    **b.** milligrams and grams
    **c.** liters and milliliters    **d.** deciliters and milliliters
    **e.** days in 1 week

**1.42** Write the equality and two conversion factors for each of the following:
    **a.** centimeters and inches    **b.** pounds and kilograms
    **c.** pounds and grams    **d.** quarts and milliliters
    **e.** dimes in 1 dollar

**1.43** Write the conversion factors for each of the following statements:
    **a.** A bee flies at an average speed of 3.5 m per second.
    **b.** The daily requirement for potassium is 3500 mg.
    **c.** An automobile traveled 46.0 km on 1.0 gal of gasoline.
    **d.** The label on a bottle reads 50 mg atenolol per tablet.
    **e.** The pesticide level in plums was 29 ppb.

**1.44** Write the conversion factors for each of the following statements:
    **a.** The label on a bottle reads 10 mg furosemide per mL.
    **b.** The daily requirement for iodine is 150 $\mu$g.
    **c.** The nitrate level in well water was 32 ppm.
    **d.** Gold jewelry contains 58% by mass gold.
    **e.** The price of a gallon of gas is $3.19.

# 1.7 PROBLEM SOLVING

### LEARNING GOAL

Use conversion factors to change from one unit to another.

The process of problem solving in chemistry often requires the conversion of an initial quantity given in one unit to the same quantity but in different units. By using one or more of the conversion factors we discussed in the previous section, the initial unit can be converted to the final unit.

Given quantity × one or more conversion factors = desired quantity

(Initial unit) $\xrightarrow{\hspace{3cm}}$ (Final unit)

You may use a sequence similar to the steps in the following guide for problem solving (GPS):

### Guide to Problem Solving (GPS) Using Conversion Factors

**Step 1**    **Given/Need**   State the initial unit given in the problem and the final unit needed.

**Step 2**    **Plan**   Write out a sequence of units that starts with the initial unit and progresses to the final unit for the answer. Be sure you can supply the equality for each unit conversion.

**Step 3**    **Equalities/Conversion Factors**   For each change of unit in your plan, state the equality and corresponding conversion factors. Recall that equalities are derived from the metric (SI) system, the U.S. system, and statements within a problem.

**Step 4**    **Set Up Problem**   Write the initial quantity and unit, and set up conversion factors that connect the units. Be sure to arrange the units in each factor so the unit in the denominator cancels the preceding unit in the numerator. Check that the units cancel properly to give the final unit. Carry out the calculations, count the significant figures in each measured number, and give a final answer with the correct number of significant figures.

Suppose a problem requires the conversion of 164 lb to kilograms. One part of this statement (164 lb) is the given quantity (initial unit), while another part (kilograms) is the final unit needed for the answer. Once you identify these units, you can determine which equalities you need to convert the initial unit to the final unit.

**Guide to Problem Solving Using Conversion Factors**

1 State the given and needed units.

2 Write a unit plan to convert the given unit to the final unit.

3 State the equalities and conversion factors to cancel units.

4 Set up problem to cancel units, calculate, and report the answer with the proper number of significant figures.

**Step 1**    **Given**    164 lb    **Need**    kg

**Step 2**    **Plan**    It is helpful to decide on a plan of units. When we look at the initial units given and the final units needed, we see that one is a metric unit and the other is a unit in the U.S. system of measurement. Therefore, the connecting conversion factor must be one that includes a metric and a U.S. unit.

$$lb \quad \boxed{\text{Metric–U.S. factor}} \quad kg$$

**Step 3**    **Equalities/Conversion Factors**    From the discussion on U.S. and metric equalities, we can write the following equality and conversion factors:

$$1 \text{ kg} = 2.20 \text{ lb}$$

$$\frac{2.20 \text{ lb}}{1 \text{ kg}} \quad \text{and} \quad \frac{1 \text{ kg}}{2.20 \text{ lb}}$$

**Step 4**    **Set Up Problem**    Now we can write the setup to solve the problem using the unit plan and a conversion factor. First, write down the initial unit, 164 lb. Then multiply by the conversion factor that has the unit lb in the denominator to cancel out the initial unit. The unit kg in the numerator (top number) gives the final unit for the answer.

Unit for answer goes here

$$164 \, \cancel{lb} \quad \times \quad \frac{1 \text{ kg}}{2.20 \, \cancel{lb}} \quad = \quad 74.5 \text{ kg}$$

Given (initial unit)       Conversion factor (cancels initial unit)       Answer (desired unit)

Take a look at how the units cancel. The unit that you want in the answer is the one that remains after all the other units have cancelled out. This is a helpful way to check that a problem is set up properly.

$$\cancel{lb} \times \frac{kg}{\cancel{lb}} = kg \quad \text{Unit needed for answer}$$

The calculation done on a calculator gives the numerical part of the answer. The calculator answer is adjusted to give a final answer with the proper number of significant figures (SFs).

$$164 \times \frac{1}{2.20} = 164 \,\boxed{\div}\, 2.20 = \boxed{74.54545455} = 74.5$$

3 SFs    3 SFs                          Calculator answer       3 SFs (rounded)

The value of 74.5 combined with the final unit, kg, gives the final answer of 74.5 kg. With few exceptions, answers to numerical problems contain a number and a unit.

## SAMPLE PROBLEM 1.11

### ■ Problem Solving Using Metric Factors

The daily recommended amount of potassium in the diet is 3500 mg. How many grams of potassium are needed each day?

**SOLUTION**

**Step 1**    **Given**    3500 mg    **Need**    g

**Step 2**    **Plan**    When we look at the initial units given and the final units needed, we see that both are metric units. Therefore, the connecting conversion factor must relate two metric units.

mg    Metric factor    g

**Step 3  Equalities/Conversion Factors**  From the discussion on prefixes and metric equalities, we can write the following equality and conversion factors:

$$1 \text{ g} = 1000 \text{ mg}$$

$$\frac{1 \text{ g}}{1000 \text{ mg}} \quad \text{and} \quad \frac{1000 \text{ mg}}{1 \text{ g}}$$

**Step 4  Set Up Problem**  We write the setup using the unit plan and a conversion factor starting with the initial unit, 3500 mg. The final answer (g) is obtained by using the conversion factor that cancels the unit mg. Round off the answer to give the proper number of significant figures.

Unit for answer goes here

$$3500 \text{ m\cancel{g}} \quad \times \quad \frac{1 \text{ g}}{1000 \text{ m\cancel{g}}} \quad = \quad 3.5 \text{ g}$$

| Given | Metric factor | Answer |
|---|---|---|
| (initial unit) | (cancels initial unit) | (desired unit) |

**STUDY CHECK**

If 1890 mL of orange juice is prepared from orange juice concentrate, how many liters of orange juice is that?

## Using Two or More Conversion Factors

In many problems, two or more conversion factors are needed to complete the change of units. In setting up these problems, one factor follows the other. Each factor is arranged to cancel the preceding unit until the final unit is obtained. Up to this point, we have used the conversion factors one at a time and calculated an answer. You can work all problems in single steps, but if you do, be sure to keep one or two extra digits in the intermediate answers and only round off the final answer to the correct number of significant figures. A more efficient way to do these problems is to use a series of two or more conversion factors set up so that the unit in the denominator of each factor cancels the unit in the preceding numerator. Both of these approaches are illustrated in the following Sample Problem.

SAMPLE PROBLEM  **1.12**

■ **Problem Solving Using Two Factors**

During a volcanic eruption on Mauna Loa, Hawaii, the lava flowed at a rate of 33 meters per minute. At this rate, how far in kilometers can the lava travel in 45 minutes?

**SOLUTION**

**Step 1  Given**    45 minutes        **Need**    kilometers

**Step 2  Plan**    min    Rate factor    m    Metric factor    km

**Step 3  Equalities/Conversion Factors**  In the problem, the information for the rate of lava flow is given as 33 m/min. We will use this rate as one of the equalities as well as the metric equality for meters and kilometers and write conversion factors for each.

$$1 \text{ min} = 33 \text{ m} \qquad\qquad 1 \text{ km} = 1000 \text{ m}$$

$$\frac{1 \text{ min}}{33 \text{ m}} \quad \text{and} \quad \frac{33 \text{ m}}{1 \text{ min}} \qquad \frac{1 \text{ km}}{1000 \text{ m}} \quad \text{and} \quad \frac{1000 \text{ m}}{1 \text{ km}}$$

**Step 4    Set Up Problem**    The problem can be set up using the rate as a conversion factor to cancel minutes, and then the metric factor to obtain kilometers in the final factor. Working in single steps we can use the rate factor to convert from minutes to meters.

$$45 \, \cancel{min} \times \frac{33 \text{ m}}{1 \, \cancel{min}} = 1500 \text{ m}$$

Then we use the metric factor to cancel meters and give kilometers as the needed unit.

$$1500 \, \cancel{m} \times \frac{1 \text{ km}}{1000 \, \cancel{m}} = 1.5 \text{ km}$$

When set up as a series, the first factor cancels minutes, and the second factor cancels meters, which gives kilometers as the final unit for the answer.

$$\cancel{min} \times \frac{\cancel{m}}{\cancel{min}} \times \frac{\text{km}}{\cancel{m}} = \text{km}$$

$$45 \, \cancel{min} \times \frac{33 \, \cancel{m}}{1 \, \cancel{min}} \times \frac{1 \text{ km}}{1000 \, \cancel{m}} = 1.485 \text{ km} = 1.5 \text{ km}$$

| Given (initial unit) | Rate factor | Metric factor | Calculator answer | Answer (desired unit) |
|---|---|---|---|---|
| 2 SFs | 2 SFs | Exact | | 2 SFs |

The calculations are done in a sequence on a calculator to give the numerical part of the answer. The calculator answer is adjusted to give a final answer with the proper number of significant figures (SF).

| 45 | $\boxed{\times}$ | 33 | $\boxed{\div}$ | 1000 | = | 1.485 | = | 1.5 |
|---|---|---|---|---|---|---|---|---|
| 2 SFs | | 2 SFs | | Exact | | Calculator answer | | 2 SFs (rounded) |

**STUDY CHECK**

One medium bran muffin contains 4.2 g of fiber. How many ounces (oz) of fiber are obtained by eating three medium bran muffins if 1 lb = 16 oz? (*Hint:* number of muffins ⟶ g of fiber ⟶ lb ⟶ oz.)

Using a sequence of two or more conversion factors is a very efficient way to set up and solve problems, especially if you are using a calculator. Once you have the problem set up, the calculations can be done without writing out the intermediate values. This process is worth practicing until you understand unit cancellation and the mathematical calculations.

## Clinical Calculations Using Conversion Factors

Conversion factors are also useful for calculating medications. For example, if an antibiotic is available in 5-mg tablets, the dosage can be written as a conversion factor, 5 mg/1 tablet. In many hospitals, the apothecary unit of grains (gr) is still in use; there are 65 mg in 1 gr. When you do a medication problem, you often start with a doctor's order that contains the quantity to give the patient. The medication dosage is used as a conversion factor.

## Veterinary Technician (VT)

"I am checking this dog's ears for foxtails and her eyes for signs of conjunctivitis," says Joyce Rhodes, veterinary assistant at the Sonoma Animal Hospital. "We always check a dog's teeth for tartar, because dental care is very important to the well-being of the animal. When I do need to give a medication to an animal, I use my chemistry to prepare the proper dose that the pet should take. Dosages may be in milligrams, kilograms, or milliliters."

As a member of the veterinary healthcare team, a veterinary technician (VT) assists a veterinarian in the care and handling of animals. A VT takes medical histories, collects specimens, performs laboratory procedures, prepares an animal for surgery, assists in surgical procedures, takes X-rays, talks with animal owners, and cleans teeth.

**SAMPLE PROBLEM   1.13**

■ **Clinical Calculations**

Synthroid is used as a replacement or supplemental therapy for diminished thyroid function. A dosage of 0.200 mg is prescribed with tablets that contain 50 μg of Synthroid. How many tablets are required to provide the prescribed medication?

## SOLUTION

**Step 1** **Given**   0.200 mg of Synthroid   **Need**   tablets

**Step 2** **Plan**   mg   Metric factor   $\mu$g   Clinical factor   tablets

**Step 3** **Equalities/Conversion Factors**   In the problem, the information for the dosage is given as 50 $\mu$g per tablet. We will use this as one of the equalities as well as the metric equality for milligrams and micrograms and write conversion factors for each.

$$1 \text{ mg} = 1000 \ \mu\text{g} \qquad\qquad 1 \text{ tablet} = 50 \ \mu\text{g}$$

$$\frac{1 \text{ mg}}{1000 \ \mu\text{g}} \quad \text{and} \quad \frac{1000 \ \mu\text{g}}{1 \text{ mg}} \qquad\qquad \frac{1 \text{ tablet}}{50 \ \mu\text{g}} \quad \text{and} \quad \frac{50 \ \mu\text{g}}{1 \text{ tablet}}$$

**Step 4** **Set Up Problem**   The problem can be set up using the metric factor to cancel "milligrams," and then the clinical factor to obtain "tablets" as the final unit.

$$0.200 \text{ mg} \times \frac{1000 \ \mu\text{g}}{1 \text{ mg}} \times \frac{1 \text{ tablet}}{50 \ \mu\text{g}} = 4 \text{ tablets}$$

### STUDY CHECK

An antibiotic dosage of 500 mg is ordered. If the antibiotic is supplied in liquid form as 250 mg in 5.0 mL, how many mL would be given?

## QUESTIONS AND PROBLEMS

### Problem Solving

**1.45** When you convert one unit to another, how do you know which unit to place in the denominator?

**1.46** When you convert one unit to another, how do you know which unit to place in the numerator?

**1.47** Use metric conversion factors to solve the following problems:
   **a.** The height of a student is 175 cm. How tall is the student in meters?
   **b.** A cooler has a volume of 5500 mL. What is the capacity of the cooler in liters?
   **c.** A hummingbird has a mass of 0.0055 kg. What is the mass of the hummingbird in grams?

**1.48** Use metric conversion factors to solve the following problems:
   **a.** The daily value of phosphorus is 800 mg. How many grams of phosphorus are recommended?
   **b.** A glass of orange juice contains 0.85 dL of juice. How many milliliters of orange juice is that?
   **c.** A package of chocolate instant pudding contains 2840 mg of sodium. How many grams of sodium is that?

**1.49** Solve the following problems using one or more conversion factors:
   **a.** A container holds 0.500 qt of liquid. How many milliliters of lemonade will it hold?
   **b.** What is the mass in kilograms of a person who weighs 145 lb?
   **c.** An athlete normally has 15% (by mass) body fat. How many lb of fat does a 74-kg athlete have?

**1.50** Solve the following problems using one or more conversion factors:
   **a.** You need 4.0 ounces of a steroid ointment. If there are 16 oz in 1 lb, how many grams of ointment does the pharmacist need to prepare?
   **b.** During surgery, a patient receives 5.0 pints of plasma. How many milliliters of plasma were given? (1 quart = 2 pints)
   **c.** Wine is 12% (by volume) alcohol. How many mL of alcohol are in a 0.750 L bottle of wine?

**1.51** Using conversion factors, solve the following clinical problems:
   **a.** You have used 250 L of distilled water for a dialysis patient. How many gallons of water is that?
   **b.** A patient needs 0.024 g of a sulfa drug. There are 8-mg tablets in stock. How many tablets should be given?
   **c.** The daily dose of ampicillin for the treatment of an ear infection is 115 mg/kg of body weight. What is the daily dose for a 34-lb child?

**1.52** Using conversion factors, solve the following clinical problems:
   **a.** The physician has ordered 1.0 g of tetracycline to be given every 6 hours to a patient. If your stock on hand is 500-mg tablets, how many will you need for 1 day's treatment?
   **b.** An intramuscular medication is given at 5.00 mg/kg of body weight. If you give 425 mg of medication to a patient, what is the patient's weight in pounds?
   **c.** A physician has ordered 0.50 mg of atropine, intramuscularly. If atropine were available as 0.10 mg/mL of solution, how many milliliters would you need to give?

**FIGURE 1.11** Objects that sink in water are more dense than water; objects float if they are less dense.
**Q** Why does a cork float and a piece of lead sink?

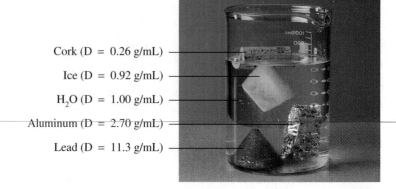

Cork (D = 0.26 g/mL)

Ice (D = 0.92 g/mL)

$H_2O$ (D = 1.00 g/mL)

Aluminum (D = 2.70 g/mL)

Lead (D = 11.3 g/mL)

## LEARNING GOAL

Calculate the density or specific gravity of a substance, and use the density or specific gravity to calculate the mass or volume of a substance.

# $1.8$ DENSITY

Differences in density determine whether an object will sink or float. In Figure 1.11, the density of lead is greater than the density of water, and the lead object sinks. The cork floats because cork is less dense than water.

The mass and volume of any object can be measured. However, the separate measurements do not tell us how tightly packed the substance might be. If we compare the mass of the object to its volume, we obtain a relationship called **density**.

$$\text{Density} = \frac{\text{mass of substance}}{\text{volume of substance}}$$

In the metric system, the densities of solids and liquids are usually expressed as grams per cubic centimeter ($g/cm^3$) or grams per milliliter (g/mL). The density of gases is usually stated as grams per liter (g/L). Table 1.11 gives the densities of some common substances.

## Density of Solids

The density of a solid is calculated from its mass and volume. When a solid is completely submerged, it displaces a volume of water that is *equal to its own volume*. In Figure 1.12, the water level rises from 35.5 mL to 45.0 mL. This means that 9.5 mL of water is displaced and that the volume of the object is 9.5 mL. The density of the zinc is calculated as follows:

$$\text{Density} = \frac{68.60 \text{ g zinc}}{9.5 \text{ mL}} = 7.2 \text{ g/mL}$$

**TABLE 1.11** Densities of Some Common Substances

| Solids (at 25 °C) | Density (g/mL) | Liquids (at 25 °C) | Density (g/mL) | Gases (at 0 °C) | Density (g/L) |
|---|---|---|---|---|---|
| Cork | 0.26 | Gasoline | 0.66 | Hydrogen | 0.090 |
| Wood (maple) | 0.75 | Ethanol | 0.79 | Helium | 0.179 |
| Ice (at 0 °C) | 0.92 | Olive oil | 0.92 | Methane | 0.714 |
| Sugar | 1.59 | Water (at 4 °C) | 1.00 | Neon | 0.90 |
| Bone | 1.80 | Plasma (blood) | 1.03 | Nitrogen | 1.25 |
| Aluminum | 2.70 | Urine | 1.003–1.030 | Air (dry) | 1.29 |
| Cement | 3.00 | Milk | 1.04 | Oxygen | 1.43 |
| Diamond | 3.52 | Mercury | 13.6 | Carbon dioxide | 1.96 |
| Silver | 10.5 | | | | |
| Lead | 11.3 | | | | |
| Gold | 19.3 | | | | |

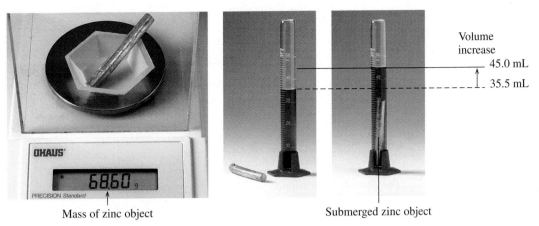

Mass of zinc object                    Submerged zinc object

Volume increase
45.0 mL
35.5 mL

**FIGURE 1.12** The density of a solid can be determined by volume displacement because a submerged object displaces a volume of water equal to its own volume.
**Q** How is the volume of zinc determined?

---

**SAMPLE PROBLEM 1.14**

■ **Calculating Density**

A copper sample has a mass of 44.65 g and a volume of 5.0 mL. What is the density of copper?

SOLUTION

Step 1 **Given** mass = 44.65 g; volume = 5.0 mL
**Need** density (g/mL)

Step 2 **Plan** To calculate density, substitute the mass (g) and the volume (mL) of the copper sample into the expression for density.

Step 3 **Equality/Conversion Factor**

$$\text{Density} = \frac{\text{mass of substance}}{\text{volume of substance}}$$

Step 4 **Set Up Problem**

$$\text{Density} = \frac{\overset{4 \text{ SFs}}{44.65 \text{ g}}}{\underset{2 \text{ SFs}}{5.0 \text{ mL}}} = \frac{\overset{2 \text{ SFs}}{8.9 \text{ g}}}{1 \text{ mL}} = 8.9 \text{ g/mL}$$

STUDY CHECK

What is the density (g/cm³) of a silver bar that has a mass of 294 g and a volume of 28.0 cm³?

---

**SAMPLE PROBLEM 1.15**

■ **Using Volume Displacement to Calculate Density**

A lead weight used in the belt of a scuba diver has a mass of 226 g. When the lead weight is carefully placed in a graduated cylinder containing 200.0 mL of water, the water level rises to 220.0 mL. What is the density of the lead weight (g/mL)?

SOLUTION

Step 1 **Given** mass = 226 g; water level before object submerged = 200.0 mL; water level after object submerged = 220.0 mL
**Need** density (g/mL)

Step 2 **Plan** To calculate density, substitute the mass (g) and the volume (mL) of the lead weight into the expression for density.

Step 3  **Equality/Conversion Factor**

$$\text{Density} = \frac{\text{mass of substance}}{\text{volume of substance}}$$

Step 4  **Set Up Problem**    The volume of the lead weight is equal to the volume of water displaced, which is calculated as follows:

| | |
|---|---|
| Water level after object submerged | = 220.0 mL |
| − Water level before object submerged | = 200.0 mL |
| Water displaced (volume of lead weight) | = 20.0 mL |

The density is calculated by dividing the mass (g) by the volume (mL). Be sure to use the volume of water the object displaced and *not* the original volume of water.

$$\text{Density} = \frac{226\ g}{20.0\ mL} = \frac{11.3\ g}{1\ mL} = 11.3\ g/mL$$
$$\quad\quad\quad\quad\ \ 3\ \text{SFs} \quad\quad\quad 3\ \text{SFs}$$

**STUDY CHECK**

A total of 0.50 lb of glass marbles is added to 425 mL of water. The water level rises to a volume of 528 mL. What is the density (g/mL) of the glass marbles?

## Problem Solving Using Density

Density can be used as a conversion factor. For example, if the volume and the density of a sample are known, the mass in grams of the sample can be calculated.

---

**SAMPLE PROBLEM  1.16**

■ **Problem Solving Using Density**

If the density of milk is 1.04 g/mL, how many grams of milk are in 0.50 qt of milk?

**SOLUTION**

Step 1  **Given**   0.50 qt      **Need**   g

Step 2  **Plan**

qt  →  U.S.–metric factor  →  L  →  Metric factor  →  mL  →  Density factor  →  g

Step 3  **Equalities/Conversion Factors**

| 1 L = 1.06 qt | 1 L = 1000 mL | 1 mL = 1.04 g |
|---|---|---|
| $\dfrac{1\ L}{1.06\ qt}$ and $\dfrac{1.06\ qt}{1\ L}$ | $\dfrac{1\ L}{1000\ mL}$ and $\dfrac{1000\ mL}{1\ L}$ | $\dfrac{1\ mL}{1.04\ g}$ and $\dfrac{1.04\ g}{1\ mL}$ |

Step 4  **Set Up Problem**

$$0.50\ \cancel{qt} \times \frac{1\ \cancel{L}}{1.06\ \cancel{qt}} \times \frac{1000\ \cancel{mL}}{1\ \cancel{L}} \times \frac{1.04\ g}{1\ \cancel{mL}} = 490\ g\ (4.9 \times 10^2\ g)$$
$$\ \ \ 2\ \text{SFs} \quad\quad 3\ \text{SFs} \quad\quad\ \text{Exact} \quad\quad\ 3\ \text{SFs} \quad 2\ \text{SFs}$$

**STUDY CHECK**

How many mL of mercury are in a thermometer that contains 20.4 g of mercury? (See Table 1.11 for the density of mercury.)

## Specific Gravity

**Specific gravity (sp gr)** is a ratio between the density of a substance and the density of water. Specific gravity is calculated by dividing the density of a sample by the density of water, which is 1.00 g/mL at 4 °C. A substance with a specific gravity of 1.00 has the same density as water. A substance with a specific gravity of 3.00 is three times as dense as water, whereas a substance with a specific gravity of 0.50 is just one-half as dense as water.

$$\text{Specific gravity} = \frac{\text{density of sample}}{\text{density of water}}$$

In the calculations for specific gravity, the units of density must match. Then all units cancel to leave only a number. *Specific gravity is one of the few unitless values you will encounter in chemistry.*

An instrument called a hydrometer is often used to measure the specific gravity of fluids such as battery fluid or a sample of urine. In Figure 1.13, a hydrometer is used to measure the specific gravity of a fluid.

### Explore Your World

#### Sink or Float?

1. Fill a large container or bucket with water. Place a can of diet soft drink and a can of nondiet soft drink in the water. What happens? Using information on the label, how might you account for your observations?
2. Design an experiment to determine the substance that is the more dense in each of the following:
   a. water and vegetable oil
   b. water and ice
   c. rubbing alcohol and ice
   d. vegetable oil, water, and ice

---

### SAMPLE PROBLEM   1.17

#### ■ Specific Gravity

What is the specific gravity of coconut oil that has a density of 0.925 g/mL?

**SOLUTION**

$$\text{sp gr oil} = \frac{\text{density of oil}}{\text{density of water}} = \frac{0.925 \ \cancel{\text{g/ml}}}{1.00 \ \cancel{\text{g/ml}}} = 0.925 \qquad \text{No units}$$

**STUDY CHECK**

What is the specific gravity of ice if 35.0 g of ice has a volume of 38.2 mL?

---

### SAMPLE PROBLEM   1.18

#### ■ Problem Solving with Specific Gravity

John took 2.0 teaspoons (tsp) of cough syrup (sp gr 1.20) for a persistent cough. If there is 5.0 mL in 1 tsp, what was the mass (in grams) of the cough syrup?

**SOLUTION**

**Step 1 Given** 2.0 tsp    **Need** grams

**Step 2 Plan**

tsp $\boxed{\text{U.S.–metric factor}}$ mL $\boxed{\text{Density factor}}$ g

**Step 3 Equalities/Conversion Factors** For problem solving, it is convenient to convert the specific gravity value (1.20) to density.

Density = (sp gr) × 1.00 g/mL = 1.20 g/mL

| 1 tsp = 5.0 mL | | 1 mL = 1.20 g | |
|---|---|---|---|
| $\dfrac{5.0 \ \text{mL}}{1 \ \text{tsp}}$ and $\dfrac{1 \ \text{tsp}}{5.0 \ \text{mL}}$ | | $\dfrac{1 \ \text{mL}}{1.20 \ \text{g}}$ and $\dfrac{1.20 \ \text{g}}{1 \ \text{mL}}$ | |

**Step 4 Set Up Problem**

$$2.0 \ \cancel{\text{tsp}} \times \frac{5.0 \ \cancel{\text{mL}}}{1 \ \cancel{\text{tsp}}} \times \frac{1.20 \ \text{g}}{1 \ \cancel{\text{mL}}} = 12 \ \text{g syrup}$$

**STUDY CHECK**

An ebony carving has a mass of 275 g. If ebony has a specific gravity of 1.33, what is the volume (mL) of the carving?

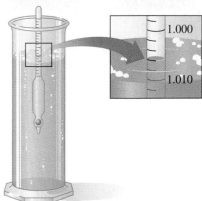

**FIGURE 1.13** When the specific gravity of beer measures 1.010 or less with a hydrometer, the fermentation process is complete.

**Q** If the hydrometer reading is 1.006, what is the density of the liquid?

## Health Note

### Determination of Percentage of Body Fat

Body mass is made up of protoplasm, extracellular fluid, bone, and adipose tissue. One way to determine the amount of adipose tissue is to measure the whole-body density. After the on-land mass of the body is determined, the underwater body mass is obtained by submerging the person in water. Because water helps support the body

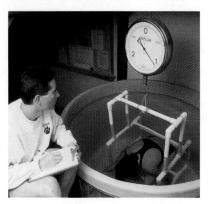

by giving it buoyancy, the underwater body mass is less. A higher percentage of body fat will make a person more buoyant, causing the underwater mass to be even lower. This occurs because fat has a lower density than the rest of the body.

The difference between the on-land mass and underwater mass, known as the buoyant force, is used to determine the body volume. Then the mass and volume of the person are used to calculate body density. For example, suppose a 70.0-kg person has a body volume of 66.7 L. The body density is calculated as

$$\frac{\text{Body mass}}{\text{Body volume}} = \frac{70.0 \text{ kg}}{66.7 \text{ L}}$$

$$= 1.05 \text{ kg/L} \quad \text{or}$$
$$1.05 \text{ g/mL}$$

When the body density is determined, it is compared with a chart that correlates the percentage of adipose tissue with body density. A person with a body density of 1.05 g/mL has 21% body fat, according to such a chart. This procedure is used by athletes to determine exercise and diet programs.

## QUESTIONS AND PROBLEMS

### Density

**1.53** What is the density (g/mL) of each of the following samples?
  **a.** A 20.0-mL sample of a salt solution that has a mass of 24.0 g.
  **b.** A solid object with a mass of 1.65 lb and a volume of 170 mL.
  **c.** A gem has a mass of 45.0 g. When the gem is placed in a graduated cylinder containing 20.0 mL of water, the water level rises to 34.5 mL.

**1.54** What is the density (g/mL) of each of the following samples?
  **a.** A medication, if 3.00 mL has a mass of 3.85 g.
  **b.** The fluid in a car battery, if it has a volume of 125 mL and a mass of 155 g.
  **c.** A 5.00-mL urine sample from a patient suffering from symptoms resembling those of diabetes mellitus. The mass of the urine sample is 5.025 g.

**1.55** Use the density value to solve the following problems:
  **a.** What is the mass, in grams, of 150 mL of a liquid with a density of 1.4 g/mL?
  **b.** What is the mass of a glucose solution that fills a 0.500-L intravenous bottle if the density of the glucose solution is 1.15 g/mL?
  **c.** A sculptor has prepared a mold for casting a bronze figure. The figure has a volume of 225 mL. If bronze has a density of 7.8 g/mL, how many ounces of bronze are needed in the preparation of the bronze figure?

**1.56** Use the density value to solve the following problems:
  **a.** A graduated cylinder contains 18.0 mL of water. What is the new water level after 35.6 g of silver metal with a density of 10.5 g/mL is submerged in the water?
  **b.** A thermometer containing 8.3 g of mercury has broken. If mercury has a density of 13.6 g/mL, what volume spilled?
  **c.** A fish tank holds 35 gal of water. Using the density of 1.0 g/mL for water, determine the number of pounds of water in the fish tank.

**1.57** Solve the following specific gravity problems:
  **a.** A urine sample has a density of 1.030 g/mL. What is the specific gravity of the sample?
  **b.** A liquid has a volume of 40.0 mL and a mass of 45.0 g. What is the specific gravity of the liquid?
  **c.** The specific gravity of a vegetable oil is 0.85. What is its density?

**1.58** Solve the following specific gravity problems:
  **a.** A 5.0% glucose solution has a specific gravity of 1.02. What is the mass of 500. mL of glucose solution?
  **b.** A bottle containing 325 g of cleaning solution is used for carpets. If the cleaning solution has a specific gravity of 0.850, what volume of solution was used?
  **c.** Butter has a specific gravity of 0.86. What is the mass, in grams, of 2.15 L of butter?

## Health Note

### Specific Gravity of Urine

The specific gravity of urine is often determined as part of a laboratory evaluation of the health of an individual. The specific gravity of urine is normally in the range 1.003–1.030. This is somewhat greater than the specific gravity of water because compounds such as salts and urea are dissolved in urine.

If the specific gravity of a person's urine is too low or too high, a physician might suspect a kidney problem. For example, if a urine sample shows a specific gravity of 1.001, significantly lower than normal, malfunctioning of the kidneys is a possibility. When water has been lost because of dehydration, the urine may have a specific gravity above normal.

## CONCEPT MAP

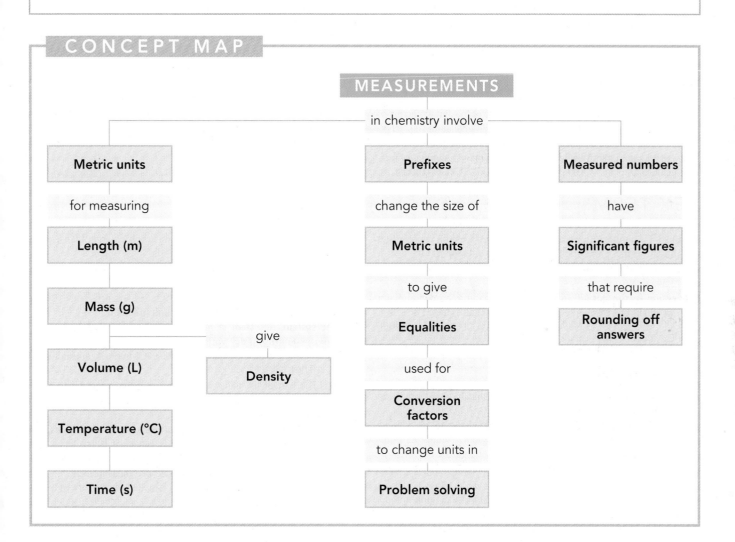

## CHAPTER REVIEW

### 1.1 Units of Measurement

**Learning Goal:** Write the names and abbreviations for the units used in measurements of length, volume, and mass.

In science, physical quantities are described in units of the metric or International System (SI). Some important units are meter (m) for length, liter (L) for volume, gram (g) and kilogram (kg) for mass, Celsius (°C) and kelvin (K) for temperature, and second (s) for time.

### 1.2 Scientific Notation

**Learning Goal:** Use scientific notation to express large and small numbers.

Large and small numbers can be written using scientific notation in which the decimal point is moved to give a coefficient from 1 to 9 and the number of decimal places moved shown as a power of 10. A large number will have a positive power of 10, while a small number will have a negative power of 10.

## 1.3 Measured Numbers and Significant Figures

**Learning Goal:** Determine the number of significant figures in measured numbers.

A measured number is any number obtained by using a measuring device. An exact number is obtained by counting items or from a definition; no measuring device is used. Significant figures are the numbers reported in a measurement including the last estimated digit. Zeros in front of a decimal number or at the end of a large number are not significant.

## 1.4 Significant Figures in Calculations

**Learning Goal:** Adjust calculated answers to the correct number of significant figures.

In multiplication or division, the final answer is written so it has the same number of significant figures as the measurement with the fewest significant figures. In addition or subtraction, the final answer is written so it has the same number of decimal places as the measurement with the fewest decimal places.

## 1.5 Prefixes and Equalities

**Learning Goal:** Use the numerical values of prefixes to write a metric equality.

Prefixes placed in front of a unit change the size of the unit by factors of 10. Prefixes such as *centi, milli,* and *micro* provide smaller units; prefixes such as *kilo* provide larger units. An equality relates two metric units that measure the same quantity of length, volume, or mass. Examples of metric equalities are 1 m = 100 cm; 1 L = 1000 mL; 1 kg = 1000 g.

## 1.6 Writing Conversion Factors

**Learning Goal:** Write a conversion factor for two units that describe the same quantity.

Conversion factors are used to express a relationship in the form of a fraction. Two factors can be written for any relationship in the metric or U.S. system.

## 1.7 Problem Solving

**Learning Goal:** Use conversion factors to change from one unit to another.

Conversion factors are useful when changing a quantity expressed in one unit to a quantity expressed in another unit. In the process, a given unit is multiplied by one or more conversion factors that cancel units until the desired answer is obtained.

## 1.8 Density

**Learning Goal:** Calculate the density or specific gravity of a substance, and use the density or specific gravity to calculate the mass or volume of a substance.

The density of a substance is a ratio of its mass to its volume, usually in units of g/mL or g/cm$^3$. Density can be used as a factor to convert between the mass and volume of a substance. Specific gravity (sp gr) compares the density of a substance to the density of water, 1.00 g/mL.

# KEY TERMS

**Celsius (°C) temperature scale** A temperature scale on which water has a freezing point of 0 °C and a boiling point of 100 °C.

**centimeter (cm)** A unit of length in the metric system; there are 2.54 cm in 1 in.

**conversion factor** A ratio in which the numerator and denominator are quantities from an equality or given relationship. For example, the conversion factors for the relationship 1 kg = 2.20 lb are written as the following:

$$\frac{2.20 \text{ lb}}{1 \text{ kg}} \quad \text{and} \quad \frac{1 \text{ kg}}{2.20 \text{ lb}}$$

**cubic centimeter (cm$^3$, cc)** The volume of a cube that has 1-cm sides, equal to 1 mL.

**density** The relationship of the mass of an object to its volume expressed as grams per cubic centimeter (g/cm$^3$), grams per milliliter (g/mL), or grams per liter (g/L).

**equality** A relationship between two units that measure the same quantity.

**exact number** A number obtained by counting or by definition.

**gram (g)** The metric unit used in measurements of mass.

**Kelvin (K) temperature scale** A temperature scale on which the lowest possible temperature is 0 K.

**kilogram (kg)** A metric mass of 1000 g, equal to 2.20 lb. The kilogram is the SI standard unit of mass.

**liter (L)** The metric unit for volume that is slightly larger than a quart.

**mass** A measure of the quantity of material in an object.

**measured number** A number obtained when a quantity is determined by using a measuring device.

**meter (m)** The metric unit for length that is slightly longer than a yard. The meter is the SI standard unit of length.

**metric system** A system of measurement used by scientists and in most countries of the world.

**milliliter (mL)** A metric unit of volume equal to one-thousandth of a L (0.001 L).

**prefix** The part of the name of a metric unit that precedes the base unit and specifies the size of the measurement. All prefixes are related on a decimal scale.

**scientific notation** A form of writing large and small numbers using a coefficient from 1 to 9, followed by a power of 10.

**second (s)** A unit of time used in both the SI and metric systems.

**significant figures** The numbers recorded in a measurement.

**specific gravity (sp gr)** A relationship between the density of a substance and the density of water:

$$\text{sp gr} = \frac{\text{density of sample}}{\text{density of water}}$$

**temperature** An indicator of the hotness or coldness of an object.

**volume** The amount of space occupied by a substance.

# UNDERSTANDING THE CONCEPTS

**1.59** In which of the following pairs do both numbers contain the same number of significant figures?
   **a.** 11.0 m and 11.00 m
   **b.** 600.0 K and 60 K
   **c.** 0.000 75 s and 75 000 s
   **d.** 255.0 L and 6.240 × 10$^{-2}$ L

**1.60** In which of the following pairs do both numbers contain the same number of significant figures?
   **a.** 5.75 × 10$^{-3}$ g and 0.002 87 g
   **b.** 0.002 50 m and 0.205 m
   **c.** 150 000 s and 1.5 × 10$^2$ s
   **d.** 0.0038 L and 75 000 mL

**1.61** Indicate if each of the following is answered with an exact number or a measured number.

a. Number of legs
b. Height of table
c. Number of chairs at the table
d. Area of tabletop

**1.62** Measure the length of each of the objects in figure **(a)**, **(b)**, and **(c)** using the metric ruler in the figure. Indicate the number of significant figures for each and the estimated digit for each.

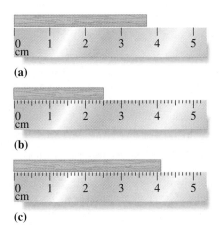

**(a)**

**(b)**

**(c)**

**1.63** Measure the length and width of the rectangle using a metric ruler.

a. What is the length and width of this rectangle measured in centimeters?
b. What is the length and width of this rectangle measured in millimeters?

c. How many significant figures are in the length measurement?
d. How many significant figures are in the width measurement?
e. What is the area of the rectangle in cm$^2$?
f. How many significant figures are in the calculated answer for area?

**1.64** Each of the following diagrams represents a container of water and a cube. Some cubes float while others sink. Match diagrams A, B, C, or D with one of the following descriptions and explain your choices.

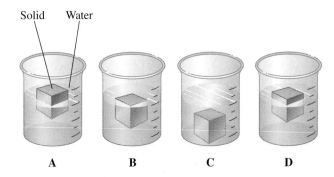

a. The cube has a greater density than water.
b. The cube has a density that is 0.60–0.80 g/mL.
c. The cube has a density that is one-half the density of water.
d. The cube has the same density as water.

**1.65** What is the density of the solid object that is weighed and submerged in water?

**1.66** Consider the following solids.
The solids A, B, and C represent gold, silver, and aluminum. If each has a mass of 10.0 g, what is the identity of each solid?

Density of aluminum = 2.70 g/mL

Density of gold = 19.3 g/mL

Density of silver = 10.5 g/mL

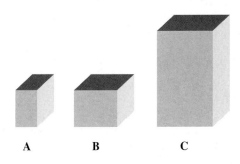

# ADDITIONAL QUESTIONS AND PROBLEMS

This group of questions and problems is related to the topics in this chapter. However, they do not all follow the chapter order and they require you to combine concepts and skills from several sections. These problems will help you increase your critical thinking skills and help you prepare for your next exam.

**1.67** Round off or add zeros to the following calculated answers to give a final answer with three significant figures:
a. 0.000 012 58 L
b. $3.528 \times 10^2$ kg
c. 125 111 m
d. 58.703 g
e. $3 \times 10^{-3}$ s
f. 0.010 826 g

**1.68** What is the total mass in grams of a dessert containing 137.25 g of vanilla ice cream, 84 g of fudge sauce, and 43.7 g of nuts?

**1.69** During a workout at the gym, you set the treadmill at a pace of 55.0 m per min. How many minutes will you walk if you cover a distance of 7500 ft?

**1.70** A fish company delivers 22 kg of salmon, 5.5 kg of crab, and 3.48 kg of oysters to your seafood restaurant.
a. What is the total mass, in kilograms, of the seafood?
b. What is the total number of pounds?

**1.71** In France, grapes are 1.75 Euros per kilogram. What is the cost of grapes in dollars per pound, if the exchange rate is 1.36 dollars per Euro?

**1.72** In Mexico, avocados are 48 pesos per kilogram. What is the cost in cents of an avocado that weighs 0.45 lb if the exchange rate is 11 pesos to the dollar?

**1.73** Bill's recipe for onion soup calls for 4.0 lb of thinly sliced onions. If an onion has an average mass of 115 g, how many onions does Bill need?

**1.74** The price of 1 pound (lb) of potatoes is $1.75. If all the potatoes sold today at the store bring in $1420, how many kilograms (kg) of potatoes did grocery shoppers buy?

**1.75** The following nutrition information is listed on a box of crackers:

Serving size 0.50 oz (6 crackers)

Fat 4 g per serving    Sodium 140 mg per serving

a. If the box has a net weight (contents only) of 8.0 oz, about how many crackers are in the box?
b. If you ate 10 crackers, how many ounces of fat are you consuming?
c. How many grams of sodium are used to prepare 50 boxes of crackers? (*Hint*: Refer to part a.)

**1.76** A dialysis unit requires 75 000 mL of distilled water. How many gallons of water are needed? (1 gal = 4 qt)

**1.77** To prevent bacterial infection, a doctor orders 4 tablets of amoxicillin per day for 10 days. If each tablet contains 250 mg of amoxicillin, how many ounces of the medication are given in 10 days?

**1.78** Celeste's diet restricts her intake of protein to 24 g per day. If she eats 1.2 oz of protein, has she exceeded her protein limit for the day?

**1.79** What is a cholesterol level of 1.85 g/L in the standard units of mg/dL?

**1.80** An object has a mass of 3.15 oz and a volume of 0.1173 L. What is the density (g/mL) of the object?

**1.81** The density of lead is 11.3 g/mL. The water level in a graduated cylinder initially at 215 mL rises to 285 mL after a piece of lead is submerged. What is the mass in grams of the lead?

**1.82** A graduated cylinder contains 155 mL of water. A 15.0-g piece of iron (density = $7.86$ g/cm$^3$) and a 20.0-g piece of lead (density = $11.3$ g/cm$^3$) are added. What is the new water level in the cylinder?

**1.83** How many cubic centimeters (cm$^3$) of olive oil have the same mass as 1.00 L of gasoline? (See Table 1.11.)

**1.84** Ethyl alcohol has a specific gravity of 0.79. What is the volume, in quarts, of 1.50 kg of alcohol?

**1.85** In a process called liposuction, a doctor removes some adipose tissue from a person's body. If body fat has a density of 0.94 g/mL and 3.0 liters of fat are removed, how many pounds of fat were removed from the patient? (See the Health Note "Determination of Percentage of Body Fat.")

**1.86** When a 50.0-kg person is immersed in a water tank for body-fat testing, she has an underwater mass of 2.0 kg. This difference in body mass is equal to the mass of water displaced by the body. The density of water is 1.00 g/mL. (See the Health Note "Determination of Percentage of Body Fat.")
a. What volume of water is displaced?
b. What is the volume, in L, of the person?
c. What is the body density, in g/mL, of that person?

**1.87** A urinometer is a hydrometer used to measure the specific gravity of a sample of urine.
a. What is the density of a urine sample that has a specific gravity of 1.012?
b. What is the mass of a 5.00-mL urine sample with a specific gravity of 1.022?

**1.88** A normal range for specific gravity of urine is 1.003 to 1.030. A 10.0-mL sample of urine has a mass of 10.31 g. What is the specific gravity of the urine? Is the urine considered normal? Why or why not?

# CHALLENGE QUESTIONS

**1.89** A balance measures mass to 0.001 g. If you determine the mass of an object that weighs about 30 g, would you record the mass as 30 g, 32.5 g, 31.25 g, 34.075 g, or 3000 g? Explain your choice by writing 2–3 complete sentences that describe your thinking.

**1.90** When three students use the same meter stick to measure the length of a paper clip, they obtain results of 5.8 cm, 5.75 cm, and 5.76 cm. If the meterstick has millimeter markings, what are some reasons for the different values?

**1.91** A car travels at 55 miles per hour and gets 11 kilometers per liter of gasoline. How many gallons of gasoline are needed for a 3.0-hour trip?

**1.92** A 50.0-g silver object and a 50.0-g gold object are both added to 75.5 mL of water contained in a graduated cylinder. What is the new water level in the cylinder?

**1.93** In the manufacturing of computer chips, cylinders of silicon are cut into thin wafers that are 3.00 inches in diameter and have a mass of 1.50 g of silicon. How thick (mm) is each wafer if silicon has a density of 2.33 $g/cm^3$? (The volume of a cylinder is $V = \pi r^2 h$.)

**1.94** A sunscreen preparation contains 2.50% by mass benzyl salicylate. If a tube contains 4.0 ounces of sunscreen, how many kilograms of benzyl salicylate are needed to manufacture 325 tubes of sunscreen?

**1.95** For a 180-lb person, calculate the quantities of each of the following that must be ingested to provide the $LD_{50}$ for caffeine given in Table 1.10:
   **a.** cups of coffee if one cup is 12 fluid ounces (fl oz) and there is 100 mg of caffeine per 6 fl oz of drip-brewed coffee.
   **b.** cans of cola if one can contains 50 mg caffeine
   **c.** tablets of No-Doz if one tablet contains 100 mg caffeine

**1.96** The label on a one-pint bottle of water lists the following components. If the density is the same as pure water and you drink 3 bottles of water in one day, how many milligrams of each component will you obtain?
   **a.** calcium 28 ppm
   **b.** fluoride 0.08 ppm
   **c.** magnesium 12 ppm
   **d.** potassium 3.2 ppm
   **e.** sodium 15 ppm

# ANSWERS

## Answers to Study Checks

**1.1** **a.** meter (m)     **b.** degree Celsius (°C)
   **c.** gram (g)

**1.2** **a.** $4.25 \times 10^5$ m     **b.** $8 \times 10^{-7}$ g

**1.3** **a.** two     **b.** one
   **c.** five

**1.4** **a.** 36 m     **b.** 0.0026 L
   **c.** $3.8 \times 10^3$ g     **d.** 1.3 kg

**1.5** **a.** 0.4924     **b.** 0.0080 or $8.0 \times 10^{-3}$
   **c.** 2.0

**1.6** **a.** 83.70 mg     **b.** 0.5 L

**1.7** **a.** giga     **b.** centi

**1.8** **a.** 1000     **b.** 0.001

**1.9** Conversion Factors: $\dfrac{1 \text{ zs}}{1 \times 10^{-21}\text{ s}}$ and $\dfrac{1 \times 10^{-21}\text{ s}}{1 \text{ zs}}$

**1.10** **a.** $\dfrac{62.2 \text{ km}}{1 \text{ h}}$, $\dfrac{1 \text{ h}}{62.2 \text{ km}}$
   **b.** $\dfrac{10\ \mu g}{1 \text{ kg}}$, $\dfrac{1 \text{ kg}}{10\ \mu g}$

**1.11** 1.89 L

**1.12** 0.44 oz

**1.13** 10 mL

**1.14** 10.5 $g/cm^3$

**1.15** 2.2 g/mL

**1.16** 1.50 mL

**1.17** 0.916

**1.18** 207 mL

## Answers to Selected Questions and Problems

**1.1** In the U.S., **a.** weight is measured in pounds (lb), **b.** height in feet and inches, **c.** gasoline in gallons, and **d.** temperature in Fahrenheit (°F). In Mexico, **a.** mass is measured in kilograms, **b.** height in meters, **c.** gasoline in liters, and **d.** temperature in Celsius (°C).

**1.3** **a.** meter, length     **b.** gram, mass
   **c.** milliliter, volume     **d.** second, time
   **e.** Celsius, temperature

**1.5** **a.** $5.5 \times 10^4$ m     **b.** $4.8 \times 10^2$ g
   **c.** $5 \times 10^{-6}$ cm     **d.** $1.4 \times 10^{-4}$ s
   **e.** $7.2 \times 10^{-3}$ L

**1.7** **a.** $7.2 \times 10^3$     **b.** $3.2 \times 10^{-2}$
   **c.** $1 \times 10^4$     **d.** $6.8 \times 10^{-2}$

**1.9** **a.** measured     **b.** exact
   **c.** exact     **d.** measured

**1.11** **a.** 6 oz meat     **b.** none
   **c.** 0.75 lb, 350 g     **d.** none (definitions are exact)

**1.13** **a.** not significant     **b.** significant
   **c.** significant     **d.** significant
   **e.** not significant

**1.15** **a.** 5     **b.** 2     **c.** 2
   **d.** 3     **e.** 4     **f.** 3

**1.17** A calculator often gives more digits than the number of significant figures allowed in the answer.

**1.19** **a.** 1.85     **b.** 184
   **c.** 0.00474     **d.** 8810
   **e.** $1.83 \times 10^5$

**1.21** **a.** $5.08 \times 10^3$ L     **b.** $3.74 \times 10^4$ g
   **c.** $1.05 \times 10^5$ m     **d.** $2.51 \times 10^{-4}$ m

**1.23**  **a.** 1.6 **b.** 0.01
**c.** 27.6 **d.** 3.5

**1.25**  **a.** 53.54 cm **b.** 127.6 g
**c.** 121.5 mL **d.** 0.50 L

**1.27**  km/h is kilometers per hour; mi/h is miles per hour

**1.29**  The prefix *kilo* means to multiply by 1000. One kg is the same mass as 1000 g.

**1.31**  **a.** mg **b.** dL
**c.** km **d.** kg
**e.** $\mu$L **f.** ng

**1.33**  **a.** 0.01 **b.** $1000 = 1 \times 10^3$
**c.** $1 \times 10^{-3}$ **d.** 0.1
**e.** $1 \times 10^6$ **f.** $1 \times 10^{-12}$

**1.35**  **a.** 100 cm **b.** 1000 m
**c.** 0.001 m **d.** 1000 mL

**1.37**  **a.** kilogram **b.** milliliter
**c.** km **d.** kL
**e.** nanometer

**1.39**  A conversion factor can be inverted to give a second conversion factor.

**1.41**  **a.** 100 cm = 1 m, $\dfrac{100 \text{ cm}}{1 \text{ m}}$ and $\dfrac{1 \text{ m}}{100 \text{ cm}}$

**b.** 1000 mg = 1 g, $\dfrac{1000 \text{ mg}}{1 \text{ g}}$ and $\dfrac{1 \text{ g}}{1000 \text{ mg}}$

**c.** 1 L = 1000 mL, $\dfrac{1000 \text{ mL}}{1 \text{ L}}$ and $\dfrac{1 \text{ L}}{1000 \text{ mL}}$

**d.** 1 dL = 100 mL, $\dfrac{100 \text{ mL}}{1 \text{ dL}}$ and $\dfrac{1 \text{ dL}}{100 \text{ mL}}$

**e.** 1 week = 7 days, $\dfrac{1 \text{ week}}{7 \text{ days}}$ and $\dfrac{7 \text{ days}}{1 \text{ week}}$

**1.43**  **a.** $\dfrac{3.5 \text{ m}}{1 \text{ s}}$ and $\dfrac{1 \text{ s}}{3.5 \text{ m}}$

**b.** $\dfrac{3500 \text{ mg potassium}}{1 \text{ day}}$ and $\dfrac{1 \text{ day}}{3500 \text{ mg potassium}}$

**c.** $\dfrac{46.0 \text{ km}}{1.0 \text{ gal}}$ and $\dfrac{1.0 \text{ gal}}{46.0 \text{ km}}$

**d.** $\dfrac{50 \text{ mg atenolol}}{1 \text{ tablet}}$ and $\dfrac{1 \text{ tablet}}{50 \text{ mg atenolol}}$

**e.** $\dfrac{29 \text{ } \mu g}{1 \text{ kg}}$ and $\dfrac{1 \text{ kg}}{29 \text{ } \mu g}$

**1.45**  The unit in the denominator must cancel with the preceding unit in the numerator.

**1.47**  **a.** 1.75 m **b.** 5.5 L **c.** 5.5 g

**1.49**  **a.** 473 mL **b.** 65.9 kg **c.** 24 lb

**1.51**  **a.** 66 gal **b.** 3 tablets **c.** 1800 mg

**1.53**  **a.** 1.20 g/mL **b.** 4.4 g/mL **c.** 3.10 g/mL

**1.55**  **a.** 210 g **b.** 575 g **c.** 62 oz

**1.57**  **a.** 1.030 **b.** 1.13 **c.** 0.85 g/mL

**1.59**  **c.** 0.000 75 s and 75 000 s
**d.** 255.0 L and $6.240 \times 10^{-2}$ L

**1.61**  **a.** exact **b.** measured
**c.** exact **d.** measured

**1.63**  **a.** length = 6.96 cm, width = 4.75 cm
**b.** length = 69.6 mm, width = 47.5 mm
**c.** 3 significant figures
**d.** 3 significant figures
**e.** 33.1 cm$^2$
**f.** 3 significant figures

**1.65**  1.8 g/mL

**1.67**  **a.** 0.000 012 6 L **b.** $3.53 \times 10^2$ kg
**c.** 125 000 m **d.** 58.7 g
**e.** $3.00 \times 10^{-3}$ s **f.** 0.0108 g

**1.69**  42 min

**1.71**  $1.08 per lb

**1.73**  16 onions

**1.75**  **a.** 96 crackers **b.** 0.2 oz of fat **c.** 110 g of sodium

**1.77**  0.35 oz

**1.79**  185 mg/dL

**1.81**  790 g

**1.83**  720 cm$^3$

**1.85**  6.2 lb

**1.87**  **a.** 1.012 g/mL **b.** 5.11 g

**1.89**  You would record the mass as 34.075 g. Since your balance will weigh to the nearest 0.001 g, the mass value would be reported to 0.001 g.

**1.91**  6.4 gal

**1.93**  0.141 mm

**1.95**  **a.** 79 cups **b.** 314 cans **c.** 157 tablets

# 2 Energy and Matter

the Chemistry place

Visit **www.chemplace.com** for extra quizzes, interactive tutorials, career resources, PowerPoint slides for chapter review, math help, and case studies.

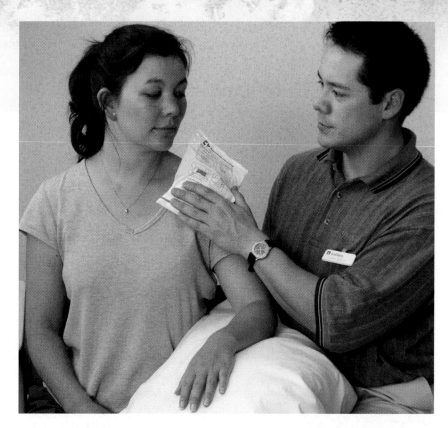

*"If you've had first aid for a sports injury," says Cort Kim, physical therapist at the Sunrise Sports Medicine Clinic, "you've likely been treated with a cold pack or hot pack. We use them for several kinds of injury. Here, I'm showing how I can use a cold pack to reduce swelling in my patient's shoulder."*

*A hot or cold pack is just a packaged chemical reaction. When you hit or open the pack to activate it, your action mixes chemicals together and thus initiates the reaction. In a cold pack, the reaction is one that absorbs heat energy, chills the pack, and draws heat from the injury. Hot packs use reactions that release energy, thus warming the pack. In both cases, the reaction proceeds at a moderate pace, so that the pack stays active for a long time and doesn't get too cold or hot.*

Most of everything we do involves energy. We use energy when we walk, play tennis, study, and breathe. We also use energy when we warm water, cook food, turn on lights, use computers, use a washing machine, or drive our cars. Of course, that energy has to come from something. In our bodies, the food we eat provides us with energy. If we don't eat for a while, we run out of energy. In our homes, schools, and automobiles, burning fossil fuels such as oil, propane, or gasoline provides energy.

When we look around, we see that matter takes the physical form of a solid, liquid, or gas. Water is a familiar example of a compound that we can observe in all three states. In an ice cube or an ice rink, water is in the solid state. Water is a liquid when it comes out of a faucet or fills a pool. Water forms a gas when it evaporates from wet clothes or boils in a pan. Substances change states by gaining or losing energy. For example, energy is added to melt ice cubes and to boil water in a teakettle. In contrast, energy is removed to freeze liquid water in an ice cube tray and to condense water vapor (gas) to liquid.

## LEARNING GOAL

Identify energy as potential or kinetic.

**FIGURE 2.1** Work is done as the rock climber moves up the cliff. At the top, the climber has more potential energy than when she started the climb.

**Q** What happens to the potential energy of the climber as she descends?

# 2.1 ENERGY

When you are running, walking, dancing, or thinking, you are using energy to do **work**. In fact, **energy** is defined as the ability to do work. Suppose you are climbing a steep hill and you become too tired to go on. You might say that you do not have the energy to do any more work. Now suppose you sit down and have lunch. In a while you will have obtained some energy from the food, and you will be able to do more work and complete the climb. (See Figure 2.1.)

## Potential and Kinetic Energy

All energy can be classified as potential energy or kinetic energy. **Potential energy** is stored energy, whereas **kinetic energy** is the energy of motion. Any object that is moving has kinetic energy. A boulder resting on top of a mountain has potential energy because of its location. If the boulder rolls down the mountain, the potential energy becomes kinetic energy. Water stored in a reservoir has potential energy. When the water goes over the dam, the potential energy is converted to kinetic energy. Even the food you eat has potential energy. When you digest food, you convert its potential energy to kinetic energy to do work in the body.

---

**SAMPLE PROBLEM 2.1**

■ **Forms of Energy**

Identify the energy in each of the following as potential or kinetic:

**a.** gasoline  **b.** skating  **c.** a candy bar

**SOLUTION**

**a.** potential energy  **b.** kinetic energy  **c.** potential energy

**STUDY CHECK**

Would the energy in a stretched rubber band be potential or kinetic?

## Heat and Units of Energy

**Heat** is the energy associated with the motion of particles in a substance. A frozen pizza feels cold because the particles in the pizza are moving very slowly. As heat is added, the motions of the particles increase, and the pizza becomes warm. Eventually the particles have enough energy to make the pizza hot and ready to eat.

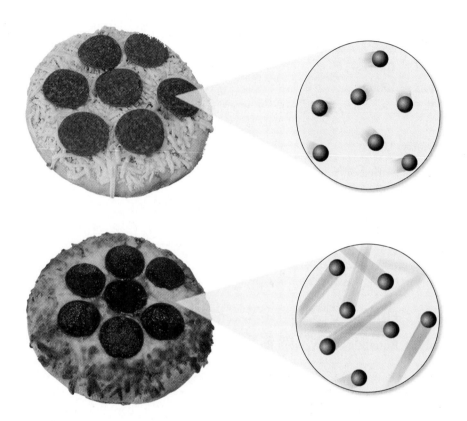

The SI unit of energy and work is the **joule (J)**, pronounced "jewel." The joule is a small amount of energy, so scientists often use the kilojoule (kJ), 1000 joules. When you heat water for one cup of tea, you use about 75 000 J or 75 kJ of heat.

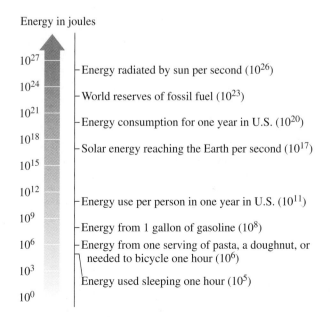

Energy in joules

$10^{27}$
  — Energy radiated by sun per second ($10^{26}$)

$10^{24}$
  — World reserves of fossil fuel ($10^{23}$)

$10^{21}$
  — Energy consumption for one year in U.S. ($10^{20}$)

$10^{18}$
  — Solar energy reaching the Earth per second ($10^{17}$)

$10^{15}$

$10^{12}$
  — Energy use per person in one year in U.S. ($10^{11}$)

$10^{9}$
  — Energy from 1 gallon of gasoline ($10^{8}$)

$10^{6}$
  — Energy from one serving of pasta, a doughnut, or needed to bicycle one hour ($10^{6}$)

$10^{3}$
  — Energy used sleeping one hour ($10^{5}$)

$10^{0}$

You may be more familiar with the older unit **calorie (cal)**, from the Latin *caloric*, meaning "heat." The calorie was originally defined as the amount of energy (heat) needed to raise the temperature of one gram of water by 1 °C (from 14.5 °C to 15.5 °C). Now one calorie is defined as exactly 4.184 J. This equality can also be written as a conversion factor.

1 cal = 4.184 J (exact)

$$\frac{4.184 \text{ J}}{1 \text{ cal}} \quad \text{and} \quad \frac{1 \text{ cal}}{4.184 \text{ J}}$$

One **kilocalorie (kcal)** is equal to 1000 calories, and one kilojoule (kJ) is equal to 1000 joules.

1 kcal = 1000 cal

1 kJ = 1000 J

---

## SAMPLE PROBLEM  2.2

■ **Energy Units**

When 1.0 g of gasoline burns in an automobile engine, 48 000 J are released. Convert this quantity of energy to the following units:

**a.** calories      **b.** kilojoules

SOLUTION

**a.** calories

Step 1  **Given**    48 000 J      **Need**    calories (cal)

Step 2  **Plan**    J    [Energy factor]    cal

Step 3  **Equalities/Conversion Factors**

$$1 \text{ cal} = 4.184 \text{ J}$$

$$\frac{1 \text{ cal}}{4.184 \text{ J}} \quad \text{and} \quad \frac{4.184 \text{ J}}{1 \text{ cal}}$$

Step 4  **Set Up Problem**

$$48\ 000\ \cancel{J} \times \frac{1 \text{ cal}}{4.184\ \cancel{J}} = 11\ 000 \text{ cal } (1.1 \times 10^4 \text{ cal})$$

**b.** kilojoules

Step 1  **Given**    48 000 J      **Need**    kilojoules

Step 2  **Plan**    J    [Energy factor]    kJ

Step 3  **Equalities/Conversion Factors**

$$1 \text{ kJ} = 1000 \text{ J}$$

$$\frac{1000 \text{ J}}{1 \text{ kJ}} \quad \text{and} \quad \frac{1 \text{ kJ}}{1000 \text{ J}}$$

Step 4    **Set Up Problem**    $48\ 000\ \cancel{J} \times \dfrac{1 \text{ kJ}}{1000\ \cancel{J}} = 48 \text{ kJ}$

STUDY CHECK

The burning of 1.0 g of coal produces 35 000 J of energy. How many kcal are produced?

# Green Chemistry Note

## Carbon Dioxide and Global Warming

Earth's climate is a product of interactions between sunlight, the atmosphere, and the oceans. The Sun provides us with energy in the form of solar radiation. Some of this radiation is reflected back into space. The rest is absorbed by the clouds, atmospheric gases including carbon dioxide, and Earth's surface. There have been fluctuations in the concentrations of carbon dioxide for millions of years. However in the last 100 years, the amount of carbon dioxide ($CO_2$) gas in our atmosphere has increased significantly. From the years 1000 to 1800, the atmospheric carbon dioxide averaged 280 ppm. But since the Industrial Revolution in 1800, the level of atmospheric carbon dioxide has risen from about 280 ppm to 375 ppm in 2000, which is about a 30% increase.

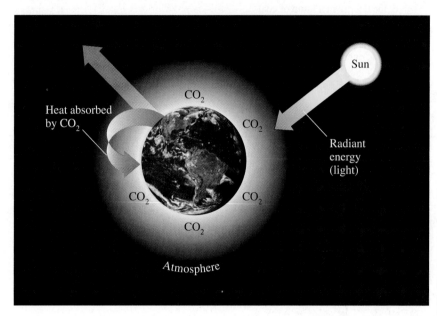

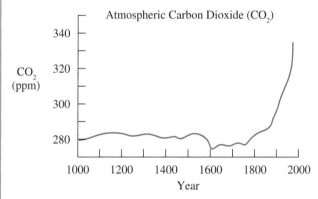

As the atmospheric $CO_2$ levels increase, more solar radiation is trapped by the atmospheric gases, which raises the temperature at the surface of Earth. Some scientists have estimated that if the carbon dioxide level doubles from its level before the Industrial Revolution, the average temperature globally could increase by 2-4.4 °C. While this may seem to be a small temperature change, it could have dramatic impact worldwide. Even now, glaciers and snow cover in much of the world have diminished. Ice sheets in Antarctica and Greenland are melting faster and breaking apart.

Although it is not known how rapidly the ice in the polar regions is melting, it will contribute to a rise in sea level. In the twentieth century, the sea level rose from 15 to 23 cm, and some scientists predict the sea level will rise 1 m in this century. Such an increase will have a major impact on coastal areas.

Until recently, the carbon dioxide levels were maintained at constant levels by algae in the oceans and the trees in the forests. However, the ability to absorb carbon dioxide is not keeping up with its increasing levels. Most scientists agree that the primary source of the increase of carbon dioxide is the burning of fossil fuels such as gasoline, coal, and natural gas. The cutting and burning of trees in the rain forests (deforestation) also reduces the amount of carbon dioxide removed from the atmosphere.

Worldwide efforts are being made to reduce the carbon dioxide produced by burning fossil fuels that heat our homes, run our cars, and provide energy for industries. Efforts are being made to explore alternative energy sources and to reduce the effects of deforestation. Meanwhile, we can reduce energy use in our homes by using appliances that are more energy efficient and replacing incandescent light bulbs with fluorescent lights. Such an effort worldwide will reduce the possible impact of global warming and at the same time save our fuel resources.

## QUESTIONS AND PROBLEMS

### Energy

**2.1** Discuss the changes in the potential and kinetic energy of a roller-coaster ride as the roller coaster climbs up a ramp and goes down the other side.

**2.2** Discuss the changes in the potential and kinetic energy of a ski jumper taking the elevator to the top of the jump and going down the ramp.

**2.3.** Indicate whether each statement describes potential or kinetic energy:
   **a.** water at the top of a waterfall
   **b.** kicking a ball
   **c.** the energy in a lump of coal
   **d.** a skier at the top of a hill

**2.4.** Indicate whether each statement describes potential or kinetic energy:
   **a.** the energy in your food
   **b.** a tightly wound spring
   **c.** an earthquake
   **d.** a car speeding down the freeway

**2.5.** A burning match releases $1.1 \times 10^3$ J. Convert the energy released by 20 matches in a box to the following energy units:
   **a.** kilojoules
   **b.** calories
   **c.** kilocalories

**2.6.** A person uses 750 kcal to run for 1.0 h. Convert the energy for running to the following energy units:
   **a.** calories
   **b.** joules
   **c.** kilojoules

## LEARNING GOAL

Use caloric values to calculate the kilocalories (Cal) in a food.

the
**Chemistry**
place

**CASE STUDY**
Calories from Hidden Sugar

# 2.2 ENERGY AND NUTRITION

When you are watching your food intake, the Calories you are counting are actually kilocalories. In the field of nutrition, the Calorie, Cal (with an uppercase C) means 1000 cal, or 1 kcal. The international unit of nutritional energy is the kilojoule (kJ). For example, a baked potato has a nutritional value of 120 Calories, which is the same as 120 kcal or about 500 kJ of energy.

$$1 \text{ Cal} = 1 \text{ kcal} = 1000 \text{ cal}$$
$$1 \text{ Cal} = 4.184 \text{ kJ} = 4184 \text{ J}$$

The number of Calories in a food is determined by using an apparatus called a calorimeter, shown in Figure 2.2.

A sample is placed in a steel container within the calorimeter, and water is added to fill a surrounding chamber. When the food burns (combustion), the heat released causes an increase in the temperature of the surrounding water. By calculating the heat in kilocalories (or kilojoules) absorbed by the water, we can determine the caloric content of the food.

## Caloric Food Values

The **caloric values** are the kilocalories per gram of the three types of food: carbohydrates, fats, and proteins. These values are listed in Table 2.1

**FIGURE 2.2** The caloric value of a food sample temperature change is determined by the combustion of a food sample in a calorimeter.
**Q** What happens to the temperature of water in a calorimeter during the combustion of a food sample?

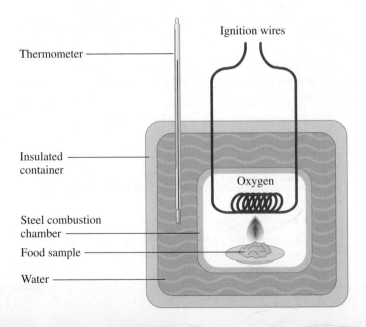

## TABLE 2.1 Caloric Values for the Three Food Types

| Food Type | Carbohydrate | Fat (lipid) | Protein |
|---|---|---|---|
| Caloric Value | $\dfrac{4\ \text{kcal}}{1\ \text{g}}$ | $\dfrac{9\ \text{kcal}}{1\ \text{g}}$ | $\dfrac{4\ \text{kcal}}{1\ \text{g}}$ |

If the composition of a food is known in terms of the mass of each food type, the caloric content (the total number of Calories) can be calculated.

$$\text{Kilocalories} = \quad \cancel{g} \quad \times \quad \frac{\text{kcal}}{\cancel{g}}$$

Mass of carbohydrate,   Caloric value
fat, or protein

The caloric content of a packaged food is listed in the nutritional information on the package, usually in terms of the number of Calories in a serving. The general composition and caloric content of some foods are given in Table 2.2.

---

### SAMPLE PROBLEM 2.3

### ■ Caloric Content for a Food

How many kilocalories are in a piece of chocolate cake that contains 34 g of carbohydrate, 8 g of fat, and 5 g of protein? (Round the final answer to the tens place.)

#### SOLUTION

Using the caloric values for carbohydrate, fat, and protein (Table 2.1), we can calculate the total number of kcal:

Carbohydrate:  $34\ \cancel{g} \times \dfrac{4\ \text{kcal}}{1\ \cancel{g}} = 136\ \text{kcal}$

Fat:  $8\ \cancel{g} \times \dfrac{9\ \text{kcal}}{1\ \cancel{g}} = 72\ \text{kcal}$

Protein:  $5\ \cancel{g} \times \dfrac{4\ \text{kcal}}{1\ \cancel{g}} = 20\ \text{kcal}$

Total caloric content: 228 kcal = 230 kcal (rounded to tens place)

#### STUDY CHECK

A 1-oz (28 g) serving of oat-bran hot cereal with half a cup of whole milk contains 22 g of carbohydrate, 7 g of fat, and 10 g of protein. If you eat two servings of the oat bran for breakfast, how many kilocalories will you obtain? (Round to the tens place.)

---

## Snack Crackers

### Nutrition Facts

Serving Size 14 crackers (31g)
Servings Per Container About 7

**Amount Per Serving**

| Calories 120 | Calories from Fat 35 |
|---|---|
| Kilojoules 500 | kJ from Fat    150 |

| | % Daily Value* |
|---|---|
| **Total Fat** 4g | **6%** |
| Saturated Fat 0.5g | **3%** |
| Trans Fat 0g | |
| Polyunsaturated Fat 0.5% | |
| Monounsaturated Fat 1.5g | |
| **Cholesterol** 0mg | **0%** |
| **Sodium** 310mg | **13%** |
| **Total Carbohydrate** 19g | **6%** |
| Dietary Fiber Less than 1g | **4%** |
| Sugars 2g | |
| **Proteins** 2g | |

| Vitamin A 0% | • | Vitamin C 0% |
|---|---|---|
| Calcium 4% | • | Iron 6% |

*Percent Daily Values are based on a 2,000 calorie diet. Your daily values may be higher or lower depending on your calorie needs.

| | | Calories: 2,000 | 2,500 |
|---|---|---|---|
| Total Fat | Less than | 65g | 80g |
| Sat Fat | Less than | 20g | 25g |
| Cholesterol | Less than | 300mg | 300mg |
| Sodium | Less than | 2,400mg | 2,400mg |
| Total Carbohydrate | | 300g | 375g |
| Dietary Fiber | | 25g | 30g |

Calories per gram:
Fat 9 • Carbohydrate 4 • Protein 4

---

## TABLE 2.2 General Composition and Caloric Content of Some Foods

| Food | Protein (g) | Fat (g) | Carbohydrate (g) | Calories (kcal[a]) |
|---|---|---|---|---|
| Banana, 1 medium | 1 | trace | 26 | 110 |
| Beans, red kidney, 1 cup | 15 | 1 | 42 | 240 |
| Beef, lean, 3 oz | 22 | 5 | trace | 130 |
| Carrots, raw, 1 cup | 1 | trace | 11 | 50 |
| Chicken, no skin, 3 oz | 20 | 3 | 0 | 110 |
| Egg, 1 large | 6 | 6 | trace | 80 |
| Milk, 4% fat, 1 cup | 9 | 9 | 12 | 165 |
| Milk, nonfat, 1 cup | 9 | trace | 12 | 85 |
| Oil, olive, 1 tbs | 0 | 14 | 0 | 130 |
| Potato, baked | 3 | trace | 23 | 105 |
| Salmon, 3 oz | 17 | 5 | 0 | 115 |
| Steak, 3 oz | 20 | 27 | 0 | 320 |
| Yogurt, lowfat, 1 cup | 8 | 4 | 13 | 120 |

[a]Values rounded to nearest 5 kcal.

---

## Explore Your World

### Counting Calories

Obtain a food item that has a nutrition label. From the information on the label, determine the number of grams of carbohydrate, fat, and protein in one serving. Using the caloric values, calculate the total Calories for one serving. (Most products round off to the tens place.)

#### QUESTION

How does your total for the Calories in one serving compare to the Calories stated on the label for a single serving?

# Health Note

## Losing and Gaining Weight

The number of kilocalories needed in the daily diet of an adult depends on gender and physical activity. Some general levels of energy needs are given in Table 2.3.

A person gains weight when food intake exceeds energy output. The amount of food a person eats is regulated by the hunger center in the hypothalamus, located in the brain. Food intake is normally proportional to the nutrient stores in the body. If these nutrient stores are low, you feel hungry; if they are high, you do not feel like eating.

A person loses weight when food intake is less than energy output. Many diet products contain cellulose, which has no nutritive value but provides bulk and makes you feel full. Some diet drugs depress the hunger center and must be used with caution because they excite the nervous system and can elevate blood pressure. Because muscular exercise is an important way to expend energy, an increase in daily exercise aids weight loss. Table 2.4 lists some activities and the amount of energy they require.

**TABLE 2.4** Energy Expended by a 70.0-kg (154-lb) Adult

| Activity | Energy Expended (kcal/h) |
|----------|--------------------------|
| Sleeping | 60 |
| Sitting | 100 |
| Walking | 200 |
| Swimming | 500 |
| Running | 750 |

**TABLE 2.3** Typical Daily Energy Requirements for 70.0-kg (154-lb) Adult

| Adult | Energy (kcal) |
|-------|---------------|
| Female | 2200 |
| Male | 3000 |

## QUESTIONS AND PROBLEMS

### Energy and Nutrition

**2.7.** Using the caloric values for foods, determine each of the following: (Round the final Cal to the tens place.)
 **a.** The total Calories for 1 cup of orange juice that has 26 g of carbohydrate, no fat, and 2 g of protein.
 **b.** The grams of carbohydrate in 1 apple if the apple has no fat and no protein and provides 72 kcal of energy.
 **c.** The number of Calories in 1 tablespoon of vegetable oil, which contains 14 g of fat and no carbohydrate or protein.
 **d.** How many Calories are in 1 breakfast roll if it has 30. g of carbohydrate, 15 g of fat, and 5 g of protein?

**2.8.** Using the caloric values for foods, determine each of the following: (Round the final Cal to the tens place.)
 **a.** The total Cal for 2 tablespoons of crunchy peanut butter that contains 6 g of carbohydrate, 16 g of fat, and 7 g of protein.

 **b.** The grams of protein in 1 cup of soup that has 110 Cal with 7 g of fat and 9 g of carbohydrate.
 **c.** How many grams of sugar (carbohydrate) are in 1 can of cola if there are 140 Cal and no fat and no protein?
 **d.** How many grams of fat are in 1 avocado if there are 405 Cal, 13 g of carbohydrate, and 5 g of protein?

**2.9.** One cup of clam chowder contains 9 g of protein, 12 g of fat, and 16 g of carbohydrate. How many kilocalories are in the clam chowder? How many kilojoules are in the clam chowder? (Round the final answers to the nearest tens place.)

**2.10.** A high-protein diet contains 70. g of carbohydrate, 150 g of protein, and 5.0 g of fat. How many kilocalories does this diet provide? How many kilojoules does this diet provide? (Round the final answers to the nearest tens place.)

# 2.3 TEMPERATURE CONVERSIONS

## LEARNING GOAL

Given a temperature, calculate a corresponding temperature on another scale.

Temperature is a measure of how hot or cold a substance is compared to another substance. Heat always flows from a substance with a higher temperature to a substance with a lower temperature until the temperatures of both are the same. When you drink hot coffee or touch a hot pan, heat flows to your mouth or hand, which is at a lower temperature. When you touch an ice cube, it feels cold because heat flows from your hand to the colder ice cube.

## Celsius and Fahrenheit Temperatures

Temperatures in science, and in most of the world, are measured and reported in *Celsius (°C)* units. In the United States, everyday temperatures are commonly reported in *Fahrenheit (°F)* units. A typical room temperature of 22 °C would be the same as 72 °F. A normal body temperature of 37.0 °C is 98.6 °F.

On the Celsius and Fahrenheit scales, the temperatures of melting ice and boiling water are used as reference points. On the Celsius scale, the freezing point of water is defined as 0 °C, and the boiling point as 100 °C. On a Fahrenheit scale, water freezes at 32 °F and boils at 212 °F. On each scale, the temperature difference between freezing and boiling is divided into smaller units, or degrees. On the Celsius scale, there are 100 degrees between the freezing and boiling of water compared with 180 degrees on the Fahrenheit scale. That makes a Celsius degree almost twice the size of a Fahrenheit degree: 1 °C = 1.8 °F. (See Figure 2.3.)

$$180 \text{ Fahrenheit degrees} = 100 \text{ Celsius degrees}$$

$$\frac{180 \text{ Fahrenheit degrees}}{100 \text{ Celsius degrees}} = \frac{1.8 \text{ °F}}{1 \text{ °C}}$$

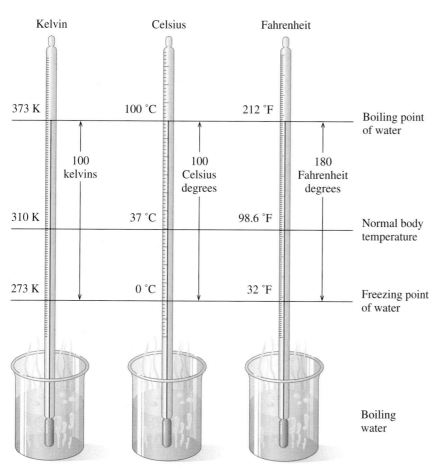

**FIGURE 2.3** A comparison of the Fahrenheit, Celsius, and Kelvin temperature scales between the freezing and boiling points of water.

**Q** What is the difference in the values for freezing on the Celsius and Fahrenheit temperature scales?

To convert to a Fahrenheit temperature, the Celsius temperature is multiplied by 1.8, and then 32 degrees is added. The 32 degrees adjusts the freezing point of 0 °C on the Celsius scale to 32 °F on the Fahrenheit scale. Both values, 1.8 and 32, are exact numbers. The equation for this conversion follows:

$$T_F = \frac{1.8\ °F(T_C)}{1\ °C} + 32°\qquad\text{or}\qquad T_F = 1.8(T_C) + 32°$$

Changes          Adjusts freezing
°C to °F         point

*Career Focus*

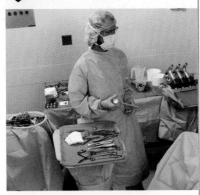

## Surgical Technologist

"As a surgical technologist, I assist the doctors during surgeries," says Christopher Ayars, surgical technologist, Kaiser Hospital. "I am there to help during general or orthopedic surgery by passing instruments, holding retractors, and maintaining the sterile field. Our equipment for surgery is sterilized by steam that is heated to 270 °F, which is the same as 130 °C."

Surgical technologists assist with surgical procedures by preparing and maintaining surgical equipment, instruments, and supplies; providing patient care in an operating room setting; preparing and maintaining a sterile field; and ensuring that there are no breaks in aseptic technique. Instruments, which have been sterilized, are wrapped and sent to surgery where they are checked again before they are opened.

---

**SAMPLE PROBLEM  2.4**

### ■ Converting Celsius to Fahrenheit

The temperature of a room is set at 22 °C. If that temperature is lowered by 1 °C, it can save as much as 5% in energy costs. What temperature in Fahrenheit degrees should be set to lower the Celsius temperature by 1 °C?

**SOLUTION**

Step 1  **Given**    22 °C − 1 °C = 21 °C    **Need**   $T_F$

Step 2  **Plan**

$$T_C \quad\boxed{\text{Temperature equation}}\quad T_F$$

Step 3  **Equality/Conversion Factor**

$$T_F = 1.8(T_C) + 32°$$

Step 4  **Set Up Problem**    Substitute the Celsius temperature into the equation and solve.

$$T_F = 1.8(21) + 32°$$

   2 SFs    Exact

$$T_F = 38° + 32°\qquad\text{1.8 is exact; 32 is exact}$$

$$= 70.\ °F\qquad\text{Answer to the ones place}$$

In the equation, *the values of 1.8 and 32 are exact numbers.* Therefore, the answer is reported to the same decimal place as the initial temperature.

**STUDY CHECK**

When making ice cream, rock salt is used to chill the mixture. If the temperature drops to −11 °C, what is it in °F?

---

To convert from Fahrenheit to Celsius, the temperature equation is rearranged for $T_C$. Start with

$$T_F = 1.8(T_C) + 32°$$

Then subtract 32 from both sides.

$$T_F - 32° = 1.8(T_C) + \cancel{32°} - \cancel{32°}$$

$$T_F - 32° = 1.8(T_C)$$

Solve the equation for $T_C$ by dividing both sides by 1.8.

$$\frac{T_F - 32°}{1.8} = \frac{\cancel{1.8}(T_C)}{\cancel{1.8}}$$

$$\frac{T_F - 32°}{1.8} = T_C$$

SAMPLE PROBLEM **2.5**

### ■ Converting Fahrenheit to Celsius

In a type of cancer treatment called thermotherapy, temperatures up to 113 °F are used to destroy cancer cells. What is that temperature in Celsius degrees?

SOLUTION

Step 1  **Given**   113 °F    **Need**   $T_C$

Step 2  **Plan**

$$T_F \quad \boxed{\text{Temperature equation}} \quad T_C$$

Step 3  **Equality/Conversion Factor**

$$T_C = \frac{T_F - 32°}{1.8}$$

Step 4  **Set Up Problem**    To solve for $T_C$, substitute the Fahrenheit temperature into the equation and solve.

$$T_C = \frac{T_F - 32°}{1.8}$$

$$T_C = \frac{(113° - 32°)}{1.8} \qquad \text{32 is exact; 1.8 is exact}$$

$$= \frac{81°}{1.8} = 45\ °C \qquad \text{Answer to the ones place}$$

STUDY CHECK

A child has a temperature of 103.6 °F. What is this temperature on a Celsius thermometer?

## Kelvin Temperature Scale

Scientists tell us that the coldest temperature possible is −273 °C (more precisely, −273.15 °C). On the Kelvin scale, this temperature, called *absolute zero,* has the value of 0 Kelvin (0 K). Units on the Kelvin scale are called kelvins (K); no degree symbol is used. Because there are no lower temperatures, the Kelvin scale has no negative numbers. Between the freezing and boiling points of water, there are 100 kelvins, which makes a kelvin equal in size to a Celsius unit.

$$1\ K = 1\ °C$$

To calculate a Kelvin temperature, add 273 to the Celsius temperature:

$$T_K = T_C + 273$$

Table 2.5 gives a comparison of some temperatures on the three scales.

**TABLE 2.5**  A Comparison of Temperatures

| Example | Fahrenheit (°F) | Celsius (°C) | Kelvin (K) |
|---|---|---|---|
| Sun | 9937 | 5503 | 5776 |
| A hot oven | 450 | 232 | 505 |
| A desert | 120 | 49 | 322 |
| A high fever | 104 | 40 | 313 |
| Room temperature | 70 | 21 | 294 |
| Water freezes | 32 | 0 | 273 |
| A northern winter | −22 | −30 | 243 |
| Helium boils | −452 | −269 | 4 |
| Absolute zero | −459 | −273 | 0 |

# Health Note

## Variation in Body Temperature

Normal body temperature is considered to be 37.0 °C, although it varies throughout the day and from person to person. Oral temperatures of 36.1 °C are common in the morning and climb to a high of 37.2 °C between 6 P.M. and 10 P.M. Temperatures above 37.2 °C for a person at rest are usually an indication of disease. Individuals involved in prolonged exercise may also experience elevated temperatures. Body temperatures of marathon runners can range from 39 °C to 41 °C as heat production during exercise exceeds the body's ability to lose heat.

Changes of more than 3.5 °C from the normal body temperature begin to interfere with bodily functions. Temperatures above 41 °C can lead to convulsions, particularly in children, which may cause permanent brain damage. Heatstroke occurs above 41.1 °C. Sweat production stops and the skin becomes hot and dry. The pulse rate is elevated, and respiration becomes weak and rapid. The person can become lethargic and lapse into a coma. Damage to internal organs is a major concern, and treatment, which must be immediate, may include immersing the person in an ice-water bath.

In hypothermia, body temperature can drop as low as 28.5 °C. The person may appear cold and pale and have an irregular heartbeat. Unconsciousness can occur if the body temperature drops below 26.7 °C. Respiration becomes slow and shallow, and oxygenation of the tissues decreases. Treatment involves providing oxygen and increasing blood volume with glucose and saline fluids. Internal temperature may be restored by injecting warm fluids (37.0 °C) into the peritoneal cavity.

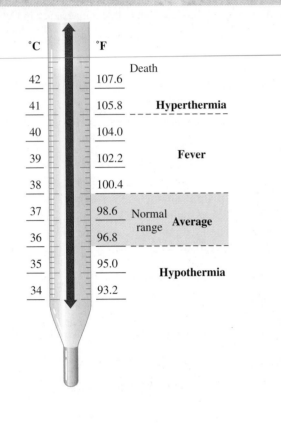

---

SAMPLE PROBLEM    2.6

## ■ Converting from Celsius to Kelvin Temperature

A dermatologist may use liquid cryogenic nitrogen at −196 °C to remove skin lesions and some skin cancers. What is the temperature of the liquid nitrogen in K?

### SOLUTION

To find the Kelvin temperature, we use the equation:

$$T_K = T_C + 273$$
$$T_K = -196 \,°C + 273$$
$$= 77 \text{ K}$$

### STUDY CHECK

On the planet Mercury, the average night temperature is 13 K, and the average day temperature is 683 K. What are these temperatures in Celsius degrees?

## QUESTIONS AND PROBLEMS

### Temperature Conversions

**2.11** Your friend who is visiting from Canada just took her temperature. When she reads 99.8, she becomes concerned that she is quite ill. How would you explain this temperature to your friend?

**2.12** You have a friend who is using a recipe for flan from a Mexican cookbook. You notice that he set your oven temperature at 175 °F. What would you advise him to do?

**2.13** Solve the following temperature conversions:
  **a.** 37.0 °C = ——— °F
  **b.** 65.3 °F = ——— °C
  **c.** −27 °C = ——— K
  **d.** 62 °C = ——— K
  **e.** 114 °F = ——— °C
  **f.** 72 °F = ——— K

**2.14** Solve the following temperature conversions:
  **a.** 25 °C = ——— °F
  **b.** 155 °C = ——— °F

  **c.** −25 °F = ——— °C
  **d.** 224 K = ——— °C
  **e.** 545 K = ——— °C
  **f.** 875 K = ——— °F

**2.15 a.** A patient with heatstroke has a temperature of 106 °F. What does this read on a Celsius thermometer?
  **b.** Because high fevers can cause convulsions in children, the doctor wants to be called if the child's temperature goes over 40.0 °C. Should the doctor be called if a child has a temperature of 103 °F?

**2.16 a.** Hot compresses are being prepared for the patient in room 32B. The water is heated to 145 °F. What is the temperature of the hot water in °C?
  **b.** During extreme hypothermia, a young woman's temperature dropped to 20.6 °C. What was her temperature on the Fahrenheit scale?

# 2.4 SPECIFIC HEAT

**LEARNING GOAL**

Use specific heat to calculate heat loss or gain, temperature change, or mass of a sample.

Every substance can absorb heat. When you want to bake a potato, you place it in a hot oven. If you are cooking pasta, you add the pasta to boiling water. You already know that adding heat to water increases its temperature until it boils. Every substance has its own characteristic ability to absorb heat. Some substances must absorb more heat than others to reach a certain temperature. These energy requirements for different substances are described in terms of a physical property called specific heat. **Specific heat** (*SH*) is the amount of heat needed to raise the temperature of exactly 1 g of a substance by exactly 1 °C. This temperature change is written as $\Delta T$ (*delta T*).

$$\text{Specific heat } (SH) = \frac{\text{heat}}{\text{grams} \times \Delta T} = \frac{\text{J (or cal)}}{1 \text{ g} \times 1 \text{ °C}}$$

Now we can write the specific heat for water using our definition of the calorie and joule.

$$\text{Specific heat } (SH) \text{ of } H_2O(l) = 4.184 \frac{\text{J}}{\text{g °C}} = 1.00 \frac{\text{cal}}{\text{g °C}}$$

The specific heat of a substance will depend on its state: solid (*s*), liquid (*l*), or gas (*g*). If we look at Table 2.6, we see that the specific heat of water is much larger than the

**TABLE 2.6** Specific Heats of Some Substances

| Substance | cal/g °C | J/g °C |
|---|---|---|
| Aluminum, Al(*s*) | 0.214 | 0.897 |
| Copper, Cu(*s*) | 0.0920 | 0.385 |
| Gold, Au(*s*) | 0.0308 | 0.129 |
| Iron, Fe(*s*) | 0.108 | 0.450 |
| Silver, Ag(*s*) | 0.0562 | 0.235 |
| Ammonia, NH₃(*g*) | 0.488 | 2.04 |
| Ethanol, C₂H₅OH(*l*) | 0.588 | 2.46 |
| Sodium chloride, NaCl(*s*) | 0.207 | 0.864 |
| Water, H₂O(*l*) | 1.00 | 4.184 |

specific heat of aluminum or copper. We see that 1 g of water requires 4.184 J to increase its temperature by 1 °C. However, adding the same amount of heat (4.184 J) will raise the temperature of 1 g of aluminum by about 5 °C and 1 g of copper by 10 °C. In the body, water can absorb or release large amounts of heat to maintain an almost constant temperature because of water's high specific heat.

## Calculations Using Specific Heat

When we know the specific heat of a substance, we can calculate the heat lost or gained by measuring the mass of the substance and the initial and final temperature. We can substitute these measurements into the specific heat expression that is rearranged to solve for heat, which we call the *heat equation*.

$$\text{Heat} = \text{mass} \times \text{temperature change} \times \text{specific heat}$$

$$\text{Heat} = \text{mass} \times \Delta T \times SH$$

$$\text{cal} = \text{grams} \times {}^\circ\text{C} \times \frac{\text{cal}}{\text{g }^\circ\text{C}}$$

$$\text{J} = \text{grams} \times {}^\circ\text{C} \times \frac{\text{J}}{\text{g }^\circ\text{C}}$$

**Guide to Calculations Using Specific Heat**

**1** List given and needed data.

**2** Calculate temperature change ($\Delta T$).

**3** Write the equation for heat.
Heat = mass $\times \Delta T \times SH$

**4** Substitute given values and solve, making sure units cancel.

---

**SAMPLE PROBLEM   2.7**

### ■ Calculating Heat with Temperature Increase

How many joules must be added to 45.2 g of aluminum to raise its temperature from 12.5 °C to 76.8 °C? (See Table 2.6.)

**SOLUTION**

**Step 1  List given and needed data.**

> **Given**    mass = 45.2 g
>
> $SH$ for aluminum = 0.897 J/g °C
>
> Initial temperature = 12.5 °C
> Final temperature = 76.8 °C
>
> **Need**    heat in joules (J)

**Step 2  Calculate the temperature change.**   The temperature change, $\Delta T$, is the difference between the two temperatures.

$$\Delta T = T_{\text{final}} - T_{\text{initial}} = 76.8\ ^\circ\text{C} - 12.5\ ^\circ\text{C} = 64.3\ ^\circ\text{C}$$

**Step 3  Write the heat equation.**

$$\text{Heat} = \text{mass} \times \Delta T \times SH$$

**Step 4  Substitute the given values into the equation and solve, making sure units cancel.**

$$\text{Heat} = 45.2\ \cancel{\text{g}} \times 64.3\ \cancel{^\circ\text{C}} \times \frac{0.897\ \text{J}}{\cancel{\text{g}}\ \cancel{^\circ\text{C}}} = 2610\ \text{J}\ (2.61 \times 10^3\ \text{J})$$

**STUDY CHECK**

Some cooking pans have a layer of copper on the bottom. How many kilojoules are needed to raise the temperature of 125 g of copper from 22 °C to 325 °C if the specific heat of copper is 0.385 J/g °C?

## QUESTIONS AND PROBLEMS

### Specific Heat

**2.17** If the same amount of heat is supplied to samples of 10.0 g each of aluminum, iron, and copper all at 15 °C, which sample would reach the highest temperature? (See Table 2.6.)

**2.18** Substances A and B are the same mass and at the same initial temperature. When they are heated, the final temperature of A is 55 °C higher than the temperature of B. What does this tell you about the specific heats of A and B?

**2.19** What is the amount of heat required in each of the following?
**a.** calories to heat 25 g of water from 15 °C to 25 °C
**b.** joules to heat 75 g of water from 22 °C to 66 °C
**c.** kilocalories to heat 150 g of water in a kettle from 15 °C to 77 °C
**d.** kilojoules to heat 175 g of copper from 28 °C to 188 °C

**2.20** What is the amount of heat involved in each of the following?
**a.** calories given off when 85 g of water cools from 45 °C to 25 °C
**b.** joules given off when 25 g of water cools from 86 °C to 61 °C
**c.** kilocalories added when 5.0 kg of water warms from 22 °C to 28 °C
**d.** kilojoules to heat 224 g of gold from 18 °C to 185 °C

**2.21** Calculate the energy in joules and calories
**a.** required to heat 25.0 g of water from 12.5 °C to 25.7 °C
**b.** required to heat 38.0 g of copper (Cu) from 122 °C to 246 °C
**c.** lost when 15.0 g of ethanol, $C_2H_5OH$, cools from 60.5 °C to −42.0 °C
**d.** lost when 125 g of iron cools from 118 °C to 55 °C

**2.22** Calculate the energy in joules and calories
**a.** required to heat 5.25 g of water, $H_2O$, from 5.5 °C to 64.8 °C
**b.** lost when 75.0 g of water, $H_2O$, cools from 86.4 °C to 2.1 °C
**c.** required to heat 10.0 g of silver (Ag) from 112 °C to 275 °C
**d.** lost when 18.0 g of gold (Au) cools from 224 °C to 118 °C

**2.23** Calculate the kilocalories each food provides when burned in a calorimeter:
**a.** one stalk of celery that produces energy to heat 505 g of water from 25.2 °C to 35.7 °C
**b.** a waffle that produces energy to heat 4980 g of water from 20.6 °C to 62.4 °C

**2.24** Calculate the kilocalories each food provides when burned in a calorimeter:
**a.** one cup of popcorn that produces energy to heat 1250 g of water from 25.5 °C to 50.8 °C
**b.** a sample of butter that produces energy to heat 357 g of water from 22.7 °C to 38.8 °C

# 2.5 STATES OF MATTER

**LEARNING GOAL**

Identify the physical state of a substance as a solid, liquid, or gas.

**Matter** is anything that occupies space and has mass. This book, the food you eat, the water you drink, your cat or dog, and the air you breathe are just a few examples of matter. On Earth, matter exists in one of three *physical states*: solid, liquid, or gas. All matter is made up of tiny particles. In a **solid**, very strong attractive forces hold the particles close together. They are arranged in such a rigid pattern they can only vibrate slowly in their fixed positions. This gives a solid a definite shape and volume. For many solids, this rigid structure produces a crystal such as seen in quartz and amethyst (see Figure 2.4).

In a **liquid**, the particles have enough energy to move freely in random directions. They are still close to each other and have enough attractions to maintain a definite volume, but there is no rigid structure. Thus, when oil, water, or vinegar is poured from one container to another, the liquid maintains its own volume but takes the shape of the new container (see Figure 2.5).

**FIGURE 2.4** The solid state of amethyst, a purple form of quartz.
**Q** Why do these crystals have a regular shape?

**FIGURE 2.5** A liquid with a volume of 100 mL takes the shape of its container.
**Q** Why does a liquid have a definite volume, but not a definite shape?

The air you breathe is made of gases, mostly nitrogen and oxygen. In a **gas**, the molecules move at high speeds, creating great distances between molecules. This behavior allows gases to fill their containers. Gases have no definite shape or volume of their own; they take the shape and volume of their container, as shown in Figure 2.6. Table 2.7 compares some of the properties of the three states of matter.

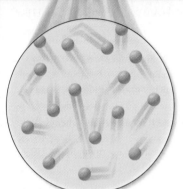

**FIGURE 2.6** A gas takes the shape and volume of its container.
**Q** Why does a gas fill the volume of a container?

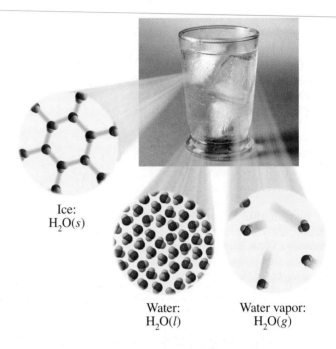

Ice:
$H_2O(s)$

Water:
$H_2O(l)$

Water vapor:
$H_2O(g)$

**TABLE 2.7** Some Properties of Solids, Liquids, and Gases

| Property | Solid | Liquid | Gas |
| --- | --- | --- | --- |
| Shape | Has a definite shape | Takes the shape of the container | Takes the shape of its container |
| Volume | Has a definite volume | Has a definite volume | Fills the volume of the container |
| Arrangement of particles | Fixed, very close | Random, close | Random, far apart |
| Interaction between particles | Very strong | Strong | Essentially none |
| Movement of particles | Very slow | Moderate | Very fast |
| Examples | Ice, salt, iron | Water, oil, vinegar | Water vapor, helium, air |

---

SAMPLE PROBLEM   2.8

■ **States of Matter**

Indicate the state(s) of matter that is (are) described by each of the following:

**a.** Volume does not change in a different container
**b.** Has a very low density
**c.** Has large distances between its particles
**d.** Shape depends on the container
**e.** Particles have a fixed arrangement

SOLUTION

**a.** solid, liquid    **b.** gas    **c.** gas    **d.** liquid, gas    **e.** solid

STUDY CHECK

What state has a definite volume but takes the shape of its container?

---

## QUESTIONS AND PROBLEMS

### States of Matter

2.25 Indicate whether each of the following describes a gas, a liquid, or a solid:
   **a.** This substance has no definite volume or shape.
   **b.** The particles in this substance do not interact strongly with each other.
   **c.** The particles of this substance are held in a definite structure.

2.26 Indicate whether each of the following describes a gas, a liquid, or a solid:
   **a.** The substance has a definite volume but takes the shape of the container.
   **b.** The particles of this substance are very far apart.
   **c.** This substance occupies the entire volume of the container.

---

# 2.6 CHANGES OF STATE

You are probably already familiar with most of these phase changes shown in Figure 2.7. Matter undergoes a **change of state** when it is converted from one state to another state.

When heat is added to a solid, the particles in the rigid structure begin to move faster. At a temperature called the **melting point (mp)**, the particles in the solid gain sufficient energy to overcome the attractive forces that hold them together. The particles in the solid separate and move about in random patterns. The substance is **melting**, changing from a solid to a liquid.

If the temperature of a liquid is lowered, the reverse process takes place. Kinetic energy is lost, the particles slow down, and attractive forces pull the particles close together. The substance is **freezing**. A liquid changes to a solid at the **freezing point (fp)**, which is the same temperature as the melting point. Every substance has its own freezing (melting) point: water freezes (melts) at 0 °C; gold freezes (melts) at 1064 °C; nitrogen freezes (melts) at −210 °C.

### LEARNING GOAL

Describe the changes of state between solids, liquids, and gases; calculate the energy involved.

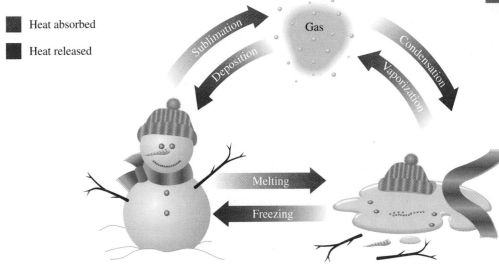

■ Heat absorbed

■ Heat released

**FIGURE 2.7**
A summary of the changes of state.
**Q** Is heat added or released when liquid water freezes?

+ Heat
Melting

Solid                          Liquid

Freezing
− Heat

During a change of state, the temperature of a substance remains constant. Suppose we have a glass containing ice and water. The ice melts when heat is added at 0 °C, forming more liquid. The liquid freezes when heat is removed at 0 °C.

## Heat of Fusion

During melting, energy called the **heat of fusion** is needed to separate the particles of a solid. For example, 80. calories of heat must be added to melt exactly 1 g of ice at its melting point.

**Heat of Fusion for Water**

$$\frac{80.\ cal}{1\ g\ ice}$$

The heat of fusion (80. cal/g) is also the heat that must be removed to freeze 1 g of water at its freezing point (0 °C). Water is sometimes sprayed in fruit orchards during very cold weather. If the air temperature drops to 0 °C, the water begins to freeze. Heat is released as the water molecules bond together, which warms the air and protects the fruit.

To determine the amount of heat needed to melt a sample of ice, multiply the mass of the ice by its heat of fusion. There is no temperature change in the calculation because temperature remains constant as long as the ice is melting.

**Calculating Heat to Melt (or Freeze) Water**

$$Heat = mass \times heat\ of\ fusion$$
$$cal = \cancel{g} \times \frac{80.\ cal}{\cancel{g}}$$

### Guide to Calculations Using Heat of Fusion/Vaporization

**1** List grams of substance and change of state.

**2** Write the plan to convert grams to heat and desired unit.

**3** Write the heat conversion factor and metric factor if needed.

**4** Set up the problem with factors.

---

**SAMPLE PROBLEM    2.9**

### ■ Heat of Fusion

Ice cubes at 0 °C with a mass of 26 g are added to your soft drink.

**a.** How much heat (cal) must be added to melt all the ice at 0 °C?
**b.** What happens to the temperature of your soft drink. Why?

SOLUTION

**a.** The heat in calories required to melt the ice is calculated as follows.

Step 1  **List grams of substance and change of state**.

    **Given**  26 g of $H_2O(s)$    **Need**  calories to melt ice

Step 2  **Write the plan to convert grams to heat and desired unit**.

    g ice  →  Heat of fusion  →  cal

Step 3  **Equalities/Conversion Factors**

$$1\ g\ H_2O(s \longrightarrow l) = 80.\ cal$$
$$\frac{80.\ cal}{1\ g\ H_2O} \quad and \quad \frac{1\ g\ H_2O}{80.\ cal}$$

Step 4  **Set Up Problem**

$$26\ \cancel{g\ H_2O} \times \frac{80.\ cal}{1\ \cancel{g\ H_2O}} = 2100\ cal$$

**b.** The soft drink will be colder because heat from the soft drink is providing the energy to melt the ice.

STUDY CHECK

In a freezer, 150 g of water at 0 °C is placed in an ice cube tray. How much heat in kilocalories must be removed to form ice cubes at 0 °C?

## Career Focus

### Histologist

"While a patient is in surgery for skin cancer, some tissue around the cancer is sent to us," says Mary Ann Pipe, histology technician. "Using the Mohs surgical technique, we place the tissue block on a glass slide, chill it to $-30\,°C$ in a machine called a cryostat, and freeze it for longer in another machine called a heat extractor. From this frozen block of tissue, we cut extremely thin slices—one-thousandth of an inch—from different depths. We prepare three separate slides from skin at three different depths up to the surface of the skin. We stain the cells pink and blue by placing the slides in hemotoxin, and then in eosin. The slices are a tissue map that the doctor can easily read to determine if the margins around the skin cancer are clear or whether more tissue must be removed."

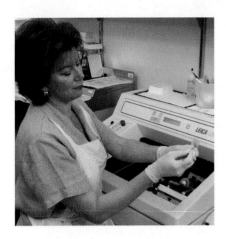

## Sublimation

In a process called **sublimation**, the particles on the surface of a solid absorb enough heat to change directly to a gas with no temperature change and without going through the liquid state. For example, dry ice, which is solid carbon dioxide, sublimes at $-78\,°C$. It is called "dry" because it does not form a liquid upon warming. In very cold areas, snow does not melt but sublimes directly to vapor. In a frost-free refrigerator, ice on the walls of the freezer sublimes when warm air is circulated through the compartment during the defrost cycle. In a process called deposition, a gas changes directly to a solid.

Freeze-dried foods prepared by sublimation are convenient for long-term storage and for camping and hiking. A food that has been frozen is placed in a vacuum chamber where it dries as the ice sublimes. The dried food retains all of its nutritional value and needs only water to be edible. A food that is freeze-dried does not need refrigeration because bacteria cannot grow without moisture.

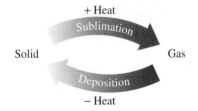

## Boiling and Condensation

Water in a mud puddle disappears, unwrapped food dries out, and clothes hung on a line dry. **Evaporation** is taking place as water molecules with sufficient energy escape from the liquid surface and enter the gas phase. (See Figure 2.8a.) The loss of the "hot" water molecules removes heat, which cools the remaining liquid water. As heat is added, more and more water molecules evaporate. At the **boiling point (bp)**, the molecules of the liquid have the energy needed to change into a gas. The **boiling** of a liquid occurs as gas bubbles form throughout the liquid, then rise to the surface and escape. (See Figure 2.8b.)

When heat is removed, a reverse process takes place. In **condensation**, water vapor is converted back to liquid as the water molecules lose kinetic energy, and slow down. Condensation occurs at the same temperature as boiling, but differs because heat is removed. You may have noticed that condensation occurs when you take a hot shower and the water vapor forms water droplets on a mirror. Because a substance loses heat as it

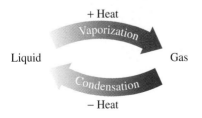

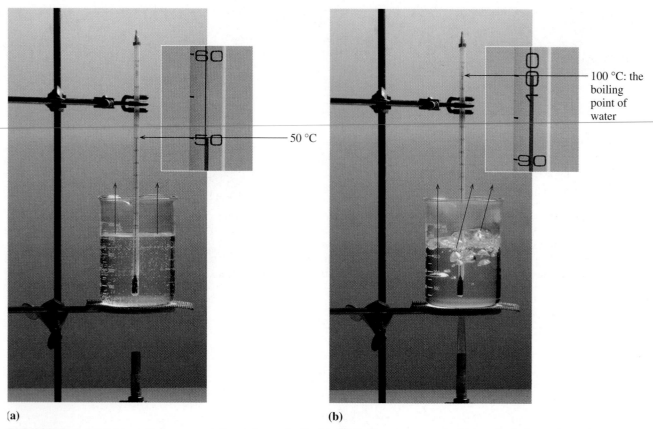

(a)                                                                    (b)

**FIGURE 2.8** **(a)** Evaporation occurs at the surface of a liquid. **(b)** Boiling occurs as bubbles of gas form throughout the liquid.

**Q** Why does water evaporate faster at 80 °C than at 20 °C?

condenses, its surroundings become warmer. That is why, when a rainstorm is approaching, we notice a warming of the air as gaseous water molecules condense to rain.

## Heat of Vaporization

The energy needed to vaporize exactly 1 g of liquid to gas at its boiling point is called the **heat of vaporization**. For water, 540 calories are needed to convert 1 g of water to vapor at 100 °C. This amount of heat (540 cal) is released when 1 g of water vapor (gas) changes to liquid at 100 °C.

### Heat of Vaporization for Water

$$\frac{540 \text{ cal}}{1 \text{ g water}}$$

To calculate the amount of heat added to vaporize (or removed to condense) a sample of water, the mass of the sample is multiplied by the heat of vaporization. As before, no temperature change occurs during a change of state.

### Calculating Heat to Vaporize (or Condense) Water

Heat = mass × heat of vaporization

$$\text{cal} = \cancel{g} \times \frac{540 \text{ cal}}{\cancel{g}}$$

# Health Note

## Steam Burns

Hot water at 100 °C will cause burns and damage to the skin. However, getting steam on the skin is even more dangerous. Let us consider 25 g of hot water at 100 °C. If this water falls on a person's skin, the temperature of the water will drop to body temperature, 37 °C. The heat released during cooling burns the skin. The amount of heat can be calculated from the temperature change, 63 °C.

$$25 \text{ g} \times 63 \text{ °C} \times \frac{1.00 \text{ cal}}{\text{g °C}} = 1600 \text{ cal heat}$$

For comparison, we can calculate the amount of heat released when 25 g of steam at 100 °C hits the skin. First, the steam condenses to water (liquid) at 100 °C:

$$25 \text{ g} \times \frac{540 \text{ cal}}{1 \text{ g}} = 14\,000 \text{ cal heat released}$$

Now the temperature of the 25 g of water drops from 100 °C to 37 °C, releasing still more heat as we saw earlier. We can calculate the total amount of heat released from the condensation and cooling of the steam as follows:

Condensation (100 °C)  =  14 000 cal

Cooling (100 °C to 37 °C)  =  1600 cal

Heat released  =  16 000 cal (rounded)

The amount of heat released from steam is 10 times greater than the heat from the same amount of hot water.

H₂O steam

Heat

H₂O liquid

---

## SAMPLE PROBLEM 2.10

### ■ Using Heat of Vaporization

In a sauna, 150 g of water is converted to steam at 100 °C. How many kilocalories of heat are needed?

SOLUTION

Step 1  **List grams of substance and change of state.**

**Given**  150 g $H_2O(l)$ to $H_2O(g)$

**Need**  kilocalories of heat to change state

Step 2  **Write the plan to convert grams to heat and desired unit.**

**Plan**  g $H_2O$  ⟩ Heat of vaporization ⟩ cal  Metric factor  kcal

**Step 3  Write the heat conversion factor and metric factor if needed.**

**Equalities/Conversion Factors**

$$1 \text{ g H}_2\text{O } (l \rightarrow g) = 540 \text{ cal} \qquad\qquad 1 \text{ kcal} = 1000 \text{ cal}$$

$$\frac{540 \text{ cal}}{1 \text{ g H}_2\text{O}} \quad \text{and} \quad \frac{1 \text{ g H}_2\text{O}}{540 \text{ cal}} \qquad \frac{1000 \text{ cal}}{1 \text{ kcal}} \quad \text{and} \quad \frac{1 \text{ kcal}}{1000 \text{ cal}}$$

**Step 4  Set up the problem with factors.**

**Problem Setup**

$$150 \text{ g H}_2\text{O} \times \frac{540 \text{ cal}}{1 \text{ g H}_2\text{O}} \times \frac{1 \text{ kcal}}{1000 \text{ cal}} = 81 \text{ kcal}$$

STUDY CHECK

When steam from a pan of boiling water reaches a cool window, it condenses. How much heat in kilocalories (kcal) is released when 25 g of steam condenses at 100 °C?

## Heating and Cooling Curves

All the changes of state during the heating of a solid can be illustrated visually. In a **heating curve**, the temperature is shown on the vertical axis and the addition of heat is shown on the horizontal axis. (See Figure 2.9a.)

## Steps on a Heating Curve

The first diagonal line indicates a warming of a solid as heat is added. When the melting temperature is reached, a horizontal line, or plateau, is drawn indicating that the solid is melting. As melting takes place, the solid is changing to liquid without any change in temperature.

   Once all the particles are in the liquid state, the heat that is added will increase the temperature of the liquid. This is drawn as a diagonal line from the melting point temperature to the boiling point temperature. Once the liquid starts to boil, another horizontal line is drawn to show that the temperature is constant as liquid changes to gas. Once all the liquid becomes a gas, any additional heat will increase the temperature of the gas.

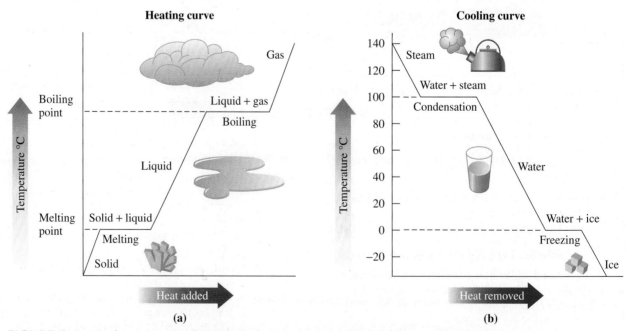

**FIGURE 2.9** **(a)** A heating curve shows the temperature increases and changes in state as heat is added. **(b)** A cooling curve shows temperature decreases and changes in state for water.
**Q** What does the plateau at 100 °C represent on the cooling curve for water?

## Steps on a Cooling Curve

All the changes of state during the cooling of a gas can also be illustrated visually. On a **cooling curve**, the temperature is plotted on the vertical axis and the removal of heat on the horizontal axis. (See Figure 2.9b.) Initially, a diagonal line to the boiling (condensation) point is drawn to show that heat is removed from a substance, cooling the gas until it begins to condense. A horizontal line (plateau) is drawn at the condensation point (same as the boiling point) to indicate the change of state as the gas condenses to form a liquid.

After all of the gas has changed into liquid, further cooling lowers the temperature. The decrease in temperature is shown as a diagonal line from the condensation point temperature to the freezing point temperature. At the freezing point, another horizontal line indicates that liquid is changing to solid at the freezing point temperature. Once all of the substance is frozen, a loss of heat decreases the temperature below its freezing point, which is shown as a diagonal line below the freezing point.

---

### SAMPLE PROBLEM 2.11

■ **Using a Cooling Curve**

Using the cooling curve for water in Figure 2.9, identify the state or change of state for water as solid (*s*), liquid (*l*), gas (*g*), condensation (C), or freezing (F).

**a.** at 120 °C          **b.** at 100 °C          **c.** at 40 °C
**d.** at 0 °C            **e.** at −10 °C

SOLUTION

**a.** *g*               **b.** C                 **c.** *l*
**d.** F                 **e.** *s*

STUDY CHECK

Describe the changes that take place when a sample of water vapor is cooled from 110 °C to 25 °C.

---

## QUESTIONS AND PROBLEMS

### Changes of State

**2.27** Identify each of the following changes of state as melting, freezing, or sublimation.
  **a.** The solid structure of a substance breaks down as liquid forms.
  **b.** Coffee is freeze-dried.
  **c.** Water on the street turns to ice during a cold wintry night.

**2.28** Identify each of the following changes of state as melting, freezing, or sublimation.
  **a.** Dry ice in an ice-cream cart disappears.
  **b.** Snow on the ground turns to liquid water.
  **c.** Heat is removed from 125 g of liquid water at 0 °C.

**2.29** Calculate the heat needed at 0 °C to make each of the following changes of state. Indicate whether heat was absorbed or released.
  **a.** calories to melt 65 g of ice
  **b.** calories to melt 17 g of ice
  **c.** kilocalories to freeze 225 g of water

**2.30** Calculate the heat needed at 0 °C to make each of the following changes of state. Indicate whether heat was absorbed or released.
  **a.** calories to freeze 35 g of water
  **b.** calories to freeze 250 g of water
  **c.** kilocalories to melt 140 g of ice

**2.31** Identify each of the following changes of state as evaporation, boiling, or condensation:
  **a.** The water vapor in the clouds changes to rain.
  **b.** Wet clothes dry on a clothesline.
  **c.** Lava flows into the ocean and steam forms.
  **d.** After a hot shower, your bathroom mirror is covered with water.

**2.32** Identify each of the following changes of state as evaporation, boiling, or condensation:
  **a.** At 100 °C, the water in a pan changes to steam.
  **b.** On a cool morning, the windows in your car fog up.
  **c.** A shallow pond dries up in the summer.
  **d.** Your teakettle whistles when the water is ready for tea.

**2.33 a.** How does perspiration during heavy exercise cool the body?
  **b.** Why do clothes dry more quickly on a hot summer day than on a cold winter day?

**2.34 a.** When a sports injury occurs during a game, a spray such as ethyl chloride may be used to numb an area of the skin. Explain how a substance like ethyl chloride that evaporates quickly can numb the skin.
  **b.** Why does water in a wide, flat, shallow dish evaporate more quickly than the same amount of water in a tall, narrow glass?

**2.35** Calculate the heat change at 100 °C in each of the following problems. Indicate whether heat was absorbed or released.
  **a.** calories to vaporize 10.0 g of water
  **b.** kilocalories to vaporize 50.0 g of water
  **c.** kilocalories to condense 8.0 kg of steam

**2.36** Calculate the heat change at 100 °C in each of the following problems. Indicate whether heat was absorbed or released.
  **a.** calories to condense 10.0 g of steam
  **b.** kilocalories to condense 75 g of steam
  **c.** kilocalories to vaporize 44 g of water

**2.37** Draw a heating curve for a sample of ice that is heated from −20 °C to 140 °C. Indicate the segment of the graph that corresponds to each of the following:
  **a.** solid          **b.** melting point
  **c.** liquid         **d.** boiling point
  **e.** gas

**2.38** Draw a cooling curve for a sample of steam that cools from 110 °C to −10 °C. Indicate the segment of the graph that corresponds to each of the following:
  **a.** solid          **b.** freezing point
  **c.** liquid         **d.** condensation point (boiling point)
  **e.** gas

## CONCEPT MAP

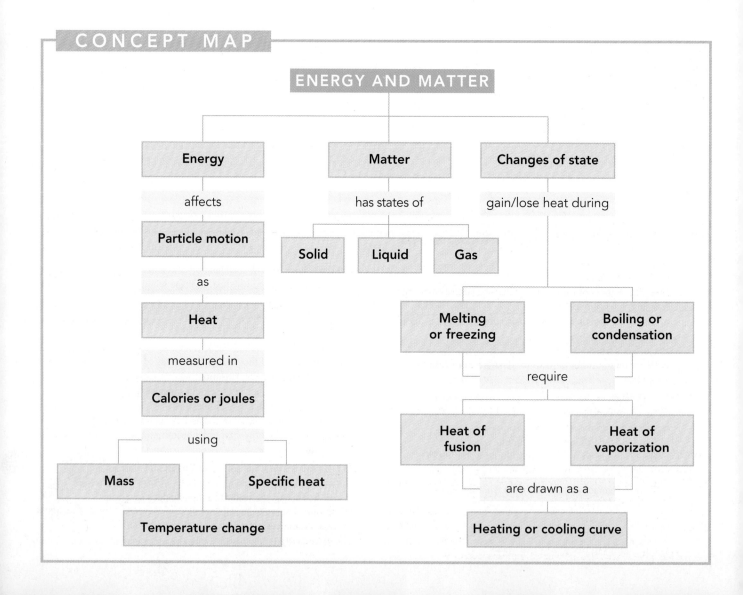

# CHAPTER REVIEW

## 2.1 Energy

**Learning Goal:** Identify energy as potential or kinetic.

Energy is the ability to do work. Potential energy is stored energy; kinetic energy is the energy of motion. Heat, a common form of energy, is measured in calories or joules. One cal is equal to 4.184 J.

## 2.2 Energy and Nutrition

**Learning Goal:** Use caloric values to calculate the kilocalories (Cal) in a food.

The nutritional Calorie is the same amount of energy as 1 kcal or 1000 calories. The caloric content of a food is the sum of kilocalories from carbohydrate, fat, and protein.

## 2.3 Temperature Conversions

**Learning Goal:** Given a temperature, calculate a corresponding temperature on another scale.

On the Celsius scale, there are 100 units between the freezing point of water (0 °C) and the boiling point (100 °C). On the Fahrenheit scale, there are 180 units between the freezing point of water (32 °F) and the boiling point (212 °F). A Fahrenheit temperature is related to its Celsius temperature by the equation $T_F = 1.8\,T_C + 32°$. The SI temperature of Kelvin is related to the Celsius temperature by the equation $T_K = T_C + 273$

## 2.4 Specific Heat

**Learning Goal:** Use specific heat to calculate heat loss or gain, temperature change, or mass of a sample.

Specific heat is the amount of energy required to raise the temperature of exactly 1 g of a substance by exactly 1 °C. The heat lost or gained by a substance is determined by multiplying its mass (g), the temperature change ($\Delta T$), and its specific heat (cal (or J)/g °C).

## 2.5 States of Matter

**Learning Goal:** Identify the physical state of a substance as a solid, liquid, or gas.

Matter is anything that has mass and occupies space. The three states of matter are solid, liquid, and gas.

## 2.6 Changes of State

**Learning Goal:** Describe the changes of state between solids, liquids, and gases; calculate the energy involved.

Melting occurs when the particles in a solid absorb enough energy to break apart and form a liquid. The amount of energy required to convert exactly 1 g of solid to liquid is called the heat of fusion. For water, 80. cal are needed to melt 1 g of ice. Boiling is the vaporization of a liquid at its boiling point. The heat of vaporization is the amount of heat needed to convert exactly 1 g of liquid to vapor. For water, 540 cal are needed to vaporize 1 g of liquid water. A heating or cooling curve illustrates the changes in temperature and state as heat is added to or removed from a substance. Plateaus on the graph indicate changes of state with no change in temperature.

# KEY TERMS

**boiling** The formation of bubbles of gas throughout a liquid.

**boiling point (bp)** The temperature at which a liquid changes to gas (boils) and gas changes to liquid (condenses).

**caloric value** The kilocalories obtained per gram of the food types: carbohydrate, fat, and protein.

**calorie (cal)** The amount of heat energy that raises the temperature of exactly 1 g of water exactly 1 °C.

**change of state** The transformation of one state of matter to another; for example, solid to liquid, liquid to solid, liquid to gas.

**condensation** The change of state of a gas to a liquid.

**cooling curve** A diagram that illustrates temperature changes and changes of state for a substance as heat is removed.

**energy** The ability to do work.

**evaporation** The formation of a gas (vapor) by the escape of high-energy molecules from the surface of a liquid.

**freezing** The change of state from liquid to solid.

**freezing point (fp)** The temperature at which a liquid changes to a solid (freezes), a solid changes to a liquid (melts).

**gas** A state of matter characterized by no definite shape or volume. Particles in a gas move rapidly.

**heat** The energy associated with the motion of particles in a substance.

**heat of fusion** The energy required to melt exactly 1 g of a substance at its melting point. For water, 80. cal are needed to melt 1 g of ice; 80. cal are released when 1 g of water freezes.

**heat of vaporization** The energy required to vaporize 1 g of a substance at its boiling point. For water, 540 calories are needed to vaporize exactly 1 g of liquid; 1 g of steam gives off 540 cal when it condenses.

**heating curve** A diagram that shows the temperature changes and changes of state of a substance as it is heated.

**joule (J)** The SI unit of heat energy; 4.184 J = 1 cal.

**kilocalorie (kcal)** An amount of heat energy equal to 1000 calories.

**kinetic energy** The energy of moving particles.

**liquid** A state of matter that takes the shape of its container but has a definite volume.

**matter** Anything that has mass and occupies space.

**melting** The change of state from a solid to a liquid.

**melting point (mp)** The temperature at which a solid becomes a liquid (melts). It is the same temperature as the freezing point.

**potential energy** An inactive type of energy that is stored for future use.

**solid** A state of matter that has its own shape and volume.

**specific heat** A quantity of heat that changes the temperature of exactly 1 g of a substance by exactly 1 °C.

**sublimation** The change of state in which a solid is transformed directly to a gas without forming a liquid first.

**work** An activity that requires energy.

# UNDERSTANDING THE CONCEPTS

**2.39** Read the temperature on each of the Celsius thermometers.

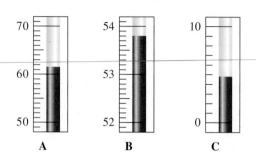

**2.40** Select the warmer temperature in each pair.
**a.** 10 °C or 10 °F      **b.** 30 °C or 15 °F
**c.** −10 °C or 32 °F     **d.** 200 °C or 200 K

**2.41** Compost can be made at home from grass clippings, some kitchen scraps, and dry leaves. As microbes break down organic matter, heat is generated and the compost can reach a temperature of 155 °F, which kills most pathogens. What is this initial temperature in Celsius degrees? In kelvins?

**2.42** After some time, biochemical reactions in compost slow, and the temperature drops to 45 °C. The dark brown organic rich mixture is ready for use in the garden. What is this temperature in Fahrenheit degrees? In kelvins?

**2.43** A 70.0-kg person has just eaten a quarter-pound cheeseburger, French fries, and a chocolate shake. Using Table 2.1, calculate the total kilocalories in this meal (round to the nearest 10 kcal).

| Item | Protein (g) | Fat (g) | Carbohydrate (g) |
|---|---|---|---|
| cheeseburger | 31 | 29 | 34 |
| French fries | 3 | 11 | 26 |
| chocolate shake | 11 | 9 | 60 |

**2.44** For the person in problem 2.43, use Table 2.4 to determine each of the following:
**a.** The number of hours of sleep needed to "burn off" the kilocalories in this meal.
**b.** The number of hours of running needed to "burn off" the kilocalories in this meal.

**2.45** Determine the energy to heat three cubes (gold, aluminum, and silver) each with a volume of 10.0 cm$^3$ from 15 °C to 25 °C. Refer to Tables 1.11 and 2.6. What do you notice about the energy needed for each?

**2.46** If you used the 8400 kilojoules you expend in energy in one day to heat 50 000 g of water at 20 °C, what would be the rise in temperature? What would be the new temperature of the water?

**2.47** The following is a heating curve for chloroform, a solvent for fats, oils, and waxes.

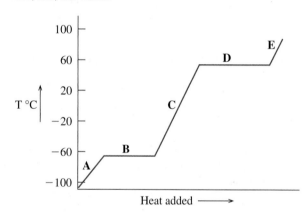

**a.** What is the melting point of chloroform?
**b.** What is the boiling point of chloroform?
**c.** On the heating curve, identify the segments A, B, C, D, and E as solid, liquid, gas, melting, or boiling.
**d.** At the following temperatures, is chloroform a solid, liquid, or gas?

   −80 °C; −40 °C; 25 °C; 80 °C

**2.48** Associate the following diagrams with a segment on the heating curve for water.

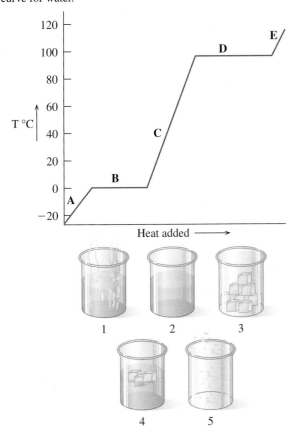

## ADDITIONAL QUESTIONS AND PROBLEMS

**2.49** Calculate the following temperatures in degrees Celsius:
  **a.** The highest recorded temperature in the continental United States was 134 °F in Death Valley, California, July 10, 1913.
  **b.** The lowest recorded temperature in the continental United States was −70. °F in Rodgers Pass, Montana, January 20, 1954.

**2.50** Calculate the following temperatures in degrees Fahrenheit:
  **a.** The highest recorded temperature in the world was 58 °C in El Azizia, Libya, September 13, 1922.
  **b.** The lowest recorded temperature in the world was −89.2 °C in Vostok, Antarctica, July 21, 1983.

**2.51** What is −15 °F in degrees Celsius and in kelvins?

**2.52** The highest recorded body temperature that a person has survived is 46.5 °C. Calculate the temperature in degrees Fahrenheit.

**2.53** On a hot day, the beach sand gets hot, but the water stays cool. Compare the specific heat of sand to that of water.

**2.54** Why do drops of liquid water form on the outside of a glass of iced tea?

**2.55** When it rains or snows, the air temperature seems warmer. Explain.

**2.56** Water is sprayed on the ground of an orchard when temperatures are near freezing to keep the fruit from freezing. Explain.

**2.57** A typical diet in the United States provides 15% of the calories from protein, 45% from carbohydrates, and the remainder from fats. Calculate the grams of protein, carbohydrate, and fat to be included each day in diets having the following caloric requirements:
  **a.** 1200 kcal
  **b.** 1900 kcal
  **c.** 2600 kcal

**2.58** Your friend has just eaten a slice of pizza, a cola soft drink, and ice cream. What is the total kilocalories your friend obtained from this meal? (Round to the nearest 10 kcal.) How many hours will your friend need to swim to "burn off" the kilocalories in this meal if your friend has a mass of 70.0 kg? (See Table 2.4.)

| Item | Protein (g) | Fat (g) | Carbohydrate (g) |
|---|---|---|---|
| Pizza | 13 | 10 | 29 |
| Cola | 0 | 0 | 51 |
| Ice cream | 8 | 28 | 44 |

**2.59** If you want to lose 1 pound of "fat," which is 15% water, how many kilocalories do you need to expend?

**2.60** Calculate the Cal (kcal) in 1 cup of whole milk: 12 g of carbohydrate, 9 g of fat, and 9 g of protein. (Round to the nearest 10 kcal.)

**2.61** Identify each of the following as solid, liquid, or gas:
  **a.** vitamin tablets in a bottle
  **b.** helium in a balloon
  **c.** milk in a glass
  **d.** the air you breathe
  **e.** charcoal briquettes on a barbecue

**2.62** Identify each of the following as solid, liquid, or gas:
   **a.** popcorn in a bag
   **b.** water in a garden hose
   **c.** a computer mouse
   **d.** air in a tire
   **e.** hot tea

**2.63** A hot-water bottle contains 725 g of water at 65 °C. If the water cools to body temperature (37 °C), how many kilocalories of heat could be transferred to sore muscles?

**2.64** A pitcher containing 0.75 L of water at 4 °C is removed from the refrigerator. How many kilojoules are needed to warm the water to a room temperature of 27 °C?

**2.65** The melting point of chloroform is −64 °C and its boiling point is 61 °C. Sketch a heating curve for chloroform from −100 °C to 100 °C.
   **a.** What is the state of chloroform at −75 °C?
   **b.** What happens on the curve at −64 °C?
   **c.** What is the state of chloroform at −18 °C?
   **d.** What is the state of chloroform at 80 °C?
   **e.** At what temperature will both solid and liquid be present?

**2.66** The melting point of benzene is 5.5 °C and its boiling point is 80.1 °C. Sketch a heating curve for benzene from 0 °C to 100 °C.
   **a.** What is the state of benzene at 15 °C?
   **b.** What happens on the curve at 5.5 °C?
   **c.** What is the state of benzene at 63 °C?
   **d.** What is the state of benzene at 98 °C?
   **e.** At what temperature will both liquid and gas be present?

# CHALLENGE QUESTIONS

**2.67** One liquid has a temperature of 140. °F and another liquid has a temperature of 60.0 °C. Are the liquids at the same temperature or at different temperatures?

**2.68** A 0.50-g sample of vegetable oil is placed in a calorimeter. When the sample is burned, 18.9 kJ are given off. What is the caloric value in kcal/g of the oil?

**2.69** A 25-g sample of an alloy at 98 °C is placed in 50. g of water at 15 °C. If the final temperature reached by the alloy sample and water is 27 °C, what is the specific heat (cal/g °C) of the alloy?

**2.70** Rearrange the heat equation to solve for each of the following:
   **a.** The mass in grams of water that absorbs 8250 J when its temperature rises from 18.3 °C to 92.6 °C.
   **b.** The mass in grams of a gold sample that absorbs 225 J when the temperature rises from 15.0 °C to 47.0 °C.
   **c.** The rise in temperature ($\Delta T$) when a 20.0 g sample of iron absorbs 1580 J.
   **d.** The specific heat of a metal when 8.50 g of the metal absorbs 28 cal and the temperature rises from 12 °C to 24 °C.

**2.71** How many kilocalories of heat are released when 75 g of steam at 100 °C is converted to ice at 0 °C? (*Hint*: The calculations include several steps.)

**2.72** A 45-g piece of ice at 0 °C is added to a sample of water at 8 °C. All of the ice melts and the temperature of the water decreases to 0 °C. How many grams of water were in the sample?

**2.73** In a large building, oil is used in a steam boiler heating system. The combustion of 1.0 lb of oil provides $2.4 \times 10^7$ J.
   **a.** How many kg of oil are needed to heat 150 kg of water from 22 °C to 100 °C?
   **b.** How many kg of oil are needed to provide steam from 150 kg of water at 100 °C?

**2.74** The combustion of 1.0 g of gasoline releases 11 kcal of heat (density of gasoline = 0.74 g/mL).
   **a.** How many megajoules are released when 1.0 gallon of gasoline burns?
   **b.** When a color television is on for 2.0 h, 300 kJ are used. How long can a color television run on the energy from 1.0 gallon of gasoline?

# ANSWERS

## Answers to Study Checks

**2.1** potential energy

**2.2** 8.4 kcal

**2.3** 380 kcal

**2.4** 12 °F

**2.5** 39.8 °C

**2.6** night −260. °C; day 410. °C

**2.7** 14.6 kJ

**2.8** liquid

**2.9** 12 kcal

**2.10** 14 kcal

**2.11** As heat is lost, the sample of water vapor cools from 110 °C to 100 °C. At 100 °C, the loss of energy causes the water vapor to condense to liquid water. Then as further cooling removes more heat, the temperature of the liquid water drops to 25 °C.

## Answers to Selected Questions and Problems

**2.1** When the car is at the top of the ramp, it has its maximum potential energy. As it descends, potential energy changes to kinetic energy. At the bottom, all the energy is kinetic.

**2.3** **a.** potential          **b.** kinetic
     **c.** potential          **d.** potential

**2.5** **a.** 22 kJ          **b.** 5300 cal
     **c.** 5.3 kcal

**2.7** **a.** 110 Cal          **b.** 18 g
     **c.** 130 Cal          **d.** 280 Cal

**2.9** 210 kcal, 880 kJ

**2.11** In the U.S., we still use the Fahrenheit temperature scale. In °F, normal body temperature is 98.6. On the Celsius scale, her temperature would be 37.7 °C, a mild fever.

**2.13** **a.** 98.6 °F          **b.** 18.5 °C
     **c.** 246 K          **d.** 335 K
     **e.** 46 °C          **f.** 295 K

**2.15 a.** 41 °C
**b.** No. The temperature is equivalent to 39 °C.

**2.17** Copper has the lowest specific heat of the samples and will reach the highest temperature.

**2.19 a.** 250 cal **b.** 14 000 J
**c.** 9.3 kcal **d.** 10.8 kJ

**2.21 a.** 1380 J, 330. cal **b.** 1810 J, 434 cal
**c.** 3780 J, 904 cal **d.** 3500 J, 850 cal

**2.23 a.** 5.30 kcal **b.** 208 kcal

**2.25 a.** gas **b.** gas
**c.** solid

**2.27 a.** melting **b.** sublimation
**c.** freezing

**2.29 a.** 5200 cal absorbed **b.** 1400 cal absorbed
**c.** 18 kcal released

**2.31 a.** condensation **b.** evaporation
**c.** boiling **d.** condensation

**2.33 a.** The heat from the skin is used to evaporate the water (perspiration). Therefore the skin is cooled.
**b.** On a hot day, there are more molecules with sufficient energy to become water vapor.

**2.35 a.** 5400 cal absorbed **b.** 27 kcal absorbed
**c.** 4300 kcal released

**2.37**

```
T °C   150
       100
        50                              Gas (e)
         0                    Boiling point (d)
       -20              Liquid (c)
                          Melting point (b)
                   Solid (a)
             Heat added ———→
```

**2.39 a.** 61.4 °C **b.** 53.80 °C
**c.** 4.8 °C

**2.41** 68 °C, 341 K

**2.43** 45 g protein $\times \dfrac{4\ \text{kcal}}{1\ \text{g protein}} = 180$ kcal

49 g fat $\times \dfrac{9\ \text{kcal}}{1\ \text{g fat}} = 440$ kcal

120 g carbohydrate $\times \dfrac{4\ \text{kcal}}{1\ \text{g carbohydrate}} = 480$ kcal

180 kcal + 441 kcal + 480 kcal = 1100 kcal

**2.45** Gold: 250 J or 59 cal; aluminum: 240 J or 58 cal; silver: 250 J or 59 cal. The heat needed for all the metal samples is almost the same.

**2.47 a.** −60 °C **b.** 60 °C
**c.** The diagonal line A represents the solid state as temperature increases. The horizontal line B represents the change from solid to liquid or melting of the substance. The diagonal line C represents the liquid state as temperature increases. The horizontal line D represents the change from liquid to gas or boiling of the liquid. The diagonal line E represents the gas state as temperature increases.
**d.** At −80 °C, solid; at −40 °C, liquid; at 25 °C, liquid; at 80 °C, gas

**2.49 a.** 56.7 °C **b.** −56.7 °C

**2.51** −26 °C; 247 K

**2.53** Sand must have a lower specific heat than water. When both substances absorb the same amount of heat, the final temperature of the sand will be higher than that of water.

**2.55** When water vapor condenses or liquid water freezes, heat is released, which warms the air.

**2.57 a.** 45 g of protein, 140 g of carbohydrate, 53 g of fat
**b.** 71 g of protein, 210 g of carbohydrate, 84 g of fat
**c.** 98 g of protein, 290 g of carbohydrate, 120 g of fat

**2.59** 3500 kcal

**2.61 a.** solid **b.** gas **c.** liquid
**d.** gas **e.** solid

**2.63** 20. kcal

**2.65**

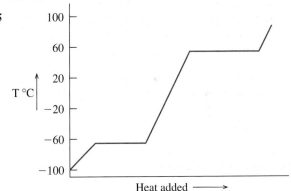

```
T °C   100
        60
        20
       −20
       −60
      −100
             Heat added ———→
```

**a.** solid **b.** solid chloroform melts
**c.** liquid **d.** gas
**e.** −64 °C

**2.67** $T_F = 1.8\,(60.0\ °C) + 32° = 140.\ °F$. The liquids are at the same temperature.

**2.69** specific heat = 0.34 cal/g °C

**2.71** 54 kcal

**2.73 a.** 0.93 kg of oil **b.** 6.4 kg of oil

# Combining Ideas from Chapters 1 and 2

**CI.1** Gold, one of the most sought-after metals in the world, has a density of 19.3 g/cm$^3$, melts at 1064 °C, and has a specific heat of 0.129 J/g °C. In 1998, a gold nugget that weighed 20.17 lb was discovered in Alaska.

A      B

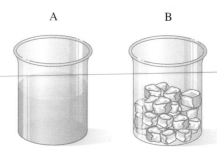

**c.** Match the diagrams (1, 2, or 3) that represent the water particles with sample A or B. Give a reason for your choice.

1      2      3

**a.** How many significant figures are in the measurement of the weight of the nugget?

**b.** Which is the mass of the nugget in kilograms?

**c.** If the nugget were pure gold, what would its volume be in cm$^3$?

**d.** What is the melting point of gold in degrees Fahrenheit and in kelvins?

**e.** How many kilocalories are required to raise the temperature of the nugget from 18 °C to 145 °C?

**CI.2** The mileage for a motorcycle with a fuel tank capacity of 22 L is 35 miles per gallon.

**a.** How long a trip, in kilometers, can be made on one full tank of gasoline?

**b.** If the price of gasoline is $3.27 per gallon, what would be the cost of fuel for the trip?

**c.** If the average speed during the trip is 44 mi/h, how many hours will it take to reach the destination?

**d.** If the density of gasoline is 0.74 g/mL, what is the mass in grams of the fuel in the tank?

**e.** When 1.00 g of gasoline burns, 46 kJ of energy is released. How many kilojoules are produced when the fuel in one full tank is burned?

**CI.3** Answer the following questions for the water samples A and B shown in the diagrams:

**a.** Which sample has its own shape?

**b.** When each sample is transferred to another container, what happens to each volume?

**d.** When the water in A changes to B, the process is called _____, which occurs at a temperature called the _____ point.
When the water in B changes to A, the process is called _____, which occurs at a temperature called the _____ point.

**e.** What happens to the solid water particles when melting occurs?

**f.** If the water in A has a mass of 19.8 g and a temperature of 45 °C, how much heat, in joules, is removed to form ice at 0 °C?

**CI.4** The label of a lemon bar lists the "nutrition facts" as 5 g of fat, 25 g of carbohydrate, and 10 g of protein.

a. Using the caloric values of carbohydrate (4 kcal/g), fat (9 kcal/g), and protein (4 kcal/g), what are the total kilocalories (Calories) listed for the lemon bar?

b. What is the energy value of the lemon bar in kilojoules?

c. If the bar has a mass of 48 g, how many kilojoules are obtained from eating 10. g of the lemon bar?

d. If you are walking (840 kJ/h), how many minutes will you need to walk to burn off the calories of two lemon bars?

**CI.5** In a box of nails, there are 75 iron nails weighing 0.25 lb. The density of iron is 7.86 g/cm³. The specific heat of iron is 0.450 J/g °C. The melting point of iron is 1535 °C.

a. What is the volume, in cm³, of all the iron nails in the box?

b. If 30 nails are added to a graduated cylinder containing 17.6 mL of water, what is the new level of water in the cylinder?

c. How many joules must be added to the nails in the box to raise the temperature from 16 °C to 125 °C?

**CI.6** A hot tub is filled with 450 gallons of water, which has a density of 1.0 g/mL.

a. What is the volume of water in the tub, in liters?

b. What is the mass, in kilograms, of water it contains?

c. How many kilocalories are needed to heat the water from 62 °F to 105 °F?

d. If the hot-tub heater provides 1400 kcal per minute, how long in hours will it take to heat the water in the hot tub from 62 °F to 105 °F?

# ANSWERS

**CI.1** a. 4 significant figures    b. 9.17 kg
c. 475 cm³                          d. 1947 °F; 1337 K
e. 35.9 kcal

**CI.3** a. B
b. The volumes of both A and B remain the same.
c. A is liquid water represented by diagram 2. In liquid water, the water particles are in a random arrangement, but close together. B is solid water represented by diagram 1. In solid water, the water particles are fixed in a definite arrangement.

d. freezing; freezing point; 0 °C; melting; melting point; 0 °C
e. The solid water particles break apart from their fixed arrangement to have a more random arrangement, but they are still close together.
f. 10 300 J

**CI.5** a. 14 cm³                  b. 23.2 mL
c. 5600 J or $5.6 \times 10^3$ J

# 3

# Atoms and Elements

the Chemistry place

Visit **www.chemplace.com** for extra quizzes, interactive tutorials, career resources, PowerPoint slides for chapter review, math help, and case studies.

*"Many of my patients have diabetes, ulcers, hypertension, and cardiovascular problems," says Sylvia Lau, registered dietitian. "If a patient has diabetes, I discuss foods that raise blood sugar such as fruit, milk, and starches. I talk about how dietary fat contributes to weight gain and complications from diabetes. For stroke patients, I suggest diets low in fat and cholesterol because high blood pressure increases the risk of another stroke."*

*If a lab test shows low levels of iron, zinc, iodine, magnesium, or calcium, a dietitian discusses foods that provide those essential elements. For instance, she may recommend more beef for an iron deficiency, whole grain for zinc, leafy green vegetables for magnesium, dairy products for calcium, and iodized table salt and seafood for iodine.*

E very day, we see around us a variety of materials with many shapes and forms. To a scientist, all of this material is *matter*. There is matter everywhere around you; the orange juice you had for breakfast, the water you put in the coffeemaker, the aluminum foil you wrap your sandwich in, your toothbrush and toothpaste, the oxygen you inhale, and the carbon dioxide you exhale are all forms of matter.

All matter is composed of *elements*, of which there are 117 different kinds. Of these, 88 elements, which occur in nature, make up all the substances in our world. Many elements are already familiar to you. You may have a ring or necklace made of gold or silver or perhaps platinum. If you play tennis or golf, your racket or clubs may be made from the elements titanium or carbon. In the body, calcium and phosphorus form the structure of bones and teeth, iron and copper are needed in the formation of red blood cells, and iodine is required for the proper functioning of the thyroid.

The amounts of certain elements are crucial to the proper growth and function of the body. Low levels of iron can lead to anemia, while low levels of iodine can cause hypothyroidism and goiter. Lab tests are used to confirm that elements such as iron, copper, zinc, or iodine are within normal ranges in a patient's blood serum. A dietitian may recommend beef for iron and zinc, whole grains and leafy green vegetables for magnesium, dairy products for calcium, and iodized table salt and seafood for iodine.

# 3.1 CLASSIFICATION OF MATTER

## LEARNING GOAL

Classify matter as pure substances or mixtures.

In Chapter 2, we defined *matter* as anything that has mass and occupies space. There are many types of matter, but we can categorize them according to their components.

## Pure Substances

A **pure substance** is a type of matter that has a definite composition. **Elements** are the simplest pure substances because they are composed of only one kind of atom. You may already know about elements such as silver, iron, and aluminum. A full list of the elements is found on the inside front cover of this text. **Compounds** are pure substances too, but they consist of two or more elements always in the same proportion. For example, the compound water, $H_2O$, always has the same proportion of the elements hydrogen and oxygen. The compound, hydrogen peroxide, $H_2O_2$, is also a combination of the elements hydrogen and oxygen, but in a different ratio.

An important difference between elements and compounds is that chemical processes can break down compounds into simpler substances such as elements. You may know that ordinary table salt consists of the compound NaCl, which can be broken down into sodium and chlorine as seen in Figure 3.1. Compounds are not broken down through physical methods such as boiling or sifting. However, elements cannot be broken down by either chemical or physical processes.

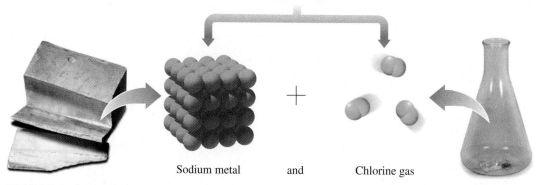

Sodium chloride

Sodium metal        and        Chlorine gas

**FIGURE 3.1** The decomposition of salt, NaCl, produces the elements sodium and chlorine.
**Q** How do elements and compounds differ?

## Mixtures

Much of the matter in our everyday lives consists of mixtures. (See Figure 3.2.) In a **mixture**, two or more substances are physically mixed, but not chemically combined. The air we breathe is a mixture of mostly oxygen and nitrogen gases. The steel in buildings and railroad tracks is a mixture of iron, nickel, carbon, and chromium. The brass in knobs and fixtures is a mixture of zinc and copper. Solutions such as tea, coffee, and ocean water are mixtures too. In any mixture, the composition can vary. For example, two sugar–water mixtures would appear the same, but one would taste sweeter because it has a higher ratio of sugar to water. Different types of brass have different properties, depending on the ratio of copper to zinc.

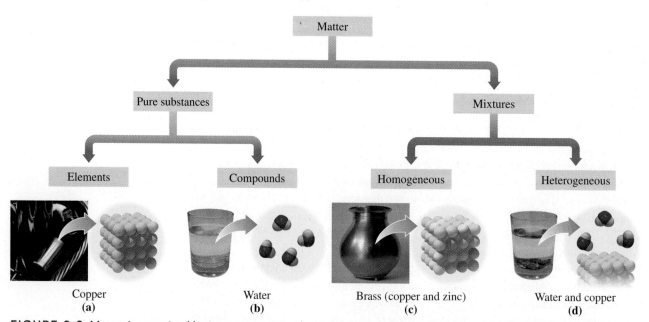

Copper
**(a)**

Water
**(b)**

Brass (copper and zinc)
**(c)**

Water and copper
**(d)**

**FIGURE 3.2** Matter is organized by its components: elements, compounds, and mixtures. **(a)** The element copper consists of copper atoms. **(b)** The compound water consists of $H_2O$ molecules. **(c)** Brass is a homogeneous mixture of copper atoms and zinc atoms. **(d)** Copper metal in water is a heterogeneous mixture of Cu atoms and $H_2O$ molecules.
**Q** Why are copper and water pure substances, but brass is a mixture?

**FIGURE 3.3** A mixture of spaghetti and water are separated using a strainer, a physical method of separation.
**Q** Why can physical methods be used to separate mixtures but not compounds?

Physical method of separation

Physical processes can be used to separate mixtures because there are no chemical interactions between the components. For example, different coins such as nickels, dimes, and quarters can be separated by size; iron particles mixed with sand can be picked up with a magnet; and water is separated from cooked spaghetti using a strainer. (See Figure 3.3.)

## Types of Mixtures

Mixtures are classified as homogeneous or heterogeneous. In a **homogeneous mixture**, also called a *solution*, the composition is uniform throughout the sample. Examples of familiar homogeneous mixtures are air, which contains oxygen and nitrogen gases; bronze, which is a mixture of copper and tin; and salt water, a solution of salt and water.

In a **heterogeneous mixture**, the components do not have a uniform composition throughout the sample. For example, a mixture of oil and water is heterogeneous because the oil floats on the surface of the water. Other examples of heterogeneous mixtures include the raisins in a cookie and a hot fudge sundae. Table 3.1 summarizes the classification of matter.

**TABLE 3.1** Classification of Matter

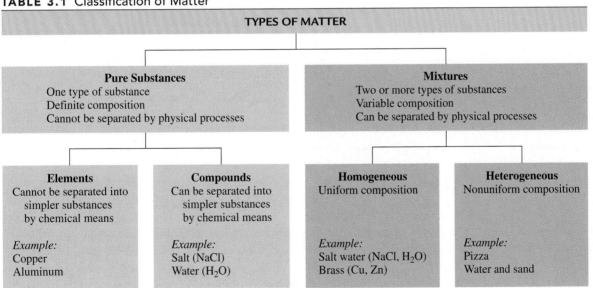

| TYPES OF MATTER | | | |
|---|---|---|---|
| **Pure Substances**<br>One type of substance<br>Definite composition<br>Cannot be separated by physical processes | | **Mixtures**<br>Two or more types of substances<br>Variable composition<br>Can be separated by physical processes | |
| **Elements**<br>Cannot be separated into simpler substances by chemical means<br><br>*Example:*<br>Copper<br>Aluminum | **Compounds**<br>Can be separated into simpler substances by chemical means<br><br>*Example:*<br>Salt (NaCl)<br>Water ($H_2O$) | **Homogeneous**<br>Uniform composition<br><br><br>*Example:*<br>Salt water (NaCl, $H_2O$)<br>Brass (Cu, Zn) | **Heterogeneous**<br>Nonuniform composition<br><br><br>*Example:*<br>Pizza<br>Water and sand |

## SAMPLE PROBLEM 3.1

### ■ Pure Substances and Mixtures

Classify each of the following as a pure substance (element or compound) or a mixture (homogeneous or heterogeneous). Give a reason for your choice of answer.

**a.** carbon dioxide ($CO_2$)

**b.** coffee with sugar

**c.** aluminum

**d.** ice cream float

SOLUTION

**a.** Pure substance; compound with a definite ratio of elements.

**b.** Mixture; homogeneous with uniform composition of coffee and sugar.

**c.** Pure substance; element with one type of matter that appears on the list of elements.

**d.** Mixture; heterogeneous with a nonuniform composition.

STUDY CHECK

A salad dressing is prepared with oil, vinegar, and chunks of blue cheese. Is this a homogeneous or heterogeneous mixture?

## QUESTIONS AND PROBLEMS

### Classification of Matter

**3.1** Classify each of the following pure substances as an element or a compound. Give a reason for your answer.
   **a.** baking soda
   **b.** oxygen
   **c.** ice
   **d.** aluminum foil

**3.2** Classify each of the following pure substances as an element or a compound. Give a reason for your answer.
   **a.** platinum in a catalytic converter
   **b.** vitamin C
   **c.** mercury in a thermometer
   **d.** carbon monoxide

**3.3** Classify each of the following mixtures as homogeneous or heterogeneous. Give a reason for your answer.
   **a.** chocolate milk
   **b.** yogurt with blackberries
   **c.** blueberry muffin
   **d.** liquid detergent

**3.4** Classify each of the following mixtures as homogeneous or heterogeneous. Give a reason for your answer.
   **a.** water and sand in an aquarium
   **b.** fruit salad
   **c.** a soft drink
   **d.** hot tea with sweetener

**3.5** Classify each of the following as a pure substance (element or compound) or mixture (homogeneous or heterogeneous). Give a reason for your answer.
   **a.** chicken noodle soup
   **b.** carbon in a tennis racket
   **c.** salt
   **d.** water with a lemon slice

**3.6** Classify each of the following as a pure substance (element or compound) or mixture (homogeneous or heterogeneous). Give a reason for your answer.
   **a.** seawater
   **b.** ham and cheese sandwich
   **c.** iodine
   **d.** propane gas

## LEARNING GOAL

Given the name of an element, write its correct symbol; from the symbol, write the correct name.

# 3.2 ELEMENTS AND SYMBOLS

*Elements* are primary substances from which all other things are built. They cannot be broken down into simpler substances. Many of the elements were named for planets, mythological figures, minerals, colors, geographic locations, and famous people. Some sources of names of elements are listed in Table 3.2.

## Chemical Symbols

**Chemical symbols** are one- or two-letter abbreviations for the names of the elements. Only the first letter of an element's symbol is capitalized; a second letter, if there is one, is lowercase. That way, we know when a different element is indicated. If both letters are capitalized, it represents the symbols of two different elements. For example, the element cobalt has the symbol Co. However, the two capital letters CO specify two elements, carbon (C) and oxygen (O).

**One-Letter Symbols**

C   carbon

S   sulfur

N   nitrogen

I   iodine

**Two-Letter Symbols**

Co   cobalt

Si   silicon

Ne   neon

Ni   nickel

Although most of the symbols use letters from the current names, some are derived from their ancient Latin or Greek names. For example, Na, the symbol for sodium, comes from the Latin word *natrium*. The symbol for iron, Fe, is derived from the Latin name *ferrum*. Table 3.3 lists the names and symbols of some common elements. Learning their names and symbols will greatly help your learning of chemistry. A complete list of all the elements and their symbols appears on the inside front cover of this text.

## Physical Properties of Elements

Suppose a friend asks you to look into a mirror and describe what you see. You might tell her the color of your hair, eyes, and skin, or the length and texture of your hair. Perhaps you have freckles or dimples. These are descriptions of some of your physical properties. Every element has **physical properties** too, which are those characteristics that can be observed or measured without affecting the identity of an element. The physical properties of a substance include its shape, color, odor, taste, density, hardness, melting point, and boiling point. For example, copper has the physical properties listed in Table 3.4.

**TABLE 3.2** Some Elements and Their Names

| Element | Source of Name |
| --- | --- |
| Uranium | The planet Uranus |
| Titanium | Titans (mythology) |
| Chlorine | *Chloros*, "greenish-yellow" (Greek) |
| Iodine | *Ioeides*, "violet" (Greek) |
| Magnesium | Magnesia, a mineral |
| Californium | California |
| Curium | Marie and Pierre Curie |

**TABLE 3.3** Names and Symbols of Some Common Elements

Aluminum

Carbon

Gold

Silver

Sulfur

| Name[a] | Symbol | Name[a] | Symbol | Name[a] | Symbol |
| --- | --- | --- | --- | --- | --- |
| Aluminum | Al | Gold (*aurum*) | Au | Oxygen | O |
| Argon | Ar | Helium | He | Phosphorus | P |
| Arsenic | As | Hydrogen | H | Platinum | Pt |
| Barium | Ba | Iodine | I | Potassium (*kalium*) | K |
| Boron | B | Iron (*ferrum*) | Fe | Radium | Ra |
| Bromine | Br | Lead (*plumbum*) | Pb | Silicon | Si |
| Cadmium | Cd | Lithium | Li | Silver (*argentum*) | Ag |
| Calcium | Ca | Magnesium | Mg | Sodium (*natrium*) | Na |
| Carbon | C | Manganese | Mn | Strontium | Sr |
| Chlorine | Cl | Mercury (*hydrargyrum*) | Hg | Sulfur | S |
| Chromium | Cr | Neon | Ne | Tin (*stannum*) | Sn |
| Cobalt | Co | Nickel | Ni | Titanium | Ti |
| Copper (*cuprum*) | Cu | Nitrogen | N | Uranium | U |
| Fluorine | F | | | Zinc | Zn |

[a]Names given in parentheses are ancient Latin or Greek words from which the symbols are derived.

**TABLE 3.4** Some Physical Properties of Copper

| Color | Reddish-orange |
|---|---|
| Odor | Odorless |
| Melting point | 1083 °C |
| Boiling point | 2567 °C |
| State at 25 °C | Solid |
| Luster | Very shiny |
| Conduction of electricity | Excellent |
| Conduction of heat | Excellent |

---

SAMPLE PROBLEM | 3.2

■ **Writing Chemical Symbols**

What are the chemical symbols for the following elements?

**a.** carbon      **b.** nitrogen      **c.** chlorine      **d.** copper

SOLUTION

**a.** C           **b.** N              **c.** Cl            **d.** Cu

STUDY CHECK

What are the chemical symbols for the elements silicon, sulfur, and silver?

---

SAMPLE PROBLEM | 3.3

■ **Names and Symbols of Chemical Elements**

Give the name of the element that corresponds to each of the following chemical symbols:

**a.** Zn      **b.** K      **c.** H      **d.** Fe

SOLUTION

**a.** zinc      **b.** potassium      **c.** hydrogen      **d.** iron

STUDY CHECK

What are the names of the elements with the chemical symbols Mg, Al, and F?

---

## *Health Note*

### Toxicity of Mercury

Mercury is a silvery, shiny element that is a liquid at room temperature. Mercury can enter the body through inhaled mercury vapor, contact with the skin, or ingestion of foods or water contaminated with mercury. In the body, mercury destroys proteins and disrupts cell function. Long-term exposure to mercury can damage the brain and kidneys, cause mental retardation, and decrease physical development. Blood, urine, and hair samples are used to test for mercury.

In fresh and salt water, bacteria convert mercury into toxic methylmercury, which primarily attacks the central nervous system (CNS). Because fish absorb methylmercury, we are exposed to mercury by consuming mercury-contaminated fish. As levels of mercury ingested from fish became a concern, the Food and Drug Administration (FDA) set a maximum level of one part mercury per million parts seafood (1 ppm), which is the same as 1 $\mu$g mercury in every gram of seafood. Fish higher in the food chain, such as swordfish and shark, can have such high levels of mercury that the Environmental Protection Agency (EPA) recommends they be consumed no more than once a week.

One of the worst incidents of mercury poisoning occurred in Minamata and Niigata, Japan, in 1950. At that time, the ocean was polluted with high levels of mercury from industrial wastes. Because fish were a major food in the diet, more than 2000 people were affected with mercury poisoning and died or developed neural damage. In the United States, industry decreased the use of mercury between 1988 and 1997 by 75% by banning mercury in paint and pesticides, reducing mercury in batteries, and regulating mercury in other products.

This mercury fountain, housed in glass, was designed by Alexander Calder for the 1937 World's Fair in Paris.

## QUESTIONS AND PROBLEMS

### Elements and Symbols

**3.7** Write the symbols for the following elements:
   **a.** copper   **b.** silicon   **c.** potassium   **d.** nitrogen
   **e.** iron   **f.** barium   **g.** lead   **h.** strontium

**3.8** Write the symbols for the following elements:
   **a.** oxygen   **b.** lithium   **c.** sulfur   **d.** aluminum
   **e.** hydrogen   **f.** neon   **g.** tin   **h.** gold

**3.9** Write the name of the element for each symbol.
   **a.** C   **b.** Cl   **c.** I   **d.** Hg   **e.** F   **f.** Ar   **g.** Zn   **h.** Ni

**3.10** Write the name of the element for each symbol.
   **a.** He   **b.** P   **c.** Na   **d.** Mg   **e.** Ca   **f.** Br   **g.** Cd   **h.** Si

**3.11** What elements are in the following substances?
   **a.** table salt, $NaCl$   **b.** plaster casts, $CaSO_4$
   **c.** Demerol, $C_{15}H_{22}ClNO_2$   **d.** antacid, $CaCO_3$

**3.12** What elements are in the following substances?
   **a.** water, $H_2O$   **b.** baking soda, $NaHCO_3$
   **c.** lye, $NaOH$   **d.** sugar, $C_{12}H_{22}O_{11}$

# 3.3 THE PERIODIC TABLE

As more and more elements were discovered, it became necessary to organize them with some type of classification system. By the late 1800s, scientists recognized that certain elements looked alike and behaved much the same way. In 1872, a Russian chemist, Dmitri Mendeleev, arranged the 60 elements known at that time into groups with similar properties and placed them in order of increasing atomic masses. Today, the arrangement of elements is known as the **periodic table**. (See Figure 3.4.)

## LEARNING GOAL

Use the periodic table to identify the group and the period of an element, and decide whether it is a metal, nonmetal, or metalloid.

**FIGURE 3.4** Groups and periods in the periodic table.
**Q** What is the symbol of the alkali metal in Period 3?

## Periodic Table of Elements

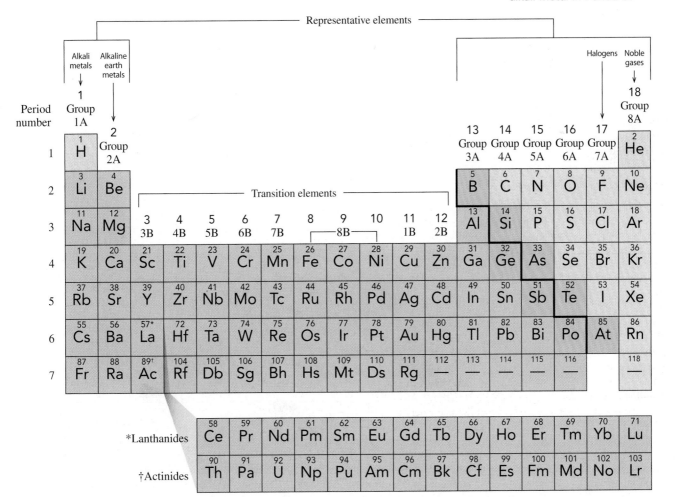

Metals    Metalloids    Nonmetals

## Health Note

### Elements Essential to Health

Many elements are essential for the well-being and survival of the human body. The four elements oxygen, carbon, hydrogen, and nitrogen are the most important elements that make up carbohydrates, fats, proteins, and DNA. Most of the hydrogen and oxygen is found in water, which makes up 55–60% of our body mass. Some examples and the amounts present in a 60-kg person are listed in Table 3.5.

**TABLE 3.5** Elements Essential to Health

| Element | Symbol | Amount in in a 60-kg Person |
|---------|--------|------------------------------|
| Oxygen | O | 39 kg |
| Carbon | C | 11 kg |
| Hydrogen | H | 6 kg |
| Nitrogen | N | 1.5 kg |
| Calcium | Ca | 1 kg |
| Phosphorus | P | 600 g |
| Potassium | K | 120 g |
| Sulfur | S | 120 g |
| Sodium | Na | 86 g |
| Chlorine | Cl | 81 g |
| Magnesium | Mg | 16 g |
| Iron | Fe | 3.6 g |
| Fluorine | F | 2.2 g |
| Zinc | Zn | 2.0 g |
| Copper | Cu | 60 mg |
| Iodine | I | 20 mg |

## Periods and Groups

Each horizontal row in the table is called a **period**. (See Figure 3.5.) The number of elements in the periods increases going down the periodic table. Each row is counted from the top of the table as Period 1 to Period 7. The first period contains only the elements hydrogen (H) and helium (He). The second period contains eight elements: lithium (Li), beryllium (Be), boron (B), carbon (C), nitrogen (N), oxygen (O), fluorine (F), and neon (Ne). The third period also contains eight elements beginning with sodium (Na) and ending with argon (Ar). The fourth period, which begins with potassium (K), and the fifth period, which begins with rubidium (Rb), have 18 elements each. The sixth period, which begins with cesium (Cs), has 32 elements. The seventh period contains the remaining elements.

Each vertical column on the periodic table contains a **group** (or family) of elements that have similar properties. At the top of each column is a number that is assigned to each group. The elements in the first two columns on the left and the last six columns on the right of the periodic table are called the *representative elements* or the *main group elements*. For many years, they have been given group numbers 1A–8A. On some periodic tables, the group numbers may be written with Roman numerals: IA–VIIIA. In the center of the periodic table is a block of elements known as the *transition elements* or *transition metals*, which are designated with the letter "B." A newer numbering system assigns group numbers of 1–18 going across the periodic table. Because both systems of group numbers are currently in use, they are both indicated on the periodic table in this text and are included in our discussions of elements and group numbers with the 1–18 group number in parentheses.

## Classification of Groups

Several groups in the periodic table have special names. (See Figure 3.6.) Group 1A (1) elements, lithium (Li), sodium (Na), potassium (K), rubidium (Rb), cesium (Cs), and francium (Fr), are a family of elements known as the **alkali metals**. (See Figure 3.7.) The elements within this group are soft, shiny metals that are good conductors of heat and

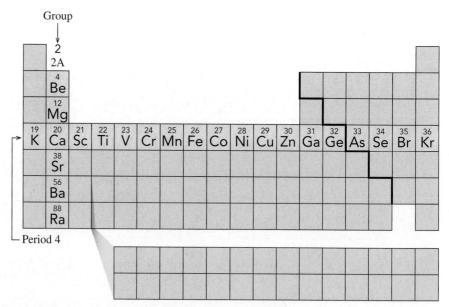

**FIGURE 3.5** On the periodic table, each vertical column represents a group of elements and each horizontal row of elements represents a period.
**Q** Are the elements Si, P, and S part of a group or a period?

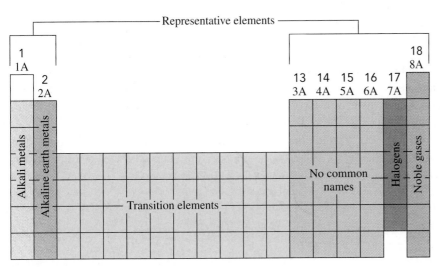

**FIGURE 3.6** Certain groups on the periodic table have common names.
**Q** What is the common name for the group of elements that includes helium and argon?

electricity and have relatively low melting points. Alkali metals react vigorously with water and form white products when they combine with oxygen.

Although hydrogen (H) is at the top of Group 1A (1), hydrogen is not an alkali metal and has very different properties than the rest of the elements in this group. Thus hydrogen is not included in the classification of alkali metals. In some periodic tables, H is sometimes placed at the top of Group 7A (17).

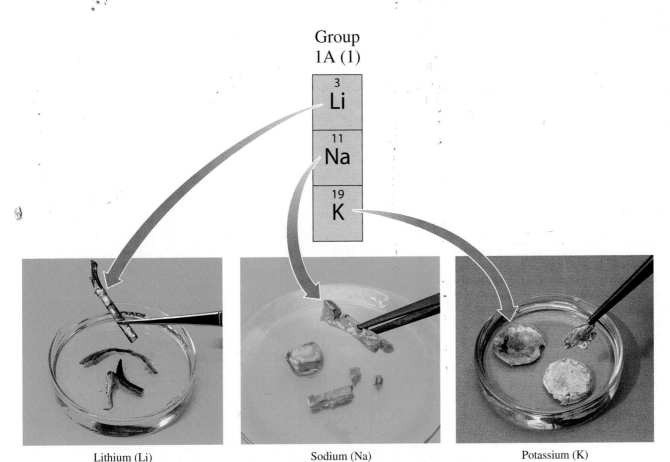

**FIGURE 3.7** Lithium (Li), sodium (Na), and potassium (K) are some alkali metals from Group 1A (1).
**Q** What physical properties do these alkali metals have in common?

## Group 7A (17)

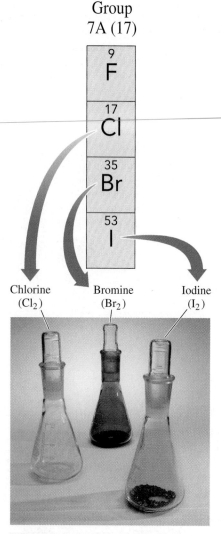

| | |
|---|---|
| 9 | **F** |
| 17 | **Cl** |
| 35 | **Br** |
| 53 | **I** |

Chlorine (Cl$_2$)    Bromine (Br$_2$)    Iodine (I$_2$)

**FIGURE 3.8** Chlorine (Cl$_2$), bromine (Br$_2$), and iodine (I$_2$) are examples of halogens from Group 7A (17).
**Q** What elements are in the halogen group?

Group 2A (2) elements, beryllium (Be), magnesium (Mg), calcium (Ca), strontium (Sr), barium (Ba), and radium (Ra), are called the **alkaline earth metals**. They are also shiny metals like those in Group 1A (1), but they are not as reactive.

The **halogens** are found on the right side of the periodic table in Group 7A (17). They include the elements fluorine (F), chlorine (Cl), bromine (Br), and iodine (I). (See Figure 3.8.) The halogens, especially fluorine and chlorine, are strongly reactive and form compounds with most of the elements.

Group 8A (18) contains the **noble gases**, helium (He), neon (Ne), argon (Ar), krypton (Kr), xenon (Xe), and radon (Rn). They are quite unreactive and are seldom found in combination with other elements.

## Metals, Nonmetals, and Metalloids

Another feature of the periodic table is the heavy zigzag line that separates the elements into the *metals* and the *nonmetals*. The metals are those elements on the left of the line *except for hydrogen*, and the nonmetals are the elements on the right. (See Figure 3.9.)

In general, most **metals** are shiny solids. They can be shaped into wires (ductile) or hammered into a flat sheet (malleable). Metals are good conductors of heat and electricity. They usually melt at higher temperatures than nonmetals. All of the metals are solids at room temperature, except for mercury (Hg), which is a liquid. Some typical metals are sodium (Na), magnesium (Mg), copper (Cu), gold (Au), silver (Ag), iron (Fe), and tin (Sn).

**Nonmetals** are not very shiny, malleable, or ductile, and they are often poor conductors of heat and electricity. They typically have low melting points and low densities. You may have heard of nonmetals such as hydrogen (H), carbon (C), nitrogen (N), oxygen (O), chlorine (Cl), and sulfur (S).

Except for aluminum, the elements located along the heavy line are called **metalloids** and include B, Si, Ge, As, Sb, Te, Po, and At. Metalloids are elements that exhibit some properties that are typical of the metals and other properties that are characteristic of the nonmetals. For example, they are better conductors of heat and electricity than the nonmetals, but not as good as the metals. The metalloids are semiconductors because they can act as conductors or insulators. Table 3.6 compares some characteristics of silver, a metal, with those of antimony, a metalloid, and sulfur, a nonmetal.

**FIGURE 3.9** Along the heavy zigzag line on the periodic table that separates the metals and nonmetals are metalloids, which exhibit characteristics of both metals and nonmetals.
**Q** On which side of the heavy zigzag line are the nonmetals located?

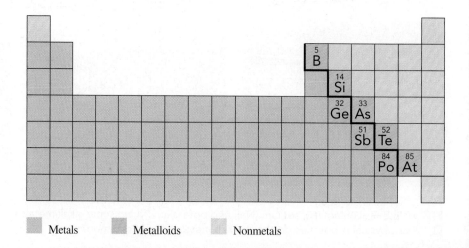

Metals         Metalloids         Nonmetals

**TABLE 3.6** Some Characteristics of a Metal, a Metalloid, and a Nonmetal

| Silver (Ag) | Antimony (Sb) | Sulfur (S) | |
|---|---|---|---|
| Metal | Metalloid | Nonmetal | |
| Shiny | Blue-grey, shiny | Dull, yellow | |
| Extremely ductile | Brittle | Brittle | |
| Can be hammered into sheets (malleable) | Shatters when hammered | Shatters when hammered | |
| Good conductor of heat and electricity | Poor conductor of heat and electricity | Poor conductor of heat and electricity, good insulator | |
| Used in coins, jewelry, tableware | Used to harden lead, color glass and plastics | Used in gunpowder, rubber, fungicides | |
| Density 10.5 g/mL | Density 6.7 g/mL | Density 2.1 g/mL | |
| Melting point 962 °C | Melting point 630 °C | Melting point 113 °C | |

## SAMPLE PROBLEM 3.4

### ■ Classification of Elements

Use the periodic table to classify each of the following elements by its group and period, group name (if any), and as a metal, nonmetal, or metalloid.

**a.** Na        **b.** Si        **c.** I        **d.** Sn

SOLUTION

**a.** Na (sodium), Group 1A (1), Period 3, is an alkali metal.
**b.** Si (silicon), Group 4A (14), Period 3, is a metalloid.
**c.** I (iodine), Group 7A (17), Period 5, halogen, is a nonmetal.
**d.** Sn (tin), Group 4A (14), Period 5, is a metal.

STUDY CHECK

Give the name and symbol of the element represented by the following:

**a.** Group 5A (15), Period 3
**b.** A noble gas in Period 2
**c.** A metalloid in Period 3

## QUESTIONS AND PROBLEMS

### The Periodic Table

**3.13** Identify the group or period number described by each of the following statements:
  **a.** contains the elements C, N, and O
  **b.** begins with helium
  **c.** the alkali metals
  **d.** ends with neon

**3.14** Identify the group or period number described by each of the following statements:
  **a.** contains Na, K, and Rb
  **b.** the row that begins with Li
  **c.** the noble gases
  **d.** contains F, Cl, Br, and I

**3.15** Classify the following as an alkali metal, alkaline earth metal, transition element, halogen, or noble gas:
  **a.** Ca        **b.** Fe        **c.** Xe
  **d.** Na        **e.** Cl

**3.16** Classify the following as an alkali metal, alkaline earth metal, transition element, halogen, or noble gas:
  **a.** Ne        **b.** Mg        **c.** Cu
  **d.** Br        **e.** Ba

**3.17** Give the symbol of the element described by the following:
  **a.** Group 4A, Period 2        **b.** a noble gas in Period 1
  **c.** an alkali metal in Period 3        **d.** Group 2, Period 4
  **e.** Group 13, Period 3

**3.18** Give the symbol of the element described by the following:
    **a.** an alkaline earth metal in Period 2
    **b.** Group 15, Period 3
    **c.** a noble gas in Period 4
    **d.** a halogen in Period 5
    **e.** Group 4A, Period 4

**3.19** Is each of the following elements a metal, nonmetal, or metalloid?
    **a.** calcium
    **b.** sulfur
    **c.** an element that is shiny
    **d.** does not conduct heat
    **e.** located in Group 7A

    **f.** phosphorus
    **g.** boron
    **h.** silver

**3.20** Is each of the following elements a metal, nonmetal, or metalloid?
    **a.** located in Group 2A
    **b.** a good conductor of electricity
    **c.** chlorine
    **d.** arsenic
    **e.** an element that is not shiny
    **f.** oxygen
    **g.** nitrogen
    **h.** aluminum

*Health Note*

## Some Important Trace Elements in the Body

Some metals and nonmetals known as trace elements are essential to the proper functioning of the body. Although they are required in very small amounts, their absence can disrupt major biological processes and cause illness. The trace elements listed in Table 3.7 are present in the body combined with other elements. The adult DV is the daily recommended amount for an adult.

**TABLE 3.7** Some Important Trace Elements in the Body

| Element | Adult DV | Biological Function | Deficiency Symptoms | Dietary Sources |
|---|---|---|---|---|
| Iron (Fe) | 10 mg (males) 18 mg (females) | Formation of hemoglobin; enzymes | Dry skin, spoon nails; decreased hemoglobin count; anemia | Liver and other organ meats, oysters, red or dark meat, green leafy vegetables, fortified breads and cereals, egg yolk |
| Copper (Cu) | 2.0–5.0 mg | Necessary in many enzyme systems; growth; aids formation of red blood cells and collagen | Uncommon; anemia; decreased white cell count; bone demineralization | Nuts, organ meats, whole grains, shellfish, eggs, poultry, green leafy vegetables |
| Zinc (Zn) | 15 mg | Amino acid metabolism; enzyme systems; energy production; collagen | Retarded growth and bone formation; skin inflammation; loss of taste and smell; poor healing | Oysters, crab, lamb, beef, organ meats, whole grains |
| Manganese (Mn) | 2.5–5.0 mg | Necessary for some enzyme systems; collagen formation; central nervous system; fat and carbohydrate metabolism; blood clotting | Abnormal skeletal growth; impairment of central nervous system | Whole grains, wheat germ, legumes, pineapple, figs |
| Iodine (I) | 150 $\mu$g | Necessary for activity of thyroid gland | Hypothyroidism; goiter; cretinism | Seafood, iodized salt |
| Fluorine (F) | 1.5–4.0 mg | Necessary for solid tooth formation and retention of calcium in bones with aging | Dental cavities | Tea, fish, water in some areas, supplementary drops, toothpaste |

# 3.4 THE ATOM

All the elements listed on the periodic table are made up of atoms. An **atom** is the smallest particle of an element that retains the characteristics of that element. You have probably seen the element aluminum. Imagine that you are tearing a piece of aluminum foil into smaller and smaller pieces. Now imagine that you have a piece so small that you cannot tear it apart any further. Then you would have an atom of aluminum, the smallest particle of an element that still retains the characteristics of that element.

Billions of atoms are packed together to build you and everything around you. The ink on this paper, even the dot over the letter *i*, contains huge numbers of atoms. There are as many atoms in that dot as there are seconds in 10 billion years.

The concept of the atom is relatively recent. Although the Greek philosophers in 500 B.C.E. reasoned that everything must contain minute particles they called *atomos*, the idea of atoms did not become a scientific theory until 1808. Then John Dalton (1766–1844) developed an atomic theory that proposed that atoms were responsible for the combinations of elements found in compounds.

## Dalton's Atomic Theory

1. All matter is made up of tiny particles called atoms.
2. All atoms of a given element are similar to one another and different from atoms of other elements.
3. Atoms of two or more different elements combine to form compounds. A particular compound is always made up of the same kinds of atoms and always has the same number of each kind of atom.
4. A chemical reaction involves the rearrangement, separation, or combination of atoms. Atoms are never created or destroyed during a chemical reaction.

Atoms are the building blocks of everything we see around us; yet, we cannot see an atom or even a billion atoms with the naked eye. However, when billions and billions of atoms are packed together, the characteristics of each atom are added to those of the next until we can see the characteristics we associate with the element. For example, a small piece of the shiny, copper-colored element we call copper consists of many, many copper atoms. A special kind of microscope called a scanning tunneling microscope (STM) produces images of individual atoms, such as the atoms of carbon in graphite shown in Figure 3.10.

## Electrical Charges in an Atom

By the end of the 1800s, experiments with electricity showed that atoms were not solid spheres, but were composed of even smaller bits of matter called *subatomic particles*. Some of these subatomic particles were discovered because they have electrical charges. An electrical charge can be positive or negative. Experiments show that like charges repel; they push away from each other. When you brush your hair on a dry day, electrical charges that are alike build up on the brush and in your hair; as a result your hair flies away from the brush. Opposite or unlike charges attract. The crackle of clothes taken from the clothes dryer indicates the presence of electrical charges. The clinginess of the clothing is due to the attraction of opposite, unlike charges, as shown in Figure 3.11.

## Structure of the Atom

In 1897, J. J. Thomson, an English physicist, showed that cathode rays were actually streams of small particles that were attracted to a positively charged electrode. Because opposite charges attract, Thomson knew that these particles must be negatively charged. In further experiments, these particles called **electrons** were found to be much smaller

Describe the electrical charge and location in an atom for a proton, a neutron, and an electron.

**WEB TUTORIAL**
Atoms and Isotopes

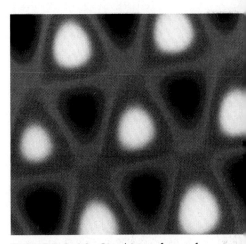

**FIGURE 3.10** Graphite, a form of carbon, magnified millions of times by a scanning tunneling microscope. This instrument generates an image of the atomic structure. The round yellow objects are atoms.
**Q** Why is a microscope with extremely high magnification needed to see these atoms?

Positive charges repel

Negative charges repel

Unlike charges attract

**FIGURE 3.11** Like charges repel and unlike charges attract.
**Q** Why are the electrons attracted to the protons in the nucleus of an atom?

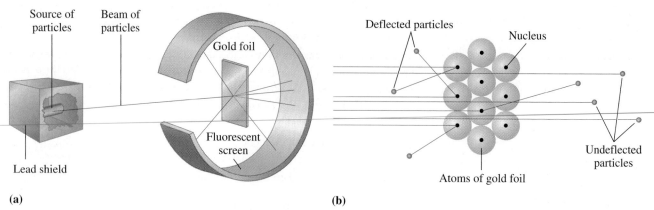

(a)                                                                          (b)

**FIGURE 3.12** **(a)** Positive particles are aimed at a piece of gold foil. **(b)** Particles that come close to the atomic nuclei are deflected from their straight path.
**Q** Why are some particles deflected while most pass through the gold foil undeflected?

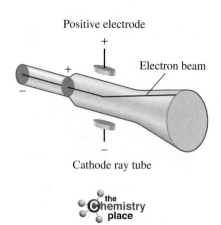

**WEB TUTORIAL**
Atoms and Isotopes

than the atom and to have an extremely small mass. Because atoms are neutral, scientists soon discovered that atoms contained positively charged particles called **protons** that were much heavier than the electrons.

Thomson proposed a model for the atom in which the electrons and protons were randomly distributed through the atom. In 1911, Ernest Rutherford worked with Thomson to test this model. In Rutherford's experiment, positively charged particles were aimed at a thin sheet of gold foil. (See Figure 3.12.) If the Thomson model were correct, the particles would travel in straight paths through the gold foil. Rutherford was greatly surprised to find that some of the particles were deflected slightly as they passed through the gold foil, and a few particles were deflected so much that they went back in the opposite direction. According to Rutherford, it was as though he had shot a cannonball at a piece of tissue paper, and it bounced back at him. Rutherford realized that the protons must be contained in a small, positively charged region at the center of the atom, which he called the **nucleus**. He proposed that the electrons in the atom occupy the space surrounding the nucleus through which most of the particles travelled undisturbed. Only the particles that came near this dense, positive center were deflected. If an atom were the size of a football stadium, the nucleus would be about the size of a golf ball placed in the center of the field.

Scientists knew that the nucleus was heavier than the mass of the protons and looked for another subatomic particle. Eventually, they discovered that the nucleus also contained a particle called a **neutron**, which is neutral. Thus, the masses of the protons and neutrons in the nucleus determine its mass. (See Figure 3.13.)

**FIGURE 3.13** In an atom, the protons and neutrons that make up almost all the mass are packed into the tiny volume of the nucleus. The electrons surround the nucleus and account for the large volume of the atom.
**Q** Why can we say that the atom is mostly empty space?

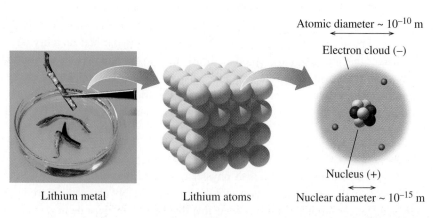

Atomic diameter ~ $10^{-10}$ m

Electron cloud (−)

Nucleus (+)

Nuclear diameter ~ $10^{-15}$ m

Lithium metal            Lithium atoms

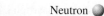

Proton ●          Neutron ○          Electron ●

**TABLE 3.8** Particles in the Atom

| Subatomic Particle | Symbol | Electrical Charge | Approximate Mass (amu) | Location in Atom |
|---|---|---|---|---|
| Proton | $p$ or $p^+$ | 1+ | 1 | Nucleus |
| Neutron | $n$ or $n^0$ | 0 | 1 | Nucleus |
| Electron | $e^-$ | 1− | 0.0005 ($^1/_{2000}$) | Outside nucleus |

## Mass of the Atom

All of the subatomic particles are extremely small compared with the things you see around you. One proton has a mass of $1.7 \times 10^{-24}$ g, and the neutron is about the same. The mass of the electron is much less. To express these masses more easily, chemists use a unit called an **atomic mass unit (amu)**. An amu is defined as one-twelfth of the mass of the carbon atom with 6 protons and 6 neutrons. In biology, the atomic mass unit is called a dalton in honor of John Dalton. On the amu scale, the proton and neutron each have a mass of about 1 amu. Because the electron mass is so small, it is usually ignored in atomic mass calculations. Table 3.8 summarizes some information about the subatomic particles in an atom.

Explore Your World

### Repulsion and Attraction

1. Obtain a tape dispenser with clear tape, a hairbrush, comb, and a piece of paper. Tear off a piece of the tape that is about 20 cm long (the length of your hand). Stick the tape to the edge of a table leaving the end hanging down. Tear off a second piece of tape and slowly bring it close to the first one. What happens? Is there an attraction or repulsion?

   Slide your thumb and finger along the tape you are holding. Bring it close to the piece that is hanging from the table. What happens? Is there an attraction or repulsion? Attach the second tape to the edge of the table. Brush your hair and bring the brush close to each piece of tape hanging from the table. What do you observe?

2. Tear a small piece of paper into bits. Brush your hair several times, and place the brush just above the bits of paper. Use your knowledge of electrical charges to give an explanation for your observations. Try the same experiment with a comb.

**QUESTIONS**

1. What happens when objects with like charges are placed close together?
2. What happens when objects with unlike charges are placed close together?

---

SAMPLE PROBLEM **3.5**

### ■ Identifying Subatomic Particles

Is each of the following statements true or false?

**a.** Protons are heavier than electrons.
**b.** Protons are attracted to neutrons.
**c.** Electrons are so small that they have no electrical charge.
**d.** The nucleus contains all the protons and neutrons of an atom.

**SOLUTION**

**a.** True
**b.** False; protons are attracted to electrons.
**c.** False; electrons have a 1− charge.
**d.** True

**STUDY CHECK**

True or false: The nucleus occupies a large volume in an atom.

## QUESTIONS AND PROBLEMS

### The Atom

**3.21** Is a proton, neutron, or electron described by each of the following?
**a.** has the smallest mass
**b.** has a 1+ charge
**c.** is found outside the nucleus
**d.** is electrically neutral

**3.22** Is a proton, neutron, or electron described by each of the following?
**a.** has a mass about the same as a proton
**b.** is found in the nucleus
**c.** is attracted to the protons
**d.** has a 1− charge

**3.23** What did Rutherford determine about the structure of the atom from his gold-foil experiment?

**3.24** Why does the nucleus in every atom have a positive charge?

**3.25** Is each of the following statements true or false?
**a.** A proton and an electron have opposite charges.
**b.** The nucleus contains most of the mass of an atom.
**c.** Electrons repel each other.
**d.** A proton is attracted to a neutron.

**3.26** Is each of the following statements true or false?
**a.** A proton is attracted to an electron.
**b.** A neutron has twice the mass of a proton.
**c.** Neutrons repel each other.
**d.** Electrons and neutrons have opposite charges.

**3.27** On a dry day, your hair flies away when you brush it. How would you explain this?

**3.28** Sometimes clothes removed from the dryer cling together. What kinds of charges are on the clothes?

---

**LEARNING GOAL**

Given the atomic number and the mass number of an atom, state the number of protons, neutrons, and electrons.

# 3.5 ATOMIC NUMBER AND MASS NUMBER

All of the atoms of the same element always have the same number of protons. This feature distinguishes atoms of one element from atoms of all the other elements.

## Atomic Number

An **atomic number**, which is equal to the number of protons in the nucleus of an atom, is used to identify each element.

Atomic number = number of protons in an atom

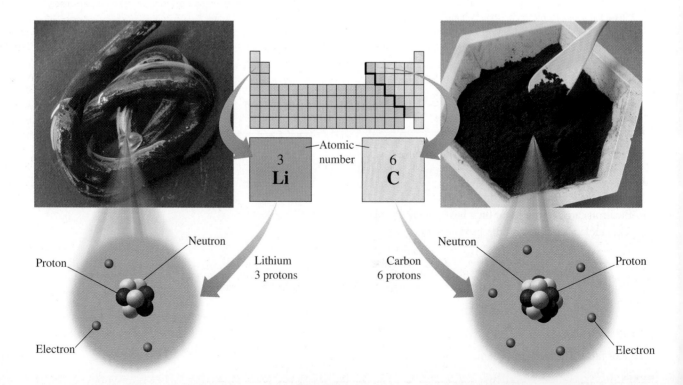

On the inside front cover of this text is a periodic table, which gives all of the elements in order of increasing atomic number. The atomic number is the whole number that appears above the symbol. For example, a hydrogen atom, with atomic number 1, has 1 proton; a lithium atom, with atomic number 3, has 3 protons; an atom of carbon, with atomic number 6, has 6 protons; and gold, with atomic number 79, has 79 protons.

An atom is electrically neutral. That means that the number of protons in an atom is equal to the number of electrons. This electrical balance gives an atom an overall charge of zero. Thus, in every atom, the atomic number also gives the number of electrons.

---

### SAMPLE PROBLEM 3.6

#### ■ Using Atomic Number to Find the Number of Protons and Electrons

Using the periodic table in Figure 3.4, state the atomic number, number of protons, and number of electrons for an atom of each of the following elements:

**a.** nitrogen
**b.** magnesium
**c.** bromine

#### SOLUTION

**a.** atomic number 7; 7 protons and 7 electrons
**b.** atomic number 12; 12 protons and 12 electrons
**c.** atomic number 35; 35 protons and 35 electrons

#### STUDY CHECK

Consider an atom that has 26 electrons.

**a.** How many protons are in its nucleus? *26*
**b.** What is its atomic number? *26*
**c.** What is its name, and what is its symbol?

---

## Mass Number

We now know that the protons and neutrons determine the mass of the nucleus. For any atom, the **mass number** is the sum of the number of protons and neutrons in the nucleus.

Mass number = number of protons + number of neutrons

For example, an atom of oxygen that contains 8 protons and 8 neutrons has a mass number of 16. An atom of iron that contains 26 protons and 30 neutrons has a mass number of 56. Table 3.9 illustrates the relationship between atomic number, mass number, and the number of protons, neutrons, and electrons in some atoms of different elements.

### Optician

"When a patient brings in a prescription, I help select the proper lenses, put them into a frame, and fit them properly on the patient's face," says Suranda Lara, optician, Kaiser Hospital. "If a prescription requires a thinner and lighter-weight lens, we formulate that lens. So we have to understand the different materials used to make lenses. Sometimes patients come in with their own glasses that they want to convert to sunglasses. We remove the lenses and put them into a tint bath, which turns them into sunglasses."

Opticians fit and adjust eyewear for patients who have had their eyesight tested by an ophthalmologist or optometrist. Optics and mathematics are used to select materials for frames and lenses that are compatible with patients' facial measurements and lifestyles.

**TABLE 3.9** Composition of Some Atoms of Different Elements

| Element | Symbol | Atomic Number | Mass Number | Number of Protons | Number of Neutrons | Number of Electrons |
|---------|--------|---------------|-------------|-------------------|--------------------|--------------------|
| Hydrogen | H | 1 | 1 | 1 | 0 | 1 |
| Nitrogen | N | 7 | 14 | 7 | 7 | 7 |
| Chlorine | Cl | 17 | 37 | 17 | 20 | 17 |
| Iron | Fe | 26 | 57 | 26 | 31 | 26 |
| Gold | Au | 79 | 197 | 79 | 118 | 79 |

SAMPLE PROBLEM 3.7

### ■ Calculating Numbers of Protons, Neutrons, and Electrons

For an atom of iron that has a mass number of 56, determine the following:

**a.** the number of protons
**b.** the number of neutrons
**c.** the number of electrons

#### SOLUTION

**a.** On the periodic table, the atomic number of iron is 26. An iron atom has 26 protons.
**b.** The number of neutrons in this atom is found by subtracting the atomic number from the mass number. The number of neutrons is 30.

$$\text{Mass number} - \text{atomic number} = \text{number of neutrons}$$
$$56 \quad - \quad 26 \quad = \quad 30$$

**c.** Because an atom is neutral, the number of electrons is equal to the number of protons. An iron atom has 26 electrons.

#### STUDY CHECK

How many neutrons are in the nucleus of a bromine atom that has a mass number of 80?

## QUESTIONS AND PROBLEMS

### Atomic Number and Mass Number

**3.29** Would you use atomic number, mass number, or both to obtain the following?
**a.** number of protons in an atom
**b.** number of neutrons in an atom
**c.** number of particles in the nucleus
**d.** number of electrons in a neutral atom

**3.30** What do you know about the subatomic particles from the following?
**a.** atomic number
**b.** mass number
**c.** mass number − atomic number
**d.** mass number + atomic number

**3.31** Write the names and symbols of the elements with the following atomic numbers:
**a.** 3 **b.** 9 **c.** 20 **d.** 30
**e.** 10 **f.** 14 **g.** 53 **h.** 8

**3.32** Write the names and symbols of the elements with the following atomic numbers:
**a.** 1 **b.** 11 **c.** 19 **d.** 26
**e.** 35 **f.** 47 **g.** 15 **h.** 2

**3.33** How many protons are in a neutral atom of the following?
**a.** magnesium **b.** zinc
**c.** iodine **d.** potassium

**3.34** How many electrons are in a neutral atom of the following?
**a.** carbon **b.** fluorine
**c.** calcium **d.** sulfur

**3.35** Complete the following table for neutral atoms.

| Name of Element | Symbol | Atomic Number | Mass Number | Number of Protons | Number of Neutrons | Number of Electrons |
|---|---|---|---|---|---|---|
| | Al | | 27 | | | |
| | | 12 | | | 12 | |
| Potassium | | | | | 20 | |
| | | | | 16 | 15 | |
| | | | 56 | | | 26 |

**3.36** Complete the following table for neutral atoms.

| Name of Element | Symbol | Atomic Number | Mass Number | Number of Protons | Number of Neutrons | Number of Electrons |
|---|---|---|---|---|---|---|
| | N | | 15 | | | |
| Calcium | | | 42 | | | |
| | | | | 38 | 50 | |
| | | 14 | | | 16 | |
| | | 56 | 138 | | | |

# 3.6 ISOTOPES AND ATOMIC MASS

We have seen that all atoms of the same element have the same number of protons and electrons. However, the atoms of any one element are not completely identical because they can have different numbers of neutrons.

## Isotopes

**Isotopes** are atoms of the same element that have different numbers of neutrons. For example, all atoms of the element magnesium (Mg) have 12 protons. However, some magnesium atoms have 12 neutrons, others have 13 neutrons, and still others have 14 neutrons. The differences in numbers of neutrons for these magnesium atoms cause their mass numbers to be different but not their chemical behavior. The three isotopes of magnesium have the same atomic number but different mass numbers.

## Atomic Symbols for the Isotopes of Magnesium

On the periodic table, the atomic number appears above the element symbol. To distinguish between the different isotopes of an element, we can write an **atomic symbol** that indicates the mass number in the upper left corner and the atomic number in the lower left corner. An isotope may be referred to by its name or symbol followed by the mass number, such as magnesium-24 or Mg-24. Magnesium has three naturally occurring isotopes, as shown in Table 3.10.

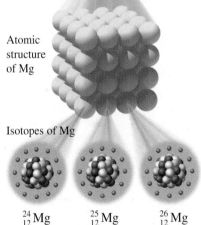

Atomic structure of Mg

Isotopes of Mg

$^{24}_{12}\text{Mg}$   $^{25}_{12}\text{Mg}$   $^{26}_{12}\text{Mg}$

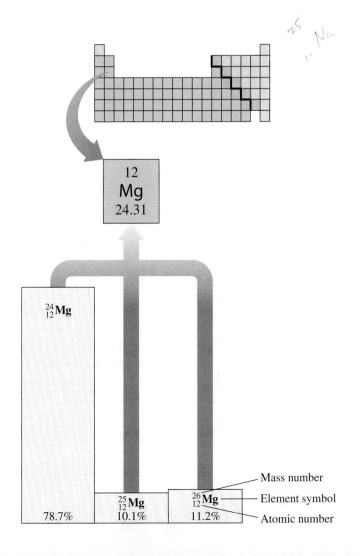

12
Mg
24.31

$^{24}_{12}\textbf{Mg}$

$^{25}_{12}\textbf{Mg}$
78.7%   10.1%

$^{26}_{12}\textbf{Mg}$   — Mass number
11.2%   — Element symbol
— Atomic number

**TABLE 3.10** Isotopes of Magnesium

| Atomic symbol | $^{24}_{12}Mg$ | $^{25}_{12}Mg$ | $^{26}_{12}Mg$ |
|---|---|---|---|
| Number of protons | 12 | 12 | 12 |
| Number of electrons | 12 | 12 | 12 |
| Mass number | **24** | **25** | **26** |
| % abundance | 78.7% | 10.1% | 11.2% |
| Number of neutrons | **12** | **13** | **14** |

## SAMPLE PROBLEM 3.8

### ■ Identifying Protons and Neutrons in Isotopes

State the number of protons and neutrons in the following isotopes of neon (Ne):

**a.** $^{20}_{10}Ne$         **b.** $^{21}_{10}Ne$         **c.** $^{22}_{10}Ne$

### SOLUTION

The atomic number of Ne is 10; each isotope has 10 protons. The number of neutrons in each isotope is found by subtracting the atomic number (10) from each mass number.

**a.** 10 protons; 10 neutrons (20 – 10)
**b.** 10 protons; 11 neutrons (21 – 10)
**c.** 10 protons; 12 neutrons (22 – 10)

### STUDY CHECK

Write an atomic symbol for each of the following isotopes:
**a.** a nitrogen atom with 8 neutrons
**b.** an atom with 20 protons and 22 neutrons
**c.** an atom with mass number 27 and 14 neutrons

## Atomic Mass

In laboratory work, a scientist generally uses samples that contain many atoms of an element. Among those atoms are all of the various isotopes with their different masses. To obtain a convenient mass to work with, chemists use the mass of an "average atom" of each element. This average atom has an **atomic mass**, which is an average of the mass of all of the naturally occurring isotopes of that element. On the periodic table, the atomic mass is given below the symbol of each element.

Most elements consist of several isotopes, which is one reason that the atomic masses on the periodic table are seldom whole numbers. For example, in a sample of chlorine atoms, there are two isotopes, $^{35}_{17}Cl$ and $^{37}_{17}Cl$. The atomic mass 35.45 amu of chlorine indicates that there will be a higher percentage of $^{35}_{17}Cl$ atoms. In fact, there are more than three atoms of $^{35}_{17}Cl$ for every atom of $^{37}_{17}Cl$ in a sample of chlorine atoms.

To determine an atomic mass of an element, the percentage of each isotope and the mass of each isotope must be determined experimentally. For example, a sample of chlorine atoms consists of 75.76% $^{35}_{17}Cl$ atoms and 24.24% $^{37}_{17}Cl$ atoms. The average atomic mass, known as a *weighted average*, is calculated using the percentage of each isotope and its mass: $^{35}_{17}Cl$ has a mass of 34.97 amu and $^{37}_{17}Cl$ has a mass of 36.97 amu.

$$\text{Atomic mass Cl} = \frac{^{35}_{17}Cl\%}{100\%} \times \text{mass } ^{35}_{17}Cl = \text{amu from } ^{35}_{17}Cl$$

$$+ \frac{^{37}_{17}Cl\%}{100\%} \times \text{mass } ^{37}_{17}Cl = \text{amu from } ^{37}_{17}Cl$$

17 — 17 protons
Cl — Symbol for chlorine
35.45 — Atomic mass 35.45 amu

17p 18n  $^{35}_{17}Cl$
17p 20n  $^{37}_{17}Cl$

75.76% of all Cl atoms     24.24% of all Cl atoms

$$\text{Atomic mass Cl} = \frac{75.76\%}{100\%} \times 34.97 \text{ amu} = 26.49 \text{ amu}$$

$$+ \frac{24.24\%}{100\%} \times 36.97 \text{ amu} = 8.962 \text{ amu}$$

$$\text{Atomic mass Cl} \qquad\qquad\qquad = 35.45 \text{ amu}$$

Table 3.11 lists the isotopes of some selected elements and their atomic masses.

**TABLE 3.11** The Atomic Mass of Some Elements

| Element | Isotopes | Atomic Mass (weighted average) |
|---------|----------|-------------------------------|
| Lithium | $^{6}_{3}\text{Li}$, $^{7}_{3}\text{Li}$ | 6.941 amu |
| Carbon | $^{12}_{6}\text{C}$, $^{13}_{6}\text{C}$, $^{14}_{6}\text{C}$ | 12.01 amu |
| Oxygen | $^{16}_{8}\text{O}$, $^{17}_{8}\text{O}$, $^{18}_{8}\text{O}$ | 16.00 amu |
| Fluorine | $^{19}_{9}\text{F}$ | 19.00 amu |
| Sulfur | $^{32}_{16}\text{S}$, $^{33}_{16}\text{S}$, $^{34}_{16}\text{S}$, $^{36}_{16}\text{S}$ | 32.07 amu |
| Copper | $^{63}_{29}\text{Cu}$, $^{65}_{29}\text{Cu}$ | 63.55 amu |

## SAMPLE PROBLEM 3.9

### ■ Average Atomic Mass

Magnesium consists of three naturally occurring isotopes, $^{24}_{12}\text{Mg}$, $^{25}_{12}\text{Mg}$, and $^{26}_{12}\text{Mg}$. Using the atomic mass on the periodic table, which isotope of magnesium is the most prevalent in a magnesium sample?

### SOLUTION

The atomic mass for magnesium is 24.3 amu, which means that $^{24}_{12}\text{Mg}$ must be the most prevalent isotope in a magnesium sample.

### STUDY CHECK

If copper consists of two isotopes, $^{63}_{29}\text{Cu}$ and $^{65}_{29}\text{Cu}$, are there more atoms of $^{63}_{29}\text{Cu}$ or $^{65}_{29}\text{Cu}$ in a sample of copper?

## QUESTIONS AND PROBLEMS

### Isotopes and Atomic Mass

**3.37** What are the number of protons, neutrons, and electrons in the following isotopes?

**a.** $^{27}_{13}\text{Al}$ **b.** $^{52}_{24}\text{Cr}$ **c.** $^{34}_{16}\text{S}$ **d.** $^{56}_{26}\text{Fe}$

**3.38** What are the number of protons, neutrons, and electrons in the following isotopes?

**a.** $^{2}_{1}\text{H}$ **b.** $^{14}_{7}\text{N}$ **c.** $^{26}_{14}\text{Si}$ **d.** $^{70}_{30}\text{Zn}$

**3.39** Write the atomic symbols for isotopes with the following:
   **a.** 15 protons and 16 neutrons
   **b.** 35 protons and 45 neutrons
   **c.** 13 electrons and 14 neutrons
   **d.** a chlorine atom with 18 neutrons
   **e.** a mercury atom with 122 neutrons

**3.40** Write the atomic symbols for isotopes with the following:
   **a.** an oxygen atom with 10 neutrons
   **b.** 4 protons and 5 neutrons
   **c.** 26 electrons and 30 neutrons
   **d.** a mass number of 24 and 13 neutrons
   **e.** a nickel atom with 32 neutrons

**3.41** There are four isotopes of sulfur with mass numbers 32, 33, 34, and 36.
   **a.** Write the atomic symbol for each of these atoms.
   **b.** How are these isotopes alike?
   **c.** How are they different?
   **d.** Why is the atomic mass of sulfur listed on the periodic table not a whole number?
   **e.** Which isotope is the most abundant in a sample of sulfur?

**3.42** There are four isotopes of strontium with mass numbers 84, 86, 87, 88.
   **a.** Write the atomic symbol for each of these atoms.
   **b.** How are these isotopes alike?
   **c.** How are they different?
   **d.** Why is the atomic mass of strontium listed on the periodic table not a whole number?
   **e.** Which isotope is the most abundant in a sample of strontium?

## LEARNING GOAL

Given the name or symbol of one of the first 18 elements in the periodic table, write the electron level arrangement.

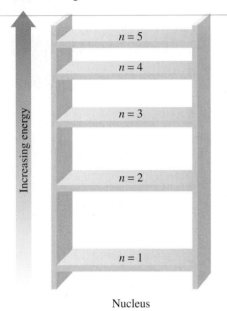

Nucleus

# 3.7 ELECTRON ENERGY LEVELS

Electrons are constantly moving within the large space of an atom, which means they possess energy. However, they do not all have the same energy. Electrons of similar energy are grouped in **energy levels**. We may think of the energy levels of an atom as similar to the rungs on a ladder. The lowest energy level would be the first rung of the ladder; the second energy level would be the second rung. As you climb up or down the ladder, you must step from one rung to the next. You cannot stop at a level between the rungs. In atoms, electrons exist only in the available energy levels. Generally, the energy levels closest to the nucleus contain electrons with the lowest energies, whereas energy levels farther away contain electrons with higher energies. Unlike the ladder, the lower energy levels are far apart compared to the higher energy levels that are closer together.

The maximum number of electrons allowed in each energy level is given by the formula $2n^2$ where $n$ is the number of the main energy level. As shown in Table 3.12, the lowest energy level, 1, can hold up to 2 electrons; level 2 can hold up to 8 electrons, level 3 can take 18 electrons, and level 4 has room for 32 electrons. In the atoms of the elements known today, electrons occupy seven energy levels.

## Orbitals

An **orbital** is a region in space around the nucleus in which an electron is most likely to be found. Each orbital can hold a maximum of two electrons. There are different types of orbitals. An $s$ orbital is spherical, with the nucleus at the center. There is just one $s$ orbital in every electron energy level.

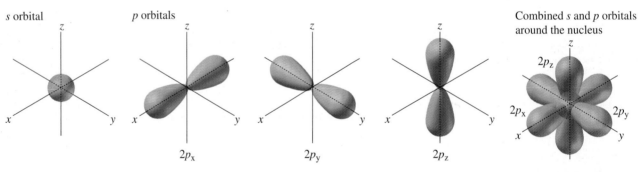

$s$ orbital

$p$ orbitals

$2p_x$

$2p_y$

$2p_z$

Combined $s$ and $p$ orbitals around the nucleus

$2p_z$

$2p_x$

$2p_y$

A $p$ orbital has two lobes. Starting with the second energy level, there is a set of three $p$ orbitals arranged in three different directions, $x$, $y$, and $z$, around the nucleus. Starting with the third energy level, there are also $d$ orbitals, and in energy level 4 and higher, there are $f$ orbitals. However, their geometry is too complex for this text.

## Electron Level Arrangements for the First 18 Elements

The electron level arrangement of an atom gives the number of electrons in each energy level. The electron level arrangements for the first 18 elements can be written by placing electrons in energy levels beginning with the lowest. The single electron of hydrogen and the 2 electrons of helium can be placed in energy level 1.

As shown in Table 3.13, the elements of the second period (lithium, Li, to neon, Ne) have enough electrons to fill the first energy level and part or all of the second energy level. For example, lithium has 3 electrons. Two of those electrons complete energy level 1. The remaining electron goes into the second energy level. As we go across Period 2, more and more electrons enter the second energy level. For example, an atom of carbon, with a total of 6 electrons, fills energy level 1 with 2 electrons, and 4 remaining electrons enter the second energy level. The last element in Period 2 is neon. The 10 electrons in an atom of neon completely fill the first and second energy levels.

In an atom of sodium, atomic number 11, the first and second energy levels are filled and the last electron enters the third energy level. The rest of the elements in the third

**TABLE 3.12** Capacity of Some Energy Levels

| Energy Level | Maximum Number of Electrons |
|:---:|:---:|
| 1 | 2 |
| 2 | 8 |
| 3 | 18 |
| 4 | 32 |

the **Chemistry** place

**WEB TUTORIAL**
Bohr's Shell Model of the Atom

**TABLE 3.13** Electron Level Arrangements for the First 18 Elements

| Element | Symbol | Atomic Number | Number of Electrons in Energy Level | | |
|---------|--------|---------------|---|---|---|
| | | | 1 | 2 | 3 |
| Hydrogen | H | 1 | 1 | | |
| Helium | He | 2 | 2 | | |
| Lithium | Li | 3 | 2 | 1 | |
| Beryllium | Be | 4 | 2 | 2 | |
| Boron | B | 5 | 2 | 3 | |
| Carbon | C | 6 | 2 | 4 | |
| Nitrogen | N | 7 | 2 | 5 | |
| Oxygen | O | 8 | 2 | 6 | |
| Fluorine | F | 9 | 2 | 7 | |
| Neon | Ne | 10 | 2 | 8 | |
| Sodium | Na | 11 | 2 | 8 | 1 |
| Magnesium | Mg | 12 | 2 | 8 | 2 |
| Aluminum | Al | 13 | 2 | 8 | 3 |
| Silicon | Si | 14 | 2 | 8 | 4 |
| Phosphorus | P | 15 | 2 | 8 | 5 |
| Sulfur | S | 16 | 2 | 8 | 6 |
| Chlorine | Cl | 17 | 2 | 8 | 7 |
| Argon | Ar | 18 | 2 | 8 | 8 |

period continue to add to the third level. For example, a sulfur atom with 16 electrons has 2 electrons in the first level, 8 electrons in the second level, and 6 electrons in the third level. At the end of Period 3, we find that argon has 8 electrons in the third level. The electron level arrangement for sulfur is written 2, 8, 6. The filling of energy levels after argon is discussed in the next section.

---

**SAMPLE PROBLEM    3.10**

■ **Writing Electron Level Arrangements**

Write the electron level arrangement for each of the following:

**a.** oxygen                    **b.** chlorine

SOLUTION

**a.** Oxygen has an atomic number of 8. Therefore, there are 8 electrons arranged with 2 electrons in energy level 1 and 6 electrons in energy level 2.

   2, 6

**b.** An atom of chlorine has 17 protons and 17 electrons. The electrons are arranged with 2 electrons in energy level 1, 8 electrons in energy level 2, and 7 electrons in energy level 3.

   2, 8, 7

STUDY CHECK

What element has an electron level arrangement of 2, 8, 2?

## Energy Level Changes

By absorbing the amount of energy equal to the difference in energy levels, an electron is raised to a higher energy level. An electron loses energy when it falls to a lower energy level and emits energy equal to the difference. (See Figure 3.14.) We see colors if the energy emitted is in the visible range. Sodium streetlights and neon lights are examples of how electrons of atoms absorb energy and emit energy as yellow light and red light.

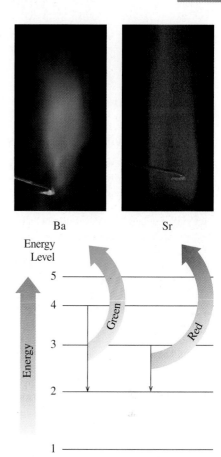

**FIGURE 3.14** The heat of a flame provides energy for electrons in atoms of strontium and barium to jump to higher energy levels. When the electrons fall to a lower energy level, energy is emitted in the visible range of red light and green light.
**Q** What causes electrons in barium and strontium to jump to higher energy levels?

# Green Chemistry Note

## Energy-Saving Light Bulbs

A compact fluorescent light (CFL) is a type of fluorescent bulb that is replacing the standard light bulb we use in our homes and workplaces. Compared to a standard light bulb, the fluorescent light bulb has a longer life and uses less electricity. Within about 20 days of use, the fluorescent bulb saves enough money in electricity costs to pay for its higher initial cost.

A standard incandescent light bulb has a thin tungsten filament inside a sealed glass bulb. When the light is switched on, electricity flows through this filament and electrical energy is converted to heat energy. When the filament reaches a temperature around 2300 °C, we see white light. We say that the light bulb is incandescent.

A fluorescent bulb produces light in a different way. When the switch is turned on, electrons move between two electrodes and collide with mercury atoms in a gas mixture of mercury and argon within the bulb. When the electrons in the mercury atoms absorb energy from the collisions, electrons are raised to higher energy levels. As electrons fall to lower energy levels, energy in the ultraviolet range is emitted. This ultraviolet light strikes the phosphor coating inside the tube, and fluorescence occurs as visible light is emitted.

The production of light in a fluorescent bulb is more efficient than in an incandescent light bulb. A 75-watt incandescent bulb can be replaced by a 20-watt fluorescent bulb that gives the same amount of light, providing an 80% reduction in electricity costs. Another saving is in the life of the bulb; a typical light bulb lasts for 1–2 months, whereas a fluorescent light bulb lasts from 1 to 2 years.

## QUESTIONS AND PROBLEMS

### Electron Energy Levels

**3.43** Electrons exist in specific energy levels. Explain.

**3.44** In what order do electrons fill the energy levels 1–3 for the first 18 elements on the periodic table?

**3.45** How many electrons are in energy level 2 of the following elements?
**a.** sodium      **b.** nitrogen      **c.** sulfur
**d.** helium      **e.** chlorine

**3.46** How many electrons are in energy level 3 of the following elements?
**a.** oxygen      **b.** sulfur      **c.** phosphorus
**d.** argon      **e.** fluorine

**3.47** Write the electron level arrangement for each of the following elements:
*Example*: sodium 2, 8, 1
**a.** carbon      **b.** argon      **c.** sulfur
**d.** silicon      **e.** an atom with 13 protons and
**f.** nitrogen             14 neutrons

**3.48** Write the electron level arrangement for each of the following atoms:
*Example*: sodium 2, 8, 1
**a.** phosphorus      **b.** neon      **c.** oxygen
**d.** an atom with atomic number 18      **e.** aluminum
**f.** fluorine

**3.49** Identify the elements that have the following electron level arrangements:

| Energy Level: | 1 | 2 | 3 |
|---|---|---|---|
| **a.** | 2 | 1 | |
| **b.** | 2 | 8 | 2 |
| **c.** | 1 | | |
| **d.** | 2 | 8 | 7 |
| **e.** | 2 | 6 | |

**3.50** Identify the elements that have the following electron level arrangements:

| Energy Level: | 1 | 2 | 3 |
|---|---|---|---|
| **a.** | 2 | 5 | |
| **b.** | 2 | 8 | 6 |
| **c.** | 2 | 4 | |
| **d.** | 2 | 8 | 8 |
| **e.** | 2 | 8 | 3 |

**3.51** Select the correct word. Electrons go to higher energy levels when they (absorb/emit) a photon.

**3.52** Select the correct word. Electrons go to lower energy levels when they (absorb/emit) a photon.

## Health Note

### Biological Reactions to UV Light

Our everyday life depends on sunlight, but exposure to sunlight can have damaging effects on living cells, and too much exposure can even cause their death. The light energy, especially ultraviolet (UV), excites electrons and may lead to unwanted chemical reactions. The list of damaging effects of sunlight includes sunburn; wrinkling; premature aging of the skin; changes in the DNA of the cells, which can lead to skin cancers; inflammation of the eyes; and perhaps cataracts. Some drugs, like the acne medications Accutane and Retin-A, as well as antibiotics, diuretics, sulfonamides, and estrogen, make the skin extremely sensitive to light.

However, medicine does take advantage of the beneficial effect of sunlight. Phototherapy can be used to treat certain skin conditions including psoriasis, eczema, and dermatitis. In the treatment of psoriasis, for example, oral drugs are given to make the skin more photosensitive; exposure to UV follows. Low-energy light is used to break down bilirubin in neonatal jaundice. Sunlight is also a factor in stimulating the immune system.

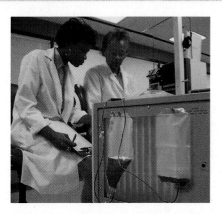

In cutaneous T-cell lymphoma, an abnormal increase in T cells causes painful ulceration of the skin. The skin is treated by photo phoresis, in which the patient receives a photosensitive chemical, and then blood is removed from the body and exposed to ultraviolet light. The blood is returned to the patient, and the treated T cells stimulate the immune system to respond to the cancer cells.

# 3.8 PERIODIC TRENDS

### LEARNING GOAL

Use the electron arrangement of elements to explain periodic trends.

The electron level arrangements of atoms are an important factor in the physical and chemical behavior of the elements. In this section, we will look at the *valence electrons* in atoms, the trends in the sizes of atoms, and *ionization energy*. Going across a period, there is a pattern of regular change in these properties from one group to the next. Known as *periodic properties*, each property increases or decreases across a period and then the trend is repeated again in each successive period. We can use the seasonal changes in temperatures as an analogy for periodic properties. In the Northern Hemisphere, temperatures in winter are usually cold and become warmer in the spring. By summer, the outdoor temperatures are hot, but begin to cool in the fall. By winter, we expect cold temperatures again as the pattern of decreasing and increasing temperatures repeats for another year.

## Group Number and Valence Electrons

If we could imagine two atoms approaching each other, the first interaction would be between the electrons in their highest energy levels. Chemists have determined that the chemical properties of representative elements are mostly due to these outermost electrons, which are known as the **valence electrons**. The **group numbers** 1A – 8A indicate the number of valence (outer) electrons for the elements in each vertical column. For example, the elements in Group 1A (1) such as lithium, sodium, and potassium, have 1 electron in the outer energy level. Elements in Group 2A (2), the alkaline earth metals, have two (2) valence electrons. The halogens in Group 7A (17) have seven (7) valence electrons. (See Table 3.14.)

**TABLE 3.14** Comparison of Electron Level Arrangements, by Group, for Some Representative Elements

| | | Number of Electrons in Energy Level | | | |
|---|---|---|---|---|---|
| Group Number | Element | 1 | 2 | 3 | 4 |
| 2A (2) | Beryllium | 2 | 2 | | |
| | Magnesium | 2 | 8 | 2 | |
| | Calcium | 2 | 8 | 8 | 2 |
| 7A (17) | Fluorine | 2 | 7 | | |
| | Chlorine | 2 | 8 | 7 | |
| | Bromine | 2 | 8 | 18 | 7 |

SAMPLE PROBLEM 3.11

### ■ Using Group Numbers

Using the periodic table, write the group number and the number of electrons in the outer electron level of the following elements:

**a.** sodium            **b.** sulfur            **c.** aluminum

#### SOLUTION

**a.** Sodium (Na) is in Group 1A (1); sodium has 1 electron in the outer energy level.
**b.** Sulfur (S) is in Group 6A (16); sulfur has 6 electrons in the outer energy level.
**c.** Aluminum (Al) is in Group 3A (13); aluminum has 3 electrons in the outer energy level.

#### STUDY CHECK

What are the group number, symbol, and name of the element with atoms that have 5 electrons in the third energy level?

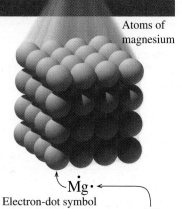

Atoms of magnesium

$\dot{M}g\cdot$

Electron-dot symbol
Electron arrangement Mg 2, 8, [2]

## Electron-Dot Symbols

An **electron-dot symbol** also known as a Lewis dot structure is a convenient way to represent the valence electrons. Valence electrons are shown as dots placed on the sides, top, or bottom of the symbol for the element. It does not matter on which of the four sides you place the dots. However, 1 to 4 valence electrons are arranged as single dots. When there are more than 4 electrons, the electrons begin to pair up. Any of the following would be an acceptable electron-dot symbol for magnesium, which has 2 valence electrons:

**Possible Electron-Dot Symbols for the 2 Valence Electrons in Magnesium**

Electron-dot symbols for selected elements are given in Table 3.15.

**TABLE 3.15** Electron-Dot Symbols for Selected Elements in Periods 1–4

| Group Number | 1A (1) | 2A (2) | 3A (13) | 4A (14) | 5A (15) | 6A (16) | 7A (17) | 8A (18) |
|---|---|---|---|---|---|---|---|---|
| **Valence Electrons** | 1 | 2 | 3 | 4 | 5 | 6 | 7 | 8 |
| **Electron-Dot Symbols** | H· | | | | | | | He: |
| | Li· | Be· | ·B· | ·Ċ· | ·N̈· | ·Ö· | ·F̈: | :N̈e: |
| | Na· | Mg· | ·Al· | ·Si· | ·P̈· | ·S̈: | ·Cl: | :Är: |
| | K· | Ca· | ·Ga· | ·Ge· | ·Äs· | ·Se: | ·Br: | :Kr: |

## SAMPLE PROBLEM 3.12

### ■ Writing Electron-Dot Symbols

Write the electron-dot symbol for each of the following elements:

**a.** bromine             **b.** aluminum

#### SOLUTION

**a.** Because the group number for bromine is 7A (17), bromine has 7 valence electrons.

·B̈r:

**b.** Aluminum, in Group 3A (3), has 3 valence electrons.

·Ȧl·

#### STUDY CHECK

What is the electron-dot symbol for phosphorus?

## Atomic Size

Although there are no fixed boundaries to atoms, scientists have a good idea of the typical volume occupied by the electrons in atoms. This volume or atomic size is determined by the *atomic radius*, which is the distance from the nucleus to the valence (outermost) electrons. Going down a group of representative elements, the outermost electrons occupy higher energy levels, which are farther from the nucleus. Therefore, the atomic radius *increases* from the top to the bottom of a group. For example, in the alkali metals, Li has a valence electron in energy level 2; Na has a valence electron in energy level 3; K has a valence electron in energy level 4; and Rb has a valence electron in energy level 5. (See Figure 3.15.)

The atomic size *decreases* going across a period. As the positive charge in the nucleus increases, there is an increase in attraction, which pulls all the electrons closer. Thus, the distance to the outermost electrons decreases and the atomic size decreases.

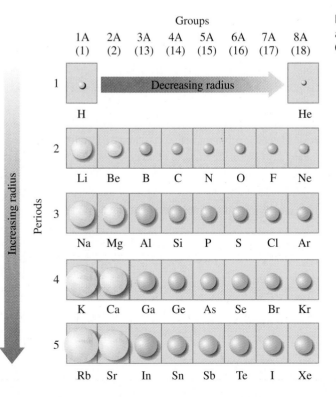

**FIGURE 3.15** The atomic radius increases going down a group, but decreases going from left to right across a period.
**Q** Why does the atomic radius increase going down a group?

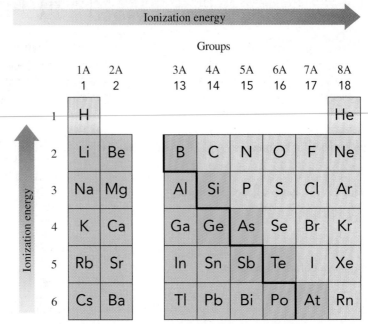

**FIGURE 3.16** Ionization energies for the representative elements tend to decrease going down a group and increase going across a period.
**Q** Why is the ionization energy for F greater than for Cl?

## Ionization Energy

Electrons are held in atoms by their attraction to the nucleus. Therefore, energy is required to remove an electron from an atom. The **ionization energy** is the energy needed to remove the least tightly bound electron from an atom in the gaseous (*g*) state. When an electron is removed from a neutral atom, a particle called a cation, with a 1+ charge, is formed.

$$Na(g) + \text{energy (ionization)} \longrightarrow Na^+(g) + e^-$$

The ionization energy *decreases* going down a group. (See Figure 3.16.) Less energy is needed to remove an electron because nuclear attraction decreases when electrons are farther from the nucleus. Going across a period from left to right, the ionization energy *increases*. As the positive charge of the nucleus increases, more energy is needed to remove an electron.

In Period 1, the valence electrons are close to the nucleus and strongly held. H and He have high ionization energies because a large amount of energy is required to remove an electron. The ionization energy for He is the highest of any element because He has a full, stable, energy level which is disrupted by removing an electron. The high ionization energies of the noble gases indicate that their electron arrangements are especially stable. In general, the ionization energy is low for the metals and high for the nonmetals.

Li atom

Na atom

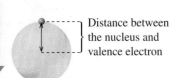

Distance between the nucleus and valence electron

K atom

Ionization energy decreases

SAMPLE PROBLEM 3.13

■ **Ionization Energy**

Indicate the element in each group that has the higher ionization energy and explain your choice.

**a.** K or Na        **b.** Mg or Cl        **c.** F, N, or C

SOLUTION

**a.** Na. In Na, the valence electron is closer to the nucleus.

**b.** Cl. Attraction for the valence electrons increases across a period, going left to right.

**c.** F. Because fluorine has more protons than nitrogen or carbon, more energy is needed to remove a valence electron from the fluorine atom.

STUDY CHECK

Arrange Sn, Sr, and I in order of increasing ionization energy.

## QUESTIONS AND PROBLEMS

### Periodic Trends

**3.53** The elements boron and aluminum are in the same group on the periodic table.
  **a.** Write the electron arrangements for B and Al.
  **b.** How many valence electrons are in the outer energy level of each atom?
  **c.** What is their group number?

**3.54** The elements fluorine and chlorine are in the same group on the periodic table.
  **a.** Write the electron arrangements for F and Cl.
  **b.** How many valence electrons are in each of their outer energy levels?
  **c.** What is their group number?

**3.55** What is the number of electrons in the outer energy level and the group number for each of the following elements?
  *Example:* fluorine $7e^-$; Group 7A (17)
  **a.** magnesium      **b.** chlorine      **c.** oxygen
  **d.** nitrogen       **e.** barium        **f.** bromine

**3.56** What is the number of electrons in the outer energy level and the group number for each of the following elements?
  *Example:* fluorine $7e^-$; Group 7A (17)
  **a.** lithium        **b.** silicon       **c.** neon
  **d.** argon          **e.** tin           **f.** cesium

**3.57** Why do Mg, Ca, and Sr have similar properties?

**3.58** Name two elements that would exhibit physical and chemical behavior similar to chlorine.

**3.59** Write the group number and electron-dot symbol for each element:
  **a.** sulfur         **b.** nitrogen      **c.** calcium
  **d.** sodium         **e.** potassium

**3.60** Write the group number and electron-dot symbol for each element:
  **a.** carbon         **b.** oxygen        **c.** fluorine
  **d.** lithium        **e.** chlorine

**3.61** Using the symbol M for a metal atom, draw the electron-dot symbol for an atom of a metal in the following groups:
  **a.** Group 1A (1)      **b.** Group 2A (2)

**3.62** Using the symbol Nm for a nonmetal atom, draw the electron-dot symbol for an atom of a nonmetal in the following groups:
  **a.** Group 5A (15)     **b.** Group 7A (17)

**3.63** The alkali metals are in the same family on the periodic table. What is their group number, and how many valence electrons does each have?

**3.64** The halogens are in the same family on the periodic table. What is their group number, and how many valence electrons does each have?

**3.65** Place the elements in each set in order of decreasing atomic radius.
  **a.** Al, Si, Mg       **b.** Cl, Br, I        **c.** I, Sb, Sr

**3.66** Place the elements in each set in order of decreasing atomic radius.
  **a.** Cl, S, P         **b.** Ge, Si, C        **c.** Ba, Ca, Sr

**3.67** Select the larger atom in each pair.
  **a.** Na or Cl         **b.** Na or Rb         **c.** Na or Mg

**3.68** Select the larger atom in each pair.
  **a.** S or Cl          **b.** S or O           **c.** S or Se

**3.69** Arrange each set of elements in order of increasing ionization energy.
  **a.** F, Cl, Br        **b.** Na, Cl, Al       **c.** Na, K, Cs

**3.70** Arrange each set of elements in order of increasing ionization energy.
  **a.** C, N, O          **b.** P, S, Cl         **c.** As, P, N

**3.71** Select the element in each pair with the higher ionization energy.
  **a.** Br or I          **b.** Mg or Sr         **c.** Si or P

**3.72** Select the element in each pair with the higher ionization energy.
  **a.** O or Ne          **b.** K or Br          **c.** Ca or Ba

CONCEPT MAP

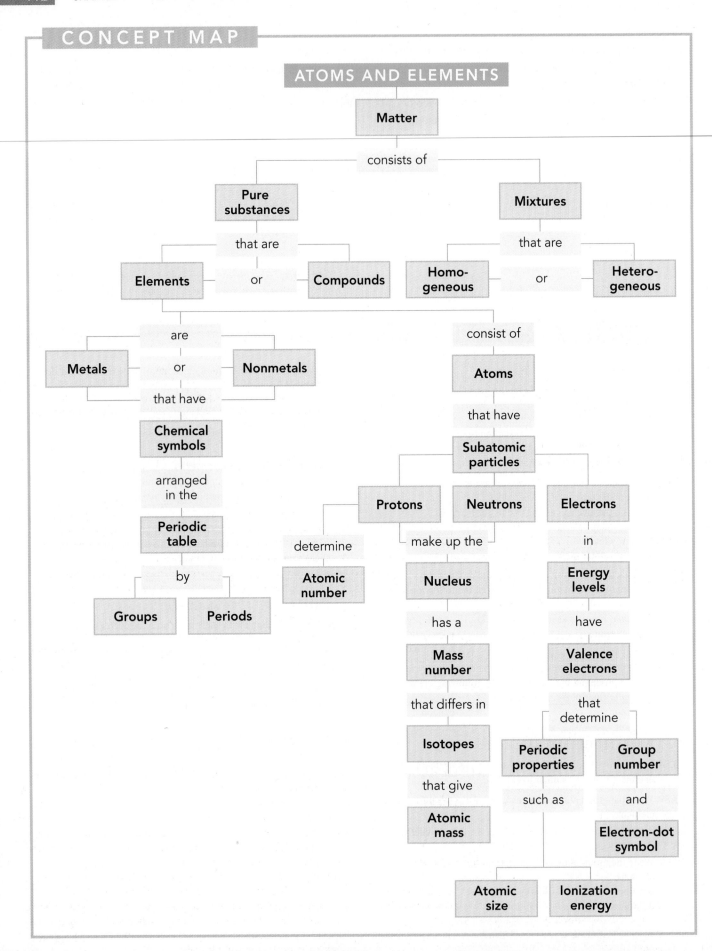

# CHAPTER REVIEW

## 3.1 Classification of Matter

**Learning Goal:** Classify matter as pure substances or mixtures.

Matter is everything that occupies space and has mass. Matter is classified as pure substances or mixtures. Pure substances, which are elements or compounds, have fixed compositions, and mixtures have variable compositions. The substances in mixtures can be separated using physical methods.

## 3.2 Elements and Symbols

**Learning Goal:** Given the name of an element, write its correct symbol; from the symbol, write the correct name.

Elements are the primary substances of matter. Chemical symbols are one- or two-letter abbreviations of the names of the elements.

## 3.3 The Periodic Table

**Learning Goal:** Use the periodic table to identify the group and the period of an element and decide whether it is a metal, nonmetal, or metalloid.

The periodic table is an arrangement of the elements by increasing atomic number. A vertical column on the periodic table containing elements with similar properties is called a group. A horizontal row is called a period. Elements in Group 1A (1) are called the alkali metals; Group 2A (2), alkaline earth metals; Group 7A (17), the halogens; and Group 8A (18), the noble gases. On the periodic table, metals are located on the left of the heavy zigzag line, and nonmetals are to the right of the heavy zigzag line. Except for aluminum, elements located on the heavy line are called metalloids.

## 3.4 The Atom

**Learning Goal:** Describe the electrical charge and location in an atom for a proton, a neutron, and an electron.

An atom is the smallest particle that retains the characteristics of an element. Atoms are composed of three subatomic particles. Protons have a positive charge (+), electrons carry a negative charge (−), and neutrons are electrically neutral. The protons and neutrons are found in the tiny, dense nucleus. Electrons are located outside the nucleus.

## 3.5 Atomic Number and Mass Number

**Learning Goal:** Given the atomic number and the mass number of an atom, state the number of protons, neutrons, and electrons.

The atomic number gives the number of protons in all the atoms of the same element. In a neutral atom, the number of protons and of electrons are equal. The mass number is the total number of protons and neutrons in an atom.

## 3.6 Isotopes and Atomic Mass

**Learning Goal:** Give the number of protons, electrons, and neutrons in the isotopes of an element.

Atoms that have the same number of protons but different numbers of neutrons are called isotopes. The atomic mass of an element is the weighted average mass of all the isotopes in a naturally occurring sample of that element.

## 3.7 Electron Energy Levels

**Learning Goal:** Given the name or symbol of one of the first 18 elements in the periodic table, write the electron level arrangement.

Every electron has a specific amount of energy. In an atom, the electrons of similar energy are grouped in specific energy levels. The first level nearest the nucleus can hold 2 electrons, the second level can hold 8 electrons, and the third level will take up to 18 electrons. Each level consists of orbitals, which represent the space where electrons of that energy are likely to be found. An $s$ orbital has a spherical shape. A $p$ orbital has two lobes like a dumbbell, and there are three $p$ orbitals in each level from $n = 2$.

## 3.8 Periodic Trends

**Learning Goal:** Use the electron level arrangement of elements to explain periodic trends.

The electron level arrangement is written by placing the number of electrons in that atom in order from the lowest energy levels and filling to higher levels. The similarity of behavior for the elements in a group is related to having the same number of valence electrons, which are the electrons in the outermost energy level. The group number for an element gives the number of electrons in its outermost energy level. With only a few minor exceptions, each group of elements has the same arrangement of valence electrons differing only in the energy level. The radius of an atom increases going down a group and decreases going across a period. The energy required to remove a valence electron is the ionization energy, which generally decreases going down a group and generally increases going across a period.

# KEY TERMS

**alkali metals** Elements of Group 1A (1) except hydrogen; these are soft, shiny metals with one outer shell electron.

**alkaline earth metals** Group 2A (2) elements, which have 2 electrons in their outer shells.

**atom** The smallest particle of an element that retains the characteristics of the element.

**atomic mass** The weighted average mass of all the naturally occurring isotopes of an element.

**atomic mass unit (amu)** A small mass unit used to describe the mass of very small particles such as atoms and subatomic particles; 1 amu is equal to one-twelfth the mass of a $^{12}C$ atom.

**atomic number** A number that is equal to the number of protons in an atom.

**atomic symbol** An abbreviation used to indicate the mass number and atomic number of an isotope.

**chemical symbol** An abbreviation that represents the name of an element.

**compound** A pure substance consisting of two or more elements, with a definite composition, that can be broken down into simpler substances by chemical methods.

**electron** A negatively charged subatomic particle having a very small mass that is usually ignored in calculations; its symbol is $e^-$.

**electron-dot symbol** The representation of an atom that shows valence electrons as dots around the symbol of the element.

**element** A pure substance that cannot be separated into any simpler substances by chemical methods.

**energy level** A group of electrons with similar energy.

**group** A vertical column in the periodic table that contains elements having similar physical and chemical properties.

**group number** A number that appears at the top of each vertical column (group) in the periodic table and indicates the number of electrons in the outermost energy level.

**halogens** Group 7A (17) elements fluorine, chlorine, bromine, iodine, and astatine.

**heterogeneous mixture** A mixture of two or more substances that are not mixed uniformly.

**homogeneous mixture** A mixture of two or more substances that are mixed uniformly.

**ionization energy** The energy needed to remove the least tightly bound electron from the outermost energy level of an atom.

**isotope** An atom that differs only in mass number from another atom of the same element. Isotopes have the same atomic number (number of protons) but different numbers of neutrons.

**mass number** The total number of neutrons and protons in the nucleus of an atom.

**metal** An element that is shiny, malleable, ductile, and a good conductor of heat and electricity. The metals are located to the left of the zigzag line in the periodic table.

**metalloid** Elements with properties of both metals and nonmetals, located along the heavy zigzag line on the periodic table.

**mixture** The physical combination of two or more substances that does not change their identities.

**neutron** A neutral subatomic particle having a mass of 1 amu and found in the nucleus of an atom; its symbol is $n$ or $n^0$.

**noble gas** An element in Group 8A (18) of the periodic table, generally unreactive and seldom found in combination with other elements.

**nonmetal** An element with little or no luster that is a poor conductor of heat and electricity. The nonmetals are located to the right of the zigzag line in the periodic table.

**nucleus** The compact, very dense center of an atom, containing the protons and neutrons of the atom.

**orbital** The region around the nucleus where electrons of a certain energy are more likely to be found. The *s* orbitals are spherical; the *p* orbitals have two lobes.

**period** A horizontal row of elements in the periodic table.

**periodic table** An arrangement of elements by increasing atomic number such that elements having similar chemical behavior are grouped in vertical columns.

**physical property** A characteristic that can be observed or measured without affecting the identity of an element, including shape, color, odor, taste, density, hardness, melting point, and boiling point.

**proton** A positively charged subatomic particle having a mass of 1 amu and found in the nucleus of an atom; its symbol is $p$ or $p^+$.

**pure substance** A type of matter with a fixed composition: elements and compounds.

**valence electrons** Electrons in the outermost energy level of an atom.

# UNDERSTANDING THE CONCEPTS

**3.73** Identify the following as an element, compound, or mixture:

a.
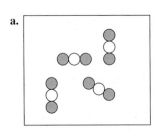

b.
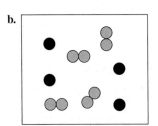

c.
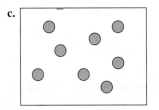

**3.74** Classify each of the following as a homogeneous or heterogeneous mixture:
   **a.** lemon-flavored water
   **b.** stuffed mushrooms
   **c.** vegetable soup
   **d.** ketchup
   **e.** hard-boiled egg
   **f.** eye drops

For Problems 3.75 and 3.76, consider the mixtures in the following diagrams:

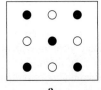
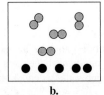
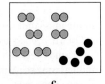

a.          b.          c.

**3.75** Which diagram(s) illustrate(s) a homogeneous mixture? Explain your choice.

**3.76** Which diagram(s) illustrate(s) a heterogeneous mixture? Explain your choice.

**3.77** According to Dalton's atomic theory, which of the following are true?
**a.** Atoms of an element are identical to atoms of other elements.
**b.** Every element is made of atoms.
**c.** Atoms of two different elements combine to form compounds.
**d.** In a chemical reaction, some atoms disappear and new atoms appear.

**3.78** Use Rutherford's gold-foil experiment to answer each of the following:
**a.** What did Rutherford expect to happen when he aimed particles at the gold foil?
**b.** How did the results differ from what he expected?
**c.** How did he use the results to propose a model of the atom?

**3.79** Match the following with the descriptions below:
**1.** protons          **2.** neutrons          **3.** electrons
**a.** atomic mass
**b.** atomic number
**c.** positive charge
**d.** negative charge
**e.** mass number – atomic number

**3.80** Match the following with the descriptions below:
**1.** protons          **2.** neutrons          **3.** electrons
**a.** mass number
**b.** surround the nucleus

**c.** nucleus
**d.** charge of 0
**e.** equal to number of electrons

**3.81** Consider the following atoms in which the chemical symbol of the element is represented by X.

$$^{16}_{8}X \quad ^{16}_{9}X \quad ^{18}_{10}X \quad ^{17}_{8}X \quad ^{18}_{8}X$$

**a.** What atoms have the same number of protons?
**b.** Which atoms are isotopes? Of what element?
**c.** Which atoms have the same mass number?
**d.** What atoms have the same number of neutrons?

**3.82** Cadmium, atomic number 48, consists of eight naturally occurring isotopes. Do you expect any of the isotopes to have the atomic mass listed on the periodic table for cadmium? Explain.

**3.83** Of the elements K, Mg, Si, S, Cl, and Ar, which
**a.** Is a metal?
**b.** Is a metalloid?
**c.** Is an alkali metal?
**d.** Has the smallest atomic size?
**e.** Has an electron arrangement 2, 8, 6?

**3.84** Of the elements K, Mg, Si, S, Cl, and Ar, which
**a.** Has the largest atomic size?
**b.** Is a halogen?
**c.** Has an electron arrangement 2, 8, 4?
**d.** Has the lowest ionization energy?
**e.** Is in Group 6A (16)?
**f.** Has the highest ionization energy?

# ADDITIONAL QUESTIONS AND PROBLEMS

**3.85** Classify each of the following as an element, compound, or mixture.
**a.** carbon in pencils
**b.** carbon dioxide ($CO_2$) we exhale
**c.** orange juice
**d.** neon gas in lights
**e.** salad dressing of oil and vinegar

**3.86** Classify each of the following as a homogenous or heterogeneous mixture:
**a.** hot fudge sundae          **b.** herbal tea
**c.** vegetable oil          **d.** water and sand
**e.** mustard

**3.87** Give the symbol and name of the element found in the following group and period on the periodic table:
**a.** Group 2A, Period 3          **b.** Group 7A, Period 4
**c.** Group 13, Period 3          **d.** Group 16, Period 2

**3.88** Give the group and period number for the following elements:
**a.** potassium          **b.** phosphorus
**c.** carbon          **d.** neon

**3.89** Give the name of the group in the periodic table for each of the following:
**a.** bromine          **b.** argon
**c.** potassium          **d.** strontium

**3.90** The following trace elements have been found to be crucial to the functions of the body. Indicate each as a metal or nonmetal.
**a.** zinc          **b.** cobalt
**c.** manganese          **d.** iodine
**e.** copper          **f.** selenium (Se)

**3.91** Indicate if each of the following statements is *true* or *false*.
**a.** The proton is a negatively charged particle.
**b.** The neutron is 2000 times as heavy as a proton.
**c.** The atomic mass unit is based on a carbon atom with 6 protons and 6 neutrons.
**d.** The nucleus is the largest part of the atom.
**e.** The electrons are located outside the nucleus.

**3.92** Indicate if each of the following statements is *true* or *false*.
**a.** The neutron is electrically neutral.
**b.** Most of the mass of an atom is due to the protons and neutrons.
**c.** The charge of an electron is equal, but opposite, to the charge of a neutron.
**d.** The proton and the electron have about the same mass.
**e.** The mass number is the number of protons.

**3.93** Complete the following statements:
**a.** The atomic number gives the number of _____ in the nucleus.

**b.** In an atom, the number of electrons is equal to the number of _____.

**c.** Sodium and potassium are examples of elements called _____.

**3.94** Complete the following statements:
   **a.** The number of protons and neutrons in an atom is also the _____ number.
   **b.** The elements in Group 7A (17) are called the _____.
   **c.** Elements that are shiny and conduct heat are called _____.

**3.95** Write the names and symbols of the elements with the following atomic numbers:
   **a.** 3      **b.** 9      **c.** 20      **d.** 33
   **e.** 50     **f.** 55     **g.** 79      **h.** 8

**3.96** Write the names and symbols of the elements with the following atomic numbers:
   **a.** 1      **b.** 11     **c.** 20      **d.** 26
   **e.** 35     **f.** 47     **g.** 83      **h.** 92

**3.97** For the following atoms, determine the number of protons, neutrons, and electrons:
   **a.** $^{27}_{13}Al$   **b.** $^{52}_{24}Cr$   **c.** $^{34}_{16}S$
   **d.** $^{56}_{26}Fe$   **e.** $^{136}_{54}Xe$

**3.98** For the following atoms, give the number of protons, neutrons, and electrons:
   **a.** $^{22}_{10}Ne$   **b.** $^{127}_{53}I$   **c.** $^{75}_{35}Br$
   **d.** $^{133}_{55}Cs$   **e.** $^{195}_{78}Pt$

**3.99** Complete the following table:

| Name | Atomic Symbol | Number of Protons | Number of Neutrons | Number of Electrons |
|---|---|---|---|---|
|  | $^{34}_{16}S$ |  |  |  |
|  |  | 30 | 40 |  |
| Magnesium |  |  | 14 |  |
|  | $^{220}_{86}Rn$ |  |  |  |

**3.100** Complete the following table:

| Name | Atomic Symbol | Number of Protons | Number of Neutrons | Number of Electrons |
|---|---|---|---|---|
| Potassium |  |  | 22 |  |
|  | $^{51}_{23}V$ |  |  |  |
|  |  | 48 | 64 |  |
| Barium |  |  | 82 |  |

**3.101** Write the atomic symbol for each of the following:
   **a.** an atom with 4 protons and 5 neutrons
   **b.** an atom with 12 protons and 14 neutrons
   **c.** a calcium atom with a mass number of 46
   **d.** an atom with 30 electrons and 40 neutrons
   **e.** a copper atom with 34 neutrons

**3.102** Write the atomic symbol for each of the following:
   **a.** an aluminum atom with 14 neutrons
   **b.** an atom with atomic number 26 and 32 neutrons
   **c.** a strontium atom with 50 neutrons
   **d.** an atom with a mass number of 72 and atomic number 33

**3.103** The most abundant isotope of lead is $^{208}_{82}Pb$.
   **a.** How many protons, neutrons, and electrons are in $^{208}_{82}Pb$?
   **b.** What is the atomic symbol of another isotope of lead with 132 neutrons?
   **c.** What is the atomic symbol and name of an atom with the same mass number as in part b and 131 neutrons?

**3.104** The most abundant isotope of silver is $^{107}_{47}Ag$.
   **a.** How many protons, neutrons, and electrons are in $^{207}_{47}Ag$?
   **b.** What is the symbol of another isotope of silver with 62 neutrons?
   **c.** What is the name and symbol of an atom with the same mass number as in part b and 49 protons?

**3.105** Write the symbol, group number, and electron arrangement for each of the following:
   **a.** nitrogen        **b.** sodium
   **c.** sulfur          **d.** boron

**3.106** Write an electron arrangement and the group number for an atom of each of the following:
   **a.** carbon          **b.** silicon
   **c.** phosphorus      **d.** argon

**3.107** Why is the ionization energy of Ca higher than K, but lower than Mg?

**3.108** Why is the ionization energy of Cl lower than F, but higher than S?

**3.109** Of the elements Na, P, Cl, and F, which
   **a.** Is a metal?
   **b.** Has the largest atomic radius?
   **c.** Has the highest ionization energy?
   **d.** Loses an electron most easily?
   **e.** Is found in Group 7A (17), Period 3?

**3.110** Of the elements: Mg, Ca, Br, Kr, which
   **a.** Is a noble gas?
   **b.** Has the smallest atomic radius?
   **c.** Has the lowest ionization energy?
   **d.** Requires the most energy to remove an electron?
   **e.** Is found in Group 2A (2), Period 4?

# CHALLENGE QUESTIONS

**3.111** For each of the following, write the symbol and name for X and the number of protons and neutrons. Which are isotopes of each other?
   **a.** $^{37}_{17}X$          **b.** $^{56}_{26}X$
   **c.** $^{116}_{50}X$         **d.** $^{124}_{50}X$
   **e.** $^{116}_{48}X$

**3.112** There are four naturally occurring isotopes of iron: $^{54}_{26}Fe$, $^{56}_{26}Fe$, $^{57}_{26}Fe$, and $^{58}_{26}Fe$.
   **a.** How many protons, neutrons, and electrons are in $^{58}_{26}Fe$?
   **b.** What is the most abundant isotope in an iron sample?
   **c.** How many neutrons are in $^{57}_{26}Fe$?
   **d.** Why don't any of the isotopes of iron have the atomic mass of 55.85 amu listed on the periodic table?

**3.113** Give the symbol of the element that has the
  **a.** smallest atomic radius in Group 6A
  **b.** smallest atomic radius in Period 3
  **c.** highest ionization energy in Group 15
  **d.** lowest ionization energy in Period 3

**3.114** Give the symbol of the element that has the
  **a.** largest atomic radius in Group 1A
  **b.** largest atomic radius in Period 4
  **c.** alkali metal with the highest ionization energy
  **d.** lowest ionization energy in Group 2

**3.115** A lead atom has a mass of $3.4 \times 10^{-22}$ g. How many lead atoms are in a cube of lead that has a volume of 2.00 cm$^3$ if the density of lead is 11.3 g/cm$^3$?

**3.116** If the diameter of a sodium atom is $3.14 \times 10^{-8}$ cm, how many sodium atoms would fit along a line exactly 1 inch long?

**3.117** Gallium has two naturally occurring isotopes: $^{69}_{31}Ga$ at 60.10% (68.926 amu) and $^{71}_{31}Ga$ at 39.90% (70.925 amu). What is the atomic mass of gallium?

**3.118** Copper has two naturally occurring isotopes: $^{63}_{29}Cu$ at 69.15% (62.930 amu) and $^{65}_{29}Cu$ at 30.85% (64.928 amu). What is the atomic mass of copper?

# ANSWERS

## Answers to Study Checks

**3.1** This is a heterogeneous mixture because it does not have a uniform composition.

**3.2** Si, S, and Ag

**3.3** magnesium, aluminum, and fluorine

**3.4** **a.** phosphorus, P    **b.** neon, Ne    **c.** silicon, Si

**3.5** False; the nucleus occupies a very small volume in an atom.

**3.6** **a.** 26    **b.** 26    **c.** iron, Fe

**3.7** 45 neutrons

**3.8** **a.** $^{15}_{7}N$    **b.** $^{42}_{20}Ca$    **c.** $^{27}_{13}Al$

**3.9** There are more atoms of $^{63}Cu$ because the atomic mass 63.55 is closer to $^{63}Cu$ than $^{65}Cu$.

**3.10** magnesium

**3.11** 5A (15); P; phosphorus

**3.12** $\cdot\ddot{P}\cdot$

**3.13** Ionization energy increases going across a period: Sr is lowest, Sn is higher, and I is the highest of this set.

## Answers to Selected Questions and Problems

**3.1** **a.** compound; combination of elements in a fixed ratio
  **b.** element; one type of substance found in the list of elements
  **c.** compound; combination of elements in a fixed ratio
  **d.** element; one type of substance found in the list of elements

**3.3** **a.** homogeneous; uniform composition
  **b.** heterogeneous; nonuniform composition
  **c.** heterogeneous; nonuniform composition
  **d.** homogeneous; uniform composition

**3.5** **a.** mixture; heterogeneous with nonuniform composition
  **b.** pure substance; element found in the list of elements
  **c.** pure substance; compound with elements in a fixed ratio
  **d.** mixture; heterogeneous with nonuniform composition

**3.7** **a.** Cu    **b.** Si    **c.** K    **d.** N
  **e.** Fe    **f.** Ba    **g.** Pb    **h.** Sr

**3.9** **a.** carbon    **b.** chlorine    **c.** iodine
  **d.** mercury    **e.** fluorine    **f.** argon
  **g.** zinc    **h.** nickel

**3.11** **a.** sodium, chlorine
  **b.** calcium, sulfur, oxygen
  **c.** carbon, hydrogen, chlorine, nitrogen, oxygen
  **d.** calcium, carbon, oxygen

**3.13** **a.** Period 2    **b.** Group 8A (18)
  **c.** Group 1A (1)    **d.** Period 2

**3.15** **a.** alkaline earth metal
  **b.** transition element    **c.** noble gas
  **d.** alkali metal    **e.** halogen

**3.17** **a.** C    **b.** He    **c.** Na    **d.** Ca    **e.** Al

**3.19** **a.** metal    **b.** nonmetal
  **c.** metal    **d.** nonmetal
  **e.** nonmetal    **f.** nonmetal
  **g.** metalloid    **h.** metal

**3.21** **a.** electron    **b.** proton
  **c.** electron    **d.** neutron

**3.23** Rutherford determined that an atom contains a small, compact nucleus that is positively charged.

**3.25** **a.** True    **b.** True
  **c.** True
  **d.** False; a proton is attracted to an electron

**3.27** In the process of brushing hair, strands of hair become charged with like charges that repel each other.

**3.29** **a.** atomic number    **b.** both
  **c.** mass number    **d.** atomic number

**3.31** **a.** lithium, Li    **b.** fluorine, F
  **c.** calcium, Ca    **d.** zinc, Zn
  **e.** neon, Ne    **f.** silicon, Si
  **g.** iodine, I    **h.** oxygen, O

**3.33** **a.** 12    **b.** 30    **c.** 53    **d.** 19

**3.35** *See Table 3.16.*

**TABLE 3.16**

| Name of Element | Symbol | Atomic Number | Mass Number | Number of Protons | Number of Neutrons | Number of Electrons |
|---|---|---|---|---|---|---|
| Aluminum | Al | 13 | 27 | 13 | 14 | 13 |
| Magnesium | Mg | 12 | 24 | 12 | 12 | 12 |
| Potassium | K | 19 | 39 | 19 | 20 | 19 |
| Sulfur | S | 16 | 31 | 16 | 15 | 16 |
| Iron | Fe | 26 | 56 | 26 | 30 | 26 |

**3.37**  **a.** 13 protons, 14 neutrons, 13 electrons
**b.** 24 protons, 28 neutrons, 24 electrons
**c.** 16 protons, 18 neutrons, 16 electrons
**d.** 26 protons, 30 neutrons, 26 electrons

**3.39**  **a.** $^{31}_{15}P$     **b.** $^{80}_{35}Br$     **c.** $^{27}_{13}Al$
**d.** $^{35}_{17}Cl$     **e.** $^{202}_{80}Hg$

**3.41**  **a.** $^{32}_{16}S$   $^{33}_{16}S$   $^{34}_{16}S$   $^{36}_{16}S$
**b.** They all have the same number of protons and electrons.
**c.** They have different numbers of neutrons, which gives them different mass numbers.
**d.** The atomic mass of S listed on the periodic table is the average atomic mass of all the isotopes.
**e.** With an atomic mass of 32.07, it is most likely that the $^{32}_{16}S$ is the most prevalent isotope of sulfur.

**3.43**  The energy of electrons is of a specific quantity for each energy level. Electrons cannot have energies that are between the energy levels.

**3.45**  **a.** 8     **b.** 5     **c.** 8     **d.** 0     **e.** 8

**3.47**  **a.** 2, 4     **b.** 2, 8, 8     **c.** 2, 8, 6
**d.** 2, 8, 4     **e.** 2, 8, 3     **f.** 2, 5

**3.49**  **a.** Li     **b.** Mg     **c.** H     **d.** Cl     **e.** O

**3.51**  absorb

**3.53**  **a.** B 2, 3; Al 2, 8, 3
**b.** 3
**c.** Group 3A (13)

**3.55**  **a.** $2e^-$, Group 2A (2)
**b.** $7e^-$, Group 7A (17)
**c.** $6e^-$, Group 6A (16)
**d.** $5e^-$, Group 5A (15)
**e.** $2e^-$, Group 2A (2)
**f.** $7e^-$, Group 7A (17)

**3.57**  They each have two electrons in their outer shells.

**3.59**  **a.** Group 6A (16)   $\ddot{\underset{\cdot\cdot}{S}}\cdot$
**b.** Group 5A (15)   $\cdot\dot{N}\cdot$
**c.** Group 2A (2)   $\cdot Ca\cdot$
**d.** Group 1A (1)   $Na\cdot$
**e.** Group 1A (1)   $K\cdot$

**3.61**  **a.** $M\cdot$     **b.** $\cdot M\cdot$

**3.63**  They are all in Group 1A; each has 1 valence electron.

**3.65**  **a.** Mg, Al, Si     **b.** I, Br, Cl     **c.** Sr, Sb, I

**3.67**  **a.** Na     **b.** Rb     **c.** Na

**3.69**  **a.** Br, Cl, F     **b.** Na, Al, Cl     **c.** Cs, K, Na

**3.71**  **a.** Br     **b.** Mg     **c.** P

**3.73**  **a.** compound     **b.** mixture     **c.** element

**3.75**  Diagrams **b** and **c** illustrate heterogeneous mixtures; they do not have uniform composition throughout the samples.

**3.77**  **a.** false     **b.** true     **c.** true     **d.** false

**3.79**  **a.** 1 + 2     **b.** 1     **c.** 1     **d.** 3     **e.** 2

**3.81**  **a.** $^{16}_{8}X$,   $^{17}_{8}X$,   $^{18}_{8}X$ all have 8 protons
**b.** $^{16}_{8}X$,   $^{17}_{8}X$,   $^{18}_{8}X$ all isotopes of O
**c.** $^{16}_{8}X$ and $^{16}_{9}X$ have mass number 16, and $^{18}_{10}X$ and $^{18}_{8}X$ have a mass number of 18.
**d.** $^{16}_{8}X$ and $^{18}_{10}X$ have 8 neutrons each.

**3.83**  **a.** K, Mg     **b.** Si     **c.** K     **d.** Ar     **e.** S

**3.85**  **a.** element     **b.** compound     **c.** mixture
**d.** element     **e.** mixture

**3.87**  **a.** Mg, magnesium     **b.** Br, bromine
**c.** Al, aluminum     **d.** O, oxygen

**3.89**  **a.** halogens     **b.** noble gases
**c.** alkali metals     **d.** alkaline earth metals

**3.91**  **a.** false     **b.** false     **c.** true
**d.** false     **e.** true

**3.93**  **a.** protons     **b.** protons     **c.** alkali metals

**3.95**  **a.** lithium, Li     **b.** fluorine, F
**c.** calcium, Ca     **d.** arsenic, As
**e.** tin, Sn     **f.** cesium, Cs
**g.** gold, Au     **h.** oxygen, O

**3.97**  **a.** 13 protons, 14 neutrons, 13 electrons
**b.** 24 protons, 28 neutrons, 24 electrons
**c.** 16 protons, 18 neutrons, 16 electrons
**d.** 26 protons, 30 neutrons, 26 electrons
**e.** 54 protons, 82 neutrons, 54 electrons

**3.99**

| Name | Atomic Symbol | Number of Protons | Number of Neutrons | Number of Electrons |
|---|---|---|---|---|
| Sulfur | $^{34}_{16}S$ | 16 | 18 | 16 |
| Zinc | $^{70}_{30}Zn$ | 30 | 40 | 30 |
| Magnesium | $^{26}_{12}Mg$ | 12 | 14 | 12 |
| Radon | $^{220}_{86}Rn$ | 86 | 134 | 86 |

**3.101** **a.** $^9_4Be$     **b.** $^{26}_{12}Mg$     **c.** $^{46}_{20}Ca$     **d.** $^{70}_{30}Zn$     **e.** $^{63}_{29}Cu$

**3.103** **a.** 82 protons, 126 neutrons, 82 electrons
**b.** $^{214}_{82}Pb$          **c.** $^{214}_{83}Bi$, bismuth

**3.105** **a.** N; Group 5A (15); 2, 5
**b.** Na; Group 1A (1); 2, 8, 1
**c.** S; Group 6A (16); 2, 8, 6
**d.** B; Group 3A (13); 2, 3

**3.107** Calcium has a greater number of protons than K. The increase in positive charge increases the attraction for electrons, which means that more energy is required to remove the valence electrons. The valence electrons are farther from the nucleus in Ca than in Mg and less energy is needed to remove them.

**3.109** **a.** Na     **b.** Na     **c.** F     **d.** Na     **e.** Cl

**3.111** **a.** Cl, chlorine; 17 protons and 20 neutrons
**b.** Fe, iron; 26 protons and 30 neutrons
**c.** Sn, tin; 50 protons and 66 neutrons
**d.** Sn, tin; 50 protons and 74 neutrons
**e.** Cd, cadmium; 48 protons and 68 neutrons
**f.** **c** and **d** are both isotopes of Sn, tin

**3.113** **a.** O     **b.** Ar     **c.** N     **d.** Na

**3.115** $6.6 \times 10^{22}$ atoms

**3.117** 69.72 amu

# 4 Compounds and Their Bonds

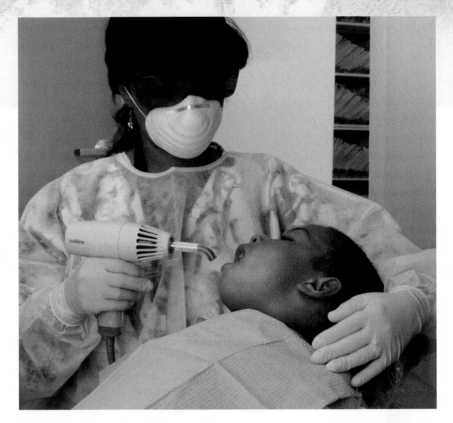

the Chemistry place

Visit **www.chemplace.com** for extra quizzes, interactive tutorials, career resources, PowerPoint slides for chapter review, math help, and case studies.

*"One way to prevent cavities in children is to apply a thin, plastic coating called a sealant to their teeth," says Dr. Pam Alston, a dentist in private practice. "We look for teeth with deep grooves and pits that trap food. We clean the teeth and apply an etching agent, which helps the sealant bind to the teeth. Then we apply the liquid sealant, which fills in the grooves and pits, and use ultraviolet light to solidify the coating."*

*The use of fluoride compounds, such as $SnF_2$ in toothpaste and $NaF$ in water, and mouth rinses have greatly reduced tooth decay. The fluoride ion replaces the hydroxide ion to form $Ca_{10}(PO_4)_6F_2$, which strengthens the enamel and makes it less susceptible to decay. Other compounds used in dentistry are the anesthetic known as laughing gas, $N_2O$, and Novocain, $C_{13}H_{20}N_2O_2$.*

I n nature, atoms of almost all the elements on the periodic table are found in combination with other atoms. Only the atoms of the noble gases—He, Ne, Ar, Kr, Xe, and Rn—do not combine in nature with other atoms. As discussed in Chapter 3, a compound is a pure substance, composed of two or more elements, with a definite composition. In a typical ionic compound, one or more electrons are transferred from the atoms of metals to atoms of nonmetals. The attraction that results is called an ionic bond.

We use many ionic compounds every day. When we cook or bake, we use ionic compounds such as salt, $NaCl$, and baking soda, $NaHCO_3$. Epsom salts, $MgSO_4$, may be used to soak sore feet. Milk of magnesia, $Mg(OH)_2$, or calcium carbonate, $CaCO_3$, may be taken to settle an upset stomach. In a mineral supplement, iron is present as iron(II) sulfate, $FeSO_4$. Certain sunscreens contain zinc oxide, $ZnO$, and the tin(II) fluoride, $SnF_2$, in toothpaste provides fluoride to help prevent tooth decay.

The structures of ionic crystals result in the beautiful facets seen in gems. Sapphires and rubies are made of aluminum oxide, $Al_2O_3$. Impurities of chromium make rubies red, and iron and titanium make sapphires blue.

In compounds of nonmetals, covalent bonding occurs by atoms sharing one or more valence electrons. There are many more covalent compounds than there are ionic ones and many simple covalent compounds are present in our everyday lives. For example, water ($H_2O$), oxygen ($O_2$), and carbon dioxide ($CO_2$) are all covalent compounds.

Covalent compounds consist of molecules, which are discrete groups of atoms. A molecule of water ($H_2O$) consists of two atoms of hydrogen and one atom of oxygen. When you have iced tea, perhaps you add molecules of sugar (sucrose), which is a covalent compound ($C_{12}H_{22}O_{11}$). Other covalent compounds include propane ($C_3H_8$), alcohol ($C_2H_6O$), the antibiotic amoxicillin ($C_{16}H_{19}N_3O_5S$), and the antidepressant Prozac ($C_{17}H_{18}F_3NO$).

# 4.1 OCTET RULE AND IONS

LEARNING GOAL

Using the octet rule, write the symbols of the simple ions for the representative elements.

Most of the elements on the periodic table combine to form compounds. However, the noble gases are so stable that they form compounds only under extreme conditions. One explanation for the stability of noble gases is that they have an **octet** of 8 valence electrons except for helium, which is stable with 2 electrons that fill its first energy level.

Compounds are the result of the formation of chemical bonds between two or more different elements. Ionic bonds occur when atoms of one element lose valence electrons and the atoms of another element gain valence electrons. Ionic compounds typically occur between metals and nonmetals. For example, atoms of sodium and chlorine form the ionic compound $NaCl$. However, atoms of nonmetals share valence electrons and form covalent compounds. For example, atoms of nitrogen and chlorine form the covalent compound $NCl_3$.

In the formation of either an ionic bond or a covalent bond, atoms lose, gain, or share valence electrons to acquire an octet. This tendency for atoms to attain a noble gas

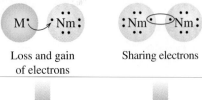

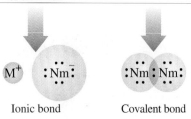

Loss and gain
of electrons

Sharing electrons

Ionic bond

Covalent bond

M is a metal
Nm is a nonmetal

electron arrangement is known as the **octet rule** and provides a key to our understanding of the ways in which atoms bond and form compounds.

## Positive Ions

In ionic bonding, the valence electrons of a metal are transferred to a nonmetal. Because the ionization energies of metals of Groups 1A (1), 2A (2), and 3A (3) are low, these metal atoms readily lose their valence electrons to nonmetals. In doing so, they acquire the electron arrangement of a noble gas (usually 8 valence electrons) and form **ions** with positive charges. For example, when a sodium atom loses its single valence electron, the remaining electrons have the noble gas arrangement of neon. By losing an electron, sodium has 10 electrons instead of 11. Because there are still 11 protons in its nucleus, the atom is no longer neutral. It has become a sodium ion and has an electrical charge, called an **ionic charge**, of 1+. In the symbol for the sodium ion, the ionic charge is written in the upper right-hand corner, as $Na^+$.

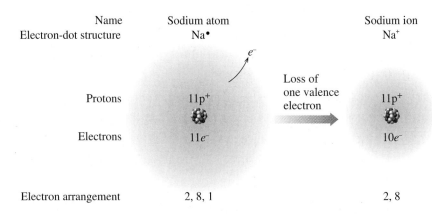

Metals in ionic compounds lose their valence electrons to form positively charged ions also called **cations** (pronounced *cat'-ions*). Magnesium, a metal in Group 2A (2), attains a noble gas electron arrangement like neon by losing 2 valence electrons to form a positive ion with a 2+ ionic charge. A metal ion is named by its element name. Thus, $Mg^{2+}$ is named the magnesium ion.

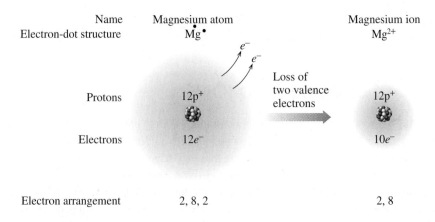

## Negative Ions

Nonmetals form negative ions when they gain valence electrons to attain an octet. For example, an atom of chlorine with 7 valence electrons obtains one electron to attain an octet and the electron arrangement of argon, a noble gas. By acquiring an electron, a chlorine atom becomes a particle called a chloride ion ($Cl^-$), which has a 1− charge. Ions with negative charges are called **anions** (pronounced *an'-ions*). The name of an anion uses *ide* to replace the last syllable of the element name.

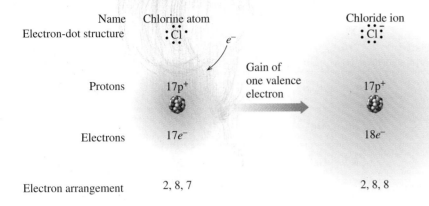

|  | Chlorine atom | | Chloride ion |
|---|---|---|---|
| Name | | | |
| Electron-dot structure | :Cl· | | :Cl: ⁻ |
| Protons | 17p⁺ | Gain of one valence electron | 17p⁺ |
| Electrons | 17e⁻ | | 18e⁻ |
| Electron arrangement | 2, 8, 7 | | 2, 8, 8 |

---

**SAMPLE PROBLEM  4.1**

### ■ Calculating Ionic Charge

Write the symbol and name of the ion described by each of the following:

**a.** a calcium atom that loses 2 electrons
**b.** an ion that has 7 protons and 10 electrons
**c.** an atom with atomic number 13 that loses 3 electrons

#### SOLUTION

**a.** A calcium atom with atomic number 20 has 20 protons and 20 electrons. A loss of 2 electrons leaves a calcium ion with 18 electrons. The sum of 20+ and 18− = 2+, which gives the calcium ion $Ca^{2+}$.

**b.** A nitrogen atom has 7 protons and 7 electrons. The nitrogen atom gains 3 electrons to attain an octet. The sum of 7+ and 10− = 3−, which gives the nitride ion $N^{3-}$.

**c.** An aluminum atom with atomic number 13 has 13 protons and 13 electrons. A loss of 3 electrons leaves an aluminum ion with 10 electrons. The sum of 13+ and 10− = 3+, which gives the aluminum ion $Al^{3+}$.

#### STUDY CHECK

How many protons and electrons are in a bromide ion, $Br^-$?

---

## Ionic Charges from Group Numbers

Group numbers can be used to determine the ionic charges for most ions of the representative elements. We have seen that metals lose electrons to form positive ions. The elements in Groups 1A (1), 2A (2), and 3A (13) lose 1, 2, and 3 electrons, respectively. Group 1A (1) metals form ions with 1+ charges, Group 2A (2) metals form ions with 2+ charges, and Group 3A (13) metals form ions with 3+ charges.

The nonmetals from Groups 5A (15), 6A (16), and 7A (17) form negative ions. Group 5A (15) nonmetals form ions with 3− charges, Group 6A (16) nonmetals form ions with 2− charges, and Group 7A (17) nonmetals form ions with 1− charges. The nonmetals of Group 4A (14) do not typically form ions. Table 4.1 lists the ionic charges for some common ions of representative elements.

**TABLE 4.1** Positive and Negative Ions Have the Same Electron Arrangement as the Nearest Noble Gases

| Noble Gases | Electron Arrangement | Metals Lose Valence Electrons | | | Nonmetals Gain Valence Electrons | | | Electron Arrangement | Noble Gases |
|---|---|---|---|---|---|---|---|---|---|
| | | 1A (1) | 2A (2) | 3A (13) | 5A (15) | 6A (16) | 7A (17) | | |
| He | ⇐ | $Li^+$ | | | | | | | |
| Ne | ⇐ | $Na^+$ | $Mg^{2+}$ | $Al^{3+}$ | $N^{3-}$ | $O^{2-}$ | $F^-$ | ⇒ | Ne |
| Ar | ⇐ | $K^+$ | $Ca^{2+}$ | | $P^{3-}$ | $S^{2-}$ | $Cl^-$ | ⇒ | Ar |
| Kr | ⇐ | $Rb^+$ | $Sr^{2+}$ | | | | $Br^-$ | ⇒ | Kr |
| Xe | ⇐ | $Cs^+$ | $Ba^{2+}$ | | | | $I^-$ | ⇒ | Xe |

## Sizes of Atoms and Their Ions

The size of ions for the representative elements compared to the size of their atoms is much smaller for metals and much larger for nonmetals.

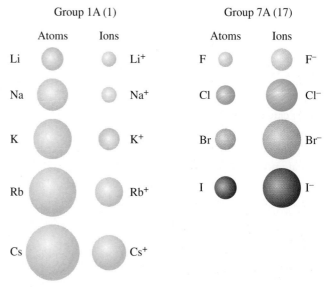

If we look at the relative sizes of the positive ions in Group 1A (1), we see that they are about half the size of their corresponding metal atoms. This occurs because metal atoms lose all of their valence electrons from their outermost energy level. For example, a sodium atom has one electron in the third energy level. When that valence electron is lost to form the sodium ion, the second energy level becomes the outermost energy level, which has an octet.

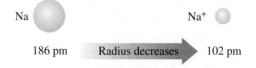

The size of nonmetal atoms increases because they gain electrons in the outermost energy level. For example, a valence electron adds to the second energy level of fluorine to

complete an octet, which increases the repulsion among the electrons and increases the size of the fluoride ion.

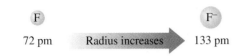

72 pm     Radius increases ⟶     133 pm

---

## SAMPLE PROBLEM 4.2

### ■ Writing Ions

Consider the elements aluminum and oxygen.

**a.** Identify each as a metal or a nonmetal.
**b.** State the number of valence electrons for each.
**c.** State the number of electrons that must be lost or gained for each to acquire an octet.
**d.** Write the symbol of each resulting ion, including its ionic charge.

#### SOLUTION

| Aluminum | Oxygen |
|---|---|
| **a.** metal | nonmetal |
| **b.** 3 valence electrons | 6 valence electrons |
| **c.** loses $3e^-$ | gains $2e^-$ |
| **d.** $Al^{3+}$ | $O^{2-}$ |

#### STUDY CHECK

What are the symbols for the ions formed by potassium and sulfur?

---

## Health Note

### Some Uses for Noble Gases

Noble gases may be used when it is necessary to have a substance that is unreactive. Scuba divers normally use a pressurized mixture of nitrogen and oxygen gases for breathing under water. However, when the air mixture is used at depths where pressure is high, the nitrogen gas is absorbed into the blood, where it can cause mental disorientation. To avoid this problem, a breathing mixture of oxygen and helium may be substituted. The diver still obtains the necessary oxygen, but the unreactive helium that dissolves in the blood does not cause mental disorientation. However, its lower density does change the vibrations of the vocal cords, and the diver will sound like Donald Duck.

Helium is also used to fill blimps and balloons. When dirigibles were first designed, they were filled with hydrogen, a very light gas. However, when they came in contact with any type of spark or heating source, they exploded violently because of the extreme reactivity of hydrogen gas with oxygen present in the air. Today blimps are filled with unreactive helium gas, which presents no danger of explosion.

Lighting tubes are generally filled with a noble gas such as neon or argon. While the electrically heated filaments that produce the light get very hot, the surrounding noble gases do not react with the hot filament. If heated in air, the elements that constitute the filament would soon burn up.

# Health Note

## Some Important Ions in the Body

A number of ions in body fluids have important physiological and metabolic functions. Some of them are listed in Table 4.2

**TABLE 4.2** Ions in the Body

| Ion | Occurrence | Function | Source | Result of Too Little | Result of Too Much |
|---|---|---|---|---|---|
| $Na^+$ | Principal cation outside the cell | Regulation and control of body fluids | Salt, cheese, pickles | Hyponatremia, anxiety, diarrhea, circulatory failure, decrease in body fluid | Hypernatremia, little urine, thirst, edema |
| $K^+$ | Principal cation inside the cell | Regulation of body fluids and cellular functions | Bananas, orange juice, milk, prunes, potatoes | Hypokalemia (hypopotassemia), lethargy, muscle weakness, failure of neurological impulses | Hyperkalemia (hyperpotassemia), irritability, nausea, little urine, cardiac arrest |
| $Ca^{2+}$ | Cation outside the cell; 90% of calcium in the body in bone as $Ca_3(PO_4)_2$ or $CaCO_3$ | Major cation of bone; muscle relaxant | Milk, yogurt, cheese, greens, and spinach | Hypocalcemia, tingling fingertips, muscle cramps, osteoporosis | Hypercalcemia, relaxed muscles, kidney stones, deep bone pain |
| $Mg^{2+}$ | Cation outside the cell; 70% of magnesium in the body in bone structure | Essential for certain enzymes, muscles, and nerve control | Widely distributed (part of chlorophyll of all green plants), nuts, whole grains | Disorientation, hypertension, tremors, slow pulse | Drowsiness |
| $Cl^-$ | Principal anion outside the cell | Gastric juice, regulation of body fluids | Salt | Same as for $Na^+$ | Same as for $Na^+$ |

## QUESTIONS AND PROBLEMS

### Octet Rule and Ions

**4.1** State the number of electrons that must be lost by atoms of each of the following elements to acquire a noble gas electron arrangement:
   **a.** Li       **b.** Mg    **c.** Al    **d.** Cs    **e.** Ba

**4.2** State the number of electrons that must be gained by atoms of each of the following elements to acquire a noble gas electron arrangement:
   **a.** Cl       **b.** O     **c.** N     **d.** I     **e.** P

**4.3** Write the symbols of the ions with the following number of protons and electrons:
   **a.** 3 protons, 2 electrons        **b.** 9 protons, 10 electrons
   **c.** 12 protons, 10 electrons      **d.** 26 protons, 23 electrons
   **e.** 30 protons, 28 electrons

**4.4** How many protons and electrons are in the following ions?
   **a.** $O^{2-}$    **b.** $K^+$    **c.** $Br^-$    **d.** $S^{2-}$    **e.** $Sr^{2+}$

**4.5** Write the symbol for the ion of each of the following:
   **a.** chlorine           **b.** potassium
   **c.** oxygen             **d.** aluminum

**4.6** Write the symbol for the ion of each of the following:
   **a.** fluorine      **b.** calcium
   **c.** sodium        **d.** lithium

**4.7** Select the larger size in each of the following pairs:
   **a.** K or $K^+$         **b.** Cl or $Cl^-$
   **c.** Ca or $Ca^{2+}$    **d.** $Li^+$ or $K^+$

**4.8** Select the smaller size in each of the following pairs:
   **a.** S or $S^{2-}$       **b.** Al or $Al^{3+}$
   **c.** $F^-$ or $Cl^-$      **d.** I or $I^-$

# 4.2 IONIC COMPOUNDS

Ionic compounds consist of positive and negative ions. The ions are held together by strong attractions between the oppositely charged ions, called ionic bonds.

## LEARNING GOAL

Using charge balance, write the correct formula for an ionic compound.

## Properties of Ionic Compounds

The physical and chemical properties of an ionic compound such as NaCl are very different from those of the original elements. For example, the original elements of NaCl were sodium, a soft, shiny metal, and chlorine, a yellow-green poisonous gas. Yet, as positive and negative ions, they form table salt, NaCl, a white, crystalline substance that is common in our diet. In ionic compounds, the attraction between the ions is very strong, which makes the melting points of ionic compounds high, often more than 500 °C. For example, the melting point of NaCl is 801 °C. At room temperature, ionic compounds are solids.

The structure of an ionic solid depends on the arrangement of the ions. In a crystal of NaCl, which has a cubic shape, the larger $Cl^-$ ions (green) are packed close together in a lattice structure. (See Figure 4.1.) The smaller $Na^+$ ions (silver) occupy the holes between the $Cl^-$ ions.

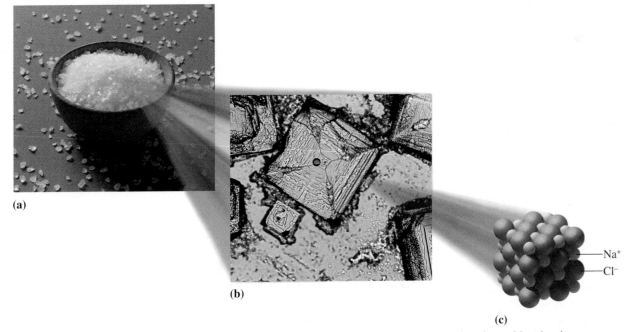

(a)

(b)

(c)

$Na^+$

$Cl^-$

**FIGURE 4.1** **(a)** The elements sodium and chlorine react to form the ionic compound sodium chloride, the compound that makes up table salt. **(b)** Crystals of NaCl under magnification. **(c)** A diagram of the arrangements of $Na^+$ and $Cl^-$ packed together in a NaCl crystal.
**Q** What is the type of bonding between $Na^+$ and $Cl^-$ ions in salt?

## Charge Balance in Ionic Compounds

The **formula** of an ionic compound indicates the number and kinds of ions that make up the ionic compound. The sum of the ionic charges in the formula is always zero. That means that the total amount of positive charge is equal to the total amount of negative charge. For example, the NaCl formula indicates that there is one sodium ion, $Na^+$, for every chloride ion, $Cl^-$, in the compound. Note that the ionic charges of the ions do not appear in the formula of the ionic compound.

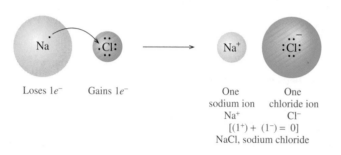

Loses $1e^-$    Gains $1e^-$

One sodium ion $Na^+$    One chloride ion $Cl^-$

$[(1^+) + (1^-) = 0]$

NaCl, sodium chloride

## Subscripts in Formulas

Consider a compound of magnesium and chlorine. To achieve an octet, a Mg atom loses its 2 valence electrons to form $Mg^{2+}$. Each Cl atom gains 1 electron to form $Cl^-$, which has a complete valence energy level. In this example, two $Cl^-$ ions are needed to balance the positive charge of $Mg^{2+}$. This gives the formula, $MgCl_2$, magnesium chloride, in which a subscript of 2 shows that two $Cl^-$ were needed for charge balance.

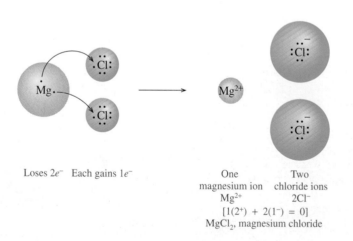

Loses $2e^-$    Each gains $1e^-$

One magnesium ion $Mg^{2+}$    Two chloride ions $2Cl^-$

$[1(2^+) + 2(1^-) = 0]$

$MgCl_2$, magnesium chloride

## Writing Ionic Formulas from Ionic Charges

The subscripts in the formula of an ionic compound represent the number of positive and negative ions that give an overall charge of zero. Thus, we can now write a formula directly from the ionic charges of the positive and negative ions. Suppose we wish to write the formula of the ionic compound containing $Na^+$ and $S^{2-}$ ions. To balance the ionic charge of the $S^{2-}$ ion, we will need to place two $Na^+$ ions in the formula. This gives the formula $Na_2S$, which has an overall charge of zero. In the formula of an ionic compound, the cation is written first followed by the anion. Appropriate subscripts are used to show the number of each of the ions.

### Physical Therapist

"Physical therapists need to understand how the body, muscles, and joints function in order to recognize when something is not working and which area needs strengthening," says Vincent Leddy, physical therapist. "Chemistry is important to understanding the body's physiology and how chemical changes in the body affect movement. I went into physical therapy because I enjoy teaching movement to children. I am guiding Maggie into her chair but allowing her to move as much as she can by herself. I help her by giving gentle pressure and reassurance that she'll be safe in the transition. I work on getting Maggie to use her body, and the occupational therapist works on Maggie's fine motor skills and how she uses her hand on the switch or picks up objects. By using both physical and occupational therapy, we enhance the child's performance."

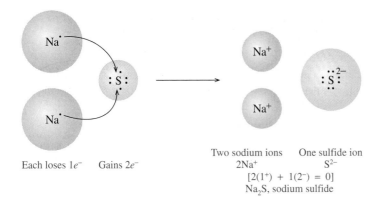

Each loses $1e^-$    Gains $2e^-$

Two sodium ions    One sulfide ion
2Na$^+$                      S$^{2-}$
$[2(1^+) + 1(2^-) = 0]$
Na$_2$S, sodium sulfide

---

### SAMPLE PROBLEM 4.3

■ **Writing Formulas from Ionic Charges**

Determine ionic charges, and write the formula for the ionic compound formed when potassium and nitrogen react.

#### SOLUTION

Potassium in Group 1A (1) forms $K^+$; nitrogen in Group 5A (15) forms $N^{3-}$. The charge of $3-$ for $N^{3-}$ is balanced by three $K^+$ ions. Writing the positive ion first gives the formula $K_3N$.

#### STUDY CHECK

Determine ionic charges, and write the formula of the compound that would form when calcium and oxygen react.

---

## QUESTIONS AND PROBLEMS

### Ionic Compounds

**4.9** Which of the following pairs of elements are likely to form an ionic compound?
 **a.** lithium and chlorine
 **b.** oxygen and chlorine
 **c.** potassium and oxygen
 **d.** sodium and neon
 **e.** sodium and magnesium

**4.10** Which of the following pairs of elements are likely to form an ionic compound?
 **a.** helium and oxygen
 **b.** magnesium and chlorine
 **c.** chlorine and bromine
 **d.** potassium and sulfur
 **e.** sodium and potassium

**4.11** Write the correct ionic formula for compounds formed between the following ions:
 **a.** $Na^+$ and $O^{2-}$
 **b.** $Al^{3+}$ and $Br^-$
 **c.** $Ba^{2+}$ and $O^{2-}$
 **d.** $Mg^{2+}$ and $Cl^-$
 **e.** $Al^{3+}$ and $S^{2-}$

**4.12** Write the correct ionic formula for compounds formed between the following ions:
 **a.** $Al^{3+}$ and $Cl^-$
 **b.** $Ca^{2+}$ and $S^{2-}$
 **c.** $Li^+$ and $S^{2-}$
 **d.** $K^+$ and $N^{3-}$
 **e.** $K^+$ and $I^-$

**4.13** Determine the ions, and write the correct formula for ionic compounds formed by the following metals and nonmetals:
 **a.** sodium and sulfur
 **b.** potassium and nitrogen
 **c.** aluminum and iodine
 **d.** lithium and oxygen

**4.14** Determine the ions, and write the correct formula for ionic compounds formed by the following metals and nonmetals:
 **a.** calcium and chlorine
 **b.** barium and bromine
 **c.** sodium and phosphorus
 **d.** magnesium and oxygen

## LEARNING GOAL

Given the formula of an ionic compound, write the correct name; given the name of an ionic compound, write the correct formula.

# 4.3 NAMING AND WRITING IONIC FORMULAS

As we mentioned in section 4.1, the name of a metal ion is the same as its elemental name. The name of a nonmetal ion is obtained by replacing the end of its elemental name with *ide*. Table 4.3 lists the names of some important metal and nonmetal ions.

**TABLE 4.3** Formulas and Names of Some Common Ions

| Group Number | Formula of Ion | Name of Ion | Group Number | Formula of Ion | Name of Ion |
|---|---|---|---|---|---|
| | **Metals** | | | **Nonmetals** | |
| 1A (1) | $Li^+$ | Lithium | 5A (15) | $N^{3-}$ | Nitride |
| | $Na^+$ | Sodium | | $P^{3-}$ | Phosphide |
| | $K^+$ | Potassium | 6A (16) | $O^{2-}$ | Oxide |
| 2A (2) | $Mg^{2+}$ | Magnesium | | $S^{2-}$ | Sulfide |
| | $Ca^{2+}$ | Calcium | 7A (17) | $F^-$ | Fluoride |
| | $Ba^{2+}$ | Barium | | $Cl^-$ | Chloride |
| 3A (3) | $Al^{3+}$ | Aluminum | | $Br^-$ | Bromide |
| | | | | $I^-$ | Iodide |

## Naming Ionic Compounds Containing Two Elements

In the name of an ionic compound made up of two elements, the metal ion is named followed by the name of the nonmetal ion. Subscripts are never mentioned; they are understood as a result of the charge balance of the ions in the compound. (See Table 4.4.)

**TABLE 4.4** Names of Some Ionic Compounds

| Compound | Metal Ion | Nonmetal Ion | Name |
|---|---|---|---|
| NaF | $Na^+$ | $F^-$ | |
| | Sodium | Fluoride | Sodium fluoride |
| $MgBr_2$ | $Mg^{2+}$ | $Br^-$ | |
| | Magnesium | Bromide | Magnesium bromide |
| $Al_2O_3$ | $Al^{3+}$ | $O^{2-}$ | |
| | Aluminum | Oxide | Aluminum oxide |

**Guide to Naming Ionic Compounds with Metals That Form a Single Ion**

1. Identify the cation and anion.

2. Name the cation by its element name.

3. Name the anion by changing the last part of its element name to *ide*.

4. Write the name of the cation first and the name of the anion second.

---

SAMPLE PROBLEM    4.4

■ **Naming Ionic Compounds**

Write the name of the ionic compound $Mg_3N_2$.

**SOLUTION**

Step 1 **Identify the cation and anion.**   The cation from Group 2A (2) is $Mg^{2+}$, and the anion from Group 5A (15) is $N^{3-}$.

Step 2 **Name the cation by its element name.**   The cation $Mg^{2+}$ is magnesium.

Step 3 **Name the anion by its element name.**   The anion $N^{3-}$ is nitride.

Step 4 **Write the name of the cation first and the name of the anion second.**   $Mg_3N_2$ is named magnesium nitride.

**STUDY CHECK**

Name the compound $CaCl_2$.

## Metals with Variable Charge

The transition metals typically form two or more kinds of positive ions because they lose their outer electrons as well as electrons from a lower energy level. For example, in some ionic compounds, iron is in the $Fe^{2+}$ form, but in other compounds, it takes the $Fe^{3+}$ form. Copper also forms two different ions: $Cu^+$ is present in some compounds and $Cu^{2+}$ in others. When a metal can form two or more ions, it is not possible to predict the ionic charge from the group number. We say that it has a variable valence.

When different ions are possible for a metal, a naming system is needed to identify the particular cation in a compound. To do this, a Roman numeral that matches the ionic charge is placed in parentheses after the elemental name of the metal. For the cations of iron, $Fe^{2+}$ is named iron(II), and $Fe^{3+}$ is named iron(III). Table 4.5 lists the ions of some common metals that produce two or more ions.

Figure 4.2 shows some common ions and their locations on the periodic table. Typically, the transition metals form more than one positive ion. However, zinc, cadmium, and silver form only one ion. The ionic charges of silver, cadmium, and zinc are fixed like Group 1A (1), 2A (2), and 3A (13) metals, so their elemental names are sufficient when naming their ionic compounds.

The selection of the correct Roman numeral depends upon the calculation of the ionic charge of the transition metal in the formula. For example, we know that in the formula $CuCl_2$ the positive charge of the copper ion must balance the negative charge of two chloride ions. Because we know that chloride ions each have a $1-$ charge, there must be a total negative charge of $2-$. Balancing the $2-$ by the positive charge gives a charge of $2+$ for Cu (a $Cu^{2+}$ ion):

**$CuCl_2$**

Cu charge $+$ $Cl^-$ charge $= 0$

(?)        $+ 2(1-)$     $= 0$

(2+)      $+ 2-$       $= 0$

To indicate the copper ion $Cu^{2+}$, we place (II) after *copper* when naming the compound:

copper(II) chloride

**TABLE 4.5** Some Metals That Form More Than One Positive Ion

| Element | Possible Ions | Name of Ion |
|---|---|---|
| Chromium | $Cr^{2+}$ | Chromium(II) |
|  | $Cr^{3+}$ | Chromium(III) |
| Copper | $Cu^+$ | Copper(I) |
|  | $Cu^{2+}$ | Copper(II) |
| Gold | $Au^+$ | Gold(I) |
|  | $Au^{3+}$ | Gold(III) |
| Iron | $Fe^{2+}$ | Iron(II) |
|  | $Fe^{3+}$ | Iron(III) |
| Lead | $Pb^{2+}$ | Lead(II) |
|  | $Pb^{4+}$ | Lead(IV) |
| Tin | $Sn^{2+}$ | Tin(II) |
|  | $Sn^{4+}$ | Tin(IV) |

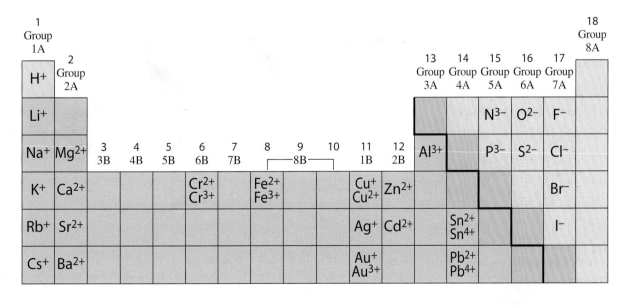

**FIGURE 4.2** On the periodic table, positive ions are produced from metals and negative ions are produced from nonmetals.

**Q** What are the typical ions produced by calcium, copper, and oxygen?

**TABLE 4.6** Some Ionic Compounds of Metals That Form Two Kinds of Positive Ions

| Compound | Systematic Name |
|---|---|
| $FeCl_2$ | Iron(II) chloride |
| $Fe_2O_3$ | Iron(III) oxide |
| $Cu_3P$ | Copper(I) phosphide |
| $CuBr_2$ | Copper(II) bromide |
| $SnCl_2$ | Tin(II) chloride |
| $PbS_2$ | Lead(IV) sulfide |

**Guide to Naming Ionic Compounds with Variable Charge Metals**

**1** Determine the charge of the cation from the anion.

**2** Name the cation by its element name and a Roman numeral in parentheses for the charge.

**3** Name the anion by changing the last part of its element name to *ide*.

**4** Write the name of the cation first and the name of the anion second.

Table 4.6 lists names of some ionic compounds in which the metals form two types of positive ion.

**SAMPLE PROBLEM   4.5**

■ **Naming Ionic Compounds with Variable Charge Metal Ions**

Write the name for $Cu_2S$.

**SOLUTION**

Step 1 **Determine the charge of the cation from the anion.** The nonmetal S in Group 6A (16) forms the $S^{2-}$ ion. Because there are two Cu ions to balance the $S^{2-}$, the charge of each Cu ion is 1+.

|  | Metal | Nonmetal |
|---|---|---|
| **Elements** | Copper | Sulfur |
| **Groups** | Transition | 6A (16) |
| **Ions** | Cu? | $S^{2-}$ |
| **Charge balance** | 2(+1)      + | (2−) = 0 |
| **Ions** | $Cu^+$ | $S^{2-}$ |

Step 2 **Name the cation by its element name, and use a Roman numeral in parentheses for the charge.**

   copper(I)

Step 3 **Name the anion by changing the last part of its element name to *ide*.**

   sulfide

Step 4 **Write the name of the cation first and the name of the anion second.**

   copper(I) sulfide

**STUDY CHECK**

Write the name of the compound whose formula is $Au_2O$.

**SAMPLE PROBLEM   4.6**

■ **Writing Formulas of Ionic Compounds**

Write the formula for iron(III) chloride.

**SOLUTION**

Step 1 **Identify the cation and anion.** The Roman numeral (III) indicates that the charge of the iron ion is 3+, $Fe^{3+}$.

|  | Metal | Nonmetal |
|---|---|---|
| **Elements** | Iron(III) | Chlorine |
| **Groups** | Transition | 7A (17) |
| **Ions** | $Fe^{3+}$ | $Cl^-$ |

**Guide to Writing Formulas from the Name of an Ionic Compound**

**1** Identify the cation and anion.

Step 2  **Balance the charges**.

$$Fe^{3+} \quad Cl^-$$
$$Cl^-$$
$$\underline{\quad\quad Cl^- \quad\quad}$$
$$\mathbf{1}(3+) + \mathbf{3}(1-) = 0$$

Becomes a subscript in the formula

Step 3  **Write the formula, cation first, using the subscripts from the charge balance**.

$$FeCl_3$$

**2** Balance the charges.

**3** Write the formula, cation first, using subscripts from charge balance.

**STUDY CHECK**

Write the correct formula for chromium(III) oxide.

---

## QUESTIONS AND PROBLEMS

### Naming and Writing Ionic Formulas

**4.15** Write names for the following ionic compounds:
  **a.** $Al_2O_3$    **b.** $CaCl_2$    **c.** $Na_2O$
  **d.** $Mg_3N_2$    **e.** $KI$

**4.16** Write names for the following ionic compounds:
  **a.** $MgCl_2$    **b.** $K_3P$    **c.** $Li_2S$
  **d.** $LiBr$    **e.** $MgO$

**4.17** Why is a Roman numeral placed after the name of most transition metal ions?

**4.18** The compound $CaCl_2$ is named calcium chloride; the compound $CuCl_2$ is named copper(II) chloride. Explain why a Roman numeral is used in one name but not the other.

**4.19** Write the names of the following Group 4A (14) and transition metal ions (include the Roman numeral when necessary):
  **a.** $Fe^{2+}$  **b.** $Cu^{2+}$  **c.** $Zn^{2+}$  **d.** $Pb^{4+}$  **e.** $Cr^{3+}$

**4.20** Write the names of the following Group 4A (14) and transition metal ions (include the Roman numeral when necessary):
  **a.** $Ag^+$  **b.** $Cu^+$  **c.** $Fe^{3+}$  **d.** $Sn^{2+}$  **e.** $Au^{3+}$

**4.21** Write names for the following ionic compounds:
  **a.** $SnCl_2$    **b.** $K_2O$    **c.** $Cu_2S$
  **d.** $CuS$    **e.** $CrBr_3$    **f.** $ZnCl_2$

**4.22** Write names for the following ionic compounds:
  **a.** $Ag_3P$    **b.** $PbS$    **c.** $Al_2O_3$
  **d.** $AuCl_3$    **e.** $FeS$    **f.** $Na_3N$

**4.23** Give the symbol of the cation in each of the following formulas:
  **a.** $AuCl_3$    **b.** $Fe_2O_3$    **c.** $PbI_4$    **d.** $SnCl_2$

**4.24** Give the symbol of the cation in each of the following formulas:
  **a.** $FeCl_2$    **b.** $CuO$    **c.** $Fe_2S_3$    **d.** $CrCl_3$

**4.25** Write formulas for the following ionic compounds:
  **a.** magnesium chloride
  **b.** sodium sulfide
  **c.** copper(I) oxide
  **d.** zinc phosphide
  **e.** gold(III) nitride
  **f.** chromium(II) chloride

**4.26** Write formulas for the following ionic compounds:
  **a.** iron(III) oxide
  **b.** barium fluoride
  **c.** tin(IV) chloride
  **d.** silver sulfide
  **e.** copper(II) chloride
  **f.** lithium nitride

---

# 4.4 POLYATOMIC IONS

**LEARNING GOAL**

Write the name and formula of a compound containing a polyatomic ion.

An ionic compound with 3 or more elements contains some type of **polyatomic ion**, which is a group of atoms that has an ionic charge. Most polyatomic ions consist of a nonmetal such as phosphorus, sulfur, carbon, or nitrogen bonded to oxygen atoms. These polyatomic ions have ionic charges of $1-$, $2-$, or $3-$ because 1, 2, or 3 electrons were gained by the atoms in the group to complete their octets. Only one of the common polyatomic ions, $NH_4^+$, is positively charged. Some models of polyatomic ions are shown in Figure 4.3.

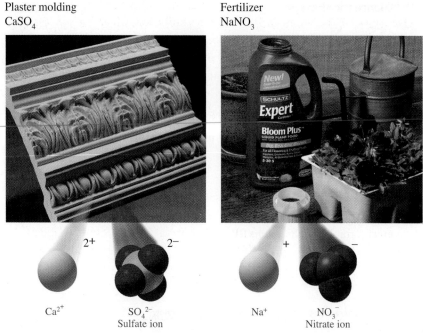

Plaster molding
$CaSO_4$

Fertilizer
$NaNO_3$

$Ca^{2+}$    $SO_4{}^{2-}$
         Sulfate ion

$Na^+$    $NO_3{}^-$
       Nitrate ion

**FIGURE 4.3** Many products contain polyatomic ions, which are groups of atoms that carry an ionic charge.

**Q** Why does the sulfate ion have a 2− charge?

## Naming Polyatomic Ions

The names of the most common polyatomic ions end in *ate*. The *ite* ending is used for the names of related ions that have one less oxygen atom. Recognizing these endings will help you identify polyatomic ions in the name of a compound. The hydroxide ion ($OH^-$) and cyanide ion ($CN^-$) are exceptions to this naming pattern. There is no easy way to learn polyatomic ions. You will need to memorize the number of oxygen atoms and the charge associated with each ion, as shown in Table 4.7. By memorizing the formulas and

**TABLE 4.7** Names and Formulas of Some Common Polyatomic Ions

| Nonmetal | Formula of Ion[a] | Name of Ion |
|---|---|---|
| Hydrogen | $OH^-$ | Hydroxide |
| Nitrogen | $NH_4{}^+$ | Ammonium |
| | $\boxed{NO_3{}^-}$ | **Nitrate** |
| | $NO_2{}^-$ | Nitrite |
| Chlorine | $\boxed{ClO_3{}^-}$ | **Chlorate** |
| | $ClO_2{}^-$ | Chlorite |
| Carbon | $\boxed{CO_3{}^{2-}}$ | **Carbonate** |
| | $HCO_3{}^-$ | Hydrogen carbonate (or bicarbonate) |
| | $CN^-$ | Cyanide |
| | $C_2H_3O_2{}^-$ | Acetate |
| Sulfur | $\boxed{SO_4{}^{2-}}$ | **Sulfate** |
| | $HSO_4{}^-$ | Hydrogen sulfate (or bisulfate) |
| | $SO_3{}^{2-}$ | Sulfite |
| | $HSO_3{}^-$ | Hydrogen sulfite (or bisulfite) |
| Phosphorus | $\boxed{PO_4{}^{3-}}$ | **Phosphate** |
| | $HPO_4{}^{2-}$ | Hydrogen phosphate |
| | $H_2PO_4{}^-$ | Dihydrogen phosphate |
| | $PO_3{}^{3-}$ | Phosphite |

[a]Boxed formulas are the most common polyatomic ion for that element.

the names of the ions shown in the boxes, you can derive the related ions. For example, the sulfate ion is $SO_4^{2-}$. We write the formula of the sulfite ion, which has one less oxygen atom, as $SO_3^{2-}$. The formula of hydrogen carbonate, or *bicarbonate*, can be written by placing a hydrogen cation ($H^+$) in front of the formula for carbonate ($CO_3^{2-}$) and decreasing the charge from 2− to 1− to give $HCO_3^-$.

## Writing Formulas for Compounds Containing Polyatomic Ions

No polyatomic ion exists by itself. Like any ion, a polyatomic ion must be associated with ions of opposite charge. The bonding between polyatomic ions and other ions is one of electrical attraction. For example, the compound sodium sulfate consists of sodium ions ($Na^+$) and sulfate ions ($SO_4^{2-}$) held together by ionic bonds.

　　To write correct formulas for compounds containing polyatomic ions, we follow the same rules of charge balance that we used for writing the formulas of simple ionic compounds. The total negative and positive charges must equal zero. For example, consider the formula for a compound containing calcium ions and carbonate ions. The ions are written as

$$Ca^{2+} \qquad CO_3^{2-}$$
　　Calcium ion　　Carbonate ion

Ionic charge: $(2+) \quad + \quad (2-) = 0$

Because one ion of each balances the charge, the formula is written as

$CaCO_3$
Calcium carbonate

When more than one polyatomic ion is needed for charge balance, parentheses are used to enclose the formula of the ion. A subscript is written outside the closing parenthesis to

## Health Note

### Polyatomic Ions in Bones and Teeth

Bone structure consists of two parts: a solid mineral material and a second phase made up primarily of collagen proteins. The mineral substance is a compound called hydroxyapatite, a solid formed from calcium ions, phosphate ions, and hydroxide ions. This material is deposited in the web of collagen to form a very durable bone material.

$Ca_{10}(PO_4)_6(OH)_2$
Calcium hydroxyapatite

　　In most individuals, bone material is continuously being absorbed and re-formed. After age 40, more bone material may be lost than formed, a condition called osteoporosis. Bone mass reduction occurs at a faster rate in women than in men and at different rates in different parts of the skeleton. The reduction in bone mass can be as much as 50% over a period of 30 to 40 years. Scanning electron

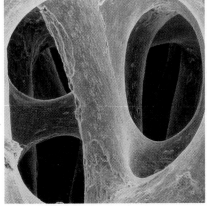

(a)

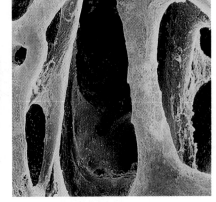

(b)

micrographs (SEMs) show (**a**) normal bone and (**b**) bone in osteoporosis because of calcium loss. It is recommended that persons over 35, especially women, include a daily calcium supplement in their diet.

indicate the number of polyatomic ions. Consider the formula for magnesium nitrate. The ions in this compound are the magnesium ion and the nitrate ion, a polyatomic ion.

$$Mg^{2+} \qquad NO_3^-$$

Magnesium ion    Nitrate ion

To balance the positive charge of 2+, two nitrate ions are needed. The formula, including the parentheses around the nitrate ion, is as follows:

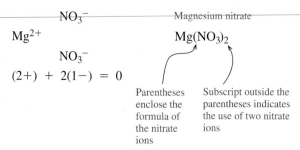

$NO_3^-$                                      Magnesium nitrate

$Mg^{2+}$                              $Mg(NO_3)_2$

$NO_3^-$

$(2+) + 2(1-) = 0$

Parentheses          Subscript outside the
enclose the          parentheses indicates
formula of           the use of two nitrate
the nitrate          ions
ions

---

## SAMPLE PROBLEM  4.7

### ■ Writing Formulas for Ionic Compounds with Polyatomic Ions

Write the formula of aluminum bicarbonate.

SOLUTION

Step 1  **Identify the cation and anion.**   The cation is the aluminum ion, $Al^{3+}$, and the anion is bicarbonate, which is a polyatomic anion, $HCO_3^-$.

|  | Cation | Anion |
|---|---|---|
| **Ions** | $Al^{3+}$ | $HCO_3^-$ |

Step 2  **Balance the charges.**

$Al^{3+}$        $HCO_3^-$
                 $HCO_3^-$
                 $HCO_3^-$
_____
$\mathbf{1}(3+) \; + \; \mathbf{3}(1-) = 0$

Becomes a subscript in the formula

Step 3  **Write the formula, cation first, using the subscripts from the charge balance.**   The formula for the compound is written by enclosing the formula of the bicarbonate ion, $HCO_3^-$, in parentheses and writing the subscript 3 outside the last parenthesis.

$Al(HCO_3)_3$

STUDY CHECK

Write the formula for a compound containing ammonium and phosphate ions.

## Naming Compounds Containing Polyatomic Ions

When naming ionic compounds containing polyatomic ions, we first write the positive ion, usually a metal, and then we write the name of the polyatomic ion. It is important that you learn to recognize the polyatomic ion in the formula and name it correctly. As with other ionic compounds, no prefixes are used.

$Na_2SO_4$          $FePO_4$            $Al_2(CO_3)_3$

$Na_2\boxed{SO_4}$        $Fe\boxed{PO_4}$          $Al_2(\boxed{CO_3})_3$

Sodium sulfate    Iron(III) phosphate    Aluminum carbonate

**TABLE 4.8 Some Compounds That Contain Polyatomic Ions**

| Formula | Name | Use |
|---|---|---|
| $BaSO_4$ | Barium sulfate | Radiopaque medium |
| $CaCO_3$ | Calcium carbonate | Antacid, calcium supplement |
| $Ca_3(PO_4)_2$ | Calcium phosphate | Calcium replenisher |
| $CaSO_3$ | Calcium sulfite | Preservative in cider and fruit juices |
| $CaSO_4$ | Calcium sulfate | Plaster casts |
| $AgNO_3$ | Silver nitrate | Topical anti-infective |
| $NaHCO_3$ | Sodium bicarbonate *or* Sodium hydrogen carbonate | Antacid |
| $Zn_3(PO_4)_2$ | Zinc phosphate | Dental cements |
| $FePO_4$ | Iron(III) phosphate | Bread/other food enrichment |
| $K_2CO_3$ | Potassium carbonate | Alkalizer, diuretic |
| $Al_2(SO_4)_3$ | Aluminum sulfate | Antiperspirant, anti-infective |
| $AlPO_4$ | Aluminum phosphate | Antacid |
| $MgSO_4$ | Magnesium sulfate | Cathartic, Epsom salts |

Table 4.8 lists the formulas and names of some ionic compounds that include polyatomic ions and also gives their uses in medicine and industry.

---

**SAMPLE PROBLEM 4.8**

■ **Naming Compounds Containing Polyatomic Ions**

Name the following ionic compounds:

**a.** $CaSO_4$    **b.** $Cu(NO_2)_2$

**SOLUTION**

| Formula | Step 1 Cation | Anion | Step 2 Name of Cation | Step 3 Name of Anion | Step 4 Name of Compound |
|---|---|---|---|---|---|
| **a.** $CaSO_4$ | $Ca^{2+}$ | $SO_4^{2-}$ | Calcium ion | Sulfate ion | Calcium sulfate |
| **b.** $Cu(NO_2)_2$ | $Cu^{2+}$ | $NO_2^{-}$ | Copper(II) ion | Nitrite ion | Copper(II) nitrite |

**STUDY CHECK**

What is the name of $Ca_3(PO_4)_2$?

**Guide to Naming Ionic Compounds with Polyatomic Ions**

**1** Identify the cation and polyatomic ion (anion).

**2** Name the cation using a Roman numeral if needed.

**3** Name the polyatomic ion usually ending with *ite* or *ate*.

**4** Write the name of the compound, cation first and the polyatomic ion second.

---

## Summary of Naming Ionic Compounds

We can now summarize the rules for naming ionic compounds. (See Figure 4.4.) Ionic compounds having two elements are named by naming the first element, followed by the second element, with an *ide* ending. If more than one positive ion is possible, write a Roman numeral for the ionic charge following the name of the metal. Ionic compounds having three or more elements include some type of polyatomic ion. They are named as ionic compounds, but usually have an *ate* or *ite* ending when the polyatomic ion has a negative charge.

**FIGURE 4.4** A flowchart for naming ionic compounds.

**Q** Why are the names of some metal ions followed by a Roman numeral in the name of a compound?

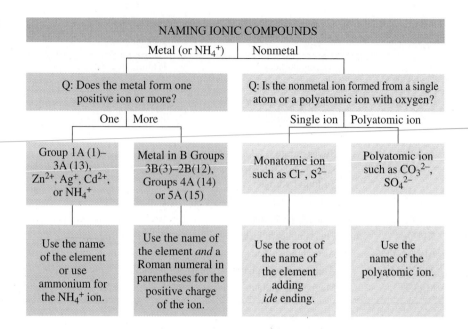

## NAMING IONIC COMPOUNDS

| Metal (or $NH_4^+$) | Nonmetal |

**Q: Does the metal form one positive ion or more?**

One | More

**Q: Is the nonmetal ion formed from a single atom or a polyatomic ion with oxygen?**

Single ion | Polyatomic ion

Group 1A (1)– 3A (13), $Zn^{2+}$, $Ag^+$, $Cd^{2+}$, or $NH_4^+$

Metal in B Groups 3B(3)–2B(12), Groups 4A (14) or 5A (15)

Monatomic ion such as $Cl^-$, $S^{2-}$

Polyatomic ion such as $CO_3^{2-}$, $SO_4^{2-}$

Use the name of the element or use ammonium for the $NH_4^+$ ion.

Use the name of the element *and* a Roman numeral in parentheses for the positive charge of the ion.

Use the root of the name of the element adding *ide* ending.

Use the name of the polyatomic ion.

## QUESTIONS AND PROBLEMS

### Polyatomic Ions

**4.27** Write the formulas, including the charge for the following polyatomic ions:
  **a.** hydrogen carbonate       **b.** ammonium
  **c.** phosphate                **d.** hydrogen sulfate

**4.28** Write the formulas, including the charge for the following polyatomic ions:
  **a.** nitrite       **b.** sulfite
  **c.** hydroxide     **d.** phosphite

**4.29** Name the following polyatomic ions:
  **a.** $SO_4^{2-}$       **b.** $CO_3^{2-}$
  **c.** $PO_4^{3-}$       **d.** $NO_3^-$

**4.30** Name the following polyatomic ions:
  **a.** $OH^-$       **b.** $HSO_3^-$
  **c.** $CN^-$       **d.** $NO_2^-$

**4.31** Complete the following table with the formula of the compound:

|          | $OH^-$ | $NO_2^-$ | $CO_3^{2-}$ | $HSO_4^-$ | $PO_4^{3-}$ |
|----------|--------|----------|-------------|-----------|-------------|
| $Li^+$   |        |          |             |           |             |
| $Cu^{2+}$|        |          |             |           |             |
| $Ba^{2+}$|        |          |             |           |             |

**4.32** Complete the following table with the formula of the compound:

|           | $OH^-$ | $NO_3^-$ | $HCO_3^-$ | $SO_3^{2-}$ | $PO_4^{3-}$ |
|-----------|--------|----------|-----------|-------------|-------------|
| $NH_4^+$  |        |          |           |             |             |
| $Al^{3+}$ |        |          |           |             |             |
| $Pb^{4+}$ |        |          |           |             |             |

**4.33** Write the formula for the polyatomic ion in each of the following, and name each compound:
  **a.** $Na_2CO_3$       **b.** $NH_4Cl$       **c.** $Li_3PO_4$
  **d.** $Cu(NO_2)_2$     **e.** $FeSO_3$

**4.34** Write the formula for the polyatomic ion in each of the following, and name each compound:
  **a.** $KOH$       **b.** $NaNO_3$       **c.** $CuCO_3$
  **d.** $NaHCO_3$   **e.** $BaSO_4$

**4.35** Write the correct formula for the following compounds:
  **a.** barium hydroxide       **b.** sodium sulfate
  **c.** iron(II) nitrate       **d.** zinc phosphate
  **e.** iron(III) carbonate

**4.36** Write the correct formula for the following compounds:
  **a.** aluminum chlorate       **b.** ammonium oxide
  **c.** magnesium bicarbonate   **d.** sodium nitrite
  **e.** copper(I) sulfate

## LEARNING GOAL

Given the formula of a covalent compound, write its correct name; given the name of a covalent compound, write its formula.

# 4.5 COVALENT COMPOUNDS

Earlier, we learned that atoms of metals and nonmetals become more stable by forming ionic compounds. However, atoms of nonmetals have high ionization energies and do not lose electrons easily. Thus, in *covalent compounds*, electrons are not transferred from one atom to another, but are shared between atoms of nonmetals to achieve stability. When atoms share electrons, they form **molecules**.

## Formation of a Hydrogen Molecule

The simplest covalent molecule is hydrogen gas, $H_2$. When two hydrogen atoms are far apart, they are not attracted to each other. As the atoms move closer, the positive charge of each nucleus attracts the electron of the other atom. This attraction pulls the atoms closer until they share a pair of valence electrons and form a *covalent bond*. In the covalent bond in $H_2$, the shared electrons give the noble gas arrangement of He to each of the H atoms. Thus, the atoms bonded in $H_2$ are more stable than two individual H atoms.

$$H\cdot \; + \; \cdot H \longrightarrow H\!:\!H \longrightarrow H\!-\!H \; = \; H_2 \;\; \text{⬤}$$

| Electrons to share | A shared pair of electrons | A covalent bond | A hydrogen molecule |

In covalent compounds, the valence electrons and the shared pairs of electrons can be shown using electron-dot formulas. In an electron-dot formula, a shared pair of electrons is written as two dots or a single line between the atomic symbols. This notation is shown in the formation of the covalent bond in the electron-dot formula for the $H_2$ molecule.

## Formation of Octets in Covalent Molecules

In most covalent compounds, atoms share electrons to achieve octets for the valence electrons. For example, a fluorine molecule, $F_2$, consists of two fluorine atoms. Each fluorine atom has 7 valence electrons. By sharing their unpaired valence electrons, each F atom achieves an octet. In the resulting $F_2$ molecule, each F atom has the noble gas arrangement of neon. In the electron-dot formula, the shared electrons, or *bonding pair*, are written between atoms with the nonbonding pairs of electrons, or *lone pairs*, on the outside. This is shown in the formation of the covalent bond for the $F_2$ molecule.

A lone pair
A bonding pair

$$:\!\ddot{F}\cdot \; + \; \cdot \ddot{F}\!: \longrightarrow \; :\!\ddot{F}\!:\!\ddot{F}\!: \quad :\!\ddot{F}\!-\!\ddot{F}\!: \; = \; F_2 \;\; \text{⬤}$$

| Electrons to share | A shared pair of electrons | A covalent bond | A fluorine molecule |

Hydrogen ($H_2$) and fluorine ($F_2$) are examples of nonmetal elements whose natural state is diatomic; that is, they contain two like atoms. The elements that exist as diatomic molecules are listed in Table 4.9.

## Sharing Electrons Between Atoms of Different Elements

In Period 2, the number of electrons that an atom shares and the number of covalent bonds it forms are usually equal to the number of electrons needed to acquire a noble gas arrangement. For example, carbon has 4 valence electrons. Because carbon needs to acquire 4 more electrons for an octet, it forms 4 covalent bonds by sharing its 4 valence electrons.

Methane, a component of natural gas, is a compound made of carbon and hydrogen. To attain an octet, each carbon shares 4 electrons, and each hydrogen shares 1 electron. In this molecule, a carbon atom forms four covalent bonds with four hydrogen atoms. The electron-dot formula for the molecule is written with the carbon atom in the center and the hydrogen atoms on the sides. Table 4.10 gives the formulas of some covalent molecules for Period 2 elements.

While the octet rule is useful, there are exceptions. We have already seen that a hydrogen ($H_2$) molecule requires just two electrons or a single bond to achieve the stability of the nearest noble gas, helium. Although the nonmetals typically form octets,

the
**C**hemistry
place

WEB TUTORIAL
Covalent Bonds

**TABLE 4.9** Elements That Exist as Diatomic, Covalent Molecules

| Element | Diatomic Molecule | Name |
|---------|-------------------|------|
| H | $H_2$ | Hydrogen |
| N | $N_2$ | Nitrogen |
| O | $O_2$ | Oxygen |
| F | $F_2$ | Fluorine |
| Cl | $Cl_2$ | Chlorine |
| Br | $Br_2$ | Bromine |
| I | $I_2$ | Iodine |

H
C H   H
H

Methane, $CH_4$

**TABLE 4.10** Electron-Dot Formulas for Some Covalent Compounds

| $CH_4$ | $NH_3$ | $H_2O$ |
|---|---|---|

**Using Electron Dots Only**

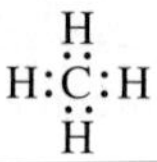

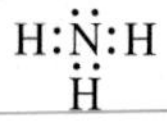

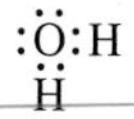

$$H:\overset{\displaystyle H}{\underset{\displaystyle H}{\overset{..}{C}}}:H \qquad H:\overset{\displaystyle \cdot\cdot}{\underset{\displaystyle H}{N}}:H \qquad :\overset{\displaystyle \cdot\cdot}{\underset{\displaystyle H}{O}}:H$$

**Using Bonds and Electron Dots**

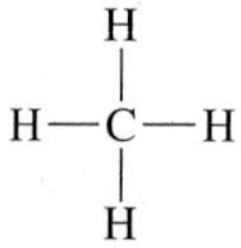

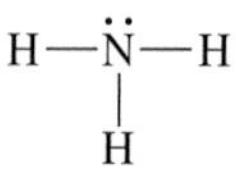

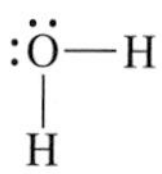

$$H-\overset{\displaystyle H}{\underset{\displaystyle H}{\overset{|}{\underset{|}{C}}}}-H \qquad H-\overset{\displaystyle \cdot\cdot}{\underset{\displaystyle H}{\underset{|}{N}}}-H \qquad :\overset{\displaystyle \cdot\cdot}{\underset{\displaystyle H}{\underset{|}{O}}}-H$$

**Molecular Models**

| Methane molecule | Ammonia molecule | Water molecule |
|---|---|---|

atoms such as P, S, Cl, Br, and I can share more valence electrons and form stable valence shells of 10 or 12 electrons. For example, in $PCl_3$, the P atom has an octet, but in $PCl_5$ the P atom has 10 valence electrons or 5 bonds. In $H_2S$, the S atom has an octet, but in $SF_6$, there are 12 valence electrons or 6 bonds to the sulfur atom.

Table 4.11 gives the bonding patterns for some nonmetals.

**TABLE 4.11** Typical Bonding Patterns of Some Nonmetals in Covalent Compounds

| 1A (1) | 3A (13) | 4A (14) | 5A (15) | 6A (16) | 7A (17) |
|---|---|---|---|---|---|
| [a]H 1 bond | | | | | |
| | [a]B 3 bonds | C 4 bonds | N 3 bonds | O 2 bonds | F 1 bond |
| | | Si 4 bonds | P 3 bonds | S 2 bonds | Cl, Br, I 1 bond |

[a]H and B do not form eight-electron octets. H atoms share one electron pair; B atoms share three electron pairs for a set of 6 electrons.

**SAMPLE PROBLEM** 4.9

■ **Writing Electron-Dot Formulas for Covalent Compounds**

Draw an electron-dot formula for $H_2S$.

SOLUTION

Sulfur, which has 6 valence electrons, shares 2 electrons to form an octet. Two atoms of hydrogen each share 1 valence electron to be stable. The shared pairs of electrons representing two covalent bonds can be written as single lines.

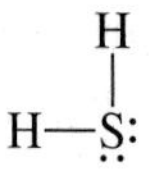

$$H-\overset{\displaystyle H}{\underset{\displaystyle ..}{\underset{|}{S}}}:$$

Electron-dot structure for $H_2S$

STUDY CHECK

Write the electron-dot structure for $PH_3$.

## Multiple Covalent Bonds

Up to now, we have looked at covalent bonding in molecules having only single bonds. In many covalent compounds, atoms share two or three pairs of electrons to complete their octets. A **double bond** occurs when two pairs of electrons are shared; in a **triple bond**, three pairs of electrons are shared. Atoms of carbon, oxygen, nitrogen, and sulfur are most likely to form multiple bonds. For example, there are double bonds in $CO_2$ because two pairs of electrons are shared between the carbon atom and each oxygen atom to give octets. Atoms of hydrogen and the halogens do not form double or triple bonds.

Octets          Molecule of
carbon dioxide

In the electron-dot structure for the covalent compound $N_2$, an octet is achieved when each nitrogen atom shares 3 electrons. Thus, three covalent bonds, or a triple bond, are formed.

Three shared   Triple bond   Nitrogen
pairs                        molecule

## Naming Covalent Compounds

When naming a covalent compound, the first nonmetal in the formula is named by its elemental name; the second nonmetal is named by its elemental name with the ending changed to *ide*. Subscripts indicating two or more atoms of an element are expressed as prefixes placed in front of each name. Table 4.12 lists prefixes used in naming covalent compounds. The names of covalent compounds need prefixes because several different compounds can be formed from the same two nonmetals. For example, carbon and oxygen can form two different compounds, carbon monoxide, CO, and carbon dioxide, $CO_2$.

When the vowels *o* and *o* or *a* and *o* appear together, the first vowel is omitted as in carbon monoxide. In the name of a covalent compound, the prefix *mono* is usually omitted, as in NO, nitrogen oxide. Traditionally, however, CO is named carbon monoxide. The prefix *mono* is not used for the first nonmetal. Table 4.13 lists the formulas, names, and commercial uses of some covalent compounds.

**TABLE 4.12** Prefixes Used in Naming Covalent Compounds

| | |
|---|---|
| 1 mono | 6 hexa |
| 2 di | 7 hepta |
| 3 tri | 8 octa |
| 4 tetra | 9 nona |
| 5 penta | 10 deca |

**TABLE 4.13** Some Common Covalent Compounds

| Formula | Name | Commercial Uses |
|---|---|---|
| $CS_2$ | Carbon disulfide | Manufacture of rayon |
| $CO_2$ | Carbon dioxide | Fire extinguishers, dry ice, propellant in aerosols, carbonation of beverages |
| NO | Nitrogen oxide | Stabilizer |
| $N_2O$ | Dinitrogen oxide | Inhalation anesthetic, "laughing gas" |
| $SiO_2$ | Silicon dioxide | Manufacture of glass |
| $SO_2$ | Sulfur dioxide | Preserving fruits, vegetables; disinfectant in breweries; bleaching textiles |
| $SF_6$ | Sulfur hexafluoride | Electrical circuits |

## Guide to Naming Covalent Compounds

**1** Name the first nonmetal by its element name.

**2** Name the second nonmetal by changing the last part of its element name to *ide*.

**3** Add prefixes to indicate the number of atoms (subscripts).

## SAMPLE PROBLEM 4.10

### ■ Naming Covalent Compounds

Name the covalent compound $NCl_3$.

**SOLUTION**

We can use the guide to naming covalent compounds to name these covalent compounds.

**Step 1  Name the first nonmetal by its element name.**  In $NCl_3$, the first nonmetal (N) is nitrogen.

**Step 2  Name the second nonmetal by changing the last part of its element name to *ide*.**  The second nonmetal (Cl) is named chloride.

**Step 3  Add prefixes to indicate the number of atoms of each nonmetal.**  Because nitrogen is the first nonmetal, the prefix *mono* is understood and not used. The subscript indicating three Cl atoms is shown as the prefix *tri*.

$NCl_3$   nitrogen trichloride

**STUDY CHECK**

Write the name of each of the following compounds:

**a.** $SiBr_4$      **b.** $Br_2O$

## Writing Formulas from the Names of Covalent Compounds

In the name of a covalent compound, the names of two nonmetals are given along with prefixes for the number of atoms of each. To obtain a formula, we write the symbol for each element and a subscript if a prefix indicates two or more atoms.

## Guide to Writing Formulas for Covalent Compounds

**1** Write the symbols in the order of the elements in the name.

**2** Write any prefixes as subscripts.

## SAMPLE PROBLEM 4.11

### ■ Writing Formulas for Covalent Compounds

Write the formula for the covalent compound diboron trioxide.

**SOLUTION**

**Step 1  Write the symbols in order of the elements in the names.**  The first nonmetal is boron, and the second nonmetal is oxygen.

B   O

**Step 2  Write prefixes as subscripts.**  The prefix *di* in *diboron* indicates that there are two atoms of boron shown as a subscript 2 in the formula. The prefix *tri* in *trioxide* indicates that there are three atoms of oxygen shown as a subscript 3 in the formula.

$B_2O_3$

**STUDY CHECK**

What is the formula of iodine pentafluoride?

## Summary of Naming Compounds

Earlier we looked at ways to name ionic compounds. Now we can include the rules for naming covalent compounds. (See Figure 4.5.) In general, compounds having two elements are named by stating the first element, followed by the second element with an *ide*

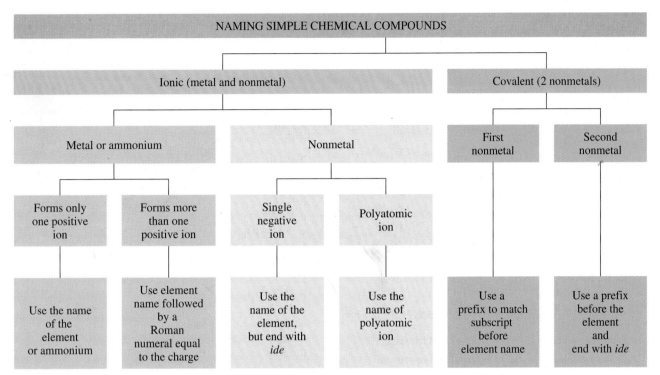

**FIGURE 4.5** A flowchart for naming ionic and covalent compounds.
**Q** Why are the names of some metal ions followed by a Roman numeral in the name of a compound?

ending. If the first element is a metal, the compound is ionic; if the first element is a non-metal, the compound is covalent. In naming covalent compounds having two elements, prefixes are used to indicate the number of atoms of each nonmetal as shown in that particular formula.

---

SAMPLE PROBLEM  **4.12**

■ **Naming Ionic and Covalent Compounds**

Name the following compounds:

**a.** $Na_3P$      **b.** $CuSO_4$      **c.** $SO_3$

SOLUTION

**a.** $Na_3P$ is an ionic compound. Na, a metal in Group 1A (1), forms $Na^+$ named sodium. Phosphorus, a nonmetal in Group 5A (15), forms $P^{3-}$, an ion named phosphide. Writing the cation first and the anion second gives the name for $Na_3P$ as sodium phosphide.

**b.** $CuSO_4$ is an ionic compound with a polyatomic ion of sulfur and oxygen. The anion $SO_4^{2-}$ has a 2− charge that is balanced by one copper ion $Cu^{2+}$. Because two cations are possible for copper, a Roman numeral in copper(II) specifies the 2+ charge. The anion $SO_4^{2-}$ is a polyatomic ion named sulfate. The compound $CuSO_4$ is named copper(II) sulfate.

**c.** $SO_3$ is covalent because it consists of two nonmetals. The first element S is named sulfur (no prefix is needed). The subscript 3 for the oxygen requires the prefix *tri*, and the element oxygen becomes oxide. The compound $SO_3$ is named sulfur trioxide.

STUDY CHECK

What is the name of $Fe(NO_3)_2$?

## QUESTIONS AND PROBLEMS

### Covalent Compounds

**4.37** Write the electron-dot formula for each of the following covalent molecules:
**a.** $Br_2$   **b.** $H_2$   **c.** HF   **d.** $OF_2$

**4.38** Write the electron-dot formula for each of the following covalent molecules:
**a.** $NCl_3$   **b.** $CCl_4$   **c.** $Cl_2$   **d.** $SiF_4$

**4.39** Name the following covalent compounds:
**a.** $PBr_3$   **b.** $CBr_4$   **c.** $SiO_2$   **d.** HF

**4.40** Name the following covalent compounds:
**a.** $CS_2$   **b.** $P_2O_5$   **c.** $Cl_2O$   **d.** $PCl_3$

**4.41** Name the following covalent compounds:
**a.** $N_2O_3$   **b.** $NCl_3$   **c.** $SiBr_4$   **d.** $PCl_5$

**4.42** Name the following covalent compounds:
**a.** $SiF_4$   **b.** $IBr_3$   **c.** $CO_2$   **d.** $SO_3$

**4.43** Write the formulas of the following covalent compounds:
**a.** carbon tetrachloride   **b.** carbon monoxide
**c.** phosphorus trichloride   **d.** dinitrogen tetroxide

**4.44** Write the formulas of the following covalent compounds:
**a.** sulfur dioxide   **b.** silicon tetrachloride
**c.** iodine pentafluoride   **d.** dinitrogen oxide

**4.45** Write the formulas of the following covalent compounds:
**a.** oxygen difluoride   **b.** boron trifluoride
**c.** dinitrogen trioxide   **d.** sulfur hexafluoride

**4.46** Write the formulas of the following covalent compounds:
**a.** sulfur dibromide   **b.** carbon disulfide
**c.** tetraphosphorus hexoxide   **d.** dinitrogen pentoxide

**4.47** Name the compounds that are found in the following:
**a.** $Al_2(SO_4)_3$     Antiperspirant
**b.** $CaCO_3$     Antacid
**c.** $N_2O$     "Laughing gas," inhaled anesthetic
**d.** $Na_3PO_4$     Cathartic
**e.** $(NH_4)_2SO_4$     Fertilizer
**f.** $Fe_2O_3$     Pigment

**4.48** Name the compounds that are found in the following:
**a.** $N_2$     Earth's atmosphere
**b.** $Mg_3(PO_4)_2$     Antacid
**c.** $FeSO_4$     Iron supplement in vitamins
**d.** $MgSO_4$     Epsom salts
**e.** $Cu_2O$     Fungicide
**f.** $SnF_2$     Prevents dental caries

# 4.6 ELECTRONEGATIVITY AND BOND POLARITY

**LEARNING GOAL**

Use electronegativity to determine the polarity of a bond.

**WEB TUTORIAL**
Electronegativity

In this chapter, we have seen that atoms form chemical bonds by gaining, losing, or sharing valence electrons. In bonds between identical nonmetal atoms, the bonding electrons are shared equally. However, in most compounds, bonds form between atoms of different elements. Then the bonding electrons are attracted to one atom more than the other.

## Electronegativity

**Electronegativity** is the attraction of an atom for valence electrons in a chemical bond. (See Figure 4.6.) Nonmetals have higher electronegativity values because they have a greater attraction for electrons than metals. The nonmetals with the highest electronegativity values are fluorine (4.0) at the top of Group 7A (17) and oxygen (3.5) at the top of Group 6A (16). The metal cesium at the bottom of Group 1A (1) has the lowest electronegativity value of 0.7. The electronegativity values for transition metals are also low, but we have not included them in our discussion. Smaller atoms tend to have higher electronegativity values because the valence electrons they share are closer to their nuclei. Electronegativity values increase from left to right across each period as well as from the bottom to the top of each group. Note that there are no electronegativity values for the noble gases because they do not typically form bonds.

**WEB TUTORIAL**
Bonds and Bond Polarity

## Types of Bonding

Earlier we discussed bonding as either *ionic*, in which electrons are transferred, or *covalent*, in which electrons were equally shared. The difference in the electronegativity

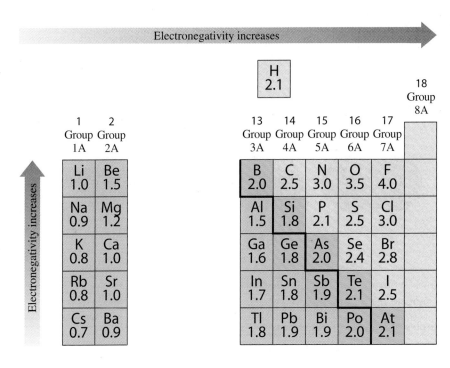

FIGURE 4.6 The electronegativity of representative elements indicates the ability of atoms to attract shared electrons. Electronegativity values increase across a period and going up a group.

**Q** What element on the periodic chart has the strongest attraction for shared electrons?

of two atoms can be used to predict the type of bond that forms. In H—H, the electronegativity difference is zero ($2.1 - 2.1 = 0$), which means the bonding electrons are shared equally. A covalent bond between atoms with identical or very similar electronegativity values is a **nonpolar covalent bond**. However, most covalent bonds are between atoms with different electronegativity values. When electrons are shared unequally, the bond is a **polar covalent bond**. In H—Cl, an electronegativity difference of $3.0\,(Cl) - 2.1(H) = 0.9$ means that the H—Cl bond is polar. (See Figure 4.7.)

In a polar covalent bond, the shared electrons are attracted to the more electronegative atom, which makes it partially negative, while the atom with the lower electronegativity becomes partially positive. A polar covalent bond that has a separation of charges is called a **dipole**. The positive and negative ends of the dipole are indicated by the lowercase Greek letter delta with a positive or negative sign, $\delta^+$ and $\delta^-$.

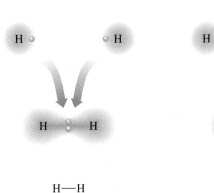

H—H

Equal sharing of electrons
in a nonpolar covalent bond

$\overset{\delta^+}{H}—\overset{\delta^-}{Cl}$

Unequal sharing of electrons
in a polar covalent bond (dipole)

FIGURE 4.7 In the nonpolar covalent bond of $H_2$, electrons are shared equally. In the polar covalent bond of HCl, electrons are shared unequally.

**Q** $H_2$ has a nonpolar covalent bond, but HCl has a polar covalent bond. Explain.

**TABLE 4.14** Electronegativity Difference and Types of Bonds

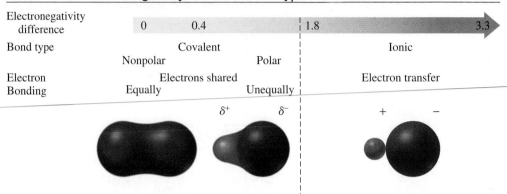

Sometimes an arrow pointing from the positive charge to the negative charge ($\longmapsto$) is used to indicate the dipole.

**Examples of Dipoles in Polar Covalent Bonds**

$$\overset{\delta^+}{C}-\overset{\delta^-}{O} \quad \overset{\delta^+}{N}-\overset{\delta^-}{O} \quad \overset{\delta^+}{Cl}-\overset{\delta^-}{F}$$
$$\longmapsto \qquad \longmapsto \qquad \longmapsto$$

## Variations in Bonding

The variations in bonding are continuous; there is no definite point at which one type of bond stops and the next starts. However, we can use some general ranges for predicting the type of bond between atoms. When electronegativity differences are from 0.0 to 0.4, the electrons are shared about equally in a nonpolar covalent bond. For example, H—H (2.1 − 2.1 = 0) and C—H (2.5 − 2.1 = 0.4) are classified as *nonpolar covalent bonds*. As the electronegativity difference increases, the shared electrons are attracted more strongly to the more electronegative atom, and the *polarity* of the bond increases. When the electronegativity difference is greater than 0.4 but less than 1.8, the bond is a *polar covalent bond*. For example, H—Cl (3.0 − 2.1 = 0.9) is classified as a *polar covalent bond*. (See Table 4.14.)

Eventually, the difference in electronegativity is great enough that the electrons are transferred from one atom to another, which results in an ionic bond. Differences in electronegativity of 1.8 or greater indicate a bond that is ionic. For example, the electronegativity difference for the ionic compound NaCl is 3.0 − 0.9 = 2.1. Thus for large differences in electronegativity, we would predict an ionic bond. (See Table 4.15.)

**TABLE 4.15** Predicting Bond Type from Electronegativity Differences

| Molecule | | Type of Electron Sharing | Electronegativity Difference[a] | Bond Type |
|---|---|---|---|---|
| $H_2$ | H—H | Shared equally | 2.1 − 2.1 = 0 | Nonpolar covalent |
| $Cl_2$ | Cl—Cl | Shared equally | 3.0 − 3.0 = 0 | Nonpolar covalent |
| HBr | $\overset{\delta^+}{H}-\overset{\delta^-}{Br}$ | Shared unequally | 2.8 − 2.1 = 0.7 | Polar covalent |
| HCl | $\overset{\delta^+}{H}-\overset{\delta^-}{Cl}$ | Shared unequally | 3.0 − 2.1 = 0.9 | Polar covalent |
| NaCl | $Na^+Cl^-$ | Electron transfer | 3.0 − 0.9 = 2.1 | Ionic |
| MgO | $Mg^{2+}O^{2-}$ | Electron transfer | 3.5 − 1.2 = 2.3 | Ionic |

[a]Values are taken from Figure 4.6.

## SAMPLE PROBLEM 4.13

### ■ Bond Polarity

Using electronegativity values, classify each bond as nonpolar covalent, polar covalent, or ionic.

N—N    O—H    Cl—As    O—K

### SOLUTION

For each bond, we calculate the difference in electronegativity.

| Bond | Electronegativity Difference | Type of Bond |
|------|------------------------------|--------------|
| N—N | $3.0 - 3.0 = 0.0$ | Nonpolar covalent |
| O—H | $3.5 - 2.1 = 1.4$ | Polar covalent |
| Cl—As | $3.0 - 2.0 = 1.0$ | Polar covalent |
| O—K | $3.5 - 0.8 = 2.7$ | Ionic |

### STUDY CHECK

Classify each bond as nonpolar covalent, polar covalent, or ionic.

**a.** Si—S    **b.** Br—Br    **c.** Na—O

## QUESTIONS AND PROBLEMS

### Electronegativity and Bond Polarity

**4.49** Describe the trend in electronegativity going from left to right across a period.

**4.50** Describe the trend in electronegativity going down a group.

**4.51** Using the periodic table, arrange the atoms in each set in order of increasing electronegativity.
**a.** Li, Na, K    **b.** Na, Cl, P    **c.** Se, Ca, O

**4.52** Using the periodic table, arrange the atoms in each set in order of increasing electronegativity.
**a.** F, Cl, Br    **b.** B, N, O    **c.** Mg, S, F

**4.53** Predict whether each of the following bonds is ionic, polar covalent, or nonpolar covalent:
**a.** Si—Br    **b.** Li—F    **c.** Br—F
**d.** Br—Br    **e.** N—P    **f.** C—O

**4.54** Predict whether each of the following bonds is ionic, polar covalent, or nonpolar covalent:
**a.** Si—O    **b.** K—Cl    **c.** S—F
**d.** P—Br    **e.** Li—O    **f.** N—P

**4.55** For each of the following bonds, indicate the positive end with $\delta^+$ and the negative end with $\delta^-$. Write an arrow to show the dipole for each.
**a.** N—F    **b.** Si—Br    **c.** C—O
**d.** P—Br    **e.** B—Cl

**4.56** For each of the following bonds, indicate the positive end with $\delta^+$ and the negative end with $\delta^-$. Write an arrow to show the dipole for each.
**a.** Si—Br    **b.** Se—F    **c.** Br—F
**d.** N—H    **e.** N—P

## 4.7 SHAPES AND POLARITY OF MOLECULES

With the information about valence electrons, electron-dot formulas, and polarity of bonds, we can now look at the three-dimensional shapes of some molecules.

To determine molecular shape, we look at the geometry of the electron groups around a central atom. The **valence-shell electron-pair repulsion (VSEPR) theory** indicates that the electron groups will move as far apart as possible to reduce the repulsion between their negative charges.

### LEARNING GOAL

Predict the three-dimensional structure of a molecule, and classify it as polar or nonpolar.

**WEB TUTORIAL**
The Shape of Molecules

The electron-dot formula is used to count the number of bonded atoms and lone pairs of electrons around the central atom. We will limit our discussion to central atoms with a total of four pairs of electrons.

In a molecule of $CH_4$, the central carbon atom is bonded to four hydrogen atoms. From the electron-dot formula, you may think that $CH_4$ is planar with 90° angles, but this is not the largest angle possible. The best arrangement for minimal repulsion is **tetrahedral**, which places the bonded atoms at the corners of a tetrahedron to give bond angles of 109°.

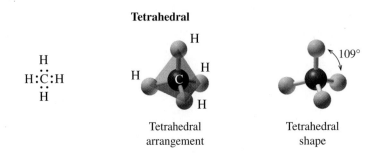

**Tetrahedral**

Tetrahedral arrangement

Tetrahedral shape

## Shapes of Molecules with Lone Pairs Around the Central Atom

In many molecules, there are one or more lone (unshared) pairs of electrons around the central atom. We predict the shape of the molecule from the number of bonded atoms. For example, in $NH_3$, the three bonded H atoms and one lone electron pair around the central N atom are arranged as a tetrahedron. A molecule with four electron pairs, but just three bonded atoms, has a **pyramidal** shape.

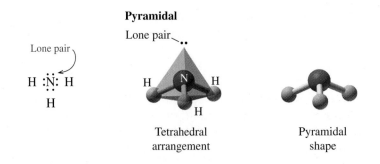

**Pyramidal**

Lone pair

Tetrahedral arrangement

Pyramidal shape

In $H_2O$, there are two bonded H atoms and two lone pairs of electrons. These four electron pairs are arranged as a tetrahedron at angles of 109°. A molecule with four electron pairs, but just two bonded atoms, has a **bent** shape.

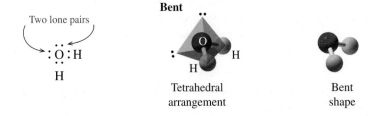

**Bent**

Two lone pairs

Tetrahedral arrangement

Bent shape

Some examples of shapes of molecules are shown in Table 4.16.

**TABLE 4.16 Examples of Shapes of Molecules**

| Molecule | Electron-Dot Formula | Bonded Atoms | Molecular Shape | |
|---|---|---|---|---|
| $CH_4$ | H:C:H (with H above and below) | 4 | Tetrahedral | 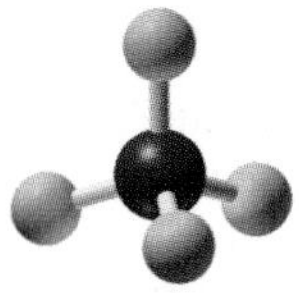 |
| $NH_3$ | H:N:H (with H below) | 3 | Pyramidal | 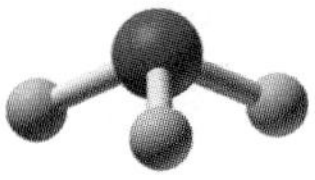 |
| $H_2O$ | :O:H (with H below) | 2 | Bent |  |

---

SAMPLE PROBLEM  **4.14**

■ **Shapes of Molecules**

Predict the shape of the following molecules:

**a.** $PH_3$        **b.** $SiCl_4$

SOLUTION

**a.** $PH_3$

Step 1  H:P:H (with H below)

Step 2  The electron-dot formula for $PH_3$ indicates four electron groups around the central P atom, which arrange themselves in a tetrahedron to minimize repulsion.

Step 3  The three bonded atoms would have a pyramidal shape.

**b.** $SiCl_4$

Step 1  :Cl:Si:Cl: (with Cl above and below)

Step 2  The electron-dot formula for $SiCl_4$ has four bonded atoms.

Step 3  With four bonded atoms and no lone pairs, the $SiCl_4$ molecule would have a tetrahedral shape.

STUDY CHECK

Predict the shape of $SCl_2$.

**Guide to Predicting Molecular Shape (VSEPR Theory)**

**1** Write the electron-dot formula for the molecule.

**2** Arrange the electron groups around the central atom to minimize repulsion.

**3** Use the atoms bonded to the central atom to determine the molecular shape.

## Polarity of Molecules

We have seen that covalent bonds in molecules can be polar or nonpolar. Molecules can also be polar or nonpolar, depending on their shape. Diatomic molecules such as $H_2$ or $Cl_2$ are nonpolar because they contain one nonpolar covalent bond.

H—H        Cl—Cl

Nonpolar

A molecule with two or more polar bonds can be a **nonpolar molecule** if the polar bonds have a symmetrical arrangement in the molecule.

In a **polar molecule**, one end of the molecule is more negatively charged than another end. Polarity in a molecule occurs when the polar bonds do not cancel each other. This cancellation depends on the type of atoms, the electron pairs around the central atom, and the shape of the molecule. For example, the HCl molecule is polar because electrons are shared unequally in a polar covalent bond.

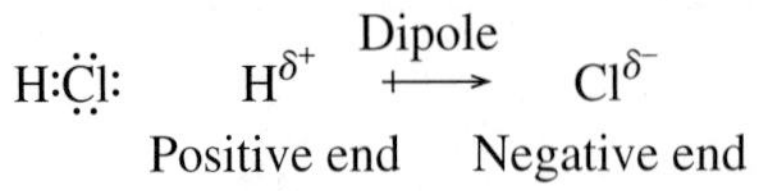

$$H:\ddot{\underset{..}{Cl}}: \qquad H^{\delta+} \xrightarrow{\text{Dipole}} Cl^{\delta-}$$

Positive end     Negative end

In polar molecules with three or more atoms, the shape of the molecule determines whether the dipoles cancel or not. Often there are lone pairs around the central atom. In $H_2O$, the dipoles do not cancel, which makes the molecule positive at one end and negative at the other end. This gives the molecule a dipole.

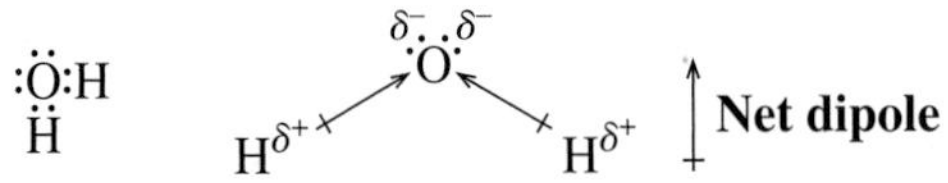

In the molecule $NH_3$, there are three dipoles, but they do not cancel.

When the polar bonds or dipoles in a molecule cancel each other, the molecule is nonpolar. For example, $CO_2$ and $CCl_4$ contain polar bonds. However, the symmetrical arrangement of the polar bonds cancels the dipoles, which makes $CO_2$ and $CCl_4$ molecules nonpolar.

### Examples of Nonpolar Molecules with Polar Bonds

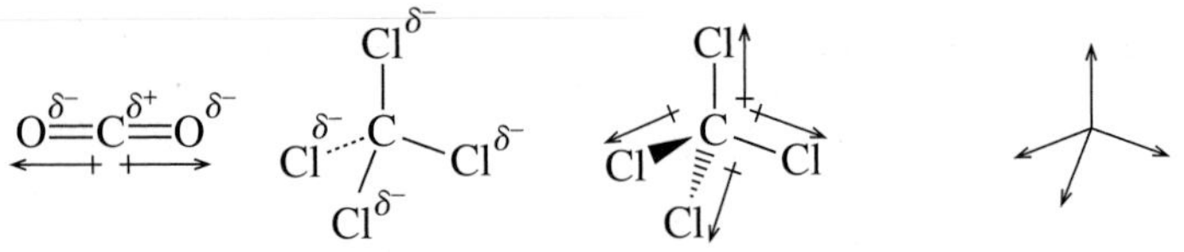

Net dipole = 0          The four individual      Net dipole = 0
                        bond polarities add up
                        to zero (they cancel)

---

## SAMPLE PROBLEM  4.15

### ■ Polarity of Molecules

Determine whether each of the following molecules is polar or nonpolar:

**a.** $CBr_4$  **b.** $OF_2$

SOLUTION

**a.** The electron-dot formula for $CBr_4$ has 4 electron pairs around the C atom bonded to Br atoms.

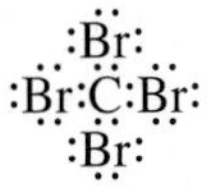

The molecule would have a tetrahedral shape. With four identical atoms bonded to the central atom and no lone pairs, the polar C-Br bonds cancel, and $CBr_4$ would be nonpolar.

**b.** The electron dot formula for $OF_2$ shows four electron groups with two bonded atoms and two lone pairs.

The molecule would have a bent shape. The two polar O-F bonds do not cancel, which makes $OF_2$ a polar molecule.

**STUDY CHECK**

Would $PCl_3$ be a polar or nonpolar molecule?

---

## QUESTIONS AND PROBLEMS

### Shapes and Polarity of Molecules

**4.57** What is the shape of a molecule if a central atom has four bonded atoms and no lone pairs?

**4.58** What is the shape of a molecule if a central atom has two bonded atoms and two lone pairs?

**4.59** In the molecule $PCl_3$, the four electron pairs around the phosphorus atom are arranged in a tetrahedral geometry. However, the shape of the molecule is called *pyramidal*. Why does the shape of the molecule have a different name from the name of the electron pair geometry?

**4.60** In the molecule $H_2S$, the four electron pairs around the sulfur atom are arranged in a tetrahedral geometry. However, the shape of the molecule is called *bent*. Why does the shape of the molecule have a different name from the name of the electron pair geometry?

**4.61** Compare the electron-dot formulas of $PH_3$ and $NH_3$. Why do these molecules have the same shape?

**4.62** Compare the electron-dot formulas $CH_4$ and $H_2O$. Why do these molecules have approximately the same angles but different shapes?

**4.63** Use the VSEPR theory to predict the shape of each molecule:
  **a.** $OF_2$     **b.** $CCl_4$

**4.64** Use the VSEPR theory to predict the shape of each molecule:
  **a.** $NCl_3$     **b.** $SCl_2$

**4.65** The molecule $Cl_2$ is nonpolar, but HCl is polar. Explain.

**4.66** The molecules $CH_4$ and $CH_3Cl$ both contain four bonds. Why is $CH_4$ nonpolar whereas $CH_3Cl$ is polar?

**4.67** Identify the following molecules as polar or nonpolar:
  **a.** HBr     **b.** $NF_3$     **c.** $CBr_4$

**4.68** Identify the following molecules as polar or nonpolar:
  **a.** $OF_2$     **b.** $PBr_3$     **c.** $SiCl_4$

---

# 4.8 ATTRACTIVE FORCES IN COMPOUNDS

**LEARNING GOAL**

Describe the attractive forces between ions, polar molecules, and nonpolar molecules.

In gases, the interactions between particles are minimal, which allows gas molecules to move far apart from each other. In solids and liquids, there are sufficient interactions between the particles to hold them close together, although some solids have low melting points while others have very high melting points. Such differences in properties are explained by looking at the various kinds of attractive forces between particles.

Ionic compound have high melting points. For example, solid NaCl melts at 801 °C. Large amounts of energy are needed to overcome the strong attractive forces between positive and negative ions. In solids containing molecules with covalent bonds, there are attractive forces too, but they are weaker than those of an ionic compound.

**WEB TUTORIAL**
Intermolecular Forces

## Dipole–Dipole Attractions and Hydrogen Bonds

For polar molecules, attractive forces called **dipole–dipole attractions** occur between the positive end of one molecule and the negative end of another. For a polar molecule with a dipole such as HCl, the partially positive H atom of one HCl molecule attracts the partially negative Cl atom in another molecule.

When a hydrogen atom is attached to highly electronegative atoms of fluorine, oxygen, or nitrogen, there are strong dipole–dipole attractions between the polar molecules. This type of attraction, called a **hydrogen bond**, occurs between the partially positive hydrogen atom of one molecule and a lone pair of electrons on a nitrogen, oxygen, or fluorine atom in another molecule. Hydrogen bonds are the strongest type of attractive forces between polar molecules. They are a major factor in the formation and structure of biological molecules such as proteins and DNA.

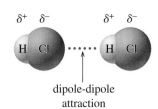

dipole-dipole attraction

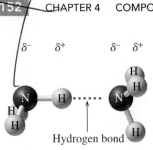

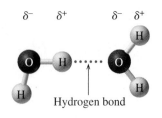

Hydrogen bond

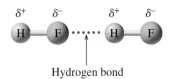

Hydrogen bond

## Dispersion Forces

Nonpolar compounds do form solids, but at low temperatures. Very weak attractions called **dispersion forces** occur between nonpolar molecules. Usually, the electrons in a nonpolar molecule are distributed symmetrically. However, electrons may accumulate more in one part of the molecule than another, which forms a temporary dipole. Although dispersion forces are very weak, they make it possible for nonpolar molecules to form liquids and solids.

The melting points of substances are related to the strength of the attractive forces within the compound. Compounds with weak attractive forces such as dispersion forces have low melting points because only a small amount of energy is needed to separate the molecules and form a liquid. Compounds with hydrogen bonds and dipole-dipole interactions require more energy to break the attractive forces between the molecules. The highest melting points are seen with ionic compounds that have the very strong attractions between ions. Table 4.17 compares the melting points of some substances with various kinds of attractive forces. The various types of attractions between particles in solids and liquids are summarized in Table 4.18.

**TABLE 4.17** Melting Points of Selected Substances

| Substance | Melting Point (°C) |
|---|---|
| **Ionic bonds** | |
| $MgF_2$ | 1248 |
| NaCl | 801 |
| **Hydrogen bonds** | |
| $H_2O$ | 0 |
| $NH_3$ | −78 |
| **Dipole–dipole interactions** | |
| HBr | −89 |
| HCl | −115 |
| **Dispersion forces** | |
| $Cl_2$ | −101 |
| $F_2$ | −220 |
| $CH_4$ | −182 |

**TABLE 4.18** Comparison of Bonding and Attractive Forces

| Type of Force | Particle Arrangement | Example | Strength |
|---|---|---|---|
| **Ionic bond** | | $Na^+$ - - - $Cl^-$ | **Strong** |
| **Hydrogen bond** (X = F, O, or N) | $\overset{\delta+}{H}\overset{\delta-}{X}$ - - $\overset{\delta+}{H}\overset{\delta-}{X}$ | $\overset{\delta+}{H}$—$\overset{\delta-}{F}$ - - - $\overset{\delta+}{H}$—$\overset{\delta-}{F}$ | |
| **Dipole–dipole** (X and Y = different nonmetals) | $\overset{\delta+}{Y}\overset{\delta-}{X}$ - - $\overset{\delta+}{Y}\overset{\delta-}{X}$ | $\overset{\delta+}{Br}$—$\overset{\delta-}{Cl}$ - - - $\overset{\delta+}{Br}$—$\overset{\delta-}{Cl}$ | |
| **Dispersion** (Temporary shift of electrons in nonpolar bonds) | $\overset{\delta+}{X}\overset{\delta-}{:X}$ - - $\overset{\delta+}{X}\overset{\delta-}{:X}$ (temporary dipoles) | $\overset{\delta+}{F}$—$\overset{\delta-}{F}$ - - - $\overset{\delta+}{F}$—$\overset{\delta-}{F}$ | **Weak** |

## SAMPLE PROBLEM    4.16

### ■ Attractive Forces Between Particles

Indicate the major type of molecular interaction expected of each of the following:

**a.** dipole–dipole        **b.** hydrogen bonding        **c.** dispersion forces
**1.** H—F    **2.** $Br_2$    **3.** $PCl_3$

SOLUTION

**1. (b)** H—F is a polar molecule that interacts with other H—F molecules by hydrogen bonding.

**2. (c)** $Br_2$ is nonpolar; only dispersion forces provide attractive forces.

**3. (a)** The polarity of the $PCl_3$ molecules provides dipole–dipole interactions.

STUDY CHECK

Why is the boiling point of $H_2S$ lower than that of $H_2O$?

## QUESTIONS AND PROBLEMS

### Attractive Forces in Compounds

**4.69** Identify the major type of interactive force in each of the following substances:
   **a.** BrF   **b.** KCl   **c.** HF   **d.** $Cl_2$

**4.70** Identify the major type of interactive force in each of the following substances:
   **a.** $OF_2$   **b.** $MgF_2$   **c.** $Br_2$   **d.** $NH_3$

**4.71** Identify the strongest attractive forces between each of the following:
   **a.** $H_2O$   **b.** $Cl_2$   **c.** HCl   **d.** $NF_3$

**4.72** Identify the strongest attractive forces between each of the following:
   **a.** $O_2$   **b.** HI   **c.** NaF   **d.** $CH_3—OH$

## CONCEPT MAP

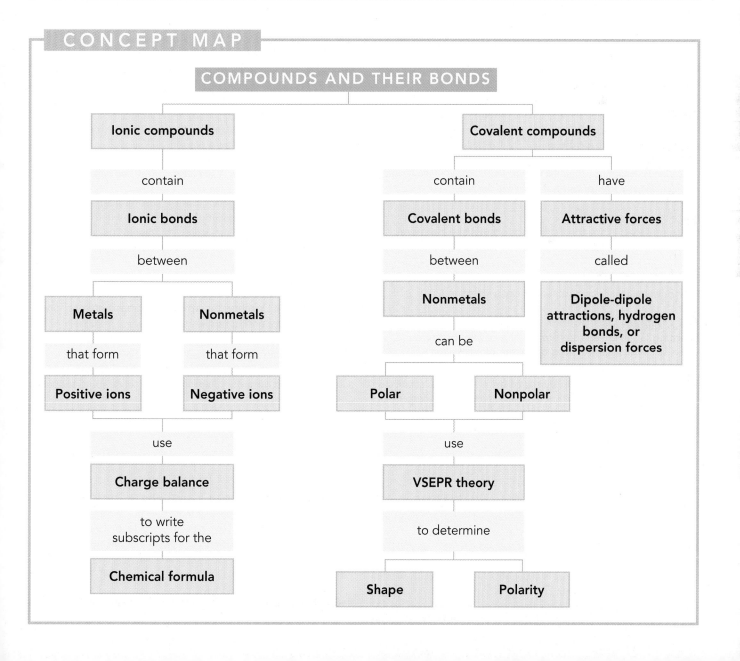

# CHAPTER REVIEW

## 4.1 Octet Rule and Ions

**Learning Goal:** Using the octet rule, write the symbols of the simple ions for the representative elements.

The stability of the noble gases is associated with an octet of 8 electrons in their valence shells; helium needs 2 electrons for stability. Atoms of elements in Groups 1A–7A (1, 2, 13–17) achieve stability by losing, gaining, or sharing their valence electrons in the formation of compounds. Metals of the representative elements form octets by losing valence electrons to form positively charged ions (cations): Group 1A (1), 1+, Group 2A (2), 2+, and Group 3A (13), 3+. When reacting with metals, nonmetals gain electrons to form octets and form negatively charged ion (anions): Group 5A (15), 3−, Group 6A (16), 2−, Group 7A (17), 1−.

## 4.2 Ionic Compounds

**Learning Goal:** Using charge balance, write the correct formula for an ionic compound.

The total positive and negative ionic charge is balanced in the formula of an ionic compound. Charge balance in a formula is achieved by using subscripts after each symbol so that the overall charge is zero.

## 4.3 Naming and Writing Ionic Formulas

**Learning Goal:** Given the formula of an ionic compound, write the correct name; given the name of an ionic compound, write the correct formula.

In naming ionic compounds, the name of the positive ion is given first, followed by the name of the negative ion. Ionic compounds containing two elements end with *ide*. Except for Ag, Cd, and Zn, transition metals form cations with two or more ionic charges. Then the charge of the cation is determined from the total negative charge in the formula and included as a Roman numeral following the name.

## 4.4 Polyatomic Ions

**Learning Goal:** Write the name and formula of a compound containing a polyatomic ion.

A polyatomic ion is a group of nonmetal atoms that carries an electrical charge; for example, the carbonate ion has the formula $CO_3{}^{2-}$. Most polyatomic ions have names that end with *ate* or *ite*.

## 4.5 Covalent Compounds

**Learning Goal:** Given the formula of a covalent compound, write its correct name; given the name of a covalent compound, write its formula.

In a covalent bond, electrons are shared by atoms of two nonmetals such that each of the atoms achieves an octet (or two for hydrogen). In a nonpolar covalent bond, electrons are shared equally by atoms. In a polar covalent bond, electrons are unequally shared due to their attraction to the more electronegative atom. In some covalent compounds, double or triple bonds are needed to provide an octet. Their names use prefixes to indicate the subscripts of each type of atom in the formula. The ending of the second nonmetal is changed to *ide*.

## 4.6 Electronegativity and Bond Polarity

**Learning Goal:** Use electronegativity to determine the polarity of a bond.

Electronegativity is the ability of an atom to attract shared pairs of electrons. The electronegativity values of the metals are low, while nonmetals have high electronegativities. If atoms share the bonding pair of electrons equally, it is called a nonpolar covalent bond. If the bonding electrons are unequally shared, it is called a polar covalent bond. In polar covalent bonds, the atom with the lower electronegativity is partially positive ($\delta^+$) and the atom with the higher electronegativity is partially negative ($\delta^-$). Atoms that form ionic bonds have large differences in electronegativity.

## 4.7 Shapes and Polarity of Molecules

**Learning Goal:** Predict the three-dimensional structure of a molecule, and classify it as polar or nonpolar.

The VSEPR theory indicates that the repulsion of electrons around a central atom pushes the electron groups as far apart as possible. The shape of a molecule is predicted from the arrangement of the bonded atoms and lone pairs around the central atom. The electron arrangement of four atoms around a central atom with no lone pairs is tetrahedral. A central atom with three atoms bonded to the central atom and one lone pair has a pyramidal shape. A central atom with two bonded atoms and two lone pairs has a bent shape.

Molecules are nonpolar if they contain nonpolar covalent bonds or have an arrangement of polar covalent bonds with dipoles that cancel out. In polar molecules, the dipoles do not cancel because there are nonidentical bonded atoms or lone pairs on the central atom.

## 4.8 Attractive Forces in Compounds

**Learning Goal:** Describe the attractive forces between ions, polar molecules, and nonpolar molecules.

Ionic bonds consist of very strong attractive forces between oppositely charged ions. Attractive forces in polar covalent compounds are weaker than ionic bonds and include dipole–dipole attractions and hydrogen bonds. Nonpolar covalent compounds form solids using temporary dipoles called dispersion forces.

# KEY TERMS

**anion** A negatively charged ion such as $Cl^-$, $O^{2-}$, or $SO_4{}^{2-}$.

**bent** The shape of a molecule with two bonded atoms and two lone pairs.

**cation** A positively charged ion such as $Na^+$, $Mg^{2+}$, $Al^{3+}$, and $NH_4{}^+$.

**dipole** The separation of positive and negative charges in a polar bond indicated by an arrow that is drawn from the more positive atom to the more negative atom.

**dipole–dipole attractions** Attractive forces between oppositely charged ends of polar molecules.

**dispersion forces** Weak dipole bonding that results from a momentary polarization of nonpolar molecules in a substance.

**double bond** A sharing of two pairs of electrons by two atoms.

**electronegativity** The relative ability of an element to attract electrons in a bond.

**formula** The group of symbols and subscripts that represents the atoms or ions in a compound.

**hydrogen bond** The attraction between a partially positive H atom and a strongly electronegative atom of F, O, or N.

**ion** An atom or group of atoms having an electrical charge because of a loss or gain of electrons.

**ionic charge** The difference between the number of protons (positive) and the number of electrons (negative) written in the upper right corner of the symbol for the element or polyatomic ion.

**molecule** The smallest unit of two or more atoms held together by covalent bonds.

**nonpolar covalent bond** A covalent bond in which the electrons are shared equally between atoms.

**nonpolar molecule** A molecule that has only nonpolar bonds or in which the bond dipoles cancel.

**octet** 8 valence electrons.

**octet rule** Elements in Groups 1A–7A (1, 2, 13–17) react with other elements by forming ionic or covalent bonds to produce a noble gas arrangement, usually 8 electrons in the outer shell.

**polar covalent bond** A covalent bond in which the electrons are shared unequally between atoms.

**polar molecule** A molecule containing bond dipoles that do not cancel.

**polyatomic ion** A group of covalently bonded nonmetal atoms that has an overall electrical charge.

**pyramidal** The shape of a molecule that has three bonded atoms and one lone pair around a central atom.

**tetrahedral** The shape of a molecule with four bonded atoms.

**triple bond** A sharing of three pairs of electrons by two atoms.

**valence-shell electron-pair repulsion (VSEPR) theory** A theory that predicts the shape of a molecule by placing the electron pairs on a central atom as far apart as possible to minimize the mutual repulsion of the electrons.

# UNDERSTANDING THE CONCEPTS

**4.73  a.** How does the octet rule explain the formation of a sodium ion?
  **b.** What noble gas has the same electron arrangement as the sodium ion?
  **c.** Why are Group 1A (1) and Group 2A (2) elements found in many compounds, but not Group 8A (18) elements?

**4.74  a.** How does the octet rule explain the formation of a chloride ion, $Cl^-$?
  **b.** What noble gas has the same electron arrangement as the chloride ion, $Cl^-$?
  **c.** Why are Group 7A (17) elements found in many compounds, but not Group 8A (18) elements?

**4.75** Identify each of the following atoms or ions:

$$15p^+ \quad 18e^-$$
$$16n$$
**A**

$$8p^+ \quad 8e^-$$
$$8n$$
**B**

$$30p^+ \quad 28e^-$$
$$35n$$
**C**

$$26p^+ \quad 23e^-$$
$$28n$$
**D**

**4.76** Consider the following bonds:

$$Ca—O \quad C—O \quad K—O \quad O—O \quad N—O$$

  **a.** Which bonds are polar covalent?
  **b.** Which bonds are nonpolar covalent?
  **c.** Which bonds are ionic?
  **d.** Arrange the covalent bonds in order of decreasing polarity.

**4.77** In the following electron-dot formulas, assume X and Y are atoms of nonmetals and all bonds are polar covalent.

  **a.** X—Y—X    **b.** :Y—X    **c.** X—Y—X

Match each of the molecules with the correct diagram of its shape, and name the shape; indicate if each molecule is polar or nonpolar.

  **1.**    **2.**    **3.**

**4.78** As discussed in the Health Note "Polyatomic Ions in Bones and Teeth," the mineral component of bone material and teeth is composed of calcium hydroxyapatite, $Ca_{10}(PO_4)_6(OH)_2$. Name the ions in the formula.

**4.79** Identify the following as
  **1.** H          **2.** Li          **3.** $Li^+$
  **4.** $H^+$       **5.** $N^{3-}$

$$3p^+ \quad 2e^-$$
$$4n$$
**A**

$$1p^+ \quad 0e^-$$
**B**

$$3p^+ \quad 3e^-$$
$$4n$$
**C**

$$7p^+ \quad 10e^-$$
$$8n$$
**D**

$$1p^+ \quad 1e^-$$
$$2n$$
**E**

**4.80** Using the electron arrangements for the elements, write the electron-dot symbols of the atoms, the cations and anions that form, and the formulas and names of their ionic compounds.

| Electron Arrangements | | Electron-Dot Symbols | | Cations | Anions | Formula of Compound | Name of Compound |
|---|---|---|---|---|---|---|---|
| 2, 8, 2 | 2, 5 | | | | | | |
| 2, 8, 8, 1 | 2, 6 | | | | | | |
| 2, 8, 3 | 2, 8, 7 | | | | | | |
| 2, 1 | 2, 8, 6 | | | | | | |

# ADDITIONAL QUESTIONS AND PROBLEMS

**4.81** Write the electron arrangement of the following:
**a.** $N^{3-}$          **b.** $Mg^{2+}$          **c.** $P^{3-}$
**d.** $Al^{3+}$          **e.** $Li^+$

**4.82** Write the electron arrangement of the following:
**a.** $K^+$          **b.** $Na^+$          **c.** $S^{2-}$
**d.** $Cl^-$          **e.** $Ca^{2+}$

**4.83** Consider an ion with the symbol $X^{2+}$ formed from a representative element.
**a.** What is the group number of the element?
**b.** What is the electron-dot symbol of the element?
**c.** If X is in Period 3, what is the element?
**d.** What is the formula of the compound formed from X and the nitride ion?

**4.84** Consider an ion with the symbol $Y^{3-}$ formed from a representative element.
**a.** What is the group number of the element?
**b.** What is the electron-dot symbol of the element?
**c.** If Y is in Period 3, what is the element?
**d.** What is the formula of the compound formed from $Ba^{2+}$ and Y?

**4.85** One of the ions of tin is tin(IV).
**a.** What is the symbol for this ion?
**b.** How many protons and electrons are in the ion?
**c.** What is the formula of tin(IV) oxide?
**d.** What is the formula of tin(IV) phosphate?

**4.86** One of the ions of gold is gold(III).
**a.** What is the symbol for this ion?
**b.** How many protons and electrons are in the ion?
**c.** What is the formula of gold(III) sulfate?
**d.** What is the formula of gold(III) nitrate?

**4.87** Write the formula of the following ionic compounds:
**a.** gold(III) chloride          **b.** lead(IV) oxide
**c.** silver chloride          **d.** calcium nitride
**e.** copper(I) phosphide          **f.** chromium(II) chloride

**4.88** Write the formula of the following ionic compounds:
**a.** tin(IV) oxide          **b.** iron(III) sulfide
**c.** lead(II) sulfate          **d.** chromium(III) iodide
**e.** lithium nitride          **f.** gold(I) oxide

**4.89** Name each of the following:
**a.** $NCl_3$          **b.** $SCl_2$          **c.** $N_2O$
**d.** $F_2$          **e.** $PCl_5$          **f.** $P_2O_5$

**4.90** Name each of the following:
**a.** $CBr_4$          **b.** $SF_6$          **c.** $Br_2$
**d.** $N_2O_4$          **e.** $SO_2$          **f.** $CS_2$

**4.91** Give the formula for each of the following:
**a.** carbon monoxide          **b.** diphosphorus pentoxide
**c.** dihydrogen sulfide          **d.** sulfur dichloride

**4.92** Give the formula for each of the following:
**a.** silicon dioxide          **b.** carbon tetrabromide
**c.** sulfur trioxide          **d.** dinitrogen oxide

**4.93** Classify each of the following as ionic or covalent, and give its name.
**a.** $FeCl_3$          **b.** $Na_2SO_4$          **c.** $N_2O$
**d.** $F_2$          **e.** $PCl_5$          **f.** $CF_4$

**4.94** Classify each of the following as ionic or covalent, and give its name.
**a.** $Al_2(CO_3)_3$          **b.** $SF_6$          **c.** $Br_2$
**d.** $Mg_3N_2$          **e.** $SO_2$          **f.** $CrPO_4$

**4.95** Write the formulas for the following:
**a.** tin(II) carbonate
**b.** lithium phosphide
**c.** silicon tetrachloride
**d.** iron(III) sulfide
**e.** bromine
**f.** calcium bromide

**4.96** Write the formulas for the following:
**a.** sodium carbonate
**b.** nitrogen dioxide
**c.** aluminum nitrate
**d.** copper(I) nitride
**e.** potassium phosphate
**f.** lead(IV) oxide

**4.97** Select the more polar bond in each of the following pairs:
**a.** C—N or C—O          **b.** N—F or N—Br
**c.** Br—Cl or S—Cl          **d.** Br—Cl or Br—I
**e.** N—F or N—O

**4.98** Select the more polar bond in each of the following pairs:
**a.** C—C or C—O  **b.** P—Cl or P—Br
**c.** Si—S or Si—Cl  **d.** F—Cl or F—Br
**e.** P—O or P—S

**4.99** Calculate the electronegativity difference and classify each of the following bonds as nonpolar covalent, polar covalent, or ionic.
**a.** Si—Cl  **b.** C—C  **c.** Na—Cl
**d.** C—H  **e.** F—F

**4.100** Calculate the electronegativity difference and classify each of the following bonds as nonpolar covalent, polar covalent, or ionic.
**a.** C—N  **b.** Cl—Cl  **c.** K—Br
**d.** H—H  **e.** N—F

**4.101** Classify the following molecules as polar or nonpolar:
**a.** $PBr_3$  **b.** $CH_3Cl$  **c.** $SiF_4$

**4.102** Classify the following molecules as polar or nonpolar:
**a.** $GeH_4$  **b.** $PCl_3$  **c.** $SCl_2$

**4.103** Predict the shape and polarity of each of the following molecules:
**a.** A central atom with three identical bonded atoms and one lone pair.
**b.** A central atom with two bonded atoms and two lone pairs.

**4.104** Predict the shape and polarity of each of the following molecules:
**a.** A central atom with four identical bonded atoms and no lone pairs.
**b.** A central atom with four bonded atoms that are not identical and no lone pairs.

**4.105** Predict the shape and polarity of each of the following molecules:
**a.** $H_2S$  **b.** $NF_3$

**4.106** Predict the shape and polarity of each of the following molecules:
**a.** $H_2O$  **b.** $CF_4$

**4.107** Indicate the major type of attractive force—(1) ionic, (2) dipole–dipole, (3) hydrogen bond, (4) dispersion forces—that occurs between particles of the following substances:
**a.** $NH_3$  **b.** HI  **c.** $Br_2$  **d.** $Cs_2O$

**4.108** Indicate the major type of attractive force—(1) ionic, (2) dipole–dipole, (3) hydrogen bond, (4) dispersion forces—that occurs between particles of the following substances:
**a.** $CHCl_3$  **b.** $H_2O$  **c.** LiCl  **d.** $Cl_2$

# CHALLENGE QUESTIONS

**4.109** Complete the following table for atoms or ions:

| Atom or Ion | Number of Protons | Number of Electrons | Electrons Lost/Gained |
|---|---|---|---|
| $K^+$ | | | |
| | $12p^+$ | $10e^-$ | |
| | $8p^+$ | | $2e^-$ gained |
| | | $10e^-$ | $3e^-$ lost |

**4.110** Complete the following table for atoms or ions:

| Atom or Ion | Number of Protons | Number of Electrons | Electrons Lost/Gained |
|---|---|---|---|
| | $30p^+$ | | $2e^-$ lost |
| | $36p^+$ | $36e^-$ | |
| | $16p^+$ | | $2e^-$ gained |
| | | $46e^-$ | $4e^-$ lost |

**4.111** Consider the following electron-dot formulas for elements X and Y.

**a.** What are the group numbers of X and Y?
**b.** Will a compound of X and Y be ionic or covalent?
**c.** What ions would be formed by X and Y?
**d.** What would be the formula of a compound of X and Y?
**e.** What would be the formula of a compound of X and sulfur?
**f.** What would be the formula of a compound of Y and chlorine?
**g.** Is the compound in **f**. ionic or covalent?

**4.112** Consider the following electron-dot formulas for elements X and Y.

$$\dot{X}\cdot \quad \cdot\ddot{Y}\cdot$$

**a.** What are the group numbers of X and Y?
**b.** Will a compound of X and Y be ionic or covalent?
**c.** What ions would be formed by X and Y?
**d.** What would be the formula of a compound of X and Y?
**e.** What would be the formula of a compound of X and sulfur?
**f.** What would be the formula of a compound of Y and chlorine?
**g.** Is the compound in **f**. ionic or covalent?

**4.113** Identify the group number in the periodic table of X, a representative element, in each of the following ionic compounds:
**a.** $XCl_3$  **b.** $Al_2X_3$  **c.** $XCO_3$

**4.114** Identify the group number in the periodic table of X, a representative element, in each of the following ionic compounds:
**a.** $X_2O_3$  **b.** $X_2SO_3$  **c.** $Na_3X$

**4.115** Classify the following as ionic or covalent, and name each.
**a.** $Li_2O$  **b.** $N_2O$  **c.** $CF_4$
**d.** $Cr(NO_3)_2$  **e.** $Mg(HCO_3)_2$  **f.** $NF_3$
**g.** $CaCl_2$  **h.** $K_3PO_4$  **i.** $Au_2(SO_3)_3$
**j.** $I_2$

**4.116** Classify the following as ionic or covalent, and name each.
**a.** $FeCl_2$  **b.** $Cl_2O_7$  **c.** $N_2$
**d.** $Ca_3(PO_4)_2$  **e.** $PCl_3$  **f.** $Al(NO_3)_2$
**g.** $PbCl_4$  **h.** $MgCO_3$  **i.** $NO_2$
**j.** $SnSO_4$

# ANSWERS

## Answers to Study Checks

**4.1** $35p^+$ and $36e^-$

**4.2** $K^+$ and $S^{2-}$

**4.3** $Ca^{2+}$ and $O^{2-}$; CaO

**4.4** calcium chloride

**4.5** gold(I) oxide

**4.6** $Cr_2O_3$

**4.7** $(NH_4)_3PO_4$

**4.8** calcium phosphate

**4.9** H:$\ddot{\text{P}}$:H      H—$\ddot{\text{P}}$—H
　　$\overset{\cdot\cdot}{\text{H}}$　　　　　|
　　　　　　　　　H

**4.10** **a.** silicon tetrabromide
**b.** dibromine oxide

**4.11** $IF_5$

**4.12** iron(II) nitrate

**4.13** **a.** polar covalent (0.7)
**b.** nonpolar covalent (0)
**c.** ionic (2.6)

**4.14** Sulfur has 4 electron pairs; two bonded atoms and two lone pairs. The geometry of $SCl_2$ would be bent.

**4.15** polar

**4.16** The attractive forces between $H_2S$ molecules are dipole–dipole attractions, whereas $H_2O$ molecules have hydrogen bonds, which are stronger and require more energy to break.

## Answers to Selected Questions and Problems

**4.1** **a.** 1    **b.** 2    **c.** 3    **d.** 1    **e.** 2

**4.3** **a.** $Li^+$    **b.** $F^-$    **c.** $Mg^{2+}$    **d.** $Fe^{3+}$    **e.** $Zn^{2+}$

**4.5** **a.** $Cl^-$    **b.** $K^+$    **c.** $O^{2-}$    **d.** $Al^{3+}$

**4.7** **a.** K    **b.** $Cl^-$    **c.** Ca    **d.** $K^+$

**4.9** **a, c**

**4.11** **a.** $Na_2O$    **b.** $AlBr_3$    **c.** BaO    **d.** $MgCl_2$    **e.** $Al_2S_3$

**4.13** **a.** $Na^+$ and $S^{2-}$, $Na_2S$
**b.** $K^+$ and $N^{3-}$, $K_3N$
**c.** $Al^{3+}$ and $I^-$, $AlI_3$
**d.** $Li^+$ and $O^{2-}$, $Li_2O$

**4.15** **a.** aluminum oxide
**b.** calcium chloride
**c.** sodium oxide
**d.** magnesium nitride
**e.** potassium iodide

**4.17** Most of the transition metals form more than one positive ion. The specific ion is indicated in a name by writing a Roman numeral that is the same as the ionic charge. For example, iron forms $Fe^{2+}$ and $Fe^{3+}$ ions, which are named iron(II) and iron(III).

**4.19** **a.** iron(II)
**b.** copper(II)
**c.** zinc
**d.** lead(IV)
**e.** chromium(III)

**4.21** **a.** tin(II) chloride
**b.** potassium oxide
**c.** copper(I) sulfide
**d.** copper(II) sulfide
**e.** chromium(III) bromide
**f.** zinc chloride

**4.23** **a.** $Au^{3+}$    **b.** $Fe^{3+}$    **c.** $Pb^{4+}$    **d.** $Sn^{2+}$

**4.25** **a.** $MgCl_2$    **b.** $Na_2S$    **c.** $Cu_2O$
**d.** $Zn_3P_2$    **e.** AuN    **f.** $CrCl_2$

**4.27** **a.** $HCO_3^-$    **b.** $NH_4^+$    **c.** $PO_4^{3-}$    **d.** $HSO_4^-$

**4.29** **a.** sulfate    **b.** carbonate    **c.** phosphate    **d.** nitrate

**4.31**

|         | $OH^-$ | $NO_2^-$ | $CO_3^{2-}$ | $HSO_4^-$ | $PO_4^{3-}$ |
|---------|--------|----------|-------------|-----------|-------------|
| $Li^+$  | LiOH   | $LiNO_2$ | $Li_2CO_3$  | $LiHSO_4$ | $Li_3PO_4$  |
| $Cu^{2+}$ | $Cu(OH)_2$ | $Cu(NO_2)_2$ | $CuCO_3$ | $Cu(HSO_4)_2$ | $Cu_3(PO_4)_2$ |
| $Ba^{2+}$ | $Ba(OH)_2$ | $Ba(NO_2)_2$ | $BaCO_3$ | $Ba(HSO_4)_2$ | $Ba_3(PO_4)_2$ |

**4.33** **a.** $CO_3^{2-}$, sodium carbonate
**b.** $NH_4^+$, ammonium chloride
**c.** $PO_4^{3-}$, lithium phosphate
**d.** $NO_2^-$, copper(II) nitrite
**e.** $SO_3^{2-}$, iron(II) sulfite

**4.35** **a.** $Ba(OH)_2$
**b.** $Na_2SO_4$
**c.** $Fe(NO_3)_2$
**d.** $Zn_3(PO_4)_2$
**e.** $Fe_2(CO_3)_3$

**4.37** **a.** :$\ddot{\text{Br}}$:$\ddot{\text{Br}}$: or :$\ddot{\text{Br}}$—$\ddot{\text{Br}}$:    **b.** H:H or H—H
**c.** H:$\ddot{\text{F}}$: or H—$\ddot{\text{F}}$:    **d.**
　　　　　　　　　　　　　　　　　　　　:$\ddot{\text{F}}$:
　　　　　　:$\ddot{\text{F}}$:$\ddot{\text{O}}$: or :$\ddot{\text{F}}$—$\ddot{\text{O}}$:

**4.39** **a.** phosphorus tribromide
**b.** carbon tetrabromide
**c.** silicon dioxide
**d.** hydrogen fluoride

**4.41** **a.** dinitrogen trioxide
**b.** nitrogen trichloride
**c.** silicon tetrabromide
**d.** phosphorus pentachloride

**4.43** **a.** $CCl_4$    **b.** CO    **c.** $PCl_3$    **d.** $N_2O_4$

**4.45** **a.** $OF_2$    **b.** $BF_3$    **c.** $N_2O_3$    **d.** $SF_6$

**4.47** **a.** aluminum sulfate
**b.** calcium carbonate
**c.** dinitrogen oxide
**d.** sodium phosphate
**e.** ammonium sulfate
**f.** iron(III) oxide

**4.49** The electronegativity increases going across a period.

**4.51** **a.** K, Na, Li    **b.** Na, P, Cl    **c.** Ca, Se, O

**4.53** **a.** polar covalent
**b.** ionic
**c.** polar covalent
**d.** nonpolar covalent
**e.** polar covalent
**f.** polar covalent

**4.55** **a.** $\overset{\delta^+ \ \ \delta^-}{N-F}$    **b.** $\overset{\delta^+ \ \ \delta^-}{Si-Br}$    **c.** $\overset{\delta^+ \ \ \delta^-}{C-O}$

     **d.** $\overset{\delta^+ \ \ \delta^-}{P-Br}$    **e.** $\overset{\delta^+ \ \ \delta^-}{B-Cl}$

**4.57** tetrahedral

**4.59** The four electron groups in $PCl_3$ have a tetrahedral arrangement, but three bonded atoms with a lone pair around a central atom give a pyramidal shape.

**4.61** In both $PH_3$ and $NH_3$, there are three bonded atoms and one lone pair on the central atoms. The shapes of both are pyramidal.

**4.63** **a.** bent
**b.** tetrahedral

**4.65** $Cl_2$ is a nonpolar molecule because there is a nonpolar covalent bond between Cl atoms, which have identical electronegativity values. In HCl, the bond is a polar bond, which makes HCl a polar molecule.

**4.67** **a.** polar
**b.** polar
**c.** nonpolar

**4.69** **a.** dipole–dipole attraction
**b.** ionic
**c.** hydrogen bond
**d.** dispersion forces

**4.71** **a.** hydrogen bond
**b.** dispersion forces
**c.** dipole–dipole attraction
**d.** dipole–dipole attraction

**4.73** **a.** By losing 1 valence electron from the third energy level, sodium achieves an octet in the second energy level.
**b.** The sodium ion $Na^+$ has the same electron arrangement as Ne (2, 8).
**c.** Group 1A (1) and 2A (2) elements acquire octets by losing electrons to form compounds. Group 8A (18) elements are stable with octets (or two electrons for helium).

**4.75** **a.** $P^{3-}$ ion
**b.** O atom
**c.** $Zn^{2+}$ ion
**d.** $Fe^{3+}$ ion

**4.77** **a.** 2. pyramidal, polar
**b.** 1. bent, polar
**c.** 3. tetrahedral, nonpolar

**4.79** **1.** H (E)    **2.** Li (C)    **3.** $Li^+$ (A)
    **4.** $H^+$ (B)    **5.** $N^{3-}$ (D)

**4.81** **a.** 2, 8    **b.** 2, 8    **c.** 2, 8, 8
    **d.** 2, 8    **e.** 2

**4.83** **a.** 2A (2)    **b.** $\overset{\bullet}{X}\bullet$
    **c.** Mg    **d.** $X_3N_2$

**4.85** **a.** $Sn^{4+}$
**b.** 50 protons and 46 electrons
**c.** $SnO_2$
**d.** $Sn_3(PO_4)_4$

**4.87** **a.** $AuCl_3$    **b.** $PbO_2$    **c.** AgCl
    **d.** $Ca_3N_2$    **e.** $Cu_3P$    **f.** $CrCl_2$

**4.89** **a.** nitrogen trichloride
**b.** sulfur dichloride
**c.** dinitrogen oxide
**d.** fluorine
**e.** phosphorus pentachloride
**f.** diphosphorus pentoxide

**4.91** **a.** CO    **b.** $P_2O_5$    **c.** $H_2S$    **d.** $SCl_2$

**4.93** **a.** ionic, iron(III) chloride
**b.** ionic, sodium sulfate
**c.** covalent, dinitrogen oxide
**d.** covalent, fluorine
**e.** covalent, phosphorus pentachloride
**f.** covalent, carbon tetrafluoride

**4.95** **a.** $SnCO_3$    **b.** $Li_3P$    **c.** $SiCl_4$
    **d.** $Fe_2S_3$    **e.** $Br_2$    **f.** $CaBr_2$

**4.97** **a.** $C-O$    **b.** $N-F$    **c.** $S-Cl$
    **d.** $Br-I$    **e.** $N-F$

**4.99** **a.** polar covalent (Cl 3.0 − Si 1.8 = 1.2)
**b.** nonpolar covalent (C 2.5 − C 2.5 = 0)
**c.** ionic (Cl 3.0 − Na 0.9 = 2.1)
**d.** nonpolar covalent (C 2.5 − H 2.1 = 0.4)
**e.** nonpolar covalent (F 4.0 − F 4.0 = 0)

**4.101** **a.** polar
**b.** polar
**c.** nonpolar

**4.103** **a.** pyramidal, polar
**b.** bent, polar

**4.105** **a.** bent, polar
**b.** pyramidal, polar

**4.107** **a.** (3) hydrogen bond
**b.** (2) dipole–dipole
**c.** (4) dispersion forces
**d.** (1) ionic

**4.109**

| Atom or Ion | Number of Protons | Number of Electrons | Electrons Lost/Gained |
|---|---|---|---|
| $K^+$ | $19p^+$ | $18e^-$ | $1e^-$ lost |
| $Mg^{2+}$ | $12p^+$ | $10e^-$ | $2e^-$ lost |
| $O^{2-}$ | $8p^+$ | $10e^-$ | $2e^-$ gained |
| $Al^{3+}$ | $13p^+$ | $10e^-$ | $3e^-$ lost |

**4.111** **a.** X = Group 1A (1), Y = Group 6A (16)
**b.** ionic
**c.** $X^+$ and $Y^{2-}$
**d.** $X_2Y$
**e.** $X_2S$

**f.** $YCl_2$
**g.** covalent

**4.113  a.** Group 3A (13)
     **b.** Group 6A (16)
     **c.** Group 2A (2)

**4.115  a.** ionic, lithium oxide
     **b.** covalent, dinitrogen oxide
     **c.** covalent, carbon tetrafluoride

**d.** ionic, chromium(II) nitrate
**e.** ionic, magnesium bicarbonate or magnesium hydrogen carbonate
**f.** covalent, nitrogen trifluoride
**g.** ionic, calcium chloride
**h.** ionic, potassium phosphate
**i.** ionic, gold(III) sulfite
**j.** covalent, iodine

# 5

# Chemical Quantities and Reactions

Visit **www.chemplace.com** for extra quizzes, interactive tutorials, career resources, PowerPoint slides for chapter review, math help, and case studies.

*"In our food science laboratory I develop a variety of food products, from cake doughnuts to energy beverages," says Anne Cristofano, senior food technologist at Mattson & Company. "When I started the doughnut project, I researched the ingredients, then weighed them out in the lab. I added water to make a batter and cooked the doughnuts in a fryer. The batter and the oil temperature make a big difference. If I don't get the right taste or texture, I adjust the ingredients, such as sugar and flour, or adjust the temperature."*

*A food technologist studies the physical and chemical properties of food and develops scientific ways to process and preserve it for extended shelf life. The food products are tested for texture, color, and flavor. The results of these tests help improve the quality and safety of food.*

I n chemistry, we calculate and measure the amounts of substances to use in the lab. Actually, measuring the amount of a substance is something you do every day. When you cook, you measure out the proper amounts of ingredients so you don't have too much of one and too little of another. At the gas station, you measure out a certain amount of fuel in your gas tank. If you paint the walls of a room, you measure the area and purchase the amount of paint that will cover the walls. In the lab, the formula of a substance tells us the number and kinds of atoms it has, which we use to determine the amount of a substance.

Chemical reactions occur everywhere. The fuel in our cars burns with oxygen to provide energy to make the car move, play the radio, and run the air conditioner. When we cook our food or bleach our hair, chemical reactions take place. In our bodies, chemical reactions convert food into molecules to build muscles and move them. In the leaves of trees and plants, carbon dioxide and water are converted into carbohydrates.

Some chemical reactions are simple, whereas others are quite complex. However, they can all be written with the chemical equations that chemists use to describe chemical reactions. In every chemical reaction, the atoms in the reacting substances, called reactants, are rearranged to give new substances called products.

In this chapter, we will see how equations are written and how we can determine the amount of reactant or product involved. We do the same thing at home when we use a recipe to make cookies. At the automotive repair shop, a mechanic does essentially the same thing when adjusting the fuel system of an engine to allow for the correct amounts of fuel and oxygen. In the body a certain amount of $O_2$ must reach the tissues for efficient metabolic reactions. If the oxygenation of the blood is low, the therapist will oxygenate the patient and recheck the blood levels.

## LEARNING GOAL

Use Avogadro's number to determine the number of particles in a given number of moles.

# 5.1 THE MOLE

At the store, you buy eggs by the dozen. In an office, pencils are ordered by the gross, and paper by the ream. In a restaurant, soda is ordered by the case. The terms *dozen, gross, ream,* and *case* are used to count the number of items present. For example, when you buy a dozen eggs, you know you will get 12 eggs in the carton.

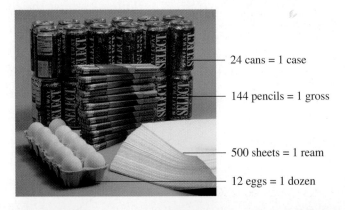

24 cans = 1 case

144 pencils = 1 gross

500 sheets = 1 ream

12 eggs = 1 dozen

# Avogadro's Number

In chemistry, particles such as atoms, molecules, and ions are counted by the **mole**, a unit that contains $6.02 \times 10^{23}$ items. This very large number, called **Avogadro's number** after Amedeo Avogadro, an Italian physicist, looks like this when written with 3 significant figures:

**Avogadro's Number**

$$602\ 000\ 000\ 000\ 000\ 000\ 000\ 000 = 6.02 \times 10^{23}$$

One mole of an element contains Avogadro's number of atoms. For example, 1 mole of carbon contains $6.02 \times 10^{23}$ carbon atoms; 1 mole of aluminum contains $6.02 \times 10^{23}$ aluminum atoms; 1 mole of sulfur contains $6.02 \times 10^{23}$ sulfur atoms.

Sulfur

1 mole of an element $= 6.02 \times 10^{23}$ atoms of that element

Avogadro's number tells us that one mole of a compound contains $6.02 \times 10^{23}$ of the particular type of particles that make up that compound. One mole of a covalent compound contains Avogadro's number of molecules. For example, 1 mole of $CO_2$ contains $6.02 \times 10^{23}$ molecules of $CO_2$. One mole of an ionic compound contains Avogadro's number of **formula units**, which are the groups of ions represented by the formula of an ionic compound. One mole of NaCl contains $6.02 \times 10^{23}$ formula units of NaCl ($Na^+$, $Cl^-$). Table 5.1 gives examples of the number of particles in some 1-mole quantities.

**Calculating Particles and Moles**

Moles of element or compound

Avogadro's number

Particles: atoms, ions, molecules, or formula units

**TABLE 5.1** Number of Particles in One-Mole Samples

| Substance | Number and Type of Particles |
|---|---|
| 1 mole of aluminum | $6.02 \times 10^{23}$ aluminum atoms |
| 1 mole of sulfur | $6.02 \times 10^{23}$ sulfur atoms |
| 1 mole of water ($H_2O$) | $6.02 \times 10^{23}$ $H_2O$ molecules |
| 1 mole of NaCl | $6.02 \times 10^{23}$ NaCl formula units |
| 1 mole of vitamin C ($C_6H_8O_6$) | $6.02 \times 10^{23}$ vitamin C molecules |

We can use Avogadro's number as a conversion factor to convert between the moles of a substance and the number of particles it contains.

$$\frac{6.02 \times 10^{23}\ \text{particles}}{1\ \text{mole}} \quad \text{and} \quad \frac{1\ \text{mole}}{6.02 \times 10^{23}\ \text{particles}}$$

For example, we use Avogadro's number to convert 4.00 moles of sulfur to atoms of sulfur.

$$4.00\ \cancel{\text{moles S atoms}} \times \frac{6.02 \times 10^{23}\ \text{S atoms}}{1\ \cancel{\text{mole S atoms}}} = 2.41 \times 10^{24}\ \text{S atoms}$$

Avogadro's number as a conversion factor

We can also use Avogadro's number to convert $3.01 \times 10^{24}$ molecules of $CO_2$ to moles of $CO_2$.

$$3.01 \times 10^{24}\ \cancel{CO_2\ \text{molecules}} \times \frac{1\ \text{mole}\ CO_2\ \text{molecules}}{6.02 \times 10^{23}\ \cancel{CO_2\ \text{molecules}}}$$

Avogadro's number as a conversion factor

$= 5.00$ moles of $CO_2$ molecules

### ■ Calculating the Number of Molecules

How many molecules are present in 1.75 moles of carbon dioxide, $CO_2$?

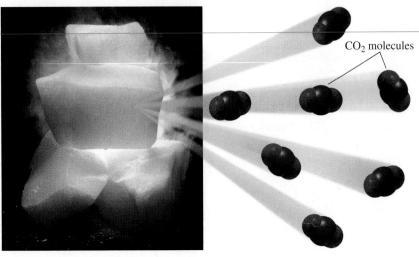

$CO_2$ molecules

Dry ice (solid $CO_2$)

**SOLUTION**

**Step 1  Given**   1.75 moles of $CO_2$      **Need**   molecules of $CO_2$

**Step 2  Plan**   moles of $CO_2$    Avogadro's number       molecules of $CO_2$

**Step 3  Equalities/Conversion Factors**

$$1 \text{ mole of } CO_2 = 6.02 \times 10^{23} \text{ molecules of } CO_2$$

$$\frac{6.02 \times 10^{23} \text{ molecules } CO_2}{1 \text{ mole } CO_2} \quad \text{and} \quad \frac{1 \text{ mole } CO_2}{6.02 \times 10^{23} \text{ molecules } CO_2}$$

**Step 4  Set Up Problem**    Calculate the number of $CO_2$ molecules:

$$1.75 \text{ moles } CO_2 \ \times \ \frac{6.02 \times 10^{23} \text{ molecules } CO_2}{1 \text{ mole } CO_2} \ = \ 1.05 \times 10^{24} \text{ molecules of } CO_2$$

**STUDY CHECK**

How many moles of water, $H_2O$, contain $2.60 \times 10^{23}$ molecules of water?

## Moles of Elements in a Formula

We have seen that the subscripts in a chemical formula of a compound indicate the number of atoms of each type of element. For example, in a molecule of aspirin, chemical formula $C_9H_8O_4$, there are 9 carbon atoms, 8 hydrogen atoms, and 4 oxygen atoms. The subscripts also state the number of moles of each element in 1 mole of aspirin: 9 moles of carbon atoms, 8 moles of hydrogen atoms, and 4 moles of oxygen atoms.

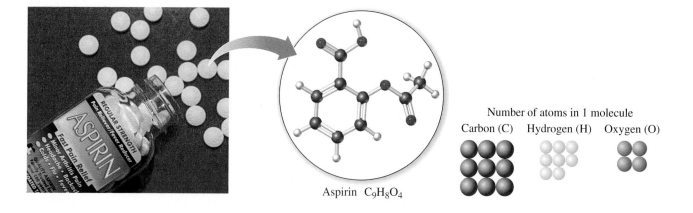

Aspirin $C_9H_8O_4$

Number of atoms in 1 molecule
Carbon (C)   Hydrogen (H)   Oxygen (O)

**The formula subscripts specify the**

$$C_9H_8O_4$$

|  | **Carbon** | **Hydrogen** | **Oxygen** |
|---|---|---|---|
| **Atoms in 1 molecule** | 9 atoms C | 8 atoms H | 4 atoms O |
| **Moles of atoms in 1 mole** | 9 moles C | 8 moles H | 4 moles O |

Using the subscripts from the formula, $C_9H_8O_6$, we can write the conversion factors for each of the elements in 1 mole of aspirin.

$$\frac{9 \text{ moles C}}{1 \text{ mole } C_9H_8O_4} \qquad \frac{8 \text{ moles H}}{1 \text{ mole } C_9H_8O_4} \qquad \frac{4 \text{ moles O}}{1 \text{ mole } C_9H_8O_4}$$

$$\frac{1 \text{ mole } C_9H_8O_4}{9 \text{ moles C}} \qquad \frac{1 \text{ mole } C_9H_8O_4}{8 \text{ moles H}} \qquad \frac{1 \text{ mole } C_9H_8O_4}{4 \text{ moles O}}$$

---

**SAMPLE PROBLEM  5.2**

■ **Calculating the Moles of an Element in a Compound**

For 1.50 moles of aspirin, $C_9H_8O_4$, how many moles of carbon are present? How many atoms of carbon is that?

SOLUTION

**Step 1  Given**    1.50 moles of $C_9H_8O_4$     **Need**   moles of C

**Step 2  Plan**    Moles of $C_9H_8O_4$   Subscript   moles of C atoms

**Step 3  Equalities/Conversion Factors**

$$1 \text{ mole of } C_9H_8O_4 = 9 \text{ moles of C}$$

$$\frac{9 \text{ moles C}}{1 \text{ mole } C_9H_8O_4} \quad \text{and} \quad \frac{1 \text{ mole } C_9H_8O_4}{9 \text{ moles C}}$$

**Step 4  Set Up Problem**

$$1.50 \; \cancel{\text{moles } C_9H_8O_4} \times \frac{9 \text{ moles C}}{1 \; \cancel{\text{mole } C_9H_8O_4}} = 13.5 \text{ moles of C}$$

The number of atoms of C is calculated from moles of C using Avogadro's number.

$$13.5 \; \text{moles C} \times \frac{6.02 \times 10^{23} \text{ atoms C}}{1 \text{ mole C}} = 8.13 \times 10^{24} \text{ C atoms}$$

The conversion factors of subscripts and Avogadro's number can be combined to calculate the number of atoms of C from moles of $C_9H_8O_4$.

$$1.50 \; \text{moles C}_9\text{H}_8\text{O}_4 \times \frac{9 \text{ moles C}}{1 \text{ mole C}_9\text{H}_8\text{O}_4} \times \frac{6.02 \times 10^{23} \text{ C atoms}}{1 \text{ mole C atoms}}$$

$$= 8.13 \times 10^{24} \text{ C atoms}$$

**STUDY CHECK**

How many moles of aspirin, $C_9H_8O_4$, contain 0.480 mole of O atoms?

## QUESTIONS AND PROBLEMS

**The Mole**

**5.1** What is a mole?

**5.2** What is Avogadro's number?

**5.3** Calculate each of the following using Avogadro's number:
**a.** number of C atoms in 0.500 mole of C
**b.** number of $SO_2$ molecules in 1.28 moles of $SO_2$
**c.** moles of Fe in $5.22 \times 10^{22}$ atoms of Fe
**d.** moles of $C_2H_5OH$ in $8.50 \times 10^{24}$ molecules of $C_2H_5OH$

**5.4** Calculate each of the following using Avogadro's number:
**a.** number of Li atoms in 4.5 moles of Li
**b.** number of $CO_2$ molecules in 0.0180 mole of $CO_2$
**c.** moles of Cu in $7.8 \times 10^{21}$ atoms of Cu
**d.** moles of $C_2H_6$ in $3.75 \times 10^{23}$ molecules of $C_2H_6$

**5.5** Calculate each of the following quantities in 2.00 moles of $H_3PO_4$:
**a.** moles of H     **b.** moles of O
**c.** atoms of P     **d.** atoms of O

**5.6** Calculate each of the following quantities in 0.185 mole of $(C_3H_5)_2O$:
**a.** moles of C     **b.** moles of O
**c.** atoms of H     **d.** atoms of C

**5.7** Consider the formula for quinine, $C_{20}H_{24}N_2O_2$.
**a.** How many moles of hydrogen are in 1.0 mole of quinine?
**b.** How many moles of carbon are in 5.0 moles of quinine?
**c.** How many moles of nitrogen are in 0.020 mole of quinine?

**5.8** Consider the formula for $Al_2(SO_4)_3$, which is used in antiperspirants.
**a.** How many moles of sulfur are present in 3.0 moles of $Al_2(SO_4)_3$?
**b.** How many moles of aluminum ions are present in 0.40 mole of $Al_2(SO_4)_3$?
**c.** How many moles of sulfate ions ($SO_4^{2-}$) are present in 1.5 moles of $Al_2(SO_4)_3$?

## LEARNING GOALS

Determine the molar mass of a substance, and use molar mass to convert between grams and moles.

**WEB TUTORIAL**
Stoichiometry

# 5.2 MOLAR MASS

A single atom or molecule is much too small to weigh, even on the most sensitive balance. In fact, it takes a huge number of atoms or molecules to make enough of a substance for you to see. An amount of water that contains Avogadro's number of water molecules is only a few sips. In the laboratory, we can use a balance to weigh out Avogadro's number of particles or 1 mole of a substance.

For any element, the quantity called **molar mass** is the quantity in grams that equals the atomic mass of that element. We are counting out $6.02 \times 10^{23}$ atoms of an element when we weigh out the number of grams equal to its molar mass. For example, if we need 1 mole of carbon (C) atoms, we would first find the atomic mass of 12.01 on the periodic table. Then to obtain 1 mole of carbon atoms, we would weigh out 12.01 g of carbon. Thus, the molar mass of carbon is found by looking at the atomic mass on the periodic table.

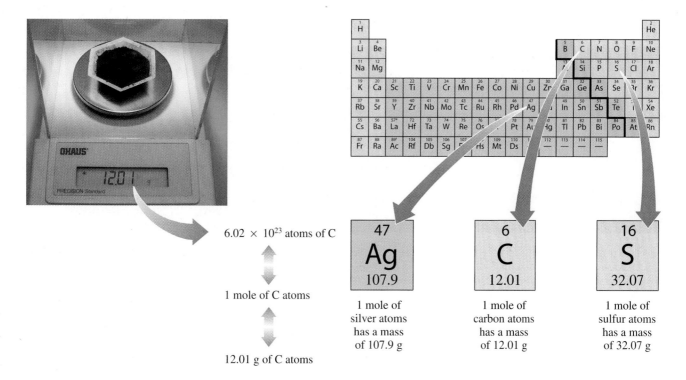

6.02 × 10²³ atoms of C

⇕

1 mole of C atoms

⇕

12.01 g of C atoms

| 47 | 6 | 16 |
|---|---|---|
| **Ag** | **C** | **S** |
| 107.9 | 12.01 | 32.07 |

1 mole of silver atoms has a mass of 107.9 g

1 mole of carbon atoms has a mass of 12.01 g

1 mole of sulfur atoms has a mass of 32.07 g

## Molar Mass of a Compound

To determine the molar mass of a compound, multiply the molar mass of each element by its subscript in the formula, and add the results. For example, the molar mass of sulfur trioxide, $SO_3$, is obtained by adding the molar masses of 1 mole of sulfur and 3 moles of oxygen. ***In this text, we round molar mass to the tenths (0.1 g) place for calculations.***

**Step 1**    Using the periodic table, obtain the molar masses of sulfur and oxygen.

$$\frac{32.1 \text{ g S}}{1 \text{ mole S}} \quad \frac{16.0 \text{ g O}}{1 \text{ mole O}}$$

**Step 2**    **Grams from 1 mole of S**

$$1 \text{ mole S} \times \frac{32.1 \text{ g S}}{1 \text{ mole S}} = 32.1 \text{ g of S}$$

**Grams from 3 moles of O**

$$3 \text{ moles O} \times \frac{16.0 \text{ g O}}{1 \text{ mole O}} = 48.0 \text{ g of O}$$

**Step 3**    Obtain the molar mass of $SO_3$ by adding the masses of 1 mole of S and 3 moles of O.

1 mole S            = 32.1 g of S

3 moles O           = 48.0 g of O

Molar mass of $SO_3$ = 80.1 g of $SO_3$

**Guide to Calculating Molar Mass**

**1** Obtain the molar mass of each element.

**2** Multiply each molar mass by the number of moles (subscript) in the formula.

**3** Calculate the molar mass by adding the masses of the elements.

Figure 5.1 shows some 1-mole quantities of substances. Table 5.2 lists the molar mass for several 1-mole samples.

**One–Mole Quantities**

|  S  |  Fe  |  NaCl  |  $K_2Cr_2O_7$  |  $C_{12}H_{22}O_{11}$  |

**FIGURE 5.1**  One-mole samples: sulfur, S (32.1 g); iron, Fe (55.9 g); salt, NaCl (58.5 g); potassium dichromate, $K_2Cr_2O_7$ (294.2 g); and sugar, sucrose, $C_{12}H_{22}O_{11}$ (342.2 g).
**Q**  How is the molar mass for $K_2Cr_2O_7$ obtained?

**TABLE 5.2**  The Molar Mass of Selected Elements and Compounds

| Substance | Molar Mass |
|---|---|
| 1 mole of carbon (C) | 12.0 g |
| 1 mole of sodium (Na) | 23.0 g |
| 1 mole of iron (Fe) | 55.9 g |
| 1 mole of NaF | 42.0 g |
| 1 mole of $CaCO_3$ (antacid) | 100.1 g |
| 1 mole of $C_6H_{12}O_6$ (glucose) | 180.0 g |
| 1 mole of $C_8H_{10}N_4O_2$ (caffeine) | 194.0 g |

■ **Calculating the Molar Mass of a Compound**

Find the molar mass of $Li_2CO_3$ used to produce red color in fireworks.

SOLUTION

Step 1  Using the periodic table, obtain the molar masses of lithium, carbon, and oxygen.

$$\frac{6.9 \text{ g Li}}{1 \text{ mole Li}} \qquad \frac{12.0 \text{ g C}}{1 \text{ mole C}} \qquad \frac{16.0 \text{ g O}}{1 \text{ mole O}}$$

Step 2  Obtain the mass of each element in the formula by multiplying each molar mass by its number of moles (subscript) in the formula.

**Grams from 2 moles of Li**

$$2 \text{ moles Li} \times \frac{6.9 \text{ g Li}}{1 \text{ mole Li}} = 13.8 \text{ g of Li}$$

**Grams from 1 mole of C**

$$1 \text{ mole C} \times \frac{12.0 \text{ g C}}{1 \text{ mole C}} = 12.0 \text{ g of C}$$

**Grams from 3 moles of O**

$$3 \text{ moles O} \times \frac{16.0 \text{ g O}}{1 \text{ mole O}} = 48.0 \text{ g of O}$$

Step 3  Obtain the molar mass of $Li_2CO_3$ by adding the masses of 2 moles of Li, 1 mole of C, and 3 moles of O.

13.8 g of Li + 12.0 g of C + 48.0 g of O = Molar mass of $Li_2CO_3$ = 73.8 g

STUDY CHECK

Calculate the molar mass of salicylic acid, $C_7H_6O_3$.

## Explore Your World

### Calculating Moles in the Kitchen

The labels on food products list the components in grams and milligrams. It is possible to convert those values to moles, which is also interesting. Read the labels of some products in the kitchen and convert the amounts given in grams or milligrams to moles using molar mass. Here are some examples.

**QUESTIONS**

1. How many moles of NaCl are in a box of salt containing 746 g of NaCl?
2. How many moles of sugar are contained in a 5-lb bag of sugar if sugar has the formula $C_{12}H_{22}O_{11}$?
3. A serving of cereal contains 90 mg of potassium. If there are 11 servings of cereal in the box, how many moles of $K^+$ are present in the cereal in the box?

## Calculations Using Molar Mass

The molar mass of an element or a compound is one of the most useful conversion factors in chemistry. Molar mass is used to change from moles of a substance to grams, or from grams to moles. To do these calculations, we use the molar mass as a conversion factor. For example, 1 mole of magnesium has a mass of 24.3 g. To express molar mass as an equality, we can write

$$1 \text{ mole of Mg} = 24.3 \text{ g of Mg}$$

From this equality, two conversion factors can be written.

$$\frac{24.3 \text{ g Mg}}{1 \text{ mole Mg}} \quad \text{and} \quad \frac{1 \text{ mole Mg}}{24.3 \text{ g Mg}}$$

Conversion factors are written for compounds in the same way. For example, the molar mass of the compound $H_2O$ is 18.0 g.

$$1 \text{ mole of } H_2O = 18.0 \text{ g of } H_2O$$

The conversion factors from the molar mass of $H_2O$ are written as

$$\frac{18.0 \text{ g } H_2O}{1 \text{ mole } H_2O} \quad \text{and} \quad \frac{1 \text{ mole } H_2O}{18.0 \text{ g } H_2O}$$

We can now change from moles to grams, or grams to moles, using the conversion factors derived from the molar mass. (Remember, you must determine the molar mass of the substance first.)

### SAMPLE PROBLEM 5.4

■ **Converting Moles of an Element to Grams**

Silver metal is used in the manufacture of tableware, mirrors, jewelry, and dental alloys. If the design for a piece of jewelry requires 0.750 mole of silver, how many grams of silver are needed?

**Guide to Converting Moles to Grams**

**1** Write a plan that converts moles to grams.

**2** Write the conversion factors for molar mass.

**3** Set up the problem to convert moles to grams.

**SOLUTION**

**Step 1 Given** 0.750 mole of Ag   **Need** grams of Ag

**Step 2 Plan** moles of Ag   Molar mass factor   grams of Ag

**Step 3 Equalities/Conversion Factors**

$$1 \text{ mole of Ag} = 107.9 \text{ g of Ag}$$

$$\frac{107.9 \text{ g Ag}}{1 \text{ mole Ag}} \quad \text{and} \quad \frac{1 \text{ mole Ag}}{107.9 \text{ g Ag}}$$

**Step 4 Set Up Problem** Calculate the grams of silver using the molar mass.

$$0.750 \text{ mole Ag} \times \frac{107.9 \text{ g Ag}}{1 \text{ mole Ag}} = 80.9 \text{ g of Ag}$$

**STUDY CHECK**

Calculate the number of grams of gold (Au) present in 0.124 mole of gold.

---

**SAMPLE PROBLEM 5.5**

■ **Converting the Mass of a Compound to Moles**

A box of salt contains 737 g NaCl. How many moles of NaCl are present in the box?

**SOLUTION**

**Step 1 Given** 737 g of NaCl   **Need** moles of NaCl

**Step 2 Plan** grams of NaCl   Molar mass factor   moles of NaCl

**Step 3 Equalities/Conversion Factors** The molar mass of NaCl is the sum of the masses of one mole $Na^+$ and one mole $Cl^-$ :

$$(1 \times 23.0 \text{ g/mole}) + (1 \times 35.5 \text{ g/mole}) = 58.5 \text{ g/mole}$$

$$1 \text{ mole of NaCl} = 58.5 \text{ g of NaCl}$$

$$\frac{58.5 \text{ g NaCl}}{1 \text{ mole NaCl}} \quad \text{and} \quad \frac{1 \text{ mole NaCl}}{58.5 \text{ g NaCl}}$$

**Step 4 Set Up Problem** We calculate the moles of NaCl using the molar mass.

$$737 \text{ g NaCl} \times \frac{1 \text{ mole NaCl}}{58.5 \text{ g NaCl}} = 12.6 \text{ moles of NaCl}$$

**STUDY CHECK**

One gel cap of an antacid contains 311 mg of $CaCO_3$ and 232 mg of $MgCO_3$. In a recommended dosage of two gel caps, how many moles each of $CaCO_3$ and $MgCO_3$ are present?

We can summarize the calculations to show the connections between the moles of a compound, its mass in grams, number of molecules (or formula units if ionic), and the moles and atoms of each element in that compound in the following flowchart.

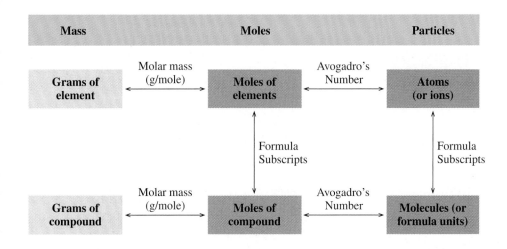

## QUESTIONS AND PROBLEMS

### Molar Mass

**5.9** Calculate the molar mass for each of the following compounds:
**a.** NaCl (table salt)
**b.** $Fe_2O_3$ (rust)
**c.** $Li_2CO_3$ (antidepressant)
**d.** $Al_2(SO_4)_3$ (antiperspirant)
**e.** $Mg(OH)_2$ (antacid)
**f.** $C_{16}H_{19}N_3O_5S$ (amoxicillin, an antibiotic)

**5.10** Calculate the molar mass for each of the following compounds:
**a.** $FeSO_4$ (iron supplement)
**b.** $Al_2O_3$ (absorbent and abrasive)
**c.** $C_7H_5NO_3S$ (saccharin)
**d.** $C_3H_8O$ (rubbing alcohol)
**e.** $(NH_4)_2CO_3$ (baking powder)
**f.** $Zn(C_2H_3O_2)_2$ (zinc dietary supplement)

**5.11** Calculate the mass in grams in each of the following:
**a.** 2.00 moles of Na       **b.** 2.80 moles of Ca
**c.** 0.125 mole of Sn       **d.** 1.76 moles of Cu

**5.12** Calculate the mass in grams in each of the following:
**a.** 1.50 moles of K       **b.** 2.5 moles of C
**c.** 0.25 mole of P       **d.** 12.5 moles of He

**5.13** Calculate the number of grams in each of the following:
**a.** 0.500 mole of NaCl       **b.** 1.75 moles of $Na_2O$
**c.** 0.225 mole of $H_2O$       **d.** 4.42 moles of $CO_2$

**5.14** Calculate the number of grams in each of the following:
**a.** 2.0 moles of $MgCl_2$       **b.** 3.5 moles of $C_3H_8$
**c.** 5.00 moles of $C_2H_6O$       **d.** 0.488 mole of $C_3H_6O_3$

**5.15 a.** The compound $MgSO_4$ is called Epsom salts. How many grams will you need to prepare a bath containing 5.00 moles of Epsom salts?
**b.** In a bottle of soda, there is 0.25 mole of $CO_2$. How many grams of $CO_2$ are in the bottle?

**5.16 a.** Cyclopropane, $C_3H_6$, is an anesthetic given by inhalation. How many grams are in 0.25 mole of cyclopropane?
**b.** The sedative Demerol hydrochloride has the formula $C_{15}H_{22}ClNO_2$. How many grams are in 0.025 mole of Demerol hydrochloride?

**5.17** How many moles are contained in each of the following?
**a.** 50.0 g of Ag       **b.** 0.200 g of C
**c.** 15.0 g of $NH_3$       **d.** 75.0 g of $SO_2$

**5.18** How many moles are contained in each of the following?
**a.** 25.0 g of Ca       **b.** 5.00 g of S
**c.** 40.0 g of $H_2O$       **d.** 100.0 g of $O_2$

**5.19** Calculate the number of moles in 25.0 g of each of the following:
**a.** Ne       **b.** $O_2$       **c.** $Al(OH)_3$       **d.** $Ga_2S_3$

**5.20** Calculate the number of moles in 4.00 g of each of the following:
**a.** He       **b.** $SnO_2$       **c.** $Cr(OH)_3$       **d.** $Ca_3N_2$

**5.21** How many moles of S are in each of the following quantities?
**a.** 25 g of S       **b.** 125 g of $SO_2$       **c.** 2.0 moles of $Al_2S_3$

**5.22** How many moles of C are in each of the following quantities?
**a.** 75 g of C       **b.** 0.25 mole of $C_2H_6$       **c.** 88 g of $CO_2$

## 5.3 CHEMICAL CHANGES

We have seen that a change of state involves a change in the physical properties of a substance. Such a change, called a **physical change**, alters the appearance of a substance, but not its formula. For example, the formula of water is $H_2O$. When liquid water boils to form a gas or freezes to a solid, it is still water with the formula $H_2O$. (See Figure 5.2.) Other physical changes can also change the appearance of a substance. When we cut an

### LEARNING GOAL

Identify a change in a substance as a chemical or a physical change.

WEB TUTORIAL
What Is Chemistry?

**A physical change:**
the boiling of water

**A chemical change:**
the tarnishing of silver

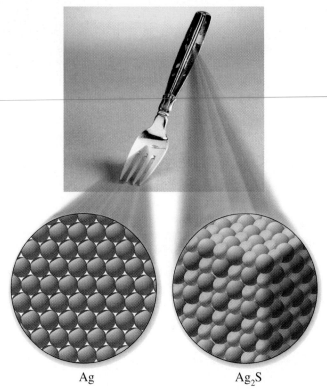

Ag              Ag$_2$S

Silver and tarnish are different substances

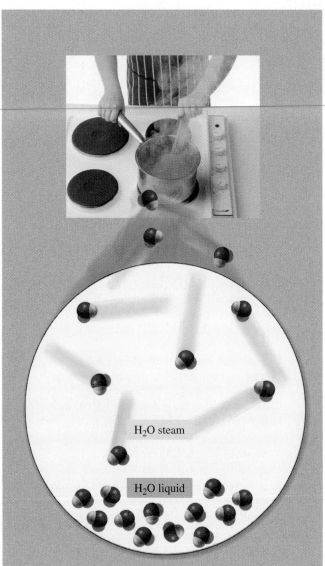

H$_2$O steam

H$_2$O liquid

Water and steam are both made
from H$_2$O molecules

**FIGURE 5.2** A chemical change produces new substances; a physical change does not.
**Q** Why is the formation of tarnish a chemical change?

apple into smaller pieces or slice a loaf of bread, the size changes. However, the smaller pieces are still apple or bread because there was no change in their composition.

In a **chemical change**, the reacting substances change into new substances that have different formulas and different properties. New properties may involve a change in color or the formation of bubbles or a solid. For instance, when silver tarnishes, the bright silver metal (Ag) reacts with sulfur (S) to become the dull, blackish substance we call tarnish (Ag$_2$S). (See Figure 5.3.) Table 5.3 gives examples of some typical physical and chemical changes.

## Changes During a Chemical Reaction

A **chemical reaction** always involves chemical change as bonds between the atoms in the original substances are broken and new bonds are formed. For example, a chemical

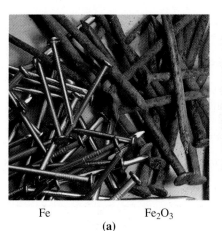

Fe          Fe$_2$O$_3$                NaHCO$_3$
      **(a)**                              **(b)**

**FIGURE 5.3** Chemical reactions involve chemical changes. (**a**) Iron (Fe) reacts with oxygen (O$_2$) to form rust (Fe$_2$O$_3$). (**b**) An antacid (NaHCO$_3$) tablet in water forms bubbles of carbon dioxide (CO$_2$).
**Q** What is the evidence for chemical change in these chemical reactions?

**TABLE 5.3** Comparison of Some Chemical and Physical Changes

| Chemical Changes | Physical Changes |
| --- | --- |
| Rusting nail | Melting ice |
| Bleaching a stain | Boiling water |
| Burning a log | Sawing a log in half |
| Tarnishing silver | Tearing paper |
| Fermenting grapes | Breaking a glass |
| Souring of milk | Pouring milk |

reaction takes place when a piece of iron combines with oxygen (O$_2$) in the air to produce a new substance, rust (Fe$_2$O$_3$), which has a reddish-brown color. When an antacid tablet is placed in a glass of water, bubbles appear as sodium hydrogen carbonate (NaHCO$_3$) and citric acid (C$_6$H$_8$O$_7$) in the tablet react to form carbon dioxide (CO$_2$) gas. (See Figure 5.3.) During each of these chemical changes, new properties become visible, which are clues that tell you a chemical reaction has taken place. Table 5.4 summarizes some types of visible evidence of a chemical reaction.

**TABLE 5.4** Types of Visible Evidence of a Chemical Reaction

1. Change in the color
2. Formation of a gas (bubbles)
3. Formation of a solid (precipitate)
4. Heat (or a flame) produced or heat absorbed

SAMPLE PROBLEM   **5.6**

■ **Evidence of a Chemical Reaction**

Identify each of the following as a physical change or a chemical reaction. If it is a chemical reaction, what is the visible evidence?

**a.** propane fuel burning in a barbecue
**b.** chopping a carrot
**c.** using peroxide to bleach hair

SOLUTION

**a.** The production of heat during burning of fuel is evidence of a chemical reaction.
**b.** Chopping a carrot into small pieces is a physical change of size, but the chopping does not change the substance.
**c.** The change in hair color is evidence of a chemical reaction.

STUDY CHECK

What evidence would you see that indicates that lighting a match is a chemical reaction?

## QUESTIONS AND PROBLEMS

### Chemical Changes

**5.23** Classify each of the following changes as chemical or physical:
  **a.** grinding coffee
  **b.** ignition of fuel in the space shuttle
  **c.** drying clothes
  **d.** neutralizing stomach acid with an antacid tablet
  **e.** formation of snowflakes
  **f.** an exploding dynamite stick

**5.24** Classify each of the following changes as chemical or physical:
  **a.** fogging the mirror during a shower
  **b.** tarnishing of a silver bracelet
  **c.** breaking a bone
  **d.** mending a broken bone
  **e.** burning paper
  **f.** slicing potatoes for fries

---

**LEARNING GOAL**

Write a balanced chemical equation from the formulas of the reactants and products for a reaction.

# 5.4 CHEMICAL EQUATIONS

When you build a model airplane, prepare a new recipe, or mix a medication, you follow a set of directions. These directions tell you what materials to use and the products you will obtain. In chemistry, a **chemical equation** tells us the materials we need and the products that will form in a chemical reaction.

## Writing a Chemical Equation

Suppose you work in a bicycle shop, assembling wheels and bodies into bicycles. You could represent this process by a simple equation.

Equation:   Wheels  +  Body  $\longrightarrow$  Bicycle

When you burn charcoal in a grill, the carbon in the charcoal combines with oxygen to form carbon dioxide. We can represent this reaction by a chemical equation that is much like the one for the bicycle:

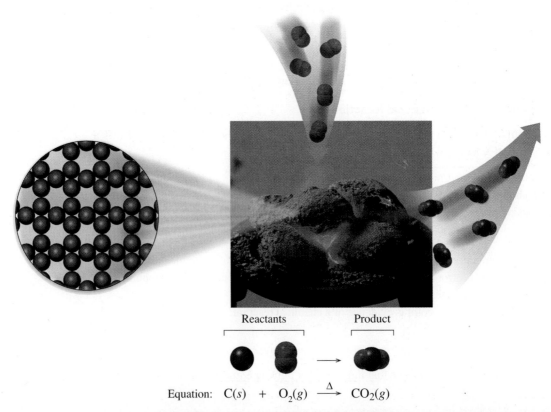

Reactants          Product

Equation:   $C(s)$  +  $O_2(g)$  $\xrightarrow{\Delta}$  $CO_2(g)$

In an equation, the formulas of the **reactants** are written on the left of the arrow and the formulas of the **products** on the right. When there are two or more formulas on the same side, they are separated by plus (+) signs. The delta sign ($\Delta$) indicates that heat was used to start the reaction.

In many equations, the formulas are followed by parentheses that contain abbreviations for the physical state of the substances: solid (*s*), liquid (*l*), or gas (*g*). If a substance is dissolved in water, it is an aqueous (*aq*) solution. Table 5.5 summarizes some of the symbols used in equations.

When a reaction takes place, the bonds between the atoms of the reactants are broken and new bonds are formed to give the products. Atoms cannot be gained, lost, or changed into other types of atoms during a chemical reaction. All the atoms in the reactants are conserved, which means the same numbers of each type of atom must appear in the products. Every reaction must be written as a **balanced equation**, which shows the same number of atoms for each element on both sides of the arrow. Let's see if the equation we wrote previously for burning carbon is balanced:

**TABLE 5.5** Some Symbols Used in Writing Equations

| Symbol | Meaning |
|---|---|
| + | Separates two or more formulas |
| $\longrightarrow$ | Reacts to form products |
| $\Delta$ | The reactants are heated |
| (*s*) | Solid |
| (*l*) | Liquid |
| (*g*) | Gas |
| (*aq*) | Aqueous |

$$C(s) + O_2(g) \longrightarrow CO_2(g)$$

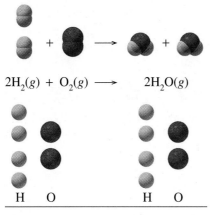

Reactant atoms  =  Product atoms

**WEB TUTORIAL**
Chemical Reactions and Equations

The answer is yes: this equation is *balanced* because there are one carbon atom and two oxygen atoms on each side of the arrow.

Now consider the reaction in which hydrogen reacts with oxygen to form water. First we write the formulas of the reactants and products:

$$H_2(g) + O_2(g) \longrightarrow H_2O(g)$$

Is the equation balanced? To find out, we add up the atoms of each element on each side of the arrow. No, the equation is *not balanced*. The number of atoms on the left side does not match the number of atoms on the right side. To balance this equation, we place whole numbers called **coefficients** in front of some of the formulas. First we write a coefficient of 2 in front of the $H_2O$ formula to represent the formation of two molecules of water. Now the product has four hydrogen atoms. That means we must also write a coefficient of 2 in front of the formula $H_2$ in the reactants to give four atoms on the reactant side. Coefficients are used to balance an equation, but the subscripts in the formulas are never changed. Now the number of hydrogen atoms and oxygen atoms are the same in the reactants as in the products. The equation is *balanced*.

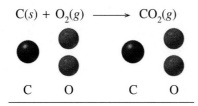

$$2H_2(g) + O_2(g) \longrightarrow 2H_2O(g)$$

H    O              H    O

Reactant atoms  =  Product atoms

## Balancing a Chemical Equation

We can now balance the equation for the reaction of the gas methane, $CH_4$, and oxygen to produce carbon dioxide and water. This is the reaction that occurs in the flame of a gas burner you use in the laboratory and in the burners of a gas stove.

**Guide to Balancing a Chemical Equation**

**1** Write an equation using the correct formulas of the reactants and products.

**2** Count the atoms of each element in reactants and products.

**3** Use coefficients to balance each element.

**4** Check the final equation for balance.

**Step 1**    **Write an equation, using the correct formulas.**    As a first step, we write the equation using the correct formulas for the reactants and products.

$$CH_4(g) + O_2(g) \longrightarrow CO_2(g) + H_2O(g)$$

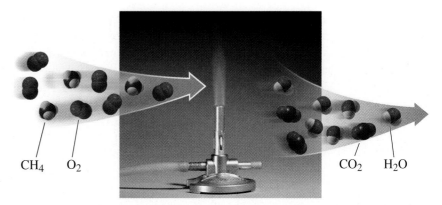

CH₄    O₂                                                    CO₂    H₂O

**Step 2**    **Determine if the equation is balanced.**    When we compare the atoms on the reactant side with the atoms on the product side, we see that there are more hydrogen atoms on the left side and more oxygen atoms on the right.

$$CH_4(g) + O_2(g) \longrightarrow CO_2(g) + H_2O(g)$$

| | |
|---|---|
| 1 C | 1 C |
| 4 H | 2 H    Not balanced |
| 2 O | 3 O    Not balanced |

**Step 3**    **Balance the equation one element at a time.**    Balance the hydrogen atoms by placing a coefficient of 2 in front of the formula for water.

$$CH_4(g) + O_2(g) \longrightarrow CO_2(g) + 2H_2O(g)$$

Then balance the oxygen atoms by placing a coefficient of 2 in front of the formula for oxygen. The oxygen atoms were balanced last because there were oxygen atoms in both products. There are now four oxygen atoms and four hydrogen atoms in both the reactants and products.

$$CH_4(g) + 2O_2(g) \longrightarrow CO_2(g) + 2H_2O(g)$$

**Step 4**    **Check to see if the equation is balanced.**    Rechecking the balanced equation shows that the numbers of atoms of carbon, hydrogen, and oxygen are the same for both the reactants and the products. The equation is balanced using the lowest possible whole numbers as coefficients.

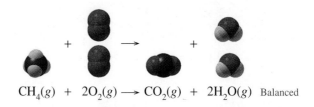

$$CH_4(g) + 2O_2(g) \longrightarrow CO_2(g) + 2H_2O(g) \quad \text{Balanced}$$

| Reactants | | Products | |
|---|---|---|---|
| 1 C atom | = | 1 C atom | Balanced |
| 4 H atoms | = | 4 H atoms | Balanced |
| 4 O atoms | = | 4 O atoms | Balanced |

Suppose you had added coefficients to the equation and obtained the following:

$$2CH_4(g) + 4O_2(g) \longrightarrow 2CO_2(g) + 4H_2O(g) \quad \text{Incorrect}$$

Although there are equal numbers of atoms on both sides of the equation, this is not written correctly. In an equation, the coefficients must be the lowest set of whole numbers that gives the same number of atoms of each element on both sides of the equation. For this *incorrectly balanced* equation, all the coefficients are divided by 2 to obtain the lowest possible whole numbers seen in Step 3.

---

SAMPLE PROBLEM  5.7

■ Balancing Equations

Balance the following equations:

**a.** $Al(s) + Cl_2(g) \longrightarrow AlCl_3(s)$
**b.** $Na_3PO_4(aq) + MgCl_2(aq) \longrightarrow Mg_3(PO_4)_2(s) + NaCl(aq)$

SOLUTION

**a. Step 1** The correct formulas are written in the equation.

$$Al(s) + Cl_2(g) \longrightarrow AlCl_3(s)$$

**Step 2** When we compare the number of atoms on the reactant and product sides, we find that the chlorine atoms are not balanced.

| Reactants | Products |
|---|---|
| $Al(s) + Cl_2(g) \longrightarrow$ | $AlCl_3(s)$ |
| 1 Al | 1 Al |
| 2 Cl | 3 Cl   Not balanced |

**Step 3** In the formulas with chlorine, there is an odd-even relationship. A 2 placed in front of $AlCl_3$ will give an even number of six Cl atoms.

$$Al(s) + Cl_2(g) \longrightarrow 2AlCl_3(s) \quad \text{Not balanced}$$

To balance the six Cl atoms in the product, place a 3 in front of $Cl_2$.

$$Al(s) + 3Cl_2(g) \longrightarrow 2AlCl_3(s) \quad \text{Not balanced}$$

Finally, the two Al atoms that are now in the product are balanced by placing a 2 in front of the Al on the reactant side.

$$2Al(s) + 3Cl_2(g) \longrightarrow 2AlCl_3(s) \quad \text{Balanced}$$

**Step 4** Check the final equation to determine that the equation is balanced.

| Reactants | Products |
|---|---|
| $2Al(s) + 3Cl_2(g) \longrightarrow$ | $2AlCl_3(s)$   Balanced |
| 2 Al | = 2 Al |
| 6 Cl | = 6 Cl |

**b. Step 1** In the equation, the correct formulas are written.

$$Na_3PO_4(aq) + MgCl_2(aq) \longrightarrow Mg_3(PO_4)_2(s) + NaCl(aq)$$

**Step 2** When we compare the number of ions on the reactant and product sides, we find that the equation is not balanced. In this equation, we can balance the phosphate polyatomic ion as a group because the phosphate group appears on both sides of the equation.

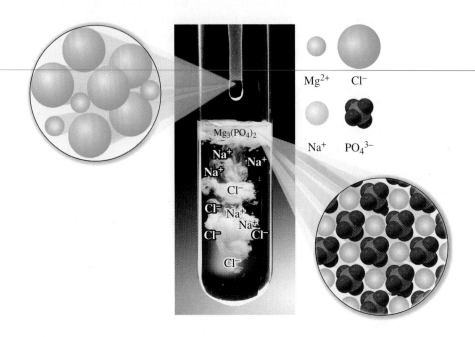

| **Reactants** | | **Products** | |
|---|---|---|---|
| $Na_3PO_4(aq) + MgCl_2(aq)$ | $\longrightarrow$ | $Mg_3(PO_4)_2(s) + NaCl(aq)$ | |
| $3\ Na^+$ | | $1\ Na^+$ | Not balanced |
| $1\ PO_4^{3-}$ | | $2\ PO_4^{3-}$ | Not balanced |
| $1\ Mg^{2+}$ | | $3\ Mg^{2+}$ | Not balanced |
| $2\ Cl^-$ | | $1\ Cl^-$ | Not balanced |

**Step 3** We begin with the formula of $Mg_3(PO_4)_2$, which is the most complex. A 3 in front of $MgCl_2$ balances magnesium, and a 2 in front of $Na_3PO_4$ balances the phosphate ion.

$$2Na_3PO_4(aq) + 3MgCl_2(aq) \longrightarrow Mg_3(PO_4)_2(s) + NaCl(aq) \quad \text{Not balanced}$$

Looking again at each of the ions in the reactants and products, we see that the sodium and chloride ions are not yet equal. A 6 in front of the NaCl balances the equation.

$$2Na_3PO_4(aq) + 3MgCl_2(aq) \longrightarrow Mg_3(PO_4)_2(s) + 6NaCl(aq) \quad \text{Balanced}$$

**Step 4** A check of the atoms indicates the equation is balanced.

| **Reactants** | | **Products** | |
|---|---|---|---|
| $2Na_3PO_4(aq) + 3MgCl_2(aq)$ | $\longrightarrow$ | $Mg_3(PO_4)_2(s) + 6NaCl(aq)$ | Balanced |
| $6\ Na^+$ | $=$ | $6\ Na^+$ | |
| $2\ PO_4^{3-}$ | $=$ | $2\ PO_4^{3-}$ | |
| $3\ Mg^{2+}$ | $=$ | $3\ Mg^{2+}$ | |
| $6\ Cl^-$ | $=$ | $6\ Cl^-$ | |

**STUDY CHECK**

Balance the following equation:

$$Sb_2S_3(s) + HCl(aq) \longrightarrow SbCl_3(s) + H_2S(g)$$

---

## QUESTIONS AND PROBLEMS

### Chemical Equations

**5.25** Determine whether each of the following equations is balanced or not balanced:
  **a.** $S(s) + O_2(g) \longrightarrow SO_3(g)$
  **b.** $2Al(s) + 3Cl_2(g) \longrightarrow 2AlCl_3(s)$
  **c.** $H_2(g) + O_2(g) \longrightarrow H_2O(g)$
  **d.** $C_3H_8(g) + 5O_2(g) \longrightarrow 3CO_2(g) + 4H_2O(g)$

**5.26** Determine whether each of the following equations is balanced or not balanced:
  **a.** $PCl_3(s) + Cl_2(g) \longrightarrow PCl_5(s)$
  **b.** $CO(g) + 2H_2(g) \longrightarrow CH_3OH(g)$
  **c.** $2KClO_3(s) \longrightarrow 2KCl(s) + O_2(g)$
  **d.** $Mg(s) + N_2(g) \longrightarrow Mg_3N_2(s)$

**5.27** Balance the following equations:
  **a.** $N_2(g) + O_2(g) \longrightarrow NO(g)$
  **b.** $HgO(s) \longrightarrow Hg(l) + O_2(g)$
  **c.** $Fe(s) + O_2(g) \longrightarrow Fe_2O_3(s)$
  **d.** $Na(s) + Cl_2(g) \longrightarrow NaCl(s)$

**5.28** Balance the following equations:
  **a.** $Ca(s) + Br_2(l) \longrightarrow CaBr_2(s)$
  **b.** $P_4(s) + O_2(g) \longrightarrow P_4O_{10}(s)$
  **c.** $Sb_2S_3(s) + HCl(aq) \longrightarrow SbCl_3(s) + H_2S(g)$
  **d.** $Fe_2O_3(s) + C(s) \longrightarrow Fe(s) + CO(g)$

**5.29** Balance the following equations:
  **a.** $Mg(s) + AgNO_3(aq) \longrightarrow Mg(NO_3)_2(aq) + Ag(s)$
  **b.** $Al(s) + CuSO_4(aq) \longrightarrow Cu(s) + Al_2(SO_4)_3(aq)$
  **c.** $Pb(NO_3)_2(aq) + NaCl(aq) \longrightarrow PbCl_2(s) + NaNO_3(aq)$
  **d.** $Al(s) + HCl(aq) \longrightarrow AlCl_3(aq) + H_2(g)$

**5.30** Balance the following equations:
  **a.** $Zn(s) + H_2SO_4(aq) \longrightarrow ZnSO_4(aq) + H_2(g)$
  **b.** $Al(s) + H_2SO_4(aq) \longrightarrow Al_2(SO_4)_3(aq) + H_2(g)$
  **c.** $K_2SO_4(aq) + BaCl_2(aq) \longrightarrow BaSO_4(s) + KCl(aq)$
  **d.** $CaCO_3(s) \longrightarrow CaO(s) + CO_2(g)$

---

# 5.5 TYPES OF REACTIONS

A great number of reactions occur in nature, in biological systems, and in the laboratory. However, there are some general patterns among all reactions that help us classify reactions. Most reactions fit into four general reaction types.

**LEARNING GOAL**

Identify a reaction as a combination, decomposition, or replacement.

## Combination Reactions

In a **combination reaction**, two or more elements or compounds bond to form one product. For example, sulfur and oxygen combine to form the product sulfur dioxide.

Two or more reactants   combine to yield   a single product

  +    $\longrightarrow$

Combination reaction

$$S(s) + O_2(g) \longrightarrow SO_2(g)$$

In Figure 5.4, the elements magnesium and oxygen combine to form a single product, magnesium oxide.

$$2Mg(s) + O_2(g) \longrightarrow 2MgO(s)$$

**FIGURE 5.4** This combination reaction is

$$2Mg(s) + O_2(g) \xrightarrow{\Delta} 2MgO(s)$$

**Q** What happens to the atoms in the reactants in a combination reaction?

## Decomposition Reactions

One reactant splits into two or more products

A B ⟶ A + B

In a **decomposition reaction**, a reactant splits into two or more simpler products. For example, when mercury(II) oxide is heated, the compound breaks apart into mercury atoms and oxygen. (See Figure 5.5.)

$$2HgO(s) \xrightarrow{\Delta} 2Hg(l) + O_2(g)$$

## Replacement Reactions

In a replacement reaction, elements in a compound are replaced by other elements. In a **single replacement reaction**, a reacting element switches place with an element in the other reacting compound.

### Single replacement

One element replaces another element

A + B C ⟶ A C + B

In the single replacement reaction shown in Figure 5.6, zinc replaces hydrogen in hydrochloric acid, HCl(*aq*).

$$Zn(s) + 2HCl(aq) \longrightarrow ZnCl_2(aq) + H_2(g)$$

In a **double replacement reaction**, the positive ions in the reacting compounds switch places.

### Double replacement

replace
Two elements each other

A B + C D ⟶ A D + C B

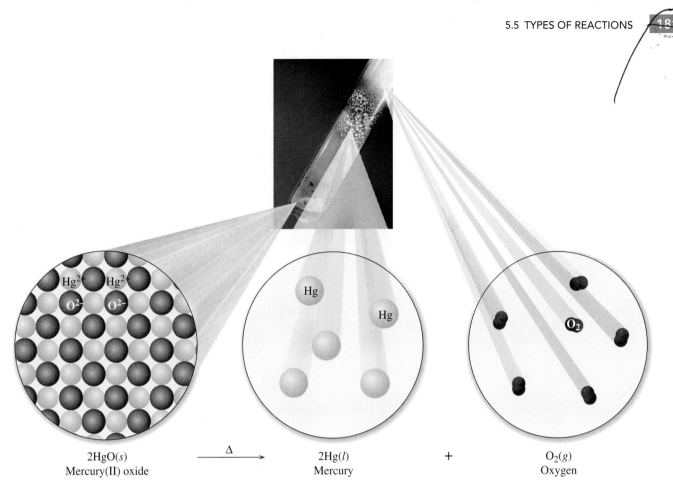

**FIGURE 5.5** In a decomposition reaction, one reactant breaks down into two or more products.
**Q** How do the differences in the reactant and products classify this as a decomposition reaction?

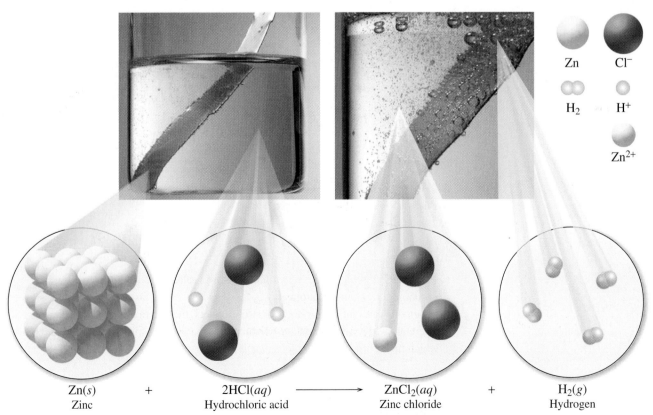

**FIGURE 5.6** In a single replacement reaction, an atom or ion replaces an atom or ion in a compound.
**Q** What changes in the formulas of the reactants identify this equation as a single replacement?

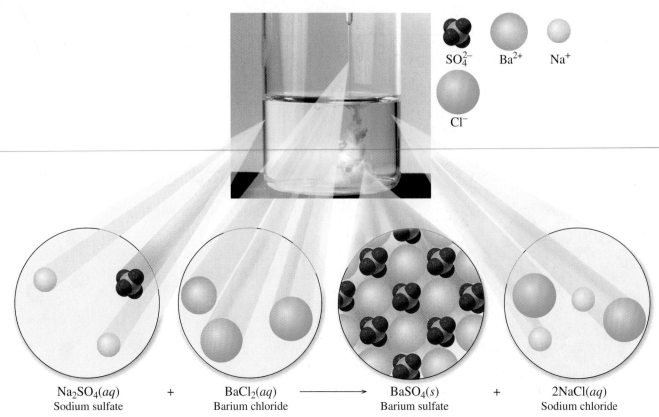

$SO_4^{2-}$    $Ba^{2+}$    $Na^+$

$Cl^-$

| $Na_2SO_4(aq)$ | + | $BaCl_2(aq)$ | $\longrightarrow$ | $BaSO_4(s)$ | + | $2NaCl(aq)$ |
| Sodium sulfate | | Barium chloride | | Barium sulfate | | Sodium chloride |

**FIGURE 5.7** In a double replacement reaction, the positive ions in the reactants replace each other.
**Q** How do the changes in the formulas of the reactants identify this equation as a double replacement reaction?

For example, in the reaction shown in Figure 5.7, barium ions change places with sodium ions in the reactants to form sodium chloride and a white solid precipitate of barium sulfate. The formulas of the products depend on the charges of the ions.

$$BaCl_2(aq) + Na_2SO_4(aq) \longrightarrow BaSO_4(s) + 2NaCl(aq)$$

### SAMPLE PROBLEM  5.8

#### ■ Identifying Reactions and Predicting Products

Classify the following reactions as combination, decomposition, or single or double replacement:

**a.** $2Fe_2O_3(s) + 3C(s) \longrightarrow 3CO_2(g) + 4Fe(s)$
**b.** $Fe_2S_3(s) \longrightarrow 2Fe(s) + 3S(s)$
**c.** $2AgNO_3(aq) + MgCl_2(aq) \longrightarrow 2AgCl(s) + Mg(NO_3)_2(aq)$

SOLUTION

**a.** In this *single replacement* reaction, a C atom replaces Fe in $Fe_2O_3$ to form the compound $CO_2$ and Fe atoms.
**b.** When one reactant breaks down to two products, the reaction is *decomposition*.
**c.** There are two reactants and two products, but the positive ions have exchanged places, which makes this a *double replacement* reaction.

STUDY CHECK

Nitrogen oxide gas and oxygen gas react to form nitrogen dioxide gas. Write the balanced equation and identify the reaction type.

# Health Note

## Smog and Health Concerns

There are two types of smog. One, photochemical smog, requires sunlight to initiate reactions that produce pollutants such as nitrogen oxides and ozone. The other type of smog, industrial or London smog, occurs in areas where coal containing sulfur is burned.

Photochemical smog is most prevalent in cities where people are dependent on cars for transportation. On a typical day in Los Angeles, for example, nitrogen oxide (NO) emissions from car exhausts increase as traffic increases on the roads. The nitrogen oxide is formed when $N_2$ and $O_2$ react at high temperatures in car and truck engines.

$$N_2(g) + O_2(g) \xrightarrow{\text{Heat}} 2NO(g)$$

Then NO reacts with oxygen in the air to produce $NO_2$, a reddish-brown gas that is irritating to the eyes and damaging to the respiratory tract.

$$2NO(g) + O_2(g) \longrightarrow 2NO_2(g)$$

When $NO_2$ is exposed to sunlight, it is converted into NO and oxygen atoms.

$$NO_2(g) \xrightarrow{\text{Sunlight}} NO(g) + \underset{\text{Oxygen atoms}}{O(g)}$$

Oxygen atoms are so reactive that they combine with oxygen molecules in the atmosphere, forming ozone.

$$O(g) + O_2(g) \longrightarrow \underset{\text{Ozone}}{O_3(g)}$$

In the upper atmosphere (the stratosphere), ozone is beneficial because it protects us from harmful ultraviolet radiation that comes from the sun. However, in the lower atmosphere, ozone irritates the eyes and respiratory tract, where it causes coughing, decreased lung function, and fatigue. It also causes deterioration of fabrics, cracks rubber, and damages trees and crops.

Industrial smog is prevalent in areas where coal with a high sulfur content is burned to produce electricity. During combustion, the sulfur is converted to sulfur dioxide:

$$S(s) + O_2(g) \longrightarrow SO_2(g)$$

The $SO_2$ is damaging to plants, suppressing growth, and it is corrosive to metals such as steel. $SO_2$ is also damaging to humans and can cause lung impairment and respiratory difficulties. The $SO_2$ in the air reacts with more oxygen to form $SO_3$. Acidic rain occurs when $SO_3$ combines with water in the air to form sulfuric acid.

$$2SO_2(g) + O_2(g) \longrightarrow 2SO_3(g)$$
$$SO_3(g) + H_2O(l) \longrightarrow \underset{\text{Sulfuric acid}}{H_2SO_4(aq)}$$

The presence of sulfuric acid in rivers and lakes causes an increase in the acidity of the water, reducing the ability of animals and plants to survive.

## QUESTIONS AND PROBLEMS

### Types of Reactions

**5.31 a.** Why is the following reaction called a decomposition reaction?

$$2Al_2O_3(s) \xrightarrow{\Delta} 4Al(s) + 3O_2(g)$$

**b.** Why is the following reaction called a single replacement reaction?

$$Br_2(g) + BaI_2(s) \longrightarrow BaBr_2(s) + I_2(g)$$

**5.32 a.** Why is the following called a combination reaction?

$$H_2(g) + Br_2(g) \longrightarrow 2HBr(g)$$

**b.** Why is the following called a double replacement reaction?

$$AgNO_3(aq) + NaCl(aq) \longrightarrow AgCl(s) + NaNO_3(aq)$$

**5.33** Classify each of the following reactions as a combination, decomposition, single replacement, or double replacement:

**a.** $4Fe(s) + 3O_2(g) \longrightarrow 2Fe_2O_3(s)$

**b.** $Mg(s) + 2AgNO_3(aq) \longrightarrow Mg(NO_3)_2(aq) + 2Ag(s)$

**c.** $CuCO_3(s) \longrightarrow CuO(s) + CO_2(g)$

**d.** $NaOH(aq) + HCl(aq) \longrightarrow NaCl(aq) + H_2O(l)$

**e.** $ZnCO_3(s) \longrightarrow CO_2(g) + ZnO(s)$

**f.** $Al_2(SO_4)_3(aq) + 6KOH(aq) \longrightarrow 2Al(OH)_3(s) + 3K_2SO_4(aq)$

**g.** $Pb(s) + O_2(g) \longrightarrow PbO_2(s)$

**5.34** Classify each of the following reactions as a combination, decomposition, single replacement, or double replacement:

**a.** $CuO(s) + 2HCl(aq) \longrightarrow CuCl_2(aq) + H_2O(l)$

**b.** $2Al(s) + 3Br_2(g) \longrightarrow 2AlBr_3(s)$

**c.** $Pb(NO_3)_2(aq) + 2NaCl(aq) \longrightarrow PbCl_2(s) + 2NaNO_3(aq)$

**d.** $2Mg(s) + O_2(g) \longrightarrow 2MgO(s)$

**e.** $Fe_2O_3(s) + 3C(s) \longrightarrow 2Fe(s) + 3CO(g)$

**f.** $C_6H_{12}O_6(aq) \longrightarrow 2C_2H_6O(aq) + 2CO_2(g)$

**g.** $BaCl_2(aq) + K_2CO_3(aq) \longrightarrow BaCO_3(s) + 2KCl(aq)$

**5.35** Try your hand at predicting the products that would result from the following types of reactions, and balance each equation.
  **a.** combination: $Mg(s) + Cl_2(g) \longrightarrow$
  **b.** decomposition: $HBr(g) \longrightarrow$
  **c.** single replacement: $Mg(s) + Zn(NO_3)_2(aq) \longrightarrow$
  **d.** double replacement: $K_2S(aq) + Pb(NO_3)_2(aq) \longrightarrow$

**5.36** Try your hand at predicting the products that would result from the following types of reactions, and balance each equation.
  **a.** combination: $Ca(s) + O_2(g) \longrightarrow$
  **b.** decomposition: $PbO_2(s) \longrightarrow$
  **c.** single replacement: $KI(s) + Cl_2(g) \longrightarrow$
  **d.** double replacement: $CuCl_2(aq) + Na_2S(aq) \longrightarrow$

## LEARNING GOAL

Define the terms *oxidation* and *reduction*.

# 5.6 OXIDATION–REDUCTION REACTIONS

Perhaps you have never heard of oxidation and reduction reactions. Yet this type of reaction has many important applications in our everyday lives. When you see a rusty nail, tarnish on a silver spoon, or corrosion on metal, you are observing oxidation.

$$4Fe(s) + 3O_2(g) \longrightarrow 2Fe_2O_3(s) \qquad \text{Fe is oxidized}$$
$$\text{Rust}$$

When you turn the lights on in your car, an oxidation–reduction reaction within your car battery provides the electricity. On a cold, wintry day, you might build a wood fire. Burning wood is an oxidation–reduction reaction. When you eat foods with starches in them, you digest the starches to give glucose, which is oxidized in your cells to give you energy along with carbon dioxide and water. Every breath you take provides oxygen to carry out oxidation in your cells.

$$C_6H_{12}O_6(aq) + 6O_2(g) \longrightarrow 6CO_2(g) + 6H_2O(l) + \text{energy}$$

## Oxidation–Reduction

In an **oxidation–reduction reaction** (*redox*), electrons are transferred from one substance to another. If one substance loses electrons, another substance must gain electrons. **Oxidation** is defined as the *loss* of electrons; **reduction** is the *gain* of electrons.

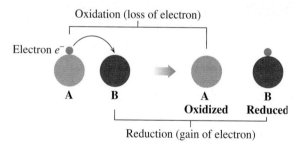

One way to remember these definitions is to use one of the following:

**LEO GER**
Loss of Electrons is Oxidation
Gain of Electrons is Reduction

**OIL RIG**
Oxidation Is Loss of electrons
Reduction Is Gain of electrons

## Oxidation–Reduction Involving Ions

In general, atoms of metals lose electrons to form positive ions, whereas nonmetals gain electrons to form negative ions. Now we can say that metals are oxidized and nonmetals are reduced.

Let's look at the formation of the ionic compound CaS.

$$Ca(s) + S(s) \longrightarrow CaS(s)$$

The calcium atom loses two electrons to form calcium ion ($Ca^{2+}$); calcium is oxidized.

$$Ca \longrightarrow Ca^{2+} + 2e^- \qquad \text{Oxidation: loss of electrons}$$

At the same time, the sulfur atom gains two electrons to form sulfide ion ($S^{2-}$); sulfur is reduced.

$$S + 2e^- \longrightarrow S^{2-} \qquad \text{Reduction: gain of electrons}$$

Therefore, the formation of CaS involves two reactions that occur simultaneously, one an oxidation and the other a reduction.

$$Ca + S \longrightarrow Ca^{2+} + S^{2-} = CaS$$

*Every time a reaction involves an oxidation and a reduction, the number of electrons lost is equal to the number of electrons gained.*

| Reduced | | Oxidized |
|---|---|---|
| Na | Oxidation: lose $e^-$ | $Na^+ + e^-$ |
| Ca | | $Ca^{2+} + 2e^-$ |
| $2Br^-$ | Reduction: gain $e^-$ | $Br_2 + 2e^-$ |
| $Fe^{2+}$ | | $Fe^{3+} + e^-$ |

Consider the reaction of zinc and copper(II) sulfate. (See Figure 5.8.)

$$Zn(s) + CuSO_4(aq) \longrightarrow ZnSO_4(aq) + Cu(s)$$

We can rewrite the equation to show the atoms and ions that react.

$$\mathbf{Zn(s)} + \mathbf{Cu^{2+}(aq)} + SO_4^{2-}(aq) \longrightarrow \mathbf{Zn^{2+}(aq)} + SO_4^{2-}(aq) + \mathbf{Cu(s)}$$

In this reaction, Zn atoms lose two electrons to form $Zn^{2+}$. At the same time, $Cu^{2+}$ gains two electrons. The $SO_4^{2-}$ ions are spectator ions and do not change.

$$Zn(s) \longrightarrow Zn^{2+}(aq) + 2e^- \qquad \text{Oxidation of Zn}$$

$$Cu^{2+}(aq) + 2e^- \longrightarrow Cu(s) \qquad \text{Reduction of } Cu^{2+}$$

In this single replacement reaction, zinc was oxidized and copper(II) ion was reduced.

**FIGURE 5.8** In this single replacement reaction, Zn(s) is oxidized to $Zn^{2+}$ when it provides two electrons to reduce $Cu^{2+}$ to Cu(s):

$$Zn(s) + Cu^{2+}(aq) \longrightarrow Cu(s) + Zn^{2+}(aq)$$

**Q** In the oxidation, does Zn(s) lose or gain electrons?

## SAMPLE PROBLEM  5.9

### ■ Oxidation–Reduction Reactions

In photographic film, the following decomposition reaction occurs in the presence of light. What is oxidized, and what is reduced?

$$2AgBr(s) \xrightarrow{\text{Light}} 2Ag(s) + Br_2(g)$$

### SOLUTION

To determine oxidation and reduction, we need to look at the ions and charges in the reactant. In AgBr, there is a silver ion ($Ag^+$) with a 1+ charge and a bromide ion ($Br^-$) with a charge of 1−. We can write the reaction as follows:

$$2Ag^+(aq) + 2Br^-(aq) \longrightarrow 2Ag(s) + Br_2(g)$$

Now we can compare $Ag^+$ with the product Ag atom. In this case, each $Ag^+$ gained an electron, which is reduction.

$$2Ag^+(aq) + 2e^- \longrightarrow 2Ag(s) \quad \text{Reduction}$$

Then we compare $Br^-$ with the Br in the product $Br_2$. In this case, each $Br^-$ lost an electron, which is oxidation.

$$2Br^-(aq) \longrightarrow Br_2(g) + 2e^- \quad \text{Oxidation}$$

In the reaction, bromide ion is oxidized, and silver ion is reduced.

### STUDY CHECK

In the following combination reaction, what reactant is oxidized and what reactant is reduced?

$$2Li(s) + F_2(g) \longrightarrow 2LiF(s)$$

## Explore Your World

### Oxidation of Fruits and Vegetables

Freshly cut surfaces of fruits and vegetables discolor when exposed to oxygen in the air. Cut three slices of a fruit or vegetable such as apple, potato, avocado, or banana. Leave one piece on the kitchen counter (uncovered). Wrap one piece in plastic wrap and leave on the kitchen counter. Dip one piece in lemon juice and leave uncovered.

#### QUESTIONS

1. What changes take place in each sample after 1–2 hours?

2. Why would wrapping fruits and vegetables slow the rate of discoloration?
3. If lemon juice contains vitamin C (an antioxidant), why would dipping a fruit or vegetable in lemon juice affect the oxidation reaction on the surface of the fruit or vegetable?
4. Other kinds of antioxidants are vitamin E, citric acid, and BHT. Look for these antioxidants on the labels of cereals, potato chips, and other foods in your kitchen. Why are antioxidants added to food products that will be stored on our kitchen shelves?

## Oxidation and Reduction in Biological Systems

Oxidation may also involve the addition of oxygen or the loss of hydrogen, and reduction may involve the loss of oxygen or the gain of hydrogen. In the cells of the body, oxidation of organic (carbon) compounds involves the transfer of hydrogen atoms (H), which are composed of electrons and protons. For example, the oxidation of a typical biochemical

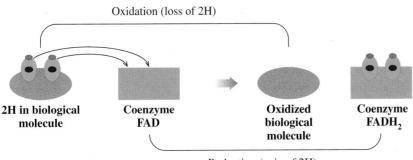

Oxidation (loss of 2H)

| 2H in biological molecule | Coenzyme FAD | Oxidized biological molecule | Coenzyme FADH$_2$ |

Reduction (gain of 2H)

molecule can involve the transfer of two hydrogen atoms (or $2H^+$ and $2e^-$) to a proton acceptor such as the coenzyme FAD (flavin adenine dinucleotide). The coenzyme is reduced to $FADH_2$. In many biochemical oxidation–reduction reactions, the transfer of hydrogen atoms is necessary for the production of energy in the cells. For example, methyl alcohol ($CH_3OH$), a poisonous substance, is metabolized in the body by the following reactions.

$$CH_3OH \longrightarrow H_2CO + 2H \quad \text{Oxidation: loss of H atoms}$$
Methyl alcohol    Formaldehyde

The formaldehyde can be oxidized further, this time by the addition of oxygen, to produce formic acid.

$$2H_2CO + O_2 \longrightarrow 2H_2CO_2 \quad \text{Oxidation: addition of O atoms}$$
Formaldehyde      Formic acid

Finally, formic acid is oxidized to carbon dioxide and water.

$$2H_2CO_2 + O_2 \longrightarrow 2CO_2 + 2H_2O \quad \text{Oxidation: addition of O atoms}$$
Formic acid

The intermediate products of the oxidation of methyl alcohol are quite toxic, causing blindness and possibly death as they interfere with key reactions in the cells of the body.

In summary, we find that the particular definition of oxidation and reduction we use depends on the process that occurs in the reaction. All of these definitions are summarized in Table 5.6. Oxidation always involves a loss of electrons, but it may also be seen as an addition of oxygen or the loss of hydrogen atoms. A reduction always involves a gain of electrons and may also be seen as the loss of oxygen, or the gain of hydrogen.

**TABLE 5.6** Characteristics of Oxidation and Reduction

| Oxidation | |
|---|---|
| **Always Involves** | **May Involve** |
| Loss of electrons | Addition of oxygen Loss of hydrogen |

| Reduction | |
|---|---|
| **Always Involves** | **May Involve** |
| Gain of electrons | Loss of oxygen Gain of hydrogen |

## QUESTIONS AND PROBLEMS

### Oxidation–Reduction Reactions

**5.37** Indicate whether each of the following describes an oxidation or a reduction:
 **a.** $Na^+(aq) + e^- \longrightarrow Na(s)$
 **b.** $Ni(s) \longrightarrow Ni^{2+}(aq) + 2e^-$
 **c.** $Cr^{3+}(aq) + 3e^- \longrightarrow Cr(s)$
 **d.** $2H^+(aq) + 2e^- \longrightarrow H_2(g)$

**5.38** Indicate whether each of the following describes an oxidation or a reduction:
 **a.** $O_2(g) + 4e^- \longrightarrow 2O^{2-}(aq)$
 **b.** $Al(s) \longrightarrow Al^{3+}(aq) + 3e^-$
 **c.** $Fe^{3+}(aq) + e^- \longrightarrow Fe^{2+}(aq)$
 **d.** $2Br^+(aq) \longrightarrow Br_2(g) + 2e^-$

**5.39** In the following reactions, identify what reactant is oxidized and what reactant is reduced:
 **a.** $Zn(s) + Cl_2(g) \longrightarrow ZnCl_2(s)$
 **b.** $Cl_2(g) + 2NaBr(aq) \longrightarrow 2NaCl(aq) + Br_2(g)$

 **c.** $2PbO(s) \longrightarrow 2Pb(s) + O_2(g)$
 **d.** $2Fe^{3+}(aq) + Sn^{2+}(aq) \longrightarrow 2Fe^{2+}(aq) + Sn^{4+}(aq)$

**5.40** In the following reactions, identify which reactant is oxidized and which reactant is reduced:
 **a.** $2Li(s) + F_2(g) \longrightarrow 2LiF(s)$
 **b.** $Cl_2(g) + 2KI(aq) \longrightarrow 2KCl(aq) + I_2(g)$
 **c.** $Zn(g) + Cu^{2+}(aq) \longrightarrow Zn^{2+}(aq) + Cu(s)$
 **d.** $Fe(s) + CuSO_4(aq) \longrightarrow FeSO_4(aq) + Cu(s)$

**5.41** In the mitochondria of human cells, energy for the production of ATP is provided by the oxidation and reduction reactions of the iron ions in the cytochromes of the electron transport chain. Identify each of the following reactions as oxidation or reduction:
 **a.** $Fe^{3+} + e^- \longrightarrow Fe^{2+}$
 **b.** $Fe^{2+} \longrightarrow Fe^{3+} + e^-$

**5.42** Chlorine ($Cl_2$) is a strong germicide used to disinfect drinking water and to kill microbes in swimming pools. If the product is $Cl^-$, was the elemental chlorine oxidized or reduced?

**5.43** When linoleic acid, an unsaturated fatty acid, reacts with hydrogen, it forms a saturated fatty acid. Is linoleic acid oxidized or reduced in the hydrogenation reaction?

$$C_{18}H_{32}O_2 + 2H_2 \longrightarrow C_{18}H_{36}O_2$$

**5.44** In one of the reactions in the citric acid cycle, which provides energy for ATP synthesis, succinic acid is converted to fumaric acid.

$$C_4H_6O_4 \longrightarrow C_4H_4O_4 + 2H$$

Succinic acid    Fumaric acid

The reaction is accompanied by a coenzyme, flavin adenine dinucleotide (FAD).

$$FAD + 2H \longrightarrow FADH_2$$

**a.** Is succinic acid oxidized or reduced?
**b.** Is FAD oxidized or reduced?
**c.** Why would the two reactions occur together?

# Green Chemistry Note

## Fuel Cells: Clean Energy for the Future

Fuel cells are of interest to scientists because they provide an alternative source of electrical energy that is more efficient, does not use up oil reserves, and generates products that do not pollute the atmosphere. Fuels cells are considered to be a clean way to produce energy.

Unlike a battery that runs down, fuel cells are provided with new reactants continually to provide an electrical current. One type of hydrogen–oxygen fuel cell has been used in automobile prototypes. In this hydrogen cell, gas enters the fuel cell and comes in contact with a platinum catalyst embedded in a plastic membrane. The catalyst assists in the oxidation of hydrogen atoms to protons and electrons.

$$2H_2(g) \longrightarrow 4H^+(aq) + 4e^-  \text{ Oxidation}$$

The electrons produce an electric current as they travel through the wire. The protons move through the plastic membrane to react with oxygen molecules. The oxygen is reduced to oxide ions that combine with the protons to form water.

$$O_2(g) + 4H^+(aq) + 4e^- \longrightarrow 2H_2O(l)  \text{ Reduction}$$

The overall hydrogen–oxygen fuel cell reaction can be written as

$$2H_2(g) + O_2(g) \longrightarrow 2H_2O(l)$$

Fuel cells have already been used to power the space shuttle and may soon be available to produce energy for cars and buses.

A major drawback to the practical use of fuel cells is the economic impact of converting cars to fuel cell operation. The storage and cost of producing hydrogen are also problems. Some manufacturers are experimenting with systems that convert gasoline or methanol to hydrogen for immediate use in fuel cells.

In homes, fuel cells may one day replace the batteries currently used to provide electrical power for cell phones, CD and DVD players, and laptop computers. Fuel cell design is still in the prototype phase, although there is much interest in development of these cells. We already know they can work, but modifications must still be made before they become reasonably priced and part of our everyday lives.

Electric current

$e^-$          $e^-$

$H_2$ gas $\longrightarrow$    $H^+$  $e^-$    $\longleftarrow$ $O_2$ gas $\longleftarrow$

$e^-$                          $O^{2-}$

$e^-$                          $e^-$

$H^+ \rightarrow H^+$           $O^{2-}$

$H^+$      $H^+ \rightarrow$

$H^+ \rightarrow H^+$           $O^{2-}$

Plastic membrane              $\rightarrow H_2O \rightarrow$

Cathode              Anode

Catalyst

Oxidation
$$2H_2(g) \longrightarrow 4H^+(aq) + 4e^-$$

Reduction
$$O_2(g) + 4H^+(aq) + 4e^- \longrightarrow 2H_2O(l)$$

# 5.7 MOLE RELATIONSHIPS IN CHEMICAL EQUATIONS

In an earlier section, we saw that equations are balanced in terms of the numbers of each type of atom in the reactants and products. However, when experiments are done in the laboratory or medications are prepared in the pharmacy, samples contain billions of atoms and molecules so it is impossible to count them individually. What can be measured conveniently is mass, using a balance. Because mass is related to the number of particles through the molar mass, measuring the mass is equivalent to counting the number of particles or moles.

**LEARNING GOAL**

Given a quantity in moles of reactant or product, use a mole–mole factor from the balanced equation to calculate the moles of another substance in the reaction.

## Conservation of Mass

In any chemical reaction, the total amount of matter in the reactants is equal to the total amount of matter in the products. If all of the reactants were weighed, they would have a total mass equal to the total mass of the products. This is known as the *law of conservation of mass,* which says that there is no change in the total mass of the substances reacting in a chemical reaction. Thus, no material is lost or gained when original substances are changed to new substances.

For example, tarnish forms when silver reacts with sulfur to form silver sulfide.

$$2Ag(s) + S(s) \longrightarrow Ag_2S(s)$$

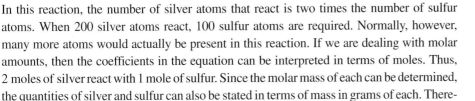

|  |  |  |
|---|---|---|
| $2Ag(s)$ | $+$ | $S(s)$ |

Mass of reactants    $=$    Mass of products

In this reaction, the number of silver atoms that react is two times the number of sulfur atoms. When 200 silver atoms react, 100 sulfur atoms are required. Normally, however, many more atoms would actually be present in this reaction. If we are dealing with molar amounts, then the coefficients in the equation can be interpreted in terms of moles. Thus, 2 moles of silver react with 1 mole of sulfur. Since the molar mass of each can be determined, the quantities of silver and sulfur can also be stated in terms of mass in grams of each. Therefore, an equation for a chemical equation can be interpreted several ways, as seen in Table 5.7.

**TABLE 5.7  Information Available from a Balanced Equation**

|  | Reactants |  | Products |
|---|---|---|---|
| **Equation** | **$2Ag(s)$** | **$+ S(s)$** | $\longrightarrow$ **$Ag_2S(s)$** |
| **Atoms** | 2 Ag atoms | + 1 S atom | $\longrightarrow$ 1 Ag$_2$S formula unit |
|  | 200 Ag atoms | + 100 S atoms | $\longrightarrow$ 100 Ag$_2$S formula units |
| **Avogadro's number of atoms** | $2(6.02 \times 10^{23})$ Ag atoms | $+ 1(6.02 \times 10^{23})$ S atoms | $\longrightarrow$ $1(6.02 \times 10^{23})$ Ag$_2$S formula units |
| **Moles** | 2 moles of Ag | + 1 mole of S | $\longrightarrow$ 1 mole of Ag$_2$S |
| **Mass (g)** | $2(107.9 \text{ g})$ Ag | $+ 1(32.1 \text{ g})$ S | $\longrightarrow$ $1(247.9 \text{ g})$ Ag$_2$S |
| **Total mass (g)** | 247.9 g |  | $\longrightarrow$ 247.9 g |

## Mole–Mole Factors from an Equation

When iron reacts with sulfur, the product is iron(III) sulfide.

$$2Fe(s) + 3S(s) \longrightarrow Fe_2S_3(s)$$

| Iron (Fe) | Sulfur (S) | Iron(III) sulfide ($Fe_2S_3$) |
|:---:|:---:|:---:|
| $2Fe(s)$     +     | $3S(s)$     $\longrightarrow$ | $Fe_2S_3(s)$ |

Because the equation is balanced, we know the proportions of iron and sulfur in the reaction. For this reaction, we see that 2 moles of iron reacts with 3 moles of sulfur to form 1 mole of iron(III) sulfide. From the coefficients, we can write **mole–mole factors** between reactants and between reactants and products.

$$\text{Fe and S:} \quad \frac{2 \text{ moles Fe}}{3 \text{ moles S}} \quad \text{and} \quad \frac{3 \text{ moles S}}{2 \text{ moles Fe}}$$

$$\text{Fe and } Fe_2S_3: \quad \frac{2 \text{ moles Fe}}{1 \text{ mole } Fe_2S_3} \quad \text{and} \quad \frac{1 \text{ mole } Fe_2S_3}{2 \text{ moles Fe}}$$

$$\text{S and } Fe_2S_3: \quad \frac{3 \text{ moles S}}{1 \text{ mole } Fe_2S_3} \quad \text{and} \quad \frac{1 \text{ mole } Fe_2S_3}{3 \text{ moles S}}$$

## Using Mole–Mole Factors in Calculations

Whenever you prepare a recipe, adjust an engine for the proper mixture of fuel and air, or prepare medicines in a pharmaceutical laboratory, you need to know the proper amounts of reactants to use and how much of the product will form. Earlier we wrote all the possible conversion factors that can be obtained from the balanced equation: $2Fe(s) + 3S(s) \longrightarrow Fe_2S_3(s)$. Now we will show how mole–mole factors are used in chemical calculations.

---

### SAMPLE PROBLEM    5.10

■ **Calculating Moles of a Reactant**

In the reaction of iron and sulfur, how many moles of sulfur are needed to react with 6.0 moles of iron?

$$2Fe(s) + 3S(s) \longrightarrow Fe_2S_3(s)$$

**SOLUTION**

Step 1  **Write the given and needed number of moles.**   In this problem, we need to find the number of moles of S that react with 6.0 moles of Fe.

**Given**   moles of Fe      **Need**   moles of S

**Step 2  Write the plan to convert the given to the needed.**

moles of Fe → Mole–mole factor → moles of S

**Step 3  Use coefficients to write relationships and mole–mole factors.**  Use coefficients to write the mole–mole factors for the given and needed substances.

2 moles Fe = 3 moles S

$$\frac{2 \text{ moles Fe}}{3 \text{ moles S}} \quad \text{and} \quad \frac{3 \text{ moles S}}{2 \text{ moles Fe}}$$

**Step 4  Set up problem using the mole–mole factor that cancels given moles.**  Use a mole–mole factor to cancel the given moles and provide needed moles.

$$6.0 \text{ moles Fe} \times \frac{3 \text{ moles S}}{2 \text{ moles Fe}} = 9.0 \text{ moles of S}$$

The answer is given with 2 significant figures because the given quantity, 6.0 moles Fe, has 2 SFs. The values in the mole–mole factor are exact.

**STUDY CHECK**

Using the equation in Sample Problem 5.10, calculate the number of moles of iron needed to completely react with 2.7 moles of sulfur.

**Guide to Using Mole Factors**

**1** Write the given and needed moles.

**2** Write a plan to convert the given to the needed moles.

**3** Use coefficients to write relationships and mole–mole factors.

**4** Set up problem using the mole factor that cancels given moles.

**SAMPLE PROBLEM  5.11**

### ■ Calculating Moles of a Product

Propane gas ($C_3H_8$), a fuel for camp stoves and specially equipped automobiles, reacts with oxygen to produce carbon dioxide and water. How many moles of $CO_2$ can be produced when 2.25 moles of $C_3H_8$ reacts?

$$C_3H_8(g) + 5O_2(g) \longrightarrow 3CO_2(g) + 4H_2O(g)$$
Propane

**SOLUTION**

**Step 1  Write the given and needed moles.**  In this problem, we need to find the number of moles of $CO_2$ that can be produced when 2.25 moles of $C_3H_8$ react.

**Given**  moles of $C_3H_8$     **Need**  moles of $CO_2$

**Step 2  Write the plan to convert the given to the needed.**

moles of $C_3H_8$ → Mole–mole factor → moles of $CO_2$

**Step 3  Use coefficients to write relationships and mole–mole factors.**  Use coefficients to write the mole–mole factors for the given and needed substances.

1 mole $C_3H_8$ = 3 moles $CO_2$

$$\frac{1 \text{ mole } C_3H_8}{3 \text{ moles } CO_2} \quad \text{and} \quad \frac{3 \text{ moles } CO_2}{1 \text{ mole } C_3H_8}$$

**Step 4  Set up problem using the mole–mole factor that cancels given moles.**  Use a mole–mole factor to cancel the given moles and provide needed moles.

$$2.25 \text{ moles } C_3H_8 \times \frac{3 \text{ moles } CO_2}{1 \text{ mole } C_3H_8} = 6.75 \text{ moles of } CO_2$$

The answer is given with 3 significant figures because the given quantity, 2.25 moles of $C_3H_8$, has 3 SFs. The values in the mole–mole factor are exact.

STUDY CHECK

Using the equation in Sample Problem 5.11, calculate the number of moles of oxygen that must react to produce 0.756 mole of water.

## QUESTIONS AND PROBLEMS

### Mole Relationships in Chemical Equations

**5.45** Give an interpretation of the following equations in terms of (1) number of particles and (2) number of moles:
  **a.** $2SO_2(g) + O_2(g) \longrightarrow 2SO_3(g)$
  **b.** $4P(s) + 5O_2(g) \longrightarrow 2P_2O_5(s)$

**5.46** Give an interpretation of the following equations in terms of (1) number of particles and (2) number of moles:
  **a.** $2Al(s) + 3Cl_2(g) \longrightarrow 2AlCl_3(s)$
  **b.** $4HCl(g) + O_2(g) \longrightarrow 2Cl_2(g) + 2H_2O(g)$

**5.47** Write all of the mole–mole factors for the equations listed in problem 5.45.

**5.48** Write all of the mole–mole factors for the equations listed in problem 5.46.

**5.49** The reaction of hydrogen with oxygen produces water.

$$2H_2(g) + O_2(g) \longrightarrow 2H_2O(g)$$

  **a.** How many moles of $O_2$ are required to react with 2.0 moles of $H_2$?
  **b.** If you have 5.0 moles of $O_2$, how many moles of $H_2$ are needed for the reaction?
  **c.** How many moles of $H_2O$ form when 2.5 moles of $O_2$ reacts?

**5.50** Ammonia is produced by the reaction of hydrogen and nitrogen.

$$N_2(g) + 3H_2(g) \longrightarrow 2NH_3(g) \quad \text{ammonia}$$

  **a.** How many moles of $H_2$ are needed to react with 1.0 mole of $N_2$?

  **b.** How many moles of $N_2$ reacted if 0.60 mole of $NH_3$ is produced?
  **c.** How many moles of $NH_3$ are produced when 1.4 moles of $H_2$ reacts?

**5.51** Carbon disulfide and carbon monoxide are produced when carbon is heated with sulfur dioxide.

$$5C(s) + 2SO_2(g) \longrightarrow CS_2(l) + 4CO(g)$$

  **a.** How many moles of C are needed to react with 0.500 mole of $SO_2$?
  **b.** How many moles of CO are produced when 1.2 moles of C reacts?
  **c.** How many moles of $SO_2$ are required to produce 0.50 mole of $CS_2$?
  **d.** How many moles of $CS_2$ are produced when 2.5 moles of C reacts?

**5.52** In the acetylene torch, acetylene gas ($C_2H_2$) burns in oxygen to produce carbon dioxide and water.

$$2C_2H_2(g) + 5O_2(g) \longrightarrow 4CO_2(g) + 2H_2O(g)$$

  **a.** How many moles of $O_2$ are needed to react with 2.00 moles of $C_2H_2$?
  **b.** How many moles of $CO_2$ are produced when 3.5 moles of $C_2H_2$ reacts?
  **c.** How many moles of $C_2H_2$ are required to produce 0.50 mole of $H_2O$?
  **d.** How many moles of $CO_2$ are produced from 0.100 mole of $O_2$?

## 5.8 MASS CALCULATIONS FOR REACTIONS

### LEARNING GOAL

Given the mass in grams of a substance in a reaction, calculate the mass in grams of another substance in the reaction.

When you perform a chemistry experiment in the laboratory, you use a laboratory balance to obtain a certain mass of reactant. From the mass in grams, you can determine the number of moles of reactant. By using mole–mole factors, you can predict the moles of product that can be produced. Then the molar mass of the product is used to convert the moles back into mass in grams as seen in the following Sample Problems.

SAMPLE PROBLEM 5.12

■ **Mass of Products from Moles of Reactant**

In the formation of smog, nitrogen reacts with oxygen to produce nitrogen oxide. Calculate the grams of NO produced when 1.50 moles of $O_2$ react.

$$N_2(g) + O_2(g) \longrightarrow 2NO(g)$$

**SOLUTION**

Step 1  **Given**   1.50 moles of $O_2$    **Need**   grams of NO

Step 2  **Plan**   moles of $O_2$  [Mole–mole factor]  moles of NO  [Molar mass]  grams of NO

Step 3  **Equalities/Conversion Factors**   The mole–mole factor that converts moles of $O_2$ to moles of NO is derived from the coefficients in the balanced equation.

$$1 \text{ mole of } O_2 = 2 \text{ moles of NO}$$

$$\frac{2 \text{ moles NO}}{1 \text{ mole } O_2} \quad \text{and} \quad \frac{1 \text{ mole } O_2}{2 \text{ moles NO}}$$

$$1 \text{ mole of NO} = 30.0 \text{ g of NO}$$

$$\frac{30.0 \text{ g NO}}{1 \text{ mole NO}} \quad \text{and} \quad \frac{1 \text{ mole NO}}{30.0 \text{ g NO}}$$

Step 4  **Set Up Problem**   First, we can change the given, 1.50 moles of $O_2$, to moles of NO.

$$1.50 \text{ moles } O_2 \times \frac{2 \text{ moles NO}}{1 \text{ mole } O_2} = 3.00 \text{ moles of NO}$$

Now, the moles of NO can be converted to grams of NO, using its molar mass.

$$3.00 \text{ moles NO} \times \frac{30.0 \text{ g NO}}{1 \text{ mole NO}} = 90.0 \text{ g of NO}$$

These two steps can also be written as a sequence of conversion factors that lead to the mass in grams of NO.

$$1.50 \text{ moles } O_2 \times \frac{2 \text{ moles NO}}{1 \text{ mole } O_2} \times \frac{30.0 \text{ g NO}}{1 \text{ mole NO}} = 90.0 \text{ g of NO}$$

**STUDY CHECK**

Using the equation in Sample Problem 5.12, calculate the grams of NO that can be produced when 0.734 mole of $N_2$ reacts.

---

## Guide to Calculating the Masses of Reactants and Products in a Chemical Reaction

**1**  Use molar mass to convert grams of given to moles (if necessary).

**2**  Write a mole–mole factor from the coefficients in the equation.

**3**  Convert moles of given to moles of needed substance using mole–mole factor.

**4**  Convert moles of needed substance to grams using molar mass.

---

## SAMPLE PROBLEM  5.13

### ■ Mass of Product from Mass of Reactant

Acetylene ($C_2H_2$), used in welding, burns with oxygen.

$$2C_2H_2(g) + 5O_2(g) \longrightarrow 4CO_2(g) + 2H_2O(g)$$

How many grams of carbon dioxide are produced when 54.6 g of $C_2H_2$ are burned?

**SOLUTION**

Step 1  **Given**   grams of $C_2H_2$    **Need**   grams of $CO_2$

Step 2  **Plan**   Once we convert grams of $C_2H_2$ to moles of $C_2H_2$ using its molar mass, we can use a mole–mole factor to find the moles of $CO_2$. Then the molar mass of $CO_2$ will give us the grams of $CO_2$.

grams of $C_2H_2$  [Molar mass]  moles of $C_2H_2$  [Mole–mole factor]  moles of $CO_2$  [Molar mass]  grams of $CO_2$

Step 3  **Equalities/Conversion Factors**   We need the factor for the molar masses of $C_2H_2$ and $CO_2$ and the mole–mole factor that converts moles of $C_2H_2$ to moles of $CO_2$ from the coefficients in the balanced equation.

$$1 \text{ mole of } C_2H_2 = 26.0 \text{ g of } C_2H_2$$

$$\frac{26.0 \text{ g } C_2H_2}{1 \text{ mole } C_2H_2} \quad \text{and} \quad \frac{1 \text{ mole } C_2H_2}{26.0 \text{ g } C_2H_2}$$

$$2 \text{ moles of } C_2H_2 = 4 \text{ moles of } CO_2$$

$$\frac{2 \text{ moles } C_2H_2}{4 \text{ moles } CO_2} \quad \text{and} \quad \frac{4 \text{ moles } CO_2}{2 \text{ moles } C_2H_2}$$

$$1 \text{ mole of } CO_2 = 44.0 \text{ g of } CO_2$$

$$\frac{44.0 \text{ g } CO_2}{1 \text{ mole } CO_2} \quad \text{and} \quad \frac{1 \text{ mole } CO_2}{44.0 \text{ g } CO_2}$$

**Step 4  Set Up Problem**  Using our plan, we first convert grams of $C_2H_2$ to moles of $C_2H_2$.

$$54.6 \text{ g } C_2H_2 \times \frac{1 \text{ mole } C_2H_2}{26.0 \text{ g } C_2H_2} = 2.10 \text{ moles of } C_2H_2$$

Then we change moles of $C_2H_2$ to moles of $CO_2$ by using the mole–mole factor.

$$2.10 \text{ moles } C_2H_2 \times \frac{4 \text{ moles } CO_2}{2 \text{ moles } C_2H_2} = 4.20 \text{ moles of } CO_2$$

Finally, we can convert moles of $CO_2$ to grams of $CO_2$.

$$4.20 \text{ moles } CO_2 \times \frac{44.0 \text{ g } CO_2}{1 \text{ mole } CO_2} = 185 \text{ g of } CO_2$$

The solution can be obtained using the conversion factors in sequence.

$$54.6 \text{ g } C_2H_2 \times \frac{1 \text{ mole } C_2H_2}{26.0 \text{ g } C_2H_2} \times \frac{4 \text{ moles } CO_2}{2 \text{ moles } C_2H_2} \times \frac{44.0 \text{ g } CO_2}{1 \text{ mole } CO_2} = 185 \text{ g of } CO_2$$

**STUDY CHECK**

Using the equation in Sample Problem 5.13, calculate the grams of $CO_2$ that can be produced when 25.0 g of $O_2$ reacts.

## QUESTIONS AND PROBLEMS

### Mass Calculations for Reactions

**5.53** Sodium reacts with oxygen to produce sodium oxide.

$$4Na(s) + O_2(g) \longrightarrow 2Na_2O(s)$$

**a.** How many grams of $Na_2O$ are produced when 2.50 moles of Na react?

**b.** If you have 18.0 g of Na, how many grams of $O_2$ are required for reaction?

**c.** How many grams of $O_2$ are needed in a reaction that produces 75.0 g of $Na_2O$?

**5.54** Nitrogen gas reacts with hydrogen gas to produce ammonia by the following equation:

$$N_2(g) + 3H_2(g) \longrightarrow 2NH_3(g)$$

**a.** If you have 1.80 moles of $H_2$, how many grams of $NH_3$ can be produced?

**b.** How many grams of $H_2$ are needed to react with 2.80 g of $N_2$?

**c.** How many grams of $NH_3$ can be produced from 12.0 g of $H_2$?

**5.55** Ammonia and oxygen react to form nitrogen and water.

$$4NH_3(g) + 3O_2(g) \longrightarrow 2N_2(g) + 6H_2O(g)$$
Ammonia

**a.** How many grams of $O_2$ are needed to react with 8.00 moles of $NH_3$?

**b.** How many grams of $N_2$ can be produced when 6.50 g of $O_2$ reacts?

**c.** How many grams of water are formed from the reaction of 34.0 g of $NH_3$?

**5.56** Iron(III) oxide reacts with carbon to give iron and carbon monoxide.

$$Fe_2O_3(s) + 3C(s) \longrightarrow 2Fe(s) + 3CO(g)$$

**a.** How many grams of C are required to react with 2.50 moles of $Fe_2O_3$?

**b.** How many grams of CO are produced when 36.0 g of C reacts?

**c.** How many grams of Fe can be produced when 6.00 g of $Fe_2O_3$ reacts?

**5.57** Nitrogen dioxide and water react to produce nitric acid, $HNO_3$, and nitrogen oxide.

$$3NO_2(g) + H_2O(l) \longrightarrow 2HNO_3(aq) + NO(g)$$

**a.** How many grams of $H_2O$ are required to react with 28.0 g of $NO_2$?

**b.** How many grams of NO are obtained from 15.8 g of $NO_2$?

**c.** How many grams of $HNO_3$ are produced from 8.25 g of $NO_2$?

**5.58** Calcium cyanamide reacts with water to form calcium carbonate and ammonia.

$$CaCN_2(s) + 3H_2O(l) \longrightarrow CaCO_3(s) + 2NH_3(g)$$

**a.** How many grams of water are needed to react with 75.0 g of $CaCN_2$?

**b.** How many grams of $NH_3$ are produced from 5.24 g of $CaCN_2$?

**c.** How many grams of $CaCO_3$ form if 155 g of water react?

**5.59** When the ore lead(II) sulfide burns in oxygen, the products are lead(II) oxide and sulfur dioxide.
**a.** Write the balanced equation for the reaction.
**b.** How many grams of oxygen are required to react with 0.125 mole of lead(II) sulfide?
**c.** How many grams of sulfur dioxide can be produced when 65.0 g of lead(II) sulfide react?
**d.** How many grams of lead(II) sulfide are used to produce 128 g of lead(II) oxide?

**5.60** When the gases dihydrogen sulfide and oxygen react, they form the gases sulfur dioxide and water vapor.
**a.** Write the balanced equation for the reaction.
**b.** How many grams of oxygen are required to react with 2.50 g of dihydrogen sulfide?
**c.** How many grams of sulfur dioxide can be produced when 38.5 g of oxygen react?
**d.** How many grams of oxygen are required to produce 55.8 g of water vapor?

# 5.9 ENERGY IN CHEMICAL REACTIONS

For a chemical reaction to take place, the molecules of the reactants must collide with each other and have the proper orientation and energy. Even when a collision has the proper orientation, there still must be sufficient energy to break the bonds of the reactants. The amount of energy needed to break apart those bonds is called the **activation energy**. If the energy of a collision is less than the activation energy, the molecules bounce apart without reacting. Many collisions occur, but only a few actually lead to the formation of product.

The concept of activation energy is analogous to climbing over a hill. To reach a destination on the other side, we must expend energy to climb to the top of the hill. Once we are at the top, we can easily run down the other side. The energy needed to get us from our starting point to the top of the hill would be the activation energy.

### Three Conditions Required for a Reaction to Occur

1. **Collision**  The reactants must collide.
2. **Orientation**  The reactants must align properly to break and form bonds.
3. **Energy**  The collision must provide the energy of activation.

## LEARNING GOAL

Describe exothermic and endothermic reactions and factors that affect the rate of a reaction.

the
Chemistry
place

**WEB TUTORIAL**
Activation Energy and Transition State

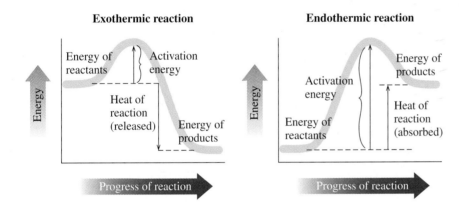

## Exothermic Reactions

The *heat of reaction* is the energy difference between the reactants and the products. In **exothermic reactions**, the energy of the products is lower than that of the reactants and heat is given off. For example, the combustion of methane ($CH_4$) is an exothermic reaction. Heat (213 kcal per mole of $CH_4$ burned) is released along with the products $CO_2$ and $H_2O$.

$$CH_4(g) + 2O_2(g) \longrightarrow CO_2(g) + 2H_2O(g) + 213 \, kcal$$
Methane                                      Heat released

## Endothermic Reactions

In **endothermic reactions**, the energy of the products is higher than the energy of the reactants and heat must be absorbed for products to form. For example, the formation of hydrogen iodide is an endothermic reaction. In the overall energy change, heat is absorbed when the hydrogen and iodine undergo reaction.

$$H_2(g) + I_2(g) + 12 \, kcal \ of \ heat \longrightarrow 2HI(g)$$
Heat absorbed

| Reaction | Energy Change | Heat in the Equation |
|---|---|---|
| Exothermic | Heat released | Product side |
| Endothermic | Heat absorbed | Reactant side |

# Health Note

## Hot Packs and Cold Packs

In a hospital, at a first-aid station, or at an athletic event, a *cold pack* may be used to reduce swelling from an injury, remove heat from inflammation, or decrease capillary size to lessen the effect of hemorrhaging. Inside the plastic container of a cold pack, there are two compartments, one containing solid ammonium nitrate ($NH_4NO_3$) and one with water. The pack is activated when it is hit or squeezed hard enough to break the walls between the compartments and cause the ammonium nitrate to mix with the water (shown as $H_2O$ over the reaction arrow). In an endothermic process, one mole of $NH_4NO_3$ that dissolves absorbs

6.3 kcal of heat from the water. The temperature drops, and the pack becomes cold and ready to use:

**Endothermic Reaction in a Cold Pack**

$$6.3 \text{ kcal} + NH_4NO_3(s) \xrightarrow{H_2O} NH_4NO_3(aq)$$

Hot packs are used to relax muscles, lessen aches and cramps, and increase circulation by expanding capillary size. Constructed in the same way as cold packs, a hot pack may contain the salt $CaCl_2$. The dissolving of the salt in water is exothermic and releases 18 kcal per mole of salt. The temperature rises, and the pack becomes hot and ready to use.

**Exothermic Reaction in a Hot Pack**

$$CaCl_2(s) \xrightarrow{H_2O} CaCl_2(aq) + 18 \text{ kcal}$$

---

## SAMPLE PROBLEM  5.14

### ■ Exothermic and Endothermic Reactions

In the reaction of one mole of solid carbon with oxygen gas, the energy of the carbon dioxide product is 94 kcal lower than the energy of the reactants.

**a.** Is the reaction exothermic or endothermic?
**b.** Write the balanced equation for the reaction, including the heat of the reaction.

#### SOLUTION

**a.** When the products have a lower energy than the reactants, the reaction is exothermic.

**b.** $C(s) + O_2(g) \longrightarrow CO_2(g) + 94 \text{ kcal}$

#### STUDY CHECK

When two moles of liquid ethanol ($C_2H_5OH$) react with oxygen gas ($O_2$), the products are carbon dioxide, water, and 326 kcal of heat. Write a balanced equation including the heat of reaction.

## Rate of Reaction

The *rate* (or speed) *of reaction* is measured by the amount of reactant used up or the amount of product formed in a certain period of time. Reactions with low activation energies go faster than reactions with high activation energies. The rate of a reaction can be affected by changes in the temperature, the amounts of reactants in the container, and the addition of catalysts.

## Temperature

At higher temperatures, the increase in kinetic energy of the reactants makes them move faster and collide more often, and it provides more collisions with the required energy of activation. Reactions almost always go faster at higher temperatures. For every 10 °C increase in temperature, most reaction rates approximately double. If we want food to cook faster, we raise the temperature. When body temperature rises, there is an increase in the pulse rate, rate of breathing, and metabolic rate. On the other hand, we slow down a reaction by lowering the temperature. In some cardiac surgeries, body temperature is lowered to 28 °C so the heart can be stopped and less oxygen is required by the brain. This is also the reason why some people have survived submersion in icy lakes for long periods of time. We refrigerate perishable foods to make them last longer.

## Concentrations of Reactants

The rate of a reaction also increases when reactants are added. Then there are more collisions between the reactants and the reaction goes faster. For example, a patient having difficulty breathing may be given a breathing mixture with a higher oxygen content than the atmosphere. The increase in the number of oxygen molecules in the lungs increases the rate at which oxygen combines with hemoglobin. The increased rate of oxygenation of the blood means that the patient can breathe more easily.

$$Hb \quad + \quad O_2 \quad \longrightarrow \quad HbO_2$$

Hemoglobin    Oxygen                Oxyhemoglobin

## Catalysts

Another way to speed up a reaction is to lower the energy of activation. This can be done by adding a **catalyst**. Earlier, we discussed the energy required to climb a hill. If instead we find a tunnel through the hill, we do not need as much energy to get to the other side. A catalyst acts by providing an alternate pathway with a lower energy requirement. As a result, more collisions form product successfully. Catalysts have found many uses in industry. In the production of margarine, the reaction of hydrogen with vegetable oils is normally very slow. However, when finely divided platinum is present as a catalyst, the reaction occurs rapidly. In the body, biocatalysts called enzymes make most metabolic reactions go at the rates necessary for proper cellular activity.

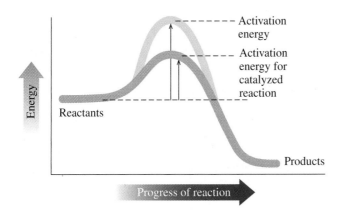

A summary of the factors affecting reaction rate is given in Table 5.8.

**TABLE 5.8** Factors That Increase Reaction Rate

| Factor | Reason |
| --- | --- |
| More reactants | More collisions |
| Higher temperature | More collisions with energy of activation |
| Adding a catalyst | Lowers energy of activation |

SAMPLE PROBLEM 5.15

■ Factors That Affect the Rate of Reaction

Indicate whether the following changes will increase, decrease, or have no effect upon the rate of reaction:

**a.** increase in temperature
**b.** decrease in the number of reactants
**c.** addition of a catalyst

SOLUTION

**a.** increase
**b.** decrease
**c.** increase

STUDY CHECK

How does lowering the temperature affect the rate of reaction?

## QUESTIONS AND PROBLEMS

### Energy in Chemical Reactions

**5.61 a.** Why do chemical reactions require energy of activation?
**b.** What is the function of a catalyst?
**c.** In an exothermic reaction, is the energy of the products higher or lower than the reactants?
**d.** Draw an energy diagram for an exothermic reaction.

**5.62 a.** What is measured by the heat of reaction?
**b.** How does the heat of reaction differ in exothermic and endothermic reactions?
**c.** In an endothermic reaction, is the energy of the products higher or lower than the reactants?
**d.** Draw an energy diagram for an endothermic reaction.

**5.63** Classify the following as exothermic or endothermic reactions:
**a.** 125 kcal is released.
**b.** In the energy diagram, the energy level of the products is higher than the reactants.
**c.** The metabolism of glucose in the body provides energy.

**5.64** Classify the following as exothermic or endothermic reactions:
**a.** In the energy diagram, the energy level of the products is lower than the reactants.
**b.** In the body, the synthesis of proteins requires energy.
**c.** 30 kcal is absorbed.

**5.65** Classify the following as exothermic or endothermic reactions:
**a.** gas burning in a Bunsen burner:

$$CH_4(g) + 2O_2(g) \longrightarrow CO_2(g) + 2H_2O(g) + 213 \text{ kcal}$$

**b.** dehydrating limestone:

$$Ca(OH)_2(s) + 15.6 \text{ kcal} \longrightarrow CaO(s) + H_2O(g)$$

**c.** formation of aluminum oxide and iron from aluminum and iron(III) oxide:

$$2Al(s) + Fe_2O_3(s) \longrightarrow Al_2O_3(s) + 2Fe(s) + 204 \text{ kcal}$$

**5.66** Classify the following as exothermic or endothermic reactions:
**a.** the combustion of propane:

$$C_3H_8(g) + 5O_2(g) \longrightarrow 3CO_2(g) + 4H_2O(g) + 531 \text{ kcal}$$

**b.** the formation of table salt:

$$2Na(s) + Cl_2(g) \longrightarrow 2NaCl(s) + 196 \text{ kcal}$$

**c.** decomposition of phosphorus pentachloride:

$$PCl_5(s) + 16 \text{ kcal} \longrightarrow PCl_3(s) + Cl_2(g)$$

**5.67 a.** What is meant by the rate of a reaction?
**b.** Why does bread grow mold more quickly at room temperature than in the refrigerator?

**5.68 a.** How does a catalyst affect the activation energy?
**b.** Why is pure oxygen used in respiratory distress?

**5.69** How would each of the following change the rate of the reaction shown here?

$$2SO_2(g) + O_2(g) \longrightarrow 2SO_3(g)$$

**a.** adding SO₂
**b.** raising the temperature
**c.** adding a catalyst
**d.** removing some SO₂

**5.70** How would each of the following change the rate of the reaction shown here?

$$2NO(g) + 2H_2(g) \longrightarrow N_2(g) + 2H_2O(g)$$

**a.** adding more NO
**b.** lowering the temperature
**c.** removing some H₂
**d.** adding a catalyst

## CONCEPT MAP

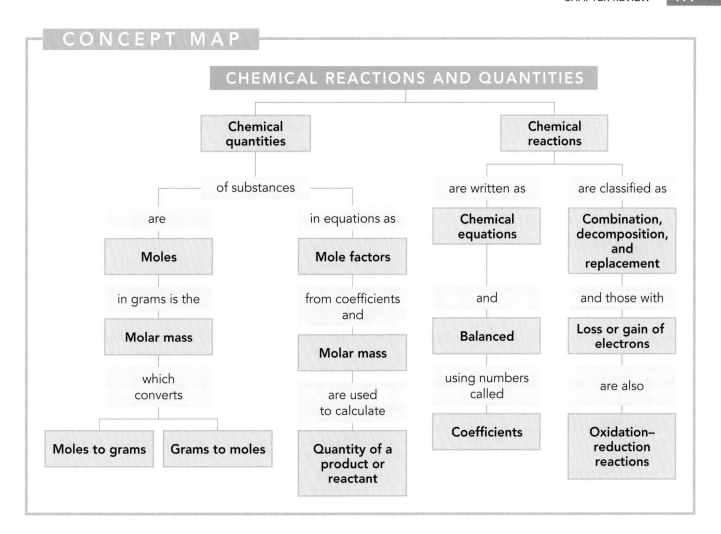

**CHEMICAL REACTIONS AND QUANTITIES**

Chemical quantities

of substances

are — **Moles**

in grams is the

**Molar mass**

which converts

**Moles to grams**    **Grams to moles**

in equations as — **Mole factors**

from coefficients and

**Molar mass**

are used to calculate

**Quantity of a product or reactant**

Chemical reactions

are written as — **Chemical equations**

and

**Balanced**

using numbers called

**Coefficients**

are classified as — **Combination, decomposition, and replacement**

and those with

**Loss or gain of electrons**

are also

**Oxidation–reduction reactions**

# CHAPTER REVIEW

## 5.1 The Mole

**Learning Goal:** Use Avogadro's number to determine the number of particles in a given number of moles.

One mole of an element contains $6.02 \times 10^{23}$ atoms; a mole of a compound contains $6.02 \times 10^{23}$ molecules or formula units.

## 5.2 Molar Mass

**Learning Goals:** Determine the molar mass of a substance, and use molar mass to convert between grams and moles.

The molar mass (g/mole) of any substance is the mass in grams equal numerically to its atomic mass, or the sum of the atomic masses, which have been multiplied by their subscripts in a formula. It becomes a conversion factor when it is used to change a quantity in grams to moles or to change a given number of moles to grams.

## 5.3 Chemical Changes

**Learning Goal:** Identify a change in a substance as a chemical or a physical change.

A chemical change occurs when the atoms of the initial substances rearrange to form new substances. When new substances form, a chemical reaction has taken place. In a physical change, the substance is the same, but its size, shape, or state changes.

## 5.4 Chemical Equations

**Learning Goal:** Write a balanced chemical equation from the formulas of the reactants and products for a reaction.

A chemical equation shows the formulas of the substances that react on the left side of a reaction arrow and the products that form on the right side of the reaction arrow. An equation is balanced by writing the smallest whole numbers (coefficients) in front of formulas to equalize the atoms of each element in the reactants and the products.

## 5.5 Types of Reactions

**Learning Goal:** Identify a reaction as a combination, decomposition, or replacement.

Many chemical reactions can be organized by reaction type: combination, decomposition, single replacement, or double replacement.

## 5.6 Oxidation–Reduction Reactions

**Learning Goal:** Define the terms *oxidation* and *reduction*.

When electrons are transferred in a reaction, it is an oxidation–reduction reaction. One reactant loses electrons, and another reactant gains electrons. Overall, the number of electrons lost and gained is equal.

## 5.7 Mole Relationships in Chemical Equations

**Learning Goal:** Given a quantity in moles of reactant or product, use a mole–mole factor from the balanced equation to calculate the moles of another substance in the reaction.

In a balanced equation, the total mass of the reactants is equal to the total mass of the products. The coefficients in an equation describing the relationship between the moles of any two components are used to write mole–mole factors. When the number of moles for one substance is known, a mole–mole factor is used to find the moles of a different substance in the reaction.

### 5.8 Mass Calculations for Reactions

**Learning Goal:** Given the mass in grams of a substance in a reaction, calculate the mass in grams of another substance in the reaction.

In calculations using equations, the molar masses of the substances and their mole factors are used to change the number of grams of one substance to the corresponding grams of a different substance.

### 5.9 Energy in Chemical Reactions

**Learning Goal:** Describe exothermic and endothermic reactions and factors that affect the rate of a reaction.

In a reaction, the reacting particles must collide with energy equal to or greater than the energy of activation. The heat of reaction is the energy difference between the initial energy of the reactants and the final energy of the products. In exothermic reactions, heat is released. In endothermic reactions, heat is absorbed. The rate of a reaction, which is the speed at which the reactants are converted to products, can be increased by adding more reactants, raising the temperature, or adding a catalyst.

## KEY TERMS

**activation energy** The energy needed upon collision to break apart the bonds of the reacting molecules.

**Avogadro's number** The number of items in a mole, equal to $6.02 \times 10^{23}$.

**balanced equation** The final form of a chemical equation that shows the same number of atoms of each element in the reactants and products.

**catalyst** A substance that increases the rate of reaction by lowering the activation energy.

**chemical change** The formation of a new substance with a different composition and different properties than the initial substance.

**chemical equation** A shorthand way to represent a chemical reaction using chemical formulas to indicate the reactants and products and coefficients to show reacting ratios.

**chemical reaction** The process by which a chemical change takes place.

**coefficients** Whole numbers placed in front of the formulas to balance the number of atoms or moles of atoms of each element on both sides of an equation.

**combination reaction** A reaction in which reactants combine to form a single product.

**decomposition reaction** A reaction in which a single reactant splits into two or more simpler substances.

**double replacement reaction** A reaction in which parts of two different reactants exchange places.

**endothermic reaction** A reaction that requires heat; the energy of the products is higher than the energy of the reactants.

**exothermic reaction** A reaction that releases heat; the energy of the products is lower than the energy of the reactants.

**formula unit** The group of ions represented by the formula of an ionic compound.

**molar mass** The mass in grams of 1 mole of an element equal numerically to its atomic mass. The molar mass of a compound is equal to the sum of the masses of the elements multiplied by their subscripts in the formula.

**mole** A group of atoms, molecules, or formula units that contains $6.02 \times 10^{23}$ of these items.

**mole–mole factor** A conversion factor that relates the number of moles of two compounds derived from the coefficients in an equation.

**oxidation** The loss of electrons by a substance. Biological oxidation may involve the addition of oxygen or the loss of hydrogen.

**oxidation–reduction reaction** A reaction in which the oxidation of one reactant is always accompanied by the reduction of another reactant.

**physical change** A change in which the physical appearance of a substance changes but the chemical composition stays the same.

**products** The substances formed as a result of a chemical reaction.

**reactants** The initial substances that undergo change in a chemical reaction.

**reduction** The gain of electrons by a substance. Biological reduction may involve the loss of oxygen or the gain of hydrogen.

**single replacement reaction** A reaction in which an element replaces a different element in a compound.

## UNDERSTANDING THE CONCEPTS

**5.71** Using the following models of the molecules (black = C, light blue = H, yellow = S, green = Cl), determine each of the following.

**1.**     **2.**

   **a.** formula
   **b.** molar mass
   **c.** number of moles in 10.0 g

**5.72** Using the following models of the molecules, (black = C, light blue = H, yellow = S, red = O), determine each of the following:

**1.**    **2.**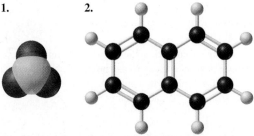

   **a.** formula    **b.** molar mass    **c.** number of moles in 10.0 g

**5.73** A dandruff shampoo contains pyrithion, $C_{10}H_8N_2O_2S_2$, an antibacterial and antifungal agent.

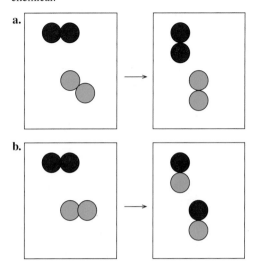

    **a.** What is the molar mass of pyrithion?
    **b.** How many moles of pyrithion are in 25.0 g?
    **c.** How many moles of carbon are in 25.0 g of pyrithion?
    **d.** How many moles of pyrithion contain $8.2 \times 10^{24}$ atoms of nitrogen?

**5.74** Ibuprofen is an anti-inflammatory with the formula $C_{13}H_{18}O_2$.

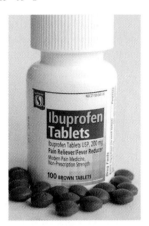

    **a.** What is the molar mass of ibuprofen?
    **b.** How many grams of ibuprofen are in 0.525 mole?
    **c.** How many moles of carbon are in 12.0 g of ibuprofen?
    **d.** How many moles of ibuprofen contain $1.22 \times 10^{23}$ atoms of carbon?

**5.75** Balance each of the following by adding coefficients, and identify the type of reaction for each:

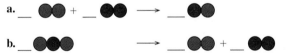

**5.76** Balance each of the following by adding coefficients, and identify the type of reaction for each:

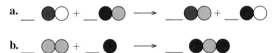

**5.77** Identify some of the physical and chemical changes of a burning wax candle.

**5.78** In each of the following, identify the change as physical or chemical:

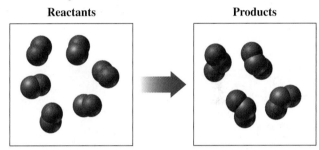

**5.79** If red spheres represent oxygen atoms and blue spheres represent nitrogen atoms,

**Reactants**                **Products**

    **a.** Write a balanced equation for the reaction.
    **b.** Indicate the type of reaction as decomposition, combination, single replacement, or double replacement.

**5.80** If purple spheres represent iodine atoms and light blue spheres represent hydrogen atoms,

**Reactants**                **Products**

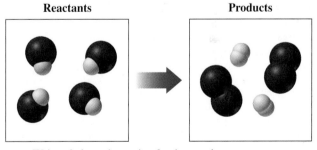

    **a.** Write a balanced equation for the reaction.
    **b.** Indicate the type of reaction as decomposition, combination, single replacement, or double replacement.

**5.81** If blue spheres represent nitrogen atoms and purple spheres represent iodine atoms,

**Reactants**                **Products**

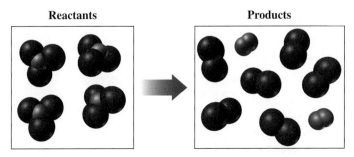

a. Write a balanced equation for the reaction.
b. Indicate the type of reaction as decomposition, combination, single replacement, or double replacement.

**5.82** If green spheres represent chlorine atoms, yellow-green spheres represent fluorine, and light blue spheres represent hydrogen atoms,

| Reactants | Products |
|---|---|

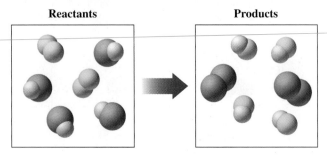

a. Write a balanced equation for the reaction.
b. Indicate the type of reaction as decomposition, combination, single replacement, or double replacement.

**5.83** If green spheres represent chlorine atoms and red spheres represent oxygen,

| Reactants | Products |
|---|---|

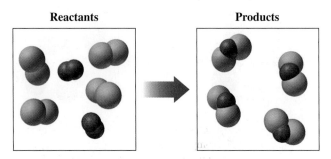

a. Write a balanced equation for the reaction.
b. Indicate the type of reaction as decomposition, combination, single replacement, or double replacement.

**5.84** If blue spheres represent nitrogen atoms and purple spheres represent iodine atoms,

| Reactants | Products |
|---|---|

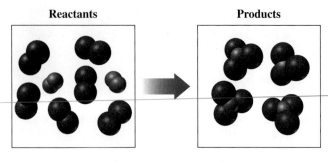

a. Write a balanced equation for the reaction.
b. Indicate the type of reaction as decomposition, combination, single replacement, or double replacement.

**5.85** Propane gas, $C_3H_8$, is used as a fuel for many barbecues.
a. How many grams of propane gas are in 1.50 moles of propane?
b. How many moles of propane gas are in 34.0 g of propane?
c. How many grams of carbon are in 34.0 g of propane?

**5.86** Allyl sulfide $(C_3H_5)_2S$ is the substance that gives garlic its characteristic odor.
a. How many moles of allyl sulfide are in 23.2 g?
b. How many hydrogen atoms are in 0.85 mole of allyl sulfide?
c. How many atoms of carbon are in $4.20 \times 10^{23}$ molecules of allyl sulfide?

# ADDITIONAL QUESTIONS AND PROBLEMS

**5.87** Calculate the molar mass of each of the following:
a. $FeSO_4$, ferrous sulfate, iron supplement
b. $Ca(IO_3)_2$, calcium iodate, iodine source in table salt
c. $C_5H_8NNaO_4$, monosodium glutamate, flavor enhancer
d. $C_6H_{12}O_2$, isoamyl formate used to make artificial fruit syrups

**5.88** Calculate the molar mass of each of the following:
a. $Mg(HCO_3)_2$, magnesium hydrogen carbonate
b. $Au(OH)_3$, gold(III) hydroxide, used in gold plating
c. $C_{18}H_{34}O_2$, oleic acid from olive oil
d. $C_{21}H_{26}O_5$, prednisone, anti-inflammatory

**5.89** How many grams are in 0.150 mole of each of the following?
a. K          b. $Cl_2$          c. $Na_2CO_3$

**5.90** How many grams are in 2.25 moles of each of the following?
a. $N_2$          b. NaBr          c. $C_6H_{14}$

**5.91** How many moles are in 25.0 g of each of the following compounds?
a. $CO_2$          b. $Al(OH)_3$          c. $MgCl_2$

**5.92** How many moles are in 4.00 g of each of the following compounds?
a. $NH_3$          b. $Ca(NO_3)_2$          c. $SO_3$

**5.93** During heavy exercise and workouts, lactic acid, $C_3H_6O_3$, accumulates in the muscles where it can cause pain and soreness.
a. How many molecules are in 0.500 mole of lactic acid?
b. How many C atoms are in 1.50 moles of lactic acid?
c. How many moles of lactic acid contain $4.5 \times 10^{24}$ atoms of O?
d. What is the molar mass of lactic acid?

**5.94** Ammonium sulfate, $(NH_4)_2SO_4$, is used in fertilizers to provide nitrogen for the soil.
a. How many formula units are in 0.200 mole of ammonium sulfate?
b. How many H atoms are in 0.100 mole of ammonium sulfate?
c. How many moles of ammonium sulfate contain $7.4 \times 10^{25}$ atoms of N?
d. What is the molar mass of ammonium sulfate?

**5.95** Balance each of the following unbalanced equations, and identify the type of reaction.

**a.** $NH_3(g) + HCl(g) \longrightarrow NH_4Cl(s)$

**b.** $Fe_3O_4(s) + H_2(g) \longrightarrow Fe(s) + H_2O(g)$

**c.** $Sb(s) + Cl_2(g) \longrightarrow SbCl_3(s)$

**d.** $NI_3(s) \longrightarrow N_2(g) + I_2(g)$

**e.** $KBr(aq) + Cl_2(aq) \longrightarrow KCl(aq) + Br_2(l)$

**f.** $Fe(s) + H_2SO_4(aq) \longrightarrow Fe_2(SO_4)_3(aq) + H_2(g)$

**g.** $Al_2(SO_4)_3(aq) + NaOH(aq) \longrightarrow Na_2SO_4(aq) + Al(OH)_3(s)$

**5.96** Balance each of the following unbalanced equations, and identify the type of reaction.

**a.** $Li_3N(s) \longrightarrow Li(s) + N_2(g)$

**b.** $Mg(s) + N_2(g) \longrightarrow Mg_3N_2(s)$

**c.** $Al(s) + HCl(aq) \longrightarrow AlCl_3(aq) + H_2(g)$

**d.** $Mg(s) + H_3PO_4(aq) \longrightarrow Mg_3(PO_4)_2(s) + H_2(g)$

**e.** $Cr_2O_3(s) + H_2(g) \longrightarrow Cr(s) + H_2O(g)$

**f.** $Al(s) + Cl_2(g) \longrightarrow AlCl_3(s)$

**g.** $MgCl_2(aq) + AgNO_3(aq) \longrightarrow Mg(NO_3)_2(aq) + AgCl(s)$

**5.97** Identify each of the following as an oxidation or a reduction reaction.

**a.** $Zn^{2+} + 2e^- \longrightarrow Zn$    **b.** $Al \longrightarrow Al^{3+} + 3e^-$

**c.** $Pb \longrightarrow Pb^{2+} + 2e^-$    **d.** $Cl_2 + 2e^- \longrightarrow 2Cl^-$

**5.98** Write a balanced chemical equation for each of the following oxidation–reduction reactions:

**a.** Sulfur reacts with molecular chlorine to form sulfur dichloride.

**b.** Molecular chlorine and sodium bromide react to form molecular bromine and sodium chloride.

**c.** Aluminum metal and iron(III) oxide react to produce aluminum oxide and elemental iron.

**d.** Copper(II) oxide reacts with carbon to form elemental copper and carbon dioxide.

**5.99** At a winery, glucose ($C_6H_{12}O_6$) in grapes undergoes fermentation to produce ethanol ($C_2H_6O$) and carbon dioxide.

$$C_6H_{12}O_6 \longrightarrow 2C_2H_6O + 2CO_2$$
$$\text{Glucose} \qquad \text{Ethanol}$$

**a.** How many moles of glucose are required to form 124 g of ethanol?

**b.** How many grams of ethanol would be formed from the reaction of 0.240 kg of glucose?

**5.100** Gasohol is a fuel containing ethanol ($C_2H_6O$) that burns in oxygen ($O_2$) to give carbon dioxide and water.

**a.** State the reactants and products for this reaction in the form of a balanced equation.

**b.** How many moles of $O_2$ are needed to completely react with 4.0 moles of $C_2H_6O$?

**c.** If a car produces 88 g of $CO_2$, how many grams of $O_2$ are used up in the reaction?

**d.** If you add 125 g of $C_2H_6O$ to your gas, how many grams of $CO_2$ and $H_2O$ can be produced?

**5.101** Balance the following equation:

$$NH_3(g) + F_2(g) \longrightarrow N_2F_4(g) + HF(g)$$

**a.** How many moles of each reactant are needed to produce 4.00 moles of HF?

**b.** How many grams of $F_2$ are required to react with 1.50 moles of $NH_3$?

**c.** How many grams of $N_2F_4$ can be produced when 3.40 g of $NH_3$ reacts?

**5.102** When nitrogen dioxide ($NO_2$) from car exhaust combines with water in the air, it forms nitric acid ($HNO_3$), which causes acid rain and nitrogen oxide.

$$3NO_2(g) + H_2O(l) \longrightarrow 2HNO_3(aq) + NO(g)$$

**a.** How many molecules of $NO_2$ are needed to react with 0.250 mole of $H_2O$?

**b.** How many grams of $HNO_3$ are produced when 60.0 g of $NO_2$ completely reacts?

**5.103** Pentane gas, $C_5H_{12}$, reacts with oxygen to produce carbon dioxide and water.

$$C_5H_{12}(g) + 8O_2(g) \longrightarrow 5CO_2(g) + 6H_2O(g)$$

**a.** How many grams of pentane must react to produce 4.0 moles of water?

**b.** How many grams of $CO_2$ are produced from 32.0 g of oxygen?

**5.104** Propane gas, $C_3H_8$, reacts with oxygen to produce water and carbon dioxide. Propane has a density of 2.02 g/L at room temperature.

$$C_3H_8(g) + 5O_2(g) \longrightarrow 3CO_2(g) + 4H_2O(l)$$
$$\text{Propane}$$

**a.** How many moles of water form when 5.00 L of propane gas ($C_3H_8$) completely react?

**b.** How many grams of $CO_2$ are produced from 18.5 g of oxygen gas?

**c.** How many grams of $H_2O$ can be produced from the reaction of $8.50 \times 10^{22}$ molecules of propane gas, $C_3H_8$?

**5.105** The equation for the formation of silicon tetrachloride from silicon and chlorine is

$$Si(s) + 2Cl_2(g) \longrightarrow SiCl_4(g) + 157 \text{ kcal}$$

**a.** Is the formation of $SiCl_4$ an endothermic or exothermic reaction?

**b.** Is the energy of the products higher or lower than the energy of the reactants?

**5.106** The equation for the formation of nitrogen monoxide is

$$N_2(g) + O_2(g) + 90.2 \text{ kJ} \longrightarrow 2NO(g)$$

**a.** Is the formation of NO an endothermic or exothermic reaction?

**b.** Is the energy of the product higher or lower than the energy of the reactants?

# CHALLENGE QUESTIONS

**5.107** A gold bar is 2.31 cm long, 1.48 cm wide, and 0.0758 cm thick.

**a.** If gold has a density of 19.3 g/mL, what is the mass of the gold bar?

**b.** How many atoms of gold are in the bar?

**c.** When the same mass of gold combines with oxygen, the oxide product has a mass of 5.61 g. How many moles of oxygen atoms are combined with the gold?

**d.** What is the formula of the oxide product?

**5.108** A toothpaste contains 0.24% by mass sodium fluoride used to prevent dental caries and 0.30% by mass triclosan $C_{12}H_7Cl_3O_2$,

a preservative and antigingivitis agent. One tube contains 119 g of toothpaste.
   **a.** How many moles of NaF are in the tube of toothpaste?
   **b.** How many fluoride atoms $F^-$ are in the tube of toothpaste?
   **c.** How many grams of sodium ion $Na^+$ are in 1.50 g of toothpaste?
   **d.** How many molecules of triclosan are in the tube of toothpaste?

**5.109** Write a balanced equation for each of the following reaction descriptions, and identify each type of reaction.
   **a.** An aqueous solution of lead(II) nitrate is mixed with aqueous sodium phosphate to produce solid lead(II) phosphate and aqueous sodium nitrate.
   **b.** Gallium metal heated in oxygen gas forms solid gallium(III) oxide.
   **c.** When solid sodium nitrate is heated, solid sodium nitrite and oxygen gas are produced.
   **d.** Solid bismuth(III) oxide and solid carbon react to form bismuth metal and carbon monoxide gas.

**5.110** The gaseous hydrocarbon acetylene, $C_2H_2$, used in welders' torches, releases a large amount of heat when it burns according to the following equation:

$$2C_2H_2(g) + 5O_2(g) \longrightarrow 4CO_2(g) + 2H_2O(g)$$

   **a.** How many moles of water are produced from the complete reaction of 64.0 g of oxygen?

   **b.** How many moles of oxygen are needed to react completely with $2.25 \times 10^{24}$ molecules of acetylene?
   **c.** How many grams of carbon dioxide are produced from the complete reaction of 78.0 g of acetylene?

**5.111** Consider the following *unbalanced* equation:

$$Al(s) + O_2(g) \longrightarrow Al_2O_3(s)$$

   **a.** Balance the equation.
   **b.** Identify the type of reaction.
   **c.** How many moles of oxygen must react with 4.50 moles of aluminum?
   **d.** How many grams of aluminum oxide are produced when 50.2 g of aluminum react?
   **e.** When aluminum is reacted in a closed container with 8.00 g of oxygen, how many grams of aluminum oxide can form?

**5.112** Methanol ($CH_3OH$), which is used as a cooking fuel, burns with oxygen gas ($O_2$) to produce carbon dioxide and water. The reaction produces 363 kJ of heat per mole of methanol.
   **a.** Write a balanced equation for the reaction, including the heat of reaction.
   **b.** Is the reaction endothermic or exothermic?
   **c.** How many moles of oxygen must react with 25.0 g of methanol?
   **d.** How many kilocalories are released when 75.0 g of methanol burns?

# ANSWERS

## Answers to Study Checks

**5.1**  0.432 mole of $H_2O$

**5.2**  0.120 mole of aspirin

**5.3**  138.1 g of salicylic acid

**5.4**  24.4 g of Au

**5.5**  0.00621 mole of $CaCO_3$, 0.00550 mole of $MgCO_3$

**5.6**  The production of heat, light, smoke, and ash is evidence of a chemical reaction.

**5.7**  $Sb_2S_3(s) + 6HCl(aq) \longrightarrow 2SbCl_3(s) + 3H_2S(g)$

**5.8**  $2NO(g) + O_2(g) \longrightarrow 2NO_2(g)$ Combination reaction

**5.9**  Lithium is oxidized: $2Li \longrightarrow 2Li^+ + 2e^-$
    Fluorine is reduced: $F_2 + 2e^- \longrightarrow 2F^-$

**5.10**  1.8 moles of Fe

**5.11**  0.945 mole of $O_2$

**5.12**  44.0 g of NO

**5.13**  27.5 g of $CO_2$

**5.14**  $C_2H_5OH(l) + 3O_2(g) \longrightarrow 2CO_2(g) + 3H_2O(g) + 163$ kcal

**5.15**  The rate of reaction will decrease because the number of collisions between reacting particles will be less, and a smaller number of the collisions that do occur will have sufficient activation energy.

## Answers to Selected Questions and Problems

**5.1**  One mole contains $6.02 \times 10^{23}$ atoms of an element, molecules of a covalent substance, or formula units of an ionic substance.

**5.3**  **a.** $3.01 \times 10^{23}$ C atoms
    **b.** $7.71 \times 10^{23}$ $SO_2$ molecules
    **c.** 0.0867 mole of Fe
    **d.** 14.1 moles of $C_2H_5OH$

**5.5**  **a.** 6.00 moles of H
    **b.** 8.00 moles of O
    **c.** $1.20 \times 10^{24}$ P atoms
    **d.** $4.82 \times 10^{24}$ O atoms

**5.7**  **a.** 24 moles of H
    **b.** $1.0 \times 10^2$ moles of C
    **c.** 0.040 mole of N

**5.9**  **a.** 58.5 g          **b.** 159.8 g          **c.** 73.8 g
    **d.** 342.3 g          **e.** 58.3 g          **f.** 365.1 g

**5.11**  **a.** 46.0 g          **b.** 112 g
    **c.** 14.8 g          **d.** 112 g

**5.13**  **a.** 29.3 g          **b.** 109 g
    **c.** 4.05 g          **d.** 194 g

**5.15**  **a.** 602 g          **b.** 11 g

**5.17**  **a.** 0.463 mole          **b.** 0.0167 mole
    **c.** 0.882 mole          **d.** 1.17 moles

**5.19**  **a.** 1.24 moles of Ne          **b.** 0.781 mole of $O_2$
    **c.** 0.321 mole of $Al(OH)_3$          **d.** 0.106 mole of $Ga_2S_3$

**5.21** **a.** 0.78 mole of S    **b.** 1.95 moles of S
**c.** 6.0 moles of S

**5.23** **a.** physical    **b.** chemical    **c.** physical
**d.** chemical    **e.** physical    **f.** chemical

**5.25** **a.** not balanced    **b.** balanced
**c.** not balanced    **d.** balanced

**5.27** **a.** $N_2(g) + O_2(g) \longrightarrow 2NO(g)$
**b.** $2HgO(s) \longrightarrow 2Hg(l) + O_2(g)$
**c.** $4Fe(s) + 3O_2(g) \longrightarrow 2Fe_2O_3(s)$
**d.** $2Na(s) + Cl_2(g) \longrightarrow 2NaCl(s)$

**5.29** **a.** $Mg(s) + 2AgNO_3(aq) \longrightarrow Mg(NO_3)_2(aq) + 2Ag(s)$
**b.** $2Al(s) + 3CuSO_4(aq) \longrightarrow 3Cu(s) + Al_2(SO_4)_3(aq)$
**c.** $Pb(NO_3)_2(aq) + 2NaCl(aq) \longrightarrow PbCl_2(s) + 2NaNO_3(aq)$
**d.** $2Al(s) + 6HCl(aq) \longrightarrow 2AlCl_3(aq) + 3H_2(g)$

**5.31** **a.** A single reactant splits into two simpler substances (elements).
**b.** One element in the reacting compound is replaced by the other reactant.

**5.33** **a.** combination    **b.** single replacement
**c.** decomposition    **d.** double replacement
**e.** decomposition    **f.** double replacement
**g.** combination

**5.35** **a.** $Mg(s) + Cl_2(g) \longrightarrow MgCl_2(s)$
**b.** $2HBr(g) \longrightarrow H_2(g) + Br_2(g)$
**c.** $Mg(s) + Zn(NO_3)_2(aq) \longrightarrow Zn(s) + Mg(NO_3)_2(aq)$
**d.** $K_2S(aq) + Pb(NO_3)_2(aq) \longrightarrow 2KNO_3(aq) + PbS(s)$

**5.37** **a.** reduction    **b.** oxidation
**c.** reduction    **d.** reduction

**5.39** **a.** Zn is oxidized, $Cl_2$ is reduced.
**b.** $Br^-$ in NaBr is oxidized, $Cl_2$ is reduced.
**c.** The $O^{2-}$ in PbO is oxidized, the $Pb^{2+}$ is reduced.
**d.** $Sn^{2+}$ is oxidized, $Fe^{3+}$ is reduced.

**5.41** **a.** reduction    **b.** oxidation

**5.43** Linoleic acid gains hydrogen atoms and is reduced.

**5.45** **a.** (1) Two molecules of sulfur dioxide react with one molecule of oxygen to produce two molecules of sulfur trioxide.
(2) Two moles of sulfur dioxide react with one mole of oxygen to produce two moles of sulfur trioxide.
**b.** (1) Four atoms of phosphorus react with five molecules of oxygen to produce two molecules of diphosphorus pentoxide.
(2) Four moles of phosphorus react with five moles of oxygen to produce two moles of diphosphorus pentoxide.

**5.47** **a.** $\dfrac{2 \text{ moles } SO_2}{1 \text{ mole } O_2}$ and $\dfrac{1 \text{ mole } O_2}{2 \text{ moles } SO_2}$

$\dfrac{2 \text{ moles } SO_2}{2 \text{ moles } SO_3}$ and $\dfrac{2 \text{ moles } SO_3}{2 \text{ moles } SO_2}$

$\dfrac{2 \text{ moles } SO_3}{1 \text{ mole } O_2}$ and $\dfrac{1 \text{ mole } O_2}{2 \text{ moles } SO_3}$

**b.** $\dfrac{4 \text{ moles } P}{5 \text{ moles } O_2}$ and $\dfrac{5 \text{ moles } O_2}{4 \text{ moles } P}$

$\dfrac{4 \text{ moles } P}{2 \text{ moles } P_2O_5}$ and $\dfrac{2 \text{ moles } P_2O_5}{4 \text{ moles } P}$

$\dfrac{5 \text{ moles } O_2}{2 \text{ moles } P_2O_5}$ and $\dfrac{2 \text{ moles } P_2O_5}{5 \text{ moles } O_2}$

**5.49** **a.** 1.0 mole of $O_2$    **b.** 10. moles of $H_2$
**c.** 5.0 moles of $H_2O$

**5.51** **a.** 1.25 moles of C    **b.** 0.96 mole of CO
**c.** 1.0 mole of $SO_2$    **d.** 0.50 mole of $CS_2$

**5.53** **a.** 77.5 g of $Na_2O$    **b.** 6.26 g of $O_2$
**c.** 19.4 g of $O_2$

**5.55** **a.** 192 g of $O_2$    **b.** 3.79 g of $N_2$
**c.** 54.0 g of $H_2O$

**5.57** **a.** 3.65 g of $H_2O$    **b.** 3.43 g of NO
**c.** 7.53 g of $HNO_3$

**5.59** **a.** $2PbS(s) + 3O_2(g) \longrightarrow 2PbO(s) + 2SO_2(g)$
**b.** 6.00 g of $O_2$    **c.** 17.4 g of $SO_2$    **d.** 137 g of PbS

**5.61** **a.** The energy of activation is the energy required to break the bonds of the reacting molecules.
**b.** A catalyst provides a pathway that lowers the activation energy and speeds up a reaction.
**c.** In exothermic reactions, the energy of the products is lower than the reactants.
**d.**

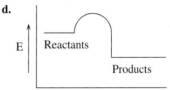

**5.63** **a.** exothermic    **b.** endothermic    **c.** exothermic

**5.65** **a.** exothermic    **b.** endothermic    **c.** exothermic

**5.67** **a.** The rate of a reaction tells how fast the products are formed or how fast the reactants are consumed.
**b.** Reactions go faster at higher temperatures.

**5.69** **a.** increase    **b.** increase
**c.** increase    **d.** decrease

**5.71** **1. a.** $S_2Cl_2$    **b.** 135.2 g/mole    **c.** 0.0740 mole
**2. a.** $C_6H_6$    **b.** 78.0 g/mole    **c.** 0.0128 mole

**5.73** **a.** 252.2 g/mole    **b.** 0.0991 mole of pyrithion
**c.** 0.991 mole of C    **d.** 6.8 moles of pyrithion

**5.75** **a.** 1,1,2 combination    **b.** 2,2,1 decomposition

**5.77** physical: solid candle wax melts (changes state), candle height is shorter, melted wax turns solid (changes state), shape of the wax changes, the wick becomes shorter, and others

chemical: wax burns in oxygen, heat and light are emitted, and wick burns in the presence of oxygen

**5.79** **a.** $2NO(g) + O_2(g) \longrightarrow 2NO_2(g)$
**b.** combination

**5.81** **a.** $2NI_3(s) \longrightarrow N_2(g) + 3I_2(g)$
**b.** decomposition

**5.83** **a.** $2Cl_2(g) + O_2(g) \longrightarrow 2OCl_2(g)$
**b.** combination

**5.85** **a.** 66.0 g of propane    **b.** 0.773 mole of propane
**c.** 27.8 g of carbon

**5.87** **a.** 152.0 g    **b.** 389.9 g
**c.** 169.0 g    **d.** 116.0 g

**5.89** **a.** 5.87 g      **b.** 10.7 g      **c.** 15.9 g

**5.91** **a.** 0.568 mole      **b.** 0.321 mole      **c.** 0.262 mole

**5.93** **a.** $3.01 \times 10^{23}$ molecules      **b.** $2.71 \times 10^{24}$ C atoms
**c.** 2.5 moles of lactic acid      **d.** 90.0 g

**5.95** **a.** $NH_3(g) + HCl(g) \longrightarrow NH_4Cl(s)$ combination
**b.** $Fe_3O_4(s) + 4H_2(g) \longrightarrow 3Fe(s) + 4H_2O(g)$ single replacement
**c.** $2Sb(s) + 3Cl_2(g) \longrightarrow 2SbCl_3(s)$ combination
**d.** $2NI_3(s) \longrightarrow N_2(g) + 3I_2(g)$ decomposition
**e.** $2KBr(aq) + Cl_2(aq) \longrightarrow 2KCl(aq) + Br_2(l)$ single replacement
**f.** $2Fe(s) + 3H_2SO_4(aq) \longrightarrow Fe_2(SO_4)_3(aq) + 3H_2(g)$ single replacement
**g.** $Al_2(SO_4)_3(aq) + 6NaOH(aq) \longrightarrow$
         $3Na_2SO_4(aq) + 2Al(OH)_3(s)$ double replacement

**5.97** **a.** reduction      **b.** oxidation
**c.** oxidation      **d.** reduction

**5.99** **a.** 1.35 moles of glucose      **b.** 123 g of ethanol

**5.101** $2NH_3 + 5F_2 \longrightarrow N_2F_4 + 6HF$
**a.** 1.33 moles of $NH_3$ and 3.33 moles of $F_2$
**b.** 143 g of $F_2$      **c.** 10.4 g of $N_2F_4$

**5.103** **a.** 48 g of pentane
**b.** 27.5 g of $CO_2$

**5.105** **a.** exothermic      **b.** lower

**5.107** **a.** 5.00 g of gold      **b.** $1.52 \times 10^{22}$ Au atoms
**c.** 0.038 moles      **d.** $Au_2O_3$

**5.109** **a.** $3Pb(NO_3)_2(aq) + 2Na_3PO_4(aq) \longrightarrow$
         $Pb_3(PO_4)_2(s) + 6NaNO_3(aq)$ double replacement
**b.** $4Ga(s) + 3O_2(g) \longrightarrow 2Ga_2O_3(s)$ combination
**c.** $2NaNO_3(s) \longrightarrow 2NaNO_2(s) + O_2(g)$ decomposition
**d.** $Bi_2O_3(s) + 3C(s) \longrightarrow 2Bi(s) + 3CO(g)$ single replacement

**5.111** **a.** $4Al(s) + 3O_2(g) \longrightarrow 2Al_2O_3(s)$
**b.** combination      **c.** 3.38 moles of oxygen
**d.** 94.9 g of $Al_2O_3$      **e.** 17.0 g of $Al_2O_3$

# Combining Ideas from Chapters 3 to 5

**CI.7** The following reaction occurs between a metal and a nonmetal.

 +  +  ⟶  +

**X      Y      Y**

a. Which spheres represent a metal? A nonmetal?
b. Which reactant has the higher electronegativity?
c. What are the ionic charges of X and Y in the product?
d. If these elements are both in period 3,
 1. Write the electron arrangement of the atoms.
 2. Write the electron arrangement of their ions.
 3. Give the names of the noble gases with the same electron arrangement as these ions.
 4. Write the formula and name of the product.
e. Match the spheres below with atoms of Li, Na, K, and Rb.

**A.      B.      C.      D.**

**CI.8** A gold bar has a volume of 728 cm³ and a density of 19.3 g/cm³.
a. What is the mass in kilograms of the gold bar?
b. How many atoms of gold are in the bar?
c. Give the protons and neutrons in each of the following isotopes of gold:

$$^{185}_{79}\text{Au} \quad ^{197}_{79}\text{Au} \quad ^{198}_{79}\text{Au}$$

Au-198 is used in liver imaging and cancer therapy.

**CI.9** Ethanol, $C_2H_5OH$, is obtained from renewable crops such as corn that use the sun as their source of energy. In the United States, automobiles can now use a fuel known as E10 that contains 10.0% ethanol and 90% unleaded gasoline by volume (%v/v). Ethanol has a melting point of −115 °C, a boiling point of 78 °C, a heat of fusion of 23.6 cal/g, and a heat of vaporization of 201 cal/g. Liquid ethanol has a density of 0.796 g/mL and a specific heat of 0.588 cal/g °C.
a. Draw a heating curve for ethanol from −150 °C to 100 °C.

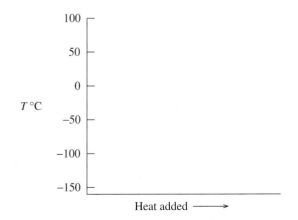

*T* °C

Heat added ⟶

b. When 20.0 g of ethanol at −55 °C is heated to 37 °C, how much energy in calories is required?
c. If the density of ethanol is 0.796 g/mL, how many kilojoules are needed to vaporize 1.00 L of ethanol at 78 °C?
d. If a 15-gallon gas tank is filled with E10, how many liters of ethanol are in the gas tank?
e. Write the balanced chemical equation for the reaction of ethanol with oxygen gas to produce carbon dioxide and water vapor.
f. How many kilograms of carbon dioxide, $CO_2$, are produced from the complete reaction of the ethanol in a full 15-gallon gas tank?

**CI.10** The active ingredient in Tums is calcium carbonate. One Tums tablet contains 500. mg of calcium carbonate.
a. What is the formula for calcium carbonate?

b. What is the molar mass of calcium carbonate?
c. How many moles of calcium carbonate are in 1 roll of Tums that contains 12 tablets?
d. If a person takes two Tums tablets a day, how many grams of calcium are obtained?
e. If the daily recommended quantity of $Ca^{2+}$ to maintain bone strength in older women is 1500 mg, how many Tums tablets are needed each day?

**CI.11** Oseltamivir, $C_{16}H_{28}N_2O_4$, is the active ingredient in Tamiflu, an antiviral drug used to treat influenza. The preparation of Tamiflu begins with the extraction of shikimic acid from the seedpods of the spice Chinese star anise. However, the star anise has no antiviral activity itself. From 2.6 g of star anise, 0.13 g of shikimic acid can be obtained and used to produce one capsule containing 75 mg of Tamiflu. The usual adult dosage for treatment of influenza is two capsules of Tamiflu daily for 5 days.

Shikimic acid

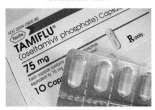

**a.** What is the formula of shikimic acid? (Black spheres are carbon, light blue spheres are hydrogen, and red spheres are oxygen.)

**b.** What is the molar mass of shikimic acid?

**c.** How many moles of shikimic acid can be obtained from 1.3 g of shikimic acid?

**d.** How many 75-mg capsules of Tamiflu could be produced from 154 g of star anise?

**e.** What is the molar mass of Tamiflu (oseltamivir)?

**f.** How many grams of carbon are in 75 mg of Tamiflu in one capsule?

**g.** How many kilograms of Tamiflu would be needed to treat all the people in a city with a population of 500 000 if each person consumes two Tamiflu capsules a day for 5 days?

**CI.12** The compound butyric acid gives rancid butter its characteristic odor.

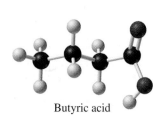

Butyric acid

**a.** If black spheres are carbon atoms, light blue spheres are hydrogen atoms, and red spheres are oxygen atoms, what is the formula of butyric acid?

**b.** What is the molar mass of butyric acid?

**c.** How many grams of butyric acid contain $3.28 \times 10^{23}$ oxygen atoms?

**d.** How many grams of carbon are in 5.28 g of butyric acid?

**e.** Butyric acid has a density of 0.959 g/mL at 20 °C. How many moles of butyric acid are contained in 1.56 mL of butyric acid?

# ANSWERS

**CI.7**  **a.** X is a metal; Y is a nonmetal.
   **b.** Y has the higher electronegativity.
   **c.** $X^{2+}$, $Y^-$
   **d.** **1.** $X = 2, 8, 2$    $Y = 2, 8, 7$
      **2.** $X^{2+} = 2, 8$    $Y^- = 2, 8, 8$
      **3.** $X^{2+}$ has the same electron arrangement as Ne.
         $Y^-$ has the same electron arrangement as Ar.
      **4.** $MgCl_2$, magnesium chloride
   **e.** Li is D; Na is A; K is C; and Rb is B.

**CI.9**  **a.**

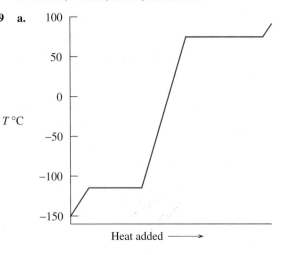

   **b.** 1100 cal
   **c.** 669 kJ
   **d.** 5.7 L
   **e** $C_2H_5OH(l) + 3O_2(g) \longrightarrow 2CO_2(g) + 3H_2O(g)$
   **f.** 8.7 kg of $CO_2$

**CI.11**  **a.** $C_7H_{10}O_5$
   **b.** 174.0 g/mole
   **c.** 0.0075 mole
   **d.** 59
   **e.** 312.0 g/mole
   **f.** 0.046 g of carbon
   **g.** 380 kg

# 6 Gases

the
Chemistry
place

Visit **www.chemplace.com** for extra
quizzes, interactive tutorials, career
resources, PowerPoint slides for
chapter review, math help, and case
studies.

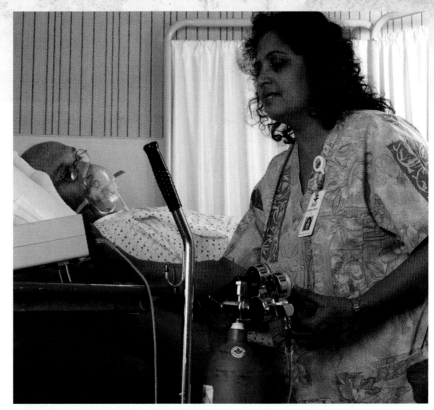

*"When oxygen levels in the blood are low, the cells in the body
don't get enough oxygen," says Sunanda Tripathi, registered
nurse, Santa Clara Valley Medical Center. "We use a nasal
cannula to give supplemental oxygen to a patient. At a flow rate
of 2 liters per minute, a patient breathes in a gaseous mixture
that is about 28% oxygen compared to 21% in ambient air."*

*When a patient has a breathing disorder, the flow and volume
of oxygen into and out of the lungs are measured. A ventilator
may be used if a patient has difficulty breathing. When pressure
is increased, the lungs expand. When the pressure of the
incoming gas is reduced, the lung volume contracts to expel
carbon dioxide. These relationships—known as gas laws—are
an important part of ventilation and breathing.*

We all live at the bottom of a sea of gases called the atmosphere. The most important of these gases is oxygen, which constitutes about 21% of the atmosphere. Without oxygen, life on this planet would be impossible—oxygen is vital to all life processes of plants and animals. Ozone ($O_3$), formed in the upper atmosphere by the interaction of oxygen with ultraviolet light, absorbs some of the harmful radiation before it can strike Earth's surface. The other gases in the atmosphere include nitrogen (78% of the atmosphere), argon, carbon dioxide ($CO_2$), and water vapor. Carbon dioxide gas, a product of combustion and metabolism, is used by plants in photosynthesis, a process that produces the oxygen that is essential for humans and animals.

The atmosphere has become a dumping ground for other gases, such as methane, chlorofluorocarbons (CFCs), and nitrogen oxides. The chemical reactions of these gases with sunlight and oxygen in the air are contributing to air pollution, ozone depletion, global warming, and acid rain. Such chemical changes can seriously affect our health and our lifestyle. An understanding of gases and some of the laws that govern gas behavior can help us understand the nature of matter and allow us to make decisions concerning important environmental and health issues.

## 6.1 PROPERTIES OF GASES

**LEARNING GOAL**

Describe the kinetic molecular theory of gases and the properties of gases.

**WEB TUTORIAL**
Properties of Gases

The behavior of gases is quite different from that of liquids and solids. Gas particles are far apart, whereas particles of both liquids and solids are held close together as strong attractive forces become more important at lower temperatures. This means that a gas has no definite shape or volume and will completely fill any container. Because there are great distances between its particles, a gas is less dense than a solid or liquid and can be compressed. A model for the behavior of a gas, called the **kinetic molecular theory of gases**, helps us understand gas behavior.

### Kinetic Molecular Theory of Gases

1. **A gas consists of small particles (atoms or molecules) that move randomly with rapid velocities**. Gas molecules moving in all directions at high speeds cause a gas to fill the entire volume of a container.
2. **The attractive forces between the particles of a gas can be neglected**. Gas particles move far apart and fill a container of any size and shape.
3. **The actual volume occupied by gas molecules is extremely small compared with the volume that the gas occupies**. The volume of the container is considered equal to the volume of the gas. Most of the volume of a gas is empty space, which allows gases to be easily compressed.
4. **The average kinetic energy of gas molecules is proportional to the Kelvin temperature**. Gas particles move faster as the temperature increases. At higher temperatures, gas particles hit the walls of the container with more force, which produces higher pressures.

5. **Gas particles are in constant motion, moving rapidly in straight paths**. When gas particles collide, they rebound and travel in new directions. Every time they hit the walls of the container, they exert gas pressure. An increase in the number or force of collisions against the walls of the container causes an increase in the pressure of the gas.

The kinetic molecular theory helps explain some of the characteristics of gases. For example, we can quickly smell perfume from a bottle that is opened on the other side of a room, because its particles move rapidly in all directions. They move faster at higher temperatures and more slowly at lower temperatures. Sometimes tires and gas-filled containers explode when temperatures are too high. From the kinetic molecular theory, we know that gas particles move faster when heated, hit the walls of a container with more force, and cause a buildup of pressure inside a container.

---

SAMPLE PROBLEM　6.1

### ■ Kinetic Molecular Theory of Gases

How would you explain the following characteristics of a gas using the kinetic molecular theory of gases?

**a.** The particles of gas are not attracted to each other.
**b.** Gas particles move faster when the temperature is increased.

#### SOLUTION

**a.** The distance between the particles of a gas is so great that the particles are not attracted to each other.
**b.** Gas particles move faster because raising the temperature increases their kinetic energy.

#### STUDY CHECK

Why can you smell food that is cooking in the kitchen when you are in a different part of the house?

---

When we talk about a gas, we describe it in terms of four properties: pressure, volume, temperature, and the amount of gas.

## Pressure (P)

Gas particles are extremely small and move rapidly. As gas particles hit against the walls of a container, they exert pressure. As more molecules strike the wall, the pressure increases. (See Figure 6.1.) If we heat the container, the molecules move faster and smash into the walls with increased force, which increases the pressure. The gas molecules of oxygen and nitrogen in the air are exerting pressure on us all the time. We call the pressure exerted by the air **atmospheric pressure**. (See Figure 6.2.)

As you go to higher altitudes, the atmospheric pressure is less because there are fewer molecules in the air. The most common units used for gas measurement are the atmosphere (atm) and millimeters of mercury (mmHg). On the TV weather report, you may hear or see the atmospheric pressure given in inches of mercury or in kilopascals in countries other than the United States. In a chemistry lab, the unit torr may be used.

## Volume (V)

The volume of gas equals the size of the container in which the gas is placed. When you inflate a tire or a basketball, you are adding more gas particles. The increase in the number

**FIGURE 6.1** Gas particles move in straight lines within a container. The gas particles exert pressure when they collide with the walls of the container.
**Q** Why does heating the container increase the pressure of the gas within it?

Molecules in air

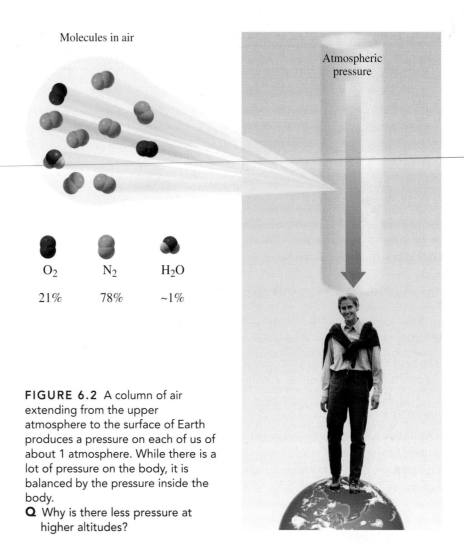

O₂     N₂     H₂O
21%    78%    ~1%

Atmospheric pressure

**FIGURE 6.2** A column of air extending from the upper atmosphere to the surface of Earth produces a pressure on each of us of about 1 atmosphere. While there is a lot of pressure on the body, it is balanced by the pressure inside the body.
**Q** Why is there less pressure at higher altitudes?

of particles hitting the walls of the tire or basketball increases its volume. Sometimes, on a cool morning, a tire looks flat. The volume of the tire has decreased because a lower temperature decreases the speed of the molecules, which reduces the force of their impacts on the walls of the tire. The most common units for volume measurement are liters (L) and milliliters (mL).

## Temperature (*T*)

The temperature of a gas is related to the kinetic energy of its particles. For example, if we have a gas at 200 K in a rigid container and heat it to a temperature of 400 K, the gas particles will have twice the kinetic energy that they did at 200 K. That also means that the gas at 400 K exerts twice the pressure of the gas at 200 K. Although you measure gas temperature using a Celsius thermometer, all comparisons of gas behavior and all calculations related to temperature must use the Kelvin temperature. No one has quite created the conditions for absolute zero (K), but we predict that the particles will have zero kinetic energy and the gas will exert zero pressure at absolute zero (0 K).

## Amount of Gas (*n*)

When you add air to a bicycle tire, you increase the amount of gas, which results in a higher pressure in the tire. Usually we measure the amount of gas by its mass (grams). In gas law calculations, we need to change the grams of gas to moles.

A summary of the four properties of a gas are given in Table 6.1.

**TABLE 6.1** Properties That Describe a Gas

| Property | Description | Units of Measurement |
|---|---|---|
| Pressure (*P*) | The force exerted by gas against the walls of the container | atmosphere (atm); mmHg; torr; pascal (Pa) |
| Volume (*V*) | The space occupied by the gas | liter (L); milliliter (mL) |
| Temperature (*T*) | Determines the kinetic energy and rate of motion of the gas particles | Celsius (°C); Kelvin (K) *required in calculations* |
| Amount (*n*) | The quantity of gas present in a container | grams (g); moles (mole) *required in calculations* |

## Forming a Gas

Obtain baking soda and a jar or a plastic bottle. You will also need an elastic glove that fits over the mouth of the jar or a balloon that fits snugly over the top of the plastic bottle. Place a cup of vinegar in the jar or bottle. Sprinkle some baking soda into the fingertips of the glove or into the balloon. Carefully fit the glove or balloon over the top of the jar or bottle. Slowly lift the fingers of the glove or the balloon so that the baking soda falls into the vinegar. Watch what happens. Squeeze the glove or balloon.

### QUESTIONS

1. Describe the properties of gas that you observe as the reaction takes place between vinegar and baking soda.
2. How do you know that a gas was formed?

---

**SAMPLE PROBLEM** 6.2

### ■ Properties of Gases

Identify the property of a gas described by each of the following:

**a.** increases the kinetic energy of gas particles
**b.** the force of the gas particles hitting the walls of the container
**c.** the space that is occupied by a gas

#### SOLUTION

**a.** temperature      **b.** pressure      **c.** volume

#### STUDY CHECK

As more helium gas is added to a balloon, the number of grams of helium increases. What property of a gas is described?

---

## QUESTIONS AND PROBLEMS

### Properties of Gases

**6.1** Use the kinetic molecular theory of gases to explain each of the following:
 **a.** Gases move faster at higher temperatures.
 **b.** Gases can be compressed much more than liquids or solids.

**6.2** Use the kinetic molecular theory of gases to explain each of the following:
 **a.** A container of nonstick cooking spray explodes when thrown into a fire.

**b.** The air in a hot-air balloon is heated to make the balloon rise.

**6.3** Identify the property of a gas that is measured in each of the following:
 **a.** 350 K   **b.** space occupied by a gas   **c.** 2.00 g of $O_2$
 **d.** force of gas particles striking the walls of the container

**6.4** Identify the property of a gas that is measured in each of the following measurements:
 **a.** 425 K        **b.** 1.0 atm
 **c.** 10.0 L        **d.** 0.50 mole of He

---

# 6.2 GAS PRESSURE

When billions and billions of gas particles hit against the walls of a container, they exert **pressure**, which is defined as a force acting on a certain area.

$$\text{Pressure}\,(P) = \frac{\text{force}}{\text{area}}$$

The air that covers the surface of Earth, the atmosphere, contains vast numbers of gas particles. Because the air particles have mass, when they collide with the surface of Earth they exert an *atmospheric pressure*.

## LEARNING GOAL

Describe the units of measurement used for pressure and change from one unit to another.

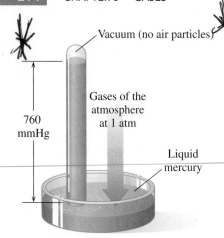

**FIGURE 6.3** A barometer: The pressure exerted by the gases in the atmosphere is equal to the downward pressure of a mercury column in a closed glass tube. The height of the mercury column measured in mmHg is called atmospheric pressure.

**Q** Why does the height of the mercury column change from day to day?

the
**Chemistry**
place

CASE STUDY
Scuba Diving and Blood Gases

**TABLE 6.2** Units for Measuring Pressure

| Unit | Abbreviation | Unit Equivalent to 1 atm |
|------|-------------|--------------------------|
| Atmosphere | atm | 1 atm |
| Millimeters of Hg | mmHg | 760 mmHg (exact) |
| Torr | torr | 760 torr (exact) |
| Pounds per square inch | lb/in.$^2$ (psi) | 14.7 lb/in.$^2$ |
| Pascal | Pa | 101 325 Pa |
| Kilopascal | kPa | 101.325 kPa |

The atmospheric pressure can be measured using a barometer. (See Figure 6.3.) At a pressure of exactly 1 atmosphere (atm), the mercury column would be exactly 760 mm high. **One atmosphere (atm)** is defined as exactly 760 mmHg. One atmosphere of pressure may also be expressed as 760 *torr*, a pressure unit named to honor Evangelista Torricelli, the inventor of the barometer. Because they are equal, units of torr and mmHg are used interchangeably.

$$1 \text{ atm} = 760 \text{ mmHg} = 760 \text{ torr}$$

$$1 \text{ mmHg} = 1 \text{ torr}$$

In SI units, pressure is measured in pascals (Pa); 1 atm is equal to 101 325 Pa. Because a pascal is a very small unit, it is likely that pressures would be reported in kilopascals.

$$1 \text{ atm} = 1.01325 \times 10^5 \text{ Pa} = 101.325 \text{ kPa}$$

The U.S. equivalent of 1 atm is 14.7 pounds per square inch (psi). When you use a pressure gauge to check the air pressure in the tires of a car, it may read 30–35 psi. Table 6.2 summarizes the various units used in the measurement of pressure.

If you have a barometer in your home, it probably gives pressure in inches of mercury. Atmospheric pressure changes with variations in weather and altitude. On a hot, sunny day, a column of air has more molecules which increases the pressure on the mercury surface. The mercury column rises indicating a higher atmospheric pressure. On a rainy day, the atmosphere exerts less pressure, which causes the mercury column to go lower. In the weather report, this type of weather is called a low-pressure system. Above sea level, the density of the gases in the air decreases, which causes lower atmospheric pressures; the atmospheric pressure is greater at the Dead Sea because it is below sea level. (See Table 6.3.)

**TABLE 6.3** Altitude and Atmospheric Pressure

| Location | Altitude (km) | Atmospheric Pressure (mmHg) |
|----------|---------------|------------------------------|
| Dead Sea | −0.40 | 800 |
| Sea level | 0 | 760 |
| Los Angeles | 0.09 | 750 |
| Las Vegas | 0.70 | 700 |
| Denver | 1.60 | 630 |
| Mount Whitney | 4.50 | 440 |
| Mount Everest | 8.90 | 250 |

Divers must be concerned about increasing pressures on their ears and lungs when they dive below the surface of the ocean. Because water is more dense than air, the pressure on a diver increases rapidly as the diver descends. At a depth of 33 ft below the surface of the ocean, an additional 1 atmosphere of pressure is exerted by the water on a diver, for a total of 2 atm. At 100 ft down, there is a total pressure of 4 atm on a diver. The air tanks a diver carries continuously adjust the pressure of the breathing mixture to match the increase in pressure.

---

## SAMPLE PROBLEM 6.3

### ■ Units of Pressure

A sample of neon gas has a pressure of 0.50 atm. Give the pressure of the neon in mmHg.

#### SOLUTION

The equality 1 atm = 760 mmHg can be written as conversion factors:

$$\frac{760 \text{ mmHg}}{1 \text{ atm}} \quad \text{or} \quad \frac{1 \text{ atm}}{760 \text{ mmHg}}$$

Using the appropriate conversion factor, the problem is set up as

$$0.50 \text{ atm} \times \frac{760 \text{ mmHg}}{1 \text{ atm}} = 380 \text{ mmHg}$$

#### STUDY CHECK

What is the pressure in atmospheres for a gas that has a pressure of 655 torr?

---

## *Health Note*

### Measuring Blood Pressure

The measurement of your blood pressure is one of the important measurements a doctor or nurse makes during a physical examination. Acting like a pump, the heart contracts to create the pressure that pushes blood through the circulatory system. During contraction, the blood pressure is called systolic and is at its highest. When the heart muscles relax, the blood pressure is called diastolic and falls. Normal range for systolic pressure is 100–120 mmHg, and for diastolic pressure, 60–80 mmHg, usually expressed as a ratio such as 100/80. These values are somewhat higher in older people. When blood pressures are elevated, such as 140/90, there is a greater risk of stroke, heart attack, or kidney damage. Low blood pressure prevents the brain from receiving adequate oxygen, causing dizziness and fainting.

The blood pressures are measured by a sphygmomanometer, an instrument consisting of a stethoscope and an inflatable cuff connected to a tube of mercury called a manometer. After the cuff is wrapped around the upper arm, it is pumped up with air until it cuts off the flow of blood through the arm. With the stethoscope over the artery, the air is slowly released from the cuff. When the pressure equals the systolic pressure, blood starts to flow again, and the noise it makes is heard through the stethoscope. As air continues to be released, the cuff deflates until no sound in the artery is heard. That second pressure reading is noted as the diastolic pressure, the pressure when the heart is not contracting.

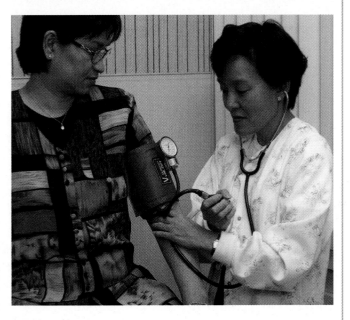

The use of digital blood pressure monitors is becoming more common. However, they have not been validated for use in all situations and can sometimes give inaccurate readings.

## QUESTIONS AND PROBLEMS

### Gas Pressure

**6.5** What units are used to measure the pressure of a gas?

**6.6** Which of the following statement(s) describes the pressure of a gas?
   **a.** the force of the gas particles on the walls of the container
   **b.** the number of gas particles in a container
   **c.** the volume of the container
   **d.** 3.00 atm      **e.** 750 torr

**6.7** An oxygen tank contains oxygen ($O_2$) at a pressure of 2.00 atm. What is the pressure in the tank in terms of the following units?
   **a.** torr      **b.** mmHg

**6.8** On a climb up Mt. Whitney, the atmospheric pressure is 467 mmHg. What is the pressure in terms of the following units?
   **a.** atm      **b.** torr

---

## LEARNING GOAL

Use the pressure–volume relationship (Boyle's law) to determine the new pressure or volume of a certain amount of gas at a constant temperature.

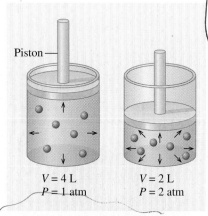

Piston

$V = 4$ L      $V = 2$ L
$P = 1$ atm      $P = 2$ atm

**FIGURE 6.4 Boyle's law:** As volume decreases, gas molecules become more crowded, which causes the pressure to increase. Pressure and volume are inversely related.

**Q** If the volume of a gas increases, what will happen to its pressure?

# 6.3 PRESSURE AND VOLUME (BOYLE'S LAW)

Imagine that you can see air particles hitting the walls inside a bicycle tire pump. What happens to the pressure inside the pump as we push down on the handle? As the volume decreases, there is a decrease in the surface area of the container. The air particles are crowded together, more collisions occur per unit area, and the air pressure increases within the container.

When a change in one property (in this case, volume) causes a change in another property (in this case, pressure), the properties are related. If the changes occur in opposite directions, the properties have an **inverse relationship**. The inverse relationship between the pressure and volume of a gas is known as **Boyle's law**. The law states that the volume ($V$) of a sample of gas changes inversely with the pressure ($P$) of the gas as long as there has been no change in the temperature ($T$) or amount of gas ($n$). (See Figure 6.4.)

If the volume or pressure of a gas sample changes without any change in the temperature or in the amount of the gas, the new pressure and volume will give the same $PV$ product as the initial pressure and volume. Then we can set the initial and final $PV$ products equal to each other.

## Boyle's Law

$$P_1V_1 = P_2V_2 \quad \text{No change in number of moles and temperature}$$

---

### SAMPLE PROBLEM 6.4

■ **Calculating Pressure When Volume Changes**

A sample of hydrogen gas ($H_2$) has a volume of 5.0 L and a pressure of 1.0 atm. What is the new pressure if the volume is decreased to 2.0 L at constant temperature?

SOLUTION

Step 1 **Organize the data in a table.**   In this problem, we want to know the final pressure ($P_2$) for the change in volume. In calculations with gas laws, it is helpful to organize the data in a table. Because we know that the volume decreases, we can predict that the pressure will increase.

| Conditions 1 | Conditions 2 | Know | Predict |
|---|---|---|---|
| $V_1 = 5.0$ L | $V_2 = 2.0$ L | $V$ decreases | |
| $P_1 = 1.0$ atm | $P_2 = ?$ | | $P$ increases |

Step 2  **Rearrange the gas law for the unknown.**   For a $PV$ relationship, we use
Boyle's law and solve for $P_2$ by dividing both sides by $V_2$.

$$P_1 V_1 = P_2 V_2$$

$$\frac{P_1 V_1}{V_2} = \frac{P_2 \cancel{V_2}}{\cancel{V_2}}$$

$$P_2 = \frac{P_1 V_1}{V_2}$$

Step 3  **Substitute values into the gas law to solve for the unknown.**   When we
substitute in the values, we see that the ratio of the volumes (volume factor)
is greater than 1. The final pressure ($P_2$) has increased as we predicted in
step 1. Note that the units of volume (L) cancel to give the final pressure in
atmospheres.

$$P_2 = 1.0 \text{ atm} \times \frac{5.0 \cancel{L}}{2.0 \cancel{L}} = 2.5 \text{ atm}$$

Volume factor increases pressure

## STUDY CHECK

A sample of helium gas has a volume of 150 mL at 750 torr. If the volume expands to
450 mL at constant temperature, what is the new pressure in torr?

---

## SAMPLE PROBLEM  6.5

### ■ Calculating Volume When Pressure Changes

The gauge on a 12-L tank of compressed oxygen reads 3800 mmHg. How many liters
would this same gas occupy at a pressure of 0.75 atm at constant temperature?

### SOLUTION

Step 1  **Organize the data in a table.**   First, we need to match the units for the initial and final pressures.

$$0.75 \cancel{\text{atm}} \times \frac{760 \text{ mmHg}}{1 \cancel{\text{atm}}} = 570 \text{ mmHg}$$

The pressure in mmHg could also be changed to atm.

$$3800 \cancel{\text{mmHg}} \times \frac{1 \text{ atm}}{760 \cancel{\text{mmHg}}} = 5.0 \text{ atm}$$

Once we place this information with the units of mmHg for pressure in a
table, we know that the pressure decreases and predict that the volume should
increase. (We could have used the two pressures in atm.)

| Conditions 1 | Conditions 2 | Know | Predict |
|---|---|---|---|
| $P_1 = 3800$ mmHg | $P_2 = 570$ mmHg | $P$ decreases | |
| $V_1 = 12$ L | $V_2 = ?$ | | $V$ increases |

1   Organize the data in a table of initial and final conditions.

2   Rearrange the gas law to solve for unknown quantity.

3   Substitute values into the gas law equation to solve for the unknown.

## Health Note

## Pressure–Volume Relationship in Breathing

The importance of Boyle's law becomes more apparent when you consider the mechanics of breathing. Our lungs are elastic, balloonlike structures contained within an airtight chamber called the thoracic cavity. The diaphragm, a muscle, forms the flexible floor of the cavity.

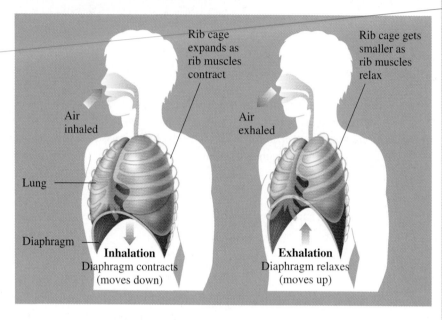

## Inspiration

The process of taking a breath of air begins when the diaphragm contracts and the rib cage expands, causing an increase in the volume of the thoracic cavity. The elasticity of the lungs allows them to expand when the thoracic cavity expands. According to Boyle's law, the pressure inside the lungs will decrease when their volume increases. This causes the pressure inside the lungs to fall below the pressure of the atmosphere. This difference in pressures produces a *pressure gradient* between the lungs and the atmosphere. In a pressure gradient, molecules flow from an area of higher pressure to an area of lower pressure. Thus, we inhale as air flows into the lungs (*inspiration*), until the pressure within the lungs becomes equal to the pressure of the atmosphere.

## Expiration

*Expiration*, or the exhalation phase of breathing, occurs when the diaphragm relaxes and moves back up into the thoracic cavity to its resting position. This reduces the volume of the thoracic cavity, which squeezes the lungs and decreases their volume. Now the pressure in the lungs is higher than the pressure of the atmosphere, so air flows out of the lungs. Thus, breathing is a process in which pressure gradients are continuously created between the lungs and the environment as a result of the changes in the volume and pressure.

**Step 2** **Rearrange the gas law for the unknown.**  Using Boyle's law, we solve for $V_2$. According to Boyle's law, a decrease in the pressure will cause an increase in the volume.

$$P_1 V_1 = P_2 V_2$$

$$\frac{P_1 V_1}{P_2} = \frac{\cancel{P_2} V_2}{\cancel{P_2}}$$

$$V_2 = \frac{P_1 V_1}{P_2}$$

**Step 3** **Substitute values into the gas law to solve for the unknown.**  When we substitute in the values with pressures in units of mmHg or atm, the ratio of pressures (pressure factor) is greater than 1, which increases the volume.

$$V_2 = 12 \text{ L} \times \frac{3800 \, \cancel{\text{mmHg}}}{570 \, \cancel{\text{mmHg}}} = 80. \text{ L}$$

Pressure factor increases volume

Or:

$$V_2 = 12 \text{ L} \times \frac{5.0 \text{ atm}}{0.75 \text{ atm}} = 80. \text{ L}$$

Pressure factor increases volume

**STUDY CHECK**

A sample of methane gas ($CH_4$) has a volume of 125 mL at 0.600 atm pressure. How many milliliters will it occupy at a pressure of 1140 mmHg at constant temperature?

## QUESTIONS AND PROBLEMS

### Pressure and Volume (Boyle's Law)

**6.9** Why do scuba divers need to exhale air when they ascend to the surface of the water?

**6.10** Why does a sealed bag of chips expand when you take it to a higher altitude?

**6.11** The air in a cylinder with a piston has a volume of 220 mL and a pressure of 650 mmHg.
**a.** To obtain a higher pressure inside the cylinder at constant temperature, should the cylinder change as shown in A or B? Explain your choice.

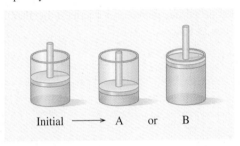

Initial ⟶ A or B

**b.** If the pressure inside the cylinder increases to 1.2 atm, what is the final volume of the cylinder? Complete the following data table:

| Property | Conditions 1 | Conditions 2 | Know | Predict |
|---|---|---|---|---|
| Pressure (P) | | | | |
| Volume (V) | | | | |

**6.12** A balloon is filled with helium gas. When the following changes are made at constant temperature, which of these diagrams (A, B, or C) shows the new volume of the balloon?

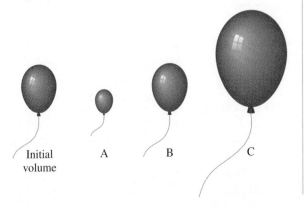

Initial volume    A    B    C

**a.** The balloon floats to a higher altitude where the outside pressure is lower.
**b.** The balloon is taken inside the house, but the atmospheric pressure remains the same.
**c.** The balloon is put in a hyperbaric chamber in which the pressure is increased.

**6.13** A gas with a volume of 4.0 L is contained in a closed container. Indicate what changes in pressure must have occurred if the volume undergoes the following changes at constant temperature:
**a.** The volume is compressed to 2.0 L.
**b.** The volume is allowed to expand to 12 L.
**c.** The volume is compressed to 400. mL.

**6.14** A gas at a pressure of 2.0 atm is contained in a closed container. Indicate the changes in its volume when the pressure undergoes the following changes at constant temperature:
**a.** The pressure increases to 6.0 atm.
**b.** The pressure drops to 1.0 atm.
**c.** The pressure drops to 0.40 atm.

**6.15** A 10.0-L balloon contains helium gas at a pressure of 655 mmHg. What is the new pressure of the helium gas at each of the following volumes if there is no change in temperature?
**a.** 20.0 L         **b.** 2.50 L         **c.** 1500 mL

**6.16** The air in a 5.00-L tank has a pressure of 1.20 atm. What is the new pressure of the air when the air is placed in tanks that have the following volumes, if there is no change in temperature?
**a.** 1.00 L         **b.** 2500. mL         **c.** 750. mL

**6.17** A sample of nitrogen ($N_2$) has a volume of 50.0 L at a pressure of 760. mmHg. What is the volume of the gas at each of the following pressures if there is no change in temperature?
**a.** 1500 mmHg      **b.** 2.0 atm         **c.** 0.500 atm

**6.18** A sample of methane ($CH_4$) has a volume of 25 mL at a pressure of 0.80 atm. What is the volume of the gas at each of the following pressures if there is no change in temperature?
**a.** 0.40 atm       **b.** 2.00 atm        **c.** 2500 mmHg

**6.19** Cyclopropane, $C_3H_6$, is a general anesthetic. A 5.0-L sample has a pressure of 5.0 atm. What is the volume of the anesthetic given to a patient at a pressure of 1.0 atm?

**6.20** The volume of air in a person's lungs is 615 mL at a pressure of 760 mmHg. Inhalation occurs as the pressure in the lungs drops to 752 mmHg. To what volume did the lungs expand?

**6.21** Use the words *inspiration* and *expiration* to describe the part of the breathing cycle that occurs as a result of each of the following:
   **a.** The diaphragm contracts (flattens out).
   **b.** The volume of the lungs decreases.
   **c.** The pressure within the lungs is less than the atmosphere.

**6.22** Use the words *inspiration* and *expiration* to describe the part of the breathing cycle that occurs as a result of each of the following:
   **a.** The diaphragm relaxes, moving up into the thoracic cavity.
   **b.** The volume of the lungs expands.
   **c.** The pressure within the lungs is greater than the atmosphere.

# 6.4 TEMPERATURE AND VOLUME (CHARLES'S LAW)

## LEARNING GOAL

Use the temperature–volume relationship (Charles's law) to determine the new temperature or volume of a certain amount of gas at a constant pressure.

Suppose that you are going to take a ride in a hot-air balloon. The captain turns on a propane burner to heat the air inside the balloon. As the temperature rises, the air particles move faster and spread out, causing the volume of the balloon to increase. As the air is heated, it becomes less dense than the air outside, causing the balloon and its passengers to lift off. In 1787, Jacques Charles, a balloonist as well as a physicist, proposed that the volume of a gas is related to the temperature. This became **Charles's law**, which states that the volume ($V$) of a gas is directly related to the temperature ($T$) when there is no change in the pressure ($P$) or amount ($n$) of gas. (See Figure 6.5.) A **direct relationship** is one in which the related properties increase or decrease together. For two conditions, we can write Charles's law as follows.

## Charles's Law

$$\frac{V_1}{T_1} = \frac{V_2}{T_2} \qquad \text{No change in number of moles and pressure}$$

*All temperatures used in gas law calculations must be converted to their corresponding Kelvin (K) temperatures.*

---

**SAMPLE PROBLEM   6.6**

■ **Calculating Volume When Temperature Changes**

A sample of neon gas has a volume of 5.40 L and a temperature of 15 °C. Find the new volume of the gas after the temperature has been increased to 42 °C at constant pressure.

**SOLUTION**

Step 1 **Organize the data in a table.**   *When the temperatures are given in degrees Celsius, they must be changed to kelvins.*

$$T_1 = 15\,°C + 273 = 288\ K$$

$$T_2 = 42\,°C + 273 = 315\ K$$

| Conditions 1 | Conditions 2 | Know | Predict |
|---|---|---|---|
| $T_1 = 288\ K$ | $T_2 = 315\ K$ | $T$ increases | |
| $V_1 = 5.40\ L$ | $V_2 = ?$ | | $V$ increases |

Step 2 **Rearrange the gas law for the unknown**. In this problem, we want to know the final volume ($V_2$) when the temperature increases. Using Charles's law, we solve for $V_2$ by multiplying both sides by $T_2$.

$$\frac{V_1}{T_1} = \frac{V_2}{T_2}$$

$$\frac{V_1}{T_1} \times T_2 = \frac{V_2}{\cancel{T_2}} \times \cancel{T_2}$$

$$V_2 = V_1 \times \frac{T_2}{T_1}$$

Step 3 **Substitute values into the gas law to solve for the unknown**. From the table, we see that the temperature has increased. Because temperature is directly related to volume, the volume must increase. When we substitute in the values, we see that the ratio of the temperatures (temperature factor) is greater than 1, which increases the volume, as predicted.

$$V_2 = 5.40 \text{ L} \times \frac{315 \cancel{K}}{288 \cancel{K}} = 5.91 \text{ L}$$

Temperature factor
increases volume

**STUDY CHECK**

A mountain climber inhales 486 mL of air at a temperature of −8 °C. What volume in mL will the air occupy in the lungs if the climber's body temperature is 37 °C?

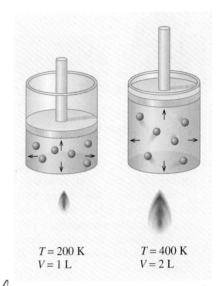

$T = 200$ K      $T = 400$ K
$V = 1$ L         $V = 2$ L

**FIGURE 6.5 Charles's law:** The Kelvin temperature of a gas is directly related to the volume of the gas when there is no change in the pressure. When the temperature increases, making the molecules move faster, the volume must increase to maintain constant pressure.
**Q** If the temperature of a gas decreases at constant pressure, how will the volume change?

---

## QUESTIONS AND PROBLEMS

### Temperature and Volume (Charles's Law)

**6.23** Select the diagram that shows the new volume of a balloon when the following changes are made at constant pressure:

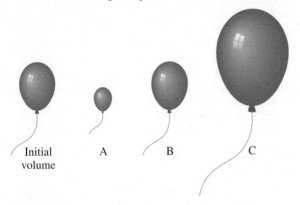

Initial volume      A      B      C

    **a.** The temperature is changed from 100 K to 300 K.
    **b.** The balloon is placed in a freezer.
    **c.** The balloon is first warmed and then returned to its starting temperature.

**6.24** Indicate whether the final volume of gas in each of the following is the same, larger, or smaller than the initial volume:

    **a.** A volume of 505 mL of air on a cold winter day at 5 °C is breathed into the lungs, where body temperature is 37 °C.
    **b.** The heater used to heat 1400 L of air in a hot-air balloon is turned off.
    **c.** A balloon filled with helium at the amusement park is left in a car on a hot day.

**6.25** A sample of neon initially has a volume of 2.50 L at 15 °C. What is the new temperature in °C when the volume of the sample is changed at constant pressure to each of the following?
    **a.** 5.00 L    **b.** 1250 mL    **c.** 7.50 L    **d.** 3550 mL

**6.26** A gas has a volume of 4.00 L at 0 °C. What final temperature in degrees Celsius is needed to cause the volume of the gas to change to the following if $n$ and $P$ are not changed?
    **a.** 100 L    **b.** 1200 mL    **c.** 250 L    **d.** 50.0 mL

**6.27** A balloon contains 2500 mL of helium gas at 75 °C. What is the new volume (mL) of the gas when the temperature changes to the following, if $n$ and $P$ are not changed?
    **a.** 55 °C    **b.** 680. K    **c.** −25 °C    **d.** 240. K

**6.28** An air bubble has a volume of 0.500 L at 18 °C. If the pressure does not change, what is the volume in liters at each of the following temperatures?
    **a.** 0 °C    **b.** 425 K    **c.** −12 °C    **d.** 575 K

# Green Chemistry Note

## Greenhouse Gases

The term *greenhouse gases* was first used in early 1800s for the gases in the atmosphere that trap heat. Among the *greenhouse gases* are carbon dioxide ($CO_2$), methane ($CH_4$), dinitrogen oxide ($N_2O$), and chlorofluorocarbons (CFCs). The molecules of greenhouse gases consist of more than two atoms that vibrate when heat is absorbed. By contrast, oxygen and nitrogen are not greenhouse gases because the two atoms in their molecules are so tightly bonded, they do not absorb heat.

Greenhouses gases are beneficial in keeping the average surface temperature for Earth at 15 °C. Without greenhouse gases, it is estimated that the average surface temperature of Earth would be −18 °C. Most scientists say that the concentration of greenhouse gases in the atmosphere, and therefore the surface temperature of Earth, is increasing as a result of human activities. As we discussed in Chapter 2, the increase in atmospheric carbon dioxide is mostly a result of the burning of fossil fuels and wood.

Methane ($CH_4$) is a colorless, odorless gas that is released by the decomposition of organic plant material in landfills; livestock; rice farming; and mining, drilling, and transport of coal and oil. The contribution from livestock comes from the digestion of organic material in the digestive tracts of cows, sheep, and camels. The level of methane in the atmosphere has increased about 150% since industrialization. In one year, as much as $5 \times 10^{11}$ kg of methane is added to the atmosphere. Livestock produces about 20% of the greenhouse gases. In one day, one cow emits about 200 g of methane. For a global population of 1.5 billion livestock, a total of $3 \times 10^8$ kg of methane is produced every day. In the past few years, methane levels have stabilized because of improvements in the recovery of methane. Methane remains in the atmosphere for about ten years, but its molecular structure causes it to trap twenty times more heat than does carbon dioxide.

Dinitrogen oxide ($N_2O$), commonly called nitrous oxide, is a colorless greenhouse gas that has a sweet odor. Most people recognize it as an anesthetic used in dentistry called "laughing gas." Although some dinitrogen oxide is released naturally from soil bacteria, the primary increases are from agricultural and industrial processes. Atmospheric dinitrogen oxide has increased by about 15% since industrialization from the extensive use of fertilizers, sewage treatment plants, and car exhaust. Each year, $1 \times 10^{10}$ kg of dinitrogen oxide is added to the atmosphere. Dinitrogen oxide released today will remain

### Percent Greenhouse Gases in the Atmosphere

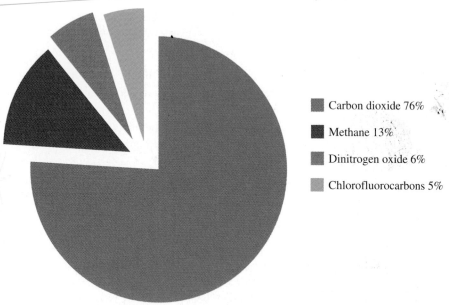

- Carbon dioxide 76%
- Methane 13%
- Dinitrogen oxide 6%
- Chlorofluorocarbons 5%

in the atmosphere for about 150–180 years, where it has a greenhouse effect that is 300 times greater than that of carbon dioxide.

Chlorofluorinated gases (CFCs) are synthetic compounds containing chlorine, fluorine, and carbon. Chlorofluorocarbons were used as propellents in aerosol cans and in refrigerants in refrigerators and air conditioners. During the 1970s, scientists determined that CFCs in the atmosphere were destroying the protective ozone layer. Since then, many countries banned the production and use of CFCs, and their levels in the atmosphere have declined slightly. Hydrofluorocarbons (HFCs), in which hydrogen atoms replace chlorine atoms, are now used as refrigerants. Although HFCs do not destroy the ozone layer, they are greenhouse gases because they trap heat in the atmosphere.

Based on current trends and climate models, scientists estimate that levels of atmospheric carbon dioxide will increase by about 2% each year up to 2025. As long as more heat is trapped by the greenhouse gases than is reflected back into space, average surface temperatures on Earth will continue to rise. Efforts are taking place around the world to slow or decrease the emissions of greenhouse gases into the atmosphere. It is anticipated that temperature will stabilize only when the amount of energy that reaches the surface of Earth is equal to the heat that is reflected back into space.

In 2007, former U.S. Vice-president Al Gore and the U.N. Panel on Climate Change were awarded the Nobel Peace Prize for increasing global awareness of the relationship between human activities and global warming.

# 6.5 TEMPERATURE AND PRESSURE (GAY–LUSSAC'S LAW)

If we could watch the molecules of a gas as the temperature rises, we would notice that they move faster and hit the sides of the container more often and with greater force. If we keep the volume of the container the same, we would observe an increase in the pressure. A temperature–pressure relationship, also known as **Gay–Lussac's law**, states that the pressure of a gas is directly related to its Kelvin temperature. This means that an increase in temperature increases the pressure of a gas and a decrease in temperature decreases the pressure of the gas, provided the volume and number of moles of the gas remain the same. (See Figure 6.6.) The ratio of pressure ($P$) to temperature ($T$) is the same under all conditions as long as volume ($V$) and amount of gas ($n$) do not change.

## Gay–Lussac's Law

$$\frac{P_1}{T_1} = \frac{P_2}{T_2} \qquad \text{No change in number of moles and volume}$$

*All temperatures used in gas law calculations must be converted to their corresponding Kelvin (K) temperatures.*

---

### SAMPLE PROBLEM  6.7

■ **Calculating Pressure When Temperature Changes**

Aerosol containers can be dangerous if they are heated, because they can explode. Suppose a container of hair spray with a pressure of 4.0 atm at a room temperature of 25 °C is thrown into a fire. If the temperature of the gas inside the aerosol can reaches 402 °C, what will be its pressure? The aerosol container may explode if the pressure inside exceeds 8.0 atm. Would you expect it to explode?

#### SOLUTION

Step 1 **Organize the data in a table.**  We must first change the temperatures to kelvins.

$$T_1 = 25\,°C + 273 = 298\ K$$

$$T_2 = 402\,°C + 273 = 675\ K$$

| Conditions 1 | Conditions 2 | Know | Predict |
|---|---|---|---|
| $P_1 = 4.0$ atm | $P_2 = ?$ | | $P$ increases |
| $T_1 = 298$ K | $T_2 = 675$ K | $T$ increases | |

Step 2 **Rearrange the gas law for the unknown.**  Using Gay–Lussac's law, we can solve for $P_2$.

$$\frac{P_1}{T_1} = \frac{P_2}{T_2}$$

$$\frac{P_1}{T_1} \times T_2 = \frac{P_2}{T_2} \times T_2$$

$$P_2 = P_1 \times \frac{T_2}{T_1}$$

Step 3 **Substitute values into the gas law to solve for the unknown.**  From the table, we see that the temperature has increased. Because pressure and

## LEARNING GOAL

Use the temperature–pressure relationship (Gay–Lussac's law) to determine the new temperature or pressure of a certain amount of gas at a constant volume.

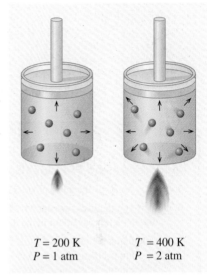

$$T = 200\ K \qquad T = 400\ K$$
$$P = 1\ atm \qquad P = 2\ atm$$

**FIGURE 6.6 Gay–Lussac's law:** The pressure of a gas is directly related to the temperature of the gas. When the Kelvin temperature of a gas is doubled, the pressure is doubled at constant volume.
**Q** How does a decrease in the temperature of a gas affect its pressure at constant volume?

temperature are directly related, the pressure must increase. When we substitute in the values, we see the ratio of the temperatures (temperature factor) is greater than 1, which increases pressure.

$$P_2 = 4.0 \text{ atm} \times \frac{675 \text{ K}}{298 \text{ K}} = 9.1 \text{ atm}$$

Temperature factor increases volume

Because the calculated pressure of 9.1 atm is greater than 8.0 atm, we expect the can to explode.

### STUDY CHECK

In a storage area where the temperature has reached 55 °C, the pressure of oxygen gas in a 15.0-L steel cylinder is 965 torr. To what temperature (°C) would the gas have to be cooled to reduce the pressure to 850. torr?

## QUESTIONS AND PROBLEMS

### Temperature and Pressure (Gay–Lussac's Law)

**6.29** Why do aerosol cans explode if heated?

**6.30** How can the tires on a car have a blowout when the car is driven on hot pavement in the desert?

**6.31** Calculate the new temperature in degrees Celsius when pressure is changed with $n$ and $V$ constant.
   **a.** A sample of xenon at 25 °C and 740 mmHg is cooled to give a pressure of 620 mmHg.
   **b.** A tank of argon gas with a pressure of 0.950 atm at −18 °C is heated to give a pressure of 1250 torr.

**6.32** Calculate the new temperature in degrees Celsius when pressure is changed with $n$ and $V$ constant.
   **a.** A tank of helium gas with a pressure of 250 torr at 0 °C is heated to give a pressure of 1500 torr.

   **b.** A sample of air at 40 °C and 740 mmHg is cooled to give a pressure of 680 mmHg.

**6.33** Solve for the new pressure in each of the following with $n$ and $V$ constant:
   **a.** The gas with a pressure of 1200 torr at 155 °C is cooled to 0 °C.
   **b.** An aerosol can with a pressure of 1.40 atm at 12 °C is heated to 35 °C.

**6.34** Solve for the new pressure in each of the following with $n$ and $V$ constant:
   **a.** A gas with a pressure of 1.20 atm at 75 °C is cooled to −32 °C.
   **b.** A sample of $N_2$ with a pressure of 780 mmHg at −75 °C is heated to 28 °C.

*Career Focus*

## Nurse Anesthetist

"During surgery, I work with the surgeon to provide a safe level of anesthetics that renders the patient free from pain," says Mark Noguchi, nurse anesthetist (CRNA), Kaiser Hospital. "We do spinal and epidural blocks as well as general anesthetics, which means the patient is totally asleep. We use a variety of pharmaceutical agents including halothane ($C_2HBrClF_3$) and bupivacain ($C_{18}H_{28}N_2O$), as well as muscle relaxants such as midazolam ($C_{18}H_{13}ClFN_3$) to achieve the results we want for the surgical situation. We also assess the patient's overall hemodynamic status. If blood is lost, we replace components such as plasma, platelets, and coagulation factors. We also monitor the heart rate and run EKGs to determine cardiac function."

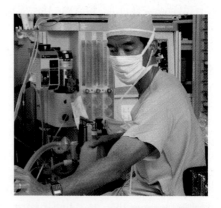

# 6.6 THE COMBINED GAS LAW

All pressure–volume–temperature relationships for gases that we have studied may be combined into a single relationship called the **combined gas law**. This expression is useful for studying the effect of changes in two of these variables on the third as long as the amount of gas (number of moles) remains constant.

**LEARNING GOAL**

Use the combined gas law to find the new pressure, volume, or temperature of a gas when changes in two of these properties are given.

## Combined Gas Law

$$\frac{P_1 V_1}{T_1} = \frac{P_2 V_2}{T_2}$$    No change in moles of gas

By using the combined gas law, we can derive any of the gas laws by omitting those properties that do not change as seen in Table 6.4.

**TABLE 6.4** Summary of Gas Laws

| Combined Gas Law | Properties Held Constant | Relationship | Name of Gas Law |
|---|---|---|---|
| $\frac{P_1 V_1}{\cancel{T_1}} = \frac{P_2 V_2}{\cancel{T_2}}$ | $T, n$ | $P_1 V_1 = P_2 V_2$ | Boyle's law |
| $\frac{\cancel{P_1} V_1}{T_1} = \frac{\cancel{P_2} V_2}{T_2}$ | $P, n$ | $\frac{V_1}{T_1} = \frac{V_2}{T_2}$ | Charles's law |
| $\frac{P_1 \cancel{V_1}}{T_1} = \frac{P_2 \cancel{V_2}}{T_2}$ | $V, n$ | $\frac{P_1}{T_1} = \frac{P_2}{T_2}$ | Gay–Lussac's law |

---

**SAMPLE PROBLEM  6.8**

### ■ Using the Combined Gas Law

A 25.0-mL bubble is released from a diver's air tank at a pressure of 4.00 atm and a temperature of 11 °C. What is the volume (mL) of the bubble when it reaches the ocean surface, where the pressure is 1.00 atm and the temperature is 18 °C?

**SOLUTION**

Step 1  **Organize the data in a table.**   We must first change the temperature to kelvins.

$$T_1 = 11\ °C + 273 = 284\ K$$
$$T_2 = 18\ °C + 273 = 291\ K$$

| Conditions 1 | Conditions 2 |
|---|---|
| $P_1 = 4.00$ atm | $P_2 = 1.00$ atm |
| $V_1 = 25.0$ mL | $V_2 = ?$ |
| $T_1 = 284$ K | $T_2 = 291$ K |

Step 2  **Rearrange the gas law for the unknown.**   Because the pressure and temperature are both changing, we must use the combined gas law to solve for $V_2$.

$$\frac{P_1 V_1}{T_1} = \frac{P_2 V_2}{T_2}$$

$$\frac{P_1 V_1}{T_1} \times \frac{T_2}{P_2} = \frac{P_2 V_2 \times \cancel{T_2}}{\cancel{T_2} \times \cancel{P_2}}$$

$$V_2 = V_1 \times \frac{P_1}{P_2} \times \frac{T_2}{T_1}$$

**Step 3  Substitute the values into the gas law to solve for the unknown.**    From the data table, we determine that the pressure decrease and the temperature increase will both increase the volume.

$$V_2 = 25.0 \text{ mL} \times \frac{4.00 \text{ atm}}{1.00 \text{ atm}} \times \frac{291 \text{ K}}{284 \text{ K}} = 102 \text{ mL}$$

Pressure factor increases volume    Temperature factor increases volume

STUDY CHECK

A weather balloon is filled with 15.0 L of helium at a temperature of 25 °C and a pressure of 685 mmHg. What is the pressure (mmHg) of the helium in the balloon in the upper atmosphere when the temperature is −35 °C and the volume becomes 34.0 L?

## QUESTIONS AND PROBLEMS

### The Combined Gas Law

**6.35** A sample of helium gas has a volume of 6.50 L at a pressure of 845 mmHg and a temperature of 25 °C. What is the pressure of the gas in atm when the volume and temperature of the gas sample are changed to the following?
**a.** 1850 mL and 325 K          **b.** 2.25 L and 12 °C
**c.** 12.8 L and 47 °C

**6.36** A sample of argon gas has a volume of 735 mL at a pressure of 1.20 atm and a temperature of 112 °C. What is the volume of the gas in milliliters when the pressure and temperature of the gas sample are changed to the following?
**a.** 658 mmHg and 281 K          **b.** 0.55 atm and 75 °C
**c.** 15.4 atm and −15 °C

**6.37** A 100.0-mL bubble of hot gases at 225 °C and 1.80 atm escapes from an active volcano. What is the new volume of the bubble outside the volcano where the temperature is −25 °C and the pressure is 0.80 atm?

**6.38** A scuba diver 40 ft below the ocean surface inhales 50.0 mL of compressed air in a scuba tank at a pressure of 3.00 atm and a temperature of 8 °C. What is the pressure of air in the lungs if the gas expands to 150.0 mL at a body temperature of 37 °C?

## LEARNING GOAL

Describe the relationship between the amount of a gas and its volume, and use this relationship in calculations.

# 6.7 VOLUME AND MOLES (AVOGADRO'S LAW)

In our study of the gas laws, we have looked at changes in properties for a specified amount ($n$) of gas. Now we will consider how the properties of a gas change when there is a change in number of moles or grams.

When you blow up a balloon, its volume increases because you add more air molecules. If a basketball gets a hole in it and some of the air leaks out, its volume decreases. **Avogadro's law** states that the volume of a gas is directly related to the number of moles of a gas when temperature and pressure are not changed. If the number of moles of a gas is doubled, then the volume will double as long as we do not change the pressure or the temperature. (See Figure 6.7.) For two conditions, we can write Avogadro's law as follows:

## Avogadro's Law

$$\frac{V_1}{n_1} = \frac{V_2}{n_2} \qquad \text{No change in pressure or temperature}$$

### ■ Calculating Volume for a Change in Moles

A weather balloon with a volume of 44 L is filled with 2.0 moles of helium. To what volume will the balloon expand if 3.0 moles of helium are added, to give a total of 5.0 moles of helium (the pressure and temperature do not change)?

#### SOLUTION

Step 1   **Organize the data in a table.**   A data table for our given information can be set up as follows:

| Conditions 1 | Conditions 2 | Know | Predict |
|---|---|---|---|
| $V_1 = 44$ L | $V_2 = ?$ | | $V$ increases |
| $n_1 = 2.0$ moles | $n_2 = 5.0$ moles | $n$ increases | |

Step 2   **Rearrange the gas law for the unknown.**   Using Avogadro's law, we can solve for $V_2$.

$$\frac{V_1}{n_1} = \frac{V_2}{n_2}$$

$$n_2 \times \frac{V_1}{n_1} = \frac{V_2}{\cancel{n_2}} \times \cancel{n_2}$$

$$V_2 = V_1 \times \frac{n_2}{n_1}$$

Step 3   **Substitute the values into the gas law to solve for the unknown.**   From the table, we see that the number of moles has increased. Because the number of moles and volume are directly related, the volume must increase at constant pressure and temperature. When we substitute in the values, we see the ratio of the moles (mole factor) is greater than 1, which increases volume.

$$V_2 = 44 \text{ L} \times \underbrace{\frac{5.0 \text{ moles}}{2.0 \text{ moles}}}_{\substack{\text{Mole factor} \\ \text{that increases volume}}} = 110 \text{ L}$$

#### STUDY CHECK

A sample containing 8.00 g of oxygen gas has a volume of 5.00 L. What is the volume after 4.00 g of oxygen gas is added to the sample if temperature and pressure do not change?

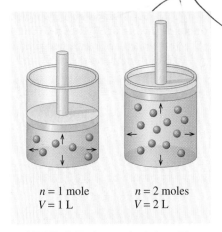

$n = 1$ mole
$V = 1$ L

$n = 2$ moles
$V = 2$ L

**FIGURE 6.7 Avogadro's law:** The volume of a gas is directly related to the number of moles of the gas. If the number of moles is doubled, the volume must double at constant temperature and pressure.
**Q** If a balloon has a leak, what happens to its volume?

## STP and Molar Volume

Using Avogadro's law, we can say that any two gases will have equal volumes if they contain the same number of moles of gas at the same temperature and pressure. To help us make comparisons between different gases, arbitrary conditions called standard temperature (273 K) and pressure (1 atm), abbreviated **STP**, were selected by scientists.

### STP Conditions

Standard temperature is 0 °C (273 K)

Standard pressure is 1 atm (760 mmHg)

At STP, it was observed that 1 mole of any gas has a volume of 22.4 L. (See Figure 6.8.) For 1 mole of a gas at STP, this is called its **molar volume**.

**FIGURE 6.8** Avogadro's law indicates that 1 mole of any gas at STP has a volume of 22.4 L.
**Q** What volume of gas is occupied by 16.0 g of methane gas, $CH_4$, at STP?

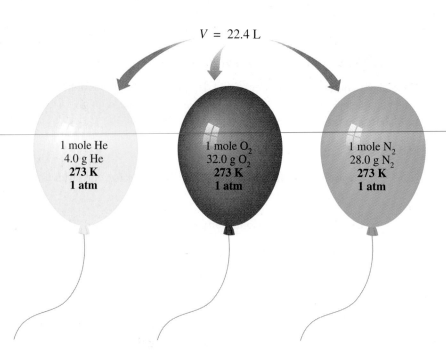

$V = 22.4$ L

1 mole He
4.0 g He
**273 K**
**1 atm**

1 mole $O_2$
32.0 g $O_2$
**273 K**
**1 atm**

1 mole $N_2$
28.0 g $N_2$
**273 K**
**1 atm**

As long as a gas is at STP conditions (0 °C and 1 atm), its molar volume can be used as a conversion factor to convert between the number of moles of gas and its volume.

**Molar Volume Conversion Factors**

Volume of 1 mole of gas (STP) $= 22.4$ L of gas

$$\frac{1 \text{ mole gas (STP)}}{22.4 \text{ L gas}} \quad \text{and} \quad \frac{22.4 \text{ L gas}}{1 \text{ mole gas (STP)}}$$

| Moles of gas | Molar volume 22.4 L/mole | Volume (L) of gas |
|---|---|---|

**SAMPLE PROBLEM 6.10**

### ■ Using Molar Volume

How many moles of helium are present in 5.25 L of helium at STP?

**SOLUTION**

The molar volume of a gas at STP can be used to calculate moles of helium.

**Guide to Using Molar Volume**

**1** Identify given and needed.

**2** Write a plan.

**3** Write conversion factors including 22.4 L/mole at STP.

**4** Set up problem with factors to cancel units.

**Step 1  Given**  5.25 L of helium at STP    **Need**  moles of helium

**Step 2  Write a plan.**  liters of He  Molar volume  moles of He

**Step 3  Write conversion factors.**

1 mole of He (STP) $= 22.4$ L of He

$$\frac{22.4 \text{ L He}}{1 \text{ mole He}} \quad \text{and} \quad \frac{1 \text{ mole He}}{22.4 \text{ L He}}$$

**Step 4  Set up problem with factors to cancel units.**

$$5.25 \text{ L He} \times \frac{1 \text{ mole He}}{22.4 \text{ L He}} = 0.234 \text{ mole of He (at STP)}$$

**STUDY CHECK**

How many moles of nitrogen ($N_2$) are present in 5.6 L of the gas at STP?

## SAMPLE PROBLEM 6.11

### ■ Using Molar Volume to Find Volume at STP

What is the volume in liters of 64.0 g of $O_2$ gas at STP?

SOLUTION

Once we convert the mass of $O_2$ to moles of $O_2$, the molar volume of a gas at STP can be used to calculate the volume (L) of $O_2$.

**Step 1 Given** 64.0 g of $O_2(g)$ at STP    **Need** volume in liters (L)

**Step 2 Write a plan.**

grams of $O_2$ | Molar mass | moles of $O_2$ | Molar volume | liters of $O_2$

**Step 3 Write conversion factors.**

$$1 \text{ mole of } O_2 = 32.0 \text{ g of } O_2 \qquad 1 \text{ mole of } O_2 \text{ (STP)} = 22.4 \text{ L of } O_2$$

$$\frac{32.0 \text{ g } O_2}{1 \text{ mole } O_2} \quad \text{and} \quad \frac{1 \text{ mole } O_2}{32.0 \text{ g } O_2} \qquad \frac{22.4 \text{ L } O_2}{1 \text{ mole } O_2} \quad \text{and} \quad \frac{1 \text{ mole } O_2}{22.4 \text{ L } O_2}$$

**Step 4 Set up problem with factors to cancel units.**

$$64.0 \text{ g } O_2 \times \frac{1 \text{ mole } O_2}{32.0 \text{ g } O_2} \times \frac{22.4 \text{ L } O_2}{1 \text{ mole } O_2} = 44.8 \text{ L of } O_2 \text{ (STP)}$$

STUDY CHECK

How many grams of $N_2(g)$ are in 5.6 L of $N_2(g)$ at STP?

## Gases in Reactions at STP

We can use the molar volume at STP to determine the moles of a gas in a reaction. Once we know the moles of gas in a reaction, we can use a mole factor to determine the moles of any other substance.

## SAMPLE PROBLEM 6.12

### ■ Gases in Chemical Reactions at STP

When potassium metal reacts with chlorine gas, the product is solid potassium chloride.

$$2K(s) + Cl_2(g) \longrightarrow 2KCl(s)$$

How many grams of potassium chloride are produced when 7.25 L of chlorine gas at STP reacts with potassium?

SOLUTION

**Step 1 Find moles of gas A using molar volume.** At STP, we can use molar volume (22.4 L/mole) to determine moles of $Cl_2$ gas.

$$7.25 \text{ L } Cl_2 \times \frac{1 \text{ mole } Cl_2}{22.4 \text{ L } Cl_2} = 0.324 \text{ mole of } Cl_2$$

**Step 2 Determine moles of substance B using mole–mole factor from the balanced equation.**

$$1 \text{ mole of } Cl_2 = 2 \text{ moles of KCl}$$

$$\frac{2 \text{ moles KCl}}{1 \text{ mole } Cl_2} \quad \text{and} \quad \frac{1 \text{ mole } Cl_2}{2 \text{ moles KCl}}$$

$$0.324 \text{ mole } Cl_2 \times \frac{2 \text{ moles KCl}}{1 \text{ mole } Cl_2} = 0.648 \text{ mole of KCl}$$

**Guide to Using Molar Volume for Reactions**

**1** Find moles of gas A using molar volume.

**2** Determine moles of substance B using mole–mole factor.

**3** Convert moles of substance B to grams or volume.

**Step 3  Convert moles of substance B to grams or volume.**  Using the molar mass of KCl, we can determine the grams of KCl.

$$1 \text{ mole of KCl} = 74.6 \text{ g of KCl}$$

$$\frac{1 \text{ mole KCl}}{74.6 \text{ g KCl}} \quad \text{and} \quad \frac{74.6 \text{ g KCl}}{1 \text{ mole KCl}}$$

$$0.648 \text{ mole KCl} \times \frac{74.6 \text{ g KCl}}{1 \text{ mole KCl}} = 48.3 \text{ g of KCl}$$

These steps can also be set up as a continuous solution.

$$7.25 \text{ L Cl}_2 \times \frac{1 \text{ mole Cl}_2}{22.4 \text{ L Cl}_2} \times \frac{2 \text{ moles KCl}}{1 \text{ mole Cl}_2} \times \frac{74.6 \text{ g KCl}}{1 \text{ moles KCl}} = 48.3 \text{ g of KCl}$$

**STUDY CHECK**

$H_2$ gas forms when zinc metal reacts with aqueous HCl.

$$Zn(s) + 2HCl(aq) \longrightarrow ZnCl_2(aq) + H_2(g)$$

How many liters of $H_2$ gas at STP are produced when 15.8 g of Zn reacts?

## QUESTIONS AND PROBLEMS

### Volume and Moles (Avogadro's Law)

**6.39** What happens to the volume of a bicycle tire or a basketball when you use an air pump to add air?

**6.40** Sometimes when you blow up a balloon and release it, it flies around the room. What is happening to the air that was in the balloon and its volume?

**6.41** A sample containing 1.50 moles of neon gas has a volume of 8.00 L. What is the new volume of the gas in liters when the following changes occur in the quantity of the gas at constant pressure and temperature?
**a.** A leak allows one-half of the neon gas to escape.
**b.** A sample of 25.0 g of neon is added to the neon gas already in the container.
**c.** A sample of 3.50 moles of $O_2$ is added to the neon gas already in the container.

**6.42** A sample containing 4.80 g of $O_2$ gas has a volume of 15.0 L. Pressure and temperature remain constant.
**a.** What is the new volume if 0.500 mole of $O_2$ gas is added?
**b.** Oxygen is released until the volume is 10.0 L. How many moles of $O_2$ are removed?

**c.** What is the volume after 4.00 g of He is added to the $O_2$ gas already in the container?

**6.43** Use the molar volume of a gas to solve the following at STP:
**a.** the number of moles of $O_2$ in 44.8 L of $O_2$ gas
**b.** the number of moles of $CO_2$ in 4.00 L of $CO_2$ gas
**c.** the volume (L) of 6.40 g of $O_2$
**d.** the volume (mL) occupied by 50.0 g of neon

**6.44** Use molar volume to solve the following problems at STP:
**a.** the volume (L) occupied by 2.50 moles of $N_2$
**b.** the volume (mL) occupied by 0.420 mole of He
**c.** the number of grams of neon contained in 11.2 L of Ne gas
**d.** the number of moles of $H_2$ in 1620 mL of $H_2$ gas

**6.45** Mg metal reacts with HCl to produce hydrogen gas.

$$Mg(s) + 2HCl(aq) \longrightarrow MgCl_2(aq) + H_2(g)$$

What volume of $H_2$ at STP is released when 8.25 g of Mg reacts?

**6.46** Aluminum oxide is formed from its elements.

$$4Al(s) + 3O_2(g) \longrightarrow 2Al_2O_3(s)$$

How many grams of Al will react with 12.0 L of $O_2$ at STP?

---

**LEARNING GOAL**

Use partial pressures to calculate the total pressure of a mixture of gases.

# 6.8 PARTIAL PRESSURES (DALTON'S LAW)

Many gas samples are a mixture of gases. For example, the air you breathe is a mixture of mostly oxygen and nitrogen gases. In gas mixtures, scientists observed that all gas particles behave in the same way. Therefore, the total pressure of the gases in a mixture is a result of the collisions of the gas particles regardless of what type of gas they are.

In a gas mixture, each gas exerts its **partial pressure**, which is the pressure it would exert if it were the only gas in the container. **Dalton's law** states that the total pressure of a gas mixture is the sum of the partial pressures of the gases in the mixture.

## Dalton's Law

$$P_{total} = P_1 + P_2 + P_3 + \cdots$$

Total pressure = Sum of the partial pressures
of a gas mixture    of the gases in the mixture

Suppose we have two separate tanks, one filled with helium at 2.0 atm and the other filled with argon at 4.0 atm. When the gases are combined in a single tank with the same volume and temperature, the number of gas molecules, not the type of gas, determines the pressure in a container. There the pressure of the gas mixture would be 6.0 atm, which is the sum of their individual or partial pressures.

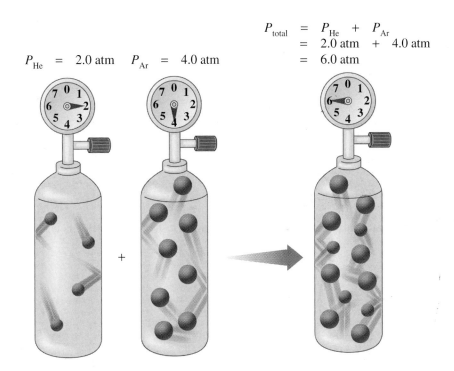

$$P_{He} = 2.0 \text{ atm} \qquad P_{Ar} = 4.0 \text{ atm}$$

$$
\begin{aligned}
P_{total} &= P_{He} + P_{Ar} \\
&= 2.0 \text{ atm} + 4.0 \text{ atm} \\
&= 6.0 \text{ atm}
\end{aligned}
$$

---

**SAMPLE PROBLEM** **6.13**

### ■ Calculating the Total Pressure of a Gas Mixture

A 10.0-L gas tank contains propane ($C_3H_8$) gas at a pressure of 300. torr. Another 10.0-L gas tank contains methane ($CH_4$) gas at a pressure of 500. torr. In preparing a gas fuel mixture, the gases from both tanks are combined in a 10.0-L container at the same temperature. What is the pressure of the gas mixture in the 10.0-L container?

#### SOLUTION

Using Dalton's law of partial pressures, we find that the total pressure of the gas mixture is the sum of the partial pressures of the gases in the mixture.

$$
\begin{aligned}
P_{total} &= P_{propane} + P_{methane} \\
&= 300. \text{ torr} + 500. \text{ torr} \\
&= 800. \text{ torr}
\end{aligned}
$$

**TABLE 6.5** Typical Composition of Air

| Gas | Partial Pressure (mmHg) | Percentage (%) |
| --- | --- | --- |
| Nitrogen, $N_2$ | 594.0 | 78 |
| Oxygen, $O_2$ | 160.0 | 21 |
| Carbon dioxide, $CO_2$ | 0.3 | 1 |
| Water vapor, $H_2O$ | 5.7 | |
| Total air | 760.0 | 100 |

Therefore, when both propane and methane are placed in the 10.0-L container, the total pressure of the mixture is 800. torr.

STUDY CHECK

A gas mixture consists of helium with a partial pressure of 315 mmHg, nitrogen with a partial pressure of 204 mmHg, and argon with a partial pressure of 422 mmHg. What is the total pressure in atmospheres?

## Air Is a Gas Mixture

The air you breathe is a mixture of gases. What we call the atmospheric pressure is actually the sum of the partial pressures of the gases in the air. Table 6.5 lists partial pressures for the gases in air on a typical day.

SAMPLE PROBLEM    6.14

■ **Partial Pressure of a Gas in a Mixture**

A mixture of oxygen and helium is prepared for a scuba diver who is going to descend 200 ft below the ocean surface. At that depth, the diver breathes a gas mixture that has a total pressure of 7.0 atm. If the partial pressure of the oxygen in the tank at that depth is 1140 mmHg, what is the partial pressure of the helium in atm?

SOLUTION

From Dalton's law of partial pressures, we know that the total pressure is equal to the sum of the partial pressures:

$$P_{total} = P_{O_2} + P_{He}$$

To solve for the partial pressure of helium ($P_{He}$), we rearrange the expression to give the following:

$$P_{He} = P_{total} - P_{O_2}$$

Convert units of pressure to match $P_{He} = P_{total} - P_{O_2}$.

$$P_{O_2} = 1140 \text{ mmHg} \times \frac{1 \text{ atm}}{760 \text{ mmHg}} = 1.50 \text{ atm}$$

Substitute known pressures, and calculate the final pressure of He.

$$P_{He} = 7.0 \text{ atm} - 1.5 \text{ atm}$$
$$= 5.5 \text{ atm}$$

Thus, in the gas mixture that the diver breathes, the partial pressure of the helium is 5.5 atm.

STUDY CHECK

An anesthetic consists of a mixture of cyclopropane gas, $C_3H_6$, and oxygen gas, $O_2$. If the mixture has a total pressure of 825 torr, and the partial pressure of the cyclopropane is 73 torr, what is the partial pressure of the oxygen in the anesthetic?

# Health Note

## Blood Gases

Our cells continuously use oxygen and produce carbon dioxide. Both gases move in and out of the lungs through the membranes of the alveoli, the tiny air sacs at the ends of the airways in the lungs. An exchange of gases occurs in which oxygen from the air diffuses into the lungs and into the blood, while carbon dioxide produced in the cells is carried to the lungs to be exhaled. In Table 6.6, partial pressures are given for the gases in air that we inhale (inspired air), air in the alveoli, and the air that we exhale (expired air).

At sea level, oxygen normally has a partial pressure of 100 mmHg in the alveoli of the lungs. Because the partial pressure of oxygen in venous blood is 40 mmHg, oxygen diffuses from the alveoli into the bloodstream. The oxygen combines with hemoglobin, which carries it to the tissues of the body where the partial pressure of oxygen can be very low, less than 30 mmHg. Oxygen diffuses from the blood where the partial pressure of $O_2$ is high into the tissues where $O_2$ pressure is low.

As oxygen is used in the cells of the body during metabolic processes, carbon dioxide is produced, so the partial pressure of $CO_2$ may be as high as 50 mmHg or more. Carbon dioxide diffuses from the tissues into the bloodstream and is carried to the lungs. There it diffuses out of the blood, where $CO_2$ has a partial pressure of 46 mmHg, into the alveoli, where the $CO_2$ is at 40 mmHg and is exhaled. Table 6.7 gives the partial pressures of blood gases in the tissues and in oxygenated and deoxygenated blood.

**TABLE 6.6** Partial Pressures of Gases During Breathing

| Gas | Partial Pressure (mmHg) | | |
| --- | --- | --- | --- |
| | Inspired Air | Alveolar Air | Expired Air |
| Nitrogen, $N_2$ | 594.0 | 573 | 569 |
| Oxygen, $O_2$ | 160.0 | 100 | 116 |
| Carbon dioxide, $CO_2$ | 0.3 | 40 | 28 |
| Water vapor, $H_2O$ | 5.7 | 47 | 47 |
| Total | 760.0 | 760 | 760 |

**TABLE 6.7** Partial Pressures of Oxygen and Carbon Dioxide in Blood and Tissues

| Gas | Partial Pressure (mmHg) | | |
| --- | --- | --- | --- |
| | Oxygenated Blood | Deoxygenated Blood | Tissues |
| $O_2$ | 100 | 40 | 30 or less |
| $CO_2$ | 40 | 46 | 50 or greater |

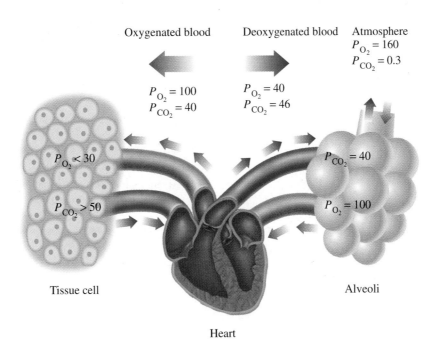

Oxygenated blood
$P_{O_2} = 100$
$P_{CO_2} = 40$

Deoxygenated blood
$P_{O_2} = 40$
$P_{CO_2} = 46$

Atmosphere
$P_{O_2} = 160$
$P_{CO_2} = 0.3$

$P_{O_2} < 30$
$P_{CO_2} > 50$

$P_{CO_2} = 40$
$P_{O_2} = 100$

Tissue cell

Alveoli

Heart

## Health Note

### Hyperbaric Chambers

A burn patient may undergo treatment for burns and infections in a hyperbaric chamber, a device in which pressures that are two to three times greater than atmospheric pressure can be obtained. A greater oxygen pressure increases the level of dissolved oxygen in the blood and tissues, where it fights bacterial infections. High levels of oxygen are toxic to many strains of bacteria. The hyperbaric chamber may also be used during surgery to help counteract carbon monoxide (CO) poisoning and to treat some cancers.

The blood is normally capable of dissolving up to 95% of the oxygen. Thus, if the partial pressure of the oxygen is 2280 mmHg (3 atm), about 2200 mmHg of oxygen can dissolve in the blood where it saturates the tissues. In the case of carbon monoxide poisoning, this oxygen can replace the carbon monoxide that has attached to the hemoglobin.

A patient undergoing treatment in a hyperbaric chamber must also undergo decompression (reduction of pressure) at a rate that slowly reduces the concentration of dissolved oxygen in the blood. If decompression is too rapid, the oxygen dissolved in the blood may form gas bubbles in the circulatory system.

If divers do not decompress slowly, they suffer a similar condition called the bends. While below the surface of the ocean, divers breathe air at higher pressures. At such high pressures, nitrogen gas will dissolve in their blood. If they ascend to the surface too quickly, the dissolved nitrogen forms bubbles in the blood that can produce life-threatening blood clots. The gas bubbles can also appear in the joints and tissues of the body and be quite painful. A diver suffering from the bends is placed immediately in a decompression chamber where pressure is first increased and then slowly decreased. The dissolved nitrogen can then diffuse through the lungs until atmospheric pressure is reached.

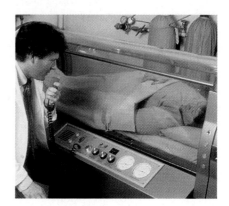

## QUESTIONS AND PROBLEMS

### Partial Pressures (Dalton's Law)

**6.47** A typical air sample in the lungs contains oxygen at 100 mmHg, nitrogen at 573 mmHg, carbon dioxide at 40 mmHg, and water vapor at 47 mmHg. Why are these pressures called partial pressures?

**6.48** Suppose a mixture contains helium and oxygen gases. If the partial pressure of helium is the same as the partial pressure of oxygen, what do you know about the number of helium atoms compared to the number of oxygen molecules? Explain.

**6.49** In a gas mixture, the partial pressures are nitrogen 425 torr, oxygen 115 torr, and helium 225 torr. What is the total pressure (torr) exerted by the gas mixture?

**6.50** In a gas mixture, the partial pressures are argon 415 mmHg, neon 75 mmHg, and nitrogen 125 mmHg. What is the total pressure (atm) exerted by the gas mixture?

**6.51** A gas mixture containing oxygen, nitrogen, and helium exerts a total pressure of 925 torr. If the partial pressures are oxygen 425 torr and helium 75 torr, what is the partial pressure (torr) of the nitrogen in the mixture?

**6.52** A gas mixture containing oxygen, nitrogen, and neon exerts a total pressure of 1.20 atm. If helium added to the mixture increases the pressure to 1.50 atm, what is the partial pressure (atm) of the helium?

**6.53** In certain lung ailments such as emphysema, there is a decrease in the ability of oxygen to diffuse into the blood.
**a.** How would the partial pressure of oxygen in the blood change?
**b.** Why does a person with severe emphysema sometimes use a portable oxygen tank?

**6.54** An accident to the head can affect the ability of a person to ventilate (breathe in and out), and so can certain drugs.
**a.** What would happen to the partial pressures of oxygen and carbon dioxide in the blood if a person cannot properly ventilate?
**b.** When a person with hypoventilation is placed on a ventilator, an air mixture is delivered at pressures that are alternately above the air pressure in the person's lung, and then below. How will this move oxygen gas into the lungs, and carbon dioxide out?

## CONCEPT MAP

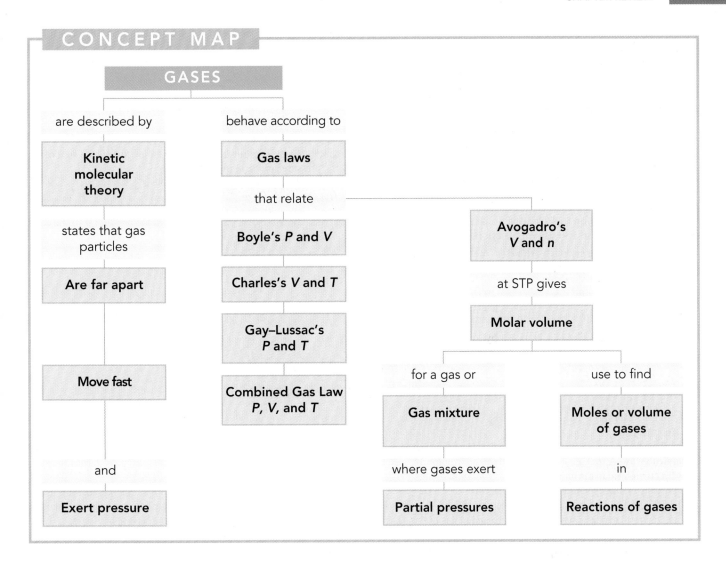

**GASES**

are described by

**Kinetic molecular theory**

states that gas particles

**Are far apart**

**Move fast**

and

**Exert pressure**

behave according to

**Gas laws**

that relate

**Boyle's *P* and *V***

**Charles's *V* and *T***

**Gay–Lussac's *P* and *T***

**Combined Gas Law *P*, *V*, and *T***

**Avogadro's *V* and *n***

at STP gives

**Molar volume**

for a gas or

**Gas mixture**

where gases exert

**Partial pressures**

use to find

**Moles or volume of gases**

in

**Reactions of gases**

# CHAPTER REVIEW

### 6.1 Properties of Gases

**Learning Goal:** Describe the kinetic molecular theory of gases and the properties of gases.

In a gas, particles are so far apart and moving so fast that their attractions are unimportant. A gas is described by the physical properties of pressure ($P$), volume ($V$), temperature ($T$), in kelvins (K) and the amount in moles ($n$).

### 6.2 Gas Pressure

**Learning Goal:** Describe the units of measurement used for pressure and change from one unit to another.

A gas exerts pressure, the force of the gas particles striking the surface of a container. Gas pressure is measured in units of torr, mmHg, atm, and Pa.

### 6.3 Pressure and Volume (Boyle's Law)

**Learning Goal:** Use the pressure–volume relationship (Boyle's law) to determine the new pressure or volume of a certain amount of gas at a constant temperature.

The volume ($V$) of a gas changes inversely with the pressure ($P$) of the gas if there is no change in the amount and temperature: $P_1V_1 = P_2V_2$. This means that the pressure increases if volume decreases; pressure decreases if volume increases.

### 6.4 Temperature and Volume (Charles's Law)

**Learning Goal:** Use the temperature–volume relationship (Charles's law) to determine the new temperature or volume of a certain amount of gas at a constant pressure.

The volume ($V$) of a gas is directly related to its Kelvin temperature ($T$) when there is no change in the amount and pressure of the gas.

$$\frac{V_1}{T_1} = \frac{V_2}{T_2}$$

Therefore, if temperature increases, the volume of the gas increases; if temperature decreases, volume decreases.

### 6.5 Temperature and Pressure (Gay–Lussac's Law)

**Learning Goal:** Use the temperature–pressure relationship (Gay–Lussac's law) to determine the new temperature or pressure of a certain amount of gas at a constant volume.

The pressure ($P$) of a gas is directly related to its Kelvin temperature ($T$).

$$\frac{P_1}{T_1} = \frac{P_2}{T_2}$$

This means that an increase in temperature increases the pressure of a gas, or a decrease in temperature decreases the pressure as long as the amount and volume stay constant.

## 6.6 The Combined Gas Law

**Learning Goal:** Use the combined gas law to find the new pressure, volume, or temperature of a gas when changes in two of these properties are given.

Gas laws combine into a relationship of pressure ($P$), volume ($V$), and temperature ($T$) for a constant amount ($n$) of gas.

$$\frac{P_1 V_1}{T_1} = \frac{P_2 V_2}{T_2}$$

This expression is used to determine the effect of changes in two of the variables on the third.

## 6.7 Volume and Moles (Avogadro's Law)

**Learning Goal:** Describe the relationship between the amount of a gas and its volume, and use this relationship in calculations.

The volume ($V$) of a gas is directly related to the number of moles ($n$) of the gas when the pressure and temperature of the gas do not change.

$$\frac{V_1}{n_1} = \frac{V_2}{n_2}$$

If the moles of gas are increased, the volume must increase; if the moles of gas are decreased, the volume must decrease. At standard temperature (273 K) and pressure (1 atm), abbreviated STP, 1 mole of any gas has a volume of 22.4 L.

## 6.8 Partial Pressures (Dalton's Law)

**Learning Goal:** Use partial pressures to calculate the total pressure of a mixture of gases.

In a mixture of two or more gases, the total pressure is the sum of the partial pressures of the individual gases.

$$P_{total} = P_1 + P_2 + P_3 + \cdots$$

The partial pressure of a gas in a mixture is the pressure it would exert if it were the only gas in the container.

# KEY TERMS

**atmosphere (atm)** The pressure exerted by a column of mercury 760 mm high.

**atmospheric pressure** The pressure exerted by the atmosphere.

**Avogadro's law** A gas law that states that the volume of gas is directly related to the number of moles of gas in the sample when pressure and temperature do not change.

**Boyle's law** A gas law stating that the pressure of a gas is inversely related to the volume when temperature and moles of the gas do not change.

**Charles's law** A gas law stating that the volume of a gas changes directly with a change in Kelvin temperature when pressure and moles of the gas do not change.

**combined gas law** A relationship that combines several gas laws relating pressure, volume, and temperature when the amount of gas does not change.

$$\frac{P_1 V_1}{T_1} = \frac{P_2 V_2}{T_2}$$

**Dalton's law** A gas law stating that the total pressure exerted by a mixture of gases in a container is the sum of the partial pressures that each gas would exert alone.

**direct relationship** A relationship in which two properties increase or decrease together.

**Gay–Lussac's law** A gas law stating that the pressure of a gas changes directly with a change in temperature when the number of moles of a gas and its volume do not change.

**inverse relationship** A relationship in which two properties change in opposite directions.

**kinetic molecular theory of gases** A model used to explain the behavior of gases.

**molar volume** A volume of 22.4 L occupied by 1 mole of a gas at STP conditions of 0 °C (273 K) and 1 atm.

**partial pressure** The pressure exerted by a single gas in a gas mixture.

**pressure** The force exerted by gas particles that hit the walls of a container.

**STP** Standard conditions of 0 °C (273 K) temperature and 1 atm pressure used for the comparison of gases.

# UNDERSTANDING THE CONCEPTS

**6.55** At 100 °C, which of the following gas samples exerts
  **a.** the lowest pressure?
  **b.** the highest pressure?

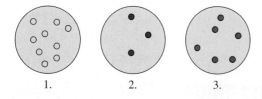

1.        2.        3.

**6.56** Indicate which diagram represents the volume of the gas sample in a flexible container when each of the following changes takes place:

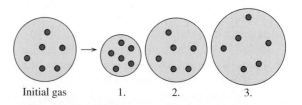

Initial gas        1.        2.        3.

**a.** Temperature increases at constant pressure.
**b.** Temperature decreases at constant pressure.
**c.** Pressure increases at constant temperature.
**d.** Pressure decreases at constant temperature.
**e.** Doubling the pressure and doubling the Kelvin temperature.

**6.57** A balloon is filled with helium gas with a pressure of 1.00 atm and neon gas with a pressure of 0.50 atm. For each of the following changes of the initial balloon, select the diagram (A, B, or C) that shows the final (new) volume of the balloon:

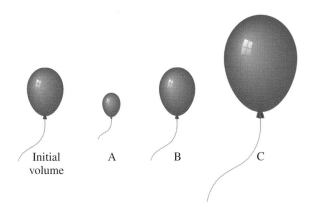

Initial     A       B       C
volume

**a.** The balloon is put in a cold storage unit ($P$ and $n$ constant).
**b.** The balloon floats to a higher altitude where the pressure is less ($n$, $T$ constant).
**c.** All of the neon gas is removed ($T$ constant).
**d.** The Kelvin temperature doubles, and one-half of the gas atoms leak out ($P$ constant).
**e.** 2.0 moles of $O_2$ gas is added at constant $T$ and $P$.

**6.58** Indicate if pressure increases, decreases, or stays the same in each of the following:

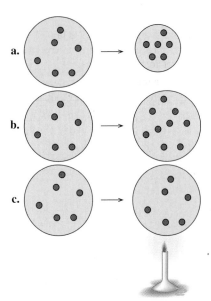

**6.59** At a restaurant, a customer chokes on a piece of food. You put your arms around the person's waist and use your fists to push up on the person's abdomen, an action called the Heimlich maneuver.
**a.** How would this action change the volume of the chest and lungs?
**b.** Why does it cause the person to expel the food item from the airway?

**6.60** An airplane is pressurized to 650 mmHg, which is the atmospheric pressure at a ski resort at 13 000 ft altitude.
**a.** If air is 21% oxygen, what is the partial pressure of oxygen on the plane?
**b.** If the partial pressure of oxygen drops below 100 mmHg, passengers become drowsy. If this happens, oxygen masks are released. What is the total cabin pressure at which oxygen masks are dropped?

# ADDITIONAL QUESTIONS AND PROBLEMS

**6.61** In 1783, Jacques Charles launched his first balloon filled with hydrogen gas because it was lighter than air. If the balloon has a volume of 31 000 L, how many grams of hydrogen would be needed to fill the balloon at STP?

**6.62** In problem 6.61, the balloon reached an altitude of 1000 m, where the pressure was 658 mmHg and the temperature was $-8\,°C$. What was the volume in liters of the balloon at these conditions?

**6.63** A fire extinguisher has a pressure of 10. atm at 25 °C. What is the pressure in atmospheres if the fire extinguisher is used at a temperature of 75 °C?

**6.64** A weather balloon has a volume of 750 L when filled with helium at 8 °C at a pressure of 380 torr. What is the new volume of the balloon, where the pressure is 0.20 atm and the temperature is $-45\,°C$?

**6.65** A sample of hydrogen ($H_2$) gas at 127 °C has a pressure of 2.00 atm. At what temperature (°C) will the pressure of the $H_2$ decrease to 0.25 atm?

**6.66** A sample of nitrogen ($N_2$) and helium has a volume of 250 mL at 30 °C and a total pressure of 745 mmHg.

**a.** If the pressure of helium is 32 mmHg, what is the partial pressure of the nitrogen?

**b.** What is the volume of the nitrogen at STP?

**6.67** A weather balloon is partially filled with helium to allow for expansion at high altitudes. At STP, a weather balloon is filled with enough helium to give a volume of 25.0 L. At an altitude of 30.0 km and −35 °C, it has expanded to 2460 L. The increase in volume causes it to burst and a small parachute returns the instruments to Earth.

**a.** How many grams of helium are added to the balloon?

**b.** What is the pressure in mmHg of the helium inside the balloon when it bursts?

**6.68** What is the total pressure in mmHg of a gas mixture containing argon gas at 0.25 atm, helium gas at 350 mmHg, and nitrogen gas at 360 torr?

**6.69** A gas mixture contains oxygen and argon at partial pressures of 0.60 atm and 425 mmHg. If nitrogen gas added to the sample increases the total pressure to 1250 torr, what is the partial pressure in torr of the nitrogen added?

**6.70** A gas mixture contains helium and oxygen at partial pressures of 255 torr and 0.450 atm. What is the total pressure in mmHg of the mixture after it is placed in a container one-half the volume of the original container?

**6.71** What is the density (g/L) of oxygen gas at STP?

**6.72** At a party, a helium-filled balloon floats close to some hot lights and bursts. What might be a reason for this?

**6.73** Compare the partial pressures of the following respiratory gases:

**a.** oxygen in the lungs and in the blood coming to the alveoli

**b.** oxygen in arterial blood and venous blood

**c.** carbon dioxide in the tissues and arterial blood

**d.** carbon dioxide in the venous blood and the lungs

**6.74** For each of the comparisons in problem 6.73, describe the direction of diffusion for each gas.

**6.75** When heated, calcium carbonate decomposes to give calcium oxide and carbon dioxide gas.

$$CaCO_3(s) \rightarrow CaO(s) + CO_2(g)$$

If 2.00 moles of $CaCO_3$ react, how many liters of $CO_2$ gas are produced at STP?

**6.76** Magnesium reacts with oxygen to form magnesium oxide. How many liters of oxygen gas at STP are needed to react completely with 8.0 g of magnesium?

$$2Mg(s) + O_2(g) \rightarrow 2MgO(s)$$

**6.77** Your spaceship has docked at a space station above Mars. The temperature inside the space station is a carefully controlled 24 °C at a pressure of 745 mmHg. A balloon with a volume of 425 mL drifts into the air lock where the temperature is −95 °C and the pressure is 0.115 atm. What is the new volume of the balloon? Assume that the balloon is very elastic.

**6.78** How many liters of $H_2$ gas can be produced at STP from 25.0 g of Zn?

$$Zn(s) + 2HCl(aq) \rightarrow ZnCl_2(aq) + H_2(g)$$

**6.79** Aluminum oxide can be formed from its elements.

$$4Al(s) + 3O_2(g) \rightarrow 2Al_2O_3(s)$$

What volume of oxygen is needed at STP to completely react 5.4 g of aluminum?

**6.80** Glucose, $C_6H_{12}O_6$, is metabolized in living systems according to the reaction

$$C_6H_{12}O_6(s) + 6O_2(g) \rightarrow 6CO_2(g) + 6H_2O(l)$$

How many grams of water can be produced when 12.5 L of $O_2$ reacts at STP?

# CHALLENGE QUESTIONS

**6.81** Two flasks of equal volume and at the same temperature contain different gases. One flask contains 1.00 g of Ne, and the other flask contains 1.00 g of He. Which of the following statements are correct? Explain your answers.

**a.** Both flasks contain the same number of atoms.

**b.** The pressures in the flasks are the same.

**c.** The flask that contains helium has a higher pressure than the flask that contains neon.

**d.** The densities of the gases are the same.

**6.82** In the fermentation of glucose (wine making), 780 mL of $CO_2$ gas was produced at 37 °C and 1.00 atm. What is the volume (L) of the gas when measured at 22 °C and 675 mmHg?

**6.83** A gas sample has a volume of 4250 mL at 15 °C and 745 mmHg. What is the new temperature (°C) after the sample is transferred to a new container with a volume of 2.50 L and a pressure of 1.20 atm?

**6.84** A weather balloon has a volume of 750 L when filled with helium at 8 °C at a pressure of 380 torr. What is the new volume of the balloon, where the pressure is 0.20 atm and the temperature is −45 °C?

**6.85** A liquid is placed in a 25.0-L flask. At 140 °C, the liquid evaporates completely to give a pressure of 0.900 atm. If the flask can withstand pressures up to 1.30 atm, calculate the maximum temperature that the gas can be heated to without breaking.

**6.86** You are doing research on planet X, and you take your favorite balloon with you. The temperature inside the space station is a carefully controlled 24 °C and the pressure is 755 mmHg. Your balloon drifts into the air lock, which cycles, sending your balloon, which has a volume of 850 mL, into the atmosphere of planet X. If the night temperature on planet X is −103 °C and the pressure is 0.150 atm, what volume will your balloon occupy assuming that it is very elastic and will not burst?

**6.87** When sensors in a car detect a collision, they cause the reaction of sodium azide, $NaN_3$, which generates nitrogen gas to fill the air bags within 0.03 second.

$$2NaN_3(s) \rightarrow 2Na(s) + 3N_2(g)$$

How many liters of $N_2$ are produced at STP if the air bag contains 132 g of $NaN_3$?

**6.88** Nitrogen dioxide reacts with water to produce oxygen and ammonia.

$$4NO_2(g) + 6H_2O(g) \rightarrow 7O_2(g) + 4NH_3(g)$$

How many liters of oxygen at STP are produced when $2.5 \times 10^{23}$ molecules of nitrogen dioxide react?

**6.89** As seen in Chapter 1, one teragram (Tg) is equal to $10^{12}$ g. In 2000, $CO_2$ emissions from the generation of electricity for use in homes in the United States was 780 Tg. In 2020, it is estimated that $CO_2$ emissions from the generation of electricity for use in homes will be 990 Tg.

**a.** Calculate the number of kilograms of $CO_2$ emitted for the years 2000 and 2020.
**b.** Calculate the number of moles of $CO_2$ emitted for the years 2000 and 2020.

**c.** What is the increase in megagrams for the $CO_2$ emissions between the years 2000 and 2020?

**6.90** As seen in Chapter 1, one teragram (Tg) is equal to $10^{12}$ g. In 2000, $CO_2$ emissions from fuels used for transportation in the United States was 1990 Tg. In 2020, it is estimated that $CO_2$ emissions from the fuels used for transportation in the United States will be 2760 Tg.

**a.** Calculate the number of kilograms of $CO_2$ emitted for the years 2000 and 2020.
**b.** Calculate the number of moles of $CO_2$ emitted for the years 2000 and 2020.
**c.** What is the increase in megagrams for the $CO_2$ emissions between the years 2000 and 2020?

# ANSWERS

## Answers to Study Checks

**6.1** The gas molecules that carry the odor of the food move throughout the house until they reach the room that you are in.

**6.2** The mass in grams gives the amount of gas.

**6.3** 0.862 atm

**6.4** 250 torr

**6.5** 50.0 mL

**6.6** 569 mL

**6.7** 16 °C

**6.8** 241 mmHg

**6.9** 7.50 L

**6.10** 0.25 mole of $N_2$

**6.11** 7.0 g of $N_2$

**6.12** 5.41 L of $H_2$

**6.13** 1.24 atm

**6.14** 752 torr

## Answers to Selected Questions and Problems

**6.1** **a.** At a higher temperature, gas particles have greater kinetic energy, which makes them move faster.
**b.** Because there are great distances between the particles of a gas, they can be pushed closer together and still remain a gas.

**6.3** **a.** temperature
**b.** volume
**c.** amount
**d.** pressure

**6.5** atmospheres (atm), mmHg, torr, lb/in.$^2$, kPa

**6.7** **a.** 1520 torr
**b.** 1520 mmHg

**6.9** As a diver ascends to the surface, external pressure decreases. If the air in the lungs were not exhaled, its volume would expand and severely damage the lungs. The pressure in the lungs must adjust to changes in the external pressure.

**6.11** **a.** The pressure is greater in cylinder A. According to Boyle's law, a decrease in volume pushes the gas particles closer together, which will cause an increase in the pressure.
**b.**

| Property | Conditions 1 | Conditions 2 | Know | Predict |
|---|---|---|---|---|
| Pressure (P) | 650 mmHg | 1.2 atm (910 mmHg) | P increase | |
| Volume (V) | 220 mL | 160 mL | | V decreases |

**6.13** **a.** The pressure doubles.
**b.** The pressure falls to one-third the initial pressure.
**c.** The pressure increases to ten times the original pressure.

**6.15** **a.** 328 mmHg    **b.** 2620 mmHg    **c.** 4400 mmHg

**6.17** **a.** 25 L    **b.** 25 L    **c.** 100. L

**6.19** 25 L

**6.21** **a.** inspiration    **b.** expiration    **c.** inspiration

**6.23** **a.** C    **b.** A    **c.** B

**6.25** **a.** 303 °C    **b.** −129 °C
**c.** 591 °C    **d.** 136 °C

**6.27** **a.** 2400 mL    **b.** 4900 mL
**c.** 1800 mL    **d.** 1700 mL

**6.29** An increase in temperature increases the pressure inside the can. When the pressure exceeds the pressure limit of the can, it explodes.

**6.31** **a.** −23°C    **b.** 168 °C

**6.33** **a.** 770 torr    **b.** 1.51 atm

**6.35** **a.** 4.26 atm    **b.** 3.07 atm    **c.** 0.606 atm

**6.37** 110 mL

**6.39** The volume increases because the number of gas particles is increased.

**6.41** **a.** 4.00 L    **b.** 14.6 L    **c.** 26.7 L

**6.43** **a.** 2.00 moles of $O_2$    **b.** 0.179 mole of $CO_2$
**c.** 4.48 L    **d.** 55 400 mL

**6.45** 7.60 L of $H_2$

**6.47** In a gas mixture, the pressure that each gas exerts as part of the total pressure is called the partial pressure of that gas. Because the air sample is a mixture of gases, the total pressure is the sum of the partial pressures of each gas in the sample.

**6.49** 765 torr

**6.51** 425 torr

**6.53** **a.** The partial pressure of oxygen will be lower than normal.
**b.** Breathing a higher concentration of oxygen will help to increase the supply of oxygen in the lungs and blood and raise the partial pressure of oxygen in the blood.

**6.55** **a.** 2      **b.** 1

**6.57** **a.** A    **b.** C    **c.** A    **d.** B    **e.** C

**6.59** **a.** The volume of the chest and lungs is decreased.
**b.** The decrease in volume increases the pressure, which can dislodge the food in the trachea.

**6.61** $2.8 \times 10^3$ g of $H_2$

**6.63** 12 atm

**6.65** $-223\,°C$

**6.67** **a.** 4.46 g of helium          **b.** 6.73 mmHg

**6.69** 370 torr

**6.71** 1.43 g/L

**6.73** **a.** The $P_{O_2}$ is high in the lungs and low in the blood coming to the lungs.
**b.** The $P_{O_2}$ is high in arterial blood and low in venous blood.
**c.** The $P_{CO_2}$ is high in the tissues and low in arterial blood.
**d.** The $P_{CO_2}$ is high in the venous blood and low in the lungs.

**6.75** 44.8 L of $CO_2$

**6.77** 2170 mL

**6.79** 3.4 L of $O_2(g)$

**6.81** **a.** False. The flask containing helium gas contains more atoms because one gram of helium contains more moles of helium and therefore more atoms than one gram of neon.
**b.** False. There are different numbers of moles of gas in the flask, which means that the pressures are different.
**c.** True. There are more moles of helium, which makes the pressure of helium greater than that of neon.
**d.** True. Density is mass divided by volume. If gases have the same mass and volume, they have the same density.

**6.83** $-66\,°C$

**6.85** $324\,°C$

**6.87** 68.2 L at STP

**6.89** **a.** $7.8 \times 10^{11}$ kg of $CO_2$ (2000); $9.9 \times 10^{11}$ kg of $CO_2$ (2020)
**b.** $1.8 \times 10^{13}$ moles of $CO_2$ (2000); $2.3 \times 10^{13}$ moles of $CO_2$ (2020)
**c.** $2.1 \times 10^8$ Mg of $CO_2$ increase

# 7 Solutions

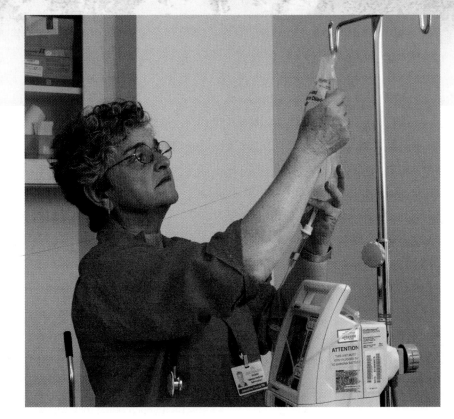

*"There is a lot of chemistry going on in the body, including drug interactions," says Josephine Firenze, registered nurse, Kaiser Hospital.*

*Normally, the body maintains a homeostasis of fluids and electrolytes. Conditions that alter the composition of body fluids can lead to convulsions, coma, or death. To halt the disease process and to establish homeostasis, a patient may be given intravenous fluid therapy. Solutions that are compatible with body fluids such as a 5% glucose or a 0.9% saline are used. An infusion pump delivers the desired number of milliliters per hour to the patient. During IV therapy, a patient is checked for fluid overload as indicated by edema, which is swelling, or a greater fluid input than output.*

**S**olutions are everywhere around us. Most consist of one substance dissolved in another. The air we breathe is a solution of oxygen and nitrogen gases. Carbon dioxide gas dissolved in water makes carbonated drinks. When we make solutions of coffee or tea, we use hot water to dissolve substances from coffee beans or tea leaves. The ocean is also a solution, consisting of many salts such as sodium chloride dissolved in water. In a hospital, the antiseptic tincture of iodine is a solution of iodine dissolved in alcohol.

Our body fluids contain water and dissolved substances such as glucose and urea and electrolytes such as $K^+$, $Na^+$, $Cl^-$, $Mg^{2+}$, $HCO_3^-$, and $HPO_4^{2-}$. Proper amounts of each of these dissolved substances and water must be maintained in the body fluids. Small changes in electrolyte levels can seriously disrupt cellular process and endanger our health. Therefore, the measurement of their concentrations is a valuable diagnostic tool.

In the processes of osmosis and dialysis, water, essential nutrients, and waste products enter and leave the cells of the body. In osmosis, water flows in and out of the cells of the body. In dialysis, small particles in solution as well as water diffuse through semipermeable membranes. The kidneys utilize osmosis and dialysis to regulate the amount of water and electrolytes that are excreted.

## LEARNING GOAL

Identify the solute and solvent in a solution. Describe the formation of a solution.

Solute: The substance present in lesser amount

— Salt

— Water

Solvent: The substance present in greater amount

## 7.1 SOLUTIONS

A **solution** is a homogeneous mixture in which one substance called the **solute** is uniformly dispersed in another substance called the **solvent**. Because the solute and the solvent do not react with each other, they can be mixed in varying proportions. A little salt dissolved in water tastes slightly salty. When more salt dissolves, the water tastes very salty. Usually, the solute (in this case, salt) is the substance present in the smaller amount, whereas the solvent (in this case, water) is present in the larger amount. In a solution, the particles of the solute are evenly dispersed among the molecules of the solvent. (See Figure 7.1.)

### Types of Solutes and Solvents

Solutes and solvents may be solids, liquids, or gases. The solution that forms has the same physical state as the solvent. When sugar crystals are dissolved in water, the resulting sugar solution is liquid. Sugar is the solute, and water is the solvent. Soda water and soft drinks are prepared by dissolving carbon dioxide gas in water. The carbon dioxide gas is the solute, and water is the solvent. Table 7.1 lists some solutes and solvents and their solutions.

---

**SAMPLE PROBLEM** 7.1

■ **Identifying a Solute and a Solvent**

Identify the solute and the solvent in each of the following solutions:

**a.** 15 g of sugar dissolved in 100 mL of water
**b.** 75 mL of water mixed with 25 mL of isopropyl alcohol (rubbing alcohol)

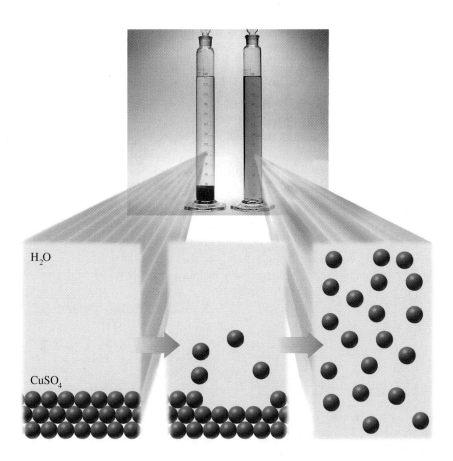

### SOLUTION

**a.** Sugar, the smaller quantity, is the solute; water is the solvent.

**b.** Isopropyl alcohol with the smaller volume is the solute. Water is the solvent.

### STUDY CHECK

A tincture of iodine is prepared with 0.10 g of $I_2$ and 10.0 mL of ethyl alcohol. What is the solute, and what is the solvent?

**TABLE 7.1** Some Examples of Solutions

| Type | Example | Solute | Solvent |
|------|---------|--------|---------|
| **Gas Solutions** | | | |
| Gas in a gas | Air | Oxygen (gas) | Nitrogen (gas) |
| **Liquid Solutions** | | | |
| Gas in a liquid | Soda water | Carbon dioxide (gas) | Water (liquid) |
| | Household ammonia | Ammonia (gas) | Water (liquid) |
| Liquid in a liquid | Vinegar | Acetic acid (liquid) | Water (liquid) |
| Solid in a liquid | Seawater | Sodium chloride (solid) | Water (liquid) |
| | Tincture of iodine | Iodine (solid) | Alcohol (liquid) |
| **Solid Solutions** | | | |
| Liquid in a solid | Dental amalgam | Mercury (liquid) | Silver (solid) |
| Solid in a solid | Brass | Zinc (solid) | Copper (solid) |
| | Steel | Carbon (solid) | Iron (solid) |

## Water as a Solvent

Water is one of the most common solvents in nature. In the $H_2O$ molecule, an oxygen atom shares electrons with two hydrogen atoms. Because the oxygen atom is much more electronegative, the O—H bonds are polar. In each of the polar bonds, the oxygen atom has a partial negative ($\delta^-$) charge, and the hydrogen atom has a partial positive ($\delta^+$) charge. Because of the arrangement of the polar bonds, water is a *polar substance*.

**Hydrogen bonds** occur between molecules where a partially positive hydrogen is attracted to the strongly electronegative atoms of O, N, or F in other molecules. In water, hydrogen bonds are formed by the attraction between the oxygen atom of one water molecule and a hydrogen atom in another water molecule. In the diagram, hydrogen bonds are shown as dotted lines between the water molecules. Although hydrogen bonds are much weaker than covalent or ionic bonds, there are many of them linking molecules together. As a result, hydrogen bonding plays an important role in the properties of water and biological compounds such as proteins, carbohydrates, and DNA.

**WEB TUTORIAL**
Hydrogen Bonding

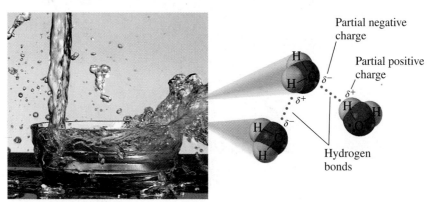

## Formation of Solutions

Ionic compounds such as sodium chloride, NaCl, are held together by ionic bonds. For NaCl, attractions occur between the positively charged $Na^+$ ions and the negatively charged $Cl^-$ ions. Water is a good solvent for many ionic compounds because water molecules are polar. When NaCl crystals are placed in water, water molecules collide with the ions on the surface of the crystal. (See Figure 7.2.) The negatively charged oxygen atom at one end of a water molecule attracts the positive $Na^+$ ions. The positively charged hydrogen atoms of other water molecules attract the negative $Cl^-$ ions. The attractive forces of several water molecules provide the energy to break the ionic bonds between the $Na^+$ and $Cl^-$ ions in the NaCl crystal. As soon as the $Na^+$ and $Cl^-$ ions are dissolved, they undergo **hydration** as water molecules surround each ion. The hydration of the ions diminishes their attraction to other ions and helps keep them in solution. In the equation for the formation of the NaCl solution, the solid and aqueous NaCl are shown along with the formula $H_2O$ over the arrow, which indicates that water is needed for the dissociation process, but it is not a reactant.

$$NaCl(s) \xrightarrow{H_2O} Na^+(aq) + Cl^-(aq)$$

## Like Dissolves Like

Gases form solutions with other gases because their particles are moving so rapidly that they are far apart and attractions to the other gas particles are not important. When solids or liquids form solutions, there must be an attraction between the solute particles and the solvent particles. Then the particles of the solute and solvent will mix together. If there is no attraction between a solute and a solvent, their particles do not mix and no solution forms.

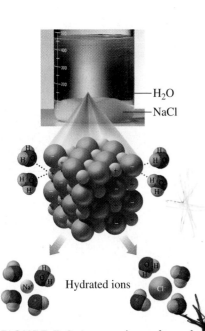

**FIGURE 7.2** Ions on the surface of a crystal of NaCl dissolve in water as they are attracted to the polar water molecules that pull the ions into solution and surround them.
**Q** What helps keep the $Na^+$ and $Cl^-$ ions in solution?

# Health Note

## Water in the Body

The average adult contains about 60% water by weight, and the average infant about 75%. About 60% of the body's water is contained within the cells as intracellular fluids; the other 40% makes up extracellular fluids, which include the interstitial fluid in tissue and the plasma in the blood. These external fluids carry nutrients and waste materials between the cells and the circulatory system.

Every day you lose between 1500 and 3000 mL of water from the kidneys as urine, from the skin as perspiration, from the lungs as you exhale, and from the gastrointestinal tract. Serious dehydration can occur in an adult if there is a 10% net loss in total body fluid, and a 20% loss of fluid can be fatal. An infant suffers severe dehydration with a 5–10% loss in body fluid.

Water loss is continually replaced by the liquids and foods in the diet and from metabolic processes that produce water in the cells of the body. Table 7.2 lists the % by mass of water contained in some foods.

### 24 Hours

| Water gain | | | Water loss | |
|---|---|---|---|---|
| Liquid | 1000 mL | | Urine | 1500 mL |
| Food | 1200 mL | | Perspiration | 300 mL |
| Metabolism | 300 mL | | Breath | 600 mL |
| | | | Feces | 100 mL |
| Total | 2500 mL | | Total | 2500 mL |

**TABLE 7.2  Percentage of Water in Some Foods**

| Food | Water (% by mass) | Food | Water (% by mass) |
|---|---|---|---|
| **Vegetables** | | **Meats/Fish** | |
| Carrot | 88 | Chicken, cooked | 71 |
| Celery | 94 | Hamburger, broiled | 60 |
| Cucumber | 96 | Salmon | 71 |
| Tomato | 94 | | |
| | | **Grains** | |
| **Fruits** | | Cake | 34 |
| Apple | 85 | French bread | 31 |
| Banana | 76 | Noodles, cooked | 70 |
| Cantaloupe | 91 | | |
| Orange | 86 | **Milk Products** | |
| Strawberry | 90 | Cottage cheese | 78 |
| Watermelon | 93 | Milk, whole | 87 |
| | | Yogurt | 88 |

A salt such as NaCl will form a solution with water because the $Na^+$ and $Cl^-$ ions in the salt are attracted to the positive and negative parts of the individual water molecules. A covalent compound such as methanol, $CH_3—OH$, dissolves in water because the molecule has a polar OH group that forms hydrogen bonds with the water.

However, compounds containing nonpolar molecules such as iodine ($I_2$), oil, or grease do not dissolve in water because water is polar. Nonpolar solutes require

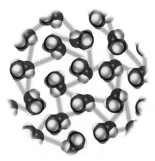

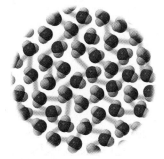

Methanol ($CH_3OH$) solute      Water solvent      Methanol–water solution with hydrogen bonding

## Explore Your World

### Like Dissolves Like

Mix together small amounts of the following substances:

**a.** oil and water
**b.** water and vinegar
**c.** salt and water
**d.** sugar and water
**e.** salt and oil

#### QUESTIONS

1. Which of the mixtures formed a solution? Which did not?
2. Why do some mixtures form solutions, but others do not?

nonpolar solvents for a solution to form. The expression "like dissolves like" is a way of saying that the polarities of a solute and a solvent must be similar in order to form a solution. Figure 7.3 illustrates the formation of some polar and nonpolar solutions.

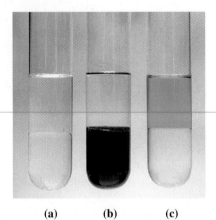

**(a)      (b)      (c)**

**FIGURE 7.3** Like dissolves like. **(a)** The test tubes contain an upper layer of water (polar) and a lower layer of $CH_2Cl_2$ (nonpolar). **(b)** The nonpolar solute $I_2$ dissolves in the nonpolar layer. **(c)** The ionic solute, $Ni(NO_3)_2$, dissolves in the water.

**Q** Which layer would dissolve polar molecules of sugar?

---

### SAMPLE PROBLEM 7.2

#### ■ Polar and Nonpolar Solutes

Indicate whether each of the following substances will dissolve in water. Explain.

**a.** KCl
**b.** octane, $C_8H_{18}$, a compound in gasoline
**c.** ethanol, $C_2H_5OH$, a substance in mouthwash

#### SOLUTION

**a.** Yes. KCl is an ionic compound.
**b.** No. $C_8H_{18}$ is a nonpolar substance.
**c.** Yes. $C_2H_5OH$ is a polar substance.

#### STUDY CHECK

Will oil, a nonpolar substance, dissolve in hexane, a nonpolar solvent?

---

## QUESTIONS AND PROBLEMS

### Solutions

**7.1** Identify the solute and the solvent in each solution composed of the following:
**a.** 10.0 g of NaCl and 100.0 g of $H_2O$
**b.** 50.0 mL of ethanol, $C_2H_5OH(l)$, and 10.0 mL of $H_2O$
**c.** 0.20 L of $O_2$ and 0.80 L of $N_2$

**7.2** Identify the solute and the solvent in each solution composed of the following:
**a.** 50.0 g of silver and 4.0 g of mercury
**b.** 100.0 mL of water and 5.0 g of sugar
**c.** 1.0 g of $I_2$ and 50.0 mL of alcohol

**7.3** Describe the formation of an aqueous KI solution.

**7.4** Describe the formation of an aqueous LiBr solution.

**7.5** Water is a polar solvent; $CCl_4$ is a nonpolar solvent. In which solvent is each of the following more likely to be soluble?
**a.** KCl, ionic         **b.** $I_2$, nonpolar
**c.** sugar, polar      **d.** gasoline, nonpolar

**7.6** Water is a polar solvent; hexane is a nonpolar solvent. In which solvent is each of the following more likely to be soluble?
**a.** vegetable oil, nonpolar
**b.** benzene, nonpolar
**c.** $LiNO_3$, ionic      **d.** $Na_2SO_4$, ionic

---

### LEARNING GOAL

Identify solutes as electrolytes or nonelectrolytes.

## 7.2 ELECTROLYTES AND NONELECTROLYTES

Solutes can be classified by their ability to conduct an electrical current. When solutes called **electrolytes** dissolve in water, they separate into ions, which are able to conduct electricity. When solutes called **nonelectrolytes** dissolve in water, they do not separate into ions and their solutions do not conduct electricity.

To test solutions for ions, we can use an apparatus that consists of a battery and a pair of electrodes connected by wires to a light bulb. The light bulb glows when electricity can flow, which only happens when electrolytes provide ions to complete the circuit.

### Strong Electrolytes

A **strong electrolyte** is a compound that dissociates completely into ions when it dissolves in water. In a process called *dissociation*, the ions separate from the solid and form a solution that conducts electricity. The equation for the dissociation of NaCl in water is written with $H_2O$ over the arrow to show that water is needed for the dissociation process.

Using a single arrow from reactant to products indicates that the reactant is a strong electrolyte and dissociates completely.

$$NaCl(s) \xrightarrow{H_2O} Na^+(aq) + Cl^-(aq)$$

In an equation for dissociation of a compound in water, the electrical charges must balance. For example, magnesium nitrate dissociates to give one magnesium ion for every two nitrate ions. However, only the ionic bonds between $Mg^{2+}$ and $NO_3^-$ are broken, not the covalent bonds within the polyatomic ion. The dissociation for $Mg(NO_3)_2$ is written as follows:

$$Mg(NO_3)_2(s) \xrightarrow{H_2O} Mg^{2+}(aq) + 2NO_3^-(aq)$$

## Weak Electrolytes

A **weak electrolyte** is a compound that dissolves in water mostly as whole molecules. A few of the dissolved molecules separate, which produces a small number of ions in solution. Thus solutions of weak electrolytes do not conduct electrical current as well as solutions of strong electrolytes. For example, an aqueous solution of HF, which is a weak electrolyte, consists of mostly HF molecules and a few $H^+$ and $F^-$ ions. Within the solution, a few HF molecules dissociate into ions. As more $H^+$ and $F^-$ ions form, some recombine to give HF molecules indicated by a backwards arrow. Eventually the rate of formation of ions is equal to the rate at which they recombine as indicated by two arrows between reactant and products.

$$HF(aq) \underset{\text{Recombination}}{\overset{\text{Dissociation}}{\rightleftharpoons}} H^+(aq) + F^-(aq)$$

## Nonelectrolytes

A nonelectrolyte is a compound that dissolves in water as molecules. They do not separate into ions, and solutions of nonelectrolytes do not conduct electricity. For example, sucrose (sugar) is a nonelectrolyte that dissolves in water as whole molecules only.

$$\underset{\text{Sucrose}}{C_{12}H_{22}O_{11}(s)} \xrightarrow{H_2O} \underset{\text{Solution of sucrose molecules}}{C_{12}H_{22}O_{11}(aq)}$$

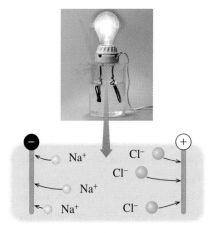

Strong electrolyte

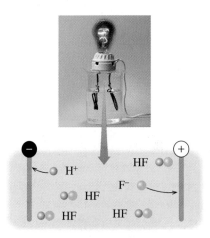

Weak electrolyte

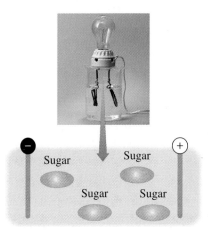

Nonelectrolyte

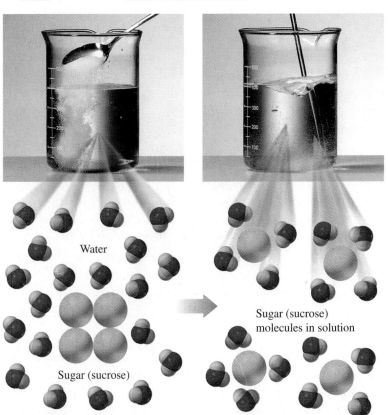

Water

Sugar (sucrose)

Sugar (sucrose) molecules in solution

Table 7.3 summarizes the classification of solutes in aqueous solutions.

**TABLE 7.3** Classification of Solutes in Aqueous Solutions

| Types of Solute | Dissociates | Contained in Solution | Conducts Electricity | Examples |
|---|---|---|---|---|
| Strong electrolyte | Completely | Ions only | Yes | Ionic compounds such as NaCl, KBr, $MgCl_2$, $NaNO_3$; bases such as NaOH, KOH; acids such as HCl, HBr, $HNO_3$, $HClO_4$, $H_2SO_4$ |
| Weak electrolyte | Partially | Mostly molecules and a few ions | Yes, but poorly | HF, $H_2O$, $NH_3$, $HC_2H_3O_2$ (acetic acid) |
| Nonelectrolyte | None | Molecules only | No | Carbon compounds such as $CH_3OH$, $C_2H_5OH$, $C_{12}H_{22}O_{11}$, $CH_4N_2O$ (urea) |

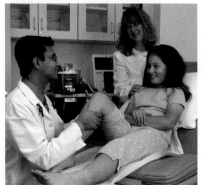

## Orthopedic Physician Assistant

"At a time when we have a shortage of health care professionals, I think of myself as a physician extender," says Pushpinder Beasley, orthopedic physician assistant, Kaiser Hospital. "We can put a significant amount of time into our patient care. Just today, I examined a child's knee. One of the most common injuries to children is disruption of either knee ligaments or the soft tissue around the knees. In this child's case, we were checking her anterior ligaments, also known as ACL. I think an important role of the health care professional is to earn the trust of young people."

As part of a health care team, physician assistants examine patients, order laboratory tests, make diagnoses, report patient progress, order therapeutic procedures and, in most states, prescribe medications.

### ■ Solutions of Electrolytes and Nonelectrolytes

Indicate whether solutions of each of the following contain only ions, only molecules, or mostly molecules and a few ions:

**a.** $Na_2SO_4$, a strong electrolyte
**b.** $CH_3OH$, a nonelectrolyte

#### SOLUTION

**a.** A solution of $Na_2SO_4$ contains only the ions $Na^+$ and $SO_4^{2-}$.
**b.** A nonelectrolyte such as $CH_3OH$ dissolves only as molecules.

#### STUDY CHECK

Boric acid, $H_3BO_3$, is a weak electrolyte. Would you expect a boric acid solution to contain only ions, only molecules, or mostly molecules and a few ions?

## Equivalents

Body fluids typically contain a mixture of several electrolytes, such as $Na^+$, $Cl^-$, $K^+$, and $Ca^{2+}$. We measure each individual ion in terms of an **equivalent (Eq)**, which is the amount of that ion equal to 1 mole of positive or negative electrical charge. For example, 1 mole of $Na^+$ ions and 1 mole of $Cl^-$ ions are each 1 equivalent because they each contain 1 mole of charge. For an ion with a charge of 2+ or 2−, there are 2 equivalents for each mole. Some examples of ions and equivalents are shown in Table 7.4.

**TABLE 7.4** Equivalents of Electrolytes

| Ion | Electrical Charge | Number of Equivalents in 1 Mole |
|---|---|---|
| $Na^+$ | 1+ | 1 Eq |
| $Ca^{2+}$ | 2+ | 2 Eq |
| $Fe^{3+}$ | 3+ | 3 Eq |
| $Cl^-$ | 1− | 1 Eq |
| $SO_4^{2-}$ | 2− | 2 Eq |

In a solution, the charge of the positive ions is always balanced by the charge of the negative ions. For example, a solution containing 25 mEq/L of $Na^+$ and 4 mEq/L of $K^+$ has a total positive charge of 29 mEq/L. If $Cl^-$ is the only anion, its concentration must be 29 mEq/L.

---

SAMPLE PROBLEM   7.4

### ■ Electrolyte Concentration

In body fluids, concentrations of electrolytes are often expressed as milliequivalents (mEq) per liter. The laboratory tests for a patient indicate a blood calcium level of 8.8 mEq/L.

**a.** How many moles of calcium ion are in 0.50 L of blood?
**b.** If chloride ion is the only other ion present, what is its concentration in mEq/L?

SOLUTION

**a.** Using the volume and the electrolyte concentration in mEq/L we can find the number of equivalents in 0.50 L of blood.

$$0.50 \; \cancel{L} \times \frac{8.8 \; \cancel{mEq}}{1 \; \cancel{L}} \times \frac{1 \; Eq}{1000 \; \cancel{mEq}} = 0.0044 \; Eq \; of \; Ca^{2+}$$

We can then convert equivalents to moles (for $Ca^{2+}$ there are 2 Eq per mole).

$$0.0044 \; \cancel{Eq \; Ca^{2+}} \times \frac{1 \; mole \; Ca^{2+}}{2 \; \cancel{Eq \; Ca^{2+}}} = 0.0022 \; mole \; of \; Ca^{2+}$$

**b.** If the concentration of $Ca^{2+}$ is 8.8 mEq/L, then the concentration of $Cl^-$ must be 8.8 mEq/L to balance the charge.

STUDY CHECK

A Ringer's solution for intravenous fluid replacement contains 155 mEq $Cl^-$ per liter of solution. If a patient receives 1250 mL of Ringer's solution, how many moles of chloride were given?

---

## QUESTIONS AND PROBLEMS

### Electrolytes and Nonelectrolytes

**7.7** KF is a strong electrolyte, and HF is a weak electrolyte. How are they different?

**7.8** NaOH is a strong electrolyte, and $CH_3OH$ is a non-electrolyte. How are they different?

**7.9** The following salts are strong electrolytes. Write a balanced equation for their dissociation in water:
   **a.** KCl    **b.** $CaCl_2$    **c.** $K_3PO_4$    **d.** $Fe(NO_3)_3$

**7.10** The following salts are strong electrolytes. Write a balanced equation for their dissociation in water:
   **a.** LiBr    **b.** $NaNO_3$    **c.** $FeCl_3$    **d.** $Mg(NO_3)_2$

**7.11** Indicate whether aqueous solutions of the following will contain ions only, molecules only, or mostly molecules and a few ions:
   **a.** acetic acid ($HC_2H_3O_2$), found in vinegar, a weak electrolyte
   **b.** NaBr, a strong electrolyte
   **c.** fructose ($C_6H_{12}O_6$), a nonelectrolyte

**7.12** Indicate whether aqueous solutions of the following will contain ions only, molecules only, or mostly molecules and a few ions:
   **a.** $Na_2SO_4$, a strong electrolyte
   **b.** ethanol, $C_2H_5OH$, a nonelectrolyte
   **c.** HCN, hydrocyanic acid, a weak electrolyte

**7.13** Indicate the type of electrolyte represented in the following equations:

a. $K_2SO_4(s) \xrightarrow{H_2O} 2K^+(aq) + SO_4^{2-}(aq)$

b. $NH_4OH(aq) \xrightleftharpoons{H_2O} NH_4^+(aq) + OH^-(aq)$

c. $C_6H_{12}O_6(s) \xrightarrow{H_2O} C_6H_{12}O_6(aq)$

**7.14** Indicate the type of electrolyte represented in the following equations:

a. $CH_3OH(l) \xrightarrow{H_2O} CH_3OH(aq)$

b. $MgCl_2(s) \xrightarrow{H_2O} Mg^{2+}(aq) + 2Cl^-(aq)$

c. $HClO(aq) \xrightleftharpoons{H_2O} H^+(aq) + ClO^-(aq)$

**7.15** Indicate the number of equivalents in each of the following:
a. 1 mole of $K^+$          b. 2 moles of $OH^-$
c. 1 mole of $Ca^{2+}$      d. 3 moles of $CO_3^{2-}$

**7.16** Indicate the number of equivalents in each of the following:
a. 1 mole of $Mg^{2+}$     b. 0.5 mole of $H^+$
c. 4 moles of $Cl^-$        d. 2 moles of $Fe^{3+}$

**7.17** A physiological saline solution contains 154 mEq/L each of $Na^+$ and $Cl^-$. How many moles each of $Na^+$ and $Cl^-$ are in 1.00 L of the saline solution?

**7.18** A solution to replace potassium loss contains 40. mEq/L each of $K^+$ and $Cl^-$. How many moles each of $K^+$ and $Cl^-$ are in 1.5 L of the solution?

**7.19** A solution contains 40. mEq/L of $Cl^-$ and 15 mEq/L of $HPO_4^{2-}$. If $Na^+$ is the only cation in the solution, what is the $Na^+$ concentration in milliequivalents per liter?

**7.20** A sample of Ringer's solution contains the following concentrations (mEq/L) of cations: $Na^+$ 147, $K^+$ 4, and $Ca^{2+}$ 4. If $Cl^-$ is the only anion in the solution, what is the $Cl^-$ concentration in milliequivalents per liter?

*Health Note*

## Electrolytes in Body Fluids

The concentrations of electrolytes present in body fluids and in intravenous fluids given to a patient are often expressed in milliequivalents per liter (mEq/L) of solution.

1 Eq = 1000 mEq

Table 7.5 gives the concentrations of some typical electrolytes in blood plasma. There is a charge balance because the total number of positive charges is equal to the total number of negative charges. The use of a specific intravenous solution depends on the nutritional, electrolyte, and fluid needs of the individual patient. Examples of various types of solutions are given in Table 7.6.

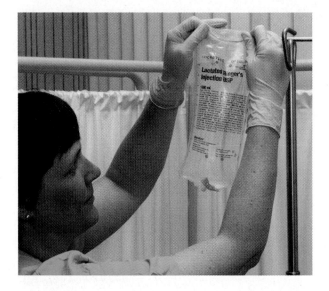

**TABLE 7.5** Some Typical Concentrations of Electrolytes in Blood Plasma

| Electrolyte | Concentration (mEq/L) |
|---|---|
| **Cations** | |
| $Na^+$ | 138 |
| $K^+$ | 5 |
| $Mg^{2+}$ | 3 |
| $Ca^{2+}$ | 4 |
| Total | 150 |
| **Anions** | |
| $Cl^-$ | 110 |
| $HCO_3^-$ | 30 |
| $HPO_4^{2-}$ | 4 |
| Proteins | 6 |
| Total | 150 |

**TABLE 7.6** Electrolyte Concentrations in Intravenous Replacement Solutions

| Solution | Electrolytes (mEq/L) | Use |
|---|---|---|
| Sodium chloride (0.9%) | $Na^+$ 154, $Cl^-$ 154 | Replacement of fluid loss |
| Potassium chloride with 5.0% dextrose | $K^+$ 40, $Cl^-$ 40 | Treatment of malnutrition (low potassium levels) |
| Ringer's solution | $Na^+$ 147, $K^+$ 4, $Ca^{2+}$ 4, $Cl^-$ 155 | Replacement of fluids and electrolytes lost through dehydration |
| Maintenance solution with 5.0% dextrose | $Na^+$ 40, $K^+$ 35, $Cl^-$ 40, lactate$^-$ 20, $HPO_4^{2-}$ 15 | Maintenance of fluid and electrolyte levels |
| Replacement solution (extracellular) | $Na^+$ 140, $K^+$ 10, $Ca^{2+}$ 5, $Mg^{2+}$ 3, $Cl^-$ 103, acetate$^-$ 47, citrate$^{3-}$ 8 | Replacement of electrolytes in extracellular fluids |

# 7.3 SOLUBILITY

The term **solubility** is used to describe the amount of a solute that can dissolve in a given amount of solvent. Many factors such as the type of solute, the type of solvent, and the temperature affect a solute's solubility. Solubility, usually expressed in grams of solute in 100 grams of solvent, is the maximum amount of solute that can be dissolved at a certain temperature. If a solute readily dissolves when added to the solvent, the solution does not contain the maximum amount of solute. We call the solution an **unsaturated solution**.

A solution that contains all the solute that can dissolve is a **saturated solution**. When a solution is saturated, the rate of the reaction that dissolves the solute becomes equal to the rate of recrystallization. Then there is no further change in the amount of dissolved solute in solution.

$$\text{Solute} + \text{Solvent} \underset{\text{Solute crystallizes}}{\overset{\text{Solute dissolves}}{\rightleftharpoons}} \text{Saturated solution}$$

We can prepare a saturated solution by adding solute greater than needed for solubility. Stirring the solution will dissolve the maximum amount of solute and leave the excess on the bottom of the container. Once we have a saturated solution, the addition of more solute will only increase the amount of undissolved solute.

## LEARNING GOAL

Define *solubility*; distinguish between an unsaturated and a saturated solution.

the Chemistry place

CASE STUDY
Kidney Stones and Saturated Solutions

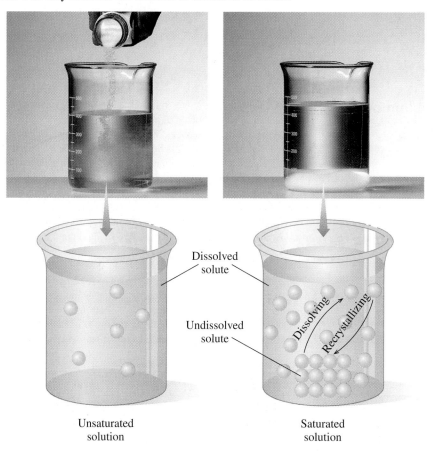

Unsaturated solution — Dissolved solute — Undissolved solute — Saturated solution — Dissolving — Recrystallizing

## SAMPLE PROBLEM 7.5

### ■ Saturated Solutions

At 20 °C, the solubility of KCl is 34 g/100 g of water. In the laboratory, a student mixes 75 g of KCl with 200. g of water at a temperature of 20 °C.

**a.** How much of the KCl can dissolve?
**b.** Is the solution saturated or unsaturated?
**c.** What is the mass of any solid KCl on the bottom of the container?

SOLUTION

**a.** KCl has a solubility of 34 g of KCl in 100 g of water. Using the solubility as a conversion factor, the maximum amount of KCl that can dissolve in 200. g of water is calculated as follows:

$$200. \text{ g } H_2O \times \frac{34 \text{ g KCl}}{100 \text{ g } H_2O} = 68 \text{ g of KCl}$$

**b.** Because 75 g of KCl exceeds the amount that can dissolve in 200. g of water, the KCl solution is saturated.

**c.** If we add 75 g of KCl to 200. g of water and only 68 g of KCl can dissolve, there is 7 g of solid (undissolved) KCl.

STUDY CHECK

At 50 °C, the solubility of $NaNO_3$ is 114 g/100 g of water. How many grams of $NaNO_3$ are needed to make a saturated $NaNO_3$ solution with 50. g of water at 50 °C?

## *Health Note*

### Gout and Kidney Stones: A Problem of Saturation in Body Fluids

The conditions of gout and kidney stones involve compounds in the body that exceed their solubility levels and form solid products. Gout affects adults, primarily men, over the age of 40. Attacks of gout may occur when the concentration of uric acid in blood plasma exceeds its solubility, which is 7 mg/100 mL of plasma at 37 °C. Insoluble deposits of needle-like crystals of uric acid can form in the cartilage, tendons, and soft tissues where they cause painful gout attacks. They may also form in the tissues of the kidneys, where they can cause renal damage. High levels of uric acid in the body can be caused by an increase in uric acid production, failure of the kidneys to remove uric acid, or by a diet with an overabundance of foods containing purines, which are metabolized to uric acid in the body. Foods in the diet that contribute to high levels of uric acid include certain meats, sardines, mushrooms, asparagus, and beans.

Drinking alcoholic beverages may also significantly increase uric acid levels and bring about gout attacks.

Treatment for gout involves diet changes and drugs. Depending on the levels of uric acid, a medication, such as probenecid, can be used to help the kidneys eliminate uric acid, or allopurinol, which blocks the production of uric acid by the body.

Kidney stones are solid materials that form in the urinary tract. Most kidney stones are composed of calcium phosphate and calcium oxalate, although they can be solid uric acid. The excessive ingestion of minerals and insufficient water intake can cause the concentration of mineral salts to exceed the solubility of the mineral salts and lead to the formation of kidney stones. When a kidney stone passes through the urinary tract, it causes considerable pain and discomfort, necessitating the use of painkillers and surgery. Sometimes ultrasound is used to break up kidney stones. Persons prone to kidney stones are advised to drink six to eight glasses of water every day to prevent saturation levels of minerals in the urine.

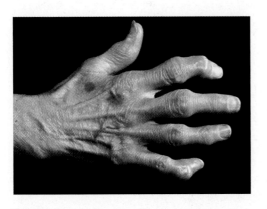

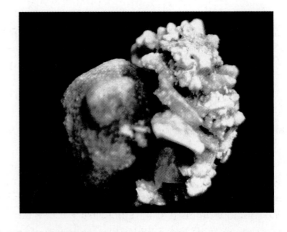

## Explore Your World

### Preparing Solutions

Using a drinking glass and water, place $\frac{1}{4}$ or $\frac{1}{2}$ cup of cold water in a glass. Begin adding 1 tablespoon of sugar at a time and stir thoroughly. Take a sip of the liquid in the glass as you proceed. As the sugar solution becomes more concentrated, you may need to stir for a few minutes until all the sugar dissolves. Each time, observe the solution after several minutes to determine when it is saturated.

Repeat the above activity with $\frac{1}{4}$ or $\frac{1}{2}$ cup of warm water. Count the number of tablespoons of sugar you need to form a saturated solution.

### Questions

1. What do you notice about how sweet each sugar solution tastes?
2. How did you know when you obtained a saturated solution?
3. How much sugar dissolved in the warm water compared to the cold water?

## Effect of Temperature on Solubility

The solubility of most solids is greater as temperature increases, which means that solutions usually contain more dissolved solute at higher temperature. A few substances show little change in solubility at higher temperatures, and a few are less soluble. (See Figure 7.4.) For example, when you add sugar to iced tea, some undissolved sugar may form on the bottom of the glass. But if you add sugar to hot tea, many teaspoons of sugar are needed before solid sugar appears. Hot tea dissolves more sugar than does cold tea because the solubility of sugar is much greater at a higher temperature. When a saturated solution is carefully cooled, it becomes a *supersaturated solution* because it contains more solute than the solubility allows. Such a solution is unstable, and if the solution is agitated or if a solute crystal is added, the excess solute will crystallize to give a saturated solution again.

The solubility of a gas in water decreases as the temperature increases. At higher temperatures, more gas molecules have the energy to escape from the solution. Perhaps you have observed the bubbles escaping from a cold carbonated soft drink as it warms. At high temperatures, bottles containing carbonated solutions may burst as more gas molecules leave the solution and increase the gas pressure inside the bottle. Biologists have found that increased temperatures in rivers and lakes cause the amount of dissolved oxygen to decrease until the warm water can no longer support a biological community. Electricity-generating plants are required to have their own ponds to use with their cooling towers to lessen the threat of thermal pollution.

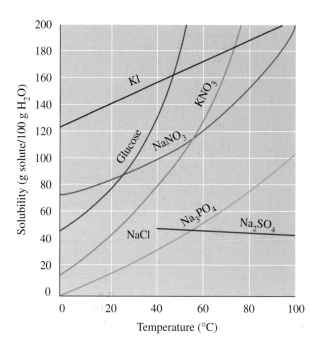

**FIGURE 7.4** In water, most common solids are more soluble as the temperature increases.
**Q** Compare the solubility of $NaNO_3$ at 20 °C and 60 °C.

### Henry's Law

**Henry's law** states that the solubility of gas in a liquid is directly related to the pressure of that gas above the liquid. At higher pressures, there are more gas molecules available to enter and dissolve in the liquid. A can of soda is carbonated by using $CO_2$ gas at high pressure to increase the solubility of the $CO_2$ in the beverage. When you open the can at atmospheric pressure, the pressure on the $CO_2$ drops, which decreases the solubility of $CO_2$. As a result, bubbles of $CO_2$ rapidly escape from the solution. The burst of bubbles is even more noticeable when you open a warm can of soda.

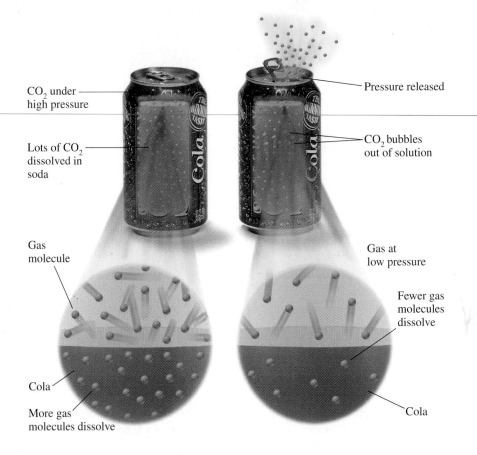

CO₂ under high pressure

Lots of CO₂ dissolved in soda

Gas molecule

Cola

More gas molecules dissolve

Pressure released

CO₂ bubbles out of solution

Gas at low pressure

Fewer gas molecules dissolve

Cola

### SAMPLE PROBLEM 7.6

■ **Factors Affecting Solubility**

Indicate whether the solubility of the solute will increase or decrease in each of the following situations:

**a.** dissolving sugar using 80 °C water instead of 25 °C water
**b.** effect on dissolved $O_2$ in a lake as it warms

SOLUTION

**a.** An increase in the temperature increases the solubility of the sugar.
**b.** An increase in the temperature decreases the solubility of $O_2$ gas.

STUDY CHECK

At 10 °C, the solubility of $KNO_3$ is 20 g/100 g $H_2O$. Would you expect the solubility of $KNO_3$ to be higher or lower at 40 °C?

## QUESTIONS AND PROBLEMS

### Solubility

**7.21** State whether each of the following refers to a saturated or unsaturated solution:

  **a.** A crystal added to a solution does not change in size.
  **b.** A sugar cube completely dissolves when added to a cup of coffee.

**7.22** State whether each of the following refers to a saturated or unsaturated solution:

  **a.** A spoonful of salt added to boiling water dissolves.
  **b.** A layer of sugar forms on the bottom of a glass of tea as ice is added.

Use this table for problems 7.23–7.26.

| Substance | Solubility (g/100 g H₂O) | |
|---|---|---|
| | 20 °C | 50 °C |
| KCl | 34 | 43 |
| NaNO₃ | 88 | 114 |
| $C_{12}H_{22}O_{11}$ (sugar) | 204 | 260 |

**7.23** Using the above table, determine whether each of the following solutions will be saturated or unsaturated at 20 °C:
  **a.** Adding 25 g of KCl to 100. g of $H_2O$
  **b.** Adding 11 g of $NaNO_3$ to 25 g of $H_2O$
  **c.** Adding 400. g of sugar to 125 g of $H_2O$

**7.24** Using the above table, determine whether each of the following solutions will be saturated or unsaturated at 50 °C:
  **a.** Adding 25 g of KCl to 50. g of $H_2O$
  **b.** Adding 150 g of $NaNO_3$ to 75 g of $H_2O$
  **c.** Adding 80. g of sugar to 25 g of $H_2O$

**7.25** A solution containing 80. g of KCl in 200. g of $H_2O$ at 50 °C is cooled to 20 °C.
  **a.** How many grams of KCl remain in solution at 20 °C?
  **b.** How many grams of solid KCl crystallized after cooling?

**7.26** A solution containing 80. g of $NaNO_3$ in 75 g of $H_2O$ at 50 °C is cooled to 20 °C.
  **a.** How many grams of $NaNO_3$ remain in solution at 20 °C?
  **b.** How many grams of solid $NaNO_3$ crystallized after cooling?

**7.27** Explain the following observations:
  **a.** More sugar dissolves in hot tea than in iced tea.
  **b.** Champagne in a warm room goes flat.
  **c.** A warm can of soda has more spray when opened than a cold one.

**7.28** Explain the following observations:
  **a.** An open can of soda loses its "fizz" faster at room temperature than in the refrigerator.
  **b.** Chlorine gas in tap water escapes as the sample warms to room temperature.
  **c.** Less sugar dissolves in iced coffee than in hot coffee.

# 7.4 PERCENT CONCENTRATION

The amount of solute dissolved in a certain amount of solution is called the **concentration** of the solution. Although there are many ways to express a concentration, they all specify a certain amount of solute in a given amount of solution.

$$\text{Concentration of a solution} = \frac{\text{amount of solute}}{\text{amount of solution}}$$

## Mass Percent

The **mass percent** (% m/m) concentration of a solution describes the mass of the solute in every 100 g of solution. The mass in grams of the solution is the sum of the mass of the solute and the mass of the solvent.

$$\text{Mass percent (\% m/m)} = \frac{\text{mass of solute (g)}}{\text{mass of solute (g)} + \text{mass of solvent (g)}} \times 100\%$$

$$= \frac{\text{mass of solute (g)}}{\text{mass of solution (g)}} \times 100\%$$

Suppose we prepared a solution by mixing 8.00 g of KCl (solute) with 42.00 g of water (solvent). Together, the mass of the solute and mass of solvent give the mass of the solution (8.00 g + 42.00 g = 50.00 g). The mass % is calculated by substituting in the values into the mass percent expression.

$$\frac{8.00 \text{ g KCl}}{50.00 \text{ g solution}} \times 100\% = 16.0\% \text{ (m/m)}$$

$$\underbrace{8.00 \text{ g KCl} + 42.00 \text{ g } H_2O}$$
(Solute  +  Solvent)

**LEARNING GOAL**

Calculate the percent concentration of a solute in a solution; use percent concentration to calculate the amount of solute or solution.

Add 8.00 g of KCl

Add water until the solution weighs 50.00 g

■ **Percent Concentration (m/m)**

What is the mass percent of a solution prepared by dissolving 30.0 g of NaOH in 120.0 g of $H_2O$?

**Guide to Calculating
Solution Concentrations**

**1** State the given and needed concentration.

**2** Write a plan to calculate needed concentration.

**3** Write equalities and conversion factors.

**4** Set up problem to calculate answer.

SOLUTION

Step 1  **Given**  30.0 g of NaOH and 120.0 g of $H_2O$
**Need**  mass percent (% m/m) of NaOH

Step 2  **Plan**  The mass percent is calculated by using the mass in grams of the solute and solution in the definition of mass percent.

Step 3  **Equalities/Conversion Factors**

$$\text{Mass percent (\% m/m)} = \frac{\text{mass of solute (g)}}{\text{mass of solute (g)} + \text{mass of solvent (g)}} \times 100\%$$

$$= \frac{\text{mass of solute (g)}}{\text{mass of solution (g)}} \times 100\%$$

Step 4  **Set Up Problem**  The mass of the solute and the solution are obtained from the data.

$$\begin{aligned}\text{Mass of solute} &= \ \ 30.0 \text{ g NaOH}\\ + \ \text{Mass of solvent} &= 120.0 \text{ g } H_2O\\ \hline \text{Mass of solution} &= 150.0 \text{ g solution}\end{aligned}$$

$$\text{Mass percent (\% m/m)} = \frac{30.0 \text{ g NaOH}}{150.0 \text{ g solution}} \times 100\%$$

$$= \ \ 20.0\% \text{ (m/m) NaOH}$$

STUDY CHECK

What is the mass percent of NaCl in a solution made by dissolving 2.0 g of NaCl in 56.0 g of $H_2O$?

## Volume Percent

Because the volumes of liquids or gases are easily measured, the concentrations of their solutions are often expressed as **volume percent** (% v/v). The units of volume used in the ratio must be the same, for example, both in milliliters or both in liters.

$$\text{Volume percent (\% v/v)} = \frac{\text{volume of solute}}{\text{volume of solution}} \times 100\%$$

We interpret a volume/volume percent as the volume of solute in 100 mL of solution. In the wine industry, a label that reads 12% (v/v) means 12 mL of alcohol in 100 mL of wine.

## Mass/Volume Percent

A **mass/volume percent** (% m/v), or weight/volume percent (% w/v), is calculated by dividing the grams of the solute by the volume (mL) of solution and multiplying by 100.

The mass/volume percent is widely used in hospitals and pharmacies for the preparation of intravenous solutions and medicines.

$$\text{Mass/volume percent (\% m/v)} = \frac{\text{grams of solute}}{\text{milliliters of solution}} \times 100\%$$

The mass/volume percent indicates the grams of a substance that are contained in 100 mL of a solution. For example, a 5% (m/v) glucose solution contains 5 g of glucose in 100 mL of solution. The volume of solution represents the combination of the glucose and $H_2O$.

---

## SAMPLE PROBLEM 7.8

### ■ Calculating Percent Concentration

A student prepared a solution by dissolving 5.0 g of KI in enough water to give a final volume of 250 mL. What is the mass/volume percent of the KI solution?

Water added to
make a solution         250 mL

5.0 g of KI   2.0% (m/v)
KI solution

SOLUTION

Step 1 **Given**   5.0 g of KI and 250 mL of solution

   **Need**   mass/volume percent (% m/v) of KI

Step 2 **Plan**   The mass/volume percent is calculated by using the mass in grams of the solute and the volume in mL of the solution in the definition of mass/volume percent.

Step 3 **Equalities/Conversion Factors**   Write the mass/volume percent expression.

$$\text{Mass/volume percent (\% m/v)} = \frac{\text{grams of solute}}{\text{milliliters of solution}} \times 100\%$$

Step 4 **Set Up Problem**   Substitute solute and solution quantities into the mass/volume percent expression.

$$\text{Mass/volume percent (\% m/v)} = \frac{\overset{\text{Mass of solute}}{5.0 \text{ g KI}}}{\underset{\text{Volume of solution}}{250 \text{ mL solution}}} \times 100\% = 2.0\% \text{ (m/v) KI}$$

STUDY CHECK

What is the mass/volume percent (% m/v) of $Br_2$ in a solution prepared by dissolving 12 g of bromine ($Br_2$) in enough carbon tetrachloride to make 250 mL of solution?

---

## Percent Concentrations as Conversion Factors

In the preparation of solutions, we often need to calculate the amount of solute or solution. Then the percent concentration is useful as a conversion factor. The value of 100 in the denominator of a percent expression is an *exact* number. Some examples of percent concentrations, their meanings, and possible conversion factors are given in Table 7.7.

**TABLE 7.7** Conversion Factors from Percent Concentrations

| Percent Concentration | Meaning | Conversion Factors | | |
|---|---|---|---|---|
| 10% (m/m) KCl | There are 10 g of KCl in 100 g of solution. | $\dfrac{10 \text{ g KCl}}{100 \text{ g solution}}$ | and | $\dfrac{100 \text{ g solution}}{10 \text{ g KCl}}$ |
| 5% (m/v) glucose | There are 5 g of glucose in 100 mL of solution. | $\dfrac{5 \text{ g glucose}}{100 \text{ mL solution}}$ | and | $\dfrac{100 \text{ mL solution}}{5 \text{ g glucose}}$ |
| 12% (v/v) ethanol | There are 12 mL of ethanol in 100 mL of solution. | $\dfrac{12 \text{ mL ethanol}}{100 \text{ mL solution}}$ | and | $\dfrac{100 \text{ mL solution}}{12 \text{ mL ethanol}}$ |

## SAMPLE PROBLEM 7.9

### ■ Using Mass/Volume Percent to Find Mass of Solute

A topical antibiotic is 1.0% (m/v) Clindamycin. How many grams of Clindamycin are in 60. mL of the 1.0% (m/v) solution?

SOLUTION

**Step 1  Given**　1.0% (m/v) Clindamycin　　**Need**　grams of Clindamycin

**Step 2  Plan**　milliliters of solution　%(m/v) factor　grams of Clindamycin

**Step 3  Equalities/Conversion Factors**　The percent (m/v) indicates the grams of a solute in every 100 mL of a solution. The 1.0% (m/v) can be written as two conversion factors.

$$100 \text{ mL of solution} = 1.0 \text{ g of Clindamycin}$$

$$\frac{1.0 \text{ g Clindamycin}}{100 \text{ mL solution}} \quad \text{and} \quad \frac{100 \text{ mL solution}}{1.0 \text{ g Clindamycin}}$$

**Step 4  Set Up Problem**　The volume of the solution is converted to mass of solute using the conversion factor.

$$60. \text{ mL solution} \times \frac{1.0 \text{ g Clindamycin}}{100 \text{ mL solution}} = 0.60 \text{ g of Clindamycin}$$

STUDY CHECK

Calculate the grams of KCl in 225 g of an 8.00% (m/m) KCl solution.

### Guide to Using Concentration to Calculate Mass or Volume

**1** State the given and needed quantities.

**2** Write a plan to calculate mass or volume.

**3** Write equalities and conversion factors including concentration.

**4** Set up problem to calculate mass or volume.

## QUESTIONS AND PROBLEMS

### Percent Concentration

**7.29** What is the difference between a 5% (m/m) glucose solution and a 5% (m/v) glucose solution?

**7.30** What is the difference between a 10% (v/v) methyl alcohol ($CH_3OH$) solution and a 10% (m/m) methyl alcohol solution?

**7.31** Calculate the mass percent, % (m/m), for the solute in each of the following solutions:
**a.** 25 g of KCl and 125 g of $H_2O$
**b.** 12 g of sugar in 225 g of tea solution with sugar

**7.32** Calculate the mass percent, % (m/m), for the solute in each of the following solutions:
**a.** 75 g of NaOH in 325 g of NaOH solution
**b.** 2.0 g of KOH in 20.0 g of $H_2O$

**7.33** Calculate the mass/volume percent, % (m/v) for the solute in each of the following solutions:
**a.** 75 g of $Na_2SO_4$ in 250 mL of $Na_2SO_4$ solution
**b.** 39 g of sucrose in 355 mL of a carbonated drink

**7.34** Calculate the mass/volume percent, % (m/v), for the solute in each of the following solutions:
  **a.** 2.50 g of KCl in 50.0 mL of solution
  **b.** 7.5 g of casein in 120 mL of low-fat milk

**7.35** Calculate the amount of solute needed to prepare the following solutions:
  **a.** 50.0 mL of a 5.0% (m/v) KCl solution
  **b.** 1250 mL of a 4.0% (m/v) NH₄Cl solution

**7.36** Calculate the amount of solute needed to prepare the following solutions:
  **a.** 150 mL of a 40.0% (m/v) LiNO₃ solution
  **b.** 450 mL of a 2.0% (m/v) KCl solution

**7.37** A mouthwash contains 22.5% alcohol by volume. If the bottle of mouthwash contains 355 mL, what is the volume in milliliters of the alcohol?

**7.38** A bottle of champagne is 11% alcohol by volume. If there are 750 mL of champagne in the bottle, how many milliliters of alcohol are present?

**7.39** A patient receives 100. mL of 20.% (m/v) mannitol solution every hour.
  **a.** How many grams of mannitol are given in 1 hour?
  **b.** How many grams of mannitol does the patient receive in 12 hours?

**7.40** A patient receives 250 mL of a 4.0% (m/v) amino acid solution twice a day.
  **a.** How many grams of amino acids are in 250 mL of solution?
  **b.** How many grams of amino acids does the patient receive in 1 day?

**7.41** A patient needs 100. g of glucose in the next 12 hours. How many liters of a 5% (m/v) glucose solution must be given?

**7.42** A patient received 2.0 g of NaCl in 8 hours. How many milliliters of a 0.90% (m/v) NaCl (saline) solution were delivered?

# 7.5 MOLARITY AND DILUTION

**LEARNING GOAL**

Calculate the molarity of a solution; use molarity to calculate the moles of solute or the volume needed to prepare a solution. Describe the dilution of a solution.

When the solutes of solutions take part in reactions, chemists are interested in the number of reacting particles. For this purpose, chemists use **molarity (M)**, a concentration that states the number of moles of solute in exactly 1 liter of solution. The molarity of a solution can be calculated knowing the moles of solute and the volume of solution in liters.

$$\text{Molarity (M)} = \frac{\text{moles of solute}}{\text{liters of solution}}$$

For example, if 1.0 mole of NaCl were dissolved in enough water to prepare 1.0 L of solution, the resulting NaCl solution has a molarity of 1.0 M. The abbreviation M indicates the units of moles per liter (moles/L).

$$M = \frac{\text{moles of solute}}{\text{liters of solution}} = \frac{1.0 \text{ mole NaCl}}{1 \text{ L}} = 1.0 \text{ M NaCl}$$

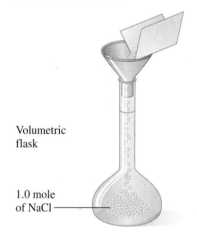

Volumetric flask

1.0 mole of NaCl

**SAMPLE PROBLEM 7.10**

### ■ Calculating Molarity

What is the molarity (M) of 60.0 g of NaOH in 0.250 L of solution?

**SOLUTION**

Step 1 **Given** 60.0 g of NaOH in 0.250 L of solution

  **Need** molarity (moles/L)

Step 2 **Plan** The calculation of molarity requires the moles of NaOH and the volume of the solution in liters.

$$\text{Molarity (M)} = \frac{\text{moles of solute}}{\text{liters of solution}}$$

$$\boxed{\text{g NaOH}} \quad \boxed{\text{Molar mass}} \quad \frac{\text{moles NaOH}}{\text{volume (L)}} = \text{M NaOH solution}$$

Add water until 1-liter mark is reached.

Mix

A 1.0 molar (M) NaCl solution

**Step 3  Equalities/Conversion Factors**

$$1 \text{ mole of NaOH} = 40.0 \text{ g of NaOH}$$

$$\frac{1 \text{ mole NaOH}}{40.0 \text{ g NaOH}} \quad \text{and} \quad \frac{40.0 \text{ g NaOH}}{1 \text{ mole NaOH}}$$

**Step 4  Set Up Problem**

$$\text{Moles of NaOH} = 60.0 \text{ g NaOH} \times \frac{1 \text{ mole NaOH}}{40.0 \text{ g NaOH}} = 1.50 \text{ moles of NaOH}$$

The molarity is calculated by dividing the moles of NaOH by the volume in liters.

$$\frac{1.50 \text{ moles NaOH}}{0.250 \text{ L}} = \frac{6.00 \text{ moles NaOH}}{1 \text{ L}} = 6.00 \text{ M NaOH}$$

STUDY CHECK

What is the molarity of a solution that contains 75.0 g of $KNO_3$ dissolved in 0.350 L of solution?

## Molarity as a Conversion Factor

When we need to calculate the moles of solute or the volume of solution, the molarity is used as a conversion factor. Examples of conversion factors from molarity are given in Table 7.8.

**TABLE 7.8**  Some Examples of Molar Solutions

| Molarity | Meaning | Conversion Factors | | |
|---|---|---|---|---|
| 6.0 M HCl | 6.0 moles of HCl in 1 liter of solution | $\dfrac{6.0 \text{ moles HCl}}{1 \text{ L}}$ | and | $\dfrac{1 \text{ L}}{6.0 \text{ moles HCl}}$ |
| 0.20 M NaOH | 0.20 mole of NaOH in 1 liter of solution | $\dfrac{0.20 \text{ mole NaOH}}{1 \text{ L}}$ | and | $\dfrac{1 \text{ L}}{0.20 \text{ mole NaOH}}$ |

To prepare a solution, we must convert the number of moles of solute needed into grams. Using the volume and the molarity of the solution with the molar mass of the solute, we can calculate the number of grams of solute necessary. This type of calculation is illustrated in Sample Problem 7.11.

## SAMPLE PROBLEM  7.11

■ **Using Molarity to Find Mass of Solute**

How many grams of KCl would you need to weigh out to prepare 0.250 L of a 2.00 M KCl solution?

SOLUTION

Step 1  **Given**  0.250 L of 2.00 M KCl solution    **Need**  grams of KCl

Step 2  **Plan**  The grams of KCl can be calculated using the volume in liters, the molarity of the KCl solution, and the molar mass of KCl.

liters of KCl  Molarity  moles of KCl  Molar mass  grams of KCl

**Step 3  Equalities/Conversion Factors**

1 L of KCl solution = 2.00 moles of KCl    1 mole of KCl = 74.6 g of KCl

$$\frac{1 \text{ L KCl}}{2.00 \text{ moles KCl}} \quad \text{and} \quad \frac{2.00 \text{ moles KCl}}{1 \text{ L KCl}} \qquad \frac{1 \text{ mole KCl}}{74.6 \text{ g KCl}} \quad \text{and} \quad \frac{74.6 \text{ g KCl}}{1 \text{ mole KCl}}$$

**Step 4  Set Up Problem**

$$\text{Moles of KCl} = 0.250 \text{ L solution} \times \frac{2.00 \text{ moles KCl}}{1 \text{ L KCl}} = 0.500 \text{ mole of KCl}$$

The grams of KCl are calculated by multiplying the moles of KCl by the molar mass.

$$\text{Grams of KCl} = 0.500 \text{ mole KCl} \times \frac{74.6 \text{ g KCl}}{1 \text{ mole KCl}} = 37.3 \text{ g of KCl}$$

Combining the calculations, we can write a complete problem setup as follows:

$$0.250 \text{ L KCl solution} \times \frac{2.00 \text{ moles KCl}}{1 \text{ L KCl}} \times \frac{74.6 \text{ g KCl}}{1 \text{ mole KCl}}$$
$$= 37.3 \text{ g of KCl}$$

**STUDY CHECK**

How many grams of $NaHCO_3$ are in 325 mL of a 4.50 M $NaHCO_3$ solution?

---

## SAMPLE PROBLEM 7.12

### ■ Using Molarity to Find Volume

How many liters of a 2.00 M NaCl solution are needed to provide 67.3 g of NaCl?

**SOLUTION**

**Step 1  Given**   67.3 g of NaCl from a 2.00 M NaCl solution
**Need**   liters of NaCl

**Step 2  Plan**   The volume of NaCl is calculated using the moles of NaCl and molarity of the NaCl solution.

grams of NaCl  Molar mass   moles of NaCl  Molarity   liters of NaCl

**Step 3  Equalities/Conversion Factors**

1 mole of NaCl  =  58.5 g of NaCl

$$\frac{1 \text{ mole NaCl}}{58.5 \text{ g NaCl}} \quad \text{and} \quad \frac{58.5 \text{ g NaCl}}{1 \text{ mole NaCl}}$$

The molarity of any solution can be written as two conversion factors.

1 L of NaCl  =  2.00 moles of NaCl

$$\frac{1 \text{ L NaCl}}{2.00 \text{ moles NaCl}} \quad \text{and} \quad \frac{2.00 \text{ moles NaCl}}{1 \text{ L NaCl}}$$

**Step 4  Set Up Problem**

$$\text{Liters of NaCl} = 67.3 \text{ g NaCl} \times \frac{1 \text{ mole NaCl}}{58.5 \text{ g NaCl}} \times \frac{1 \text{ L NaCl}}{2.00 \text{ moles NaCl}}$$
$$= 0.575 \text{ L of NaCl}$$

**STUDY CHECK**

How many milliliters of a 6.0 M HCl solution will provide 4.5 moles of HCl?

## Dilution

In chemistry, we often need to prepare a dilute solution from a more concentrated solution. In a process called **dilution**, a solvent, usually water, is added to a solution, which increases the volume and decreases the concentration. In an everyday example, you are making a dilution when you add three cans of water to a can of concentrated orange juice.

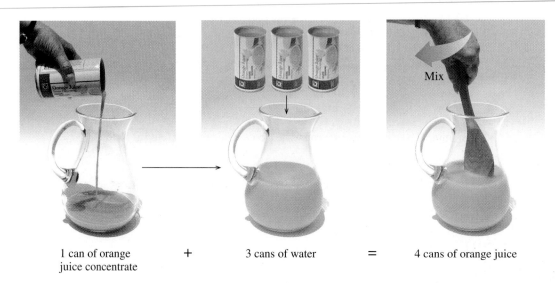

| 1 can of orange juice concentrate | + | 3 cans of water | = | 4 cans of orange juice |

When a solution is diluted, the amount of solute before dilution is equal to the amount of solute in the diluted solution. (See Figure 7.5.)

Grams or moles of solute = grams or moles of solute
    Concentrated solution    Diluted solution

Amount of solute in the = amount of solute in the
concentrated solution         diluted solution

We can write this equality in terms of the concentration, $C$, and the volume, $V$.

$$C_1V_1 = C_2V_2$$
Concentrated    Diluted
solution         solution

We know from the discussion of percent concentration that the grams of solute are obtained from the volume and the percent concentration.

Grams of solute = percent (grams/100 mL) × volume (mL)

**FIGURE 7.5** When water is added to a concentrated solution, there is no change in the number of particles, but the solute particles can spread out as the volume of the diluted solution increases.
**Q** What is the concentration of the diluted solution after an equal volume of water is added to a sample of 6 M HCl?

We can express the number of grams for the concentrated solution as $\%_1V_1$ and the number of grams in the diluted solution as $\%_2V_2$.

Grams of solute = grams of solute
Concentrated solution    Diluted solution

$$\%_1V_1 = \%_2V_2$$

When the concentration is given as molarity (M), the moles of solute are obtained from the volume (liters) and the molarity.

Moles of solute = molarity (moles/L) × volume (L)

Expressing the number of moles in the concentrated solution as $M_1V_1$ and the number of moles in the diluted solution as $M_2V_2$, the equality is written as follows:

Moles of solute = moles of solute
Concentrated solution    Diluted solution

$$M_1V_1 = M_2V_2$$

If we are given any 3 of the 4 variables, we can rearrange the dilution expression to solve for the unknown quantity as seen in Sample Problem 7.13.

SAMPLE PROBLEM 7.13

### Volume of a Diluted Solution

What volume (mL) of a 2.5% (m/v) KOH solution can be prepared by diluting 50.0 mL of a 12% (m/v) KOH solution?

SOLUTION

Step 1 **Give Data in a Table**  We make a table of the concentrations and volumes of the initial and diluted solutions. For the calculation, units must be the same.

**Initial:**  $C_1 = 12\%$ (m/v)    $V_1 = 50.0$ mL

**Diluted:**  $C_2 = 2.5\%$ (m/v)    $V_2 = ?$ mL

Step 2 **Plan**  The volume of the dilute solution can be calculated by solving the dilution expression for $V_2$.

$$C_1V_1 = C_2 V_2$$

$$\frac{C_1V_1}{C_2} = \frac{\cancel{C_2}}{\cancel{C_2}} V_2$$

$$V_2 = \frac{C_1V_1}{C_2}$$

Step 3 **Set Up Problem**  The values from the table are now used in the dilution expression to solve for $V_2$.

$$V_2 = \frac{C_1V_1}{C_2}$$

$$V_2 = \frac{12\% \times 50.0 \text{ mL}}{2.5\%} = 240 \text{ mL (diluted KOH solution)}$$

STUDY CHECK

What is the percent (% m/v) of the dilute solution when 25.0 mL of 15% (m/v) HCl is diluted to 125 mL?

**Guide to Calculating Dilution Quantities**

1 Prepare a table of the initial and diluted volumes and concentrations.

2 Write a plan that solves the dilution expression for the unknown quantity.

3 Set up problem by placing known quantities in dilution expression.

SAMPLE PROBLEM 7.14

### ■ Molarity of a Diluted Solution

What is the molarity of a solution prepared when 75.0 mL of a 4.00 M KCl solution is diluted to a volume of 0.500 L?

SOLUTION

Step 1 **Give Data in a Table**   We make a table of the molar concentrations and volumes of the initial and diluted solutions.

**Initial:**   $M_1$ = 4.00 M KCl    $V_1$ = 75.0 mL = 0.0750 L

**Diluted:**  $M_2$ = ? M KCl        $V_2$ = 0.500 L

Step 2 **Plan**   The unknown molarity can be calculated by solving the dilution expression for $M_2$.

$$M_1 V_1 = M_2 V_2$$
$$\frac{M_1 V_1}{V_2} = M_2 \frac{\cancel{V_2}}{\cancel{V_2}}$$
$$M_2 = M_1 \times \frac{V_1}{V_2}$$

Step 3 **Set Up Problem**   The diluted concentration is calculated by placing the values from the table into the dilution expression.

$$M_2 = 4.00 \text{ M} \times \frac{0.075 \cancel{L}}{0.500 \cancel{L}} = 0.600 \text{ M KCl (diluted solution)}$$

STUDY CHECK

You need to prepare 600. mL of 2.00 M NaOH solution from a 10.0 M NaOH solution. What volume of the 10.0 M NaOH solution do you use?

---

## QUESTIONS AND PROBLEMS

### Molarity and Dilution

**7.43** Calculate the molarity (M) of the following solutions:
 **a.** 2.0 moles of glucose in 4.0 L of solution
 **b.** 4.0 g of KOH in 2.0 L of solution
 **c.** 5.85 g NaCl in 400. mL of solution

**7.44** Calculate the molarity (M) of the following solutions:
 **a.** 0.50 mole of glucose in 0.200 L of solution
 **b.** 36.5 g of HCl in 1.0 L of solution
 **c.** 30.0 g of NaOH in 350. mL of solution

**7.45** Calculate the moles of solute needed to prepare each of the following:
 **a.** 1.0 L of a 3.0 M NaCl solution
 **b.** 0.40 L of a 1.0 M KBr solution
 **c.** 125 mL of a 2.0 M $MgCl_2$ solution

**7.46** Calculate the moles of solute needed to prepare each of the following:
 **a.** 5.0 L of a 2.0 M $CaCl_2$ solution
 **b.** 4.0 L of a 0.10 M NaOH solution
 **c.** 215 mL of a 4.0 M $HNO_3$ solution

**7.47** Calculate the grams of solute needed to prepare each of the following solutions:
 **a.** 2.0 L of a 1.5 M NaOH solution
 **b.** 4.0 L of a 0.20 M KCl solution
 **c.** 25.0 mL of a 6.0 M HCl solution

**7.48** Calculate the grams of solute needed to prepare each of the following solutions:
 **a.** 2.0 L of a 6.0 M NaOH solution
 **b.** 5.0 L of a 0.10 M $CaCl_2$ solution
 **c.** 175 mL of a 3.00 M $NaNO_3$ solution

**7.49** What volume is needed to obtain each of the following amounts of solute?
 **a.** liters of a 2.0 M NaOH solution to obtain 3.0 moles of NaOH
 **b.** liters of a 1.5 M NaCl solution to obtain 15 moles of NaCl
 **c.** milliliters of a 0.800 M $Ca(NO_3)_2$ solution to obtain 0.0500 moles of $Ca(NO_3)_2$

**7.50** What volume is needed to obtain each of the following amounts of solute?
**a.** liters of 4.0 M KCl solution to obtain 0.100 mole of KCl
**b.** liters of a 6.0 M HCl solution to obtain 5.0 moles of HCl
**c.** milliliters of a 2.5 M $K_2SO_4$ solution to obtain 1.2 moles of $K_2SO_4$

**7.51** To make tomato soup, you add one can of water to the condensed soup. Why is this a dilution?

**7.52** A can of frozen lemonade calls for the addition of three cans of water to make a pitcher of the beverage. Why is this a dilution?

**7.53** Calculate the final concentration of each of the following diluted solutions:
**a.** 2.0 L of a 6.0 M HCl solution is added to water so that the final volume is 6.0 L.
**b.** Water is added to 0.50 L of a 12 M NaOH solution to make 3.0 L of a diluted NaOH solution.
**c.** A 10.0-mL sample of 25% (m/v) KOH solution is diluted with water so that the final volume is 100.0 mL.
**d.** A 50.0-mL sample of 15% (m/v) $H_2SO_4$ solution is added to water to give a final volume of 250 mL.

**7.54** Calculate the final concentration of each of the following diluted solutions:
**a.** 1.0 L of a 4.0 M $HNO_3$ solution is added to water so that the final volume is 8.0 L.

**b.** Water is added to 0.25 L of a 6.0 M KOH solution to make 2.0 L of a diluted KOH solution.
**c.** A 50.0-mL sample of an 8.0% (m/v) NaOH is diluted with water so that the final volume is 200.0 mL.
**d.** A 5.0-mL sample of 50.0% (m/v) acetic acid ($HC_2H_3O_2$) solution is added to water to give a final volume of 25 mL.

**7.55** What is the final volume of each of the following diluted solutions?
**a.** liters of a 0.20 M HCl solution prepared from 20.0 mL of a 6.0 M HCl solution
**b.** milliliters of a 2.0% (m/v) NaOH solution prepared from 50.0 mL of a 10.0% (m/v) NaOH solution
**c.** liters of a 0.50 M $H_3PO_4$ solution prepared from 0.500 L of a 6.0 M $H_3PO_4$ solution
**d.** milliliters of a 5.0% (m/v) glucose solution prepared from 75 mL of a 12% (m/v) glucose solution

**7.56** What is the final volume (mL) of each of the following diluted solutions?
**a.** a 1.0% (m/v) KOH solution prepared from 10.0 mL of a 20.0% KOH solution
**b.** a 0.10 M HCl solution prepared from 25 mL of a 6.0 M HCl solution
**c.** a 1.0 M NaOH solution prepared from 50.0 mL of a 12 M NaOH solution
**d.** a 1.0% (m/v) NaCl solution prepared from 18 mL of a 4.0% (m/v) NaCl solution

# 7.6 SOLUTIONS IN CHEMICAL REACTIONS

When chemical reactions involve aqueous solutions, we use molarity and volume to determine the moles of the substances required or produced. Using the balanced equation, we can determine the volume of a solution from the molarity and moles of a solute as seen in Sample Problem 7.15.

## LEARNING GOAL

Given the volume and molarity of a solution, calculate the amount of another reactant or product in the reaction.

---

SAMPLE PROBLEM   **7.15**

■ **Volume of a Solution in a Reaction**

Zinc reacts with HCl to produce $ZnCl_2$ and hydrogen gas, $H_2$.

$$Zn(s) + 2HCl(aq) \longrightarrow ZnCl_2(aq) + H_2(g)$$

How many liters of a 1.50 M HCl solution completely react with 5.32 g of zinc?

**SOLUTION**

Step 1  **Given**   5.32 g of Zn and a 1.50 M HCl solution

**Need**   L of HCl solution

Step 2  **Plan**   We start the problem with the grams of Zn given and use its molar mass to calculate moles. Then we can use the mole–mole factor from the equation and the molarity of the HCl as conversion factors.

g of Zn   Molar mass   moles of Zn   Mole–mole factor   moles of HCl   Molarity   L of HCl

## Guide to Calculations Involving Solutions in Chemical Reactions

**1** State the given and needed quantities.

**2** Write a plan to calculate needed quantity or concentration.

**3** Write equalities and conversion factors including mole–mole and concentration factors.

**4** Set up problem to calculate needed quantity or concentration.

### Step 3 Equalities/Conversion Factors

**Molar mass of Zn**

1 mole of Zn = 65.4 g of Zn

$$\frac{1 \text{ mole Zn}}{65.4 \text{ g Zn}} \quad \text{and} \quad \frac{65.4 \text{ g Zn}}{1 \text{ mole Zn}}$$

**Mole–mole factor**

1 mole of Zn = 2 moles of HCl

$$\frac{1 \text{ mole Zn}}{2 \text{ moles HCl}} \quad \text{and} \quad \frac{2 \text{ moles HCl}}{1 \text{ mole Zn}}$$

**Molarity of HCl solution**

1 L of HCl = 1.50 moles of HCl

$$\frac{1 \text{ L HCl}}{1.50 \text{ moles HCl}} \quad \text{and} \quad \frac{1.50 \text{ moles HCl}}{1 \text{ L HCl}}$$

**Step 4 Set Up Problem** We can write the problem setup as seen in our plan.

$$5.32 \text{ g Zn} \times \frac{1 \text{ mole Zn}}{65.4 \text{ g Zn}} \times \frac{2 \text{ moles HCl}}{1 \text{ mole Zn}} \times \frac{1 \text{ L HCl}}{1.50 \text{ moles HCl}} = 0.108 \text{ L of HCl}$$

### STUDY CHECK

Using the reaction in Sample Problem 7.15, how many grams of zinc can react with 225 mL of 0.200 M HCl?

### SAMPLE PROBLEM 7.16

#### ■ Volume of a Reactant

How many liters of 0.250 M $BaCl_2$ are needed to react with 0.0325 L of 0.160 M $Na_2SO_4$ solution?

$$Na_2SO_4(aq) + BaCl_2(aq) \longrightarrow BaSO_4(s) + 2NaCl(aq)$$

#### SOLUTION

We list the given volumes and molarities for the components in the chemical equation.

**Step 1 Given** 0.0325 L of 0.160 M $Na_2SO_4$ and 0.250 M $BaCl_2$

**Need** L of $BaCl_2$

**Step 2 Plan** We start with the volume of 0.0325 L and use the molarities of the solutions as conversion factors. Then we use the volume and molarity of the $Na_2SO_4$ to calculate the moles of $Na_2SO_4$, which is converted to moles of $BaCl_2$ using a mole–mole factor. Then the molarity of $BaCl_2$ is used as a conversion factor to calculate the volume of $BaCl_2$.

L $Na_2SO_4$ — Molarity → moles $Na_2SO_4$ — Mole–mole factor → moles $BaCl_2$ — Molarity → L $BaCl_2$

**Step 3 Equalities/Conversion Factors**

**Molarity of $Na_2SO_4$**

1 L of $Na_2SO_4$ solution = 0.160 mole of $Na_2SO_4$

$$\frac{1 \text{ L Na}_2SO_4}{0.160 \text{ mole Na}_2SO_4} \quad \text{and} \quad \frac{0.160 \text{ mole Na}_2SO_4}{1 \text{ L Na}_2SO_4}$$

**Mole–mole factor**

1 mole of $Na_2SO_4$ = 1 mole of BaC

$$\frac{1 \text{ mole Na}_2SO_4}{1 \text{ mole BaCl}_2} \quad \text{and} \quad \frac{1 \text{ mole BaC}}{1 \text{ mole Na}_2S}$$

**Molarity of $BaCl_2$**

1 L of $BaCl_2$ solution = 0.250 mole of $BaCl_2$

$$\frac{1 \text{ L BaCl}_2}{0.250 \text{ mole BaCl}_2} \quad \text{and} \quad \frac{0.250 \text{ mole BaCl}_2}{1 \text{ L BaCl}_2}$$

**Step 4 Set Up Problem**

$$0.0325 \text{ L Na}_2\text{SO}_4 \times \frac{0.160 \text{ moles Na}_2\text{SO}_4}{1 \text{ L Na}_2\text{SO}_4} \times \frac{1 \text{ mole BaCl}_2}{1 \text{ mole Na}_2\text{SO}_4} \times \frac{1 \text{ L BaCl}_2}{0.250 \text{ mole BaCl}_2} = 0.0208 \text{ L of BaCl}_2$$

**STUDY CHECK**

For the reaction in Sample Problem 7.16, how many milliliters of 0.330 M $Na_2SO_4$ are needed to react with 26.8 mL of a 0.216 M $BaCl_2$ solution?

## QUESTIONS AND PROBLEMS

### Solutions in Chemical Reactions

**7.57** Given the reaction

$$Pb(NO_3)_2(aq) + 2KCl(aq) \longrightarrow PbCl_2(s) + 2KNO_3(aq)$$

**a.** How many grams of $PbCl_2$ will be formed from 50.0 mL of 1.50 M KCl and excess $Pb(NO_3)_2$?

**b.** How many milliliters of 2.00 M $Pb(NO_3)_2$ will react with 50.0 mL of 1.50 M KCl?

**7.58** In the reaction

$$NiCl_2(aq) + 2NaOH(aq) \longrightarrow Ni(OH)_2(s) + 2NaCl(aq)$$

**a.** How many milliliters of 0.200 M NaOH are needed to react with 18.0 mL of 0.500 M $NiCl_2$?

**b.** How many grams of $Ni(OH)_2$ are produced from the reaction of 35.0 mL of 1.75 M NaOH?

**7.59** In the reaction

$$Mg(s) + 2HCl(aq) \longrightarrow MgCl_2(aq) + H_2(g)$$

**a.** How many milliliters of a 6.00 M HCl solution are required to react with 15.0 g of magnesium?

**b.** How many moles of hydrogen gas form when 0.500 L of 2.00 M HCl reacts?

**7.60** The calcium carbonate in limestone reacts with HCl to produce a calcium chloride solution and carbon dioxide gas.

$$CaCO_3(s) + 2HCl(aq) \longrightarrow CaCl_2(aq) + H_2O(l) + CO_2(g)$$

**a.** How many milliliters of 0.200 M HCl can react with 8.25 g of $CaCO_3$?

**b.** How many moles of $CO_2$ form when 15.5 mL of 3.00 M HCl react with excess $CaCO_3$?

# 7.7 PROPERTIES OF SOLUTIONS

The solute particles in a solution play an important role in determining the properties of that solution. In most of the solutions discussed so far, the solute is dissolved as small particles that are uniformly dispersed throughout the solvent to give a homogeneous solution. When you observe a solution, such as salt water, you cannot visually distinguish the solute from the solvent. The solution appears transparent. The particles are so small that they go through filters and through semipermeable membranes. A **semipermeable membrane** allows solvent molecules such as water and very small solute particles to pass through, but not large solute molecules.

## LEARNING GOAL

Identify a mixture as a solution, a colloid, or a suspension. Describe osmosis and dialysis.

## Colloids

The particles in colloidal dispersions, or **colloids**, are much larger than solute particles in a solution. Colloidal particles are large molecules, such as proteins, or groups of molecules or ions. Colloids are homogeneous mixtures that do not separate or settle out. Colloidal particles are small enough to pass through filters, but too large to pass through semipermeable membranes. Table 7.9 lists several examples of colloids.

## Suspensions

**Suspensions** are heterogeneous, nonuniform mixtures that are very different from solutions or colloids. The particles of a suspension are so large that they can often be seen with the naked eye. They are trapped by filters and semipermeable membranes.

## Colloids and Solutions in the Body

In the body, colloids are retained by semipermeable membranes. For example, the intestinal lining allows solution particles to pass into the blood and lymph circulatory systems. However, the colloids from foods are too large to pass through the membrane, and they remain in the intestinal tract. Digestion breaks down large colloidal particles, such as starch and protein, into smaller particles, such as glucose and amino acids that can pass through the intestinal membrane and enter the circulatory system. Certain foods, such as bran, a fiber, cannot be broken down by human digestive processes, and they move through the intestine intact.

Because large proteins, such as enzymes, are colloids, they remain inside cells. However, many of the substances that must be obtained by cells, such as oxygen, amino acids, electrolytes, glucose, and minerals, can pass through cellular membranes. Waste products, such as urea and carbon dioxide, pass out of the cell to be excreted.

**TABLE 7.9** Examples of Colloids

|  | Substance Dispersed | Dispersing Medium |
|---|---|---|
| Fog, clouds, sprays | Liquid | Gas |
| Dust, smoke | Solid | Gas |
| Shaving cream, whipped cream, soapsuds | Gas | Liquid |
| Styrofoam, marshmallows | Gas | Solid |
| Mayonnaise, butter, homogenized milk, hand lotions | Liquid | Liquid |
| Cheese, butter | Liquid | Solid |
| Blood plasma, paints (latex), gelatin | Solid | Liquid |

The weight of the suspended solute particles causes them to settle out soon after mixing. If you stir muddy water, it mixes but then quickly separates as the suspended particles settle to the bottom and leave clear liquid at the top. You can find suspensions among the medications in a hospital or in your medicine cabinet. These include Kaopectate, calamine lotion, antacid mixtures, and liquid penicillin. It is important to shake well before using to suspend all the particles before giving a medication that is a suspension.

Water-treatment plants make use of the properties of suspensions to purify water. When flocculants such as aluminum sulfate or ferric sulfate are added to untreated water, they react with impurities to form large suspended particles called floc. In the water-treatment plant, a system of filters traps the suspended particles but clean water passes through.

Table 7.10 compares the different types of mixtures and Figure 7.6 illustrates some properties of solutions, colloids, and suspensions.

**TABLE 7.10** Comparison of Solutions, Colloids, and Suspensions

| Type of Mixture | Type of Particle | Settling | Separation |
|---|---|---|---|
| Solution | Small particles such as atoms, ions, or small molecules | Particles do not settle | Particles cannot be separated by filters or semipermeable membranes |
| Colloid | Larger molecules or groups of molecules or ions | Particles do not settle | Particles can be separated by semipermeable membranes but not by filters |
| Suspension | Very large particles that may be visible | Particles settle rapidly | Particles can be separated by filters |

SAMPLE PROBLEM 7.17

### ■ Classifying Types of Mixtures

Classify each of the following as a solution, colloid, or suspension:

**a.** a mixture that settles rapidly upon standing
**b.** a mixture whose solute particles pass through both filters and membranes

SOLUTION

**a.** suspension        **b.** solution

STUDY CHECK

Enzymes are large protein molecules that catalyze chemical reactions inside the cells of the body. If an aqueous mixture of enzymes cannot pass through the cell membrane, is the mixture a solution or a colloid?

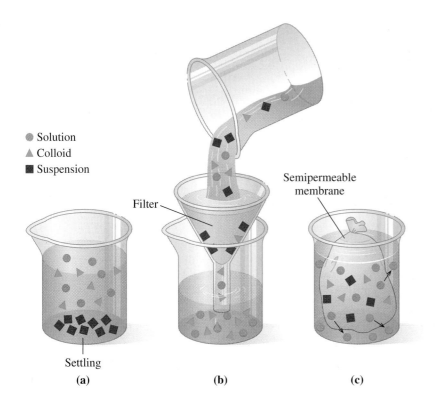

• Solution
▲ Colloid
■ Suspension

Filter

Semipermeable
membrane

Settling

**(a)**          **(b)**          **(c)**

**FIGURE 7.6** Properties of different types of mixtures: **(a)** suspensions settle out; **(b)** suspensions are separated by a filter; **(c)** solution particles go through a semipermeable membrane, but colloids and suspensions do not.

**Q** A filter can be used to separate suspension particles from a solution, but a semipermeable membrane is needed to separate colloids from a solution. Explain.

## Osmosis and Dialysis

The movement of water into and out of the cells of plants as well as our own bodies is an important biological process. In a process called **osmosis**, water molecules move through a semipermeable membrane from the solution with the lower solute concentration into a solution with the higher solute concentration. In an osmosis apparatus, water is placed on one side of a semipermeable membrane and a sucrose (sugar) solution on the other side. The semipermeable membrane allows small water molecules to flow back and forth, but blocks the sucrose molecules because they are too large to pass through the membrane. Because the sucrose solution has a higher solute concentration, there are more water molecules flowing into the sucrose solution. The volume of the sucrose solution increases, while the volume on the water side decreases. The movement of water dilutes the sucrose solution in order to equalize (or attempt to equalize) the concentrations on both sides of the membrane.

Eventually the height of the sucrose solution creates pressure that pushes water molecules back to the water side so that the flow of water between the two compartments becomes equal. This pressure, called **osmotic pressure**, prevents the flow of additional water into the more concentrated solution. When the flow of water molecules between the two compartments is equal, there is no further change in the volumes of the two solutions. The osmotic pressure of a solution depends on the concentration of solute particles in the solution. The greater the number of particles dissolved in a solution, the higher its osmotic pressure. In this example, the sucrose solution has a higher osmotic pressure than pure water, which has an osmotic pressure of zero.

In a process called reverse osmosis, a pressure greater than the osmotic pressure is applied to a solution. The flow of water is reversed so that water flows out of the solution with the higher solute concentration. The process of reverse osmosis is used in desalination plants to obtain pure water from sea (salt) water.

**WEB TUTORIAL**
Diffusion
Osmosis

## Explore Your World

### Everyday Osmosis

1. Place a few pieces of dry fruit such as raisins, prunes, or banana chips in water. Observe them after 1 hour or more. Look at them again the next day.
2. Place some grapes in a concentrated salt-water solution. Observe them after 1 hour or more. Look at them again the next day.
3. Place one potato slice in water and another slice in a concentrated salt-water solution. After 1–2 hours, observe the shapes and size of the slices. Look at them again the next day.

#### QUESTIONS

1. How did the shape of the dried fruit change after being in water? Explain.
2. How did the appearance of the grapes change after being in a concentrated salt solution? Explain.
3. How does the appearance of the potato slice that was placed in water compare to the appearance of the potato slice placed in salt water? Explain.
4. At the grocery store, why are sprinklers used to spray water on fresh produce such as lettuce, carrots, and cucumbers?

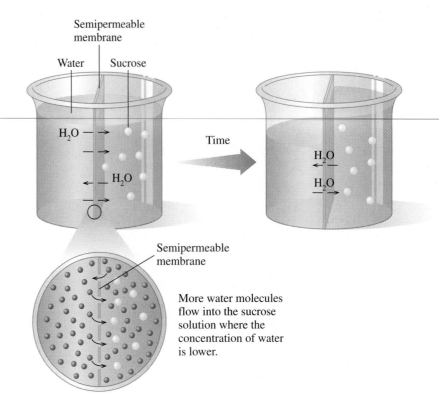

More water molecules flow into the sucrose solution where the concentration of water is lower.

### SAMPLE PROBLEM 7.18

#### ■ Osmotic Pressure

A 2% (m/v) sucrose solution and an 8% (m/v) sucrose solution are separated by a semipermeable membrane.

a. Which sucrose solution exerts the greater osmotic pressure?
b. In what direction does water flow initially?
c. Which solution will have the higher level of liquid at equilibrium?

#### SOLUTION

a. The 8% (m/v) sucrose solution has the higher solute concentration, more solute particles, and the greater osmotic pressure.
b. Initially, water will flow out of the 2% (m/v) solution into the more concentrated 8% (m/v) solution.
c. The level of the side initially containing the 8% (m/v) solution will be higher.

#### STUDY CHECK

If a 10% (m/v) glucose solution is separated from a 5% (m/v) glucose solution by a semipermeable membrane, which solution will decrease in volume?

## Isotonic Solutions

Because the cell membranes in biological systems are semipermeable, osmosis is an ongoing process. The solutes in body solutions such as blood, tissue fluids, lymph, and plasma all exert osmotic pressure. Most intravenous solutions are **isotonic solutions**, which exert the same osmotic pressure as body fluids. *Iso* means "equal to," and *tonic* refers to the osmotic pressure of the solution in the cell. In the hospital, isotonic solutions or **physiological**

**solutions** include 0.90% (m/v) NaCl solution and 5.0% (m/v) glucose solution. Although they do not contain the same particles as the body fluids, they exert the same osmotic pressure.

## Hypotonic and Hypertonic Solutions

A red blood cell placed in an isotonic solution retains its normal volume because there is an equal flow of water into and out of the cell. (See Figure 7.7a.) However, if a red blood cell is placed in a solution that is not isotonic, the differences in osmotic pressure inside and outside the cell can drastically alter the volume of the cell. When a red blood cell is placed in pure water, a **hypotonic solution** (*hypo* means "lower than"), water flows into the cell by osmosis. (See Figure 7.7b.) The increase in fluid causes the cell to swell and possibly burst, a process called **hemolysis**. A similar process occurs when you place dehydrated food, such as raisins or dried fruit, in water. The water enters the cells and the food becomes plump and smooth.

If a red blood cell is placed in a **hypertonic solution**, which has a higher solute concentration (*hyper* means "greater than"), water leaves the cell by osmosis. Suppose a red blood cell is placed in a 10% (m/v) NaCl solution. Because the osmotic pressure in the red blood cell is equal to that of a 0.90% (m/v) NaCl solution, the 10% (m/v) NaCl solution has a much greater osmotic pressure. As water is lost, the cell shrinks, a process called **crenation**. (See Figure 7.7c.) A similar process occurs when making pickles; a hypertonic salt solution causes the cucumbers to shrivel as they lose water.

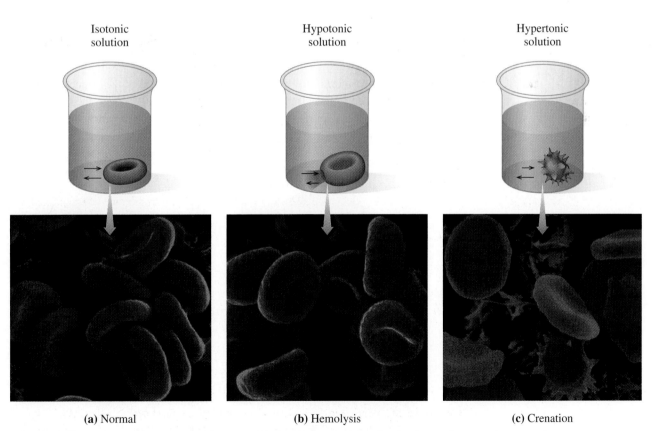

Isotonic solution    Hypotonic solution    Hypertonic solution

**(a)** Normal          **(b)** Hemolysis          **(c)** Crenation

**FIGURE 7.7** (**a**) In an isotonic solution, a red blood cell retains its normal volume. (**b**) Hemolysis: In a hypotonic solution, water flows into a red blood cell, causing it to swell and burst. (**c**) Crenation: In a hypertonic solution, water leaves the red blood cell, causing it to shrink.

**Q** What happens to a red blood cell placed in a 4% (m/v) NaCl solution?

### ■ Isotonic, Hypotonic, and Hypertonic Solutions

Describe each of the following solutions as isotonic, hypotonic, or hypertonic. Indicate whether a red blood cell placed in each solution will undergo hemolysis, crenation, or no change.

**a.** a 5.0% (m/v) glucose solution
**b.** a 0.2% (m/v) NaCl solution

#### SOLUTION

**a.** A 5.0% (m/v) glucose solution is isotonic. A red blood cell will not undergo any change.
**b.** A 0.2% (m/v) NaCl solution is hypotonic. A red blood cell will undergo hemolysis.

#### STUDY CHECK

What is the effect of a 10% (m/v) glucose solution on a red blood cell?

## Dialysis

**Dialysis** is a process that is similar to osmosis. In dialysis, a semipermeable membrane, called a dialyzing membrane, permits small solute molecules and ions as well as solvent water molecules to pass through, but it retains large particles, such as colloids. Dialysis is a way to separate solution particles from colloids.

Suppose we fill a cellophane bag with a solution containing NaCl, glucose, starch, and protein and place it in pure water. Cellophane is a dialyzing membrane, and the sodium ions, chloride ions, and glucose molecules will pass through it into the surrounding water. However, starch and protein remain inside because they are colloids. Water molecules will flow by osmosis into the cellophane bag. Eventually the concentrations of sodium ions, chloride ions, and glucose molecules inside and outside the dialysis bag become equal. To remove more NaCl or glucose, the cellophane bag must be placed in a fresh sample of pure water.

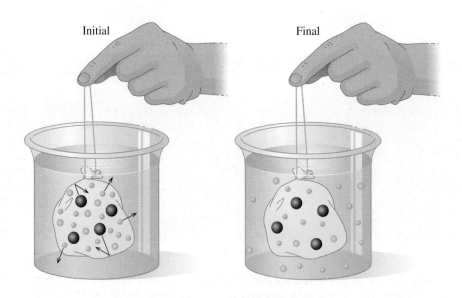

Solution particles such as Na+, Cl−, glucose
Colloidal particles such as protein, starch

# Health Note

## Dialysis by the Kidneys and the Artificial Kidney

The fluids of the body undergo dialysis by the membranes of the kidneys, which remove waste materials, excess salts, and water. In an adult, each kidney contains about 2 million nephrons. At the top of each nephron, there is a network of arterial capillaries called the glomerulus.

As blood flows into the glomerulus, small particles, such as amino acids, glucose, urea, water, and certain ions, will move through the capillary membranes into the nephron. As this solution moves through the nephron, substances still of value to the body (such as amino acids, glucose, certain ions, and 99% of the water) are reabsorbed. The major waste product, urea, is excreted in the urine.

### HEMODIALYSIS

If the kidneys fail to dialyze waste products, increased levels of urea can become life-threatening in a relatively short time. A person with kidney failure must use an artificial kidney, which cleanses the blood by **hemodialysis**.

A typical artificial kidney machine contains a large tank filled with about 100 L of water containing selected electrolytes. In the center of this dialyzing bath (dialysate), there is a dialyzing coil or membrane made of cellulose tubing. As the patient's blood flows through the dialyzing coil, the highly concentrated waste products dialyze out of the blood. No blood is lost because the membrane is not permeable to large particles such as red blood cells.

Dialysis patients do not produce much urine. As a result, they retain large amounts of water between dialysis treatments, which produces a strain on the heart. The intake of fluids for a dialysis patient may be restricted to as little as a few teaspoons of water a day. In the dialysis procedure, the pressure of the blood is increased as it circulates through the dialyzing coil so water can be squeezed out of the blood. For some dialysis patients, 2–10 L of water may be removed during one treatment. Dialysis patients have from two to three treatments a week, each treatment requiring about 5–7 hours. Some of the newer treatments require less time. For many patients, dialysis is done at home with a home dialysis unit.

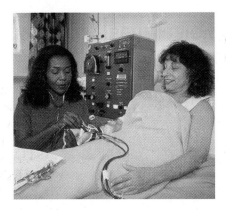

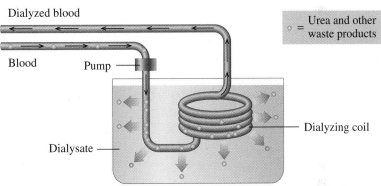

Dialyzed blood

Blood

Pump

Dialysate

○ = Urea and other waste products

Dialyzing coil

## QUESTIONS AND PROBLEMS

### Properties of Solutions

**7.61** Identify the following as characteristic of a solution, colloid, or suspension:
   **a.** a mixture that cannot be separated by a semipermeable membrane
   **b.** a mixture that settles out upon standing

**7.62** Identify the following as characteristic of a solution, colloid, or suspension:
   **a.** Particles of this mixture remain inside a semipermeable membrane, but pass through filters.
   **b.** The particles of solute in this solution are very large and visible.

**7.63 a.** How do plants obtain water from the ground?
   **b.** How does a pickle get all shriveled up?

**7.64 a.** Why should you not drink large quantities of seawater?
   **b.** Why does salt preserve foods?

**7.65** A 10% (m/v) starch solution is separated from pure water by an osmotic membrane.
   **a.** Which has the higher osmotic pressure?
   **b.** In which direction will water flow initially?
   **c.** In which compartment will the volume level rise?

**7.66** Two solutions, a 0.1% (m/v) albumin solution and a 2% (m/v) albumin solution, are separated by a semipermeable membrane. (Albumins are colloidal proteins.)
   **a.** Which compartment has the higher osmotic pressure?
   **b.** In which direction will water flow initially?
   **c.** In which compartment will the volume level rise?

**7.67** Indicate the compartment (A or B) that will increase in volume for each of the following pairs of solutions separated by semipermeable membranes:

| **A** | **B** |
|---|---|
| **a.** 5.0% (m/v) glucose | 10% (m/v) glucose |
| **b.** 4% (m/v) albumin | 8% (m/v) albumin |
| **c.** 0.1% (m/v) NaCl | 10% (m/v) NaCl |

**7.68** Indicate the compartment (A or B) that will increase in volume for each of the following pairs of solutions separated by semipermeable membranes:

| **A** | **B** |
|---|---|
| **a.** 20% (m/v) glucose | 10% (m/v) glucose |
| **b.** 10% (m/v) albumin | 2% (m/v) albumin |
| **c.** 0.5% (m/v) NaCl | 5% (m/v) NaCl |

**7.69** Are the following solutions isotonic, hypotonic, or hypertonic compared with a red blood cell?
**a.** distilled $H_2O$
**b.** 1% (m/v) glucose
**c.** 0.90% (m/v) NaCl
**d.** 5.0% (m/v) glucose

**7.70** Will a red blood cell undergo crenation, hemolysis, or no change in each of the following solutions?
**a.** 1% (m/v) glucose
**b.** 2% (m/v) NaCl
**c.** 5% (m/v) NaCl
**d.** 0.1% (m/v) NaCl

**7.71** Each of the following mixtures is placed in a dialyzing bag and immersed in distilled water. Which substances will be found outside the bag in the distilled water?
**a.** NaCl solution
**b.** starch (colloid) and alanine (amino acid solution)
**c.** NaCl solution and starch (colloid)
**d.** urea solution

**7.72** Each of the following mixtures is placed in a dialyzing bag and immersed in distilled water. Which substances will be found outside the bag in the distilled water?
**a.** KCl solution and glucose solutions
**b.** an albumin solution (colloid)
**c.** an albumin solution (colloid), KCl solution, and glucose solution
**d.** urea solution and NaCl solution

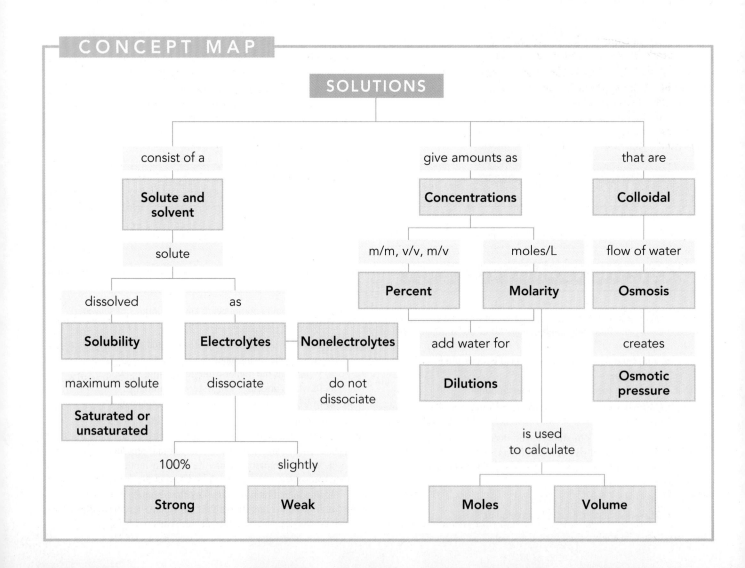

CONCEPT MAP

# CHAPTER REVIEW

## 7.1 Solutions

**Learning Goal:** Identify the solute and solvent in a solution. Describe the formation of a solution.

A solution forms when a solute dissolves in a solvent. In a solution, the particles of solute are evenly distributed in the solvent. The solute and solvent may be solid, liquid, or gas. The polar O—H bond leads to hydrogen bonding between water molecules. An ionic solute dissolves in water, a polar solvent, because the polar water molecules attract and pull the ions into solution, where they become hydrated. The expression "like dissolves like" means that a polar or an ionic solute dissolves in a polar solvent while a nonpolar solute dissolves in a nonpolar solvent.

## 7.2 Electrolytes and Nonelectrolytes

**Learning Goal:** Identify solutes as electrolytes or nonelectrolytes.

Substances that release ions in water are called electrolytes because their solutions will conduct an electrical current. Strong electrolytes are completely ionized, whereas weak electrolytes are only partially ionized. Nonelectrolytes are substances that dissolve in water to produce molecules and cannot conduct electrical currents. An equivalent is the amount of an electrolyte that carries one mole of positive or negative charge. One mole of $Na^+$ is 1 equivalent. One mole of $Ca^{2+}$ has 2 equivalents. In fluid replacement solutions, the concentrations of electrolytes are expressed as mEq/L of solution.

## 7.3 Solubility

**Learning Goal:** Define solubility; distinguish between an unsaturated and a saturated solution.

The solubility of a solute is the maximum amount of a solute that can dissolve in 100 g of solvent. A solution that contains the maximum amount of dissolved solute is a saturated solution. A solution containing less than the maximum amount of dissolved solute is unsaturated. An increase in temperature increases the solubility of most solids in water, but decreases the solubility of gases in water.

## 7.4 Percent Concentration

**Learning Goal:** Calculate the percent concentration of a solute in a solution; use percent concentration to calculate the amount of solute or solution.

The concentration of a solution is the amount of solute dissolved in a certain amount of solution. Mass percent expresses the ratio of the mass of solute to the mass of solution multiplied by 100. Percent concentration is also expressed as volume/volume and mass/volume ratios. In calculations of grams or milliliters of solute or solution, the percent concentration is used as a conversion factor.

## 7.5 Molarity and Dilution

**Learning Goal:** Calculate the molarity of a solution; use molarity to calculate the moles of solute or the volume needed to prepare a solution. Describe the dilution of a solution.

Molarity is the moles of solute per liter of solution. Units of molarity, moles/liter, are used in conversion factors to solve for moles of solute or volume of solution. In dilution, the volume of solvent increases and the solute concentration decreases.

## 7.6 Solutions in Chemical Reactions

**Learning Goal:** Given the volume and molarity of a solution, calculate the amount of another reactant or product in the reaction.

When solutions are involved in chemical reactions, the moles of a substance in solution can be determined from the volume and molarity of the solution. If the mass or solution volume and molarity of substances in a reaction are given, the balanced equation can be used to determine the quantities or concentrations of any of the other substances in the reaction.

## 7.7 Properties of Solutions

**Learning Goal:** Identify a mixture as a solution, a colloid, or a suspension. Describe osmosis and dialysis.

Colloids contain particles that do not settle out; they pass through filters but not through semipermeable membranes. Suspensions have very large particles that settle out of solution.

In osmosis, solvent (water) passes through a semipermeable membrane from a solution of a lower solute concentration to a solution of a higher concentration. Isotonic solutions have osmotic pressures equal to that of body fluids. A red blood cell maintains its volume in an isotonic solution, but swells and may burst (hemolyze) in a hypotonic solution and shrinks (crenates) in a hypertonic solution. In dialysis, water and small solute particles pass through a dialyzing membrane, while larger particles are retained.

# KEY TERMS

**colloid** A mixture having particles that are moderately large. Colloids pass through filters but cannot pass through semipermeable membranes.

**concentration** A measure of the amount of solute that is dissolved in a specified amount of solution.

**crenation** The shriveling of a cell because of water's leaving the cell when the cell is placed in a hypertonic solution.

**dialysis** A process in which water and small solute particles pass through a semipermeable membrane.

**dilution** A process by which water (solvent) is added to a solution to increase the volume and decrease (dilute) the concentration of the solute.

**electrolyte** A substance that produces ions when dissolved in water; its solution conducts electricity.

**equivalent (Eq)** The amount of a positive or negative ion that supplies 1 mole of electrical charge.

**hemodialysis** A mechanical cleansing of the blood by an artificial kidney using the principle of dialysis.

**hemolysis** A swelling and bursting of red blood cells in a hypotonic solution because of an increase in fluid volume.

**Henry's law** The solubility of a gas in a liquid is directly related to the pressure of that gas above the liquid.

**hydration** The process of surrounding dissolved ions by water molecules.

**hydrogen bond** The attraction between a partially positive hydrogen atom in one molecule and a highly electronegative atom such as oxygen in another molecule.

**hypertonic solution** A solution that has a higher particle concentration and higher osmotic pressure than the cells of the body.

**hypotonic solution** A solution that has a lower particle concentration and lower osmotic pressure than the cells of the body.

**isotonic solution** A solution that has the same particle concentration and osmotic pressure as that of the cells of the body.

**mass percent** The grams of solute in exactly 100 g of solution.

**mass/volume percent** The grams of solute in exactly 100 mL of solution.

**molarity (M)** The number of moles of solute in exactly 1 L of solution.

**nonelectrolyte** A substance that dissolves in water as molecules; its solution will not conduct an electrical current.

**osmosis** The flow of a solvent, usually water, through a semipermeable membrane into a solution of higher solute concentration.

**osmotic pressure** The pressure that prevents the flow of water into the more concentrated solution.

**physiological solution** A solution that is isotonic with and exerts the same osmotic pressure as normal body fluids.

**saturated solution** A solution containing the maximum amount of solute that can dissolve at a given temperature. Any additional solute will remain undissolved in the container.

**semipermeable membrane** A membrane that permits the passage of certain substances while blocking or retaining others.

**solubility** The maximum amount of solute that can dissolve in exactly 100 g of solvent, usually water, at a given temperature.

**solute** The component in a solution that is present in the smaller quantity.

**solution** A homogeneous mixture in which the solute is made up of small particles (ions or molecules) that can pass through filters and semipermeable membranes.

**solvent** The substance in which the solute dissolves; usually the component present in greatest amount.

**strong electrolyte** A polar or ionic compound that ionizes completely when it dissolves in water. Its solution is a good conductor of electricity.

**suspension** A mixture in which the solute particles are large enough and heavy enough to settle out and be retained by both filters and semipermeable membranes.

**unsaturated solution** A solution that contains less solute than can be dissolved.

**volume percent** A percent concentration that relates the volume of the solute in exactly 100 ml of the solution.

**weak electrolyte** A substance that produces only a few ions along with many molecules when it dissolves in water. Its solution is a weak conductor of electricity.

# UNDERSTANDING THE CONCEPTS

**7.73** Match the diagrams with
   **a.** a polar solute and a polar solvent
   **b.** a nonpolar solute and a polar solvent
   **c.** a nonpolar solute and a nonpolar solvent

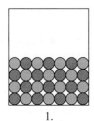 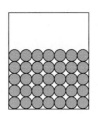

1.                                    2.

**7.74** Do you think solution (1) has undergone heating or cooling to give the solid shown in (2) and (3)?

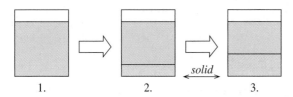

1.                      2.          *solid*          3.

**7.75** Select the diagram that represents the solution formed by a solute 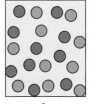 that is a
   **a.** nonelectrolyte
   **b.** weak electrolyte
   **c.** strong electrolyte

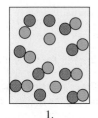 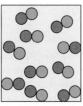

1.                      2.                      3.

**7.76** Why do the lettuce leaves in a salad wilt after a vinaigrette dressing containing salt is added?

**7.77** A pickle is made by soaking a cucumber in *brine*, a salt-water solution. What makes the smooth cucumber become wrinkled like a prune?

**7.78** Select the container that represents the dilution of a 4% (m/v) KCl solution to each of the following:
   **a.** 2% (m/v) KCl
   **b.** 1% (m/v) KCl

4% (m/v) NaCl          1.          2.          3.

**7.79** A semipermeable membrane separates compartments A and B. If the levels of the following solutions in A and B are equal initially, select the diagram that illustrates the final levels:

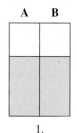
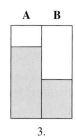

1.  2.  3.

| Solution in A | Solution in B |
|---|---|
| **a.** 2% (m/v) starch | 8% (m/v) starch |
| **b.** 1% (m/v) starch | 1% (m/v) starch |
| **c.** 5% (m/v) sucrose | 1% (m/v) sucrose |
| **d.** 0.1% (m/v) sucrose | 1% (m/v) sucrose |

**7.80** Select the diagram that represents the shape of a red blood cell when placed in each of the following solutions:

1.  2.  3.

Normal red blood cell

**a.** 0.90% (m/v) NaCl
**b.** 10% (m/v) glucose
**c.** 0.01% (m/v) NaCl
**d.** 5.0% (m/v) glucose
**e.** 1% (m/v) glucose

# ADDITIONAL QUESTIONS AND PROBLEMS

**7.81** Why does iodine dissolve in hexane, but not in water?

**7.82** How does temperature and pressure affect the solubility of solids and gases in water?

**7.83** Calculate the mass percent (% m/m) of a solution containing 15.5 g of $Na_2SO_4$ and 75.5 g of $H_2O$.

**7.84** How many grams of $K_2CO_3$ are in 750 mL of a 3.5% (m/v) $K_2CO_3$ solution?

**7.85** A patient receives all her nutrition from fluids given through the vena cava. Every 12 hours, 500 mL of a solution that is 5.0% (m/v) amino acids (protein) and 20% (m/v) glucose (carbohydrate) is given along with 500 mL of a 10% (m/v) lipid (fat).
  **a.** In 1 day, how many grams each of amino acids, glucose, and lipid are given to the patient?
  **b.** How many kilocalories does she obtain in 1 day?

**7.86** An 80-proof brandy is 40.0% (v/v) ethyl alcohol. The "proof" is twice the percent concentration of alcohol in the beverage. How many milliliters of alcohol are present in 750 mL of brandy?

**7.87** How many milliliters of a 12% (v/v) propyl alcohol solution would you need to obtain 4.5 mL of propyl alcohol?

**7.88** How many liters of a 5.0% (m/v) glucose solution would you need to obtain 75 g of glucose?

**7.89** If you were in the laboratory, how would you prepare 0.250 L of a 2.00 M KCl solution?

**7.90** What is the molarity of a solution containing 15.6 g of KCl in 274 mL of solution?

**7.91** A solution is prepared with 70.0 g of $HNO_3$ and 130.0 g of $H_2O$. It has a density of 1.21 g/mL.
  **a.** What is the mass percent (% m/m) of the $HNO_3$ solution?
  **b.** What is the total volume of the solution?
  **c.** What is the mass/volume (% m/v) percent?
  **d.** What is its molarity (M)?

**7.92** What is the molarity of a 15% (m/v) NaOH solution?

**7.93** How many grams of solute are in each of the following solutions?
  **a.** 2.5 L of 3.0 M $Al(NO_3)_3$
  **b.** 75 mL of 0.50 M $C_6H_{12}O_6$
  **c.** 235 mL of 1.80 M LiCl

**7.94** How many milliliters of each of the following solutions will provide 25.0 g of KOH?
  **a.** 2.50 M KOH
  **b.** 0.750 M KOH
  **c.** 5.60 M KOH

**7.95** The antacid Amphogel contains aluminum hydroxide $Al(OH)_3$. How many milliliters of 6.00 M HCl are required to react with 60.0 mL of 1.00 M $Al(OH)_3$?

$$Al(OH)_3(s) + 3HCl(aq) \longrightarrow AlCl_3(aq) + 3H_2O(aq)$$

**7.96** Why would a dialysis unit (artificial kidney) use isotonic concentrations of NaCl, KCl, $NaHCO_3$, and glucose in the dialysate?

**7.97** Why would solutions with high salt content be used to prepare dried flowers?

**7.98** A patient on dialysis has a high level of urea, a high level of sodium, and a low level of potassium in the blood. Why is the dialyzing solution prepared with a high level of potassium but no sodium or urea?

**7.99** Why can't you drink seawater even if you are stranded on a desert island?

**7.100** Calcium carbonate $CaCO_3$ reacts with stomach acid (HCl, hydrochloric acid) according to the following equation:

$$CaCO_3(s) + 2HCl(aq) \longrightarrow CaCl_2(aq) + H_2O(l) + CO_2(g)$$

Tums®, an antacid, contains $CaCO_3$. If Tums is added to 20.0 mL of 0.400 M HCl, how many grams of $CO_2$ gas are produced?

# CHALLENGE QUESTIONS

**7.101**  In a laboratory experiment, a 10.0-mL sample of NaCl solution is poured into an evaporating dish with a mass of 24.10 g. The combined mass of the evaporating dish and NaCl is 36.15 g. After heating, the evaporating dish and dry NaCl have a combined mass of 25.50 g.
  **a.** What is the % (m/m) of the NaCl solution?
  **b.** What is the molarity (M) of the NaCl solution?
  **c.** If water is added to 10.0 mL of the initial NaCl solution to give a final volume of 60.0 mL, what is the molarity of the dilute NaCl solution?

**7.102**  A solution contains 4.56 g of KCl in 175 mL of solution. If the density of the KCl solution is 1.12 g/mL, what are the percent (m/m) and molarity, M, for the potassium chloride solution?

**7.103**  Potassium fluoride has a solubility of 92 g of KF in 100 g of $H_2O$ at 18 °C. State if each of the following mixtures forms an unsaturated or saturated solution at 18 °C.
  **a.** 35 g of KF and 25 g of $H_2O$
  **b.** 42 g of KF and 50. g of $H_2O$
  **c.** 145 g of KF and 150. g of $H_2O$

**7.104**  A solution is prepared by dissolving 22.0 g of NaOH in 118.0 g of water. The NaOH solution has a density of 1.15 g/mL.
  **a.** What is the % (m/m) concentration of the NaOH solution?
  **b.** What is the total volume (mL) of the solution?
  **c.** What is the molarity (M) of the solution?

**7.105**  How many milliliters of a 1.75 M LiCl solution contain 15.2 g of LiCl?

**7.106**  How many grams of NaBr are contained in 75.0 mL of a 1.50 M NaBr solution?

**7.107**  Magnesium reacts with HCl to produce magnesium chloride and hydrogen gas:

$$Mg(s) + 2HCl(aq) \longrightarrow MgCl_2(aq) + H_2(g)$$

What is the molarity of the HCl solution if 250. mL of the HCl solution reacts with magnesium to produce 4.20 L of $H_2$ gas measured at STP?

**7.108**  How many grams of NO gas can be produced from 80.0 mL of 4.00 M $HNO_3$ and excess Cu?

$$3Cu(s) + 8HNO_3(aq) \longrightarrow 3Cu(NO_3)_2(aq) + 4H_2O(l) + 2NO(g)$$

# ANSWERS

## Answers to Study Checks

**7.1**  Iodine is the solute, and ethyl alcohol is the solvent.

**7.2**  Yes. Both the solute and solvent are nonpolar substances; "like dissolves like."

**7.3**  A solution of a weak electrolyte will contain mostly molecules and a few ions.

**7.4**  0.194 mole of $Cl^-$

**7.5**  57 g of $NaNO_3$

**7.6**  A higher solubility is more likely because the solubility of most solids increases when the temperature increases.

**7.7**  3.4% (m/m) NaCl solution

**7.8**  4.8% (m/v) $Br_2$ in $CCl_4$

**7.9**  18.0 g of KCl

**7.10**  2.12 M $KNO_3$

**7.11**  123 g of $NaHCO_3$

**7.12**  750 mL of HCl

**7.13**  3.0% (m/v) HCl

**7.14**  120. mL

**7.15**  1.47 g of Zn

**7.16**  17.5 mL

**7.17**  colloid

**7.18**  5% (m/v) glucose

**7.19**  The red blood cell will shrink (crenate).

## Answers to Selected Questions and Problems

**7.1**  **a.** NaCl, solute; water, solvent
  **b.** water, solute; ethanol, solvent
  **c.** oxygen, solute; nitrogen, solvent

**7.3**  The polar water molecules pull the $K^+$ and $I^-$ ions away from the solid and into solution, where they are hydrated.

**7.5**  **a.** water       **b.** $CCl_4$       **c.** water       **d.** $CCl_4$

**7.7**  In a solution of KF, only the ions of $K^+$ and $F^-$ are present in the solvent. In an HF solution, there are a few ions of $H^+$ and $F^-$ present but mostly dissolved HF molecules.

**7.9**  **a.** $KCl(s) \xrightarrow{H_2O} K^+(aq) + Cl^-(aq)$
  **b.** $CaCl_2(s) \xrightarrow{H_2O} Ca^{2+}(aq) + 2Cl^-(aq)$
  **c.** $K_3PO_4(s) \xrightarrow{H_2O} 3K^+(aq) + PO_4^{3-}(aq)$
  **d.** $Fe(NO_3)_3(s) \xrightarrow{H_2O} Fe^{3+}(aq) + 3NO_3^-(aq)$

**7.11**  **a.** mostly molecules and a few ions
  **b.** ions only                    **c.** molecules only

**7.13**  **a.** strong electrolyte       **b.** weak electrolyte
  **c.** nonelectrolyte

**7.15**  **a.** 1 Eq       **b.** 2 Eq       **c.** 2 Eq       **d.** 6 Eq

**7.17**  0.154 mole of $Na^+$, 0.154 mole of $Cl^-$

**7.19**  55 mEq/L

**7.21**  **a.** saturated              **b.** unsaturated

**7.23**  **a.** unsaturated       **b.** unsaturated       **c.** saturated

**7.25**  **a.** 68 g of KCl       **b.** 12 g of KCl

**7.27** **a.** The solubility of solid solutes typically increases as temperature increases.
**b.** The solubility of a gas is less at a higher temperature.
**c.** Gas solubility is less at a higher temperature and the $CO_2$ pressure in the can is increased.

**7.29** 5% (m/m) is 5 g of glucose in 100 g of solution, whereas 5% (m/v) is 5 g of glucose in 100 mL of solution.

**7.31** **a.** 17% (m/m)      **b.** 5.3% (m/m)

**7.33** **a.** 30.% (m/v)      **b.** 11% (m/v)

**7.35** **a.** 2.5 g of KCl      **b.** 50. g of $NH_4Cl$

**7.37** 79.9 mL of alcohol

**7.39** **a.** 20. g of mannitol      **b.** 240 g of mannitol

**7.41** 2 L

**7.43** **a.** 0.50 M glucose      **b.** 0.036 M KOH
**c.** 0.250 M NaCl

**7.45** **a.** 3.0 moles of NaCl      **b.** 0.40 mole of KBr
**c.** 0.25 mole of $MgCl_2$

**7.47** **a.** 120 g of NaOH      **b.** 60. g of KCl
**c.** 5.5 g of HCl

**7.49** **a.** 1.5 L      **b.** 10. L      **c.** 62.5 mL

**7.51** Adding water (solvent) to the soup increases the volume and dilutes the tomato concentration.

**7.53** **a.** 2.0 M HCl      **b.** 2.0 M NaOH
**c.** 2.5% (m/v) KOH      **d.** 3.0% (m/v) $H_2SO_4$

**7.55** **a.** 0.60 L      **b.** 250 mL
**c.** 6.0 L      **d.** 180 mL

**7.57** **a.** 10.4 g of $PbCl_2$
**b.** 18.8 mL of $Pb(NO_3)_2$ solution

**7.59** **a.** 206 mL of HCl solution
**b.** 0.500 mole of $H_2$ gas

**7.61** **a.** solution      **b.** suspension

**7.63** **a.** Water flows from wet soil through plant membranes toward the higher solute concentration in the plants.
**b.** Because a brine solution (salts in water) has a higher solute concentration, water flows from the cucumber to the brine and the cucumber becomes a pickle.

**7.65** **a.** starch solution
**b.** from pure water into the starch
**c.** starch solution

**7.67** **a.** B 10% (m/v) glucose solution
**b.** B 8% (m/v) albumin solution
**c.** B 10% (m/v) NaCl solution

**7.69** **a.** hypotonic      **b.** hypotonic
**c.** isotonic      **d.** isotonic

**7.71** **a.** NaCl      **b.** alanine
**c.** NaCl      **d.** urea

**7.73** **a.** 1      **b.** 2
**c.** 1

**7.75** **a.** 3      **b.** 1      **c.** 2

**7.77** The skin of the cucumber acts like a semipermeable membrane, and the more dilute solution inside flows into the brine solution.

**7.79** **a.** 2      **b.** 1      **c.** 3      **d.** 2

**7.81** Because iodine is a nonpolar molecule, it will dissolve in hexane, a nonpolar solvent. Iodine does not dissolve in water because water is a polar solvent.

**7.83** 17.0% (m/m)

**7.85** **a.** 50 g of amino acids, 200 g of glucose, and 100 g of lipids
**b.** 1900 kcal

**7.87** 38 mL of solution

**7.89** To make a 2.00 M KCl solution, weigh out 37.3 g of KCl (0.500 mole) and place into a volumetric flask. Add water to dissolve the KCl and give a final volume of 0.250 liter.

**7.91** **a.** 35.0% (m/m) $HNO_3$      **b.** 165 mL
**c.** 42.4% (m/v) $HNO_3$      **d.** 6.73 M

**7.93** **a.** 1600 g of $Al(NO_3)_3$      **b.** 6.8 g of $C_6H_{12}O_6$
**c.** 17.9 g of LiCl

**7.95** 30.0 mL of HCl solution

**7.97** The solution will dehydrate the flowers because water will flow out of the cells of the flowers into the more concentrated (hypertonic) salt solution.

**7.99** Drinking seawater will cause water to flow out of the body cells and further dehydrate a person.

**7.101** **a.** 11.6% (m/m)      **b.** 2.39 M      **c.** 0.398 M

**7.103** **a.** saturated      **b.** unsaturated      **c.** saturated

**7.105** 205 mL

**7.107** 1.50 M

# 8 Acids and Bases

Visit **www.chemplace.com** for extra quizzes, interactive tutorials, career resources, PowerPoint slides for chapter review, math help, and case studies.

*"In a stat lab, we are sent blood samples of patients in emergency situations," says Audrey Trautwein, clinical laboratory technician, Stat Lab, Santa Clara Valley Medical Center. "We may need to assess the status of a trauma patient in ER or a patient who is in surgery. For example, an acidic blood pH diminishes cardiac function and affects the actions of certain drugs. In a stat situation, it is critical that we obtain our results fast. This is done using a blood gas analyzer. As I put a blood sample into the analyzer, a small probe draws out a measured volume, which is tested simultaneously for pH, $P_{O_2}$, and $P_{CO_2}$ as well as electrolytes, glucose, and hemoglobin. In about one minute, we have our test results, which are sent to the doctor's computer."*

Lemons, grapefruit, and vinegar taste sour because they contain acids. We have acid in our stomach that helps us digest food; we produce lactic acid in our muscles when we exercise. Acid from bacteria turns milk sour to make cottage cheese or yogurt. Bases are solutions that neutralize acids. Sometimes we take antacids such as milk of magnesia to offset the effects of too much stomach acid.

The pH of a solution describes its acidity. The pH of body fluids, including blood and urine, is regulated primarily by the lungs and the kidneys. Major changes in the pH of the body fluids can severely affect biological activities within the cells. Buffers are present to prevent large fluctuations in pH.

# 8.1 ACIDS AND BASES

The term *acid* comes from the Latin word *acidus*, which means "sour." We are familiar with the sour tastes of vinegar and lemons and other common acids in foods.

In 1887, the Swedish chemist Svante Arrhenius was the first to describe **acids** as substances that produce hydrogen ions ($H^+$) when they dissolve in water. For example, hydrogen chloride ionizes in water to give hydrogen ions, $H^+$, and chloride ions, $Cl^-$. The hydrogen ions, $H^+$, give acids a sour taste, change blue litmus indicator to red, and corrode some metals.

$$HCl(g) \xrightarrow{\;H_2O\;} H^+(aq) + Cl^-(aq)$$

Polar covalent compound — Ionization in water

## Naming Acids

When an acid dissolves in water to produce hydrogen ion and a simple nonmetal anion, the prefix *hydro* is used before the name of the nonmetal and its *ide* ending is changed to *ic acid*. For example, hydrogen chloride (HCl) dissolves in water to form HCl(*aq*), which is named hydrochloric acid.

When an acid contains a polyatomic ion, the name of the acid comes from the name of the polyatomic ion. The *ate* in the name is replaced with *ic acid*. If the acid contains a polyatomic ion with an *ite* ending, its name ends with *ous acid*. The names of some common acids and their anions are listed in Table 8.1.

**TABLE 8.1** Naming Common Acids

| Acid | Name of Acid | Anion | Name of Anion |
|------|--------------|-------|----------------|
| HCl | **Hydro**chlor**ic acid** | $Cl^-$ | Chlor**ide** |
| HBr | **Hydro**brom**ic acid** | $Br^-$ | Brom**ide** |
| $HNO_3$ | Nitr**ic acid** | $NO_3^-$ | Nit**rate** |
| $HNO_2$ | Nitr**ous acid** | $NO_2^-$ | Nit**rite** |
| $H_2SO_4$ | Sulfur**ic acid** | $SO_4^{2-}$ | Sul**fate** |
| $H_2SO_3$ | Sulfur**ous acid** | $SO_3^{2-}$ | Sul**fite** |
| $H_2CO_3$ | Carbon**ic acid** | $CO_3^{2-}$ | Carbon**ate** |
| $H_3PO_4$ | Phosphor**ic acid** | $PO_4^{3-}$ | Phosph**ate** |
| $HClO_3$ | Chlor**ic acid** | $ClO_3^-$ | Chlor**ate** |
| $HClO_2$ | Chlor**ous acid** | $ClO_2^-$ | Chlor**ite** |
| $HC_2H_3O_2$ | Acet**ic acid** | $C_2H_3O_2^-$ | Acet**ate** |

LEARNING GOAL

Describe and name acids and bases; identify Brønsted–Lowry acids and bases.

the
**Chemistry**
place

**WEB TUTORIAL**
Nature of Acids and Bases

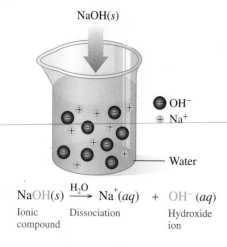

NaOH(s)

● OH⁻
⊕ Na⁺

Water

$$NaOH(s) \xrightarrow{H_2O} Na^+(aq) + OH^-(aq)$$

Ionic        Dissociation        Hydroxide
compound                        ion

## Bases

You may be familiar with some bases such as antacids, drain openers, and oven cleaners. According to the Arrhenius theory, **bases** are ionic compounds that dissociate into a metal ion and hydroxide ions ($OH^-$) when they dissolve in water. For example, sodium hydroxide is an Arrhenius base that dissociates in water to give sodium ions, $Na^+$, and hydroxide ions, $OH^-$.

Most Arrhenius bases are formed from Groups 1A (1) and 2A (2) metals, such as NaOH, KOH, LiOH, and $Ca(OH)_2$. The hydroxide ions ($OH^-$) give Arrhenius bases common characteristics such as a bitter taste and soapy, slippery feel. A base turns litmus indicator blue and phenolphthalein indicator pink.

## Naming Bases

Typical Arrhenius bases are named as hydroxides.

| Bases | Name |
|---|---|
| NaOH | Sodium **hydroxide** |
| KOH | Potassium **hydroxide** |
| $Ca(OH)_2$ | Calcium **hydroxide** |
| $Al(OH)_3$ | Aluminum **hydroxide** |

---

### SAMPLE PROBLEM  8.1

#### ■ Names and Formulas of Acids and Bases

**a.** Name each of the following as an acid or a base:
   **1.** $H_3PO_4$          **2.** NaOH
**b.** Write the formula of each of the following acid or base:
   **1.** nitrous acid          **2.** hydrobromic acid

**SOLUTION**

**a. 1.** phosphoric acid          **2.** sodium hydroxide
**b. 1.** $HNO_2$          **2.** HBr

**STUDY CHECK**

**a.** Give the name for $H_2SO_4$.
**b.** Write the formula of potassium hydroxide.

---

### SAMPLE PROBLEM  8.2

#### ■ Dissociation of an Arrhenius Base

Write an equation for the dissociation of $Ca(OH)_2(s)$ in water.

**SOLUTION**

$Ca(OH)_2$ dissolves in water to give a solution of calcium ions ($Ca^{2+}$) and twice as many hydroxide ions ($OH^-$).

$$Ca(OH)_2(s) \xrightarrow{H_2O} Ca^{2+}(aq) + 2OH^-(aq)$$

**STUDY CHECK**

Write an equation for dissociation of lithium hydroxide.

## Brønsted–Lowry Acids and Bases

In 1923, J.N. Brønsted in Denmark and T.M. Lowry in Great Britain expanded the definition of acids and bases. A **Brønsted–Lowry acid** donates a proton (hydrogen ion, $H^+$) to another substance, and a **Brønsted–Lowry base** accepts a proton.

A Brønsted–Lowry acid is a proton ($H^+$) donor.

A Brønsted–Lowry base is a proton ($H^+$) acceptor.

A free, dissociated proton ($H^+$) does not actually exist in water. Its attraction to polar water molecules is so strong that the proton bonds to the water molecule and forms a **hydronium ion, $H_3O^+$**.

$$H{-}\ddot{O}{:} + H^+ \longrightarrow \left[ H{-}\ddot{O}{-}H \right]^+$$

Water        Proton        Hydronium ion

We can write the formation of a hydrochloric acid solution as a transfer of a proton from hydrogen chloride to water. By accepting a proton in the reaction, water is acting as a base, according to the Brønsted–Lowry concept.

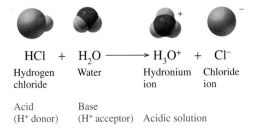

$$HCl + H_2O \longrightarrow H_3O^+ + Cl^-$$

| Hydrogen chloride | Water | Hydronium ion | Chloride ion |
|---|---|---|---|
| Acid ($H^+$ donor) | Base ($H^+$ acceptor) | Acidic solution | |

In another reaction, ammonia ($NH_3$) reacts with water. Because the nitrogen atom of $NH_3$ has a stronger attraction for a proton, water acts as an acid by donating a proton.

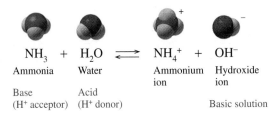

$$NH_3 + H_2O \rightleftharpoons NH_4^+ + OH^-$$

| Ammonia | Water | Ammonium ion | Hydroxide ion |
|---|---|---|---|
| Base ($H^+$ acceptor) | Acid ($H^+$ donor) | | Basic solution |

Table 8.2 compares some characteristics of acids and bases.

**TABLE 8.2** Some Characteristics of Acids and Bases

| Characteristic | Acids | Bases |
|---|---|---|
| Reaction: Arrhenius | Produce $H^+$ | Produce $OH^-$ |
| Reaction: Brønsted–Lowry | Donate $H^+$ | Accept $H^+$ |
| Electrolytes | Yes | Yes |
| Taste | Sour | Bitter, chalky |
| Feel | May sting | Soapy, slippery |
| Litmus | Red | Blue |
| Phenolphthalein | Colorless | Pink |
| Neutralization | Neutralize bases | Neutralize acids |

SAMPLE PROBLEM  **8.3**

### ■ Acids and Bases

In each of the following equations, identify the reactant that is a Brønsted–Lowry acid and the reactant that is a Brønsted–Lowry base:

**a.** $HBr(aq) + H_2O(l) \longrightarrow H_3O^+(aq) + Br^-(aq)$
**b.** $H_2O(l) + CN^-(aq) \rightleftharpoons HCN(aq) + OH^-(aq)$

#### SOLUTION

**a.** HBr, acid; $H_2O$, base          **b.** $H_2O$, acid; $CN^-$, base

#### STUDY CHECK

When $HNO_3$ reacts with water, water acts as a Brønsted–Lowry base. Write the equation for the reaction.

## Conjugate Acid–Base Pairs

Conjugate acid–base pair

Donates H+

HF          F⁻

Conjugate acid–base pair

Accepts H+

$H_2O$          $H_3O^+$

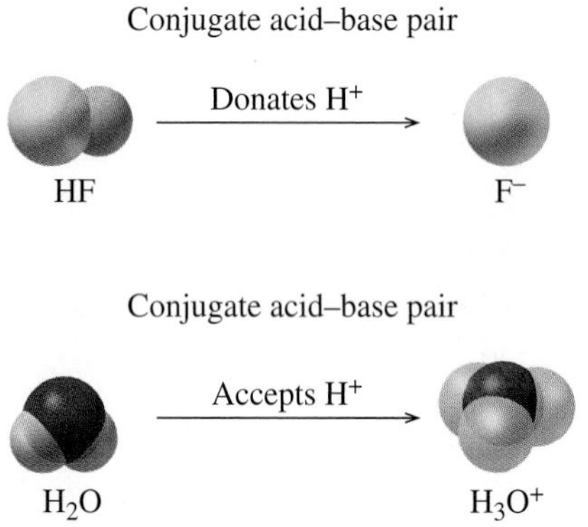

According to the Brønsted–Lowry theory, a **conjugate acid–base pair** consists of molecules or ions related by the loss or gain of one $H^+$. Every acid–base reaction contains two conjugate acid–base pairs because protons are transferred in both the forward and the reverse reaction. When the acid HA donates one $H^+$, the conjugate base $A^-$ forms. When the base B accepts the $H^+$, the conjugate acid $BH^+$ forms. We can write this as a general equation for a Brønsted–Lowry acid–base reaction.

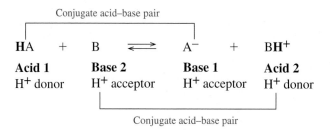

Now we can identify the conjugate acid–base pairs in a reaction such as hydrofluoric acid and water. Because the reaction is reversible, the conjugate acid $H_3O^+$ can transfer a proton to the conjugate base $F^-$ and re-form the acid HF. Using the relationship of loss and gain of one $H^+$, we identify the conjugate acid–base pairs as HF and $F^-$ along with $H_3O^+$ and $H_2O$.

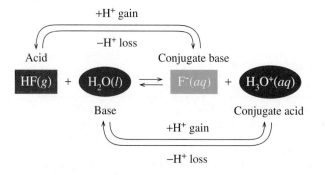

In another proton-transfer reaction, ammonia, $NH_3$, accepts $H^+$ from $H_2O$ to form the conjugate acid $NH_4^+$ and conjugate base $OH^-$. Each of these conjugate acid–base pairs, $NH_4^+$ and $NH_3$ as well as $H_2O$ and $OH^-$, are related by the loss and gain of one $H^+$. In these two examples, we see that water can act as an acid when it donates one $H^+$ or a base when it accepts $H^+$. Substances that can act as both acids and bases are *amphoteric*. For water, the most common amphoteric substance, the acidic or basic behavior depends on the other reactant.

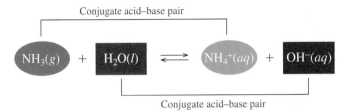

---

## SAMPLE PROBLEM 8.4

### ■ Conjugate Acid–Base Pairs

Write the formula of the conjugate base of each of the following Brønsted–Lowry acids.

**a.** $HClO_3$      **b.** $H_2CO_3$

SOLUTION

**a.** $ClO_3^-$ is the conjugate base that forms when $HClO_3$ donates one $H^+$.
**b.** $HCO_3^-$ is the conjugate base that forms when $H_2CO_3$ donates one $H^+$.

STUDY CHECK

Write the conjugate acid of each of the following Brønsted–Lowry bases.

**a.** $HS^-$          **b.** $NO_2^-$

---

## SAMPLE PROBLEM 8.5

### ■ Identifying Conjugate Acid–Base Pairs

Identify the conjugate acid–base pairs in the following equation:

$$HBr(aq) + NH_3(aq) \longrightarrow Br^-(aq) + NH_4^+(aq)$$

SOLUTION

Acting as a Brønsted–Lowry acid, HBr donates $H^+$ to form $Br^-$ as its conjugate base. The $NH_3$ acting as a Brønsted–Lowry base accepts $H^+$ to form its conjugate acid, $NH_4^+$. One conjugate acid–base pair is HBr and $Br^-$, and the other is $NH_4^+$ and $NH_3$.

STUDY CHECK

In the following reaction, identify the conjugate acid–base pairs.

$$HCN(aq) + SO_4^{2-}(aq) \rightleftharpoons CN^-(aq) + HSO_4^-(aq)$$

---

## QUESTIONS AND PROBLEMS

### Acids and Bases

**8.1** Indicate whether each of the following statements is characteristic of an acid or a base:
 **a.** has a sour taste
 **b.** neutralizes bases
 **c.** produces $H^+$ ions in water
 **d.** is named potassium hydroxide

**8.2** Indicate whether each of the following statements is characteristic of an acid or a base:
 **a.** neutralizes acids
 **b.** produces $OH^-$ in water
 **c.** has a soapy feel
 **d.** turns litmus red

**8.3**  Name each of the following as an acid or a base:
  **a.** HCl          **b.** $Ca(OH)_2$        **c.** $H_2CO_3$
  **d.** $HNO_3$      **e.** $H_2SO_3$          **f.** LiOH

**8.4**  Name each of the following as an acid or a base:
  **a.** $Al(OH)_3$     **b.** HBr             **c.** $H_2SO_4$
  **d.** KOH            **e.** $HNO_2$         **f.** $H_3PO_4$

**8.5**  Write formulas for the following acids and bases:
  **a.** magnesium hydroxide          **b.** hydrofluoric acid
  **c.** phosphoric acid              **d.** lithium hydroxide
  **e.** copper(II) hydroxide         **f.** sulfuric acid

**8.6**  Write formulas for the following acids and bases:
  **a.** barium hydroxide
  **b.** hydroiodic acid
  **c.** nitric acid
  **d.** iron(III) hydroxide
  **e.** sodium hydroxide
  **f.** hydrobromic acid

**8.7**  In each of the following, identify the Brønsted–Lowry acid and Brønsted–Lowry base:
  **a.** $HI(aq) + H_2O(l) \longrightarrow H_3O^+(aq) + I^-(aq)$
  **b.** $F^-(aq) + H_2O(l) \rightleftarrows HF(aq) + OH^-(aq)$

**8.8**  In each of the following, identify the Brønsted–Lowry acid and the Brønsted–Lowry base:
  **a.** $CO_3^{2-}(aq) + H_2O(l) \rightleftarrows HCO_3^-(aq) + OH^-(aq)$
  **b.** $H_2SO_4(aq) + H_2O(l) \longrightarrow H_3O^+(aq) + HSO_4^-(aq)$

**8.9**  Write the formula and name of the conjugate base for each of the following acids:
  **a.** HF          **b.** $H_2O$          **c.** $H_2CO_3$          **d.** $HSO_4^-$

**8.10**  Write the formula and name of the conjugate base for each of the following acids:
  **a.** $HCO_3^-$     **b.** $H_3O^+$        **c.** $HPO_4^{2-}$        **d.** $HNO_2$

**8.11**  Write the formula and name of the conjugate acid for each of the following bases:
  **a.** $CO_3^{2-}$     **b.** $H_2O$        **c.** $H_2PO_4^-$        **d.** $Br^-$

**8.12**  Write the formula and name of the conjugate acid for each of the following bases:
  **a.** $SO_4^{2-}$
  **b.** $CN^-$
  **c.** $OH^-$
  **d.** $ClO_2^-$, chlorite ion

**8.13**  Identify the Brønsted–Lowry acid–base pairs in the following equations:
  **a.** $H_2CO_3(aq) + H_2O(l) \rightleftarrows H_3O^+(aq) + HCO_3^-(aq)$
  **b.** $NH_4^+(aq) + H_2O(l) \rightleftarrows H_3O^+(aq) + NH_3(aq)$
  **c.** $HCN(aq) + NO_2^-(aq) \rightleftarrows CN^-(aq) + HNO_2(aq)$

**8.14**  Identify the Brønsted–Lowry acid–base pairs in the following equations:
  **a.** $H_3PO_4(aq) + H_2O(l) \rightleftarrows H_3O^+(aq) + H_2PO_4^-(aq)$
  **b.** $CO_3^{2-} + H_2O(l) \rightleftarrows OH^-(aq) + HCO_3^-(aq)$
  **c.** $H_3PO_4(aq) + NH_3(aq) \rightleftarrows NH_4^+(aq) + H_2PO_4^-(aq)$

---

**LEARNING GOAL**

Write equations for the dissociation of strong and weak acids.

# $8.2$ STRENGTHS OF ACIDS AND BASES

The *strength* of an acid or a base in water is determined by its ability to donate or accept protons. A strong acid donates protons easily, and a strong base accepts protons easily. Strong acids and strong bases dissociate completely in water. Only a few of the weak acids donate their protons, and only a few of the weak bases accept protons.

## Strong and Weak Acids

**Strong acids** are examples of strong electrolytes because they donate protons so easily that their dissociation in water is virtually complete. For example, when HCl, a strong acid, dissociates in water, $H^+$ is transferred to $H_2O$ and the resulting solution contains only the ions $H_3O^+$ and $Cl^-$. We consider the reaction of HCl in $H_2O$ as going nearly 100% to products. Therefore, the equation for a strong acid such as HCl is written with a single arrow to the product.

$$HCl(g) + H_2O(l) \longrightarrow H_3O^+(aq) + Cl^-(aq)$$

The common strong acids along with the more numerous weak acids are listed in Table 8.3.

Most acids are weak acids, which means they are also weak electrolytes. **Weak acids** dissociate slightly in water, which means that only a small percentage of the dissolved molecules donate $H^+$ to $H_2O$. Thus, a weak acid reacts with water to form only a small amount of $H_3O^+$ ions. Even at high concentrations, weak acids produce low concentrations of $H_3O^+$ ions. (See Figure 8.1.) Many of the products you drink or use at home contain weak

**TABLE 8.3** Common Strong and Weak Acids

| Strong Acids | |
|---|---|
| Perchloric acid | $HClO_4$ |
| Sulfuric acid | $H_2SO_4$ |
| Hydroiodic acid | HI |
| Hydrobromic acid | HBr |
| Hydrochloric acid | HCl |
| Nitric acid | $HNO_3$ |
| **Weak Acids** | |
| Hydronium ion | $H_3O^+$ |
| Hydrogen sulfate ion | $HSO_4^-$ |
| Nitrous acid | $HNO_2$ |
| Phosphoric acid | $H_3PO_4$ |
| Acetic acid | $HC_2H_3O_2$ |
| Hydrofluoric acid | HF |
| Carbonic acid | $H_2CO_3$ |
| Hydrosulfuric acid | $H_2S$ |
| Ammonium ion | $NH_4^+$ |
| Hydrocyanic acid | HCN |
| Bicarbonate ion | $HCO_3^-$ |
| Hydrogen sulfide ion | $HS^-$ |
| Water | $H_2O$ |

Increasing acid strength ↑

acids. In carbonated soft drinks, $CO_2$ dissolves in water to form carbonic acid, $H_2CO_3$, a weak acid.

$$H_2CO_3\,(aq) + H_2O(l) \rightleftharpoons H_3O^+(aq) + HCO_3^-(aq)$$

Carbonic acid        Bicarbonate ion

**FIGURE 8.1** A strong acid such as HCl is completely dissociated (≈100%), whereas a weak acid such as $HC_2H_3O_2$ contains mostly molecules and a few ions.

**Q** What is the difference between a strong acid and a weak acid?

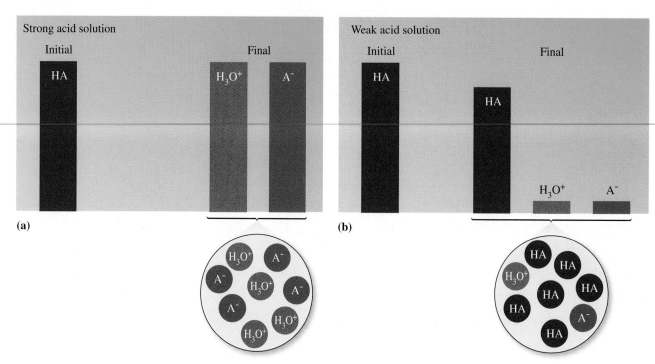

**FIGURE 8.2** **(a)** A strong acid dissociates in water to give $H_3O^+$ and $A^-$ ions. **(b)** A weak acid in water dissociates only slightly, to form a solution containing only a few $H_3O^+$ and $A^-$ ions and mostly undissociated HA molecules.

**Q** How does the height of the $H_3O^+$ and $A^-$ in the bar diagram change for a strong acid compared to a weak acid?

Citric acid is a weak acid found in fruits and fruit juices such as lemons, oranges, and grapefruit. Vinegar contains another weak acid known as acetic acid, $HC_2H_3O_2$. In the vinegar used on salads, acetic acid is present as a 5% (m/v) solution.

$$HC_2H_3O_2(l) + H_2O(l) \rightleftharpoons H_3O^+(aq) + C_2H_3O_2^-(aq)$$

Acetic acid                                             Acetate ion

In summary, if HA is a strong acid in water, the solution consists of the ions $H_3O^+$ and $A^-$. However, if HA is a weak acid, the aqueous solution consists mostly of undissociated HA and only a few $H_3O^+$ and $A^-$ ions. (See Figure 8.2.)

Strong acid: $HA(aq) + H_2O(l) \longrightarrow H_3O^+(aq) + A^-(aq)$ (~100% dissociated)

Weak acid: $HA(aq) + H_2O(l) \rightleftharpoons H_3O^+(aq) + A^-(aq)$ (small % dissociated)

## Strong and Weak Bases

As strong electrolytes, the Arrhenius bases are **strong bases** that dissociate virtually completely in water. Because these strong bases are ionic compounds, they dissociate in water to give an aqueous solution of a metal ion and hydroxide ion. The Group 1A (1) hydroxides are very soluble in water, which can give high concentrations of $OH^-$ ions. The other strong bases are much less soluble in water, but they dissolve completely as ions.

$$KOH(s) \xrightarrow{H_2O} K^+(aq) + OH^-(aq)$$

## Strong Bases

LiOH
NaOH
KOH
$Ca(OH)_2$
$Sr(OH)_2$
$Ba(OH)_2$

Strong bases, such as NaOH (also known as lye), are used in household products to remove grease in ovens and to clean drains. Because high concentrations of hydroxide ions cause severe damage to the skin and eyes, directions must be followed carefully when such products are used in the home, and use in the chemistry laboratory should be carefully supervised. If you spill an acid or a base on your skin or get some in your eyes, be sure to flood the area immediately with water.

**Weak bases** are weak electrolytes that are poor acceptors of protons and produce very few ions in solution. A typical weak base, ammonia, $NH_3$, is used in cleaning products. In an aqueous solution of $NH_3$, only a few molecules accept protons to form ammonium hydroxide.

$$NH_3(g) + H_2O(l) \rightleftharpoons NH_4^+(aq) + OH^-(aq)$$

Ammonia              Ammonium hydroxide

---

## SAMPLE PROBLEM 8.6

### ■ Strengths of Acids and Bases

For the following questions, select from one of the following:

$H_2CO_3$       $H_2SO_4$       $HNO_2$

**a.** Which is the strongest acid?
**b.** Which acid is the weakest?

### SOLUTION

As derived from the information in Table 8.3:

**a.** The strongest acid in this group is $H_2SO_4$.
**b.** The weakest acid in this group is carbonic acid, $H_2CO_3$.

### STUDY CHECK

Which is the stronger base: KOH or $NH_3$?

---

## QUESTIONS AND PROBLEMS

### Strengths of Acids and Bases

**8.15** Identify the stronger acid in each pair.
    **a.** HBr or $HNO_2$
    **b.** $H_3PO_4$ or $HSO_4^-$
    **c.** HCN or $H_2CO_3$

**8.16** Identify the stronger acid in each pair.
    **a.** $NH_4^+$ or $H_3O^+$
    **b.** $H_2SO_4$ or HCN
    **c.** $H_2O$ or $H_2CO_3$

**8.17** Identify the weaker acid in each pair.
    **a.** HCl or $HSO_4^-$
    **b.** $HNO_2$ or HF
    **c.** $HCO_3^-$ or $NH_4^+$

**8.18** Identify the weaker acid in each pair.
    **a.** $HNO_3$ or $HCO_3^-$
    **b.** $HSO_4^-$ or $H_2O$
    **c.** $H_2SO_4$ or $H_2CO_3$

---

# 8.3 IONIZATION OF WATER

We have seen that in some acid–base reactions, water acts as an acid and in other reactions, as a base. Does that mean water can be both an acid and a base? Yes, this is exactly what happens with water molecules in pure water. Let's see how this happens. One water molecule acts as an acid by donating $H^+$ to another water molecule, which acts as

### LEARNING GOAL

Use the ion product of water to calculate the $[H_3O^+]$ and $[OH^-]$ in an aqueous solution.

a base. The products are the conjugate acid $H_3O^+$ and conjugate base $OH^-$. Let's take a look at the conjugate acid–base pairs in water.

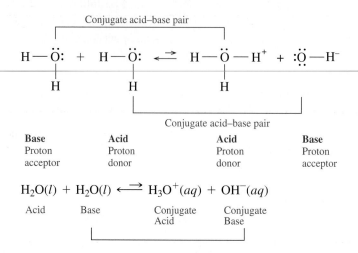

Every time a $H^+$ is transferred between two water molecules, the products are one $H_3O^+$ and one $OH^-$. Experiments have determined that, in pure water, the concentrations of $H_3O^+$ and $OH^-$ at 25 °C are each $1.0 \times 10^{-7}$ M. Square brackets around the symbols indicate their concentrations in moles per liter (M).

Pure water    $[H_3O^+] = [OH^-] = 1.0 \times 10^{-7}$ M

When we multiply these concentrations, we obtain the **ion-product constant of water**, $K_w$, which is $1.0 \times 10^{-14}$. The concentration units are omitted in the $K_w$ value.

$$K_w = [H_3O^+] \times [OH^-]$$
$$= (1.0 \times 10^{-7}\,\text{M})(1.0 \times 10^{-7}\,\text{M}) = 1.0 \times 10^{-14}$$

The $K_w$ value of $1.0 \times 10^{-14}$ is important because it applies to any aqueous solution: all aqueous solutions have $H_3O^+$ and $OH^-$.

When the $[H_3O^+]$ and $[OH^-]$ in a solution are equal, the solution is **neutral**. However, most solutions are not neutral; they have different concentrations of $[H_3O^+]$ and $[OH^-]$. If acid is added to water, there is an increase in $[H_3O^+]$ and a decrease in $[OH^-]$, which makes an acidic solution. If base is added, $[OH^-]$ increases and $[H_3O^+]$ decreases, which makes a basic solution. (See Figure 8.3.) However, for any aqueous solution, whether it is neutral, acidic, or basic, the product $[H_3O^+] \times [OH^-]$ is equal to

**FIGURE 8.3** In a neutral solution, $[H_3O^+]$ and $[OH^-]$ are equal. In acidic solutions, the $[H_3O^+]$ is greater than the $[OH^-]$. In basic solutions, the $[OH^-]$ is greater than the $[H_3O^+]$.

**Q** Is a solution that has a $[H_3O^+]$ of $1.0 \times 10^{-3}$ M acidic, basic, or neutral?

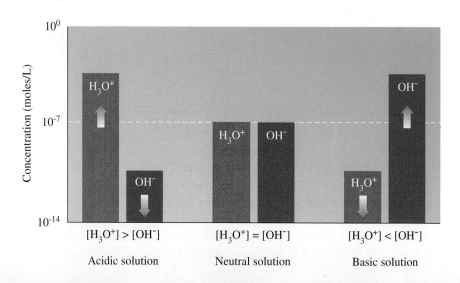

**TABLE 8.4** Examples of $[H_3O^+]$ and $[OH^-]$ in Neutral, Acidic, and Basic Solutions

| Type of Solution | $[H_3O^+]$ | $[OH^-]$ | $K_w$ |
|---|---|---|---|
| Neutral | $1.0 \times 10^{-7}$ M | $1.0 \times 10^{-7}$ M | $1.0 \times 10^{-14}$ |
| Acidic | $1.0 \times 10^{-2}$ M | $1.0 \times 10^{-12}$ M | $1.0 \times 10^{-14}$ |
| Acidic | $2.5 \times 10^{-5}$ M | $4.0 \times 10^{-10}$ M | $1.0 \times 10^{-14}$ |
| Basic | $1.0 \times 10^{-8}$ M | $1.0 \times 10^{-6}$ M | $1.0 \times 10^{-14}$ |
| Basic | $5.0 \times 10^{-11}$ M | $2.0 \times 10^{-4}$ M | $1.0 \times 10^{-14}$ |

$K_w (1.0 \times 10^{-14})$. Therefore, if the $[H_3O^+]$ is given, $K_w$ can be used to calculate the $[OH^-]$. Or if the $[OH^-]$ is given, $K_w$ can be used to calculate the $[H_3O^+]$. (See Table 8.4.)

$$K_w = [H_3O^+] \times [OH^-]$$

$$[OH^-] = \frac{K_w}{[H_3O^+]} \qquad [H_3O^+] = \frac{K_w}{[OH^-]}$$

To illustrate these calculations, let us calculate the $[H_3O^+]$ for a solution that has an $[OH^-] = 1.0 \times 10^{-6}$ M.

**Step 1    Write the $K_w$ for water.**

$$K_w = \boxed{[H_3O^+]} [OH^-] = 1.0 \times 10^{-14}$$

**Step 2    Arrange the $K_w$ to solve for the unknown.** Dividing through by the $[OH^-]$ gives

$$\frac{K_w}{[OH^-]} = \frac{[H_3O^+] \times [\cancel{OH^-}]}{[\cancel{OH^-}]}$$

$$\boxed{[H_3O^+]} = \frac{1.0 \times 10^{-14}}{[OH^-]}$$

**Step 3    Substitute the $[OH^-]$, and calculate the $[H_3O^+]$.**

$$\boxed{[H_3O^+]} = \frac{1.0 \times 10^{-14}}{1.0 \times 10^{-6}} = 1.0 \times 10^{-8} \text{ M}$$

Note that the square brackets around $H_3O^+$ and $OH^-$ indicate the molarity (moles/liter). Because the $[OH^-]$ of $1.0 \times 10^{-6}$ M is larger than the $[H_3O^+]$ of $1.0 \times 10^{-8}$ M, the solution is basic.

**Guide to Calculating $[H_3O^+]$ and $[OH^-]$ in Aqueous Solutions**

**1** Write the $K_w$ constant for water.

**2** Solve the $K_w$ for the unknown $[H_3O^+]$ or $[OH^-]$.

**3** Substitute the known $[H_3O^+]$ or $[OH^-]$ and calculate.

## SAMPLE PROBLEM 8.7

■ **Calculating $[H_3O^+]$ and $[OH^-]$ in Solution**

A vinegar solution has a $[H_3O^+] = 2.0 \times 10^{-3}$ M at 25 °C. What is the $[OH^-]$ of the vinegar solution? Is the solution acidic, basic, or neutral?

**SOLUTION**

Step 1  **Write the $K_w$ for water.**

$$K_w = [H_3O^+] \times [OH^-] = 1.0 \times 10^{-14}$$

Step 2  **Arrange the $K_w$ to solve for the unknown.** Rearranging the $K_w$ for $[OH^-]$ gives

$$\frac{K_w}{[H_3O^+]} = \frac{[\cancel{H_3O^+}] \times [OH^-]}{[\cancel{H_3O^+}]}$$

$$[OH^-] = \frac{1.0 \times 10^{-14}}{[H_3O^+]}$$

**Step 3 Substitute the known [$H_3O^+$] or [$OH^-$] and calculate.**

$$[OH^-] = \frac{1.0 \times 10^{-14}}{2.0 \times 10^{-3}} = 5.0 \times 10^{-12} \text{ M}$$

Because the [$H_3O^+$] of $2.0 \times 10^{-3}$ M is much larger than the [$OH^-$] of $5.0 \times 10^{-12}$ M, the solution is acidic.

**STUDY CHECK**

What is the [$H_3O^+$] of an ammonia cleaning solution with a [$OH^-$] = $4.0 \times 10^{-4}$ M? Is the solution acidic, basic, or neutral?

## QUESTIONS AND PROBLEMS

### Ionization of Water

**8.19** Why are the concentrations of $H_3O^+$ and $OH^-$ equal in pure water?

**8.20** What is the meaning and value of $K_w$?

**8.21** In an acidic solution, how does the concentration of $H_3O^+$ compare to the concentration of $OH^-$?

**8.22** If a base is added to pure water, why does the concentration of $H_3O^+$ decrease?

**8.23** Indicate whether the following are acidic, basic, or neutral solutions:
a. [$H_3O^+$] = $2.0 \times 10^{-5}$ M
b. [$H_3O^+$] = $1.4 \times 10^{-9}$ M
c. [$OH^-$] = $8.0 \times 10^{-3}$ M
d. [$OH^-$] = $3.5 \times 10^{-10}$ M

**8.24** Indicate whether the following are acidic, basic, or neutral solutions:
a. [$H_3O^+$] = $6.0 \times 10^{-12}$ M
b. [$H_3O^+$] = $1.4 \times 10^{-4}$ M
c. [$OH^-$] = $5.0 \times 10^{-12}$ M
d. [$OH^-$] = $4.5 \times 10^{-2}$ M

**8.25** Calculate the [$OH^-$] of each aqueous solution with the following [$H_3O^+$]:

a. coffee, $1.0 \times 10^{-5}$ M
b. soap, $1.0 \times 10^{-8}$ M
c. cleanser, $5.0 \times 10^{-10}$ M
d. lemon juice, $2.5 \times 10^{-2}$ M

**8.26** Calculate the [$OH^-$] of each aqueous solution with the following [$H_3O^+$]:
a. NaOH, $1.0 \times 10^{-12}$ M
b. aspirin, $6.0 \times 10^{-4}$ M
c. milk of magnesia, $1.0 \times 10^{-9}$ M
d. stomach acid, $5.2 \times 10^{-2}$ M

**8.27** Calculate the [$OH^-$] of each aqueous solution with the following [$H_3O^+$]:
a. vinegar, $1.0 \times 10^{-3}$ M
b. urine, $5.0 \times 10^{-6}$ M
c. ammonia, $1.8 \times 10^{-12}$ M
d. NaOH, $4.0 \times 10^{-13}$ M

**8.28** Calculate the [$OH^-$] of each aqueous solution with the following [$H_3O^+$]:
a. baking soda, $1.0 \times 10^{-8}$ M
b. orange juice, $2.0 \times 10^{-4}$ M
c. milk, $5.0 \times 10^{-7}$ M
d. bleach, $4.8 \times 10^{-12}$ M

## LEARNING GOAL

Calculate pH from [$H_3O^+$]; given the pH, calculate [$H_3O^+$] and [$OH^-$] of a solution.

# 8.4 THE pH SCALE

Many kinds of careers such as respiratory therapy, wine and beer making, medicine, agriculture, spa cleaning, and soap manufacturing require personnel to measure the [$H_3O^+$] and [$OH^-$] of solutions. The proper levels of acidity are necessary for soil to support plant growth and prevent algae in swimming pool water. Measuring the acidity levels of blood and urine checks the function of the kidneys.

On the pH scale, a number between 0 and 14 represents the $H_3O^+$ concentration for most solutions. A neutral solution has a pH of 7.0. An acidic solution has a pH less than 7.0, and a basic solution has a pH greater than 7.0. (See Figure 8.4.)

| Neutral solution | pH = 7.0 | [$H_3O^+$] = $1 \times 10^{-7}$ M |
| Acidic solution | pH < 7.0 | [$H_3O^+$] > $1 \times 10^{-7}$ M |
| Basic solution | pH > 7.0 | [$H_3O^+$] < $1 \times 10^{-7}$ M |

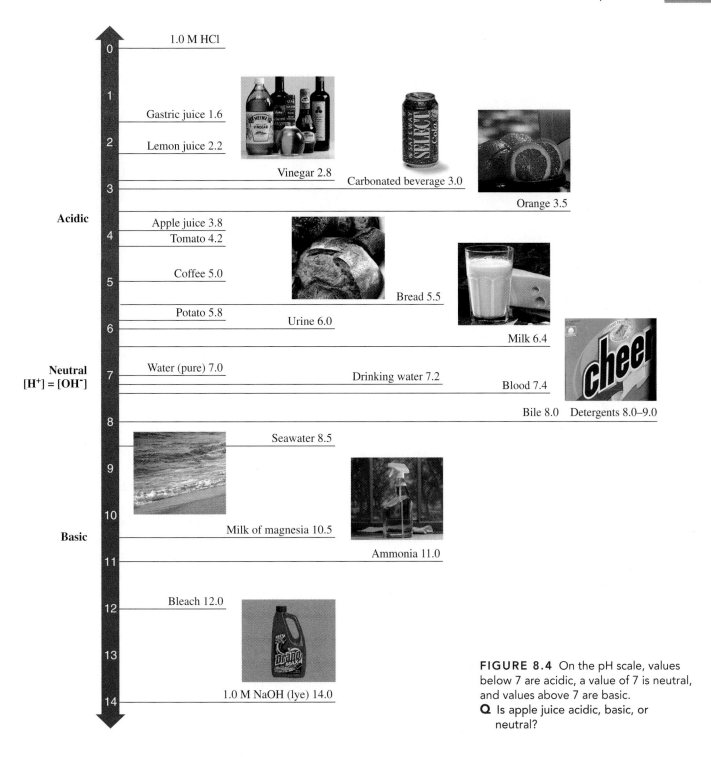

**FIGURE 8.4** On the pH scale, values below 7 are acidic, a value of 7 is neutral, and values above 7 are basic.
**Q** Is apple juice acidic, basic, or neutral?

In the laboratory, a pH meter is commonly used to determine the pH of a solution. There are also indicators and pH papers that turn specific colors when placed in solutions of different pH values. The pH is found by comparing the colors to a color chart. (See Figure 8.5.)

## Calculating the pH of Solutions

The pH scale is a log scale that corresponds to the hydronium-ion concentrations of aqueous solutions. Mathematically, **pH** is the negative logarithm (base 10) of the $H_3O^+$ concentration.

$$pH = -\log[H_3O^+]$$

WEB TUTORIAL
The pH Scale

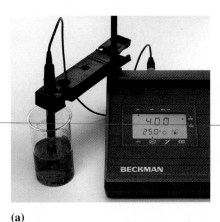

(a)

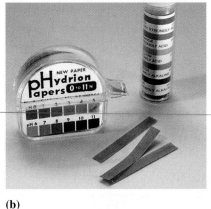

(b)

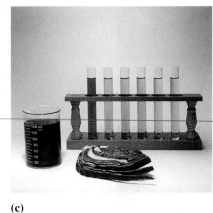

(c)

**FIGURE 8.5** The pH of a solution can be determined using **(a)** a pH meter, **(b)** pH paper, and **(c)** indicators that turn different colors corresponding to different pH values.
**Q** If a pH meter reads 4.00, is the solution acidic, basic, or neutral?

Essentially, the negative powers of 10 in the molar concentrations are converted to positive numbers. For example, a lemon juice solution with $[H_3O^+] = 1 \times 10^{-2}$ M has a pH of 2.0. This can be calculated using the pH equation.

$$pH = -\log[1 \times 10^{-2}]$$
$$pH = -(-2.0)$$
$$= 2.0$$

For molar concentrations of $[H_3O^+]$ that are whole numbers, significant zeros must be added to the resulting pH obtained on your calculator.

Let's look at how we determine the number of significant figures in the pH. For a logarithm, the number of decimal places in the pH value is equal to the number of significant figures in the coefficient of $[H_3O^+]$. The number to the left of the decimal point is the power of 10.

$$[H_3O^+] = \mathbf{1.0} \times 10^{-2} \qquad pH = \mathbf{2.00}$$

Two significant figures          Two decimal places

## Steps for a pH Calculation

The pH of a solution is determined using the *log* key and *changing sign*. For example, to calculate the pH of a vinegar solution with $[H_3O^+] = 2.4 \times 10^{-3}$ M you can use the following steps:

| Operation | | Display Shows |
|---|---|---|
| Step 1 | **Enter the $[H_3O^+]$ value.** 2.4 (EE or EXP) 3 | $2.4^{\,03}$  or  2.4 E03 |
| | Press the (+/-) to change the power to −3. (For calculators without a change sign key, consult the instructions for the calculator.) | $2.4^{-03}$ or  2.4 E–03 |
| Step 2 | **Press the (log) key.** | −2.619789 |
| | Change the sign by pressing the (+/-) key. | 2.619789 |

The steps can be combined to give the calculator sequence as follows:

$$pH = -\log[2.4 \times 10^{-3}] = 2.4 \boxed{\text{EE or EXP}} \; 3 \; \boxed{+/-} \; \boxed{\log} \; \boxed{+/-}$$
$$= 2.619789$$

Be sure to check the instructions for your calculator. On some calculators, the log key is used first, followed by the concentration.

**Step 3** **Adjust significant figures.** In a pH value, the number to the *left* of the decimal point is an *exact* number derived from the power of 10. The number of digits to the *right* of the decimal point is equal to the number of significant figures in the coefficient.

| Coefficient | Power of 10 |
|---|---|
| $[H_3O^+] = \mathbf{2.4}$ | $\times \quad 10^{-3}$ M |

Two significant figures (2 SFs) — Exact

$$pH = -\log[2.4 \times 10^{-3}] = \mathbf{2.62}$$

Exact — Two decimal places

Because pH is a log scale, a change of one pH unit corresponds to a ten-fold change in $[H_3O^+]$. It is important to note that the pH decreases as the $[H_3O^+]$ increases. For example, a solution with a pH of 2.00 has a $[H_3O^+]$ 10 times higher than a solution with a pH of 3.00 and 100 times higher than a solution with a pH of 4.00.

---

## SAMPLE PROBLEM 8.8

### ■ Calculating pH

Determine the pH for the following solutions:

**a.** $[H_3O^+] = 1.0 \times 10^{-5}$ M    **b.** $[H_3O^+] = 5 \times 10^{-8}$ M

SOLUTION

**a. Step 1** **Enter the concentration of $[H_3O^+]$ using the *change sign* key.**

Display

$$pH = -\log[1.0 \times 10^{-5}] = 1.0 \boxed{\text{EE or EXP}} \; 5 \boxed{+/-} \qquad 1.0^{-05} \quad \text{or} \quad 1.0\,E{-}05$$

**Step 2** **Press the *log* key and *change sign* key.**

$\boxed{\log} \; \boxed{+/-}$          5

**Step 3** **Adjust significant figures to the *right* of the decimal point to equal the number of significant figures in the coefficient.**

$1.0 \times 10^{-5}$ M    pH = 5.00

2 SFs ———→ 2 SFs to the *right* of the decimal point

**b. Step 1** **Enter the concentration using the *change sign* key.**

$$pH = -\log[5 \times 10^{-8}] = 5 \boxed{\text{EE or EXP}} \; 8 \boxed{+/-} \qquad 5^{-08} \quad \text{or} \quad 5\,E{-}08$$

**Step 2** **Press the *log* key, and then the *change sign* key.**

$\boxed{\log} \; \boxed{+/-}$          7.301029

**Step 3** **Adjust significant figures to the *right* of the decimal point to equal the significant figures in the coefficient.**

$5 \times 10^{-8}$ M    pH = 7.3

1 SF ———→ 1 SF to the *right* of the decimal point

**Guide to Calculating pH of an Aqueous Solution**

**1** Enter the concentration of $[H_3O^+]$.

**2** Press the *log* key and change the sign.

**3** Adjust significant figures to the *right* of the decimal point to equal SFs in the coefficient.

---

STUDY CHECK

What is the pH of bleach with $[H_3O^+] = 4.2 \times 10^{-12}$ M?

■ **Calculating pH from [OH⁻]**

What is the pH of an ammonia solution with $[OH^-] = 3.7 \times 10^{-3}$ M?

SOLUTION

~~Step 1~~ **Enter the [H₃O⁺] and press the change sign key.** Because $[OH^-]$ is given for the ammonia solution, we have to calculate $[H_3O^+]$ using the ion product of water, $K_w$. Dividing through by $[OH^-]$ gives $[H_3O^+]$.

$$\frac{K_w}{[OH^-]} = \frac{[H_3O^+] \times [\cancel{OH^-}]}{[\cancel{OH^-}]}$$

$$[H_3O^+] = \frac{1.0 \times 10^{-14}}{3.7 \times 10^{-3}} = 2.7 \times 10^{-12} \, M$$

**Display**

$$pH = -\log[2.7 \times 10^{-12}] = 2.7 \boxed{\text{EE or EXP}} \; 12 \boxed{+/-} \qquad 2.7^{-12} \text{ or } 2.7 \, E{-}12$$

**Step 2** **Press the log key, and then the change sign key.**

$$\boxed{\text{log}} \; \boxed{+/-} \qquad\qquad\qquad\qquad 11.56863$$

**Step 3** **Adjust significant figures to the right of the decimal point to equal the SFs in the coefficient.**

$$\mathbf{2.7} \times 10^{-12} \, M \qquad pH = 11.\mathbf{57}$$

2 SFs ⟶ 2 SFs *after* the decimal point

STUDY CHECK

Calculate the pH of a sample of acid rain that has $[OH^-] = 2 \times 10^{-10}$ M.

## Calculating the [H₃O⁺] from pH

In another type of calculation, we are given the pH of a solution and asked to determine the $[H_3O^+]$. This is a reverse of the pH calculation. For whole number pH values, the negative pH value is the power of 10 in the $H_3O^+$ concentration.

$$[H_3O^+] = 1 \times 10^{-pH}$$

For pH values that are not whole numbers, the calculation requires the use of the $10^x$ key, which is usually a $2^{nd}$ function key. On some calculators, this operation is done using the inverse key and the log key.

■ **Calculating [H₃O⁺] from pH**

Determine $[H_3O^+]$ for solutions having the following pH values:

**a.** pH = 3.0      **b.** pH = 8.2

SOLUTION

**a.** For pH values that are whole numbers, the $[H_3O^+]$ can be written $1 \times 10^{-pH}$.

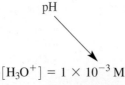

$$[H_3O^+] = 1 \times 10^{-3} \, M$$

**b.** For pH values that are not whole numbers, the $[H_3O^+]$ is calculated as follows:

Step 1  **Enter the pH value and press the *change sign* key.**

                                                    **Display**
        8.2  [+/−]                                     −8.2

Step 2  **Convert −pH to concentration.**  Press the *2nd function* key and then the
        $10^x$ key. Or press the *inverse* key and then the *log* key.

        [2nd] [10ˣ]  or  [inv] [log]              $6.3095^{-09}$  or  $6.3095\,E{-}09$

Step 3  **Adjust the significant figures in the coefficient.**  Because the pH value of
        8.2 has one digit to the *right* of the decimal point, the $[H_3O^+]$ is written with
        one significant figure.

        $[H_3O^+] = 6 \times 10^{-9}$ M

STUDY CHECK

What is the $[H_3O^+]$ of a beer that has a pH of 4.5?

A comparison of $[H_3O^+]$, $[OH^-]$, and their corresponding pH values is given in Table 8.5.

**TABLE 8.5**  A Comparison of $[H_3O^+]$, $[OH^-]$, and Corresponding pH Values

| $[H_3O^+]$ | pH | $[OH^-]$ | |
|---|---|---|---|
| $10^0$ | 0 | $10^{-14}$ | |
| $10^{-1}$ | 1 | $10^{-13}$ | |
| $10^{-2}$ | 2 | $10^{-12}$ | |
| $10^{-3}$ | 3 | $10^{-11}$ | Acidic |
| $10^{-4}$ | 4 | $10^{-10}$ | |
| $10^{-5}$ | 5 | $10^{-9}$ | |
| $10^{-6}$ | 6 | $10^{-8}$ | |
| $10^{-7}$ | 7 | $10^{-7}$ | Neutral |
| $10^{-8}$ | 8 | $10^{-6}$ | |
| $10^{-9}$ | 9 | $10^{-5}$ | |
| $10^{-10}$ | 10 | $10^{-4}$ | |
| $10^{-11}$ | 11 | $10^{-3}$ | Basic |
| $10^{-12}$ | 12 | $10^{-2}$ | |
| $10^{-13}$ | 13 | $10^{-1}$ | |
| $10^{-14}$ | 14 | $10^0$ | |

## QUESTIONS AND PROBLEMS

### The pH Scale

**8.29** Why does a neutral solution have a pH of 7.00?

**8.30** If you know the $[OH^-]$, how can you determine the pH of a solution?

**8.31** State whether each of the following solutions is acidic, basic, or neutral:
   **a.** blood, pH 7.38          **b.** vinegar, pH 2.8
   **c.** drain cleaner, pH 11.2  **d.** coffee, pH 5.5
   **e.** tomatoes, pH 4.2        **f.** chocolate cake, pH 7.6

**8.32** State whether each of the following solutions is acidic, basic, or neutral:
   **a.** soda, pH 3.2            **b.** shampoo, pH 5.7

   **c.** laundry detergent, pH 9.4   **d.** rain, pH 5.8
   **e.** honey, pH 3.9          **f.** cheese, pH 7.4

**8.33** Calculate the pH of each solution given the following $[H_3O^+]$ or $[OH^-]$ values:
   **a.** $[H_3O^+] = 1 \times 10^{-4}$ M       **b.** $[H_3O^+] = 3 \times 10^{-9}$ M
   **c.** $[OH^-] = 1 \times 10^{-5}$ M         **d.** $[OH^-] = 2.5 \times 10^{-11}$ M
   **e.** $[H_3O^+] = 6.7 \times 10^{-8}$ M     **f.** $[OH^-] = 8.2 \times 10^{-4}$ M

**8.34** Calculate the pH of each solution given the following $[H_3O^+]$ or $[OH^-]$ values:
   **a.** $[H_3O^+] = 1 \times 10^{-8}$ M       **b.** $[H_3O^+] = 5 \times 10^{-6}$ M
   **c.** $[OH^-] = 4 \times 10^{-2}$ M         **d.** $[OH^-] = 8 \times 10^{-3}$ M
   **e.** $[H_3O^+] = 4.7 \times 10^{-2}$ M     **f.** $[OH^-] = 3.9 \times 10^{-6}$ M

**8.35** Complete the following table:

| $[H_3O^+]$ | $[OH^-]$ | pH | Acidic, Basic, or Neutral? |
|---|---|---|---|
| | $1 \times 10^{-6}$ M | | |
| | | 3.0 | |
| $2 \times 10^{-5}$ M | | | |
| $1 \times 10^{-12}$ M | | | |
| | | 4.62 | |

**8.36** Complete the following table:

| $[H_3O^+]$ | $[OH^-]$ | pH | Acidic, Basic, or Neutral? |
|---|---|---|---|
| | | 10.0 | |
| | | | Neutral |
| | $1 \times 10^{-5}$ M | | |
| $1 \times 10^{-2}$ M | | | |
| | | 11.3 | |

## Using Vegetables and Flowers as pH Indicators

Many flowers and vegetables with strong color, especially reds and purples, contain compounds that change color with changes in pH. Some examples are red cabbage, cranberry juice, and cranberry drinks.

### MATERIALS NEEDED

Red cabbage, or cranberry juice or drinks, water, and a saucepan.

Several glasses or small glass containers and some tape and a pen or pencil to mark the containers.

Several colorless household solutions such as vinegar, lemon juice, other fruit juices, baking soda, antacids, aspirin, window cleaners, soaps, shampoos, and detergents.

### PROCEDURE

1. Obtain a bottle of cranberry juice or cranberry drink, or use a red cabbage to prepare the red cabbage pH indicator, as follows: Tear up several red cabbage leaves and place them in a saucepan and cover with water. Boil for about 5 minutes. Cool and collect the purple solution.

2. Place small amounts of each household solution into separate clear glass containers and mark what each one is. If the sample is a solid or a thick liquid, add a small amount of water. Add some cranberry juice or some red cabbage indicator until you obtain a color.

3. Observe the colors of the various samples. The colors that indicate acidic solutions are the pink and orange colors (pH 1–4) and the pink to lavender colors (pH 5–6). A neutral solution has about the same purple color as the indicator. Bases will give blue to green color (pH 8–11) or a yellow color (pH 12–13).

4. Arrange your samples by color and pH. Classify each of the solutions as acidic (1–6), neutral (7), or basic (8–13).

5. Try to make an indicator using other colorful fruits or flowers.

### QUESTIONS

1. Which products that tested acidic listed an acid on their labels?
2. Which products that tested basic listed a base on their labels?
3. How many products were neutral?
4. Which flowers or vegetables behaved as indicators?

## LEARNING GOAL

Write balanced equations for reactions of acids and bases; calculate the molarity or volume of an acid or a base from titration information.

# 8.5 REACTIONS OF ACIDS AND BASES

Typical reactions of acids and bases include the reactions of acids with metals, bases, and carbonate or bicarbonate ions. For example, when you drop an antacid tablet in water, the bicarbonate ion and citric acid in the tablet react to produce carbon dioxide bubbles, a salt, and water.

## Acids and Metals

Acids react with certain metals to produce hydrogen gas ($H_2$) and a salt, which is an ionic compound that does not contain $H^+$ or $OH^-$. Metals that react with acids include

# Green Chemistry Note

## Acid Rain

Natural rain is slightly acidic, with a pH of 5.6. In the atmosphere, carbon dioxide combines with water to form carbonic acid, a weak acid that dissociates to give hydronium ions and bicarbonate.

$$CO_2(g) + H_2O(l) \rightleftharpoons H_2CO_3(aq)$$
$$H_2CO_3(aq) + H_2O(l) \rightleftharpoons H_3O^+(aq) + HCO_3^-(aq)$$

However, in many parts of the world, rain has become considerably more acidic. *Acid rain* is a term given to precipitation such as rain, snow, hail, or fog when the water has a pH that is less than 5.6. In the United States, pH values of rain have decreased to about 4–4.5. In some parts of the world, pH values of rain have been reported as low as 2.6, which is about as acidic as lemon juice or vinegar. Because the calculation of pH involves powers of 10, a pH value of 2.6 would be 1000 times more acidic than natural rain.

Although natural sources such as volcanoes and forest fires release $SO_2$, the primary sources of acid rain today are from the burning of fossil fuels in automobiles and coal in industrial plants. When coal and oil are burned, the sulfur impurities combine with the oxygen in the air to produce $SO_2$ and $SO_3$. The reaction of $SO_3$ with water forms sulfuric acid, $H_2SO_4$, a strong acid.

$$S(s) + O_2(g) \longrightarrow SO_2(g)$$
$$2SO_2(g) + O_2(g) \longrightarrow 2SO_3(g)$$
$$SO_3(g) + H_2O(l) \longrightarrow H_2SO_4(aq)$$

In an effort to decrease the formation of acid rain, legislation has required a reduction in $SO_2$ emissions. Coal-burning plants have installed equipment called "scrubbers" that absorb $SO_2$ before it is emitted. In a smokestack, "scrubbing" involves passing the flue gases containing $SO_2$ through limestone ($CaCO_3$) and water where 95% of the $SO_2$ is removed. The end product, $CaSO_4$, also called "gypsum," is used in agriculture and to prepare cement products.

Nitrogen oxide forms at high temperatures in the engines of automobiles when air containing nitrogen and oxygen gases is burned. As nitrogen oxide is emitted into the air, it combines with more oxygen to form nitrogen dioxide, which is responsible for the brown color of smog. When nitrogen dioxide dissolves in water in the atmosphere, nitric acid, a strong acid, forms.

$$N_2(g) + O_2(g) \longrightarrow 2NO(g)$$
$$2NO(g) + O_2(g) \longrightarrow 2NO_2(g)$$
$$3NO_2(g) + H_2O(g) \longrightarrow 2HNO_3(aq) + NO(g)$$

The sulfuric acid and nitric acid in the atmosphere are carried by the air currents many thousands of kilometers before they

1994
Marble statue in Washington Square Park

precipitate in areas far away from the site of the contamination. The acids in acid rain have detrimental effects on marble and limestone structures, lakes, and forests. Throughout the world, monuments made of marble (a form of $CaCO_3$) are deteriorating as acid rain dissolves the marble.

$$CaCO_3(s) + H_2SO_4(aq) \longrightarrow CaSO_4(aq) + H_2O(l) + CO_2(g)$$

Acid rain is changing the pH of many lakes and streams in parts of the United States and Europe. When the pH of a lake falls below 4.5–5, most fish and plant life cannot survive. As the soil near a lake becomes more acidic, aluminum becomes more soluble. Increased levels of aluminum ions in lakes are toxic to fish and water animals.

Trees and forests are susceptible to acid rain too. Acid rain breaks down the protective waxy coating on leaves and interferes with photosynthesis. Tree growth is impaired as nutrients and minerals in the soil dissolve and wash away. In Eastern Europe, acid rain is causing an environmental disaster. Nearly 70% of the forests in the Czech Republic have been severely damaged, and some parts of the land are so acidic that crops will not grow.

potassium, sodium, calcium, magnesium, aluminum, zinc, iron, and tin. Such reactions are single replacement reactions in which the metal ion replaces the hydrogen in the acid.

$$Mg(s) + 2HCl(aq) \longrightarrow H_2(g) + MgCl_2(aq)$$
Metal        Acid                    Hydrogen  Salt

$$Zn(s) + 2HCl(aq) \longrightarrow H_2(g) + ZnCl_2(aq)$$
Metal        Acid                    Hydrogen  Salt

## Acids, Carbonates, and Bicarbonates

When an acid is added to a carbonate or bicarbonate (hydrogen carbonate), the products are carbon dioxide gas, water, and an ionic compound (salt). The acid reacts with $CO_3^{2-}$ or $HCO_3^-$ to produce carbonic acid, $H_2CO_3$, which breaks down rapidly to $CO_2$ and $H_2O$.

$$HCl(aq) + NaHCO_3(aq) \longrightarrow CO_2(g) + H_2O(l) + NaCl(aq)$$
Acid          Bicarbonate              Carbon dioxide   Water      Salt

$$2HCl(aq) + Na_2CO_3(aq) \longrightarrow CO_2(g) + H_2O(l) + 2NaCl(aq)$$
Acid          Carbonate               Carbon dioxide   Water      Salt

## Acids and Hydroxides: Neutralization

**Neutralization** is a reaction between an acid and a base to produce a salt and water. In the reaction, the $H^+$ of an acid that can be strong or weak and the $OH^-$ of a strong base combine to form water as one product. The salt that is usually soluble is the cation from the base and the anion from the acid. We can write the following equation for the neutralization reaction between HCl and NaOH.

$$HCl(aq) + NaOH(aq) \longrightarrow H_2O(l) + NaCl(aq)$$
Acid          Base                Water        Salt

If we write HCl and NaOH as ions, we see that $H^+$ reacts with $OH^-$ to form water, leaving the ions $Na^+$ and $Cl^-$ in solution.

$$\mathbf{H^+}(aq) + Cl^-(aq) + Na^+(aq) + \mathbf{OH^-}(aq) \longrightarrow \mathbf{H_2O}(l) + Na^+(aq) + Cl^-(aq)$$

When we omit the ions that do not change ($Na^+$ and $Cl^-$) from the equation, we see that the reaction for neutralization is the formation of $H_2O$ from $H^+$ and $OH^-$.

$$\mathbf{H^+}(aq) + \cancel{Cl^-}(aq) + \cancel{Na^+}(aq) + \mathbf{OH^-}(aq) \longrightarrow \mathbf{H_2O}(l) + \cancel{Na^+}(aq) + \cancel{Cl^-}(aq)$$
$$H^+(aq) + OH^-(aq) \longrightarrow H_2O(l)$$

**Guide to Balancing
an Equation for Neutralization**

**1** Write the reactants and products.

**2** Balance the $H^+$ in the acid with the $OH^-$ in the base.

**3** Balance the $H_2O$ with the $H^+$ and the $OH^-$.

**4** Write the salt from the remaining ions.

## Balancing Neutralization Equations

In a neutralization reaction, one $H^+$ always combines with one $OH^-$. Therefore, coefficients are used to balance $H^+$ in the acid with the $OH^-$ in the base. We balance the neutralization of HCl and $Ba(OH)_2$ as follows:

Step 1 **Write the reactants and products.**

$$HCl(aq) + Ba(OH)_2(s) \longrightarrow H_2O(l) + salt$$

Step 2 **Balance the $H^+$ in the acid with the $OH^-$ in the base.**   Place a coefficient of 2 in front of HCl to balance $2OH^-$ in $Ba(OH)_2$.

$$\mathbf{2}HCl(aq) + Ba(OH)_2(s) \longrightarrow H_2O(l) + salt$$

**Step 3  Balance the $H_2O$ with the $H^+$ and $OH^-$.**  Use a coefficient of 2 in front of $H_2O$ to balance $2H^+$ and $2OH^-$.

$$2HCl(aq) + Ba(OH)_2(s) \longrightarrow 2H_2O(l) + \text{salt}$$

**Step 4  Write the salt from the remaining ions.**  The remaining ions $Ba^{2+}$ and $2Cl^-$ are used to write the formula of the salt as $BaCl_2$.

$$2HCl(aq) + Ba(OH)_2(s) \longrightarrow 2H_2O(l) + BaCl_2(aq)$$

---

## SAMPLE PROBLEM  8.11

### ■ Reactions of Acids

Write a balanced equation for the reaction of $HCl(aq)$ with each of the following:

**a.** $Al(s)$          **b.** $K_2CO_3(aq)$          **c.** $Mg(OH)_2(s)$

#### SOLUTION

**a.** Al

**Step 1  Write the reactants and products.**  When a metal reacts with an acid, the products are $H_2$ gas and a salt.

$$Al(s) + HCl(aq) \longrightarrow H_2(g) + \text{salt}$$

**Step 2  Determine the formula of the salt.**  When $Al(s)$ dissolves, it forms $Al^{3+}$, which is balanced by $3\ Cl^-$ from HCl.

$$Al(s) + HCl(aq) \longrightarrow H_2(g) + AlCl_3(aq)$$

**Step 3  Balance the equation.**

$$2Al(s) + 6HCl(aq) \longrightarrow 3H_2(g) + 2AlCl_3(aq)$$

**b.** $K_2CO_3$

**Step 1  Write the reactants and products.**  When a carbonate reacts with an acid, the products are $CO_2(g)$, $H_2O(l)$, and a salt.

$$K_2CO_3(s) + HCl(aq) \longrightarrow CO_2(g) + H_2O(l) + \text{salt}$$

**Step 2  Determine the formula of the salt.**  When $K_2CO_3(s)$ dissolves, it forms $K^+$, which is balanced by $1\ Cl^-$ from HCl.

$$K_2CO_3(s) + HCl(aq) \longrightarrow CO_2(g) + H_2O(l) + KCl(aq)$$

**Step 3  Balance the equation.**  A coefficient of 2 in front of KCl balances the 2K in $K_2CO_3$. A coefficient of 2 in front of HCl balances the 2KCl.

$$K_2CO_3(aq) + 2HCl(aq) \longrightarrow CO_2(g) + H_2O(l) + 2KCl(aq)$$

**c.** $Mg(OH)_2$

**Step 1  Write the reactants and products.**  When a base reacts with an acid, the products are $H_2O(l)$ and a salt.

$$Mg(OH)_2(s) + HCl(aq) \longrightarrow H_2O(l) + \text{salt}$$

**Step 2  Balance the $H^+$ in the acid with the $OH^-$ in the base.**  Placing a 2 in front of the HCl balances $2OH^-$ in $Mg(OH)_2$.

$$Mg(OH)_2(s) + 2HCl(aq) \longrightarrow H_2O(l) + \text{salt}$$

**Step 3  Balance the $H_2O$ with the $H^+$ and $OH^-$.**  Use a coefficient of 2 in front of $H_2O$ to balance $2H^+$ and $2OH^-$.

$$Mg(OH)_2(s) + 2HCl(aq) \longrightarrow 2H_2O(l) + \text{salt}$$

**Step 4  Write the salt from the remaining ions in the acid and base.**   The remaining ions, $Mg^{2+}$ and $2Cl^-$, are used to write the formula of the salt as $MgCl_2$.

$$Mg(OH)_2(s) + 2HCl(aq) \longrightarrow 2H_2O(l) + MgCl_2(aq)$$

STUDY CHECK

Write the balanced equation for the reaction between $H_2SO_4$ and $NaHCO_3$.

## Acid–Base Titration

Suppose we need to find the molarity of a HCl solution of unknown concentration. We can do this by a laboratory procedure called **titration** in which we neutralize an acid sample with a known amount of base. In our titration, we first place a measured volume of the acid in a flask and add a few drops of an *indicator* such as phenolphthalein. In an acidic solution, phenolphthalein is colorless. Then we fill a buret with a NaOH solution of known molarity and carefully add NaOH to the acid in the flask. (See Figure 8.6.)

In the titration, we neutralize the acid by adding a volume of base that contains a matching number of moles of $OH^-$. We know that neutralization has taken place when the phenolphthalein in the solution changes from colorless to pink. This is called the neutralization *end point*. From the volume added and molarity of the NaOH, we can calculate the number of moles of NaOH and then the concentration of the acid.

Health Note

### Antacids

Antacids are substances used to neutralize excess stomach acid (HCl). Some antacids are mixtures of aluminum hydroxide and magnesium hydroxide. These hydroxides are not very soluble in water, so the levels of available $OH^-$ are not damaging to the intestinal tract. However, aluminum hydroxide has the side effects of producing constipation and binding phosphate in the intestinal tract, which may cause weakness and loss of appetite. Magnesium hydroxide has a laxative effect. These side effects are less likely when a combination of the antacids is used.

$$Al(OH)_3(aq) + 3HCl(aq) \longrightarrow 3H_2O(l) + AlCl_3(aq)$$
$$Mg(OH)_2(s) + 2HCl(aq) \longrightarrow 2H_2O(l) + MgCl_2(aq)$$

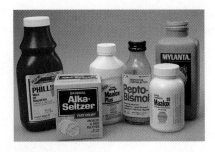

Some antacids use calcium carbonate to neutralize excess stomach acid. About 10% of the calcium is absorbed into the bloodstream, where it elevates the levels of serum calcium. Calcium carbonate is not recommended for patients who have peptic ulcers or a tendency to form kidney stones.

$$CaCO_3(s) + 2HCl(aq) \longrightarrow CO_2(g) + H_2O(l) + CaCl_2(aq)$$

Still other antacids contain sodium bicarbonate. This type of antacid has a tendency to increase blood pH and elevate sodium levels in the body fluids. It also is not recommended in the treatment of peptic ulcers.

$$NaHCO_3(s) + HCl(aq) \longrightarrow CO_2(g) + H_2O(l) + NaCl(aq)$$

The neutralizing substances in some antacid preparations are given in Table 8.6.

**TABLE 8.6** Basic Compounds in Some Antacids

| Antacid | Base(s) |
| --- | --- |
| Amphojel | $Al(OH)_3$ |
| Milk of magnesia | $Mg(OH)_2$ |
| Mylanta, Maalox, Di-Gel, Gelusil, Riopan | $Mg(OH)_2$, $Al(OH)_3$ |
| Bisodol | $CaCO_3$, $Mg(OH)_2$ |
| Titralac, Tums, Pepto-Bismol | $CaCO_3$ |
| Alka-Seltzer | $NaHCO_3$, $KHCO_3$ |

**FIGURE 8.6** The titration of an acid. A known volume of an acid is placed in a flask with an indicator and titrated with a measured volume of a base, such as NaOH, to the neutralization point.

**Q** What data is needed to determine the molarity of the acid in the flask?

---

## SAMPLE PROBLEM 8.12

### ■ Titration of an Acid

A 25.0-mL sample of a HCl solution is placed in a flask with a few drops of phenolphthalein (indicator). If 32.6 mL of 0.185 M NaOH is needed to reach the end point, what is the concentration (M) of the HCl solution?

$$NaOH(aq) + HCl(aq) \longrightarrow NaCl(aq) + H_2O(l)$$

**SOLUTION**

**Step 1  Given:** 32.6 mL of 0.185 M NaOH; 25.0 mL of HCl = 0.0250 L of HCl
**Need:** Molarity of HCl

**Step 2  Plan**

32.6 mL → | Metric factor | L → | Molarity factor | moles NaOH → | Mole factor | moles HCl → | Divide by liters | M HCl

**Step 3  Equalities/Conversion Factors**

$$1 \text{ L of NaOH} = 1000 \text{ mL of NaOH}$$
$$\frac{1 \text{ L}}{1000 \text{ mL}} \text{ and } \frac{1000 \text{ mL}}{1 \text{ L}}$$

$$0.185 \text{ mole of NaOH} = 1 \text{ L of NaOH}$$
$$\frac{0.185 \text{ mole NaOH}}{1 \text{ L}} \text{ and } \frac{1 \text{ L}}{0.185 \text{ mole NaOH}}$$

$$1 \text{ mole of HCl} = 1 \text{ mole of NaOH}$$
$$\frac{1 \text{ mole HCl}}{1 \text{ mole NaOH}} \text{ and } \frac{1 \text{ mole NaOH}}{1 \text{ mole HCl}}$$

**Step 4  Set up problem**

$$32.6 \text{ mL NaOH} \times \frac{1 \text{ L NaOH}}{1000 \text{ mL NaOH}} \times \frac{0.185 \text{ mole NaOH}}{1 \text{ L NaOH}}$$
$$\times \frac{1 \text{ mole HCl}}{1 \text{ mole NaOH}} = 0.00603 \text{ mole of HCl}$$

$$\text{Molarity of HCl} = \frac{0.00603 \text{ mole HCl}}{0.0250 \text{ HCl}} = 0.241 \text{ M}$$

**Guide to Calculations for an Acid-Base Titration**

**1** State the given and needed quantities and concentrations.

**2** Write a plan to calculate molarity or volume.

**3** State equalities and conversion factors including concentration.

**4** Set up problem to calculate needed quantity.

**STUDY CHECK**

What is the molarity of a HCl solution, if 28.6 mL of a 0.175 M NaOH solution is needed to neutralize a 25.0-mL sample of the HCl solution?

## QUESTIONS AND PROBLEMS

### Reactions of Acids and Bases

**8.37** Complete and balance the equation for each of the following reactions:
a. $ZnCO_3(s) + HBr(aq) \longrightarrow$
b. $Zn(s) + HCl(aq) \longrightarrow$
c. $HCl(aq) + NaHCO_3(s) \longrightarrow$
d. $H_2SO_4(aq) + Mg(OH)_2(s) \longrightarrow$

**8.38** Complete and balance the equations for each of the following reactions:
a. $KHCO_3(s) + HCl(aq) \longrightarrow$
b. $Ca(s) + H_2SO_4(aq) \longrightarrow$
c. $H_2SO_4(aq) + Al(OH)_3(s) \longrightarrow$
d. $Na_2CO_3(s) + H_2SO_4(aq) \longrightarrow$

**8.39** Balance each of the following neutralization reactions:
a. $HCl(aq) + Mg(OH)_2(s) \longrightarrow H_2O(l) + MgCl_2(aq)$
b. $H_3PO_4(aq) + LiOH(aq) \longrightarrow H_2O(l) + Li_3PO_4(aq)$

**8.40** Balance each of the following neutralization reactions:
a. $HNO_3(aq) + Ba(OH)_2(s) \longrightarrow H_2O(l) + Ba(NO_3)_2(aq)$
b. $H_2SO_4(aq) + Al(OH)_3(aq) \longrightarrow H_2O(l) + Al_2(SO_4)_3(aq)$

**8.41** Write a balanced equation for the neutralization of each of the following:
a. $H_2SO_4(aq)$ and $NaOH(aq)$
b. $HCl(aq)$ and $Fe(OH)_3(aq)$
c. $H_2CO_3(aq)$ and $Mg(OH)_2(s)$

**8.42** Write a balanced equation for the neutralization of each of the following:
a. $H_3PO_4(aq)$ and $NaOH(aq)$
b. $HI(aq)$ and $LiOH(aq)$
c. $HNO_3(aq)$ and $Ca(OH)_2(s)$

**8.43** What is the molarity of a HCl solution if 5.00 mL of a HCl solution is titrated with 28.6 mL of a 0.145 M NaOH solution?

$$HCl(aq) + NaOH(aq) \longrightarrow H_2O(l) + NaCl(aq)$$

**8.44** If 29.7 mL of 0.205 M KOH is required to neutralize completely 25.0 mL of a $HC_2H_3O_2$ solution, what is the molarity of the acetic acid solution?

$$HC_2H_3O_2(aq) + KOH(aq) \longrightarrow H_2O(l) + KC_2H_3O_2(aq)$$

**8.45** If 38.2 mL of 0.163 M KOH is required to neutralize completely 25.0 mL of a $H_2SO_4$ solution, what is the molarity of the acid?

$$H_2SO_4(aq) + 2KOH(aq) \longrightarrow 2H_2O(l) + K_2SO_4(aq)$$

**8.46** A solution of 0.162 M NaOH is used to neutralize 25.0 mL of $H_2SO_4$ solution. If 32.8 mL of the NaOH solution is required, what is the molarity of the $H_2SO_4$ solution?

$$H_2SO_4(aq) + 2NaOH(aq) \longrightarrow 2H_2O(l) + Na_2SO_4(aq)$$

---

## LEARNING GOAL

Describe the role of buffers in maintaining the pH of a solution.

**WEB TUTORIAL**
pH and Buffers

# 8.6 BUFFERS

The pH of water and most solutions changes drastically when a small amount of acid or base is added. However, if a solution is buffered, there is little change in pH. A **buffer** is a solution that maintains pH by neutralizing added acid or base. For example, blood contains buffers that maintain a consistent pH of about 7.4. If the pH of the blood goes slightly above or below 7.4, changes in oxygen uptake and metabolic processes can be drastic enough to cause death. Even though we obtain acids and bases from foods and cellular reactions, the buffers in the body absorb those compounds so effectively that the pH of the blood remains essentially unchanged. (See Figure 8.7.)

Buffers consist of nearly equal concentrations of a weak acid and its conjugate base or a weak base and its conjugate acid. In a buffer, the acid neutralizes added $OH^-$, and the base neutralizes added $H_3O^+$. For example, a typical buffer contains acetic acid ($HC_2H_3O_2$) and a salt such as sodium acetate ($NaC_2H_3O_2$). As a weak acid, acetic acid dissociates slightly in water to form $H_3O^+$ and a very small amount of $C_2H_3O_2^-$, its conjugate base. The salt is needed to provide more acetate ion ($C_2H_3O_2^-$) than is obtained from the weak acid.

$$HC_2H_3O_2(aq) + H_2O(l) \Longleftrightarrow H_3O^+(aq) + C_2H_3O_2^-(aq)$$

Large amount ............................................ Large amount

When a small amount of acid is added to this buffer, it combines with $C_2H_3O_2^-$ (conjugate base) in a reverse reaction that produces more $HC_2H_3O_2$. There will be a slight decrease in the $[C_2H_3O_2^-]$ and a slight increase in $[HC_2H_3O_2]$, but the $[H_3O^+]$, and therefore the pH, will not change very much.

$$HC_2H_3O_2(aq) + H_2O(l) \longleftarrow H_3O^+(aq) + C_2H_3O_2^-(aq)$$

**CASE STUDY**
Hyperventilation and Blood pH

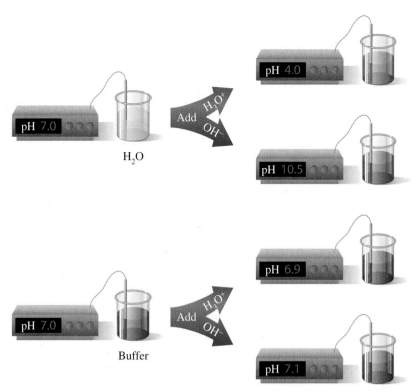

**FIGURE 8.7** Adding an acid or a base to water changes the pH drastically, but a buffer resists pH change when small amounts of acid or base are added.
**Q** Why does the pH change several pH units when acid is added to water, but not when acid is added to a buffer?

When a small amount of base is added to this buffer, it is neutralized by the $[HC_2H_3O_2]$ to produce water and $C_2H_3O_2^-$, the conjugate base. Now the $[HC_2H_3O_2]$ decreases slightly and the $[C_2H_3O_2^-]$ increases slightly, but again the $[H_3O^+]$, and therefore the pH, does not change very much. (See Figure 8.8.)

$$HC_2H_3O_2(aq) + OH^-(aq) \longleftarrow H_2O(aq) + C_2H_3O_2^-(aq)$$

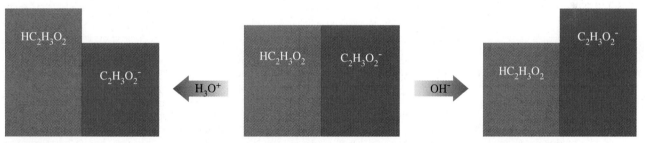

**FIGURE 8.8** The buffer described here consists of about equal concentrations of acetic acid ($HC_2H_3O_2$) and its conjugate base, acetate ion ($C_2H_3O_2^-$). Adding $H_3O^+$ to the buffer uses up some $C_2H_3O_2^-$, whereas adding $OH^-$ neutralizes some $HC_2H_3O_2$. The pH of the solution is maintained as long as the added amounts of acid or base are small compared to the concentrations of the buffer components.
**Q** How does this acetic acid–acetate ion buffer maintain pH?

SAMPLE PROBLEM    8.13

■ **Identifying Buffer Solutions**

Indicate whether each of the following would make a buffer solution:

**a.** HCl, a strong acid, and NaCl
**b.** $H_3PO_4$, a weak acid
**c.** HF, a weak acid, and NaF

SOLUTION

**a.** No. A solution of a strong acid and its salt is completely ionized.

**b.** No. A weak acid is not sufficient for a buffer; the salt of the weak acid is also needed.

**c.** Yes. This mixture contains a weak acid and its salt.

STUDY CHECK

Will a mixture of NaCl and $Na_2CO_3$ make a buffer solution? Explain.

## Buffers in the Blood

The arterial blood has a normal pH of 7.35–7.45. If changes in $H_3O^+$ lower the pH below 6.8 or raise it above 8.0, cells cannot function properly and death may result. In our cells, $CO_2$ is continually produced as an end product of cellular metabolism. Some $CO_2$ is carried to the lungs for elimination, and the rest dissolves in body fluids such as plasma and saliva, forming carbonic acid. As a weak acid, carbonic acid dissociates to give bicarbonate and $H_3O^+$. More of the anion $HCO_3^-$ is supplied by the kidneys to give an important buffer system in the body fluid: the $H_2CO_3/HCO_3^-$ buffer.

$$CO_2 + H_2O \rightleftharpoons H_2CO_3 \overset{H_2O}{\rightleftharpoons} H_3O^+ + HCO_3^-$$

Excess $H_3O^+$ entering the body fluids reacts with the $HCO_3^-$ and excess $OH^-$ reacts with the carbonic acid.

$$H_2CO_3 + H_2O \longleftarrow H_3O^+ + HCO_3^-$$

$$H_2CO_3 + OH^- \longrightarrow H_2O + HCO_3^-$$

In the body, the concentration of carbonic acid is closely associated with the partial pressure of $CO_2$. Table 8.7 lists the normal values for arterial blood. If the $CO_2$ increases, it produces more $H_2CO_3$ and

more $H_3O^+$, lowering the pH. This condition is called **acidosis**. Difficulty with ventilation or gas diffusion can lead to respiratory acidosis, which can happen in emphysema or when the medulla of the brain is affected by an accident or depressive drugs.

A decrease in the $CO_2$ leads to a high blood pH, a condition called **alkalosis**. Excitement, trauma, or a high temperature may cause a person to hyperventilate, which expels large amounts of $CO_2$. As the partial pressure of $CO_2$ in the blood falls below normal, $H_2CO_3$ forms $CO_2$ and $H_2O$, decreasing the $[H_3O^+]$ and raising the pH. Table 8.8 lists some of the conditions that lead to changes in the blood pH and some possible treatments. The kidneys also regulate $H_3O^+$ and $HCO_3^-$ components, but more slowly than the adjustment made by the lungs through ventilation.

**TABLE 8.7** Normal Values for Blood Buffer in Arterial Blood

| | |
|---|---|
| $P_{CO_2}$ | 40 mmHg |
| $H_2CO_3$ | 2.4 mmoles/L of plasma |
| $HCO_3^-$ | 24 mmoles/L of plasma |
| pH | 7.35–7.45 |

**TABLE 8.8** Acidosis and Alkalosis: Symptoms, Causes, and Treatments

| Respiratory Acidosis: $CO_2 \uparrow$ pH $\downarrow$ | | Metabolic Acidosis: $H^+ \uparrow$ pH $\downarrow$ | |
|---|---|---|---|
| Symptoms: | Failure to ventilate, suppression of breathing, disorientation, weakness, coma | Symptoms: | Increased ventilation, fatigue, confusion |
| Causes: | Lung disease blocking gas diffusion (e.g., emphysema, pneumonia, bronchitis, and asthma); depression of respiratory center by drugs, cardiopulmonary arrest, stroke, poliomyelitis, or nervous system disorders | Causes: | Renal disease, including hepatitis and cirrhosis; increased acid production in diabetes mellitus, hyperthyroidism, alcoholism, and starvation; loss of alkali in diarrhea; acid retention in renal failure |
| Treatment: | Correction of disorder, infusion of bicarbonate | Treatment: | Sodium bicarbonate given orally, dialysis for renal failure, insulin treatment for diabetic ketosis |
| **Respiratory Alkalosis: $CO_2 \downarrow$ pH $\uparrow$** | | **Metabolic Alkalosis: $H^+ \downarrow$ pH $\uparrow$** | |
| Symptoms: | Increased rate and depth of breathing, numbness, light-headedness, tetany | Symptoms: | Depressed breathing, apathy, confusion |
| Causes: | Hyperventilation because of anxiety, hysteria, fever, exercise; reaction to drugs such as salicylate, quinine, and antihistamines; conditions causing hypoxia (e.g., pneumonia, pulmonary edema, and heart disease) | Causes: | Vomiting, diseases of the adrenal glands, ingestion of excess alkali |
| Treatment: | Elimination of anxiety-producing state, rebreathing into a paper bag | Treatment: | Infusion of saline solution, treatment of underlying diseases |

## QUESTIONS AND PROBLEMS

### Buffers

**8.47** Which of the following represent a buffer system? Explain.
  **a.** NaOH and NaCl          **b.** $H_2CO_3$ and $NaHCO_3$
  **c.** HF and KF              **d.** KCl and NaCl

**8.48** Which of the following represent a buffer system? Explain.
  **a.** $H_3PO_4$             **b.** $NaNO_3$
  **c.** $HC_2H_3O_2$ and $NaC_2H_3O_2$   **d.** HCl and NaOH

**8.49** Consider the buffer system of hydrofluoric acid, HF, and its salt, NaF.

$$HF(aq) \rightleftharpoons H^+(aq) + F^-(aq)$$

**a.** What is the purpose of the buffer system?
**b.** Why is a salt of the acid needed?
**c.** How does the buffer react when some $H^+$ is added?
**d.** How does the buffer react when some $OH^-$ is added?

**8.50** Consider the buffer system of nitrous acid, $HNO_2$, and its salt, $NaNO_2$.

$$HNO_2(aq) \rightleftharpoons H^+(aq) + NO_2^-(aq)$$

**a.** What is the purpose of a buffer system?
**b.** What is the purpose of $NaNO_2$ in the buffer?
**c.** How does the buffer react when some $H^+$ is added?
**d.** How does the buffer react when some $OH^-$ is added?

## CONCEPT MAP

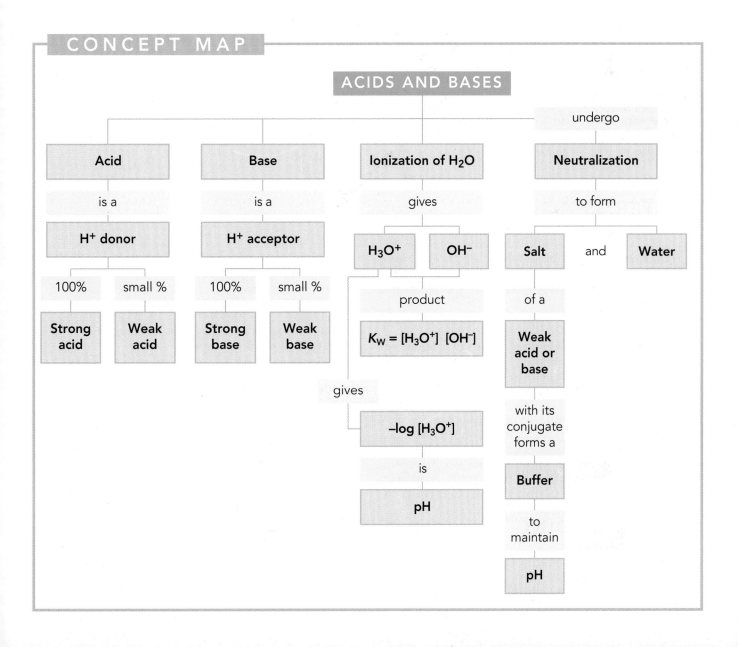

# CHAPTER REVIEW

## 8.1 Acids and Bases

**Learning Goal:** Describe and name acids and bases; identify Brønsted–Lowry acids and bases.

An Arrhenius acid produces $H^+$ and an Arrhenius base produces $OH^-$ in aqueous solutions. According to the Brønsted–Lowry theory, acids are proton $(H^+)$ donors and bases are proton acceptors. Two conjugate acid–base pairs are present in an acid–base reaction. Each acid–base pair is related by the loss or gain of one $H^+$. For example, when the acid HF donates a $H^+$, the $F^-$ it forms is its conjugate base because $F^-$ is capable of accepting a $H^+$. The other acid–base pair would be $H_2O$ and $H_3O^+$.

$$HF(aq) + H_2O(l) \rightleftharpoons H_3O^+(aq) + F^-(aq)$$

## 8.2 Strengths of Acids and Bases

**Learning Goal:** Write equations for the dissociation of strong and weak acids.

In strong acids, all the $H^+$ in the acid is donated to $H_2O$; in a weak acid, only a small percentage of acid molecules produce $H_3O^+$. Strong bases are hydroxides of Groups 1A (1) and 2A (2) that dissociate completely in water. An important weak base is ammonia, $NH_3$.

## 8.3 Ionization of Water

**Learning Goal:** Use the ion product of water to calculate the $[H_3O^+]$ and $[OH^-]$ in an aqueous solution.

In pure water, a few molecules transfer protons to other water molecules, producing small but equal amounts of $[H_3O^+]$ and $[OH^-]$, such that each has a concentration of $1.0 \times 10^{-7}$ mole/L. The ion product, $K_w$, $[H_3O^+][OH^-] = 1.0 \times 10^{-14}$, applies to all aqueous solutions at 25 °C. In acidic solutions, the $[H_3O^+]$ is greater than the $[OH^-]$. In basic solutions, the $[OH^-]$ is greater than the $[H_3O^+]$.

## 8.4 The pH Scale

**Learning Goal:** Calculate pH from $[H_3O^+]$; given the pH, calculate $[H_3O^+]$ and $[OH^-]$ of a solution.

The pH scale is a range of numbers from 0 to 14 related to the $[H_3O^+]$ of the solution. A neutral solution has a pH of 7. In acidic solutions, the pH is below 7, and in basic solutions, the pH is above 7. Mathematically, pH is the negative logarithm of the hydronium ion concentration $(-\log[H_3O^+])$.

## 8.5 Reactions of Acids and Bases

**Learning Goal:** Write balanced equations for reactions of acids and bases; calculate the molarity or volume of an acid or a base from titration information.

When an acid reacts with a metal, hydrogen gas and a salt are produced. The reaction of an acid with a carbonate or bicarbonate produces carbon dioxide, a salt, and water. In neutralization, an acid reacts with a base to produce a salt and water. In titration, an acid sample is neutralized with a known amount of a base. From the volume and molarity of the base, the concentration of the acid is calculated.

## 8.6 Buffers

**Learning Goal:** Describe the role of buffers in maintaining the pH of a solution.

A buffer solution resists changes in pH when small amounts of acid or base are added. A buffer contains either a weak acid and its salt or a weak base and its salt. The weak acid picks up added $OH^-$, and the anion of the salt picks up added $H^+$. Buffers are important in maintaining the pH of the blood.

# KEY TERMS

**acid** A substance that dissolves in water and produces hydrogen ions $(H^+)$, according to the Arrhenius theory. All acids are proton donors, according to the Brønsted–Lowry theory.

**acidosis** A physiological condition in which the blood pH is lower than 7.35.

**alkalosis** A physiological condition in which the blood pH is higher than 7.45.

**base** A substance that dissolves in water and produces hydroxide ions $(OH^-)$, according to the Arrhenius theory. All bases are proton acceptors, according to the Brønsted–Lowry theory.

**Brønsted–Lowry acids and bases** An acid is a proton donor, and a base is a proton acceptor.

**buffer** A solution of a weak acid and its conjugate base or a weak base and its conjugate acid that maintains the pH by neutralizing added acid or base.

**conjugate acid–base pair** An acid and base that differ by one $H^+$. When an acid donates a proton, the product is its conjugate

base, which is capable of accepting a proton in the reverse reaction.

**hydronium ion, $H_3O^+$** The ion formed by the attraction of a proton $(H^+)$ to a $H_2O$ molecule.

**ion-product constant of water, $K_w$** The product of $[H_3O^+]$ and $[OH^-]$ in solution; $K_w = [H_3O^+][OH^-]$.

**neutral** The term that describes a solution with equal concentrations of $[H_3O^+]$ and $[OH^-]$.

**neutralization** A reaction between an acid and a base to form a salt and water.

**pH** A measure of the $[H_3O^+]$ in a solution; pH $= -\log[H_3O^+]$.

**strong acid** An acid that completely ionizes in water.

**strong base** A base that completely ionizes in water.

**titration** The addition of base to an acid sample to determine the concentration of the acid.

**weak acid** An acid that donates only a few $H^+$.

**weak base** A base that accepts only a few $H^+$.

# UNDERSTANDING THE CONCEPTS

**8.51** Identify each of the following as an acid, a base, or a salt, and give its name.
   **a.** LiOH     **b.** $Ca(NO_3)_2$     **c.** HBr
   **d.** $Ba(OH)_2$     **e.** $H_2CO_3$

**8.52** Identify each of the following as an acid, a base, or a salt, and give its name.
   **a.** $H_3PO_4$     **b.** $MgBr_2$     **c.** $NH_3$
   **d.** $H_2SO_4$     **e.** NaCl

**8.53**  Complete the following table:

| Acid | Conjugate base |
|------|----------------|
| $H_2O$ | |
| | $CN^-$ |
| $HNO_2$ | |
| | $H_2PO_4^-$ |

**8.54**  Complete the following table:

| Base | Conjugate acid |
|------|----------------|
| | $HS^-$ |
| | $H_3O^+$ |
| $NH_3$ | |
| $HCO_3^-$ | |

**8.55**  In each of the following diagrams of acid solutions, determine if the diagram represents a strong acid or a weak acid. The acid has the formula HX.

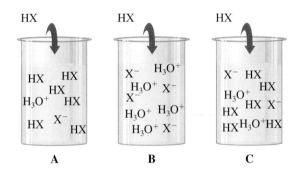

**8.56**  Adding a few drops of a strong acid to water will lower the pH appreciably. However, adding the same number of drops to a buffer does not appreciably alter the pH. Why?

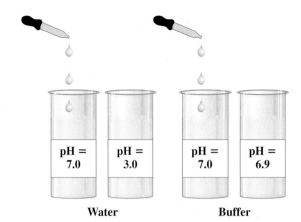

**8.57**  Sometimes, during stress or trauma, a person can start to hyperventilate. Then the person might breathe into a paper bag to avoid fainting.
**a.** What changes occur in the blood pH during hyperventilation?
**b.** How does breathing into a paper bag help return blood pH to normal?

**8.58**  In the blood plasma, pH is maintained by the carbonic acid–bicarbonate buffer system.
**a.** How is pH maintained when acid is added to the buffer system?
**b.** How is pH maintained when base is added to the buffer system?

# ADDITIONAL QUESTIONS AND PROBLEMS

**8.59**  Name each of the following:
**a.** $H_2SO_4$
**b.** KOH
**c.** $Ca(OH)_2$
**d.** HCl
**e.** $HNO_2$

**8.60**  Name each of the following:
**a.** $Sr(OH)_2$
**b.** $H_3PO_4$
**c.** $HClO_3$
**d.** LiOH
**e.** $H_2CO_3$

**8.61**  Are each of the following examples acidic, basic, or neutral?
**a.** rain, pH 5.2
**b.** tears, pH 7.5
**c.** tea, pH 3.8
**d.** cola, pH 2.5
**e.** photo developer, pH 12.0

**8.62**  Are the following examples of body fluids acidic, basic, or neutral?
**a.** saliva, pH 6.8
**b.** urine, pH 5.9
**c.** pancreatic juice, pH 8.0
**d.** bile, pH 8.4
**e.** blood, pH 7.45

**8.63**  What are some similarities and differences between strong and weak acids?

**8.64**  What are some ingredients found in antacids? What do they do?

**8.65**  One ingredient in some antacids is $Mg(OH)_2$.
**a.** If the base is not very soluble in water, why is it considered a strong base?

**b.** What is the neutralization reaction of $Mg(OH)_2$ with stomach acid, HCl?

**8.66** Acetic acid, $HC_2H_3O_2$, used to prepare vinegar, is a weak acid. Why?

**8.67** Determine the pH for the following solutions:
**a.** $[H_3O^+] = 1.0 \times 10^{-8}$ M
**b.** $[H_3O^+] = 5.0 \times 10^{-2}$ M
**c.** $[OH^-] = 3.5 \times 10^{-4}$ M
**d.** $[OH^-] = 0.005$ M

**8.68** Identify each of the solutions in problem 8.67 as acidic, basic, or neutral.

**8.69** Determine the pH for the following solutions:
**a.** $[OH^-] = 1.0 \times 10^{-7}$ M
**b.** $[H_3O^+] = 4.2 \times 10^{-3}$ M
**c.** $[H_3O^+] = 0.0001$ M
**d.** $[OH^-] = 8.5 \times 10^{-9}$ M

**8.70** Identify each of the solutions in problem 8.69 as acidic, basic, or neutral.

**8.71** What are the $[H_3O^+]$ and $[OH^-]$ for a solution with the following pH values?
**a.** 3.0  **b.** 6.00  **c.** 8.0
**d.** 11.0  **e.** 9.20

**8.72** What are the $[H_3O^+]$ and $[OH^-]$ for a solution with the following pH values?
**a.** 10.0
**b.** 5.0
**c.** 7.00
**d.** 6.5
**e.** 1.82

**8.73** Solution A has a pH of 4.0, and solution B has a pH of 6.0.
**a.** Which solution is more acidic?
**b.** What is the $[H_3O^+]$ in each?
**c.** What is the $[OH^-]$ in each?

**8.74** Solution X has a pH of 9.5, and solution Y has a pH of 7.5.
**a.** Which solution is more acidic?
**b.** What is the $[H_3O^+]$ in each?
**c.** What is the $[OH^-]$ in each?

**8.75** A buffer is made by dissolving $H_3PO_4$ and $NaH_2PO_4$ in water.
**a.** Write an equation that shows how this buffer neutralizes added acid.
**b.** Write an equation that shows how this buffer neutralizes added base.

**8.76** A buffer is made by dissolving acetic acid, $HC_2H_3O_2$, and sodium acetate, $NaC_2H_3O_2$, in water.
**a.** Write an equation that shows how this buffer neutralizes added acid.
**b.** Write an equation that shows how this buffer neutralizes added base.

**8.77** Calculate the volume (mL) of a 0.150 M NaOH solution needed to neutralize each of the following:
**a.** 25.0 mL of a 0.288 M HCl solution
**b.** 10.0 mL of a 0.560 M $H_2SO_4$ solution
**c.** 5.00 mL of a 0.618 M HBr solution

**8.78** Calculate the volume (mL) of a 0.215 M KOH solution that will completely neutralize each of the following:
**a.** 2.50 mL of a 0.825 M $H_2SO_4$ solution
**b.** 18.5 mL of a 0.560 M $HNO_3$ solution
**c.** 5.00 mL of a 3.18 M $H_2SO_4$ solution

**8.79** A solution of 0.205 M NaOH is used to neutralize 20.0 mL of $H_2SO_4$. If 45.6 mL of a NaOH solution is required, what is the molarity of the $H_2SO_4$ solution?

$$H_2SO_4(aq) + 2NaOH(aq) \longrightarrow Na_2SO_4(aq) + 2H_2O(l)$$

**8.80** A 10.0-mL sample of vinegar, which is an aqueous solution of acetic acid, $HC_2H_3O_2$, requires 16.5 mL of 0.500 M NaOH to reach the end point in a titration. What is the molarity of the acetic acid solution?

$$HC_2H_3O_2(aq) + NaOH(aq) \longrightarrow NaC_2H_3O_2(aq) + H_2O(l)$$

# CHALLENGE QUESTIONS

**8.81** Consider the following:
**1.** $H_2S$  **2.** $H_3PO_4$  **3.** $HCO_3^-$
**a.** For each, write the formula of the conjugate base.
**b.** Write the formula of the weakest acid.
**c.** Write the formula of the strongest acid.

**8.82** Identify the conjugate acid–base pairs in each of the following equations:
**a.** $NH_3(aq) + HNO_3(aq) \rightleftharpoons NH_4^+(aq) + NO_3^-(aq)$

**b.** $H_2O(l) + HBr(aq) \rightleftharpoons H_3O^+(aq) + Br^-(aq)$

**c.** $HNO_2(aq) + HS^-(aq) \rightleftharpoons H_2S(g) + NO_2^-(aq)$

**d.** $Cl^-(aq) + H_2O(l) \rightleftharpoons OH^-(aq) + HCl(aq)$

**8.83** Complete and balance each of the following:
**a.** $ZnCO_3(s) + H_2SO_4(aq) \longrightarrow$
**b.** $Al(s) + HCl(aq) \longrightarrow$
**c.** $H_3PO_4(aq) + Ca(OH)_2(s) \longrightarrow$
**d.** $KHCO_3(s) + HNO_3(aq) \longrightarrow$

**8.84** Determine each of the following for a 0.050 M KOH solution.
**a.** $[H_3O^+]$
**b.** pH
**c.** products when allowed to react with $H_3PO_4$
**d.** milliliters required to neutralize 40.0 mL of 0.035 M $H_2SO_4$

**8.85** What is the pH of a solution prepared by dissolving 2.5 g of HCl in water to make 425 mL of HCl solution?

**8.86** Consider the reaction of KOH and $HNO_2$.
**a.** Write the balanced chemical equation.
**b.** Calculate the milliliters of 0.122 M KOH required to neutralize 36.0 mL of 0.250 M $HNO_2$.

**8.87** A solution of 0.204 M NaOH is used to titrate 50.0 mL of a $H_3PO_4$ solution.

**a.** Write the balanced chemical equation.
**b.** What is the molarity of the $H_3PO_4$ solution if 16.4 mL of the NaOH solution is required?

**8.88** A solution of 0.312 M KOH is used to titrate 15.0 mL of a $H_2SO_4$ solution.
**a.** Write the balanced chemical equation.
**b.** What is the molarity of the $H_2SO_4$ solution if 28.2 mL of the KOH solution is required?

**8.89** One of the most acidic lakes in the United States is Little Echo Pond in the Adirondacks in New York. Recently, this lake had a pH of 4.2, well below the recommended pH of 6.5.

**a.** What are the $[H_3O^+]$ and $[OH^-]$ of Little Echo Pond?
**b.** What are the $[H_3O^+]$ and $[OH^-]$ of a lake that has a pH of 6.5?
**c.** One way to raise the pH and restore aquatic life of an acidic lake is to add limestone ($CaCO_3$). How many g of $CaCO_3$ are needed to neutralize 1.0 kL of the acidic water from Little Echo Pond if the acid is written as HA?

$$2HA + CaCO_3(s) \longrightarrow CaA_2 + CO_2(g) + H_2O(l)$$

**8.90** The daily output of stomach acid (gastric juices) is 1000 mL to 1400 mL. Prior to a meal, stomach acid (HCl) typically has a pH of 1.42.
**a.** What is the $[H_3O^+]$ of stomach acid?
**b.** The antacid Maalox contains 200. mg of $Al(OH)_3$ per tablet. Write the neutralization equation, and calculate the milliliters of stomach acid neutralized by two tablets of Maalox.
**c.** The antacid milk of magnesia contains 400. mg of $Mg(OH)_2$ per teaspoon. Write the neutralization equation, and calculate the number of milliliters of stomach acid neutralized by 1 tablespoon of Maalox (1 tablespoon = 3 teaspoons).

# ANSWERS

## Answers to Study Checks

**8.1 a.** Sulfuric acid
**b.** KOH

**8.2** $LiOH(s) \xrightarrow{H_2O} Li^+(aq) + OH^-(aq)$

**8.3** $HNO_3(aq) + H_2O(l) \longrightarrow H_3O^+(aq) + NO_3^-(aq)$

**8.4 a.** $H_2S$ is the conjugate acid that forms when $HS^-$ accepts one $H^+$.
**b.** $HNO_2$ is the conjugate acid that forms when $NO_2^-$ accepts one $H^+$.

**8.5** One of the conjugate acid–base pairs is HCN and $CN^-$; the other pair is $SO_4^{2-}$ and $HSO_4^-$.

**8.6** KOH

**8.7** $[H_3O^+] = 2.5 \times 10^{-11}$ M, basic

**8.8** 11.38

**8.9** 4.3

**8.10** $[H_3O^+] = 3 \times 10^{-5}$ M

**8.11** $H_2SO_4(aq) + 2NaHCO_3(s) \longrightarrow$
$$Na_2SO_4(aq) + 2CO_2(g) + 2H_2O(l)$$

**8.12** 0.200 M HCl

**8.13** No. Both substances are salts; the mixture has no weak acid present.

## Answers to Selected Questions and Problems

**8.1 a.** acid     **b.** acid     **c.** acid     **d.** base

**8.3 a.** hydrochloric acid
**b.** calcium hydroxide
**c.** carbonic acid
**d.** nitric acid
**e.** sulfurous acid
**f.** lithium hydroxide

**8.5 a.** $Mg(OH)_2$
**b.** HF
**c.** $H_3PO_4$
**d.** LiOH
**e.** $Cu(OH)_2$
**f.** $H_2SO_4$

**8.7 a.** HI is the acid (proton donor) and $H_2O$ is the base (proton acceptor).
**b.** $H_2O$ is the acid (proton donor) and $F^-$ is the base (proton acceptor).

**8.9 a.** $F^-$, fluoride ion
**b.** $OH^-$, hydroxide ion
**c.** $HCO_3^-$, bicarbonate ion *or* hydrogen carbonate ion
**d.** $SO_4^{2-}$, sulfate ion

**8.11 a.** $HCO_3^-$, bicarbonate ion *or* hydrogen carbonate ion
**b.** $H_3O^+$, hydronium ion
**c.** $H_3PO_4$, phosphoric acid
**d.** HBr, hydrobromic acid

**8.13 a.** $H_2CO_3$ and its conjugate base $HCO_3^-$; $H_2O$ and its conjugate acid $H_3O^+$
**b.** $NH_4^+$ and its conjugate base $NH_3$; $H_2O$ and its conjugate acid $H_3O^+$
**c.** HCN and its conjugate base $CN^-$; $NO_2^-$ and its conjugate acid $HNO_2$

**8.15 a.** HBr
**b.** $HSO_4^-$
**c.** $H_2CO_3$

**8.17 a.** $HSO_4^-$
**b.** HF
**c.** $HCO_3^-$

**8.19** In pure water, $[H_3O^+] = [OH^-]$ because one of each is produced every time a proton transfers from one water molecule to another.

**8.21** In an acidic solution, the $[H_3O^+]$ is greater than the $[OH^-]$.

**8.23** **a.** acidic
**b.** basic
**c.** basic
**d.** acidic

**8.25** **a.** $1.0 \times 10^{-9}$ M
**b.** $1.0 \times 10^{-6}$ M
**c.** $2.0 \times 10^{-5}$ M
**d.** $4.0 \times 10^{-13}$ M

**8.27** **a.** $1.0 \times 10^{-11}$ M
**b.** $2.0 \times 10^{-9}$ M
**c.** $5.6 \times 10^{-3}$ M
**d.** $2.5 \times 10^{-2}$ M

**8.29** In a neutral solution, the $[H_3O^+]$ is $1.0 \times 10^{-7}$ M and the pH is 7.00, which is the negative value of the power of 10.

**8.31** **a.** basic
**b.** acidic
**c.** basic
**d.** acidic
**e.** acidic
**f.** basic

**8.33** **a.** 4.0
**b.** 8.5
**c.** 9.0
**d.** 3.40
**e.** 7.17
**f.** 10.92

**8.35**

| $[H_3O^+]$ | $[OH^-]$ | pH | Acidic, Basic, or Neutral? |
|---|---|---|---|
| $1 \times 10^{-8}$ M | $1 \times 10^{-6}$ M | 8.0 | Basic |
| $1 \times 10^{-3}$ M | $1 \times 10^{-11}$ M | 3.0 | Acidic |
| $2 \times 10^{-5}$ M | $5 \times 10^{-10}$ M | 4.7 | Acidic |
| $1 \times 10^{-12}$ M | $1 \times 10^{-2}$ M | 12.0 | Basic |
| $2.4 \times 10^{-5}$ M | $4.2 \times 10^{-10}$ M | 4.62 | Acidic |

**8.37** **a.** $ZnCO_3(s) + 2HBr(aq) \longrightarrow$
$ZnBr_2(aq) + CO_2(g) + H_2O(l)$
**b.** $Zn(s) + 2HCl(aq) \longrightarrow ZnCl_2(aq) + H_2(g)$
**c.** $HCl(g) + NaHCO_3(s) \longrightarrow$
$NaCl(aq) + H_2O(l) + CO_2(g)$
**d.** $H_2SO_4(aq) + Mg(OH)_2(s) \longrightarrow MgSO_4(aq) + 2H_2O(l)$

**8.39** **a.** $2HCl(aq) + Mg(OH)_2(s) \longrightarrow 2H_2O(l) + MgCl_2(aq)$
**b.** $H_3PO_4(aq) + 3LiOH(aq) \longrightarrow 3H_2O(l) + Li_3PO_4(aq)$

**8.41** **a.** $H_2SO_4(aq) + 2NaOH(aq) \longrightarrow Na_2SO_4(aq) + 2H_2O(l)$
**b.** $3HCl(aq) + Fe(OH)_3(aq) \longrightarrow FeCl_3(aq) + 3H_2O(l)$
**c.** $H_2CO_3(aq) + Mg(OH)_2(s) \longrightarrow MgCO_3(s) + 2H_2O(l)$

**8.43** 0.829 M HCl

**8.45** 0.125 M $H_2SO_4$

**8.47** (b) and (c) are buffer systems. (b) contains the weak acid $H_2CO_3$ and its salt $NaHCO_3$. (c) contains HF, a weak acid, and its salt KF.

**8.49** **a.** A buffer system keeps the pH constant.
**b.** To neutralize any $H^+$ added.
**c.** The added $H^+$ reacts with $F^-$ from NaF.
**d.** The added $OH^-$ is neutralized by the HF.

**8.51** **a.** base, lithium hydroxide
**b.** salt, calcium nitrate
**c.** acid, hydrobromic acid
**d.** base, barium hydroxide
**e.** acid, carbonic acid

**8.53**

| Acid | Conjugate base |
|---|---|
| $H_2O$ | $OH^-$ |
| HCN | $CN^-$ |
| $HNO_2$ | $NO_2^-$ |
| $H_3PO_4$ | $H_2PO_4^-$ |

**8.55** **a.** weak acid
**b.** strong acid
**c.** weak acid

**8.57** **a.** Hyperventilation will lower the $CO_2$ level in the blood, which lowers the $H_2CO_3$ concentration, which decreases the $H_3O^+$ and increases the blood pH.
**b.** Breathing into a bag will increase the $CO_2$ level, increase the $H_2CO_3$, increase $H_3O^+$, and lower the blood pH.

**8.59** **a.** sulfuric acid
**b.** potassium hydroxide
**c.** calcium hydroxide
**d.** hydrochloric acid
**e.** nitrous acid

**8.61** **a.** acidic
**b.** basic
**c.** acidic
**d.** acidic
**e.** basic

**8.63** Both strong and weak acids produce $H_3O^+$ in water. Weak acids are only slightly dissociated, to give an aqueous solution of only a few ions; whereas a strong acid is completely dissociated, to give only ions in solution.

**8.65** **a.** The $Mg(OH)_2$ that dissolves is completely ionized, making it a strong base.
**b.** $Mg(OH)_2(aq) + 2HCl(aq) \longrightarrow MgCl_2(aq) + 2H_2O(l)$

**8.67** **a.** 8.00
**b.** 1.30
**c.** 10.54
**d.** 11.7

**8.69** **a.** 7.00
**b.** 2.38
**c.** 4.0
**d.** 5.93

**8.71** **a.** $[H_3O^+] = 1 \times 10^{-3}$ M, $[OH^-] = 1 \times 10^{-11}$ M
**b.** $[H_3O^+] = 1.0 \times 10^{-6}$ M, $[OH^-] = 1.0 \times 10^{-8}$ M
**c.** $[H_3O^+] = 1 \times 10^{-8}$ M, $[OH^-] = 1 \times 10^{-6}$ M
**d.** $[H_3O^+] = 1 \times 10^{-11}$ M, $[OH^-] = 1 \times 10^{-3}$ M
**e.** $[H_3O^+] = 6.3 \times 10^{-10}$ M, $[OH^-] = 1.6 \times 10^{-5}$ M

**8.73** **a.** Solution A with a pH of 4.0 is more acidic than solution B.
**b.** Solution A: $[H_3O^+] = 1 \times 10^{-4}$ M,
Solution B: $[H_3O^+] = 1 \times 10^{-6}$ M
**c.** Solution A: $[OH^-] = 1 \times 10^{-10}$ M,
Solution B: $[OH^-] = 1 \times 10^{-8}$ M

**8.75** **a.** acid: $H_2PO_4^-(aq) + H_3O^+(aq) \longrightarrow$
$$H_3PO_4(aq) + H_2O(l)$$
**b.** base: $H_3PO_4(aq) + OH^-(aq) \longrightarrow$
$$H_2PO_4^-(aq) + H_2O(l)$$

**8.77** **a.** 48.0 mL
**b.** 74.7 mL
**c.** 20.6 mL

**8.79** 0.234 M $H_2SO_4$

**8.81** **a.** **1.** $HS^-$ **2.** $H_2PO_4^-$ **3.** $CO_3^{2-}$
**b.** $H_2S$
**c.** $H_3PO_4$

**8.83** **a.** $ZnCO_3(s) + H_2SO_4(aq) \longrightarrow$
$$ZnSO_4(aq) + CO_2(g) + H_2O(l)$$
**b.** $2Al(s) + 6HCl(aq) \longrightarrow 2AlCl_3(aq) + 3H_2(g)$
**c.** $2H_3PO_4(aq) + 3Ca(OH)_2(s) \longrightarrow$
$$Ca_3(PO_4)_2(aq) + 6H_2O(l)$$
**d.** $KHCO_3(s) + HNO_3(aq) \longrightarrow$
$$KNO_3(aq) + CO_2(g) + H_2O(l)$$

**8.85** 0.80

**8.87** **a.** $H_3PO_4(aq) + 3NaOH(aq) \longrightarrow Na_3PO_4(aq) + 3H_2O(l)$
**b.** 0.0223 M

**8.89** **a.** $[H_3O^+] = 6 \times 10^{-5}$ M; $[OH^-] = 2 \times 10^{-10}$ M
**b.** $[H_3O^+] = 3 \times 10^{-7}$ M; $[OH^-] = 3 \times 10^{-8}$ M
**c.** 3 g of $CaCO_3$

**CI.13** Methane is a major component of purified natural gas used for heating and cooking. When 1 mole of methane gas burns with oxygen to produce carbon dioxide and water, 211 kcal of heat is produced. Methane gas has a density of 0.715 g/L at STP. For transport, the volume of natural gas is decreased by cooling it to −163 °C, which gives liquefied natural gas (LNG) with a density of 0.45 g/mL. A tank on a ship can hold 7.0 million gallons of LNG.

a. Write the electron-dot formula and the molecular formula of methane if it consists of one carbon atom bonded to hydrogen atoms.
b. What is the mass in kilograms of LNG (assume that LNG is all methane) transported in one tank on a ship?
c. What is the volume in liters of methane gas when the LNG in one tank is converted to gas at STP?
d. Write the balanced equation for the reaction of methane and oxygen in a gas burner.

e. How many kilograms of oxygen are needed to react with all of the methane provided by one tank of LNG?
f. How much heat, in kilocalories, is released from burning all of the LNG in one tank of methane?

**CI.14** Automobile exhaust is a major cause of air pollution. The pollutants formed from gasoline include nitrogen oxide that is produced at high temperatures in an automobile engine from nitrogen and oxygen gases in the air. Once emitted into the air, nitrogen oxide reacts with oxygen to produce nitrogen dioxide, a reddish gas with a sharp, pungent odor that makes up smog. A major component of gasoline is octane, $C_8H_{18}$, which has a density of 0.803 g/cm$^3$. In 1 year, a typical automobile uses 550 gal of gasoline and produces 41 lb of nitrogen oxide.

a. Write balanced equations for the production of nitrogen oxide and nitrogen dioxide.
b. If all the nitrogen oxide emitted by one automobile is converted to nitrogen dioxide in the atmosphere, how many kilograms of nitrogen dioxide are produced in 1 year by a single automobile?
c. Write a balanced equation for the reaction of octane with oxygen gas to give carbon dioxide and water vapor.
d. How many moles of $C_8H_{18}$ are present in 15.2 gal of octane?
e. How many kilograms of carbon dioxide would this car produce in 1 year? (Assume complete reaction of octane.)

**CI.15** When clothes have stains, bleach is often added to the wash to react with the soil and make the stains colorless. The active ingredient in bleach is sodium hypochlorite (NaClO). A bleach solution can be prepared by bubbling chlorine gas into a solution of sodium hydroxide to produce sodium hypochlorite, sodium chloride, and water. A typical bottle of bleach contains 1.42 gal of bleach solution with 282 g of NaClO and has a density of 1.08 g/mL.

a. Is sodium hypochlorite an ionic or a covalent compound?
b. What is the percent concentration (% m/v) of sodium hypochlorite in bleach?
c. Write the equation for the preparation of a bleach solution.
d. How many liters of chlorine gas at STP are required to produce 1.42 gal of bleach for one bottle of bleach?
e. If the pH of the bleach solution is 10.3, what is the $[H_3O^+]$ and $[OH^-]$?

**CI.16** In wine making, glucose $(C_6H_{12}O_6)$ from grapes undergoes fermentation in the absence of oxygen to produce ethanol and carbon dioxide. A bottle of vintage port wine has a volume of 750 mL and contains 135 mL of ethanol $(C_2H_6O)$. Ethanol has a density of 0.789 g/mL. In 1.5 lb of grapes, there is 26 g of glucose.

a. Calculate the percent concentration ethanol by volume (% v/v).
b. What is the molarity (M) of ethanol in the port wine?
c. Write the balanced equation for the fermentation reaction of glucose.
d. How many grams of glucose from grapes are required to produce one bottle of port wine?
e. How many bottles of port wine can be produced from 1.0 ton of grapes? (1 ton = 2000 lb)

**CI.17** A metal completely reacts with 34.8 mL of 0.520 M HCl.

a. Write a balanced equation for the reaction of the metal M and HCl(*aq*) to form $MCl_3(aq)$ and $H_2$ gas.
b. What volume in milliliters of $H_2$ at STP is produced?
c. How many moles of metal M reacted?
d. If the metal has a mass of 0.420 g, use your results from part **c.** to determine the atomic mass of the metal M.
e. What are the name and symbol of metal M in part **d**?
f. Write the balanced equation using the symbol of the metal from part **e.**

**CI.18** In a teaspoon (5.0 mL) of a common liquid antacid, there are 200. mg $Ca(OH)_2$ and 200. mg $Al(OH)_3$. A 0.080 M HCl, which is similar to stomach acid, is used to neutralize 5.0 mL of the liquid antacid.

a. Write the equation for the neutralization of HCl and $Ca(OH)_2$ .
b. Write the equation for the neutralization of HCl and $Al(OH)_3$ .
c. What is the pH of the HCl solution?
d. How many milliliters of the HCl solution are needed to neutralize the $Ca(OH)_2$?
e. How many milliliters of the HCl solution are needed to neutralize the $Al(OH)_3$?

## ANSWERS

**CI.13  a.**

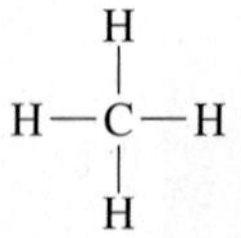

H—C—H   $CH_4$

with H above and H below the C

**b.** $1.2 \times 10^7$ kg of LNG (methane)

**c.** $1.7 \times 10^{10}$ L of methane at STP

**d.** $CH_4(g) + 2O_2(g) \longrightarrow CO_2(g) + 2H_2O(g)$

**e.** $4.8 \times 10^7$ kg of $O_2$

**f.** $1.6 \times 10^{11}$ kcal

**CI.15  a.** ionic

**b.** 5.26% (m/v)

**c.** $2NaOH(aq) + Cl_2(g) \longrightarrow$
$$NaCl(aq) + NaClO(aq) + H_2O(l)$$

**d.** 84.8 L of chlorine gas

**e.** $[H_3O^+] = 5 \times 10^{-11}$ M; $[OH^-] = 2 \times 10^{-4}$ M

**CI.17  a.** $2M(s) + 6HCl(aq) \longrightarrow 2MCl_3(aq) + 3H_2(g)$

**b.** 203 mL of $H_2$

**c.** $6.03 \times 10^{-3}$ mole of M

**d.** 69.7 g/mole

**e.** Gallium; Ga

**f.** $2Ga(s) + 6HCl(aq) \longrightarrow 2GaCl_3(aq) + 3H_2(g)$

# 9 Nuclear Radiation

Visit **www.chemplace.com** for extra quizzes, interactive tutorials, career resources, PowerPoint slides for chapter review, math help, and case studies.

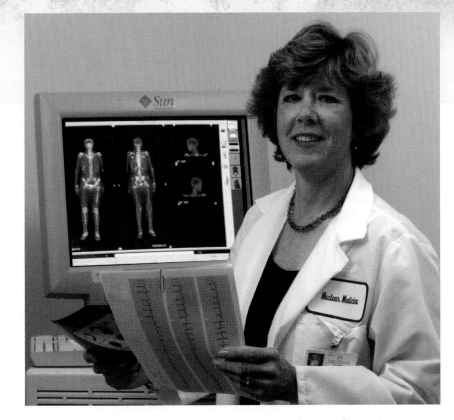

*"Everything we do in this department involves radioactive materials," says Julie Goudak, nuclear medicine technologist at Kaiser Hospital. "The radioisotopes are given in several ways. The patient may ingest an isotope, breathe it in, or receive it by an IV injection. We do many diagnostic tests, particularly of the heart function, to determine if a patient needs a cardiac CAT scan."*

*A nuclear medicine technologist administers isotopes that emit radiation to determine the level of function of an organ such as the thyroid or heart, to detect the presence and size of a tumor, or to treat disease. A radioisotope locates in a specific organ, and its radiation is used by a computer to create an image of that organ. From this data, a physician can make a diagnosis and design a treatment program.*

A female patient, age 50, complains of nervousness, irritability, increased perspiration, brittle hair, and muscle weakness. Her hands are shaky at times and her heart often beats rapidly. She has been experiencing weight loss. The doctor decides to test for hyperthyroidism. To get a detailed look at the thyroid, a thyroid scan is ordered. The patient is given a small amount of an iodine radioisotope, which will be taken up by the thyroid. The scan shows a higher than normal rate of uptake of the radioactive iodine, indicating an overactive thyroid gland, a condition called hyperthyroidism. Treatment for hyperthyroidism includes the use of drugs to lower the level of thyroid hormone, the use of radioactive iodine to destroy thyroid cells, or surgical removal of part of or the entire thyroid. In our case, the nuclear physician decides to use radioactive iodine. To begin treatment, the patient drinks a solution containing radioactive iodine. In the following few weeks, the cells that take up the radioactive iodine are destroyed by the radiation. Tests show that the patient's thyroid is smaller and the blood level of thyroid hormone is normal once again.

With the production of artificial radioactive substances in 1934, the field of nuclear medicine was established. In 1937, the first radioactive isotope was used to treat a patient with leukemia at the University of California at Berkeley. Major strides in the use of radioactivity in medicine occurred in 1946, when a radioactive iodine isotope was successfully used to diagnose thyroid function and to treat hyperthyroidism and thyroid cancer. In the 1970s and 1980s, a variety of radioactive substances were used to produce an image of an organ such as liver, spleen, thyroid gland, kidney, and brain, and to detect heart disease. Today, procedures in nuclear medicine provide information about the function and structure of every organ in the body, allowing the nuclear physician to diagnose and treat diseases early.

# 9.1 NATURAL RADIOACTIVITY

Most naturally occurring isotopes of elements up to atomic number 19 have stable nuclei. In a stable nucleus, the repulsions between the positively charged protons are balanced by other nuclear forces. Elements with atomic numbers 20 and higher usually have one or more isotopes that have unstable nuclei, in which the nuclear forces cannot offset the repulsions between the protons. An unstable nucleus is **radioactive**, which means that it spontaneously emits small particles or energy, called **radiation**, to become more stable. Radiation may take the form of particles such as alpha ($\alpha$) and beta ($\beta$) particles, positrons ($\beta^+$), or pure energy such as gamma ($\gamma$) rays. An isotope that emits radiation is called a *radioisotope*. Elements with atomic numbers of 93 and higher are produced artificially in nuclear laboratories and consist only of radioactive isotopes.

In Chapter 3, we wrote symbols for the different isotopes of an element that had the mass number written in the upper left corner and the atomic number written in the lower left corner. Recall that the mass number is equal to the number of protons and neutrons in the nucleus, and atomic number is equal to the number of protons. For example, a radioactive isotope of iodine used in the diagnosis and treatment of thyroid conditions has a symbol with a mass number of 131 and an atomic number of 53, as shown below.

Mass number (protons and neutrons)
Element
Atomic number (protons)

$$^{131}_{53}\text{I}$$

Radioactive isotopes are named by writing the mass number after the element's name or symbol. This isotope is named iodine-131 or I-131. Table 9.1 compares some stable, nonradioactive isotopes with some radioactive isotopes.

**TABLE 9.1** Stable and Radioactive Isotopes of Some Elements

| Magnesium | Iodine | Uranium |
| --- | --- | --- |
| **Stable Isotopes** | | |
| $^{24}_{12}\text{Mg}$ | $^{127}_{53}\text{I}$ | None |
| Magnesium-24 | Iodine-127 | |
| **Radioactive Isotopes** | | |
| $^{23}_{12}\text{Mg}$ | $^{125}_{53}\text{I}$ | $^{235}_{92}\text{U}$ |
| Magnesium-23 | Iodine-125 | Uranium-235 |
| $^{27}_{12}\text{Mg}$ | $^{131}_{53}\text{I}$ | $^{238}_{92}\text{U}$ |
| Magnesium-27 | Iodine-131 | Uranium-238 |

## Types of Radiation

Different types of radiation are emitted from an unstable nucleus to form a more stable, lower energy nucleus. One type of radiation consists of alpha particles. An **alpha particle** is identical to a helium (He) nucleus, which has 2 protons and 2 neutrons. An alpha particle has a mass number of 4, an atomic number of 2, and a charge of 2+. The symbol for an alpha particle is the Greek letter alpha ($\alpha$) or the symbol of a helium nucleus, but the 2+ charge is omitted.

$^{4}_{2}\text{He}$

Alpha ($\alpha$) particle

A **beta particle** is an electron that is emitted when a neutron in an unstable nucleus changes to a proton and electron. A beta particle has a charge of 1– and a mass number of 0. It is represented by the Greek letter beta ($\beta$) or by the symbol for the electron ($e$) including the mass number and the charge.

$^{0}_{-1}e$     ●

Beta ($\beta$) particle

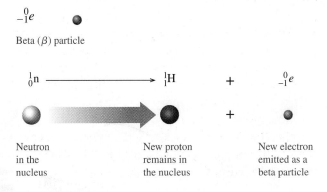

| $^{1}_{0}\text{n}$ | $\longrightarrow$ | $^{1}_{1}\text{H}$ | + | $^{0}_{-1}e$ |

Neutron in the nucleus    New proton remains in the nucleus    New electron emitted as a beta particle

A **positron** represented as $\beta^+$ has a positive (1+) charge with a mass number of 0, which makes it similar to a beta particle. We write the symbols of a beta particle and a positron as follows.

|  | **Electron** | **Positron** |
|---|---|---|
| Mass number<br>Charge | $_{-1}^{0}e$ | $_{+1}^{0}e$ |

A positron is produced by an unstable nucleus when a proton is transformed into a neutron and a positron.

$$_{1}^{1}\text{H} \longrightarrow \ _{0}^{1}\text{n} \ + \ _{+1}^{0}e \ (\text{or} \ \beta^+)$$

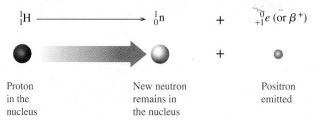

| Proton<br>in the<br>nucleus | + | New neutron<br>remains in<br>the nucleus | + | Positron<br>emitted |

A positron is an example of *antimatter*, a term physicists use to describe a particle that is the exact opposite of another particle, in this case, an electron. When an electron and a positron collide, their minute masses are completely converted to energy in the form of gamma rays.

$$_{-1}^{0}e \ + \ _{+1}^{0}e \ \longrightarrow \ 2_{0}^{0}\gamma$$

When the symbol $\beta$ is used with no charge, it is a beta particle rather than a positron.

**Gamma rays** are high-energy radiation, released when an unstable nucleus undergoes a rearrangement of its particles to give a more stable, lower energy nucleus. A gamma ray is written as the Greek letter gamma ($\gamma$). Because gamma rays are energy only, zeros are used to show that a gamma ray has no mass or charge.

$$_{0}^{0}\gamma$$

Gamma ($\gamma$) ray

Table 9.2 summarizes the types of radiation we will use in nuclear equations.

**TABLE 9.2** Some Common Forms of Radiation

| Type of Radiation | Symbol | | Mass Number | Charge |
|---|---|---|---|---|
| Alpha particle | $\alpha$ | $_{2}^{4}\text{He}$ | 4 | 2+ |
| Beta particle | $\beta$ | $_{-1}^{0}e$ | 0 | 1− |
| Positron | $\beta^+$ | $_{+1}^{0}e$ | 0 | 1+ |
| Gamma ray | $\gamma$ | $_{0}^{0}\gamma$ | 0 | 0 |
| Proton | $p$ | $_{1}^{1}\text{H}$ | 1 | 1+ |
| Neutron | $n$ | $_{0}^{1}\text{n}$ | 1 | 0 |

SAMPLE PROBLEM    9.1

■ **Writing Symbols for Radiation Particles**

Write the symbol for an alpha particle.

SOLUTION

The alpha particle contains 2 protons and 2 neutrons. It has a mass number of 4 and an atomic number of 2.

$$\alpha \quad \text{or} \quad _{2}^{4}\text{He}$$

STUDY CHECK

What is the symbol used for beta radiation?

When radiation strikes molecules in its path, electrons may be knocked away, forming unstable ions. For example, when radiation passes through the human body, it may interact with water molecules, removing electrons and producing $H_2O^+$, which can cause undesirable chemical reactions.

The cells most sensitive to radiation are the ones undergoing rapid division—those of the bone marrow, skin, reproductive organs, and intestinal lining, as well as all cells of growing children. Damaged cells may lose their ability to produce necessary materials. For example, if radiation damages cells of the bone marrow, red blood cells may no longer be produced. If sperm cells, ova, or the cells of a fetus are damaged, birth defects may result. In contrast, cells of the nerves, muscles, liver, and adult bones are much less sensitive to radiation because they undergo little or no cellular division.

Cancer cells are another example of rapidly dividing cells. Because cancer cells are highly sensitive to radiation, large doses of radiation are used to destroy them. The normal tissue that surrounds cancer cells divides at a slower rate and suffers less damage from radiation. However, radiation may cause malignant tumors, leukemia, anemia, and genetic mutations.

## Radiation Protection

It is essential that the radiologist, doctor, and nurse working with radioactive isotopes use proper radiation protection. Proper **shielding** is necessary to prevent exposure. Alpha particles, the heaviest of the radiation particles, travel only a few centimeters in the air before they collide with air molecules, acquire electrons, and become helium atoms. A piece of paper, clothing, and our skin are protection against alpha particles. Lab coats and gloves will also provide sufficient shielding. However, if ingested or inhaled, alpha particles can bring about serious internal damage because their mass and high charge causes much ionization in a short distance.

Beta particles move much faster and farther than alpha particles, traveling as much as several meters through air. They can pass through paper and penetrate as far as 4–5 mm into body tissue. External exposure to beta particles can burn the surface of the skin, but they are stopped before they reach the internal organs. Heavy clothing such as lab coats and gloves are needed to protect the skin from beta particles.

Gamma rays travel great distances through the air and pass through many materials, including body tissues. Because gamma rays can penetrate so deeply, exposure to these rays can be extremely hazardous. Only very dense shielding, such as lead or concrete, will stop them. Syringes used for injections of radioactive isotopes use shielding made of lead or heavy-weight materials such as tungsten and plastic composites.

When preparing radioactive materials, the radiologist wears special gloves and works behind lead glass windows. Long tongs are used within the work area to pick up vials of radioactive material, keeping them away from the hands and body. (See Figure 9.1.) Table 9.3 summarizes the shielding materials required for the various types of radiation.

Radiologists keep the time they spend in a radioactive area to a minimum. A certain amount of radiation is emitted every minute. Remaining in a radioactive area twice as long exposes a person to twice as much radiation.

**WEB TUTORIAL**
Radiation and Its Biological Effects

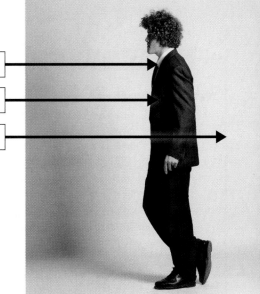

**TABLE 9.3** Properties of Ionizing Radiation and Shielding Required

| Property | Alpha (α) particle | Beta (β) particle | Gamma (γ) ray |
|---|---|---|---|
| Travel distance in air | 2–4 cm | 200–300 cm | 500 m |
| Tissue depth | 0.05 mm | 4–5 mm | 50 cm or more |
| Shielding | Paper, clothing | Heavy clothing, lab coats, gloves | Lead, thick concrete |
| Typical source | Radium-226 | Carbon-14 | Technetium-99m |

**FIGURE 9.1** A person working with radioisotopes wears protective clothing and gloves and stands behind a lead shield.

**Q** What types of radiation does the lead shield block?

Keep your distance! The greater the distance from the radioactive source, the lower the intensity of radiation received. If you double your distance from the radiation source, the intensity of radiation drops to $\left(\frac{1}{2}\right)^2$ or one-fourth of its previous value.

SAMPLE PROBLEM  9.2

■ Radiation Protection

How does the type of shielding for alpha radiation differ from that used for gamma radiation?

SOLUTION

Alpha radiation is stopped by paper and clothing. However, lead or concrete is needed for protection from gamma radiation.

STUDY CHECK

Besides shielding, what other methods help reduce exposure to radiation?

## QUESTIONS AND PROBLEMS

### Natural Radioactivity

9.1  a. How are an alpha particle and a helium nucleus similar?
   b. What symbols are used for alpha particles?
   c. What is the source of an alpha particle?

9.2  a. How are a beta particle and an electron similar?
   b. What symbols are used for beta particles?
   c. What is the source of a beta particle?

9.3  Naturally occurring potassium consists of three isotopes: potassium-39, potassium-40, and radioactive potassium-41.
   a. Write the atomic symbols for each isotope.
   b. In what ways are the isotopes similar, and in what ways do they differ?

9.4  Naturally occurring iodine is iodine-127. Medically, radioactive isotopes of iodine-125 and iodine-130 are used.
   a. Write the atomic symbols for each isotope.
   b. In what ways are the isotopes similar, and in what ways do they differ?

9.5  Supply the missing information in the following table:

| Medical Use | Atomic Symbol | Mass Number | Number of Protons | Number of Neutrons |
|---|---|---|---|---|
| Heart imaging | $^{201}_{81}\text{Tl}$ | | | |
| Radiation therapy | | 60 | 27 | |
| Abdominal scan | | | 31 | 36 |
| Hyperthyroidism | $^{131}_{53}\text{I}$ | | | |
| Leukemia treatment | | 32 | | 17 |

9.6  Supply the missing information in the following table:

| Medical Use | Atomic Symbol | Mass Number | Number of Protons | Number of Neutrons |
|---|---|---|---|---|
| Cancer treatment | $^{60}_{27}\text{Co}$ | | | |
| Brain scan | | 99 | 43 | |
| Blood flow | | 141 | 58 | |
| Bone scan | | 85 | | 47 |
| Lung function | $^{133}_{54}\text{Xe}$ | | | |

9.7  Write a symbol for each of the following:
   a. alpha particle       b. neutron
   c. beta particle        d. nitrogen-15
   e. iodine-125

9.8  Write a symbol for each of the following:
   a. proton               b. gamma ray
   c. electron             d. positron
   e. cobalt-60

9.9  Identify each of the following:
   a. $^{0}_{-1}X$         b. $^{4}_{2}X$
   c. $^{1}_{0}X$          d. $^{24}_{11}X$
   e. $^{14}_{6}X$

9.10 Identify each of the following:
   a. $^{1}_{1}X$    b. $^{32}_{15}X$    c. $^{0}_{0}X$
   d. $^{59}_{26}X$    e. $^{0}_{+1}X$

9.11 a. Why does beta radiation penetrate farther in solid material than alpha radiation?
   b. How does radiation cause damage to cells of the body?
   c. Why does the radiation technician leave the room when you receive an X-ray?
   d. What is the purpose of wearing gloves when handling radioactive isotopes?

9.12 a. As a nurse in an oncology unit, you may give an injection of a radioactive isotope. What are three ways you can minimize your exposure to radiation?
   b. Why are cancer cells more sensitive to radiation than nerve cells?
   c. What is the purpose of placing a lead apron on a patient who is receiving routine dental X-rays?
   d. Why are the walls in a radiology office built of thick concrete blocks?

# 9.2 NUCLEAR REACTIONS

When a nucleus spontaneously breaks down by emitting radiation, the process is called **radioactive decay**. It can be shown as a *nuclear equation* using the symbols for the original radioactive nucleus, the new nucleus, and the radiation emitted.

$$\text{Radioactive nucleus} \longrightarrow \text{new nucleus} + \text{radiation } (\alpha, \beta, \beta^+, \gamma)$$

In Chapter 4, we balanced chemical equations to give the same number of atoms in the reactants and products. In a nuclear equation, the mass numbers and the atomic numbers must balance so the number of protons and neutrons are equal on both sides. However, in a nuclear equation, there is often a change in the number of protons, which gives a different element.

The changes in mass number and atomic number of an unstable nucleus that undergoes radioactive decay are shown in Table 9.4.

**TABLE 9.4** Mass Number and Atomic Number Changes due to Radiation

| Decay Process | Radiation Symbol | Change in Mass Number | Change in Atomic Number | Change in Neutron Number |
|---|---|---|---|---|
| Alpha emission | $^{4}_{2}\text{He}$ | −4 | −2 | −2 |
| Beta emission | $^{0}_{-1}e$ | 0 | +1 | −1 |
| Positron emission | $^{0}_{+1}e$ | 0 | −1 | +1 |
| Gamma emission | $^{0}_{0}\gamma$ | 0 | 0 | 0 |

## Alpha Decay

An unstable nucleus undergoes alpha decay by emitting an alpha particle. Because an alpha particle consists of 2 protons and 2 neutrons, the mass number decreases by 4, and the atomic number decreases by 2. For example, uranium-238 emits an alpha particle to form a nucleus with a mass number of 234. Compared to uranium with 92 protons, the new nucleus has 90 protons, which makes it thorium.

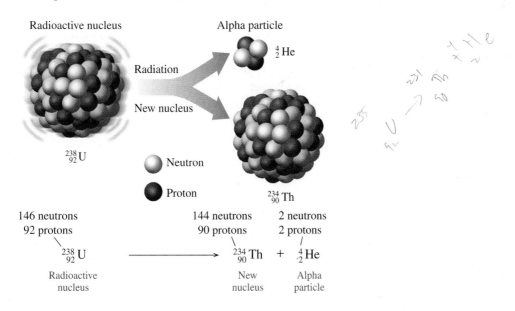

Radioactive nucleus

Alpha particle

$^{4}_{2}\text{He}$

Radiation

New nucleus

$^{238}_{92}\text{U}$

○ Neutron

● Proton

$^{234}_{90}\text{Th}$

| 146 neutrons | 144 neutrons | 2 neutrons |
| 92 protons | 90 protons | 2 protons |

$$^{238}_{92}\text{U} \longrightarrow \ ^{234}_{90}\text{Th} + \ ^{4}_{2}\text{He}$$

Radioactive nucleus        New nucleus        Alpha particle

## Guide to Completing a Nuclear Equation

In another example of radioactive decay, radium-226 emits an alpha particle to form a new isotope whose mass number, atomic number, and identity we must determine.

## LEARNING GOAL

Write an equation showing mass numbers and atomic numbers for radioactive decay.

**WEB TUTORIAL**
Radiation and Its Biological Effects

**Guide to Completing
a Nuclear Equation**

**1** Write the incomplete
nuclear equation.

**2** Determine the missing
mass number.

**3** Determine the missing
atomic number.

**4** Determine the symbol
of the new nucleus.

**5** Complete the
nuclear equation.

**Step 1** **Write the incomplete nuclear equation.**

$$^{226}_{88}\text{Ra} \longrightarrow ? + {}^{4}_{2}\text{He}$$

**Step 2** **Determine the missing mass number.** In the equation, the mass number, 226, of the radium is equal to the combined mass numbers of the alpha particle and the new nucleus.

$$226 \quad = ? + 4$$
$$226 - 4 = ?$$
$$222 \quad = ? \text{ (mass number of new nucleus)}$$

**Step 3** **Determine the missing atomic number.** The atomic number of radium, 88, must equal the sum of the atomic numbers of the alpha particle and the new nucleus.

$$88 \quad = ? + 2$$
$$88 - 2 = ?$$
$$86 \quad = ? \text{ (atomic number of new nucleus)}$$

**Step 4** **Determine the symbol of the new nucleus.** On the periodic table, the element that has atomic number 86 is radon, Rn. The nucleus of this isotope of Rn is written as

$$^{222}_{86}\text{Rn}$$

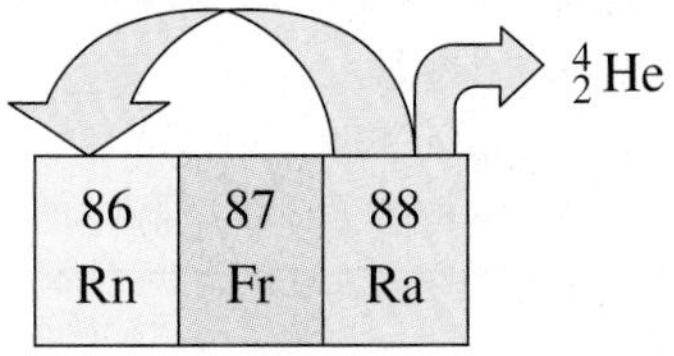

| 86 | 87 | 88 |
|----|----|----|
| Rn | Fr | Ra |

$${}^{4}_{2}\text{He}$$

**Step 5** **Complete the nuclear equation.**

$$^{226}_{88}\text{Ra} \longrightarrow {}^{222}_{86}\text{Rn} + {}^{4}_{2}\text{He}$$

In this nuclear reaction, a radium-226 nucleus decays by releasing an alpha particle and producing a radon-222 nucleus.

---

**SAMPLE PROBLEM 9.3**

### ■ Writing an Equation for Alpha Decay

Smoke detectors, required in homes and apartments, contain americium-241, which undergoes alpha decay. The alpha particles ionize air molecules, producing a constant stream of electrical current. However, when smoke particles enter the detector, they interfere with the formation of ions in the air, and the electric current is interrupted. This causes the alarm to sound, and it warns the occupants of the danger of fire. Complete the following nuclear equation for the decay of americium-241:

$$^{241}_{95}\text{Am} \longrightarrow ? + {}^{4}_{2}\text{He}$$

**SOLUTION**

**Step 1** **Write the incomplete nuclear equation.**

$$^{241}_{95}\text{Am} \longrightarrow ? + {}^{4}_{2}\text{He}$$

**Step 2** **Determine the missing mass number.** In the equation, the mass number, 241, of the americium is equal to the sum of the mass numbers of the alpha particle and the new nucleus.

$$241 \quad = ? + 4$$
$$241 - 4 = ?$$
$$237 \quad = ? \text{ (mass number of new nucleus)}$$

**Step 3** **Determine the missing atomic number.** The atomic number of americium-95 must equal the sum of the atomic numbers of the alpha particle and the new nucleus.

$$95 = ? + 2$$
$$95 - 2 = ?$$
$$93 = ? \text{ (atomic number of new nucleus)}$$

**Step 4** **Determine the symbol of the new nucleus.** On the periodic table, the element that has atomic number 93 is neptunium, Np. The symbol is written

$$^{237}_{93}\text{Np}.$$

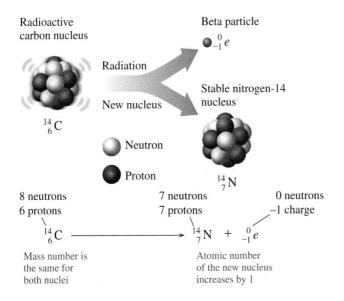

| 93 | 94 | 95 |
|----|----|----|
| Np | Pu | Am |

$$^{4}_{2}\text{He}$$

**Step 5** **Complete the nuclear equation.**

$$^{241}_{95}\text{Am} \longrightarrow {}^{237}_{93}\text{Np} + {}^{4}_{2}\text{He}$$

In this nuclear reaction, an Am-241 nucleus decays by releasing an alpha particle and producing a Np-237 nucleus.

STUDY CHECK

Write a balanced nuclear equation for the alpha decay of polonium Po-214.

## Beta Decay

When an unstable nucleus emits a beta particle, the newly formed proton increases the atomic number by 1, but the mass number stays the same. For example, when carbon-14 decays by beta emission, it becomes nitrogen-14.

Radioactive
carbon nucleus

Beta particle
• $^{0}_{-1}e$

Radiation

Stable nitrogen-14
nucleus

New nucleus

$^{14}_{6}\text{C}$

○ Neutron
● Proton

$^{14}_{7}\text{N}$

8 neutrons        7 neutrons        0 neutrons
6 protons         7 protons         −1 charge

$$^{14}_{6}\text{C} \longrightarrow {}^{14}_{7}\text{N} + {}^{0}_{-1}e$$

Mass number is          Atomic number
the same for            of the new nucleus
both nuclei             increases by 1

## Green Chemistry Note

### Radon in Our Homes

The presence of radon has become a much publicized environmental and health issue because of radiation danger. Radioactive isotopes such as uranium-238 and radium-226, are naturally present in many types of rocks and soils. Uranium-238 has been found in particularly high levels in an area between Pennsylvania and New England. When uranium-238 decays, it forms radium-226. Radium-226 emits an alpha particle and is converted into radon gas, which diffuses out of the rocks and soil.

$$^{226}_{88}\text{Ra} \longrightarrow {}^{222}_{86}\text{Rn} + {}^{4}_{2}\text{He}$$

Outdoors, radon gas poses little danger because it disperses in the air. However, if the radioactive source is under a house or building, the radon gas can enter the house through cracks in the foundation or other openings. Then the radon gas may be inhaled by those living or working there. Inside the lungs, radon-222 emits alpha particles to form polonium-218, which is known to cause lung cancer.

$$^{222}_{86}\text{Rn} \longrightarrow {}^{218}_{84}\text{Po} + {}^{4}_{2}\text{He}$$

Some researchers have estimated that 10% of all lung cancer deaths in the United States are due to radon. The Environmental Protection Agency (EPA) recommends that the maximum level of radon not exceed 4 picocuries (pCi) per liter of air in a home. One (1) picocurie (pCi) is equal to $10^{-12}$ curies (Ci); curies are described in Section 9.3. In California, 1% of all the houses surveyed exceeded the EPA's recommended maximum radon level.

## Beta Emitters in Medicine

The radioactive isotopes of several biologically important elements are beta emitters. When a radiologist wants to treat a malignancy within the body, a beta emitter may be used. The short range of penetration into the tissue by beta particles is advantageous for certain conditions. For example, some malignant tumors increase the fluid within the body tissues. A compound containing phosphorus-32, a beta emitter, is injected into the body cavity where the tumor is located. The beta particles travel only a few millimeters through the tissue, so only the malignancy and any tissue within that range are affected. The growth of the tumor is slowed or stopped, and the production of fluid decreases. Phosphorus-32 is also used to treat leukemia, polycythemia vera (excessive production of red blood cells), and lymphomas.

$$^{32}_{15}P \longrightarrow {}^{32}_{16}S + {}^{0}_{-1}e$$

Another beta emitter, iron-59, is used in blood tests to determine the level of iron in the blood and the rate of production of red blood cells by the bone marrow.

$$^{59}_{26}Fe \longrightarrow {}^{59}_{27}Co + {}^{0}_{-1}e$$

### ■ Writing an Equation for Beta Decay

Write the nuclear equation for the beta decay of cobalt-60.

SOLUTION

**Step 1** **Write the incomplete nuclear equation.**

$$^{60}_{27}Co \longrightarrow ? + {}^{0}_{-1}e$$

**Step 2** **Determine the missing mass number.** In the equation, the mass number of cobalt-60 is equal to the sum of the mass numbers of the beta particle and the new nucleus.

$$60 \quad = ? + 0$$
$$60 - 0 = ?$$
$$60 \quad = ? \text{ (mass number of new nucleus)}$$

**Step 3** **Determine the missing atomic number.** The atomic number of cobalt-60 must equal the sum of the atomic number of the beta particle and the new nucleus.

$$27 \quad = ? - 1$$
$$27 + 1 = ?$$
$$28 \quad = ? \text{ (atomic number of new nucleus)}$$

**Step 4** **Determine the symbol of the new nucleus.** On the periodic table, the element that has atomic number 28 is nickel (Ni). The symbol of this isotope is

$$^{60}_{28}Ni$$

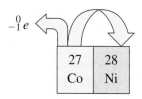

**Step 5** **Complete the nuclear equation.**

$$^{60}_{27}Co \longrightarrow {}^{60}_{28}Ni + {}^{0}_{-1}e$$

In this nuclear reaction, cobalt-60 undergoes beta decay to produce nickel-60.

STUDY CHECK

Write the nuclear equation for iodine-131, a beta emitter.

## Positron Emission

When a radioactive isotope emits a positron, the mass number does not change. However, the atomic number of the new nucleus decreases by 1. For example, manganese-49 undergoes positron emission to produce chromium-49. The atomic number of chromium (24) and the charge of the positron ($+1$) added together give the atomic number of manganese (25).

$$^{49}_{25}Mn \longrightarrow {}^{49}_{24}Cr + {}^{0}_{+1}e$$

# Gamma Emission

There are very few pure gamma emitters, although gamma radiation accompanies most alpha and beta radiation. In radiology, one of the most commonly used gamma emitters is technetium (Tc). The symbol $m$ is used to indicate an unstable or metastable isotope, shown here as technetium-99m, Tc-99m, or $^{99m}_{43}\text{Tc}$. By emitting energy in the form of gamma rays, the unstable nucleus becomes more stable.

$$^{99m}_{43}\text{Tc} \longrightarrow ^{99}_{43}\text{Tc} + ^{0}_{0}\gamma$$

Figure 9.2 summarizes the changes in the nucleus for alpha, beta, positron, and gamma radiation.

# Producing Radioactive Isotopes

Today, many radioisotopes are produced in small amounts by converting stable, nonradioactive isotopes into radioactive ones. In a process called *transmutation*, a stable nucleus is bombarded by high-speed particles such as alpha particles, protons, neutrons, and small nuclei. When one of these particles is absorbed, the nucleus becomes a radioactive isotope.

When boron-10, a nonradioactive isotope is bombarded by an alpha particle, it is converted to nitrogen-13, and a neutron is emitted.

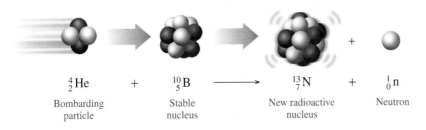

$$\underset{\text{Bombarding particle}}{^{4}_{2}\text{He}} + \underset{\text{Stable nucleus}}{^{10}_{5}\text{B}} \longrightarrow \underset{\text{New radioactive nucleus}}{^{13}_{7}\text{N}} + \underset{\text{Neutron}}{^{1}_{0}\text{n}}$$

All elements that have an atomic number greater than 92 have been produced by bombardment; none of these elements occurs naturally. Most have been produced in small amounts and exist for such a short time that it is difficult to study their properties. An example is element 105, dubnium (Db), which is produced when californium-249 is bombarded with nitrogen-15.

$$^{249}_{98}\text{Cf} + ^{15}_{7}\text{N} \longrightarrow ^{260}_{105}\text{Db} + 4^{1}_{0}\text{n}$$

Technetium-99m is a radioisotope used in nuclear medicine for several diagnostic procedures, including the detection of brain tumors and examinations of the liver and spleen. The source of technetium-99m is molybdenum-99, which is produced in a nuclear reactor by neutron bombardment of molybdenum-98.

$$^{98}_{42}\text{Mo} + ^{1}_{0}\text{n} \longrightarrow ^{99}_{42}\text{Mo}$$

Many radiology laboratories have a small generator containing molybdenum-99, which decays to give technetium-99m.

$$^{99}_{42}\text{Mo} \longrightarrow ^{99m}_{43}\text{Tc} + ^{0}_{-1}e$$

The technetium-99m radioisotope decays by emitting gamma rays. Gamma emission is desirable for diagnostic work because the gamma rays pass through the body to the detection equipment.

$$^{99m}_{43}\text{Tc} \longrightarrow ^{99}_{43}\text{Tc} + ^{0}_{0}\gamma$$

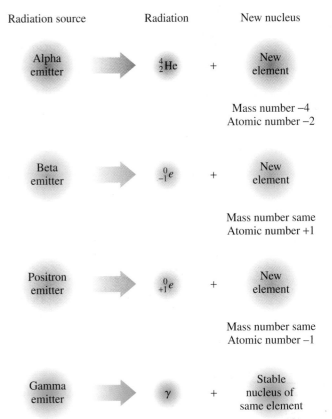

| Radiation source | Radiation | New nucleus |
|---|---|---|
| Alpha emitter | $^{4}_{2}\text{He}$ + | New element |
| | | Mass number −4 Atomic number −2 |
| Beta emitter | $^{0}_{-1}e$ + | New element |
| | | Mass number same Atomic number +1 |
| Positron emitter | $^{0}_{+1}e$ + | New element |
| | | Mass number same Atomic number −1 |
| Gamma emitter | $\gamma$ + | Stable nucleus of same element |
| | | Mass number same Atomic number same |

**FIGURE 9.2** When the nuclei of alpha, beta, positron, and gamma emitters emit radiation, new and more stable nuclei are produced.
**Q** What changes occur in the number of protons and neutrons when an alpha emitter gives off radiation?

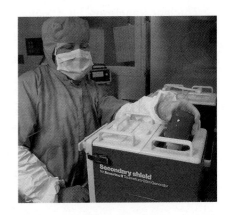

## SAMPLE PROBLEM 9.5

### ■ Producing Radioactive Isotopes

Write the equation when zinc-66 absorbs one proton $\left(^{1}_{1}H\right)$ during bombardment to form a radioactive isotope.

SOLUTION

**Step 1** **Write the incomplete nuclear equation.**

$$^{66}_{30}Zn + ^{1}_{1}H \longrightarrow ?$$

**Step 2** **Determine the missing mass number.** In the equation, the sum of the mass numbers of zinc (66) and hydrogen (1) must equal the mass number of the new nucleus.

$$66 + 1 = ?$$
$$67\ \ \ \ = ? \text{ (mass number of new nucleus)}$$

**Step 3** **Determine the missing atomic number.** The sum of the atomic numbers of zinc (30) and hydrogen (1) must equal the atomic number of the new nucleus.

$$30 + 1 = ?$$
$$31\ \ \ \ = ? \text{ (atomic number of new nucleus)}$$

**Step 4** **Determine the symbol of the new nucleus.** On the periodic table, the element that has atomic number 31 is gallium, Ga. The symbol of this isotope is

$$^{67}_{31}Ga$$

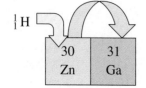

**Step 5** **Complete the nuclear equation.**

$$^{66}_{30}Zn + ^{1}_{1}H \longrightarrow ^{67}_{31}Ga$$

STUDY CHECK

The first radioactive isotope was produced in 1934 by the bombardment of aluminum-27 by an alpha particle to produce a radioactive isotope and one neutron. What is the balanced nuclear equation for this transmutation?

---

## QUESTIONS AND PROBLEMS

### Nuclear Reactions

**9.13** Write a balanced nuclear equation for the alpha decay of each of the following:
  **a.** $^{208}_{84}Po$    **b.** $^{232}_{90}Th$    **c.** $^{251}_{102}No$    **d.** $^{220}_{86}Rn$

**9.14** Write a balanced nuclear equation for the alpha decay of each of the following:
  **a.** $^{243}_{96}Cm$    **b.** $^{252}_{99}Es$    **c.** $^{251}_{98}Cf$    **d.** $^{261}_{107}Bh$

**9.15** Write a balanced nuclear equation for the beta decay of each of the following:
  **a.** $^{25}_{11}Na$    **b.** $^{20}_{8}O$
  **c.** strontium-92    **d.** potassium-42

**9.16** Write a balanced nuclear equation for the beta decay of each of the following:
  **a.** $^{42}_{19}K$
  **b.** iron-59
  **c.** iron-60
  **d.** $^{141}_{56}Ba$

**9.17** Complete each of the following nuclear equations, and describe the type of radiation:
  **a.** $^{28}_{13}Al \longrightarrow ? + ^{0}_{-1}e$
  **b.** $^{135}_{60}Nd \longrightarrow ^{135}_{59}Pr + ?$
  **c.** $^{66}_{29}Cu \longrightarrow ^{66}_{30}Zn + ?$
  **d.** $^{228}_{90}Th \longrightarrow ^{224}_{88}Ra + ?$

**9.18** Complete each of the following nuclear equations, and describe the type of radiation:
  **a.** $^{11}_{6}C \longrightarrow ^{7}_{4}Be + ?$
  **b.** $^{35}_{16}S \longrightarrow ? + ^{0}_{-1}e$
  **c.** $? \longrightarrow ^{90}_{39}Y + ^{0}_{-1}e$
  **d.** $^{20}_{12}Mg \longrightarrow ? + ^{0}_{+1}e$

**9.19** Complete each of the following bombardment reactions:
  **a.** $^{9}_{4}Be + ^{1}_{0}n \longrightarrow ?$
  **b.** $^{32}_{16}S + ? \longrightarrow ^{32}_{15}P$

  **c.** $? + ^{1}_{0}n \longrightarrow ^{24}_{11}Na + ^{4}_{2}He$
  **d.** $^{27}_{13}Al + ^{4}_{2}He \longrightarrow ? + ^{1}_{0}n$

**9.20** Complete each of the following bombardment reactions:
  **a.** $^{40}_{18}Ar + ? \longrightarrow ^{43}_{19}K + ^{1}_{1}H$
  **b.** $^{238}_{92}U + ^{1}_{0}n \longrightarrow ?$
  **c.** $? + ^{1}_{0}n \longrightarrow ^{14}_{6}C + ^{1}_{1}H$
  **d.** $? + ^{64}_{28}Ni \longrightarrow ^{272}_{111}Rg + ^{1}_{0}n$

# 9.3  RADIATION MEASUREMENT

**LEARNING GOAL**

Describe the detection and measurement of radiation.

One of the most common instruments for detecting beta and gamma radiation is the Geiger counter. It consists of a metal tube filled with a gas such as argon. When radiation enters a window on the end of the tube, argon atoms form ions, which produce an electrical current. Each burst of current is amplified to give a click and a reading on a meter.

$$Ar + radiation \longrightarrow Ar^+ + e^-$$

Radiation is measured in several different ways. We can measure the activity of a radioactive sample or determine the impact of radiation on biological tissue.

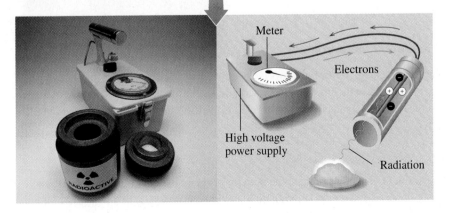

## Measuring Radiation

When a radiology laboratory obtains a radioisotope, the *activity* of the sample is measured in terms of the number of nuclear disintegrations per second. The **curie (Ci)**, the original unit of activity, was defined as the number of disintegrations that occur in

WEB TUTORIAL
Nuclear Chemistry

1 second for 1 g of radium, which is equal to $3.7 \times 10^{10}$ disintegrations per second. The curie was named for Marie Curie, a Polish scientist, who along with her husband, Pierre, discovered the radioactive elements radium and polonium. A newer unit of radiation activity is the **becquerel (Bq)**, which is one disintegration per second.

After we measure the activity of a radioisotope, we often want to know how much radiation the tissues in the body absorb. The **rad (radiation absorbed dose)** is a unit that measures the amount of radiation absorbed by a gram of material such as body tissue. The newer unit for absorbed dose is the **gray (Gy)**, which is equal to 100 rads.

The **rem (radiation equivalent in humans)** measures the biological effects of different kinds of radiation. Alpha particles do not penetrate the skin. But if they should enter the body by some other route, they cause a lot of damage even though the particles travel a short distance in tissue. High-energy radiation such as beta particles and high-energy protons and neutrons that penetrate the skin and travel into tissue cause more damage. Gamma rays are damaging because they travel a long way through tissue and create a great deal of ionization.

To determine the **equivalent dose** or rem dose, the absorbed dose (rads) is multiplied by a factor that adjusts for biological damage caused by a particular form of radiation. For beta and gamma radiation the factor is 1, so the biological damage in rems is the same as the absorbed radiation (rads). For high-energy protons and neutrons, the factor is about 10, and for alpha particles it is 20.

Biological damage (rem) = absorbed dose (rad) × factor

Often the measurement for an equivalent dose is in units of millirems (mrem). One rem is equal to 1000 mrem. The newer unit is the **sievert (Sv)**. One sievert is equal to 100 rems. Table 9.5 summarizes the units used to measure radiation.

CASE STUDY
Food Irradiation

**TABLE 9.5** Some Units of Radiation Measurement

| Measurement | Common Unit | SI Unit |
|---|---|---|
| Activity | curie (Ci) = $3.7 \times 10^{10}$ disintegrations/s | becquerel (Bq) = 1 disintegration/s |
| Absorbed dose | rad | gray (Gy) |
| Biological damage | rem = rad × factor | sievert (Sv) |

SAMPLE PROBLEM 9.6

■ **Radiation Measurement**

One treatment of bone pain involves intravenous administration of the radioisotope phosphorus-32, which is primarily incorporated into bone. A typical dose of 7 mCi can produce up to 450 rads in the bone. What is the difference between the units of mCi and rads?

SOLUTION

The millicuries (mCi) indicate the activity of the phosphorus-32 in terms of nuclei that break down in 1 second. The radiation absorbed dose (rads) is a measure of amount of radiation absorbed by the bone and the tissue.

STUDY CHECK

If phosphorus-32 is a beta emitter, how do the number of rems compare to the rads?

## Health Note

### Radiation and Food

Food-borne illnesses caused by pathogenic bacteria such as *Salmonella, Listeria*, and *Escherichia coli* have become a major health concern in the United States. The Centers for Disease Control and Prevention estimates that each year *E. coli* in contaminated foods infects 20 000 people in the United States, and that 500 people die. *E. coli* has been responsible for outbreaks of illness from contaminated ground beef, fruit juices, lettuce, spinach, and alfalfa sprouts.

The Food and Drug Administration (FDA) has approved the use of 0.3 kilogray (0.3 kGy) to 1 kGy of ionizing radiation produced by cobalt-60 or cesium-137 for the treatment of foods. The irradiation technology is much like that used to sterilize medical supplies. Cobalt pellets are placed in stainless steel tubes, which are arranged in racks. When food moves through the series of racks, the gamma rays pass through the food and kill the bacteria.

It is important for consumers to understand that when food is irradiated, it never comes in contact with the radioactive source. The gamma rays pass through the food to kill bacteria, but that does not make the food radioactive. The radiation kills bacteria because it stops their ability to divide and grow. We cook or heat food thoroughly for the same purpose. Radiation, as well as heat, has little effect on the food itself because its cells are no longer dividing or growing. Thus, irradiated food is not harmed although a small amount of vitamin B$_1$ and C may be lost.

Currently, tomatoes, blueberries, strawberries, and mushrooms are irradiated to allow them to be harvested when completely ripe and extend their shelf life. (See Figure 9.3.) The FDA has also approved the irradiation of pork, poultry, and beef in order to decrease potential infections and to extend shelf life. Currently, irradiated vegetable and meat products are available in retail markets in South Africa. Apollo 17 astronauts ate irradiated foods on the moon, and some U.S. hospitals and nursing homes now use irradiated poultry to reduce the possibility of infections among patients. The extended shelf life of irradiated food also makes it useful for campers and military personnel. Soon consumers concerned about food safety will have a choice of irradiated meats, fruits, and vegetables at the market.

**(a)**

**(b)**

**FIGURE 9.3** **(a)** The FDA requires this symbol to appear on irradiated retail foods. **(b)** After 2 weeks, the irradiated strawberries on the right show no spoilage. Mold is growing on the nonirradiated ones on the left.
**Q** Why are irradiated foods used on spaceships and in nursing homes?

## Exposure to Radiation

Every day, we are exposed to low levels of *background* radiation from naturally occurring radioactive isotopes in the buildings where we live and work, in our food and water, and in the air we breathe. For example, potassium-40 is a naturally occurring isotope that is present in any potassium-containing food. Other naturally occurring radioisotopes in air and food are carbon-14, radon-222, strontium-90, and iodine-131. The average person in the United States is exposed to about 360 mrem of radiation annually. Table 9.6 lists some common sources of radiation.

Another source of background radiation is cosmic radiation produced in space by the sun. People who live at high altitudes or travel by airplane receive a greater amount of cosmic radiation because there are fewer molecules in the atmosphere to absorb the radiation. For example, a person living in Denver receives about twice the cosmic radiation as a person living in Los Angeles.

A person living close to a nuclear power plant normally does not receive much additional radiation, perhaps 0.1 millirem (mrem) in 1 year. (One rem equals 1000 mrem.) However, in the accident at the Chernobyl nuclear power plant in 1986, it is estimated that people in a nearby town received as much as 1 rem/h.

Medical sources of radiation including dental, hip, spine and chest X-rays, and mammograms add to our radiation exposure.

**TABLE 9.6** Average Annual Radiation Received by a Person in the United States

| Source | Dose (mrem) |
| --- | --- |
| **Natural** | |
| The ground | 20 |
| Air, water, food | 30 |
| Cosmic rays | 40 |
| Wood, concrete, brick | 50 |
| **Medical** | |
| Chest X-ray | 20 |
| Dental X-ray | 20 |
| Hip X-ray | 60 |
| Lumbar spine X-ray | 70 |
| Mammogram | 40 |
| Upper gastrointestinal tract X-ray | 200 |
| **Other** | |
| Television | 20 |
| Air travel | 10 |
| Radon | 200[a] |

[a]Varies widely.

# Health Note

## Brachytherapy

The process called brachytherapy, or seed implantation, is an internal form of radiation therapy. The prefix *brachy* is from the Greek word for short distance. With internal radiation, a high dose of radiation is delivered to a cancerous area, while normal tissue sustains minimal damage. Because higher doses are used, fewer treatments of shorter duration are needed. Conventional external treatment delivers a lower dose per treatment, but requires 6 to 8 weeks of treatments.

### PERMANENT BRACHYTHERAPY

One of the most common forms of cancer in males is prostate cancer. In addition to surgery and chemotherapy, one treatment option is to place 40 or more titanium capsules or "seeds" in the malignant area. Each seed, which is the size of a small grain of rice, contains radioactive iodine-125, palladium-103, or cesium-131. The radiation from the seeds destroys the cancer by interfering with the reproduction of cancer cells. Because the radiation targets the cancer cells, there is minimal damage to normal tissues. Ninety percent (90%) of the radioisotopes decay within a few months because they have short half-lives.

| Isotope | I-125 | Pd-103 | Cs-131 |
|---|---|---|---|
| **Half-life** | 60 days | 17 days | 10 days |
| **Time to deliver 90% of radiation** | 7 months | 2 months | 1 month |

Very little radiation passes out of the patient's body. The amount of radiation received by a family member is no greater than that received on a long plane flight. The titanium capsules are left in the body permanently, but the products of decay are not radioactive and cause no further damage.

### TEMPORARY BRACHYTHERAPY

In another type of treatment for prostate cancer, long needles containing iridium-192 are placed in the tumor. However, the needles are removed after 5 to 10 minutes depending on the activity of the iridium isotope. Compared to permanent brachytherapy, temporary brachytherapy can deliver a higher dose of radiation over a shorter time. The procedure may be repeated in a few days.

Brachytherapy is also used following breast cancer lumpectomy. An iridium-192 isotope is inserted into the catheter implanted in the space left by the removal of the tumor. The isotope is removed after 5 to 10 minutes depending on the activity of the iridium source. Radiation is delivered primarily to the tissue surrounding the cavity that contained the tumor and where the cancer is most likely to reoccur. The procedure is repeated twice a day for 5 days to give an absorbed dose of 34 Gy (3400 rads). The catheter is removed and no radioactive material remains in the body.

In conventional external beam therapy for breast cancer, a patient receives 2 Gy/treatment once a day for 35 days or about 7 weeks, which gives a total absorbed dose of about 70 Gy or 7000 rads. The external beam therapy irradiates the entire breast including the tumor cavity.

## Radiation Sickness

The larger the dose of radiation received at one time, the greater the effect on the body. Exposure to radiation below 25 rem usually cannot be detected. Whole-body exposure of 100 rem produces a temporary decrease in the number of white blood cells. If the exposure to radiation is 100 rem or higher, the person suffers the symptoms of radiation sickness: nausea, vomiting, fatigue, and a reduction in white-cell count. A whole-body dosage greater than 300 rem can decrease the white-cell count to zero. The victim suffers diarrhea, hair loss, and infection. Exposure to radiation of about 500 rem is expected to cause death in 50% of the people receiving that dose. This amount of radiation is called the lethal dose for one-half the population, or the $LD_{50}$. The $LD_{50}$ varies for different life-forms, as Table 9.7 shows. Radiation dosages of about 600 rem would be fatal to all humans within a few weeks.

**TABLE 9.7** Lethal Doses of Radiation for Some Life-Forms

| Life-Form | $LD_{50}$ (rem) |
|---|---|
| Insect | 100 000 |
| Bacterium | 50 000 |
| Rat | 800 |
| Human | 500 |
| Dog | 300 |

## QUESTIONS AND PROBLEMS

### Radiation Measurement

**9.21 a.** How does a Geiger counter detect radiation?
 **b.** What are the SI unit and the older unit that describe the activity of a radioactive sample?
 **c.** What are the SI unit and the older unit that describe the radiation dose absorbed by tissue?
 **d.** What is meant by the term kilogray?

**9.22 a.** What is background radiation?
 **b.** What are the SI unit and the older unit that describe the biological effect of radiation?
 **c.** What is meant by the terms mCi and mrem?
 **d.** Why is a factor used to determine the dose equivalent?

**9.23** The recommended dosage of iodine-131 is 4.20 $\mu$Ci/kg of body weight. How many microcuries of iodine-131 are needed for a 70.0-kg patient with hyper-thyroidism?

**9.24 a.** The dosage of technetium-99m for a lung scan is 20 $\mu$Ci/kg of body weight. How many millicuries should be given to a 50.0-kg patient? (1 mCi = 1000 $\mu$Ci)
 **b.** Suppose a person absorbed 50 mrads of alpha radiation. What would be the dose equivalent in mrems?

**9.25** Why would an airline pilot be exposed to more background radiation than the person who works at the ticket counter?

**9.26** In radiation therapy, a patient receives high doses of radiation. What symptoms of radiation sickness might the patient exhibit?

# 9.4 HALF-LIFE OF A RADIOISOTOPE

**LEARNING GOAL**

Given the half-life of a radioisotope, calculate the amount of radioisotope remaining after one or more half-lives.

The **half-life** of a radioisotope is the amount of time it takes for one-half of a sample to decay. For example, $^{131}_{53}I$ has a half-life of 8.0 days. As $^{131}_{53}I$ decays, it produces a beta particle and the nonradioactive isotope $^{131}Xe$.

$$^{131}_{53}I \longrightarrow ^{131}_{54}Xe + ^{0}_{-1}\beta$$

Suppose we have a sample that contains 20.0 g of $^{131}_{53}I$. We do not know which specific nucleus will emit radiation, but we do know that in 8 days, one-half of all the nuclei in the sample will decay to give $^{131}_{54}Xe$. That means that after 8.0 days, there are one-half of the number of $^{131}_{53}I$ atoms in the sample, or 10. g of $^{131}_{53}I$, remaining. The decay process has also produced 10. g of the product $^{131}_{54}Xe$. After another half-life or 8.0 days passes, 5.0 g of the 10. g of $^{131}_{53}I$ will decay to $^{131}_{54}Xe$. Now there are 5.0 g of $^{131}_{53}I$ left, while there is a total of 15 g of $^{131}_{54}Xe$. A third half-life, or another 8.0 days, results in 2.5 g of the $^{131}_{53}I$ decaying to give $^{131}_{54}Xe$, which leaves 2.5 g of $^{131}_{53}I$ still capable of producing radiation.

**WEB TUTORIAL**
Nuclear Chemistry

20.0 g of $^{131}_{53}I$ $\xrightarrow{\text{1 half-life}}$ 10.0 g of $^{131}_{53}I$ $\xrightarrow{\text{2 half-lives}}$ 5.0 g of $^{131}_{53}I$ $\xrightarrow{\text{3 half-lives}}$ 2.5 g of $^{131}_{53}I$

| $^{131}_{53}I$ | | | |
|---|---|---|---|
| | $^{131}_{54}Xe$ | $^{131}_{54}Xe$ | $^{131}_{54}Xe$ |

1 Half-life     2 Half-lives     3 Half-lives

10. g
$^{131}_{53}I$

15 g

$^{131}_{53}I$

17.5 g
$^{131}_{53}I$

20. g        10. g        5.0 g        2.5 g

A **decay curve** is a diagram of the decay of a radioactive isotope. Figure 9.4 shows such a curve for the $^{131}_{53}I$ we have discussed. This information can be summarized in Table 9.8.

**FIGURE 9.4** The decay curve for iodine-131 shows that one-half of the radioactive sample decays and one-half remains radioactive after each half-life of 8 days.

**Q** How many grams of the 20-g sample remain radioactive after 2 half-lives?

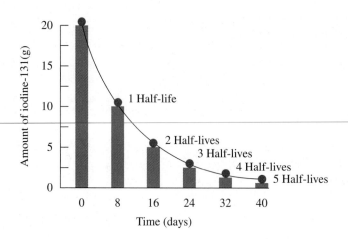

**TABLE 9.8** Activity of an $^{131}_{53}I$ Sample with Time

| Time elapsed | 0 days | 8.0 days | 16 days | 24 days |
|---|---|---|---|---|
| Half-lives | 0 | 1 | 2 | 3 |
| $^{131}_{53}I$ remaining | 20. g | 10. g | 5.0 g | 2.5 g |
| $^{131}_{54}Xe$ produced | 0 g | 10. g | 15 g | 17.5 g |

---

**SAMPLE PROBLEM  9.7**

■ **Using Half-Lives of a Radioisotope**

Phosphorus-32, a radioisotope used in the treatment of leukemia, has a half-life of 14 days. If a sample contains 8.0 g of phosphorus-32, how many grams of phosphorus-32 remain after 42 days?

SOLUTION

Step 1  **Given**  8.0 g of $^{32}_{15}P$; 42 days; 14 days/half-life
      **Need**  g of $^{32}_{15}P$ remaining

Step 2  **Plan**

42 days  →  Half-life  →  Number of half-lives

8.0 g of $^{32}_{15}P$  →  Number of half-lives  →  g of $^{32}_{15}P$ remaining

Step 3  **Equalities/Conversion Factors**

$$1 \text{ half-life} = 14 \text{ days}$$

$$\frac{14 \text{ days}}{1 \text{ half-life}} \quad \text{and} \quad \frac{1 \text{ half-life}}{14 \text{ days}}$$

Step 4  **Set Up Problem**  We can do this problem with two calculations. First, we determine the number of half-lives in the amount of time that has elapsed.

$$\text{Number of half-lives} = 42 \text{ days} \times \frac{1 \text{ half-life}}{14 \text{ days}} = 3 \text{ half-lives}$$

Now, we determine how much of the sample decays in 3 half-lives and how many grams of the phosphorus remain.

$$8.0 \text{ g of } ^{32}_{15}P \xrightarrow{\text{1 half-life}} 4.0 \text{ g of } ^{32}_{15}P \xrightarrow{\text{2 half-lives}} 2.0 \text{ g of } ^{32}_{15}P \xrightarrow{\text{3 half-lives}} 1.0 \text{ g of } ^{32}_{15}P$$

**STUDY CHECK**

Iron-59 has a half-life of 46 days. If the laboratory received a sample of 8.0 g iron-59, how many grams are still active after 184 days?

Naturally occurring isotopes of the elements usually have long half-lives, as shown in Table 9.9. They disintegrate slowly and produce radiation over a long period of time, even hundreds or millions of years. In contrast, many of the radioisotopes used in nuclear medicine have much shorter half-lives. They disintegrate rapidly and produce almost all their radiation in a short period of time. For example, technetium-99m emits half of its radiation in the first 6 h. This means that a small amount of the radioisotope given to a patient is essentially gone within 2 days. The decay products of technetium-99m are totally eliminated from the body.

**TABLE 9.9** Half-Lives of Some Radioisotopes

| Element | Radioisotope | Half-Life |
|---------|--------------|-----------|
| **Naturally Occurring Radioisotopes** | | |
| Carbon | $^{14}_{6}C$ | 5730 yr |
| Potassium | $^{40}_{19}K$ | $1.3 \times 10^9$ yr |
| Radium | $^{226}_{88}Ra$ | 1600 yr |
| Uranium | $^{238}_{92}U$ | $4.5 \times 10^9$ yr |
| **Some Medical Radioisotopes** | | |
| Chromium | $^{51}_{24}Cr$ | 28 days |
| Iodine | $^{131}_{53}I$ | 8 days |
| Iron | $^{59}_{26}Fe$ | 46 days |
| Technetium | $^{99m}_{43}Tc$ | 6.0 h |

*Explore Your World*

**Modeling Half-Lives**

Obtain a piece of paper and a licorice stick or celery stalk. Draw a vertical and a horizontal axis on the paper. Label the vertical axis as radioactive atoms and the horizontal axis as minutes. Place the licorice stick or celery against the vertical axis and mark its height for zero minutes. In the next minute, cut the licorice stick or celery in two. (You can eat the half if you are hungry.) Place the shortened licorice stick or celery at 1 minute on the horizontal axis and mark its height. Every minute cut the licorice stick or celery in half again and mark the shorter height at the corresponding time. Keep reducing the length by half until you cannot divide the licorice or celery in half any more. Connect the points you made for each minute. What does the curve look like? How does this curve represent the concept of a half-life for a radioisotope?

---

**SAMPLE PROBLEM 9.8**

■ **Dating Using Half-Lives**

In Los Angeles, the remains of ancient animals have been unearthed at La Brea tar pits. Suppose a bone sample from the tar pits is subjected to the carbon-14 dating method. If the sample shows that two half-lives have passed, when did the animal live?

**SOLUTION**

We can calculate the age of the bone sample by using the half-life of carbon-14 (5730 years).

$$2 \text{ half-lives} \times \frac{5730 \text{ yr}}{1 \text{ half-life}} = 11\,500 \text{ yr}$$

We would estimate that the animal lived 11 500 years ago, or about 9500 B.C.E.

**STUDY CHECK**

Suppose that a piece of wood found in a tomb had $\frac{1}{8}$ (3 half-lives) of its original carbon-14 activity. About how many years ago was the wood part of a living tree?

# *Environmental Note*

## Dating Ancient Objects

A technique known as radiological dating is used by geologists, archaeologists, and historians as a way to determine the age of ancient objects. The age of an object derived from plants or animals (such as wood, fiber, natural pigments, bone, and cotton or woolen clothing) is determined by measuring the amount of carbon-14, a naturally occurring radioactive form of carbon. In 1960, Willard Libby received the Nobel Prize in Chemistry for the work he did developing carbon-14 dating techniques during the 1940s. Carbon-14 is produced in the upper atmosphere by the bombardment of $^{14}_{7}N$ by high-energy neutrons from cosmic rays.

$$^{1}_{0}n \quad + \quad ^{14}_{7}N \quad \longrightarrow \quad ^{14}_{6}C \quad + \quad ^{1}_{1}H$$

Neutron from cosmic rays    Nitrogen in atmosphere    Radioactive carbon-14    Proton

The carbon-14 reacts with oxygen to form radioactive carbon dioxide: $^{14}CO_2$. Carbon dioxide is continuously absorbed by living plants, incorporating carbon-14 into the plant material. The uptake of carbon-14 stops when the plant dies.

$$^{14}_{6}C \longrightarrow ^{14}_{7}N + ^{0}_{-1}e$$

As the carbon-14 decays, the amount of radioactive carbon-14 in the plant material steadily decreases. In a process called **carbon dating**, scientists use the half-life of carbon-14 (5730 years) to calculate the length of time since the plant died. As the plant material ages, there is less radioactive carbon-14 remaining, and the approximate age of the sample can be determined. For example, a wooden beam found in an ancient Indian dwelling might have one-half of the carbon-14 found in living plants. Because one half-life of carbon-14 is 5730 years, the dwelling was constructed about 5730 years ago. Carbon-14 dating was used to determine that the Dead Sea Scrolls are about 2000 years old.

A radiological dating method used for determining the age of rocks is based on the radioisotope uranium-238, which decays through a series of reactions to lead-206. The uranium-238 isotope has a very long half-life, about $4 \times 10^9$ (4 billion) years. Measurements of the amounts of uranium-238 and lead-206 enable geologists to determine the age of rock samples. The older rocks will have a higher percentage of lead-206 because more of the uranium-238 has decayed. The age of rocks brought back from the moon by the *Apollo* missions was determined using uranium-238. They were found to be about $4 \times 10^9$ years old, approximately the same age calculated for Earth.

## QUESTIONS AND PROBLEMS

### Half-Life of a Radioisotope

**9.27** What is meant by the term half-life?

**9.28** Why are radioisotopes with short half-lives used for diagnosis in nuclear medicine?

**9.29** Technetium-99m is an ideal radioisotope for scanning organs because it has a half-life of 6.0 hours and is a pure gamma emitter. Suppose that 80.0 mg were prepared in the technetium generator this morning. How many milligrams would remain after the following intervals?
**a.** one half-life
**b.** two half-lives
**c.** 18 hours
**d.** 24 hours

**9.30** A sample of sodium-24 with an activity of 12 mCi is used to study the rate of blood flow in the circulatory system. If sodium-24 has a half-life of 15 hours, what is the activity of the sodium after 2.5 d?

**9.31** Strontium-85, used for bone scans, has a half-life of 64 d. How long will it take for the radiation level of strontium-85 to drop to one-fourth of its original level? To one-eighth?

**9.32** Fluorine-18, which has a half-life of 110 min, is used in PET scans. If 100 mg of fluorine-18 is shipped at 8:00 A.M., how many milligrams of the radioisotope are still active if the sample arrives at the radiology laboratory at 1:30 P.M.?

# 9.5 MEDICAL APPLICATIONS USING RADIOACTIVITY

LEARNING GOAL

Describe the use of radioisotopes in medicine.

To determine the condition of an organ in the body, a radiologist may give a patient a radioisotope that concentrates in that organ. The cells in the body cannot differentiate between a nonradioactive atom and a radioactive one. However, radioactive atoms can be detected because they emit radiation. Some radioisotopes used in nuclear medicine are listed in Table 9.10.

**TABLE 9.10** Medical Applications of Radioisotopes

| Isotope | Half-Life | Medical Application |
|---------|-----------|---------------------|
| Ce-141 | 32.5 d | Gastrointestinal tract diagnosis; measuring myocardial blood flow |
| Ga-67 | 78 h | Abdominal imaging; tumor detection |
| Ga-68 | 68 min | Detect pancreatic cancer |
| P-32 | 4.3 d | Treatment of leukemia, polycythemia vera (excess red blood cells), pancreatic cancer |
| I-125 | 60 d | Treatment of brain cancer; osteoporosis detection |
| I-131 | 8 d | Imaging thyroid; treatment of Graves' disease, goiter, and hyperthyroidism; treatment of thyroid and prostate cancer |
| Sr-85 | 65 d | Detection of bone lesions; brain scans |
| Tc-99m | 6.0 h | Imaging of skeleton and heart muscle, brain, liver, heart, lungs, bone, spleen, kidney, and thyroid; *most widely used radioisotope in nuclear medicine* |

## Scans with Radioisotopes

After a patient receives a radioisotope, the radiologist determines the level and location of radioactivity emitted by the radioisotope. An apparatus called a scanner is used to produce an image of the organ. The scanner moves slowly across the patient's body above the region where the organ containing the radioisotope is located. The gamma rays emitted from the radioisotope in the organ can be used to expose a photographic plate, producing a **scan** of the organ. On a scan, an area of decreased or increased radiation can indicate such conditions as a disease of the organ, a tumor, a blood clot, or edema.

A common method of determining thyroid function is the use of radioactive iodine uptake (RAIU). Taken orally, the radioisotope iodine-131 mixes with the iodine already present in the thyroid. Twenty-four hours later, the amount of iodine taken up by the thyroid is determined. A detection tube held up to the area of the thyroid gland detects the radiation coming from the iodine-131 that has located there. (See Figure 9.5.)

A patient with a hyperactive thyroid will have a higher than normal level of radioactive iodine, whereas a patient with a hypoactive thyroid will record low values. If the patient has hyperthyroidism, treatment is begun to lower the activity of the thyroid. One treatment involves giving the patient a therapeutic dosage of radioactive iodine, which has a higher radiation count than the diagnostic dose. The radioactive iodine goes to the thyroid where its radiation destroys some of the thyroid cells. The thyroid produces less thyroid hormone, bringing the hyperthyroid condition under control.

## Positron Emission Tomography (PET)

Positron emitters with short half-lives, such as carbon-11, oxygen-15, nitrogen-13, and fluorine-18, are used in an imaging method called positron emission tomography (PET).

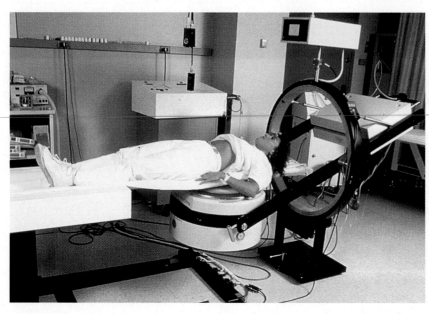

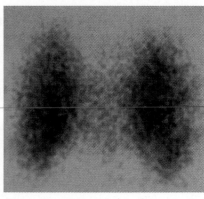

**(b)**

**(a)**

**FIGURE 9.5** (a) A scanner is used to detect radiation from a radioisotope that has accumulated in an organ. (b) A scan of the thyroid shows the accumulation of radioactive iodine-131 in the thyroid.
**Q** What type of radiation would move through body tissues to create a scan?

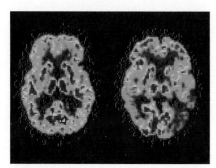

**FIGURE 9.6** These PET scans of the brain show a normal brain on the left and a brain affected by Alzheimer's disease on the right.
**Q** When positrons collide with electrons, what type of radiation is produced that gives an image of an organ?

A positron-emitting isotope such as fluorine-18 combined with substances in the body such as glucose is used to study brain function, metabolism, and blood flow.

$$^{18}_{9}\text{F} \longrightarrow {}^{18}_{8}\text{O} + {}^{0}_{+1}e$$

As positrons are emitted, they combine with electrons to produce gamma rays that are detected by computerized equipment to create a three-dimensional image of the organ. (See Figure 9.6.)

## SAMPLE PROBLEM 9.9

### ■ Medical Application of Radioactivity

In the determination of thyroid function, a patient receives an oral dose of sodium iodide that contains 10 $\mu$Ci of iodine-131, which is a beta emitter. Write the nuclear equation for the beta decay of iodine-131.

#### SOLUTION

We can write the incomplete nuclear equation starting with iodine-131, which has the atomic number 53.

$$^{131}_{53}\text{I} \longrightarrow ? + {}^{0}_{-1}e$$

In beta decay, the mass number (131) does not change, but the atomic number of the new nucleus increases by 1. The new atomic number is 54, which is xenon (Xe).

$$^{131}_{53}\text{I} \longrightarrow {}^{131}_{54}\text{Xe} + {}^{0}_{-1}e$$

#### STUDY CHECK

In an experimental treatment, a patient is given boron-10, which is taken up by malignant tumors. When bombarded with neutrons, boron-10 decays by emitting alpha particles that destroy the surrounding tumor cells. Write the equation for the nuclear reaction for this experimental procedure.

# Health Note

## Radiation Doses in Diagnostic and Therapeutic Procedures

We can compare the levels of radiation exposure commonly used during diagnostic and therapeutic procedures in nuclear medicine. In diagnostic procedures, the radiologist minimizes radiation damage by using the minimum amount of radioisotope needed to evaluate the condition of an organ or tissue. The doses used in radiation therapy are much greater than those used for diagnostic procedures. For example, a therapeutic dose would be used to destroy the cells in a malignant tumor. Although there will be some damage to surrounding tissue, the healthy cells are more resistant to radiation and can repair themselves. (See Table 9.11.)

**TABLE 9.11** Radiation Doses Used for Diagnostic and Therapeutic Procedures

| Organ/Condition | Dose (rem) |
|---|---|
| **Diagnostic** | |
| Liver | 0.3 |
| Thyroid | 50.0 |
| Lung | 2.0 |
| **Therapeutic** | |
| Lymphoma | 4500 |
| Skin cancer | 5000–6000 |
| Lung cancer | 6000 |
| Brain tumor | 6000–7000 |

# Health Note

## Other Imaging Methods

### COMPUTED TOMOGRAPHY (CT)

Another imaging method used to detect changes within the body is computed tomography (CT). A computer monitors the degree of absorption of 30 000 X-ray beams directed at the brain at successive layers. The differences in absorption based upon the densities of the tissues and fluids in the brain provide a series of images of the brain. This technique is successful in the identification of brain hemorrhages, tumors, and atrophy. (See Figure 9.7.)

### MAGNETIC RESONANCE IMAGING (MRI)

Magnetic resonance imaging (MRI) is a powerful imaging technique that does not involve X-ray radiation. It is the least invasive imaging method available. MRI is based on the absorption of energy when the protons in hydrogen atoms are excited by a strong magnetic field. Hydrogen atoms make up 63% of all the atoms in the body. In the nucleus, the spin of the single proton acts like a tiny magnet. With no external field, the spins of the protons have random orientations. However, when placed within a large magnet, the spins of the protons align with the magnetic field. A magnet aligned with the field has a lower energy than one that is aligned against the field. As the MRI scan proceeds, radiofrequency pulses of energy are applied. When a nucleus absorbs certain energy, its proton "flips" and becomes aligned against the field. Because hydrogen atoms in the body are in different chemical environments, frequencies of different energies are absorbed. The energies absorbed are calculated and converted to color images of the body. MRI is particularly useful to image soft tissues because soft tissues contain large amounts of water. (See Figure 9.8.)

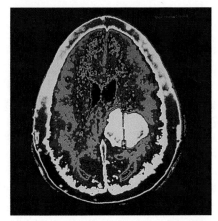

**FIGURE 9.7** A CT scan shows a brain tumor (yellow area) in the center of the right side of the brain.
**Q** What is the type of radiation used to give a CT scan?

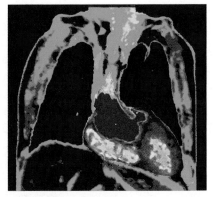

**FIGURE 9.8** An MRI scan of the heart and lungs, with the left ventricle shown in red.
**Q** What is the source of energy in MRI?

## QUESTIONS AND PROBLEMS

### Medical Applications Using Radioactivity

**9.33** Bone and bony structures contain calcium and phosphorus.
   **a.** Why would the radioisotopes of calcium-47 and phosphorus-32 be used in the diagnosis and treatment of bone diseases?
   **b.** The radioisotope strontium-89, a beta emitter, is used to treat bone cancer. Write the nuclear equation, and explain why a strontium radioisotope would be used to treat bone cancer.

**9.34 a.** Technetium-99m emits only gamma radiation. Why would this type of radiation be used in diagnostic imaging rather than an isotope that also emits beta or alpha radiation?

   **b.** A patient with polycythemia vera (excess production of red blood cells) receives radioactive phosphorus-32. Why would this treatment reduce the production of red blood cells in the bone marrow of the patient?

**9.35** In a diagnostic test for leukemia, a patient receives 4.0 mL of a solution containing selenium-75. If the activity of the selenium-75 is 45 $\mu$Ci/mL, what is the dose received by the patient?

**9.36** A vial contains radioactive iodine-131 with an activity of 2.0 mCi per milliliter. If the thyroid test requires 3.0 mCi in an "atomic cocktail," how many milliliters are used to prepare the iodine-131 solution?

---

## LEARNING GOAL

Describe the processes of nuclear fission and fusion.

# 9.6 NUCLEAR FISSION AND FUSION

In the 1930s, scientists bombarding uranium-235 with neutrons discovered that the U-235 nucleus splits into two medium-weight nuclei and produces a great amount of energy. This was the discovery of nuclear **fission**. The energy generated by splitting the atom was called atomic energy. When uranium-235 absorbs a neutron, it breaks apart to form two smaller nuclei, several neutrons, and a great amount of energy. A typical equation for nuclear fission is

$$\,_{0}^{1}n + \,_{92}^{235}U \longrightarrow \,_{36}^{91}Kr + \,_{56}^{142}Ba + 3\,_{0}^{1}n + Energy$$

If we could weigh these products with great accuracy, we would find that their total mass is slightly less than the mass of the starting materials. The missing mass has been converted into energy, consistent with the famous equation derived by Albert Einstein:

$$E = mc^2$$

$E$ is the energy released, $m$ is the mass lost, and $c$ is the speed of light, $3 \times 10^8$ m/s. Even though the mass loss is very small, when it is multiplied by the speed of light squared, the result is a large value for the energy released. The fission of 1 g of uranium-235 produces about as much energy as the burning of 3 tons of coal.

## Chain Reaction

Fission begins when a neutron collides with the nucleus of a uranium atom. The resulting nucleus is unstable and splits into smaller nuclei. This fission process also releases several neutrons and large amounts of gamma radiation and energy. The neutrons emitted have high energies and bombard more uranium-235 nuclei. As fission continues, there is a rapid increase in the number of high-energy neutrons capable of splitting more uranium atoms, a process called a **chain reaction**. To sustain a nuclear chain reaction, sufficient quantities of uranium-235 must be brought together to provide a critical mass in which almost all the neutrons immediately collide with more uranium-235 nuclei. So much heat and energy are released that an atomic explosion occurs. (See Figure 9.9.)

## Nuclear Fusion

In **fusion**, two small nuclei such as those in hydrogen combine to form a larger nucleus. Mass is lost, and a tremendous amount of energy is released, even more than the energy

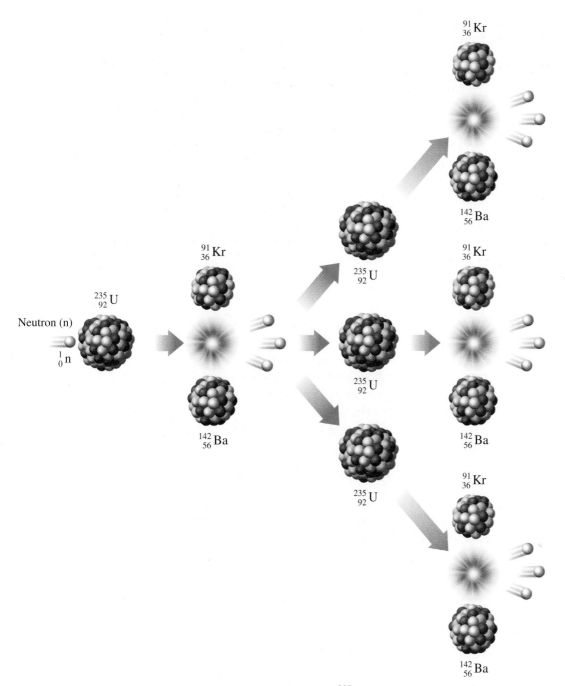

**FIGURE 9.9** In a nuclear chain reaction, the fission of each $^{235}$U atom produces three neutrons that cause the nuclear fission of more and more $^{235}$U atoms.

**Q** Why is the fission of $^{235}$U called a chain reaction?

released from nuclear fission. However, a very high temperature (100 000 000 °C) is required to overcome the repulsion of the hydrogen nuclei and cause them to undergo fusion. Fusion reactions occur continuously in the sun and other stars, providing us with heat and light. The huge amounts of energy produced by our sun come from the fusion of $6 \times 10^{11}$ kg of hydrogen every second. The following fusion reaction involves the combination of two isotopes of hydrogen.

$$^{3}_{1}\text{H} \quad + \quad ^{2}_{1}\text{H} \quad \longrightarrow \quad ^{4}_{2}\text{He} \quad + \quad ^{1}_{0}\text{n} \quad + \quad \text{Energy}$$

# Green Chemistry Note

## Nuclear Power Plants

In a nuclear power plant, the quantity of uranium-235 is held below a critical mass, so it cannot sustain a chain reaction. The fission reactions are slowed by placing control rods, which absorb some of the fast-moving neutrons, among the uranium samples. In this way, less fission occurs, and there is a slower, controlled production of energy. The heat from the controlled fission is used to produce steam. The steam drives a generator, which produces electricity. Approximately 10% of the electrical energy produced in the United States is generated in nuclear power plants.

Although nuclear power plants help meet some of our energy needs, there are some problems. One of the most serious problems is the production of radioactive by-products that have very long half-lives. It is essential that these waste products be stored safely for a very long time in a place where they do not contaminate the environment. Early in 1990, the EPA gave its approval for the storage of radioactive hazardous wastes in chambers 2150 ft underground. In 1998, the Waste Isolation Pilot Plant (WIPP) repository site in New Mexico was ready to receive plutonium waste from former U.S. bomb factories. Although authorities claim the caverns are safe, some people are concerned with the safe transport of the radioactive waste by trucks on the highways.

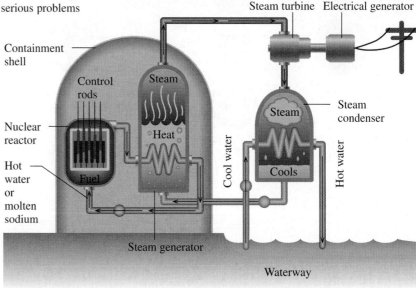

The fusion reaction has tremendous potential as a possible source for future energy needs. Scientists expect less radioactive waste with shorter half-lives from fusion reactors. However, fusion is still in the experimental stage because the extremely high temperatures needed have been difficult to reach and even more difficult to maintain. Research groups around the world are attempting to develop the technology needed to make the fusion reaction a reality in our lifetime.

## SAMPLE PROBLEM 9.10

### ■ Identifying Fission and Fusion

Classify the following as pertaining to nuclear fission, nuclear fusion, or both:

**a.** Small nuclei combine to form larger nuclei.
**b.** Large amounts of energy are released.
**c.** Very high temperatures are needed for reaction.

SOLUTION

**a.** fusion
**b.** both fusion and fission
**c.** fusion

STUDY CHECK

Would the following reaction be an example of a fission or fusion reaction?

$$^{2}_{1}\text{H} + ^{1}_{1}\text{H} \longrightarrow ^{3}_{2}\text{He}$$

## QUESTIONS AND PROBLEMS

### Nuclear Fission and Fusion

**9.37** What is nuclear fission?

**9.38** How does a chain reaction occur in nuclear fission?

**9.39** Complete the following fission reaction:

$$^{235}_{92}\text{U} + ^{1}_{0}\text{n} \longrightarrow ^{131}_{50}\text{Sn} + ? + 2^{1}_{0}\text{n} + \text{energy}$$

**9.40** In another fission reaction, U-235 bombarded with a neutron produces Sr-94, another small nucleus, and 3 neutrons. Write the complete equation for the fission reaction.

**9.41** Indicate whether each of the following are characteristic of the fission or fusion process or both:

**a.** Neutrons bombard a nucleus.
**b.** The nuclear process occurs in the sun.
**c.** A large nucleus splits into smaller nuclei.
**d.** Small nuclei combine to form larger nuclei.

**9.42** Indicate whether each of the following are characteristic of the fission or fusion process or both:

**a.** Very high temperatures are required to initiate the reaction.
**b.** Less radioactive waste is produced.
**c.** Hydrogen nuclei are the reactants.
**d.** Large amounts of energy are released when the nuclear reaction occurs.

## CONCEPT MAP

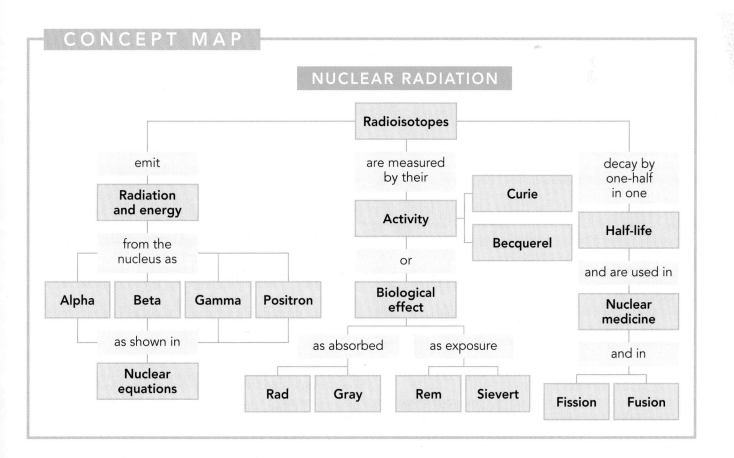

# CHAPTER REVIEW

## 9.1 Natural Radioactivity

**Learning Goal:** Describe alpha, beta, positron, and gamma radiation.

Radioactive isotopes have unstable nuclei that break down (decay), spontaneously emitting alpha ($\alpha$), beta ($\beta$), positron ($\beta^+$), and gamma ($\gamma$) radiation. Because radiation can damage the cells in the body, proper protection must be used: shielding, limiting the time of exposure, and distance.

## 9.2 Nuclear Reactions

**Learning Goal:** Write an equation showing mass numbers and atomic numbers for radioactive decay.

A balanced equation is used to represent the changes that take place in the nuclei of the reactants and products. The new isotopes and the type of radiation emitted can be determined from the symbols that show the mass numbers and atomic numbers of the isotopes in the nuclear reaction. A radioisotope is produced artificially when a nonradioactive isotope is bombarded by a small particle. Many radioactive isotopes used in nuclear medicine are produced in this way.

## 9.3 Radiation Measurement

**Learning Goal:** Describe the detection and measurement of radiation.

In a Geiger counter, radiation ionizes gas in a metal tube, which produces an electrical current. The curie (Ci) measures the number of nuclear transformations of a radioactive sample. Activity is also measured in becquerel (Bq) units. The amount of radiation absorbed by a substance is measured in rads or the gray (Gy). The rem and the sievert (Sv) are units used to determine the biological damage from the different types of radiation.

## 9.4 Half-Life of a Radioisotope

**Learning Goal:** Given the half-life of a radioisotope, calculate the amount of radioisotope remaining after one or more half-lives.

Every radioisotope has its own rate of emitting radiation. The time it takes for one-half of a radioactive sample to decay is called its half-life. For many medical radioisotopes, such as Tc-99m and I-131, half-lives are short. For other isotopes, usually naturally occurring ones such as C-14, Ra-226, and U-238, half-lives are extremely long.

## 9.5 Medical Applications Using Radioactivity

**Learning Goal:** Describe the use of radioisotopes in medicine.

In nuclear medicine, radioisotopes are given that go to specific sites in the body. By detecting the radiation they emit, an evaluation can be made about the location and extent of an injury, disease, tumor, or the level of function of a particular organ. Higher levels of radiation are used to treat or destroy tumors.

## 9.6 Nuclear Fission and Fusion

**Learning Goal:** Describe the processes of nuclear fission and fusion.

In fission, a large nucleus breaks apart into smaller pieces, releasing one or more types of radiation and a great amount of energy. In fusion, small nuclei combine to form a larger nucleus, while great amounts of energy are released.

# KEY TERMS

**alpha particle** A nuclear particle identical to a helium nucleus with symbol $\alpha$ or $^4_2\text{He}$.

**becquerel (Bq)** A unit of activity of a radioactive sample equal to one disintegration per second.

**beta particle** A particle identical to an electron with symbol $\beta$ or $^0_{-1}e$ that forms in the nucleus when a neutron changes to a proton and an electron.

**carbon dating** A technique used to date ancient specimens that contain carbon. The age is determined by the amount of active carbon-14 that remains in the sample.

**chain reaction** A fission reaction that will continue once it has been initiated by a high-energy neutron bombarding a heavy nucleus such as U-235.

**curie (Ci)** A unit of radiation equal to $3.7 \times 10^{10}$ disintegrations/s.

**decay curve** A diagram of the decay of a radioactive element.

**equivalent dose** The measure of biological damage from an absorbed dose that has been adjusted for the type of radiation.

**fission** A process in which large nuclei are split into smaller pieces, releasing large amounts of energy.

**fusion** A reaction in which large amounts of energy are released when small nuclei combine to form larger nuclei.

**gamma ray** High-energy radiation with symbol $^0_0\gamma$ emitted by an unstable nucleus.

**gray (Gy)** A unit of absorbed dose equal to 100 rads.

**half-life** The length of time it takes for one-half of a radioactive sample to decay.

**positron** A particle with no mass and a positive charge produced when a proton is transformed into a neutron and a positron.

**rad (radiation absorbed dose)** A measure of an amount of radiation absorbed by the body.

**radiation** Energy or particles released by radioactive atoms.

**radioactive decay** The process by which an unstable nucleus breaks down with the release of high-energy radiation.

**rem (radiation equivalent in humans)** A measure of the biological damage caused by the various kinds of radiation (rad $\times$ radiation biological factor).

**scan** The image of a site in the body created by the detection of radiation from radioactive isotopes that have accumulated in that site.

**shielding** Materials used to provide protection from radioactive sources.

**sievert (Sv)** A unit of biological damage (equivalent dose) equal to 100 rems.

# UNDERSTANDING THE CONCEPTS

**9.43** Consider the following nucleus of a radioactive isotope.

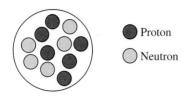

- Proton
- Neutron

**a.** What is the nuclear symbol for this isotope?
**b.** If this isotope decays by emitting a positron, what does the resulting nucleus look like?

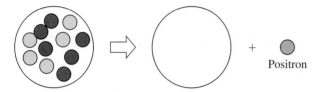

+ Positron

**9.44** Draw in the radioactive nucleus that emits a beta particle to form the following nucleus.

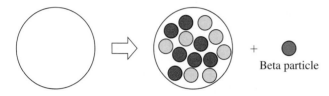

+ Beta particle

**9.45** Draw in the nucleus of the atom to complete the following nuclear reaction.

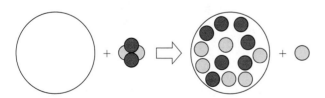

**9.46** Complete the following equation by drawing the nucleus of the atom produced.

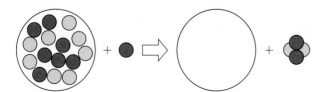

**9.47** Carbon dating of small bits of charcoal used in cave paintings has determined that some of the paintings are from 10 000 to 30 000 years old. Carbon-14 has a half-life of 5730 years. In a 1-$\mu$g sample of carbon from a live tree, the activity of carbon-14 is 6.4 $\mu$Ci. If researchers determine that 1 $\mu$g of charcoal from a prehistoric cave painting in France has an activity of 0.80 $\mu$Ci, what is the age of the painting?

**9.48** Using the decay curve for $^{131}_{53}I$, determine the following:

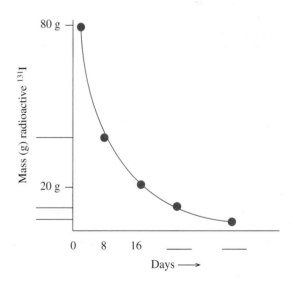

**a.** the missing values for the mass of radioactive $^{131}_{53}I$ on the vertical axis
**b.** the number of days on the horizontal axis
**c.** the half-life in days of $^{131}_{53}I$

# ADDITIONAL QUESTIONS AND PROBLEMS

**9.49** Give the number of protons and the number of neutrons in the nucleus of each the following:
  **a.** sodium-25
  **b.** nickel-61
  **c.** rubidium-84
  **d.** silver-110

**9.50** Give the number of protons and the number of neutrons in the nucleus of each the following:
  **a.** boron-10
  **b.** zinc-72
  **c.** iron-59
  **d.** gold-198

**9.51** Describe alpha, beta, and gamma radiation in terms of the following:
  **a.** type of radiation
  **b.** symbols

**9.52** Describe alpha, beta, and gamma radiation in terms of the following:
  **a.** depth of tissue penetration
  **b.** type of shielding needed for protection

**9.53** Identify each of the following as alpha decay, beta decay, positron emission, or gamma radiation:
  **a.** $^{27m}_{13}\text{Al} \longrightarrow {}^{27}_{13}\text{Al} + {}^{0}_{0}\gamma$
  **b.** $^{8}_{5}\text{B} \longrightarrow {}^{8}_{4}\text{Be} + {}^{0}_{+1}e$
  **c.** $^{220}_{86}\text{Rn} \longrightarrow {}^{216}_{84}\text{Po} + {}^{4}_{2}\text{He}$

**9.54** Identify each of the following as alpha decay, beta decay, positron emission, or gamma radiation:
  **a.** $^{127}_{55}\text{Cs} \longrightarrow {}^{127}_{54}\text{Xe} + {}^{0}_{+1}e$
  **b.** $^{90}_{38}\text{Sr} \longrightarrow {}^{90}_{39}\text{Y} + {}^{0}_{-1}e$
  **c.** $^{218}_{85}\text{At} \longrightarrow {}^{214}_{83}\text{Bi} + {}^{4}_{2}\text{He}$

**9.55** Write a balanced nuclear equation for each of the following:
  **a.** Th-225 ($\alpha$ decay)
  **b.** Bi-210 ($\alpha$ decay)
  **c.** Cs-137 ($\beta$ decay)
  **d.** Sn-126 ($\beta$ decay)
  **e.** N-13 ($\beta^+$ emission)

**9.56** Write a balanced nuclear equation for each of the following:
  **a.** potassium-40 ($\beta$ decay)
  **b.** sulfur-35 ($\beta$ decay)
  **c.** platinum-190 ($\alpha$ decay)
  **d.** Ra-210 ($\alpha$ decay)
  **e.** In-113m ($\gamma$ emission)

**9.57** Complete each of the following nuclear equations:
  **a.** $^{14}_{7}\text{N} + {}^{4}_{2}\text{He} \longrightarrow ? + {}^{1}_{1}\text{H}$
  **b.** $^{27}_{13}\text{Al} + {}^{4}_{2}\text{He} \longrightarrow {}^{30}_{14}\text{Si} + ?$
  **c.** $^{235}_{92}\text{U} + {}^{1}_{0}n \longrightarrow {}^{90}_{38}\text{Sr} + 3{}^{1}_{0}n + ?$

**9.58** Complete each of the following nuclear equations:
  **a.** $^{59}_{27}\text{Co} + ? \longrightarrow {}^{56}_{25}\text{Mn} + {}^{4}_{2}\text{He}$
  **b.** $? \longrightarrow {}^{14}_{7}\text{N} + {}^{0}_{-1}e$
  **c.** $^{76}_{36}\text{Kr} + {}^{0}_{-1}e \longrightarrow ?$

**9.59** Write the symbols and a balanced nuclear equation for the following:
  **a.** When two oxygen-16 atoms collide, one of the products is an alpha particle.
  **b.** When californium-249 is bombarded by oxygen-18, a new isotope and four neutrons are produced.
  **c.** Radon-222 undergoes alpha decay.
  **d.** The product from c. undergoes alpha decay.

**9.60** Write the symbols and a balanced nuclear equation for the following:
  **a.** Polonium-210 decays to give lead-206.
  **b.** Bismuth-211 decays by emitting an alpha particle.
  **c.** The product from b. emits a beta particle.
  **d.** When an alpha particle bombards aluminum-27, one product is silicon-30.

**9.61** Write a balanced nuclear equation for the positron decay of each of the following:
  **a.** $^{26}_{14}\text{Si}$     **b.** $^{54}_{27}\text{Co}$
  **c.** $^{77}_{37}\text{Rb}$     **d.** $^{93}_{45}\text{Rh}$

**9.62** Write a balanced nuclear equation for the positron decay of each of the following:
  **a.** $^{8}_{5}\text{B}$     **b.** $^{13}_{7}\text{N}$
  **c.** $^{49}_{19}\text{K}$     **d.** $^{118}_{54}\text{Xe}$

**9.63** If the amount of radioactive phosphorus-32 in a sample decreases from 1.2 g to 0.30 g in 28 days, what is the half-life of phosphorus-32?

**9.64** If the amount of radioactive iodine-123 in a sample decreases from 0.4 g to 0.1 g in 26.2 hours, what is the half-life of iodine-123?

**9.65** Iodine-131, a beta emitter, has a half-life of 8.0 days.
  **a.** Write the nuclear equation for the beta decay of iodine-131.
  **b.** How many grams of a 12.0-g sample of iodine-131 would remain after 40 days?
  **c.** How many days have passed if 48 g of iodine-131 decayed to 3.0 g of iodine-131?

**9.66** Cesium-137, a beta emitter, has a half-life of 30 years.
  **a.** Write the nuclear equation for the beta decay of cesium-137.
  **b.** How many grams of a 16-g sample of cesium-137 would remain after 90 years?
  **c.** How many years will be needed for 28 g of cesium-137 to decay to 3.5 g of cesium-137?

**9.67** A nurse was accidentally exposed to potassium-42 while doing some brain scans for possible tumors. The error was not discovered until 36 hours later when the activity of the potassium-42 sample was 2.0 $\mu$Ci. If potassium-42 has a half-life of 12 hours, what was the activity of the sample at the time the nurse was exposed?

**9.68** A wooden object from the site of an ancient temple has a carbon-14 activity of 10 counts per minute, compared with a reference piece of wood cut today that has an activity of 40 counts per minute. If the half-life for carbon-14 is 5730 years, what is the age of the ancient wood object?

**9.69** A 120-mg sample of technetium-99m is used for a diagnostic test. If technetium-99m has a half-life of 6.0 hours, how much of the technetium-99m sample remains 24 hours after the test?

**9.70** The half-life of oxygen-15 is 124 seconds. If a sample of oxygen-15 has an activity of 4000 becquerels, how many minutes will elapse before it reaches an activity of 500 becquerels?

**9.71** What is the purpose of irradiating meats, fruits, and vegetables?

**9.72** The irradiation of foods was approved in the United States in the 1980s.
  **a.** Why have we not seen many irradiated products in our markets?
  **b.** Would you buy foods that have been irradiated? Why or why not?

**9.73** What is the difference between fission and fusion?

**9.74**  **a.** What are the products in the fission of uranium-235 that make possible a nuclear chain reaction?

  **b.** What is the purpose of placing control rods among uranium samples in a nuclear reactor?

**9.75**  Where does fusion occur naturally?

**9.76**  Why are scientists continuing to try to build a fusion reactor even though very high temperatures have been difficult to reach and maintain?

# CHALLENGE QUESTIONS

**9.77**  Uranium-238 decays in a series of nuclear changes until stable $^{206}$Pb is produced. Complete the following nuclear equations that are part of the $^{238}$U decay series:

  **a.** $^{238}_{92}\text{U} \longrightarrow\ ^{234}_{90}\text{Th} + ?$

  **b.** $^{234}_{90}\text{Th} \longrightarrow\ ? +\ ^{0}_{-1}e$

  **c.** $? \longrightarrow\ ^{222}_{86}\text{Rn} +\ ^{4}_{2}\text{He}$

**9.78**  The iceman known as "Ötzi" was discovered in a high mountain pass on the Austrian–Italian border. Samples of his hair and bones had carbon-14 activity that was about 50% of that present in new hair or bone. Carbon-14 is a beta emitter.

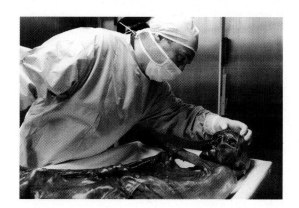

  **a.** How long ago did "Ötzi" live if the half-life for C-14 is 5730 years?

  **b.** Write a nuclear equation for the decay of $^{14}$C.

**9.79**  Complete and balance each of the following nuclear equations:

  **a.** $^{23m}_{12}\text{Mg} \longrightarrow\ ? +\ ^{0}_{0}\gamma$

  **b.** $^{61}_{30}\text{Zn} \longrightarrow\ ^{61}_{29}\text{Cu} + ?$

  **c.** $\text{Cf-249} + \text{C-12} \longrightarrow\ ? + 4 \text{ neutrons}$

**9.80**  Complete and balance each of the following nuclear equations:

  **a.** $\text{Bi-209} + \text{Cr-54} \longrightarrow\ ? + 1 \text{ neutron}$

  **b.** Np-237 undergoes alpha decay.

  **c.** $^{241}_{95}\text{Am} +\ ^{4}_{2}\text{He} \longrightarrow\ ? + 2\ ^{0}_{1}n$

**9.81**  The half-life for the radioactive decay of calcium-47 is 4.5 days. If a sample has an activity of 4.0 $\mu$Ci after 18 days, what was the initial activity of the sample?

**9.82**  A 16-$\mu$g sample of sodium-24 decays to 2.0 $\mu$g in 45 h. What is the half-life of sodium-24?

# ANSWERS

### Answers to Study Checks

**9.1**  $\beta$ or $^{0}_{-1}e$

**9.2**  Limiting the time one spends near a radioactive source and staying as far away as possible will reduce exposure to radiation.

**9.3**  $^{214}_{84}\text{Po} \longrightarrow\ ^{210}_{82}\text{Pb} +\ ^{4}_{2}\text{He}$

**9.4**  $^{131}_{53}\text{I} \longrightarrow\ ^{131}_{54}\text{Xe} +\ ^{0}_{-1}e$

**9.5**  $^{27}_{13}\text{Al} +\ ^{4}_{2}\text{He} \longrightarrow\ ^{30}_{15}\text{P} +\ ^{1}_{0}n$

**9.6**  For $\beta$, the factor is 1; rads and rems are equal.

**9.7**  0.50 g

**9.8**  17 200 years

**9.9**  $^{10}_{5}\text{B} +\ ^{1}_{0}n \longrightarrow\ ^{7}_{3}\text{Li} +\ ^{4}_{2}\text{He}$

**9.10**  fusion

### Answers to Selected Questions and Problems

**9.1**  **a.** Both an alpha particle and a helium nucleus have 2 protons and 2 neutrons.

  **b.** $\alpha$, $^{4}_{2}\text{He}$

  **c.** An alpha particle is emitted from an unstable nucleus during radioactive decay.

**9.3**  **a.** $^{39}_{19}\text{K}$,   $^{40}_{19}\text{K}$,   $^{41}_{19}\text{K}$

  **b.** They all have 19 protons and 19 electrons, but they differ in the number of neutrons.

**9.5**

| Medical Use | Isotope Symbol | Mass Number | Number of Protons | Number of Neutrons |
| --- | --- | --- | --- | --- |
| Heart imaging | $^{201}_{81}\text{Tl}$ | 201 | 81 | 120 |
| Radiation therapy | $^{60}_{27}\text{Co}$ | 60 | 27 | 33 |
| Abdominal scan | $^{67}_{31}\text{Ga}$ | 67 | 31 | 36 |
| Hyperthyroidism | $^{131}_{53}\text{I}$ | 131 | 53 | 78 |
| Leukemia treatment | $^{32}_{15}\text{P}$ | 32 | 15 | 17 |

**9.7**  **a.** $\alpha$, $^{4}_{2}\text{He}$   **b.** $^{1}_{0}n$   **c.** $\beta$, $^{0}_{-1}e$
  **d.** $^{15}_{7}\text{N}$   **e.** $^{125}_{53}\text{I}$

**9.9**  **a.** $\beta$, $^{0}_{-1}e$   **b.** $\alpha$, $^{4}_{2}\text{He}$   **c.** $^{1}_{0}n$
  **d.** $^{24}_{11}\text{Na}$   **e.** $^{14}_{6}\text{C}$

**9.11** **a.** Because $\beta$ particles have less mass and move faster than $\alpha$ particles, they can penetrate farther into tissue.

**b.** Ionizing radiation produces ions that cause undesirable reactions in the cells.

**c.** Radiation technicians leave the room to increase the distance between themselves and the radiation. Also a wall that contains lead shields them.

**d.** Wearing gloves shields the skin from $\alpha$ and $\beta$ radiation.

**9.13** **a.** $^{208}_{84}Po \longrightarrow ^{204}_{82}Pb + ^{4}_{2}He$

**b.** $^{232}_{90}Th \longrightarrow ^{228}_{88}Ra + ^{4}_{2}He$

**c.** $^{251}_{102}No \longrightarrow ^{247}_{100}Fm + ^{4}_{2}He$

**d.** $^{220}_{86}Rn \longrightarrow ^{216}_{84}Po + ^{4}_{2}He$

**9.15** **a.** $^{25}_{11}Na \longrightarrow ^{25}_{12}Mg + ^{0}_{-1}e$

**b.** $^{20}_{8}O \longrightarrow ^{20}_{9}F + ^{0}_{-1}e$

**c.** $^{92}_{38}Sr \longrightarrow ^{92}_{39}Y + ^{0}_{-1}e$

**d.** $^{42}_{19}K \longrightarrow ^{42}_{20}Ca + ^{0}_{-1}e$

**9.17** **a.** $^{28}_{14}Si$, beta decay

**b.** $^{0}_{+1}e$, positron emission

**c.** $^{0}_{-1}e$, beta decay

**d.** $^{4}_{2}He$, alpha decay

**9.19** **a.** $^{10}_{4}Be$    **b.** $^{0}_{-1}e$    **c.** $^{27}_{13}Al$    **d.** $^{30}_{15}P$

**9.21** **a.** When radiation enters the Geiger counter, charged particles are produced that create a burst of current that is detected by the instrument.

**b.** becquerel (Bq), curie (Ci)

**c.** gray (Gy), rad

**d.** 1000 Gy

**9.23** 294 $\mu$Ci

**9.25** When pilots are flying at high altitudes, there is less atmosphere to protect them from cosmic radiation.

**9.27** A half-life is the time it takes for one-half of a radioactive sample to decay.

**9.29** **a.** 40.0 mg    **b.** 20.0 mg    **c.** 10.0 mg    **d.** 5.00 mg

**9.31** 128 days, 192 days

**9.33** **a.** Since the elements Ca and P are part of bone, their radioactive isotopes will also become part of the bony structures of the body where their radiation can be used to diagnose or treat bone diseases.

**b.** $^{89}_{38}Sr \longrightarrow ^{89}_{39}Y + ^{0}_{-1}e$

Strontium (Sr) acts much like calcium (Ca) because both are Group 2A (2) elements. The body will accumulate radioactive strontium in bones in the same way that it incorporates calcium. Once the strontium isotope is absorbed by the bone, the beta radiation will destroy cancer cells.

**9.35** 180 $\mu$Ci

**9.37** Nuclear fission is the splitting of a large atom into smaller fragments with the release of large amounts of energy.

**9.39** $^{103}_{42}Mo$

**9.41** **a.** fission    **b.** fusion

**c.** fission    **d.** fusion

**9.43** **a.** $^{11}_{6}C$

**b.**

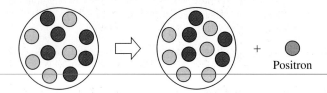

Positron

**9.45**

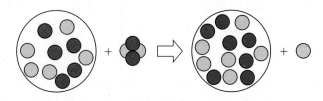

**9.47** 17 200 years old

**9.49** **a.** 11 protons and 14 neutrons

**b.** 28 protons and 33 neutrons

**c.** 37 protons and 47 neutrons

**d.** 47 protons and 63 neutrons

**9.51** **a.** In alpha decay, a helium nucleus is emitted from a radioisotope. In beta decay, a neutron in an unstable nucleus is converted to a proton and an electron, which is emitted as a beta particle. In gamma emission, high-energy radiation is emitted from the nucleus of a radioisotope.

**b.** alpha radiation: $\alpha$ or $^{4}_{2}He$

beta radiation: $\beta$ or $^{0}_{-1}e$

gamma radiation: $\gamma$ or $^{0}_{0}\gamma$

**9.53** **a.** gamma radiation

**b.** positron emission

**c.** alpha decay

**9.55** **a.** $^{225}_{90}Th \longrightarrow ^{221}_{88}Ra + ^{4}_{2}He$

**b.** $^{210}_{83}Bi \longrightarrow ^{206}_{81}Tl + ^{4}_{2}He$

**c.** $^{137}_{55}Cs \longrightarrow ^{137}_{56}Ba + ^{0}_{-1}e$

**d.** $^{126}_{50}Sn \longrightarrow ^{126}_{51}Sb + ^{0}_{-1}e$

**e.** $^{13}_{7}N \longrightarrow ^{13}_{6}C + ^{0}_{+1}e$

**9.57** **a.** $^{17}_{8}O$

**b.** $^{1}_{1}H$

**c.** $^{143}_{54}Xe$

**9.59** **a.** $^{16}_{8}O + ^{16}_{8}O \longrightarrow ^{4}_{2}He + ^{28}_{14}Si$

**b.** $^{249}_{98}Cf + ^{18}_{8}O \longrightarrow ^{263}_{106}Sg + 4^{1}_{0}n$

**c.** $^{222}_{86}Rn \longrightarrow ^{218}_{84}Po + ^{4}_{2}He$

**d.** $^{218}_{84}Po \longrightarrow ^{214}_{82}Pb + ^{4}_{2}He$

**9.61** **a.** $^{26}_{14}Si \longrightarrow ^{26}_{13}Al + ^{0}_{+1}e$

**b.** $^{54}_{27}Co \longrightarrow ^{54}_{26}Fe + ^{0}_{+1}e$

**c.** $^{77}_{37}Rb \longrightarrow ^{77}_{36}Kr + ^{0}_{+1}e$

**d.** $^{93}_{45}Rh \longrightarrow ^{93}_{44}Ru + ^{0}_{+1}e$

**9.63** 14 days

**9.65** **a.** $^{131}_{53}I \longrightarrow ^{0}_{-1}e + ^{131}_{54}Xe$

**b.** 0.375 g

**c.** 32 days

**9.67** 16 $\mu$Ci

**9.69** 7.5 mg

**9.71** The irradiation of meats, fruits, and vegetables kills bacteria such as *E. coli* that can cause food-borne illnesses. In addition, spoilage is deterred, and shelf life is extended.

**9.73** In the fission process, an atom splits into smaller nuclei. In fusion, small nuclei combine (fuse) to form a larger nucleus.

**9.75** Fusion occurs naturally in the sun and other stars.

**9.77** a. $^{238}_{92}U \longrightarrow \, ^{234}_{90}Th + \, ^{4}_{2}He$

b. $^{234}_{90}Th \longrightarrow \, ^{234}_{91}Pa + \, ^{0}_{-1}e$

c. $^{226}_{88}Ra \longrightarrow \, ^{222}_{86}Rn + \, ^{4}_{2}He$

**9.79** a. $^{23}_{12}Mg$

b. $^{0}_{+1}e$

c. $^{257}_{104}Rf$

**9.81** 64 $\mu$Ci

# 10

# Introduction to Organic Chemistry: Alkanes

Visit **www.chemplace.com** for extra quizzes, interactive tutorials, career resources, PowerPoint slides for chapter review, math help, and case studies.

*"When we have a hazardous materials spill, the first thing we do is isolate it," says Don Dornell, assistant fire chief, Burlingame Fire Station. "Then our technicians and a county chemist identify the product from its flammability and solubility in water so we can use the proper materials to clean up the spill. We use different methods for alcohol, which mixes with water, than for gasoline, which floats. Because hydrocarbons are volatile, we use foam to cover them and trap the vapors. At oil refineries, we will use foams, but many times we squirt water on the tanks to cool the contents below their boiling points, too. By knowing the boiling point of the product and its density and vapor density, we know if it floats or sinks in water and where its vapors will go."*

Organic chemistry is the chemistry of carbon compounds that contain, primarily, carbon and hydrogen. The element carbon has a special role in chemistry because it bonds with other carbon atoms to give a vast array of molecules. The variety of molecules is so great that we find organic compounds in many common products we use, such as gasoline, medicine, shampoos, plastic bottles, and perfumes. The food we eat is composed of different organic compounds that supply us with fuel for energy and the carbon atoms needed to build and repair the cells of our bodies.

Although many organic compounds occur in nature, chemists have synthesized even more. The cotton, wool, or silk in your clothes contain naturally occurring organic compounds, whereas materials such as polyester, nylon, and plastic have been synthesized through organic reactions. Sometimes it is convenient to synthesize a molecule in the lab even though that molecule is also found in nature. For example, vitamin C synthesized in a laboratory has the same structure and properties as the vitamin C in oranges and lemons. In these chapters, you will learn about the structures and reactions of organic molecules, which will provide a foundation for understanding the more complex molecules of biochemistry.

# 10.1 ORGANIC COMPOUNDS

At the beginning of the nineteenth century, scientists classified chemical compounds as inorganic and organic. An inorganic compound was a substance that was composed of minerals, and an organic compound was a substance that came from an organism, thus the use of the word "organic." It was thought that some type of "vital force," which could only be found in living cells, was required to synthesize an organic compound. This perception was shown to be incorrect in 1828 when the German chemist Friedrick Wöhler synthesized urea, a product of protein metabolism, by heating an inorganic compound, ammonium cyanate.

$$NH_4CNO \xrightarrow{\text{Heat}} \underset{\text{Urea (organic)}}{H_2N-\overset{\overset{\displaystyle O}{\|}}{C}-NH_2}$$

Ammonium
cyanate
(inorganic)

We now define organic chemistry as the study of carbon compounds. **Organic compounds** always contain carbon (C), usually hydrogen (H), and may also have other nonmetallic elements such as oxygen (O), sulfur (S), nitrogen (N), or chlorine (Cl). In any organic compound, there are always four bonds to every carbon. Organic compounds are usually nonpolar molecules with weak attractions between molecules, which accounts for their low melting and boiling points. Many organic compounds burn vigorously in air. Typically, organic compounds are not soluble in water. For example, vegetable oil, which is a mixture of organic compounds, does not dissolve in water, but floats on top.

**TABLE 10.1** Some Properties of Organic and Inorganic Compounds

| Property | Organic | Example: $C_3H_8$ | Inorganic | Example: NaCl |
|---|---|---|---|---|
| Elements | C and H, sometimes O, S, N, or Cl | C and H | Most metals and nonmetals | Na and Cl |
| Particles | Molecules | $C_3H_8$ | Mostly ions | $Na^+$ and $Cl^-$ |
| Bonding | Mostly covalent | Covalent (4 bonds to each C) | Many are ionic, some covalent | Ionic |
| Polarity of bonds | Nonpolar, unless a more electronegative atom is present | Nonpolar | Most are ionic or polar covalent, a few are nonpolar covalent | Ionic |
| Melting point | Usually low | $-188\ °C$ | Usually high | $801\ °C$ |
| Boiling point | Usually low | $-42\ °C$ | Usually high | $1413\ °C$ |
| Flammability | High | Burns in air | Low | Does not burn |
| Solubility in water | Not soluble, unless a polar group is present | No | Most are soluble, unless nonpolar | Yes |

In contrast, many of the inorganic compounds are ionic, which leads to high melting and boiling points. Inorganic compounds that are ionic or polar covalent are usually soluble in water. Most inorganic substances do not burn in air. Table 10.1 contrasts some of the properties associated with organic and inorganic compounds such as propane, $C_3H_8$, and sodium chloride, NaCl. (See Figure 10.1.)

SAMPLE PROBLEM 10.1

■ **Properties of Organic Compounds**

Indicate whether the following properties are most typical of organic or inorganic compounds:

**a.** not soluble in water     **b.** high melting point     **c.** burns in air

**FIGURE 10.1** Propane, $C_3H_8$, is an organic compound, whereas sodium chloride, NaCl, is an inorganic compound.
**Q** Why is propane used as a fuel?

SOLUTION

**a.** Many organic compounds are not soluble in water.
**b.** Inorganic compounds are most likely to have high melting points.
**c.** Organic compounds are most likely to be flammable.

STUDY CHECK

Octane is not soluble in water. What type of compound is octane?

## Bonding in Organic Compounds

The **hydrocarbons** are organic compounds that consist of only carbon and hydrogen. In the simplest hydrocarbon, methane ($CH_4$), the carbon atom forms an octet by sharing four valence electrons with four hydrogen atoms. In the electron-dot formula, each shared pair of electrons represents a single bond. In all organic molecules, every carbon atom has four bonds. An **expanded structural formula** is written when we show the bonds between all of the atoms.

$$\cdot\overset{\cdot}{\underset{\cdot}{C}}\cdot + 4H\cdot \longrightarrow H\overset{\overset{H}{\cdot\cdot}}{\underset{\underset{H}{\cdot\cdot}}{C}}H \quad = \quad H-\overset{\overset{H}{|}}{\underset{\underset{H}{|}}{C}}-H$$

Methane

## The Tetrahedral Structure of Carbon

The VSEPR theory (Chapter 4) predicts that a molecule with four atoms bonded to a central atom has a tetrahedral shape. In $CH_4$, the bonds from the carbon atom to the four hydrogen atoms are directed to the corners of a tetrahedron with bond angles of 109.5°. The structure of methane is illustrated in Figure 10.2 as a ball-and-stick model and as a space-filling model.

In ethane, $C_2H_6$, each carbon atom is bonded to another carbon and three hydrogen atoms. As in methane, each carbon retains the tetrahedral shape, with bond angles close to 109.5°. (See Figure 10.3.)

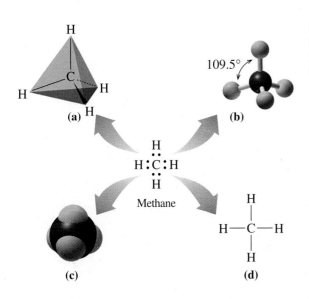

**FIGURE 10.2** Representations of methane, $CH_4$: **(a)** tetrahedron, **(b)** ball-and-stick model, **(c)** space-filling model, **(d)** expanded structural formula.
**Q** Why does methane have a tetrahedral shape and not a flat shape?

**FIGURE 10.3** Representations of ethane, $C_2H_6$: **(a)** tetrahedral shape of each carbon, **(b)** ball-and-stick model, **(c)** space-filling model, **(d)** expanded structural formula.

**Q** How is the tetrahedral shape maintained in a molecule with two carbon atoms?

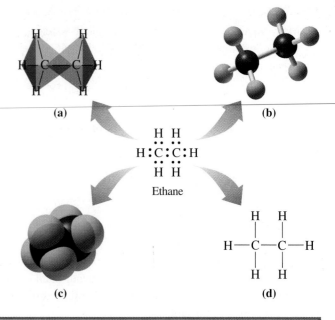

Ethane

(a)   (b)   (c)   (d)

## QUESTIONS AND PROBLEMS

### Organic Compounds

**10.1** Identify the following as formulas of organic or inorganic compounds:
  **a.** KCl      **b.** $C_4H_{10}$      **c.** $CH_3CH_2OH$
  **d.** $H_2SO_4$      **e.** $CaCl_2$      **f.** $CH_3CH_2Cl$

**10.2** Identify the following as formulas of organic or inorganic compounds:
  **a.** $C_6H_{12}O_6$      **b.** $Na_2SO_4$      **c.** $I_2$
  **d.** $C_2H_5Cl$      **e.** $C_{10}H_{22}$      **f.** $CH_4$

**10.3** Identify the following properties as most typical of organic or inorganic compounds:
  **a.** soluble in water      **b.** low boiling point
  **c.** burns in air      **d.** high melting point

**10.4** Identify the following properties as most typical of organic or inorganic compounds:
  **a.** contains Na      **b.** boils at $-50$ °C
  **c.** covalent bonds      **d.** produces ions in water

**10.5** Match the following physical and chemical properties with the compounds ethane, $C_2H_6$, or sodium bromide, NaBr.
  **a.** boils at $-89$ °C      **b.** burns vigorously
  **c.** solid at 250 °C      **d.** dissolves in water

**10.6** Match the following physical and chemical properties with the compounds cyclohexane, $C_6H_{12}$, or calcium nitrate, $Ca(NO_3)_2$.
  **a.** melts at 500 °C      **b.** insoluble in water
  **c.** produces ions in water      **d.** is a liquid at room temperature

**10.7** Why is the shape of a $CH_4$ molecule three-dimensional rather than two-dimensional?

**10.8** In a propane molecule with three carbon atoms, what is the shape around each carbon atom?

Propane

## 10.2 ALKANES

**LEARNING GOAL**

Write the IUPAC names and structural formulas for alkanes.

**WEB TUTORIAL**
Introduction to Organic Molecules

More than 90% of the compounds in the world are organic compounds. The large number of carbon compounds is possible because the covalent bond between carbon atoms (C—C) is very strong, allowing carbon atoms to form long, stable chains. To help us study this large group of compounds, we organize them into classes that have similar structures and chemical properties.

    The **alkanes** are a class of hydrocarbons in which the atoms are connected only by single bonds. One of the most common uses of alkanes is as fuels. Methane, used in gas heaters and gas cooktops, is an alkane with one carbon atom. The alkanes ethane, propane, and butane contain two, three, and four carbon atoms, respectively, connected in a row or a *continuous* chain. These names are part of the **IUPAC** (International Union of Pure and Applied Chemistry) **system**, which chemists use to name organic compounds. Alkanes with five or more carbon atoms in a chain are named using Greek prefixes: *pent*(5), *hex*(6), *hept*(7), *oct*(8), *non*(9), and *dec*(10). (See Table 10.2.)

**TABLE 10.2** IUPAC Names for the First Ten Alkanes

| Number of Carbon Atoms | Prefix | Name | Molecular Formula | Condensed Structural Formula |
|---|---|---|---|---|
| 1 | Meth | Methane | $CH_4$ | $CH_4$ |
| 2 | Eth | Ethane | $C_2H_6$ | $CH_3-CH_3$ |
| 3 | Prop | Propane | $C_3H_8$ | $CH_3-CH_2-CH_3$ |
| 4 | But | Butane | $C_4H_{10}$ | $CH_3-CH_2-CH_2-CH_3$ |
| 5 | Pent | Pentane | $C_5H_{12}$ | $CH_3-CH_2-CH_2-CH_2-CH_3$ |
| 6 | Hex | Hexane | $C_6H_{14}$ | $CH_3-CH_2-CH_2-CH_2-CH_2-CH_3$ |
| 7 | Hept | Heptane | $C_7H_{16}$ | $CH_3-CH_2-CH_2-CH_2-CH_2-CH_2-CH_3$ |
| 8 | Oct | Octane | $C_8H_{18}$ | $CH_3-CH_2-CH_2-CH_2-CH_2-CH_2-CH_2-CH_3$ |
| 9 | Non | Nonane | $C_9H_{20}$ | $CH_3-CH_2-CH_2-CH_2-CH_2-CH_2-CH_2-CH_2-CH_3$ |
| 10 | Dec | Decane | $C_{10}H_{22}$ | $CH_3-CH_2-CH_2-CH_2-CH_2-CH_2-CH_2-CH_2-CH_2-CH_3$ |

## Condensed Structural Formulas

In a **condensed structural formula**, each carbon atom and its attached hydrogen atoms are written as a group. A subscript indicates the number of hydrogen atoms bonded to each carbon atom.

$$
\begin{array}{ccc}
H-\underset{\underset{H}{|}}{\overset{\overset{H}{|}}{C}}- & = & CH_3- \\
\text{Expanded} & & \text{Condensed}
\end{array}
\qquad
\begin{array}{ccc}
-\underset{\underset{H}{|}}{\overset{\overset{H}{|}}{C}}- & = & -CH_2- \\
\text{Expanded} & & \text{Condensed}
\end{array}
$$

By contrast, the molecular formula gives the total number of each kind of atom, but does not indicate their arrangement in the molecule. Table 10.3 shows the molecular, expanded, and condensed structural formulas for alkanes with one, two, and three carbon atoms.

When a molecule consists of a chain of three or more carbon atoms, the carbon atoms do not lie in a straight line. The tetrahedral shape of carbon arranges the carbon bonds in a zigzag pattern, which is seen in the ball-and-stick model of hexane. (See Figure 10.4.)

Because an alkane has only C—C single bonds, the groups attached to each C are not in fixed positions. They can rotate freely about the bond connecting the carbon atoms. This motion is analogous to the independent rotation of the wheels

**FIGURE 10.4** A ball-and-stick model of hexane.

**Q** Why do the carbon atoms in hexane appear to be arranged in a zigzag chain?

**TABLE 10.3** Writing Structural Formulas for Some Alkanes

| Alkane | Methane | Ethane | Propane |
|---|---|---|---|
| Molecular formula | $CH_4$ | $C_2H_6$ | $C_3H_8$ |
| Structural formulas | | | |
| Expanded | $H-\underset{\underset{H}{|}}{\overset{\overset{H}{|}}{C}}-H$ | $H-\underset{\underset{H}{|}}{\overset{\overset{H}{|}}{C}}-\underset{\underset{H}{|}}{\overset{\overset{H}{|}}{C}}-H$ | $H-\underset{\underset{H}{|}}{\overset{\overset{H}{|}}{C}}-\underset{\underset{H}{|}}{\overset{\overset{H}{|}}{C}}-\underset{\underset{H}{|}}{\overset{\overset{H}{|}}{C}}-H$ |
| Condensed | $CH_4$ | $CH_3-CH_3$ | $CH_3-CH_2-CH_3$ |

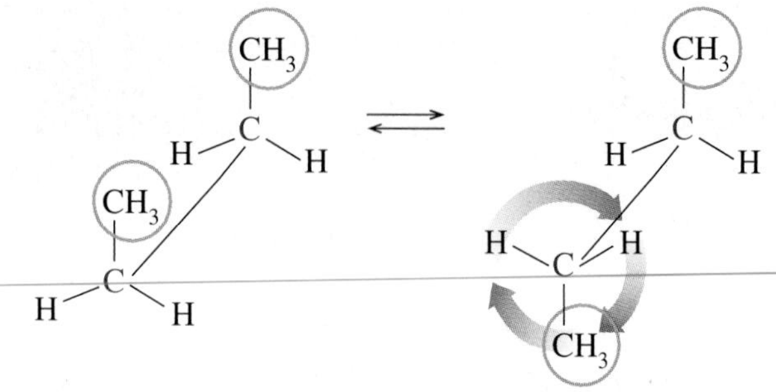

of a toy car. Thus, different arrangements occur during the rotation about a single bond.

Suppose we could look at butane, $C_4H_{10}$, as it rotates. Sometimes the $CH_3$ groups line up in front of each other, and at other times they are opposite each other. As the $CH_3$ groups turn around the single bond, many arrangements are possible. Butane can be depicted by a variety of two-dimensional condensed structural formulas as shown in Table 10.4. All of these condensed structural formulas represent the same compound with four carbon atoms.

**TABLE 10.4** Some Structural Formulas for Butane, $C_4H_{10}$

**Expanded structural formula**

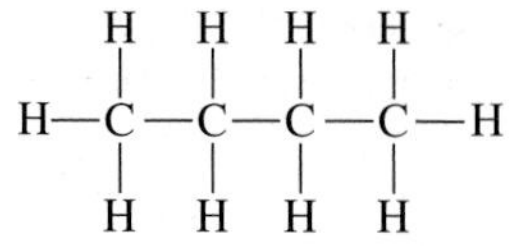

$$
\begin{array}{cccc}
H & H & H & H \\
| & | & | & | \\
H-C-C-C-C-H \\
| & | & | & | \\
H & H & H & H
\end{array}
$$

**Condensed structural formulas**

$CH_3-CH_2-CH_2-CH_3$

$CH_3-CH_2$
　　　$|$
　　$CH_2-CH_3$

$CH_2-CH_2$
$|$　　$|$
$CH_3$　$CH_3$

$CH_3$
$|$
$CH_2-CH_2-CH_3$

$CH_3$
$|$
$CH_2-CH_2$
　　　$|$
　　　$CH_3$

$CH_3$
$|$
$CH_2$
$|$
$CH_2$
$|$
$CH_3$

$CH_3$　　$CH_2$
　$CH_2$　　$CH_3$

---

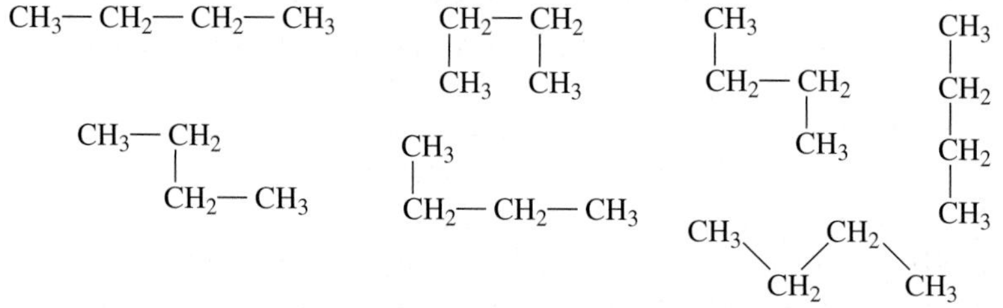

## SAMPLE PROBLEM 10.2

### ■ Drawing Expanded and Condensed Structural Formulas for Alkanes

A molecule of butane, $C_4H_{10}$, has four carbon atoms in a row. What are its expanded and condensed structural formulas?

**SOLUTION**

In the expanded structural formula, four carbon atoms are connected to each other and to hydrogen atoms using single bonds to give each carbon atom a total of four bonds. In the condensed structural formula, each carbon atom and its attached hydrogen atoms are written as $CH_3-$, or $-CH_2-$.

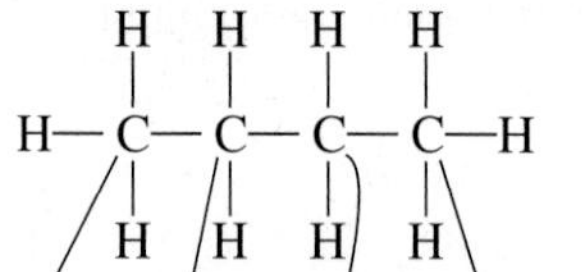

$$
\begin{array}{cccc}
H & H & H & H \\
| & | & | & | \\
H-C-C-C-C-H \\
| & | & | & | \\
H & H & H & H
\end{array}
$$
Expanded structural formula

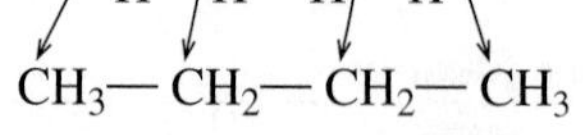

$CH_3-CH_2-CH_2-CH_3$   Condensed structural formula

**STUDY CHECK**

Write the expanded and condensed structural formulas of pentane, $C_5H_{12}$.

## Cycloalkanes

Hydrocarbons can also form cyclic structures called **cycloalkanes**, which have two fewer hydrogen atoms than the corresponding alkanes. Thus, the simplest cycloalkane, cyclopropane ($C_3H_6$) has a ring of three carbon atoms bonded to six hydrogen atoms.

# Career Focus

## Geologist

"Chemistry underpins geology," says Vic Abadie, consulting geologist. "I am a self-employed geologist consulting in exploration for petroleum and natural gas. Predicting the occurrence of an oil reservoir depends in part on understanding chemical reactions of minerals. This is because over geologic time, such reactions create or destroy pore spaces that host crude oil or gas in a reservoir rock formation. Chemical analysis can match oil in known reservoir formations with distant formations that generated oil and from which oil migrated into the reservoirs in the geologic past. This can help identify target areas to explore for new reservoirs.

"I evaluate proposals to drill for undiscovered oil and gas. I recommend that my clients invest in proposed wells that my analysis suggests have strong geologic and economic merit. This is a commercial application of the scientific method: the proposal to drill is

the hypothesis, and the drill bit tests it. A successful well validates the hypothesis and generates oil or gas production and revenue for my clients and me. I do this and other consulting for private and corporate clients. The risk is high, the work is exciting, and my time is flexible."

A simplified formula, which omits the hydrogen atoms and looks like a geometric figure, is a convenient way to show cyclic structures. Each corner of the triangle represents a carbon atom with four bonds to other carbon and hydrogen atoms.

The ball-and-stick models, condensed structural formulas, and geometric formulas for several cycloalkanes are shown in Table 10.5. A cycloalkane is named by adding the prefix *cyclo* to the name of the alkane with the same number of carbon atoms.

**TABLE 10.5**  Formulas of Some Common Cycloalkanes

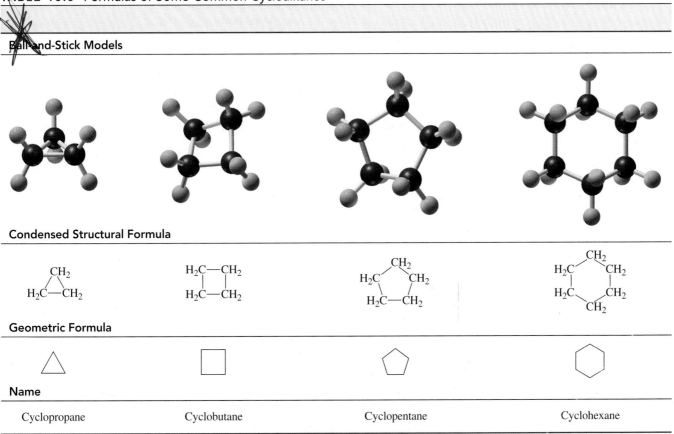

| Ball-and-Stick Models | | | |
|---|---|---|---|
| **Condensed Structural Formula** | | | |
| $H_2C-CH_2$ (with $CH_2$) | $H_2C-CH_2$ / $H_2C-CH_2$ | $CH_2$ / $H_2C$—$CH_2$ / $H_2C-CH_2$ | $CH_2$ / $H_2C$—$CH_2$ / $H_2C$—$CH_2$ / $CH_2$ |
| **Geometric Formula** | | | |
| △ | □ | ⬠ | ⬡ |
| **Name** | | | |
| Cyclopropane | Cyclobutane | Cyclopentane | Cyclohexane |

■ **Naming Alkanes**

Give the IUPAC name for each of the following:

**a.** $CH_3-CH_2-CH_2-CH_2-CH_3$

**b.**

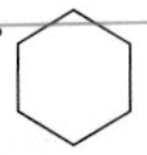

**c.** $CH_3-CH_2-CH_3$

### SOLUTION

**a.** A chain with five carbon atoms is pentane.
**b.** The ring of six carbon atoms is named cyclohexane.
**c.** This alkane is named propane because it has three carbon atoms.

### STUDY CHECK

What is the IUPAC name of the following compound?

---

## QUESTIONS AND PROBLEMS

### Alkanes

**10.9** Give the IUPAC name for each of the following alkanes:

  **a.** 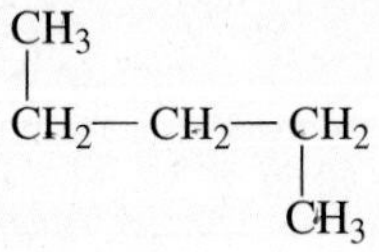 $CH_3$
       |
    $CH_2-CH_2-CH_2$
                |
                $CH_3$

  **b.** $CH_3-CH_3$
               $CH_2-CH_3$
               |
               $CH_2$
               |
  **c.** 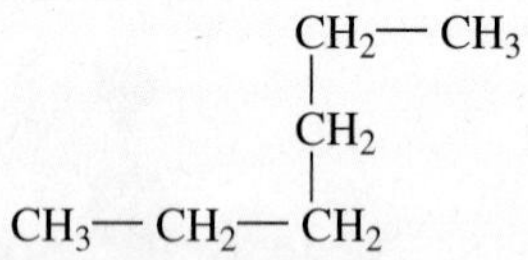 $CH_3-CH_2-CH_2$

**10.10** Give the IUPAC name for each of the following alkanes:

  **a.** $CH_4$
  **b.** $CH_3-CH_2-CH_2-CH_3$
  **c.** 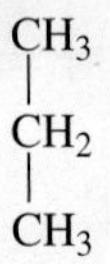 $CH_3$
      |
    $CH_2$
      |
    $CH_3$

**10.11** Write the condensed structural formula or geometric figure for each of the following:
  **a.** methane
  **b.** ethane
  **c.** pentane
  **d.** cyclopropane

**10.12** Write the condensed structural formula or geometric figure for each of the following:
  **a.** propane
  **b.** hexane
  **c.** heptane
  **d.** cyclopentane

---

## LEARNING GOAL

Write the IUPAC names for alkanes with substituents.

# 10.3 ALKANES WITH SUBSTITUENTS

When an alkane has four or more carbon atoms, the atoms can be arranged so that a side group called a **branch** or **substituent** is attached to a carbon chain. For example, there are different ball-and-stick models for two compounds that have the molecular formula $C_4H_{10}$. One model is shown as a chain of four carbon atoms. In the other model, a carbon atom is attached as a branch or substituent to a carbon in a chain of three atoms.

(See Figure 10.5.) An alkane with at least one branch is called a **branched alkane**. When two compounds have the same molecular formula but different arrangements of atoms, they are called **isomers**.

In another example, we can write the condensed structural formulas of three different isomers with the molecular formula $C_5H_{12}$ as follows:

**Isomers of $C_5H_{12}$**

| Alkane | Branched Alkanes |
|---|---|
| $CH_3$—$CH_2$—$CH_2$—$CH_2$—$CH_3$ | $\begin{array}{c} CH_3 \\ | \\ CH_3-CH-CH_2-CH_3 \end{array}$ $\quad$ $\begin{array}{c} CH_3 \\ | \\ CH_3-C-CH_3 \\ | \\ CH_3 \end{array}$ |

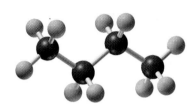

**FIGURE 10.5** The isomers of $C_4H_{10}$ have the same number and type of atoms, but bonded in a different order.

**Q** What makes these molecules isomers?

---

## SAMPLE PROBLEM  10.4

### ■ Isomers

Identify each pair of condensed structural formulas as isomers or the same molecule.

**a.** $\begin{array}{cc} CH_3 & CH_3 \\ | & | \\ CH_2 - CH_2 \end{array}$ and $\begin{array}{c} CH_2-CH_2-CH_3 \\ | \\ CH_3 \end{array}$

**b.** $\begin{array}{c} CH_3 \\ | \\ CH_3-CH-CH_2-CH_2-CH_3 \end{array}$ and $\begin{array}{c} CH_3 \quad CH_3 \\ | \quad\; | \\ CH_3-CH-CH-CH_3 \end{array}$

### SOLUTION

**a.** The condensed structural formulas represent the same molecule because they both have four C atoms in a chain with no substituents.

**b.** These are isomers because the molecular formula $C_6H_{14}$ is identical, but the C atoms are bonded in a different order. One has a $CH_3$ group attached to a five-carbon chain, and the other has two $CH_3$ groups attached to a four-carbon chain.

### STUDY CHECK

Is the following condensed structural formula an isomer or identical to one of the molecules in Sample Problem 10.4 b?

$\begin{array}{c} CH_3 \\ | \\ CH_3-CH_2-CH-CH_2-CH_3 \end{array}$

---

## Substituents in Alkanes

In the IUPAC names for alkanes, a carbon branch is named as an **alkyl group**, which is an alkane that is missing one hydrogen atom. The alkyl group is named by replacing the *ane* ending of the corresponding alkane name with *yl*. Alkyl groups cannot exist on their own: they must be attached to a carbon chain. When a halogen atom is attached to a carbon chain, it is named as a *halo* group: *fluoro* (F), *chloro* (Cl), *bromo* (Br), or *iodo* (I). Some of the common groups attached to carbon chains are illustrated in Table 10.6.

**TABLE 10.6** Names and Formulas of Some Common Substituents

| Substituent | Name |
|---|---|
| $CH_3$— | Methyl |
| $CH_3$—$CH_2$— | Ethyl |
| $CH_3$—$CH_2$—$CH_2$— | Propyl |
| $\begin{array}{c} | \\ CH_3-CH-CH_3 \end{array}$ | Isopropyl |
| F—, Cl—, Br—, I— | Fluoro, chloro, bromo, iodo |

## Rules for Naming Alkanes with Substituents

In the IUPAC system, a carbon chain with a substituent is numbered to give the location of that substituent. Let's take a look at how we use the IUPAC system to name the following alkane:

$$CH_3-\overset{\overset{\displaystyle CH_3}{|}}{CH}-CH_2-CH_2-CH_3$$

**Step 1**  **Write the alkane name of the longest chain of carbon atoms.**  In this alkane, the longest chain has five carbon atoms, which is *pentane*.

$$CH_3-\overset{\overset{\displaystyle CH_3}{|}}{CH}-CH_2-CH_2-CH_3 \qquad \text{pentane}$$

**Step 2**  **Number the carbon atoms starting from the end nearest a substituent.**  Once you start numbering, continue in that same direction.

$$\underset{1\quad\;\; 2\qquad 3\qquad\; 4\qquad 5}{CH_3-\overset{\overset{\displaystyle CH_3}{|}}{CH}-CH_2-CH_2-CH_3} \qquad \text{pentane}$$

**Step 3**  **Give the location and name of each substituent as a prefix to the alkane name.**  Place a hyphen between the number and the substituent name.

$$\underset{1\quad\;\; 2\qquad 3\qquad\; 4\qquad 5}{CH_3-\overset{\overset{\displaystyle CH_3}{|}}{CH}-CH_2-CH_2-CH_3} \qquad \text{2-methylpentane}$$

Below are more examples of giving the location of substituents and using prefixes for two or more of the same substituent. Different substituents are listed in alphabetical order.

$$\underset{1\quad\;\; 2\qquad 3\qquad\; 4\qquad 5}{CH_3-\overset{\overset{\displaystyle CH_3}{|}}{CH}-\overset{\overset{\displaystyle Cl}{|}}{CH}-CH_2-CH_3} \qquad \text{3-chloro-2-methylpentane}$$

Use a prefix (*di*, *tri*, *tetra*) to indicate a group that appears more than once. Use commas to separate two or more numbers.

$$\underset{1\quad\;\; 2\qquad 3\qquad\; 4\qquad 5}{CH_3-\overset{\overset{\displaystyle CH_3}{|}}{CH}-\overset{\overset{\displaystyle CH_3}{|}}{CH}-CH_2-CH_3} \qquad \text{2,3-dimethylpentane}$$

When there are two or more substituents, the main chain is numbered in the direction that gives the lowest set of numbers.

$$\underset{5\qquad\; 4\qquad 3\qquad\;\; 2\qquad 1}{CH_3-\overset{\overset{\displaystyle Br}{|}}{CH}-CH_2-\underset{\underset{\displaystyle Br}{|}}{\overset{\overset{\displaystyle CH_3}{|}}{C}}-CH_3} \qquad \text{2,4-dibromo-2-methylpentane}$$

No number is necessary for a compound with one or two carbon atoms and one substituent.

$$CH_3-Br \qquad\qquad\qquad\qquad\quad \text{bromomethane}$$
$$CH_3-CH_2-Cl \qquad\qquad\qquad \text{chloroethane}$$

When a substituent is attached to a cycloalkane, its name is listed in front of the cycloalkane name. No number in needed for a single substituent.

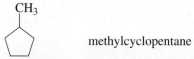

methylcyclopentane

SAMPLE PROBLEM **10.5**

■ **Writing IUPAC Names**

Give the IUPAC name for the following alkane:

$$CH_3—CH_2—\overset{\overset{\displaystyle CH_3}{|}}{\underset{\underset{\displaystyle CH_3}{|}}{C}}—CH_3$$

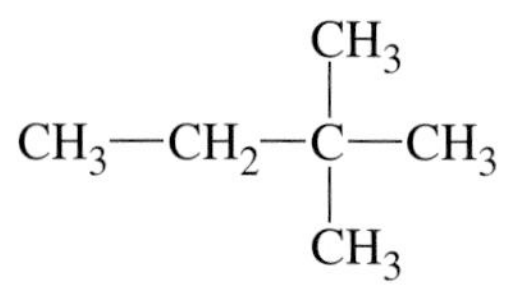

SOLUTION

**Step 1** **Write the alkane name of the longest continuous chain of carbon atoms**. In this alkane, the longest chain has four carbon atoms, which is *butane*.

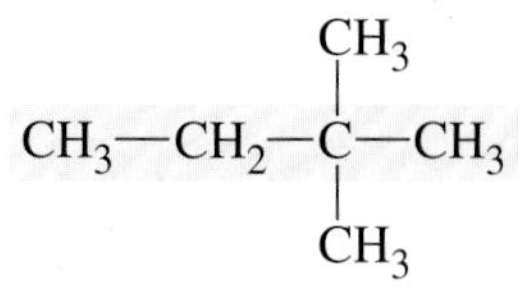

$$CH_3—CH_2—\overset{\overset{\displaystyle CH_3}{|}}{\underset{\underset{\displaystyle CH_3}{|}}{C}}—CH_3 \qquad \text{butane}$$

**Step 2** **Number the carbon atoms starting from the end nearest a substituent.**

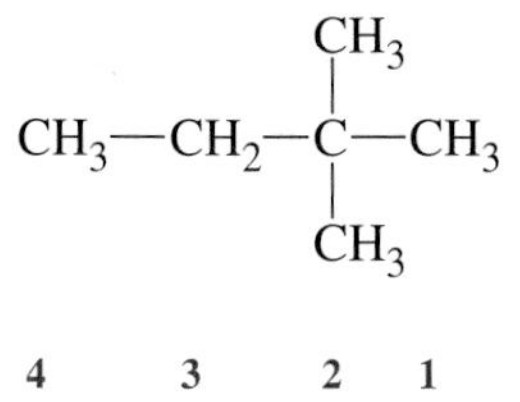

$$CH_3—CH_2—\overset{\overset{\displaystyle CH_3}{|}}{\underset{\underset{\displaystyle CH_3}{|}}{C}}—CH_3 \qquad \text{butane}$$

   4    3    2   1

**Step 3** **Give the location and name of each substituent in front of the name of the longest chain. List the names of different substituents in alphabetical order.**   Place a hyphen between the number and the substituent names, and use commas to separate two or more numbers. A prefix (*di, tri, tetra*) indicates a group that appears more than once.

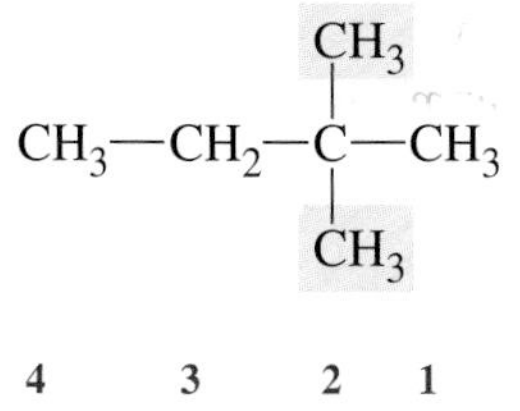

$$CH_3—CH_2—\overset{\overset{\displaystyle CH_3}{|}}{\underset{\underset{\displaystyle CH_3}{|}}{C}}—CH_3 \qquad \text{2,2-dimethylbutane}$$

   4    3    2   1

STUDY CHECK

Give the IUPAC name for the following compound:

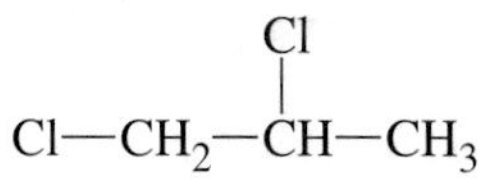

$$Cl—CH_2—\overset{\overset{\displaystyle Cl}{|}}{CH}—CH_3$$

**Guide to Naming Alkanes**

**1** Write the alkane name of the longest chain of carbon atoms.

**2** Number the carbon atoms starting from the end nearest a substituent.

**3** Give the location and name of each substituent (alphabetical order) as a prefix to the name of the main chain.

SAMPLE PROBLEM **10.6**

■ **Writing IUPAC Names for Different Substituents**

Give the IUPAC name for the following alkane:

$$CH_3—\overset{\overset{\displaystyle CH_3}{|}}{CH}—CH_2—\overset{\overset{\displaystyle Br}{|}}{\underset{\underset{\displaystyle CH_3}{|}}{C}}—CH_2—CH_3$$

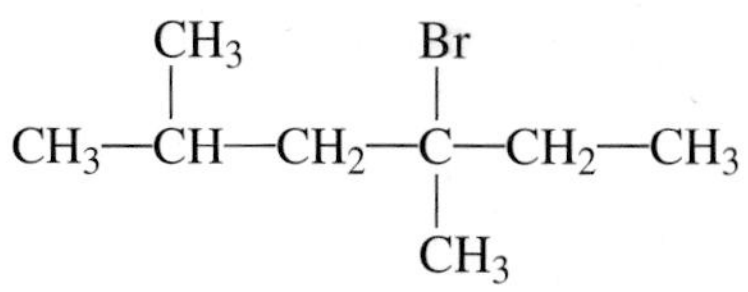

SOLUTION

Step 1 **Write the alkane name of the longest continuous chain of carbon atoms.**   In this alkane, the longest chain has six carbon atoms, which is *hexane*.

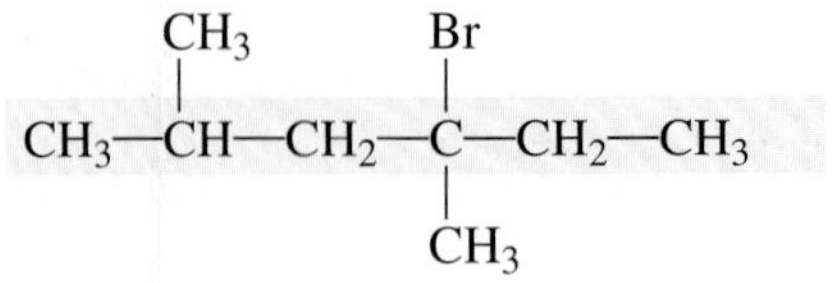

$$CH_3-CH-CH_2-C-CH_2-CH_3 \qquad \text{hexane}$$

with $CH_3$ above the $CH$, and $Br$ above the $C$, and $CH_3$ below the $C$.

Step 2 **Number the carbon atoms starting from the end nearest a substituent.**

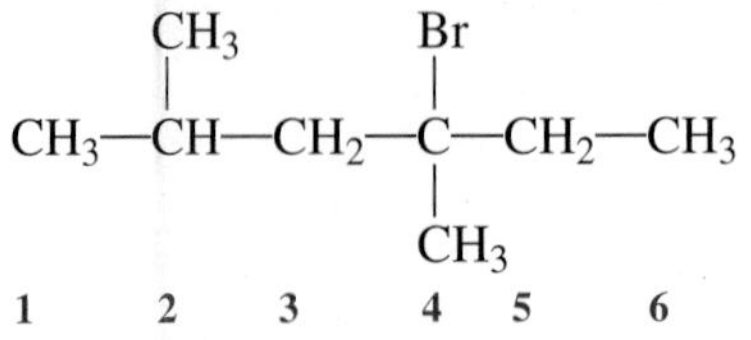

$$CH_3-CH-CH_2-C-CH_2-CH_3 \qquad \text{hexane}$$

with $CH_3$ above the second carbon, $Br$ above the fourth carbon, and $CH_3$ below the fourth carbon.

1    2    3    4    5    6

Step 3 **Give the location and name of each substituent in front of the name of the longest chain. List the names of different substituents in alphabetical order.**   Place a hyphen between the number and the substituent names and commas to separate two or more numbers. A prefix (*di*, *tri*, *tetra*) indicates a group that appears more than once.

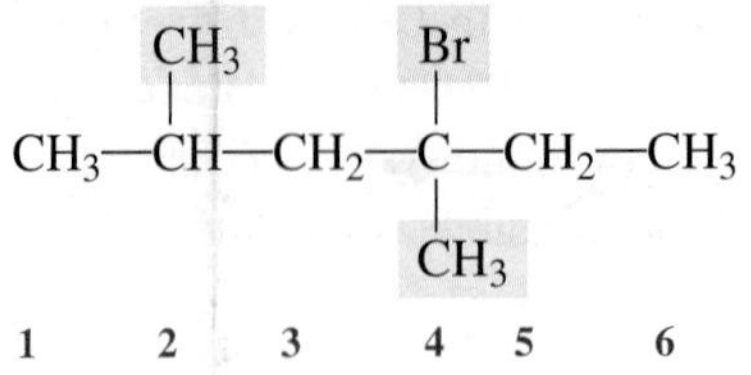

$$CH_3-CH-CH_2-C-CH_2-CH_3 \qquad \text{4-bromo-2,4-dimethylhexane}$$

with $CH_3$ above the second carbon, $Br$ above the fourth carbon, and $CH_3$ below the fourth carbon.

1    2    3    4    5    6

STUDY CHECK

Give the IUPAC name for the following compound:

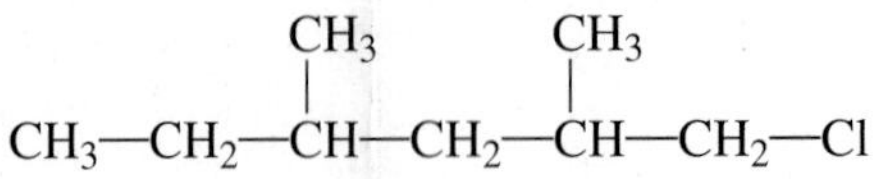

$$CH_3-CH_2-CH-CH_2-CH-CH_2-Cl$$

with $CH_3$ above the third carbon and $CH_3$ above the fifth carbon.

## Drawing Condensed Structural Formulas for Alkanes

The IUPAC name gives all the information needed to draw the condensed structural formula of an alkane. Suppose you are asked to draw the condensed structural formula of 2,3-dimethylbutane. The alkane name gives the number of carbon atoms in the longest chain. The names in the beginning indicate the substituents and where they are attached. We can break down the name in the following way.

**2,3-Dimethylbutane**

| 2,3- | Di | methyl | but | ane |
|---|---|---|---|---|
| Substituents on carbons 2 and 3 | Two identical groups | $CH_3$— alkyl groups | Four carbon atoms in the main chain | Single C—C bonds |

SAMPLE PROBLEM **10.7**

■ **Drawing Condensed Structures from IUPAC Names**

Write the condensed structural formula for 2,3-dimethylbutane.

SOLUTION

We can use the following guide to draw the condensed structural formula.

Step 1 **Draw the main chain of carbon atoms.**   For butane, we draw a chain of four carbon atoms.

$$C—C—C—C$$

Step 2 **Number the chain and place the substituents on the carbons indicated by the numbers.**   The first part of the name indicates two methyl groups ($CH_3—$), one on carbon 2 and one on carbon 3.

$$
\begin{array}{c}
\text{Methyl Methyl}\\
CH_3\,CH_3\\
|\quad\;\;|\\
C—C—C—C\\
1\quad 2\quad 3\quad 4
\end{array}
$$

Step 3 **Add the correct number of hydrogen atoms to give four bonds to each C atom.**

$$
\begin{array}{c}
CH_3\;\;CH_3\\
|\quad\;\;|\\
CH_3—CH—CH—CH_3\\
\text{2,3-Dimethylbutane}
\end{array}
$$

STUDY CHECK

What is the condensed structural formula for 2,4-dimethylpentane?

### Guide to Drawing Alkane Formulas

**1** Draw the main chain of carbon atoms.

**2** Number the chain and place the substituents on the carbons indicated by the numbers.

**3** Add the correct number of hydrogen atoms to give four bonds to each C atom.

## QUESTIONS AND PROBLEMS

### Alkanes with Substituents

**10.13** Indicate whether each of the following pairs of structural formulas represent isomers or the same molecule.

$$
\textbf{a.}\;\; CH_3—\underset{\underset{CH_3}{|}}{CH}—CH_3 \quad\text{and}\quad \underset{\underset{CH_3}{|}}{CH}—CH_3
$$

$$
\textbf{b.}\;\; CH_3—\underset{\underset{CH_3}{|}}{CH}—CH_2—CH_3 \quad\text{and}\quad \underset{\underset{CH_3}{|}}{CH_2}—CH_2—\underset{\underset{CH_3}{|}}{CH_2}
$$

$$
\textbf{c.}\;\; \underset{\underset{CH_3}{|}}{CH_2}—\underset{\underset{CH_3}{|}}{CH}—CH_2—CH_3 \quad\text{and}\quad CH_3—\underset{\underset{CH_3}{|}}{CH}—\underset{\underset{CH_3}{|}}{CH}—CH_3
$$

**10.14** Indicate whether each of the following pairs of structural formulas represent isomers or the same molecule.

$$
\textbf{a.}\;\; CH_3—\underset{\underset{CH_3}{|}}{\overset{\overset{CH_3}{|}}{C}}—CH_3 \quad\text{and}\quad \underset{\underset{CH_3}{|}}{CH}—CH_2—CH_3
$$

$$
\textbf{b.}\;\; CH_3—\underset{\underset{CH_3}{|}}{\overset{\overset{CH_3}{|}}{CH}}—\overset{\overset{CH_3}{|}}{CH}—CH_2 \quad\text{and}\quad CH_3—\underset{\underset{CH_3}{|}}{CH}—CH_2—\underset{\underset{CH_3}{|}}{CH}—CH_3
$$

$$
\textbf{c.}\;\; CH_3—\underset{\underset{CH_3}{|}}{CH}—CH_2—CH_3 \quad\text{and}\quad CH_3—CH_2—\underset{\underset{CH_3}{|}}{CH}—CH_3
$$

**10.15** Give the IUPAC name for each of the following alkanes:

$$
\textbf{a.}\;\; CH_3—\underset{\underset{CH_3}{|}}{CH}—CH_2—CH_3 \qquad \textbf{b.}\;\; CH_3—\underset{\underset{CH_3}{|}}{\overset{\overset{CH_3}{|}}{C}}—CH_3
$$

$$
\textbf{c.}\;\; CH_3—CH_2—\underset{\underset{CH_3}{|}}{CH}—\underset{\underset{CH_3}{|}}{CH}—CH_3
$$

$$
\textbf{d.}\;\; CH_3—\underset{\underset{CH_3}{|}}{\overset{\overset{CH_3}{|}}{C}}—CH_2—\overset{\overset{CH_2—CH_3}{|}}{CH}—CH_2—CH_3
$$

**10.16** Give the IUPAC name for each of the following alkanes:

$$
\textbf{a.}\;\; CH_3—\underset{\underset{CH_3}{|}}{CH}—CH_2—CH_2—CH_3
$$

$$
\textbf{b.}\;\; CH_3—\underset{\underset{CH_3}{|}}{CH}—\underset{\underset{CH_3}{|}}{CH}—CH_3
$$

$$
\textbf{c.}\;\; CH_3—CH_2—\underset{\underset{CH_3}{|}}{CH}—CH_2—\underset{\underset{CH_3}{|}}{CH}—CH_3
$$

$$
\textbf{d.}\;\; CH_3—CH_2—\underset{\underset{CH_2—CH_3}{|}}{CH}—\overset{\overset{CH_2—CH_3}{|}}{CH}—CH_2—CH_3
$$

**10.17** Give the IUPAC name for each of the following cycloalkanes:

a. [square]     b. CH₃ [cyclopentane with CH₃]     c. CH₂—CH₃ [cyclohexane with CH₂—CH₃]

**10.18** Give the IUPAC name for each of the following cycloalkanes:

a. CH₃ [cyclopropane with CH₃]     b. [cyclopentane]     c. [cyclohexane with CH₃]

**10.19** Draw a condensed structural formula for each of the following alkanes:
a. 2-methylbutane
b. 3,3-dimethylpentane
c. 2,3,5-trimethylhexane
d. 3-ethyl-2,5-dimethyloctane

**10.20** Draw a condensed structural formula for each of the following alkanes:
a. 3-ethylpentane
b. 3-ethyl-2-methylpentane
c. 2,2,3,5-tetramethylhexane
d. 4-ethyl-2,2-dimethyloctane

**10.21** Draw the geometric formula for each of the following cycloalkanes:
a. methylcyclopentane
b. cyclobutane

**10.22** Draw the geometric formula for each of the following cycloalkanes:
a. cyclopropane
b. ethylcyclohexane

**10.23** Give the IUPAC name for each of the following compounds:
a. $CH_3—CH_2—Br$
b. $CH_3—CH_2—CH_2—F$
c. $CH_3—\overset{\displaystyle CH_3}{\underset{\displaystyle }{CH}}—Cl$
d. $CHCl_3$

**10.24** Give the IUPAC name for each of the following compounds:
a. $CH_3—CH_2—\overset{\displaystyle Cl}{\underset{\displaystyle }{CH}}—CH_3$     b. $CCl_4$
c. $CH_3—\overset{\displaystyle CH_3}{\underset{\displaystyle CH_3}{C}}—I$     d. $CH_3F$

**10.25** Write the condensed structural formula for each of the following compounds:
a. 2-chloropropane
b. 2-bromo-3-chlorobutane
c. bromomethane
d. tetrabromomethane

**10.26** Write the condensed structural formula for each of the following compounds:
a. 1,1,2,2-tetrabromopropane
b. 2-bromopropane
c. 2,3-dichloro-2-methylbutane
d. dibromodichloromethane

## Health Note

### Common Uses of Halogenated Alkanes

Halogenated alkanes are commonly used as solvents and anesthetics. For many years, carbon tetrachloride was widely used in dry cleaners and in home spot removers to take oils and grease out of clothes. However, this use was discontinued when carbon tetrachloride was found to be toxic to the liver, where it can cause cancer. Today, dry cleaners use other halogenated compounds such as dichloromethane, 1,1,1-trichloroethane, and 1,1,2-trichloro-1,2,2-trifluoroethane.

| $CH_2Cl_2$ | $Cl_3C—CH_3$ | $FCl_2C—CClF_2$ |
|---|---|---|
| Dichloromethane | 1,1,1-Trichloroethane | 1,1,2-Trichloro-1,2,2-trifluoroethane |

General anesthetics are compounds that are inhaled or injected to cause a loss of sensation so that surgery or other procedures can be done without causing pain to the patient. As nonpolar compounds, they are soluble in the nonpolar nerve membranes, where they decrease the ability of the nerve cells to conduct the sensation of pain. Trichloromethane, commonly called chloroform ($CHCl_3$),

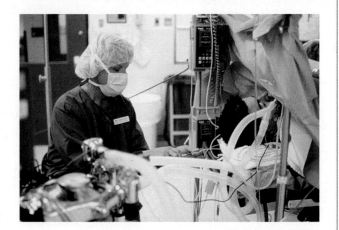

was once used as an anesthetic, but it is toxic and may be carcinogenic. One of the most widely used general anesthetics is halothane (2-bromo-2-chloro-1,1,1-trifluoroethane), also called Fluothane. It has a pleasant odor, is nonexplosive, has few side effects, undergoes few reactions within the body, and is eliminated quickly.

$$F-\overset{\overset{\displaystyle F}{|}}{\underset{\underset{\displaystyle F}{|}}{C}}-\overset{\overset{\displaystyle Cl}{|}}{\underset{\underset{\displaystyle H}{|}}{C}}-Br \qquad \text{Halothane (Fluothane)}$$

For minor surgeries, a local anesthetic such as chloroethane $CH_3-CH_2-Cl$ is applied to an area of the skin. Chloroethane evaporates quickly, cooling the skin and causing a loss of sensation.

# 10.4 PROPERTIES OF ALKANES

Many types of alkanes are the components of fuels that power our cars and oil that heats our homes. You may have used a mixture of hydrocarbons such as mineral oil as a laxative or petrolatum to soften your skin. The differences in uses of many of the alkanes result from their physical properties including solubility, density, and boiling point.

## Solubility and Density

Alkanes are nonpolar, which makes them insoluble in water. However, they are soluble in nonpolar solvents such as other alkanes. Alkanes have densities from 0.62 g/mL to about 0.79 g/mL, which is less dense than water (1.0 g/mL). If there is an oil spill in the ocean, the alkanes in the crude oil remain on the surface and spread over a large area. In the *Exxon Valdez* oil spill in 1989, 40 million liters of oil covered over 25 000 square kilometers of water in Prince William Sound, Alaska. (See Figure 10.6.) If the crude oil reaches the beaches and inlets, there can be considerable damage to beaches, shellfish, fish, birds, and wildlife habitats. Cleanup includes both mechanical and chemical methods. In one method, a nonpolar compound that is "oil attracting" is used to pick up oil, which is then scraped off into recovery tanks.

## Some Uses of Alkanes

The first four alkanes—methane, ethane, propane, and butane—are gases at room temperature and are widely used as heating fuels.

Alkanes having 5–8 carbon atoms (pentane, hexane, heptane, and octane) are liquids at room temperature. They are highly volatile, which makes them useful in fuels such as gasoline.

Liquid alkanes with 9–17 carbon atoms have higher boiling points and are found in kerosene, diesel, and jet fuels. Motor oil is a mixture of high-molecular-weight liquid hydrocarbons and is used to lubricate the internal components of engines. Mineral oil is a mixture of liquid hydrocarbons and is used as a laxative and a lubricant. Alkanes with 18 or more carbon atoms are waxy solids at room temperature. The high-molecular-weight alkanes, known as paraffins, are used to coat fruits and vegetables to retain moisture, inhibit mold growth, and enhance appearance. (See Figure 10.7.) Petrolatum, or Vaseline, is a mixture of liquid hydrocarbons that have low boiling points and are

## LEARNING GOAL

Identify the properties of alkanes and write a balanced equation for combustion.

CASE STUDY
Hazardous Materials

**FIGURE 10.6** In oil spills, large quantities of oil spread over the water.

**Q** What physical properties cause oil to remain on the surface of water?

**FIGURE 10.7** The solid alkanes that make up waxy coatings on fruits and vegetables help retain moisture, inhibit mold, and enhance appearance.

**Q** Why does the waxy coating help the fruits and vegetables retain moisture?

encapsulated in solid hydrocarbons. It is used in ointments and cosmetics and as a lubricant and a solvent.

## Combustion of Alkanes

An alkane undergoes **combustion** when it completely reacts with oxygen to produce carbon dioxide, water, and energy. Carbon–carbon single bonds are difficult to break, which makes alkanes the least reactive family of organic compounds. However, alkanes burn readily in oxygen.

$$\text{Alkane} + O_2 \longrightarrow CO_2 + H_2O + \text{energy}$$

Methane is the gas we use to cook our foods and heat our homes.

$$CH_4 + 2O_2 \longrightarrow CO_2 + 2H_2O + \text{energy}$$

Propane is the gas used in portable heaters and gas barbecues. (See Figure 10.8.)

$$C_3H_8 + 5O_2 \longrightarrow 3CO_2 + 4H_2O + \text{energy}$$

In the cells of our bodies, energy is produced by the combustion of glucose. Although a series of reactions is involved, we can write the overall combustion of glucose in our cells as follows:

$$C_6H_{12}O_6 + 6O_2 \xrightarrow{\text{Enzymes}} 6CO_2 + 6H_2O + \text{energy}$$

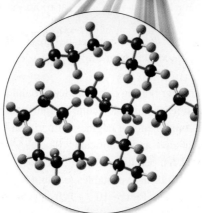

**FIGURE 10.8** The propane fuel in the tank undergoes combustion, which provides energy.

**Q** What is the balanced equation for the combustion of propane?

## Health Note

## Toxicity of Carbon Monoxide

When a propane heater, fireplace, or woodstove is used in a closed room, there must be adequate ventilation. If the supply of oxygen is limited, *incomplete combustion* from burning gas, oil, or wood produces carbon monoxide. The incomplete combustion of methane in natural gas is written as:

$$2CH_4(g) + 3O_2(g) \longrightarrow 2CO(g) + 4H_2O(g) + heat$$

       Limited           Carbon
       oxygen          monoxide
       supply

Carbon monoxide (CO) is a colorless, odorless, poisonous gas. When inhaled, CO passes into the bloodstream, where it attaches to hemoglobin. When CO binds to the hemoglobin, it reduces the amount of oxygen ($O_2$) reaching the organs and cells. As a result, a healthy person can experience a reduction in exercise capability, visual perception, and manual dexterity.

When the amount of hemoglobin (Hb) bound to CO as COHb is 10% or less, a person may experience shortness of breath, mild headache, and drowsiness. Heavy smokers can have as high as 9% COHb in their blood. When as much as 30% of the hemoglobin is COHb, a person may experience more severe symptoms including dizziness, mental confusion, severe headache, and nausea. If 50% or more of the hemoglobin is bound to CO, a person could become unconscious and die if not treated immediately with oxygen.

---

## SAMPLE PROBLEM 10.8

### ■ Combustion

Write a balanced equation for the complete combustion of butane.

#### SOLUTION

The balanced equation for the complete combustion of butane can be written

$$2C_4H_{10} + 13O_2 \longrightarrow 8CO_2 + 10H_2O$$

#### STUDY CHECK

Write a balanced equation for the complete combustion of the following:

$$\begin{array}{c} CH_3 \\ | \\ CH_3-CH-CH_2-CH_3 \end{array}$$

---

## QUESTIONS AND PROBLEMS

### Properties of Alkanes

**10.27** Heptane has a density of 0.68 g/mL and boils at 98 °C.
    **a.** What is the condensed structural formula of heptane?
    **b.** Is it a solid, liquid, or gas at room temperature?
    **c.** Is it soluble in water?
    **d.** Will it float on water or sink?

**10.28** Nonane has a density of 0.79 g/mL and boils at 151 °C.
    **a.** What is the condensed structural formula of nonane?
    **b.** Is it a solid, liquid, or gas at room temperature?
    **c.** Is it soluble in water?
    **d.** Will it float on water or sink?

**10.29** Write a balanced equation for the complete combustion of each of the following compounds:
    **a.** ethane
    **b.** cyclopropane, $C_3H_6$
    **c.** octane
    **d.** cyclohexane, $C_6H_{12}$

**10.30** Write a balanced equation for the complete combustion of each of the following compounds:
    **a.** hexane
    **b.** cyclopentane, $C_5H_{10}$
    **c.** nonane
    **d.** 2-methylbutane

# Green Chemistry Note

## Crude Oil

Crude oil or petroleum contains a wide variety of hydrocarbons. At an oil refinery, the components in crude oil are separated by fractional distillation, a process that removes groups or fractions of hydrocarbons by continually heating the mixture to higher temperatures. (See Table 10.7.) Fractions containing alkanes with longer carbon chains require higher temperatures before they reach their boiling temperature and form gases. The gases are removed and passed through a distillation column where they cool and condense back to liquids. The major use of crude oil is to obtain gasoline, but a barrel of crude oil is only about 35% gasoline. To increase the production of gasoline, heating oils are broken down to give the lower-weight alkanes.

**TABLE 10.7** Typical Alkane Mixtures Obtained by Distillation of Crude Oil

| Distillation Temperatures (°C) | Number of Carbon Atoms | Product |
|---|---|---|
| Below 30 | 1–4 | Natural gas |
| 30–200 | 5–12 | Gasoline |
| 200–250 | 12–16 | Kerosene, jet fuel |
| 250–350 | 15–18 | Diesel fuel, heating oil |
| 350–450 | 18–25 | Lubricating oil |
| Nonvolatile residue | Over 25 | Asphalt, tar |

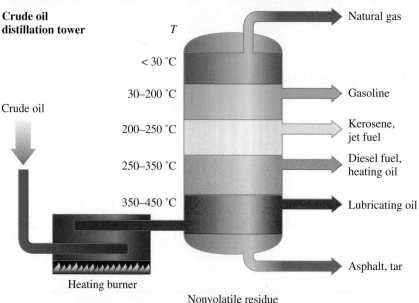

**Crude oil distillation tower**

Crude oil

Heating burner

Nonvolatile residue

*T*

< 30 °C → Natural gas

30–200 °C → Gasoline

200–250 °C → Kerosene, jet fuel

250–350 °C → Diesel fuel, heating oil

350–450 °C → Lubricating oil

→ Asphalt, tar

# 10.5 FUNCTIONAL GROUPS

In organic compounds, carbon atoms are most likely to bond with hydrogen, oxygen, nitrogen, sulfur, and halogens. Table 10.8 lists the number of covalent bonds most often formed by elements found in organic compounds to achieve a complete set of valence electrons. Hydrogen and the halogens form one covalent bond, and carbon forms four covalent bonds. Nitrogen forms three covalent bonds, whereas oxygen and sulfur each form two covalent bonds.

Organic compounds number in the millions, and more are synthesized every day. Within this vast number of compounds, there are specific groups of atoms called **functional groups** that give compounds similar properties. The identification of functional groups allows us to classify organic compounds according to their structure, to name compounds within each family, and to predict their chemical reactions.

## LEARNING GOAL

Classify organic molecules according to their functional groups.

## Alkenes, Alkynes, and Aromatic Compounds

In the hydrocarbon family, there are also alkenes, alkynes, and aromatics. The **alkenes** contain one or more double bonds between carbon atoms; **alkynes** contain triple bonds. **Aromatic compounds** contain benzene, a molecule that has a ring of six carbon atoms with one hydrogen atom attached to each carbon. The benzene structure is represented as a hexagon with a circle in the center.

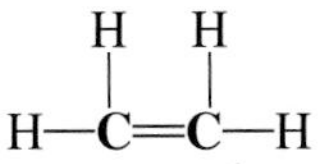

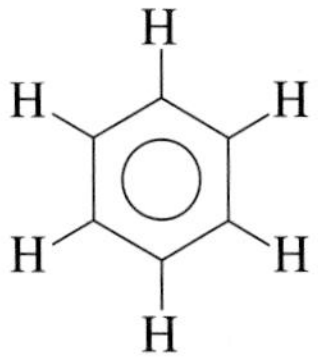

|  | An alkene | An alkyne | An aromatic |
|---|---|---|---|
| Functional group | —C=C— | —C≡C— | (hexagon with circle) |
| Condensed structural formula | CH₂=CH₂ | HC≡CH | |

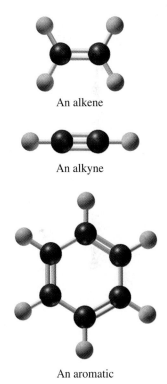

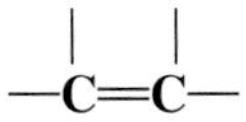

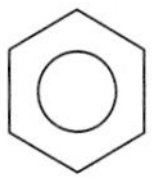

An alkene

An alkyne

An aromatic

## TABLE 10.8  Covalent Bonds for Elements in Organic Compounds

| Element | Group | Covalent Bonds | Structure of Atoms |
|---|---|---|---|
| H | 1A (1) | 1 | H— |
| C | 4A (14) | 4 | —C— |
| N | 5A (15) | 3 | —N— |
| O, S | 6A (16) | 2 | —O—   —S— |
| F, Cl, Br, I | 7A (17) | 1 | —X:   (X = F, Cl, Br, I) |

## Alcohols, Thiols, and Ethers

The characteristic functional group in **alcohols** is the **hydroxyl** (—OH) **group** bonded to a carbon atom. In **thiols**, the functional group —SH is bonded to a carbon atom. In **ethers**, the characteristic feature is an oxygen atom bonded to two carbon atoms. The oxygen atom also has two unshared pairs of electrons, but they are not shown in the condensed structural formulas. In the abbreviation for a functional group, a single line (—) is assumed to be attached to a carbon atom.

An alcohol

$$CH_3\text{---}CH_2\text{---}\mathbf{OH} \qquad CH_3\text{---}O\text{---}CH_3 \qquad CH_3\text{---}CH_2\text{---}\mathbf{SH}$$

An alcohol · An ether · A thiol

Functional group    —O—H    —O—    —S—H

A thiol

## Aldehydes and Ketones

The aldehydes and ketones contain a **carbonyl group** (C=O), which is a carbon with a double bond to oxygen. In an **aldehyde**, the carbon atom of the carbonyl group is bonded to another carbon and one hydrogen atom. Only the simplest aldehyde, $CH_2O$, has a carbonyl group attached to two hydrogen atoms. In a **ketone**, the carbonyl group is bonded to two other carbon atoms.

An ether

$$
\begin{array}{cc}
\overset{\displaystyle O}{\underset{\|}{}} & \overset{\displaystyle O}{\underset{\|}{}} \\
CH_3\text{---}C\text{---}H & CH_3\text{---}C\text{---}CH_3 \\
\text{An aldehyde} & \text{A ketone}
\end{array}
$$

$$
\begin{array}{cc}
\overset{\displaystyle O}{\underset{\|}{}} & \overset{\displaystyle O}{\underset{\|}{}} \\
\text{---}C\text{---}H & \text{---}C\text{---}
\end{array}
$$

Functional group

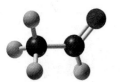

An aldehyde

When we look at a molecule with functional groups, we need to isolate the functional group before we can classify the compound. In the following examples, we have highlighted the functional groups and classified the compound.

$$CH_3\text{---}CH_2\text{---}O\text{---}CH_3 \qquad\qquad CH_3\text{---}CH_2\text{---}CH_2\text{---}OH$$

An ether    An alcohol

A ketone

$$
\begin{array}{cc}
\overset{\displaystyle O}{\underset{\|}{}} & \\
CH_3\text{---}C\text{---}CH_2\text{---}CH_3 & CH_3\text{---}CH{=}CH\text{---}CH_3 \\
\text{A ketone} & \text{An alkene}
\end{array}
$$

---

### SAMPLE PROBLEM   10.9

#### ■ Classifying Organic Compounds

Classify the following organic compounds according to their functional groups:

**a.** $CH_3\text{---}CH_2\text{---}CH_2\text{---}OH$     **b.** $CH_3\text{---}CH{=}CH\text{---}CH_3$

**c.** $CH_3\text{---}CH_2\overset{\displaystyle O}{\overset{\|}{\text{---}C\text{---}}}CH_2\text{---}CH_3$

SOLUTION

**a.** alcohol     **b.** alkene     **c.** ketone

STUDY CHECK

Why is $CH_3\text{---}CH_2\text{---}O\text{---}CH_3$ an ether, but $CH_3\overset{\displaystyle OH}{\overset{|}{\text{---}CH\text{---}}}CH_3$ is an alcohol?

## Carboxylic Acids and Esters

In **carboxylic acids**, the functional group is the *carboxyl group*, which is a combination of the *carb*onyl and hydr*oxyl* groups.

$$CH_3\overset{\overset{\displaystyle O}{\|}}{-C}-O-H \quad \text{or} \quad CH_3COOH \quad \text{or} \quad CH_3CO_2H$$

A carboxylic acid

Functional group   $-\overset{\overset{\displaystyle O}{\|}}{C}-O-H$   or   $-COOH$   or   $-CO_2H$

A carboxylic acid

An **ester** is similar to a carboxylic acid, except that the oxygen of the carboxyl group is attached to a carbon and not to hydrogen.

$$CH_3\overset{\overset{\displaystyle O}{\|}}{-C}-O-CH_3 \quad \text{or} \quad CH_3COOCH_3 \quad \text{or} \quad CH_3CO_2CH_3$$

An ester

Functional group   $-\overset{\overset{\displaystyle O}{\|}}{C}-O-$   or   $-COO-$   or   $-CO_2-$

An ester

## Amines and Amides

In **amines**, the central atom is a nitrogen atom. Amines are derivatives of ammonia, $NH_3$, in which carbon atoms replace one, two, or three of the hydrogen atoms.

$NH_3$        $CH_3-NH_2$        $CH_3-\underset{\underset{\displaystyle CH_3}{|}}{NH}$        $CH_3-\underset{\underset{\displaystyle CH_3}{|}}{N}-CH_3$

Ammonia       Examples of amines

An amine

In an **amide**, the hydroxyl group of a carboxylic acid is replaced by a nitrogen group.

$$CH_3\overset{\overset{\displaystyle O}{\|}}{-C}-NH_2$$

An amide

A list of the common functional groups in organic compounds is shown in Table 10.9.

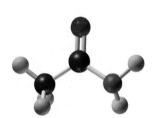

An amide

---

SAMPLE PROBLEM **10.10**

### ■ Identifying Functional Groups

Classify the following organic compounds according to their functional groups:

**a.** $CH_3-CH_2-NH-CH_3$

**b.** $CH_3\overset{\overset{\displaystyle O}{\|}}{-C}-O-CH_2-CH_3$

**c.** $CH_3-CH_2\overset{\overset{\displaystyle O}{\|}}{-C}-OH$

SOLUTION

**a.** amine        **b.** ester        **c.** carboxylic acid

STUDY CHECK

How does a carboxylic acid differ from an ester?

**TABLE 10.9** Classification of Organic Compounds

| Class | Example | Functional Group | Characteristic |
|---|---|---|---|
| Alkene | $H_2C{=}CH_2$ | C=C | Carbon–carbon double bond |
| Alkyne | $HC{\equiv}CH$ | $-C{\equiv}C-$ | Carbon–carbon triple bond |
| Aromatic | 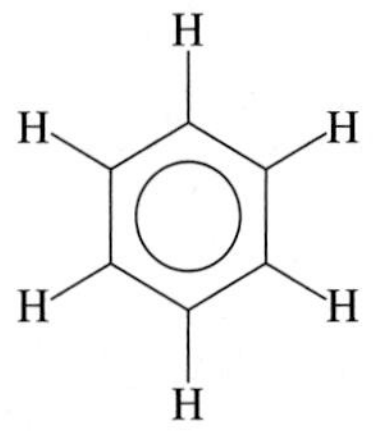 (benzene ring structure) | 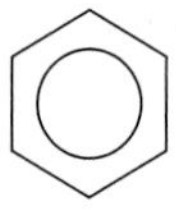 (benzene ring symbol) | Benzene ring (six carbon atoms and six hydrogen atoms) |
| Alcohol | $CH_3-CH_2-OH$ | $-OH$ | Hydroxyl group ($-OH$) |
| Ether | $CH_3-O-CH_3$ | $-O-$ | Oxygen atom bonded to two carbons |
| Thiol | $CH_3-SH$ | $-SH$ | A $-SH$ group bonded to carbon |
| Aldehyde | $CH_3-\overset{O}{\underset{\|}{C}}-H$ | $-\overset{O}{\underset{\|}{C}}-H$ | Carbonyl group (carbon–oxygen double bond) with $-H$ |
| Ketone | $CH_3-\overset{O}{\underset{\|}{C}}-CH_3$ | $-\overset{O}{\underset{\|}{C}}-$ | Carbonyl group (carbon–oxygen double bond) between carbon atoms |
| Carboxylic acid | $CH_3-\overset{O}{\underset{\|}{C}}-O-H$ | $-\overset{O}{\underset{\|}{C}}-O-H$ | Carboxyl group (carbon–oxygen double bond and $-OH$) |
| Ester | $CH_3-\overset{O}{\underset{\|}{C}}-O-CH_3$ | $-\overset{O}{\underset{\|}{C}}-O-$ | Carboxyl group with $-H$ replaced by a carbon |
| Amine | $CH_3-NH_2$ | $-N-$ | Nitrogen atom with one or more carbon groups |
| Amide | $CH_3-\overset{O}{\underset{\|}{C}}-NH_2$ | $-\overset{O}{\underset{\|}{C}}-N-$ | Carbonyl group bonded to nitrogen |

## QUESTIONS AND PROBLEMS

### Functional Groups

**10.31** Identify the class of compounds that contains each of the following functional groups:
  **a.** hydroxyl group attached to a carbon chain
  **b.** carbon–carbon double bond
  **c.** carbonyl group attached to a hydrogen atom
  **d.** carboxyl group attached to two carbon atoms

**10.32** Identify the class of compounds that contains each of the following functional groups:
  **a.** a nitrogen atom attached to one or more carbon atoms
  **b.** carboxyl group
  **c.** oxygen atom bonded to two carbon atoms
  **d.** a carbonyl group between two carbon atoms

**10.33** Classify the following molecules according to their functional groups. The possibilities are alcohol, ether, ketone, carboxylic acid, or amine.
  **a.** $CH_3-CH_2-O-CH_2-CH_3$
  **b.** $CH_3-\overset{OH}{\underset{\|}{CH}}-CH_3$
  **c.** $CH_3-\overset{O}{\underset{\|}{C}}-CH_2-CH_3$
  **d.** $CH_3-CH_2-CH_2-COOH$
  **e.** $CH_3-CH_2-NH_2$

**10.34** Classify the following molecules according to their functional groups. The possibilities are alkene, aldehyde, carboxylic acid, ester, or amide.

**a.** CH$_3$—C(=O)—O—CH$_3$

**b.** CH$_3$—C(=O)—NH$_2$

**c.** CH$_3$—CH$_2$—CH$_2$—C(=O)—H

**d.** CH$_3$—CH$_2$—CH$_2$—CH$_2$—CH$_2$—COOH

**e.** CH$_3$—CH=CH—CH$_3$

## Environmental Note

### Functional Groups in Familiar Compounds

The flavors and odors of foods and many household products can be attributed to the functional groups of organic compounds. As we discuss these familiar products, look for the functional groups we have described.

Ethyl alcohol is the alcohol found in alcoholic beverages. Isopropyl alcohol is another alcohol commonly used to disinfect skin before giving injections and to treat cuts.

CH$_3$—CH$_2$—OH
Ethyl alcohol

CH$_3$—CH(OH)—CH$_3$
Isopropyl alcohol

Acetone or dimethyl ketone is produced in great amounts commercially. Acetone is used as an organic solvent because it dissolves a wide variety of organic substances.

CH$_3$—C(=O)—CH$_3$
Acetone

Ketones and aldehydes are found in flavorings such as vanilla, cinnamon, and spearmint. When we buy a small bottle of liquid flavorings, the aldehyde or ketone is dissolved in alcohol because the compounds are not very soluble in water. The aldehyde butyraldehyde adds a buttery taste to foods and margarine.

CH$_3$—CH$_2$—CH$_2$—C(=O)—H
Butyraldehyde — butter flavoring

The sour tastes of vinegar and fruit juices and the pain from ant stings are all due to carboxylic acids. Acetic acid is the carboxylic acid that makes up vinegar. Aspirin also contains a carboxylic acid group. Esters found in fruits produce the pleasant aromas and tastes of bananas, oranges, pears, and pineapples. Esters are also used as solvents in many household cleaners, polishes, and glues.

CH$_3$—C(=O)—OH
Acetic acid (in vinegar)

CH$_3$—C(=O)—O—CH$_2$—CH$_2$—CH$_3$
Propyl acetate (pears)

CH$_3$—NH$_2$
Methyl amine

CH$_3$—C(=O)—O—CH$_2$—CH$_2$—CH$_2$—CH$_2$—CH$_3$
Pentyl acetate (bananas)

One of the characteristics of fish is their odor, which is due to amines. Amines produced when proteins decay have a particularly pungent and offensive odor.

H$_2$N—CH$_2$—CH$_2$—CH$_2$—CH$_2$—NH$_2$
Putrescine

H$_2$N—CH$_2$—CH$_2$—CH$_2$—CH$_2$—CH$_2$—NH$_2$
Cadaverine

Alkaloids are biologically active amines synthesized by plants to ward off insects and animals. Some typical alkaloids include caffeine, nicotine, histamine, and the decongestant epinephrine. Many are painkillers and hallucinogens such as morphine, LSD, marijuana, and cocaine. Certain parts of our neurons have receptor sites that respond to the various alkaloids. By modifying the structures of certain alkaloids to eliminate side effects, chemists have synthesized painkillers and drugs such as Novocain, codeine, and Valium.

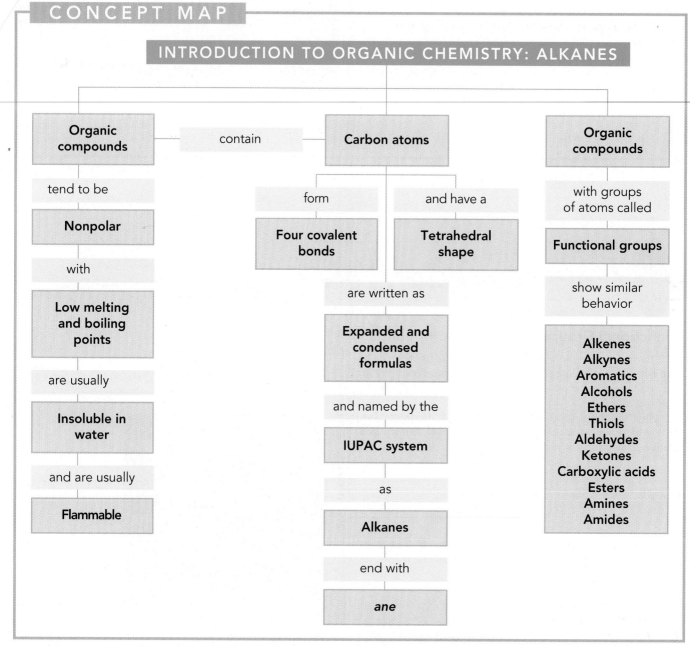

# CHAPTER REVIEW

## 10.1 Organic Compounds

**Learning Goal:** Identify properties characteristic of organic or inorganic compounds.

Most organic compounds have covalent bonds and form nonpolar molecules. Often they have low melting points and low boiling points, are not very soluble in water, produce molecules in solutions, and burn vigorously in air. In contrast, many inorganic compounds are ionic or contain polar covalent bonds and form polar molecules. Many have high melting and boiling points, are usually soluble in water, produce ions in water, and do not burn in air. Carbon atoms share four valence electrons to form four covalent bonds. In the simplest organic molecule, methane ($CH_4$), the four bonds that bond hydrogen to the carbon atom are directed o the corners of a tetrahedron with bond angles of $109.5°$.

## 10.2 Alkanes

**Learning Goal:** Write the IUPAC names and structural formulas for alkanes.

Alkanes are hydrocarbons that have only C—C single bonds. In the expanded structural formula, a separate line is drawn for every bonded atom. A condensed structural formula depicts groups composed of each carbon atom and its attached hydrogen atoms. In cycloalkanes, the carbon atoms form a ring or cyclic structure. The name is written by placing the prefix *cyclo* before the alkane name with the same number of carbon atoms. The IUPAC system is used to name organic compounds in a systematic manner. The IUPAC name indicates the number of carbon atoms.

## 10.3 Alkanes with Substituents

**Learning Goal:** Write the IUPAC names for alkanes with substituents.

In an alkane, the carbon atoms are connected in a chain and bonded to hydrogen atoms. Substituents such as alkyl groups can replace hydrogen atoms on an alkane. In the IUPAC system, halogen atoms are named as fluoro, chloro, bromo, or iodo substituents attached to the main chain.

## 10.4 Properties of Alkanes

**Learning Goal:** Identify the properties of alkanes and write a balanced equation for combustion.

As nonpolar molecules, alkanes are not soluble in water. They are less dense than water. With only weak attractions, they have low melting and boiling points. Although the $C-C$ bonds in alkanes resist most reactions, alkanes undergo combustion. In combustion, or burning, alkanes react with oxygen to produce carbon dioxide, water, and energy.

## 10.5 Functional Groups

**Learning Goal:** Classify organic molecules according to their functional groups.

An organic molecule contains a characteristic group of atoms called a functional group that determines the molecule's family name and chemical reactivity. Functional groups are used to classify organic compounds, act as reactive sites in the molecule, and provide a system of naming organic compounds. Some common functional groups include the hydroxyl group ($-OH$) in alcohols, the carbonyl group ($C=O$) in aldehydes and ketones, and a nitrogen atom $\left(-\overset{|}{N}-\right)$ in amines, and a nitrogen and carbonyl group in amides.

# SUMMARY OF NAMING

| Type | Example | Characteristic | Structure |
|------|---------|----------------|-----------|
| Alkane | Propane | Single $C-C$, $C-H$ bonds | $CH_3-CH_2-CH_3$ |
| | 2-Methylbutane | | $CH_3-\overset{\overset{\displaystyle CH_3}{|}}{CH}-CH_2-CH_3$ |
| | 1-Chloropropane | Halogen atom | $CH_3-CH_2-CH_2-Cl$ |
| Cycloalkane | Cyclobutane | Carbon ring | □ |

# SUMMARY OF REACTIONS

### COMBUSTION

$$\text{Alkane} + O_2 \longrightarrow CO_2 + H_2O + \text{energy}$$

# KEY TERMS

**alcohols** A class of organic compounds that contains the hydroxyl ($-OH$) group bonded to a carbon atom.

**aldehydes** A class of organic compounds that contains a carbonyl group ($C=O$) bonded to at least one hydrogen atom.

**alkanes** Hydrocarbons containing only single bonds between carbon atoms.

**alkenes** Hydrocarbons that contain carbon–carbon double bonds ($C=C$).

**alkyl group** An alkane minus one hydrogen atom. Alkyl groups are named like the alkanes except a *yl* ending replaces *ane*.

**alkynes** Hydrocarbons that contain carbon–carbon triple bonds ($C\equiv C$).

**amides** A class of organic compounds in which the hydroxyl group of a carboxylic acid is replaced by a nitrogen group.

**amines** A class of organic compounds that contains a nitrogen atom bonded to one or more carbon atoms.

**aromatic compound** A compound that contains benzene. Benzene has a six-carbon ring with a hydrogen atom attached to each carbon.

**branch** A carbon group or halogen bonded to the main carbon chain.

**branched alkane** A hydrocarbon containing a substituent bonded to the main chain.

**carbonyl group** A functional group that contains a double bond between a carbon atom and an oxygen atom ($C=O$).

**carboxylic acids** A class of organic compounds that contains the functional group $-COOH$.

**combustion** A chemical reaction in which an alkane reacts with oxygen to produce $CO_2$, $H_2O$, and energy.

**condensed structural formula** A structural formula that shows the arrangement of the carbon atoms in a molecule but groups each carbon atom with its bonded hydrogen atoms ($CH_3$, $CH_2$, or $CH$).

**cycloalkane** An alkane that is a ring or cyclic structure.

**esters** A class of organic compounds that contains a $-COO-$ group with an oxygen atom bonded to carbon.

**ethers** A class of organic compounds that contains an oxygen atom bonded to two carbon atoms.

**expanded structural formula** A type of structural formula that shows the arrangement of the atoms by drawing each bond in the hydrocarbon as $C-H$ or $C-C$.

**functional group** A group of atoms that determine the physical and chemical properties and naming of a class of organic compounds.

**hydrocarbons** Organic compounds consisting of only carbon and hydrogen.

**hydroxyl group** The group of atoms ($-OH$) characteristic of alcohols.

**isomers** Organic compounds in which identical molecular formulas have different arrangements of atoms.

**IUPAC system** A system for naming organic compounds devised by the International Union of Pure and Applied Chemistry.

**ketones** A class of organic compounds in which a carbonyl group is bonded to two carbon atoms.

**organic compounds** Compounds made of carbon that typically have covalent bonds, nonpolar molecules, low melting and boiling points, are insoluble in water, and are flammable.

**substituent** Groups of atoms such as an alkyl group or a halogen bonded to the main chain or ring of carbon atoms.

**thiol** A class of organic molecules that contains the —SH functional group bonded to a carbon atom.

# UNDERSTANDING THE CONCEPTS

**10.35** Sunscreens contain compounds that absorb UV light such as oxybenzone and 2-ethylhexyl *p*-methoxycinnamate.

Identify the functional groups in each of the following UV-absorbing compounds used in sunscreens:
**a.** oxybenzone

$$CH_3—O—\text{(ring)}—\overset{\displaystyle O}{\overset{\|}{C}}—\text{(ring)}$$
$$\text{OH}$$

**b.** 2-ethylhexyl *p*-methoxycinnamate

$$CH_3—O—\text{(ring)}—CH{=}CH—\overset{\displaystyle O}{\overset{\|}{C}}—CH—(CH_2)_3—CH_3$$
$$\overset{|}{CH_2—CH_3}$$

**10.36** Oxymetazoline is a vasoconstrictor used in nasal decongestant sprays such as Afrin.

What functional groups are in oxymetazoline?

$$(CH_3)_3C—\text{(ring)}—\overset{\text{OH}}{}\;CH_3$$

**10.37** Decimemide is used as an anticonvulsant.

What functional groups are in decimemide?

$$CH_3—O$$
$$CH_3(CH_2)_9O—\text{(ring)}—\overset{\displaystyle O}{\overset{\|}{C}}—NH_2$$
$$CH_3—O$$

**10.38** The odor and taste of pineapples is from ethyl butyrate.

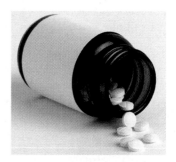

What functional group is in ethyl butyrate?

$$CH_3—CH_2—CH_2—\overset{\displaystyle O}{\overset{\|}{C}}—O—CH_2—CH_3$$

# ADDITIONAL QUESTIONS AND PROBLEMS

**10.39** Compare organic and inorganic compounds in terms of:
**a.** types of bonds   **b.** solubility in water
**c.** melting points   **d.** flammability

**10.40** Identify each of the following compounds as organic or inorganic:
**a.** $Na_2SO_4$   **b.** $CH_2{=}CH_2$
**c.** $Cr_2O_3$   **d.** $C_{12}H_{22}O_{11}$

**10.41** Match the following physical and chemical properties with the compounds butane ($C_4H_{10}$) used in lighters or potassium chloride (KCl) in salt substitutes.

**a.** melts at $-138\ °C$
**b.** burns vigorously in air
**c.** melts at $770\ °C$
**d.** produces ions in water
**e.** is a gas at room temperature

**10.42** Match the following physical and chemical properties with the compounds cyclohexane, $C_6H_{12}$, used as a paint remover, or calcium nitrate, $Ca(NO_3)_2$, used in fertilizers, fireworks, and explosives.
**a.** contains only covalent bonds
**b.** soluble in water
**c.** density is 0.78 g/mL
**d.** flammable
**e.** strong electrolyte

**10.43** Identify the functional groups in each of the following compounds:

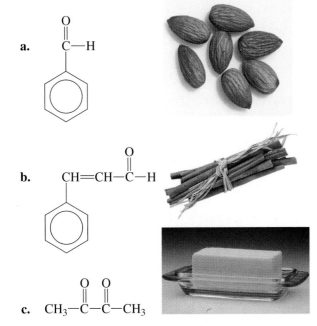

**a.**

**b.** CH=CH—C—H

**c.** $CH_3{-}\overset{O}{\underset{}{C}}{-}\overset{O}{\underset{}{C}}{-}CH_3$

**10.44** Identify the functional groups in each of the following compounds:
**a.** BHA, an antioxidant used as a preservative in foods such as baked goods, butter, meats, and snack foods.

**b.** Vanillin, a flavoring, obtained from the seeds of the vanilla bean.

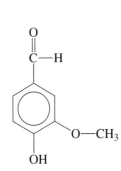

**10.45** The sweetener aspartame is made from two amino acids: aspartic acid and phenylalanine. Identify the functional groups in aspartame.

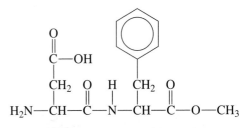

**10.46** Some aspirin substitutes contain phenacetin to reduce fever. Identify the functional groups in phenacetin.

$$CH_3-CH_2-O-\bigcirc-NH-\overset{\overset{\displaystyle O}{\|}}{C}-CH_3$$

**10.47** Write the name of each of the following substituents:
  **a.** $CH_3-$
  **b.** $CH_3-CH_2-CH_2-$
  **c.** $Cl-$

**10.48** Write the name of each of the following substituents:
  **a.** $Br-$
  **b.** $CH_3-\overset{\overset{\displaystyle CH_3}{|}}{CH}-$
  **c.** $CH_3-CH_2-$

**10.49** Give the IUPAC names for each of the following molecules:
  **a.** $CH_3-CH_2-\overset{\overset{\displaystyle CH_3}{|}}{\underset{\underset{\displaystyle CH_3}{|}}{C}}-CH_3$
  **b.** $CH_3-CH_2-Cl$
  **c.** $CH_3-CH_2-\overset{\overset{\displaystyle CH_3-CH_2}{|}}{CH}-CH_2-\overset{\overset{\displaystyle Br}{|}}{CH}-CH_3$
  **d.** ⬡

**10.50** Give the IUPAC names for each of the following molecules:
  **a.** (cyclopentane with $CH_3$)
  **b.** $Cl-CH_2-\overset{\overset{\displaystyle Br}{|}}{CH}-CH_2-Br$
  **c.** $CH_3-\overset{\overset{\displaystyle CH_3}{|}}{CH}-\overset{\overset{\displaystyle CH_3}{|}}{\underset{\underset{\underset{\underset{\displaystyle CH_3}{|}}{CH_2}}{|}}{CH}}-CH_3$

**d.** $CH_3-CH_2-\overset{\overset{\displaystyle Cl}{|}}{\underset{\underset{\underset{\underset{\displaystyle CH_3}{|}}{CH_2}}{|}}{C}}-CH_2-CH_3$

**10.51** Write the condensed structural formula for each of the following molecules:
  **a.** 3-ethylhexane
  **b.** 2,3-dimethylpentane
  **c.** 1,3-dichloro-3-methylheptane

**10.52** Write the condensed structural formula for each of the following molecules:
  **a.** ethylcyclopropane
  **b.** methylcyclohexane
  **c.** isopropylcyclopentane

**10.53** Write a balanced equation for the complete combustion of each of the following:
  **a.** propane
  **b.** $C_5H_{12}$
  **c.** cyclobutane, $C_4H_8$
  **d.** octane
  **e.** $CH_3-CH=CH_2$

**10.54** Write a balanced equation for the complete combustion of each of the following:
  **a.** hexane
  **b.** $HC\equiv C-CH_2-CH_3$
  **c.** cyclopentane, $C_5H_{10}$
  **d.** 2-methylpropane
  **e.** $CH_3-CH_2-CH_2-CH_3$

**10.55** Identify the functional group in each:
  **a.** $CH_3-NH_2$
  **b.** $CH_3-\overset{\overset{\displaystyle OH}{|}}{\underset{\underset{\displaystyle CH_3}{|}}{C}}-CH_2-CH_3$
  **c.** $CH_3-\overset{\overset{\displaystyle CH_3}{|}}{CH}-\overset{\overset{\displaystyle O}{\|}}{C}-O-CH_3$
  **d.** $CH_3-\overset{\overset{\displaystyle CH_3}{|}}{CH}-CH=CH_2$

**10.56** Identify the functional group in each:
  **a.** $CH_3-C\equiv CH$
  **b.** $CH_3-O-CH_2-CH_3$
  **c.** $CH_3-CH_2-CH_2-SH$
  **d.** $CH_3-CH_2-\overset{\overset{\displaystyle O}{\|}}{C}-H$

**10.57** For each definition, find a corresponding term from the following list: alkane, alkene, alkyne, alcohol, ether, aldehyde, ketone, carboxylic acid, ester, amine, functional group, isomers.
  **a.** An organic compound that contains a hydroxyl group bonded to a carbon.
  **b.** A hydrocarbon that contains one or more carbon–carbon double bonds.

c. An organic compound in which the carbon of a carbonyl group is bonded to a hydrogen.

d. A hydrocarbon that contains only carbon–carbon single bonds.

e. An organic compound in which the carbon of a carbonyl group is bonded to a hydroxyl group.

f. An organic compound that contains a nitrogen atom bonded to one or more carbon atoms.

**10.58** For each definition, find a corresponding term from the following list: alkane, alkene, alkyne, alcohol, ether, aldehyde, ketone, carboxylic acid, ester, amine, functional group, isomers.

a. Organic compounds with identical molecular formulas that differ in the order the atoms are connected.

b. An organic compound in which the hydrogen atom of a carboxyl group is replaced by a carbon atom.

c. An organic compound that contains an oxygen atom bonded to two carbon atoms.

d. A hydrocarbon that contains a carbon–carbon triple bond.

e. A characteristic group of atoms that make compounds behave and react in a particular way.

f. An organic compound in which the carbonyl group is bonded to two carbon atoms.

**10.59** A tank on an outdoor heater contains 5.0 lb of propane.

a. Write the equation for the complete combustion of propane.

b. How many kilograms of $CO_2$ are produced by the complete combustion of all the propane?

**10.60** A butane fireplace lighter contains 56.0 g of butane.

a. Write the equation for the complete combustion of butane.

b. How many grams of oxygen are needed for the complete combustion of the butane in the lighter?

# CHALLENGE QUESTIONS

**10.61** The density of pentane, a component of gasoline, is 0.63 g/mL. The heat of combustion for pentane is 845 kcal per mole.

a. Write an equation for the complete combustion of pentane.

b. What is the molar mass?

c. How much heat is produced when 1 gallon of pentane is burned (1 gallon = 3.78 liters)?

d. How many liters of $CO_2$ at STP are produced from the complete combustion of 1 gallon of pentane?

**10.62** Write condensed structural formulas of two esters and a carboxylic acid that each have molecular formula $C_3H_6O_2$.

**10.63** Draw all the possible condensed structural formulas for the organic compounds with 6 carbon atoms that have a 4-carbon chain.

**10.64** Draw all the possible condensed structural formulas for the organic compounds with 4 carbon atoms that have a 3-carbon ring and a hydroxyl group.

**10.65** Consider the compound propane.

a. Draw the condensed structural formula.

b. Write the equation for the complete combustion of propane.

c. How many grams $O_2$ are needed to react with 12.0 L propane at STP?

d. How many grams of $CO_2$ would be produced from the reaction in part **c.**?

**10.66** Consider the compound ethylcyclopentane.

a. Draw the geometric formula.

b. Write the equation for the complete combustion of ethylcyclopentane.

c. Calculate the grams of $O_2$ required for the reaction of 25.0 g ethylcyclopentane.

d. How many liters of $CO_2$ would be produced at STP from the reaction in part **c.**?

**10.67** In an automobile engine, "knocking" occurs when the combustion of gasoline occurs too rapidly. The octane number of gasoline represents the ability of a gasoline mixture to reduce knocking. A sample of gasoline is compared with heptane, rated 0 because it reacts with severe knocking, and 2,2,4-trimethylpentane, which has a rating of 100 because of its low knocking. Write the condensed structural formula, molecular formula, and equation for the complete combustion of 2,2,4-trimethylpentane.

# ANSWERS

## Answers to Study Checks

**10.1** Octane is not soluble in water; it is an organic compound.

**10.2**

$$H-\underset{\underset{H}{|}}{\overset{\overset{H}{|}}{C}}-\underset{\underset{H}{|}}{\overset{\overset{H}{|}}{C}}-\underset{\underset{H}{|}}{\overset{\overset{H}{|}}{C}}-\underset{\underset{H}{|}}{\overset{\overset{H}{|}}{C}}-\underset{\underset{H}{|}}{\overset{\overset{H}{|}}{C}}-H$$

$$CH_3-CH_2-CH_2-CH_2-CH_3$$

**10.3** cyclobutane

**10.4** This is another isomer. There is a five-carbon chain with a carbon group bonded to the middle (third) carbon.

**10.5** 1,2-dichloropropane

**10.6** 1-chloro-2,4-dimethylhexane

**10.7** $CH_3-\underset{\underset{CH_3}{|}}{CH}-CH_2-\underset{\underset{CH_3}{|}}{CH}-CH_3$

**10.8** $CH_3-\underset{\underset{CH_3}{|}}{CH}-CH_2-CH_3 = C_5H_{12}$

$$C_5H_{12} + 8O_2 \longrightarrow 5CO_2 + 6H_2O$$

**10.9** $CH_3—CH_2—O—CH_3$ contains the functional group $C—O—C$; it is an ether.

$$CH_3—\overset{\overset{\displaystyle OH}{|}}{CH}—CH_3$$

contains the $—OH$ functional group; it is an alcohol.

**10.10** A carboxylic acid has a carboxyl group COOH. In an ester, the oxygen atom of the carboxyl group is attached to a carbon atom, not hydrogen.

## Answers to Selected Questions and Problems

**10.1**  **a.** inorganic  **b.** organic
  **c.** organic  **d.** inorganic
  **e.** inorganic  **f.** organic

**10.3**  **a.** inorganic  **b.** organic
  **c.** organic  **d.** inorganic

**10.5**  **a.** ethane  **b.** ethane
  **c.** NaBr  **d.** NaBr

**10.7**  VSEPR theory predicts that the four bonds in $CH_4$ will be as far apart as possible, which means that the hydrogen atoms are at the corners of a tetrahedron.

**10.9**  **a.** pentane  **b.** ethane  **c.** hexane

**10.11**  **a.** $CH_4$  **b.** $CH_3—CH_3$
  **c.** $CH_3—CH_2—CH_2—CH_2—CH_3$
  **d.** △

**10.13**  **a.** same molecule
  **b.** isomers of $C_5H_{12}$
  **c.** isomers of $C_6H_{14}$

**10.15**  **a.** 2-methylbutane
  **b.** 2,2-dimethylpropane
  **c.** 2,3-dimethylpentane
  **d.** 4-ethyl-2,2-dimethylhexane

**10.17**  **a.** cyclobutane
  **b.** methylcyclopentane
  **c.** ethylcyclohexane

**10.19**  **a.** $CH_3—\overset{\overset{\displaystyle CH_3}{|}}{CH}—CH_2—CH_3$

  **b.** $CH_3—CH_2—\overset{\overset{\displaystyle CH_3}{|}}{\underset{\underset{\displaystyle CH_3}{|}}{C}}—CH_2—CH_3$

  **c.** $CH_3—\overset{\overset{\displaystyle CH_3}{|}}{CH}—\overset{\overset{\displaystyle CH_3}{|}}{CH}—CH_2—\overset{\overset{\displaystyle CH_3}{|}}{CH}—CH_3$

  **d.** $CH_3—\overset{\overset{\displaystyle CH_3}{|}}{CH}—\overset{\overset{\displaystyle CH_2—CH_3}{|}}{CH}—CH_2—\overset{\overset{\displaystyle CH_3}{|}}{CH}—CH_2—CH_2—CH_3$

**10.21**  **a.** 
  **b.** 

**10.23**  **a.** bromoethane  **b.** 1-fluoropropane
  **c.** 2-chloropropane  **d.** trichloromethane

**10.25**  **a.** $CH_3—\overset{\overset{\displaystyle Cl}{|}}{CH}—CH_3$

  **b.** $CH_3—\overset{\overset{\displaystyle Br}{|}}{CH}—\overset{\overset{\displaystyle Cl}{|}}{CH}—CH_3$
  **c.** $CH_3Br$
  **d.** $CBr_4$

**10.27**  **a.** $CH_3—CH_2—CH_2—CH_2—CH_2—CH_2—CH_3$
  **b.** liquid
  **c.** insoluble in water
  **d.** float

**10.29**  **a.** $2C_2H_6 + 7O_2 \longrightarrow 4CO_2 + 6H_2O$
  **b.** $2C_3H_6 + 9O_2 \longrightarrow 6CO_2 + 6H_2O$
  **c.** $2C_8H_{18} + 25O_2 \longrightarrow 16CO_2 + 18H_2O$
  **d.** $C_6H_{12} + 9O_2 \longrightarrow 6CO_2 + 6H_2O$

**10.31**  **a.** alcohol
  **b.** alkene
  **c.** aldehyde
  **d.** ester

**10.33**  **a.** ether
  **b.** alcohol
  **c.** ketone
  **d.** carboxylic acid
  **e.** amine

**10.35**  **a.** aromatic, ether, alcohol, ketone
  **b.** aromatic, ether, alkene, ketone

**10.37**  aromatic, ether, amide

**10.39**  **a.** Organic compounds have mostly covalent bonds; inorganic compounds have ionic as well as polar covalent bonds, and a few have nonpolar covalent bonds.
  **b.** Most organic compounds are insoluble in water; many inorganic compounds are soluble in water.
  **c.** Most organic compounds have low melting points; inorganic compounds have high melting points.
  **d.** Most organic compounds are flammable; inorganic compounds are not usually flammable.

**10.41**  **a.** butane
  **b.** butane
  **c.** potassium chloride
  **d.** potassium chloride
  **e.** butane

**10.43**  **a.** aldehyde, aromatic
  **b.** alkene, aldehyde, aromatic
  **c.** ketone

**10.45**  carboxylic acid, aromatic, amine, amide, ester

**10.47**  **a.** methyl
  **b.** propyl
  **c.** chloro

**10.49**  **a.** 2,2-dimethylbutane
  **b.** chloroethane
  **c.** 2-bromo-4-ethylhexane
  **d.** cyclohexane

**10.51** **a.** 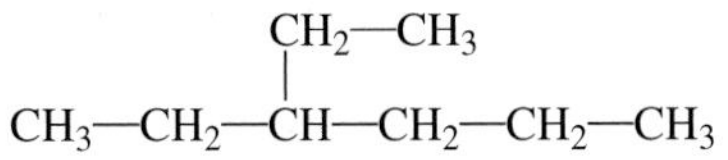 CH$_3$—CH$_2$—$\overset{\displaystyle \text{CH}_2\text{—CH}_3}{\underset{|}{\text{CH}}}$—CH$_2$—CH$_2$—CH$_3$

**b.** 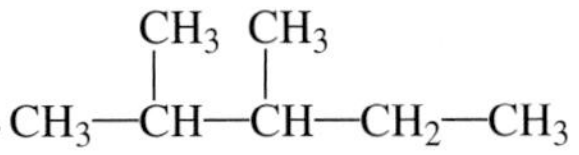 CH$_3$—$\overset{\displaystyle \text{CH}_3}{\underset{|}{\text{CH}}}$—$\overset{\displaystyle \text{CH}_3}{\underset{|}{\text{CH}}}$—CH$_2$—CH$_3$

**c.** 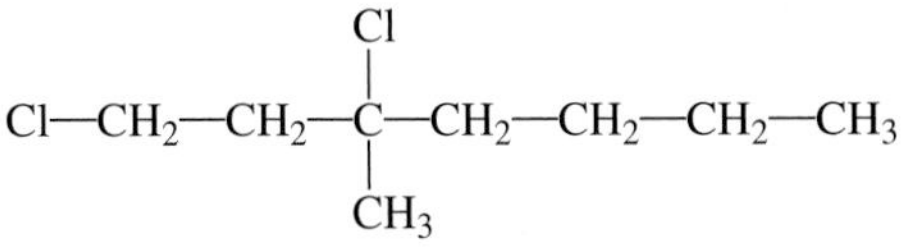 Cl—CH$_2$—CH$_2$—$\overset{\displaystyle \text{Cl}}{\underset{\displaystyle \text{CH}_3}{\overset{|}{\underset{|}{\text{C}}}}}$—CH$_2$—CH$_2$—CH$_2$—CH$_3$

**10.53** **a.** C$_3$H$_8$ + 5O$_2$ $\longrightarrow$ 3CO$_2$ + 4H$_2$O
**b.** C$_5$H$_{12}$ + 8O$_2$ $\longrightarrow$ 5CO$_2$ + 6H$_2$O
**c.** C$_4$H$_8$ + 6O$_2$ $\longrightarrow$ 4CO$_2$ + 4H$_2$O
**d.** 2C$_8$H$_{18}$ + 25O$_2$ $\longrightarrow$ 16CO$_2$ + 18H$_2$O
**e.** 2C$_3$H$_6$ + 9O$_2$ $\longrightarrow$ 6CO$_2$ + 6H$_2$O

**10.55** **a.** amine
**b.** alcohol
**c.** ester
**d.** alkene

**10.57** **a.** alcohol
**b.** alkene
**c.** aldehyde
**d.** alkane
**e.** carboxylic acid
**f.** amine

**10.59** **a.** C$_3$H$_8$ + 5O$_2$ $\longrightarrow$ 3CO$_2$ + 4H$_2$O
**b.** 6.8 kg of CO$_2$

**10.61** **a.** C$_5$H$_{12}$ + 8O$_2$ $\longrightarrow$ 5CO$_2$ + 6H$_2$O
**b.** 72.0 g/mole
**c.** 2.8 $\times$ 10$^4$ kcal
**d.** 3.7 $\times$ 10$^3$ L of CO$_2$ at STP

**10.63** CH$_3$—$\overset{\displaystyle \text{CH}_3}{\underset{|}{\text{CH}}}$—$\overset{\displaystyle \text{CH}_3}{\underset{|}{\text{CH}}}$—CH$_3$     CH$_3$—$\overset{\displaystyle \text{CH}_3}{\underset{\displaystyle \text{CH}_3}{\overset{|}{\underset{|}{\text{C}}}}}$—CH$_2$—CH$_3$

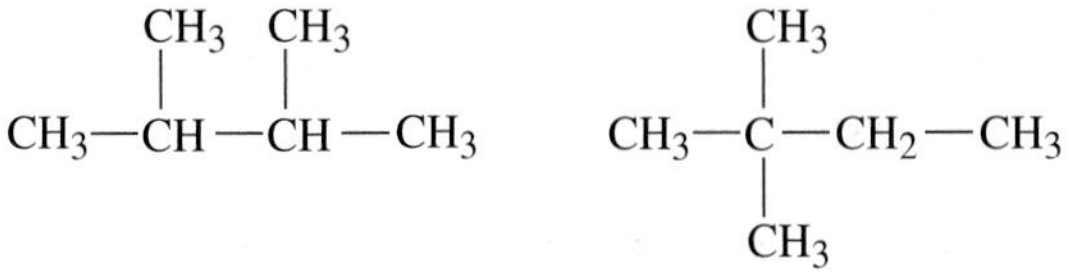

**10.65** **a.** CH$_3$—CH$_2$—CH$_3$
**b.** C$_3$H$_8$ + 5O$_2$ $\longrightarrow$ 3CO$_2$ + 4H$_2$O
**c.** 85.7 g of O$_2$
**d.** 70.0 g of CO$_2$

**10.67** Condensed structural formula

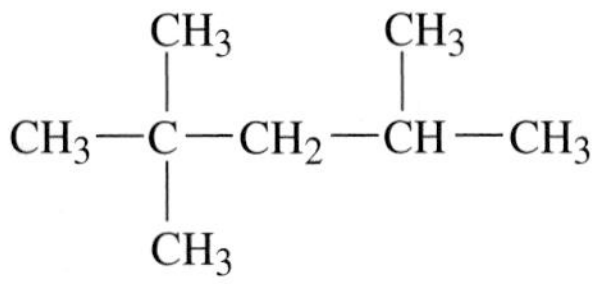

CH$_3$—$\overset{\displaystyle \text{CH}_3}{\underset{\displaystyle \text{CH}_3}{\overset{|}{\underset{|}{\text{C}}}}}$—CH$_2$—$\overset{\displaystyle \text{CH}_3}{\underset{|}{\text{CH}}}$—CH$_3$

Molecular formula C$_8$H$_{18}$

2C$_8$H$_{18}$ + 25O$_2$ $\longrightarrow$ 16CO$_2$ + 18H$_2$O

# 11 Unsaturated Hydrocarbons

**the Chemistry place**

Visit **www.chemplace.com** for extra quizzes, interactive tutorials, career resources, PowerPoint slides for chapter review, math help, and case studies.

*"Dentures replace natural teeth that are extracted due to cavities, bad gums, or trauma," says Dr. Irene Hilton, dentist, La Clinica De La Raza. "I make an impression of teeth using alginate, which is a polysaccharide extracted from seaweed. I mix the compound with water and place the gel-like material in the patient's mouth, where it becomes a hard, cement-like substance. I fill this mold with gypsum (CaSO₄) and water, which form a solid to which I add teeth made of plastic or porcelain. When I get a good match to the patient's own teeth, I prepare a preliminary wax denture. This is placed in the patient's mouth to check the bite and adjust the position of the replacement teeth. Then a permanent denture is made using a hard plastic polymer (methyl methacrylate)."*

In Chapter 10, we looked at alkanes, the hydrocarbons that contain only single bonds. Now we will investigate hydrocarbons that contain double bonds or triple bonds between carbon atoms. When we cook with vegetable oils such as corn oil, safflower oil, or olive oil, we are using organic compounds called lipids, which have one or more double bonds in their long carbon chains. Animal fats also contain long chains of carbon atoms, but with fewer double bonds. If we compare the two types of fats, we find considerable differences in their physical and chemical properties. Vegetable oils are liquid at room temperature, whereas animal fats are solid. Because double bonds are very reactive, oils are more easily oxidized by oxygen in the air, especially at warm temperatures, forming products that have rancid, unpleasant odors. The fats containing mostly single bonds are more resistant to reactions.

# 11.1 ALKENES AND ALKYNES

Alkenes and alkynes are families of hydrocarbons that contain double and triple bonds, respectively. They are called **unsaturated hydrocarbons** because they do not contain the maximum number of hydrogen atoms as do alkanes. They react with hydrogen gas to increase the number of hydrogen atoms to become alkanes, which are **saturated hydrocarbons** because they do have the maximum number of hydrogen atoms.

**LEARNING GOAL**

Identify structural formulas as alkenes, cycloalkenes, and alkynes, and write their IUPAC or common names.

## Identifying Alkenes and Alkynes

**Alkenes** contain one or more carbon–carbon double bonds formed when adjacent carbon atoms share two pairs of valence electrons. Recall that a carbon atom always forms four covalent bonds. In the simplest alkene, ethene, $C_2H_4$, two carbon atoms are connected by a double bond and each is also attached to two H atoms. The ethene molecule is flat because the carbon and hydrogen atoms all lie in the same plane. (See Figure 11.1.)

Ethene, more commonly called ethylene, is an important plant hormone involved in promoting the ripening of fruit. Commercially grown fruit, such as avocados, bananas, and tomatoes, are often picked before they are ripe. Before the fruit is brought to market, it is exposed to ethylene to accelerate the ripening process. Ethylene also accelerates the breakdown of cellulose in plants, which causes flowers to wilt and leaves to fall from trees.

In an **alkyne**, a triple bond forms when two carbon atoms share three pairs of valence electrons. In the simplest alkyne, ethyne ($C_2H_2$), the two carbon atoms in the triple bond are each attached to one hydrogen atom to give a linear geometry. Ethyne, commonly called acetylene, is used in welding where it reacts with oxygen to produce flames with temperatures above 3300 °C.

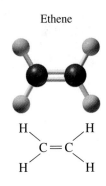

Ethene

Ethyne

H—C≡C—H

**FIGURE 11.1** Ball-and-stick models of ethene and ethyne show the functional groups of double or triple bonds.

**Q** Why are these compounds called unsaturated hydrocarbons?

SAMPLE PROBLEM   11.1

### ■ Identifying Unsaturated Compounds

Classify each of the following condensed structural formulas as an alkane, alkene, or alkyne:

**a.** $CH_3—C≡C—CH_3$          **b.** $CH_3—CH_2—CH_3$

**c.** $CH_3—CH_2—\overset{\overset{\displaystyle CH_3}{|}}{C}=CH—CH_2—CH_3$

#### SOLUTION

The condensed structural formula with a double bond is an alkene, and the one with a triple bond is an alkyne.

**a.** alkyne          **b.** alkane          **c.** alkene

#### STUDY CHECK

Classify each of the following as an alkane, alkene, or alkyne:

**a.** $CH_3—CH_2—CH=CH_2$

**b.** $CH_3—C≡C—CH_3$

**WEB TUTORIAL**
Organic Molecules and Isomers
Introduction to Organic Molecules

## Ripening Fruit

Obtain two unripe green bananas. Place one in a plastic bag and seal it. Leave both bananas on the counter. Check the bananas twice a day to observe any difference in the ripening process.

### QUESTIONS

1. What compound helps ripen the bananas?
2. What are some possible reasons for any difference in the ripening rate?
3. If you wish to ripen an avocado, what procedure might you use?

## Naming Alkenes and Alkynes

The IUPAC names for alkenes and alkynes are similar to those of alkanes.

When naming alkenes and alkynes with substituents, the longest carbon chain must contain the double or triple bond.

**Step 1**   **Name the longest carbon chain that contains the double or triple bond.** Replace the corresponding alkane ending with *ene* for an alkene and *yne* for an alkyne. Cyclic alkenes are named as *cycloalkenes*.

**Step 2**   **Number the longest chain from the end nearest the double or triple bond.** Indicate the position of the double or triple bond, using the lowest number.

$$\underset{4\quad\ 3\quad\ 2\quad\ 1}{CH_3—CH_2—CH=CH_2}\qquad\underset{1\quad\ 2\quad\ 3\quad\ 4}{CH_3—CH=CH—CH_3}\qquad\underset{1\quad\ 2\quad\ 3\quad\ 4}{CH_3—C≡C—CH_3}$$

$$\text{1-Butene}\qquad\qquad\text{2-Butene}\qquad\qquad\text{2-Butyne}$$

Alkenes or alkynes with two or three carbons do not need numbers. For example, the double bond in propene must be between carbon 1 and carbon 2, which can be written $CH_2=CH—CH_3$ or $CH_3—CH=CH_2$.

**Step 3**   **Give the location and name of each substituent (alphabetical order) as a prefix to the alkene or alkyne name.** For *cycloalkenes*, the double bond is always between carbons 1 and 2, and the ring is numbered to give the lowest number to a substituent.

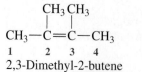

$$\underset{1\quad\ 2\quad\ 3\quad\ 4}{CH_3—\overset{\overset{\displaystyle CH_3}{|}}{C}=\overset{\overset{\displaystyle CH_3}{|}}{C}—CH_3}$$
2,3-Dimethyl-2-butene

$$\underset{4\quad\ 3\quad\ 2\ \ 1}{CH_3—\overset{\overset{\displaystyle Cl}{|}}{CH}—C≡CH}$$
3-Chloro-1-butyne

Cyclobutene

3-Methylcyclopentene

Table 11.1 compares the names of alkanes, alkenes, and alkynes.

**TABLE 11.1 Comparison of Names for Alkanes, Alkenes, and Alkynes**

| Alkane | Alkene | Alkyne |
|---|---|---|
| $H_3C-CH_3$ | $H_2C=CH_2$ | $HC\equiv CH$ |
| Ethane | Ethene (ethylene) | Ethyne (acetylene) |
| $CH_3-CH_2-CH_3$ | $CH_3-CH=CH_2$ | $CH_3-C\equiv CH$ |
| Propane | Propene | Propyne |

SAMPLE PROBLEM  **11.2**

### ■ Naming Alkenes and Alkynes

Write the IUPAC name for each of the following:

$$\begin{matrix} & CH_3 & \\ & | & \\ \textbf{a. } CH_3 & -CH-CH=CH-CH_3 & \end{matrix}$$

**b.** $CH_3-CH_2-C\equiv C-CH_2-CH_3$

#### SOLUTION

**a. Step 1  Name the longest carbon chain that contains the double or triple bond.**
There are five carbon atoms in the longest carbon chain containing the double bond. Replacing the corresponding alkane ending with *ene* gives pentene.

**Step 2  Number the longest chain from the end nearest the double or triple bond.**  The number of the first carbon in the double bond is used to give the location of the double bond.

$$\begin{matrix} & CH_3 & & \\ & | & & \\ CH_3 & -CH-CH=CH-CH_3 & & \text{2-pentene} \\ 5 & 4 \quad 3 \quad 2 \quad 1 & & \end{matrix}$$

**Step 3  Give the location and name of each substituent (alphabetical order) as a prefix to the alkene or alkyne name.**  The methyl group is located on carbon 4.

$$\begin{matrix} & CH_3 & & \\ & | & & \\ CH_3 & -CH-CH=CH-CH_3 & & \text{4-methyl-2-pentene} \\ 5 & 4 \quad 3 \quad 2 \quad 1 & & \end{matrix}$$

**b. Step 1  Name the longest carbon chain that contains the double or triple bond.**
There are six carbon atoms in the longest chain containing the triple bond. Replacing the corresponding alkane ending with *yne* gives hexyne.

**Step 2  Number the main chain from the end nearest the double or triple bond.**
The number of the first carbon in the triple bond is used to give the location of the triple bond.

$$CH_3-CH_2-C\equiv C-CH_2-CH_3 \qquad \text{3-hexyne}$$
$$1 \quad \ \ 2 \quad \ \ 3 \quad \ 4 \quad \ 5 \quad \ \ 6$$

**Step 3  Give the location and name of each substituent (alphabetical order) as a prefix to the alkene or alkyne name.**  There are no substituents in this formula.

#### STUDY CHECK

Draw the condensed structural formulas for each of the following:

**a.** 2-pentyne          **b.** 5-chloro-1-pentene

**Guide to Naming Alkenes and Alkynes**

**1**  Name the longest carbon chain with a double or triple bond.

**2**  Number the carbon chain starting from the end nearest a double or triple bond.

**3**  Give the location and name of each substituent (alphabetical order) as a prefix to the name, if needed.

# *Environmental Note*

## Fragrant Alkenes

The odors you associate with lemons, oranges, roses, and lavender are due to volatile compounds that are synthesized by the plants. Often it is unsaturated compounds that are responsible for the pleasant flavors and fragrances of many fruits and flowers. They were some of the first kinds of compounds to be extracted from natural plant material. In ancient times, they were highly valued in their pure forms. Limonene and myrcene give the characteristic odors and flavors to lemons and bay leaves, respectively. Geraniol and citronellal give roses and lemon grass their distinct aromas, respectively. In the food and perfume industries, these compounds are extracted or synthesized and used as perfumes and flavorings.

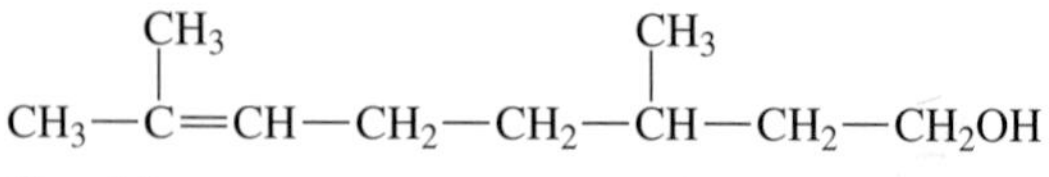

Geraniol, roses

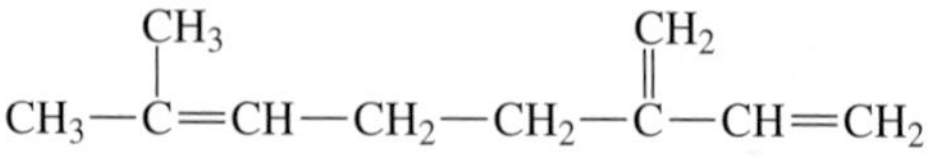

Myrcene, bay leaves

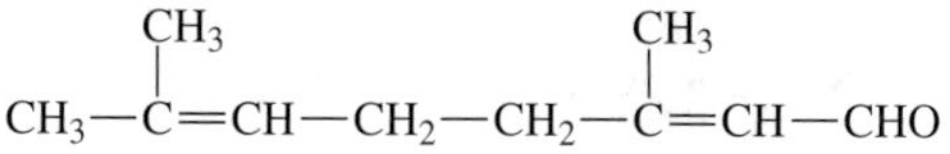

Citronellal, lemon grass

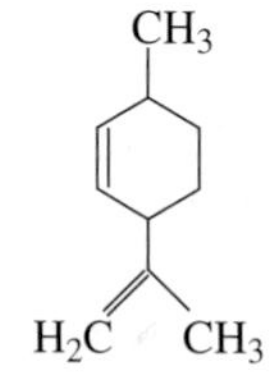

Limonene, lemons

---

## QUESTIONS AND PROBLEMS

### Alkenes and Alkynes

**11.1** Identify the following as alkanes, alkenes, cycloalkenes, or alkynes:

**a.** 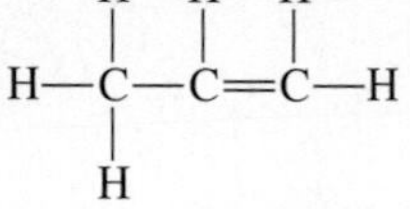 H—C—C=C—H (with H's)

**b.** $CH_3$—$CH_2$—C≡C—H

**c.** 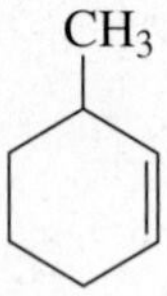 (cyclohexene with $CH_3$)

**d.** $CH_2$=CH—$CH_2$—$CH_2$—$CH_2$—$CH_3$

**11.2** Identify the following as alkanes, alkenes, cycloalkenes, or alkynes:

**a.** 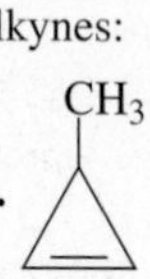 (cyclopropane with $CH_3$)

**b.** 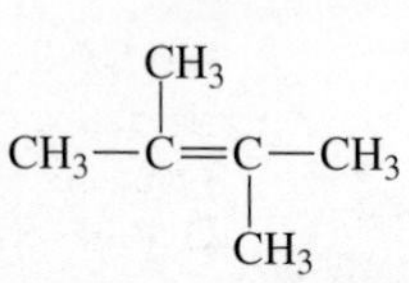 $CH_3$—C=C—$CH_3$ (with $CH_3$ groups)

**c.** (cyclopentane with C≡CH)

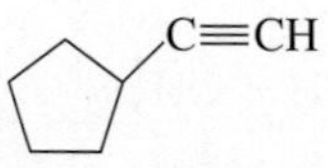

**d.** $CH_3$—$CH_2$—$CH_2$—$CH_2$—$CH_3$

**11.3** Compare the condensed structural formulas of propene and propyne.

**11.4** Compare the condensed structural formulas of 1-butyne and 2-butyne.

**11.5** Give the IUPAC name for each of the following:

**a.** $CH_2$=$CH_2$

**b.** 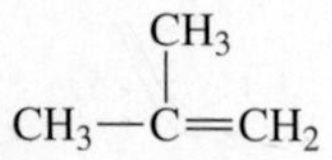 $CH_3$—C=$CH_2$ (with $CH_3$)

**c.** $CH_3$—$CH_2$—C≡C—$CH_3$

**d.**  (square)

**11.6** Give the IUPAC name for each of the following:

**a.** $CH_2$=CH—$CH_2$—$CH_3$

**b.** 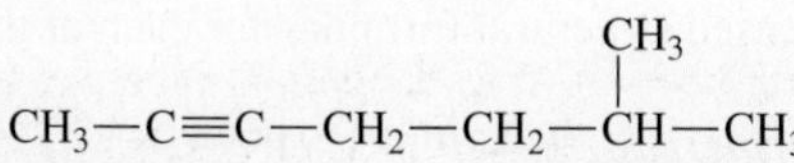 $CH_3$—C≡C—$CH_2$—$CH_2$—CH—$CH_3$ (with $CH_3$)

**c.**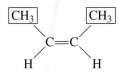

**d.** $CH_3—CH_2—CH\text{=}CH—CH_3$

**11.7** Draw the condensed structural formula for each of the following:
  **a.** propene
  **b.** 1-pentene
  **c.** 2-methyl-1-butene
  **d.** cyclohexene

**e.** 1-butyne
**f.** 1-bromo-3-hexyne

**11.8** Draw the condensed structural formula for each of the following:
  **a.** cyclopentene
  **b.** 3-methyl-1-butyne
  **c.** 3,4-dimethyl-1-pentene
  **d.** cyclobutene
  **e.** propyne
  **f.** 2-methyl-2-hexene

# 11.2 CIS–TRANS ISOMERS

In alkenes, there is no rotation around the carbons in the double bond. (See Explore Your World, "Modeling Cis–Trans Isomers.") Because the double bond is rigid, groups attached to the carbon atoms in the double bond are on one side or the other. In a **cis isomer**, the hydrogen atoms are on the same side of the double bond. In the **trans isomer**, the hydrogen atoms are attached on opposite sides. For example, cis–trans isomers can be written for 2-butene. (See Figure 11.2.) In the cis isomer, the hydrogen atoms are on the same side. This also means that the end carbons of the chain appear as $CH_3$ groups on the other side. In the trans isomer, the hydrogen atoms appear on the opposite sides of the double bond as do the $CH_3$ groups. Because the groups attached to the double bond do not rotate, this molecule has two possible structures or isomers indicated in the name by using the prefix of *cis* or *trans*.

In general, trans isomers are more stable than their cis counterparts because the large groups attached to the double bond are farther apart. The cis–trans isomers of 2-butene are different compounds with different physical properties, such as melting and boiling points, as well as different chemical properties.

$$CH_3 \quad CH_3$$
$$\diagdown \quad \diagup$$
$$C\text{=}C$$
$$\diagup \quad \diagdown$$
$$H \quad H$$

*cis*-2-Butene
(mp −139 °C; bp 3.7 °C)

$$H \quad CH_3$$
$$\diagdown \quad \diagup$$
$$C\text{=}C$$
$$\diagup \quad \diagdown$$
$$CH_3 \quad H$$

*trans*-2-Butene
(mp −106 °C; bp 0.3 °C)

As long as the groups attached to the double bond are different, an alkene will show cis–trans isomers. Another example of cis–trans isomers is the following:

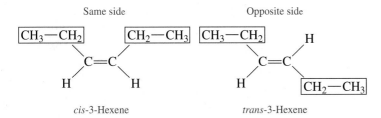

Same side

$$CH_3—CH_2 \quad CH_2—CH_3$$
$$C\text{=}C$$
$$H \quad H$$

*cis*-3-Hexene

Opposite side

$$CH_3—CH_2 \quad H$$
$$C\text{=}C$$
$$H \quad CH_2—CH_3$$

*trans*-3-Hexene

An alkene does not have cis–trans isomers if identical groups are attached to either of the carbon atoms in the double bond. For example, in 1-butene there are two hydrogen atoms on carbon 1. In 2-methyl-1-propene, there are identical groups on both carbon atoms in

## LEARNING GOAL

Write the condensed structural formulas and names for cis–trans isomers of alkenes.

**WEB TUTORIAL**
Geometric Isomers

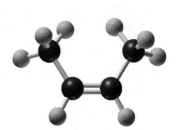

*cis*-2-Butene

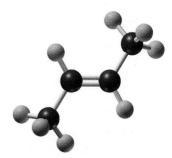

*trans*-2-Butene

**FIGURE 11.2** Ball-and-stick models of the cis and trans isomers of 2-butene.
**Q** What feature in 2-butene accounts for the cis and trans isomers?

## Modeling Cis–Trans Isomers

Because cis–trans isomerism is not easy to imagine, here are some things you can do to understand the difference in rotation around a single bond compared to a double bond and how it affects groups that are attached to the carbon atoms in the double bond.

Put the fingertips of your index fingers together. This is a model of a single bond. Consider the index fingers as a pair of carbon atoms, and think of your thumbs and other fingers as other parts of a carbon chain. While your index fingers are touching, twist your hands and change the position of the thumbs relative to each other. Notice how the relationship of your other fingers changes.

Now place the tips of your index fingers and middle fingers together in a model of a double bond. As you did before, twist your hands to move the thumbs apart. What happens? Can you change the location of your thumbs relative to each other without breaking the double bond? The difficulty of moving your hands with two fingers touching represents the lack of rotation about a double bond. You have

made a model of a cis isomer when both thumbs point in the same direction. If you turn one hand over so one thumb points down and the other thumb points up, you have made a model of a trans isomer.

Cis-hands (cis-thumbs/fingers)

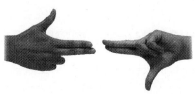

Trans-hands (trans-thumbs/fingers)

the double bond. Alkynes do not have cis–trans isomers because the carbons in the triple bond are each attached to only one group.

$$
\text{Identical atoms} \quad
\begin{array}{c}
\text{H} \qquad\qquad \text{CH}_2-\text{CH}_3 \\
\diagdown \qquad\qquad \diagup \\
\text{C}=\text{C} \\
\diagup \qquad\qquad \diagdown \\
\text{H} \qquad\qquad\qquad \text{H}
\end{array}
\qquad\qquad
\begin{array}{c}
\text{H} \qquad\qquad \text{CH}_3 \\
\diagdown \qquad\qquad \diagup \\
\text{C}=\text{C} \\
\diagup \qquad\qquad \diagdown \\
\text{H} \qquad\qquad\qquad \text{CH}_3
\end{array}
\quad \text{Identical groups}
$$

1-Butene             2-Methyl-1-propene

---

### SAMPLE PROBLEM    11.3

#### ■ Identifying Cis–Trans Isomers

Identify the following as cis or trans isomers.

**a.**
$$
\begin{array}{c}
\text{Br} \qquad\qquad \text{Cl} \\
\diagdown \qquad\qquad \diagup \\
\text{C}=\text{C} \\
\diagup \qquad\qquad \diagdown \\
\text{H} \qquad\qquad \text{H}
\end{array}
$$

**b.**
$$
\begin{array}{c}
\text{CH}_3 \qquad\qquad \text{H} \\
\diagdown \qquad\qquad \diagup \\
\text{C}=\text{C} \\
\diagup \qquad\qquad \diagdown \\
\text{H} \qquad\qquad \text{CH}_2-\text{CH}_3
\end{array}
$$

#### SOLUTION

**a.** This is a cis isomer because the two H atoms in the double bond are on the same side.

**b.** This is a trans isomer because the two H atoms are on the opposite sides of the double bond.

#### STUDY CHECK

Is the following compound *cis*-3-hexene or *trans*-3-hexene?

$$
\begin{array}{c}
\text{CH}_3-\text{CH}_2 \qquad\qquad \text{H} \\
\diagdown \qquad\qquad\qquad \diagup \\
\text{C}=\text{C} \\
\diagup \qquad\qquad\qquad \diagdown \\
\text{H} \qquad\qquad \text{CH}_2-\text{CH}_3
\end{array}
$$

## QUESTIONS AND PROBLEMS

### Cis–Trans Isomers

**11.9** Write the IUPAC name of each of the following, using cis or trans prefixes:

**a.**

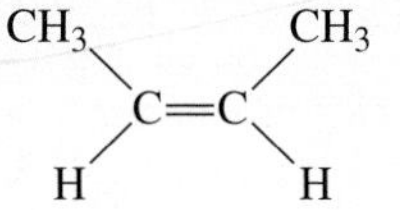

$$\underset{H}{\overset{CH_3}{\diagdown}}C=C\underset{H}{\overset{CH_3}{\diagup}}$$

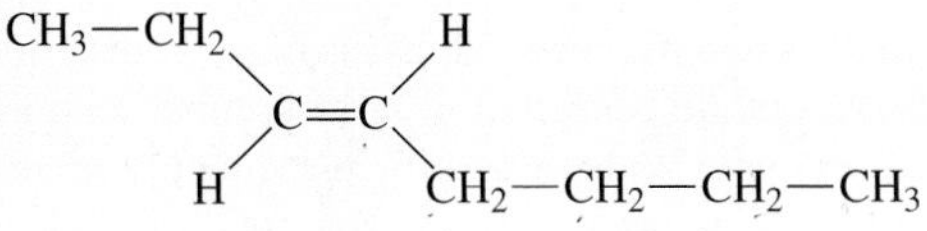

**b.**

$$\underset{H}{\overset{CH_3-CH_2}{\diagdown}}C=C\underset{CH_2-CH_2-CH_2-CH_3}{\overset{H}{\diagup}}$$

**c.**

$$\underset{H}{\overset{CH_3-CH_2-CH_2}{\diagdown}}C=C\underset{H}{\overset{CH_2-CH_3}{\diagup}}$$

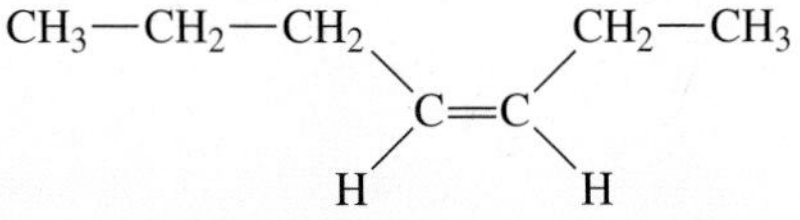

**11.10** Write the IUPAC name of each of the following, using cis or trans prefixes:

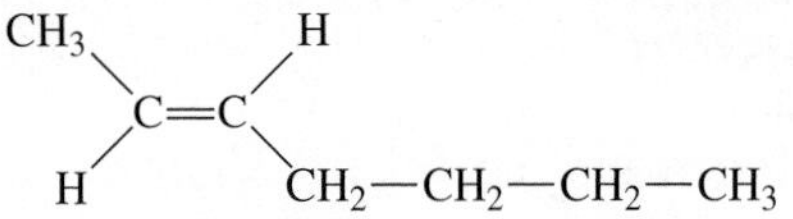

**a.**

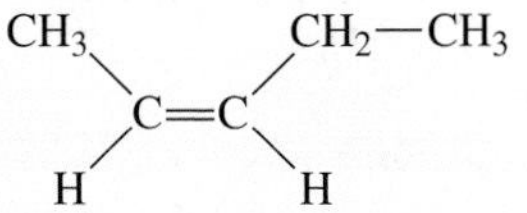

$$\underset{H}{\overset{CH_3}{\diagdown}}C=C\underset{H}{\overset{CH_2-CH_3}{\diagup}}$$

**b.**

$$\underset{H}{\overset{CH_3}{\diagdown}}C=C\underset{CH_2-CH_2-CH_2-CH_3}{\overset{H}{\diagup}}$$

**c.**

$$\underset{H}{\overset{CH_3-CH_2-CH_2}{\diagdown}}C=C\underset{CH_2-CH_3}{\overset{H}{\diagup}}$$

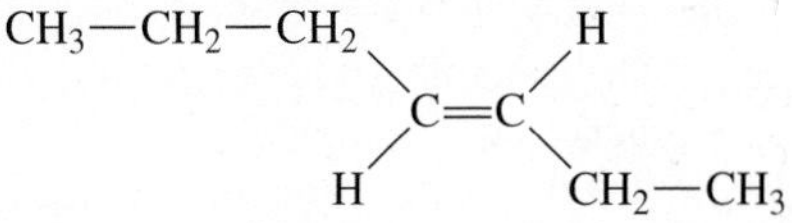

**11.11** Draw the condensed structural formula for each of the following:
**a.** *trans*-2-butene
**b.** *cis*-2-pentene
**c.** *trans*-3-heptene

**11.12** Draw the condensed structural formula for each of the following:
**a.** *cis*-3-hexene
**b.** *trans*-2-pentene
**c.** *cis*-4-octene

---

## Pheromones in Insect Communication

Insects and many other organisms emit minute quantities of chemicals called pheromones. Insects use pheromones to send messages to individuals of the same species. Some pheromones warn of danger, others call for defense, mark a trail, or attract the opposite sex. In the last 40 years, the structures of many pheromones have been chemically determined. One of the most studied is bombykol, the sex pheromone produced by the female of the silkworm moth species. The bombykol molecule is a 16-carbon chain with one cis double bond, one trans double bond, and an alcohol group. A few molecules of synthetic bombykol will attract male silkworm moths from distances of over one kilometer. The effectiveness of many of these pheromones depends on the cis or trans configuration of the double bonds in the molecules. A certain species will respond to one isomer but not the other.

Scientists are interested in synthesizing pheromones for use as nontoxic alternatives to pesticides. When used in a trap, bombykol can be used to isolate male silkworm moths. When a synthetic pheromone is released in several areas of a field or crop, the males cannot locate the females, which disrupts the reproductive cycle. This technique has been successful with controlling the oriental fruit moth, the grapevine moth, and the pink bollworm.

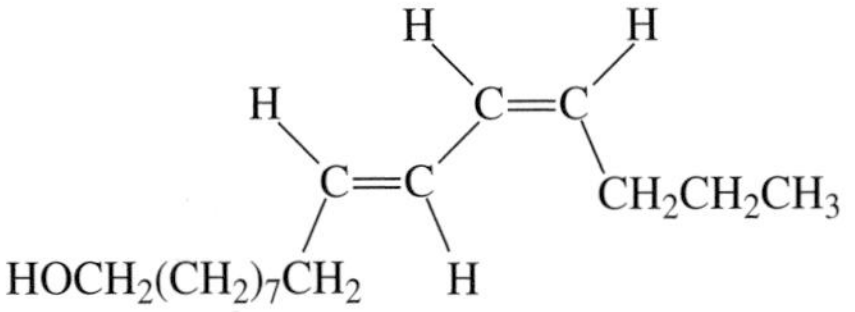

$$\underset{HOCH_2(CH_2)_7CH_2}{\overset{H}{\diagdown}}C=C\underset{H}{\overset{\overset{\displaystyle H \qquad\qquad H}{|\qquad\qquad|}}{\diagup}}$$

Bombykol, sex attractant for the silkworm moth

## Cis–Trans Isomers for Night Vision

The retinas of the eyes consist of two types of cells: rods and cones. The rods on the edge of the retina allow us to see in dim light, and the cones, in the center, produce our vision in bright light. In the rods, there is a substance called rhodopsin that absorbs light. Rhodopsin is composed of *cis*-11-retinal, an unsaturated compound, attached to a protein. When rhodopsin absorbs light, the *cis*-11-retinal isomer is converted to its trans isomer, which changes its shape. The trans form no longer fits the protein, and it separates from the protein. The change from the cis to trans isomer and its separation from the protein generate an electrical signal that the brain converts into an image.

An enzyme (isomerase) converts the trans isomer back to the *cis*-11-retinal isomer and the rhodopsin re-forms. If there is a deficiency of rhodopsin in the rods of the retina, night blindness may occur. One common cause is a lack of vitamin A in the diet. In our diet, we obtain vitamin A from plant pigments containing β-carotene, which is found in foods such as carrots, squash, and spinach. In the small intestine, the β-carotene is converted to vitamin A, which can be converted to *cis*-11-retinal or stored in the liver for future use. Without a sufficient quantity of retinal, not enough rhodopsin is produced to enable us to see adequately in dim light.

### CIS-TRANS ISOMERS OF RETINAL

*cis*-11-Retinal → Light → *trans*-11-Retinal

## LEARNING GOAL

Write the condensed structural formulas and names for the organic products of addition reactions of alkenes and alkynes.

# 11.3 ADDITION REACTIONS

For alkenes and alkynes, the most characteristic reaction is the **addition** of atoms or groups of atoms to the carbons of the double or triple bond. Addition occurs because double and triple bonds are easily broken, providing electrons for new single bonds.

The addition reactions have different names that depend on the type of reactant we add to the alkene, as Table 11.2 shows.

**TABLE 11.2** Summary of Addition Reactions

| Name of Addition Reaction | Reactants | Catalysts | Products |
|---|---|---|---|
| Hydrogenation | Alkene + $H_2$ | Pt, Ni, Pd | Alkane |
|  | Alkyne + $2H_2$ | Pt, Ni, Pd | Alkane |
| Hydration | Alkene + $H_2O$ | $H^+$ (strong acid) | Alcohol |

## Hydrogenation

In a reaction called **hydrogenation**, two atoms of hydrogen add to the carbons in a double bond of an alkene to form an alkane. During hydrogenation, double bonds are converted to single bonds. A catalyst such as platinum (Pt), nickel (Ni), or palladium (Pd) is used to speed up the reaction. The general equation for hydrogenation can be written as follows:

Double bond (alkene)    Single bond (alkane)

Some examples of the hydrogenation of alkenes follow:

$$CH_3-CH=CH-CH_3 + H-H \xrightarrow{Pt} CH_3-\overset{\overset{\displaystyle H}{|}}{C}H-\overset{\overset{\displaystyle H}{|}}{C}H-CH_3$$

2-Butene                                                    Butane

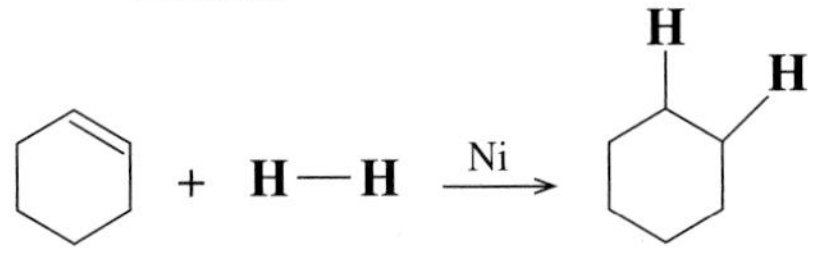

Cyclohexene                     Cyclohexane

The hydrogenation of alkynes requires two molecules of hydrogen ($H_2$) to form the alkane product.

$$CH_3-C\equiv C-CH_3 + 2\,H-H \xrightarrow{Pt} CH_3-\overset{\overset{\displaystyle H}{|}}{\underset{\underset{\displaystyle H}{|}}{C}}-\overset{\overset{\displaystyle H}{|}}{\underset{\underset{\displaystyle H}{|}}{C}}-CH_3$$

2-Butyne                                       Butane

---

## SAMPLE PROBLEM 11.4

### ■ Writing Equations for Hydrogenation

Write the condensed structural formula for the product of the following hydrogenation reactions:

**a.** $CH_3-CH=CH_2 + H_2 \xrightarrow{Pt}$

**b.** 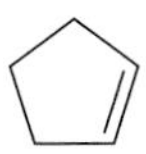 (pentagon ring with double bond) $+ H_2 \xrightarrow{Pt}$

**c.** $HC\equiv CH + 2H_2 \xrightarrow{Ni}$

#### SOLUTION

In an addition reaction, hydrogen adds to the double or triple bond to give an alkane.

**a.** $CH_3-CH_2-CH_3$    **b.** 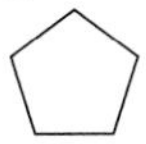 (pentagon ring)    **c.** $H_3C-CH_3$

#### STUDY CHECK

Draw the condensed structural formula of the product of the hydrogenation of 2-methyl-1-butene, using a platinum catalyst.

---

---

## *Health Note*

### Hydrogenation of Unsaturated Fats

Vegetable oils such as corn oil or safflower oil are unsaturated fats composed of fatty acids that contain double bonds. The process of hydrogenation is used commercially to convert the double bonds in the unsaturated fats in vegetable oils to saturated fats such as margarine, which are more solid. Adjusting the amount of added hydrogen produces partially hydrogenated fats such as soft margarine, solid margarine in sticks, and shortenings, which are used in cooking. For example, oleic acid is a typical unsaturated fatty acid in olive oil and has a cis double bond at carbon 9. When oleic acid is hydrogenated, it is converted to stearic acid, a saturated fatty acid.

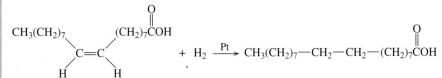

Oleic acid (found in olive oil and other unsaturated fats)

Stearic acid (found in saturated fats)

# Hydration

In **hydration**, an alkene reacts with water (H—OH). A hydrogen atom (H) forms a bond with one carbon atom in the double bond, and the oxygen atom in OH forms a bond with the other carbon. The reaction is catalyzed by a strong acid such as $H_2SO_4$. Hydration is used to prepare alcohols, which have the hydroxyl (—OH) functional group. In the general equation for hydration, the acid is represented by $H^+$.

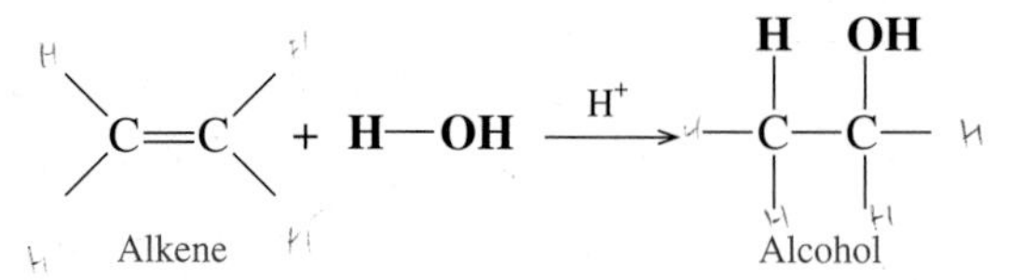

$$\underset{\text{Alkene}}{\overset{H}{\underset{}{\text{C}}}=\overset{}{\text{C}} + \text{H—OH} \xrightarrow{H^+} \underset{\text{Alcohol}}{\overset{H \quad OH}{-\text{C}-\text{C}-}}}$$

$$\underset{\text{Ethene}}{CH_2=CH_2} + \text{H—OH} \xrightarrow{H^+} \underset{\text{Ethanol (ethyl alcohol)}}{\overset{H \quad\quad OH \leftarrow \text{Functional group of alcohols}}{CH_2-CH_2}}$$

When water adds to a double bond in which the carbon atoms are attached to a different number of H atoms, the H from HOH attaches to the carbon that already has the most H atoms. In the following example, the $CH_2$ in the double bond has more H atoms than CH. Thus, the H from HOH attaches to the $CH_2$.

$$\underset{\text{Propene}}{CH_3-CH=CH_2} + \text{H—OH} \xrightarrow{H^+} \underset{\text{2-Propanol}}{\overset{OH \quad H}{CH_3-CH-CH_2}}$$

---

## SAMPLE PROBLEM 11.5

### ■ Writing Products of Hydration

Write the condensed structural formulas for the products that form in the following hydration reactions:

**a.** $CH_3-CH_2-CH_2-CH=CH_2 + \text{HOH} \xrightarrow{H^+}$

**b.** 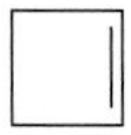 ☐ + HOH $\xrightarrow{H^+}$

### SOLUTION

**a.** The H— and —OH from water (HOH) add to the carbon atoms in the double bond. The H— from water adds to the $CH_2$, which has more H atoms, and the —OH bonds to the CH.

$$CH_3-CH_2-CH_2-\overset{OH}{\underset{\downarrow}{C}}H=\overset{H}{\underset{\downarrow}{C}}H_2 \xrightarrow{H^+} CH_3-CH_2-CH_2-\overset{OH}{\underset{}{C}}H-CH_3$$

**b.** In cyclobutene, each carbon atom in the double bond has one H. The H— from water adds to one carbon atom in the double bond, and the —OH adds to the other.

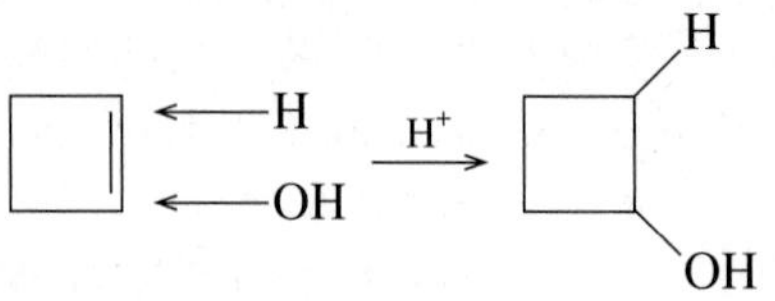

### STUDY CHECK

Draw the condensed structural formula for the alcohol obtained by the hydration of 2-methyl-2-butene.

## QUESTIONS AND PROBLEMS

### Addition Reactions

**11.13** Give the condensed structural formulas of the products in each of the following reactions:

**a.** $CH_3—CH_2—CH_2—CH{=}CH_2 + H_2 \xrightarrow{Pt}$

**b.** $CH_3—CH{=}CH—CH_3 + HOH \xrightarrow{H^+}$

**c.** ⬚ $+ HOH \xrightarrow{H^+}$

**d.** cyclopentene $+ H_2 \xrightarrow{Pt}$

**e.** 2-pentyne $+ 2H_2 \xrightarrow{Pt}$

**11.14** Give the condensed structural formulas of the products in each of the following reactions:

**a.** $CH_3—CH_2—CH{=}CH_2 + HOH \xrightarrow{H^+}$

**b.** cyclohexene $+ H_2 \xrightarrow{Pt}$

**c.** *cis*-2-butene $+ H_2 \xrightarrow{Pt}$

**d.** $CH_3—\overset{\overset{\displaystyle CH_3}{|}}{C}{=}CH—CH_2—CH_3 + H_2 \xrightarrow{Pt}$

**e.** $CH_3—\overset{\overset{\displaystyle CH_3}{|}}{C}H—C{\equiv}CH + 2H_2 \xrightarrow{Pt}$

# 11.4 POLYMERS OF ALKENES

**Polymers** are large molecules that consist of small repeating units called **monomers**. In the past hundred years, the plastics industry has made synthetic polymers that are in many of the materials we use every day, such as carpeting, plastic wrap, nonstick pans, plastic cups, and rain gear. In medicine, synthetic polymers are used to replace diseased or damaged body parts such as hip joints, teeth, heart valves, and blood vessels. (See Figure 11.3.) There are about 100 billion kg of plastics now produced every year, which is over 15 kg for every person on Earth.

## Addition Polymers

Many of the synthetic polymers are made by addition reactions of monomers that are small alkenes. The conditions for many polymerization reactions require high temperatures and very high pressure (over 1000 atm). In polymerization, a series of addition reactions joins one monomer to the next to form a long carbon chain that contains as many as 1000 monomers. Polyethylene, a polymer made from ethylene monomers, is used in plastic bottles, film, and plastic dinnerware. More polyethylene is produced worldwide than any other polymer.

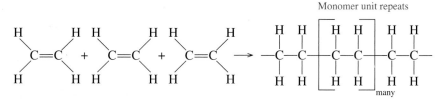

Ethene (ethylene) monomers    Polyethylene section

Table 11.3 lists several alkene monomers that are used to produce common synthetic polymers, and Figure 11.4 shows examples of each.

The alkane-like nature of these plastic synthetic polymers makes them unreactive. Thus, they do not decompose easily (they are non-biodegradable) and have become contributors to pollution. Efforts are being made to make them more degradable. It is becoming increasingly important to recycle plastic material, rather than add to our growing landfills. You

### LEARNING GOAL

Draw structural formulas of monomers that form a polymer or a three-monomer section of a polymer.

**WEB TUTORIAL**
Polymers

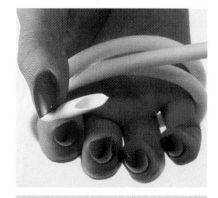

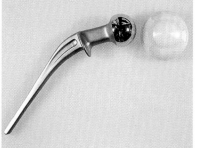

**FIGURE 11.3** Synthetic polymers are used to replace diseased veins and arteries. A broken hip is repaired using a metal piece that fits into an artificial plastic cup socket.
**Q** Why are the substances in these plastic devices called polymers?

## Explore Your World

### Polymers and Recycling Plastics

1. Make a list of the items you use or have in your room or home that are made of polymers.
2. Recycling information on the bottom or side of a plastic bottle includes a triangle with a code number that identifies the type of polymer used to make the plastic. Make a collection of several different kinds of plastic bottles. Try to find plastic items with each type of polymer.

#### QUESTIONS

1. What are the most common types of plastics among the plastic containers in your collection?
2. What are the monomer units of some of the plastics you looked at?

**TABLE 11.3** Some Alkenes and Their Polymers

| Monomer | Polymer Section | Common Uses |
|---|---|---|
| $CH_2$=$CH_2$ <br> Ethene (ethylene) | 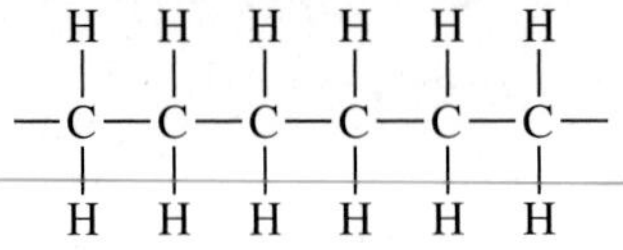 Polyethylene | Plastic bottles, film, insulation materials |
| 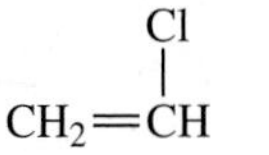 $CH_2$=CH <br> Cl <br> Chloroethene (vinyl chloride) | 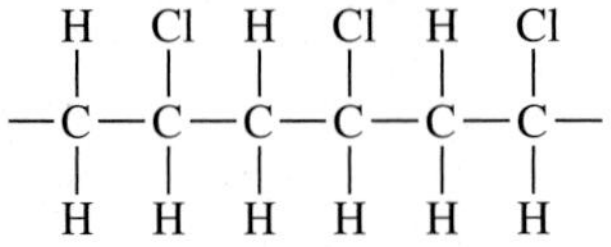 Polyvinyl chloride (PVC) | Plastic pipes and tubing, garden hoses, garbage bags |
| 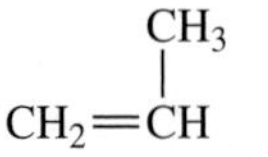 $CH_2$=CH <br> $CH_3$ <br> Propene (propylene) | 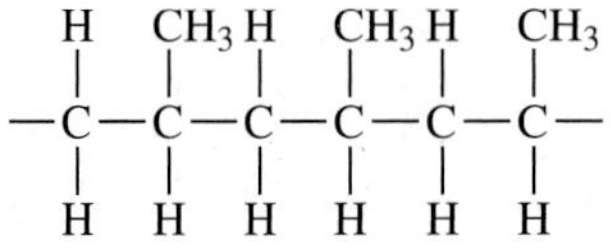 Polypropylene | Ski and hiking clothing, carpets, artificial joints |
| 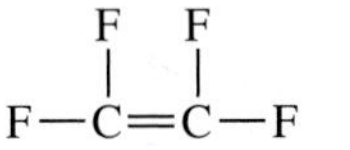 F—C=C—F <br> F   F <br> Tetrafluoroethene | 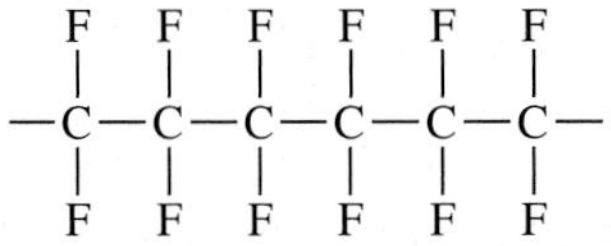 Polytetrafluoroethylene (Teflon) | Nonstick coatings |
| 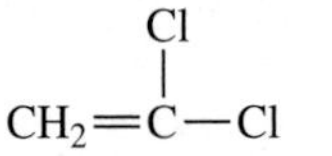 $CH_2$=C—Cl <br> Cl <br> 1,1-Dichloroethene | 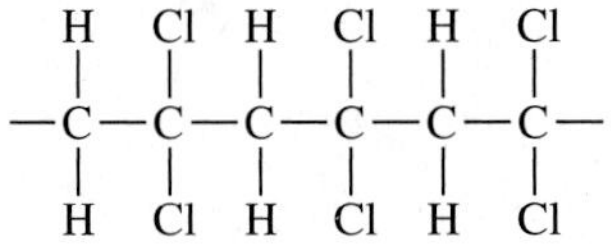 Polydichloroethylene (Saran) | Plastic film and wrap |
|  $H_2C$=CH <br> Phenylethene (styrene) | 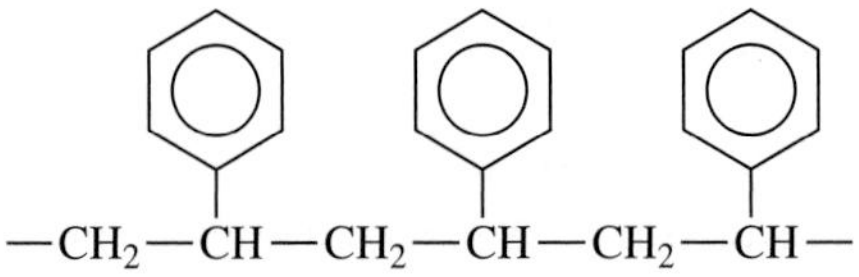 Polystyrene | Plastic coffee cups, and cartons, insulation |

can identify the type of polymer used to manufacture a plastic item by looking for the recycle symbol (arrows in a triangle) found on the label or on the bottom of the plastic container. For example, the number 5 or the letters PP inside the triangle is a code for a polypropylene plastic.

| 1 | 2 | 3 | 4 | 5 | 6 |
|---|---|---|---|---|---|
| PETE | HDPE | PVC | LDPE | PP | PS |
| Polyethylene terephthalate | High-density polyethylene | Polyvinyl chloride | Low-density polyethylene | Polypropylene | Polystyrene |

### SAMPLE PROBLEM 11.6

#### ■ Polymers

What are the starting monomers for the following polymers?

**a.** polypropylene

**b.** Saran

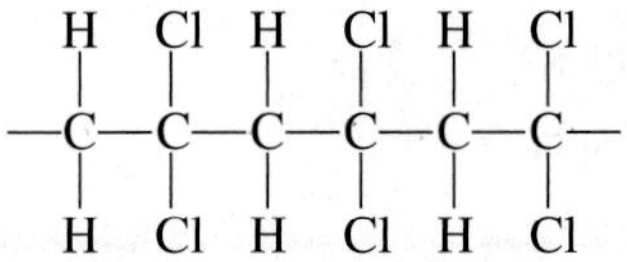

```
    H   Cl  H   Cl  H   Cl
    |   |   |   |   |   |
 —C — C — C — C — C — C—
    |   |   |   |   |   |
    H   Cl  H   Cl  H   Cl
```

Polyethylene

Polyvinyl chloride

Polypropylene

Polytetrafluoroethylene (Teflon)

Polydichloroethylene (Saran)

Polystyrene

**FIGURE 11.4** Synthetic polymers provide a wide variety of items that we use every day.
**Q** What are some alkenes used to make the polymers in these plastic items?

SOLUTION

**a.** propene (propylene), $\text{CH}_2{=}\overset{\overset{\displaystyle \text{CH}_3}{\mid}}{\text{CH}}$

**b.** 1,1-dichloroethene, $\text{CH}_2{=}\overset{\overset{\displaystyle \text{Cl}}{\mid}}{\text{C}}{-}\text{Cl}$

STUDY CHECK

What is the monomer for PVC?

## QUESTIONS AND PROBLEMS

### Polymers of Alkenes

**11.15** What is a polymer?

**11.16** What is a monomer?

**11.17** Write an equation that represents the formation of a part of the Teflon polymer from three of the monomer units.

**11.18** Write an equation that represents the formation of a part of the polystyrene polymer from three of the monomer units.

## LEARNING GOAL

Describe the bonding in benzene; name aromatic compounds, and write their structural formulas.

# 11.5 AROMATIC COMPOUNDS

In 1825, Michael Faraday isolated a hydrocarbon called benzene, which had the molecular formula $C_6H_6$. Because many compounds containing benzene had fragrant odors, the family of benzene compounds became known as **aromatic compounds**. A **benzene** molecule consists of a ring of six carbon atoms with one hydrogen atom attached to each carbon. Each carbon atom uses three valence electrons to bond to the hydrogen atom and two adjacent carbons. That leaves one valence electron to share in a double bond with an adjacent carbon. When it was first discovered, scientists expected benzene to be very reactive like alkenes, but it was found to be much less reactive. It behaved more like an alkane. In 1865, August Kekulé proposed that the carbon atoms in benzene were arranged in a flat ring with alternating single and double bonds between the carbon atoms. This idea led to two ways of writing the benzene structure, as follows:

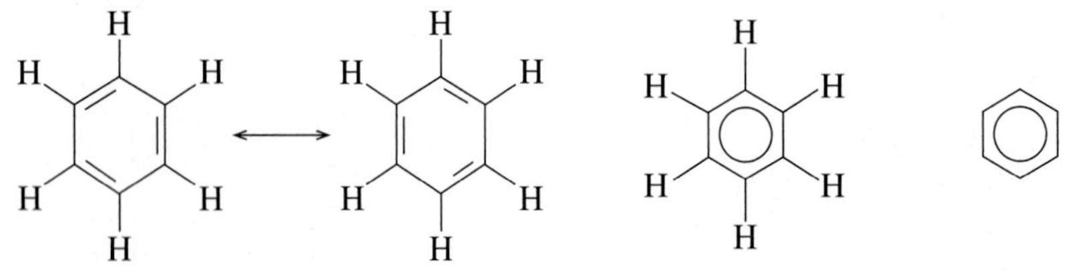

Structures for benzene

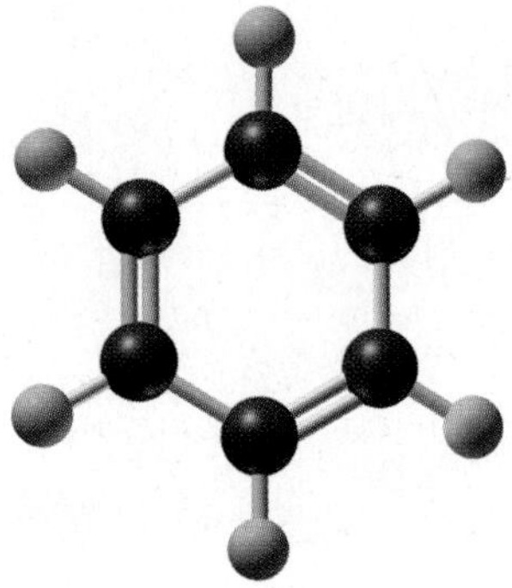

However, there is only one structure of benzene. Today we know that all the bonds in benzene are identical and the electrons are shared equally. This is the unique feature that makes aromatic compounds especially stable. Today the benzene structure is also represented as a hexagon with a circle in the center.

## Naming Aromatic Compounds

Aromatic compounds that contain a benzene ring with a single substituent are usually named as benzene derivatives. However, many of these compounds have been important in chemistry for many years and still use their common names. Some widely used names such as toluene, aniline, and phenol are allowed by IUPAC rules.

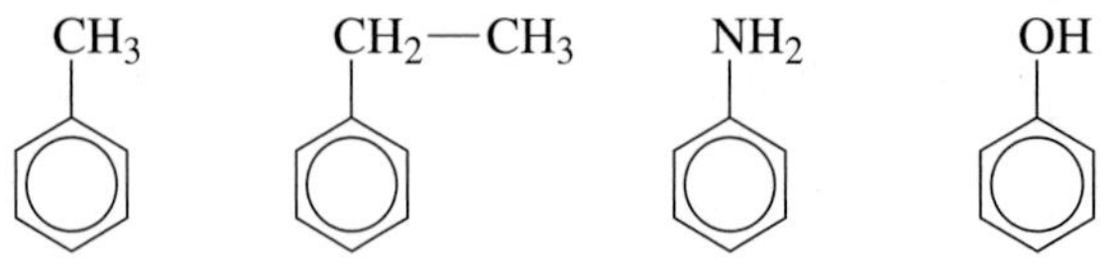

Toluene (methylbenzene)    Ethylbenzene    Aniline (benzenamine)    Phenol (hydroxybenzene)

When a benzene ring is a substituent, $C_6H_5$—, it is named as a phenyl group.

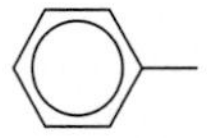

$H_3C-CH-CH=CH_2$

Phenyl group

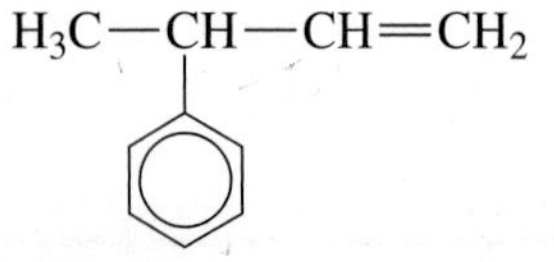

3-Phenyl-1-butene

When there are two or more substituents, the benzene ring is numbered to give the lowest numbers to the substituents.

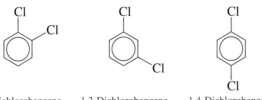

1,2-Dichlorobenzene

1,3-Dichlorobenzene

1,4-Dichlorobenzene

When a common name can be used such as toluene, phenol, or aniline, the carbon atom attached to the methyl, hydroxyl, or amine group is numbered as carbon 1. The group and benzene ring that represent the common name are highlighted in each.

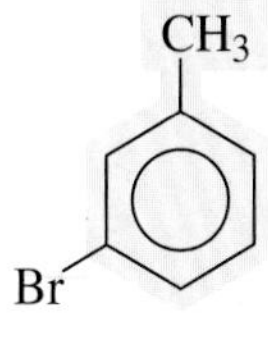

3-Bromotoluene

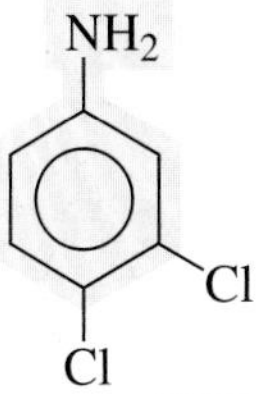

3,4-Dichloroaniline

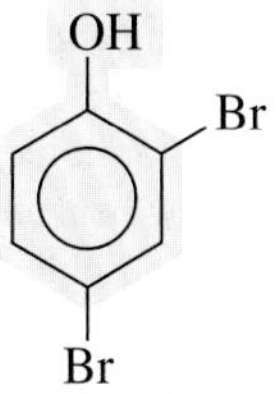

2,4-Dibromophenol

The substituents are named alphabetically.

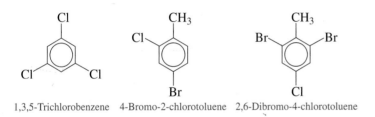

1,3,5-Trichlorobenzene

4-Bromo-2-chlorotoluene

2,6-Dibromo-4-chlorotoluene

SAMPLE PROBLEM 11.7

## Naming Aromatic Compounds

Give the IUPAC name for each of the following aromatic compounds:

**a.**

**b.**

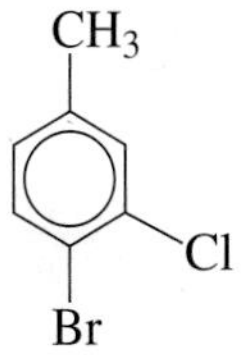

**c.**

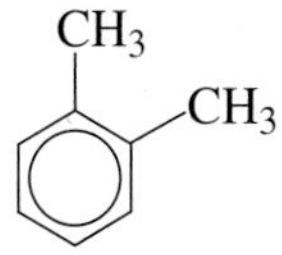

SOLUTION

**a.** chlorobenzene
**b.** 4-bromo-3-chlorotoluene
**c.** 1,2-dimethylbenzene

STUDY CHECK

Name the following compound:

## Health Note

### Some Common Aromatic Compounds

Aromatic compounds are common in nature and in medicine. Toluene is used as a reactant to make drugs, dyes, and explosives such as TNT (trinitrotoluene). The benzene ring is found in some amino acids (the building blocks of proteins); in pain relievers such as aspirin, acetaminophen, and ibuprofen; and in flavorings such as vanillin.

TNT (2,4,6-trinitrotoluene)

Aspirin

Vanillin

Ibuprofen

Acetaminophen

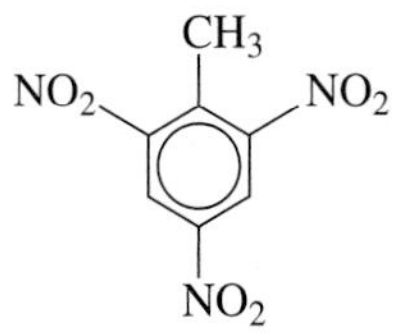

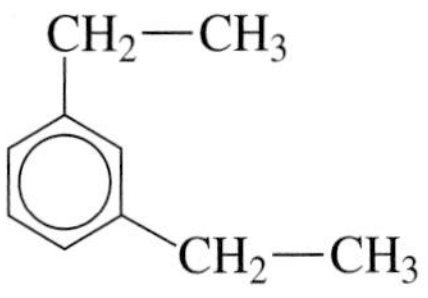

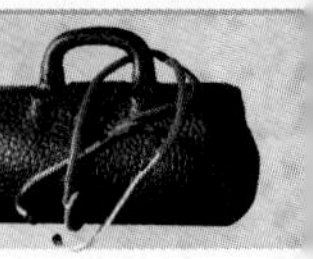

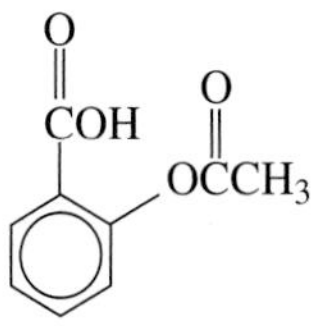

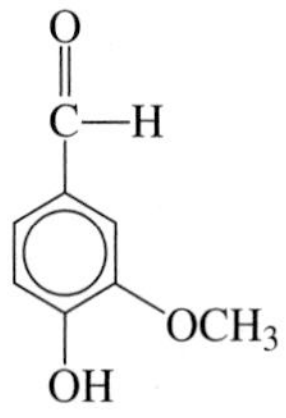

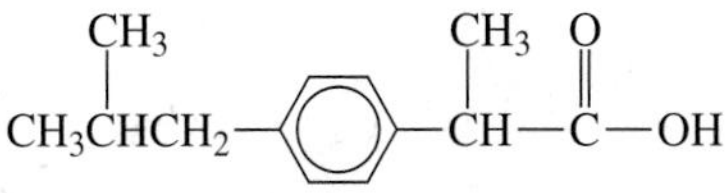

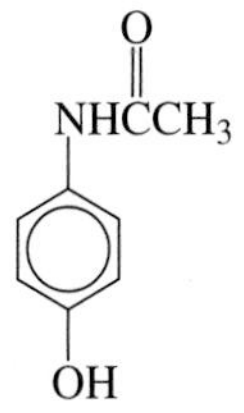

## QUESTIONS AND PROBLEMS

### Aromatic Compounds

**11.19** Cyclohexane and benzene each have six carbon atoms. How are they different?

**11.20** In the Health Note "Some Common Aromatic Compounds," what part of each molecule is the aromatic portion?

**11.21** Give the IUPAC name for each of the following:

**a.**

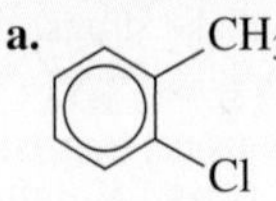

**b.**

**c.**

**d.**

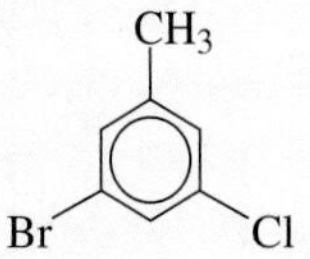

**11.22** Give the IUPAC name for each of the following:

**a.**

**b.**

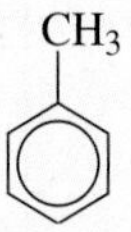

**c.**

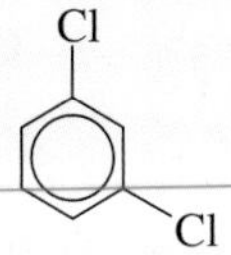

**d.**

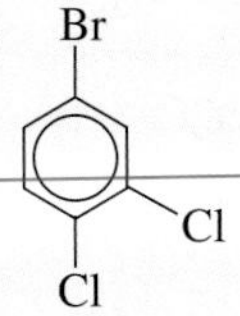

**11.23** Draw the condensed structural formula for each of the following compounds:
**a.** toluene
**b.** 1,3-dichlorobenzene
**c.** 4-ethyltoluene
**d.** 4-chlorotoluene

**11.24** Draw the condensed structural formula for each of the following compounds:
**a.** benzene
**b.** 2-chlorotoluene
**c.** propylbenzene
**d.** 1,2,4-trichlorobenzene

---

## *Health Note*

### Polycyclic Aromatic Hydrocarbons (PAHs)

Large aromatic compounds known as polycyclic aromatic hydrocarbons are formed by fusing together two or more benzene rings edge to edge. In a fused-ring compound, neighboring benzene rings share two carbon atoms. Naphthalene with two benzene rings is well known for its use in mothballs. Anthracene with three rings is used in the manufacture of dyes.

Compounds containing five or more fused benzene rings such as benzo[*a*]pyrene are potent carcinogens. The molecules interact with the DNA in the cells, causing abnormal cell growth and cancer. Increased exposure to carcinogens increases the chance of DNA alterations in the cells.

Naphthalene          Anthracene           Phenanthrene

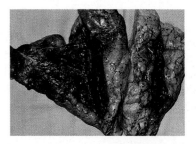

Benzo[*a*]pyrene

When a polycyclic compound contains phenanthrene, it may act as a carcinogen, a substance known to cause cancer. For example, some aromatic compounds in cigarette smoke cause cancer, as seen in the lung tissue of a heavy smoker. Benzo[*a*]pyrene, a product of combustion, has been identified in coal tar, tobacco smoke, barbecued meats, and automobile exhaust.

## CONCEPT MAP

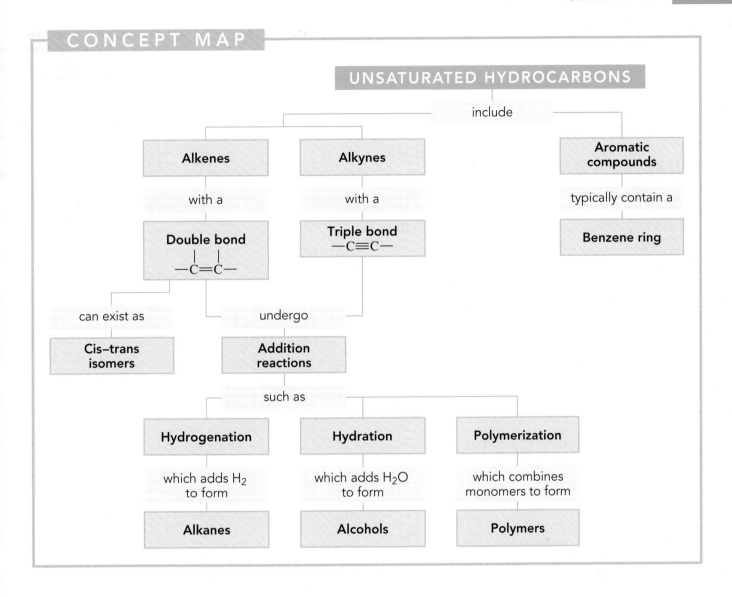

UNSATURATED HYDROCARBONS

include

Alkenes

Alkynes

Aromatic compounds

with a

with a

typically contain a

Double bond
—C=C—

Triple bond
—C≡C—

Benzene ring

can exist as                    undergo

Cis–trans isomers

Addition reactions

such as

Hydrogenation

Hydration

Polymerization

which adds $H_2$ to form

which adds $H_2O$ to form

which combines monomers to form

Alkanes

Alcohols

Polymers

# CHAPTER REVIEW

## 11.1 Alkenes and Alkynes

**Learning Goal:** Identify structural formulas as alkenes, cycloalkenes, and alkynes, and write their IUPAC or common names.

Alkenes are unsaturated hydrocarbons that contain carbon–carbon double bonds (C=C). Alkynes contain a triple bond (—C≡C—). The IUPAC names of alkenes end with *ene*, while alkyne names end with *yne*. The main chain is numbered from the end nearest the double or triple bond.

## 11.2 Cis–Trans Isomers

**Learning Goal:** Write the condensed structural formulas and names for cis–trans isomers of alkenes.

Isomers of alkenes occur when the carbon atoms in the double bond are connected to different atoms or groups. In the cis isomer, the similar groups are on the same side of the double bond, whereas in the trans isomer they are connected on the opposite sides of the double bond.

## 11.3 Addition Reactions

**Learning Goal:** Write the condensed structural formulas and names for the organic products of addition reactions of alkenes and alkynes.

The addition of small molecules to the double bond is a characteristic reaction of alkenes. Hydrogenation adds hydrogen atoms to the double bond of an alkene to yield an alkane. Water can also add to a double bond. When there are a different number of groups attached to the carbons in the double bond, the H from the HOH adds to the carbon with the greater number of hydrogen atoms, and OH adds to the other carbon.

## 11.4 Polymers of Alkenes

**Learning Goal:** Draw structural formulas of monomers that form a polymer or a three-monomer section of a polymer.

Polymers are long-chain molecules that consist of many repeating units of smaller carbon molecules called monomers. Many materials that we use every day are synthetic polymers, including carpeting, plastic wrap, nonstick pans, and nylon. These synthetic materials are often made by addition reactions in which a catalyst links the carbon atoms from various kinds of alkene molecules.

## 11.5 Aromatic Compounds

**Learning Goal:** Describe the bonding in benzene; name aromatic compounds, and write their condensed structural formulas.

Most aromatic compounds contain benzene, $C_6H_6$, a cyclic structure represented as a hexagon with a circle in the center. Aromatic compounds containing benzene are named using the parent name benzene, although common names such as toluene are retained. The benzene ring is numbered, and the substituents are listed in alphabetical order.

# SUMMARY OF NAMING

| Type | Example | Characteristic | Structure |
|------|---------|----------------|-----------|
| Alkene | Propene (propylene) | Double bond | $CH_3—CH{=}CH_2$ |
| Cycloalkene | Cyclopropene | Double bond in a carbon ring | △ |
| Alkyne | Propyne | Triple bond | $CH_3—C{\equiv}CH$ |
| Aromatic | Benzene | Aromatic ring of six carbons | ⬡ |
| | Methylbenzene, or toluene | | $CH_3$ ⬡ |

# SUMMARY OF REACTIONS

### HYDROGENATION

Alkene + $H_2$ $\xrightarrow{Pt}$ alkane

$CH_2{=}CH—CH_3 + H_2 \xrightarrow{Pt} CH_3—CH_2—CH_3$

Alkyne + $2H_2$ $\xrightarrow{Pt}$ alkane

$CH_3—C{\equiv}CH + 2H_2 \xrightarrow{Pt} CH_3—CH_2—CH_3$

### HYDRATION OF ALKENES

Alkene + H—OH $\xrightarrow{H^+}$ alcohol

$$CH_2{=}CH—CH_3 + H—OH \xrightarrow{H^+} CH_3—\overset{\overset{\displaystyle OH}{|}}{CH}—CH_3$$

# KEY TERMS

**addition** A reaction in which atoms or groups of atoms bond to a double bond or triple bond. Addition reactions include the addition of hydrogen (hydrogenation) and water (hydration).

**alkene** An unsaturated hydrocarbon containing a carbon–carbon double bond.

**alkyne** An unsaturated hydrocarbon containing a carbon–carbon triple bond.

**aromatic compounds** Compounds that usually have fragrant odors and often contain the ring structure of benzene.

**benzene** A ring of six carbon atoms, each of which is attached to a hydrogen atom, $C_6H_6$.

**cis isomer** An isomer of an alkene in which the hydrogen atoms in the double bond are on the same side.

**hydration** An addition reaction in which the components of water, H— and —OH, bond to the carbon–carbon double bond to form an alcohol.

**hydrogenation** The addition of hydrogen ($H_2$) to the double bond of alkenes or alkynes to yield alkanes.

**monomer** The small organic molecule that is repeated many times in a polymer.

**polymer** A very large molecule that is composed of many small, repeating structural units that are identical.

**saturated hydrocarbon** A compound of carbon and hydrogen that contains the maximum number of hydrogen atoms.

**trans isomer** An isomer of an alkene in which the hydrogen atoms in the double bond are on opposite sides.

**unsaturated hydrocarbon** A compound of carbon and hydrogen in which the carbon chain contains at least one double (alkene) or triple (alkyne) carbon–carbon bond. An unsaturated compound is capable of an addition reaction with hydrogen, which converts the double or triple bonds to single carbon–carbon bonds.

# UNDERSTANDING THE CONCEPTS

**11.25** Draw a part of the polymer (use four monomers) of Teflon made from 1,1,2,2-tetrafluoroethene.

**11.26** A garden hose is made of polyvinylchloride (PVC) from chloroethene (vinyl chloride). Draw a part of the polymer (use four monomers) for PVC.

**11.27** Explosives used in mining contain TNT, or trinitrotoluene.

**a.** If the functional group *nitro* is —NO$_2$, what is the structural formula of 2,4,6-trinitrotoluene, one isomer of TNT?
**b.** TNT is actually a mixture of isomers of trinitrotoluene. Draw two other possible isomers.

**11.28** Margarine is produced from the hydrogenation of vegetable oils, which contain unsaturated fatty acids. How many grams of hydrogen are required to completely saturate 75.0 g of oleic acid, C$_{18}$H$_{34}$O$_2$, which has one double bond?

# ADDITIONAL QUESTIONS AND PROBLEMS

**11.29** Compare the formulas and bonding in propane, cyclopropane, propene, and propyne.

**11.30** Compare the formulas and bonding in butane, cyclobutane, cyclobutene, and 2-butyne.

**11.31** Give the IUPAC name for each of the following compounds:

**a.** 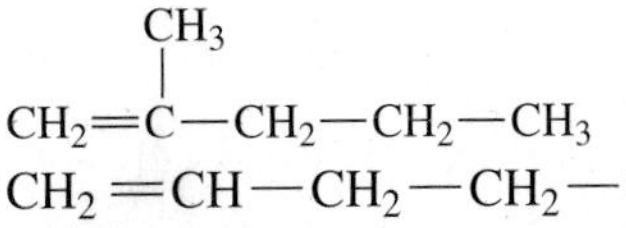 CH$_2$=C(CH$_3$)—CH$_2$—CH$_2$—CH$_3$
**b.** CH$_2$=CH—CH$_2$—CH$_2$—CH$_2$—Cl
**c.**

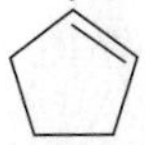

**11.32** Give the IUPAC name for each of the following compounds:
**a.** 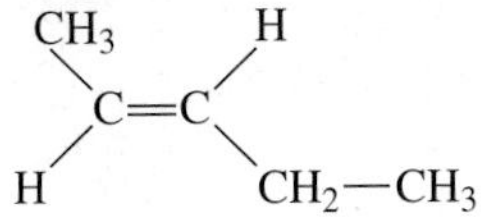
**b.** 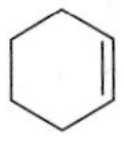
**c.** CH$_3$—CH$_2$—C≡C—CH$_3$

**11.33** Indicate if the following pairs of structures represent isomers, cis–trans isomers, or the same molecule.
**a.** 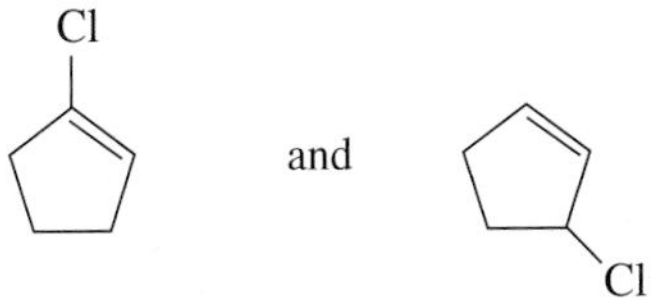
**b.** 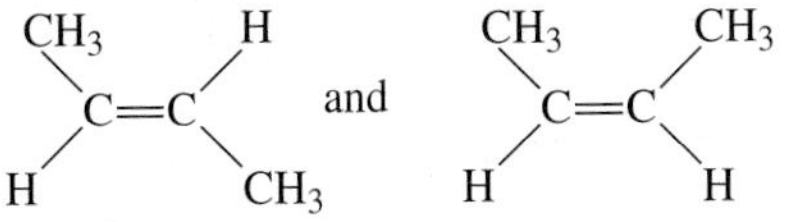
**c.** 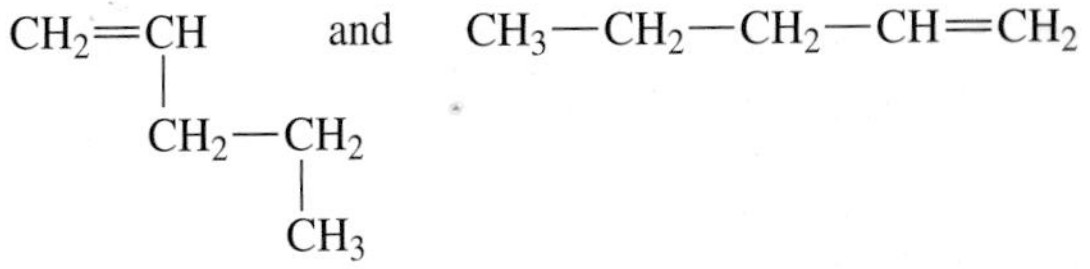 CH$_2$=CH—CH$_2$—CH$_2$—CH$_3$ and CH$_3$—CH$_2$—CH$_2$—CH=CH$_2$
**d.** 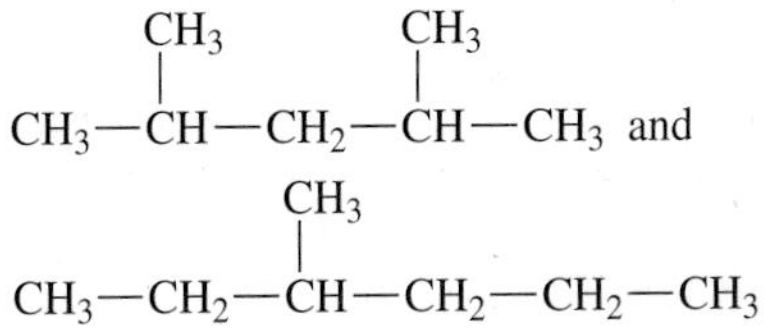 CH$_3$—CH(CH$_3$)—CH$_2$—CH(CH$_3$)—CH$_3$ and CH$_3$—CH$_2$—CH(CH$_3$)—CH$_2$—CH$_2$—CH$_3$

**11.34** Write the condensed structural formula of each of the following compounds:
  **a.** 2-pentyne
  **b.** *cis*-2-heptene
  **c.** *trans*-3-hexene
  **d.** 2,3-dichloro-1-butene

**11.35** Write the cis and trans isomers for each of the following:
  **a.** 2-pentene
  **b.** 3-hexene

**11.36** Write the cis and trans isomers for each of the following:
  **a.** 2-butene
  **b.** 2-hexene

**11.37** Give the name of the products from hydrogenation of each of the following:
  **a.** 2-butene
  **b.** 3-methyl-2-pentene
  **c.** cyclohexene
  **d.** 2-pentyne

**11.38** Give the name of the products from hydrogenation of each of the following:
  **a.** 3-hexene
  **b.** 2-methyl-2-butene
  **c.** propyne
  **d.** methylcyclopropene

**11.39** Write the condensed structural formula of the products for the following.
  **a.** $CH_3-CH=CH-CH_3 + H_2 \xrightarrow{Ni}$
  **b.**

$$\text{(cyclopentene)} + H_2 \xrightarrow{Ni}$$

  **c.** $CH_3-CH=CH-CH_3 + HOH \xrightarrow{H^+}$

**11.40** Write the condensed structural formula of the products for the following.
  **a.**

$$\text{(cyclopentene)} + HOH \xrightarrow{H^+}$$

  **b.**

$$\text{(cyclohexene)} + HOH \xrightarrow{H^+}$$

  **c.** $CH_3-CH_2-C\equiv C-CH_3 + 2H_2 \xrightarrow{Pt}$

**11.41** A plastic called polyvinylidene difluoride, PVDF, is made from monomers of 1,1-difluoroethene. Write the structure of the polymer formed from the addition of three monomers of 1,1-difluoroethene.

**11.42** An alkene called acrylonitrile is the monomer used to form the polymer used in the fabric material called Orlon. Write an equation that represents the formation of a part of the polyacrylonitrile polymer from three of the monomer units. The structure of acrylonitrile is $CH_2=CH$ with a CN group attached.

**11.43** Name each of the following aromatic compounds:
  **a.** $CH_3$ (benzene ring)
  **b.** $CH_3$, $Cl$ (benzene ring)
  **c.** $CH_3$, $CH_2-CH_3$ (benzene ring)
  **d.** $CH_2-CH_3$, $CH_2-CH_3$ (benzene ring)

**11.44** Draw the condensed structural formula for each of the following:
  **a.** ethylbenzene
  **b.** 1,3-dichlorobenzene
  **c.** 1,2,4-trimethylbenzene
  **d.** 1,4-dimethylbenzene

# CHALLENGE QUESTIONS

**11.45** How many grams of hydrogen are needed to hydrogenate 30.0 g of 2-butene?

**11.46** Using each of the following carbon chains for $C_5H_{10}$, write and name all the possible alkenes, including those with cis and trans isomers.

$$C-C-C-C-C$$

$$C-\underset{\underset{C}{|}}{C}-C-C$$

**11.47** If a female silkworm moth secretes 50 ng of bombykol, a sex attractant, how many molecules does she secrete? (See Environmental Note "Pheromones in Insect Communication.")

**11.48** Acetylene gas reacts with oxygen and burns at 3300 °C in an acetylene torch.
  **a.** Write the balanced equation for the complete combustion of acetylene.
  **b.** How many grams of oxygen are needed to react with 8.5 L of acetylene at STP?

  **c.** How many liters of $CO_2$ at STP are produced when 30.0 g of acetylene undergoes combustion?

# ANSWERS

## Answers to Study Checks

**11.1 a.** alkene
**b.** alkyne

**11.2 a.** $CH_3-C\equiv C-CH_2-CH_3$
**b.** $CH_2=CH-CH_2-CH_2-CH_2-Cl$

**11.3** *trans*-3-hexene

**11.4**

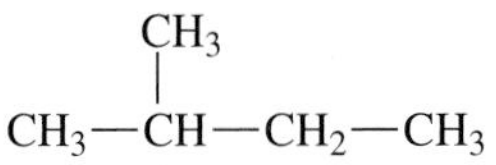

$$CH_3-\overset{\overset{\textstyle CH_3}{|}}{CH}-CH_2-CH_3$$

**11.5**

$$CH_3-\overset{\overset{\textstyle CH_3}{|}}{\underset{\underset{\textstyle OH}{|}}{C}}-CH_2-CH_3$$

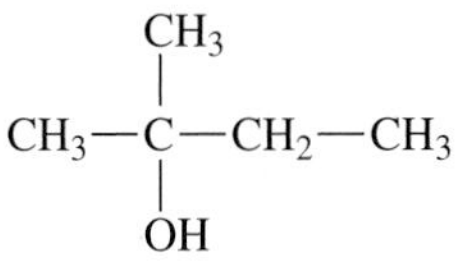

**11.6** The monomer of PVC, polyvinyl chloride, is chloroethene.

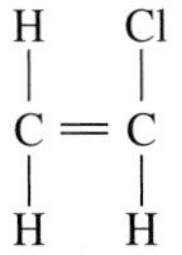

**11.7** 1,3-diethylbenzene

## Answers to Selected Questions and Problems

**11.1 a.** An alkene has a double bond.
**b.** An alkyne has a triple bond.
**c.** A cycloalkene has a double bond in a ring.
**d.** An alkene has a double bond.

**11.3** Propene contains a double bond, and propyne has a triple bond. Propene has four hydrogen atoms, and propyne has only two hydrogen atoms.

**11.5 a.** ethene       **b.** methylpropene
**c.** 2-pentyne       **d.** cyclobutene

**11.7 a.** $CH_3-CH=CH_2$
**b.** $CH_2=CH-CH_2-CH_2-CH_3$
**c.**

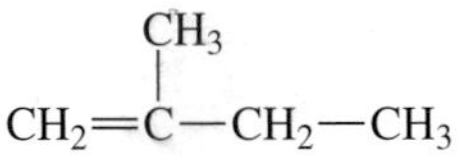

$$CH_2=\overset{\overset{\textstyle CH_3}{|}}{C}-CH_2-CH_3$$

**d.**

**e.** $CH_3-CH_2-C\equiv CH$
**f.** $Br-CH_2-CH_2-C\equiv C-CH_2-CH_3$

**11.9 a.** *cis*-2-butene
**b.** *trans*-3-octene
**c.** *cis*-3-heptene

**11.11 a.**

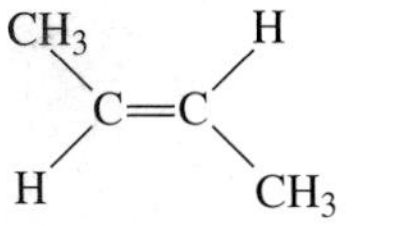

**b.**

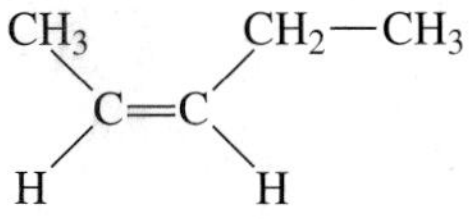

**c.**

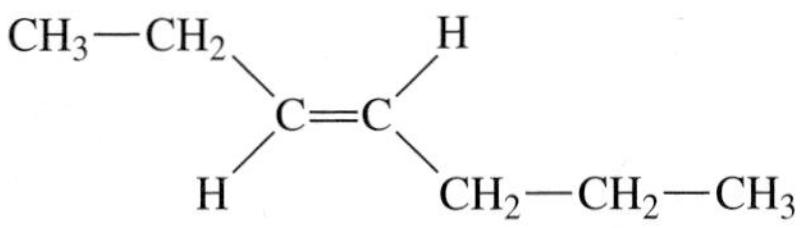

**11.13 a.** $CH_3-CH_2-CH_2-CH_2-CH_3$
**b.**

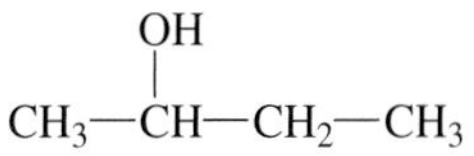

$$CH_3-\overset{\overset{\textstyle OH}{|}}{CH}-CH_2-CH_3$$

**c.**

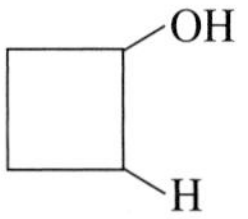

**d.**

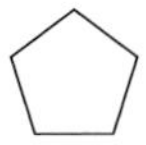

**e.** $CH_3-CH_2-CH_2-CH_2-CH_3$

**11.15** A polymer is a very large molecule composed of small units that are repeated many times.

**11.17**

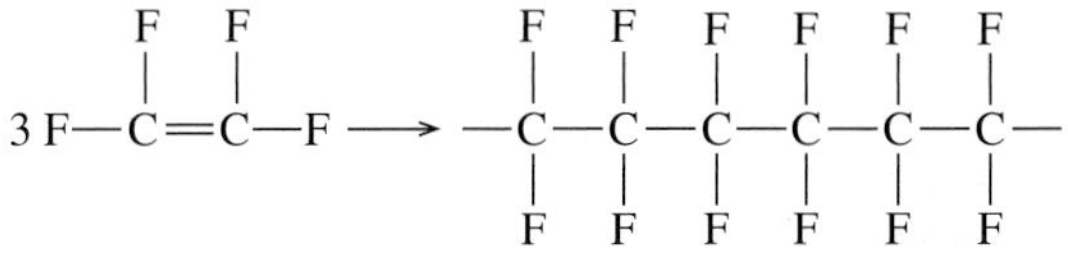

**11.19** Cyclohexane, $C_6H_{12}$, is a cycloalkane with 6 carbon atoms and 12 hydrogen atoms. The carbon atoms are connected in a ring by single bonds. Benzene, $C_6H_6$, is an aromatic compound with 6 carbon atoms and 6 hydrogen atoms. The carbon atoms are connected in a ring where the electrons are equally shared among the six carbon atoms.

**11.21 a.** 2-chlorotoluene       **b.** ethylbenzene
**c.** 1,3,5-trichlorobenzene    **d.** 3-bromo-5-chlorotoluene

**11.23 a.**

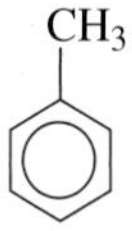

**b.**

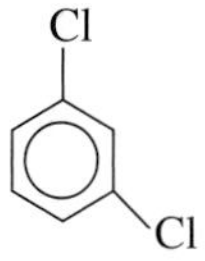

**c.**

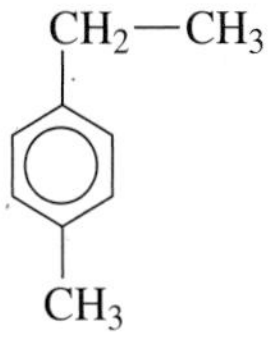

**d.**

**11.25**

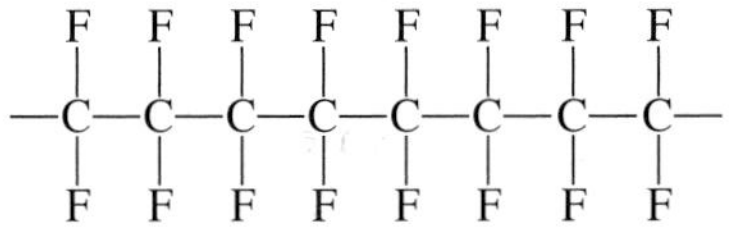

**11.27 a.**

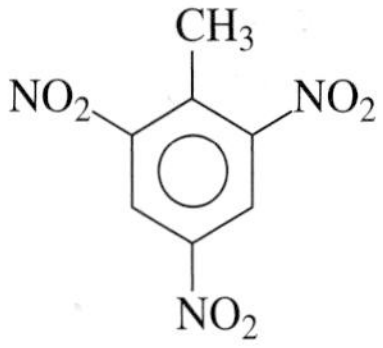

**b.**

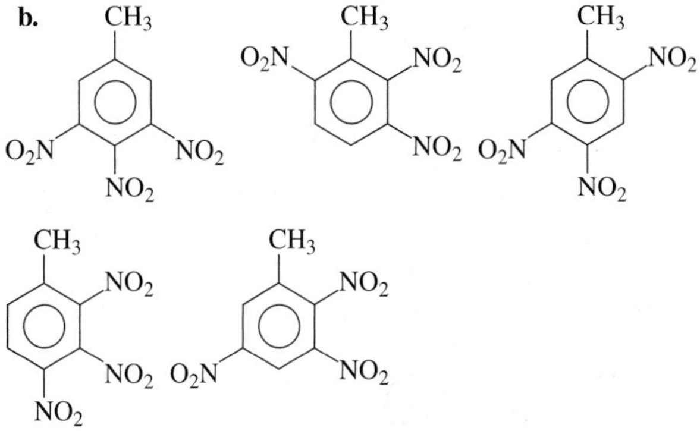

**11.29** All the compounds have three carbon atoms: propane has eight hydrogen atoms, cyclopropane has six hydrogen atoms, propene has six hydrogen atoms, and propyne has four hydrogen atoms. Propane is a saturated alkane, and cyclopropane is a saturated cyclic hydrocarbon. Both propene and propyne are unsaturated hydrocarbons, but propene has a double bond and propyne has a triple bond.

**11.31** **a.** 2-methyl-1-pentene
**b.** 5-chloro-1-pentene
**c.** cyclopentene

**11.33** **a.** isomers
**b.** cis–trans isomers
**c.** identical
**d.** isomers

**11.35** **a.**

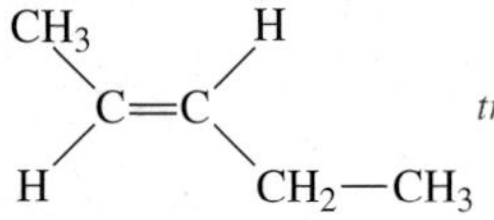

trans-2-Pentene

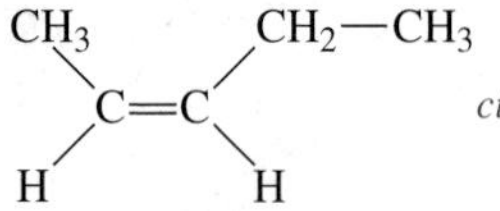

cis-2-Pentene

**b.**

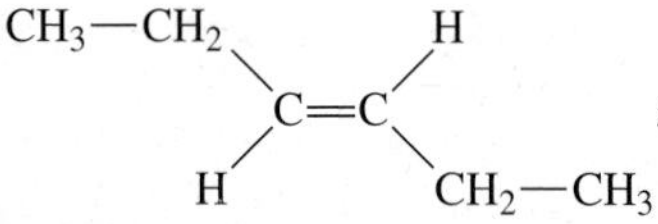

trans-3-Hexene

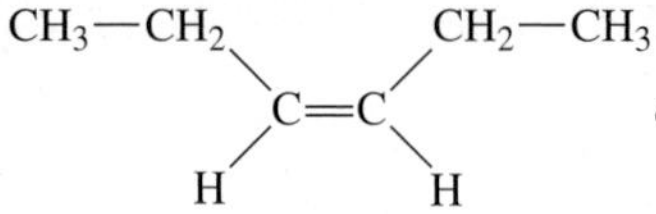

cis-3-Hexene

**11.37** **a.** butane
**b.** 3-methylpentane
**c.** cyclohexane
**d.** pentane

**11.39** **a.** $CH_3-CH_2-CH_2-CH_3$
**b.**

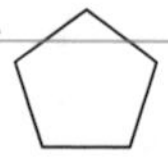

**c.**

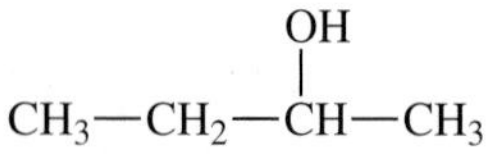

$$CH_3-CH_2-\overset{\overset{\displaystyle OH}{|}}{CH}-CH_3$$

**11.41**

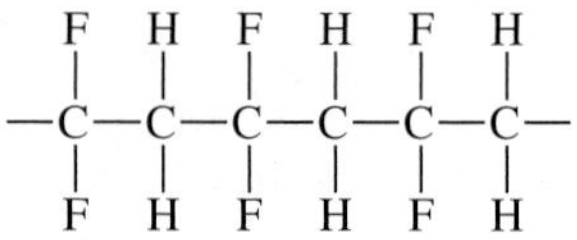

**11.43** **a.** toluene
**b.** 2-chlorotoluene
**c.** 4-ethyltoluene
**d.** 1,3-diethylbenzene

**11.45** 1.07 g of $H_2$

**11.47** $1 \times 10^{14}$ molecules

# 12

# Organic Compounds with Oxygen and Sulfur

the **Chemistry** place

Visit **www.chemplace.com** for extra quizzes, interactive tutorials, career resources, PowerPoint slides for chapter review, math help, and case studies.

*"The purpose of our research was to create a way to make Taxol," says Paul Wender, Francis W. Bergstrom Professor of organic chemistry and head of the Wender research group at Stanford University. "Taxol is a chemotherapy drug originally derived from the bark of the Pacific yew tree. However, removing the bark from yew trees destroys them, so we need a renewable resource. We worked out a synthesis that began with turpentine, which is both renewable and inexpensive. Initially, Taxol was used with patients who did not respond to chemotherapy. The first person to be treated was a woman who was diagnosed with terminal ovarian cancer and given three to six months to live. After a few treatments with Taxol, she was declared 98% disease–free. A drug like Taxol can save many lives, which is one reason that a study of organic chemistry is so important."*

I n this chapter, we will look at organic compounds that contain oxygen or sulfur atoms. Alcohols, which contain the hydroxyl group (—OH), are commonly found in nature and are used in industry and at home. For centuries, grains, vegetables, and fruits have been fermented to produce the ethanol present in alcoholic beverages. The hydroxyl group is important in biomolecules such as sugars and starches as well as in steroids such as cholesterol and estradiol. Menthol is a cyclic alcohol with a minty odor and flavor that is used in cough drops, shaving creams, and ointments. Ethers are compounds that contain an oxygen atom connected to two carbon atoms (—O—). Ethers are important solvents in chemistry and medical laboratories. Beginning in 1842, diethyl ether was used for about 100 years as a general anesthetic. Thiols, which contain a —SH group, give the strong odors we associate with garlic and onions.

We will also study two other families of organic compounds: aldehydes and ketones. Many of the odors and flavors that you associate with flavorings and perfumes are due to a carbon–oxygen double bond called a **carbonyl group** (C=O). Aldehydes in foods and perfumes provide the odors and flavors of vanilla, almond, and cinnamon. In biology, you may have seen specimens preserved in a solution of formaldehyde. You probably notice the odor of a ketone when you use paint or nail-polish remover.

## LEARNING GOAL

Identify and name alcohols, thiols, and ethers; classify alcohols as primary, secondary, or tertiary.

# 12.1 ALCOHOLS, THIOLS, AND ETHERS

As we learned in Chapter 10, alcohols and ethers are two classes of organic compounds that contain an oxygen (O) atom, shown in red in the ball-and-stick models. In an alcohol, the oxygen atom is part of a *hydroxyl* group (—OH) that is attached to a carbon atom. In an **ether**, the oxygen atom is attached to two carbon atoms. Both alcohols and ethers have bent structures similar to water. One hydrogen atom of water is replaced by an alkyl group in an alcohol and by a benzene ring in a **phenol**.

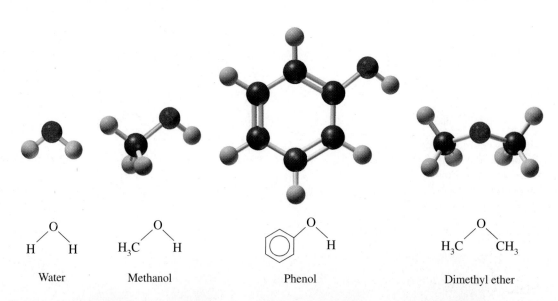

| Water | Methanol | Phenol | Dimethyl ether |

## Naming Alcohols

In the IUPAC system, the alcohol family is indicated by the *ol* ending.

**Step 1**  **Name the longest carbon chain containing the —OH group.**  Replace the *e* in the alkane name with *ol*. Consider the following alcohol.

$$CH_3—CH_2—CH_2—OH \qquad Propanol$$

**Step 2**  **Number the longest chain, starting at the end closest to the —OH group.**  For simple alcohols, the common name (shown in parentheses) gives the name of the carbon chain as an alkyl group followed by *alcohol*.

$$\underset{3}{CH_3}—\underset{2}{CH_2}—\underset{1}{CH_2}—OH \qquad \begin{array}{l} \text{1-Propanol} \\ \text{(Propyl alcohol)} \end{array}$$

Alcohols with one or two carbon atoms do not require a number for the hydroxyl group.

$$CH_3—OH \qquad CH_3—CH_2—OH \qquad Cl—CH_2—CH_2—OH$$

Methanol      Ethanol           2-Chloroethanol
(methyl alcohol)   (ethyl alcohol)

**Step 3**  **Name and number other substituents relative to the —OH group.**  Substituents are listed in alphabetical order.

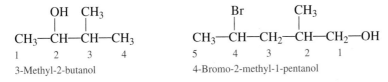

3-Methyl-2-butanol        4-Bromo-2-methyl-1-pentanol

**Step 4**  **Name a cyclic alcohol as a *cycloalkanol*.**  For other substituents, the ring is numbered with the —OH group on carbon 1.

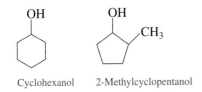

Cyclohexanol     2-Methylcyclopentanol

**Step 5**  **When the —OH group is attached to a benzene ring, it is named *phenol*.**  When there is a second substituent on the benzene ring, the ring is numbered from carbon 1, which is attached to the —OH group, to give the lowest possible number to the substituent.

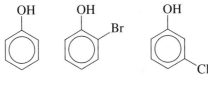

Phenol     2-Bromophenol     3-Chlorophenol

---

**SAMPLE PROBLEM  12.1**

**■ Naming Alcohols**

Give the IUPAC name for the following:

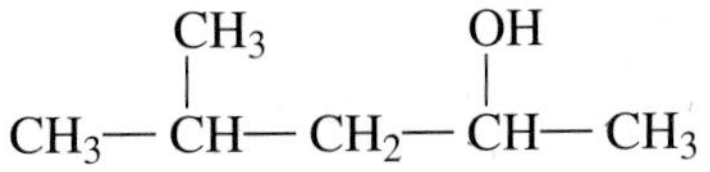

$$CH_3—\underset{\underset{CH_3}{|}}{CH}—CH_2—\underset{\underset{OH}{|}}{CH}—CH_3$$

---

WEB TUTORIAL
Alcohols, Thiols, Aldehydes, and Ketones

*Explore Your World*

## Alcohols in Household Products

Read the labels on household products such as mouthwashes, cold remedies, rubbing alcohol, and flavoring extracts. Look for names of alcohols such as ethyl alcohol, isopropyl alcohol, thymol, and menthol.

### QUESTIONS

1. What part of the name tells you that it is an alcohol?
2. What alcohol is usually meant by the term "alcohol"?
3. What is the percent of alcohol in the products?
4. Write out the structures of the alcohols you find listed on the labels. You may need to use a reference book for some structures.

SOLUTION

The parent chain is pentane; the alcohol is named pentanol. The carbon chain is numbered to give the position of the —OH group on carbon 2 and the methyl group on carbon 4. The compound is named 4-methyl-2-pentanol.

STUDY CHECK

Give the IUPAC name for the following:

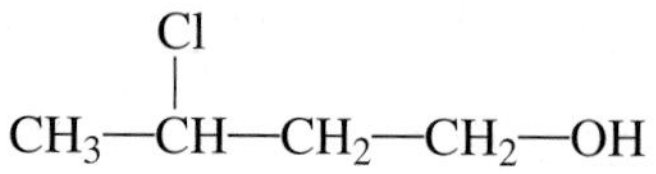

$$\underset{\text{Cl}}{\overset{\text{Cl}}{|}}$$

$$\text{CH}_3\text{—CH—CH}_2\text{—CH}_2\text{—OH}$$

## Classification of Alcohols

Alcohols are classified by the number of carbon groups attached to the carbon atom bonded to the hydroxyl (—OH) group. A **primary (1°) alcohol** has one alkyl group attached to the carbon atom bonded to the —OH, a **secondary (2°) alcohol** has two alkyl groups, and a **tertiary (3°) alcohol** has three alkyl groups.

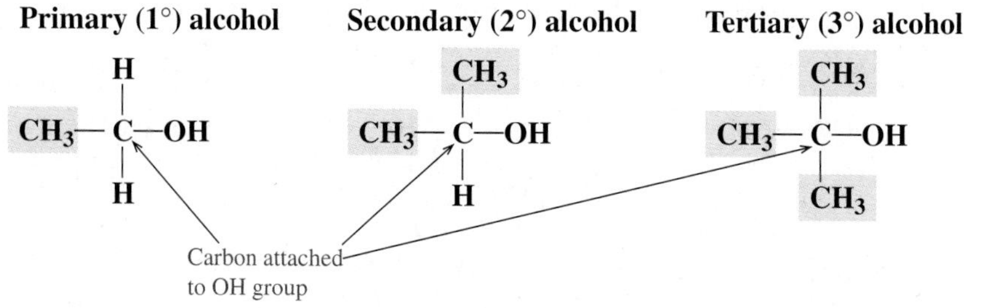

| Primary (1°) alcohol | Secondary (2°) alcohol | Tertiary (3°) alcohol |
|---|---|---|

$$\text{CH}_3\text{—}\overset{\overset{\text{H}}{|}}{\underset{\underset{\text{H}}{|}}{\text{C}}}\text{—OH} \qquad \text{CH}_3\text{—}\overset{\overset{\text{CH}_3}{|}}{\underset{\underset{\text{H}}{|}}{\text{C}}}\text{—OH} \qquad \text{CH}_3\text{—}\overset{\overset{\text{CH}_3}{|}}{\underset{\underset{\text{CH}_3}{|}}{\text{C}}}\text{—OH}$$

Carbon attached to OH group

---

## SAMPLE PROBLEM 12.2

### ■ Classifying Alcohols

Classify each of the following alcohols as primary (1°), secondary (2°), or tertiary (3°).

**a.** $\text{CH}_3\text{—CH}_2\text{—CH}_2\text{—OH}$

**b.** $\text{CH}_3\text{—CH}_2\text{—}\overset{\overset{\text{OH}}{|}}{\underset{\underset{\text{CH}_3}{|}}{\text{C}}}\text{—CH}_3$

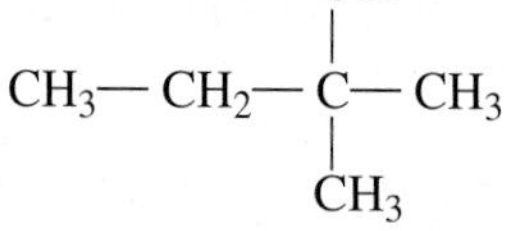

SOLUTION

**a.** One alkyl group attached to the carbon atom bonded to the —OH makes this a primary (1°) alcohol.

**b.** Three alkyl groups attached to the carbon atom bonded to the —OH makes this a tertiary (3°) alcohol.

STUDY CHECK

Classify the following as primary (1°), secondary (2°), or tertiary (3°):

$$\text{CH}_3\text{—}\overset{\overset{\text{OH}}{|}}{\text{CH}}\text{—CH}_3$$

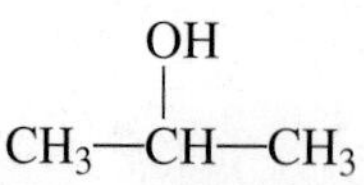

# *Health Note*

## Some Important Alcohols

*Methanol (methyl alcohol)*, the simplest alcohol, is found in many solvents and paint removers. If ingested, methanol is oxidized to formaldehyde, which can cause headaches, blindness, and death. Methanol is used to make plastics, medicines, and fuels. In car racing, it is used as a fuel because it is less flammable and has a higher octane rating than does gasoline.

*Ethanol (ethyl alcohol)* has been known since prehistoric times as an intoxicating product formed by the fermentation of grains and starches.

$$C_6H_{12}O_6 \xrightarrow{\text{Fermentation}} 2CH_3-CH_2-OH + 2CO_2$$

Today, ethanol for commercial uses is produced by allowing ethene and water to react at high temperatures and pressures. It is used as a solvent for perfumes, varnishes, and some medicines, such as tincture of iodine. "Gasohol" is a mixture of ethanol and gasoline used as a fuel.

$$H_2C=CH_2 + H_2O \xrightarrow[\text{Catalyst}]{300\ °C,\ 200\ atm} CH_3-CH_2-OH$$

Several of the essential oils of plants, which produce the odor or flavor of the plant, are derivatives of phenol. Eugenol is found in cloves, vanillin in vanilla bean, isoeugenol in nutmeg, and thymol in thyme and mint. Thymol has a pleasant, minty taste and is used in mouthwashes and by dentists to disinfect a cavity before adding a filling compound.

*1,2,3-Propanetriol (glycerol or glycerin)*, a trihydroxy alcohol, is a viscous liquid obtained from oils and fats during the production of soaps. The presence of several polar —OH groups makes it strongly attracted to water, a feature that makes glycerin useful as a skin softener in products such as skin lotions, cosmetics, shaving creams, and liquid soaps.

$$HO-CH_2-\overset{\displaystyle OH}{\underset{\displaystyle |}{CH}}-CH_2-OH$$
1,2,3-Propanetriol (glycerol)

*1,2-Ethanediol (ethylene glycol)* is used as antifreeze in heating and cooling systems. It is also a solvent for paints, inks, and plastics, and is used in the production of synthetic fibers such as Dacron. If ingested, it is extremely toxic. In the body, it is oxidized to oxalic acid, which forms insoluble salts in the kidneys that cause renal damage, convulsions, and death. Because its sweet taste is attractive to pets and children, ethylene glycol solutions must be carefully stored.

$$HO-CH_2-CH_2-OH \xrightarrow{[O]} HO-\overset{\displaystyle O}{\overset{\displaystyle \|}{C}}-\overset{\displaystyle O}{\overset{\displaystyle \|}{C}}-OH$$
1,2-Ethanediol (ethylene glycol)          Oxalic acid

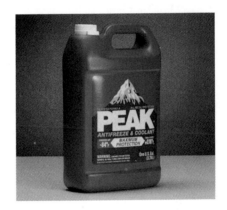

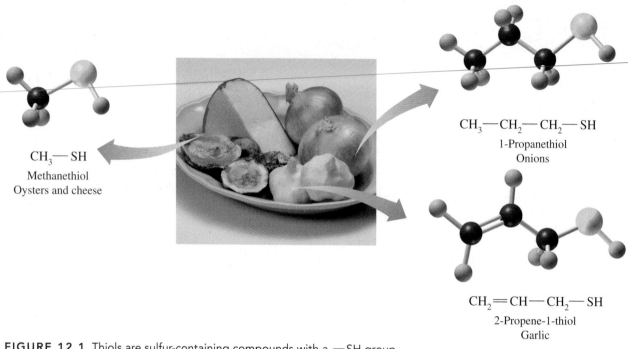

$CH_3—SH$
Methanethiol
Oysters and cheese

$CH_3—CH_2—CH_2—SH$
1-Propanethiol
Onions

$CH_2{=}CH—CH_2—SH$
2-Propene-1-thiol
Garlic

**FIGURE 12.1** Thiols are sulfur-containing compounds with a —SH group.
**Q** Why do thiols have structures similar to alcohols?

$CH_3—CH_2—SH$
Ethanethiol

## Thiols

**Thiols** are a family of sulfur-containing organic compounds that have a *thiol* (—SH) group. They have structures similar to alcohols except that a —SH group takes the place of the —OH group. The sulfur atom is shown in yellow in the ball-and-stick models. In the IUPAC system, thiols are named by adding *thiol* to the longest carbon chain connected to the —SH group. (See Figure 12.1.)

An important property of thiols is their strong, sometimes disagreeable, odor, which is characteristic of oysters, cheddar cheese, onions, and garlic. To help us detect natural gas (methane) leaks, a small amount of ethanethiol is added to the gas supply.

##  Ethers

An *ether* contains an oxygen atom that is attached by single bonds to two carbon groups that are alkyls or aromatic rings. Ethers have a bent structure like water and alcohols, except both hydrogen atoms are replaced by alkyl groups.

Most ethers are named by their common names. Write the name of each alkyl or aromatic group attached to the oxygen atom, in alphabetical order followed by the word *ether*. The IUPAC names are used only when the ether is more complex.

Methyl group    Propyl group
$CH_3—O—CH_2—CH_2—CH_3$
Common name: Methyl propyl ether

Some examples of the common names for ethers follow:

$CH_3—O—CH_3$     $CH_3—CH_2—O—CH_2—CH_3$     $CH_3—CH_2—O—\langle\bigcirc\rangle$

Dimethyl ether          Diethyl ether                    Ethyl phenyl ether

# Health Note

## Ethers as Anesthetics

Anesthesia is the loss of sensation and consciousness. A general anesthetic is a substance that blocks signals to the awareness centers in the brain so the person has a loss of memory, a loss of feeling pain, and an artificial sleep. The term *ether* has been associated with anesthesia because diethyl ether was the most widely used anesthetic for more than a hundred years. Although it is easy to administer, ether is very volatile and highly flammable. A small spark in the operating room could cause an explosion. Since the 1950s, anesthetics such as Forane (isoflurane), Ethrane (enflurane), and Penthrane (methoxyflurane) have been developed that are not as flammable and do not cause nausea. Most of these anesthetics retain the ether group, but the addition of halogen atoms reduces the volatility and flammability of the ethers. More recently, they have been replaced by halothane (1-bromo-1-chloro-2,2,2-trifluoroethane) because of the side effects of the ether-type inhalation anesthetics.

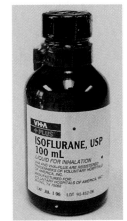

Forane
(isoflurane)

Ethrane
(enflurane)

Penthrane
(methoxyflurane)

---

## SAMPLE PROBLEM 12.3

### ■ Ethers

Give the common name for the following ether:

$$CH_3—CH_2—O—CH_2—CH_2—CH_3$$

**SOLUTION**

The groups attached to the oxygen are an ethyl group and a propyl group. The common name is ethyl propyl ether.

**STUDY CHECK**

Draw the structure of methyl phenyl ether.

---

## QUESTIONS AND PROBLEMS

### Alcohols, Thiols, and Ethers

**12.1** Give the IUPAC name for each of the following:

**a.** $CH_3—CH_2—OH$

**b.** $CH_3—CH_2—\overset{\overset{\displaystyle OH}{|}}{CH}—CH_3$

**c.** $CH_3—\overset{\overset{\displaystyle OH}{|}}{CH}—CH_2—CH_2—CH_3$

**d.** cyclohexanol with $CH_3$ group (4-methylcyclohexanol structure)

**e.** $CH_3—CH_2—CH_2—SH$

**12.2** Give the IUPAC name for each of the following:

**a.** cyclobutane with $CH_2—CH_3$ and $OH$ groups

**b.** $CH_3—CH_2—\overset{\overset{\displaystyle CH_3}{|}}{CH}—CH_2—OH$

**c.** $CH_3—CH_2—\overset{\overset{\displaystyle CH_3}{|}}{CH}—\overset{\overset{\displaystyle CH_3}{|}}{CH}—CH_2—OH$

**d.** $CH_3—CH_2—CH_2—\overset{\overset{\displaystyle OH}{|}}{CH}—CH_3$

**e.** cyclopentane with $SH$ group

**12.3** Write the condensed structural formula of each of the following alcohols:
  **a.** 1-propanol      **b.** methyl alcohol
  **c.** 3-pentanol      **d.** 2-methyl-2-butanol

**12.4** Write the condensed structural formula of each of the following alcohols:
  **a.** ethyl alcohol      **b.** 3-methyl-1-butanol
  **c.** 2,4-dichlorocyclohexanol      **d.** propyl alcohol

**12.5** Name each of the following phenols:

**a.**  OH

**b.** 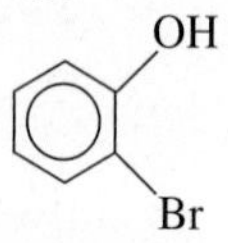 OH
Br

**c.** 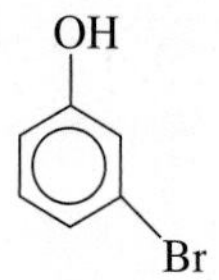 OH
Br

**12.6** Name each of the following phenols:

**a.** 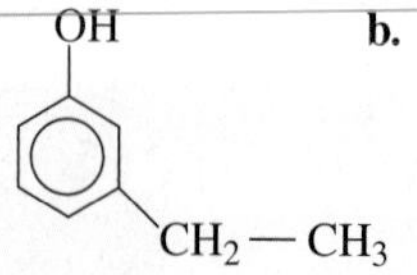 OH
$CH_2-CH_3$

**b.** 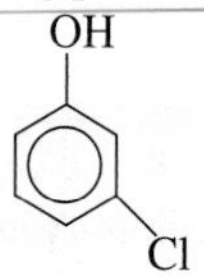 OH
Cl

**c.** 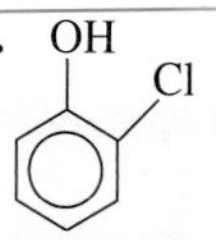 OH
Cl

**12.7** Classify each of the following as a primary (1°), secondary (2°), or tertiary (3°) alcohol:

**a.** 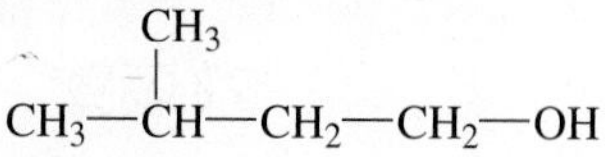 $CH_3-\overset{\overset{\displaystyle CH_3}{|}}{CH}-CH_2-CH_2-OH$

**b.** $CH_3-CH_2-CH_2-CH_2-OH$

**c.** 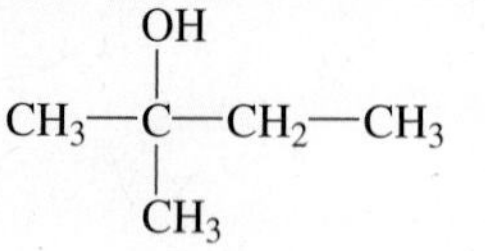 $CH_3-\overset{\overset{\displaystyle OH}{|}}{\underset{\underset{\displaystyle CH_3}{|}}{C}}-CH_2-CH_3$

**d.** 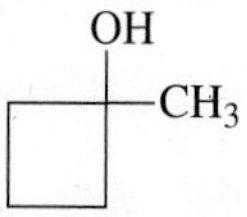 OH—CH₃ (cyclobutane ring with OH and CH₃)

**12.8** Classify each of the following as a primary (1°), secondary (2°), or tertiary (3°) alcohol:

**a.** 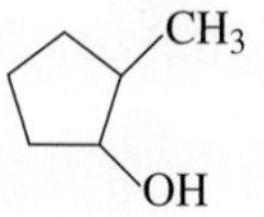 (cyclopentane ring with CH₃ and OH)

**b.** 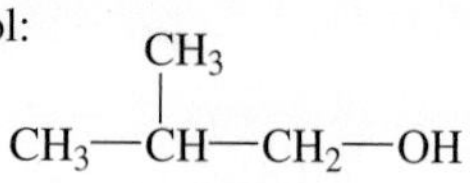 $CH_3-\overset{\overset{\displaystyle CH_3}{|}}{CH}-CH_2-OH$

**c.** 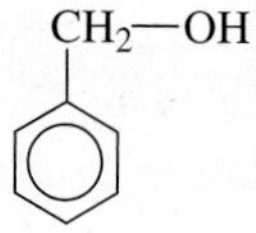 $CH_2-OH$ (benzene ring)

**d.** 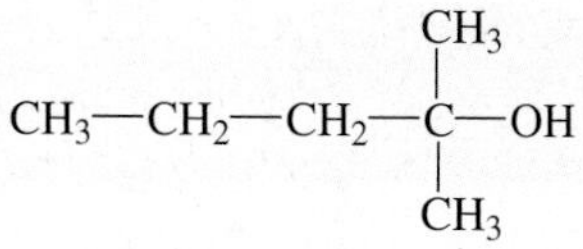 $CH_3-CH_2-CH_2-\overset{\overset{\displaystyle CH_3}{|}}{\underset{\underset{\displaystyle CH_3}{|}}{C}}-OH$

**12.9** Give a common name for each of the following ethers:

**a.** $CH_3-O-CH_2-CH_3$

**b.** $CH_3-CH_2-CH_2-O-CH_2-CH_2-CH_3$

**c.** 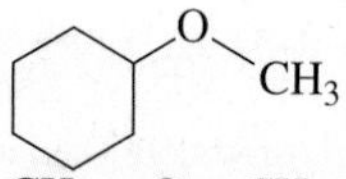 (cyclohexane ring)—O—CH₃

**d.** $CH_3-O-CH_2-CH_2-CH_3$

**12.10** Give a common name for each of the following ethers:

**a.** $CH_3-CH_2-O-CH_2-CH_2-CH_3$

**b.** 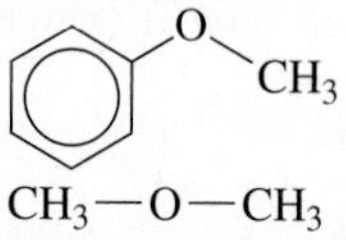 (benzene ring)—O—CH₃

**c.** 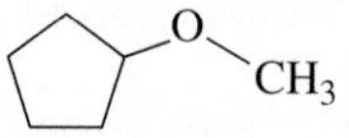 (cyclopentane ring)—O—CH₃

**d.** $CH_3-O-CH_3$

**12.11** Write the condensed structural formula for each of the following ethers:

**a.** ethyl propyl ether    **b.** cyclopropyl ethyl ether
**c.** ethyl methyl ether

**12.12** Write the condensed structural formula for each of the following ethers:

**a.** diethyl ether    **b.** diphenyl ether
**c.** cyclohexyl methyl ether

---

# 12.2 PROPERTIES OF ALCOHOLS AND ETHERS

In Chapter 10, we learned that hydrocarbons, which are composed of only carbon and hydrogen, are nonpolar. In this chapter, we look at compounds containing the element oxygen. The high electronegativity of oxygen increases the boiling points and solubility in water of alcohols and ethers.

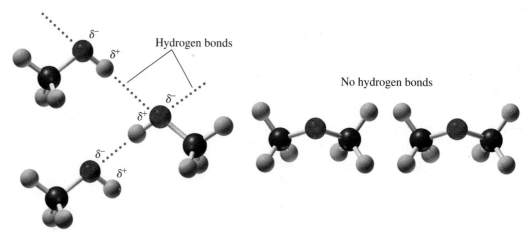

Methyl alcohol                    Dimethyl ether

## Boiling Points

In alcohols, the oxygen and hydrogen atoms in the hydroxyl group, —OH, form hydrogen bonds with each other. Ethers contain an oxygen atom, but without a hydrogen atom attached. Thus, ethers cannot hydrogen bond with each other.

Alcohols have higher boiling points than do alkanes and ethers of similar mass. Higher temperatures are required to provide the energy needed to break the hydrogen bonds between alcohol molecules. Because ethers and alkanes do not have hydrogen bonds between molecules, their boiling points are lower than alcohols and are similar to each other.

## Solubility in Water

The oxygen atom in alcohols and ethers influences their solubility in water. In alcohols, the polar —OH group forms hydrogen bonds with water. However, as the number of carbon atoms increases, the effect of the —OH group is diminished. Only an alcohol with one to four carbon atoms is soluble in water, but an alcohol with five or more carbon atoms is not.

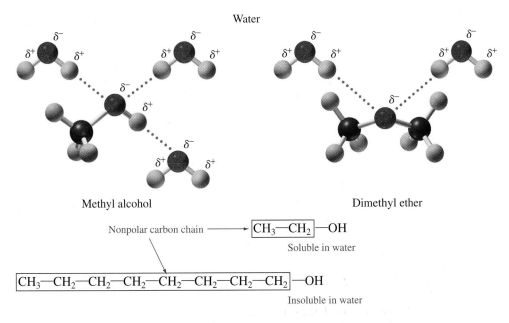

Ethers with small alkyl groups are also soluble in water because the oxygen atom forms hydrogen bonds with water. However, ethers do not form as many hydrogen bonds with water as do the alcohols. Thus, ethers are more soluble in water than are alkanes, but not as soluble as alcohols.

Table 12.1 compares the boiling points and solubility of some alkanes, alcohols, and ethers by mass.

**TABLE 12.1** Solubility and Boiling Points of Some Typical Alkanes, Alcohols, and Ethers of Similar Molar Mass

| Compound | Structural Formula | Molar Mass (g/mole) | Boiling Point (°C) | Soluble in Water |
|---|---|---|---|---|
| Propane | $CH_3$—$CH_2$—$CH_3$ | 44 | −42 | No |
| Dimethyl ether | $CH_3$—O—$CH_3$ | 46 | −23 | Yes |
| Ethanol | $CH_3$—$CH_2$—OH | 46 | 78 | Yes |
| Butane | $CH_3$—$CH_2$—$CH_2$—$CH_3$ | 58 | 0 | No |
| Ethyl methyl ether | $CH_3$—O—$CH_2$—$CH_3$ | 60 | 8 | Yes |
| 1-Propanol | $CH_3$—$CH_2$—$CH_2$—OH | 60 | 97 | Yes |

## Solubility of Phenols

Phenol is soluble in water because the hydroxyl group ionizes slightly as a weak acid. In fact, an early name for phenol was *carbolic acid*. A concentrated solution of phenol is very corrosive and highly irritating to the skin; it can cause severe burns and ingestion can be fatal. Dilute solutions of phenol were previously used in hospitals as antiseptics, but they have generally been replaced.

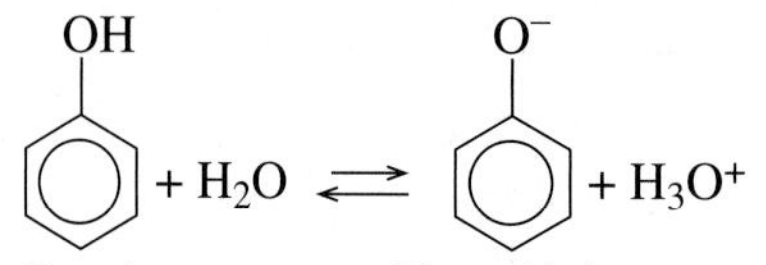

Phenol    + $H_2O$  ⇌    Phenoxide ion    + $H_3O^+$

---

## SAMPLE PROBLEM 12.4

### ■ Properties of Alcohols and Ethers

Predict which compounds in each pair will be more soluble in water.

**a.** propane or ethanol
**b.** 1-propanol or 1-hexanol

### SOLUTION

**a.** The alcohol ethanol is more soluble because it can form hydrogen bonds with water.
**b.** The 1-propanol is more soluble because it has a shorter carbon chain.

### STUDY CHECK

Dimethyl ether and ethanol both have molar masses of 46. However, ethanol has a much higher boiling point than dimethyl ether. How would you explain this difference in boiling points?

---

## QUESTIONS AND PROBLEMS

### Properties of Alcohols and Ethers

**12.13** Predict the compound with the higher boiling point in the following pairs:
**a.** ethane or methanol
**b.** diethyl ether or 1-butanol
**c.** 1-butanol or pentane

**12.14** Glycerol (1,2,3-propanetriol) has a boiling point of 290 °C. 1-Pentanol, which has about the same molar mass as glycerol, boils at 138 °C. Why is the boiling point of glycerol so much higher?

**12.15** Are each of the following soluble in water? Explain.
**a.** $CH_3-CH_2-OH$
**b.** $CH_3-O-CH_3$
**c.** $CH_3-CH_2-CH_2-CH_2-CH_2-CH_2-OH$
**d.** $CH_3-CH_2-CH_3$

**12.16** Give an explanation for the following observations:
**a.** Ethanol is soluble in water, but propane is not.
**b.** Dimethyl ether is soluble in water, but pentane is not.
**c.** 1-Propanol is soluble in water, but 1-hexanol is not.

---

## LEARNING GOAL

Write equations for the combustion, dehydration, and oxidation of alcohols.

# 12.3 REACTIONS OF ALCOHOLS AND THIOLS

In Chapter 10, we learned that hydrocarbons undergo combustion in the presence of oxygen. Alcohols burn with oxygen too. For example, in a restaurant, a dessert may be prepared by pouring a liquor on fruit or ice cream and lighting it. (See Figure 12.2.) The combustion of the ethanol in the liquor proceeds as follows:

$$CH_3-CH_2-OH + 3O_2 \longrightarrow 2CO_2 + 3H_2O + energy$$

# Dehydration of Alcohols to Form Alkenes

Earlier, we saw that alkenes can add water to yield alcohols. In a reverse reaction, alcohols lose a water molecule when they are heated with an acid catalyst such as $H_2SO_4$. During the **dehydration** of an alcohol, H— and —OH are removed from *adjacent carbon atoms of the same alcohol* to produce a water molecule. A double bond forms between the same two carbon atoms to produce an alkene product.

**FIGURE 12.2** A flaming dessert is prepared using a liquor that undergoes combustion.
**Q** What is the equation for the combustion of the ethanol in the liquor?

$$\underset{\text{Alcohol}}{\overset{\text{H  OH}}{\underset{|\quad|}{-\text{C}-\text{C}-}}} \xrightarrow[\text{Heat}]{H^+} \underset{\text{Alkene}}{\text{C}=\text{C}} + \underset{\text{Water}}{H_2O}$$

**Examples**

$$\underset{\text{Ethanol}}{\overset{\text{H  OH}}{\text{H}-\underset{\underset{\text{H}}{|}}{\overset{\overset{}{|}}{\text{C}}}-\underset{\underset{\text{H}}{|}}{\overset{\overset{}{|}}{\text{C}}}-\text{H}}} \xrightarrow[\text{Heat}]{H^+} \underset{\text{Ethene}}{\text{H}-\underset{\underset{\text{H}}{|}}{\text{C}}=\underset{\underset{\text{H}}{|}}{\text{C}}-\text{H}} + H_2O$$

Cyclopentanol → Cyclopentene + $H_2O$

---

**SAMPLE PROBLEM  12.5**

## ■ Dehydration of Alcohols

Draw the condensed structural formulas for the alkenes produced by the dehydration of the following alcohols:

**a.** $CH_3-CH_2-\overset{\overset{\text{OH}}{|}}{CH}-CH_2-CH_3 \xrightarrow[\text{Heat}]{H^+}$

**b.** (cyclohexanol with OH) $\xrightarrow[\text{Heat}]{H^+}$

### SOLUTION

**a.** $CH_3-CH_2-CH=CH-CH_3 + H_2O$

**b.** The —OH of this alcohol is removed along with a H from an adjacent carbon. Remember that the hydrogens are not drawn in this type of geometric figure.

### STUDY CHECK

What is the name of the alkene produced by the dehydration of cyclopentanol?

## Oxidation of Alcohols

From Chapter 5, we will use the term **oxidation** as a loss of hydrogen atoms or the addition of oxygen. In organic chemistry, we find that an oxidation reaction occurs when there is an increase in the number of carbon–oxygen bonds. In a reduction reaction, the product has fewer bonds between carbon and oxygen.

$$
\underset{\text{Alkane}}{CH_3-CH_3}
\underset{\text{Reduction}}{\overset{\text{Oxidation}}{\rightleftharpoons}}
\underset{\text{Alcohol (1°)}}{CH_3-\overset{\overset{\text{1 Bond to O}}{\displaystyle OH}}{\underset{\displaystyle |}{CH_2}}}
\underset{\text{Reduction}}{\overset{\text{Oxidation}}{\rightleftharpoons}}
\underset{\text{Aldehyde}}{CH_3-\overset{\overset{\text{2 Bonds to O}}{\displaystyle O}}{\underset{\displaystyle \|}{C}}-H}
\underset{\text{Reduction}}{\overset{\text{Oxidation}}{\rightleftharpoons}}
\underset{\text{Carboxylic acid}}{CH_3-\overset{\overset{\text{3 Bonds to O}}{\displaystyle O}}{\underset{\displaystyle \|}{C}}-OH}
$$

## Oxidation of Primary and Secondary Alcohols

The oxidation of a primary alcohol produces an aldehyde, which contains a double bond between carbon and oxygen. The oxidation occurs by removing two hydrogen atoms, one from the —OH group and another from the carbon that is bonded to the —OH. To indicate the presence of an oxidizing agent, reactions are often written with the symbol [O].

$$
\underset{\text{Methyl alcohol}}{H-\overset{\overset{\displaystyle OH}{|}}{\underset{\underset{\displaystyle H}{|}}{C}}-H}
\xrightarrow{[O]}
\underset{\text{Formaldehyde}}{H-\overset{\overset{\displaystyle O}{\|}}{C}-H} + \mathbf{H_2O}
$$

$$
\underset{\text{Ethyl alcohol}}{CH_3-\overset{\overset{\displaystyle OH}{|}}{CH_2}}
\xrightarrow{[O]}
\underset{\text{Acetaldehyde}}{CH_3-\overset{\overset{\displaystyle O}{\|}}{C}-H} + H_2O
$$

Aldehydes oxidize further by the addition of oxygen to form a carboxylic acid. This step occurs so readily that it is often difficult to isolate the aldehyde product during oxidation. We will learn more about carboxylic acids in Chapter 13.

$$
\underset{\substack{\text{Ethanal}\\\text{(acetaldehyde)}}}{CH_3-\overset{\overset{\displaystyle O}{\|}}{C}-H}
\xrightarrow{[O]}
\underset{\substack{\text{Ethanoic acid}\\\text{(acetic acid)}}}{CH_3-\overset{\overset{\displaystyle O}{\|}}{C}-OH}
$$

In the oxidation of secondary alcohols, the products are ketones. Two hydrogen atoms are removed; one from the —OH group and another from the carbon bonded to the —OH group. The result is a ketone that has the carbon–oxygen double bond attached to alkyl groups on both sides. There is no further oxidation of a ketone because there are no hydrogen atoms attached to the carbonyl group.

$$
\underset{\substack{\text{2-Propanol}\\\text{(isopropyl alcohol)}}}{CH_3-\overset{\overset{\displaystyle OH}{|}}{\underset{\underset{\displaystyle H}{|}}{C}}-CH_3}
\xrightarrow{[O]}
\underset{\substack{\text{Propanone}\\\text{(dimethyl ketone; acetone)}}}{CH_3-\overset{\overset{\displaystyle O}{\|}}{C}-CH_3} + \mathbf{H_2O}
$$

the **C**hemistry place

CASE STUDY
Alcohol Toxicity

*Career Focus*

### Pharmacist

"The pharmacy is one of the many factors in the final integration of chemistry and medicine in patient care," says Dorothea Lorimer, pharmacist, Kaiser Hospital. "If someone is allergic to a medication, I have to find out if a new medication has similar structural features. For instance, some people are allergic to sulfur. If there is sulfur in the new medication, there is a chance it will cause a reaction."

A prescription indicates a specific amount of a medication. At the pharmacy, the chemical name, formula, and quantity in milligrams or micrograms are checked. Then the prescribed number of capsules is prepared and placed in a container. If it is a liquid medication, a specific volume is measured and poured into a bottle for liquid prescriptions.

Tertiary alcohols do not oxidize readily because there are no hydrogen atoms on the carbon bonded to the —OH group. Because C—C bonds are usually too strong to oxidize, tertiary alcohols resist oxidation.

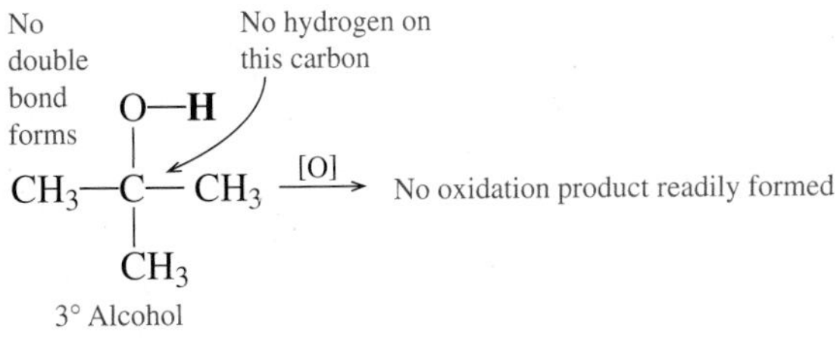

No double bond forms

No hydrogen on this carbon

$$CH_3—\underset{\underset{CH_3}{|}}{\overset{\overset{O—H}{|}}{C}}—CH_3 \xrightarrow{[O]}$$ No oxidation product readily formed

3° Alcohol

---

## SAMPLE PROBLEM 12.6

### ■ Oxidation of Alcohols

Draw the structural formula of the aldehyde or ketone formed by the oxidation of each of the following:

**a.** $CH_3—CH_2—\overset{\overset{OH}{|}}{CH}—CH_3$

**b.** $CH_3—CH_2—CH_2—OH$

### SOLUTION

**a.** A secondary (2°) alcohol oxidizes to a ketone.

$$CH_3—CH_2—\overset{\overset{O}{\|}}{C}—CH_3$$

**b.** A primary (1°) alcohol oxidizes to an aldehyde.

$$CH_3—CH_2—\overset{\overset{O}{\|}}{C}—H$$

### STUDY CHECK

Draw the condensed structural formula of the product of the oxidation of 2-propanol.

## Oxidation of Thiols

Thiols also undergo oxidation by a loss of hydrogen atoms from the —SH groups. The oxidized product is called a **disulfide**.

Much of the protein in hair is cross-linked by disulfide bonds, which occur mostly between the thiol groups of the amino acid cysteine.

Protein chain—$CH_2$—**SH** + **HS**—$CH_2$—Protein chain $\xrightarrow{[O]}$

Cysteine side groups

Protein chain—$CH_2$—**S**—**S**—$CH_2$—Protein chain + **$H_2O$**

Disulfide bond

When a person is given a "perm," a reducing substance is used to break the disulfide bonds. While the hair is still wrapped around the curlers, an oxidizing substance is applied that causes new disulfide bonds to form between different parts of the protein hair strands, which gives the hair a new shape.

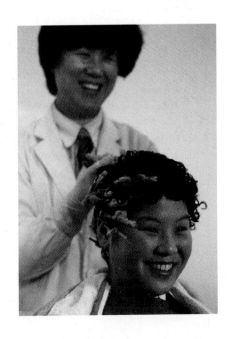

# Health Note

## Oxidation of Alcohol in the Body

Ethanol is the most commonly abused drug in the United States. When ingested in small amounts, ethanol may produce a feeling of euphoria in the body although it is a depressant. In the liver, enzymes such as alcohol dehydrogenases oxidize ethanol to acetaldehyde, a substance that impairs mental and physical coordination. If the blood alcohol concentration exceeds 0.4%, coma or death may occur. Table 12.2 gives some of the typical behaviors exhibited at various levels of blood alcohol.

$$CH_3-CH_2-OH \xrightarrow{[O]} CH_3-\overset{\displaystyle O}{\overset{\|}{C}}-H \xrightarrow{[O]} 2CO_2 + H_2O$$

Ethanol (ethyl alcohol)     Ethanal (acetaldehyde)

The acetaldehyde produced from ethanol in the liver is further oxidized to acetic acid, which is converted to carbon dioxide and water in the citric acid (Krebs) cycle. Thus, the enzymes in the liver

can eventually break down ethanol, but the aldehyde and carboxylic acid intermediates can cause considerable damage while they are present within the cells of the liver.

A person weighing 150 lb requires about one hour to metabolize 10 ounces of beer. However, the rate of metabolism of ethanol varies between nondrinkers and drinkers. Typically, nondrinkers and social drinkers can metabolize 12–15 mg of ethanol/dL of blood in one hour, but an alcoholic can metabolize as much as 30 mg of ethanol/dL in one hour. Some effects of alcohol metabolism include an increase in liver lipids (fatty liver), an increase in serum triglycerides, gastritis, pancreatitis, ketoacidosis, alcoholic hepatitis, and psychological disturbances.

When the Breathalyzer test is used for suspected drunken drivers, the driver exhales a volume of breath into a solution containing the orange $Cr^{6+}$ ion. If there is ethyl alcohol present in the exhaled air, the alcohol is oxidized, and the $Cr^{6+}$ is reduced to give a green solution of $Cr^{3+}$.

$$CH_3-CH_2-OH + Cr^{6+} \xrightarrow{[O]} CH_3-\overset{\displaystyle O}{\overset{\|}{C}}-OH + Cr^{3+}$$

Ethanol          Orange          Acetic acid          Green

Sometime an alcoholic is treated with a drug called Antabuse (disulfiram), which prevents the oxidation of acetaldehyde to acetic acid. If that person drinks alcohol, acetaldehyde accumulates in the blood, which causes nausea, profuse sweating, headache, dizziness, vomiting, and respiratory difficulties. Because of these unpleasant side effects, the person is less likely to use alcohol.

**TABLE 12.2** Typical Behaviors Exhibited by a 150-lb Person Consuming Alcohol

| Number of Beers (12 oz) or Glasses of Wine (5 oz) | Blood Alcohol Level (m/v%) | Typical Behavior |
|---|---|---|
| 1 | 0.025 | Slightly dizzy, talkative |
| 2 | 0.05 | Euphoria, loud talking and laughing |
| 4 | 0.10 | Loss of inhibition, loss of coordination, drowsiness, legally drunk in most states |
| 8 | 0.20 | Intoxicated, quick to anger, exaggerated emotions |
| 12 | 0.30 | Unconscious |
| 16–20 | 0.40–0.50 | Coma and death |

## QUESTIONS AND PROBLEMS

### Reactions of Alcohols and Thiols

**12.17** Draw the condensed structural formula of the alkene produced by each of the following dehydration reactions:

**a.** $CH_3-CH_2-CH_2-CH_2-OH \xrightarrow[\text{Heat}]{H^+}$

**b.** (cyclopentanol structure) $\xrightarrow[\text{Heat}]{H^+}$

**c.** $CH_3-CH_2-\underset{\underset{OH}{|}}{CH}-CH_2-CH_3 \xrightarrow[\text{Heat}]{H^+}$

**12.18** Draw the condensed structural formula of the alkene produced by each of the following dehydration reactions:

**a.** $CH_3-CH_2-OH \xrightarrow[\text{Heat}]{H^+}$

**b.** $CH_3-\underset{\underset{CH_3}{|}}{CH}-CH_2-OH \xrightarrow[\text{Heat}]{H^+}$

**c.** (cyclohexanol structure) $\xrightarrow[\text{Heat}]{H^+}$

**12.19** Draw the condensed structural formula of the aldehyde or ketone when each of the following alcohols is oxidized [O] (if no reaction, write *none*):

a. $CH_3—CH_2—CH_2—CH_2—CH_2—OH$

b. $CH_3—CH_2—\overset{\displaystyle OH}{\underset{\displaystyle |}{CH}}—CH_3$

c.

d. $CH_3—\overset{\displaystyle OH}{\underset{\displaystyle |}{CH}}—CH_2—\overset{\displaystyle CH_3}{\underset{\displaystyle |}{CH}}—CH_3$

**12.20** Draw the condensed structural formula of the organic product when each of the following alcohols is oxidized [O] (if no reaction, write *none*):

a. $CH_3—\overset{\displaystyle CH_3}{\underset{\displaystyle |}{CH}}—CH_2—CH_2—OH$

b. $CH_3—CH_2—\overset{\displaystyle OH}{\underset{\displaystyle |}{\underset{\displaystyle CH_3}{\overset{\displaystyle |}{C}}}}—CH_3$

c. $CH_3—CH_2—\overset{\displaystyle OH}{\underset{\displaystyle |}{CH}}—CH_2—CH_3$

d.

# 12.4 ALDEHYDES AND KETONES

As we learned in Chapter 10, the carbonyl group consists of a carbon–oxygen double bond. The double bond in the carbonyl group is similar to that of alkenes, except the carbonyl group has a dipole. The oxygen atom with two lone pairs of electrons is much more electronegative than the carbon atom. Therefore, the carbonyl group has a strong dipole with a partial negative charge ($\delta^-$) on the oxygen and a partial positive charge ($\delta^+$) on the carbon. The polarity of the carbonyl group strongly influences the physical and chemical properties of aldehydes and ketones.

$$\overset{\displaystyle O^{\delta-}}{\underset{\displaystyle /\ \backslash}{\overset{\displaystyle ||}{C^{\delta+}}}}$$

In an **aldehyde**, the carbon of the carbonyl group is bonded to at least one hydrogen atom. That carbon may also be bonded to another hydrogen atom, a carbon of an alkyl group, or an aromatic ring. (See Figure 12.3.) In a **ketone**, the carbonyl group is bonded to two alkyl groups or aromatic rings.

In the condensed structural formula, the aldehyde group may be drawn as separate atoms or it may be written as —CHO, with the double bond understood. The keto group (C=O) is sometimes written as CO.

**Aldehyde**

$$CH_3—CH_2—\overset{\displaystyle O}{\overset{\displaystyle ||}{C}}—H \quad = \quad CH_3—CH_2—CHO$$

**Ketone**

$$CH_3—\overset{\displaystyle O}{\overset{\displaystyle ||}{C}}—CH_3 \quad = \quad CH_3—CO—CH_3$$

## LEARNING GOAL

Identify compounds with the carbonyl group as aldehydes and ketones; write their IUPAC names.

**the Chemistry place**

**WEB TUTORIAL**
Aldehydes and Ketones

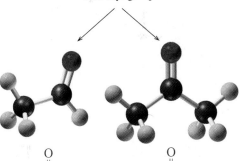

Carbonyl group

Aldehyde        Ketone

**FIGURE 12.3** The carbonyl group found in aldehydes and ketones.
**Q** If aldehydes and ketones both contain a carbonyl group, how can you differentiate between compounds from each family?

SAMPLE PROBLEM 12.7

■ **Identifying Aldehydes and Ketones**

Identify each of the following compounds as an aldehyde or a ketone:

a. $CH_3—\overset{\displaystyle CH_3}{\underset{\displaystyle CH_3}{\overset{\displaystyle |}{\underset{\displaystyle |}{C}}}}—CH_2—\overset{\displaystyle O}{\overset{\displaystyle ||}{C}}—H$

b.

c.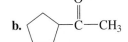

SOLUTION

**a.** aldehyde
**b.** ketone
**c.** aldehyde

STUDY CHECK

Draw the condensed structural formula of a ketone that has a carbonyl group bonded to two ethyl groups.

## Naming Aldehydes

In the IUPAC names of aldehydes, the *e* of the alkane name is replaced with *al*.

**Step 1    Name the longest carbon chain containing the carbonyl group by replacing the *e* in the corresponding alkane name by *al*.** No number is needed for the aldehyde group because it always appears at the end of the chain.
    The IUPAC system names the aldehyde of benzene as benzaldehyde.

Benzaldehyde

The first four unbranched aldehydes are often referred to by their common names, which end in *aldehyde*. (See Figure 12.4.) The roots of these common names are derived from Latin or Greek words that indicate the source of the corresponding carboxylic acid. We will look at carboxylic acids in the next section.

**Step 2    Name and number any substituents on the carbon chain by counting the carbonyl carbon as carbon 1.**

2-Methylpropanal                    4-Methylpentanal

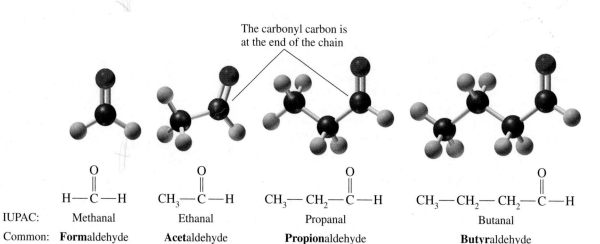

The carbonyl carbon is at the end of the chain

| IUPAC: | Methanal | Ethanal | Propanal | Butanal |
| Common: | **Form**aldehyde | **Acet**aldehyde | **Propion**aldehyde | **Butyr**aldehyde |

**FIGURE 12.4** In the structures of aldehydes, the carbonyl group is always the end carbon.
**Q** Why is the carbon in the carbonyl group in aldehydes always at the end of the chain?

SAMPLE PROBLEM 12.8

## ■ Naming Aldehydes

Give the IUPAC names for the following aldehydes:

**a.** $CH_3$—$CH_2$—$CH_2$—$CH_2$—$\overset{\displaystyle O}{\overset{\|}{C}}$—H
**b.** 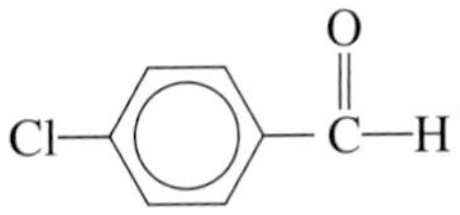 Cl—⟨benzene ring⟩—$\overset{\displaystyle O}{\overset{\|}{C}}$—H

**c.** $CH_3$—$\overset{\displaystyle CH_3}{\underset{\displaystyle |}{CH}}$—$CH_2$—$\overset{\displaystyle O}{\overset{\|}{C}}$—H

### SOLUTION

**a.** pentanal
**b.** 4-chlorobenzaldehyde
**c.** The longest unbranched chain has four atoms, with a methyl group on the third carbon. The IUPAC name is 3-methylbutanal.

### STUDY CHECK

What are the IUPAC and common names of the aldehyde with three carbon atoms?

## *Environmental Note*

### Vanilla

Vanilla has been used as a flavoring for over a thousand years. After drinking a beverage made from powdered vanilla and cocoa beans with Emperor Montezuma in Mexico, Cortez took vanilla back to Europe, where it became popular for flavoring and for scenting perfumes and tobacco. Thomas Jefferson introduced vanilla to the United States in the late 1700s. Today, much of the vanilla we use in the world is grown in Mexico, Madagascar, Réunion, Seychelles, Tahiti, Sri Lanka, Java, the Philippines, and Africa.

The vanilla plant is a member of the orchid family. There are many species of *Vanilla*, but *Vanilla planifolia* (or *V. fragrans*) is considered to produce the best flavor. The vanilla plant grows like a vine and can grow to 100 feet in length. Its flowers are hand-pollinated to produce a green fruit that is picked in 8 to 9 months. It is sun-dried to form a long, dark brown pod, which is called "vanilla bean" because it looks like a string bean. The flavor and fragrance of the vanilla bean come from the black seeds found inside the dried bean.

The seeds and pod are used to flavor desserts such as custards and ice cream. The extract of vanilla is made by chopping up vanilla beans and mixing them with a 35% alcohol–water mixture. The liquid, which contains the aldehyde vanillin, is drained from the bean residue and used for flavoring.

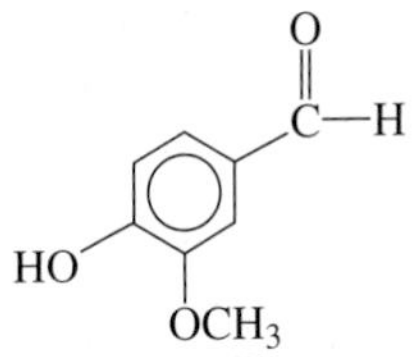

Vanillin

# Naming Ketones

Aldehydes and ketones are some of the most important classes of organic compounds. Because they have played a major role in organic chemistry for more than a century, the common names for unbranched ketones are still in use. In the common names, the alkyl groups bonded to the carbonyl group are named as substituents and are listed alphabetically, followed by *ketone*. Acetone, which is another name for propanone, has been retained by the IUPAC system.

In the IUPAC system, the name of a ketone is obtained by replacing the *e* in the corresponding alkane name with *one*.

**Step 1**  **Name the longest carbon chain containing the carbonyl group by replacing the *e* in the corresponding alkane name by *one*.**

**Step 2**  **Number the main chain starting from the end nearest the carbonyl group.**  Place the number of the carbonyl carbon in front of the ketone name. (Propanone and butanone do not require numbers.)

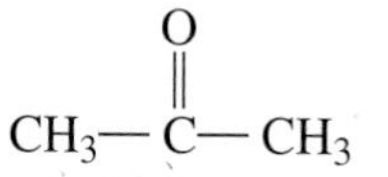

$$CH_3-\overset{\overset{\displaystyle O}{\|}}{C}-CH_3 \qquad CH_3-CH_2-\overset{\overset{\displaystyle O}{\|}}{C}-CH_3 \qquad CH_3-CH_2-\overset{\overset{\displaystyle O}{\|}}{C}-CH_2-CH_3$$

| Propanone | Butanone | 3-Pentanone |
|---|---|---|
| (dimethyl ketone: acetone) | (ethyl methyl ketone) | (diethyl ketone) |

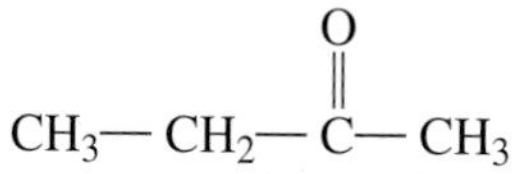

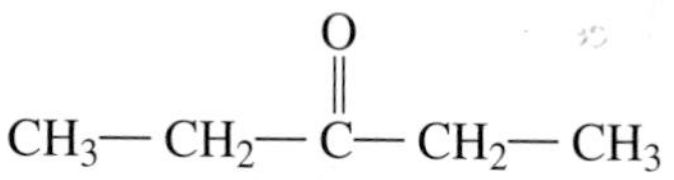

**Step 3**  **Name and number any substituents on the carbon chain.**

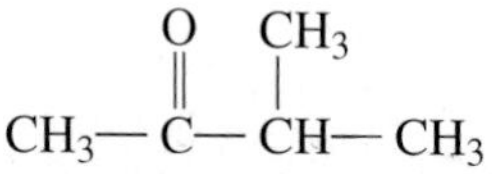

$$CH_3-\overset{\overset{\displaystyle O}{\|}}{C}-\overset{\overset{\displaystyle CH_3}{|}}{CH}-CH_3 \qquad CH_3-\overset{\overset{\displaystyle Br}{|}}{CH}-\overset{\overset{\displaystyle O}{\|}}{C}-CH_2-CH_3$$

3-Methylbutanone 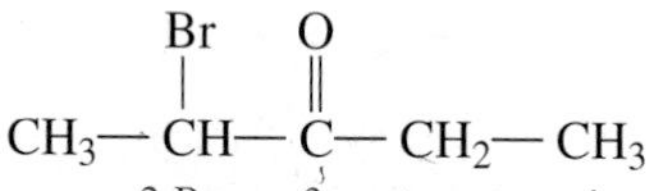            2-Bromo-3-pentanone

**Step 4**  **For cyclic ketones, the prefix *cyclo* is used in front of the ketone name.** Any substituents are located by numbering the ring starting with the carbonyl carbon as carbon 1. The ring is numbered so that the substituents have the lowest possible number.

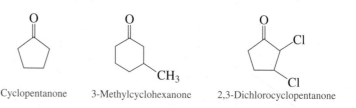

Cyclopentanone    3-Methylcyclohexanone    2,3-Dichlorocyclopentanone

---

**SAMPLE PROBLEM 12.9**

### ■ Names of Ketones

Give the IUPAC name for the following ketone:

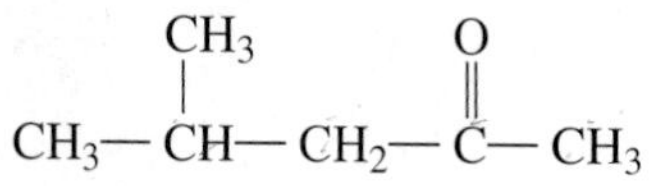

$$CH_3-\overset{\overset{\displaystyle CH_3}{|}}{CH}-CH_2-\overset{\overset{\displaystyle O}{\|}}{C}-CH_3$$

SOLUTION

The longest chain is five carbon atoms. Counting from the right, the carbonyl group is on carbon 2 and a methyl group is on carbon 4. The IUPAC name is 4-methyl-2-pentanone.

STUDY CHECK

What is the common name of 3-hexanone?

# *Health Note*

## Some Important Aldehydes and Ketones

*Formaldehyde*, the simplest aldehyde, is a colorless gas with a pungent odor. Industrially, it is a reactant in the synthesis of polymers used to make fabrics, insulation materials, carpeting, pressed-wood products such as plywood, and plastics for kitchen counters. An aqueous solution called formalin, which contains 40% formaldehyde, is used as a germicide and to preserve biological specimens. Exposure to formaldehyde fumes can irritate eyes, nose, upper respiratory tract, and can cause skin rashes, headaches, dizziness, and general fatigue. Formaldehyde is classified as a carcinogen.

*Acetone*, or propanone (dimethyl ketone), which is the simplest ketone, is a colorless liquid with a mild odor that has wide use as a solvent in cleaning fluids, paint and nail-polish removers, and rubber cement. (See Figure 12.5.) It is extremely flammable and care must be taken when using acetone. In the body, acetone may be produced in uncontrolled diabetes, fasting, and high-protein diets when large amounts of fats are metabolized for energy.

Several naturally occurring aromatic aldehydes are used to flavor food and as fragrances in perfumes. Benzaldehyde is found in almonds, vanillin in vanilla beans, and cinnamaldehyde in cinnamon.

Benzaldehyde (almond)    Vanillin (vanilla)    Cinnamaldehyde (cinnamon)

The flavor of butter or margarine is from butanedione, muscone is used to make musk perfumes, and oil of spearmint contains carvone.

Butanedione (butter flavor)    Muscone (musk)    Carvone (spearmint oil)

$$CH_3 - \overset{\overset{\displaystyle O}{\|}}{C} - \overset{\overset{\displaystyle O}{\|}}{C} - CH_3$$

Butanedione

**FIGURE 12.5** Acetone is used as a solvent in paint and nail-polish removers.
**Q** What is the IUPAC name of acetone?

## QUESTIONS AND PROBLEMS

### Aldehydes and Ketones

**12.21** Identify the following compounds as aldehydes or ketones:

**a.** CH$_3$—CH$_2$—$\overset{\overset{\text{O}}{\|}}{\text{C}}$—CH$_3$

**b.**

**c.**

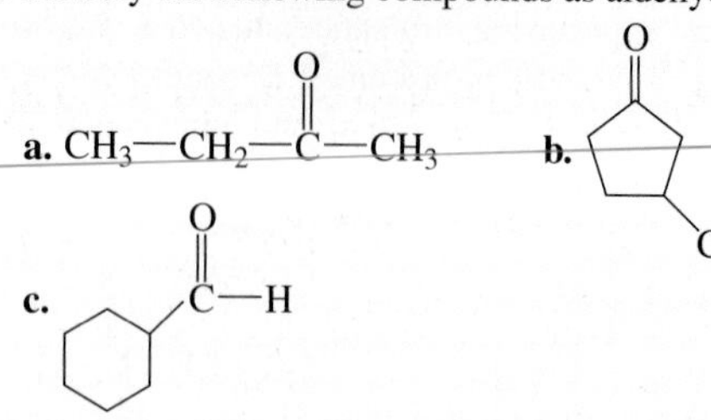

**12.22** Identify the following compounds as aldehydes or ketones:

**a.**

**b.** CH$_3$—$\overset{\overset{\text{CH}_3}{|}}{\text{CH}}$—$\overset{\overset{\text{O}}{\|}}{\text{C}}$—H

**c.**

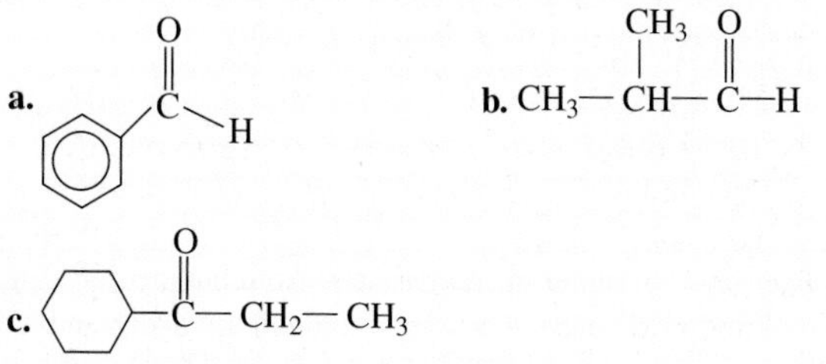

**12.23** Give a common name for each of the following compounds:

**a.** CH$_3$—$\overset{\overset{\text{O}}{\|}}{\text{C}}$—H

**b.** CH$_3$—$\overset{\overset{\text{O}}{\|}}{\text{C}}$—CH$_2$—CH$_2$—CH$_3$

**c.** H—$\overset{\overset{\text{O}}{\|}}{\text{C}}$—H

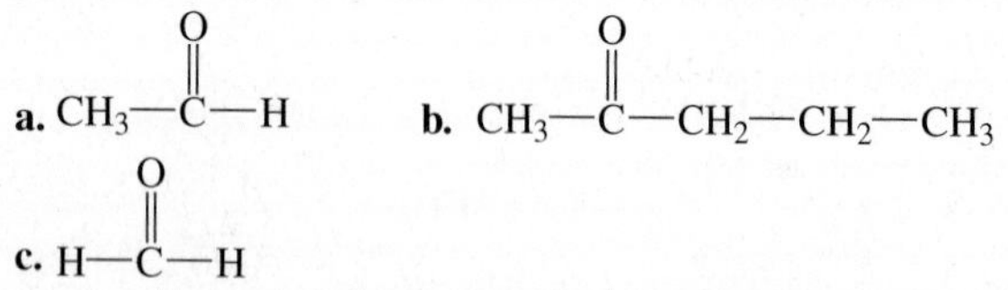

**12.24** Give the common name for each of the following compounds:

**a.** CH$_3$—$\overset{\overset{\text{O}}{\|}}{\text{C}}$—CH$_2$—CH$_3$

**b.** CH$_3$—CH$_2$—$\overset{\overset{\text{O}}{\|}}{\text{C}}$—CH$_2$—CH$_3$

**c.** CH$_3$—CH$_2$—$\overset{\overset{\text{O}}{\|}}{\text{C}}$—H

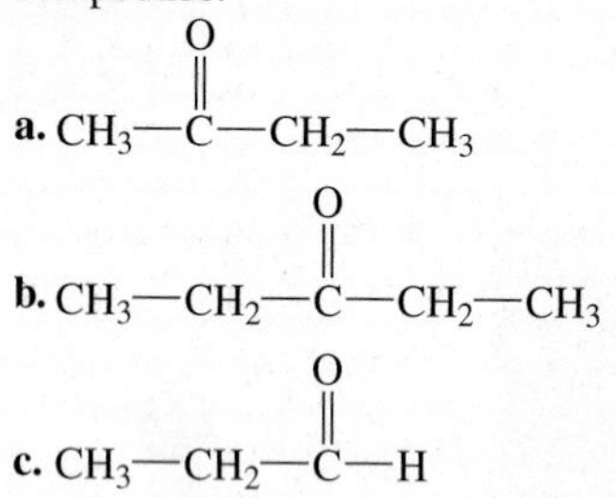

**12.25** Give the IUPAC name for each of the following compounds:

**a.** CH$_3$—CH$_2$—$\overset{\overset{\text{O}}{\|}}{\text{C}}$—H

**b.** CH$_3$—CH$_2$—$\overset{\overset{\text{O}}{\|}}{\text{C}}$—$\overset{\overset{\text{CH}_3}{|}}{\text{CH}}$—CH$_3$

**c.**

**d.**

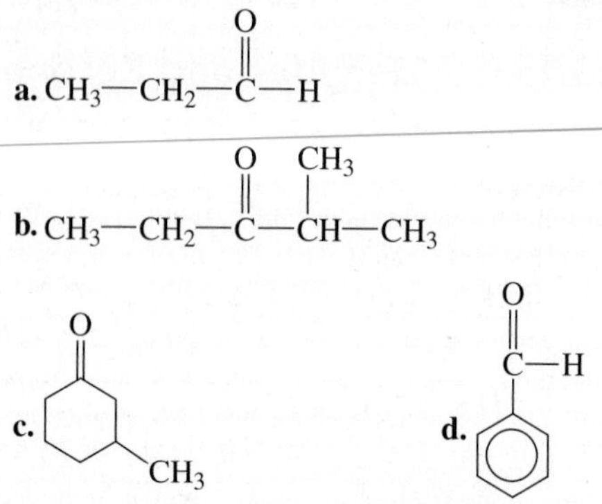

**12.26** Give the IUPAC name for each of the following compounds:

**a.** CH$_3$—CH$_2$—$\overset{\overset{\text{CH}_3}{|}}{\text{CH}}$—CH$_2$—$\overset{\overset{\text{O}}{\|}}{\text{C}}$—H

**b.** CH$_3$—CH$_2$—CH$_2$—$\overset{\overset{\text{O}}{\|}}{\text{C}}$—CH$_3$

**c.**

**d.**

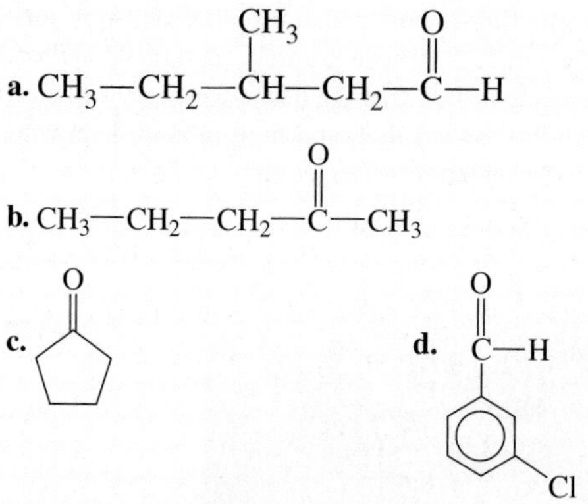

**12.27** Write the condensed structural formula for each of the following compounds:
- **a.** acetaldehyde
- **b.** 2-pentanone
- **c.** butyl methyl ketone
- **d.** 3-methylpentanal

**12.28** Write the condensed structural formula for each of the following compounds:
- **a.** propionaldehyde
- **b.** butanal
- **c.** 4-bromobutanone
- **d.** acetone

# 12.5 PROPERTIES OF ALDEHYDES AND KETONES

**LEARNING GOAL**

Compare the boiling points and solubility of aldehydes and ketones to those of alkanes and alcohols.

At room temperature, formaldehyde (bp −21 °C) and acetaldehyde (bp 21 °C) are gases. Aldehydes containing from 3 to 10 carbon atoms are liquids. The polar carbonyl group with a partially negative oxygen atom and a partially positive carbon atom has an influence on the boiling points and the solubility of aldehydes and ketones in water.

## Boiling Points

The polar carbonyl group gives aldehydes and ketones higher boiling points than alkanes and ethers of similar mass. The increase in boiling points is due to dipole–dipole interactions.

Dipole–dipole interaction

$\overset{}{>}\text{C}^{\delta+}\!\!=\!\text{O}^{\delta-}\bullet\!\bullet\!\bullet\!\bullet\!\bullet\!\bullet\!\bullet\!\bullet\overset{}{>}\text{C}^{\delta+}\!\!=\!\text{O}^{\delta-}\bullet\!\bullet\!\bullet\!\bullet\!\bullet\!\bullet\!\bullet\!\bullet\overset{}{>}\text{C}^{\delta+}\!\!=\!\text{O}^{\delta-}$

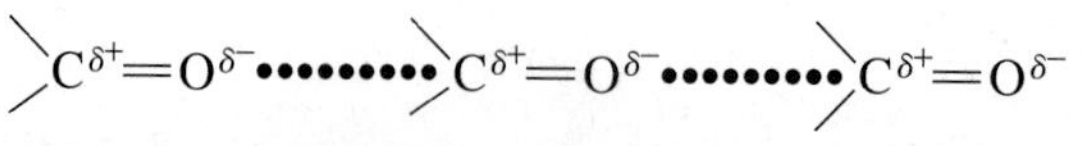

However, because there is no hydrogen on the oxygen atom, aldehydes and ketones cannot form hydrogen bonds with each other. Thus, they have boiling points that are lower than alcohols.

| | $CH_3-CH_2-CH_2-CH_3$ | $CH_3-CH_2-O-CH_3$ | $CH_3-CH_2-\overset{\overset{O}{\|\|}}{C}-H$ | $CH_3-\overset{\overset{O}{\|\|}}{C}-CH_3$ | $CH_3-CH_2-CH_2-OH$ |
|---|---|---|---|---|---|
| Name | Butane | Ethyl methyl ether | Propanal | Propanone | 1-Propanol |
| Molar Mass | 58 | 60 | 58 | 58 | 60 |
| Family | Alkane | Ether | Aldehyde | Ketone | Alcohol |
| bp | 0 °C | 8 °C | 49 °C | 56 °C | 97 °C |

Increasing boiling point

*btc of Hydrogen bonding*

## Solubility of Aldehydes and Ketones in Water

Although aldehydes and ketones do not hydrogen bond with each other, the electronegative oxygen atom does hydrogen bond with water molecules. Carbonyl compounds with one to four carbons are very soluble in water. However, those with five or more carbon atoms are not very soluble because the alkyl portions diminish the effect of the polar carbonyl group.

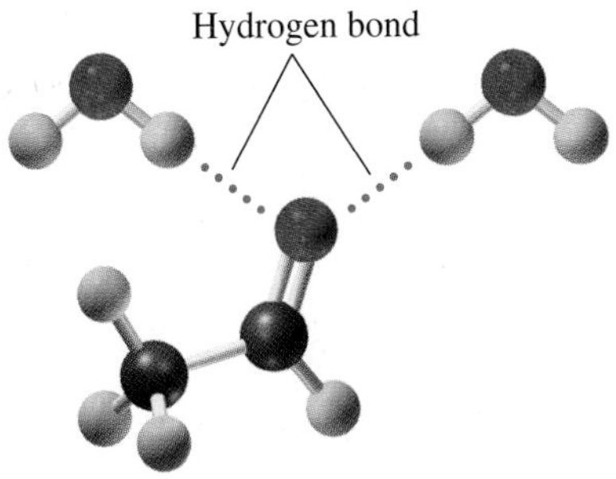

Hydrogen bond

Acetaldehyde

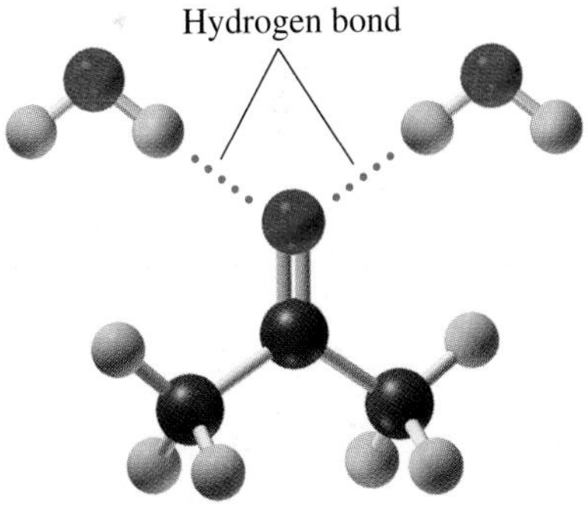

---

### SAMPLE PROBLEM 12.10

#### ■ Boiling Point and Solubility

Would you expect the boiling point of ethanol, $CH_3-CH_2-OH$, to be higher or lower than that of ethanal, $CH_3CHO$? Explain.

#### SOLUTION

Ethanol would have a higher boiling point because its molecules can hydrogen bond with each other, while molecules of ethanal cannot.

#### STUDY CHECK

If acetone molecules cannot hydrogen bond with each other, why is acetone soluble in water?

Hydrogen bond

Acetone

## Oxidation

Aldehydes oxidize readily to carboxylic acids. In contrast, ketones do not undergo further oxidation.

$$CH_3-\overset{\overset{O}{\|\|}}{C}-H \xrightarrow{[O]} CH_3-\overset{\overset{O}{\|\|}}{C}-OH$$

Acetaldehyde          Acetic acid

$$CH_3-\overset{\overset{O}{\|\|}}{C}-CH_3 \xrightarrow{[O]} \text{no reaction}$$

Propanone

**FIGURE 12.6** In Tollens' test, a silver mirror forms when the oxidation of an aldehyde reduces silver ion to metallic silver. The silvery surface of a mirror is formed in a similar way.

**Q** What is the product of the oxidation of an aldehyde?

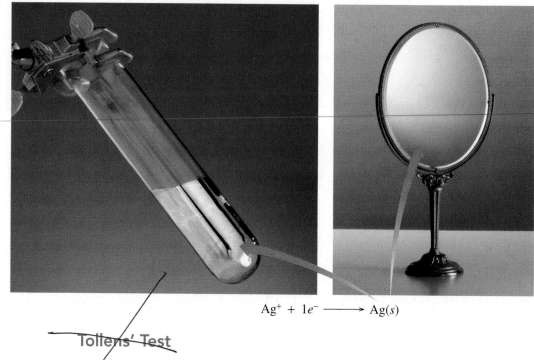

$$Ag^+ + 1e^- \longrightarrow Ag(s)$$

## Tollens' Test

**Tollens' test**, which uses a solution of $Ag^+$ ($AgNO_3$) and ammonia, oxidizes aldehydes but not ketones. The silver ion is reduced and forms a "silver mirror" on the inside of the container. Commercially, a similar process is used to make mirrors by applying a mixture of $AgNO_3$ and ammonia on glass with a spray gun. (See Figure 12.6.)

$$\underset{\text{Acetaldehyde}}{CH_3-\overset{\displaystyle O}{\overset{\|}{C}}-H} + \underset{\substack{\text{Tollens'}\\\text{reagent}}}{2Ag^+} \longrightarrow \underset{\substack{\text{Silver}\\\text{mirror}}}{2Ag(s)} + \underset{\text{Acetic acid}}{CH_3-\overset{\displaystyle O}{\overset{\|}{C}}-OH}$$

Another test, called **Benedict's test**, gives a positive result with compounds that have an aldehyde functional group and an adjacent hydroxyl group. When Benedict's reagent containing $Cu^{2+}$($CuSO_4$) ions is added to this type of aldehyde, a brick-red solid of $Cu_2O$ forms. (See Figure 12.7.) The test is negative with simple aldehydes and ketones.

$$\underset{\text{2-Hydroxypropanal}}{CH_3-\overset{\overset{\displaystyle OH}{|}}{CH}-\overset{\overset{\displaystyle O}{\|}}{C}-H} + \underset{\substack{\text{Benedict's}\\\text{reagent}}}{2Cu^{2+}} \longrightarrow \underset{\text{Brick-red solid}}{Cu_2O(s)} + \underset{\text{2-Hydroxypropanoic acid}}{CH_3-\overset{\overset{\displaystyle OH}{|}}{CH}-\overset{\overset{\displaystyle O}{\|}}{C}-OH}$$

Because many sugars, such as glucose, contain this type of aldehyde grouping, Benedict's reagent can be used to determine the presence of glucose in blood or urine.

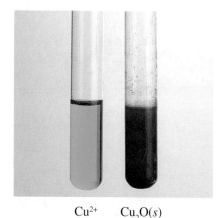

Cu²⁺        Cu₂O(s)

**FIGURE 12.7** The blue $Cu^{2+}$ in Benedict's solution forms a brick-red solid of $Cu_2O$ in a positive test for many sugars and aldehydes with adjacent hydroxyl groups.

**Q** Which test tube indicates that glucose is present?

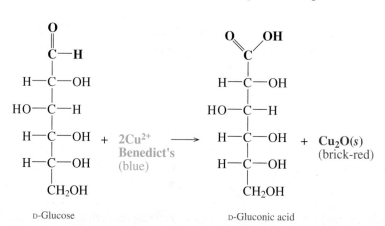

## QUESTIONS AND PROBLEMS

### Properties of Aldehydes and Ketones

**12.29** Which compound in each of the following pairs would have the higher boiling point? Explain.

**a.** $CH_3-CH_2-CH_3$  or  $CH_3-CH_2-\overset{\displaystyle O}{\overset{\displaystyle \|}{C}}-H$

**b.** propanal or pentanal

**c.** butanal or 1-butanol

**12.30** Which compound in each of the following pairs would have the higher boiling point? Explain.

**a.** $CH_3-CH_2-\overset{\displaystyle OH}{\underset{\displaystyle |}{CH}}-CH_3$  or  $CH_3-CH_2-\overset{\displaystyle O}{\overset{\displaystyle \|}{C}}-CH_3$

**b.** butane or butanone

**c.** propanone or pentanone

**12.31** Which compound in each of the following pairs would be more soluble in water? Explain.

**a.** $CH_3-\overset{\displaystyle O}{\overset{\displaystyle \|}{C}}-CH_2-CH_2-CH_3$  or

$CH_3-\overset{\displaystyle O}{\overset{\displaystyle \|}{C}}-\overset{\displaystyle O}{\overset{\displaystyle \|}{C}}-CH_2-CH_3$

**b.** acetone or 2-pentanone

**12.32** Which compound in each of the following pairs would be more soluble in water? Explain.

**a.** $CH_3-CH_2-CH_3$ or $CH_3-CH_2-CHO$

**b.** propanone or 3-hexanone

**12.33** Draw the condensed structural formula of the organic product when each of the following alcohols is oxidized [O] (if no reaction, write *none*):

**a.** $CH_3-CH_2-CH_2-CH_2-CH_2-OH$

**b.** $CH_3-CH_2-\overset{\displaystyle OH}{\underset{\displaystyle |}{CH}}-CH_3$

**c.** cyclohexanol with OH

**d.** $CH_3-\overset{\displaystyle OH}{\underset{\displaystyle |}{CH}}-CH_2-\overset{\displaystyle CH_3}{\underset{\displaystyle |}{CH}}-CH_3$

**e.** $CH_3-\overset{\displaystyle CH_3}{\underset{\displaystyle |}{CH}}-CH_2-CH_2-OH$

**12.34** Draw the condensed structural formula of the organic product when each of the following alcohols is oxidized [O] (if no reaction, write *none*):

**a.** cyclobutyl$-CH_2-OH$

**b.** $CH_3-\overset{\displaystyle CH_3}{\underset{\displaystyle |}{CH}}-CH_2-\overset{\displaystyle CH_3}{\underset{\displaystyle |}{CH}}-OH$

**c.** $CH_3-CH_2-\overset{\displaystyle OH}{\underset{\displaystyle |}{\underset{\displaystyle CH_3}{C}}}-CH_3$

**d.** $CH_3-\overset{\displaystyle OH}{\underset{\displaystyle |}{CH}}-\overset{\displaystyle OH}{\underset{\displaystyle |}{CH}}-CH_2-CH_3$

**e.** cyclohexanol with OH

## 12.6 CHIRAL MOLECULES

In the preceding chapters, we have looked at isomers. Let's review those now. Molecules are structural isomers when they have the same molecular formula, but different bonding arrangements.

**LEARNING GOAL**

Identify chiral and achiral carbon atoms in an organic molecule.

### Isomers

$C_2H_6O$    $CH_3-CH_2-OH$    $CH_3-O-CH_3$
    Ethanol    Dimethyl ether

$C_3H_6O$    $CH_3-CH_2-\overset{\displaystyle O}{\overset{\displaystyle \|}{C}}-H$    $CH_3-\overset{\displaystyle O}{\overset{\displaystyle \|}{C}}-CH_3$
    Propanal    Propanone

Another group of isomers called **stereoisomers** have identical molecular formulas, too, but they are not structural isomers. In stereoisomers, the atoms are bonded in the same sequence, but differ in the way they are arranged in space.

### Chirality

If you hold your right hand up to a mirror, you see the mirror image, which matches your left hand. (See Figure 12.8.) If you turn your palms toward each other, you also have mirror images. If you look at the palms of your hands, your thumbs are on opposite sides. If you place your right hand over your left hand, you cannot match up all the parts of the hands: palms, backs, thumbs and little fingers. The thumbs and little fingers can be matched, but then the palms and backs of your hands are facing each other. Your hands are

**FIGURE 12.8** The left and right hands are chiral because they have mirror images that cannot be superimposed on each other.
**Q** Why are your shoes chiral objects?

Left hand                                                          Right hand

Mirror image of right hand

mirror images that cannot be superimposed on each other. When organic molecules have mirror images that cannot be completely matched, we say they are nonsuperimposable.

Objects such as hands that have nonsuperimposable mirror images are **chiral**. Left and right shoes are chiral; left- and right-handed golf clubs are chiral. If we try to put a left-hand glove on our right hand or a right shoe on our left foot, or use left-handed scissors if we are right-handed, we realize that mirror images that are chiral do not match. However, sometimes a mirror image can be superimposed on the original. For example, all parts of the mirror image of a plain drinking glass can be matched to the glass. Then we say that the object is *achiral*. (See Figure 12.9.)

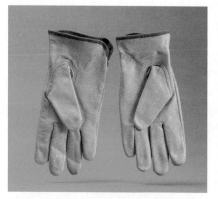

Chiral

Golf club, chiral

Achiral

Achiral

Chiral

Right-handed scissors, chiral

**FIGURE 12.9** Everyday objects can be chiral or achiral.
**Q** Why are some of the above objects chiral and others achiral?

Molecules in nature also have mirror images, and often one stereoisomer has a different biological effect than the other one. In some cases, one isomer has a certain odor, and the mirror image has a completely different odor. For example, one stereoisomer of limonene smells like oranges, while its mirror image has the odor of lemons.

---

### SAMPLE PROBLEM 12.11

#### ■ Chiral and Achiral Objects

Identify each of the following objects as chiral or achiral:

**a.** white tennis sock
**b.** flip-flop beach shoe
**c.** golf ball

SOLUTION

**a.** Achiral. The mirror image of a tennis sock is superimposable.
**b.** Chiral. The mirror image of a left flip flop is the right flip flop; they are not superimposable.
**c.** Achiral. The mirror image of a golf ball is superimposable.

STUDY CHECK

To repair a desk, a student uses a screwdriver and a wood screw. Identify the screwdriver and wood screw as chiral or achiral objects.

---

## Chiral Carbon Atoms

A carbon compound is chiral if it has at least one carbon atom bonded to four different atoms or groups. This type of carbon atom is called a **chiral carbon** because there are two different ways that it can bond to four atoms or groups of atoms. The resulting structures are mirror images of each other. Let's look at the mirror images of a carbon bonded to four different atoms. (See Figure 12.10.) If we line up the hydrogen and iodine atoms in the mirror images, the bromine and chlorine atoms appear on opposite sides. No matter how we turn the models, we cannot align all four atoms at the same time. When stereoisomers cannot be superimposed, they are called **enantiomers**. If two or more atoms are the same, the atoms can be aligned (superimposed) and the mirror images represent the same structure. (See Figure 12.11.)

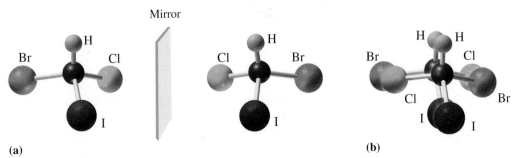

**FIGURE 12.10 (a)** The enantiomers of a chiral molecule are mirror images.
**(b)** The enantiomers of a chiral molecule cannot be superimposed on each other.
**Q** Why is the carbon atom in this compound chiral?

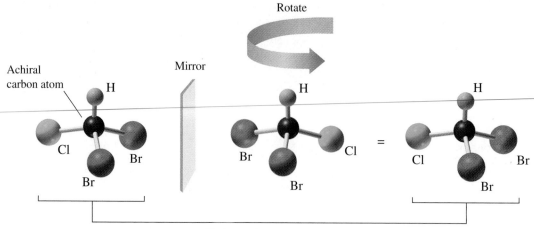

These are the same structures

**FIGURE 12.11** The mirror images of an achiral compound can be superimposed on each other. **Q** Why do the mirror images have the same structure?

---

SAMPLE PROBLEM 12.12

■ **Chiral Carbons**

Indicate whether the carbon in red is chiral or achiral.

a. Cl—C—CH₃ (with Cl above and H below)

b. CH₃—C—CH₂—CH₃ (with OH above and H below)

c. CH₃—C—CH₂—CH₃ (with O double bonded above)

SOLUTION

**a.** Achiral. Two of the substituents on the carbon in red are the same (Cl). A chiral carbon must be bonded to four different groups or atoms.

**b.** Chiral. The carbon in red is bonded to four different groups: one OH, one CH₃, one CH₂—CH₃, and one H.

**c.** Achiral. The carbon in red is bonded to only three groups, not four.

STUDY CHECK

Circle the two chiral carbons in the structural formula of the carbohydrate erythrose.

HO—CH₂—CH—CH—C—H (with OH, OH, O labels)

Erythrose

## Drawing Fischer Projections

Emil Fischer, who received the Nobel Prize in Chemistry in 1902, devised a simplified system for drawing stereoisomers, showing the arrangements of the atoms. In a **Fischer projection**, the carbon chain is written vertically with the most highly oxidized carbon at the top. Horizontal lines represent bonds that project forward in the three-dimensional structure. Vertical lines represent bonds that project back. Each intersection of lines represents a carbon atom that is usually chiral.

For glyceraldehyde, the simplest sugar, the carbonyl group (which is the most highly oxidized group in the molecule) is written at the top. The letter L is assigned to the stereoisomer that has the —OH group on the left of the chiral carbon. The letter D is assigned to the structure where the —OH is on the right of the chiral carbon. Let's look at how glyceraldehyde is converted from a three-dimensional view to a Fischer projection. (See Figure 12.12.)

Fischer projections can also be written for compounds that have two or more chiral carbons. For example, in the mirror images shown, the chiral carbon atom at each

Dash–wedge structures of glyceraldehyde

Mirror
Fischer projections of glyceraldehyde

Extend forward
(wedge)

Project back
(dash line)

L-Glyceraldehyde

D-Glyceraldehyde

**FIGURE 12.12** In a Fischer projection, the chiral carbon atom is at the center with horizontal lines for bonds that extend toward the viewer and vertical lines for bonds that point away.

**Q** Why does glyceraldehyde have only one chiral carbon atom?

intersection is bonded to four different groups. To draw the mirror image, the positions of the substituents on the horizontal lines are reversed.

We can also draw the mirror image of the carbohydrate erythrose by changing the sides of the —OH groups on the two chiral carbons.

L-Erythrose

D-Erythrose

---

## SAMPLE PROBLEM 12.13

### ■ Fischer Projections

Determine if each Fischer projection is chiral or achiral. If chiral, identify it as the D or L isomer and draw the mirror image.

### SOLUTION

**a.** Chiral. The carbon at the intersection is attached to four different substituents. It is the L isomer because the —OH is on the left. The mirror image is written by reversing the H and OH.

**b.** Achiral. The carbon atom at the intersection is attached to two identical groups ($CH_3$).

**c.** Chiral. The carbon atom at the intersection is attached to four different substituents. It is the D isomer because the —OH is on the right. The mirror image is written by reversing the H and OH.

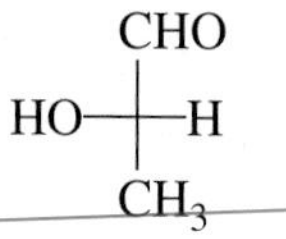

$$\text{HO}-\overset{\displaystyle \text{CHO}}{\underset{\displaystyle \text{CH}_3}{|}}-\text{H}$$

### STUDY CHECK

Draw the Fischer projections for the D and L stereoisomers of 2-hydroxypropanal.

## QUESTIONS AND PROBLEMS

### Chiral Molecules

**12.35** Identify each of the following structures as chiral or achiral. If chiral, indicate the chiral carbon.

a. 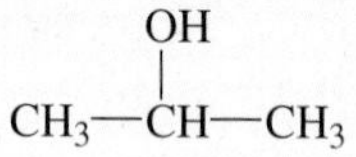 CH₃—CH—CH₃ with OH above
b. 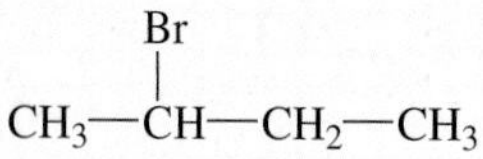 CH₃—CH—CH₂—CH₃ with Br above
c. CH₃—CH—C—H with Br above and O (double bond) above the C 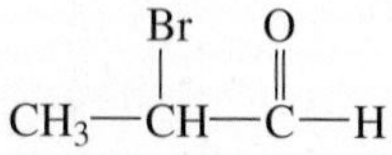
d. 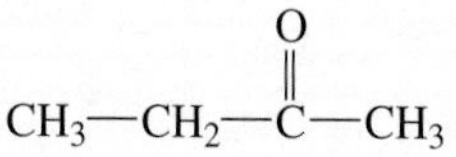 CH₃—CH₂—C—CH₃ with O (double bond)

**12.36** Identify each of the following structures as chiral or achiral. If chiral, indicate the chiral carbon.

a. 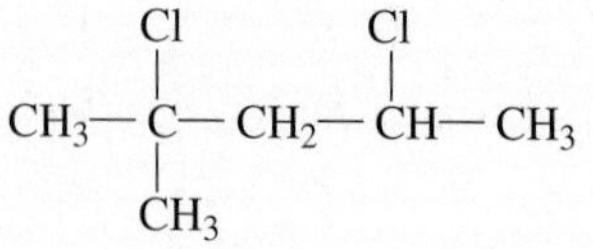 CH₃—C—CH₂—CH—CH₃ with Cl above first C, Cl above CH, CH₃ below first C
b. CH₃—C=CH—CH₃ with Br above
c. CH₃—C—CH—CH₃ with OH OH above, CH₃ below
d. 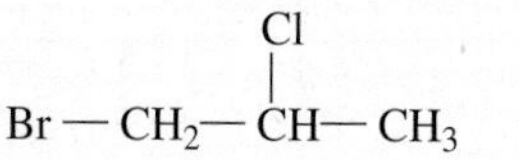 Br—CH₂—CH—CH₃ with Cl above

**12.37** Identify the chiral carbon in each of the following naturally occurring compounds:

**a.** citronellol; one enantiomer has the geranium odor

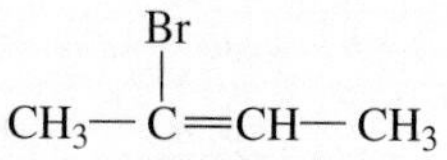

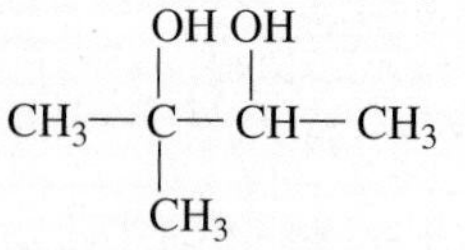

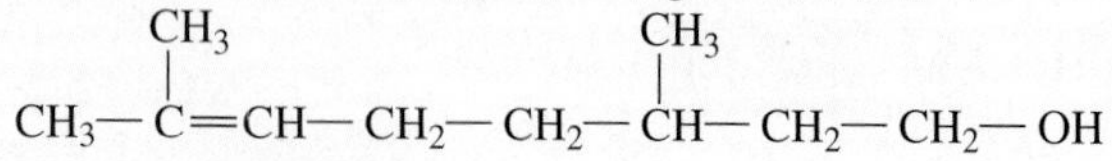

CH₃—C=CH—CH₂—CH₂—CH—CH₂—CH₂—OH with CH₃ above the C and CH₃ above the CH

**b.** alanine, amino acid

H₂N—CH—C—OH with CH₃ above the CH and O (double bond) above the C

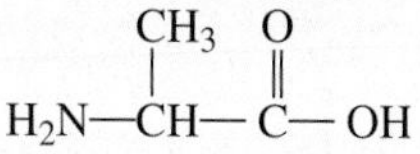

**12.38** Identify the chiral carbon in each of the following naturally occurring compounds:

**a.** amphetamine (benzedrine), stimulant, used in the treatment of hyperactivity

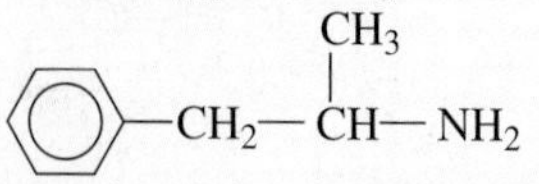

(benzene ring)—CH₂—CH—NH₂ with CH₃ above the CH

**b.** norepinephrine, increases blood pressure and nerve transmission

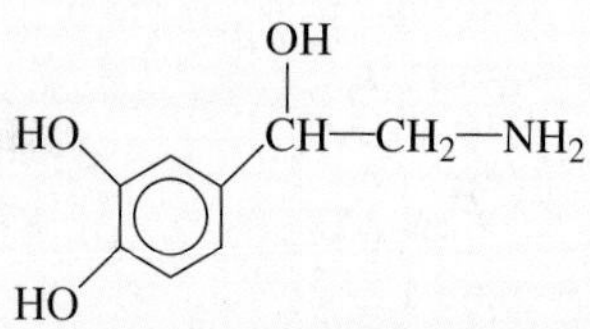

HO, HO (on benzene ring)—CH—CH₂—NH₂ with OH above the CH

**12.39** Draw Fischer projections for each of the following dash-wedge structures:

a. 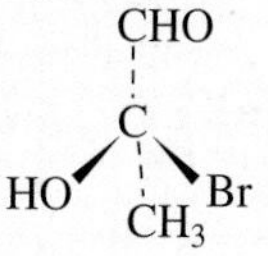 CHO, C with HO and Br and CH₃
b. CHO, C with Cl and Br and OH
c. CHO, C with HO and H and CH₂CH₃

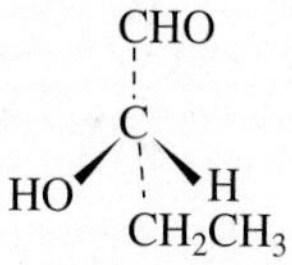

**12.40** Draw Fischer projections for each of the following dash-wedge structures:

a. 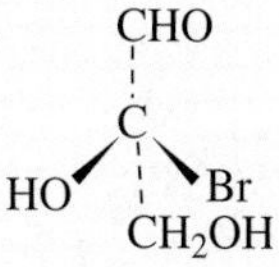 CHO, C with HO and Br and CH₂OH
b. CHO, C with H and OH and CH₂OH
c. CHO, C with HO and H and CH₂OH

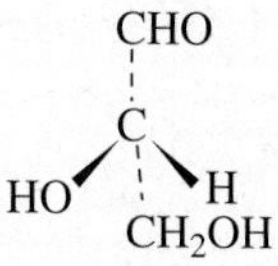

**12.41** Indicate whether each pair of Fischer projections represents enantiomers or identical structures.

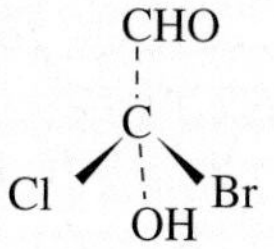

a. 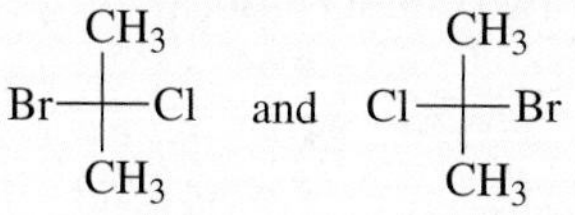 Br—|—Cl and Cl—|—Br with CH₃ top and CH₃ bottom
b. 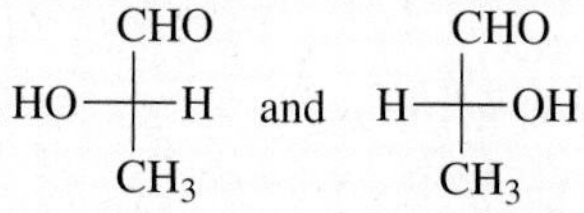 HO—|—H and H—|—OH with CHO top and CH₃ bottom
c. 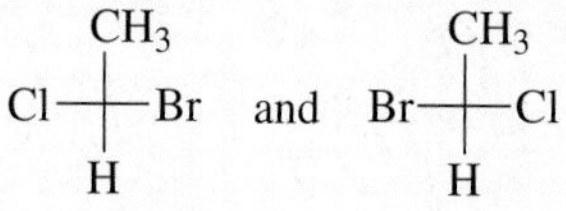 Cl—|—Br and Br—|—Cl with CH₃ top and H bottom
d. 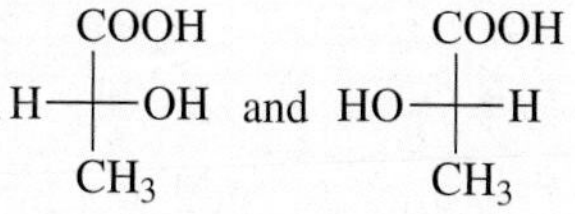 H—|—OH and HO—|—H with COOH top and CH₃ bottom

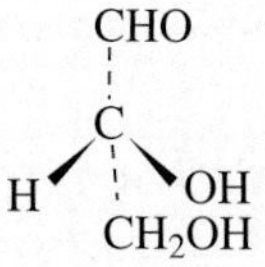

**12.42** Indicate whether each pair of Fischer projections represents enantiomers or identical structures:

a. 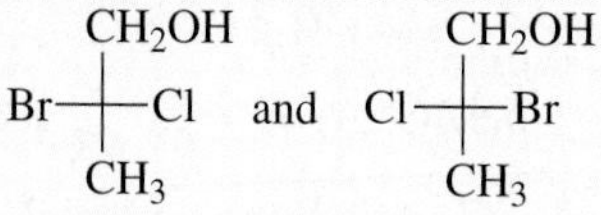 Br—|—Cl and Cl—|—Br with CH₂OH top and CH₃ bottom
b. 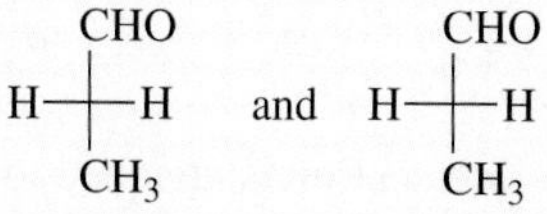 H—|—H and H—|—H with CHO top and CH₃ bottom
c. 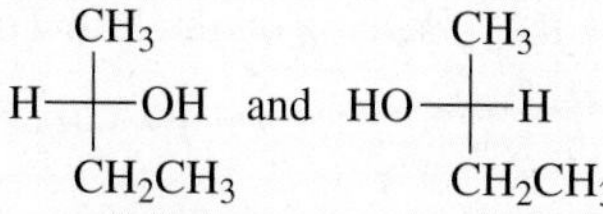 H—|—OH and HO—|—H with CH₃ top and CH₂CH₃ bottom
d. 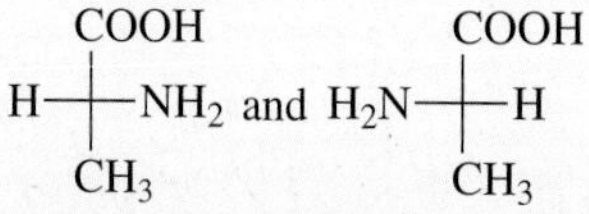 H—|—NH₂ and H₂N—|—H with COOH top and CH₃ bottom

# Health Note

## Enantiomers in Biological Systems

Most compounds that are active in biological systems consist of only one enantiomer. Rarely are both enantiomers of biological molecules active. This happens because the enzymes and cell surface receptors on which metabolic reactions take place are themselves chiral. Thus, only one enantiomer interacts with its enzymes or receptors; the other is inactive. The chiral receptor fits the arrangement of the substituents in only one enantiomer; its mirror image does not fit properly. (See Figure 12.13.)

A substance called carvone has two enantiomers. One enantiomer gives the odor of spearmint oil, while the other enantiomer produces the odor of caraway seeds. The chiral receptor sites in the olfactory cells in the nasal cavity and the gustatory cells in the taste buds on the tongue fit the shape of only one enantiomer. Thus, our senses of smell and taste are responsive to the chirality of molecules.

## Enantiomers of Carvone

From spearmint oil    From caraway seeds

In the brain, one enantiomer of LSD affects the production of serotonin, which influences sensory perception and may lead to hallucinations. However, its enantiomer produces little effect in the brain. The behavior of nicotine and epinephrine (adrenaline) also depends upon only one of their enantiomers. For example, one enantiomer of nicotine is more toxic than the other. Only one enantiomer of epinephrine is responsible for the constriction of blood vessels.

Nicotine — **Chiral carbon**    Adrenalin (epinephrine)

A substance used to treat Parkinson's disease is L-dopa, which is converted to dopamine in the brain, where it raises the serotonin level. However, the D-dopa enantiomer is not effective for the treatment of Parkinson's disease.

L-Dopa    D-Dopa

For many drugs, only one of the enantiomers is biologically active. However, for many years, drugs have been produced that were mixtures of their enantiomers. Today, drug researchers are using *chiral technology* to produce the active enantiomers of chiral drugs. Chiral catalysts are being designed that direct the formation of just one enantiomer rather than both. The benefits of producing only the active enantiomer include using a lower dose, enhancing activity, reducing interactions with other drugs, and eliminating possible harmful side effects from the nonactive enantiomer. Several active enantiomers are now being produced such as L-dopa and the active enantiomer of the popular analgesic ibuprofen used in Advil, Motrin, and Nuprin.

Ibuprofen

**FIGURE 12.13** (a) The substituents on the biologically active enantiomer bind to all the sites on a chiral receptor; (b) its enantiomer does not bind properly and is not active biologically.

**Q** Why don't all the substituents of the mirror image of the active enantiomer fit into a chiral receptor site?

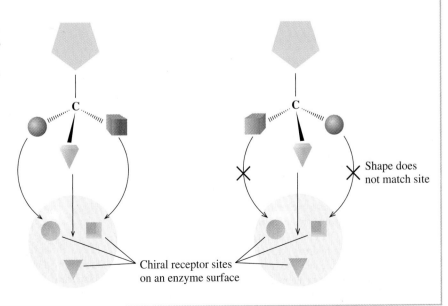

Shape does not match site

Chiral receptor sites on an enzyme surface

# CONCEPT MAP

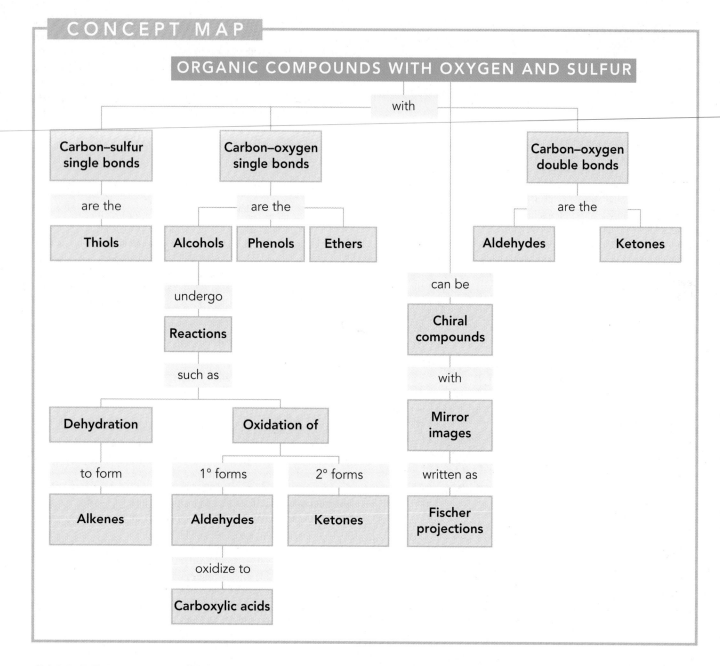

# CHAPTER REVIEW

## 12.1 Alcohols, Thiols, and Ethers

**Learning Goal:** Identify and name alcohols, thiols, and ethers; classify alcohols as primary, secondary, or tertiary.

The functional group of an alcohol is the hydroxyl group —OH bonded to a carbon chain. In a phenol, the hydroxyl group is bonded to an aromatic ring. In the IUPAC system, the names of alcohols have *ol* endings, and the location of the —OH group is given by numbering the carbon chain. Simple alcohols are generally named by their common names, with the alkyl name preceding the term *alcohol*. A cyclic alcohol is named as a cycloalkanol. Alcohols are classified according to the number of alkyl or aromatic groups bonded to the carbon that holds the —OH group. In a primary (1°) alcohol, one alkyl group is attached to the hydroxyl carbon. In a secondary (2°) alcohol, two alkyl groups are attached; and in a tertiary (3°), there are three alkyl groups bonded to the hydroxyl carbon. In thiols, the functional group is —SH, which is analogous to the —OH group of alcohols.

In an ether, an oxygen atom is connected by single bonds to two alkyl or aromatic groups. In the common names of ethers, the alkyl groups are listed alphabetically, followed by the name *ether*.

## 12.2 Properties of Alcohols and Ethers

**Learning Goal:** Describe some properties of alcohols and ethers.

The —OH group allows alcohols to hydrogen bond, which causes alcohols to have higher boiling points than alkanes and ethers of similar mass. Short-chain alcohols and ethers can hydrogen bond with water, which makes them soluble in water.

## 12.3 Reactions of Alcohols and Thiols

**Learning Goal:** Write equations for the combustion, dehydration, and oxidation of alcohols.

At high temperatures, alcohols dehydrate in the presence of an acid to yield alkenes. Primary alcohols are oxidized to aldehydes, which can oxidize further to carboxylic acids. Secondary alcohols are oxidized to ketones. Tertiary alcohols do not oxidize. Thiols oxidize to form disulfides.

## 12.4 Aldehydes and Ketones
**Learning Goal:** Identify compounds with the carbonyl group as aldehydes and ketones; write their IUPAC names.

Aldehydes and ketones contain a carbonyl group (C=O), which consists of a double bond between a carbon and an oxygen atom. In aldehydes, the carbonyl group appears at the end of carbon chains attached to at least one hydrogen atom. In ketones, the carbonyl group occurs between two alkyl or aromatic groups. In the IUPAC system, the *e* in the corresponding alkane is replaced with *al* for aldehydes and *one* for ketones. For ketones with more than four carbon atoms in the main chain, the carbonyl group is numbered to show its location. Many of the simple aldehydes and ketones use common names.

## 12.5 Properties of Aldehydes and Ketones
**Learning Goal:** Compare the boiling points and solubility of aldehydes and ketones to those of alkanes and alcohols.

Because they contain a polar carbonyl group, aldehydes and ketones have higher boiling points than alkanes and ethers. Their boiling points are lower than alcohols because aldehydes and ketones cannot hydrogen bond with each other. However, aldehydes and ketones can hydrogen bond with water molecules, which makes carbonyl compounds with one to four carbon atoms soluble in water. Aldehydes are easily oxidized to carboxylic acids, but ketones do not oxidize further. Aldehydes, but not ketones, react with Tollens' reagent to give silver mirrors. In Benedict's test, aldehydes with adjacent hydroxyl groups reduce blue $Cu^{2+}$ to give a brick-red $Cu_2O$ solid.

## 12.6 Chiral Molecules
**Learning Goal:** Identify chiral and achiral carbon atoms in an organic molecule.

Chiral molecules are molecules with mirror images that cannot be superimposed on each other. These types of stereoisomers are called enantiomers. A chiral molecule must have at least one chiral carbon, which is a carbon bonded to four different atoms or groups of atoms. The Fischer projection is a simplified way to draw the arrangements of atoms by placing the carbon atoms at the intersection of vertical and horizontal lines. The names of the mirror images are labeled D or L to differentiate between the enantiomers.

# SUMMARY OF NAMING

| Structure | Family | IUPAC Name | Common Name |
|---|---|---|---|
| $CH_3-OH$ | Alcohol | Methanol | Methyl alcohol |
| (benzene ring)—OH | Phenol | Phenol | Phenol |
| $CH_3-SH$ | Thiol | Methanethiol | |
| $CH_3-O-CH_3$ | Ether | | Dimethyl ether |
| H—C(=O)—H | Aldehyde | Methanal | Formaldehyde |
| $CH_3-C(=O)-CH_3$ | Ketone | Propanone | Acetone; dimethyl ketone |

# SUMMARY OF REACTIONS

**COMBUSTION OF ALCOHOLS**

$CH_3-CH_2-OH + 3O_2 \longrightarrow 2CO_2 + 3H_2O$
Ethanol    Oxygen   Carbon dioxide  Water

**DEHYDRATION OF ALCOHOLS TO FORM ALKENES**

$-\overset{H}{\underset{|}{C}}-\overset{OH}{\underset{|}{C}}- \xrightarrow[Heat]{H^+} -C=C- + H_2O$
Alcohol          Alkene

$CH_3-CH_2-CH_2-OH \xrightarrow[Heat]{H^+} CH_3-CH=CH_2 + H_2O$
1-Propanol          Propene

**OXIDATION OF PRIMARY ALCOHOLS TO FORM ALDEHYDES**

$CH_3-\overset{OH}{\underset{|}{CH_2}} \xrightarrow{[O]} CH_3-C(=O)-H + H_2O$
Ethanol       Acetaldehyde

**OXIDATION OF SECONDARY ALCOHOLS TO FORM KETONES**

$CH_3-\overset{OH}{\underset{|}{CH}}-CH_3 \xrightarrow{[O]} CH_3-C(=O)-CH_3 + H_2O$
2-Propanol          Propanone

**OXIDATION OF THIOLS TO FORM DISULFIDES**

$CH_3-S-H + H-S-CH_3 \xrightarrow{[O]} CH_3-S-S-CH_3 + H_2O$
Methanethiol          Dimethyl disulfide

**OXIDATION OF ALDEHYDES TO CARBOXYLIC ACIDS**

$CH_3-C(=O)-H \xrightarrow{[O]} CH_3-C(=O)-OH$
Acetaldehyde      Acetic acid

# KEY TERMS

**alcohols** Organic compounds that contain the hydroxyl (—OH) functional group, attached to a carbon chain.

**aldehyde** An organic compound with a carbonyl functional group bonded to at least one hydrogen.

$$\overset{\overset{\displaystyle O}{\|}}{-C}-H \;=\; -CHO$$

**Benedict's test** A test for aldehydes with adjacent hydroxyl groups in which $Cu^{2+}$ ($CuSO_4$) ions in Benedict's reagent are reduced to a brick-red solid of $Cu_2O$.

**carbonyl group** A functional group that contains a carbon–oxygen double bond (C=O).

**chiral** Having nonsuperimposable mirror images.

**chiral carbon** A carbon atom that is bonded to four different atoms or groups.

**dehydration** A reaction that removes water from an alcohol in the presence of an acid, to form alkenes.

**disulfides** Compounds formed from thiols that contain the —S—S— functional group.

**enantiomers** Stereoisomers that are mirror images that cannot be superimposed.

**ether** An organic compound, —O—, in which an oxygen atom is bonded to two alkyl or two aromatic groups, or one of each.

**Fischer projection** A system for drawing stereoisomers that shows horizontal lines for bonds projecting forward and vertical lines for bonds projecting back, with a carbon atom represented by each intersection.

**ketone** An organic compound in which the carbonyl functional group is bonded to two alkyl or aromatic groups.

$$\overset{\overset{\displaystyle O}{\|}}{-C}- \;=\; -CO-$$

**oxidation** The loss of two hydrogen atoms from a reactant to give a more oxidized compound. For example, primary alcohols oxidize to aldehydes, and secondary alcohols oxidize to ketones. An oxidation can also be the addition of an oxygen atom or an increase in the number of carbon-to-oxygen bonds.

**phenol** An organic compound that has a hydroxyl group attached to a benzene ring.

**primary (1°) alcohol** An alcohol that has one alkyl group bonded to the carbon atom with the —OH.

**secondary (2°) alcohol** An alcohol that has two alkyl groups bonded to the carbon atom with the —OH group.

**stereoisomers** Isomers that have atoms bonded in the same order, but with different arrangements in space.

**tertiary (3°) alcohol** An alcohol that has three alkyl groups bonded to the carbon atom with the —OH.

**thiols** Organic compounds that contain a —SH group.

**Tollens' test** A test for aldehydes in which $Ag^+$ in Tollens' reagent is reduced to metallic silver, which forms a "silver mirror" on the walls of the container.

# UNDERSTANDING THE CONCEPTS

**12.43** A compound called butylated hydroxytoluene, or BHT, is a common preservative added to cereals.

$(CH_3)_3C$ —〔ring: OH, C(CH_3)_3, CH_3〕

**a.** Identify the functional groups in BHT.
**b.** What parts of the molecule make up the toluene part of the compound?

**12.44** The compound 1,3-dihydroxy-2-propanone or dihydroxyacetone (DHA) is used in tanning lotions that darken the skin without sun. What is its structural formula?

**12.45** Identify the functional groups in each of the following:
   **a.** Urushiol, a substance in poison ivy and poison oak that causes itching and blistering of the skin.

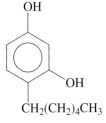

   **b.** Menthol gives a peppermint taste and odor used in candy and throat lozenges.

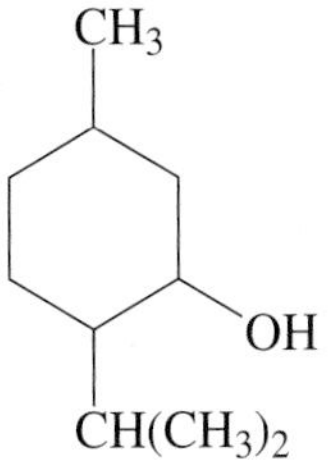

**12.46** Which of the following will give a positive Tollens' test?
   **a.** propenal
   **b.** ethanol
   **c.** ethyl methyl ether
   **d.** 1-propanol
   **e.** 2-propanol
   **f.** hexanal

**12.47** Citronellal, a constituent of oil of citronella as well as lemon and lemon grass, is used in perfumes and as an insect repellent.

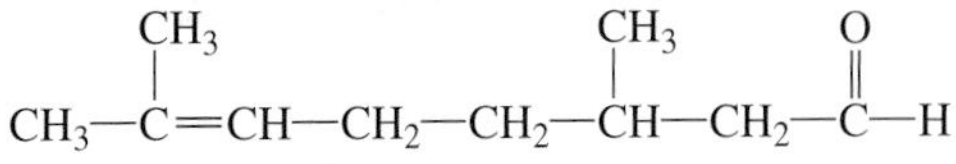

$$CH_3-\overset{\overset{\displaystyle CH_3}{|}}{C}=CH-CH_2-CH_2-\overset{\overset{\displaystyle CH_3}{|}}{CH}-CH_2-\overset{\overset{\displaystyle O}{\|}}{C}-H$$

   **a.** Complete the IUPAC name
      ____, ____ -di_____-__ -octenal
   **b.** What does the *en* in octenal signify?
   **c.** What does the *al* in octenal signify?
   **d.** Write the balanced equation for the combustion of citronellal when burned in a candle to repel insects.

**12.48** Draw the condensed structural formulas for each of the following:

   **a.** 2-heptanone, an alarm pheromone of bees
   **b.** 2,6-dimethyl-3-heptanone, a communication pheromone of bees
   **c.** 2-nonanol, an alarm pheromone released when a bee stings

# ADDITIONAL QUESTIONS AND PROBLEMS

**12.49** Classify each of the following as primary (1°), secondary (2°), or tertiary (3°) alcohols:

   **a.** 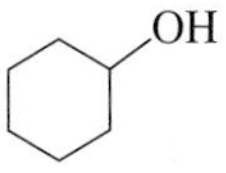 (cyclohexane with OH)

   **b.** 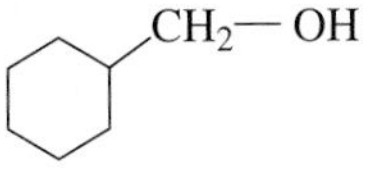 (cyclohexane with CH₂—OH)

   **c.** 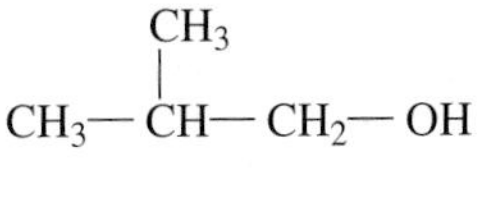 $CH_3-\overset{\overset{\displaystyle CH_3}{|}}{CH}-CH_2-OH$

   **d.** 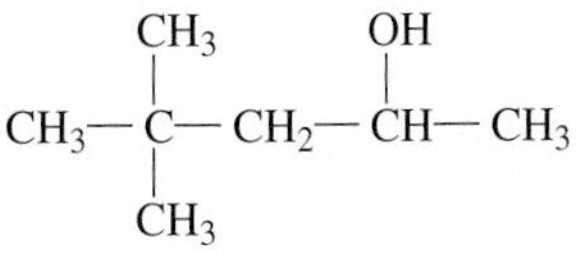 $CH_3-\overset{\overset{\displaystyle CH_3}{|}}{\underset{\underset{\displaystyle CH_3}{|}}{C}}-CH_2-\overset{\overset{\displaystyle OH}{|}}{CH}-CH_3$

   **e.** $HO-CH_2-CH_2-CH_3$

   **f.** 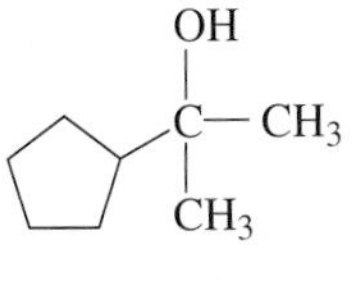 (cyclopentane with $\overset{\overset{\displaystyle OH}{|}}{\underset{\underset{\displaystyle CH_3}{|}}{C}}-CH_3$)

**12.50** Classify each of the following as primary (1°), secondary (2°), or tertiary (3°) alcohols:

a. 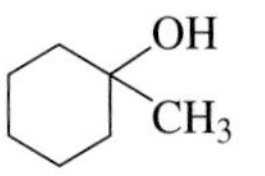   b.

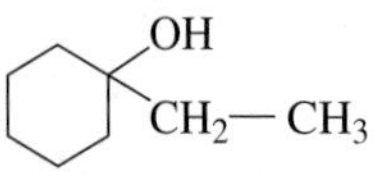

c. 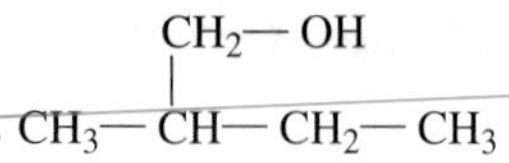 $CH_3-CH-CH_2-CH_3$ with $CH_2-OH$ branch

d. 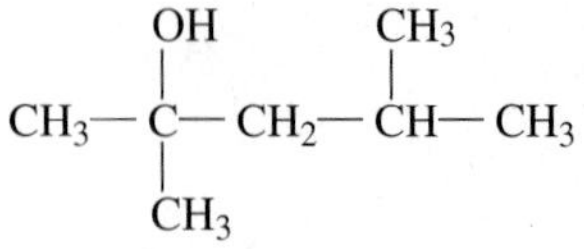 $CH_3-\underset{\underset{CH_3}{|}}{\overset{\overset{OH}{|}}{C}}-CH_2-\underset{\underset{}{\overset{\overset{CH_3}{|}}{CH}}}-CH_3$

e. $CH_3-CH_2-CH_2-CH_2-OH$

f.

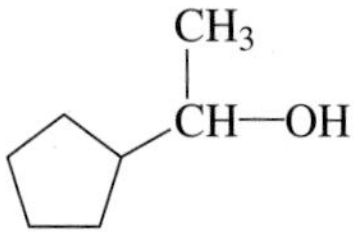

**12.51** Identify each of the following as an alcohol, a phenol, an ether, or a thiol:

a. 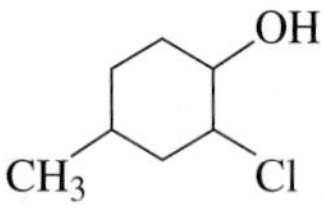   b.

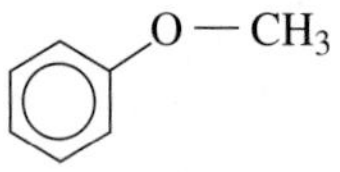

c. 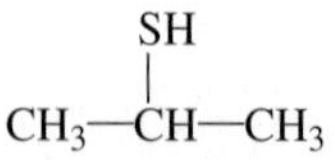 $CH_3-\overset{\overset{SH}{|}}{CH}-CH_3$

d. 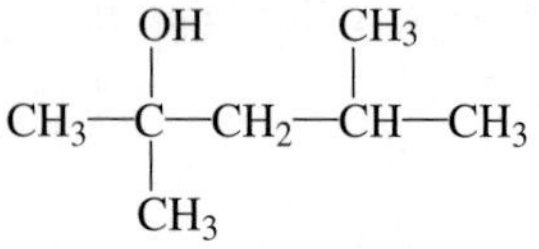 $CH_3-\overset{\overset{OH}{|}}{\underset{\underset{CH_3}{|}}{C}}-CH_2-\overset{\overset{CH_3}{|}}{CH}-CH_3$

e. $CH_3-CH_2-CH_2-O-CH_3$

f. 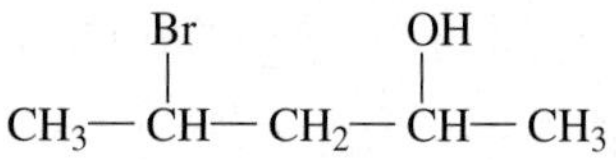 $CH_3-\overset{\overset{Br}{|}}{CH}-CH_2-\overset{\overset{OH}{|}}{CH}-CH_3$   g.

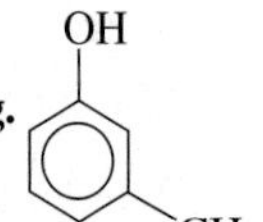

**12.52** Identify each of the following as an alcohol, a phenol, an ether, or a thiol:

a.    b. $CH_3-CH_2-CH_2-SH$

c.

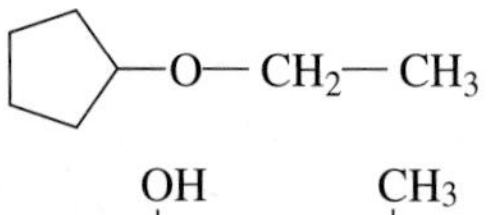

d. 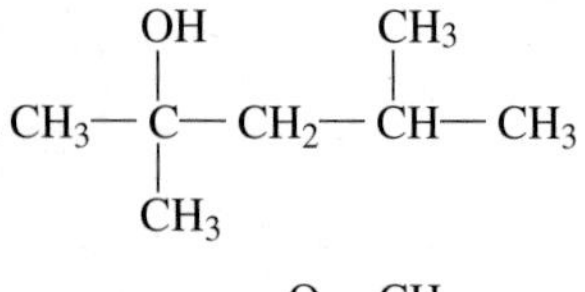 $CH_3-\overset{\overset{OH}{|}}{\underset{\underset{CH_3}{|}}{C}}-CH_2-\overset{\overset{CH_3}{|}}{CH}-CH_3$

e. 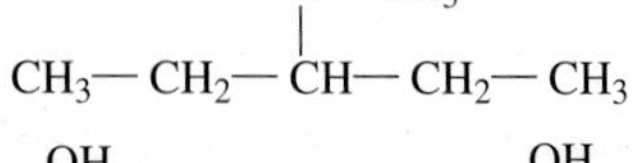 $CH_3-CH_2-\overset{\overset{O-CH_3}{|}}{CH}-CH_2-CH_3$

f. 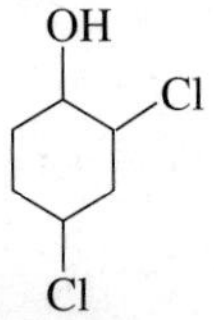   g.

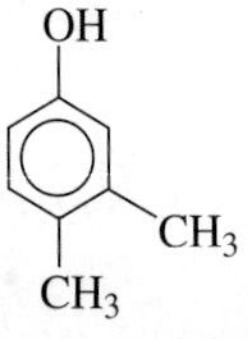

**12.53** Draw the condensed structural formula of each of the following compounds:
a. 3-methylcyclopentanol    b. 4-chlorophenol
c. 2-methyl-3-pentanol    d. ethyl phenyl ether
e. 3-pentanone    f. 2,4-dibromophenol

**12.54** Draw the condensed structural formula of each of the following compounds:
a. 3-pentanol
b. 2-pentanol
c. methyl propyl ether
d. 3-methyl-2-butanol
e. cyclohexanol

**12.55** Which compound in each pair would you expect to have the higher boiling point? Why?
a. butane or 1-propanol
b. 1-propanol or ethyl methyl ether
c. ethanol or 1-butanol

**12.56** Which compound in each pair would you expect to have the higher boiling point? Why?
a. propane or ethyl alcohol
b. 2-propanol or 2-pentanol
c. diethyl ether or 1-butanol

**12.57** Explain why each of the following compounds would be soluble or insoluble in water:
a. 2-propanol    b. dimethyl ether    c. 1-hexanol

**12.58** Explain why each of the following compounds would be soluble or insoluble in water:
a. glycerol    b. butane    c. 1,3-hexanediol

**12.59** Draw the condensed structural formula for the product of each of the following reactions:

a. $CH_3-CH_2-CH_2-OH \xrightarrow[heat]{H^+}$

b. $CH_3-CH_2-CH_2-OH \xrightarrow{[O]}$

c. 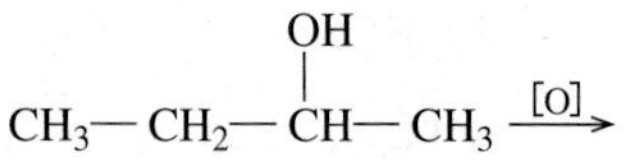 $CH_3-CH_2-\overset{\overset{OH}{|}}{CH}-CH_3 \xrightarrow{[O]}$

d. 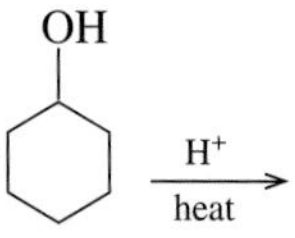 $\xrightarrow[heat]{H^+}$   e. 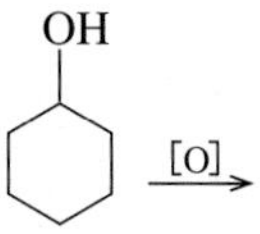 $\xrightarrow{[O]}$

**12.60** Draw the condensed structural formula for the product of each of the following reactions:

a. 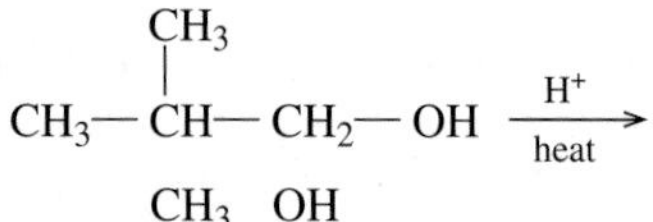 $CH_3-\overset{\overset{CH_3}{|}}{CH}-CH_2-OH \xrightarrow[heat]{H^+}$

b. 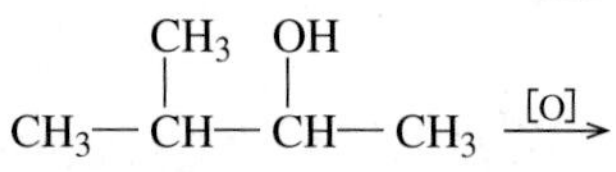 $CH_3-\overset{\overset{CH_3}{|}}{CH}-\overset{\overset{OH}{|}}{CH}-CH_3 \xrightarrow{[O]}$

c. 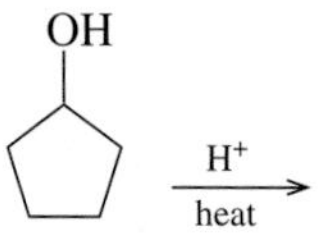 $\xrightarrow[heat]{H^+}$   d. 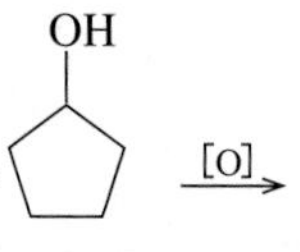 $\xrightarrow{[O]}$

e. 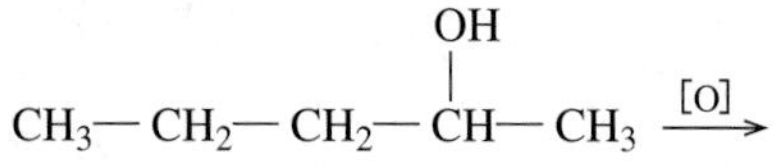 $CH_3-CH_2-CH_2-\overset{\overset{OH}{|}}{CH}-CH_3 \xrightarrow{[O]}$

**12.61** Identify the functional groups in the following molecule:

Testosterone

**12.62** Identify the functional groups in the following molecule:

Tetrahydrocannabinol (THC)

**12.63** Write the condensed structural formulas for each of the following naturally occurring compounds:
   **a.** 2,5-dichlorophenol, a defense pheromone of a grasshopper
   **b.** 3-methyl-1-butanethiol, which occurs in skunk scent
   **c.** pentachlorophenol, which is a wood preservative

**12.64** Dimethyl ether and ethyl alcohol both have the molecular formula $C_2H_6O$. One has a boiling point of $-23$ °C, and the other, 79 °C. Draw the condensed structural formulas of each compound. Decide which boiling point goes with which compound and explain. Check the boiling points in a chemistry handbook.

**12.65** Give the IUPAC and common names (if any) for each of the following compounds:

**a.** 

**b.** 

**c.** $Cl-CH_2-CH_2-\overset{\overset{\displaystyle O}{\|}}{C}-H$

**d.** $CH_3-\overset{\overset{\displaystyle Cl}{|}}{CH}-\overset{\overset{\displaystyle O}{\|}}{C}-CH_2-CH_3$   **e.** 

**12.66** Give the IUPAC and common names (if any) for each of the following compounds:

**a.** $CH_3-CH_2-\overset{\overset{\displaystyle O}{\|}}{C}-CH_3$

**b.** 

**c.** 

**d.** $CH_3-\overset{\overset{\displaystyle CH_3}{|}}{CH}-\overset{\overset{\displaystyle OH}{|}}{CH}-CH_2-\overset{\overset{\displaystyle O}{\|}}{C}-H$

**e.** 

**12.67** Draw the condensed structural formulas of each of the following:
   **a.** 3-methylcyclopentanone
   **b.** 4-chlorobenzaldehyde
   **c.** 3-chloropropionaldehyde
   **d.** ethyl methyl ketone
   **e.** 3-methylhexanal
   **f.** 2-heptanone, an alarm pheromone of bees

**12.68** Draw the condensed structural formulas of each of the following:
   **a.** propionaldehyde
   **b.** 2-chlorobutanal
   **c.** 2-methylcyclohexanone
   **d.** 3,5-dimethylhexanal
   **e.** 3-bromocyclopentanone
   **f.** *trans*-2-hexenal, an alarm pheromone of an ant

**12.69** Which of the following compounds are soluble in water?
   **a.** $CH_3-CH_2-CH_2-CH_3$

   **b.** $CH_3-CH_2-\overset{\overset{\displaystyle O}{\|}}{C}-H$   **c.** $CH_3-\overset{\overset{\displaystyle O}{\|}}{C}-CH_3$

   **d.** $CH_3-CH_2-CH_2-OH$

   **e.** $CH_3-CH_2-\overset{\overset{\displaystyle O}{\|}}{C}-CH_2-CH_2-CH_3$

**12.70** Which of the following compounds are soluble in water?

   **a.** $CH_3-CH_2-\overset{\overset{\displaystyle O}{\|}}{C}-CH_3$   **b.** $H-\overset{\overset{\displaystyle O}{\|}}{C}-H$

   **c.** $CH_3-\overset{\overset{\displaystyle O}{\|}}{C}-H$   **d.** $CH_3-CH_2-CH_3$

   **e.** $CH_3-CH_2-\overset{\overset{\displaystyle CH_3}{|}}{CH}-CH_2-CH_2-\overset{\overset{\displaystyle O}{\|}}{C}-H$

**12.71** In each of the following pairs of compounds, select the compound with the higher boiling point:

   **a.** $CH_3-CH_2-OH$  or  $CH_3-\overset{\overset{\displaystyle O}{\|}}{C}-H$

   **b.** $CH_3-CH_2-CH_2-CH_3$  or  $CH_3-CH_2-\overset{\overset{\displaystyle O}{\|}}{C}-H$

**12.72** In each of the following pairs of compounds, select the compound with the higher boiling point:

   **a.** $CH_3-CH_2-CH_2-OH$  or  $CH_3-\overset{\overset{\displaystyle O}{\|}}{C}-CH_3$

   **b.** $CH_3-CH_2-CH_2-CH_3$  or  $CH_3-\overset{\overset{\displaystyle O}{\|}}{C}-CH_3$

**12.73** Identify the chiral carbons, if any, in each of the following compounds:

a. H—C—C—O—H  (with Cl, Cl above; Cl, H below)

b. CH₃—C=C—CH₃  (with H, CH₃ above)

c. HO—CH₂—CH—CH₂—OH  (with OH above)

d. CH₃—CH—C—H  (with NH₂ and O)

e. CH₃—CH₂—CH—CH₂—CH₂—CH₃  (with Br above)

**12.74** Identify the chiral carbons, if any, in each of the following compounds:

a. CH₃—CH—CH₃  (with OCH₃ above)

b. CH₃—CH—C—CH₃  (with OH and O)

c. CH₃—C—CH₃  (with OH above and OH below)

d. CH₃—CH—C—CH₃  (with CH₃ and O)

e. CH₃—C—CH₂—CH₃  (with Br above and OH below)

**12.75** Identify each of the following pairs of Fischer projections as enantiomers or identical compounds:

a. CH₂OH / H—OH / CH₂OH   and   CH₂OH / HO—H / CH₂OH

b. CHO / H—OH / CH₂OH   and   CHO / HO—H / CH₂OH

c. CH₂OH / Cl—H / CH₃   and   CH₂OH / H—Cl / CH₃

d. OH / H—OH / CH₃   and   OH / HO—H / CH₃

**12.76** Identify each of the following pairs of Fischer projections as enantiomers or identical compounds:

a. CH₂OH / H—Cl / CH₂CH₃   and   CH₂OH / Cl—H / CH₂CH₃

b. CH₂OH / H—OH / CH₂OH   and   CH₂OH / HO—H / CH₂OH

c. CH₂OH / H—Cl / CH₃   and   CH₂OH / H—Cl / CH₃

d. CH₂OH / H—OH / CH₃   and   CH₂OH / HO—H / CH₃

**12.77** Draw the structural formula of the organic product when each of the following is oxidized:
a. CH₃—CH₂—CH₂—OH

b. CH₃—CH—CH₂—CH₂—CH₃  (with OH above)

c. CH₃—CH₂—CH₂—C—H  (with O)    d. (cyclohexane ring with OH)

**12.78** Draw the structural formula of the organic product when each of the following is oxidized:

a. CH₃—CH₂—CH—CH₂OH  (with OH above)

b. CH₃—CH₂—CH—CH₃  (with OH above)

c. CH₃—CH—CH₂—C—H  (with CH₃ and O)

d. (cyclohexane ring with CH—CH₃ and OH)

# CHALLENGE QUESTIONS

**12.79** Draw the structures and give the IUPAC names of all the alcohols that have the formula $C_5H_{12}O$.

**12.80** Draw the structures and give the IUPAC names of all the aldehydes and ketones that have the formula $C_5H_{10}O$.

**12.81** A compound with the formula $C_4H_8O$ is synthesized from 2-methyl-1-propanol and oxidizes easily to give a carboxylic acid. What are the structure and the name of the compound?

**12.82** Methyl *tert*-butyl ether (MTBE) or methyl 2-methyl-2-propyl ether is a fuel additive for gasoline to boost the octane rating. It

increases the oxygen content, which reduces CO emissions to an acceptable level determined by the Clean Air Act.

**a.** If fuel mixtures are required to contain 2.7% oxygen by mass, how many grams of MTBE must be added to each 100 g gasoline?

**b.** How many liters of liquid MTBE would be in a liter of fuel if the density of both gasoline and MTBE is 0.740 g/mL?

**c.** Write the equation for the complete combustion of MTBE.

**d.** How many liters of air containing 21% (v/v) $O_2$ are required at STP to completely react (combust) 1.00 L of liquid MTBE?

**12.83** Compound A is 1-propanol. When compound A is heated with strong acid, it dehydrates to form compound B ($C_3H_6$). When compound A is oxidized, compound C ($C_3H_6O$) forms. What are the structures and names of compounds A, B, and C?

**12.84** Compound X is 2-propanol. When compound X is heated with strong acid, it dehydrates to form compound Y ($C_3H_6$). When compound X is oxidized, compound Z ($C_3H_6O$) forms, which cannot be oxidized further. What are the structures and names of compounds X, Y, and Z?

# ANSWERS

## Answers to Study Checks

**12.1** 3-chloro-l-butanol

**12.2** secondary (2°)

**12.3**

![structure: benzene ring with O-CH3]

**12.4** Ethanol molecules can hydrogen bond with each other, but dimethyl ether molecules cannot. Thus, a higher temperature is required to break the hydrogen bonds between ethanol molecules.

**12.5** cyclopentene

**12.6** 2-Propanol loses two hydrogen atoms to form a ketone.

$$CH_3-\underset{\underset{OH}{|}}{CH}-CH_3 \xrightarrow{[O]} CH_3-\underset{\underset{O}{||}}{C}-CH_3$$
2-Propanol                Propanone

**12.7** $CH_3-CH_2-\underset{\underset{O}{||}}{C}-CH_2-CH_3$

**12.8** propanal (IUPAC), propionaldehyde (common)

**12.9** ethyl propyl ketone

**12.10** The oxygen atom in the carbonyl group of acetone hydrogen bonds with water molecules.

**12.11** A screwdriver is achiral and a wood screw is chiral.

**12.12** $HO-CH_2-\underset{\underset{OH}{|}}{CH}-\underset{\underset{OH}{|}}{CH}-\underset{\underset{O}{||}}{C}-H$

**12.13**

![Fischer projections: left D-2-Hydroxypropanal with CHO top, H—OH, CH3 bottom; right L-2-Hydroxypropanal with CHO top, HO—H, CH3 bottom]

D-2-Hydroxypropanal    L-2-Hydroxypropanal

## Answers to Selected Questions and Problems

**12.1** **a.** ethanol
  **b.** 2-butanol
  **c.** 2-pentanol
  **d.** 4-methylcyclohexanol
  **e.** 1-propanethiol

**12.3** **a.** $CH_3-CH_2-CH_2-OH$
  **b.** $CH_3-OH$
  **c.** $CH_3-CH_2-\underset{\underset{OH}{|}}{CH}-CH_2-CH_3$
  **d.** $CH_3-\underset{\underset{CH_3}{|}}{\overset{\overset{OH}{|}}{C}}-CH_2-CH_3$

**12.5 a.** phenol
**b.** 2-bromophenol
**c.** 3-bromophenol

**12.7 a.** 1°
**b.** 1°
**c.** 3°
**d.** 3°

**12.9 a.** ethyl methyl ether
**b.** dipropyl ether
**c.** cyclohexyl methyl ether
**d.** methyl propyl ether

**12.11 a.** $CH_3-CH_2-O-CH_2-CH_2-CH_3$

**b.** $CH_3-CH_2-O-\triangleleft$

**c.** $CH_3-CH_2-O-CH_3$

**12.13 a.** methanol
**b.** 1-butanol
**c.** 1-butanol

**12.15 a.** yes; hydrogen bonding
**b.** yes; hydrogen bonding
**c.** no; long carbon chain diminishes effect of —OH group
**d.** no; alkanes are nonpolar

**12.17 a.** $CH_3-CH_2-CH=CH_2$

**b.** (cyclopentene structure)

**c.** $CH_3-CH=CH-CH_2-CH_3$

**12.19 a.** $CH_3-CH_2-CH_2-CH_2-\overset{\displaystyle O}{\overset{\|}{C}}-H$

**b.** $CH_3-CH_2-\overset{\displaystyle O}{\overset{\|}{C}}-CH_3$

**c.** (cyclohexanone structure)

**d.** $CH_3-\overset{\displaystyle O}{\overset{\|}{C}}-CH_2-\overset{CH_3}{\overset{|}{C}H}-CH_3$

**12.21 a.** ketone    **b.** ketone    **c.** aldehyde

**12.23 a.** acetaldehyde    **b.** methyl propyl ketone
**c.** formaldehyde

**12.25 a.** propanal    **b.** 2-methyl-3-pentanone
**c.** 3-methylcyclohexanone    **d.** benzaldehyde

**12.27 a.** $CH_3-\overset{\displaystyle O}{\overset{\|}{C}}-H$

**b.** $CH_3-\overset{\displaystyle O}{\overset{\|}{C}}-CH_2-CH_2-CH_3$

**c.** $CH_3-\overset{\displaystyle O}{\overset{\|}{C}}-CH_2-CH_2-CH_2-CH_3$

**d.** $CH_3-CH_2-\overset{CH_3}{\overset{|}{C}H}-CH_2-\overset{\displaystyle O}{\overset{\|}{C}}-H$

**12.29 a.** $CH_3-CH_2-\overset{\displaystyle O}{\overset{\|}{C}}-H$ has a higher boiling point because it has a polar carbonyl group.
**b.** Pentanal has a higher molar mass and thus a higher boiling point.
**c.** 1-Butanol has a higher boiling point because it can hydrogen bond with other 1-butanol molecules.

**12.31 a.** $CH_3-\overset{\displaystyle O}{\overset{\|}{C}}-\overset{\displaystyle O}{\overset{\|}{C}}-CH_2-CH_3$; more hydrogen bonding
**b.** acetone; lower number of carbon atoms

**12.33 a.** $CH_3-CH_2-CH_2-CH_2-\overset{\displaystyle O}{\overset{\|}{C}}-H$

**b.** $CH_3-CH_2-\overset{\displaystyle O}{\overset{\|}{C}}-CH_3$

**c.** (cyclohexanone structure)

**d.** $CH_3-\overset{\displaystyle O}{\overset{\|}{C}}-CH_2-\overset{CH_3}{\overset{|}{C}H}-CH_3$

**e.** $CH_3-\overset{CH_3}{\overset{|}{C}H}-CH_2-\overset{\displaystyle O}{\overset{\|}{C}}-H$

**12.35 a.** achiral
**b.** chiral    $CH_3-\overset{\overset{Br}{|}}{C}H-CH_2CH_3$ (Br on chiral carbon)
**c.** chiral    $CH_3-\overset{\overset{Br}{|}}{C}H-\overset{\displaystyle O}{\overset{\|}{C}}-H$ (chiral carbon)
**d.** achiral

**12.37**
**a.** $CH_3-\overset{CH_3}{\overset{|}{C}}=CH-CH_2-CH_2-\overset{CH_3}{\overset{|}{C}H}-CH_2-CH_2-OH$ (chiral carbon)

**b.** $H_2N-\overset{CH_3}{\overset{|}{C}H}-\overset{\displaystyle O}{\overset{\|}{C}}-OH$ (chiral carbon)

**12.39 a.** (Fischer projection) CHO top, HO—C—Br, CH$_3$ bottom    **b.** (Fischer projection) CHO top, Cl—C—Br, OH bottom

**c.** (Fischer projection) CHO top, HO—C—H, CH$_2$CH$_3$ bottom

**12.41 a.** identical    **b.** enantiomers
**c.** enantiomers    **d.** enantiomers

**12.43** **a.** aromatic, alcohol
 **b.** Toluene is the benzene ring attached to the methyl ($CH_3$) group.

**12.45** **a.** aromatic, alcohol
 **b.** cycloalkane, alcohol

**12.47** **a.** 3,7-dimethyl-6-octenal
 **b.** The *en* in octenal signifies a double bond. The -6- indicates the double bond is between carbon 6 and carbon 7, counting from the aldehyde as carbon 1.
 **c.** The *al* in octenal signifies that an aldehyde group is carbon 1.
 **d.** $C_{10}H_{18}O + 14O_2 \longrightarrow 10CO_2 + 9H_2O$

**12.49** **a.** 2° **b.** 1° **c.** 1°
 **d.** 2° **e.** 1° **f.** 3°

**12.51** **a.** alcohol **b.** ether **c.** thiol
 **d.** alcohol **e.** ether **f.** alcohol
 **g.** phenol

**12.53** **a.** OH

      $CH_3$

  **b.** OH

      Cl

  **c.** $CH_3$ OH
 $CH_3-CH-CH-CH_2-CH_3$

  **d.** O $CH_2-CH_3$

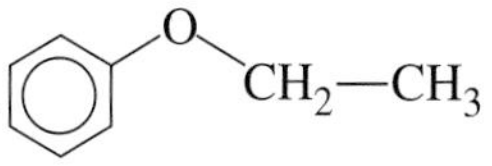

  **e.** O
 $CH_3-CH_2-C-CH_2-CH_3$

  **f.** OH Br

      Br

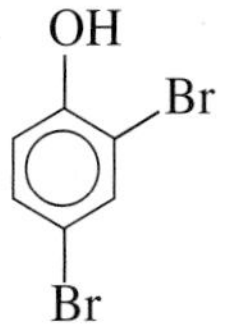

**12.55** **a.** 1-propanol; hydrogen bonding
 **b.** 1-propanol; hydrogen bonding
 **c.** 1-butanol; higher molar mass

**12.57** **a.** soluble; hydrogen bonding
 **b.** soluble; hydrogen bonding
 **c.** insoluble; long carbon chain diminishes the effect of polar —OH on hydrogen bonding

**12.59** **a.** $CH_3-CH=CH_2$

  **b.** O
 $CH_3-CH_2-C-H$

  **c.** O
 $CH_3-CH_2-C-CH_3$

  **d.**         **e.** O

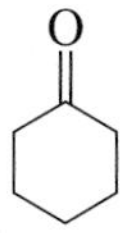

**12.61** Testosterone contains cycloalkene, alcohol, and ketone functional groups.

**12.63** **a.** OH Cl

    Cl

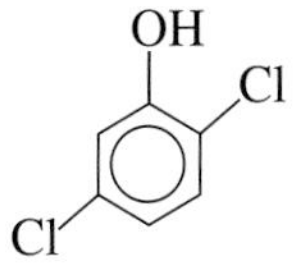

  **b.** $CH_3$
 $CH_3-CH-CH_2-CH_2-SH$

  **c.** OH
 Cl    Cl
 Cl    Cl
     Cl

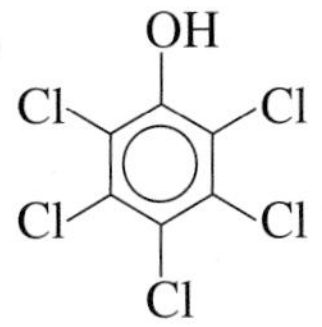

**12.65** **a.** 2-bromo-4-chlorocyclopentanone
 **b.** 3-bromo-4-chlorobenzaldehyde
 **c.** 3-chloropropanal
 **d.** 2-chloro-3-pentanone
 **e.** 3-methylcyclohexanone

**12.67** **a.** O

      $CH_3$

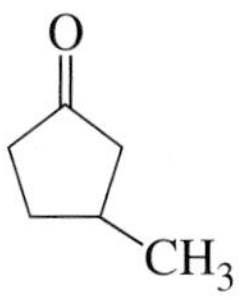

  **b.** CHO

      Cl

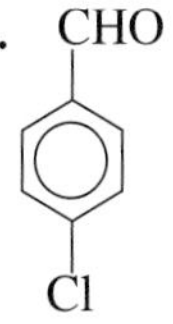

  **c.** O
 $Cl-CH_2-CH_2-C-H$

  **d.** O
 $CH_3-CH_2-C-CH_3$

  **e.** $CH_3$ O
 $CH_3-CH_2-CH_2-CH-CH_2-C-H$

  **f.** O
 $CH_3-CH_2-CH_2-CH_2-CH_2-C-CH_3$

**12.69** **b**, **c**, and **d**

**12.71** **a.** $CH_3-CH_2-OH$     **b.** O
 $CH_3-CH_2-C-H$

**12.73** **a.** Cl  Cl
 $H-C-C-O-H$
 Cl  H

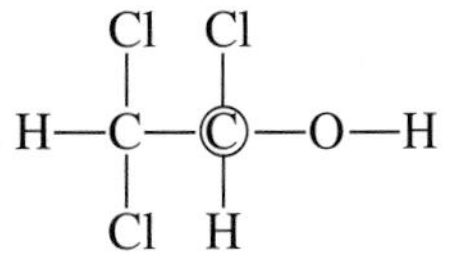

  **b.** none
 **c.** none

  **d.** $NH_2$ O
 $CH_3-CH-C-H$

  **e.** Br
 $CH_3-CH_2-CH-CH_2-CH_2-CH_3$

**12.75** **a.** identical
 **b.** enantiomers
 **c.** enantiomers
 **d.** identical

**12.77  a.**

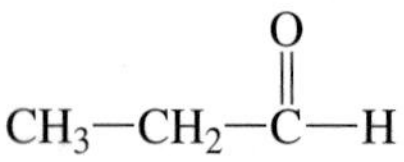

$$CH_3-CH_2-\overset{\overset{\displaystyle O}{\|}}{C}-H \quad \text{that oxidizes further to} \quad CH_3-CH_2-\overset{\overset{\displaystyle O}{\|}}{C}-OH$$

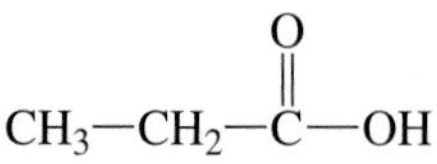

**b.** $CH_3-\overset{\overset{\displaystyle O}{\|}}{C}-CH_2-CH_2-CH_3$

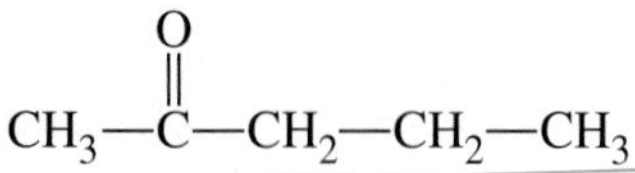

**c.** $CH_3-CH_2-CH_2-\overset{\overset{\displaystyle O}{\|}}{C}-OH$

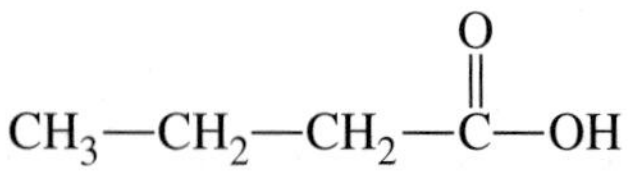

**d.**

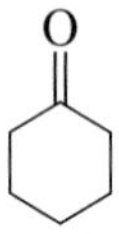

**12.79**

$CH_3-CH_2-CH_2-CH_2-CH_2-OH$    1-Pentanol

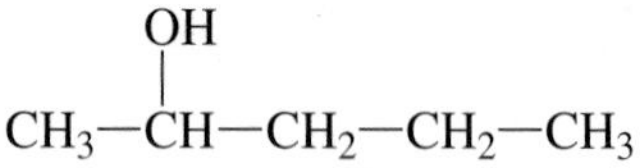

$CH_3-\overset{\overset{\displaystyle OH}{|}}{CH}-CH_2-CH_2-CH_3$    2-Pentanol

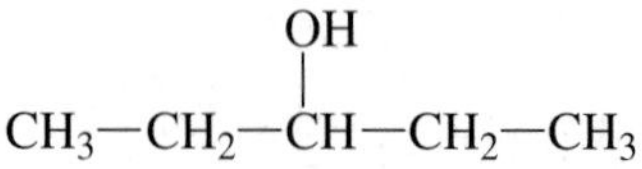

$CH_3-CH_2-\overset{\overset{\displaystyle OH}{|}}{CH}-CH_2-CH_3$    3-Pentanol

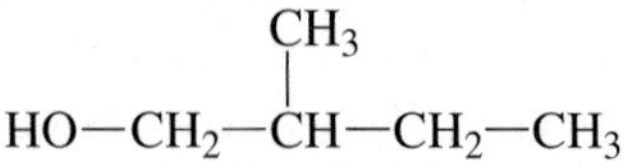

$HO-CH_2-\overset{\overset{\displaystyle CH_3}{|}}{CH}-CH_2-CH_3$    2-Methyl-1-butanol

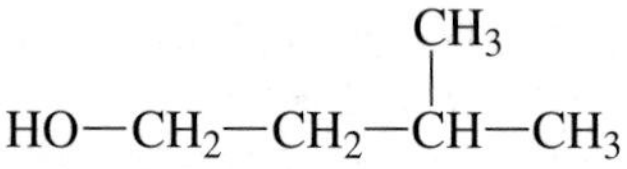

$HO-CH_2-CH_2-\overset{\overset{\displaystyle CH_3}{|}}{CH}-CH_3$    3-Methyl-1-butanol

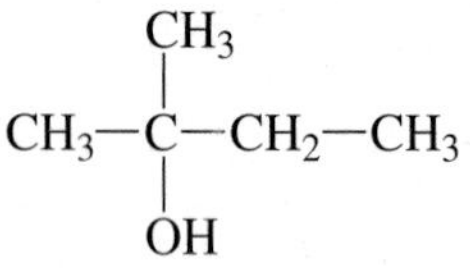

$CH_3-\overset{\overset{\displaystyle CH_3}{|}}{\underset{\underset{\displaystyle OH}{|}}{C}}-CH_2-CH_3$    2-Methyl-2-butanol

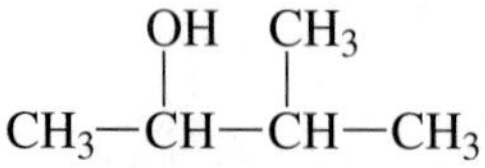

$CH_3-\overset{\overset{\displaystyle OH}{|}}{CH}-\overset{\overset{\displaystyle CH_3}{|}}{CH}-CH_3$    3-Methyl-2-butanol

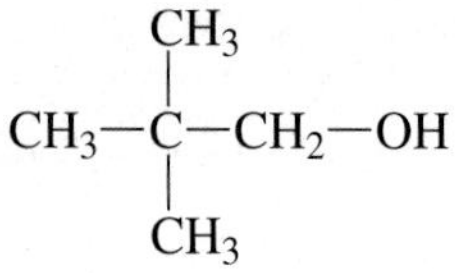

$CH_3-\overset{\overset{\displaystyle CH_3}{|}}{\underset{\underset{\displaystyle CH_3}{|}}{C}}-CH_2-OH$    2,2-Dimethyl-1-propanol

**12.81**    $CH_3-\overset{\overset{\displaystyle CH_3}{|}}{CH}-\overset{\overset{\displaystyle O}{\|}}{C}-H$    2-Methylpropanal

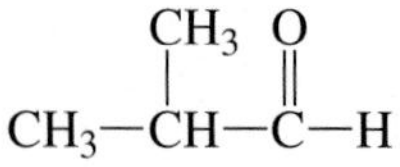

**12.83**    $CH_3-CH_2-CH_2-OH$    1-Propanol
           **A**

$CH_3-CH=CH_2$    Propene
     **B**

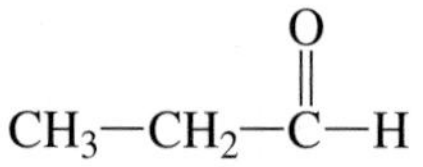

$CH_3-CH_2-\overset{\overset{\displaystyle O}{\|}}{C}-H$    Propanal
     **C**

# Combining Ideas from Chapters 9 to 12

**CI.19** Some isotopes of silicon are listed in the following table:

| Isotope | % Natural Abundance | Atomic Mass | Half-Life | Radiation Emitted |
|---|---|---|---|---|
| $^{27}Si$ | | 26.99 | 4.2 s | Positron |
| $^{28}Si$ | 92.23 | 27.99 | Stable | None |
| $^{29}Si$ | 4.67 | 28.99 | Stable | None |
| $^{30}Si$ | 3.10 | 29.98 | Stable | None |
| $^{31}Si$ | | 30.99 | 2.6 h | Beta |

**a.** Indicate the number of protons, neutrons, and electrons for each isotope listed.

| Isotope | Number of Protons | Number of Neutrons | Number of Electrons |
|---|---|---|---|
| $^{27}Si$ | | | |
| $^{28}Si$ | | | |
| $^{29}Si$ | | | |
| $^{30}Si$ | | | |
| $^{31}Si$ | | | |

**b.** What is the electron arrangement of silicon?
**c.** Calculate the atomic mass for silicon, using the isotopes that have a natural abundance.
**d.** Write the nuclear equations for $^{27}Si$ and $^{31}Si$.
**e.** Write the electron-dot formula, and predict the shape of $SiCl_4$.
**f.** How many hours are needed for a sample of $^{31}Si$ with an activity of 16 $\mu Ci$ to decay to 2.0 $\mu Ci$?

**CI.20** $K^+$ is an electrolyte required by the human body and is found in many foods, as well as salt substitutes. One of the isotopes of potassium is $^{40}K$, which has a natural abundance of 0.012% and a half-life of $1.30 \times 10^9$ years. The isotope $^{40}K$ decays to $^{40}Ca$ or to $^{40}Ar$. A typical activity for $^{40}K$ is 7.0 $\mu Ci$ per gram.

**a.** Write a nuclear equation for each type of decay.
**b.** Identify the particle emitted for each type of decay.
**c.** How many $K^+$ ions are in 3.5 oz of KCl?
**d.** What is the activity of 25 g of KCl, in bequerels?

**CI.21** Of much concern to environmentalists is the radioactive noble gas radon-222, which can seep from the ground into basements of buildings. Radon-222 is a product of the decay of radium-226 that occurs naturally in rocks and soil in much of the United States. Radon-222, which has a half-life of 3.8 days, decays by emitting an alpha particle. Because radon-222 is a gas, it can be inhaled into the lungs where it is associated with lung cancer. Environmental agencies have set the maximum level of radon-222 in a home at 4 picocuries per liter (pCi/L) of air in a home.

**a.** Write the equation for the decay of Ra-226.
**b.** Write the equation for the decay of Rn-222.
**c.** If a room contains 24 000 atoms of radon-222, how many atoms of radon-222 remain after 15.2 days?
**d.** Suppose a room in a home has a volume of 72 000 liters ($7.2 \times 10^4$ L). If the radon level is 2.5 picocuries/liter, how many alpha particles are emitted in one day? (1 Ci = $3.7 \times 10^{10}$ disintegrations per second.)

**CI.22** Ionone is a compound that gives violets their aroma. The small edible purple flowers of violets are used on salads and to make teas. An antioxidant called anthocyanin produces the blue and purple colors of violets. Liquid ionone has a density of 0.935 g/mL.

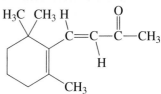

Ionone

**a.** What functional groups are present in ionone?
**b.** Is the double bond on the side chain cis or trans?

c. What is the molecular formula and molar mass of ionone?

d. How many moles are in 2.00 mL of ionone?

e. When ionone reacts with hydrogen in the presence of a platinum catalyst, hydrogen adds to the double bonds and converts the ketone group to an alcohol. What is the condensed structural formula and molecular formula of the product?

f. How many mL of hydrogen gas are needed at STP to completely react 5.0 mL of ionone?

**CI.23** Acetone (propanone), a clear liquid solvent with an acrid odor, is used to remove nail polish, paints, and resins. It has a low boiling point and is highly flammable.

a. Write the condensed structural formula of propanone.

b. What is the molecular formula and the molar mass of propanone?

c. Write the condensed structural formula of the alcohol that can be oxidized to produce propanone.

**CI.24** Acetone (propanone) has a density of 0.786 g/mL and a heat of combustion of 428 kcal/mole. Using your answers to problem CI.23, solve the following:

a. Write the equation for the complete combustion reaction of propanone.

b. How much heat in kilojoules is released if 2.58 g of propanone reacts with oxygen?

c. How many grams of oxygen gas are needed to react with 15.0 mL of propanone?

d. How many liters of carbon dioxide gas are produced at STP in part **c**?

## ANSWERS

**CI.19 a.**

| Isotope | Number of Protons | Number of Neutrons | Number of Electrons |
|---------|-------------------|--------------------|--------------------|
| $^{27}Si$ | 14 | 13 | 14 |
| $^{28}Si$ | 14 | 14 | 14 |
| $^{29}Si$ | 14 | 15 | 14 |
| $^{30}Si$ | 14 | 16 | 14 |
| $^{31}Si$ | 14 | 17 | 14 |

b. 2, 8, 4

c. Atomic mass calculated from the three stable isotopes is 28.09 amu.

d. $^{27}_{14}Si \longrightarrow {}^{27}_{13}Al + {}^{0}_{+1}e$

$^{31}_{14}Si \longrightarrow {}^{31}_{15}P + {}^{0}_{-1}e$

e. :C̈l—Si—C̈l:    Tetrahedral
with :C̈l: above and :C̈l: below

f. $16 \, \mu Ci \longrightarrow 8.0 \, \mu Ci \longrightarrow 4.0 \, \mu Ci \longrightarrow 2.0 \, \mu Ci$ is three half-lives.

$$3 \, \text{half-lives} \times \frac{2.6 \, h}{\text{half-life}} = 7.8 \, h$$

**CI.21 a.** $^{226}_{88}Ra \longrightarrow {}^{222}_{86}Rn + {}^{4}_{2}He$

b. $^{222}_{86}Rn \longrightarrow {}^{218}_{84}Po + {}^{4}_{2}He$

c. 1500 atoms of radon-222 remain

d. $5.8 \times 10^{8}$ alpha particles/day

**CI.23 a.** CH$_3$—C(=O)—CH$_3$

b. $C_3H_6O$, 58.0 g/mole

c. CH$_3$—CH(OH)—CH$_3$

# 13

# Carboxylic Acids, Esters, Amines, and Amides

Visit **www.chemplace.com** for extra quizzes, interactive tutorials, career resources, PowerPoint slides for chapter review, math help, and case studies.

*"There are many carboxylic acids, including the alpha hydroxy acids, that are found today in skin products," says Dr. Ken Peterson, pharmacist and cosmetic chemist in Oakland, California. "When you take a carboxylic acid called a fatty acid and react it with a strong base, you get a salt called soap. Soap has a high pH because the weak fatty acid and the strong base won't have a neutral pH of 7. If you take soap and drop its pH down to 7, you will convert the soap to the fatty acid. When I create fragrances, I use my nose and my chemistry background to identify and break down the reactions that produce good scents. Many fragrances are esters, which form when an alcohol reacts with a carboxylic acid. For example, the ester that smells like pineapple is made from ethanol and butyric acid."*

Carboxylic acids are similar to the weak acids we studied in Chapter 8. They have a sour or tart taste, produce hydronium ions in water, and neutralize bases. You encounter carboxylic acids when you use a vinegar salad dressing, which is a solution of acetic acid and water, or experience the sour taste of citric acid in a grapefruit or lemon. When a carboxylic acid reacts with an alcohol, an ester is produced. Fats in oils and meats are esters of glycerol and fatty acids, which are long-chain carboxylic acids. Many fruits and flavorings including bananas, oranges, and strawberries contain esters, which produce their pleasant aromas and flavors.

Amines and amides are organic compounds that contain nitrogen. Many nitrogen-containing compounds are important to life as components of amino acids, proteins, and the nucleic acids, DNA and RNA. Many amines that exhibit strong physiological activity are used in medicine as decongestants, anesthetics, and sedatives. Examples include dopamine, histamine, epinephrine, and amphetamine.

Alkaloids such as caffeine, nicotine, cocaine, and digitalis, which demonstrate powerful physiological activity, are naturally occurring amines obtained from plants. Amides, which are derived from carboxylic acids, are important in biology in proteins. Some amides used medically include acetaminophen (Tylenol) used to reduce fever; phenobarbital, a sedative and anticonvulsant medication; and penicillin, an antibiotic.

## LEARNING GOAL

Give the common names, IUPAC names, and condensed structural formulas of carboxylic acids.

# 13.1 CARBOXYLIC ACIDS

In Chapters 10 and 12, we described the carbonyl group ($C=O$) as the functional group in aldehydes and ketones. In a **carboxylic acid**, a hydroxyl group is attached to the carbonyl group, forming a **carboxyl group**. The carboxyl functional group may be attached to an alkyl group or an aromatic group.

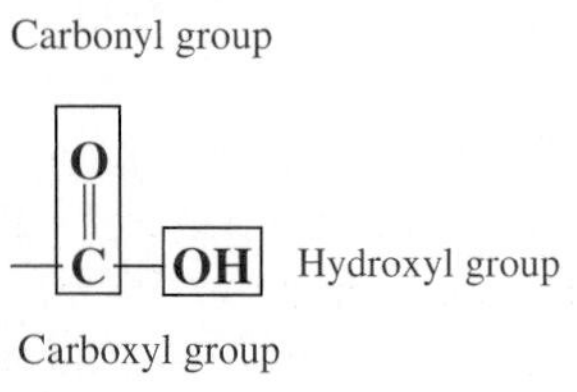

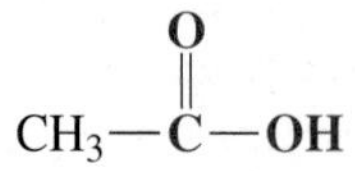

Carbonyl group

Carboxyl group    Hydroxyl group

$$CH_3-\overset{O}{\overset{\|}{C}}-OH$$
Acetic acid

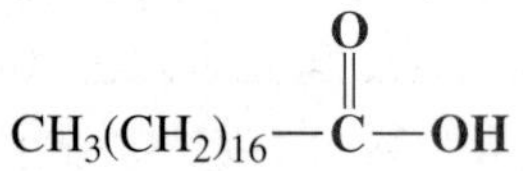

$$CH_3(CH_2)_{16}-\overset{O}{\overset{\|}{C}}-OH$$
Stearic acid (a fatty acid in fats and oils)

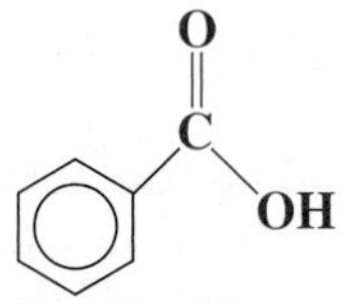

Benzoic acid (aromatic acid)

The carboxyl group can be written in several different ways. For example, the condensed structural formula for propanoic acid can be written as follows:

$$CH_3-CH_2-\overset{\overset{\displaystyle O}{\|}}{C}-OH \quad CH_3-CH_2-COOH \quad CH_3-CH_2-CO_2H$$

Some condensed structural formulas for propanoic acid

## Naming Carboxylic Acids

the
**Chemistry place**

**WEB TUTORIAL**
Carboxylic Acids

The IUPAC names of carboxylic acids use the alkane names of the corresponding carbon chains.

**Step 1**  **Identify the longest carbon chain containing the carboxyl group, and replace the _e_ of the alkane name by _oic acid_.**

**Step 2**  **Number the carbon chain, beginning with the carboxyl carbon as carbon 1.**

**Step 3**  **Give the location and names of substituents on the main chain.**

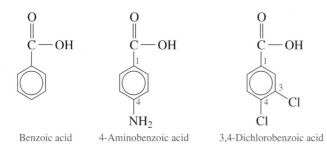

Methanoic acid    2-Methylpropanoic acid    3-Hydroxybutanoic acid

**Step 4**  **The aromatic carboxylic acid is called benzoic acid.**  With the carboxyl carbon bonded to carbon 1, the ring is numbered in the direction that gives substituents the smallest possible numbers.

Benzoic acid    4-Aminobenzoic acid    3,4-Dichlorobenzoic acid

Many carboxylic acids are still named by their common names, which are derived from their natural sources. In Chapter 12, we named aldehydes using the prefixes that represent the typical sources of carboxylic acids.

Formic acid is injected under the skin from bee or red ant stings and other insect bites. Acetic acid is the oxidation product of the ethanol in wines and apple cider. The resulting solution of acetic acid and water is known as vinegar. Butyric acid gives the foul odor to rancid butter. (See Table 13.1.)

**TABLE 13.1** Names and Natural Sources of Carboxylic Acids

| Condensed Structural Formulas | IUPAC Name | Common Name | |
|---|---|---|---|
| H—C—OH (with =O) | Methanoic acid | Formic acid | 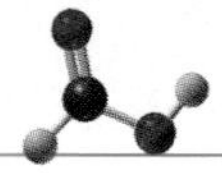 |
| $CH_3$—C—OH (with =O) | Ethanoic acid | Acetic acid | 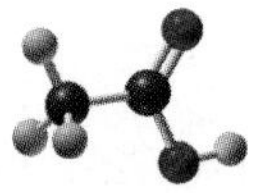 |
| $CH_3$—$CH_2$—C—OH (with =O) | Propanoic acid | Propionic acid | 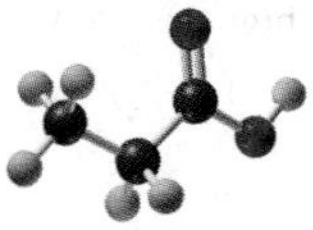 |
| $CH_3$—$CH_2$—$CH_2$—C—OH (with =O) | Butanoic acid | Butyric acid | 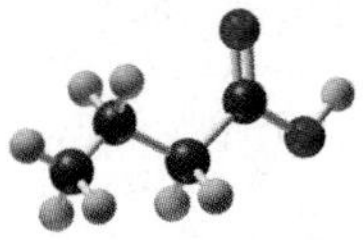 |

SAMPLE PROBLEM 13.1

### ■ Naming Carboxylic Acids

Give the IUPAC and common name, if any, for each of the following carboxylic acids:

**a.** $CH_3$—$CH_2$—C—OH (with =O)

**b.** $CH_3$—$CH_2$—CH—C—OH (with $CH_3$ on CH, and =O)

**c.** C—OH (with =O, attached to benzene ring)

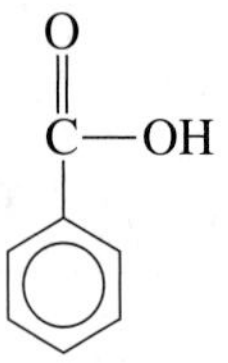

#### SOLUTION

**a.** This carboxylic acid has 3 carbon atoms. In the IUPAC system, the *e* in propane is replaced by *oic acid*, to give the name propanoic acid. Its common name is propionic acid.

**b.** This carboxylic acid has a methyl group on the second carbon. It has the IUPAC name 2-methylbutanoic acid.

**c.** An aromatic carboxylic acid is named benzoic acid.

#### STUDY CHECK

Write the condensed structural formula of pentanoic acid.

## Health Note

### Alpha Hydroxy Acids

Alpha hydroxy acids (AHAs), carboxylic acids with a hydroxy (—OH) on the alpha carbon next to the carboxyl group, are found in fruits, milk, and sugarcane. Cleopatra reportedly bathed in sour milk to smooth her skin. Dermatologists have been using products with a high concentration of AHAs to remove acne scars and reduce irregular pigmentation and age spots. Now, lower concentrations (8–10%) of AHAs have been added to skin-care products for the purpose of smoothing fine lines, improving skin texture, and cleansing pores. Several alpha hydroxy acids may be found in skin-care products singly or in combination. Glycolic acid and lactic acid are most frequently used.

Recent studies indicate that products with AHAs increase sensitivity of the skin to sun and UV radiation. It is recommended that a sunscreen with a sun protection factor (SPF) of at least 15 be used when treating the skin with products that include AHAs. Products containing AHAs at concentrations under 10% and pH values greater than 3.5 are generally considered safe. However, the Food and Drug Administration (FDA) has reports of AHAs

causing skin irritation including blisters, rashes, and discoloration of the skin. The FDA does not require product safety reports from cosmetic manufacturers, although they are responsible for marketing safe products. The FDA advises that you test any product containing AHAs on a small area of skin before you use it on a large area.

| Alpha Hydroxy Acid (Source) | Structure |
|---|---|
| Glycolic acid (sugarcane, sugar beet) | 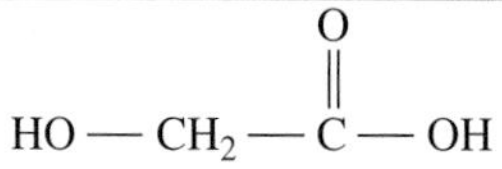 $HO-CH_2-\overset{\overset{\displaystyle O}{\|\|}}{C}-OH$ |
| Lactic acid (sour milk) | 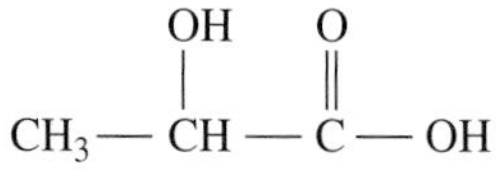 $CH_3-\overset{\overset{\displaystyle OH}{\|}}{CH}-\overset{\overset{\displaystyle O}{\|\|}}{C}-OH$ |
| Tartaric acid (grapes) | 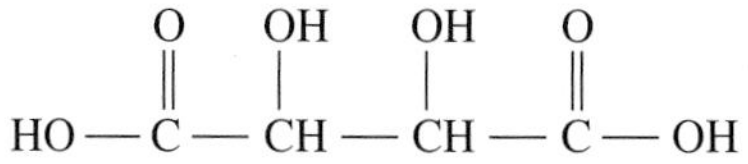 $HO-\overset{\overset{\displaystyle O}{\|\|}}{C}-\overset{\overset{\displaystyle OH}{\|}}{CH}-\overset{\overset{\displaystyle OH}{\|}}{CH}-\overset{\overset{\displaystyle O}{\|\|}}{C}-OH$ |
| Malic acid (apples, grapes) | 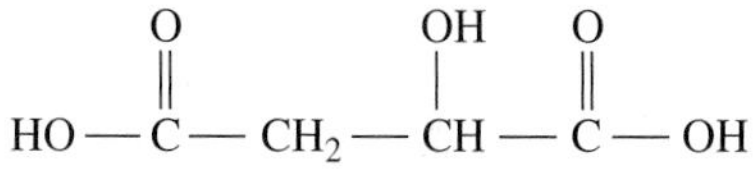 $HO-\overset{\overset{\displaystyle O}{\|\|}}{C}-CH_2-\overset{\overset{\displaystyle OH}{\|}}{CH}-\overset{\overset{\displaystyle O}{\|\|}}{C}-OH$ |
| Citric acid (citrus fruits: lemons, oranges, grapefruit) | 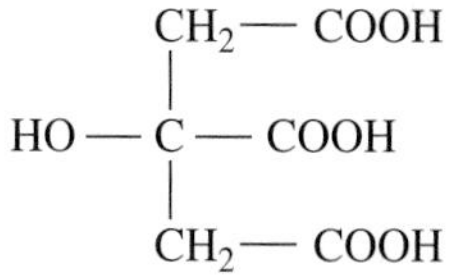 $HO-\overset{\overset{\displaystyle CH_2-COOH}{\|}}{\underset{\underset{\displaystyle CH_2-COOH}{\|}}{C}}-COOH$ |

## QUESTIONS AND PROBLEMS

### Carboxylic Acids

**13.1** What carboxylic acid is responsible for the pain of an ant sting?

**13.2** What carboxylic acid is found in vinegar?

**13.3** Explain the differences in the condensed structural formulas of propanal and propanoic acid.

**13.4** Explain the differences in the condensed structural formulas of benzaldehyde and benzoic acid.

**13.5** Give the IUPAC and common names (if any) for the following carboxylic acids:

**a.** 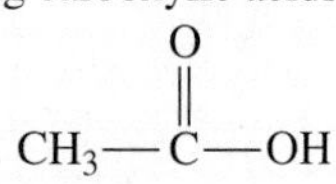 $CH_3-\overset{\overset{\displaystyle O}{\|\|}}{C}-OH$    **b.** $CH_3-CH_2-\overset{\overset{\displaystyle O}{\|\|}}{C}-OH$

**c.** (benzene ring with $\overset{\overset{\displaystyle O}{\|\|}}{C}-OH$ and OH substituents)

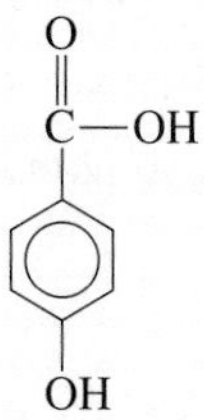

**d.** 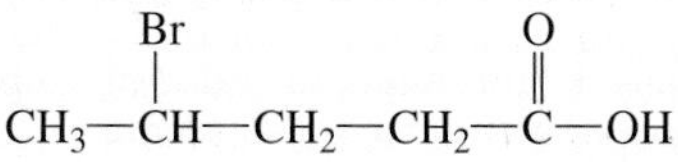 $CH_3-\overset{\overset{\displaystyle Br}{\|}}{CH}-CH_2-CH_2-\overset{\overset{\displaystyle O}{\|\|}}{C}-OH$

**13.6** Give the IUPAC and common names (if any) for the following carboxylic acids:

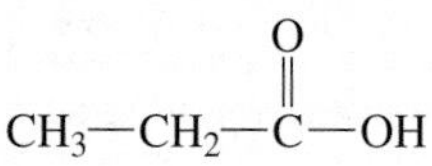

**a.** 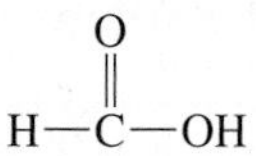 $H-\overset{\overset{\displaystyle O}{\|\|}}{C}-OH$    **b.** (benzene ring with $\overset{\overset{\displaystyle O}{\|\|}}{C}-OH$)    **c.** (benzene ring with $\overset{\overset{\displaystyle O}{\|\|}}{C}-OH$ and Cl)

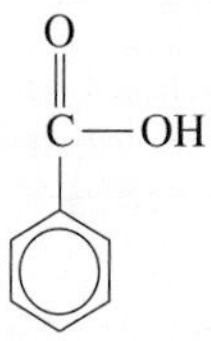

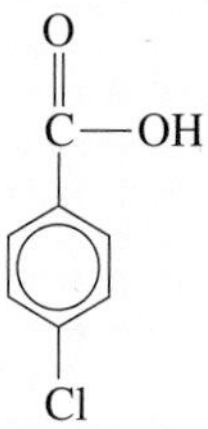

**d.** 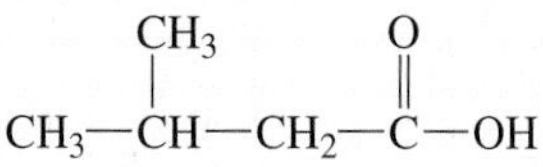 $CH_3-\overset{\overset{\displaystyle CH_3}{\|}}{CH}-CH_2-\overset{\overset{\displaystyle O}{\|\|}}{C}-OH$

**13.7** Draw the condensed structural formulas of each of the following carboxylic acids:
**a.** propionic acid          **b.** benzoic acid
**c.** chloroethanoic acid     **d.** 3-hydroxypropanoic acid
**e.** butyric acid            **f.** heptanoic acid

**13.8** Draw the condensed structural formulas of each of the following carboxylic acids:
**a.** pentanoic acid          **b.** 3-ethylbenzoic acid
**c.** 2-hydroxyacetic acid    **d.** 2,4-dibromobutanoic acid
**e.** 3-methylbenzoic acid    **f.** hexanoic acid

Describe the boiling points, solubility, and ionization of carboxylic acids in water.

**WEB TUTORIAL**
Carboxylic Acids

# $13.2$ PROPERTIES OF CARBOXYLIC ACIDS

Carboxylic acids are among the most polar organic compounds because the functional group consists of two polar groups: a hydroxyl (—OH) group and a carbonyl (C=O) group. The —OH group is similar to the functional group in alcohols, and the C=O double bond is similar to that of aldehydes and ketones.

## Boiling Points

The polar —OH group allows carboxylic acids to form several hydrogen bonds with other carboxylic acid molecules, as well as water. This effect of hydrogen bonds gives carboxylic acids higher boiling points than alcohols, ketones, and aldehydes of similar mass.

| Compound | Propanal | 1-Propanol | Acetic acid |
|---|---|---|---|
| Molar mass | 58 | 60 | 60 |
| bp | 49 °C | 97 °C | 118 °C |

An important reason for the higher boiling points is that two carboxylic acids form hydrogen bonds to give a dimer. The mass of the carboxylic acid is effectively doubled, which means that higher temperatures are required to reach the boiling point and form a gas.

A dimer of two acetic acid molecules

## Solubility in Water

Carboxylic acids with one to four carbons are very soluble in water because the carboxyl group forms hydrogen bonds with several water molecules. (See Figure 13.1.) However, as the length of the carbon chain increases, the nonpolar portion reduces solubility. Carboxylic acids having five or more carbons are not very soluble in water. Table 13.2 lists boiling points and solubilities for some selected carboxylic acids.

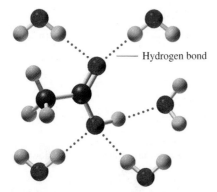

**FIGURE 13.1** Acetic acid forms hydrogen bonds with water molecules.
**Q** Why do the atoms in the carboxyl group hydrogen bond with water molecules?

**TABLE 13.2** Properties of Selected Carboxylic Acids

| IUPAC name | bp (°C) | Soluble in water? |
|---|---|---|
| Methanoic acid | 101 | Yes |
| Ethanoic acid | 118 | Yes |
| Propanoic acid | 141 | Yes |
| Butanoic acid | 164 | Yes |
| Pentanoic acid | 187 | Slightly |
| Hexanoic acid | 205 | Slightly |
| Benzoic acid | 250 | Slightly |

**SAMPLE PROBLEM 13.2**

### ■ Properties of Carboxylic Acids

Place the following organic compounds in order of increasing boiling points: butanoic acid, pentane, and 2-butanol.

**SOLUTION**

The boiling point increases when molecules with similar molar masses can form hydrogen bonds. The alkane has the lowest boiling point because alkanes cannot hydrogen bond. Alcohols and carboxylic acids have higher boiling points because they hydrogen bond via the —OH. The higher boiling points of carboxylic acids are due to the formation of stable dimers, which increase the effective mass and therefore the boiling point.

Pentane < 2-butanol < butanoic acid

STUDY CHECK

Why would methanoic acid (bp 101 °C) have a higher boiling point than ethanol (bp 78 °C) if they have about the same mass?

## Acidity of Carboxylic Acids

One of the most important properties of carboxylic acids is their ionization in water as weak acids. In the ionization, a carboxylic acid donates a proton to a water molecule to produce an anion called a **carboxylate ion** and a hydronium ion. Carboxylic acids are more acidic than other organic compounds including phenols, but only a small percentage (~1%) of the carboxylic acid molecules are ionized.

**Carboxylic acid**            **Carboxylate ion**

$$CH_3-\overset{\overset{\text{O}}{\|}}{C}-OH + H_2O \rightleftharpoons CH_3-\overset{\overset{\text{O}}{\|}}{C}-O^- + H_3O^+$$

Ethanoic acid        Ethanoate ion      Hydronium
(acetic acid)        (acetate ion)      ion

---

SAMPLE PROBLEM   13.3

### ■ Ionization of Carboxylic Acids in Water

Write the equation for the ionization of propionic acid in water.

SOLUTION

The ionization of propionic acid produces a carboxylate ion and a hydronium ion.

$$CH_3-CH_2-\overset{\overset{\text{O}}{\|}}{C}-OH + H_2O \rightleftharpoons CH_3-CH_2-\overset{\overset{\text{O}}{\|}}{C}-O^- + H_3O^+$$

STUDY CHECK

Write an equation for the ionization of formic acid in water.

## Neutralization of Carboxylic Acids

Carboxylic acids, like all weak acids, are completely neutralized by strong bases such as NaOH and KOH. The products are water and a **carboxylic acid salt**, which is a carboxylate ion and the metal ion from the base. The carboxylate ion is named by replacing the *oic acid* ending of the acid name with *ate*.

$$H-\overset{\overset{\text{O}}{\|}}{C}-OH + NaOH \longrightarrow H-\overset{\overset{\text{O}}{\|}}{C}-O^-Na^+ + H_2O$$

Formic acid                Sodium formate

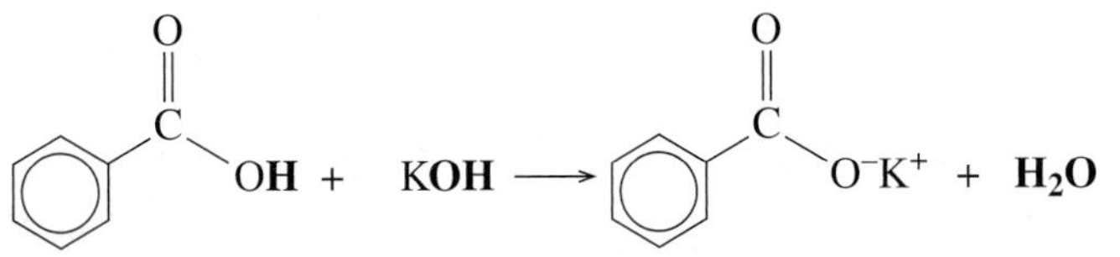

Benzoic acid                Potassium benzoate

Sodium propionate, a preservative, is added to cheeses, bread, and other bakery items to inhibit the spoilage of the food by microorganisms. Sodium benzoate, an inhibitor of mold and bacteria, is added to juices, margarine, relishes, salads, and jams. Monosodium glutamate (MSG) is added to meats, fish, vegetables, and bakery items to enhance flavor, although it causes headaches in some people.

$$CH_3-CH_2-\overset{\displaystyle O}{\overset{\|}{C}}-O^-Na^+$$
Sodium propionate

Sodium benzoate

$$HO-\overset{\displaystyle O}{\overset{\|}{C}}-\overset{NH_2}{\underset{|}{CH}}-CH_2-CH_2-\overset{\displaystyle O}{\overset{\|}{C}}-O^-Na^+$$
Monosodium glutamate

Carboxylic acid salts are ionic compounds with strong attractions between ions of metals such as $Li^+$, $Na^+$, and $K^+$ and the negatively charged carboxylate ion. Like most salts, the carboxylic acid salts are solids at room temperature, have high melting points, and are usually soluble in water.

---

## SAMPLE PROBLEM 13.4

### ■ Neutralization of a Carboxylic Acid

Write the equation for the neutralization of propanoic acid (propionic acid) with sodium hydroxide.

#### SOLUTION

The neutralization of an acid with a base produces the salt of the acid and water.

$$CH_3-CH_2-\overset{\displaystyle O}{\overset{\|}{C}}-OH + NaOH \longrightarrow CH_3-CH_2-\overset{\displaystyle O}{\overset{\|}{C}}-O^-Na^+ + H_2O$$

Propionic acid    Sodium propionate

#### STUDY CHECK

What carboxylic acid will produce potassium butanoate (potassium butyrate) when it is neutralized by KOH?

---

## QUESTIONS AND PROBLEMS

### Properties of Carboxylic Acids

**13.9** Identify the compound in each pair that has the higher boiling point. Explain.
   **a.** ethanoic acid (acetic acid) or butanoic acid
   **b.** 1-propanol or propanoic acid
   **c.** butanone or butanoic acid

**13.10** Identify the compound in each pair that has the higher boiling point. Explain.
   **a.** propanone (acetone) or propanoic acid
   **b.** propanoic acid or hexanoic acid
   **c.** ethanol or ethanoic acid (acetic acid)

**13.11** Identify the compound in each group that is the most soluble in water. Explain.
   **a.** propanoic acid, hexanoic acid, benzoic acid
   **b.** pentane, 1-butanol, propanoic acid

**13.12** Identify the compound in each group that is the most soluble in water. Explain.
   **a.** butanone, butanoic acid, butane
   **b.** ethanoic acid (acetic acid), pentanoic acid, octanoic acid

**13.13** Write equations for the ionization of each of the following carboxylic acids in water:

   **a.** $H-\overset{\displaystyle O}{\overset{\|}{C}}-OH$    **b.** $CH_3-CH_2-\overset{\displaystyle O}{\overset{\|}{C}}-OH$

   **c.** acetic acid

**13.14** Write equations for the ionization of each of the following carboxylic acids in water:

   **a.** $CH_3-\overset{CH_3}{\underset{|}{CH}}-\overset{\displaystyle O}{\overset{\|}{C}}-OH$
   **b.** hydroxyethanoic acid
   **c.** butanoic acid

**13.15** Write equations for the reaction of each of the following carboxylic acids with NaOH:
   **a.** formic acid    **b.** propanoic acid    **c.** benzoic acid

**13.16** Write equations for the reaction of each of the following carboxylic acids with KOH:
   **a.** acetic acid    **b.** 2-methylbutanoic acid
   **c.** 4-chlorobenzoic acid

# 13.3 ESTERS

LEARNING GOAL

Name an ester; write equations for the formation and hydrolysis of an ester.

A carboxylic acid reacts with an alcohol to form an **ester**. In an ester, the —H of the carboxylic acid is replaced by an alkyl group. Fats and oils in our diets contain esters of glycerol and fatty acids, which are long-chain carboxylic acids. The aromas and flavors of many fruits, including bananas, oranges, and strawberries, are due to esters.

**Carboxylic acid**     **Ester**

$$
\begin{array}{cc}
\underset{\substack{\displaystyle \text{Ethanoic acid} \\ \text{(acetic acid)}}}{CH_3-\overset{\displaystyle O}{\overset{\|}{C}}-O-H} &
\underset{\substack{\displaystyle \text{Methyl ethanoate} \\ \text{(methyl acetate)}}}{CH_3-\overset{\displaystyle O}{\overset{\|}{C}}-O-CH_3}
\end{array}
$$

## Esterification

In a reaction called **esterification**, an ester is produced when a carboxylic acid and an alcohol react in the presence of an acid catalyst (usually $H_2SO_4$). In this reaction, the —OH from the carboxylic acid and the —H from the alcohol combine to form water.

$$
\underset{\text{Acetic acid}}{CH_3-\overset{\displaystyle O}{\overset{\|}{C}}-O-H} + \underset{\text{Methyl alcohol}}{H-O-CH_3} \overset{H^+,\ heat}{\rightleftharpoons} \underset{\text{Methyl acetate}}{CH_3-\overset{\displaystyle O}{\overset{\|}{C}}-O-CH_3} + H-O-H
$$

For example, the ester responsible for the flavor and odor of pears can be prepared using acetic acid and 1-propanol. The equation for the esterification is written as

$$
\underset{\text{Acetic acid}}{CH_3-\overset{\displaystyle O}{\overset{\|}{C}}-OH} + \underset{\text{1-Propanol}}{H-O-CH_2-CH_2-CH_3} \overset{H^+,\ heat}{\rightleftharpoons} \underset{\substack{\text{Propyl acetate} \\ \text{(pears)}}}{CH_3-\overset{\displaystyle O}{\overset{\|}{C}}-O-CH_2-CH_2-CH_3} + \mathbf{H_2O}
$$

---

**SAMPLE PROBLEM  13.5**

### ■ Writing Esterification Equations

The ester that gives the flavor and odor of apples can be synthesized from butyric acid and methyl alcohol. What is the equation for the formation of the ester in apples?

**SOLUTION**

$$
\underset{\text{Butyric acid}}{CH_3-CH_2-CH_2-\overset{\displaystyle O}{\overset{\|}{C}}-OH} + \underset{\text{Methyl alcohol}}{H-O-CH_3} \overset{H^+,\ heat}{\rightleftharpoons}
$$

$$
\underset{\text{Methyl butyrate}}{CH_3-CH_2-CH_2-\overset{\displaystyle O}{\overset{\|}{C}}-O-CH_3} + \mathbf{H_2O}
$$

**STUDY CHECK**

What carboxylic acid and alcohol are needed to form the following ester, which gives the flavor and odor to apricots? (*Hint*: Separate the O and C═O of the ester group and add H and OH to give the original alcohol and carboxylic acid.)

$$
CH_3-CH_2-\overset{\displaystyle O}{\overset{\|}{C}}-O-CH_2-CH_2-CH_2-CH_2-CH_3
$$

## Health Note

### Salicylic Acid and Aspirin

Chewing on a piece of willow bark was used as a way of relieving pain for many centuries. By the 1800s, chemists discovered that salicylic acid was the agent in the bark responsible for the relief of pain. However, salicylic acid, which has both a carboxylic group and a hydroxyl group, irritates the stomach lining. A less irritating ester of salicylic acid and acetic acid, called acetylsalicylic acid or "aspirin," was prepared in 1899 by the Bayer chemical company in Germany. In some aspirin preparations, a buffer is added to neutralize the carboxylic acid group and lessen its irritation of the stomach. Aspirin is used as an analgesic (pain reliever), antipyretic (fever reducer), and anti-inflammatory agent.

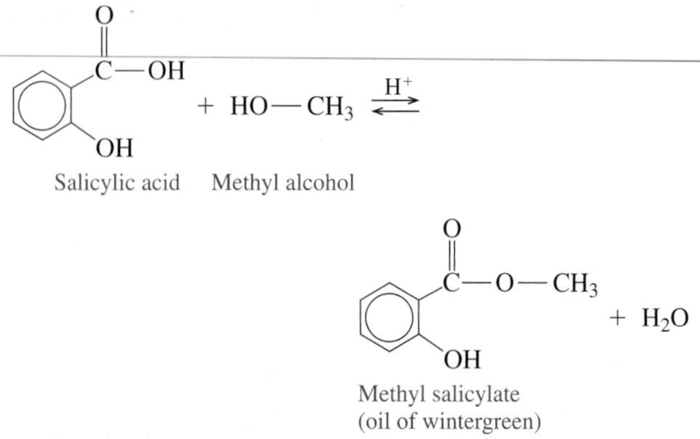

Salicylic acid    Methyl alcohol

Methyl salicylate
(oil of wintergreen)

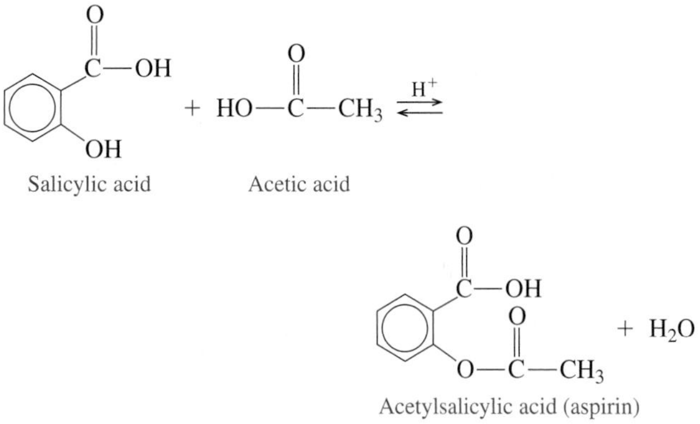

Salicylic acid          Acetic acid

Acetylsalicylic acid (aspirin)

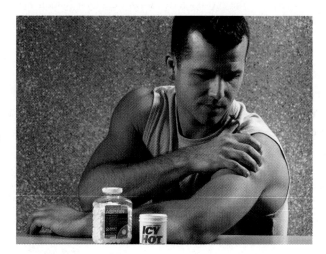

Oil of wintergreen, or methyl salicylate, has a spearmint odor and flavor. Because it can pass through the skin, methyl salicylate is used in skin ointments where it acts as a counterirritant, producing heat to soothe sore muscles.

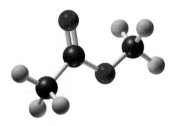

Methyl ethanoate
(methyl acetate)

## Naming Esters

The name of an ester consists of two words taken from the names of the alcohol and the acid. The first word indicates the *alkyl* part of the alcohol. The second word is the *carboxylate* name of the carboxylic acid. The IUPAC names of esters use the IUPAC names for the acid, while the common names of esters use the common names for the acid. Let's take a look at the following ester and break it into two parts, one from the alcohol and one from the acid. By writing and naming the alcohol and carboxylic acid that produced the ester, we can determine the name of the ester.

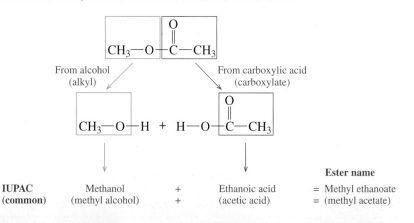

$CH_3$—O—C—$CH_3$

From alcohol          From carboxylic acid
(alkyl)               (carboxylate)

$CH_3$—O—H  +  H—O—C—$CH_3$

| **IUPAC** | Methanol | + | Ethanoic acid | = Methyl ethanoate |
| **(common)** | (methyl alcohol) | + | (acetic acid) | = (methyl acetate) |

                                                    **Ester name**

The following examples of some typical esters show the IUPAC, as well as the common names, of esters.

$$CH_3-CH_2-O-\overset{\overset{\displaystyle O}{\|}}{C}-CH_3$$

Ethyl ethanoate
(ethyl acetate)

$$CH_3-O-\overset{\overset{\displaystyle O}{\|}}{C}-CH_2-CH_3$$

Methyl propanoate
(methyl propionate)

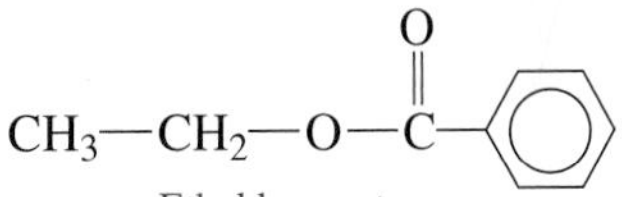

$$CH_3-CH_2-O-\overset{\overset{\displaystyle O}{\|}}{C}-\bigcirc$$

Ethyl benzoate

---

## SAMPLE PROBLEM  13.6

### ■ Naming Esters

Write the IUPAC and common names of the following ester:

$$CH_3-CH_2-\overset{\overset{\displaystyle O}{\|}}{C}-O-CH_2-CH_2-CH_3$$

SOLUTION

The alcohol part of the ester is propyl, and the carboxylic acid part is propanoic (propionic) acid.

From propanoic acid → propanoate
(or propionic acid → propionate)

From propyl alcohol → propyl

$$CH_3-CH_2-\overset{\overset{\displaystyle O}{\|}}{C}-O-CH_2-CH_2-CH_3$$

IUPAC name:    Propyl propanoate
Common name:    Propyl propionate

STUDY CHECK

Draw the condensed structural formula of pentyl ethanoate (pentyl acetate) that gives the odor and flavor to apricots.

## Esters in Plants

Many of the fragrances of perfumes and flowers and the flavors of fruits are due to esters. Small esters are volatile, so we can smell them, and they are soluble in water, so we can taste them. Several of these are listed in Table 13.3.

### TABLE 13.3  Some Esters in Fruits and Flavorings

| Condensed Structural Formula and Name | Flavor/Odor |
|---|---|
| $CH_3-\overset{\overset{\displaystyle O}{\|}}{C}-O-CH_2-CH_2-CH_3$  <br> Propyl ethanoate (propyl acetate) | Pears |
| $CH_3-\overset{\overset{\displaystyle O}{\|}}{C}-O-CH_2-CH_2-CH_2-CH_2-CH_3$  <br> Pentyl ethanoate (pentyl acetate) | Bananas |
| $CH_3-\overset{\overset{\displaystyle O}{\|}}{C}-O-CH_2-CH_2-CH_2-CH_2-CH_2-CH_2-CH_2-CH_3$  <br> Octyl ethanoate (octyl acetate) | Oranges |
| $CH_3-CH_2-CH_2-\overset{\overset{\displaystyle O}{\|}}{C}-O-CH_2-CH_3$  <br> Ethyl butanoate (ethyl butyrate) | Pineapples |
| $CH_3-CH_2-CH_2-\overset{\overset{\displaystyle O}{\|}}{C}-O-CH_2-CH_2-CH_2-CH_2-CH_3$  <br> Pentyl butanoate (pentyl butyrate) | Apricots |

## Acid Hydrolysis of Esters

In **hydrolysis**, esters are split apart with water when heated in the presence of a strong acid, usually $H_2SO_4$ or HCl. The products of acid hydrolysis are the carboxylic acid and alcohol. Therefore, hydrolysis is the reverse of the esterification reaction. When hydrolysis of biological compounds occurs in the cells, an enzyme replaces the acid as the catalyst. In the hydrolysis reaction, the —OH from a water molecule bonds to the carbonyl group of the ester to form the carboxylic acid.

| **Ester** | | **Carboxylic Acid** | **Alcohol** |
|---|---|---|---|

$$CH_3-\overset{\overset{\displaystyle O}{\|}}{C}-O-CH_3 \ + \ H-OH \ \underset{}{\overset{H^+}{\rightleftharpoons}} \ CH_3-\overset{\overset{\displaystyle O}{\|}}{C}-O-H \ + \ CH_3-OH$$

Methyl acetate　　　　　Water　　　　　　　Acetic acid　　　Methyl alcohol

---

### SAMPLE PROBLEM 13.7

#### ■ Acid Hydrolysis of Esters

Aspirin that has been stored for a long time may undergo hydrolysis in the presence of water and heat. What are the hydrolysis products of aspirin? Why does a bottle of old aspirin smell like vinegar?

$$\text{Aspirin (acetylsalicylic acid)}$$

#### SOLUTION

To write the hydrolysis products, separate the compound at the ester bond. Complete the formula of the carboxylic acid by adding —OH (from water) to the carbonyl group and a —H to complete the alcohol. The acetic acid in the products gives the vinegar odor to a sample of aspirin that has hydrolyzed.

Aspirin　　　+ H—OH　$\overset{H^+}{\rightleftharpoons}$　Salicylic acid　+ HO—C—CH$_3$　Acetic acid

#### STUDY CHECK

What are the names of the products from the acid hydrolysis of ethyl propanoate (ethyl propionate)?

## Base Hydrolysis of Esters

When an ester undergoes hydrolysis with a strong base such as NaOH or KOH, the products are the carboxylic acid salt and the corresponding alcohol. The base hydrolysis reaction is also called **saponification**, which refers to the reaction of a long-chain fatty acid with NaOH to make soap. The carboxylic acid, which is produced in acid hydrolysis, is converted to its carboxylate ion by the strong base.

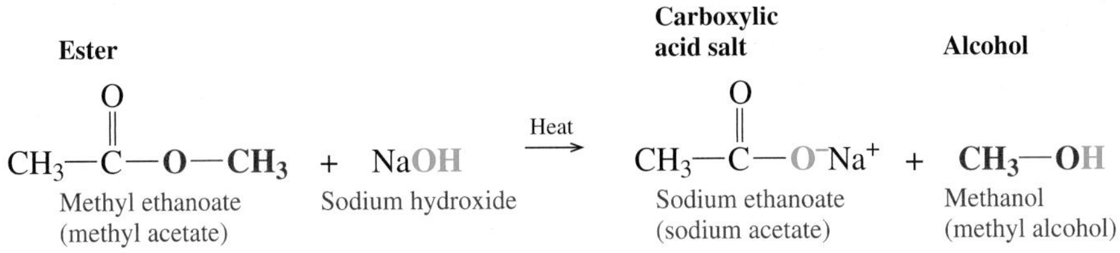

| Ester | | Carboxylic acid salt | Alcohol |
|---|---|---|---|

$$CH_3-\overset{\overset{\text{O}}{\|}}{C}-O-CH_3 + NaOH \xrightarrow{\text{Heat}} CH_3-\overset{\overset{\text{O}}{\|}}{C}-O^-Na^+ + CH_3-OH$$

Methyl ethanoate (methyl acetate)　Sodium hydroxide　　Sodium ethanoate (sodium acetate)　Methanol (methyl alcohol)

---

## SAMPLE PROBLEM  13.8

### ■ Base Hydrolysis of Esters

Ethyl acetate is a solvent widely used for fingernail polish, plastics, and lacquers. Write the equation of the hydrolysis of ethyl acetate by NaOH.

**SOLUTION**

The hydrolysis of ethyl acetate by NaOH gives the salt of acetic acid and ethyl alcohol.

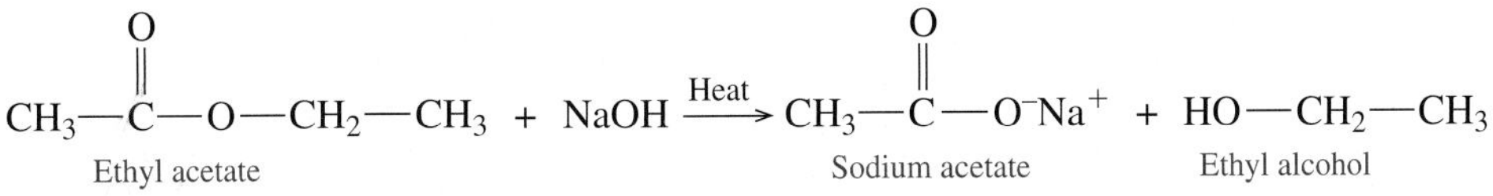

$$CH_3-\overset{\overset{\text{O}}{\|}}{C}-O-CH_2-CH_3 + NaOH \xrightarrow{\text{Heat}} CH_3-\overset{\overset{\text{O}}{\|}}{C}-O^-Na^+ + HO-CH_2-CH_3$$

Ethyl acetate　　　　　　　　Sodium acetate　　Ethyl alcohol

**STUDY CHECK**

Write the condensed structural formulas of the products from the hydrolysis of methyl benzoate by KOH.

---

## QUESTIONS AND PROBLEMS

### Esters

**13.17** Identify each of the following as an aldehyde, a ketone, a carboxylic acid, or an ester:

**a.** 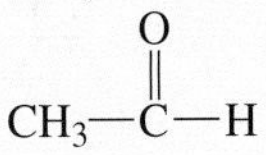 $CH_3-\overset{\overset{\text{O}}{\|}}{C}-H$

**b.** 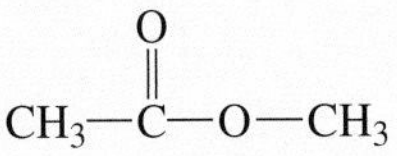 $CH_3-\overset{\overset{\text{O}}{\|}}{C}-O-CH_3$

**c.** $CH_3-CH_2-\overset{\overset{\text{O}}{\|}}{C}-CH_3$

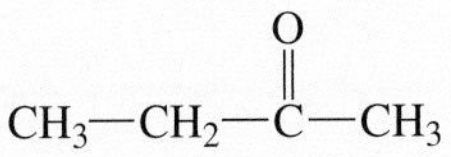

**d.** 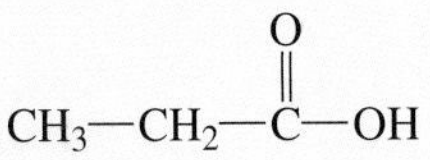 $CH_3-CH_2-\overset{\overset{\text{O}}{\|}}{C}-OH$

**13.18** Identify each of the following as an aldehyde, a ketone, a carboxylic acid, or an ester:

**a.** 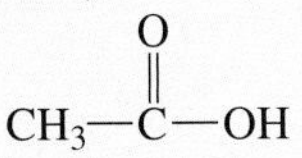 $CH_3-\overset{\overset{\text{O}}{\|}}{C}-OH$

**b.** 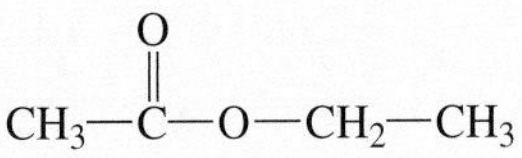 $CH_3-\overset{\overset{\text{O}}{\|}}{C}-O-CH_2-CH_3$

**c.** $CH_3-CH_2-\overset{\overset{\text{O}}{\|}}{C}-H$

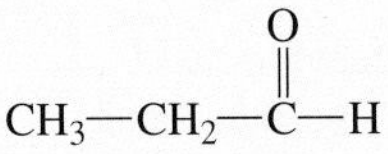

**d.** 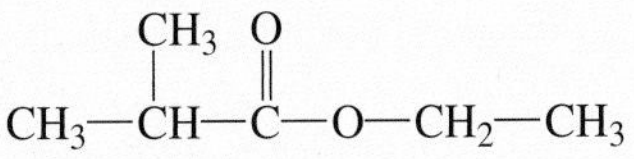 $CH_3-\overset{\overset{CH_3}{|}}{CH}-\overset{\overset{\text{O}}{\|}}{C}-O-CH_2-CH_3$

**13.19** Write the condensed structural formula of the ester formed when each of the following reacts with methyl alcohol:
**a.** acetic acid
**b.** butyric acid
**c.** benzoic acid

**13.20** Write the condensed structural formula of the ester formed when each of the following reacts with methyl alcohol:
**a.** formic acid
**b.** propionic acid
**c.** 2-methylpentanoic acid

**13.21** Draw the condensed structural formulas of the ester formed in each of the following reactions:

a. $CH_3-CH_2-\overset{\overset{\displaystyle O}{\|}}{C}-OH + HO-CH_2-CH_2-CH_3 \underset{\longleftarrow}{\overset{H^+}{\longrightarrow}}$

b. $CH_3-CH_2-CH_2-CH_2-\overset{\overset{\displaystyle O}{\|}}{C}-OH + HO-\overset{\overset{\displaystyle CH_3}{|}}{CH}-CH_3 \underset{\longleftarrow}{\overset{H^+}{\longrightarrow}}$

**13.22** Draw the condensed structural formula of the ester formed in each of the following reactions:

a. $CH_3-CH_2-\overset{\overset{\displaystyle O}{\|}}{C}-OH + HO-CH_3 \underset{\longleftarrow}{\overset{H^+}{\longrightarrow}}$

b. $\overset{\overset{\displaystyle O}{\|}}{C}-OH + HO-CH_2-CH_2-CH_2-CH_3 \underset{\longleftarrow}{\overset{H^+}{\longrightarrow}}$ (with benzene ring)

**13.23** Name each of the following esters:

a. $CH_3-O-\overset{\overset{\displaystyle O}{\|}}{C}-H$

b. $CH_3-O-\overset{\overset{\displaystyle O}{\|}}{C}-CH_3$

c. $CH_3-O-\overset{\overset{\displaystyle O}{\|}}{C}-CH_2-CH_2-CH_3$

d. $CH_3-CH_2-CH_2-\overset{\overset{\displaystyle O}{\|}}{C}-O-CH_2-CH_3$

**13.24** Name each of the following esters:

a. $CH_3-CH_2-O-\overset{\overset{\displaystyle O}{\|}}{C}-CH_2-CH_2-CH_3$

b. $CH_3-O-\overset{\overset{\displaystyle O}{\|}}{C}-CH_2-CH_2-CH_2-CH_2-CH_3$

c. $CH_3-O-\overset{\overset{\displaystyle O}{\|}}{C}-CH_2-CH_2-CH_3$

d. $CH_3-CH_2-\overset{\overset{\displaystyle O}{\|}}{C}-O-CH_2-CH_2-CH_2-CH_3$

**13.25** Draw the condensed structural formulas of each of the following esters:
a. methyl acetate          b. butyl formate
c. ethyl pentanoate        d. propyl propanoate

**13.26** Draw the condensed structural formulas of each of the following esters:
a. hexyl acetate           b. propyl propionate
c. ethyl butanoate         d. methyl benzoate

**13.27** What is the ester responsible for the flavor and odor of the following fruit?
a. banana
b. orange
c. apricot

**13.28** What flavor would you notice if you smelled or tasted the following?
a. ethyl butanoate
b. propyl acetate
c. pentyl acetate

**13.29** What are the products of the acid hydrolysis of an ester?

**13.30** What are the products of the base hydrolysis of an ester?

**13.31** Draw the condensed structural formulas of the products from the acid- or base-catalyzed hydrolysis of each of the following compounds:

a. $CH_3-CH_2-\overset{\overset{\displaystyle O}{\|}}{C}-O-CH_3 + NaOH \longrightarrow$

b. $CH_3-\overset{\overset{\displaystyle O}{\|}}{C}-O-CH_2-CH_2-CH_3 + H_2O \overset{H^+}{\longrightarrow}$

c. $CH_3-CH_2-CH_2-\overset{\overset{\displaystyle O}{\|}}{C}-O-CH_2-CH_3 + H_2O \overset{H^+}{\longrightarrow}$

d. $\overset{\overset{\displaystyle O}{\|}}{C}-O-CH_2-CH_3 + NaOH \longrightarrow$ (with benzene ring)

**13.32** Draw the condensed structural formulas of the products from the acid- or base-catalyzed hydrolysis of each of the following compounds:

a.
$CH_3-CH_2-\overset{\overset{\displaystyle O}{\|}}{C}-O-CH_2-CH_2-CH_2-CH_3 + H_2O \overset{H^+}{\longrightarrow}$

b. $H-\overset{\overset{\displaystyle O}{\|}}{C}-O-CH_2-CH_3 + NaOH \longrightarrow$

c. $CH_3-CH_2-\overset{\overset{\displaystyle O}{\|}}{C}-O-CH_3 + H_2O \overset{H^+}{\longrightarrow}$

d. $CH_3-CH_2-\overset{\overset{\displaystyle O}{\|}}{C}-O-\bigcirc + H_2O \overset{H^+}{\longrightarrow}$ (with benzene ring)

## Environmental Note

### Cleaning Action of Soaps

For many centuries, soaps were made by heating a mixture of animal fats (tallow) with lye, a basic solution obtained from wood ashes. In the soap-making process, fatty acids, which are long-chain carboxylic acids, undergo saponification with the strong base in lye.

Fatty acid

$$CH_3CH_2CH_2CH_2CH_2CH_2CH_2CH_2CH_2CH_2CH_2CH_2CH_2CH_2CH_2CH_2 - \overset{\overset{\textstyle O}{\|}}{C} - OH \ + \ NaOH \longrightarrow$$

Carboxylic acid salt (soap)

$$CH_3CH_2CH_2CH_2CH_2CH_2CH_2CH_2CH_2CH_2CH_2CH_2CH_2CH_2CH_2CH_2 - \overset{\overset{\textstyle O}{\|}}{C} - O^- Na^+$$

Nonpolar tail
(hydrophobic)

Polar head
(hydrophilic)

Today, soaps are also prepared from fats such as coconut oil. Perfumes are added to give a pleasant-smelling soap. Because a soap is the salt of a long-chain fatty acid, the two ends of a soap molecule have different polarities. The long carbon chain end is nonpolar and *hydrophobic* (water-fearing). It is soluble in nonpolar substances such as oil or grease; but it is not soluble in water. The carboxylate salt end is ionic and *hydrophilic* (water-loving). It is very soluble in water but not in oils or grease.

When soap is used to clean grease or oil, the nonpolar ends of the soap molecules dissolve in the nonpolar fats and oils that accompany dirt. The water-loving salt ends of the soap molecules extend outside, where they can dissolve in water. The soap molecules coat the oil or grease, forming clusters called *micelles*. The ionic ends of the soap molecules provide polarity to the micelles, which makes them soluble in water. As a result, small globules of oil and fat coated with soap molecules are pulled into the water and rinsed away.

One of the problems of using soaps is that the carboxylate end reacts with ions in water, such as $Ca^{2+}$ and $Mg^{2+}$, and forms insoluble substances.

$$2CH_3(CH_2)_{16}COO^- \ + \ Mg^{2+} \longrightarrow [CH_3(CH_2)_{16}COO]_2Mg^{2+}$$

Stearate ion    Magnesium ion    Magnesium stearate
(insoluble)

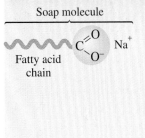

Soap molecule

Fatty acid
chain

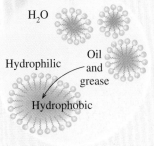

$H_2O$

Hydrophilic

Oil
and
grease

Hydrophobic

## 13.4 AMINES

**Amines** are considered as derivatives of ammonia ($NH_3$), in which one or more hydrogen atoms attached to the nitrogen atom are replaced with alkyl or aromatic groups. From Table 10.8 in Chapter 10, we know that a nitrogen atom typically forms three bonds. For example, in methylamine, a methyl group replaces one hydrogen atom in ammonia. The bonding of two methyl groups gives dimethylamine, and the three methyl groups in trimethylamine replace all the hydrogen atoms in ammonia.

### LEARNING GOAL

Classify amines as primary (1°), secondary (2°), or tertiary (3°); name amines using common names. Describe some properties of amines.

**WEB TUTORIAL**
Amine and Amide Functional
Groups

## Classification of Amines

In the classification of amines, we determine if the nitrogen atom is bonded to one, two, or three alkyl or aromatic groups. In a *primary (1°) amine*, a nitrogen atom is bonded to one alkyl group. In a *secondary (2°) amine*, a nitrogen atom is bonded to two alkyl groups, and in a *tertiary (3°) amine*, a nitrogen atom is bonded to three alkyl groups.

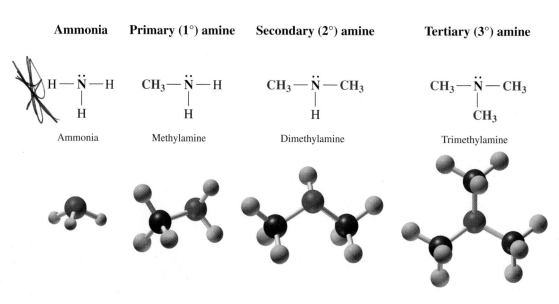

|  Ammonia  |  Primary (1°) amine  |  Secondary (2°) amine  |  Tertiary (3°) amine  |

| Ammonia | Methylamine | Dimethylamine | Trimethylamine |

---

**SAMPLE PROBLEM  13.9**

### ■ Classifying Amines

Classify the following amines as primary (1°), secondary (2°), or tertiary (3°):

**a.**  NH$_2$ (on cyclohexane ring)

**b.**
$$CH_3-\overset{\overset{\displaystyle H}{|}}{N}-CH_2-CH_3$$

**c.**  (phenyl)—N—CH$_3$ with H below N

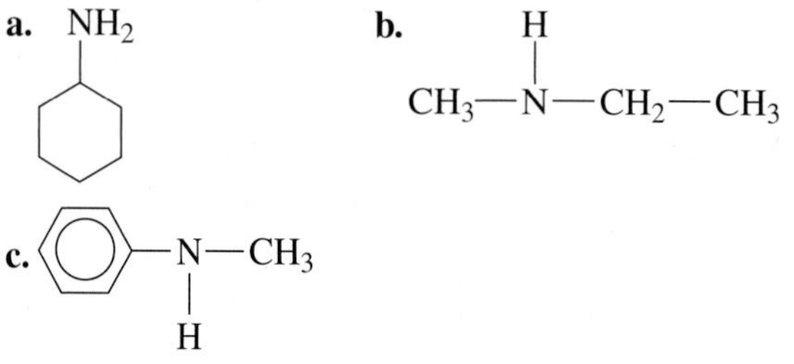

### SOLUTION

**a.** This is a primary (1°) amine because there is one alkyl group (cyclohexyl) attached to a nitrogen atom.

**b.** This is a secondary (2°) amine. There are two alkyl groups (methyl and ethyl) attached to the nitrogen atom.

**c.** This is a secondary (2°) amine with two carbon groups, methyl and phenyl, bonded to the nitrogen atom.

### STUDY CHECK

Classify the following amine as primary (1°), secondary (2°), or tertiary (3°):

$$CH_3-CH_2-\overset{\overset{\displaystyle }{|}}{N}-CH_2-CH_3$$
$$\overset{|}{CH_3}$$

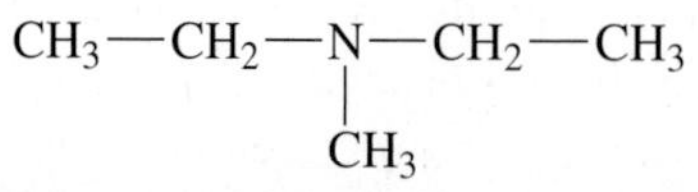

## Naming Amines

For simple amines, the common names are often used. In the common name, the alkyl groups bonded to the nitrogen atom are listed in alphabetical order. The prefixes *di* and *tri* are used to indicate two and three identical substituents.

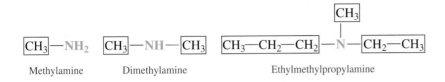

$CH_3—NH_2$     $CH_3—NH—CH_3$     Ethylmethylpropylamine

Methylamine     Dimethylamine

## Aromatic Amines

The aromatic amines use the name *aniline*. Alkyl groups attached to the nitrogen of aniline are named with the prefix *N-* followed by the alkyl name.

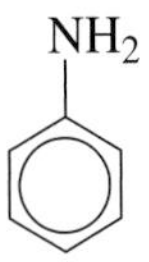

Aniline 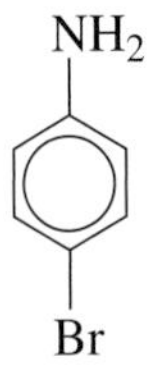        4-Bromoaniline 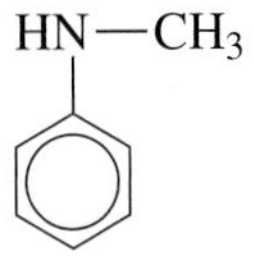        *N*-Methylaniline

---

SAMPLE PROBLEM 13.10

■ **Naming Amines**

Give the common name for each of the following amines:

**a.** $CH_3—CH_2—NH_2$

**b.** 
$$CH_3—N—CH_3$$
with $CH_3$ above N

**c.** aromatic ring with $NH_2$

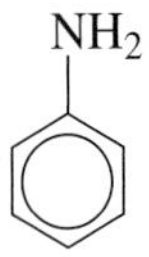

SOLUTION

**a.** This amine has one ethyl group attached to the nitrogen atom; its name is ethylamine.

**b.** This amine has three methyl groups attached to the nitrogen atom; its name is trimethylamine.

**c.** This aromatic amine is aniline.

STUDY CHECK

Draw the structure of ethylpropylamine.

# Health Note

## Amines in Health and Medicine

In response to allergic reactions or injury to cells, the body increases the production of histamine, which causes blood vessels to dilate and increases the permeability of the cells. Redness and swelling occur in the area. Administering an antihistamine such as diphenhydramine helps block the effects of histamine.

Histamine

Diphenhydramine

In the body, hormones called biogenic amines carry messages between the central nervous system and nerve cells. Epinephrine (adrenaline) and norepinephrine (noradrenaline) are released by the adrenal medulla in "fight or flight" situations to raise the blood glucose level and move the blood to the muscles. Used in remedies for colds, hay fever, and asthma, norepinephrine contracts the capillaries in the mucous membranes of the respiratory passages. The prefix *nor* in a drug name means there is one less $CH_3$— group on the nitrogen atom. Parkinson's disease is a result of a deficiency in another biogenic amine called dopamine.

Epinephrine (adrenaline)

Norepinephrine (noradrenaline)

Dopamine

Produced synthetically, amphetamines (known as "uppers") are stimulants of the central nervous system much like epinephrine, but they also increase cardiovascular activity and depress the appetite. They are sometimes used to bring about weight loss, but they can cause chemical dependency. Benzedrine and Neo-Synephrine (phenylephrine) are used in medications to reduce respiratory congestion from colds, hay fever, and asthma. Sometimes, benzedrine is taken to combat the desire to sleep, but it has side effects. Methedrine is used to treat depression and in the illegal form is known as "speed" or "crank." The prefix *meth* means that there is one more methyl group on the nitrogen atom.

Benzedrine (amphetamine)

Neo-Synephrine (phenylephrine)

Methamphetamine (methedrine)

## Properties of Amines

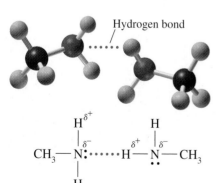

Hydrogen bond

Amines have higher boiling points than hydrocarbons of similar mass, but lower than the alcohols. Because amines contain a polar N—H bond, they form hydrogen bonds. However, the nitrogen atom in amines is not as electronegative as the oxygen in alcohols, making the hydrogen bonds weaker in amines. In primary (1°) amines, —$NH_2$ can form more hydrogen bonds, which gives them higher boiling points than the secondary (2°) amines of the same mass. It is not possible for tertiary (3°) amines to hydrogen bond with each other (no N—H bonds), making their boiling points much lower and similar to those of alkanes and ethers.

$CH_3—CH_2—CH_2—OH$
Propanol
bp 97 °C

$CH_3—CH_2—CH_2—NH_2$
Propylamine (1°)
bp 48 °C

$CH_3—CH_2—NH—CH_3$
Ethylmethylamine (2°)
bp 36 °C

$CH_3—N—CH_3$ (with $CH_3$ on N)
Trimethylamine (3°)
bp 3 °C

## Solubility in Water

Like alcohols, the smaller amines, including tertiary ones, are soluble because they form hydrogen bonds with water. However, in amines with more than six carbon atoms, the effect of hydrogen bonding is diminished. Then the nonpolar alkyl part of the molecules decreases the solubility of an amine in water.

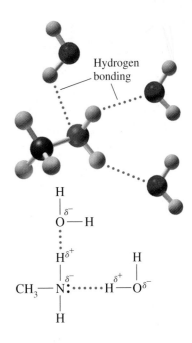

Hydrogen bonding

### SAMPLE PROBLEM 13.11

### ■ Properties of Amines

If the compounds trimethylamine and ethylmethylamine have the same molar mass, why is the boiling point of trimethylamine (3 °C) lower than that of ethylmethylamine (36 °C)?

#### SOLUTION

With a polar N—H bond, hydrogen bonds form between ethylmethylamine molecules. Thus, a higher temperature is required to break the hydrogen bonds and form a gas. However, trimethylamine, which is a tertiary amine, has no N—H bond and cannot form hydrogen bonds between the amine molecules.

#### STUDY CHECK

Why is $CH_3$—$CH_2$—$CH_2$—$CH_2$—$NH_2$ soluble in water?

## Amines React as Bases

In Chapter 8, we saw that ammonia ($NH_3$) acts as a Brønsted–Lowry base because it accepts a proton ($H^+$) from water to produce an ammonium ion ($NH_4^+$) and a hydroxide ion ($OH^-$). Let's review that equation:

$$\ddot{N}H_3 + H_2O \rightleftharpoons NH_4^+ + OH^-$$

Ammonia　　　Ammonium ion　Hydroxide ion

In water, amines act as Brønsted–Lowry bases because the lone electron pair on the nitrogen atom accepts a proton from water and produces hydroxide ions.

$$CH_3-\ddot{N}H_2 + H_2O \rightleftharpoons CH_3-\overset{+}{N}H_3 + OH^-$$

Methylamine　　　　Methylammonium ion　Hydroxide ion

$$CH_3-\underset{\underset{CH_3}{|}}{\ddot{N}H} + H_2O \rightleftharpoons CH_3-\underset{\underset{CH_3}{|}}{\overset{+}{N}H_2} + OH^-$$

Dimethylamine　　Dimethylammonium ion　Hydroxide ion

WEB TUTORIAL
Amines as Bases

## Amine Salts

When you squeeze lemon juice on fish, the "fishy odor" of the amines is removed by converting them to amine salts. In a neutralization reaction, an amine acts as a base and reacts with an acid to form an **amine salt**. The lone pair of electrons on the nitrogen atom accepts a proton $H^+$ from an acid to give an amine salt; no water is formed. An amine salt is named by replacing the *amine* part of the amine name with *ammonium*, followed by the name of the negative ion.

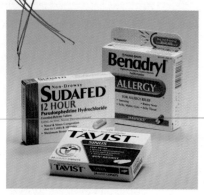

**FIGURE 13.2** Decongestants and products that relieve itch and skin irritations can contain ammonium salts.

**Q** Why are the ammonium salts used rather than the biologically active amines?

CASE STUDY
Death By Chocolate?

**Neutralization of an Amine**

| Amine | Acid | | Amine salt |
|---|---|---|---|

$$CH_3\text{—}\overset{..}{N}H_2 + HCl \longrightarrow CH_3\text{—}\overset{+}{N}H_3\ Cl^-$$

Methylamine                    Methylammonium chloride

$$CH_3\text{—}\overset{..}{N}H + HCl \longrightarrow CH_3\text{—}\overset{+}{N}H_2\ Cl^-$$
$$\quad\quad |\qquad\qquad\qquad\qquad\qquad |$$
$$\quad\quad CH_3\qquad\qquad\qquad\qquad CH_3$$

Dimethylamine                    Dimethylammonium chloride

## Properties of Amine Salts

Amine salts are ionic compounds with strong attractions between the positively charged ammonium ion and an anion, usually chloride. Like most salts, amine salts are solids at room temperature, odorless, and soluble in water and body fluids. For this reason, amines used as drugs are converted to their amine salts. The amine salt of ephedrine is used as a bronchodilator and in decongestant products such as Sudafed. The amine salt of diphenhydramine is used in products such as Benadryl for relief of itching and pain from skin irritations and rashes. (See Figure 13.2.) In pharmaceuticals, the naming of the amine salt follows an older method of giving the amine name followed by the name of the acid.

Ephedrine hydrochloride
Ephedrine HCl
Sudafed

Diphenhydramine hydrochloride
Diphenhydramine HCl
Benadryl

When an amine salt reacts with a strong base such as NaOH, it is converted back to the amine, which is also called the free amine or free base.

$$CH_3\text{—}NH_3{}^+Cl^- + NaOH \longrightarrow CH_3\text{—}NH_2 + NaCl + H_2O$$

The narcotic cocaine is typically extracted from coca leaves, using an acidic HCl solution to give a white, solid amine salt, which is cocaine hydrochloride. This is the form in which cocaine is smuggled and sold illegally on the street to be snorted or injected. "Crack cocaine" is the free amine or free base of the amine obtained by treating the cocaine hydrochloride with NaOH and ether, a process known as "free-basing." The solid product is known as "crack cocaine" because it makes a crackling noise when heated. The free amine is rapidly absorbed when smoked and gives stronger highs than the cocaine hydrochloride. Unfortunately, these effects of crack cocaine have caused a rise in addiction to cocaine.

Cocaine hydrochloride

Cocaine (free base)

# Health Note

## Alkaloids: Amines in Plants

**Alkaloids** are physiologically active nitrogen-containing compounds produced by plants. The term *alkaloid* refers to the alkali-like or basic characteristics we have seen for amines. Certain alkaloids are used in anesthetics, in antidepressants, and as stimulants, although many are habit forming.

As a stimulant, nicotine increases the level of adrenaline in the blood, which increases the heart rate and blood pressure. It is well known that smoking cigarettes can damage the lungs and that exposure to tars and other carcinogens in cigarette smoke can lead to lung cancer. However, nicotine is responsible for the addiction of smoking. Coniine, which is obtained from hemlock, is an extremely toxic alkaloid.

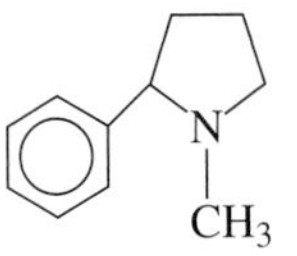

Nicotine

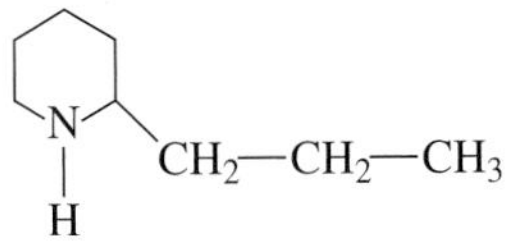

Coniine

Caffeine is a stimulant of the central nervous system. Present in coffee, tea, soft drinks, chocolate, and cocoa, caffeine increases alertness, but may cause nervousness and insomnia. Caffeine is also used in certain pain relievers to counteract the drowsiness caused by an antihistamine.

Several alkaloids are used in medicine. Quinine obtained from the bark of the cinchona tree has been used in the treatment of malaria since the 1600s. Atropine from belladonna is used in low concentrations to accelerate unhealthy slow heart rates and as an anesthetic for eye examinations.

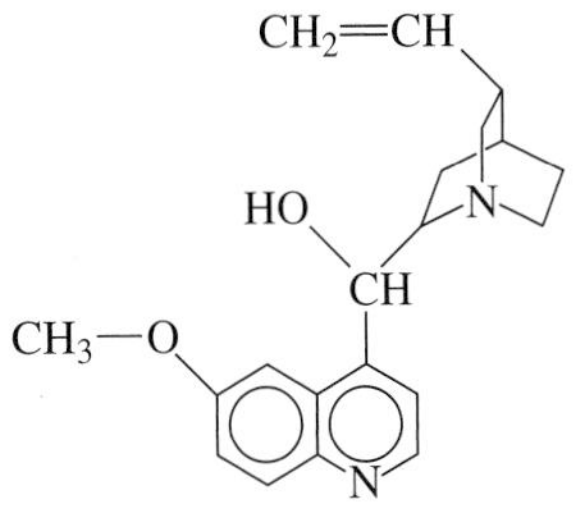

Quinine

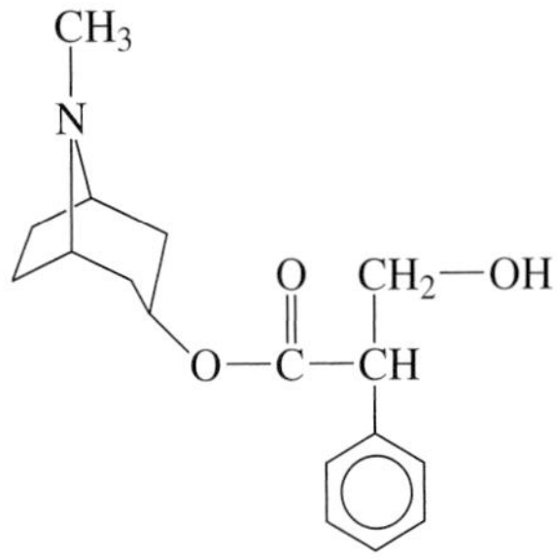

Atropine

For many centuries, morphine and codeine, alkaloids found in the oriental poppy plant, have been used as effective painkillers. Codeine, which is structurally similar to morphine, is used in some prescription painkillers and cough syrups. Heroin, obtained by a chemical modification of morphine, is strongly addicting and is not used medically.

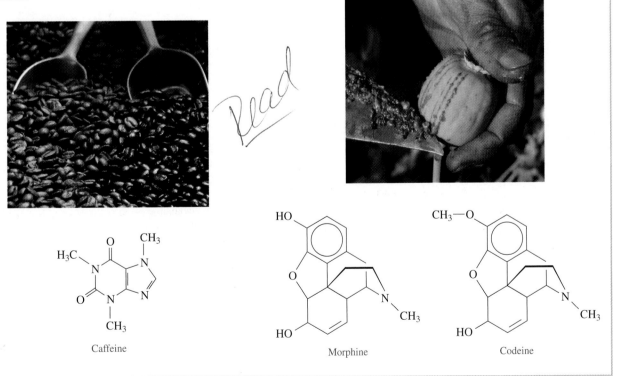

Caffeine

Morphine

Codeine

SAMPLE PROBLEM  13.12

■ **Reactions of Amines**

Write an equation that shows ethylamine

**a.** ionizing as a weak base in water
**b.** neutralized by HCl

SOLUTION

**a.** In water, ethylamine acts as a weak base by accepting a proton from water to produce ethylammonium hydroxide.

$$CH_3-CH_2-NH_2 + H-OH \rightleftharpoons CH_3-CH_2-NH_3^+ + OH^-$$

**b.** $CH_3-CH_2-NH_2 + HCl \longrightarrow CH_3-CH_2-NH_3^+Cl^-$

STUDY CHECK

What is the condensed structural formula of the salt formed by the reaction of trimethyl-amine and HCl?

## QUESTIONS AND PROBLEMS

### Amines

**13.33** Classify each of the following amines as primary, secondary, or tertiary:

**a.** $CH_3-CH_2-CH_2-NH_2$

**b.** $CH_3-\overset{\overset{\displaystyle H}{|}}{N}-CH_2-CH_3$

**c.**

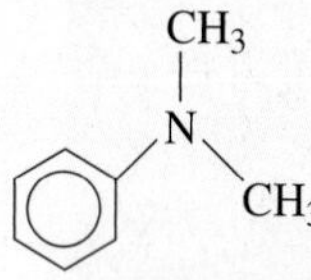

**d.** $CH_3-\overset{\overset{\displaystyle CH_3}{|}}{CH}-\overset{\overset{\displaystyle CH_3}{|}}{N}-CH_2-CH_3$

**13.34** Classify each of the following amines as primary, secondary, or tertiary:

**a.** $CH_3-CH_2-\overset{\overset{\displaystyle NH_2}{|}}{CH}-CH_3$

**b.** $CH_3-CH_2-\overset{\overset{\displaystyle CH_3}{|}}{N}-CH_2-CH_3$

**c.** $\overset{\overset{\displaystyle CH_3}{|}}{CH}-NH_2$

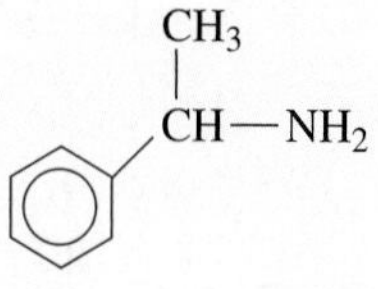

**d.** $CH_3-\overset{\overset{\displaystyle H}{|}}{N}-\overset{\overset{\displaystyle CH_3}{|}}{\underset{\underset{\displaystyle CH_3}{|}}{C}}-CH_3$

**13.35** Write the common names for each of the following:
**a.** $CH_3-CH_2-NH_2$
**b.** $CH_3-NH-CH_2-CH_2-CH_3$

**c.** $CH_3-CH_2-\overset{\overset{\displaystyle CH_3}{|}}{N}-CH_2-CH_3$    **d.** $CH_3-\overset{\overset{\displaystyle NH_2}{|}}{CH}-CH_3$

**13.36** Write the common names for each of the following:
**a.** $CH_3-CH_2-CH_2-NH_2$
**b.** $CH_3-NH-CH_2-CH_3$
**c.** $CH_3-CH_2-CH_2-CH_2-NH_2$

**d.** $CH_3-CH_2-\overset{\overset{\displaystyle CH_2-CH_3}{|}}{N}-CH_2-CH_3$

**13.37** Draw the condensed structural formulas for each of the following amines:
**a.** ethylamine
**b.** *N*-methylaniline
**c.** butylpropylamine

**13.38** Draw the condensed structural formulas for each of the following amines:
**a.** dimethylamine
**b.** 4-chloroaniline
**c.** *N,N*-diethylaniline

**13.39** Identify the compound in each pair that has the higher boiling point.
**a.** $CH_3-CH_2-NH_2$ or $CH_3-CH_2-OH$
**b.** $CH_3-NH_2$ or $CH_3-CH_2-CH_2-NH_2$

**c.** $CH_3-\overset{\overset{\displaystyle CH_3}{|}}{N}-CH_3$ or $CH_3-CH_2-CH_2-NH_2$

**13.40** Identify the compound in each pair that has the higher boiling point.
 **a.** $CH_3-CH_2-CH_2-CH_3$ or $CH_3-CH_2-CH_2-NH_2$
 **b.** $CH_3-NH_2$ or $CH_3-CH_2-NH_2$
 **c.** $CH_3-CH_2-CH_2-OH$ or $CH_3-\overset{\overset{\displaystyle NH_2}{|}}{CH}-CH_3$

**13.41** Indicate if each of the following is soluble in water. Explain.
 **a.** $CH_3-CH_2-NH_2$
 **b.** $CH_3-NH-CH_3$
 **c.** $CH_3-CH_2-CH_2-\overset{\overset{\displaystyle CH_2-CH_2-CH_3}{|}}{N}-CH_2-CH_2-CH_3$
 **d.** $CH_3-\overset{\overset{\displaystyle NH_2}{|}}{CH}-CH_2-CH_3$

**13.42** Indicate if each of the following is soluble in water. Explain.
 **a.** $CH_3-CH_2-CH_2-NH_2$
 **b.** $CH_3-CH_2-CH_2-NH-CH_2-CH_3$
 **c.** $CH_3-\overset{\overset{\displaystyle CH_3}{|}}{N}-CH_3$   **d.**

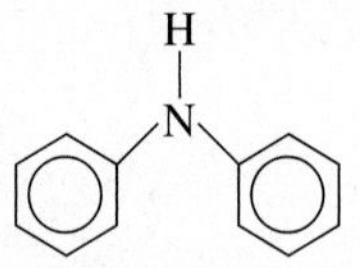

**13.43** Write an equation for the ionization of each of the following amines in water:
 **a.** methylamine   **b.** dimethylamine   **c.** aniline

**13.44** Write an equation for the ionization of each of the following amines in water:
 **a.** ethylamine
 **b.** propylamine
 **c.** N-methylaniline

# 13.5  AMIDES

The **amides** are derivatives of carboxylic acids in which a nitrogen group replaces the hydroxyl group. An amide is produced when a carboxylic acid reacts with ammonia or an amine. A molecule of water is eliminated, and the fragments of the carboxylic acid and amine molecules join to form the amide, much like the formation of ester.

Write common and IUPAC names of amides and the products of formation and hydrolysis.

$$CH_3-\overset{\overset{\displaystyle O}{||}}{C}-OH + H-\overset{\overset{\displaystyle H}{|}}{N}-H \xrightarrow{\text{Heat}} CH_3-\boxed{\overset{\overset{\displaystyle O}{||}}{C}-\overset{\overset{\displaystyle H}{|}}{N}}-H + H_2O$$

Ethanoic acid        Ammonia        Ethanamide
(acetic acid)                        (acetamide)

Ethanamide
(acetamide)

$$CH_3-\overset{\overset{\displaystyle O}{||}}{C}-OH + H-\overset{\overset{\displaystyle H}{|}}{N}-CH_3 \xrightarrow{\text{Heat}} CH_3-\boxed{\overset{\overset{\displaystyle O}{||}}{C}-\overset{\overset{\displaystyle H}{|}}{N}}-CH_3 + H_2O$$

Ethanoic acid        Methylamine        N-Methylethanamide
(acetic acid)                        (N-methylacetamide)

**SAMPLE PROBLEM  13.13**

■ **Formation of Amides**

Give the condensed structural formula of the amide product in each of the following reactions:

**a.**

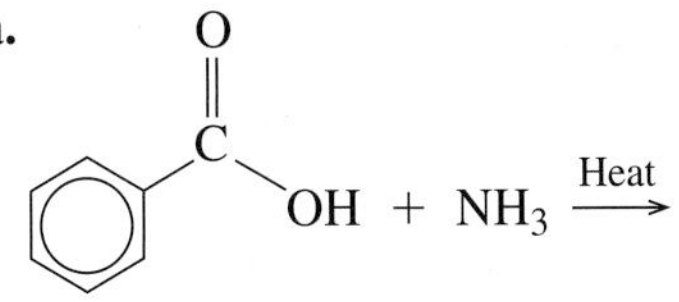

$$\overset{\overset{\displaystyle O}{||}}{C}-OH + NH_3 \xrightarrow{\text{Heat}}$$

**b.** $CH_3-\overset{\overset{\displaystyle O}{||}}{C}-OH + NH_2-CH_2-CH_3 \xrightarrow{\text{Heat}}$

## Clinical Laboratory Technologist

"We use mass spectrometry to analyze and confirm the presence of drugs," says Valli Vairavan, clinical lab technologist—Mass Spectrometry, Santa Clara Valley Medical Center. "A mass spectrometer separates and identifies compounds, including drugs, by mass. When we screen a urine sample, we look for metabolites, which are the products of drugs that have metabolized in the body. If the presence of one or more drugs such as heroin and cocaine is indicated, we confirm it by using mass spectrometry."

Drugs or their metabolites are detected in urine 24–48 hours after use. Cocaine metabolizes to benzoylecgonine and hydroxycocaine, morphine to morphine-3-glucuronide, and heroin to acetylmorphine. Amphetamines and methamphetamines are detected unchanged.

### SOLUTION

**a.** The condensed structural formula of the amide product can be written by attaching the carbonyl group from the acid to the nitrogen atom of the amine. —OH is removed from the acid and —H from the amine to form water.

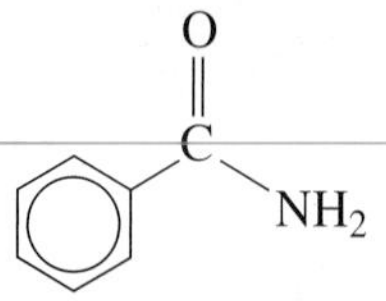

**b.**

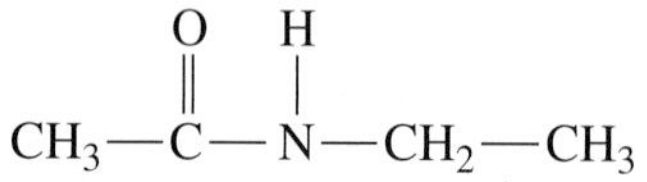

### STUDY CHECK

What are the condensed structural formulas of the carboxylic acid and amine needed to prepare the following amide? (*Hint*: Separate the N and C=O of the amide group, and add H and OH to give the original amine and carboxylic acid.)

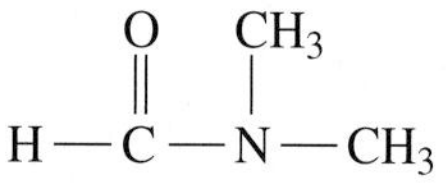

## Naming Simple Amides

In both the common and IUPAC names, simple amides are named by dropping the *ic acid* or *oic acid* from the carboxylic acid names and adding the suffix *amide*. An alkyl group attached to the nitrogen of an amide is named with the prefix *N-*, followed by the alkyl name.

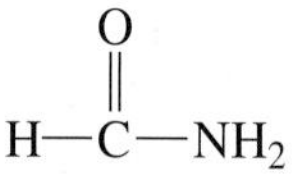

Methanamide
(formamide)

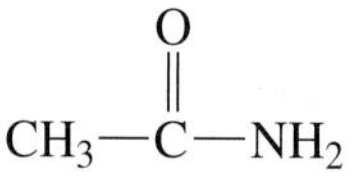

Ethanamide
(acetamide)

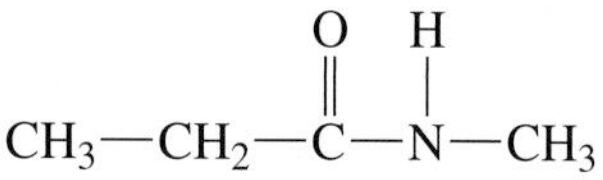

*N*-Methylpropanamide
(*N*-methylpropionamide)

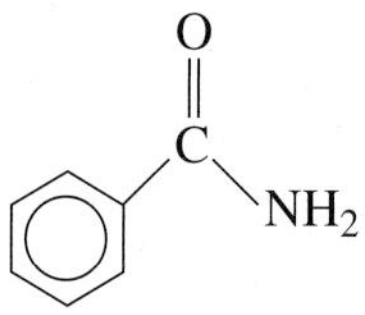

Benzamide

### SAMPLE PROBLEM 13.14

#### ■ Naming Amides

Give the common and IUPAC names for

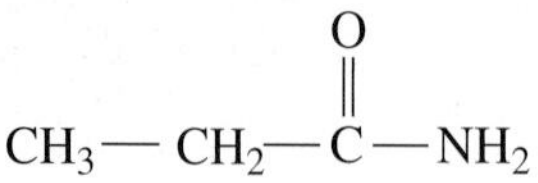

$$CH_3-CH_2-\overset{\overset{\displaystyle O}{\|}}{C}-NH_2$$

### SOLUTION

The IUPAC name of the carboxylic acid is propanoic acid; the common name is propionic acid. Replacing the *oic acid* or *ic acid* ending with *amide* gives the IUPAC name of *propanamide* and the common name of *propionamide*.

### STUDY CHECK

Draw the condensed structural formula of benzamide.

## Physical Properties of Amides

The amides do not have the properties of bases that we saw for the amines. Only for-mamide is a liquid at room temperature, while the other amides are solids. For primary amides, the —$NH_2$ group can form several hydrogen bonds, which gives primary amides high melting points. The melting points of the secondary amides are lower because the number of hydrogen bonds decreases. Tertiary amides have even lower melting points because they cannot form hydrogen bonds with other tertiary amides.

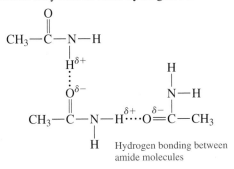

Hydrogen bonding between amide molecules

The amides with one to five carbon atoms are soluble in water because they can hydrogen bond with water molecules.

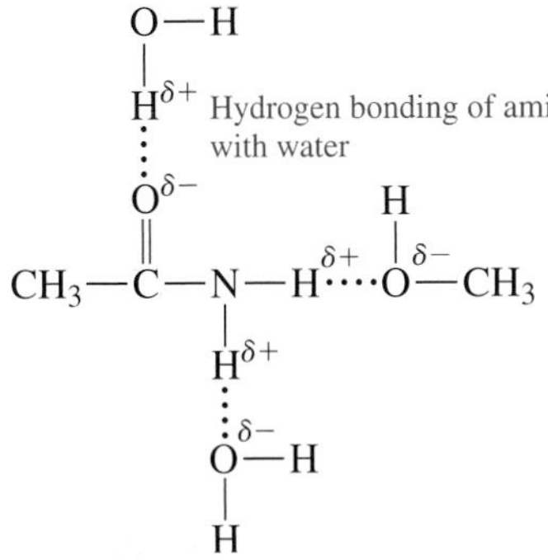

Hydrogen bonding of amides with water

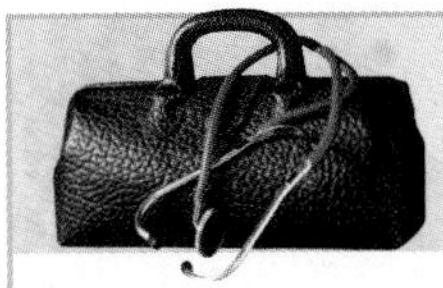

---

# Health Note

## Amides in Health and Medicine

The simplest natural amide is urea, an end product of protein metabo-lism in the body. The kidneys remove urea from the blood and pro-vide for its excretion in urine. If the kidneys malfunction, urea is not removed and builds to a toxic level, a condition called uremia. Urea is also used as a component of fertilizer, to increase nitrogen in the soil.

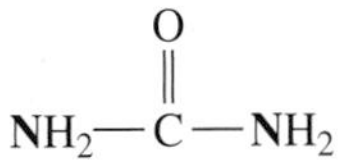

$$NH_2—\overset{\overset{O}{\|}}{C}—NH_2 \quad \text{Urea}$$

Many barbiturates are cyclic amides of barbituric acid that act as sedatives in small dosages or sleep inducers in larger dosages. They are often habit forming. Barbiturate drugs include phenobarbital (Luminal™) and pentobarbital (Nembutal™).

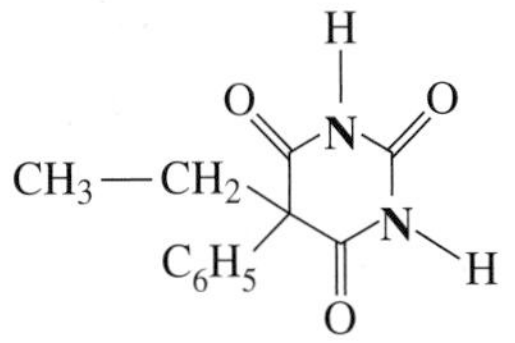

Luminal (phenobarbital)

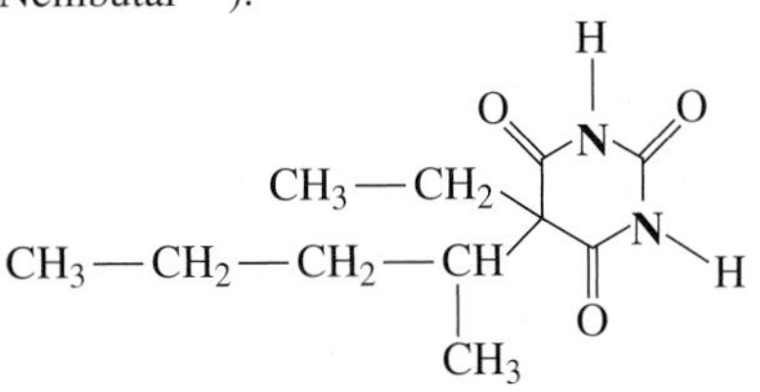

Nembutal (pentobarbital)

Aspirin substitutes contain phenacetin or acetaminophen, which is used in Tylenol. Like aspirin, acetaminophen reduces fever and pain, but it has little anti-inflammatory effect.

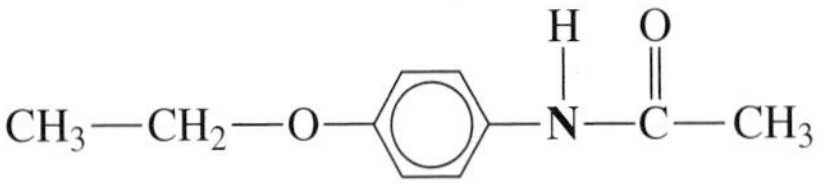

Phenacetin

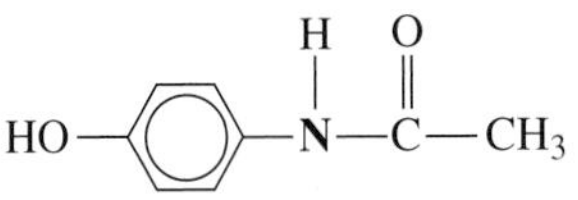

Acetaminophen

## Hydrolysis of Amides

Amides undergo hydrolysis when water is added to the amide bond to split the molecule. When an acid is used, the hydrolysis products of an amide are the carboxylic acid and the ammonium salt. In base hydrolysis, the amide produces the salt of the carboxylic acid and ammonia or an amine.

**Acid Hydrolysis of Amides**

$$CH_3-\overset{\overset{\displaystyle O}{\|}}{C}-NH_2 \; + \; HOH \; + \; HCl \; \longrightarrow \; CH_3-\overset{\overset{\displaystyle O}{\|}}{C}-OH \; + \; NH_4^+Cl^-$$

Ethanamide                  Ethanoic acid     Ammonium
(acetamide)               (acetic acid)      chloride

**Base Hydrolysis of Amides**

$$CH_3-CH_2-\overset{\overset{\displaystyle O}{\|}}{C}-NH-CH_3 \; + \; NaOH \longrightarrow CH_3-CH_2-\overset{\overset{\displaystyle O}{\|}}{C}-O^-Na^+ \; + \; NH_2-CH_3$$

*N*-Methylpropanamide            Sodium propanoate, a salt      Methanamine
(*N*-methylpropionamide)        (sodium propionate)         (methylamine)

---

**SAMPLE PROBLEM 13.15**

### ■ Hydrolysis of Amides

Write the structural formulas for the products for the hydrolysis of *N*-methylpentanamide with NaOH.

**SOLUTION**

Hydrolysis of the amide with a base produces a carboxylate salt (sodium pentanoate) and the corresponding amine (methylamine).

$$CH_3-CH_2-CH_2-CH_2-\overset{\overset{\displaystyle O}{\|}}{C}-O^-Na^+ \; + \; NH_2-CH_3$$

**STUDY CHECK**

What are the structures of the products from the hydrolysis of *N*-methylbutyramide with HBr?

---

## QUESTIONS AND PROBLEMS

### Amides

**13.45** Give the IUPAC and common names (if any) for each of the following amides:

**a.** $CH_3-\overset{\overset{\displaystyle O}{\|}}{C}-NH_2$

**b.** $CH_3-CH_2-CH_2-\overset{\overset{\displaystyle O}{\|}}{C}-NH_2$

**c.** $H-\overset{\overset{\displaystyle O}{\|}}{C}-NH_2$

**13.46** Give the IUPAC and common names (if any) for each of the following amides:

**a.** $CH_3-CH_2-\overset{\overset{\displaystyle O}{\|}}{C}-NH_2$

**b.** $CH_3-CH_2-CH_2-CH_2-CH_2-\overset{\overset{\displaystyle O}{\|}}{C}-NH_2$

**c.** 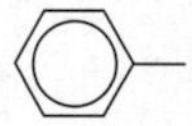 ⬡$-\overset{\overset{\displaystyle O}{\|}}{C}-NH_2$

**13.47** Draw the condensed structural formulas for each of the following amides:

 **a.** propionamide          **b.** 2-methylpentanamide
 **c.** methanamide

**13.48** Draw the condensed structural formulas for each of the following amides:

 **a.** formamide            **b.** benzamide
 **c.** 3-methylbutyramide

**13.49** Write the condensed structural formulas for the products of the acid hydrolysis of each of the following amides with HCl:

 **a.** $CH_3-\overset{\displaystyle O}{\overset{\|}{C}}-NH_2$

 **b.** $CH_3-CH_2-\overset{\displaystyle O}{\overset{\|}{C}}-NH_2$

 **c.** $CH_3-CH_2-CH_2-\overset{\displaystyle O}{\overset{\|}{C}}-NH-CH_3$

 **d.** 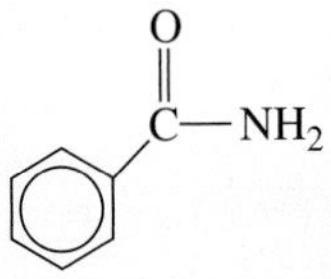 (benzene ring)$-\overset{\displaystyle O}{\overset{\|}{C}}-NH_2$

 **e.** N-ethylpentanamide

**13.50** Write the condensed structural formulas for the products of the base hydrolysis of each of the following amides with NaOH:

 **a.** $CH_3-CH_2-\overset{\displaystyle CH_3}{\underset{}{CH}}-\overset{\displaystyle O}{\overset{\|}{C}}-NH_2$

 **b.** $CH_3-CH_2-CH_2-\overset{\displaystyle O}{\overset{\|}{C}}-\overset{\displaystyle CH_2-CH_3}{\underset{}{N}}-CH_2-CH_3$

 **c.** 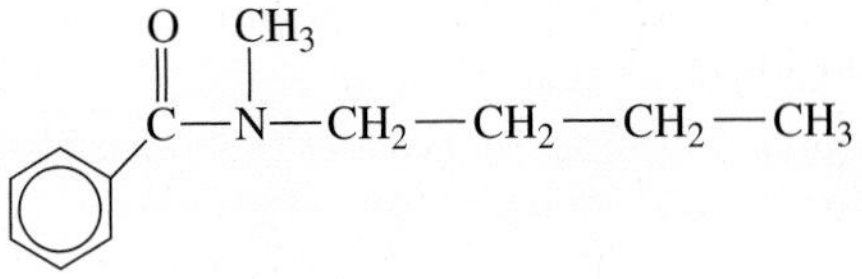 (benzene ring)$-\overset{\displaystyle O}{\overset{\|}{C}}-\overset{\displaystyle CH_3}{\underset{}{N}}-CH_2-CH_2-CH_2-CH_3$

 **d.** $CH_3-\overset{\displaystyle Cl}{\underset{}{CH}}-\overset{\displaystyle O}{\overset{\|}{C}}-\overset{\displaystyle CH_3}{\underset{}{N}}-CH_2-CH_3$

 **e.** 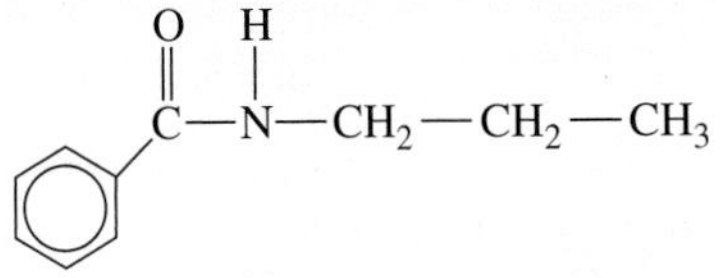 (benzene ring)$-\overset{\displaystyle O}{\overset{\|}{C}}-\overset{\displaystyle H}{\underset{}{N}}-CH_2-CH_2-CH_3$

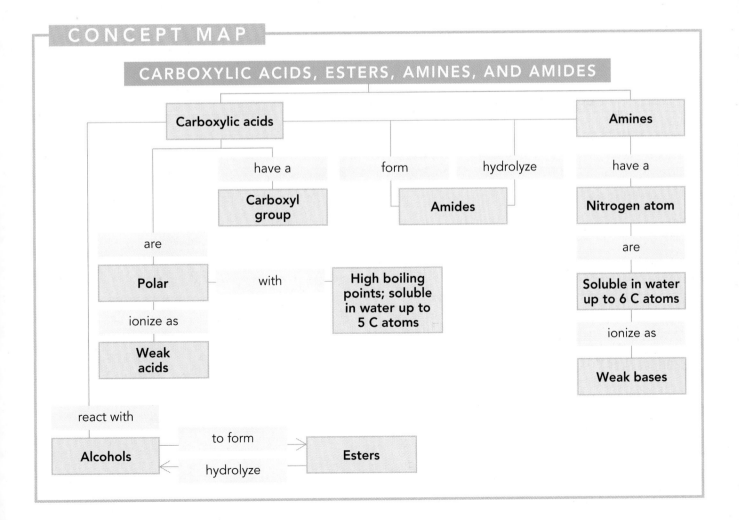

## CONCEPT MAP

### CARBOXYLIC ACIDS, ESTERS, AMINES, AND AMIDES

Carboxylic acids

Amines

have a → Carboxyl group

form → Amides ← hydrolyze

have a → Nitrogen atom

are → Polar

with → High boiling points; soluble in water up to 5 C atoms

are → Soluble in water up to 6 C atoms

ionize as → Weak acids

ionize as → Weak bases

react with → Alcohols

to form → Esters

hydrolyze

# CHAPTER REVIEW

## 13.1 Carboxylic Acids

**Learning Goal:** Give the common names, IUPAC names, and condensed structural formulas of carboxylic acids.

A carboxylic acid contains the carboxyl functional group, which is a hydroxyl group connected to the carbonyl group.

## 13.2 Properties of Carboxylic Acids

**Learning Goal:** Describe the boiling points, solubility, and ionization of carboxylic acids in water.

The carboxyl group contains polar bonds of O—H and C=O, which makes a carboxylic acid with one to four carbon atoms very soluble in water. As weak acids, carboxylic acids ionize slightly by donating a proton to water to form carboxylate and hydronium ions. Carboxylic acids are neutralized by base, producing the carboxylate salt and water.

## 13.3 Esters

**Learning Goal:** Name an ester; write equations for the formation and hydrolysis of an ester.

In an ester, an alkyl or aromatic group has replaced the H of the hydroxyl group of a carboxylic acid. In the presence of a strong acid, a carboxylic acid reacts with an alcohol to produce an ester. A molecule of water is removed: —OH from the carboxylic acid and —H from the alcohol molecule. The names of esters consist of two words, one from the alcohol and the other from the carboxylic acid with the *ic* ending replaced by *ate*. Esters undergo acid hydrolysis by adding water to yield the carboxylic acid and alcohol (or phenol). Base

hydrolysis, or saponification, of an ester produces the carboxylate salt and an alcohol.

## 13.4 Amines

**Learning Goal:** Classify amines as primary (1°), secondary (2°), or tertiary (3°); name amines using common names. Describe some properties of amines.

A nitrogen atom attached to one, two, or three alkyl or aromatic groups forms a primary, secondary, or tertiary amine. Many amines, synthetic or naturally occurring, have physiological activity. In the common names of simple amines, the alkyl groups are listed alphabetically followed by the suffix *amine*. Primary and secondary amines form hydrogen bonds, which make their boiling points higher than alkanes of similar mass. Small amines are soluble in water. In water, amines act as weak bases to produce ammonium and hydroxide ions. When amines react with acids, they form amine salts. As ionic compounds, amine salts are solids, soluble in water, and odorless compared to the amines.

## 13.5 Amides

**Learning Goal:** Write common and IUPAC names of amides and the products of formation and hydrolysis.

Amides are derivatives of carboxylic acids in which the hydroxyl group is replaced by a nitrogen group. Amides are named by replacing the *ic acid* or *oic acid* with *amide*. Hydrolysis of an amide by an acid produces an amine salt. Hydrolysis by a base produces the salt of the carboxylic acid.

# SUMMARY OF NAMING

| Family | Condensed Structural Formula | IUPAC Name | Common Name |
|---|---|---|---|
| Carboxylic acid | $CH_3-\overset{\displaystyle O}{\overset{\displaystyle \|}{C}}-OH$ | Ethanoic acid | Acetic acid |
| Ester | $CH_3-\overset{\displaystyle O}{\overset{\displaystyle \|}{C}}-O-CH_3$ | Methyl ethanoate | Methyl acetate |
| Amine | $CH_3-CH_2-NH_2$ | | Ethylamine |
| Amide | $CH_3-\overset{\displaystyle O}{\overset{\displaystyle \|}{C}}-NH_2$ | Ethanamide | Acetamide |

# SUMMARY OF REACTIONS

### IONIZATION OF A CARBOXYLIC ACID IN WATER

$$CH_3-\overset{\displaystyle O}{\overset{\displaystyle \|}{C}}-OH \ + \ H_2O \ \rightleftharpoons \ CH_3-\overset{\displaystyle O}{\overset{\displaystyle \|}{C}}-O^- \ + \ H_3O^+$$

Ethanoic acid (acetic acid)      Ethanoate ion (acetate ion)      Hydronium ion

## NEUTRALIZATION OF A CARBOXYLIC ACID

$$CH_3-CH_2-\overset{\overset{\displaystyle O}{\|}}{C}-OH \ + \ NaOH \ \longrightarrow \ CH_3-CH_2-\overset{\overset{\displaystyle O}{\|}}{C}-O^-Na^+ \ + \ H_2O$$

Propanoic acid      Sodium         Sodium propanoate
(propionic acid)     hydroxide       (sodium propionate)

## ESTERIFICATION: CARBOXYLIC ACID AND AN ALCOHOL

$$CH_3-\overset{\overset{\displaystyle O}{\|}}{C}-OH \ + \ HO-CH_3 \ \overset{H^+}{\underset{}{\rightleftharpoons}} \ CH_3-\overset{\overset{\displaystyle O}{\|}}{C}-O-CH_3 \ + \ H_2O$$

Ethanoic acid    Methanol         Methyl ethanoate
(acetic acid)     (methyl alcohol)     (methyl acetate)

## ACID HYDROLYSIS OF AN ESTER

$$CH_3-\overset{\overset{\displaystyle O}{\|}}{C}-O-CH_3 \ + \ H-OH \ \overset{H^+}{\underset{}{\rightleftharpoons}} \ CH_3-\overset{\overset{\displaystyle O}{\|}}{C}-OH \ + \ HO-CH_3$$

Methyl ethanoate           Ethanoic acid    Methanol
(methyl acetate)           (acetic acid)    (methyl alcohol)

## BASE HYDROLYSIS OF AN ESTER

$$CH_3-CH_2-\overset{\overset{\displaystyle O}{\|}}{C}-O-CH_3 \ + \ NaOH \ \overset{Heat}{\longrightarrow} \ CH_3-CH_2-\overset{\overset{\displaystyle O}{\|}}{C}-O^-Na^+ \ + \ HO-CH_3$$

Methyl propanoate    Sodium        Sodium propanoate    Methanol
(methyl propionate)   hydroxide     (sodium propionate)    (methyl
                                                                         alcohol)

## IONIZATION OF AMINES IN WATER

$$CH_3-\overset{\overset{\displaystyle H}{|}}{\underset{\underset{\displaystyle H}{|}}{N}} \ + \ HOH \ \rightleftharpoons \ CH_3-\overset{\overset{\displaystyle H}{|}}{\underset{\underset{\displaystyle H}{|}}{\overset{+}{N}}}-H \ + \ OH^-$$

Methylamine         Methylammonium hydroxide

## FORMATION OF AMINE SALTS

$$CH_3-\overset{\overset{\displaystyle H}{|}}{\underset{\underset{\displaystyle H}{|}}{N}} \ + \ HCl \ \longrightarrow \ CH_3-\overset{\overset{\displaystyle H}{|}}{\underset{\underset{\displaystyle H}{|}}{\overset{+}{N}}}-H \ Cl^-$$

Methylamine         Methylammonium chloride

**FORMATION OF AMIDES**

$$CH_3-CH_2-\overset{\overset{\displaystyle O}{\|}}{C}-OH \;+\; H-\overset{\overset{\displaystyle H}{|}}{N}-H \;\xrightarrow{\text{Heat}}\; CH_3-CH_2-\overset{\overset{\displaystyle O}{\|}}{C}-\overset{\overset{\displaystyle H}{|}}{N}-H \;+\; H_2O$$

Propanoic acid          Ammonia                      Propanamide
(propionic acid)                                       (propionamide)

**ACID HYDROLYSIS OF AMIDES**

$$CH_3-\overset{\overset{\displaystyle O}{\|}}{C}-NH_2 \;+\; HOH \;+\; HCl \;\longrightarrow\; CH_3-\overset{\overset{\displaystyle O}{\|}}{C}-OH \;+\; NH_4^+Cl^-$$

Ethanamide                          Ethanoic acid        Ammonium
(acetamide)                         (acetic acid)        chloride

**BASE HYDROLYSIS OF AMIDES**

$$CH_3-CH_2-\overset{\overset{\displaystyle O}{\|}}{C}-NH-CH_3 \;+\; NaOH \longrightarrow CH_3-CH_2-\overset{\overset{\displaystyle O}{\|}}{C}-O^-Na^+ \;+\; NH_2-CH_3$$

*N*-methylpropanamide                Sodium propanoate, a salt        Methylamine
(*N*-methylpropionamide)             (sodium propionate)

# KEY TERMS

**alkaloids** Physiologically active amines that are produced in plants.

**amides** Organic compounds containing the carbonyl group attached to an amino group or a substituted nitrogen atom.

$$-\overset{\overset{\displaystyle O}{\|}}{C}-NH_2 \qquad -\overset{\overset{\displaystyle O}{\|}}{C}-\overset{\overset{\displaystyle |}{}}{N}-$$

**amines** Organic compounds containing a nitrogen atom attached to one, two, or three hydrocarbon groups.

**amine salt** An ionic compound produced from an amine and an acid.

**carboxyl group** A functional group found in carboxylic acids composed of carbonyl and hydroxyl groups.

$$-\overset{\overset{\displaystyle O}{\|}}{C}-OH \qquad \text{Carboxyl group}$$

**carboxylate ion** The anion produced when a carboxylic acid donates a proton to water.

**carboxylic acids** A family of organic compounds containing the carboxyl group.

$$-\overset{\overset{\displaystyle O}{\|}}{C}-OH \qquad \text{Carboxylic acid}$$

**carboxylic acid salt** A carboxylate ion and the metal ion from the base that is the product of neutralization of a carboxylic acid.

**esterification** The formation of an ester from a carboxylic acid and an alcohol, with the elimination of a molecule of water in the presence of an acid catalyst.

**esters** A family of organic compounds in which an alkyl group replaces the hydrogen atom in a carboxylic acid.

$$-\overset{\overset{\displaystyle O}{\|}}{C}-O-\overset{\overset{\displaystyle |}{}}{\underset{|}{C}}- \qquad \text{Ester}$$

**hydrolysis** The splitting of a molecule by the addition of water. Esters hydrolyze to produce a carboxylic acid and an alcohol. Amides yield the corresponding carboxylic acid and amine or their salts.

**saponification** The hydrolysis of an ester with a strong base to produce a salt of the carboxylic acid and an alcohol.

# UNDERSTANDING THE CONCEPTS

**13.51** Propyl acetate is the ester that gives the odor and smell of pears.

**a.** What is the structure of propyl acetate?
**b.** Write an equation for the formation of propyl acetate.
**c.** Write an equation for the acid hydrolysis of propyl acetate.
**d.** Write an equation for the base hydrolysis of propyl acetate with NaOH.
**e.** How many mL of 0.208 M NaOH are needed to completely hydrolyze (saponify) 1.58 g of propyl acetate?

**13.52** Ethyl octanoate is a flavor component of mangoes.

**a.** What is the structure of ethyl octanoate?
**b.** Write an equation for the formation of ethyl octanoate.
**c.** Write an equation for the acid hydrolysis of ethyl octanoate.
**d.** Write an equation for the base hydrolysis of ethyl octanoate with NaOH.
**e.** How many mL of 0.315 M NaOH are needed to completely hydrolyze (saponify) 2.84 g of ethyl octanoate?

**13.53** Neo-Synephrine is the active ingredient in some nose sprays used to reduce swelling of nasal membranes. What functional groups are in the structure of Neo-Synephrine?

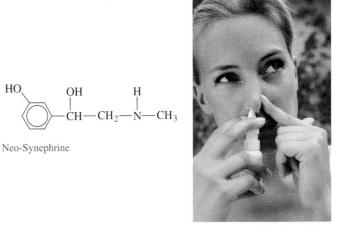

Neo-Synephrine

**13.54** Atovaquone (Malarone) is a medication used in the treatment of malaria. What functional groups are in the structure of Atovaquone?

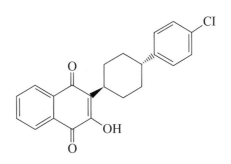

Atovaquone (malarone)

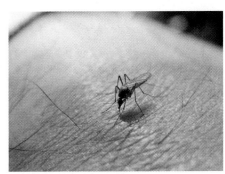

# ADDITIONAL QUESTIONS AND PROBLEMS

**13.55** Give the IUPAC and common names (if any) for each of the following compounds:

**a.** $CH_3-CH(CH_3)-CH_2-C(=O)-OH$

**b.** 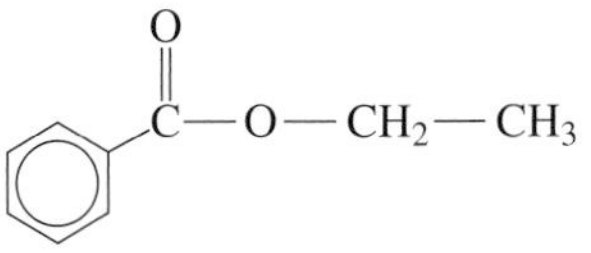 (benzene ring)$-C(=O)-O-CH_2-CH_3$

**c.** $CH_3-CH_2-O-C(=O)-CH_2-CH_3$

**d.** 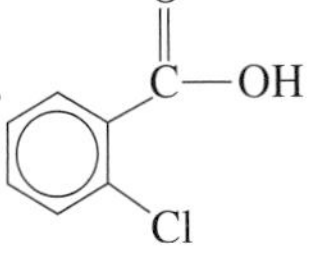 (benzene ring with Cl)$-C(=O)-OH$

**e.** $CH_3-CH_2-CH_2-CH_2-C(=O)-OH$

**13.56** Give the IUPAC and common names (if any) for each of the following compounds:

a. $CH_3$—$\overset{\overset{\displaystyle CH_3}{|}}{CH}$—$CH_2$—$CH_2$—$\overset{\overset{\displaystyle O}{\|}}{C}$—OH

b. $\overset{\overset{\displaystyle O}{\|}}{C}$—OH on benzene ring with Cl and Cl

c. benzene—$\overset{\overset{\displaystyle O}{\|}}{C}$—O—$CH_3$

d. $CH_3$—$CH_2$—$CH_2$—$\overset{\overset{\displaystyle O}{\|}}{C}$—O—$CH_3$

e. $CH_3$—$CH_2$—$\overset{\overset{\displaystyle O}{\|}}{C}$—O—$CH_2$—$CH_2$—$CH_3$

**13.57** Draw the condensed structural formula of each of the following:
a. methyl acetate
b. ethyl butanoate
c. 3-methylpentanoic acid
d. ethyl benzoate

**13.58** Draw the condensed structural formula of each of the following:
a. ethyl butyrate
b. 2-methylpentanoic acid
c. 3,5-dimethylhexanoic acid
d. propyl acetate

**13.59** For each of the following pairs, identify the compound that has the higher boiling point. Explain.

a. $CH_3$—$CH_2$—$CH_2$—OH  or  $CH_3$—$\overset{\overset{\displaystyle O}{\|}}{C}$—OH

b. $CH_3$—$CH_2$—$CH_2$—$CH_3$  or  $CH_3$—$CH_2$—$\overset{\overset{\displaystyle O}{\|}}{C}$—OH

c. $CH_3$—$\overset{\overset{\displaystyle O}{\|}}{C}$—OH  or  $CH_3$—$CH_2$—$\overset{\overset{\displaystyle O}{\|}}{C}$—OH

**13.60** For each of the following pairs, identify the compound that has the higher boiling point. Explain.

a. $CH_3$—$CH_2$—$CH_2$—OH  or  $CH_3$—$\overset{\overset{\displaystyle O}{\|}}{C}$—O—$CH_3$

b. $CH_3$—O—$\overset{\overset{\displaystyle O}{\|}}{C}$—$CH_3$  or  $CH_3$—$CH_2$—$\overset{\overset{\displaystyle O}{\|}}{C}$—OH

c. $CH_3$—$\overset{\overset{\displaystyle O}{\|}}{C}$—O—$CH_3$  or  $CH_3$—$CH_2$—$CH_2$—$CH_3$

**13.61** Write the products of the following reactions:

a. $CH_3$—$CH_2$—$\overset{\overset{\displaystyle O}{\|}}{C}$—OH  +  KOH  $\longrightarrow$

b. $CH_3$—$CH_2$—$\overset{\overset{\displaystyle O}{\|}}{C}$—OH  +  $CH_3OH$  $\overset{H^+}{\rightleftharpoons}$

c. benzene—$\overset{\overset{\displaystyle O}{\|}}{C}$—OH  +  $CH_3$—$CH_2$—OH  $\overset{H^+}{\rightleftharpoons}$

**13.62** Write the products of the following reactions:

a. $CH_3$—$\overset{\overset{\displaystyle O}{\|}}{C}$—OH  +  NaOH  $\longrightarrow$

b. $CH_3$—$\overset{\overset{\displaystyle CH_3}{|}}{CH}$—$\overset{\overset{\displaystyle O}{\|}}{C}$—OH  +  KOH  $\longrightarrow$

c. $CH_3$—$\overset{\overset{\displaystyle CH_3}{|}}{CH}$—$\overset{\overset{\displaystyle O}{\|}}{C}$—OH  +  $CH_3OH$  $\overset{H^+}{\rightleftharpoons}$

**13.63** Write the products of the following reactions:

a. $CH_3$—$CH_2$—$\overset{\overset{\displaystyle O}{\|}}{C}$—O—$\overset{\overset{\displaystyle CH_3}{|}}{CH}$—$CH_3$  +  $H_2O$  $\overset{H^+}{\rightleftharpoons}$

b. $CH_3$—$\overset{\overset{\displaystyle CH_3}{|}}{CH}$—$\overset{\overset{\displaystyle O}{\|}}{C}$—O—$CH_2$—$CH_2$—$CH_3$  +  $H_2O$  $\overset{H^+}{\rightleftharpoons}$

**13.64** Write the products of the following reactions:

a. $CH_3$—$CH_2$—$\overset{\overset{\displaystyle O}{\|}}{C}$—O—$\overset{\overset{\displaystyle CH_3}{|}}{CH}$—$CH_3$  +  NaOH $\longrightarrow$

b. $CH_3$—$\overset{\overset{\displaystyle CH_3}{|}}{CH}$—$\overset{\overset{\displaystyle O}{\|}}{C}$—O—$CH_2$—$CH_2$—$CH_3$  +  NaOH $\longrightarrow$

**13.65** Draw the structure of each of the following compounds:
a. dimethylamine
b. cyclohexylamine
c. dimethylammonium chloride
d. triethylamine
e. N-methylaniline

**13.66** In each pair, indicate the compound that has the higher boiling point. Explain.
a. 1-butanol or butylamine
b. trimethylamine or propylamine
c. butylamine or diethylamine
d. butane or propylamine

**13.67** In each pair, indicate the compound that is more soluble in water. Explain.
a. ethylamine or dipentylamine
b. trimethylamine or tripropylamine
c. butylamine or pentane
d. butyramide or hexane

**13.68** Give the IUPAC name for each of the following amides:

a. H—$\overset{\overset{\displaystyle O}{\|}}{C}$—$NH_2$

b. $CH_3$—$CH_2$—$\overset{\overset{\displaystyle O}{\|}}{C}$—$NH_2$

**c.** $CH_3-\overset{\overset{\displaystyle O}{\|}}{C}-NH_2$    **d.** $CH_3-CH_2-CH_2-\overset{\overset{\displaystyle O}{\|}}{C}-NH_2$

**13.69** Write the structure of each product of the following reactions:
**a.** $CH_3-CH_2-NH_2 + H_2O \rightleftharpoons$
**b.** $CH_3-CH_2-NH_2 + HCl \longrightarrow$
**c.** $CH_3-CH_2-NH-CH_3 + H_2O \rightleftharpoons$
**d.** $CH_3-CH_2-NH-CH_3 + HCl \longrightarrow$
**e.** $CH_3-CH_2-CH_2-NH_3^+Cl^- + NaOH \longrightarrow$

$\qquad\qquad\qquad\overset{\displaystyle CH_3}{\underset{\displaystyle |}{}}$
**f.** $CH_3-CH_2-NH_2^+Cl^- + NaOH \longrightarrow$

**13.70** Toradol is used in dentistry to relieve pain. Name the functional groups in this molecule.

**13.71** Voltaren is indicated for acute and chronic treatment of the symptoms of rheumatoid arthritis. Name the functional groups in this molecule.

**13.72** Using a reference book such as *The Merck Index* or *Physicians' Desk Reference*, look up the condensed structural formula of the following medicinal drugs, and list the functional groups in the compounds. You may need to refer to the cross-index of names in the back of the reference book.
**a.** Keflex$^{TM}$, an antibiotic
**b.** Inderal$^{TM}$, a beta-blocker used to treat heart irregularities
**c.** ibuprofen, an anti-inflammatory agent
**d.** Aldomet$^{TM}$ (methyldopa)
**e.** Percodan$^{TM}$, a narcotic pain reliever
**f.** triamterene, a diuretic

## CHALLENGE QUESTIONS

**13.73** For the following:
**a.** Give the common name of each.
**b.** Identify isomers.
**c.** Explain the following order of increasing boiling points:

$CH_3-O-CH_3$      $CH_3-CH_2-NH_2$      $CH_3-CH_2-OH$      $CH_3-\overset{\overset{\displaystyle O}{\|}}{C}-OH$

$\quad-25\,°C\qquad\qquad\quad 17\,°C\qquad\qquad\quad 79\,°C\qquad\qquad 118\,°C$

**13.74** Why is the boiling point of trimethylamine (3 °C) lower than propylamine (48 °C) when the two compounds are isomers and have the same molar mass?

**13.75** Novocain, a local anesthetic, is the amine salt of procaine.

Procaine

**a.** What is the condensed structural formula of the amine salt (procaine hydrochloride) formed when procaine reacts with HCl? (*Hint*: The tertiary amine reacts with HCl.)
**b.** Why is procaine hydrochloride used rather than procaine?

**13.76** Lidocaine (xylocaine) is used as a local anesthetic and cardiac depressant.

Lidocaine (xylocaine)

**a.** What is the formula of the amine salt formed when lidocaine reacts with HCl?
**b.** Why is the amine salt of lidocaine used rather than the amine?

## ANSWERS

### Answers to Study Checks

**13.1** $CH_3-CH_2-CH_2-CH_2-\overset{\overset{\displaystyle O}{\|}}{C}-OH$

**13.2** Methanoic acid has a higher boiling point than ethanol because two molecules of methanoic acid can hydrogen bond to form a dimer, which gives twice the mass and requires a higher temperature to reach the boiling point.

**13.3** $H-\overset{\overset{\displaystyle O}{\|}}{C}-OH + H_2O \rightleftharpoons H-\overset{\overset{\displaystyle O}{\|}}{C}-O^- + H_3O^+$

**13.4** butanoic acid (butyric acid)

**13.5** propanoic acid (propionic acid) and 1-pentanol

**13.6** $CH_3-\overset{\overset{\displaystyle O}{\|}}{C}-O-CH_2-CH_2-CH_2-CH_2-CH_3$

**13.7** propanoic (propionic) acid and ethanol

**13.8**

$$\underset{\text{(benzene ring)}}{\bigcirc}\!\!-\!\!\overset{\overset{\displaystyle O}{\|}}{C}\!\!-\!\!O^-K^+ \quad + \quad CH_3\!\!-\!\!OH$$

**13.9** tertiary (3°)

**13.10**  $CH_3\!-\!CH_2\!-\!NH\!-\!CH_2\!-\!CH_2\!-\!CH_3$

**13.11** Hydrogen bonding makes amines with six or fewer carbon atoms soluble in water.

**13.12**  $CH_3\!-\!\overset{\overset{\displaystyle CH_3}{|}}{\underset{\underset{\displaystyle CH_3}{|}}{N^+}}\!-\!H\ Cl^-$

**13.13**  $H\!-\!\overset{\overset{\displaystyle O}{\|}}{C}\!-\!OH$   and   $H\!-\!\overset{\overset{\displaystyle CH_3}{|}}{N}\!-\!CH_3$

**13.14**

$$\underset{\text{(benzene ring)}}{\bigcirc}\!\!-\!\!\overset{\overset{\displaystyle O}{\|}}{C}\!\!\underset{\diagdown}{}\!NH_2$$

**13.15**  $CH_3\!-\!CH_2\!-\!CH_2\!-\!\overset{\overset{\displaystyle O}{\|}}{C}\!-\!OH$   and   $CH_3\!-\!NH_3{}^+Br^-$

## Answers to Selected Problems

**13.1** methanoic acid (formic acid)

**13.3** Each compound contains three carbon atoms. They differ because propanal, an aldehyde, contains a carbonyl group bonded to a hydrogen. In propanoic acid, the carbonyl group connects to a hydroxyl group, forming a carboxyl group.

**13.5**  **a.** ethanoic acid (acetic acid)
**b.** propanoic acid (propionic acid)
**c.** 4-hydroxybenzoic acid
**d.** 4-bromopentanoic acid

**13.7**  **a.**  $CH_3\!-\!CH_2\!-\!\overset{\overset{\displaystyle O}{\|}}{C}\!-\!OH$

**b.**  $\underset{\text{(benzene ring)}}{\bigcirc}\!\!-\!\!\overset{\overset{\displaystyle O}{\|}}{C}\!-\!OH$

**c.**  $Cl\!-\!CH_2\!-\!\overset{\overset{\displaystyle O}{\|}}{C}\!-\!OH$

**d.**  $HO\!-\!CH_2\!-\!CH_2\!-\!\overset{\overset{\displaystyle O}{\|}}{C}\!-\!OH$

**e.**  $CH_3\!-\!CH_2\!-\!CH_2\!-\!\overset{\overset{\displaystyle O}{\|}}{C}\!-\!OH$

**f.**  $CH_3\!-\!CH_2\!-\!CH_2\!-\!CH_2\!-\!CH_2\!-\!CH_2\!-\!\overset{\overset{\displaystyle O}{\|}}{C}\!-\!OH$

**13.9**  **a.** Butanoic acid has a higher molar mass and has a higher boiling point.
**b.** Propanoic acid can form more hydrogen bonds (dimers) and has a higher boiling point.
**c.** Butanoic acid can form hydrogen bonds (dimers) and has a higher boiling point.

**13.11**  **a.** Propanoic acid is the most soluble of the group because it has the fewest number of carbon atoms in its alkyl chain. Solubility of carboxylic acids decreases as the number of carbon atoms in the alkyl chain increases.
**b.** Propanoic acid is more soluble than 1-butanol because a carboxyl group forms more hydrogen bonds with water than with a hydroxyl group. An alkane is not soluble in water.

**13.13**  **a.**  $H\!-\!\overset{\overset{\displaystyle O}{\|}}{C}\!-\!OH + H_2O \rightleftharpoons H\!-\!\overset{\overset{\displaystyle O}{\|}}{C}\!-\!O^- + H_3O^+$

**b.**  $CH_3\!-\!CH_2\!-\!\overset{\overset{\displaystyle O}{\|}}{C}\!-\!OH + H_2O \rightleftharpoons$

$CH_3\!-\!CH_2\!-\!\overset{\overset{\displaystyle O}{\|}}{C}\!-\!O^- + H_3O^+$

**c.**  $CH_3\!-\!\overset{\overset{\displaystyle O}{\|}}{C}\!-\!OH + H_2O \rightleftharpoons CH_3\!-\!\overset{\overset{\displaystyle O}{\|}}{C}\!-\!O^- + H_3O^+$

**13.15**  **a.**  $H\!-\!\overset{\overset{\displaystyle O}{\|}}{C}\!-\!OH + NaOH \longrightarrow H\!-\!\overset{\overset{\displaystyle O}{\|}}{C}\!-\!O^-Na^+ + H_2O$

**b.**  $CH_3\!-\!CH_2\!-\!\overset{\overset{\displaystyle O}{\|}}{C}\!-\!OH + NaOH \longrightarrow$

$CH_3\!-\!CH_2\!-\!\overset{\overset{\displaystyle O}{\|}}{C}\!-\!O^-Na^+ + H_2O$

**c.**  $\underset{\text{(benzene ring)}}{\bigcirc}\!\!-\!\!\overset{\overset{\displaystyle O}{\|}}{C}\!-\!OH + NaOH \longrightarrow \underset{\text{(benzene ring)}}{\bigcirc}\!\!-\!\!\overset{\overset{\displaystyle O}{\|}}{C}\!-\!O^-Na^+ + H_2O$

**13.17**  **a.** aldehyde
**b.** ester
**c.** ketone
**d.** carboxylic acid

**13.19**  **a.**  $CH_3\!-\!\overset{\overset{\displaystyle O}{\|}}{C}\!-\!O\!-\!CH_3$

**b.**  $CH_3\!-\!CH_2\!-\!CH_2\!-\!\overset{\overset{\displaystyle O}{\|}}{C}\!-\!O\!-\!CH_3$

**c.**  $\underset{\text{(benzene ring)}}{\bigcirc}\!\!-\!\!\overset{\overset{\displaystyle O}{\|}}{C}\!-\!O\!-\!CH_3$

**13.21**  **a.**  $CH_3\!-\!CH_2\!-\!\overset{\overset{\displaystyle O}{\|}}{C}\!-\!O\!-\!CH_2\!-\!CH_2\!-\!CH_3$

**b.**  $CH_3\!-\!CH_2\!-\!CH_2\!-\!CH_2\!-\!\overset{\overset{\displaystyle O}{\|}}{C}\!-\!O\!-\!\overset{\overset{\displaystyle CH_3}{|}}{CH}\!-\!CH_3$

**13.23** **a.** methyl formate (methyl methanoate)
**b.** methyl acetate (methyl ethanoate)
**c.** methyl butyrate (methyl butanoate)
**d.** ethyl butyrate (ethyl butanoate)

**13.25** **a.** $CH_3-\overset{\overset{\displaystyle O}{\|}}{C}-O-CH_3$

**b.** $H-\overset{\overset{\displaystyle O}{\|}}{C}-O-CH_2-CH_2-CH_2-CH_3$

**c.** $CH_3-CH_2-CH_2-CH_2-\overset{\overset{\displaystyle O}{\|}}{C}-O-CH_2-CH_3$

**d.** $CH_3-CH_2-\overset{\overset{\displaystyle O}{\|}}{C}-O-CH_2-CH_2-CH_3$

**13.27** **a.** pentyl ethanoate (pentyl acetate)
**b.** octyl ethanoate (octyl acetate)
**c.** pentyl butanoate (pentyl butyrate)

**13.29** The products of the acid hydrolysis of an ester are an alcohol and a carboxylic acid.

**13.31** **a.** $CH_3-CH_2-\overset{\overset{\displaystyle O}{\|}}{C}-O^-Na^+$ and $CH_3-OH$

**b.** $CH_3-\overset{\overset{\displaystyle O}{\|}}{C}-OH$ and $CH_3-CH_2-CH_2-OH$

**c.** $CH_3-CH_2-CH_2-\overset{\overset{\displaystyle O}{\|}}{C}-OH$ and $CH_3-CH_2-OH$

**d.** benzene ring–$\overset{\overset{\displaystyle O}{\|}}{C}-O^-Na^+$ and $CH_3-CH_2-OH$

**13.33** **a.** primary **b.** secondary
**c.** tertiary **d.** tertiary

**13.35** **a.** ethylamine
**b.** methylpropylamine
**c.** diethylmethylamine
**d.** isopropylamine

**13.37** **a.** $CH_3-CH_2-NH_2$
**b.** benzene ring–$NH-CH_3$
**c.** $CH_3-CH_2-CH_2-CH_2-\overset{\overset{\displaystyle H}{|}}{N}-CH_2-CH_2-CH_3$

**13.39** Amines have higher boiling points than hydrocarbons, but lower boiling points than alcohols of similar mass.
**a.** $CH_3-CH_2-OH$
**b.** $CH_3-CH_2-CH_2-NH_2$
**c.** $CH_3-CH_2-CH_2-NH_2$

**13.41** **a.** Yes; amines with fewer than six carbon atoms are soluble in water.
**b.** Yes; amines with fewer than six carbon atoms are soluble in water.
**c.** No; an amine with nine carbon atoms is not soluble in water.
**d.** Yes; amines with fewer than six carbon atoms are soluble in water.

**13.43** **a.** $CH_3-NH_2 + H_2O \rightleftharpoons CH_3-NH_3^+ + OH^-$
**b.** $CH_3-NH-CH_3 + H_2O \rightleftharpoons CH_3-\overset{+}{N}H_2-CH_3 + OH^-$
**c.** benzene ring–$NH_2$ + $H_2O \rightleftharpoons$ benzene ring–$NH_3^+$ + $OH^-$

**13.45** **a.** ethanamide (acetamide)
**b.** butanamide (butyramide)
**c.** methanamide (formamide)

**13.47** **a.** $CH_3-CH_2-\overset{\overset{\displaystyle O}{\|}}{C}-NH_2$

**b.** $CH_3-CH_2-CH_2-\overset{\overset{\displaystyle CH_3}{|}}{C}H-\overset{\overset{\displaystyle O}{\|}}{C}-NH_2$

**c.** $H-\overset{\overset{\displaystyle O}{\|}}{C}-NH_2$

**13.49** **a.** $CH_3-\overset{\overset{\displaystyle O}{\|}}{C}-OH + NH_4^+Cl^-$
**b.** $CH_3-CH_2-\overset{\overset{\displaystyle O}{\|}}{C}-OH + NH_4^+Cl^-$
**c.** $CH_3-CH_2-CH_2-\overset{\overset{\displaystyle O}{\|}}{C}-OH + CH_3-NH_3^+Cl^-$
**d.** benzene ring–$\overset{\overset{\displaystyle O}{\|}}{C}-OH + NH_4^+Cl^-$
**e.** $CH_3-CH_2-CH_2-CH_2-\overset{\overset{\displaystyle O}{\|}}{C}-OH$
    $+ CH_3-CH_2-NH_3^+Cl^-$

**13.51** **a.** $CH_3-CH_2-CH_2-O-\overset{\overset{\displaystyle O}{\|}}{C}-CH_3$
**b.** $CH_3-CH_2-CH_2-OH + HO-\overset{\overset{\displaystyle O}{\|}}{C}-CH_3 \overset{H^+, heat}{\rightleftharpoons}$
    $CH_3-CH_2-CH_2-O-\overset{\overset{\displaystyle O}{\|}}{C}-CH_3 + H_2O$
**c.** $CH_3-CH_2-CH_2-O-\overset{\overset{\displaystyle O}{\|}}{C}-CH_3 + H_2O \overset{H^+}{\rightleftharpoons}$
    $CH_3-CH_2-CH_2-OH + HO-\overset{\overset{\displaystyle O}{\|}}{C}-CH_3$
**d.** $CH_3-CH_2-CH_2-O-\overset{\overset{\displaystyle O}{\|}}{C}-CH_3 + NaOH \overset{Heat}{\rightarrow}$
    $CH_3-CH_2-CH_2-OH + Na^+\ ^-O-\overset{\overset{\displaystyle O}{\|}}{C}-CH_3$
**e.** 74.5 mL of 0.208 M NaOH

**13.53** phenol, (aromatic alcohol), alcohol, and amine

**13.55** **a.** 3-methylbutanoic acid
**b.** ethyl benzoate
**c.** ethyl propanoate; ethyl propionate

**d.** 2-chlorobenzoic acid
**e.** pentanoic acid

**13.57  a.** $CH_3-O-\overset{\displaystyle O}{\overset{\|}{C}}-CH_3$

**b.** $CH_3-CH_2-O-\overset{\displaystyle O}{\overset{\|}{C}}-CH_2-CH_2-CH_3$

**c.** $CH_3-CH_2-\overset{\displaystyle CH_3}{\overset{|}{CH}}-CH_2-\overset{\displaystyle O}{\overset{\|}{C}}-OH$

**d.** $\overset{\displaystyle O}{\overset{\|}{C}}-O-CH_2-CH_3$ (attached to benzene ring)

**13.59  a.** Ethanoic acid has a higher boiling point than 1-propanol because two molecules of ethanoic acid can hydrogen bond to form a dimer, which gives twice the mass and requires a higher temperature to reach the boiling point.
**b.** Propanoic acid forms hydrogen bonds, but butane does not.
**c.** Propanoic acid has a higher mass than ethanoic acid and requires a higher temperature to reach the boiling point.

**13.61  a.** $CH_3-CH_2-\overset{\displaystyle O}{\overset{\|}{C}}-O^-K^+ \ + \ H_2O$

**b.** $CH_3-CH_2-\overset{\displaystyle O}{\overset{\|}{C}}-O-CH_3 \ + \ H_2O$

**c.** $\overset{\displaystyle O}{\overset{\|}{C}}-O-CH_2-CH_3 \ + \ H_2O$ (attached to benzene ring)

**13.63  a.** $CH_3-CH_2-\overset{\displaystyle O}{\overset{\|}{C}}-OH$ and $HO-\overset{\displaystyle CH_3}{\overset{|}{CH}}-CH_3$

**b.** $CH_3-\overset{\displaystyle CH_3}{\overset{|}{CH}}-\overset{\displaystyle O}{\overset{\|}{C}}-OH$ and $HO-CH_2-CH_2-CH_3$

**13.65  a.** $CH_3-\overset{\displaystyle CH_3}{\overset{|}{NH}}$

**b.** (cyclohexane ring with $NH_2$)

**c.** $CH_3-\overset{\displaystyle CH_3}{\overset{|}{NH_2^+}} \ Cl^-$

**d.** $CH_3-CH_2=\overset{\displaystyle CH_2-CH_3}{\overset{|}{N}}=CH_2-CH_3$

**e.** (benzene ring with $HN-CH_3$)

**13.67  a.** Ethylamine is a small amine that is soluble because it can hydrogen bond with water. Dipentylamine has two large nonpolar alkyl groups that decrease the solubility in water.
**b.** Trimethylamine is a small tertiary amine that is soluble because it hydrogen bonds with water.
**c.** Butylamine is soluble because it hydrogen bonds with water. Pentane cannot form hydrogen bonds.
**d.** Butylamide is soluble because it hydrogen bonds with water. Hexane cannot form hydrogen bonds.

**13.69  a.** $CH_3-CH_2-NH_3^+ \ + \ OH^-$
**b.** $CH_3-CH_2-NH_3^+Cl^-$
**c.** $CH_3-CH_2-\overset{+}{N}H_2-CH_3 \ + \ OH^-$
**d.** $CH_3-CH_2-\overset{+}{N}H_2-CH_3 \quad Cl^-$
**e.** $CH_3-CH_2-CH_2-NH_2 \ + \ NaCl \ + \ H_2O$
**f.** $CH_3-CH_2-\overset{\displaystyle CH_3}{\overset{|}{NH}} + NaCl + H_2O$

**13.71** carboxylic acid salt, aromatic, amine

**13.73  a.** dimethyl ether, ethylamine, ethyl alcohol, acetic acid
**b.** Dimethyl ether and ethyl alcohol are isomers.
**c.** Acetic acid has the highest boiling point because it forms a dimer and many hydrogen bonds. Ethanol also forms hydrogen bonds. Ethylamine forms weaker hydrogen bonds than ethanol. Dimethyl ether does not form hydrogen bonds; it has the lowest boiling point.

**13.75  a.** $H_2N-\bigcirc-\overset{\displaystyle O}{\overset{\|}{C}}-O-CH_2-CH_2-\overset{\displaystyle CH_2-CH_3}{\underset{\displaystyle CH_2-CH_3}{\overset{|}{\underset{|}{N^+}}}}-H \quad Cl^-$

**b.** The amine salt (Novocain) is more soluble in body fluids than procaine.

# 14 Carbohydrates

the **Chemistry place**

Visit **www.chemplace.com** for extra quizzes, interactive tutorials, career resources, PowerPoint slides for chapter review, math help, and case studies.

*"We use a refractometer to measure sugar content in a small sample of juices from the grapes in different areas of the vineyard," says Leslie Bucher, laboratory director at Bouchaine Winery. "We also measure the alcohol content during fermentation and run tests for sulfur, pH, and total acid."*

*As grapes ripen, there is an increase in the sugars, which are the monosaccharides fructose and glucose. The sugar content is affected by soil conditions and the amount of sun and water. When the grapes are ripe and sugar content is at a desirable level, they are harvested. During fermentation, enzymes from yeast convert about half the sugar to ethanol, and half to carbon dioxide. Grapes harvested with 22.5% sugar will ferment to give a wine with 12.5–13.5% alcohol content.*

Of all the organic compounds in nature, carbohydrates are the most abundant. In plants, energy from the sun converts carbon dioxide and water into the carbohydrate glucose. Many of the glucose molecules are made into long-chain polymers of starch that store energy. About 65% of the foods in our diet consist of carbohydrates. Each day you enjoy carbohydrates in foods such as bread, pasta, potatoes, and rice. During digestion and cellular metabolism, the starches are broken down into glucose, which is oxidized further in our cells to provide our bodies with energy and to provide the cells with carbon atoms for building molecules of protein, lipids, and nucleic acids. In plants, a polymer of glucose called cellulose builds the structural framework. Cellulose has other important uses, too. The wood in our furniture, the pages in this book, and the cotton in our clothing are made of cellulose.

## LEARNING GOAL

Classify a monosaccharide as an aldose or ketose, and indicate the number of carbon atoms.

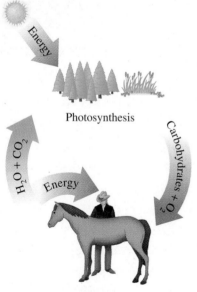

Photosynthesis

Respiration

**FIGURE 14.1** During photosynthesis, energy from the sun combines $CO_2$ and $H_2O$ to form glucose ($C_6H_{12}O_6$) and $O_2$. During respiration in the body, carbohydrates are oxidized to $CO_2$ and $H_2O$, while energy is produced.

**Q** What are the reactants and products of respiration?

# 14.1 CARBOHYDRATES

**Carbohydrates** such as table sugar, lactose in milk, and cellulose are all made of carbon, hydrogen, and oxygen. Simple sugars, which have formulas of $C_n(H_2O)_n$, were once thought to be hydrates of carbon, thus the name *carbohydrate*. In a series of reactions called photosynthesis, energy from the sun is used to combine the carbon atoms from carbon dioxide ($CO_2$) and the hydrogen and oxygen atoms of water into the carbohydrate glucose.

$$6CO_2 + 6H_2O + energy \underset{\text{Respiration}}{\overset{\text{Photosynthesis}}{\rightleftharpoons}} C_6H_{12}O_6 + 6O_2$$

Glucose

In the body, glucose is oxidized in a series of metabolic reactions known as respiration, which releases chemical energy to do work in the cells. Carbon dioxide and water are produced and returned to the atmosphere. The combination of photosynthesis and respiration is called the carbon cycle, in which energy from the sun is stored in plants by photosynthesis and made available to us when the carbohydrates in our diets are metabolized. (See Figure 14.1.)

## Types of Carbohydrates

The simplest carbohydrates are the **monosaccharides**. A monosaccharide cannot be split or hydrolyzed into smaller carbohydrates. One of the most common carbohydrates, glucose ($C_6H_{12}O_6$), is a monosaccharide. **Disaccharides** consist of two monosaccharide units joined together. Likewise, a disaccharide can be split into two monosaccharide units. For example, ordinary table sugar, sucrose ($C_{12}H_{22}O_{11}$), is a disaccharide that hydrolyzes in the presence of an acid or an enzyme to give one molecule of glucose and one molecule of another monosaccharide, fructose.

$$C_{12}H_{22}O_{11} + H_2O \xrightarrow{\text{H}^+ \text{ or enzyme}} C_6H_{12}O_6 + C_6H_{12}O_6$$

Sucrose                          Glucose      Fructose

**Polysaccharides** are carbohydrates that are naturally occurring polymers containing many monosaccharide units. In the presence of an acid or an enzyme, a polysaccharide can be completely hydrolyzed to yield many molecules of monosaccharides.

the
**Chemistry**
place

WEB TUTORIAL
Carbohydrates

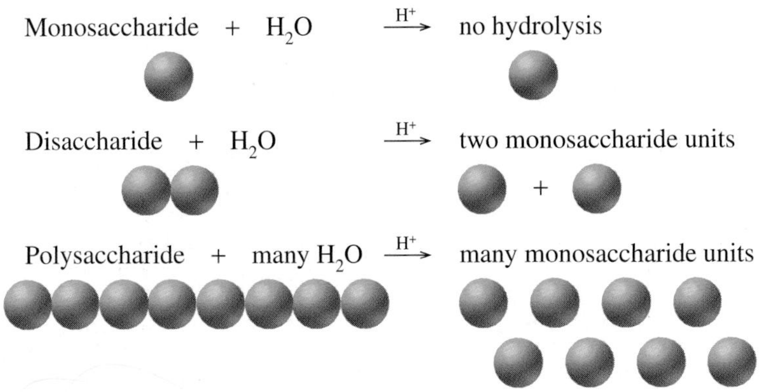

---

## SAMPLE PROBLEM 14.1

### ■ Carbohydrates

Classify the following carbohydrates as mono-, di-, or polysaccharides:

**a.** When lactose, milk sugar, is hydrolyzed, two monosaccharide units are produced.
**b.** Cellulose, a carbohydrate in cotton, yields thousands of monosaccharide units when completely hydrolyzed.

#### SOLUTION

**a.** A disaccharide contains two monosaccharide units.
**b.** A polysaccharide contains many monosaccharide units.

#### STUDY CHECK

Fructose found in fruits does not undergo hydrolysis. What type of carbohydrate is it?

---

## Monosaccharides

Monosaccharides are simple sugars that have a chain of three to eight carbon atoms, one in a carbonyl group and the rest attached to hydroxyl groups. There are two types of monosaccharide structures. In an **aldose**, the carbonyl group is on the first carbon (—CHO); a **ketose** contains the carbonyl group on the second carbon atom as a ketone (C=O).

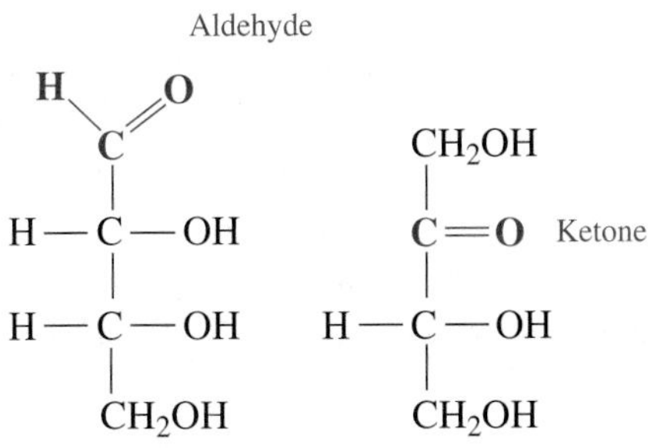

Erythrose, a
polyhydroxy
aldehyde

Erythrulose, a
polyhydroxy
ketone

A monosaccharide with three carbon atoms is a *triose*, one with four carbon atoms is a *tetrose*, a *pentose* has five carbons, and a *hexose* contains six carbons. We can use both classification systems to indicate the type of carbonyl group and the number of carbon atoms. An aldopentose is a five-carbon monosaccharide that is an aldehyde; a ketohexose is a six-carbon monosaccharide that is a ketone. Some examples:

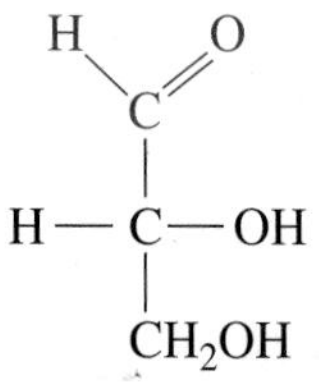

Glyceraldehyde
(aldotriose)

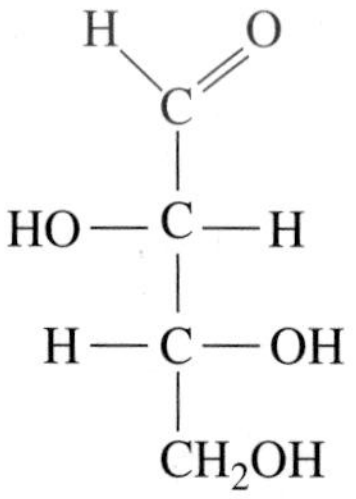

Threose
(aldotetrose)

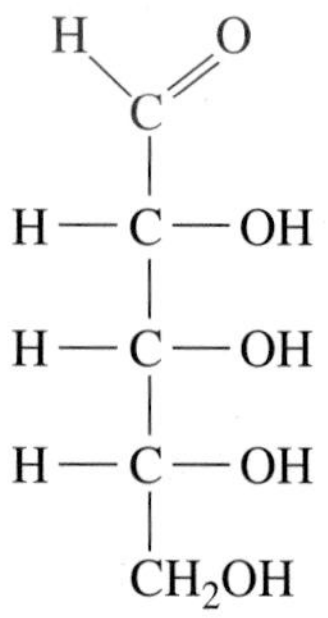

Ribose
(aldopentose)

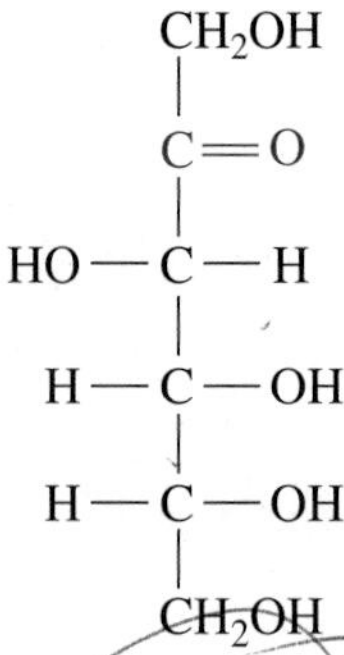

Fructose
(ketohexose)

## SAMPLE PROBLEM 14.2

### ■ Monosaccharides

Classify each of the following monosaccharides to indicate their carbonyl group and number of carbon atoms:

a. Ribulose

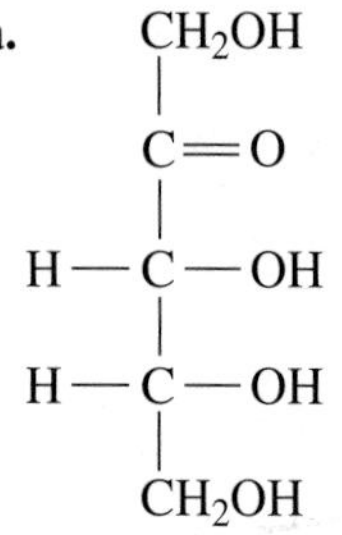

b. Glucose

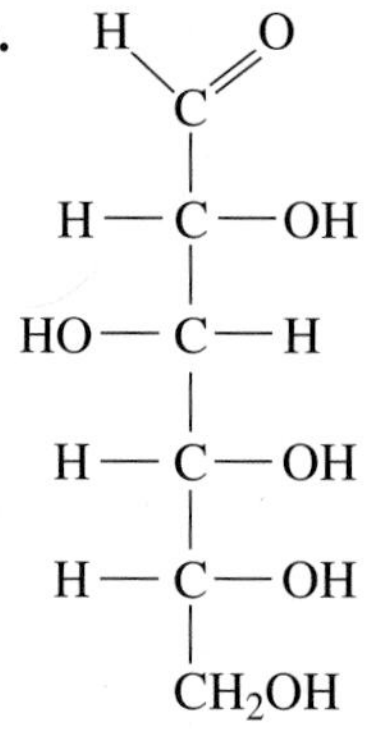

### SOLUTION

**a.** Ribulose has five carbon atoms (pentose) with a ketone group (keto), which makes it a ketopentose.

**b.** Glucose has six carbon atoms (hexose) with an aldehyde (aldo), which makes it an aldohexose.

### STUDY CHECK

Draw the condensed structural formula for dihydroxyacetone, a ketotriose.

## QUESTIONS AND PROBLEMS

### Carbohydrates

**14.1** What reactants are needed for photosynthesis and respiration?

**14.2** What is the relationship between photosynthesis and respiration?

**14.3** What is a monosaccharide? A disaccharide?

**14.4** What is a polysaccharide?

**14.5** What functional groups are found in all monosaccharides?

**14.6** What is the difference between an aldose and a ketose?

**14.7** What are the functional groups and number of carbons in a ketopentose?

**14.8** What are the functional groups and number of carbons in an aldohexose?

**14.9** Classify each of the following monosaccharides as an aldose or a ketose.

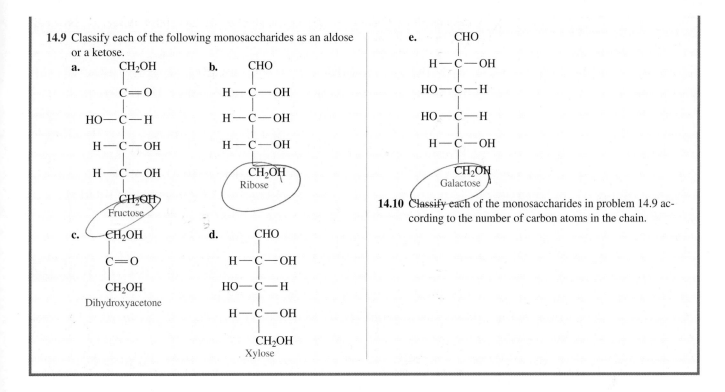

a.

CH₂OH
|
C=O
|
HO—C—H
|
H—C—OH
|
H—C—OH
|
CH₂OH
Fructose

b.

CHO
|
H—C—OH
|
H—C—OH
|
H—C—OH
|
CH₂OH
Ribose

c.

CH₂OH
|
C=O
|
CH₂OH
Dihydroxyacetone

d.

CHO
|
H—C—OH
|
HO—C—H
|
H—C—OH
|
CH₂OH
Xylose

e.

CHO
|
H—C—OH
|
HO—C—H
|
HO—C—H
|
H—C—OH
|
CH₂OH
Galactose

**14.10** Classify each of the monosaccharides in problem 14.9 according to the number of carbon atoms in the chain.

# 14.2 FISCHER PROJECTIONS OF MONOSACCHARIDES

In Chapter 12, we learned that chiral compounds exist as mirror images that cannot be superimposed. The monosaccharides, which contain chiral carbons, exist as mirror images.

## LEARNING GOAL

Draw the D or L isomers of glucose, galactose, and fructose.

**WEB TUTORIAL**
Forms of Carbohydrates

## Fischer Projections

Let's take a look again at the Fischer projection for the simplest aldose, glyceraldehyde. By convention, the carbon chain is written vertically with the aldehyde group (most oxidized carbon) at the top. The letter L is assigned to the stereoisomer if the —OH group is on the left of the chiral carbon. In D-glyceraldehyde, the —OH is on the right. The carbon atom in the CH₂OH group at the bottom of the Fischer projection is not chiral because it does not have four different groups bonded to it.

CHO
|
HO——H
|
CH₂OH
L-Glyceraldehyde

CHO
|
H——OH
|
CH₂OH
D-Glyceraldehyde

Most of the carbohydrates we will study have carbon chains with five or six carbon atoms. Because there are several chiral carbons, the chiral carbon farthest from the carbonyl group is used to determine the D or L isomer. The following are the Fischer projections for the D and L isomers of ribose, a five-carbon monosaccharide, and the D and L isomers of glucose, a six-carbon monosaccharide. In each of the mirror images, it is important to understand that the —OH groups on all the chiral carbon atoms are reversed from one side to the other. For example, in L-ribose, the —OH groups are all written on the left sides of the horizontal lines. In the mirror image, D-ribose, the —OH groups are all written on the right sides of the horizontal lines.

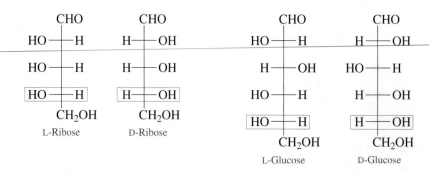

L-Ribose    D-Ribose

L-Glucose    D-Glucose

### ■ Identifying D and L Isomers of Sugars

Identify the following Fischer projection as D- or L-ribose.

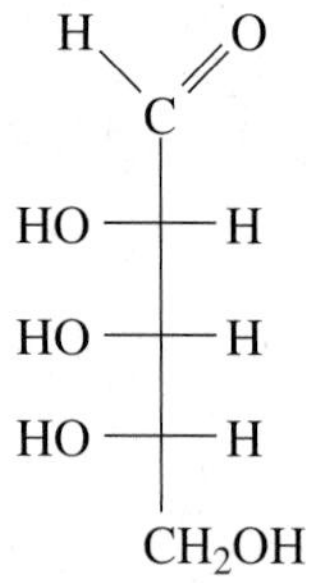

$$
\begin{array}{c}
\underset{\text{H}}{\phantom{}}\ \overset{\displaystyle O}{\underset{\displaystyle C}{\diagup\diagup}} \\
\text{HO}-\!\!\!\!\!\begin{array}{|c|}\hline\end{array}\!\!\!\!\!-\text{H} \\
\text{HO}-\!\!\!\!\!\begin{array}{|c|}\hline\end{array}\!\!\!\!\!-\text{H} \\
\text{HO}-\!\!\!\!\!\begin{array}{|c|}\hline\end{array}\!\!\!\!\!-\text{H} \\
\text{CH}_2\text{OH}
\end{array}
$$

**SOLUTION**

In ribose, carbon 4 is the chiral atom farthest from the carbonyl group. Because the hydroxyl group on carbon 4 is on the left, this is L-ribose.

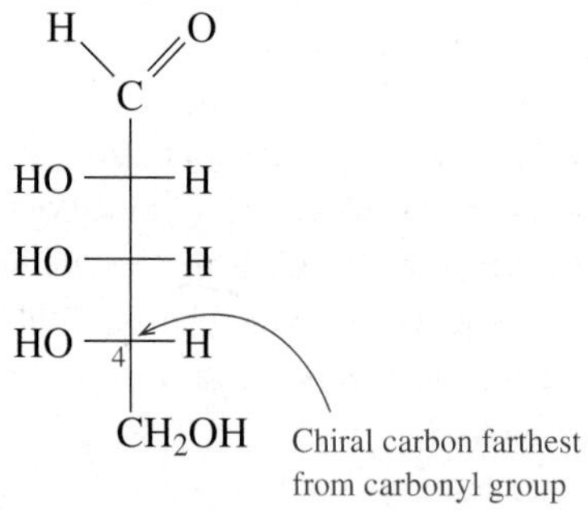

$$
\begin{array}{c}
\underset{\text{H}}{\phantom{}}\ \overset{\displaystyle O}{\underset{\displaystyle C}{\diagup\diagup}} \\
\text{HO}-\!\!\!\!\!\begin{array}{|c|}\hline\end{array}\!\!\!\!\!-\text{H} \\
\text{HO}-\!\!\!\!\!\begin{array}{|c|}\hline\end{array}\!\!\!\!\!-\text{H} \\
\text{HO}-\!\!\!\!\!\underset{4}{\begin{array}{|c|}\hline\end{array}}\!\!\!\!\!-\text{H} \\
\text{CH}_2\text{OH}
\end{array}
$$
Chiral carbon farthest from carbonyl group

**STUDY CHECK**

Draw the Fischer projection for D-ribose.

## Some Important Monosaccharides

The hexoses glucose, galactose, and fructose are important monosaccharides. Although we show Fischer projections for D and L isomers, the D isomers are more commonly

found in nature and used in the cells of the body. The Fischer projections for the D isomers are written as follows:

D-Glucose    D-Galactose    D-Fructose

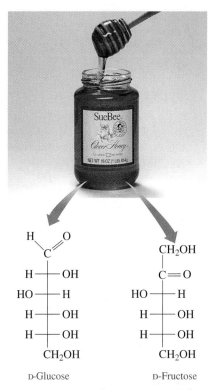

The most common hexose, D-**glucose**, $C_6H_{12}O_6$, also known as dextrose and blood
sugar, is found in fruits, vegetables, corn syrup, and honey. (See Figure 14.2.) It is a
building block of the disaccharides sucrose, lactose, and maltose and polysaccharides
such as starch, cellulose, and glycogen.

**Galactose** is an aldohexose that does not occur in the free form in nature. It is
obtained as a hydrolysis product of the disaccharide lactose, a sugar found in milk and
milk products. Galactose is important in the cellular membranes of the brain and nervous
system. The only difference in the Fischer projections of D-glucose and D-galactose is the
arrangement of the —OH group on carbon 4.

D-Glucose    D-Fructose

**FIGURE 14.2** The sweet taste of
honey is due to the monosaccharides D-glucose and D-fructose.
**Q** What are some differences in
the Fischer projections of D-
glucose and D-fructose?

D-Glucose    D-Galactose

In a condition called *galactosemia*, an enzyme that is needed to convert galactose to
glucose is missing. The accumulation of galactose in the blood and tissues can lead to
cataracts, mental retardation, and cirrhosis. The treatment for galactosemia is the removal
of all galactose-containing foods, mainly milk and milk products, from the diet. If this is
done for an infant immediately after birth, the damaging effects of galactose accumulation can be avoided.

In contrast to glucose and galactose, **fructose** is a ketohexose. The structure of fructose differs from glucose by the location of the carbonyl group.

D-Glucose    D-Fructose

# Health Note

## Hyperglycemia and Hypoglycemia

In the body, glucose normally occurs at a concentration of 70–90 mg/dL (1 dL = 100 mL) of blood. However, the amount of glucose depends on the time that has passed since eating. In the first hour after a meal, the level of glucose rises to about 130 mg/dL of blood, and then decreases over the next 2–3 hours as it is used in the tissues.

A doctor may order a glucose tolerance test to evaluate the body's ability to return to normal glucose concentration in response to the ingestion of a specified amount of glucose. The patient fasts for 12 hours and then drinks a solution containing glucose. A blood sample is taken immediately, followed by more blood samples each half hour for 2 hours, and then every hour, for a total of 5 hours. If the blood glucose exceeds 140 mg/dL in plasma and remains high, hyperglycemia may be indicated. The term *glyc* or *gluco* refers to "sugar." The prefix *hyper* means "above" or "over," and *hypo* is "below" or "under." Thus the blood sugar level in *hyperglycemia* is above normal, and it is below normal in *hypoglycemia*.

An example of a disease that can cause hyperglycemia is diabetes mellitus, which occurs when the pancreas is unable to produce sufficient quantities of insulin. As a result, glucose levels in the body fluids can rise as high as 350 mg/dL plasma. Symptoms of

diabetes in people under the age of 40 include thirst, excessive urination, increased appetite, and weight loss. In older persons, diabetes is sometimes a consequence of excessive weight gain.

When a person is hypoglycemic, the blood glucose level rises and then decreases rapidly to levels as low as 40 mg/dL plasma. In some cases, hypoglycemia is caused by overproduction of insulin by the pancreas. Low blood glucose can cause dizziness, general weakness, and muscle tremors. A diet may be prescribed that consists of several small meals high in protein and low in carbohydrate. Some hypoglycemic patients are finding success with diets that include more complex carbohydrates rather than simple sugars.

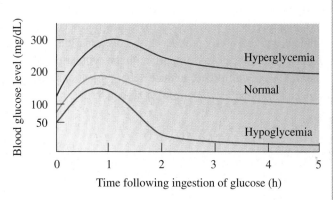

Fructose is the sweetest of the carbohydrates, twice as sweet as sucrose. This makes fructose popular with dieters because less fructose, and therefore fewer calories, are needed to provide a pleasant taste. After fructose enters the bloodstream, it is converted to its isomer, glucose. (See Figure 14.2.) Fructose is found in fruit juices and honey; it is also called levulose and fruit sugar. Fructose is also obtained as one of the hydrolysis products of sucrose, the disaccharide known as table sugar.

## SAMPLE PROBLEM   14.4

### ■ Monosaccharides

Ribulose has the following Fischer projection.

$$
\begin{array}{c}
CH_2OH \\
| \\
C{=}O \\
H{-}\!\!-\!\!OH \\
H{-}\!\!-\!\!OH \\
CH_2OH
\end{array}
$$

**a.** Identify the compound as D- or L-ribulose.

**b.** Write the Fischer projection of its mirror image.

## SOLUTION

**a.** The compound is D-ribulose because the —OH is on the right side of the chiral carbon farthest from the carbonyl group.

**b.** To write the mirror image, all the —OH groups on the chiral carbon atoms are written on the opposite side, as shown in the D isomer. L-Ribulose has the following Fischer projection:

$$
\begin{array}{c}
\text{CH}_2\text{OH} \\
|\\
\text{C}{=}\text{O} \\
\text{HO}{-}\!\!{-}\text{H} \\
\text{HO}{-}\!\!{-}\text{H} \\
\text{CH}_2\text{OH}
\end{array}
$$

## STUDY CHECK

What type of carbohydrate is ribulose?

---

## QUESTIONS AND PROBLEMS

### Fischer Projections of Monosaccharides

**14.11** How is a Fischer projection identified as a D or an L isomer?

**14.12** Write the Fischer projection for D-glyceraldehyde and L-glyceraldehyde.

**14.13** Identify each of the following as the D or L isomer:

**a.**
$$
\begin{array}{c}
\text{CHO} \\
\text{HO}{-}\!\!{-}\text{H} \\
\text{H}{-}\!\!{-}\text{OH} \\
\text{CH}_2\text{OH}
\end{array}
$$
Threose

**b.**
$$
\begin{array}{c}
\text{CH}_2\text{OH} \\
\text{C}{=}\text{O} \\
\text{HO}{-}\!\!{-}\text{H} \\
\text{H}{-}\!\!{-}\text{OH} \\
\text{CH}_2\text{OH}
\end{array}
$$
Xylulose

**c.**
$$
\begin{array}{c}
\text{CHO} \\
\text{H}{-}\!\!{-}\text{OH} \\
\text{H}{-}\!\!{-}\text{OH} \\
\text{HO}{-}\!\!{-}\text{H} \\
\text{HO}{-}\!\!{-}\text{H} \\
\text{CH}_2\text{OH}
\end{array}
$$
Mannose

**d.**
$$
\begin{array}{c}
\text{CHO} \\
\text{H}{-}\!\!{-}\text{OH} \\
\text{H}{-}\!\!{-}\text{OH} \\
\text{H}{-}\!\!{-}\text{OH} \\
\text{H}{-}\!\!{-}\text{OH} \\
\text{CH}_2\text{OH}
\end{array}
$$
Allose

**14.14** Identify each of the following as the D or L isomer:

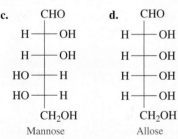

**a.**
$$
\begin{array}{c}
\text{CH}_2\text{OH} \\
\text{C}{=}\text{O} \\
\text{H}{-}\!\!{-}\text{OH} \\
\text{H}{-}\!\!{-}\text{OH} \\
\text{CH}_2\text{OH}
\end{array}
$$
Ribulose

**b.**
$$
\begin{array}{c}
\text{CH}_2\text{OH} \\
\text{C}{=}\text{O} \\
\text{HO}{-}\!\!{-}\text{H} \\
\text{H}{-}\!\!{-}\text{OH} \\
\text{HO}{-}\!\!{-}\text{H} \\
\text{CH}_2\text{OH}
\end{array}
$$
Sorbose

**c.**
$$
\begin{array}{c}
\text{CHO} \\
\text{H}{-}\!\!{-}\text{OH} \\
\text{HO}{-}\!\!{-}\text{H} \\
\text{H}{-}\!\!{-}\text{OH} \\
\text{H}{-}\!\!{-}\text{OH} \\
\text{CH}_2\text{OH}
\end{array}
$$
Glucose

**d.**
$$
\begin{array}{c}
\text{CHO} \\
\text{HO}{-}\!\!{-}\text{H} \\
\text{HO}{-}\!\!{-}\text{H} \\
\text{HO}{-}\!\!{-}\text{H} \\
\text{CH}_2\text{OH}
\end{array}
$$
Ribose

**14.15** Write the Fischer projections for the mirror images for a–d in problem 14.13.

**14.16** Write the Fischer projections for the mirror images for a–d in problem 14.14.

**14.17** Draw the Fischer projections for D-glucose and L-glucose.

**14.18** Draw the Fischer projections for D-fructose and L-fructose.

**14.19** How does the Fischer projection of D-galactose differ from D-glucose?

**14.20** How does the Fischer projection of D-fructose differ from D-glucose?

**14.21** Identify the monosaccharide that fits each of the following descriptions:
**a.** also called blood sugar
**b.** not metabolized in galactosemia
**c.** also called fruit sugar

**14.22** Identify a monosaccharide that fits each of the following descriptions:
**a.** high blood levels in diabetes
**b.** obtained as a hydrolysis product of lactose
**c.** the sweetest of the monosaccharides

## LEARNING GOAL

Draw and identify the Haworth structures of monosaccharides.

# 14.3 HAWORTH STRUCTURES OF MONOSACCHARIDES

Most of the time, monosaccharides exist as cyclic structures formed when a carbonyl group and a hydroxyl group in the *same* molecule react. The products are ring structures, also known as Haworth structures, that represent the most stable form of aldopentoses and aldohexoses. While the carbonyl group in the open-chain form of an aldohexose could react with several of the —OH groups, the aldohexoses typically form six-atom rings. Let's look at how we draw the Haworth structures for some D-isomers, starting with the open-chain structure of D-glucose.

## Drawing Haworth Structures for Glucose

Step 1    Turn the open-chain structure of D-glucose to the right. This places the —OH groups on the right of the vertical open chain below the carbon atoms and the —OH group on the left of the open chain above the carbon atom.

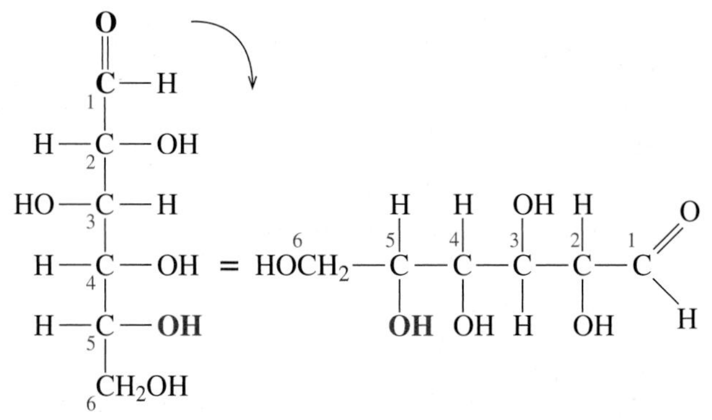

D-Glucose (open chain)

Step 2    Fold the carbon chain into a hexagon (move carbons 4, 5, and 6 clockwise). Write the —CH$_2$OH above carbon 5 and the —OH group on carbon 5 next to the carbonyl carbon. Complete the Haworth structure by bonding the oxygen from the —OH group to the carbonyl carbon.

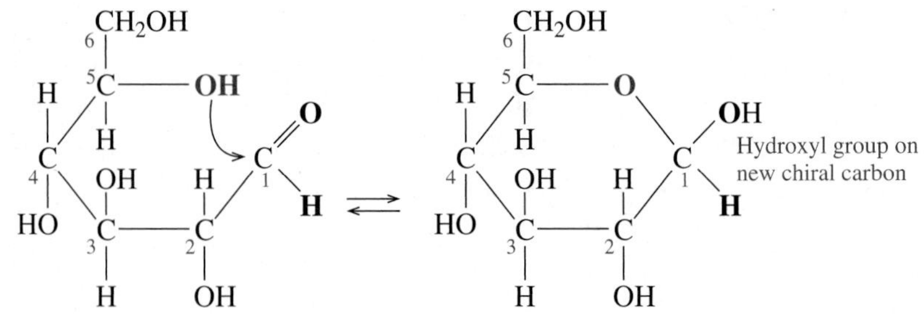

Carbon-5 oxygen bonds to carbonyl          Cyclic structure

Step 3    In a Haworth structure, a new —OH group forms on carbon 1. There are two ways to draw the —OH, either up or down. The —OH group is written *down* in the α (alpha) form and *up* in the β (beta) form.

Sometimes, the Haworth structure is simplified to show only the hydroxyl groups on the six-atom ring structure.

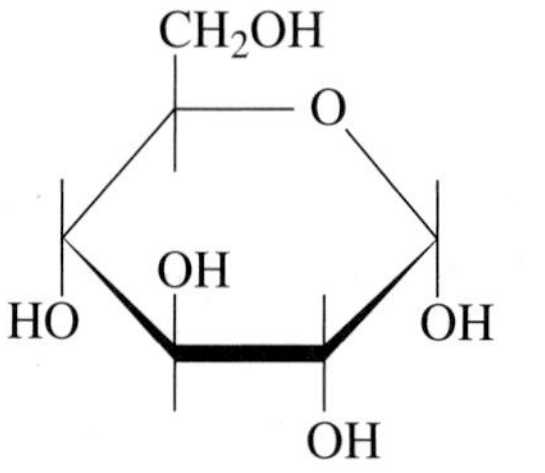

α-D-Glucose (simplified structure)

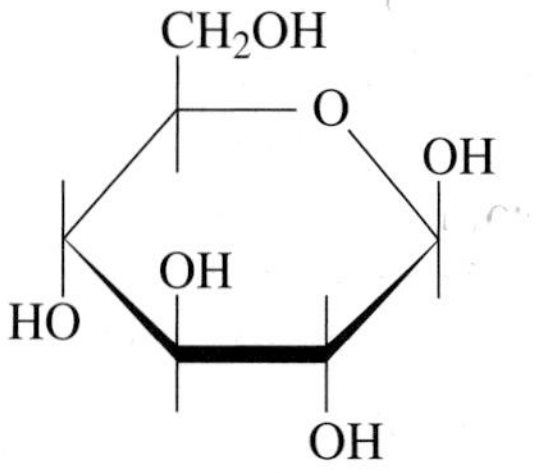

β-D-Glucose

In aqueous solution, the cyclic form of α-D-glucose opens and closes to form β-D-glucose. When the ring opens, the structure is the open chain of D-glucose with an aldehyde group. At any given time, there is only a trace amount of the open chain.

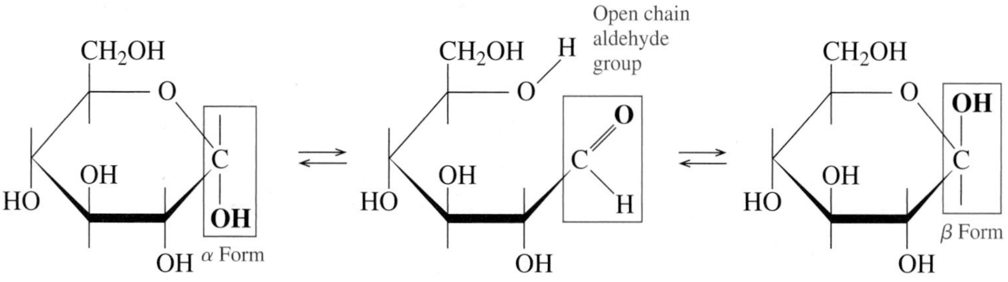

α-D-Glucose
(36% in mixture)

D-Glucose
open-chain (trace)

β-D-Glucose
(64% in mixture)

## Haworth Structures of Galactose

Galactose is an aldohexose that differs from glucose only in the arrangement of the —OH group on carbon 4. Thus, its Haworth structure is similar to glucose, except that in galactose the —OH on carbon 4 is up. With a new hydroxyl group on carbon 1, galactose also exists in α and β forms.

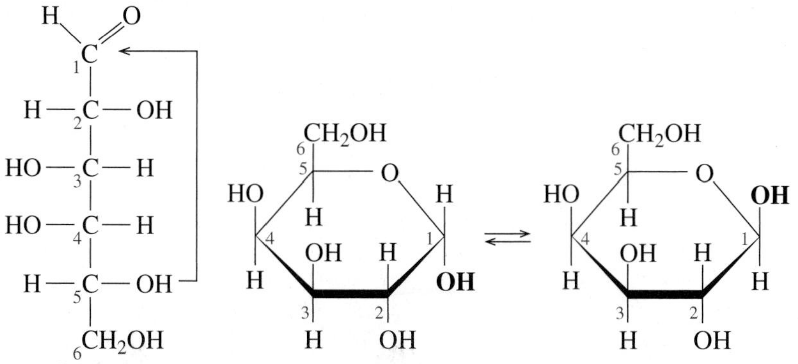

D-Galactose

α-D-Galactose

β-D-Galactose

## Haworth Structures of Fructose

In contrast to glucose and galactose, fructose is a ketohexose. The Haworth structure for fructose is a five-atom ring with carbon 2 at the right corner. The cyclic structure forms when the hydroxyl group on carbon 5 reacts with the ketone group on carbon 2. The new hydroxyl group on carbon 2 gives the α and β forms of fructose.

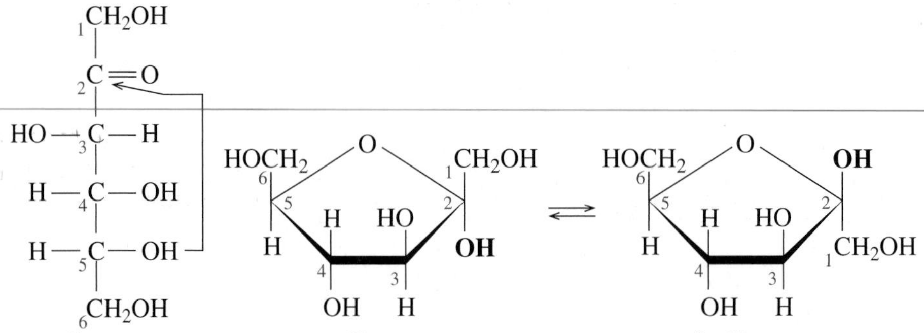

D-Fructose          α-D-Fructose          β-D-Fructose

---

## SAMPLE PROBLEM 14.5

### ■ Haworth Structures for Sugars

Identify the following as α- or β-D-mannose, a carbohydrate found in immunoglobulins.

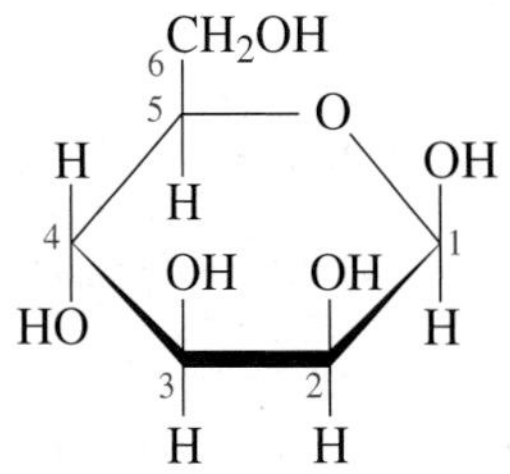

#### SOLUTION

The Haworth structure is β-D-mannose because the hydroxyl group on carbon 1 is written above carbon 1.

#### STUDY CHECK

Draw the Haworth structure for α-D-mannose.

---

## QUESTIONS AND PROBLEMS

### Haworth Structures of Monosaccharides

**14.23** What are the kind and number of atoms in the ring portion of the Haworth structure of glucose?

**14.24** What are the kind and number of atoms in the ring portion of the Haworth structure of fructose?

**14.25** Draw the Haworth structures for α- and β-D-glucose.

**14.26** Draw the Haworth structures for α- and β-D-fructose.

**14.27** Identify each of the following as the α or β form.

a.

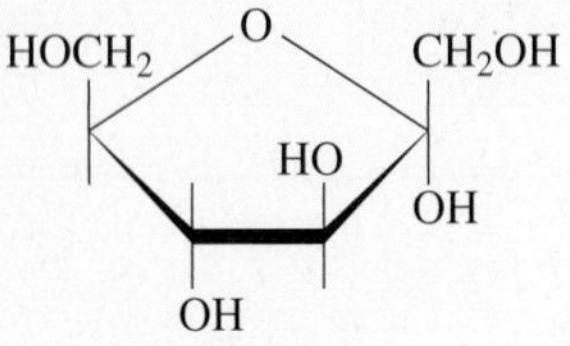

b.

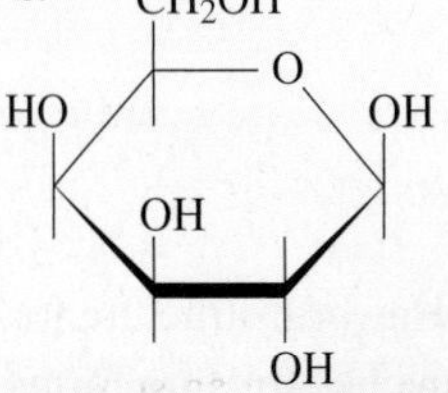

**14.28** Identify each of the following as the α or β form.

a.

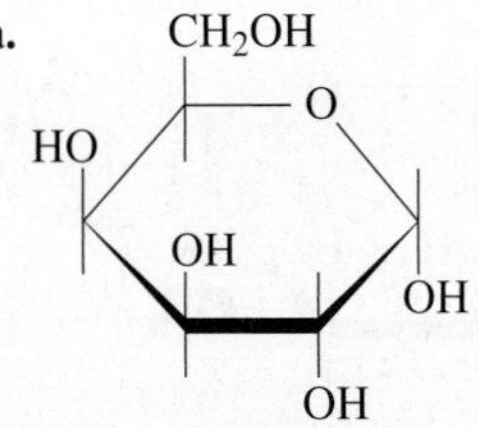

b.

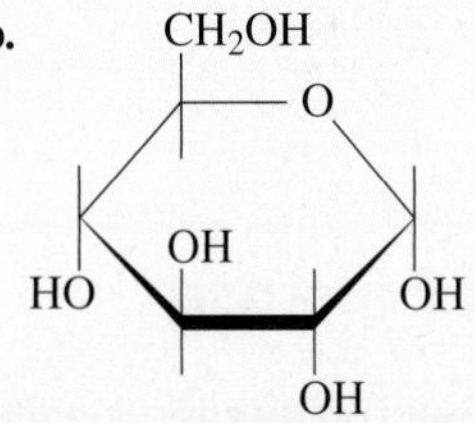

# 14.4 CHEMICAL PROPERTIES OF MONOSACCHARIDES

Monosaccharides contain functional groups that can undergo chemical reactions. In an aldose, the aldehyde group can be oxidized to a carboxylic acid. The carbonyl group in both an aldose and a ketose can be reduced to give a hydroxyl group. The hydroxyl groups can react with other compounds to form a variety of derivatives that are important in biological structures.

## Oxidation of Monosaccharides

Although monosaccharides exist mostly in cyclic forms, we have seen that a small amount of the open-chain form is always present, which provides an aldehyde group. When Benedict's reagent is added, the aldehyde group is oxidized, and $Cu^{2+}$ is reduced to $Cu^+$, which forms a brick-red precipitate of $Cu_2O$. Monosaccharides that reduce another substance are called **reducing sugars**.

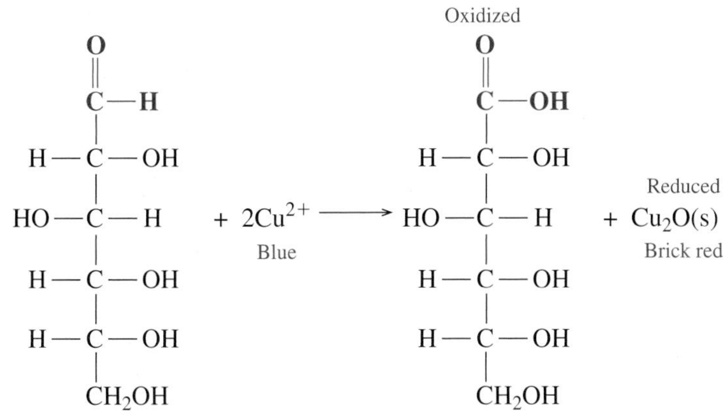

Oxidized

$$C-H + 2Cu^{2+} \longrightarrow C-OH + Cu_2O(s)$$

Open chain of D-glucose,
a reducing sugar          Blue        D-Gluconic acid      Reduced
Brick red

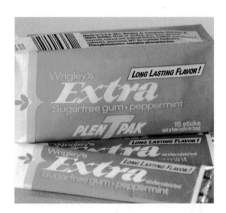

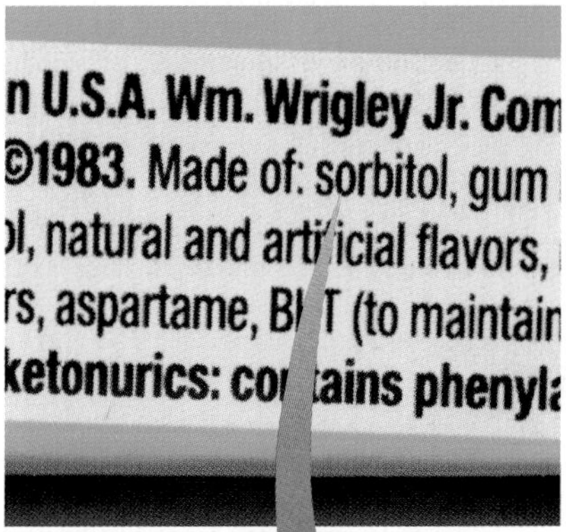

Fructose is also a reducing sugar. In the open-chain form, a rearrangement between the hydroxyl group on carbon 1 and the ketone group provides an aldehyde group that can be oxidized.

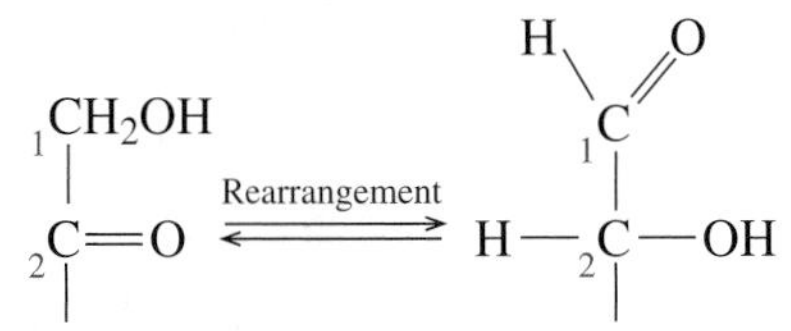

D-Fructose          D-Glucose
(ketose)            (aldose)

Rearrangement

## Reduction of Monosaccharides

The reduction of the carbonyl group in monosaccharides produces sugar alcohols, which are also called *alditols*. D-Glucose is reduced to D-glucitol, better known as sorbitol. D-Mannose is reduced to give D-mannitol.

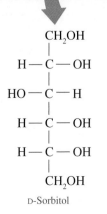

$CH_2OH$

D-Sorbitol

the **C**hemistry place

CASE STUDY
Diabetes and Blood Glucose

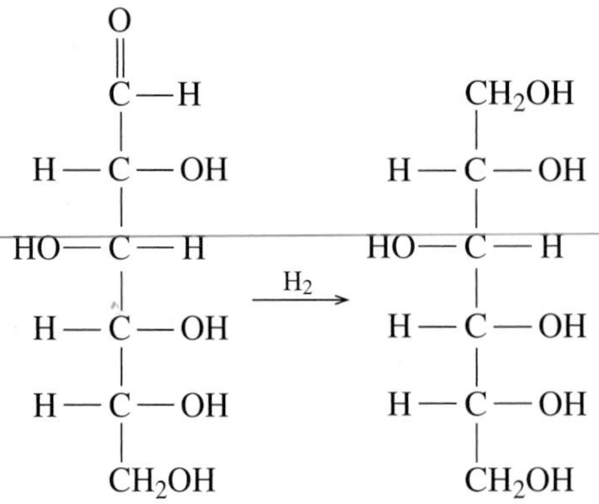

$$
\begin{array}{ccc}
\underset{\text{D-Glucose}}{
\begin{array}{c}
\overset{O}{\underset{\|}{C}}\!-\!H \\
H-C-OH \\
HO-C-H \\
H-C-OH \\
H-C-OH \\
CH_2OH
\end{array}
}
&
\xrightarrow{H_2}
&
\underset{\text{D-Glucitol or D-Sorbitol}}{
\begin{array}{c}
CH_2OH \\
H-C-OH \\
HO-C-H \\
H-C-OH \\
H-C-OH \\
CH_2OH
\end{array}
}
\end{array}
$$

Sugar alcohols such as sorbitol, xylitol from xylose, and mannitol from mannose are used as sweeteners in many sugar-free products such as diet drinks and sugarless gum as well as products for people with diabetes. However, there are some side effects of these sugar substitutes. Some people experience some discomfort such as gas and diarrhea from the ingestion of sugar alcohols. The development of cataracts in diabetics is attributed to the accumulation of sorbitol in the lens of the eye.

## *Health Note*

### Testing for Glucose in Urine

Normally, blood glucose flows through the kidneys and is reabsorbed into the bloodstream. However, if the blood level exceeds about 160 mg of glucose/dL of blood, the kidneys cannot reabsorb all of the glucose, and it spills over into the urine, a condition known as glucosuria. A symptom of diabetes mellitus is a high level of glucose in the urine.

Benedict's test can be used to determine the presence of glucose in urine. The amount of cuprous oxide ($Cu_2O$) formed is proportional to the amount of reducing sugar present in the urine. Low to moderate levels of reducing sugar turn the solution green; solutions with high glucose levels turn Benedict's reagent yellow or brick-red. Table 14.1 lists some colors associated with the concentration of glucose in the urine.

In another clinical test that is more specific for glucose, the enzyme glucose oxidase is used. The oxidase enzyme converts glucose to gluconic acid and hydrogen peroxide, $H_2O_2$. The peroxide produced reacts with a dye in the test strip to give different colors. The level of glucose present in the urine is found by matching the color produced to a color chart on the test strip container.

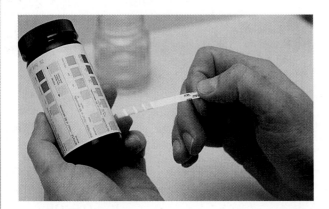

**TABLE 14.1** Glucose Test Results

| Color | Glucose Present in Urine | |
|---|---|---|
| | % | mg/dL |
| Blue | 0 | 0 |
| Blue-green | 0.25 | 250 |
| Green | 0.50 | 500 |
| Yellow | 1.00 | 1000 |
| Brick-red | 2.00 | 2000 |

## SAMPLE PROBLEM 14.6

### ■ Reducing Sugars

Why is D-glucose a reducing sugar?

#### SOLUTION

The aldehyde group of D-Glucose is easily oxidized by Benedict's reagent. A carbohydrate that reduces $Cu^{2+}$ to $Cu^{+}$ is called a reducing sugar.

#### STUDY CHECK

A test using Benedict's reagent turns brick-red with a urine sample. According to Table 14.1, what might this result indicate?

## QUESTIONS AND PROBLEMS

### Chemical Properties of Monosaccharides

**14.29** Draw the structure of xylitol produced when D-xylose is reduced.

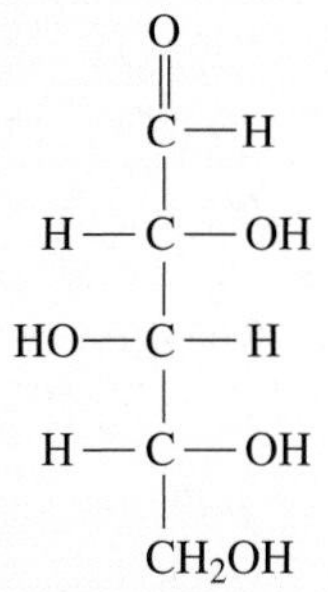

$$
\begin{array}{c}
\text{O} \\
\parallel \\
\text{C}-\text{H} \\
\text{H}-\text{C}-\text{OH} \\
\text{HO}-\text{C}-\text{H} \\
\text{H}-\text{C}-\text{OH} \\
\text{CH}_2\text{OH}
\end{array}
$$
D-Xylose

**14.30** Draw the structure of mannitol produced when D-mannose is reduced.

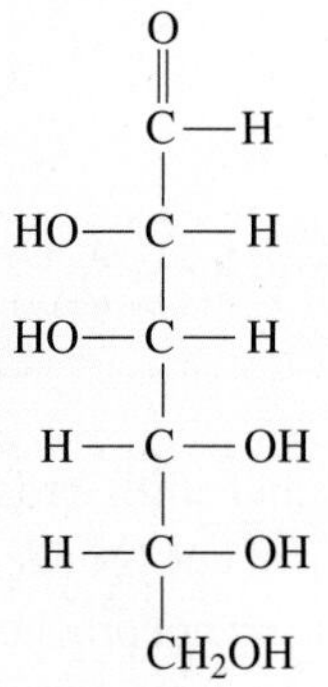

$$
\begin{array}{c}
\text{O} \\
\parallel \\
\text{C}-\text{H} \\
\text{HO}-\text{C}-\text{H} \\
\text{HO}-\text{C}-\text{H} \\
\text{H}-\text{C}-\text{OH} \\
\text{H}-\text{C}-\text{OH} \\
\text{CH}_2\text{OH}
\end{array}
$$
D-Mannose

**14.31** Write the oxidation and reduction products of D-arabinose. What is the name of the sugar alcohol produced?

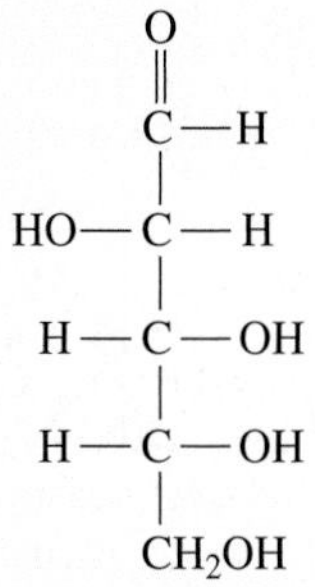

$$
\begin{array}{c}
\text{O} \\
\parallel \\
\text{C}-\text{H} \\
\text{HO}-\text{C}-\text{H} \\
\text{H}-\text{C}-\text{OH} \\
\text{H}-\text{C}-\text{OH} \\
\text{CH}_2\text{OH}
\end{array}
$$
D-Arabinose

**14.32** Write the oxidation and reduction products of D-ribose. What is the name of the sugar alcohol produced?

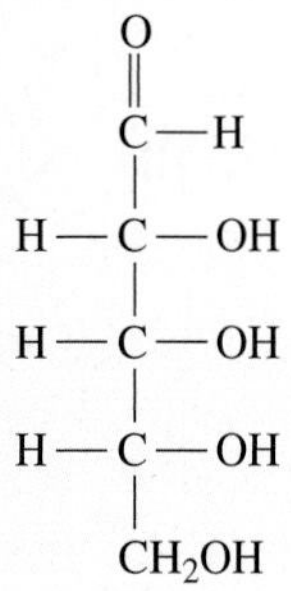

$$
\begin{array}{c}
\text{O} \\
\parallel \\
\text{C}-\text{H} \\
\text{H}-\text{C}-\text{OH} \\
\text{H}-\text{C}-\text{OH} \\
\text{H}-\text{C}-\text{OH} \\
\text{CH}_2\text{OH}
\end{array}
$$
D-Ribose

**LEARNING GOAL**

Describe the monosaccharide units and linkages in disaccharides.

# 14.5 DISACCHARIDES

A disaccharide is composed of two monosaccharides linked together. The most common disaccharides are maltose, lactose, and sucrose. Their hydrolysis, by an acid or an enzyme, gives the following monosaccharides.

$$\text{Maltose} + H_2O \xrightarrow{H^+} \text{glucose} + \text{glucose}$$

$$\text{Lactose} + H_2O \xrightarrow{H^+} \text{glucose} + \text{galactose}$$

$$\text{Sucrose} + H_2O \xrightarrow{H^+} \text{glucose} + \text{fructose}$$

**Maltose**, or malt sugar, is a disaccharide obtained from starch. When maltose in barley and other grains is hydrolyzed by yeast enzymes, glucose is obtained, which can undergo fermentation to give ethanol. Maltose is used in cereals, candies, and the brewing of beverages.

When a hydroxyl group in one monosaccharide reacts with a hydroxyl group in another monosaccharide, a **glycosidic bond** forms and the product is a disaccharide. In maltose, the glycosidic bond that joins the two glucose molecules is designated an $\alpha$-1,4 linkage to show that —OH on carbon 1 of $\alpha$-D-glucose bonds to carbon 4 of the second glucose. In the second glucose molecule, the —OH group on carbon 1 gives the $\alpha$ and $\beta$ forms. Maltose is a reducing sugar because the —OH on carbon 1 of the second glucose opens to give an aldehyde group.

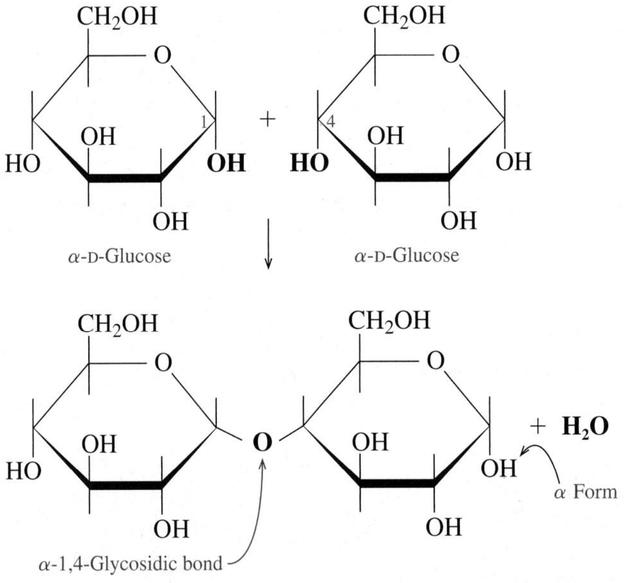

$\alpha$-D-Glucose          $\alpha$-D-Glucose

$\alpha$-1,4-Glycosidic bond

$\alpha$-Maltose, a disaccharide

**Lactose**, milk sugar, is a disaccharide found in milk and milk products. (See Figure 14.3.) It makes up 6–8% of human milk and about 4–5% of cow's milk and is used in products that attempt to duplicate mother's milk. Some people do not produce sufficient quantities of the enzyme needed to hydrolyze lactose, and the sugar remains undigested, causing abdominal cramps and diarrhea. In some commercial milk products, an enzyme called lactase is added to break down lactose. The bond in lactose is a $\beta$-1,4-glycosidic bond because the $\beta$ form of galactose forms a bond with a hydroxyl group on carbon 4 of glucose. The hydroxyl group on carbon 1 of glucose gives both $\alpha$- and $\beta$-lactose.

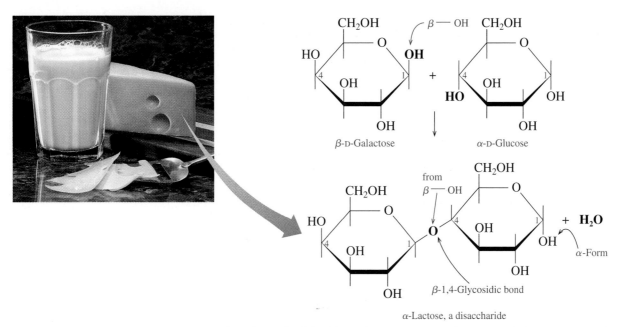

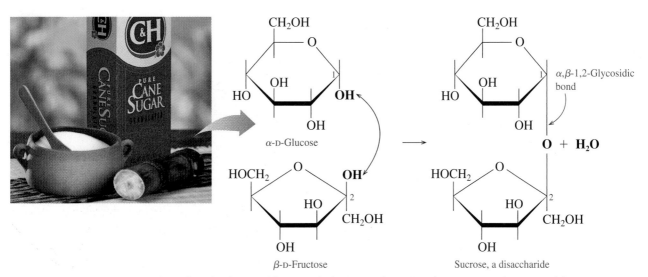

FIGURE 14.3 Lactose, a disaccharide found in milk and milk products, contains galactose and glucose.
Q What type of glycosidic bond links galactose and glucose in lactose?

Because the open chain has an aldehyde group that can be oxidized, lactose is a reducing sugar.

**Sucrose** consists of an $\alpha$-D-glucose and $\beta$-D-fructose molecule joined by an $\alpha$, $\beta$-1,2-glycosidic bond. (See Figure 14.4.) Unlike the other disaccharides, the glycosidic bond in sucrose cannot open to give an aldehyde. Sucrose does not react with Benedict's reagent, which means it is not a reducing sugar.

The sugar we use to sweeten our cereal, coffee, or tea is sucrose. Most of the sucrose for table sugar comes from sugar cane (20% by mass) or sugar beets (15% by mass). Both the raw and refined forms of sugar are sucrose. Some estimates in 2003 indicate that each person in the United States consumes an average of 68 kg (150 lb) of sucrose every year, either by itself or in a variety of food products.

FIGURE 14.4 Sucrose, a disaccharide obtained from sugar beets and sugar cane, contains glucose and fructose.
Q Why is sucrose a nonreducing sugar?

## Explore Your World

### Sugar and Sweeteners

Add a tablespoon of sugar to a glass of water and stir. Taste. Add more sugar, stir, and taste. If you have other carbohydrates such as fructose, honey, cornstarch, arrowroot, or flour, add some of each to separate glasses of water and stir. If you have some artificial sweeteners, add a few drops of the sweetener or a package, if solid, to a glass of water. Taste each.

### QUESTIONS

1. Which substance is the most soluble in water?
2. Place the substances in order from the one that tastes least sweet to the sweetest.
3. How does your list compare to the values in Table 14.2?
4. How does the sweetness of sucrose compare with the artificial sweeteners?
5. Check the labels of food products in your kitchen. Look for sugars such as sucrose or fructose, or artificial sweeteners such as aspartame or sucralose on the label. How many grams of sugar are in a serving of the food?

---

### ■ Glycosidic Bonds in Disaccharides

Melebiose is a disaccharide that is 30 times sweeter than sucrose.

Melebiose

**a.** What are the monosaccharide units in melebiose?
**b.** What type of glycosidic bond links the monosaccharides?
**c.** Identify the structure as $\alpha$- or $\beta$-melebiose.

#### SOLUTION

**a.** The monosaccharide on the left side is $\alpha$-D-galactose; on the right is $\alpha$-D-glucose.
**b.** The monosaccharide units are linked by an $\alpha$-1,6-glycosidic bond.
**c.** The downward position of the hydroxyl group on carbon 1 of the D-glucose end that can open makes it $\alpha$-melebiose.

#### STUDY CHECK

Cellobiose is a disaccharide composed of two $\beta$-D-glucose molecules linked by a $\beta$-1,4-glycosidic linkage. Draw a structural formula for $\beta$-cellobiose.

---

## Career Focus

### Phlebotomist

As part of the medical team, phlebotomists collect and process blood for laboratory tests. They work directly with patients, calming them if necessary prior to the collection of blood. Phlebotomists are trained to collect blood in a safe manner and to provide patient care if fainting occurs. Blood is drawn through venipuncture methods such as syringe, vacutainer, blood culture, and finger stick. They also prepare patients for procedures such as glucose tolerance tests. In the preparation of specimens for analysis, a phlebotomist determines media, inoculation method, and reagents for culture setup.

# Health Note

## How Sweet Is My Sweetener?

Although many of the monosaccharides and disaccharides taste sweet, they differ considerably in their degree of sweetness. Dietetic foods contain sweeteners that are non-carbohydrate or carbohydrates that are sweeter than sucrose. Some examples of sweeteners compared with sucrose are shown in Table 14.2.

Sucralose is made from sucrose by replacing some of the hydroxyl groups with chlorine atoms.

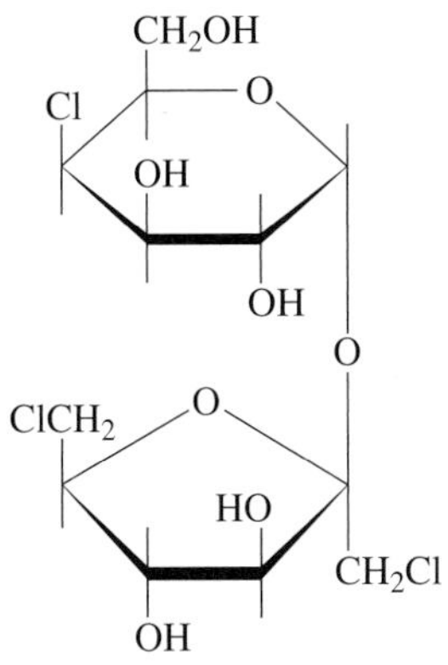

Sucralose

Aspartame, which is marketed as Nutra-Sweet™, is used in a large number of sugar-free products. It is a non-carbohydrate sweetener made of aspartic acid and a methyl ester of phenylalanine. It does have some caloric value, but it is so sweet that a very small quantity is needed. However, phenylalanine, one of the breakdown products, poses a danger to anyone who cannot metabolize it properly, a condition called phenylketonuria (PKU).

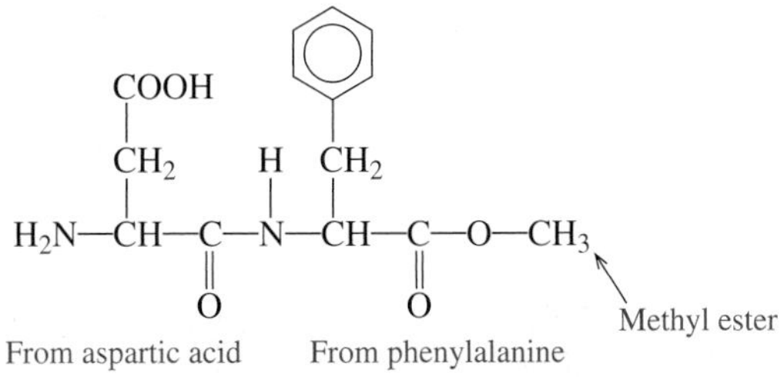

From aspartic acid    From phenylalanine
Aspartame (Nutra-Sweet)

A new artificial sweetener, Neotame™, is a modification of the aspartame structure. The addition of a large alkyl group to the amine group prevents enzymes from breaking the amide bond between aspartic acid and phenylalanine. Thus, phenylalanine is not produced when Neotame is used as a sweetener. Very small amounts of Neotame are needed because it is about 10 000 times sweeter than sucrose.

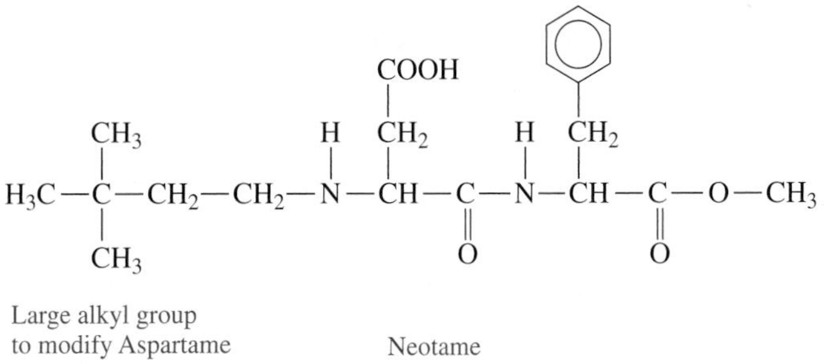

Large alkyl group
to modify Aspartame          Neotame

**TABLE 14.2** Relative Sweetness of Sugars and Artificial Sweeteners

| | Sweetness Relative to Sucrose (= 100) |
|---|---|
| **Monosaccharides** | |
| Galactose | 30 |
| Glucose | 75 |
| Fructose | 175 |
| **Disaccharides** | |
| Lactose | 16 |
| Maltose | 33 |
| Sucrose | 100  ← reference standard |
| **Sugar Alcohols (Polyols)** | |
| Sorbitol | 60 |
| Malitol | 80 |
| Xylitol | 100 |
| **Artificial Sweeteners (Non-carbohydrate)** | |
| Aspartame | 18 000 |
| Saccharin | 45 000 |
| Sucralose | 60 000 |
| Neotame | 1 000 000 |

Saccharin has been used as a non-carbohydrate artificial sweetener for the past 25 years. The use of saccharin has been banned in Canada because studies indicate that it may cause bladder tumors. However, it is still approved for use by the FDA in the United States.

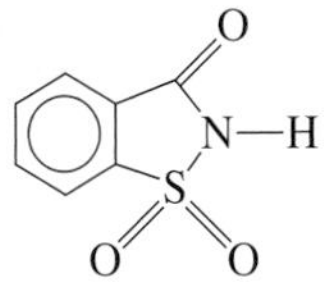

Saccharin

*Sweet + low not a carbohydrate*

## Blood Types and Carbohydrates

Every individual's blood can be typed as one of four blood groups, A, B, AB, and O. Although there is some variation among ethnic groups in the United States, the incidence of blood types in the general population is about 43% O, 40% A, 12% B, and 5% AB.

The blood types are determined by three or four monosaccharides that are attached to the red blood cells. All the blood types include *N*-acetylglucosamine, galactose, and fucose. In type A blood, a fourth monosaccharide, *N*-acetylgalactosamine, is bonded to the galactose. The structures of these monosaccharides are as follows:

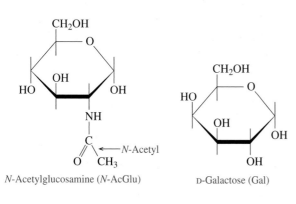

*N*-Acetylglucosamine (*N*-AcGlu)    D-Galactose (Gal)

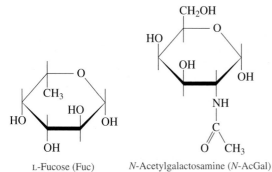

L-Fucose (Fuc)    *N*-Acetylgalactosamine (*N*-AcGal)

In type B blood, there is a second molecule of galactose. Type AB blood contains both A and B blood types. A person with type A blood produces antibodies against type B, whereas a person with type B blood produces antibodies against A. Type AB blood produces no antibodies, whereas type O produces both. Thus, if a person with type A blood receives a transfusion of type B blood, factors in the recipient's blood will agglutinate the donor's red blood cells. If you become a blood donor, your blood is screened to make sure that an exact match is made with the blood type of the recipient. Table 14.3 summarizes the compatibility of blood groups for transfusion.

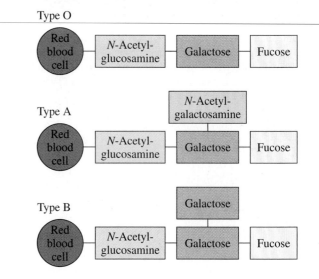

Type O

Type A

Type B

### TABLE 14.3  Compatibility of Blood Groups

| Blood Group | Can Receive Blood Types | Cannot Receive Blood Types |
|---|---|---|
| A | A, O | B, AB |
| B | B, O | A, AB |
| AB[a] | A, B, AB, O | Can receive all blood types |
| O[b] | O | A, B, AB |

[a] Universal recipient
[b] Universal donor

## QUESTIONS AND PROBLEMS

### Disaccharides

**14.33** For each of the following, give the monosaccharide units produced by hydrolysis, the type of glycosidic bond, and the identity of the disaccharide including the α or β form:

**a.**

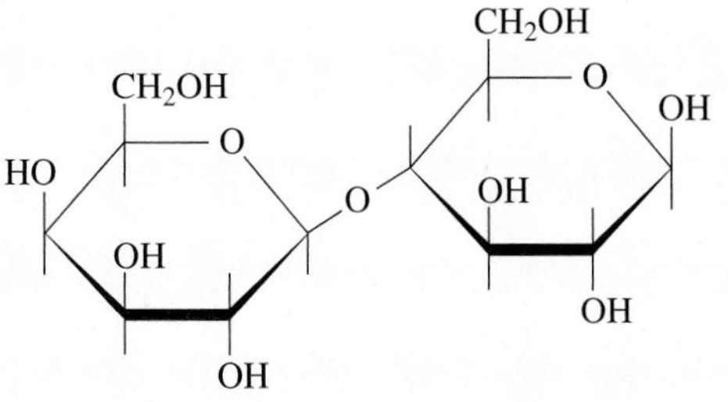

**b.**

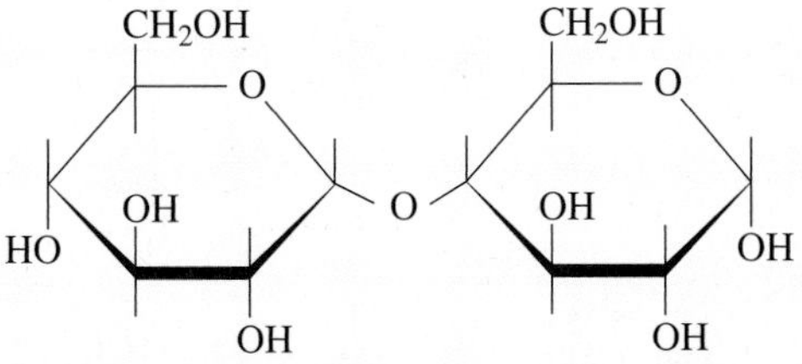

**14.34** For each of the following, give the monosaccharide units produced by hydrolysis, the type of glycosidic bond, and the identity of the disaccharide including the α or β form:

**a.**

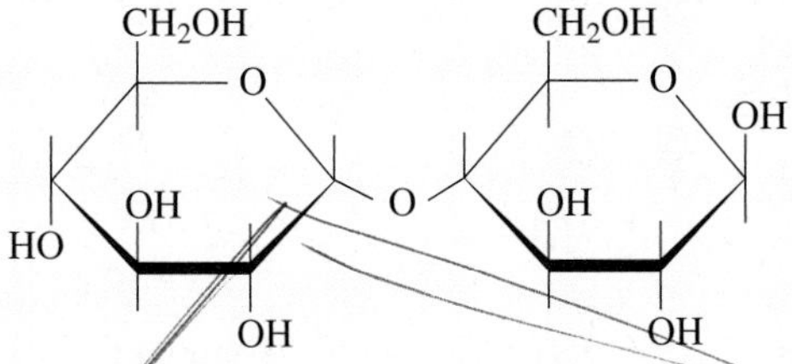

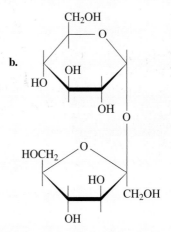

**b.**

**14.35** Indicate whether the disaccharides in problem 14.33 will undergo oxidation.

**14.36** Indicate whether the disaccharides in problem 14.34 will undergo oxidation.

**14.37** Identify disaccharides that fit each of the following descriptions:
   **a.** ordinary table sugar
   **b.** found in milk and milk products
   **c.** also called malt sugar
   **d.** hydrolysis gives galactose and glucose

**14.38** Identify disaccharides that fit each of the following descriptions:
   **a.** not a reducing sugar
   **b.** composed of two glucose units
   **c.** also called milk sugar
   **d.** hydrolysis gives glucose and fructose

# 14.6 POLYSACCHARIDES

A polysaccharide is a polymer of many monosaccharides joined together. Four biologically important polysaccharides—amylose, amylopectin, cellulose, and glycogen—are all polymers of D-glucose that differ only in the type of glycosidic bonds and the amount of branching in the molecule.

Starch, a storage form of glucose in plants, is found as insoluble granules in rice, wheat, potatoes, beans, and cereals. Starch is composed of two kinds of polysaccharides, amylose and amylopectin. **Amylose**, which makes up about 20% of starch, consists of 250–4000 α-D-glucose molecules connected by α-1,4-glycosidic bonds in a continuous chain. Sometimes called a straight-chain polymer, polymers of amylose are actually coiled in helical fashion.

**Amylopectin**, which makes up as much as 80% of plant starch, is a branched-chain polysaccharide. Like amylose, the glucose molecules are connected by α-1,4-glycosidic bonds. However, at about every 25 glucose units, there is a branch of glucose molecules attached by an α-1,6-glycosidic bond between carbon 1 of the branch and carbon 6 in the main chain. (See Figure 14.5.)

Starches hydrolyze easily in water and acid to give dextrins, which then hydrolyze to maltose and finally glucose. In our bodies, these complex carbohydrates are digested

## LEARNING GOAL

Describe the structural features of amylose, amylopectin, glycogen, and cellulose.

**WEB TUTORIAL**
Polymers

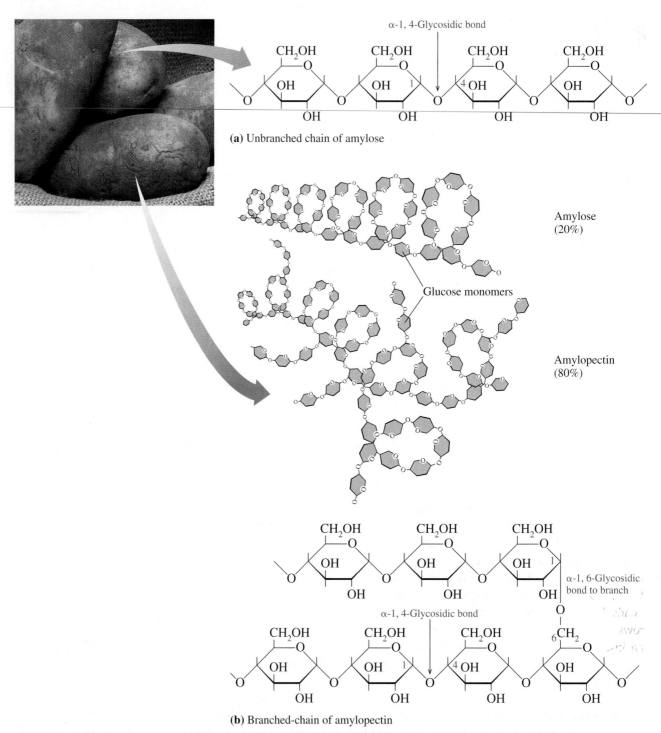

**(a)** Unbranched chain of amylose

Amylose (20%)

Glucose monomers

Amylopectin (80%)

**(b)** Branched-chain of amylopectin

**FIGURE 14.5** The structure of **(a)** amylose is a straight-chain polysaccharide of glucose units, and **(b)** amylopectin is a branched chain of glucose.

**Q** What are the two types of glycosidic bonds that link glucose molecules in amylopectin?

by the enzymes amylase (in saliva) and maltase. The glucose obtained provides about 50% of our nutritional calories.

$$\text{Amylose, amylopectin} \xrightarrow{\text{H}^+ \text{ or amylase}} \text{dextrins} \xrightarrow{\text{H}^+ \text{ or amylase}} \text{maltose} \xrightarrow{\text{H}^+ \text{ or maltase}} \text{many D-glucose units}$$

**Glycogen**, or animal starch, is a polymer of glucose that is stored in the liver and muscle of animals. It is hydrolyzed in our cells at a rate that maintains the blood level of glucose and provides energy between meals. The structure of glycogen is very similar to

**FIGURE 14.6** The polysaccharide cellulose is composed of $\beta$-1,4-glycosidic bonds.
**Q** Why are humans unable to digest cellulose?

that of amylopectin found in plants, except that glycogen is more highly branched. In glycogen, the glucose units are joined by $\alpha$-1,4-glycosidic bonds, and branches occurring about every 10–15 glucose units are attached by $\alpha$-1,6-glycosidic bonds.

**Cellulose** is the major structural material of wood and plants. Cotton is almost pure cellulose. In cellulose, glucose molecules form a long unbranched chain similar to that of amylose. However, the glucose units in cellulose are linked by $\beta$-1,4-glycosidic bonds. The $\beta$ isomers do not form coils like the $\alpha$ isomers but are aligned in parallel rows that are held in place by hydrogen bonds between hydroxyl groups in adjacent chains, making cellulose insoluble in water. This gives a rigid structure to the cell walls in wood and fiber that is more resistant to hydrolysis than are the starches. (See Figure 14.6.)

Humans have enzymes called $\alpha$-amylase in saliva and pancreatic juices that hydrolyze the $\alpha$-1,4-glycosidic bonds of starches, but not the $\beta$-1,4-glycosidic bonds of cellulose. Thus, humans cannot digest cellulose. Animals such as horses, cows, and goats can obtain glucose from cellulose because their digestive systems contain bacteria that provide enzymes such as cellulase to hydrolyze $\beta$-1,4-glycosidic bonds.

---

**SAMPLE PROBLEM 14.8**

### ■ Structures of Polysaccharides

Identify the polysaccharide described by each of the following:

**a.** a polysaccharide that is stored in the liver and muscle tissues
**b.** an unbranched polysaccharide containing $\beta$-1,4-glycosidic bonds
**c.** a starch containing $\alpha$-1,4- and $\alpha$-1,6-glycosidic bonds

---

*Explore Your World*

### Polysaccharides

Read the nutrition label on a box of crackers, cereal, bread, chips, or pasta. The major ingredient in crackers is flour, a starch. Chew on a single cracker for 4–5 minutes. An enzyme (amylase) in your saliva breaks apart the bonds in starch.

**QUESTIONS**

1. How are carbohydrates listed on the label?
2. What other carbohydrates are listed?
3. How did the taste of the cracker change during the time that you chewed it?
4. What happened to the starches in the cracker as the amylase enzyme in your saliva reacted with the amylose or amylopectin?

SOLUTION

**a.** glycogen    **b.** cellulose    **c.** amylopectin, glycogen

STUDY CHECK

Cellulose and amylose are both unbranched glucose polymers. How do they differ?

## QUESTIONS AND PROBLEMS

### Polysaccharides

**14.39** Describe the similarities and differences in the following:
   **a.** amylose and amylopectin
   **b.** amylopectin and glycogen

**14.40** Describe the similarities and differences in the following:
   **a.** amylose and cellulose    **b.** cellulose and glycogen

**14.41** Give the name of one or more polysaccharides that matches each of the following descriptions:
   **a.** not digestible by humans
   **b.** the storage form of carbohydrates in plants

   **c.** contains only $\alpha$-1,4-glycosidic bonds
   **d.** the most highly branched polysaccharide

**14.42** Give the name of one or more polysaccharides that matches each of the following descriptions:
   **a.** the storage form of carbohydrates in animals
   **b.** contains only $\beta$-1,4-glycosidic bonds
   **c.** contains both $\alpha$-1,4- and $\alpha$-1,6-glycosidic bonds
   **d.** produces maltose during digestion

## CONCEPT MAP

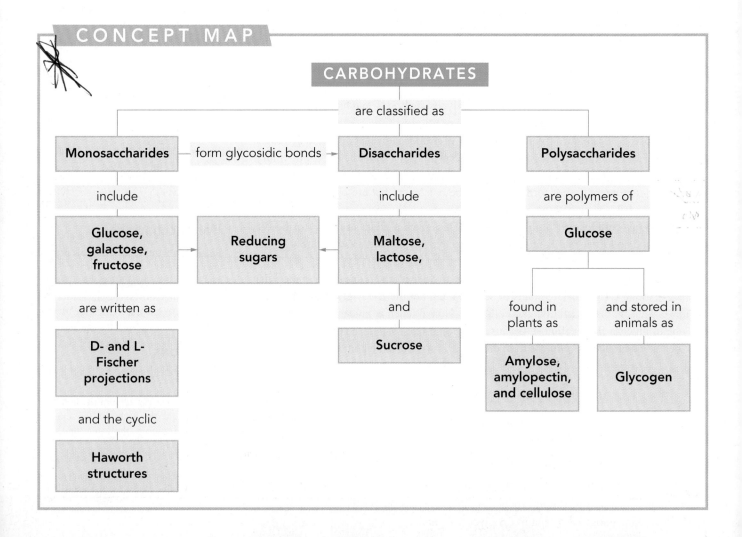

# CHAPTER REVIEW

## 14.1 Carbohydrates
**Learning Goal:** Classify a monosaccharide as an aldose or a ketose, and indicate the number of carbon atoms.

Carbohydrates are classified as monosaccharides (simple sugars), disaccharides (two monosaccharide units), and polysaccharides (many monosaccharide units). Monosaccharides are polyhydroxy aldehydes (aldoses) or ketones (ketoses). Monosaccharides are also classified by their number of carbon atoms: *triose, tetrose, pentose,* or *hexose.*

## 14.2 Fischer Projections of Monosaccharides
**Learning Goal:** Draw the D or L isomers of glucose, galactose, and fructose.

In a Fischer projection, the prefixes D- and L- are used to distinguish between mirror images. In a D isomer, the —OH is on the right of the chiral carbon farthest from the carbonyl carbon; it is on the left in the L isomer. Important monosaccharides are the aldohexoses, glucose and galactose, and the ketohexose, fructose.

## 14.3 Haworth Structures of Monosaccharides
**Learning Goal:** Draw and identify the Haworth structures of monosaccharides.

The predominant form of monosaccharides is a ring of five or six atoms. The cyclic structure forms when an —OH (usually the one on carbon 5 in hexoses) reacts with the carbonyl group of the same molecule. The formation of a new hydroxyl group on carbon 1 gives $\alpha$ and $\beta$ forms of the cyclic monosaccharide.

## 14.4 Chemical Properties of Monosaccharides
**Learning Goal:** Identify the products of oxidation or reduction of monosaccharides; determine whether a carbohydrate is a reducing sugar.

The aldehyde group in an aldose can be oxidized to a carboxylic acid, while the carbonyl group in an aldose or a ketose can be reduced to give a hydroxyl group. Monosaccharides that are reducing sugars have an aldehyde group in the open chain that is oxidized.

## 14.5 Disaccharides
**Learning Goal:** Describe the monosaccharide units and linkages in disaccharides.

Disaccharides are two monosaccharide units joined together by a glycosidic bond. In the common disaccharides maltose, lactose, and sucrose, there is at least one glucose unit.

## 14.6 Polysaccharides
**Learning Goal:** Describe the structural features of amylose, amylopectin, glycogen, and cellulose.

Polysaccharides are polymers of monosaccharide units. Amylose is an unbranched chain of glucose with $\alpha$-1,4-glycosidic bonds, and amylopectin is a branched polymer of glucose with $\alpha$-1,4- and $\alpha$-1,6-glycosidic bonds. Glycogen, the storage form of glucose in animals, is similar to amylopectin with more branching. Cellulose is also a polymer of glucose, but in cellulose the glycosidic bonds are $\beta$-1,4-bonds rather than $\alpha$-1,4-bonds in amylose.

# SUMMARY OF CARBOHYDRATES

| Carbohydrate | Food Sources | Monosaccharides |
|---|---|---|
| **Monosaccharides** | | |
| Glucose | Fruit juices, honey, corn syrup | |
| Galactose | Lactose hydrolysis | |
| Fructose | Fruit juices, honey, sucrose hydrolysis | |
| **Disaccharides** | | **Monosaccharides** |
| Maltose | Germinating grains, starch hydrolysis | Glucose + glucose |
| Lactose | Milk, yogurt, ice cream | Glucose + galactose |
| Sucrose | Sugar cane, sugar beets | Glucose + fructose |
| **Polysaccharides** | | |
| Amylose | Rice, wheat, grains, cereals | Unbranched polymer of glucose joined by $\alpha$-1,4-glycosidic bonds |
| Amylopectin | Rice, wheat, grains, cereals | Branched polymer of glucose joined by $\alpha$-1,4- and $\alpha$-1,6-glycosidic bonds |
| Glycogen | Liver, muscles | Highly branched polymer of glucose joined by $\alpha$-1,4- and $\alpha$-1,6-glycosidic bonds |
| Cellulose | Plant fiber, bran, beans, celery | Unbranched polymer of glucose joined by $\beta$-1,4-glycosidic bonds |

# SUMMARY OF REACTIONS

## FORMATION OF DISACCHARIDES

Glycosidic bond

Monosaccharide + Monosaccharide → Disaccharide + $H_2O$

## OXIDATION AND REDUCTION OF MONOSACCHARIDES

D-Glucitol ← Reduction — D-Glucose — Oxidation → D-Gluconic acid

## HYDROLYSIS OF DISACCHARIDES

Sucrose + $H_2O$ ⟶ glucose + fructose
Lactose + $H_2O$ ⟶ glucose + galactose
Maltose + $H_2O$ ⟶ glucose + glucose

# KEY TERMS

**aldose** A monosaccharide that contains an aldehyde group.

**amylopectin** A branched-chain polymer of starch composed of glucose units joined by α-1,4- and α-1,6-glycosidic bonds.

**amylose** An unbranched polymer of starch composed of glucose units joined by α-1,4-glycosidic bonds.

**carbohydrate** A simple or complex sugar composed of carbon, hydrogen, and oxygen.

**cellulose** An unbranched polysaccharide composed of glucose units linked by β-1,4-glycosidic bonds that cannot be hydrolyzed by the human digestive system.

**disaccharides** Carbohydrates composed of two monosaccharides joined by a glycosidic bond.

**fructose** A monosaccharide that is also called levulose and fruit sugar and is found in honey and fruit juices; it is combined with glucose in sucrose.

**galactose** A monosaccharide that occurs combined with glucose in lactose.

**glucose** The most prevalent monosaccharide in the diet. An aldohexose found in fruits, vegetables, corn syrup, and honey that is also known as blood sugar and dextrose. Most polysaccharides are polymers of glucose.

**glycogen** A polysaccharide formed in the liver and muscles for the storage of glucose as an energy reserve. It is composed of glucose

in a highly branched polymer joined by α-1,4- and α-1,6-glycosidic bonds.

**glycosidic bond** The bond that forms when the hydroxyl group of one monosaccharide reacts with the hydroxyl group of another monosaccharide. It is the type of bond that links monosaccharide units in di- or polysaccharides.

**Haworth structure** The ring structure of a monosaccharide.

**ketose** A monosaccharide that contains a ketone group.

**lactose** A disaccharide consisting of glucose and galactose found in milk and milk products.

**maltose** A disaccharide consisting of two glucose units; it is obtained from the hydrolysis of starch and in germinating grains.

**monosaccharide** A polyhydroxy compound that contains an aldehyde or ketone group.

**polysaccharides** Polymers of many monosaccharide units, usually glucose. Polysaccharides differ in the types of glycosidic bonds and the amount of branching in the polymer.

**reducing sugar** A carbohydrate with an aldehyde group capable of reducing the $Cu^{2+}$ in Benedict's reagent.

**sucrose** A disaccharide composed of glucose and fructose; a nonreducing sugar, commonly called table sugar or sugar.

# UNDERSTANDING THE CONCEPTS

**14.43** Isomaltose, obtained from the breakdown of starch, has the following Haworth structure:

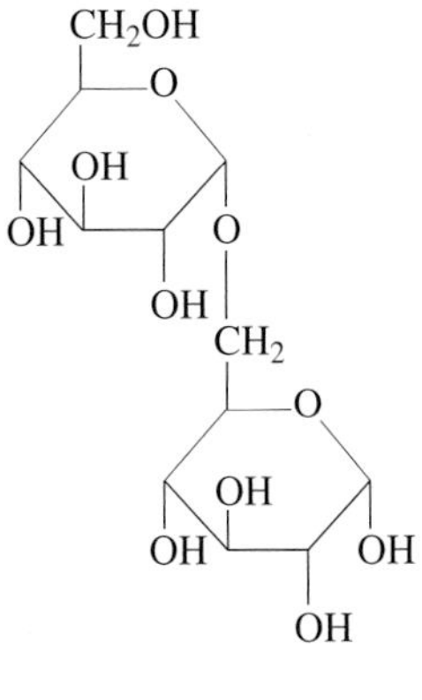

**a.** Is isomaltose a mono-, di-, or polysaccharide?
**b.** What are the monosaccharides in isomaltose?
**c.** What is the glycosidic link in isomaltose?
**d.** Is this the $\alpha$ or $\beta$ form of isomaltose?
**e.** Is isomaltose a reducing sugar?

**14.44** Sophorose, a carbohydrate found in certain types of beans, has the following Haworth structure:

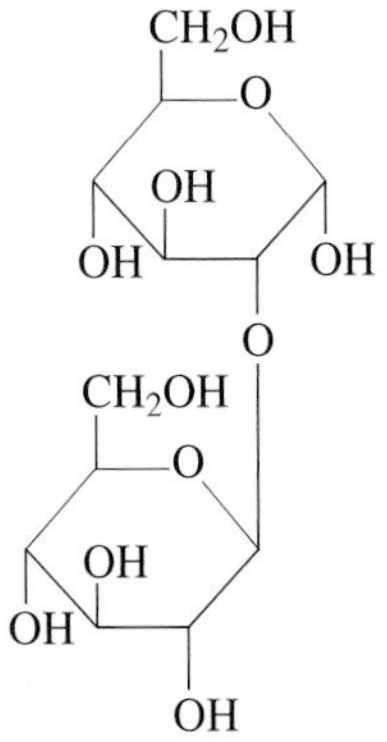

**a.** Is sophorose a mono-, di-, or polysaccharide?
**b.** What are the monosaccharides in sophorose?
**c.** What is the glycosidic link in sophorose?
**d.** Is this the $\alpha$ or $\beta$ form of sophorose?
**e.** Is sophorose a reducing sugar?

**14.45** Melezitose, a carbohydrate found in tree sap, has the following Haworth structure:

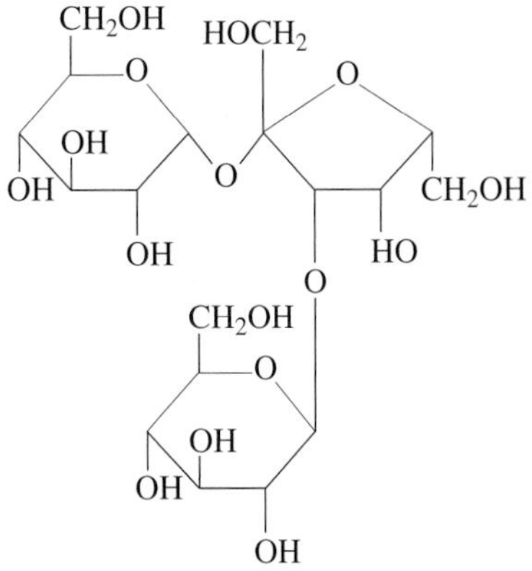

**a.** Is melezitose a mono-, di-, tri-, or polysaccharide?
**b.** What aldohexose is part of melezitose?

**14.46** What are the disaccharides and polysaccharides present in each of the following?

(a)  (b)

(c)  (d)

# ADDITIONAL QUESTIONS AND PROBLEMS

**14.47** What are the differences in the Fischer projections of D-fructose and D-galactose?

**14.48** What are the differences in the Fischer projections of D-glucose and D-fructose?

**14.49** What are the differences in the Fischer projections of D-galactose and L-galactose?

**14.50** What are the differences in the Haworth structures of $\alpha$-D-glucose and $\beta$-D-glucose?

**14.51** Consider the sugar D-gulose.

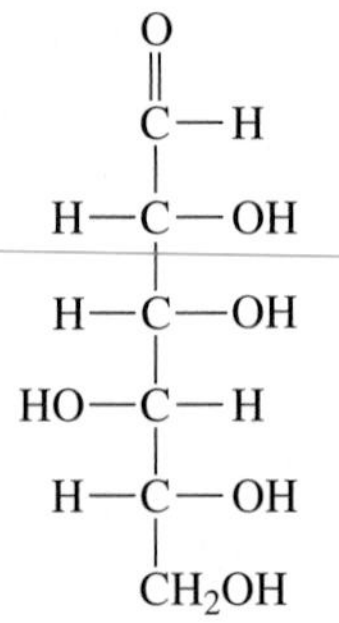

D-Gulose

**a.** Draw the Fisher projection for L-gulose.
**b.** Draw the Haworth structures for $\alpha$- and $\beta$-D-gulose.

**14.52** Consider the open-chain structure for D-gulose in question 14.51.
**a.** Draw the structure and name the product formed by the reduction of D-gulose.
**b.** Draw the structure and name the product formed by the oxidation of D-gulose.

**14.53** D-Sorbitol, a sweetener found in seaweed and berries, contains only hydroxyl functional groups. When D-sorbitol is oxidized, it forms D-glucose. Draw the Fischer projection of D-sorbitol.

**14.54** Raffinose is a trisaccharide found in Australian manna and in cottonseed meal. It is composed of three different monosaccharides. Identify the monosaccharides in raffinose.

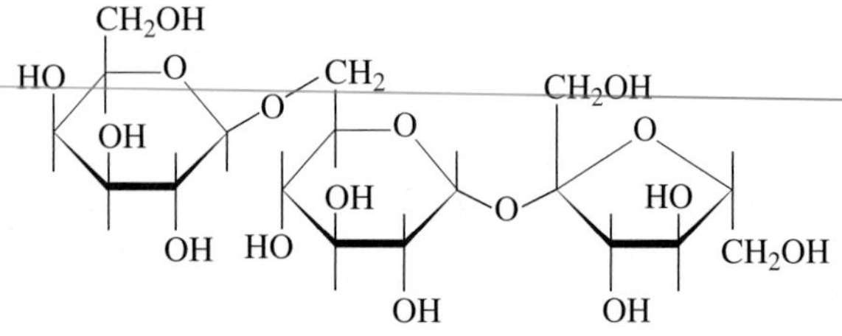

**14.55** If $\alpha$-galactose is dissolved in water, $\beta$-galactose is eventually present. Explain how this occurs.

**14.56** Why are lactose and maltose reducing sugars, but sucrose is not?

**14.57** $\beta$-Cellobiose is a disaccharide obtained from the hydrolysis of cellulose. It is quite similar to maltose except it has a $\beta$-1,4-glycosidic bond. Draw the Haworth structure of $\beta$-cellobiose.

**14.58** The disaccharide trehalose found in mushrooms is composed of two $\alpha$-D-glucose molecules joined by an $\alpha$-1,1-glycosidic bond. Draw the Haworth structure of trehalose.

## CHALLENGE QUESTIONS

**14.59** Gentiobiose is found in saffron.
**a.** Gentiobiose contains two glucose molecules linked by a $\beta$-1,6-glycosidic bond. Draw the Haworth structure of $\beta$-gentiobiose.
**b.** Is gentiobiose a reducing sugar? Explain.

**14.60** Identify the open-chain formulas that match each of the following:
**a.** the L-isomer of mannose
**b.** a ketopentose
**c.** an aldopentose
**d.** a ketohexose

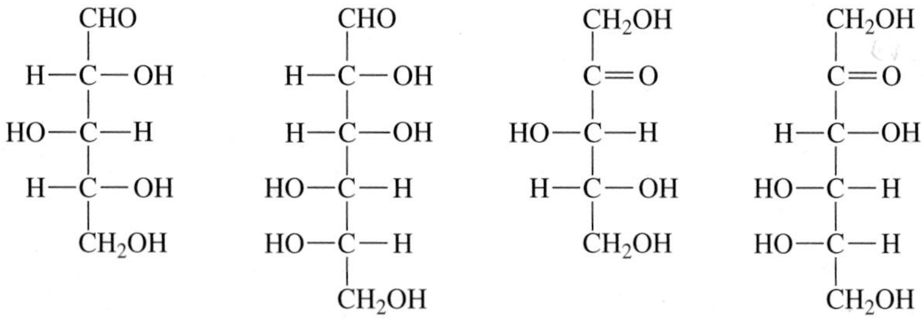

## ANSWERS

### Answers to Study Checks

**14.1**  a monosaccharide

**14.2**

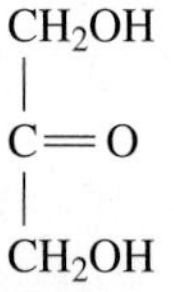

**14.3**

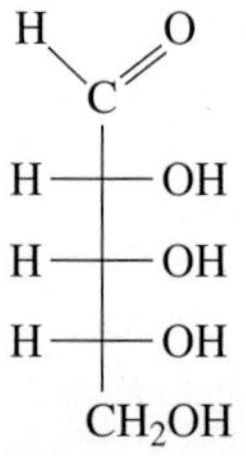

**14.4** Ribulose is a ketopentose.

**14.5**

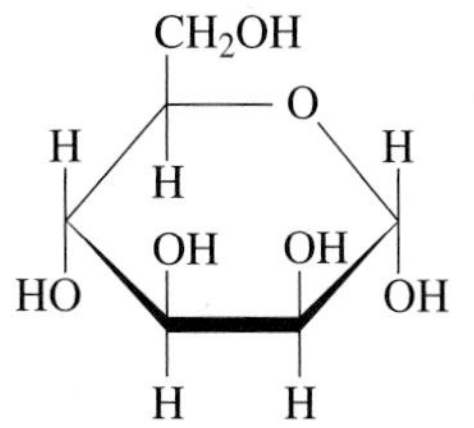

**14.6** This indicates a high level of reducing sugar (probably glucose) in the urine. One common cause of this condition is diabetes mellitus.

**14.7**

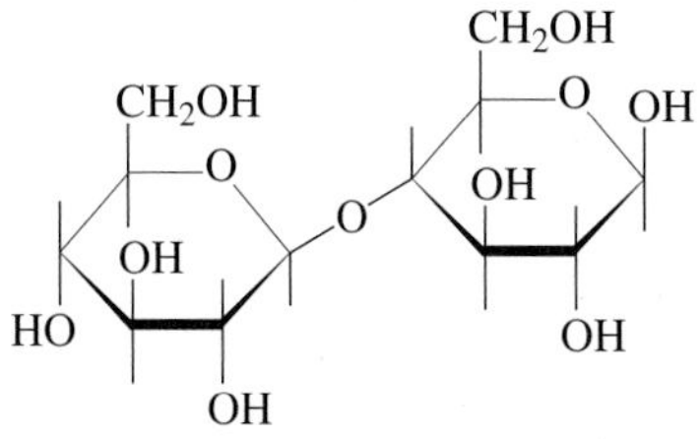

**14.8** Cellulose contains glucose units connected by $\beta$-1,4-glycosidic bonds, whereas the glucose units in amylose are connected by $\alpha$-1,4-glycosidic bonds.

## Answers to Selected Questions and Problems

**14.1** Photosynthesis requires $CO_2$, $H_2O$, and the energy from the sun. Respiration requires $O_2$ from the air and glucose from our foods.

**14.3** Monosaccharides can be a chain of three to eight carbon atoms, one in a carbonyl group as an aldehyde or ketone, and the rest attached to hydroxyl groups. A monosaccharide cannot be split or hydrolyzed into smaller carbohydrates. A disaccharide consists of two monosaccharide units joined together that can be split.

**14.5** Hydroxyl groups are found in all monosaccharides along with a carbonyl on the first or second carbon.

**14.7** A ketopentose contains hydroxyl and ketone functional groups and has five carbon atoms.

**14.9** **a.** ketose **b.** aldose **c.** ketose **d.** aldose **e.** aldose

**14.11** In the D isomer, the —OH on the chiral carbon atom at the bottom of the chain is on the right side, whereas in the L isomer the —OH appears on the left side.

**14.13** **a.** D **b.** D **c.** L **d.** D

**14.15** **a.**

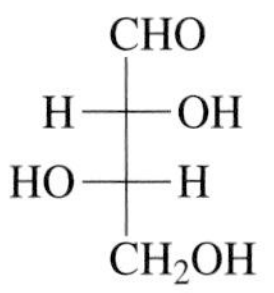

**b.**

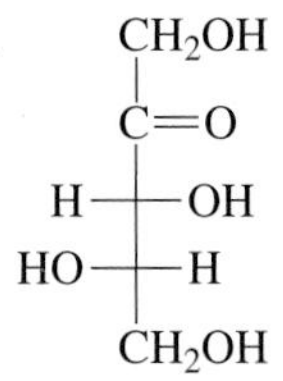

**c.**

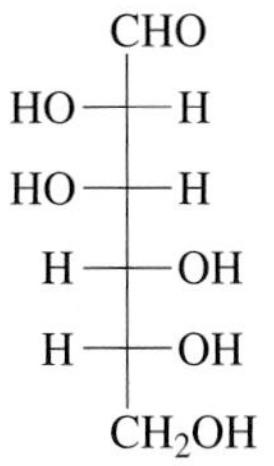

**d.**

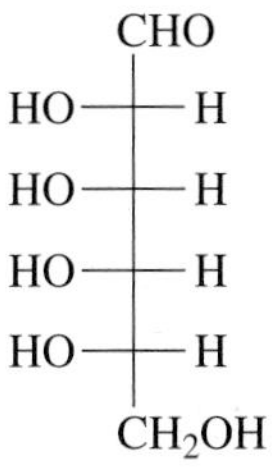

**14.17**

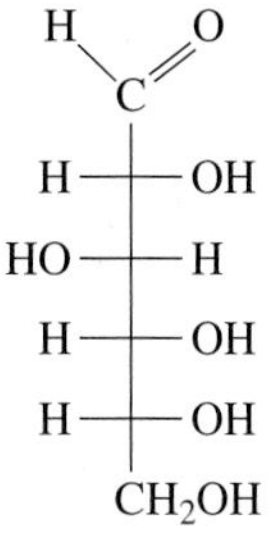

D-Glucose

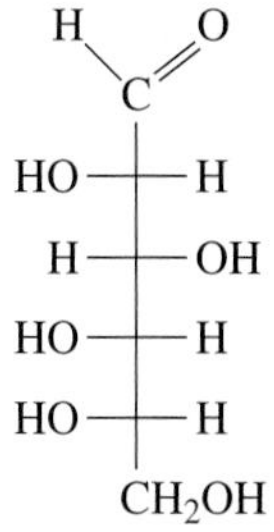

L-Glucose

**14.19** In D-galactose, the hydroxyl on carbon 4 extends to the left. In D-glucose, this hydroxyl goes to the right.

**14.21** **a.** glucose **b.** galactose **c.** fructose

**14.23** In the cyclic structure of glucose, there are five carbon atoms and an oxygen atom.

**14.25**

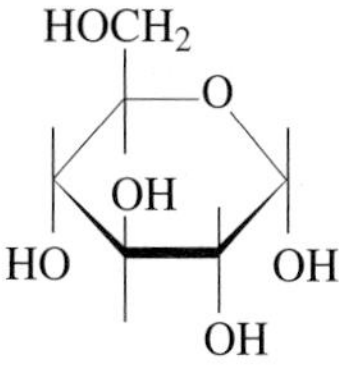

$\alpha$-D-Glucose

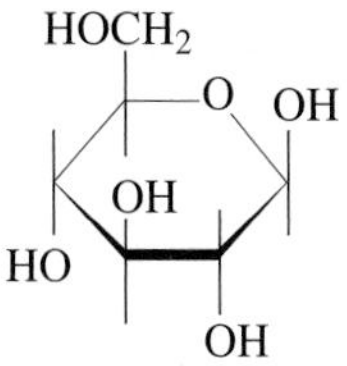

$\beta$-D-Glucose

**14.27** **a.** $\alpha$ form **b.** $\beta$ form

**14.29**

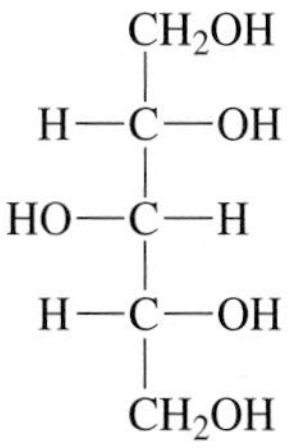

Xylitol

**14.31** Oxidation product:

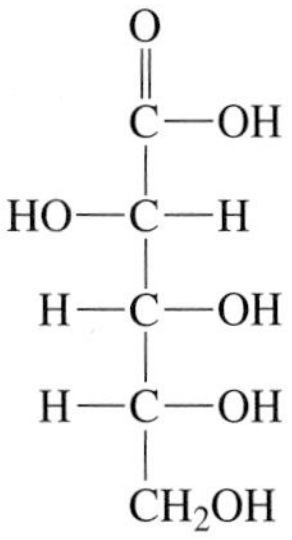

Reduction product (sugar alcohol):

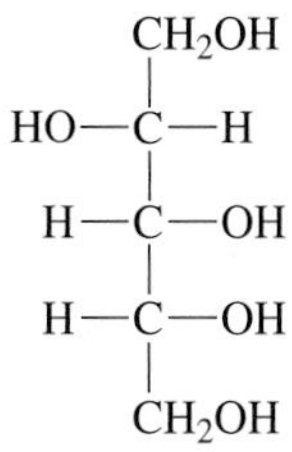

D-Arabitol

**14.33** **a.** galactose and glucose; $\beta$-1,4 bond; $\beta$-lactose
**b.** glucose and glucose; $\alpha$-1,4 bond; $\alpha$-maltose

**14.35** **a.** can be oxidized
**b.** can be oxidized

**14.37** **a.** sucrose
**b.** lactose
**c.** maltose
**d.** lactose

**14.39** **a.** Amylose is an unbranched polymer of glucose units joined by $\alpha$-1,4 bonds; amylopectin is a branched polymer of glucose joined by $\alpha$-1,4 and $\alpha$-1,6 bonds.
**b.** Amylopectin, which is produced in plants, is a branched polymer of glucose, joined by $\alpha$-1,4 and $\alpha$-1,6 bonds. Glycogen, which is produced in animals, is a highly branched polymer of glucose, joined by $\alpha$-1,4 and $\alpha$-1,6 bonds.

**14.41** **a.** cellulose
**b.** amylose, amylopectin
**c.** amylose
**d.** glycogen

**14.43** **a.** disaccharide
**b.** $\alpha$-glucose
**c.** $\alpha$-1,6
**d.** $\alpha$
**e.** Yes.

**14.45** **a.** trisaccharide
**b.** glucose

**14.47** D-Fructose is a ketohexose, whereas D-galactose is an aldohexose. In galactose, the —OH on carbon 4 is on the left; in fructose the —OH is on the right.

**14.49** D-Galactose is the mirror image of L-galactose. In D-galactose, the —OH on carbons 2 and 5 is on the right side, but on the left for carbons 3 and 4. In L-galactose, the —OHs are reversed; carbons 2 and 5 have —OH on the left, and carbons 3 and 4 have —OH on the right.

**14.51** **a.**

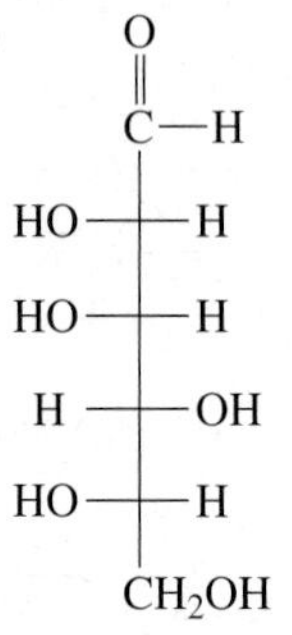

$$
\begin{array}{c}
\text{O} \\
\parallel \\
\text{C—H} \\
\text{HO}\!-\!\!-\!\text{H} \\
\text{HO}\!-\!\!-\!\text{H} \\
\text{H}\!-\!\!-\!\text{OH} \\
\text{HO}\!-\!\!-\!\text{H} \\
\text{CH}_2\text{OH}
\end{array}
$$
L-Gulose

**b.**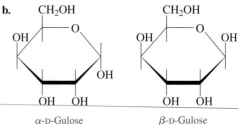

$\alpha$-D-Gulose          $\beta$-D-Gulose

**14.53**

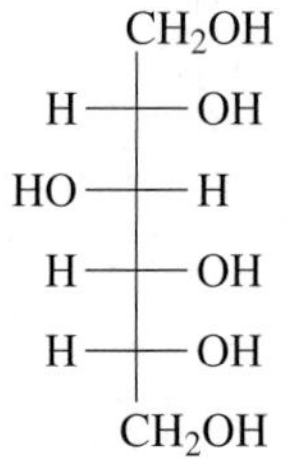

$$
\begin{array}{c}
\text{CH}_2\text{OH} \\
\text{H}\!-\!\!-\!\text{OH} \\
\text{HO}\!-\!\!-\!\text{H} \\
\text{H}\!-\!\!-\!\text{OH} \\
\text{H}\!-\!\!-\!\text{OH} \\
\text{CH}_2\text{OH}
\end{array}
$$

**14.55** When $\alpha$-galactose forms an open-chain structure, it can close to form either $\alpha$- or $\beta$-galactose.

**14.57**

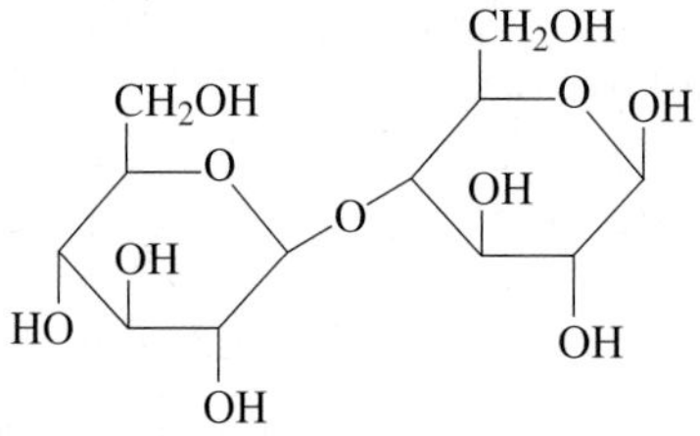

**14.59** **a.**

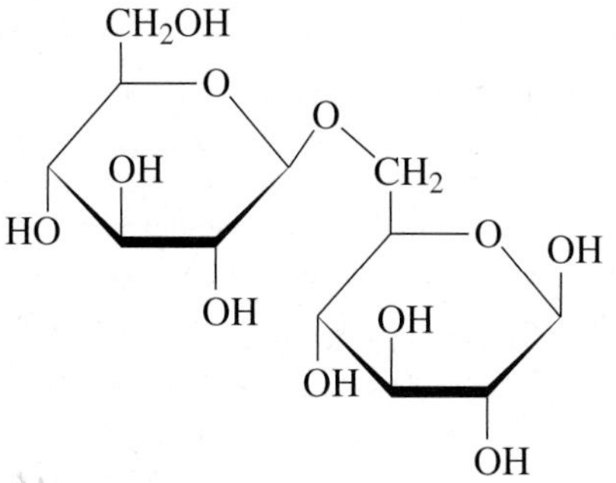

**b.** Yes. Gentiobiose is a reducing sugar. The ring on the right side can open up to form an aldehyde that can be oxidized.

# 15

# Lipids

Visit **www.chemplace.com** for extra quizzes, interactive tutorials, career resources, PowerPoint slides for chapter review, math help, and case studies.

*"In our toxicology lab, we measure the drugs in samples of urine or blood," says Penny Peng, assistant supervisor of chemistry, Toxicology Lab, Santa Clara Valley Medical Center. "But first we extract the drugs from the fluid and concentrate them so they can be detected in the machine we use. We extract the drugs by using different organic solvents such as methanol, ethyl acetate, or methylene chloride, and by changing the pH. We evaporate most of the organic solvent to concentrate any drugs it may contain. A small sample of the concentrate is placed into a machine called a gas chromatograph. As the gas moves over a column, the drugs in it are separated. From the results, we can identify as many as 10 to 15 different drugs from one urine sample."*

**W**hen we talk of fats and oils, waxes, steroids, cholesterol, and fat-soluble vitamins, we are discussing lipids. All the lipids are naturally occurring compounds that vary considerably in structure but share a common feature of being soluble in nonpolar solvents, but not in water. Fats, which are one family of lipids, have many functions in the body; they store energy and protect and insulate internal organs. Other types of lipids are found in nerve fibers and in hormones, which act as chemical messengers. Because they are not soluble in water, a major function of lipids as components of cell membranes is to separate the internal contents of cells from the external environment.

Many people are concerned about the amounts of saturated fats and cholesterol in our diets. Researchers suggest that saturated fats and cholesterol are associated with diseases such as diabetes; cancers of the breast, pancreas, and colon; and atherosclerosis, a condition in which deposits of lipid materials called *plaque* accumulate in the coronary blood vessels. In atherosclerosis, plaque restricts the flow of blood to the tissue, causing necrosis (death). An accumulation of plaque in the heart could result in a *myocardial infarction* (heart attack).

The American Institute for Cancer Research has recommended that we increase the fiber and starch content of our diets by adding more vegetables, fruits, and whole grains with moderate amounts of foods with low levels of fat and cholesterol such as fish, poultry, lean meats, and low-fat dairy products. The AICR has also suggested that we limit our intake of foods high in fat and cholesterol such as eggs, nuts, fatty or organ meats, cheeses, butter, and coconut and palm oil.

## LEARNING GOAL

Describe the classes of lipids.

WEB TUTORIAL
Fats

# 15.1 LIPIDS

**Lipids** are a family of biomolecules that have the common property of being soluble in organic solvents but not in water. The word "lipid" comes from the Greek word *lipos*, meaning "fat" or "lard." Typically, the lipid content of a cell can be extracted using a nonpolar solvent such as ether or chloroform. Lipids are an important feature in cell membranes, fat-soluble vitamins, and steroid hormones.

## Types of Lipids

Within the lipid family, there are distinct structures that distinguish the different types of lipids. Lipids such as waxes, fats, oils, and glycerophospholipids are esters that can be hydrolyzed to give fatty acids along with other products including an alcohol. Steroids are characterized by the steroid nucleus of four fused carbon rings. They do not contain fatty acids and cannot be hydrolyzed. Figure 15.1 illustrates the types and general structure of lipids we will discuss in this chapter.

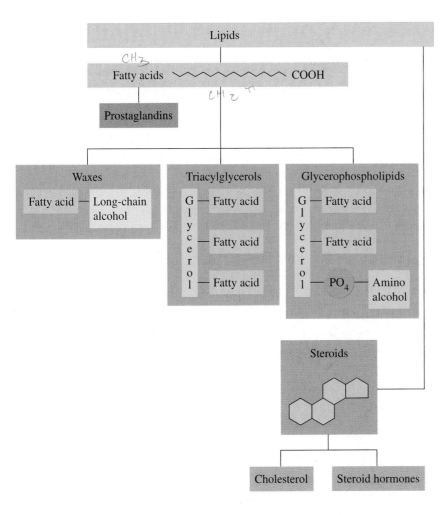

**FIGURE 15.1** Structures for some classes of lipids that are naturally occurring compounds in cells and tissues.

**Q** What chemical property do waxes, triacylglycerols, and steroids have in common?

## Explore Your World

### Solubility of Fats and Oils

Place some water in a small bowl. Add a drop of a vegetable oil. Then add a few more drops of the oil. Now add a few drops of liquid soap and mix. Record your observations.

Place a small amount of fat such as margarine, butter, shortening, or vegetable oil on a dish or plate. Run water over it. Record your observations. Mix some soap with the fat substance and run water over it again. Record your observations.

#### QUESTIONS

1. Do the drops of oil in the water separate or do they come together? Explain.
2. How does the soap affect the oil layer?
3. Why don't the fats on the dish or plate wash off with water?
4. In general, what is the solubility of lipids in water?
5. Why does soap help to wash the fats off the plate?

---

**SAMPLE PROBLEM** 15.1

### ■ Classes of Lipids

What type of lipid does not contain fatty acids?

**SOLUTION**

The steroids are a group of lipids with no fatty acids.

**STUDY CHECK**

What are the components of a glycerophospholipid?

---

## QUESTIONS AND PROBLEMS

### Lipids

**15.1** What are some functions of lipids in the body?

**15.2** What are some of the different kinds of lipids?

**15.3** Lipids are not soluble in water. Are lipids polar or nonpolar molecules?

**15.4** Which of the following solvents might be used to dissolve an oil stain?
   **a.** water                 **b.** $CCl_4$
   **c.** diethyl ether          **d.** benzene
   **e.** NaCl solution

**LEARNING GOAL**

Write structures of fatty acids, and identify each as saturated or unsaturated.

# 15.2 FATTY ACIDS

The fatty acids are the simplest type of lipids and are found as components in more complex lipids. A **fatty acid** contains a long, unbranched carbon chain attached to a carboxylic acid group at one end. Although the carboxylic acid part is hydrophilic, the long hydrophobic carbon chain makes long-chain fatty acids insoluble in water.

Naturally occurring fatty acids have an even number of carbon atoms, usually between 10 and 20. An example of a fatty acid is lauric acid, a 12-carbon acid found in coconut oil. In a simplified structure of a fatty acid called a line–bond formula, the carbon chain is written as a zigzag line that indicates the bonds between carbon atoms. In a zigzag line representation of a fatty acid, the ends and bends of the zigzag line are the carbon atoms. The structural formula of lauric acid can be written in several forms as follows.

## Writing Formulas for Lauric Acid

$$CH_3{-}(CH_2)_{10}{-}\overset{\overset{\displaystyle O}{\|}}{C}{-}OH \qquad CH_3{-}(CH_2)_{10}{-}COOH$$

$$CH_3{-}CH_2{-}CH_2{-}CH_2{-}CH_2{-}CH_2{-}CH_2{-}CH_2{-}CH_2{-}CH_2{-}CH_2{-}C\begin{smallmatrix}\diagup\!\!\diagup O\\ \diagdown OH\end{smallmatrix}$$

Condensed structural formula

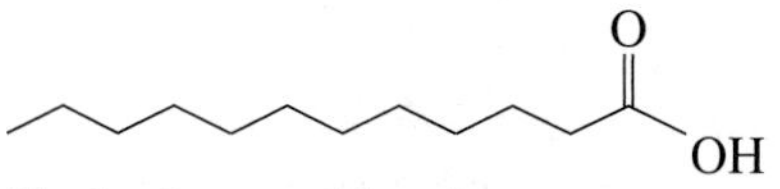

Line-bond structural formula

In a **saturated fatty acid**, the long carbon chain is like an alkane because there are only single carbon–carbon bonds. In a **monounsaturated fatty acid**, the long carbon chain has one double bond, which makes its properties similar to an alkene. In a **polyunsaturated fatty acid**, there are at least two carbon–carbon double bonds. Table 15.1 lists some of the typical fatty acids in lipids.

## Cis and Trans Isomers of Unsaturated Fatty Acids

Unsaturated fatty acids can be written as cis and trans isomers in the same way as the cis and trans alkene structures we looked at in Chapter 11. For example, oleic acid, a monounsaturated fatty acid found in olives and corn, has one double bond at carbon 9. We can show its cis and trans structural formulas using the line–bond notation. The cis structure is the most prevalent isomer found in naturally occurring unsaturated fatty acids. In the cis isomer, the carbon chain has a "kink" at the double bond site. As we will see, the cis bond has a major impact on the physical properties and uses of unsaturated fatty acids.

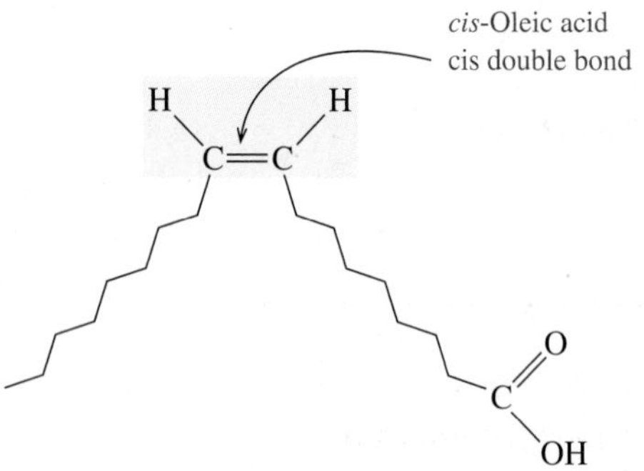

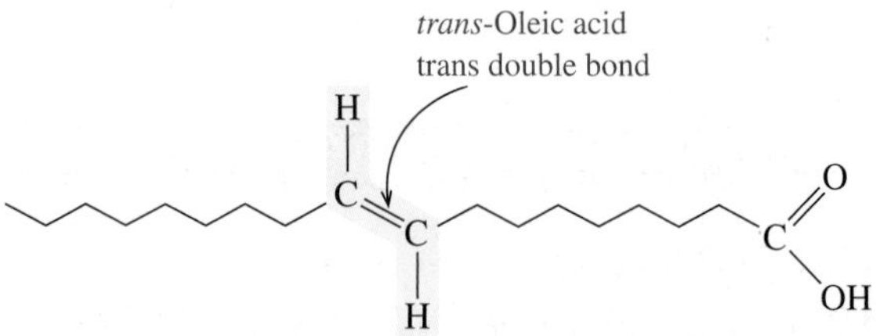

The human body is capable of synthesizing most fatty acids from carbohydrates or other fatty acids. However, humans do not synthesize sufficient amounts of polyunsaturated fatty acids such as linoleic acid, linolenic acid, and arachidonic acid. Because they must be obtained from the diet, they are known as *essential fatty acids*. In infants, a deficiency of

**TABLE 15.1** Structures and Melting Points of Common Fatty Acids

| Name | Carbon Atoms | Source | Melting Point (°C) | Structures |
|------|------|------|------|------|
| **Saturated Fatty Acids** | | | | |
| Lauric acid | 12 | Coconut | 43 | $CH_3-(CH_2)_{10}-COOH$ 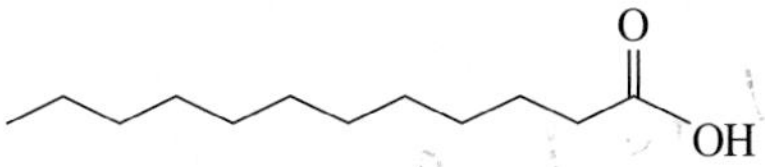 |
| Myristic acid | 14 | Nutmeg | 54 | $CH_3-(CH_2)_{12}-COOH$ 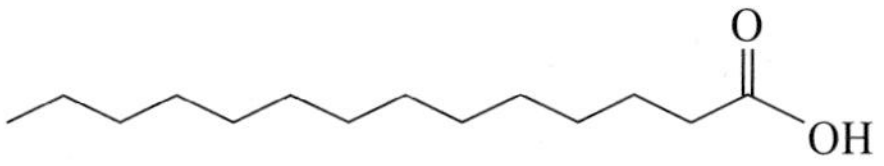 |
| Palmitic acid | 16 | Palm | 62 | $CH_3-(CH_2)_{14}-COOH$ 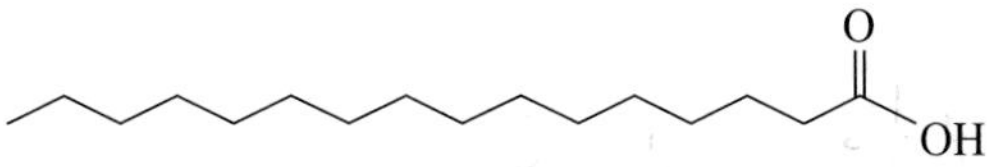 |
| Stearic acid | 18 | Animal fat | 69 | $CH_3-(CH_2)_{16}-COOH$ 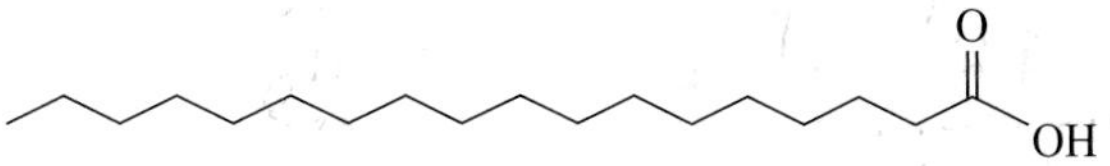 |
| **Monounsaturated Fatty Acids** | | | | |
| Palmitoleic acid | 16 | Butter | 0 | $CH_3-(CH_2)_5-CH=CH-(CH_2)_7-COOH$ 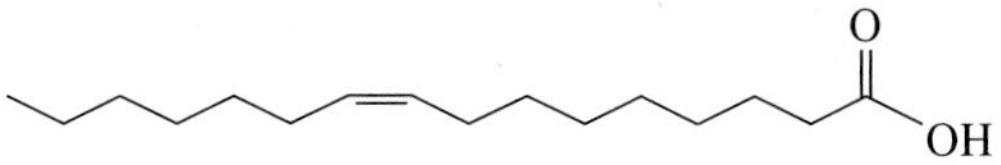 |
| Oleic acid | 18 | Olives, corn | 13 | $CH_3-(CH_2)_7-CH=CH-(CH_2)_7-COOH$ 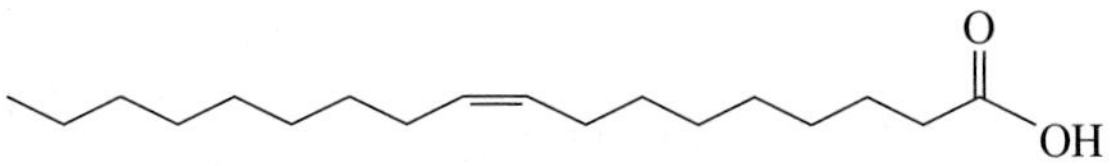 |
| **Polyunsaturated Fatty Acids** | | | | |
| Linoleic acid | 18 | Soybean, safflower, sunflower | −9 | $CH_3-(CH_2)_4-CH=CH-CH_2-CH=CH-(CH_2)_7-COOH$ 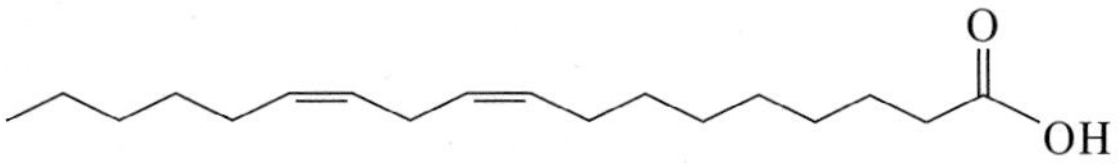 |
| Linolenic acid | 18 | Corn | −17 | $CH_3-CH_2-CH=CH-CH_2-CH=CH-CH_2-CH=CH-(CH_2)_7-COOH$ 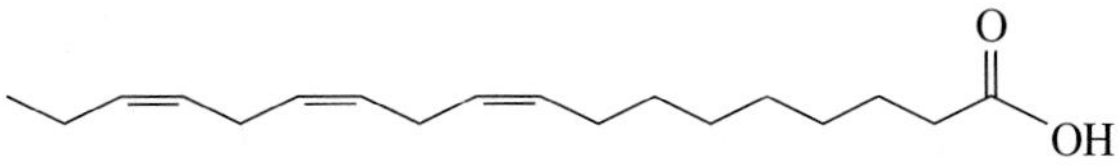 |
| Arachidonic acid | 20 | Meat, eggs, fish | −50 | $CH_3-(CH_2)_3-(CH_2-CH=CH)_4-(CH_2)_3-COOH$ 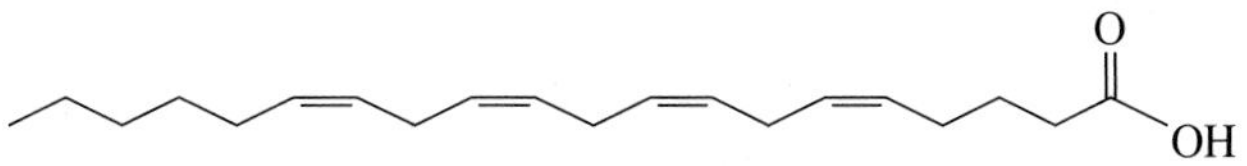 |

essential fatty acids can cause skin dermatitis. However, the role of fatty acids in adult nutrition is not well understood. Adults do not usually have a deficiency of essential fatty acids.

## Physical Properties of Fatty Acids

The saturated fatty acids fit close together in a regular pattern, which allows strong attractions to occur between the carbon chains. As a result, a significant amount of energy and high temperatures are required to separate the fatty acids and melt the fat. As the length of the carbon chain increases, more interactions occur between the carbon chains, requiring higher melting points. Saturated fatty acids are usually solids at room temperature.

In unsaturated fatty acids, the cis double bonds cause the carbon chain to bend or kink, giving the molecules an irregular shape. As a result, fewer interactions occur between carbon chains. Consequently, less energy is required to separate the molecules, making the melting points of unsaturated fats lower than those of saturated fats. (See Figure 15.2.) Most unsaturated fats are liquid oils at room temperature.

We might think of saturated fatty acids as chips with matching shapes that stack close together in a can. Then irregularly shaped chips would be like unsaturated fatty acids that do not pack close together.

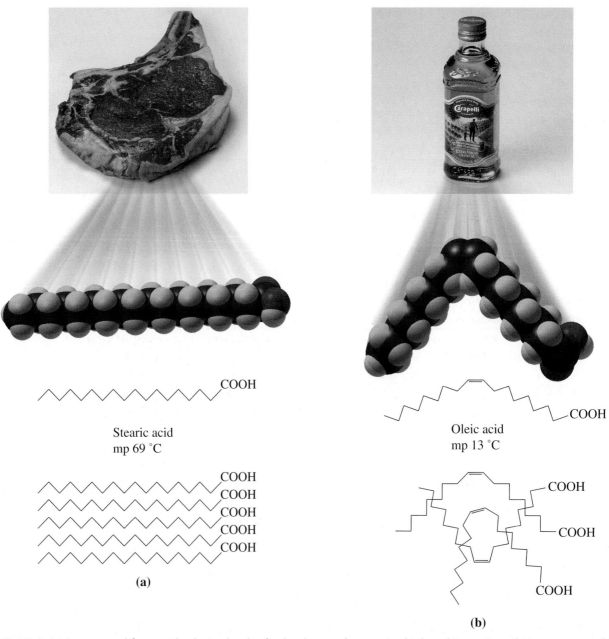

Stearic acid
mp 69 °C

Oleic acid
mp 13 °C

(a)

(b)

**FIGURE 15.2** **(a)** In saturated fatty acids, the molecules fit closely together to give high melting points. **(b)** In unsaturated fatty acids, molecules cannot pack closely together, resulting in lower melting points.
**Q** Why does the cis double bond affect the melting points of unsaturated fatty acids?

## SAMPLE PROBLEM 15.2

### ■ Structures and Properties of Fatty Acids

Consider the condensed structural formula of oleic acid.

$$CH_3-(CH_2)_7-CH=CH-(CH_2)_7-\overset{\overset{\displaystyle O}{\|}}{C}-OH$$

**a.** Why is the substance an acid?
**b.** How many carbon atoms are in oleic acid?
**c.** Is it a saturated or unsaturated fatty acid?
**d.** Is it most likely to be solid or liquid at room temperature?
**e.** Would it be soluble in water?

### SOLUTION

**a.** Oleic acid contains a carboxylic acid group.
**b.** It contains 18 carbon atoms.
**c.** It is an unsaturated fatty acid.
**d.** It is liquid at room temperature.
**e.** No, its long hydrocarbon chain makes it insoluble in water.

### STUDY CHECK

Palmitoleic acid is a fatty acid with the following formula:

$$CH_3-(CH_2)_5-CH=CH-(CH_2)_7-\overset{\overset{\displaystyle O}{\|}}{C}-OH$$

**a.** How many carbon atoms are in palmitoleic acid?
**b.** Is it a saturated or unsaturated fatty acid?
**c.** Is it most likely to be solid or liquid at room temperature?

## Prostaglandins

**Prostaglandins** are hormone-like substances produced in low amounts in most cells of the body. The various kinds of prostaglandins are formed from arachidonic acid, the polyunsaturated fatty acid with 20 carbon atoms. Prostaglandins are sometimes referred to as eicosanoids (*eicos* is the Greek word for 20). A prostaglandin with a ketone group on carbon 9 is designated as prostaglandin E (PGE), and as prostaglandin F (PGF) if carbon 9 is bonded to a hydroxyl group. Most prostaglandins have a hydroxyl group on carbon 11 and carbon 15. The number of double bonds is shown as a subscript 1 or 2.

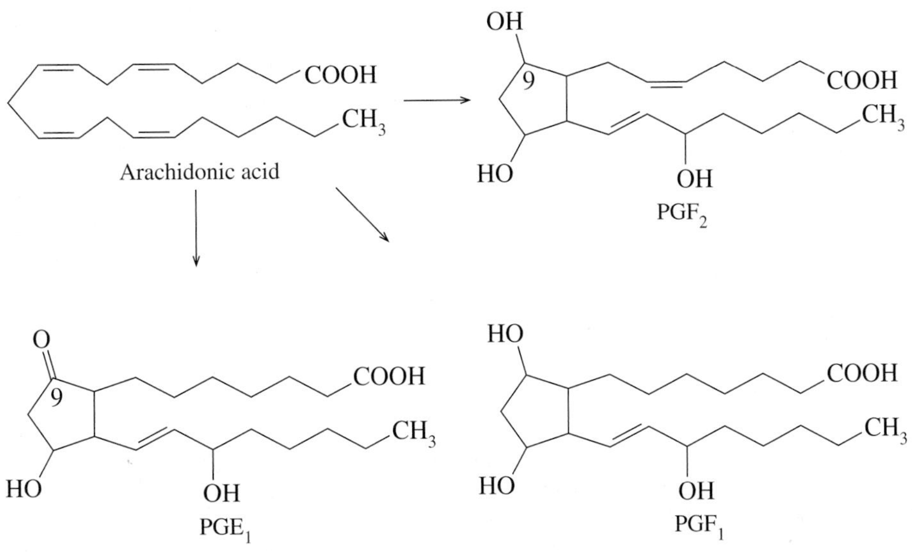

Arachidonic acid

$PGF_2$

$PGE_1$

$PGF_1$

# Health Note

## Omega-3 Fatty Acids in Fish Oils

Because unsaturated fats are now recognized as being more beneficial to health than saturated fats, Americans have been changing their diets to include more unsaturated fats. Unsaturated fats contain two types of unsaturated fatty acids, omega-3 and omega-6. Counting from the $CH_3$— end, the first double bond occurs at carbon 6 in an omega-6 fatty acid, whereas in the omega-3 type, the first double bond occurs at the third carbon.

The omega-6 fatty acids are mostly found in grains, oils from plants, and eggs, and the omega-3 fatty acids are mostly found in cold-water fish such as tuna and salmon. The benefits of omega-3 fatty acids were first recognized when a study was conducted of the diets of the Inuit peoples of Alaska who have a high-fat diet and high levels of blood cholesterol, but a very low occurrence of coronary heart disease. The fats in the Inuit diet are omega-3 fatty acids obtained primarily from fish.

In coronary heart disease, cholesterol forms plaque that adheres to the walls of the blood vessels. Blood pressure rises as blood has to squeeze through a smaller opening in the blood vessel. As more plaque forms, there is also a possibility of blood clots blocking the blood vessels and causing a heart attack. Omega-3 fatty acids lower the tendency of blood platelets to stick together, thereby reducing the possibility of blood clots. However, high levels of omega-3 fatty acids can increase bleeding if the ability of the platelets to form blood clots is reduced too much. It does seem that a diet that includes fish such as salmon, tuna, and herring can provide higher amounts of the omega-3 fatty acids, which help lessen the possibility of developing heart disease.

**Omega-6 Fatty Acids**

Linoleic acid    $CH_3$—$(CH_2)_4$—$CH{=}CH$—$CH_2$—$CH{=}CH$—$(CH_2)_7$—$COOH$
                      1                    6

Arachidonic acid    $CH_3$—$(CH_2)_4$—$(CH{=}CH$—$CH_2)_4$—$(CH_2)_2$—$COOH$
                          1                          6

**Omega-3 Fatty Acids**

Linolenic acid    $CH_3$—$CH_2$—$(CH{=}CH$—$CH_2)_3$—$(CH_2)_6$—$COOH$
                       1        2           3

Eicosapentaenoic acid (EPA)    $CH_3$—$CH_2$—$(CH{=}CH$—$CH_2)_5$—$(CH_2)_2$—$COOH$
                                   1        2           3

Docosahexaenoic acid (DHA)    $CH_3$—$CH_2$—$(CH{=}CH$—$CH_2)_6$—$CH_2$—$COOH$
                                  1        2           3

Although prostaglandins are broken down quickly, they have potent physiological effects. Some prostaglandins increase blood pressure, and others lower blood pressure. Other prostaglandins stimulate contraction and relaxation in the smooth muscle of the uterus. When tissues are injured, arachidonic acid is converted to prostaglandins such as PGE and PGF that produce inflammation and pain in the area.

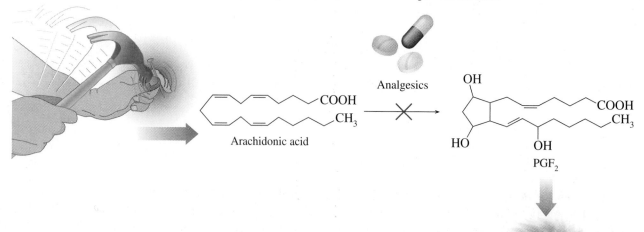

The treatment of pain, fever, and inflammation is based on inhibiting the enzymes that convert arachidonic acid to prostaglandins. Several nonsteroidal anti-inflammatory drugs (NSAIDs), such as aspirin, block the production of prostaglandins, and in doing so decrease pain and inflammation and reduce fever (antipyretics). Ibuprofen has similar anti-inflammatory and analgesic effects. Other NSAIDs include naproxen (Aleve and Naprosyn), ketoprofen (Actron), and nabumetone (Relafen). Long-term use of such products can result in liver, kidney, and gastrointestinal damage. Some forms of PGE are being tested as inhibitors of gastric secretion for use in the treatment of stomach ulcers.

Aspirin (acetylsalicylic acid)    Ibuprofen (Advil, Motrin)    Naproxen (Aleve, Naprosyn)

## QUESTIONS AND PROBLEMS

### Fatty Acids

**15.5** Describe some similarities and differences in the structures of a saturated fatty acid and an unsaturated fatty acid.

**15.6** Stearic acid and linoleic acid both have 18 carbon atoms. Why does stearic acid melt at 69 °C, but linoleic acid melts at −9 °C?

**15.7** Write the line–bond structure of the following fatty acids:
a. palmitic acid      b. oleic acid

**15.8** Write the line–bond structure of the following fatty acids:
a. stearic acid      b. linoleic acid

**15.9** Which of the following fatty acids are saturated, and which are unsaturated?
a. lauric acid      b. linolenic acid
c. palmitoleic acid      d. stearic acid

**15.10** Which of the following fatty acids are saturated, and which are unsaturated?
a. linoleic acid      b. palmitic acid
c. myristic acid      d. oleic acid

**15.11** How does the structure of a fatty acid with a cis double bond differ from the structure of a fatty acid with a trans double bond?

**15.12** In each pair, identify the fatty acid with the lower melting point. Explain.
a. myristic acid and stearic acid
b. stearic acid and linoleic acid
c. oleic acid and linolenic acid

**15.13** What is the difference in the location of the first double bond in an omega-3 and an omega-6 fatty acid? (See Health Note "Omega-3 Fatty Acids in Fish Oils.")

**15.14** a. What are some sources of omega-3 and omega-6 fatty acids? (See Health Note "Omega-3 Fatty Acids in Fish Oils.")
b. How may omega-3 fatty acids help in lowering the risk of heart disease?

**15.15** Compare the structures and functional groups between arachidonic acid and prostaglandin PGE$_1$?

**15.16** Compare the structures and functional groups between PGF$_1$ and PGF$_2$?

**15.17** What are some effects of prostaglandins in the body?

**15.18** How does an anti-inflammatory drug reduce inflammation?

# 15.3 WAXES, FATS, AND OILS

Waxes are found in many plants and animals. Coatings of carnauba wax on fruits and the leaves and stems of plants help to prevent loss of water and damage from pests. Waxes on the skin, fur, and feathers of animals provide a waterproof coating. A **wax** is an ester of a saturated fatty acid and a long-chain alcohol, each containing from 14 to 30 carbon atoms.

## LEARNING GOAL

Write the structural formula of a wax, fat, or oil produced by the reaction of a fatty acid and an alcohol or glycerol.

Ester bond

$$Fatty\ acid—C—O—Long\text{-}chain\ alcohol$$

Wax

**TABLE 15.2** Some Typical Waxes

| Type | Structural Formula | Source | Uses |
|---|---|---|---|
| Beeswax | $CH_3(CH_2)_{14}—C—O—(CH_2)_{29}CH_3$ (C=O) | Honeycomb | Candles, shoe polish, wax paper |
| Carnauba wax | $CH_3(CH_2)_{24}—C—O—(CH_2)_{29}CH_3$ (C=O) | Brazilian palm tree | Waxes for furniture, cars, floors, shoes |
| Jojoba wax | $CH_3(CH_2)_{18}—C—O—(CH_2)_{19}CH_3$ (C=O) | Jojoba | Candles, soaps, cosmetics |

The formulas of some common waxes are given in Table 15.2. Beeswax obtained from honeycombs and carnauba wax from palm trees are used to give a protective coating to furniture, cars, and floors. Jojoba wax is used in making candles and cosmetics such as lipstick. Lanolin, a mixture of waxes obtained from wool, is used in hand and facial lotions to aid retention of water, softening the skin.

## Fats and Oils: Triacylglycerols

In the body, fatty acids are stored as fats and oils known as **triacylglycerols**. These substances, also called *triglycerides*, are triesters of glycerol (a trihydroxy alcohol) and fatty acids. The general formula of a triacylglycerol follows:

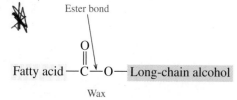

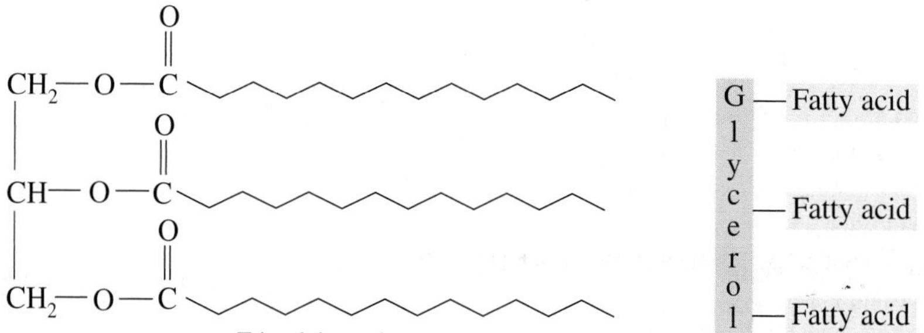

$$\begin{array}{l} CH_2—O—C(=O)\sim\sim\sim\sim \\ CH—O—C(=O)\sim\sim\sim\sim \\ CH_2—O—C(=O)\sim\sim\sim\sim \end{array}$$

Triacylglycerol

Glycerol — Fatty acid / Fatty acid / Fatty acid

In Chapter 13, we saw that esters are produced from a reaction between an alcohol and a carboxylic acid. In a triacylglycerol, three hydroxyl groups of glycerol form ester bonds with the carboxyl groups of fatty acid. For example, glycerol and three molecules of stearic acid form glyceryl tristearate which is commonly named tristearin.

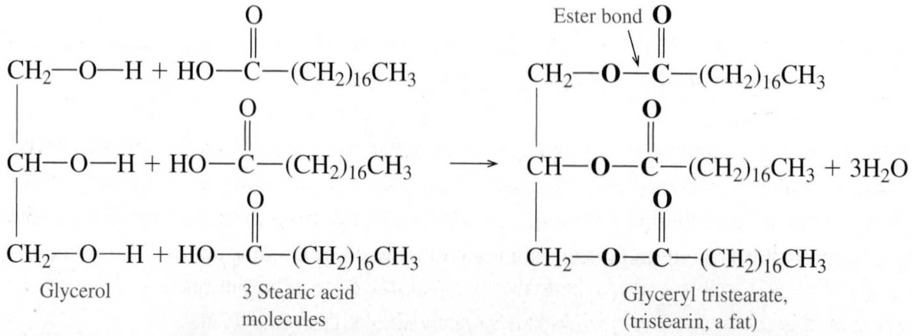

$$\begin{array}{l} CH_2—O—H\ +\ HO—C(=O)—(CH_2)_{16}CH_3 \\ CH—O—H\ +\ HO—C(=O)—(CH_2)_{16}CH_3 \\ CH_2—O—H\ +\ HO—C(=O)—(CH_2)_{16}CH_3 \end{array}$$

Glycerol          3 Stearic acid molecules

$$\longrightarrow$$

Ester bond

$$\begin{array}{l} CH_2—O—C(=O)—(CH_2)_{16}CH_3 \\ CH—O—C(=O)—(CH_2)_{16}CH_3\ +\ 3H_2O \\ CH_2—O—C(=O)—(CH_2)_{16}CH_3 \end{array}$$

Glyceryl tristearate, (tristearin, a fat)

Most fats and oils are mixed triacylglycerols that contain two or three different fatty acids. For example, a mixed triacylglycerol might be made from lauric acid, myristic acid, and palmitic acid. One possible structure for the mixed triacylglycerol follows:

$$CH_2-O-\overset{\displaystyle O}{\overset{\displaystyle \|}{C}}-(CH_2)_{10}CH_3 \quad \text{Lauric acid}$$

$$CH-O-\overset{\displaystyle O}{\overset{\displaystyle \|}{C}}-(CH_2)_{12}CH_3 \quad \text{Myristic acid}$$

$$CH_2-O-\overset{\displaystyle O}{\overset{\displaystyle \|}{C}}-(CH_2)_{14}CH_3 \quad \text{Palmitic acid}$$

A mixed triacylglycerol

Triacylglycerols are the major form of energy storage for animals. Animals that hibernate eat large quantities of plants, seeds, and nuts that contain high levels of fats and oils. They gain as much as 14 kilograms a week. As the external temperature drops, the animal goes into hibernation. The body temperature drops to nearly freezing, and there is a dramatic reduction in cellular activity and in respiration and heart rates. Animals that live in extremely cold climates hibernate for 4–7 months. During this time, stored fat is the only source of energy.

## SAMPLE PROBLEM

### ■ Writing Structures for a Triacylglycerol

Draw the condensed structural formula of glyceryl trioleate (triolein).

SOLUTION

Glyceryl trioleate (triolein) is the triacylglycerol that contains ester bonds between glycerol and three oleic acid molecules.

$$CH_2-O-\overset{\displaystyle O}{\overset{\displaystyle \|}{C}}-(CH_2)_7CH=CH(CH_2)_7CH_3$$

$$CH-O-\overset{\displaystyle O}{\overset{\displaystyle \|}{C}}-(CH_2)_7CH=CH(CH_2)_7CH_3$$

$$CH_2-O-\overset{\displaystyle O}{\overset{\displaystyle \|}{C}}-(CH_2)_7CH=CH(CH_2)_7CH_3$$

Glyceryl trioleate (triolein)

STUDY CHECK

Write the condensed structural formula of the triacylglycerol containing 3 molecules of myristic acid.

## Melting Points of Fats and Oils

A **fat** is a triacylglycerol that is solid at room temperature, and usually comes from animal sources such as meat, whole milk, butter, and cheese.

An **oil** is a triacylglycerol that is usually a liquid at room temperature and is obtained from a plant source. Olive oil and peanut oil are monounsaturated because they contain large amounts of oleic acid. Oils from corn, cottonseed, safflower, and sunflower are polyunsaturated because they contain large amounts of fatty acids with two or more double bonds. (See Figure 15.3.) Palm oil and coconut oil are solids at room temperature because they consist mostly of saturated fatty acids. The amounts of saturated, monounsaturated, and polyunsaturated fatty acids in some typical fats and oils are shown in Figure 15.4.

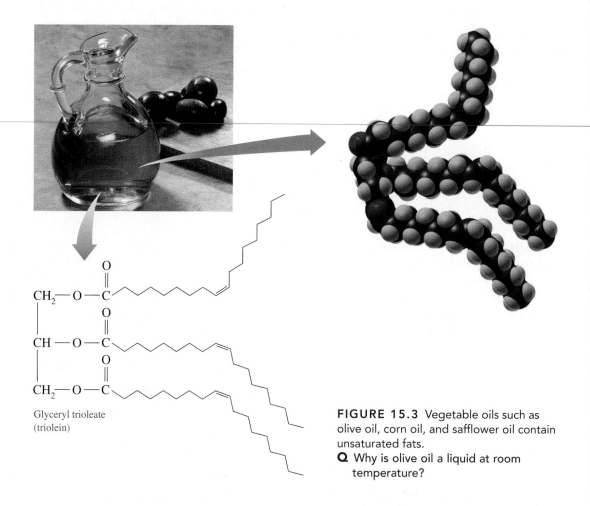

Glyceryl trioleate
(triolein)

**FIGURE 15.3** Vegetable oils such as olive oil, corn oil, and safflower oil contain unsaturated fats.
**Q** Why is olive oil a liquid at room temperature?

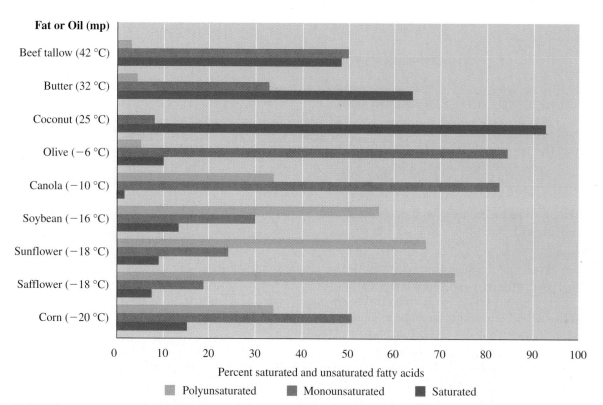

**Fat or Oil (mp)**

Beef tallow (42 °C)

Butter (32 °C)

Coconut (25 °C)

Olive (−6 °C)

Canola (−10 °C)

Soybean (−16 °C)

Sunflower (−18 °C)

Safflower (−18 °C)

Corn (−20 °C)

0   10   20   30   40   50   60   70   80   90   100

Percent saturated and unsaturated fatty acids

■ Polyunsaturated   ■ Monounsaturated   ■ Saturated

**FIGURE 15.4** Vegetable oils have low melting points because they have a higher percentage of unsaturated fatty acids than do animal fats.
**Q** Why is the melting point of butter higher than olive oil or canola oil?

Saturated fatty acids have higher melting points than unsaturated fatty acids because they pack together more tightly. Animal fats usually contain more saturated fatty acids than do vegetable oils. Therefore the melting points of animal fats are higher than those of vegetable oils.

## QUESTIONS AND PROBLEMS

### Waxes, Fats, and Oils

**15.19** Draw the structure of an ester in beeswax formed from myricyl alcohol, $CH_3(CH_2)_{29}OH$, and palmitic acid.

**15.20** Draw the structure of an ester in jojoba wax formed from arachidic acid, a 20-carbon saturated fatty acid, and 1-docosanol, $CH_3(CH_2)_{21}OH$.

**15.21** Draw the structure of a triacylglycerol that contains stearic acid and glycerol.

**15.22** A mixed triacylglycerol contains two palmitic acid molecules and one oleic acid molecule. Write two possible structures (isomers) for the compound.

**15.23** Draw the structure of glyceryl tripalmitate (tripalmitin).

**15.24** Draw the structure of glyceryl trioleate (triolein).

**15.25** Safflower oil is polyunsaturated, whereas olive oil is monounsaturated. Explain.

**15.26** Why does olive oil have a lower melting point than butter fat?

**15.27** Why does coconut oil, a vegetable oil, have a melting point similar to that of fats from animal sources?

**15.28** A label on a bottle of 100% sunflower seed oil states that it is lower in saturated fats than all the leading oils.
   **a.** How does the percentage of saturated fats in sunflower seed oil compare to that of safflower, corn, and canola oils? (See Figure 15.4.)
   **b.** Is the claim valid?

# 15.4 CHEMICAL PROPERTIES OF TRIACYLGLYCEROLS

**LEARNING GOAL**

Draw the structure of the product from the reaction of a triacylglycerol with hydrogen, or an acid or base.

The chemical reactions of the triacylglycerols (fats and oils) are the same as those we discussed for alkenes and esters. We will look at the hydrogenation (Chapter 11) and the hydrolysis and saponification (Chapter 13) of fats and oils.

## Hydrogenation

In the **hydrogenation** of an unsaturated fat, hydrogen is added to carbon–carbon double bonds to form carbon–carbon single bonds. The hydrogen gas is bubbled through the heated oil in the presence of a nickel catalyst.

$$-CH=CH- + H_2 \xrightarrow{Ni} -\overset{\displaystyle H}{\underset{\displaystyle H}{C}}-\overset{\displaystyle H}{\underset{\displaystyle H}{C}}-$$

For example, when hydrogen adds to all of the double bonds of glyceryl trioleate (triolein) using a nickel catalyst, the product is the saturated fat glyceryl tristearate (tristearin).

$$
\begin{array}{l}
CH_2-O-\overset{O}{\overset{\|}{C}}-(CH_2)_7CH=CH(CH_2)_7CH_3 \\
\quad\quad\quad\quad O \\
CH-O-\overset{\|}{C}-(CH_2)_7CH=CH(CH_2)_7CH_3 \;+\; 3H_2 \\
\quad\quad\quad\quad O \\
CH_2-O-\overset{\|}{C}-(CH_2)_7CH=CH(CH_2)_7CH_3
\end{array}
\xrightarrow{Ni}
\begin{array}{l}
CH_2-O-\overset{O}{\overset{\|}{C}}-(CH_2)_{16}CH_3 \\
\quad\quad\quad\quad O \\
CH-O-\overset{\|}{C}-(CH_2)_{16}CH_3 \\
\quad\quad\quad\quad O \\
CH_2-O-\overset{\|}{C}-(CH_2)_{16}CH_3
\end{array}
$$

Glyceryl trioleate
(triolein)

Glyceryl tristearate
(tristearin)

In commercial hydrogenation, the addition of hydrogen is stopped before all the double bonds in a liquid vegetable oil become completely saturated. Complete hydrogenation

## Health Note

### Olestra: A Fat Substitute

In 1968, food scientists designed an artificial fat called *olestra* as a source of nutrition for premature babies. However, olestra could not be digested and was never used for that purpose. Then scientists realized that olestra had the flavor and texture of a fat without the calories.

Olestra is manufactured by obtaining the fatty acids from the fats in cottonseed or soybean oils and bonding the fatty acids with the hydroxyl groups on sucrose. Chemically, olestra is composed of six to eight long-chain fatty acids attached by ester links to a sugar (sucrose) rather than to a glycerol molecule found in fats. This makes olestra a very large molecule that cannot be absorbed through the intestinal walls. The enzymes and bacteria in the intestinal tract are unable to break down the olestra molecule and it travels through the intestinal tract undigested.

In 1996, the Food and Drug Administration (FDA) approved olestra for use in potato chips, tortilla chips, crackers, and fried snacks. In 1996, olestra snack products were test marketed in parts of Iowa, Wisconsin, Indiana, and Ohio. By 1997, there were reports of some adverse reactions, including diarrhea, abdominal cramps, and anal leakage, indicating that olestra may act as a laxative in some people. However, the manufacturers contend there is no direct proof that olestra is the cause of those effects.

The large molecule of olestra also combines with fat-soluble vitamins (A, D, E, and K) as well as the carotenoids from the foods we eat before they can be absorbed through the intestinal wall. Carotenoids are plant pigments in fruits and vegetables that protect against cancer, heart disease, and macular degeneration, a form of blindness in the elderly. The FDA now requires manufacturers to add the four vitamins, but not the carotenoids, to olestra products. Snack foods made with olestra are now in supermarkets nationwide. Since there are already low-fat snacks on the market, it remains to be seen whether olestra will have any significant effect on reducing the problem of obesity.

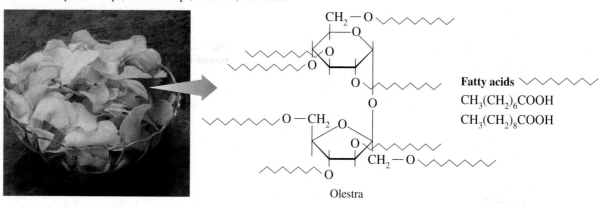

Fatty acids

$CH_3(CH_2)_6COOH$

$CH_3(CH_2)_8COOH$

Olestra

gives a very brittle product, whereas the partial hydrogenation of a liquid vegetable oil changes it to a soft, semisolid fat. As it becomes more saturated, the melting point increases, and the substance becomes more solid at room temperature. By controlling the amount of hydrogen, manufacturers can produce various types of products such as soft margarines, solid stick margarines, and solid shortenings. (See Figure 15.5.) Although these products now contain more saturated fatty acids than the original oils, they contain no cholesterol, unlike similar products from animal sources, such as butter and lard.

**FIGURE 15.5** Many soft margarines, stick margarines, and solid shortenings are produced by the partial hydrogenation of vegetable oils.

**Q** How does hydrogenation change the structure of the fatty acids in the vegetable oils?

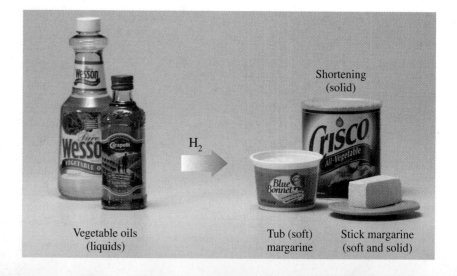

Vegetable oils (liquids)

$H_2$

Shortening (solid)

Tub (soft) margarine

Stick margarine (soft and solid)

# Health Note

## Trans Fatty Acids and Hydrogenation

In the early 1900s, margarine became a popular replacement for the highly saturated fats such as butter and lard. Margarine is produced by partially hydrogenating the unsaturated fats in vegetable oils such as safflower oil, corn oil, canola oil, cottonseed oil, and sunflower oil. Fats that are more saturated are more resistant to oxidation.

In vegetable oils, the unsaturated fats usually contain cis double bonds. As hydrogenation occurs, double bonds are converted to single bonds. However, some of the cis double bonds are converted to trans double bonds, which causes a change in the overall structure of the fatty acids. If the label on a product states that the oils have been "partially" or "fully hydrogenated," that product will also contain trans fatty acids. In the United States, it is estimated that 2–4% of our total calories come from trans fatty acids.

The concern about trans fatty acids is that their altered structure may make them behave like saturated fatty acids in the body. In the 1980s, research indicated that trans fatty acids have an effect on blood cholesterol similar to that of saturated fats, although study results vary. Several studies reported that trans fatty acids raise the levels of LDL-cholesterol, low-density lipoproteins containing cholesterol that can accumulate in the arteries. (LDLs and HDLs are described in the section on lipoproteins later in this chapter.) Some studies also report that trans fatty acids lower HDL-cholesterol, high-density lipoproteins that carry cholesterol to the liver to be excreted. But other studies did not report any decrease in HDL-cholesterol. In some American and European studies, an increased risk of breast cancer was associated with increased intake of trans fatty acids. However, these studies are not conclusive and not all studies have supported such findings. Current evidence does not yet indicate that the intake of trans fatty acids is a significant risk factor for heart disease. The trans fatty acids controversy will continue to be debated as more research is done.

Foods containing trans fatty acids include milk, bread, fried foods, ground beef, baked goods, stick margarine, butter, soft margarine, cookies, crackers, and vegetable shortening. The American Heart Association recommends that margarine should have no more than 2 grams of saturated fat per tablespoon and a liquid vegetable oil should be the first ingredient. They recommend the use of soft margarine, which is lower in trans fatty acids because soft margarine is only slightly hydrogenated, and diet margarine also, because it has less fat and therefore fewer trans fatty acids.

Many health organizations agree that fat should account for less than 30% of daily calories (the current average for Americans is 34%) and saturated fat should be less than 10% of total calories. Lowering the overall fat intake would also decrease the amount of trans fatty acids. The FDA and the U.S. Department of Agriculture are encouraging the use of new food labels to inform consumers of the fat content of food. The best advice may be to reduce total fat in the diet by using fats and oils sparingly, cooking with little or no fat, substituting olive oil or canola oil for other oils, and limiting the use of coconut oil and palm oil, which are high in saturated fatty acids.

There are several products including peanut butter and butter-like spreads on the market that have 0% trans fatty acids, as indicated on the labels. However, in the list of natural vegetable oils, such as soy and canola oil, there is also palm oil. Because palm oil has a melting point of 30 °C, it increases the overall melting point of the spread and gives a product that is solid at room temperature. However, palm oil contains high amounts of saturated fatty acids and has a similar effect in the body as do stearic acid (18 carbons) and fats derived from animal sources. Health experts recommend that we limit the amount of saturated fats, including palm oil, in our diets.

cis-Oleic acid

$H_2/Ni$

COOH    Double bond opens

Ni catalyst

$H_2$  Isomerization

Addition of $H_2$

COOH

COOH

Undesired side product (*trans*-oleic acid)

Desired saturated product (stearic acid)

## Hydrolysis

Triacylglycerols are hydrolyzed (split by water) in the presence of strong acids or digestive enzymes called *lipases*. The products of hydrolysis of the ester bonds are glycerol and three fatty acids. The polar glycerol is soluble in water, but the fatty acids with their long hydrocarbon chains are not.

Water adds to ester bonds

$$
\begin{array}{c}
\underset{\substack{\text{Glyceryl tripalmitate}\\ \text{(tripalmitin)}}}{
\begin{array}{l}
CH_2{-}O{-}\overset{\displaystyle O}{\overset{\|}{C}}{-}(CH_2)_{14}CH_3 \\
CH{-}O{-}\overset{\displaystyle O}{\overset{\|}{C}}{-}(CH_2)_{14}CH_3 \; + \; 3H_2O \\
CH_2{-}O{-}\overset{\displaystyle O}{\overset{\|}{C}}{-}(CH_2)_{14}CH_3
\end{array}}
\xrightarrow[\text{lipase}]{\substack{H^+ \\ \text{or}}}
\underset{\text{Glycerol}}{
\begin{array}{l}
CH_2{-}OH \\
CH{-}OH \; + \\
CH_2{-}OH
\end{array}}
\underset{\substack{3 \text{ Palmitic acid}\\ \text{molecules}}}{3HO{-}\overset{\displaystyle O}{\overset{\|}{C}}{-}(CH_2)_{14}CH_3}
\end{array}
$$

## Saponification

When a fat is heated with a strong base such as sodium hydroxide, saponification of the fat gives glycerol and the sodium salts of the fatty acids, which are soaps. When NaOH is used, a solid soap is produced that can be molded into a desired shape; KOH produces a softer, liquid soap. Oils that are polyunsaturated produce softer soaps. Names like "coconut" or "avocado shampoo" tell you the sources of the oil used in the reaction.

Fat or oil + strong base $\longrightarrow$ glycerol + salts of fatty acids (soaps)

$$
\begin{array}{c}
\underset{\substack{\text{Glyceryl tripalmitate}\\ \text{(tripalmitin)}}}{
\begin{array}{l}
CH_2{-}O{-}\overset{\displaystyle O}{\overset{\|}{C}}{-}(CH_2)_{14}CH_3 \\
CH{-}O{-}\overset{\displaystyle O}{\overset{\|}{C}}{-}(CH_2)_{14}CH_3 \; + \; 3NaOH \\
CH_2{-}O{-}\overset{\displaystyle O}{\overset{\|}{C}}{-}(CH_2)_{14}CH_3
\end{array}}
\longrightarrow
\underset{\text{Glycerol}}{
\begin{array}{l}
CH_2{-}OH \\
CH{-}OH \; + \\
CH_2{-}OH
\end{array}}
\underset{\substack{3 \text{ Sodium palmitate}\\ \text{(soap)}}}{3Na^+{}^-O{-}\overset{\displaystyle O}{\overset{\|}{C}}{-}(CH_2)_{14}CH_3}
\end{array}
$$

## *Explore Your World*

### Types of Fats

Read the labels on food products that contain fats such as butter, margarine, vegetable oils, peanut butter, and potato chips. Look for terms such as saturated, monounsaturated, polyunsaturated, and partially or fully hydrogenated.

#### QUESTIONS

1. How many grams of saturated, monounsaturated, and polyunsaturated fat are in one serving of the product?
2. What type(s) of fats or oils are in the product?
3. What percent of the total fat is saturated fat? Unsaturated fat?
4. The label on a container of peanut butter states that the cottonseed and canola oils used to make the peanut butter have been fully hydrogenated. What are the typical products that would form when hydrogen is added?
5. For each packaged food, determine the following:
   a. How many grams of fat are in one serving of the food?
   b. Using the caloric value for fat (9 kcal/gram of fat), how many Calories (kilocalories) come from the fat in one serving?
   c. What is the percentage of fat in one serving?

# *Green Chemistry Note*

## Biodiesel as an Alternative Fuel

**Biodiesel** is a name of a nonpetroleum fuel that can be used in place of diesel fuel. Biodiesel is produced from renewable biological resources such as vegetable oils (primarily soybean), waste vegetable oils from restaurants, and some animal fats. Biodiesel is nontoxic and biodegradable.

Biodiesel is prepared from triacylglycerols and alcohols (usually ethanol) to form ethyl esters and glycerol. The glycerol that separates from the fat is used in soaps and other products. The reaction of triacylglycerols is catalyzed by a base such as NaOH or KOH at low temperatures to give a very high percentage of the fatty acid esters, which make up the biodiesel product.

In many cases, diesel engines need no or only slight modification to use biodiesel. Manufacturers of diesel cars, trucks, boats, and tractors have different suggestions for the percentage of biodiesel to use, ranging from 2% (B2) blended with standard diesel fuel to using 100% pure biodiesel (B100). For example, B20 is 20 percent by volume of biodiesel blended with 80 percent by volume petroleum diesel. In 2006, $9.8 \times 10^8$ million liters of biodiesel were used in the United States. Some fuel stations in Europe and the United States are now stocking biodiesel fuel.

Compared to diesel fuel from petroleum, biodiesel burns in an engine to produce much lower levels of carbon dioxide emissions, particulates, unburned hydrocarbons, and polycyclic aromatic hydrocarbons that cause lung cancer. Because biodiesel has extremely low sulfur content, it does not contribute to the formation of the sulfur oxides that produce acid rain. The energy output from the combustion of biodiesel is almost the same as energy produced by petroleum diesel.

$$H_2C-O-\overset{\overset{\displaystyle O}{\|}}{C}-(CH_2)_{12}-CH_3$$
$$H-C-O-\overset{\overset{\displaystyle O}{\|}}{C}-(CH_2)_7-CH=CH-(CH_2)_7-CH_3 + 3\ CH_3-CH_2-OH$$
$$H_2C-O-\overset{\overset{\displaystyle O}{\|}}{C}-(CH_2)_{16}-CH_3$$

Triacylglycerol from vegetable oil                          Ethanol

$$\xrightarrow{\text{NaOH catalyst}}$$

$$H_2C-OH \quad CH_3-CH_2-O-\overset{\overset{\displaystyle O}{\|}}{C}-(CH_2)_{12}-CH_3$$
$$H-C-OH + CH_3-CH_2-O-\overset{\overset{\displaystyle O}{\|}}{C}-(CH_2)_7-CH=CH-(CH_2)_7-CH_3$$
$$H_2C-OH \quad CH_3-CH_2-O-\overset{\overset{\displaystyle O}{\|}}{C}-(CH_2)_{16}-CH_3$$

Glycerol                    Ethyl esters used for biodiesel

## SAMPLE PROBLEM 15.4

### ■ Reactions of Lipids

Write the equation for the reaction catalyzed by the enzyme lipase that hydrolyzes glyceryl trilaurate (trilaurin) during the digestion process.

SOLUTION

$$CH_2-O-\overset{\displaystyle O}{\overset{\displaystyle \|}{C}}-(CH_2)_{10}CH_3$$

$$CH-O-\overset{\displaystyle O}{\overset{\displaystyle \|}{C}}-(CH_2)_{10}CH_3 + 3\ H_2O \xrightarrow{\text{Lipase}}$$

$$CH_2-O-\overset{\displaystyle O}{\overset{\displaystyle \|}{C}}-(CH_2)_{10}CH_3$$

Glyceryl trilaurate
(trilaurin)

$$CH_2-OH$$
$$CH-OH + 3HO\overset{\displaystyle O}{\overset{\displaystyle \|}{C}}-(CH_2)_{10}CH_3$$
$$CH_2-OH$$

Glycerol          3 Lauric acid
                  molecules

STUDY CHECK

What is the name of the product formed when a triacylglycerol containing oleic acid and linoleic acid is completely hydrogenated?

## QUESTIONS AND PROBLEMS

### Chemical Properties of Triacylglycerols

**15.29** Write an equation for the hydrogenation of glyceryl tri-oleate, a fat containing glycerol and three oleic acid units.

**15.30** Write an equation for the hydrogenation of glyceryl trilinole-nate, a fat containing glycerol and three linolenic acid units.

**15.31 a.** Write an equation for the acid hydrolysis of glyceryl trimyristate (trimyristin).
**b.** Write an equation for the NaOH saponification of glyceryl trimyristate (trimyristin).

**15.32 a.** Write an equation for the acid hydrolysis of glyceryl trioleate (triolein).
**b.** Write an equation for the NaOH saponification of glyceryl trioleate (triolein).

**15.33** Compare the structure of a triacylglycerol to the structure of olestra. (See Health Note "Olestra: A Fat Substitute.")

**15.34** A vegetable oil is partially hydrogenated.
**a.** Are all or just some of the double bonds converted to single bonds?
**b.** What happens to many of the cis double bonds during hydrogenation?

**c.** How can you reduce the amount of trans fatty acids in your diet?

**15.35** Write the product of the hydrogenation of the following triacylglycerol:

$$CH_2-O-\overset{\displaystyle O}{\overset{\displaystyle \|}{C}}-(CH_2)_{16}CH_3$$

$$CH-O-\overset{\displaystyle O}{\overset{\displaystyle \|}{C}}-(CH_2)_7CH=CH(CH_2)_7CH_3$$

$$CH_2-O-\overset{\displaystyle O}{\overset{\displaystyle \|}{C}}-(CH_2)_{16}CH_3$$

**15.36** Write all the products that would be obtained when the triacylglycerol in problem 15.35 undergoes complete hydrolysis.

# 15.5 GLYCEROPHOSPHOLIPIDS

The **glycerophospholipids** are a family of lipids similar to triacylglycerols except that one hydroxyl group of glycerol is replaced by the ester of phosphoric acid and an amino alcohol, bonded through a phosphodiester bond. We can compare the general structures of a triacylglycerol and a glycerophospholipid as follows:

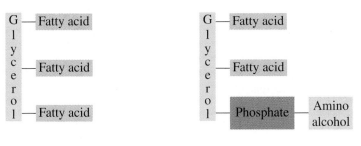

Triacylglycerol (triglyceride)                    Glycerophospholipid

Three amino alcohols found in glycerophospholipids are choline, serine, and ethanolamine. In the body, at physiological pH of 7.4, they are ionized.

$$HO-CH_2-CH_2-\overset{+}{\underset{\underset{CH_3}{|}}{\overset{\overset{CH_3}{|}}{N}}}-CH_3 \qquad HO-CH_2-\underset{\underset{}{}}{\overset{\overset{+}{NH_3}}{CH}}-COO^- \qquad HO-CH_2-CH_2-\overset{+}{N}H_3$$

Choline                          Serine                    Ethanolamine

**Lecithins** and **cephalins** are two types of glycerophospholipids that are particularly abundant in brain and nerve tissues as well as in egg yolks, wheat germ, and yeast. Lecithins contain choline, and cephalins contain ethanolamine and sometimes serine. In the following structural formulas, palmitic acid is used as an example of a fatty acid.

$$
\begin{array}{l}
CH_2-O-\overset{\overset{O}{\|}}{C}-(CH_2)_{14}CH_3 \\
| \qquad\quad O \\
| \qquad\quad \| \\
CH-O-C-(CH_2)_{14}CH_3 \\
| \qquad\quad O \\
| \qquad\quad \| \\
CH_2-O-\overset{}{\underset{\underset{O^-}{|}}{P}}-O-CH_2CH_2\overset{+}{N}(CH_3)_3 \\
\end{array}
\qquad
\begin{array}{l}
CH_2-O-\overset{\overset{O}{\|}}{C}-(CH_2)_{14}CH_3 \\
| \qquad\quad O \\
| \qquad\quad \| \\
CH-O-C-(CH_2)_{14}CH_3 \\
| \qquad\quad O \\
| \qquad\quad \| \\
CH_2-O-\overset{}{\underset{\underset{O^-}{|}}{P}}-O-CH_2CH_2\overset{+}{N}H_3 \\
\end{array}
$$

Nonpolar fatty acids   Polar

A lecithin          Choline          A cephalin          Ethanolamine

Glycerophospholipids contain both polar and nonpolar regions, which allow them to interact with both polar and nonpolar substances. The ionized alcohol and phosphate portion called "the head" is polar and strongly attracted to water. (See Figure 15.6.) The two fatty acids connected to the glycerol molecule represent the nonpolar "tails" of the glycerophospholipid. The hydrocarbon chains that make up the tails are soluble in other nonpolar substances only, mostly lipids.

Glycerophospholipids are the most abundant lipids in cell membranes, where they play an important role in cellular permeability. They make up much of the myelin sheath that protects nerve cells. In the body fluids, they combine with the less polar triglycerides and cholesterol to make them more soluble as they are transported in the body.

**LEARNING GOAL**

Describe the characteristics of glycerophospholipids.

**WEB TUTORIAL**
Phospholipids

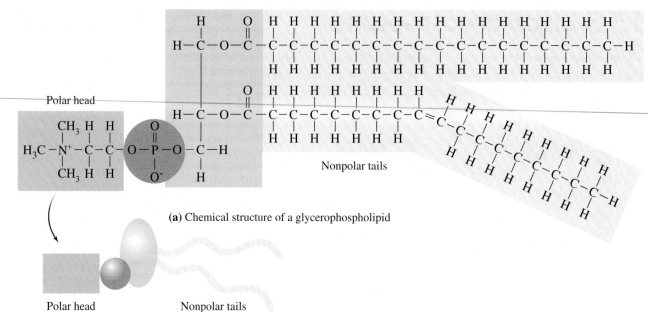

(a) Chemical structure of a glycerophospholipid

Polar head                    Nonpolar tails

(b) Simplified way to draw a glycerophospholipid

**FIGURE 15.6** (a) In a glycerophospholipid, a polar head contains the ionized amino alcohol and phosphoric acid groups, while the two fatty acids make up the nonpolar tails. (b) A simplified drawing indicates the polar region and the nonpolar region.

**Q** Why are glycerophospholipids polar?

---

**SAMPLE PROBLEM** 15.5

### ■ Drawing Glycerophospholipid Structures

Draw the condensed structural formula of a cephalin that contains stearic acid and serine. Describe each component in the glycerophospholipid.

**SOLUTION**

In general, glycerophospholipids are composed of a glycerol molecule in which two carbon atoms are attached to fatty acids such as stearic acid. The third carbon atom is attached by an ester bond to phosphate linked to an amino alcohol. In this example, the amino alcohol is serine.

$$
\begin{array}{l}
CH_2-O-\overset{\overset{\textstyle O}{\|}}{C}-(CH_2)_{16}CH_3 \\[2pt]
\quad\;\; \overset{\overset{\textstyle O}{\|}}{} \\
CH-O-\overset{\overset{\textstyle O}{\|}}{C}-(CH_2)_{16}CH_3 \\[2pt]
\qquad\qquad\qquad\;\; \overset{+}{N}H_3 \\
CH_2-O-\overset{\overset{\textstyle O}{\|}}{\underset{\underset{\textstyle O^-}{|}}{P}}-O-CH_2-CH-COO^-
\end{array}
$$

Stearic acids

Serine

**STUDY CHECK**

Draw the structure of a lecithin, using myristic acid for the fatty acids and choline for the amino alcohol.

## Career Focus

### Physical Therapist

"We do all kinds of activities that are typical of the things kids would do," says Helen Tong, physical therapist. "For example, we play with toys to help improve motor control. When a child can't do something, we use adaptive equipment to help him or her do that activity in a different way, which still allows participation. In school, we learn how the body works, and why it does not work. Then we can figure out what to do to help a child learn new skills. For example, this child with Rett syndrome has motor difficulties. Although she has difficulty talking, she does amazing work at a computer using the switch. There has been a real growth in assisted technology for children, and it has changed our work tremendously."

## QUESTIONS AND PROBLEMS

### Glycerophospholipids

**15.37** Describe the differences between triacylglycerols and glycerophospholipids.

**15.38** Describe the differences between lecithins and cephalins.

**15.39** Draw the structure of a glycerophospholipid containing two molecules of palmitic acid and ethanolamine. What is another name for this type of glycerophospholipid?

**15.40** Draw the structure of a glycerophospholipid that contains choline and palmitic acids.

**15.41** Identify the type of the following glycerophospholipid, and list its components:

$$\begin{array}{l} CH_2{-}O{-}\overset{\displaystyle O}{\overset{\|}{C}}{-}(CH_2)_7CH{=}CH(CH_2)_7CH_3 \\ \quad| \\ CH{-}O{-}\overset{\displaystyle O}{\overset{\|}{C}}{-}(CH_2)_{16}CH_3 \\ \quad| \\ CH_2{-}O{-}\overset{\displaystyle O}{\overset{\|}{P}}{-}O{-}CH_2{-}CH_2{-}\overset{+}{N}H_3 \\ \qquad| \\ \qquad O^- \end{array}$$

**15.42** Identify the type of the following glycerophospholipid, and list its components:

$$\begin{array}{l} CH_2{-}O{-}\overset{\displaystyle O}{\overset{\|}{C}}{-}(CH_2)_{14}CH_3 \\ \quad| \\ CH{-}O{-}\overset{\displaystyle O}{\overset{\|}{C}}{-}(CH_2)_{16}CH_3 \\ \quad| \qquad\qquad\qquad\qquad CH_3 \\ CH_2{-}O{-}\overset{\displaystyle O}{\overset{\|}{P}}{-}O{-}CH_2{-}CH_2{-}\overset{+}{N}{-}CH_3 \\ \qquad| \qquad\qquad\qquad\qquad | \\ \qquad O^- \qquad\qquad\qquad\qquad CH_3 \end{array}$$

# 15.6 STEROIDS: CHOLESTEROL AND STEROID HORMONES

**LEARNING GOAL**

Describe the structures of steroids.

**Steroids** are compounds containing the steroid nucleus, which consists of three cyclohexane rings and one cyclopentane ring fused together. The four rings in the steroid nucleus are designated A, B, C, and D. The carbon atoms are numbered beginning with the carbons in ring A, and in steroids like cholesterol, ending with two methyl groups.

Steroid nucleus

Steroid numbering system

## Cholesterol

Attaching other atoms and groups of atoms to the steroid structure forms a wide variety of steroid compounds. **Cholesterol**, which is one of the most important and abundant steroids in the body, is a *sterol* because it contains an oxygen atom as a hydroxyl (—OH) group on carbon 3. Like many steroids, cholesterol has methyl groups at carbons 10 and 13, a carbon chain at carbon 17, and a double bond between carbons 5 and 6. In other steroids, the oxygen atom typically at carbon 3 forms a carbonyl (C=O) group.

Cholesterol

## Cholesterol in the Body

Cholesterol is a component of cellular membranes, myelin sheath, and brain and nerve tissue. It is also found in the liver, bile salts, and skin, where it forms vitamin D. In the adrenal gland, it is used to synthesize steroid hormones. Cholesterol in the body is obtained from eating meats, milk, and eggs, and it is also synthesized by the liver from fats, carbohydrates, and proteins. There is no cholesterol in vegetable and plant products.

If a diet is high in cholesterol, the liver produces less. A typical daily American diet includes 400–500 mg of cholesterol, one of the highest in the world. The American Heart Association has recommended that we consume no more than 300 mg of cholesterol a day. The cholesterol contents of some typical foods are listed in Table 15.3.

When cholesterol exceeds its saturation level in the bile, gallstones may form. Gallstones are composed of almost 100% cholesterol, with some calcium salts, fatty acids, and glycerophospholipids. (See Figure 15.7.) High levels of cholesterol are also associated with the accumulation of lipid deposits (plaque) that line and narrow the coronary arteries. (See Figure 15.8.) Clinically, cholesterol levels are considered elevated if the total plasma cholesterol level exceeds 200 mg/dL.

A diet that is low in foods containing cholesterol and saturated fats appears to be helpful in reducing the serum cholesterol level. Other factors that may also increase the risk of heart disease are family history, lack of exercise, smoking, obesity, diabetes, gender, and age.

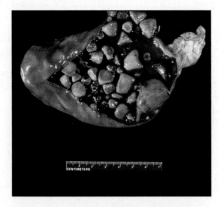

**FIGURE 15.7** Gallstones form in the gallbladder when cholesterol levels are high.

**Q** What type of steroid is stored in the gallbladder?

**TABLE 15.3** Cholesterol Content of Some Foods

| Food | Serving Size | Cholesterol (mg) |
|---|---|---|
| Liver (beef) | 3 oz | 370 |
| Egg | 1 | 250 |
| Lobster | 3 oz | 175 |
| Fried chicken | $3\frac{1}{2}$ oz | 130 |
| Hamburger | 3 oz | 85 |
| Chicken (no skin) | 3 oz | 75 |
| Fish (salmon) | 3 oz | 40 |
| Butter | 1 tablespoon | 30 |
| Whole milk | 1 cup | 35 |
| Skim milk | 1 cup | 5 |
| Margarine | 1 tablespoon | 0 |

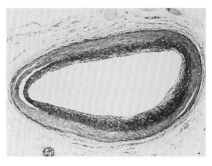
(a)

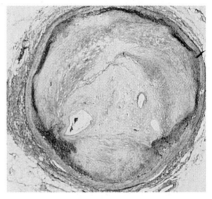
(b)

**SAMPLE PROBLEM 15.6**

### ■ Cholesterol

Observe the structure of cholesterol to answer the following questions:

**a.** What part of cholesterol is the steroid nucleus?
**b.** What features have been added to the steroid nucleus in cholesterol?
**c.** What classifies cholesterol as a sterol?

#### SOLUTION

**a.** The four fused rings form the steroid nucleus.
**b.** The cholesterol molecule contains an alcohol group (—OH) on the first ring, one double bond in the second ring, methyl groups (—CH₃) at carbons 10 and 13, and a branched carbon chain.
**c.** The alcohol group determines the sterol classification.

#### STUDY CHECK

Why is cholesterol in the lipid family?

**FIGURE 15.8** Excess cholesterol forms plaque that can block an artery, resulting in a heart attack. **(a)** A normal, open artery shows no buildup of plaque. **(b)** An artery that is almost completely clogged by atherosclerotic plaque.

**Q** What property of cholesterol would cause it to form deposits along the coronary arteries?

## Lipoproteins: Transport of Lipids

In the body, lipids must be transported through the bloodstream to tissues where they are stored, used for energy, or to make hormones. However, most lipids are nonpolar and insoluble in the aqueous environment of blood. They are made more soluble by combining them with glycerophospholipids and proteins to form water-soluble complexes called **lipoproteins**. In general, lipoproteins are spherical particles with an outer surface of polar proteins and glycerophospholipids that surround hundreds of nonpolar molecules of triacylglycerols and cholesteryl esters. (See Figure 15.9.)

Cholesteryl esters are the prevalent form of cholesterol in the blood. They are formed by the esterification of the hydroxyl group in cholesterol with a fatty acid.

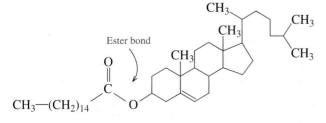
Cholesteryl ester

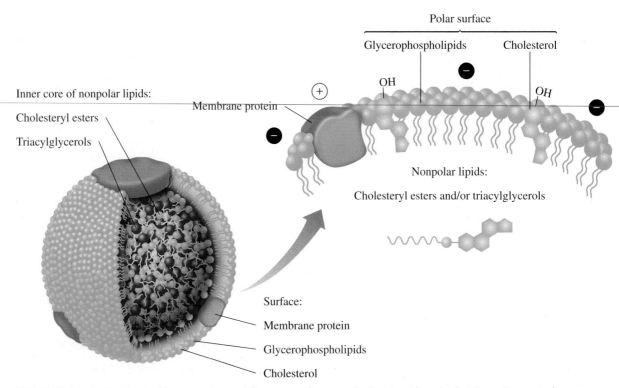

**FIGURE 15.9** A spherical lipoprotein particle surrounds nonpolar lipids with polar lipids and protein for transport to body cells.

**Q** Why are the polar components on the surface of a lipoprotein particle and the nonpolar components at the center?

There are several types of lipoproteins that differ in density, lipid composition, and function. They include chylomicrons, very-low-density lipoprotein (VLDL), low-density lipoprotein (LDL), and high-density lipoprotein (HDL). The LDLs form when the triacylglycerol portion is removed from VLDLs. The density of the lipoproteins increases as the percentage of protein in each type also increases. (See Table 15.4.)

**TABLE 15.4** Composition and Properties of Plasma Lipoproteins

|  | Chylomicron | VLDL | LDL | HDL |
|---|---|---|---|---|
| **Density (g/mL)** | 0.94 | 0.950–1.006 | 1.006–1.063 | 1.063–1.210 |
| **Composition (% by mass)** | | | | |
| Triacylglycerol | 86 | 55 | 6 | 4 |
| Phospholipids | 7 | 18 | 22 | 24 |
| Cholesterol | 2 | 7 | 8 | 2 |
| Cholesteryl esters | 3 | 12 | 42 | 15 |
| Protein | 2 | 8 | 22 | 55 |

The chylomicrons and the VLDLs transport triacylglycerols, glycerophospholipids, and cholesterol to the tissues for storage or to the muscles for energy. (See Figure 15.10.) The LDLs transport cholesterol to tissues where it is used for the synthesis of cell membranes, steroid hormones, and bile salts. When the level of LDL exceeds the amount of cholesterol needed by the tissues, the LDLs deposit cholesterol in the arteries, restricting blood flow and increasing the risk of developing heart disease and/or myocardial infarctions (heart attacks). This is why LDL cholesterol is called "bad" cholesterol.

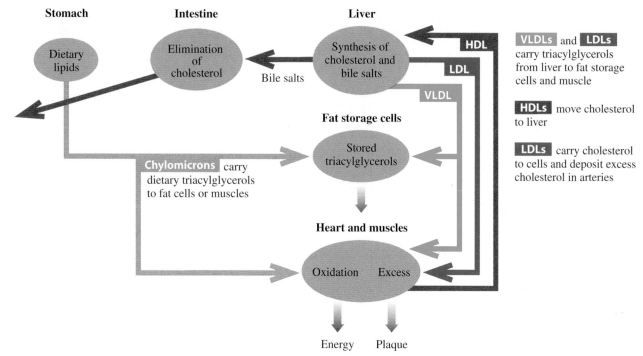

**FIGURE 15.10** Lipoproteins such as HDLs and LDLs transport nonpolar lipids and cholesterol to cells and the liver.
**Q** What type of lipoprotein transports cholesterol to the liver?

The HDLs remove excess cholesterol from the tissues and carry it to the liver, where it is converted to bile salts and eliminated. When HDL levels are high, cholesterol that is not needed by the tissues is carried to the liver for elimination rather than deposited in the arteries, which gives the HDLs the name of "good" cholesterol. Most of the cholesterol in the body is synthesized in the liver, although some comes from the diet. However, a person on a high-fat diet reabsorbs cholesterol from the bile salts, causing less cholesterol to be eliminated. In addition, higher levels of saturated fats stimulate the synthesis of cholesterol by the liver.

Because high cholesterol levels are associated with the onset of arteriosclerosis and heart disease, the serum levels of LDL and HDL are generally determined as part of a medical examination. For adults, recommended levels for total cholesterol are less than 200 mg/dL, with LDL less than 130 mg/dL and HDL higher than 40 mg/dL. A lower level of serum cholesterol decreases the risk of heart disease. Increased HDL levels are found in people who exercise regularly and eat less saturated fat.

## Steroid Hormones

The word *hormone* comes from the Greek "to arouse" or "to excite." Hormones are chemical messengers that serve as a kind of communication system from one part of the body to another. The *steroid* hormones, which include the sex hormones and the adrenocortical hormones, are closely related in structure to cholesterol and depend on cholesterol for their synthesis.

Two of the male sex hormones, *testosterone* and *androsterone*, promote the growth of muscle and of facial hair and the maturation of the male sex organs and of sperm.

The *estrogens*, a group of female sex hormones, direct the development of female sexual characteristics: the uterus increases in size, fat is deposited in the breasts, and the pelvis broadens. *Progesterone* prepares the uterus for the implantation of a fertilized egg. If an egg is not fertilized, the levels of progesterone and estrogen drop sharply, and menstruation follows. Synthetic forms of the female sex hormones are used in birth-control pills. As with other kinds of steroids, side effects can include weight gain and a greater risk of forming blood clots. The structures of some steroid hormones follow:

| Hormone | Biological Effects |
|---|---|

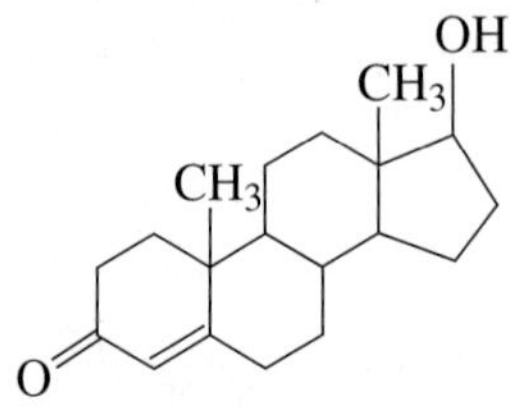

Testosterone (androgen)
(produced in testes)

Development of male organs; male sexual characteristics including muscles and facial hair; sperm formation

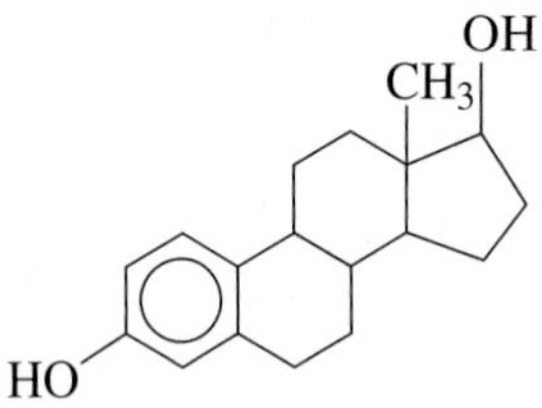

Estradiol (estrogen)
(produced in ovaries)

Development of female sexual characteristics; ovulation

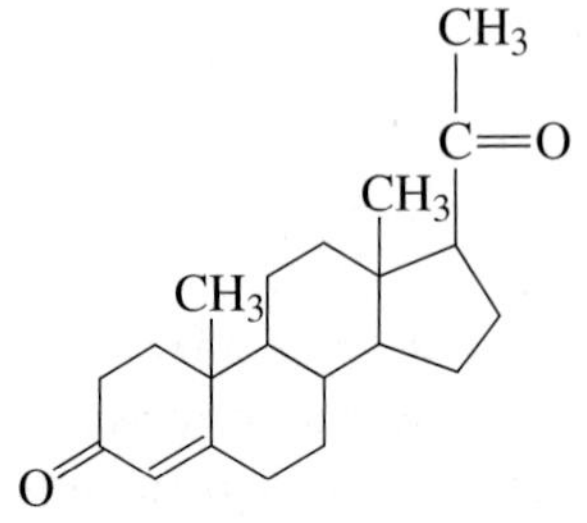

Progesterone
(produced in ovaries)

Prepares uterus for fertilized egg

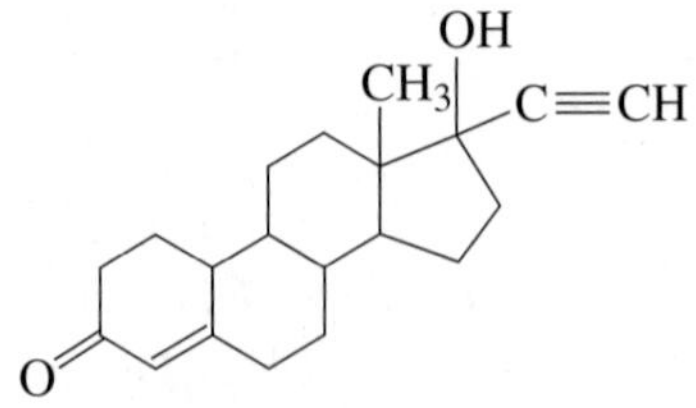

Norethindrone
(synthetic progestin)

Contraceptive (birth-control) pill

## Anabolic Steroids

Some of the physiological effects of testosterone are to increase muscle mass and decrease body fat. Derivatives of testosterone called *anabolic steroids* that enhance these effects have been synthesized. Although they have some medical uses, anabolic steroids have been used in rather high dosages by some athletes in an effort to increase muscle mass. Such use is illegal.

Use of anabolic steroids in attempting to improve athletic strength can cause side effects including hypertension, fluid retention, increased hair growth, sleep disturbances, and acne. Over a long period of time, their use may cause irreversible liver damage and decreased sperm production.

**SOME ANABOLIC STEROIDS**

Methandienone    Oxandrolone    Nandrolone    Stanozolol

## Adrenal Corticosteroids

The adrenal glands, located on the top of each kidney, produce the corticosteroids. *Aldosterone*, a mineralocorticoid, is responsible for electrolyte and water balance by the kidneys. *Cortisone*, a glucocorticoid, increases the blood glucose level and stimulates the synthesis of glycogen in the liver from amino acids. Synthetic corticoids such as *prednisone* are derived from cortisone and used medically for reducing inflammation and treating asthma and rheumatoid arthritis, although precautions are given for long-term use.

**Corticosteroids**

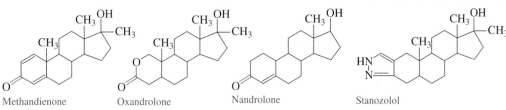

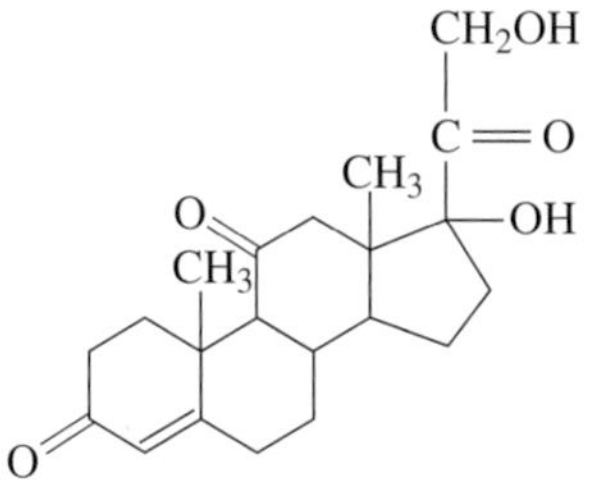

Cortisone
(produced in adrenal gland)

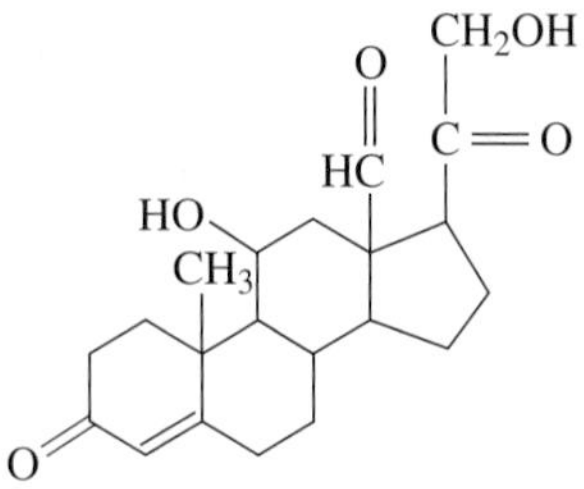

Aldosterone (mineralocorticoid)
(produced in adrenal gland)

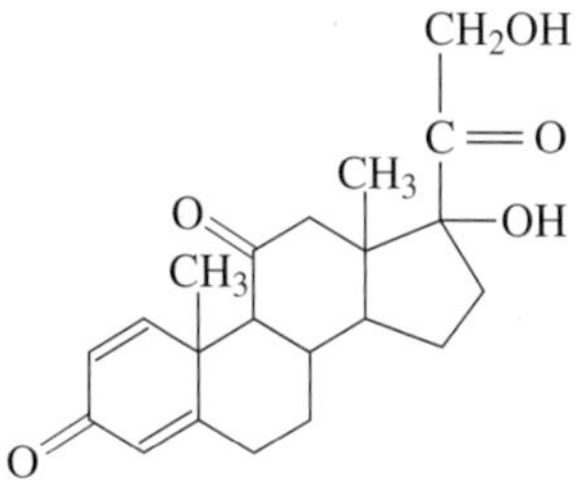

Prednisone
(synthetic corticoid)

**Biological Effects**

Increases the blood glucose and glycogen levels from fatty acids and amino acids

Increases the reabsorption of $Na^+$ in kidneys; retention of water

Reduces inflammation; treatment of asthma and rheumatoid arthritis

■ **Steroid Hormones**

What are the groups on the steroid nucleus in the sex hormones estradiol and testosterone?

SOLUTION

Estradiol contains a benzene ring, one methyl group, and two hydroxyl groups. Testosterone contains a ketone group, a double bond, two methyl groups, and a hydroxyl group.

STUDY CHECK

What are the similarities and differences in the structures of testosterone and the anabolic steroid nandrolone? (See Health Note "Anabolic Steroids.")

## QUESTIONS AND PROBLEMS

**Steroids: Cholesterol and Steroid Hormones**

**15.43** Draw the structure for the steroid nucleus.

**15.44** Which of the following compounds are derived from cholesterol?
   a. glyceryl tristearate          b. cortisone
   c. testosterone                  d. estradiol

**15.45** What is the general structure of lipoproteins?

**15.46** Why are lipoproteins needed to transport lipids in the bloodstream?

**15.47** How do chylomicrons differ from VLDLs (very-low-density lipoproteins)?

**15.48** How do LDLs differ from HDLs?

**15.49** Why are LDLs called "bad" cholesterol?

**15.50** Why are HDLs called "good" cholesterol?

**15.51** What are the similarities and differences between the sex hormones estradiol and testosterone?

**15.52** What are the similarities and differences between the adrenal hormone cortisone and the synthetic corticoid prednisone?

**15.53** Which of the following are male sex hormones?
   a. cholesterol          b. aldosterone
   c. estrogen             d. testosterone
   e. choline

**15.54** Which of the following are adrenal steroids?
   a. cholesterol          b. aldosterone
   c. estrogen             d. testosterone
   e. choline

## LEARNING GOAL

Describe the composition and function of the lipid bilayer in cell membranes.

WEB TUTORIAL
Membrane Structure

# 15.7 CELL MEMBRANES

The membrane of a cell separates the contents of a cell from the external fluids. It is semipermeable so that nutrients can enter the cell and waste products can leave. The main components of a cell membrane are the glycerophospholipids. Earlier in this chapter, we saw that the structures of such glycerophospholipids consist of a nonpolar region, or tail, with long-chain fatty acids and a polar region, or head, from phosphoric acid and amino alcohols that ionize at physiological pH.

   In a cell membrane, two rows of glycerophospholipids are arranged like a sandwich. Their nonpolar tails, which are hydrophobic ("water-fearing"), move to the center, while their polar heads, which are hydrophilic ("water-loving"), align on the outer edges of the membrane. This double-row arrangement of glycerophospholipids is called a **lipid bilayer**. (See Figure 15.11.) One row of glycerophospholipids forms the outside surface of the membrane, which is in contact with the external fluids, and the other row forms the inside surface or edge of the membrane, which is in contact with the internal contents of the cell.

   Most of the glycerophospholipids in the lipid bilayer contain unsaturated fatty acids. Because of the kinks in the carbon chains at the cis double bonds, the glycerophospholipids do not fit closely together. As a result, the lipid bilayer is not a rigid, fixed structure,

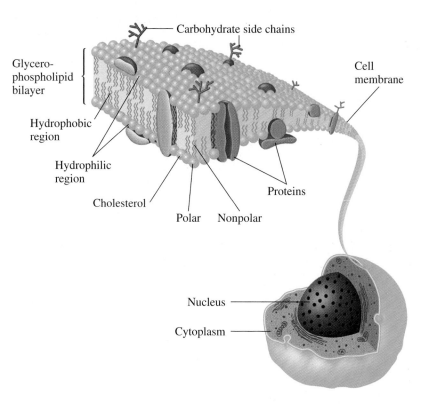

**FIGURE 15.11** The fluid mosaic model of a cell membrane. Proteins and cholesterol are embedded in a lipid bilayer of glycerophospholipids. The bilayer forms a membrane-type barrier with polar heads at the membrane surfaces and the nonpolar tails in the center away from the water.

**Q** What types of fatty acids are found in the glycerophospholipids of the lipid bilayer?

but one that is dynamic and fluid-like. In this liquid-like bilayer, there are also proteins, carbohydrates, and cholesterol molecules. For this reason, the model of biological membranes is referred to as the **fluid mosaic model** of membranes.

In the fluid mosaic model, proteins known as peripheral proteins emerge on just one of the surfaces, outer or inner. The integral proteins extend through the entire lipid bilayer and appear on both surfaces of the membrane. Some proteins and lipids on the outer surface of the cell membrane are attached to carbohydrates. These carbohydrate chains project into the surrounding fluid environment where they are responsible for cell recognition and communication with chemical messengers such as hormones and neurotransmitters. In animals, cholesterol molecules embedded among the glycerophospholipids make up 20–25% of the lipid bilayer. Because cholesterol molecules are large and rigid, they reduce the flexibility of the lipid bilayer and add strength to the cell membrane.

## SAMPLE PROBLEM 15.8

### ■ Lipid Bilayer in the Cell Membranes

Describe the role of glycerophospholipids in the lipid bilayer.

#### SOLUTION

Glycerophospholipids consist of polar and nonpolar parts. In a cell membrane, an alignment of the nonpolar sections toward the center with the polar sections on the outside produces a barrier that prevents the contents of a cell from mixing with the fluids on the outside of the cell.

#### STUDY CHECK

What is the function of cholesterol in the cell membrane?

# Health Note

## Transport Through Cell Membranes

Ions and molecules flow in and out of the cell in several ways. In the simplest transport mechanism called diffusion or passive transport, ions and small molecules migrate from a higher concentration to a lower concentration. For example, some ions as well as small molecules such as $O_2$, urea, and water diffuse through cell membranes. If their concentration is greater outside the cell than inside, they diffuse into the cell. If water has a high concentration in the cell, it diffuses out of the cell.

Another type of transport called facilitated transport increases the rate of diffusion for substances that diffuse too slowly by passive diffusion to meet cell needs. This process utilizes the integral proteins that extend from one edge of the cell membrane to the other. Groups of integral proteins provide channels to transport chloride

ion ($Cl^-$), bicarbonate ion ($HCO_3^-$), and glucose molecules in and out of the cell more rapidly.

Certain ions such as $K^+$, $Na^+$, and $Ca^{2+}$, move across a cell membrane against a concentration gradient. For example, the $K^+$ concentration is greater inside a cell, and the $Na^+$ concentration is greater outside. However, in the conduction of nerve impulses and contraction of muscles, $K^+$ moves into the cell, and $Na^+$ moves out. To move an ion from a lower to a higher concentration requires energy, which is accomplished by a process known as active transport. In active transport, a protein complex called a $Na^+/K^+$ pump breaks down ATP to ADP (Chapter 18), which releases energy to move $Na^+$ and $K^+$ against their concentration gradients.

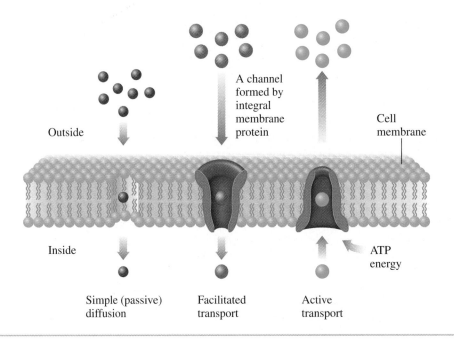

## QUESTIONS AND PROBLEMS

### Cell Membranes

**15.55** What types of lipids are found in cell membranes?

**15.56** Describe the structure of a lipid bilayer.

**15.57** What is the function of the lipid bilayer in a cell membrane?

**15.58** How do the unsaturated fatty acids in the glycerophospholipids affect the structure of cell membranes?

**15.59** Where are proteins located in cell membranes?

**15.60** What is the difference between peripheral and integral protein?

**15.61** What is the function of the carbohydrates on a cell membrane surface?

**15.62** Why is a cell membrane semipermeable?

## CONCEPT MAP

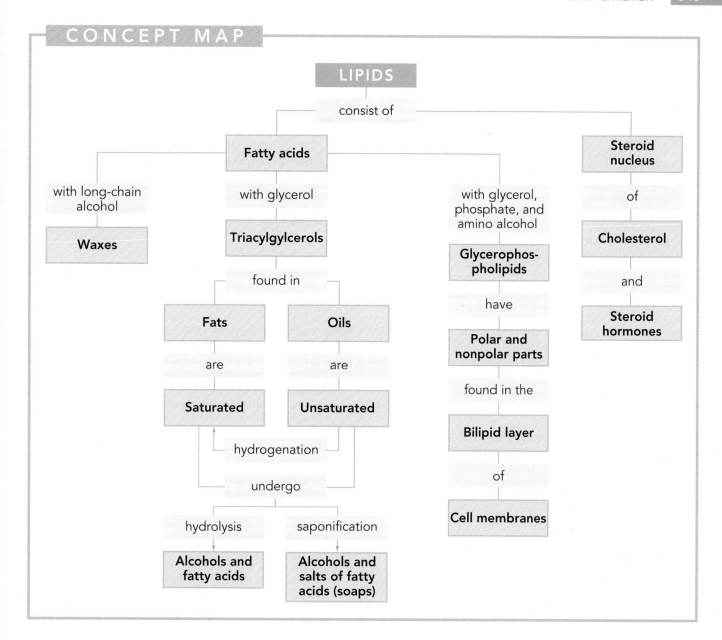

# CHAPTER REVIEW

### 15.1 Lipids

**Learning Goal: Describe the classes of lipids.**

Lipids are nonpolar compounds that are not soluble in water. Classes of lipids include waxes, fats and oils, glycerophospholipids, and steroids.

### 15.2 Fatty Acids

**Learning Goal: Write structures of fatty acids, and identify each as saturated or unsaturated.**

Fatty acids are unbranched carboxylic acids that typically contain an even number (10–20) of carbon atoms. Fatty acids may be saturated, monounsaturated with one double bond, or polyunsaturated with two or more double bonds. The double bonds in unsaturated fatty acids are almost always cis.

### 15.3 Waxes, Fats, and Oils

**Learning Goal: Write the structural formula of a wax, fat, or oil produced by the reaction of a fatty acid and an alcohol or glycerol.**

A wax is an ester of a long-chain fatty acid and a long-chain alcohol. The triacylglycerols of fats and oils are esters of glycerol with three long-chain fatty acids. Fats contain more saturated fatty acids and have higher melting points than most oils.

### 15.4 Chemical Properties of Triacylglycerols

**Learning Goal: Draw the structure of the product from the reaction of a triacylglycerol with hydrogen, or an acid or base.**

The hydrogenation of unsaturated fatty acids converts double bonds to single bonds. The hydrolysis of the ester bonds in fats or oils produces glycerol and fatty acids. In saponification, a fat heated with a strong base produces glycerol and the salts of the fatty acids or soaps.

## 15.5 Glycerophospholipids

**Learning Goal:** Describe the characteristics of glycerophospholipids.

Glycerophospholipids are esters of glycerol with two fatty acids and a phosphate group attached to an amino alcohol.

## 15.6 Steroids: Cholesterol and Steroid Hormones

**Learning Goal:** Describe the structures of steroids.

Steroids are lipids containing the steroid nucleus, which is a fused structure of four rings. Lipids, which are nonpolar, are transported through the aqueous environment of the blood by forming lipoproteins. Lipoproteins such as chylomicrons and LDL transport triacylglycerols from the intestines and the liver to fat cells for storage and to muscles for energy. HDLs transport cholesterol from the tissues to the liver for elimination. The steroid hormones are closely related in structure to cholesterol and depend on cholesterol for their synthesis. The sex hormones, such as estrogen and testosterone, are responsible for sexual characteristics and reproduction. The adrenal corticosteroids, such as aldosterone and cortisone, regulate water balance and glucose levels in the cells.

## 15.7 Cell Membranes

**Learning Goal:** Describe the composition and function of the lipid bilayer in cell membranes.

All animal cells are surrounded by a semipermeable membrane that separates the cellular contents from the external fluids. The membrane is composed of two rows of glycerophospholipids in a lipid bilayer.

# KEY TERMS

**cephalins**  Glycerophospholipids found in brain and nerve tissues that incorporate the amino alcohol serine or ethanolamine.

**cholesterol**  The most prevalent of the steroid compounds; needed for cellular membranes, the synthesis of vitamin D, hormones, and bile salts.

**fat**  A triacylglycerol obtained from an animal source; usually a solid at room temperature.

**fatty acids**  Long-chain carboxylic acids found in many lipids.

**fluid mosaic model**  The concept that cell membranes are lipid bilayer structures that contain an assortment of polar lipids and proteins in a dynamic, fluid arrangement.

**glycerophospholipids**  Polar lipids of glycerol attached to two fatty acids and a phosphate group connected to an amino alcohol such as choline, serine, or ethanolamine.

**hydrogenation**  The addition of hydrogen to unsaturated fats.

**lecithins**  Glycerophospholipids containing choline as the amino alcohol.

**lipid bilayer**  A model of a cell membrane in which glycerophospholipids are arranged in two rows.

**lipids**  A family of compounds that is nonpolar in nature and not soluble in water; includes fats, waxes, glycerophospholipids, and steroids.

**lipoprotein**  A combination of nonpolar lipids with glycerophospholipids and proteins to form a polar complex that can be transported through body fluids.

**monounsaturated fatty acids**  Fatty acids with one double bond.

**oil**  A triacylglycerol obtained from a plant source; usually liquid at room temperature.

**polyunsaturated fatty acids**  Fatty acids that contain two or more double bonds.

**prostaglandins**  A number of compounds derived from arachidonic acid that regulate several physiological processes.

**saturated fatty acids**  Fatty acids that have no double bonds; they have higher melting points than unsaturated lipids and are usually solid at room temperatures.

**steroids**  Types of lipid composed of a multicyclic ring system.

**triacylglycerols**  A family of lipids composed of three fatty acids bonded through ester bonds to glycerol, a trihydroxy alcohol.

**wax**  The ester of a long-chain alcohol and a long-chain saturated fatty acid.

# UNDERSTANDING THE CONCEPTS

**15.63**  Palmitic acid is obtained from palm oil as glyceryl tripalmitate (tripalmitin). Draw the structure of glyceryl tripalmitate.

**15.64**  Jojoba wax in candles consists of stearic acid and a 22-carbon saturated alcohol. Write the structure of jojoba wax.

**15.65** Sunflower oil can be used to make margarine. A triacylglycerol in sunflower oil consists of 2 linoleic acids and 1 oleic acid.

    **a.** Write two isomers for the triacylglycerol in sunflower oil.

    **b.** Using one of the isomers, write the reaction that would be used when sunflower oil is used to make solid margarine.

**15.66** Identify each of the following as saturated, monounsaturated, polyunsaturated, omega-3, or omega-6 fatty acids:

    **a.**

$$CH_3-(CH_2)_4-CH=CH-CH_2-CH=CH-(CH_2)_7-COOH$$

    **b.** linolenic acid

    **c.** $CH_3-(CH_2)_{14}-COOH$

    **d.** $CH_3-(CH_2)_7-CH=CH-(CH_2)_7-COOH$

# ADDITIONAL QUESTIONS AND PROBLEMS

**15.67** Among the ingredients in lipstick are beeswax, carnauba wax, hydrogenated vegetable oils, and glyceryl tricaprate (tricaprin). What types of lipids have been used? Draw the condensed structural formula of glyceryl tricaprate (tricaprin). Capric acid is a saturated 10-carbon fatty acid.

**15.68** Because peanut oil floats on the top of peanut butter, many brands of peanut butter are hydrogenated. A solid product then forms that is mixed into the peanut butter and does not separate. If a triacylglycerol in peanut oil that contains one palmitic acid, one oleic acid, and one linoleic acid is completely hydrogenated, what is the product?

**15.69** Trans fats are produced during the hydrogenation of polyunsaturated oils.

    **a.** What is the typical configuration of the double bond in a monounsaturated fatty acid?

    **b.** How does a trans fatty acid differ from a cis fatty acid?

    **c.** Draw the structure of *trans*-oleic acid.

**15.70** One mole of glyceryl trioleate (triolein) is completely hydrogenated. Draw the condensed structural formula of the product. How many moles of hydrogen are required? How many grams of hydrogen? How many liters of hydrogen are needed if the reaction is run at STP?

**15.71** On the list of ingredients of a cosmetic product are glyceryl tristearate and a lecithin. Draw the structure of glyceryl tristearate and a lecithin with palmitic acids and choline.

**15.72** Some typical meals at fast-food restaurants are listed here. Calculate the number of kilocalories from fat and the percentage of total kilocalories that are due to fat (1 gram of fat = 9 kcal). Would you expect the fats to be mostly saturated or unsaturated? Why?

    **a.** a chicken dinner, 830 kcal, 46 g of fat

    **b.** a quarter-pound cheeseburger, 518 kcal, 29 g of fat

    **c.** pepperoni pizza (three slices), 560 kcal, 18 g of fat

    **d.** beef burrito, 470 kcal, 21 g of fat

    **e.** deep-fried fish (three pieces), 480 kcal, 28 g of fat

**15.73** Identify the following as fatty acids, soaps, triacylglycerols, waxes, glycerophospholipids, or steroids:

    **a.** beeswax

    **b.** cholesterol

    **c.** lecithin

    **d.** glyceryl tripalmitate (tripalmitin)

    **e.** sodium stearate

    **f.** safflower oil

    **g.** whale blubber

    **h.** adipose tissue

    **i.** progesterone

    **j.** cortisone

    **k.** stearic acid

**15.74** Why would an animal that lives in a cold climate have more unsaturated triacylglycerols in its body fat than an animal that lives in a warm climate?

**15.75** Identify the components (1–5) contained in each of the lipids a–e.

    **1.** glycerol    **2.** fatty acid    **3.** phosphate

    **4.** amino alcohol    **5.** steroid nucleus

    **a.** estrogen

    **b.** cephalin

    **c.** wax

    **d.** triacylglycerol

    **e.** glycerophospholipid

**15.76** Which of the following are found in cell membranes?

    **a.** cholesterol

    **b.** triacylglycerols

    **c.** carbohydrates

    **d.** proteins

    **e.** waxes

    **f.** glycerophospholipids

# CHALLENGE QUESTIONS

**15.77** Identify the lipoprotein (1–4) in each description (a–h).

   **1.** chylomicrons    **2.** VLDL

   **3.** LDL           **4.** HDL

   **a.** "good" cholesterol

   **b.** transports most of the cholesterol to the cells

   **c.** carries triacylglycerols from the intestine to the fat cells

   **d.** transports cholesterol to the liver

   **e.** has the greatest abundance of protein

   **f.** "bad" cholesterol

   **g.** carries triacylglycerols synthesized in the liver to the muscles

   **h.** has the lowest density

**15.78**  **a.** Which of the following fatty acids has the lowest melting point? Explain.

   **b.** Which of the following fatty acids has the highest melting point? Explain.

$$CH_3-(CH_2)_{16}-COOH$$
Stearic acid

$$CH_3-(CH_2)_4-CH=CH-CH_2-CH=CH-(CH_2)_7-COOH$$
Linoleic acid

$$CH_3-(CH_2)_7-CH=CH-(CH_2)_7-COOH$$
Oleic acid

**15.79** Draw the structure of a glycerophospholipid that is made from 2 stearic acids, and a phosphate bonded to ethanolamine.

**15.80** Olive oil consists of a high percentage of glyceryl trioleate (triolein).

   **a.** Draw the structure for glyceryl trioleate (triolein).

   **b.** How many liters of $H_2$ gas at STP are needed to completely saturate 100. g of glyceryl trioleate (triolein)?

   **c.** How many mL of 0.250 M NaOH are needed to completely saponify 100. g of glyceryl trioleate (triolein)?

**15.81** A sink drain can become clogged with solid fat such as glyceryl tristearate (tristearin).

   **a.** How would adding lye (NaOH) to the sink drain remove the blockage?

   **b.** Write an equation for the reaction that occurs.

# ANSWERS

## Answers to Study Checks

**15.1** Glycerol, two fatty acids, phosphate, and an amino alcohol.

**15.2**  **a.** 16      **b.** unsaturated     **c.** liquid

**15.3**

$$CH_2-O-\overset{\overset{\displaystyle O}{\|}}{C}-(CH_2)_{12}-CH_3$$
$$CH-O-\overset{\overset{\displaystyle O}{\|}}{C}-(CH_2)_{12}-CH_3$$
$$CH_2-O-\overset{\overset{\displaystyle O}{\|}}{C}-(CH_2)_{12}-CH_3$$

**15.4** glyceryl tristearate (tristearin)

**15.5**

$$CH_2-O-\overset{\overset{\displaystyle O}{\|}}{C}-(CH_2)_{12}-CH_3$$
$$CH-O-\overset{\overset{\displaystyle O}{\|}}{C}-(CH_2)_{12}-CH_3$$
$$CH_2-O-\overset{\overset{\displaystyle O}{\|}}{P}-O-CH_2-CH_2-\overset{\overset{\displaystyle CH_3}{|}}{\underset{\underset{\displaystyle CH_3}{|}}{N^+}}-CH_3$$
$$\qquad\qquad O^-$$

**15.6** Cholesterol is not soluble in water; it is part of the lipid family.

**15.7** Testosterone and nandrolone both contain a steroid nucleus with one double bond and a ketone group in the first ring, and a methyl and an alcohol group on the five-carbon ring. Nandrolone does not have a second methyl group at the first and second ring fusion point, which is seen in the structure of testosterone.

**15.8** Cholesterol adds strength and rigidity to the cell membrane.

## Answers to Selected Questions and Problems

**15.1** Lipids provide energy and protection and insulation for the organs in the body.

**15.3** Because lipids are not soluble in water, a polar solvent, they are nonpolar molecules.

**15.5** All fatty acids contain a long chain of carbon atoms with a carboxylic acid group. Saturated fatty acids contain only carbon–carbon single bonds; unsaturated fatty acids contain one or more double bonds.

**15.7**  **a.** palmitic acid

      COOH

   **b.** oleic acid

      COOH

**15.9**  **a.** saturated  **b.** unsaturated
 **c.** unsaturated  **d.** saturated

**15.11** In a cis fatty acid, the hydrogen atoms are on the same side of the double bond, which produces a bend in the carbon chain. In a trans fatty acid, the hydrogen atoms are on opposite sides of the double bond, which gives a carbon chain without any bend.

**15.13** In an omega-3 fatty acid, there is a double bond on carbon 3, counting from the methyl group, whereas in an omega-6 fatty acid, there is a double bond beginning at carbon 6, counting from the methyl group.

**15.15** Arachidonic acid contains four double bonds and no side groups. In $PGE_1$, a part of the chain forms cyclopentane and there are hydroxyl and ketone functional groups.

**15.17** Prostaglandins raise blood pressure, stimulate contraction and relaxation of smooth muscle, and may cause inflammation and pain.

**15.19**

$$CH_3(CH_2)_{14}\overset{\overset{\displaystyle O}{\|}}{C}O(CH_2)_{29}CH_3$$

**15.21**

$$CH_2O\overset{\overset{\displaystyle O}{\|}}{C}(CH_2)_{16}CH_3$$
$$| \quad \overset{\displaystyle O}{\|}$$
$$CHO\overset{}{C}(CH_2)_{16}CH_3$$
$$| \quad \overset{\displaystyle O}{\|}$$
$$CH_2O\overset{}{C}(CH_2)_{16}CH_3$$

**15.23**

$$CH_2O\overset{\overset{\displaystyle O}{\|}}{C}(CH_2)_{14}CH_3$$
$$| \quad \overset{\displaystyle O}{\|}$$
$$CHO\overset{}{C}(CH_2)_{14}CH_3$$
$$| \quad \overset{\displaystyle O}{\|}$$
$$CH_2O\overset{}{C}(CH_2)_{14}CH_3$$

**15.25** Safflower oil contains fatty acids with two or three double bonds; olive oil contains a large amount of oleic acid, which has only one (monounsaturated) double bond.

**15.27** Although coconut oil comes from a vegetable, it has large amounts of saturated fatty acids and small amounts of unsaturated fatty acids.

**15.29**

$$CH_2O\overset{\overset{\displaystyle O}{\|}}{C}(CH_2)_7CH=CH(CH_2)_7CH_3$$
$$| \quad \overset{\displaystyle O}{\|}$$
$$CHO\overset{}{C}(CH_2)_7CH=CH(CH_2)_7CH_3 + 3\ H_2 \xrightarrow{Ni}$$
$$| \quad \overset{\displaystyle O}{\|}$$
$$CH_2O\overset{}{C}(CH_2)_7CH=CH(CH_2)_7CH_3$$

$$CH_2O\overset{\overset{\displaystyle O}{\|}}{C}(CH_2)_{16}CH_3$$
$$| \quad \overset{\displaystyle O}{\|}$$
$$CHO\overset{}{C}(CH_2)_{16}CH_3$$
$$| \quad \overset{\displaystyle O}{\|}$$
$$CH_2O\overset{}{C}(CH_2)_{16}CH_3$$

**15.31  a.**

$$CH_2O\overset{\overset{\displaystyle O}{\|}}{C}(CH_2)_{12}CH_3$$
$$| \quad \overset{\displaystyle O}{\|}$$
$$CHO\overset{}{C}(CH_2)_{12}CH_3 + 3\ H_2O \xrightarrow{H^+}$$
$$| \quad \overset{\displaystyle O}{\|}$$
$$CH_2O\overset{}{C}(CH_2)_{12}CH_3$$

$$CH_2OH$$
$$|$$
$$CHOH + 3CH_3(CH_2)_{12}\overset{\overset{\displaystyle O}{\|}}{C}OH$$
$$|$$
$$CH_2OH$$

**b.**

$$CH_2O\overset{\overset{\displaystyle O}{\|}}{C}(CH_2)_{12}CH_3$$
$$| \quad \overset{\displaystyle O}{\|}$$
$$CHO\overset{}{C}(CH_2)_{12}CH_3 + 3\ NaOH \longrightarrow$$
$$| \quad \overset{\displaystyle O}{\|}$$
$$CH_2O\overset{}{C}(CH_2)_{12}CH_3$$

$$CH_2OH$$
$$|$$
$$CHOH + 3CH_3(CH_2)_{12}\overset{\overset{\displaystyle O}{\|}}{C}O^-Na^+$$
$$|$$
$$CH_2OH$$

**15.33** A triacylglycerol is composed of glycerol with three hydroxyl groups that form ester links with three long-chain fatty acids. In olestra, six to eight long-chain fatty acids form ester links with the hydroxyl groups on sucrose, a sugar. The olestra cannot be digested because our enzymes cannot break down the large olestra molecule.

**15.35**

$$CH_2O-\overset{\overset{\displaystyle O}{\|}}{C}-(CH_2)_{16}CH_3$$
$$| \quad \overset{\displaystyle O}{\|}$$
$$CHO-C-(CH_2)_{16}CH_3$$
$$| \quad \overset{\displaystyle O}{\|}$$
$$CH_2O-C-(CH_2)_{16}CH_3$$

**15.37** A triacylglycerol consists of glycerol and three fatty acids. A glycerophospholipid consists of glycerol, two fatty acids, a phosphate group, and an amino alcohol.

**15.39**

$$CH_2O\overset{\overset{\displaystyle O}{\|}}{C}(CH_2)_{14}CH_3$$
$$| \quad \overset{\displaystyle O}{\|}$$
$$CHO\overset{}{C}(CH_2)_{14}CH_3$$
$$| \quad \overset{\displaystyle O}{\|}$$
$$CH_2O\overset{}{P}OCH_2CH_2\overset{+}{N}H_3$$
$$|$$
$$O^-$$

This is a cephalin

**15.41** This glycerophospholipid is a cephalin. It contains glycerol, oleic acid, stearic acid, a phosphate, and ethanolamine.

**15.43**

**15.45** Lipoproteins are large, spherically shaped molecules that transport lipids in the bloodstream. They consist of an outside layer of glycerophospholipids and proteins surrounding an inner core of hundreds of nonpolar lipids and cholesteryl esters.

**15.47** Chylomicrons have a lower density than VLDLs. They pick up triacylglycerols from the intestine, whereas VLDLs transport triacylglycerols synthesized in the liver.

**15.49** "Bad" cholesterol is the cholesterol carried by LDLs that can form deposits in the arteries called plaque, which narrows the arteries.

**15.51** Both estradiol and testosterone contain the steroid nucleus and a hydroxyl group. Testosterone has a ketone group, a double bond, and two methyl groups. Estradiol has a benzene ring, a hydroxyl group in place of the ketone, and a methyl group.

**15.53** **d.** Testosterone is a male sex hormone.

**15.55** The lipids in a cell membrane are glycerophospholipids with smaller amounts of cholesterol.

**15.57** The lipid bilayer in a cell membrane surrounds the cell and separates the contents of the cell from the external fluids.

**15.59** The peripheral proteins in the membrane emerge on the inner or outer surface only, whereas the integral proteins extend through the membrane to both surfaces.

**15.61** The carbohydrates attached to proteins and lipids on the surface of cells act as receptors for cell recognition and chemical messengers such as neurotransmitters.

**15.63**

$$H_2C-O-\overset{\overset{\displaystyle O}{\|}}{C}-(CH_2)_7-CH=CH-CH_2-CH=CH-(CH_2)_4-CH_3$$

$$H-C-O-\overset{\overset{\displaystyle O}{\|}}{C}-(CH_2)_7-CH=CH-(CH_2)_7-CH_3$$

$$H_2C-O-\overset{\overset{\displaystyle O}{\|}}{C}-(CH_2)_7-CH=CH-CH_2-CH=CH-(CH_2)_4-CH_3$$

**b.**

$$H_2C-O-\overset{\overset{\displaystyle O}{\|}}{C}-(CH_2)_7-CH=CH-CH_2-CH=CH-(CH_2)_4-CH_3$$

$$H-C-O-\overset{\overset{\displaystyle O}{\|}}{C}-(CH_2)_7-CH=CH-(CH_2)_7-CH_3 + 5H_2 \xrightarrow{Ni}$$

$$H_2C-O-\overset{\overset{\displaystyle O}{\|}}{C}-(CH_2)_7-CH=CH-CH_2-CH=CH-(CH_2)_4-CH_3$$

$$H_2C-O-\overset{\overset{\displaystyle O}{\|}}{C}-(CH_2)_{16}-CH_3$$

$$H-C-O-\overset{\overset{\displaystyle O}{\|}}{C}-(CH_2)_{16}-CH_3$$

$$H_2C-O-\overset{\overset{\displaystyle O}{\|}}{C}-(CH_2)_{16}-CH_3$$

**15.67** Beeswax and carnauba are waxes. Vegetable oil and glyceryl tricaprate (tricaprin) are triacylglycerols.

$$CH_2O\overset{\overset{\displaystyle O}{\|}}{C}(CH_2)_8CH_3$$

$$CHO\overset{\overset{\displaystyle O}{\|}}{C}(CH_2)_8CH_3$$

$$CH_2O\overset{\overset{\displaystyle O}{\|}}{C}(CH_2)_8CH_3$$

Glyceryl tricaprate (tricaprin)

**15.69** **a.** A typical unsaturated fatty acid has a cis double bond.
  **b.** A cis unsaturated fatty acid contains hydrogen atoms on the same side of each double bond. A trans unsaturated fatty acid has hydrogen atoms on the opposite sides of each double bond that forms during hydrogenation.

**c.**

$$\underset{CH_3(CH_2)_6CH_2}{\overset{H}{\diagdown}} C=C \underset{H}{\overset{CH_2(CH_2)_6\overset{\overset{\displaystyle O}{\|}}{C}OH}{\diagup}}$$

**15.65** **a.**

$$H_2C-O-\overset{\overset{\displaystyle O}{\|}}{C}-(CH_2)_7-CH=CH-CH_2-CH=CH-(CH_2)_4-CH_3$$

$$H-C-O-\overset{\overset{\displaystyle O}{\|}}{C}-(CH_2)_7-CH=CH-CH_2-CH=CH-(CH_2)_4-CH_3$$

$$H_2C-O-\overset{\overset{\displaystyle O}{\|}}{C}-(CH_2)_7-CH=CH-(CH_2)_7-CH_3$$

**15.71**

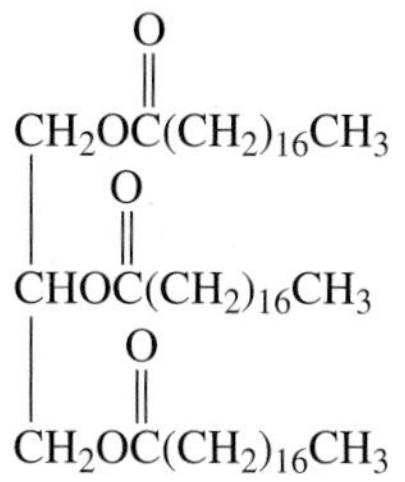

$$CH_2OC(CH_2)_{16}CH_3$$
$$CHOC(CH_2)_{16}CH_3$$
$$CH_2OC(CH_2)_{16}CH_3$$
Glyceryl tristearate

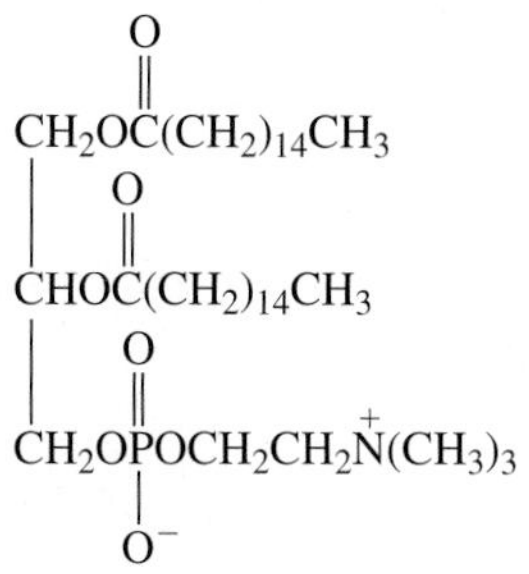

$$CH_2OC(CH_2)_{14}CH_3$$
$$CHOC(CH_2)_{14}CH_3$$
$$CH_2OPOCH_2CH_2\overset{+}{N}(CH_3)_3$$
$$O^-$$
Lecithin

**15.73**
**a.** wax
**b.** steroid
**c.** glycerophospholipid
**d.** triacylglycerol
**e.** soap
**f.** triacylglycerol
**g.** triacylglycerol
**h.** triacylglycerol
**i.** steroid
**j.** steroid
**k.** fatty acid

**15.75**
**a.** 5
**b.** 1, 2, 3, 4
**c.** 2
**d.** 1, 2
**e.** 1, 2, 3, 4

**15.77**
**a.** 4
**b.** 3
**c.** 1
**d.** 4
**e.** 4
**f.** 3
**g.** 2
**h.** 1

**15.79**

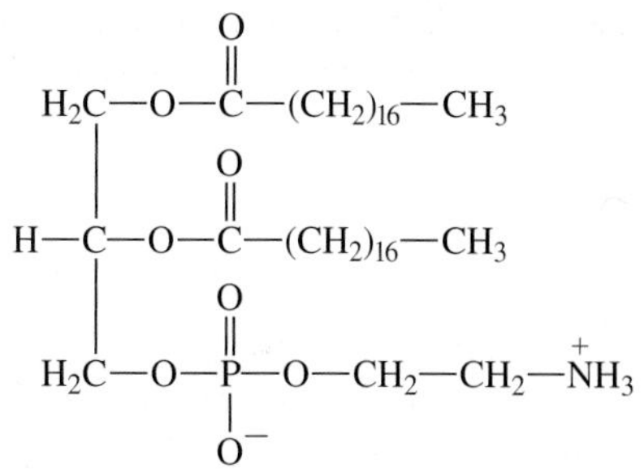

$$H_2C-O-C-(CH_2)_{16}-CH_3$$
$$H-C-O-C-(CH_2)_{16}-CH_3$$
$$H_2C-O-P-O-CH_2-CH_2-\overset{+}{N}H_3$$
$$O^-$$

**15.81** **a.** Adding NaOH would hydrolyze lipids such as glyceryl tristearate (tristearin), forming glycerol and salts of the fatty acids that are soluble in water and wash down the drain.

**b.**

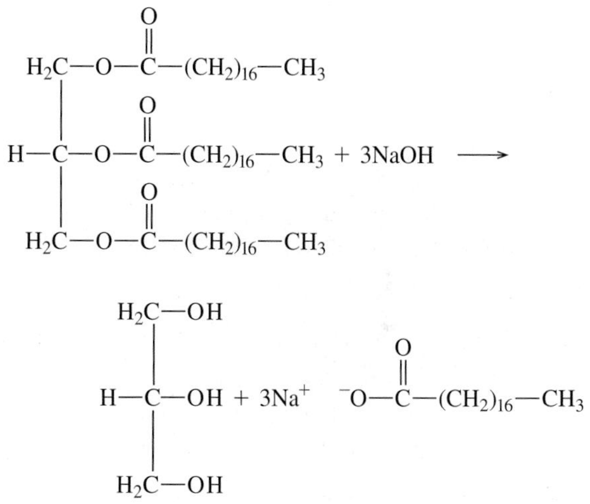

$$H_2C-O-C-(CH_2)_{16}-CH_3$$
$$H-C-O-C-(CH_2)_{16}-CH_3 + 3NaOH \longrightarrow$$
$$H_2C-O-C-(CH_2)_{16}-CH_3$$

$$H_2C-OH$$
$$H-C-OH + 3Na^+ \quad {}^-O-C-(CH_2)_{16}-CH_3$$
$$H_2C-OH$$
Glycerol                Sodium stearate

# Combining Ideas from Chapters 13 to 15

**CI.25** The plastic known as PETE (**p**ol**y**ethylene**te**rephthalate) is used to make plastic soft drink bottles and containers for salad dressing, shampoos, and dishwashing liquids. Today, PETE is the most widely recycled of all the plastics. PETE is a polymer of terephthalic acid and ethylene glycol. In one year, 1.7 billon pounds of PETE are recycled. After it is separated from other plastics, PETE can be used in polyester fabric, door mats, tennis ball containers, and fill for sleeping bags. The density of PETE is 1.38 g/mL.

$$HO-\overset{\displaystyle O}{\overset{\|}{C}}-\underset{}{\bigcirc}-\overset{\displaystyle O}{\overset{\|}{C}}-OH \qquad HO-CH_2-CH_2-OH$$

Terephthalic acid          Ethylene glycol

**a.** Draw the formula of the ester formed from one molecule of terephthalic acid and one molecule of ethylene glycol.

**b.** Draw the formula of the product formed when a second molecule of ethylene glycol reacts with the ester you drew in part **a**.

**c.** How many kilograms of PETE are recycled in one year?

**d.** What volume, in liters, of PETE is recycled in one year?

**e.** Suppose a landfill with an area of a football field and a depth of 5.0 m holds $2.7 \times 10^7$ L of recycled PETE. If all of the PETE that is recycled in a year were placed instead in landfills, how many would it fill?

**CI.26** Using the Internet or a reference book such as *The Merck Index* or *Physicians' Desk Reference,* look up the structural formulas of the following medicinal drugs and list the functional groups in the compounds. You may need to refer to the cross-index of names at the back of the reference book.

**a.** baclofen, a muscle relaxant

**b.** anethole, a licorice flavoring agent in anise and fennel

**c.** alibendol, antispasmodic drug

**d.** pargyline, antihypertensive drug

**e.** naproxen, non-steroidal anti-inflammatory drug

**CI.27** The insect repellent DEET is an amide that can be made from 3-methylbenzoic acid and diethylamine. A 6.0-fl oz can of DEET repellent contains 25% DEET by mass (1 qt = 32 fluid ounces). Assume that the density of DEET solution in a can is 1.0 g/mL.

**a.** What is the structure of DEET?

**b.** What is the molecular formula of DEET?

**c.** What is the molar mass, with 3 significant figures, of DEET?

**d.** How many grams of DEET are in one spray can?

**e.** How many molecules of DEET are in one spray can?

**CI.28** Glyceryl trimyristate (trimyristin) is found in the seeds of the nutmeg (*Myristica fragrans*). The oil known as nutmeg butter contains 75% trimyristin. Ground nutmeg, which is sweet, is used to flavor many foods. It is used as a lubricant and fragrance in soaps and shaving creams. Isopropyl myristate is used to increase absorption of skin creams. Draw the structures for each of the following:

**a.** myristic acid

**b.** glyceryl trimyristate (trimyristin)

**c.** isopropyl myristate

**d.** products of the hydrolysis of glyceryl trimyristate with an acid catalyst

**e.** products of the saponication of glyceryl trimyristate with KOH

**f.** reactant and product for oxidation of myristyl alcohol to myristic acid

**CI.29** Panose is a trisaccharide that is being considered as a possible sweetener by the food industry.

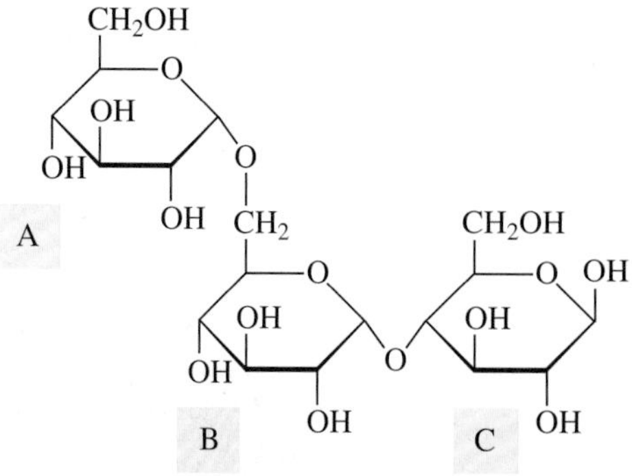

a. What are the monosaccharides units of A, B, and C in panose?
b. What type of bond connects monosaccharides A and B?
c. What type of bond connects monosaccharides B and C?
d. Is the structure drawn as $\alpha$- or $\beta$-panose?
e. Why would panose be a reducing sugar?

**CI.30** Hyaluronic acid (HA), a polymer of about 25 000 disaccharide units, is a natural component of eye and joint fluid as well as skin and cartilage. Because of the ability of HA to absorb water, it is used in skin-care products and injections to smooth wrinkles and for treatment of arthritis. The repeating disaccharide units in HA consist of D-gluconic acid and D-acetylglucosamine. D-Acetylglucosamine is an amide derived from acetic acid and D-glucosamine, in which an amine group ($-NH_2$) replaces the hydroxyl on carbon 2 of D-glucose. Another natural polymer called chitin is found in the shells of lobsters and crabs. Chitin is made of repeating units of D-acetylglucosamine connected by $\beta$-1,4-glycosidic bonds.

a. Draw the Haworth structures for the oxidation reaction of hydroxyl group on carbon 6 in $\beta$-glucose to form $\beta$-gluconic acid.
b. Draw the Haworth structure of D-glucosamine.
c. Draw the Haworth structure for the amide of D-glucosamine and acetic acid.
d. What are the two types of glycosidic bonds that link the monosaccharides?
e. Draw the structure of a section of chitin with three D-acetylglucosamine units linked by $\beta$-1,4-glycosidic bonds.

## ANSWERS

**CI.25 a.**

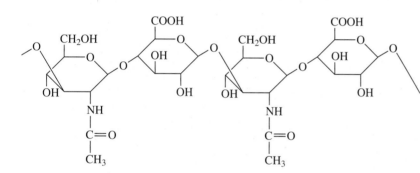

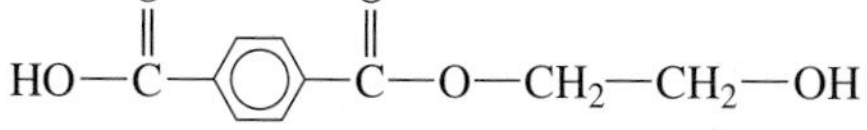

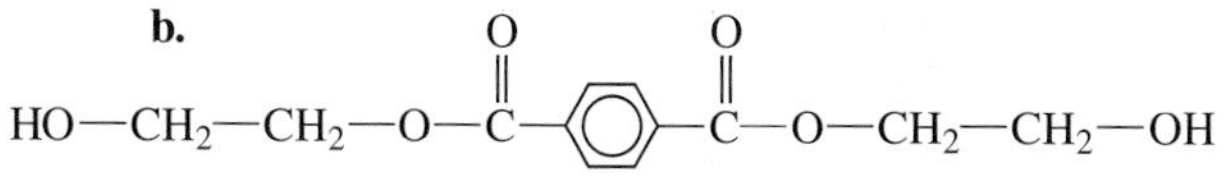

c. $7.7 \times 10^8$ kg   d. $5.6 \times 10^8$ L   e. 21 landfills

**CI.27 a.**

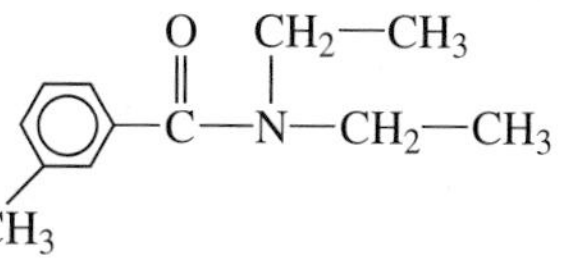

b. $C_{12}H_{17}NO$

c. 191 g/mole
d. 44 g of DEET
e. $1.4 \times 10^{23}$ molecules

**CI.29 a.** A, B, and C are all glucose.
b. An $\alpha$-1,6-glycosidic bond links A and B.
c. An $\alpha$-1,4-glycosidic bond links B and C.
d. $\beta$-panose
e. Panose is a reducing sugar because it has a free hydroxyl on carbon 1 of structure C, which allows glucose (C) to form the aldehyde.

# 16 Amino Acids, Proteins, and Enzymes

**the Chemistry place**

Visit **www.chemplace.com** for extra quizzes, interactive tutorials, career resources, PowerPoint slides for chapter review, math help, and case studies.

*"This lamb is fed with Lamb Lac, which is a chemically formulated replacement for ewe's milk,"* says part-time farmer Dennis Samuelson. *"Its mother had triplets and didn't have enough milk to feed them all, so they weren't thriving the way the other lambs were. The Lamb Lac includes dried skim milk, dried whey, milk proteins, egg albumin, the amino acids methionine and lysine, vitamins, and minerals."*

*A veterinary technician diagnoses and treats diseases of animals, takes blood and tissue samples, and administers drugs and vaccines. Agricultural technologists assist in the study of farm crops to increase productivity and ensure a safe food supply. They look for ways to improve crop yields, develop safer methods of weed and pest control, and design methods to conserve soil and water.*

The word "protein" is derived from the Greek word *proteios*, meaning "first." All proteins in humans are polymers made up from 20 different amino acids. Each kind of protein is composed of amino acids arranged in a specific order that determines the characteristics of the protein and its biological action. Proteins provide structure in membranes, build cartilage and connective tissue, transport oxygen in blood and muscle, direct biological reactions as enzymes, defend the body against infection, and control metabolic processes as hormones. They can even be a source of energy.

The different functions of proteins depend on the structures and chemical behavior of amino acids, the building blocks of proteins. We will see how peptide bonds link amino acids and how the order of the amino acids in these protein polymers directs the formation of unique three-dimensional structures.

Every second, thousands of chemical reactions occur in the cells of the human body at rates that meet our physiological and metabolic needs. To make this happen, enzymes catalyze the chemical reactions in our cells, with a different enzyme for every reaction. Digestive enzymes in the mouth, stomach, and small intestine catalyze the hydrolysis of carbohydrates, fats, and proteins. Enzymes in the mitochondria extract energy from biomolecules to give us energy.

# 16.1 FUNCTIONS OF PROTEINS

The many kinds of proteins perform different functions in the body. There are proteins that form structural components such as cartilage, muscles, hair, and nails. Wool, silk, feathers, and horns are some other proteins made by animals. Proteins that function as enzymes regulate biological reactions such as digestion and cellular metabolism. Some proteins, such as hemoglobin and myoglobin, transport oxygen in the blood and muscle. (See Figure 16.1.) Table 16.1 gives examples of proteins that are classified by their functions in biological systems.

## LEARNING GOAL

Classify proteins by their functions in the cells.

**WEB TUTORIAL**
Functions of Proteins

**TABLE 16.1** Classification of Some Proteins and Their Functions

| Class of Protein | Function in the Body | Examples |
|---|---|---|
| Structural | Provide structural components | *Collagen* is in tendons and cartilage. *Keratin* is in hair, skin, wool, and nails. |
| Contractile | Movement of muscles | *Myosin* and *actin* contract muscle fibers. |
| Transport | Carry essential substances throughout the body | *Hemoglobin* transports oxygen. *Lipoproteins* transport lipids. |
| Storage | Store nutrients | *Casein* stores protein in milk. *Ferritin* stores iron in the spleen and liver. |
| Hormone | Regulate body metabolism and nervous system | *Insulin* regulates blood glucose level. *Growth hormone* regulates body growth. |
| Enzyme | Catalyze biochemical reactions in the cells | *Sucrase* catalyzes the hydrolysis of sucrose. *Trypsin* catalyzes the hydrolysis of proteins. |
| Protection | Recognize and destroy foreign substances | *Immunoglobulins* stimulate immune responses. |

**FIGURE 16.1** The horns of animals are made of proteins.
**Q** What class of protein would be in horns?

### ■ Classifying Proteins by Function

Give the class of protein that would perform each of the following functions.

**a.** catalyzes metabolic reactions of lipids
**b.** carries oxygen in the bloodstream
**c.** stores amino acids in milk

SOLUTION

**a.** Enzymes catalyze metabolic reactions.
**b.** Transport proteins carry substances such as oxygen through the bloodstream.
**c.** A storage protein stores nutrients such as amino acids in milk.

STUDY CHECK

What class of proteins helps regulate metabolism?

## QUESTIONS AND PROBLEMS

### Functions of Proteins

**16.1** Classify each of the following proteins according to its function:
   **a.** hemoglobin, oxygen carrier in the blood
   **b.** collagen, a major component of tendons and cartilage
   **c.** keratin, a protein found in hair
   **d.** amylases that hydrolyze starch

**16.2** Classify each of the following proteins according to its function:
   **a.** insulin, a protein needed for glucose utilization
   **b.** antibodies, proteins that disable foreign proteins
   **c.** casein, milk protein
   **d.** lipases that hydrolyze lipids

---

LEARNING GOAL

Draw the zwitterion for an amino acid.

# 16.2 AMINO ACIDS

Proteins are composed of molecular building blocks called amino acids. An **amino acid** contains two functional groups, an amine group ($-NH_2$) and a carboxylic acid group ($-COOH$). In all of the 20 amino acids found in proteins, the amine group, the carboxylic acid group, and a hydrogen atom are bonded to a central carbon atom called the $\alpha$ (alpha) carbon. Only 20 $\alpha$-amino acids are present in human proteins, and they have unique characteristics because of different side chains (R) attached to the $\alpha$-carbon.

## General Structure of an $\alpha$-Amino Acid

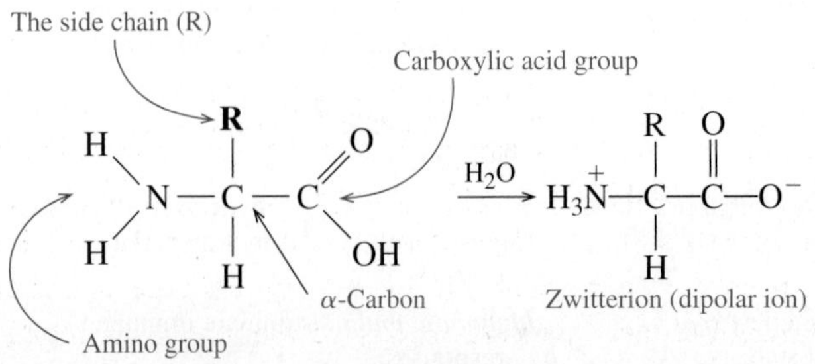

The side chain (R)
Carboxylic acid group
$\alpha$-Carbon
Amino group
Zwitterion (dipolar ion)

Although it is convenient to write amino acids with neutral groups, they are ionized at the pH of most body fluids. Under physiological conditions, the —COOH group loses $H^+$ to give —COO⁻, and the —NH$_2$ group gains $H^+$ to give —NH$_3^+$. With a positively charged end and a negatively charged end, an α-amino acid is a dipolar ion called a **zwitterion**.

## Classification of Amino Acids

**Nonpolar amino acids**, which have alkyl or aromatic side chains, are *hydrophobic* ("water-fearing"). **Polar amino acids**, which have polar side chains such as hydroxyl (—OH), thiol (—SH), and amide (—CONH$_2$) that interact with water are *hydrophilic* ("water attracting"). The side chains of the **acidic amino acids** contain carboxylic acid groups (—COOH) that ionize as weak acids. The side chains of the **basic amino acids** contain amine groups that ionize as weak bases. The structures of the side chains (R), common names, and the three-letter abbreviations of the 20 α-amino acids in proteins are listed in Table 16.2. The isoelectric points known as pI values are discussed in Section 16.3.

---

### SAMPLE PROBLEM 16.2

■ **Structural Formulas of Amino Acids**

Draw the zwitterion and write the abbreviation for each of the following amino acids:

**a.** alanine (R = —CH$_3$)          **b.** serine (R = —CH$_2$OH)

SOLUTION

**a.** The zwitterion of an amino acid is written by attaching a side group (R) to the central carbon atom of the dipolar ion of an amino acid.

alanine (Ala)

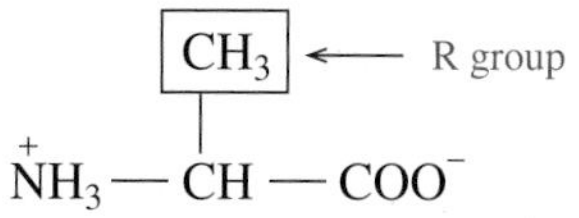

$$\boxed{CH_3} \longleftarrow \text{R group}$$
$$\overset{+}{NH_3} - CH - COO^-$$

**b.** serine (Ser)

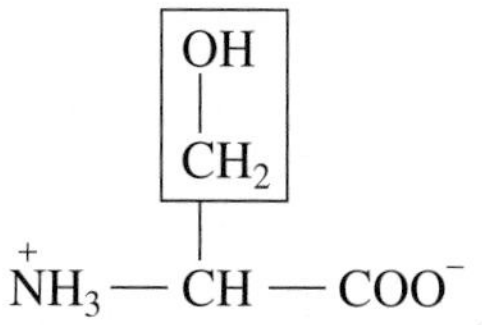

$$\boxed{\begin{array}{c} OH \\ | \\ CH_2 \end{array}}$$
$$\overset{+}{NH_3} - CH - COO^-$$

STUDY CHECK

Classify the amino acids in the sample problem as polar or nonpolar.

---

## Amino Acid Stereoisomers

All of the α-amino acids except for glycine are chiral because the α carbon is attached to four different atoms. Thus amino acids can exist as D and L isomers. We can write Fischer projections for α-amino acids as we did in Chapter 12 for aldehydes by placing the carboxylate group at the top and the side chain at the bottom. In the L isomer, the —NH$_3^+$, is on the left, and in the D isomer, it is on the right. In biological systems, the only amino acids incorporated into proteins are the L isomers. There are D amino acids found in

**TABLE 16.2** The 20 Amino Acids in Proteins

### Nonpolar Amino Acids

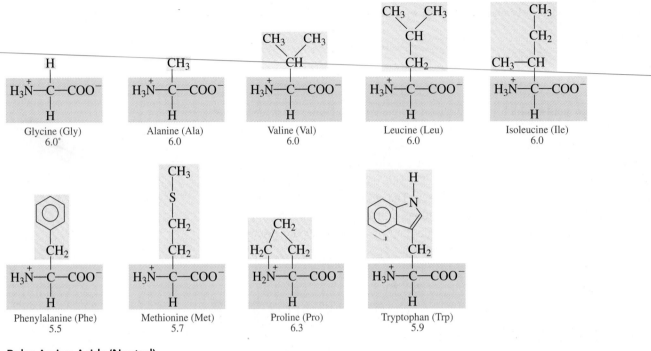

Glycine (Gly)
6.0*

Alanine (Ala)
6.0

Valine (Val)
6.0

Leucine (Leu)
6.0

Isoleucine (Ile)
6.0

Phenylalanine (Phe)
5.5

Methionine (Met)
5.7

Proline (Pro)
6.3

Tryptophan (Trp)
5.9

### Polar Amino Acids (Neutral)

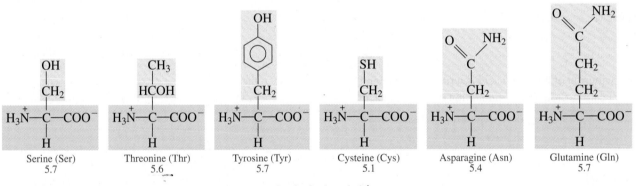

Serine (Ser)
5.7

Threonine (Thr)
5.6

Tyrosine (Tyr)
5.7

Cysteine (Cys)
5.1

Asparagine (Asn)
5.4

Glutamine (Gln)
5.7

### Acidic Amino Acids                    Basic Amino Acids

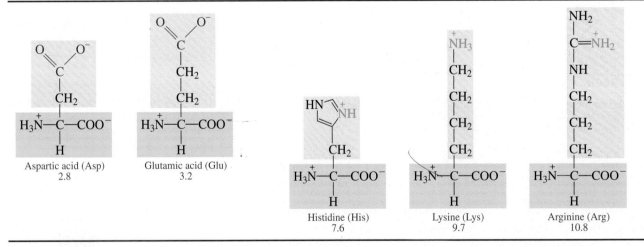

Aspartic acid (Asp)
2.8

Glutamic acid (Glu)
3.2

Histidine (His)
7.6

Lysine (Lys)
9.7

Arginine (Arg)
10.8

*Isoelectric points (pI)

nature, but not in proteins. Let's take a look at the isomers for L- and D-glyceraldehyde and the isomers of L- and D-alanine and L- and D-cysteine.

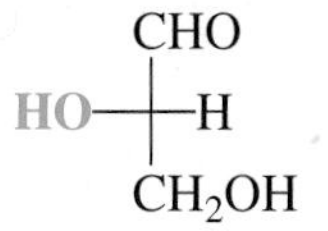

$$CHO$$
$$HO \text{---} H$$
$$CH_2OH$$
L-Glyceraldehyde

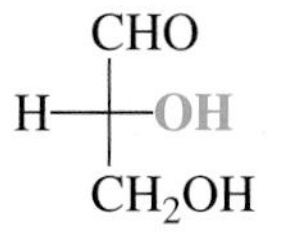

$$CHO$$
$$H \text{---} OH$$
$$CH_2OH$$
D-Glyceraldehyde

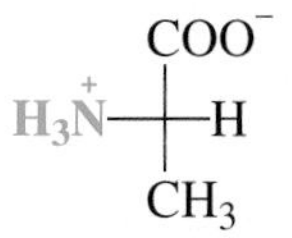

$$COO^-$$
$$H_3\overset{+}{N} \text{---} H$$
$$CH_3$$
L-Alanine

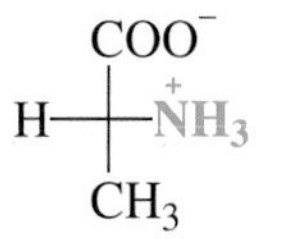

$$COO^-$$
$$H \text{---} \overset{+}{N}H_3$$
$$CH_3$$
D-Alanine

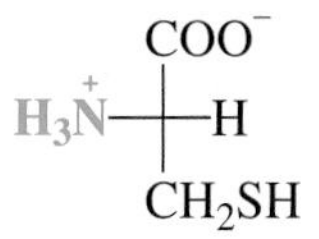

$$COO^-$$
$$H_3\overset{+}{N} \text{---} H$$
$$CH_2SH$$
L-Cysteine

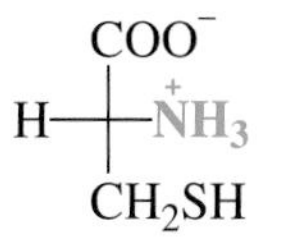

$$COO^-$$
$$H \text{---} \overset{+}{N}H_3$$
$$CH_2SH$$
D-Cysteine

## QUESTIONS AND PROBLEMS

### Amino Acids

**16.3** Describe the functional groups found in all α-amino acids.

**16.4** How does the polarity of the side chain in leucine compare to the side chain in serine?

**16.5** Draw the zwitterion form for each of the following amino acids:
a. alanine
b. threonine
c. glutamic acid
d. phenylalanine

**16.6** Draw the zwitterion form for each of the following amino acids:
a. lysine
b. aspartic acid
c. leucine
d. tyrosine

**16.7** Classify the amino acids in problem 16.5 as hydrophobic (nonpolar), hydrophilic (polar, neutral), acidic, or basic.

**16.8** Classify the amino acids in problem 16.6 as hydrophobic (nonpolar), hydrophilic (polar, neutral), acidic, or basic.

**16.9** Give the name of the amino acid represented by each of the following three-letter abbreviations:
a. Ala
b. Val
c. Lys
d. Cys

**16.10** Give the name of the amino acid represented by each of the following three-letter abbreviations:
a. Trp
b. Met
c. Pro
d. Gly

**16.11** Draw the Fischer projections for the following amino acids:
a. L-valine
b. D-cysteine

**16.12** Draw the Fischer projections for the following amino acids:
a. L-threonine
b. D-valine

# 16.3 AMINO ACIDS AS ACIDS AND BASES

**LEARNING GOAL**

Draw the ionized structure of an amino acid at pH values above or below pI.

At a certain pH known as the **isoelectric point (pI)**, the positive and negative charges of an amino acid are equal, which gives an overall charge of zero. In a solution that is more acidic (lower pH) than the pI, the —$COO^-$ group accepts a $H^+$ to form —COOH. Because of the —$NH_3^+$, the amino acid has an overall positive charge (1+). In a solution that is more basic (higher pH) than the pI, the —$NH_3^+$ loses a $H^+$. Because there is a carboxylate group —$COO^-$, the amino acid has an overall negative charge (1−).

| Condition | pH < pI | pH = pI | pH > pI |
|---|---|---|---|
| **Change in H⁺** | $[H^+] \Uparrow$ | None | $[H^+] \Downarrow$ |
| **Change in ionized groups** | —COOH<br>—$NH_3^+$ | —$COO^-$<br>—$NH_3^+$ | —$COO^-$<br>—$NH_2$ |
| **Charge of amino acid** | 1+ | 0 | 1− |

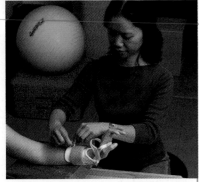

## Rehabilitation Specialist

"I am interested in the biomechanical part of rehabilitation, which involves strengthening activities that help people return to the activities of daily living," says Minna Robles, rehabilitation specialist. "Here I am fitting a patient with a wrist extension splint that allows her to lift her hand. This exercise will also help the muscles and soft tissues in that wrist area to heal. An understanding of the chemicals of the body, how they interact, and how we can affect the body on a chemical level is important in understanding our work. One technique we use is called myofacial release. We apply pressure to a part of the body, which helps to increase circulation. By increasing circulation, we can move the soft tissues better, which improves movement and range of motion."

Let's take a look at the changes in all the ionic forms of alanine from its zwitterion (pI = 6.0), to the positive ion in a more acidic solution, and the negative ion in basic solution.

$$\overset{+}{NH_3}-CH-\overset{\overset{O}{\|}}{C}-OH \underset{H_3O^+}{\overset{OH^-}{\rightleftharpoons}} \overset{+}{NH_3}-CH-\overset{\overset{O}{\|}}{C}-O^- \underset{H_3O^+}{\overset{OH^-}{\rightleftharpoons}} NH_2-CH-\overset{\overset{O}{\|}}{C}-O^-$$

| | | |
|---|---|---|
| $CH_3$ | $CH_3$ | $CH_3$ |
| Alanine at a pH < 6 (charge = 1+) | Zwitterion of alanine pH = 6.0 (charge = 0) | Alanine at a pH > 6 (charge = 1−) |

The pI values for polar and nonpolar amino acids are in the range of pH 5.1 to 6.0. The pI values of the acidic amino acids are at a lower pH (about pH 3) because the carboxyl groups in their side chains also accept $H^+$. The pI values of the basic amino acids are at a high pH (about pH 7.6 to 10.8) because the amine groups in their side chains also donate $H^+$. The pI values are included below the names of the amino acids in Table 16.2.

### SAMPLE PROBLEM 16.3

#### ■ Amino Acids in Acid or Base

Draw the structural formula for serine at pH 5.7, which is equal to the pI of serine.

SOLUTION

At the pI of 5.7, serine exists as a zwitterion with both the carboxylic acid and the amine groups ionized.

$$\overset{+}{NH_3}-\overset{\overset{\displaystyle OH}{\displaystyle |}}{\underset{\displaystyle CH}{\overset{\displaystyle |}{\underset{\displaystyle |}{CH_2}}}}-\overset{\overset{O}{\|}}{C}-O^- \quad \text{Zwitterion of serine}$$

STUDY CHECK

Draw the structure of serine at a pH below the pI of 5.7.

## QUESTIONS AND PROBLEMS

### Amino Acids as Acids and Bases

**16.13** Draw the zwitterion of each of the following amino acids:
 **a.** glycine   **b.** cysteine   **c.** serine   **d.** alanine

**16.14** Draw the zwitterion of each of the following amino acids:
 **a.** phenylalanine   **b.** methionine   **c.** leucine   **d.** valine

**16.15** Draw the positive ion (acidic ion) of each of the amino acids in problem 16.13 at a pH below 1.0.

**16.16** Draw the negative ion (basic ion) of each of the amino acids in problem 16.13 at a pH above 12.0.

**16.17** Would the following ions of valine exist at a pH above, below, or at pI?
 **a.** $H_2N-\underset{\underset{CH_3\,\,CH_3}{\overset{|}{CH}}}{\overset{|}{CH}}-COO^-$   **b.** $\overset{+}{H_3N}-\underset{\underset{CH_3\,\,CH_3}{\overset{|}{CH}}}{\overset{|}{CH}}-COOH$

 **c.** $\overset{+}{H_3N}-\underset{\underset{CH_3\,\,CH_3}{\overset{|}{CH}}}{\overset{|}{CH}}-COO^-$

**16.18** Would the following ions of serine exist at a pH above, below, or at pI?
 **a.** $\overset{+}{H_3N}-\underset{\overset{|}{CH_2OH}}{\overset{|}{CH}}-COO^-$   **b.** $\overset{+}{H_3N}-\underset{\overset{|}{CH_2OH}}{\overset{|}{CH}}-COOH$

 **c.** $H_2N-\underset{\overset{|}{CH_2OH}}{\overset{|}{CH}}-COO^-$

# 16.4 FORMATION OF PEPTIDES

The linking of two or more amino acids forms a **peptide**. A **peptide bond** is an amide bond that forms when the $-COO^-$ group of one amino acid reacts with the $-NH_3^+$ group of the next amino acid. Two amino acids linked together by a peptide bond form a *dipeptide*. We can write the formation of the dipeptide between glycine and alanine as follows. (See Figure 16.2.) In a peptide, the amino acid written on the left with the free $-NH_3^+$ group is called the **N terminal** amino acid. The **C terminal** amino acid is the last amino acid in the chain with the free $-COO^-$ group.

## Naming Peptides

In naming a peptide, each amino acid beginning from the N terminal is named with an *yl* ending followed by the full name of the amino acid at the C terminal. For example, a tripeptide consisting of alanine, glycine, and serine is named as alan**yl**glyc**yl**serine. For convenience, the order of amino acids in the peptide is often written as the sequence of three-letter abbreviations.

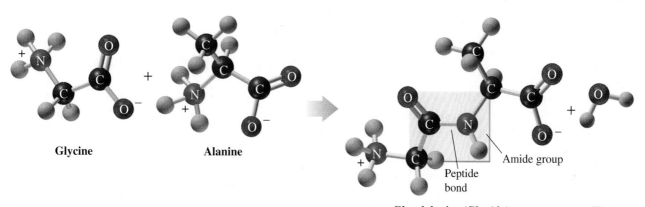

**FIGURE 16.2** A peptide bond between zwitterions of glycine and alanine form the dipeptide glycylalanine.
**Q** What functional groups in glycine and alanine form the peptide bond?

■ **Identifying a Tripeptide**

Consider the following tripeptide.

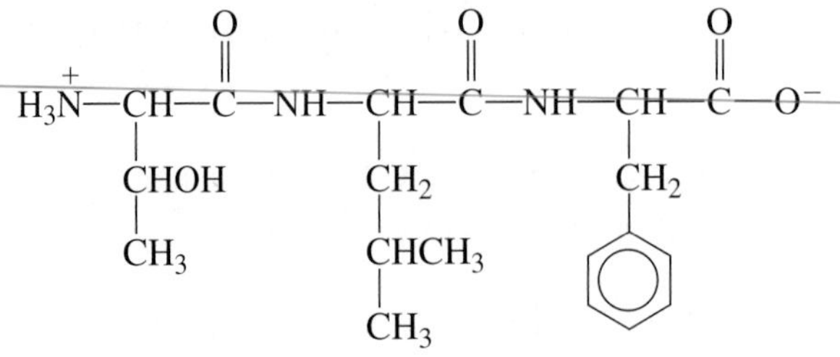

$$\overset{+}{H_3N}-CH-\overset{\overset{\displaystyle O}{\|}}{C}-NH-CH-\overset{\overset{\displaystyle O}{\|}}{C}-NH-CH-\overset{\overset{\displaystyle O}{\|}}{C}-O^-$$

with side chains:
- CHOH—CH₃
- CH₂—CHCH₃—CH₃
- CH₂—(phenyl ring)

**a.** What amino acid is the N terminal? What amino acid is the C terminal?

**b.** What is the three-letter abbreviation for the order of amino acids in the tripeptide?

SOLUTION

**a.** Threonine is the N terminal; phenylalanine is the C terminal.

**b.** Thr-Leu-Phe

STUDY CHECK

What is the full name of the tripeptide in Sample Problem 16.4?

---

## QUESTIONS AND PROBLEMS

**Formation of Peptides**

**16.19** Draw the condensed structural formula of each of the following peptides, and give the abbreviation for their names:
   **a.** alanylcysteine
   **b.** serylphenylalanine
   **c.** glycylalanylvaline
   **d.** valylisoleucyltryptophan

**16.20** Draw the condensed structural formula of each of the following peptides, and give the abbreviation for their names:
   **a.** methionylaspartic acid
   **b.** alanyltryptophan
   **c.** methionylglutaminyllysine
   **d.** histidylglycylglutamylalanine

---

**LEARNING GOAL**

Identify the structural levels of a protein.

**WEB TUTORIAL**
Structure of Proteins
Primary and Secondary Structure

# 16.5  LEVELS OF PROTEIN STRUCTURE

When there are more than 50 amino acids in a chain, the polypeptide is usually called a **protein**. Each protein in our cells has a unique sequence of amino acids that determines its biological function.

## Primary Structure

The **primary structure** of a protein is the particular sequence of amino acids held together by peptide bonds. For example, a hormone that stimulates the thyroid to release thyroxin is a tripeptide with the amino acid sequence Glu-His-Pro. Although other amino acid sequences are possible, such as His-Pro-Glu or Pro-His-Glu, they do not produce hormonal activity. Thus the biological function of peptides and proteins depends on the sequence of the amino acids,

The first protein to have its primary structure determined was insulin, which is a hormone that regulates the glucose level in the blood. In the primary structure of human insulin, there are two polypeptide chains. In chain A, there are 21 amino acids, and chain B has 30 amino acids. The polypeptide chains are held together by disulfide bonds formed by the side chains of the cysteine amino acids in each of the chains. (See Figure 16.3.) Today, human insulin produced through genetic engineering is used in the treatment of diabetes.

## Secondary Structure

The **secondary structure** of a protein describes the type of structure that forms when amino acids form hydrogen bonds within a polypeptide or between polypeptides. The three most common types of secondary structure are the *alpha helix*, the *beta-pleated sheet*, and the *triple helix*.

In an **alpha helix ($\alpha$ helix)**, hydrogen bonds form between each N—H group and the oxygen of a C=O group of an amino acid farther along the polypeptide. (See Figure 16.4.) Because there are many hydrogen bonds along the polypeptide, it has the helical shape of a coiled telephone cord or a Slinky toy. The side chains (R groups) of the different $\alpha$-amino acids extend to the outside of the helix.

The secondary structure in silk is beta-pleated sheet.

## Health Note

### Natural Opiates in the Body

Painkillers known as enkephalins and endorphins are produced naturally in the body. They are polypeptides that bind to receptors in the brain to give relief from pain. This effect appears to be responsible for the runner's high, for the temporary loss of pain when severe injury occurs, and for the analgesic effects of acupuncture.

The *enkephalins*, found in the thalamus and the spinal cord, are pentapeptides, the smallest molecules with opiate activity. The

amino acid sequence of an enkephalin is found in the longer amino acid sequence of the endorphins.

Four groups of *endorphins* have been identified: $\alpha$-endorphin contains 16 amino acids, $\beta$-endorphin contains 31 amino acids, $\gamma$-endorphin has 17 amino acids, and $\delta$-endorphin has 27 amino acids. Endorphins may produce their sedating effects by preventing the release of substance P, a polypeptide with 11 amino acids, which has been found to transmit pain impulses to the brain.

$\alpha$-Endorphin

Tyr — Gly — Gly — Phe — Met — Thr — Ser — Glu — Lys — Ser — Glu — Thr — Pro — Leu — Val — Thr

Enkephalin

Leu

Glu — Gly — Lys — Lys — Tyr — Ala — Asn — Lys — Ile — Ile — Ala — Asn — Lys — Phe

$\beta$-Endorphin

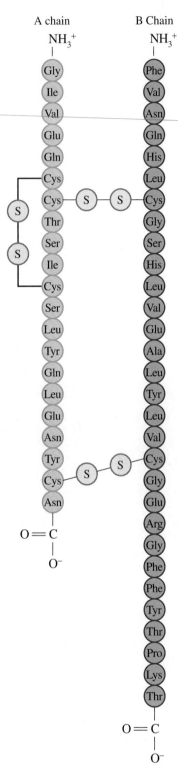

**A chain**

NH$_3^+$
|
Gly
Ile
Val
Glu
Gln
Cys
Cys
Thr
Ser
Ile
Cys
Ser
Leu
Tyr
Gln
Leu
Glu
Asn
Tyr
Cys
Asn
|
O=C
|
O$^-$

**B Chain**

NH$_3^+$
|
Phe
Val
Asn
Gln
His
Leu
Cys
Gly
Ser
His
Leu
Val
Glu
Ala
Leu
Tyr
Leu
Val
Cys
Gly
Glu
Arg
Gly
Phe
Phe
Tyr
Thr
Pro
Lys
Thr
|
O=C
|
O$^-$

**FIGURE 16.3** The sequence of amino acids in human insulin is its primary structure.

**Q** What kinds of bonds occur in the primary structure of a protein?

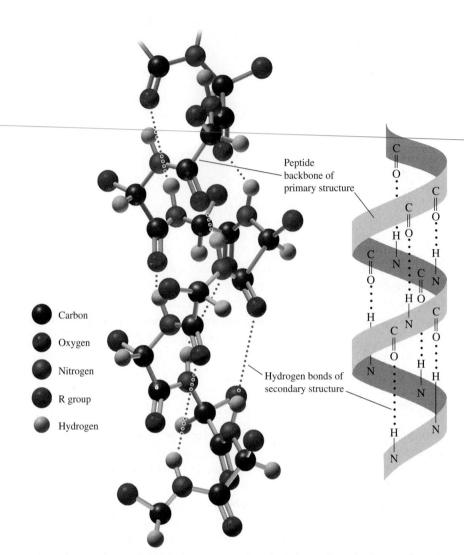

Carbon
Oxygen
Nitrogen
R group
Hydrogen

Peptide backbone of primary structure

Hydrogen bonds of secondary structure

**FIGURE 16.4** The $\alpha$ (alpha) helix acquires a coiled shape from hydrogen bonds between the N—H of the peptide bond in one loop and the C=O of the peptide bond in the next loop.

**Q** What are the partial charges of the H in N—H and the O in C=O that permit hydrogen bonds to form?

Another type of secondary structure is the **beta-pleated sheet ($\beta$-pleated sheet)**. In a $\beta$-pleated sheet, hydrogen bonds form between polypeptides so they are held together side by side like folded or pleated sheets. In silk fibers, the predominance of glycine, alanine, and serine with small side chains allows the $\beta$-pleated sheets to stack closely together. This compact arrangement gives strength, flexibility, and durability to proteins such as silk. (See Figure 16.5.)

**Collagen**, the most abundant protein, makes up as much as one-third of all the protein in vertebrates. It is found in connective tissue, blood vessels, skin, tendons, ligaments, the cornea of the eye, and cartilage. The strong structure of collagen is a result of three polypeptides woven together like a braid to form a **triple helix**. (See Figure 16.6.) Collagen has a high content of glycine (33%), proline (22%), alanine (12%), and smaller amounts of hydroxyproline and hydroxylysine. The hydroxy forms of proline and lysine

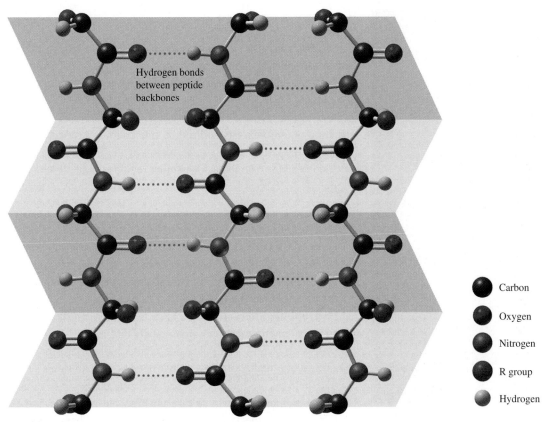

**FIGURE 16.5** In a β (beta)-pleated sheet secondary structure, hydrogen bonds form between the peptide chains.

**Q** How do the hydrogen bonds differ in a β-pleated sheet from the α helix?

Carbon

Oxygen

Nitrogen

R group

Hydrogen

contain —OH groups that form hydrogen bonds across the peptide chains and give strength to the collagen triple helix.

Hydroxyproline

Hydroxylysine

When a diet is deficient in vitamin C, collagen fibrils are weakened because the enzymes needed to form hydroxyproline and hydroxylysine require vitamin C. Because there are fewer —OH groups, there is less hydrogen bonding between collagen fibrils. As a person ages, more cross-links form between the fibrils, which make collagen less elastic. Bones, cartilage, and tendons become more brittle, and wrinkles are seen as the skin loses elasticity.

Triple helix                         3 $\alpha$-Helix peptide chains

**FIGURE 16.6** Hydrogen bonds between polar side chains of three polypeptide chains form the triple helices that form fibers of collagen.

**Q** What are some of the amino acids in collagen that form hydrogen bonds between the polypeptide chains?

## Tertiary Structure

The **tertiary structure** of a protein involves attractions and repulsions between the side chain groups of the amino acids in the polypeptide chain. As interactions occur between different parts of the peptide chain, segments of the chain twist and bend until the protein acquires a specific three-dimensional shape. The tertiary structure of a protein is stabilized by interactions between the side chains of the amino acids in one region of the polypeptide chain, with side chains of amino acids in other regions of the protein. (See Figure 16.7.) Table 16.3 lists the stabilizing interactions of tertiary structures.

1. **Hydrophobic interactions** are interactions between two nonpolar side chains. Within the protein, the amino acids with nonpolar side chains push as far away from the aqueous environment as possible, which forms a hydrophobic center at the interior of the protein molecule.

2. **Hydrophilic interactions** are attractions between the external aqueous environment and amino acids that have polar, or ionized, side chains. The polar side chains pull toward the outer surface of globular proteins to hydrogen bond with water.

3. **Salt bridges** are ionic bonds between side chains of basic and acidic amino acids, which have positive and negative charges. For example, the ionized side chain of lysine has a positive charge, and the ionized side chain of aspartic acid has a negative charge. The attraction of these oppositely charged side chains forms a strong ionic bond or salt bridge.

4. **Hydrogen bonds** form between H of one side group and O or N of another amino acid. For example, a hydrogen bond can form between the —OH groups of two serines or between the —OH of serine and the —$NH_2$ in the side chain of glutamine.

5. **Disulfide bonds** (—S—S—) are covalent bonds that form between the —SH groups of cysteines in the polypeptide chain.

**TABLE 16.3** Some Cross-Links in Tertiary Structures

| | Nature of Bonding |
|---|---|
| **Hydrophobic interactions** | Attractions between nonpolar groups |
| **Hydrophilic interactions** | Attractions between polar or ionized groups and water on the surface of the tertiary structure |
| **Salt bridges** | Ionic interactions between ionized acidic and basic amino acids |
| **Hydrogen bonds** | Occur between H and O or N |
| **Disulfide bonds** | Strong covalent links between sulfur atoms of two cysteine amino acids |

**WEB TUTORIAL**
Tertiary and Quaternary Structure

**SAMPLE PROBLEM 16.5**

■ **Cross-Links in Tertiary Structures**

What type of attraction would you expect between the side chains of the following?

**a.** cysteine and cysteine

**b.** glutamic acid and lysine

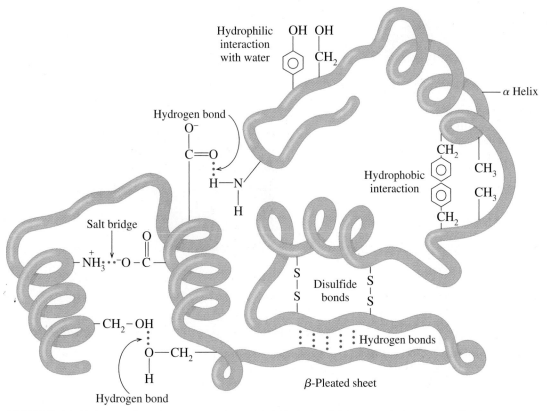

**FIGURE 16.7** Interactions between amino acid side chains fold a protein into a specific three-dimensional shape called its tertiary structure.

**Q** Why would one section of a protein move to the center while another section remains on the surface of the tertiary structure?

### SOLUTION

**a.** Because cysteine has a side chain containing —SH, a disulfide bond could form.

**b.** A salt bridge can form by the attraction of the —COO⁻ in the side chain of glutamic acid and the —NH₃⁺ in the side chain of lysine.

### STUDY CHECK

Would you expect to find valine and leucine on the outside or the inside of the tertiary structure? Why?

## Globular and Fibrous Proteins

A group of proteins known as **globular proteins** have compact, spherical shapes because sections of the polypeptide chain fold over on top of each other. It is the globular proteins that carry out the work of the cells: functions such as synthesis, transport, and metabolism.

Myoglobin is a globular protein that stores oxygen in skeletal muscle. High concentrations of myoglobin are found in the muscles of sea mammals, such as seals and whales, allowing them to stay under the water for long periods. Myoglobin contains 153 amino acids in a single polypeptide chain with about three-fourths of the chain in the α-helix secondary structure. The polypeptide chain, including its helical regions, forms a compact tertiary structure by folding upon itself. (See Figure 16.8.) Within the tertiary structure, a pocket of amino acids and a heme group binds and stores oxygen ($O_2$).

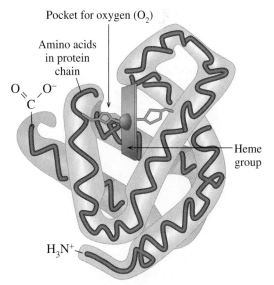

**FIGURE 16.8** Myoglobin is a globular protein with a heme pocket in its tertiary structure that binds oxygen to be carried to the tissues.

**Q** Would hydrophilic amino acids be found on the outside or inside of the myoglobin structure?

## Health Note

### Essential Amino Acids

Of the 20 amino acids used to build the proteins in the body, only 10 can be synthesized in the body. The other 10 amino acids, listed in Table 16.4, are *essential amino acids* that cannot be synthesized and must be obtained from the proteins in the diet.

**TABLE 16.4** Essential Amino Acids

| | |
|---|---|
| Arginine (Arg)* | Methionine (Met) |
| Histidine (His)* | Phenylalanine (Phe) |
| Isoleucine (Ile) | Threonine (Thr) |
| Leucine (Leu) | Tryptophan (Trp) |
| Lysine (Lys) | Valine (Val) |

\* Required in diets of children, not adults.

*Complete proteins*, which contain all of the essential amino acids, are found in most animal products such as eggs, milk, meat, fish, and poultry. However, gelatin and plant proteins such as grains, beans, and nuts are *incomplete proteins* because they are deficient in one or more of the essential amino acids. Diets that rely on plant foods for protein must contain a variety of protein sources to obtain all the essential amino acids. For example, a diet of rice and beans contains all the essential amino acids because they are complementary protein sources. Rice contains the methionine and tryptophan that are deficient in beans, while beans contain the lysine that is lacking in rice. (See Table 16.5.)

**TABLE 16.5** Amino Acid Deficiency in Selected Vegetables and Grains

| Food Source | Amino Acids Missing |
|---|---|
| Eggs, milk, meat, fish, poultry | None |
| Wheat, rice, oats | Lysine |
| Corn | Lysine, tryptophan |
| Beans | Methionine, tryptophan |
| Peas | Methionine |
| Almonds, walnuts | Lysine, tryptophan |
| Soy | Low in methionine |

The **fibrous proteins** are proteins that consist of long, thin, fiber-like shapes. They are typically involved in the structure of cells and tissues. Two types of fibrous protein are the $\alpha$- and $\beta$-keratins. The $\alpha$-keratins are the proteins that make up hair, wool, skin, and nails. In hair, three $\alpha$-helices coil together like a braid to form a fibril. Within the fibril, the $\alpha$-helices are held together by disulfide ($-S-S-$) linkages between the side chains of the many cysteine amino acids in hair. Several fibrils bind together to form a strand of hair. (Figure 16.9.) The $\beta$-keratins are the type of proteins found in the feathers of birds and scales of reptiles. In $\beta$-keratins, the proteins consist of large amounts of $\beta$-pleated sheet structure.

### Quaternary Structure: Hemoglobin

When a biologically active protein consists of two or more polypeptide subunits, the structural level is referred to as a **quaternary structure**. Hemoglobin, a globular protein that transports oxygen in blood, consists of four polypeptide chains or subunits, two $\alpha$ chains, and two $\beta$ chains. (See Figure 16.10.) The subunits are held together in the quaternary structure by the same interactions that stabilize the tertiary structure, such as hydrogen bonds and salt bridges between side groups, disulfide links, and hydrophobic attractions. Each subunit of the hemoglobin contains a heme group that binds oxygen. In the adult hemoglobin molecule, all four subunits must be combined for the hemoglobin to properly function as an oxygen carrier. Therefore, the complete quaternary structure of hemoglobin can bind and transport four molecules of oxygen. Table 16.6 and Figure 16.11 summarize the structural levels of proteins.

$\alpha$ Helix

$\alpha$-Keratin

**FIGURE 16.9** The fibrous proteins of $\alpha$-keratin wrap together to form fibrils of hair and wool.

**Q** Why does hair have a large amount of cysteine?

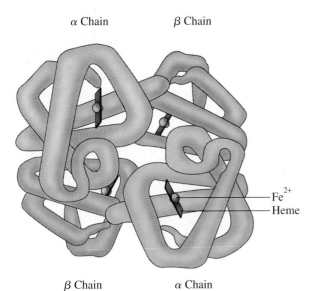

α Chain    β Chain

β Chain    α Chain

**FIGURE 16.10** The quaternary structure of hemoglobin consists of four polypeptide subunits, each containing a heme group that binds an oxygen molecule.
**Q** What is the difference between a tertiary structure and a quaternary structure?

**TABLE 16.6** Summary of Structural Levels in Proteins

| Structural Level | Characteristics |
| --- | --- |
| Primary | The sequence of amino acids |
| Secondary | The coiled α helix, β-pleated sheet, or a triple helix formed by hydrogen bonding between peptide bonds along the chain |
| Tertiary | A folding of the protein into a compact, three-dimensional shape stabilized by interactions between side chains of amino acids |
| Quaternary | A combination of two or more protein subunits to form a larger, biologically active protein |

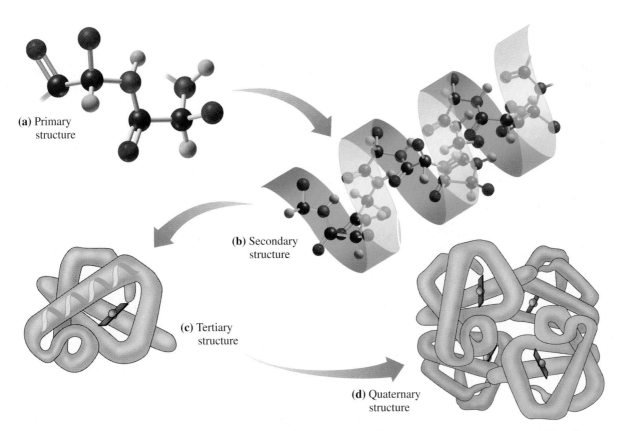

**(a)** Primary structure

**(b)** Secondary structure

**(c)** Tertiary structure

**(d)** Quaternary structure

**FIGURE 16.11** Proteins consist of primary, secondary, tertiary, and sometimes quaternary structural levels.
**Q** What is the difference between a primary structure and a tertiary structure?

## Health Note

### Protein Structure and Mad Cow Disease

Until recently, researchers thought that only viruses or bacteria were responsible for transmitting diseases. Now a group of diseases have been found in which the infectious agents are proteins called *prions*. Bovine spongiform encephalopathy (BSE), or "mad cow disease," is a fatal brain disease of cattle in which the brain fills with cavities resembling a sponge. In the noninfectious form of the prion PrP$^c$, the N-terminal portion is a random coil (see structure at right). Although the noninfectious form may be ingested from meat products, its structure can change to what is known as PrP$^{sc}$ or *prion-related protein scrapie*. In this infectious form, the end of the peptide chain folds into a $\beta$-pleated sheet, which has disastrous effects on the brain and spinal cord. The conditions that cause this structural change are not yet known.

The human variant is called Creutzfeldt–Jakob (CJD) disease. Around 1955, Dr. Carleton Gajdusek was studying the Fore tribe in Papua New Guinea where many were dying of neurological disease known as kuru. Within this Fore tribe, it was a custom to cannibalize members of the tribe upon their death. Because of the long incubation period, it took years to determine that this practice was responsible for transmitting the infectious agent from one tribe member to another.

BSE was diagnosed in Great Britain in 1986. The protein is present in nerve tissue but is not found in meat. Control measures that exclude brain and spinal cord from animal feed are now in place to reduce the incidence of BSE. In 2003, the first case of BSE was identified in the United States.

---

### ■ Identifying Protein Structure

Indicate whether the following conditions are responsible for primary, secondary, tertiary, or quaternary protein structures:

**a.** Disulfide bonds form between portions of a protein chain.
**b.** Peptide bonds form a chain of amino acids.

#### SOLUTION

**a.** Disulfide bonds help to stabilize the tertiary structure of a protein.
**b.** The sequence of amino acids in a polypeptide is a primary structure.

#### STUDY CHECK

What structural level is represented by the interactions of the two subunits in insulin?

## Denaturation of Proteins

**Denaturation** of a protein occurs when there is a disruption of any of the bonds that stabilize the secondary, tertiary, or quaternary structure. However, the covalent amide bonds of the primary structure are not affected.

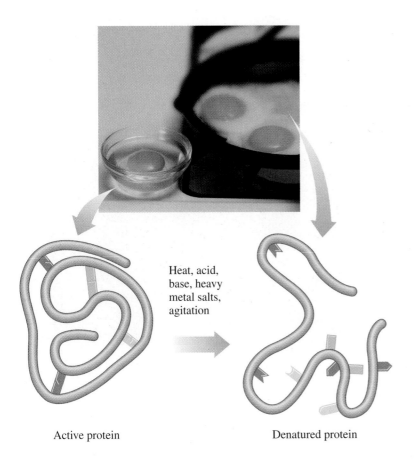

Active protein

Heat, acid, base, heavy metal salts, agitation

Denatured protein

## Explore Your World

### Denaturation of Milk Protein

Place some milk in five glasses. Add the following to the milk samples in glasses 1–4. The fifth glass of milk is a reference sample.

1. Vinegar, drop by drop. Stir.
2. One-half teaspoon of meat tenderizer. Stir.
3. One teaspoon of fresh pineapple juice. (Canned juice has been heated and cannot be used.)
4. One teaspoon of fresh pineapple juice after the juice is heated to boiling.

#### QUESTIONS

1. How did the appearance of the milk change in each of the samples?
2. What enzyme is listed on the package label of the tenderizer?
3. How does the effect of the heated pineapple juice compare with that of the fresh juice? Explain.
4. Why is cooked pineapple used when making gelatin (a protein) desserts?

The loss of secondary and tertiary structures occurs when conditions change, such as increasing the temperature or making the pH very acidic or basic. If the pH changes, the basic and acidic side chains lose their ionic charges and cannot form salt bridges, which causes a change in the shape of the protein. Denaturation can also occur by adding certain organic compounds or heavy metal ions or by mechanical agitation. (See Table 16.7.)

When the interactions between side chains are disrupted, a globular protein unfolds to become like a loose piece of spaghetti. With the loss of its overall shape, the protein is no longer biologically active.

**TABLE 16.7** Examples of Protein Denaturation

| Denaturing Agent | Bonds Disrupted | Examples |
|---|---|---|
| Heat above 50 °C | Hydrogen bonds; hydrophobic attractions between nonpolar side chains | Cooking food and autoclaving surgical items |
| Acids and bases | Hydrogen bonds between polar side chains; salt bridges | Lactic acid from bacteria, which denatures milk protein in the preparation of yogurt and cheese |
| Organic compounds | Hydrophobic interactions | Ethanol and isopropyl alcohol, which disinfect wounds and prepare the skin for injections |
| Heavy metal ions $Ag^+$, $Pb^{2+}$, and $Hg^{2+}$ | Disulfide bonds in proteins, by forming ionic bonds | Mercury and lead poisoning |
| Agitation | Hydrogen bonds and hydrophobic interactions, by stretching polypeptide chains and disrupting stabilizing interactions | Whipped cream, meringue made from egg whites |

## Health Note

### Sickle-Cell Anemia

Sickle-cell anemia is a disease caused by an abnormality in the shape of one of the subunits of the hemoglobin protein. In the $\beta$ chain, the sixth amino acid, glutamic acid, which is polar, is replaced by valine, a nonpolar amino acid.

Because valine has a nonpolar side chain, it is attracted to the nonpolar regions of other beta hemoglobin chains. The affected red blood cells change from a rounded shape to a crescent shape, like a sickle, which interferes with their ability to transport adequate quantities of oxygen. Hydrophobic attractions also cause sickle-cell hemoglobin molecules to stick together. They form insoluble fibers of sickle-cell hemoglobin that clog capillaries, where they cause inflammation, pain, and organ damage. Critically low oxygen levels may occur in the affected tissues.

In sickle-cell anemia, both genes for the altered hemoglobin must be inherited. However, a few sickled cells are found in persons who carry one gene for sickle-cell hemoglobin, a condition that is also known to provide protection from malaria.

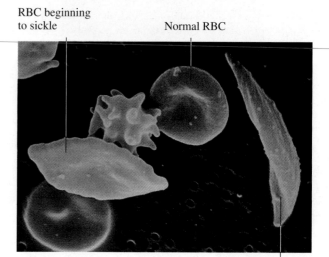

RBC beginning to sickle

Normal RBC

Sickled RBC

Normal $\beta$ chain:    Val—His—Leu—Thr—Pro—|Glu|—Glu—Lys—
Sickled $\beta$ chain:    Val—His—Leu—Thr—Pro—|Val|—Glu—Lys—

Polar amino acid

Nonpolar amino acid

---

## QUESTIONS AND PROBLEMS

### Levels of Protein Structure

**16.21** What type of bonding occurs in the primary structure of a protein?

**16.22** How can two proteins with exactly the same number and type of amino acids have different primary structures?

**16.23** Two peptides each contain one molecule of valine and two molecules of serine. What are their possible primary structures?

**16.24** What are three different types of secondary protein structure?

**16.25** What happens to the primary structure of a protein when a protein forms a secondary structure?

**16.26** In an $\alpha$ helix, how does bonding occur between the amino acids in the polypeptide chain?

**16.27** What is the difference in bonding between an $\alpha$ helix and a $\beta$-pleated sheet?

**16.28** How is the secondary structure of a $\beta$-pleated sheet different from that of a triple helix?

**16.29** What type of interaction would you expect from the side chains of the following amino acids in a tertiary structure?
**a.** cysteine and cysteine    **b.** glutamic acid and lysine
**c.** serine and aspartic acid    **d.** leucine and leucine

**16.30** In myoglobin, about one-half of the 153 amino acids have nonpolar side chains.
**a.** Where would you expect those amino acids to be located in the tertiary structure?
**b.** Where would you expect the polar side chains to be?
**c.** Why is myoglobin more soluble in water than is silk or wool?

**16.31** A portion of a polypeptide chain contains the following sequence of amino acids:

-Leu-Val-Cys-Asp-

**a.** Which amino acids can form a disulfide cross-link?
**b.** Which amino acids are likely to be found on the inside of the protein structure? Why?

c. Which amino acids would be found on the outside of the protein? Why?

d. How does the primary structure of a protein affect its tertiary structure?

**16.32** State whether the following statements describe primary, secondary, tertiary, or quaternary protein structure:

a. Side chains interact to form disulfide bonds or ionic bonds.

b. Peptide bonds join amino acids in a polypeptide chain.

c. Several polypeptides are held together by hydrogen bonds between adjacent chains.

d. Hydrogen bonding between amino acids in the same polypeptide gives a coiled shape to the protein.

e. Hydrophobic side chains seeking a nonpolar environment move toward the inside of the folded protein.

f. Protein chains of collagen form a triple helix.

g. An active protein contains four tertiary subunits.

**16.33** Indicate the changes in protein structure for each of the following:

a. An egg placed in water at 100 °C is soft boiled in about 3 minutes.

b. Prior to giving an injection, the skin is wiped with an alcohol swab.

c. Surgical instruments are placed in a 120 °C autoclave.

d. During surgery, a wound is closed by cauterization (heat).

**16.34** Indicate the changes in protein structure for each of the following:

a. Tannic acid is placed on a burn.

b. Milk is heated to 60 °C to make yogurt.

c. To avoid spoilage, seeds are treated with a solution of $HgCl_2$.

d. Hamburger is cooked at high temperatures to destroy *E. coli* bacteria that may cause intestinal illness.

# 16.6 ENZYMES

Biological catalysts known as **enzymes** are needed for most chemical reactions that take place in the body. As we discussed in Chapter 5, a catalyst increases the reaction rate by changing the way a reaction takes place, but is itself not changed at the end of the reaction. An uncatalyzed reaction in a cell may take place eventually, but not at a rate fast enough for survival. For example, the hydrolysis of proteins in our diet would eventually occur without a catalyst, but not fast enough to meet the body's requirements for amino acids. The chemical reactions in our cells must occur at incredibly fast rates under the mild conditions of pH 7.4 and a body temperature of 37 °C.

As catalysts, enzymes lower the activation energy for a chemical reaction. (See Figure 16.12.) Less energy is required to convert reactant molecules to products, which increases the rate of a biochemical reaction compared to the rate of the uncatalyzed reaction. For example, an enzyme in the blood called carbonic anhydrase converts carbon dioxide and water to carbonic acid. In 1 minute, one molecule of carbonic anhydrase catalyzes the reaction of about 1 million molecules of carbon dioxide.

$$CO_2 + H_2O \xrightarrow{\text{Carbonic anhydrase}} H_2CO_3$$

## Names and Classification of Enzymes

The names of enzymes describe the compound or the reaction that is catalyzed. The actual names of enzymes are derived by replacing the end of the name of the reaction or reacting compound with the suffix *ase*. For example, an *oxidase* catalyzes an oxidation reaction, and a *dehydrogenase* removes hydrogen atoms. The compound sucrose is hydrolyzed by the enzyme *sucrase*, and a lipid is hydrolyzed by a *lipase*. Some early known enzymes use names that end in the suffix *in*, such as *papain* found in papaya, *rennin* found in milk, and *pepsin* and *trypsin*, enzymes that catalyze the hydrolysis of proteins.

More recently, a systematic method of classifying and naming enzymes has been established. The name and class of each indicates the type of reaction it catalyzes. There are six main classes of enzymes, as described in Table 16.8.

## LEARNING GOAL

Describe how enzymes function as catalysts, and give their names.

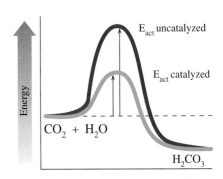

**FIGURE 16.12** The enzyme carbonic anhydrase lowers the activation energy needed for the reaction of $CO_2$ and $H_2O$.

**Q** Why are enzymes used in biological reactions?

**TABLE 16.8** Classes of Enzymes

| Class | Reaction Catalyzed | Examples |
|---|---|---|
| Oxidoreductases | Oxidation–reduction reactions | *Oxidases* oxidize a substance. |
| | | *Reductases* reduce a substance. |
| | | *Dehydrogenases* remove 2 H atoms to form a double bond. |
| Transferases | Transfer a group between two compounds | *Transaminases* transfer amino groups. |
| | | *Kinases* transfer phosphate groups. |
| Hydrolases | Hydrolysis reactions | *Proteases* hydrolyze peptide bonds in proteins. |
| | | *Lipases* hydrolyze ester bonds in lipids. |
| | | *Carbohydrases* hydrolyze glycosidic bonds in carbohydrates. |
| | | *Phosphatases* hydrolyze phosphate–ester bonds. |
| | | *Nucleases* hydrolyze nucleic acids. |
| Lyases | Add or remove groups involving a double bond without hydrolysis | *Carboxylases* add $CO_2$. |
| | | *Deaminases* remove $NH_3$. |
| Isomerases | Rearrange atoms in a molecule to form an isomer | *Isomerases* convert cis to trans or trans to cis. |
| | | *Epimerases* convert D to L isomers or L to D. |
| Ligases | Form bonds between molecules using ATP energy | *Synthetases* combine two molecules. |

---

## SAMPLE PROBLEM 16.7

### ■ Naming Enzymes

What chemical reaction does each of the following enzymes catalyze?

**a.** amino transferase     **b.** lactate dehydrogenase

SOLUTION

**a.** catalyzes the transfer of an amino group
**b.** catalyzes the removal of hydrogen from lactate

STUDY CHECK

What is the name of the enzyme that catalyzes the hydrolysis of lipids?

---

## QUESTIONS AND PROBLEMS

### Enzymes

**16.35** Why do chemical reactions in the body require enzymes?

**16.36** How do enzymes make chemical reactions in the body proceed at faster rates?

**16.37** What is the reactant for each of the following enzymes?
  **a.** galactase
  **b.** lipase
  **c.** aspartase

**16.38** What is the reactant for each of the following enzymes?
  **a.** peptidase
  **b.** cellulase
  **c.** lactase

**16.39** What is the name of the class of enzymes that would catalyze each of the following reactions?
  **a.** hydrolysis of sucrose
  **b.** addition of oxygen
  **c.** converting glucose ($C_6H_{12}O_6$) to fructose ($C_6H_{12}O_6$)
  **d.** moving an amino group from one molecule to another

**16.40** What is the name of the class of enzymes that would catalyze each of the following reactions?
  **a.** addition of water to a double bond
  **b.** removing hydrogen atoms
  **c.** splitting peptide bonds in proteins
  **d.** removing $CO_2$ from pyruvate

# 16.7 ENZYME ACTION

Nearly all enzymes are globular proteins. Each has a unique three-dimensional shape that recognizes and binds a small group of reacting molecules, which are called **substrates**. The tertiary structure of an enzyme plays an important role in how that enzyme catalyzes reactions.

## Active Site

In a catalyzed reaction, an enzyme must first bind to a substrate in a way that favors catalysis. A typical enzyme is much larger than its substrate. However, within its large tertiary structure, there is a region called the **active site** where the enzyme binds one or more substrates and catalyzes the reaction. This active site is often a small pocket that closely fits the structure of the substrate. (See Figure 16.13.) Within the active site, the side chains of amino acids bind the substrate with hydrogen bonds, salt bridges, or hydrophobic attractions. The active site of a particular enzyme fits the shape of only a few types of substrates, which makes enzymes very specific about the type of substrate they bind.

## Enzyme Catalyzed Reaction

The proper alignment of a substrate within the active site forms an **enzyme–substrate (ES) complex**. This combination of enzyme and substrate provides an alternative pathway for the reaction that has a lower activation energy. Within the active site, the amino acid side chains take part in catalyzing the chemical reaction. As soon as the catalyzed reaction is complete, the products are quickly released from the enzyme so it can bind to a new substrate molecule. We can write the catalyzed reaction of an enzyme (E) with a substrate (S) to form product (P) as follows:

Step 1   $E + S \rightleftharpoons ES$

Step 2   $ES \longrightarrow E + P$
_____
$E + S \rightleftharpoons ES \longrightarrow E + P$

Enzyme + substrate    ES complex    Enzyme + product

Let's consider the hydrolysis of sucrose by sucrase. When sucrose binds to the active site of sucrase, the glycosidic bond of sucrose is placed in a position most favorable for reaction. The amino acid side chains catalyze the cleavage of the glycosidic bond to give glucose and fructose.

Sucrase + sucrose $\rightleftharpoons$ sucrase–sucrose complex $\longrightarrow$ sucrase + glucose + fructose

E   +   S              ES complex              E   +   $P_1$   +   $P_2$

Because the active site does not bond to the products, they are released, allowing the sucrase to catalyze the hydrolysis of another sucrose substrate. (See Figure 16.14.)

## Lock-and-Key and Induced-Fit Models

In an early theory of enzyme action called the **lock-and-key model**, the active site is described as having a rigid, nonflexible shape. Thus only those substrates with shapes that fit exactly into the active site are able to bind with that enzyme. The shape of the active site is analogous to a lock, and the proper substrate is the key that fits into the lock. (See Figure 16.15a.)

## LEARNING GOAL

Describe the role of an enzyme in an enzyme-catalyzed reaction.

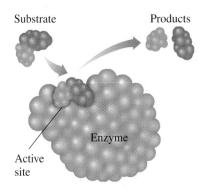

**FIGURE 16.13** On the surface of an enzyme, a small region called an active site binds a substrate and catalyzes a reaction of that substrate.
**Q** Why does an enzyme catalyze a reaction of only certain substrates?

**WEB TUTORIAL**
How Enzymes Work

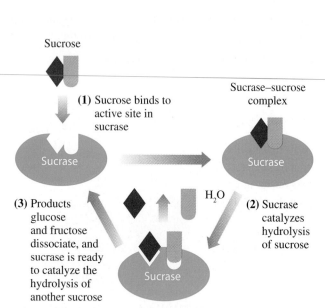

Sucrose

**(1)** Sucrose binds to active site in sucrase

Sucrose

Sucrase–sucrose complex

Sucrase

$H_2O$

Sucrase

**(3)** Products glucose and fructose dissociate, and sucrase is ready to catalyze the hydrolysis of another sucrose

**(2)** Sucrase catalyzes hydrolysis of sucrose

**FIGURE 16.14** After sucrose binds to sucrase at the active site, it is properly aligned for the hydrolysis reaction by the active site. The monosaccharide products dissociate from the active site, and the enzyme is ready to bind to another sucrose molecule.

**Q** Why does the enzyme-catalyzed hydrolysis of sucrose go faster than the hydrolysis of sucrose in the chemistry laboratory?

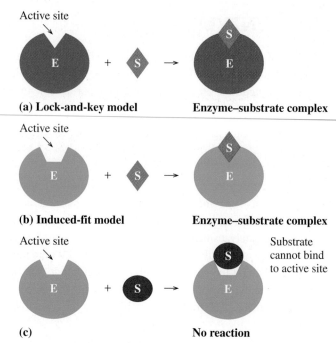

Active site

E + S → S E

**(a) Lock-and-key model**    **Enzyme–substrate complex**

Active site

E + S → S E

**(b) Induced-fit model**    **Enzyme–substrate complex**

Active site

E + S → S E

Substrate cannot bind to active site

**(c)**    **No reaction**

**FIGURE 16.15** **(a)** In the lock-and-key model, a substrate fits the shape of the active site and forms an enzyme–substrate complex. **(b)** In the induced-fit model, a flexible active site and substrate adjust shape to provide the best fit for the reaction. **(c)** A substrate that does not fit or induce a fit in the active site cannot undergo catalysis by the enzyme.

**Q** How does the induced-fit model differ from the lock-and-key model?

While the lock-and-key model explains the binding of substrates for many enzymes, certain enzymes have a broader range of activity than the lock-and-key model allows. In the **induced-fit model**, there is an interaction between both the enzyme and the substrate. (See Figure 16.15b.) The active site adjusts to fit the shape of the substrate more closely. At the same time, the substrate adjusts its shape to better adapt to the geometry of the active site. As a result, the reacting section of the substrate becomes aligned exactly with the groups in the active site that catalyze the reaction. A different substrate could not induce these structural changes, and no catalysis would occur. (See Figure 16.15c.)

SAMPLE PROBLEM 16.8

■ **Enzyme Action**

What is the function of the active site in an enzyme?

SOLUTION

The active site in an enzyme binds the substrate and contains the amino acid side chains that catalyze the reaction.

STUDY CHECK

How do the lock-and-key and the induced-fit models differ in their description of the active site in an enzyme?

# Health Note

## Isoenzymes as Diagnostic Tools

**Isoenzymes** are different forms of an enzyme that catalyze the same reaction in different cells or tissues of the body. They consist of quaternary structures with slight variations in the amino acids in the polypeptide subunits. For example, there are five isoenzymes of *lactate dehydrogenase (LDH)* that catalyze the conversion between lactate and pyruvate.

$$CH_3-\underset{\underset{\text{Lactate}}{|}}{\overset{\overset{OH}{|}}{C}H}-COO^- \underset{\underset{\text{dehydrogenase}}{\text{Lactate}}}{\rightleftharpoons} CH_3-\underset{\underset{\text{Pyruvate}}{}}{\overset{\overset{O}{\|}}{C}}-COO^- + 2H$$

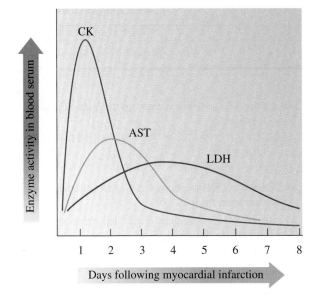

Days following myocardial infarction

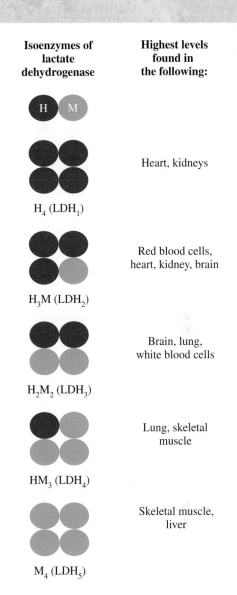

| Isoenzymes of lactate dehydrogenase | Highest levels found in the following: |
|---|---|
| $H_4$ (LDH$_1$) | Heart, kidneys |
| $H_3M$ (LDH$_2$) | Red blood cells, heart, kidney, brain |
| $H_2M_2$ (LDH$_3$) | Brain, lung, white blood cells |
| $HM_3$ (LDH$_4$) | Lung, skeletal muscle |
| $M_4$ (LDH$_5$) | Skeletal muscle, liver |

Each LDH isoenzyme contains a mix of polypeptide subunits, M and H. In the liver and muscle, lactate is converted to pyruvate by a LDH$_5$ isoenzyme with four M subunits designated $M_4$. In the heart, the same reaction is catalyzed by a LDH$_1$ isoenzyme ($H_4$) containing four H subunits. Different combinations of the M and H subunits are found in the LDH isoenzymes of the brain, red blood cells, kidney, and white blood cells.

The different forms of an enzyme allow a medical diagnosis of damage or disease to a particular organ or tissue. In healthy tissues, isoenzymes function within the cells. However, when a disease damages a particular organ, cells die, releasing cell contents including the isoenzymes into the blood. Measurements of the elevated levels of specific isoenzymes in the blood serum help to identify the disease and its location in the body. For example, an elevation in serum LDH$_5$, which is the $M_4$ isoenzyme of lactate dehydrogenase, indicates liver damage or disease. When a myocardial infarction (MI), or heart attack, damages the cells in heart muscle, an increase in the level of LDH$_1$ ($H_4$) isoenzyme is detected in the blood serum. (See Table 16.9.)

**TABLE 16.9** Isoenzymes of Lactate Dehydrogenase and Creatinine Kinase

| Isoenzyme | Abundant in | Subunits |
|---|---|---|
| **Lactate Dehydrogenase (LDH)** | | |
| LDH$_1$ | Heart, kidneys | $H_4$ |
| LDH$_2$ | Red blood cells, heart, kidney, brain | $MH_3$ |
| LDH$_3$ | Brain, lung, white blood cells | $M_2H_2$ |
| LDH$_4$ | Lung, skeletal muscle | $M_3H$ |
| LDH$_5$ | Skeletal muscle, liver | $M_4$ |
| **Creatinine Kinase (CK)** | | |
| CK$_1$ | Brain, lung | BB |
| CK$_2$ | Heart muscle | MB |
| CK$_3$ | Skeletal muscle, red blood cells | MM |

### Isoenzymes as Diagnostic Tools (*Continued*)

Another isoenzyme used diagnostically is creatine kinase (CK), which consists of two types of polypeptide subunits. Subunit B is prevalent in the brain, and subunit M predominates in muscle. Normally $CK_3$ (subunits MM) is present in low amounts in the blood serum. However, in a patient who has suffered a myocardial infarction, the level of $CK_2$ (subunits MB) is elevated within 4 to 6 h and reaches a peak in about 24 h. Table 16.10 lists some enzymes used to diagnose tissue damage and diseases of certain organs.

**TABLE 16.10** Serum Enzymes Used in Diagnosis of Tissue Damage

| Condition | Diagnostic Enzymes Elevated |
|---|---|
| Heart attack or liver disease (cirrhosis, hepatitis) | Lactate dehydrogenase (LDH) Aspartate transaminase (AST) |
| Heart attack | Creatinine kinase (CK) |
| Hepatitis | Alanine transaminase (ALT) |
| Liver (carcinoma) or bone disease (rickets) | Alkaline phosphatase (ALP) |
| Pancreatic disease | Amylase, cholinesterase, lipase (LPS) |
| Prostate carcinoma | Acid phosphatase (ACP) Prostate specific antigen (PSA) |

## QUESTIONS AND PROBLEMS

### Enzyme Action

**16.41** Match the terms, (1) enzyme–substrate complex, (2) enzyme, and (3) substrate, with each of the following:
   **a.** has a tertiary structure that recognizes the substrate
   **b.** the combination of an enzyme with the substrate
   **c.** has a structure that fits the active site of an enzyme

**16.42** Match the terms, (1) active site, (2) lock-and-key model, and (3) induced-fit model, with each of the following:
   **a.** the portion of an enzyme where catalytic activity occurs
   **b.** an active site that adapts to the shape of a substrate
   **c.** an active site that has a rigid shape

**16.43** **a.** Write an equation that represents an enzyme-catalyzed reaction.
   **b.** How is the active site different from the whole enzyme structure?

**16.44** **a.** How does an enzyme speed up the reaction of a substrate?
   **b.** After the products have formed, what happens to the enzyme?

**16.45** What are isoenzymes?

**16.46** How is the LDH isoenzyme in the heart different from LDH isoenzyme in the liver?

**16.47** A patient arrives in an emergency department complaining of chest pains. What enzymes would you test for in the blood serum?

**16.48** A patient who is an alcoholic has elevated levels of LDH and AST. What condition might be indicated?

## LEARNING GOAL

Describe the effect of temperature, pH, concentration of substrate, and inhibitors on enzyme activity.

# 16.8 FACTORS AFFECTING ENZYME ACTIVITY

The **activity** of an enzyme describes how fast an enzyme catalyzes the reaction that converts a substrate to product. This activity is strongly affected by reaction conditions, which include temperature, pH, concentration of the substrate, and the presence of inhibitors.

## Temperature

Enzymes are very sensitive to temperature. At low temperatures, most enzymes show little activity because there is not a sufficient amount of energy for the catalyzed reaction to take place. At higher temperatures, enzyme activity increases as reacting molecules move faster to cause more collisions with enzymes. Enzymes are most active at **optimum temperature**, which is 37 °C or body temperature for most enzymes. (See Figure 16.16.) At temperatures above 50 °C, the tertiary structure—and thus the shape of most proteins—is destroyed, causing a loss in enzyme activity. For this reason, equipment in hospitals and laboratories is sterilized in autoclaves, where the high temperatures denature the enzymes in harmful bacteria.

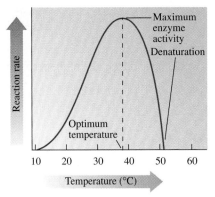

**FIGURE 16.16** An enzyme attains maximum activity at its optimum temperature, usually 37 °C. Lower temperatures slow the rate of reaction, and temperatures above 50 °C denature an enzyme with a loss of catalytic activity.
**Q** Why is 37 °C the optimum temperature for many enzymes?

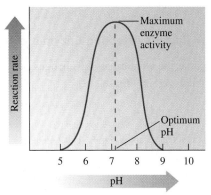

**FIGURE 16.17** Enzymes are most active at their optimum pH. At a higher or lower pH, denaturation of the enzyme causes a loss of catalytic activity.
**Q** Why does the digestive enzyme pepsin have an optimum pH of 2?

## Explore Your World

### Enzyme Activity

The enzymes on the surface of a freshly cut apple, avocado, or banana react with oxygen in the air to turn the surface brown. An antioxidant, such as vitamin C in lemon juice, prevents the oxidation reaction. Cut an apple, an avocado, or a banana into several slices. Place one slice in a plastic zipper lock bag, squeeze out all the air, and close the zipper lock. Dip another slice in lemon juice and place it on a plate. Sprinkle another slice with a crushed vitamin C tablet. Leave another slice alone as a control. Observe the surface of each of your samples. Record your observations immediately, then every hour for 6 hours or longer.

#### QUESTIONS

1. Which slice(s) shows the most oxidation (turns brown)?
2. Which slice(s) shows little or no oxidation?
3. How was the oxidation reaction on each slice affected by treatment with an antioxidant?

## pH

Enzymes are most active at their **optimum pH**, the pH that maintains the proper tertiary structure of the protein. (See Figure 16.17.) A pH value that is above or below the optimum pH causes a change in the three-dimensional structure of the enzyme, disrupting the active site. As a result, the enzyme cannot bind substrate properly and no reaction occurs.

Enzymes in most cells have optimum pH values around 7.4. However, enzymes in the stomach have a low optimum pH because they hydrolyze proteins at the acidic pH in the stomach. For example, pepsin, a digestive enzyme in the stomach, has an optimum pH of 2. Between meals, the pH in the stomach is 4 or 5 and pepsin shows little or no digestive activity. When food enters the stomach, the secretion of HCl lowers the pH to about 2, which activates pepsin.

If small changes in pH are corrected, an enzyme can regain its structure and activity. However, large variations from optimum pH permanently destroy the structure of the enzyme. Table 16.11 lists the optimum pH values for selected enzymes.

**TABLE 16.11** Optimum pH for Selected Enzymes

| Enzyme | Location | Substrate | Optimum pH |
|---|---|---|---|
| Pepsin | Stomach | Peptide bonds | 2 |
| Urease | Liver | Urea | 5 |
| Sucrase | Small intestine | Sucrose | 6.2 |
| Pancreatic amylase | Pancreas | Amylose | 7 |
| Trypsin | Small intestine | Peptide bonds | 8 |
| Arginase | Liver | Arginine | 9.7 |

## Substrate Concentration

In any catalyzed reaction, the substrate must first bind with the enzyme to form the substrate–enzyme complex. When enzyme concentration is kept constant, increasing the substrate concentration increases the rate of the catalyzed reaction until the substrate saturates the enzyme. With all the available enzyme molecules bonded to substrate, the rate of the catalyzed reaction reaches its maximum. Adding more substrate molecules cannot increase the rate further. (See Figure 16.18.)

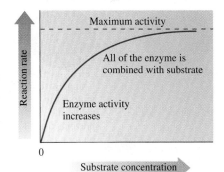

**FIGURE 16.18** Increasing the substrate concentration increases the rate of reaction until the enzyme molecules are saturated with substrate.
**Q** What happens to the rate of reaction when the substrate saturates the enzyme?

### ■ Factors Affecting Enzymatic Activity

Describe what effect the changes in parts a and b would have on the rate of the reaction catalyzed by urease.

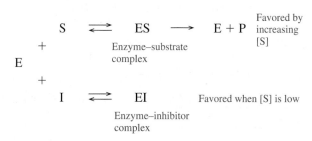

$$H_2N-\overset{\overset{\displaystyle O}{\|}}{C}-NH_2 \ + \ H_2O \ \xrightarrow{\text{Urease}} \ 2NH_3 \ + \ CO_2$$

Urea

**a.** increasing the urea concentration     **b.** lowering the temperature to 10 °C

SOLUTION

**a.** An increase in urea concentration will increase the rate of reaction until all the enzyme molecules are bound to the urea substrate.

**b.** Because 10 °C is lower than the optimum temperature of 37 °C, the lower temperature will decrease the rate of the reaction.

STUDY CHECK

If urease has an optimum pH of 5, what is the effect of lowering the pH to 3?

**WEB TUTORIAL**
Enzyme Inhibition

## Enzyme Inhibition

Many kinds of molecules called **inhibitors** cause enzymes to lose catalytic activity. Although inhibitors act differently, they all prevent the active site from binding with a substrate. Inhibition can be competitive or noncompetitive. In competitive inhibition, an inhibitor competes for the active site, whereas in noncompetitive inhibition, the inhibitor acts on a site that is not the active site.

A **competitive inhibitor** has a structure that is so similar to the substrate it competes for the active site on the enzyme. As long as the inhibitor occupies the active site, the substrate cannot bind to the enzyme and no reaction takes place. (See Figure 16.19.)

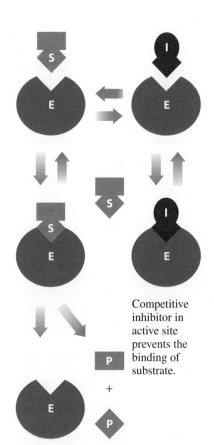

Competitive inhibitor in active site prevents the binding of substrate.

**FIGURE 16.19** With a structure similar to the substrate for an enzyme, a competitive inhibitor also fits the active site and competes with the substrate when both are present.

**Q** Why does increasing the substrate concentration reverse the inhibition by a competitive inhibitor?

$$
\begin{array}{llll}
& S & \rightleftharpoons & ES & \longrightarrow \quad E + P & \text{Favored by} \\
& + & & \text{Enzyme–substrate} & & \text{increasing} \\
E & & & \text{complex} & & [S] \\
& + & & & & \\
& I & \rightleftharpoons & EI & & \text{Favored when [S] is low} \\
& & & \text{Enzyme–inhibitor} & & \\
& & & \text{complex} & & \\
\end{array}
$$

As long as the concentration of the inhibitor is substantial, there is a loss of enzyme activity. However, increasing the substrate concentration displaces more of the inhibitor molecules. As more enzyme molecules bind to substrate (ES), enzyme activity is regained.

The structure of a **noncompetitive inhibitor** does not resemble the substrate and does not compete for the active site. Instead, a noncompetitive inhibitor binds to a site on the enzyme that is not the active site. When the noncompetitive inhibitor is bonded to the enzyme, the shape of the enzyme is distorted. Inhibition occurs because the substrate cannot fit in the active site or it does not fit properly. Without the proper alignment of substrate with the amino acid side groups, no catalysis can take place. (See Figure 16.20.)

Because a noncompetitive inhibitor is not competing for the active site, the addition of more substrate does not reverse this type of inhibition. Examples of noncompetitive inhibitors are the heavy metal ions $Pb^{2+}$, $Ag^+$, and $Hg^{2+}$ that bond with amino acid side groups such as $-COO^-$, or $-OH$. Catalytic activity is restored when chemical reagents remove the inhibitors.

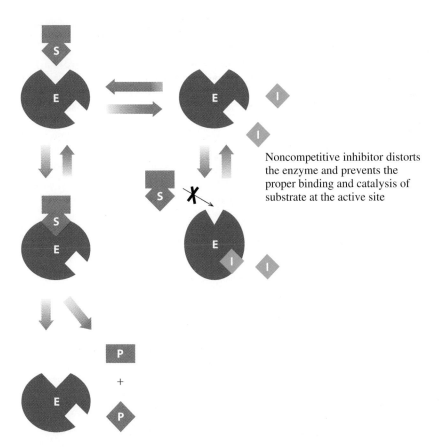

Noncompetitive inhibitor distorts the enzyme and prevents the proper binding and catalysis of substrate at the active site

Antibiotics produced by bacteria, mold, or yeast are inhibitors used to stop bacterial growth. For example, penicillin inhibits an enzyme needed for the formation of cell walls in bacteria, but not human cell membranes. With an incomplete cell wall, bacteria cannot survive and the infection is stopped. However, some bacteria are resistant to penicillin because they produce penicillinase, an enzyme that breaks down penicillin. Over the years, derivatives of penicillin to which bacteria have not yet become resistant have been produced.

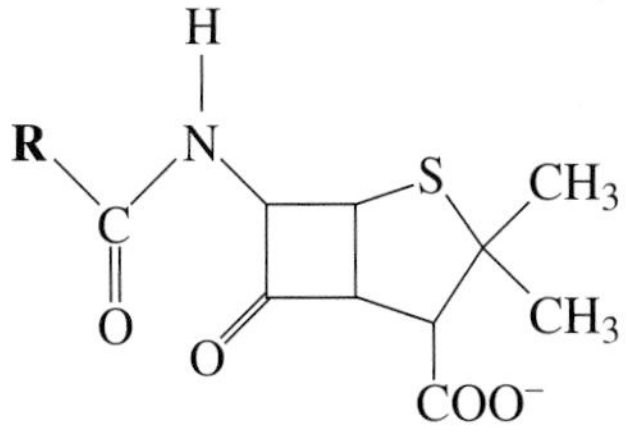

Penicillin

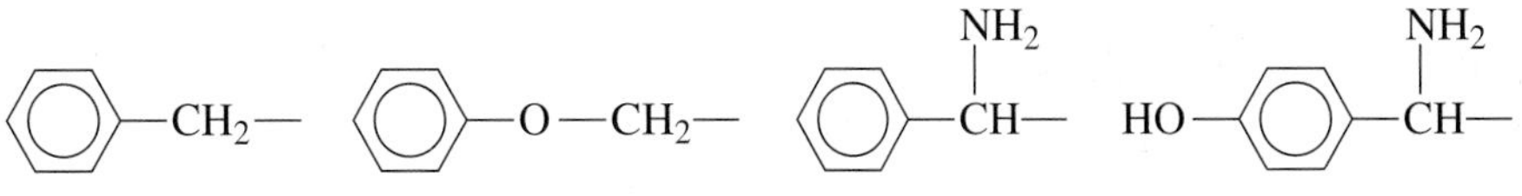

R groups for penicillin derivatives

## SAMPLE PROBLEM 16.10

### ■ Enzyme Inhibition

State the type of inhibition in the following:

**a.** The inhibitor has a structure that is similar to the substrate.
**b.** This inhibitor binds to the surface of the enzyme, changing its shape in such a way that it cannot bind to substrate.

#### SOLUTION

**a.** competitive inhibition          **b.** noncompetitive inhibition

#### STUDY CHECK

What type of enzyme inhibition can be reversed by adding more substrate?

## QUESTIONS AND PROBLEMS

### Factors Affecting Enzyme Activity

**16.49** Trypsin, a peptidase that hydrolyzes polypeptides, functions in the small intestine at an optimum pH of 8. How is the rate of a trypsin-catalyzed reaction affected by each of the following conditions?
  **a.** lowering the concentration of polypeptides
  **b.** changing the pH to 3
  **c.** running the reaction at 75 °C

**16.50** Pepsin, a peptidase that hydrolyzes proteins, functions in the stomach at an optimum pH of 2. How is the rate of a pepsin-catalyzed reaction affected by each of the following conditions?
  **a.** increasing the concentration of proteins
  **b.** changing the pH to 5
  **c.** running the reaction at 0 °C

**16.51** The following graph shows the curves for pepsin, urease, and trypsin. Estimate the optimum pH for each.

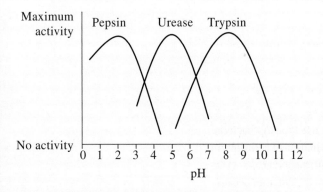

**16.52** Refer to the graph in problem 16.51 to determine if the reaction rate in each condition will be at the optimum rate or not.
  **a.** trypsin, pH 5      **b.** urease, pH 5
  **c.** pepsin, pH 4       **d.** trypsin, pH 8
  **e.** pepsin, pH 2

**16.53** Indicate whether the following describe a competitive or a noncompetitive enzyme inhibitor:
  **a.** The inhibitor has a structure similar to the substrate.
  **b.** The effect of the inhibitor cannot be reversed by adding more substrate.
  **c.** The inhibitor competes with the substrate for the active site.
  **d.** The structure of the inhibitor is not similar to the substrate.
  **e.** The addition of more substrate reverses the inhibition.

**16.54** Oxaloacetate is an inhibitor of succinate dehydrogenase:

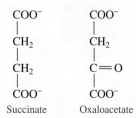

  **a.** Would you expect oxaloacetate to be a competitive or a noncompetitive inhibitor? Why?
  **b.** Would oxaloacetate bind to the active site or elsewhere on the enzyme?
  **c.** How would you reverse the effect of the inhibitor?

**16.55** Methanol and ethanol are oxidized by alcohol dehydrogenase. In methanol poisoning, ethanol is given intravenously to prevent the formation of formaldehyde that has toxic effects.
  **a.** Compare the structure of methanol and ethanol.
  **b.** Would ethanol compete for the active site or bind to a different site?
  **c.** Would ethanol be a competitive or noncompetitive inhibitor of methanol oxidation?

**16.56** In humans, the antibiotic amoxycillin (a type of penicillin) is used to treat certain bacterial infections.
  **a.** Does the antibiotic inhibit enzymes in humans?
  **b.** Why does the antibiotic kill bacteria, but not humans?

## LEARNING GOAL

Describe the types of cofactors found in enzymes.

# 16.9 ENZYME COFACTORS

Enzymes known as **simple enzymes** consist only of proteins. However, many enzymes require small molecules or metal ions called **cofactors** to catalyze reactions properly. When the cofactor is a small organic molecule, it is known as a **coenzyme**. If an enzyme requires a cofactor, neither the protein structure nor the cofactor alone has catalytic activity.

Forms of Active Enzymes

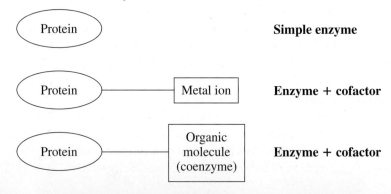

## Metal Ions

Many enzymes must contain a metal ion to carry out their catalytic activity. The metal ions are bonded to one or more of the amino acid side chains. The metal ions from the minerals that we obtain from foods in our diet have various functions in catalysis. Ions such as $Fe^{2+}$ and $Cu^{2+}$ are used by oxidases because they lose or gain electrons in oxidation and reduction reactions. Other metals ions such as $Zn^{2+}$ stabilize the amino acid side chains during hydrolysis reactions. Some metal cofactors required by enzymes are listed in Table 16.12.

**TABLE 16.12** Enzymes and the Metal Ions Required as Cofactors

| Metal Ion Cofactor | Function | Enzyme |
|---|---|---|
| $Cu^{2+}$ | Oxidation–reduction | Cytochrome oxidase |
| $Fe^{2+}/Fe^{3+}$ | Oxidation–reduction | Catalase |
| | Oxidation–reduction | Cytochrome oxidase |
| $Zn^{2+}$ | Used with $NAD^+$ | Alcohol dehydrogenase |
| | | Carbonic anhydrase |
| | | Carboxypeptidase A |
| $Mg^{2+}$ | Hydrolyzes phosphate esters | Glucose-6-phosphatase |
| $Mn^{2+}$ | Removes electrons | Arginase |
| $Ni^{2+}$ | Hydrolyzes amides | Urease |

## Vitamins and Coenzymes

**Vitamins** are organic molecules that are essential for normal health and growth. They are required in trace amounts and must be obtained from the diet because they are not synthesized in the body. Before vitamins were discovered, it was known that lime juice prevented the disease scurvy in sailors and that cod liver oil could prevent rickets. In 1912, scientists found that, in addition to carbohydrates, fats, and proteins, certain other factors called vitamins must be obtained from the diet.

Vitamins are classified into two groups by solubility: water-soluble and fat-soluble. **Water-soluble vitamins** have polar groups such as —OH and —COOH, which make them soluble in the aqueous environment of the cells. The **fat-soluble vitamins** are nonpolar compounds that are soluble in the fat (lipid) components of the body such as fat deposits and cell membranes.

Most water-soluble enzymes are not stored in the body, and excess amounts are eliminated in the urine each day. Therefore, the water-soluble vitamins must be in the foods of our daily diets. (See Figure 16.21.) Because many water-soluble vitamins are easily destroyed by heat, oxygen, and ultraviolet light, care must be taken in food preparation, processing, and storage. In the 1940s, a Committee on Food and Nutrition of the National Research Council began to recommend dietary enrichment of cereal grains. It was known that refining grains such as wheat caused a loss of vitamins. Thiamine ($B_1$), riboflavin ($B_2$), niacin, and iron were in the first group of added nutrients recommended. We now see the Recommended Daily Allowance (RDA) for many vitamins and minerals on most food product labels.

The water-soluble vitamins are required by many enzymes as cofactors to carry out certain aspects of catalytic action. (See Table 16.13.) The coenzymes do not remain bonded to a particular enzyme, but are used over and over again by different enzymes to facilitate an enzyme-catalyzed reaction. (See Figure 16.22.) Thus, only small amounts of coenzymes are required in the cells.

The fat-soluble vitamins A, D, E, and K are not involved as coenzymes, but they are important in processes such as vision, formation of bone, protection from oxidation, and proper blood clotting. Because the fat-soluble vitamins are stored in the body and not eliminated, it is possible to take too much, which could be toxic. (See Figure 16.23.)

**FIGURE 16.21** Oranges, lemons, peppers, and tomatoes contain vitamin C, or ascorbic acid.

**Q** What happens to the excess vitamin C that may be consumed in a day?

**TABLE 16.13** Vitamins and Function

| Water-Soluble Vitamins | Coenzyme | Function |
|---|---|---|
| Thiamine (vitamin $B_1$) | Thiamine pyrophosphate | Decarboxylation |
| Riboflavin (vitamin $B_2$) | Flavin adenine dinucleotide (FAD); flavin mononucleotide (FMN) | Electron transfer |
| Niacin (vitamin $B_3$) | Nicotinamide adenine dinucleotide ($NAD^+$); nicotinamide adenine dinucleotide phosphate ($NADP^+$) | Oxidation–reduction |
| Pantothenic acid (vitamin $B_5$) | Coenzyme A | Acetyl group transfer |
| Pyridoxine (vitamin $B_6$) | Pyridoxal phosphate | Transamination |
| Cobalamin (vitamin $B_{12}$) | Methylcobalamin | Methyl group transfer |
| Ascorbic acid (vitamin C) | Vitamin C | Collagen synthesis, healing of wounds |
| Biotin | Biocytin | Carboxylation |
| Folic acid | Tetrahydrofolate | Methyl group transfer |
| **Fat-Soluble Vitamins** | | |
| Vitamin A | | Formation of visual pigments; development of epithelial cells |
| Vitamin D | | Absorption of calcium and phosphate; deposition of calcium and phosphate in bone |
| Vitamin E | | Antioxidant; prevents oxidation of vitamin A and unsaturated fatty acids |
| Vitamin K | | Synthesis of prothrombin for blood clotting |

**FIGURE 16.22** The active forms of many enzymes require the combination of the protein with a coenzyme.

**Q** What is the function of water-soluble vitamins in enzymes?

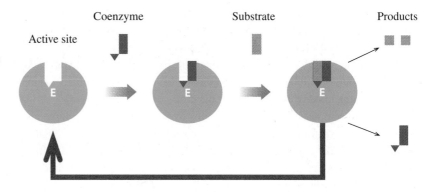

Coenzyme        Substrate        Products

Active site

E        E        E

**FIGURE 16.23** Yellow and green fruits and vegetables contain vitamin A.

**Q** Why is vitamin A called a fat-soluble vitamin?

---

SAMPLE PROBLEM    16.11

■ **Cofactors**

Indicate whether each of the following enzymes is active as a simple enzyme or requires a cofactor.

**a.** a polypeptide that needs $Mg^{2+}$ for catalytic activity
**b.** an active enzyme composed only of a polypeptide chain
**c.** an enzyme that consists of a quaternary structure attached to vitamin $B_6$

SOLUTION

**a.** The enzyme requires a cofactor.
**b.** An active enzyme that consists of only a polypeptide chain is a simple enzyme.
**c.** The enzyme requires a cofactor.

STUDY CHECK

Which of the nonprotein portions of the enzymes in Sample Problem 16.11 is a coenzyme?

## QUESTIONS AND PROBLEMS

### Enzyme Cofactors

**16.57** Is the enzyme described in each of the following statements a simple enzyme or one that requires a cofactor?
**a.** requires vitamin $B_1$ (thiamine)
**b.** needs $Zn^{2+}$ for catalytic activity
**c.** its active form consists of two polypeptide chains

**16.58** Is the enzyme described in each of the following statements a simple enzyme or one that requires a cofactor?
**a.** requires vitamin $B_2$ (riboflavin)
**b.** its active form is composed of 155 amino acids
**c.** uses $Cu^{2+}$ during catalysis

**16.59** Identify a vitamin that is a component of each of the following coenzymes:
**a.** coenzyme A
**b.** tetrahydrofolate (THF)
**c.** $NAD^+$

**16.60** Identify a vitamin that is a component of each of the following coenzymes:
**a.** thiamine pyrophosphate
**b.** FAD
**c.** pyridoxal phosphate

## CONCEPT MAP

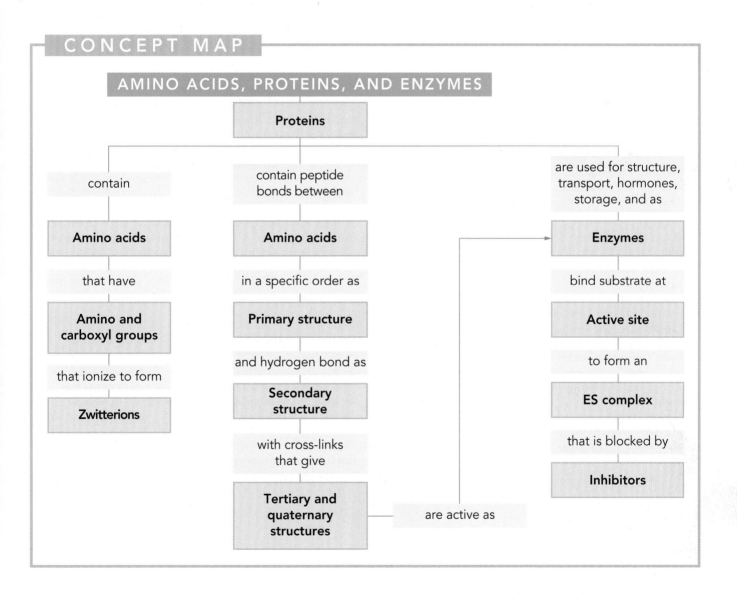

**AMINO ACIDS, PROTEINS, AND ENZYMES**

Proteins

contain — Amino acids — that have — Amino and carboxyl groups — that ionize to form — Zwitterions

contain peptide bonds between — Amino acids — in a specific order as — Primary structure — and hydrogen bond as — Secondary structure — with cross-links that give — Tertiary and quaternary structures — are active as

are used for structure, transport, hormones, storage, and as — Enzymes — bind substrate at — Active site — to form an — ES complex — that is blocked by — Inhibitors

# CHAPTER REVIEW

## 16.1 Functions of Proteins

**Learning Goal:** Classify proteins by their functions in the cells.

Some proteins are enzymes or hormones, whereas others are important in structure, transport, protection, storage, and muscle contraction.

## 16.2 Amino Acids

**Learning Goal:** Draw the zwitterion for an amino acid.

A group of 20 amino acids provides the molecular building blocks of proteins. Attached to the central (alpha) carbon of each amino acid are an amino group, a carboxyl group, and a unique side chain (R). The side chain gives an amino acid the property of being nonpolar, polar, acidic, or basic. At physiological pH, most amino acids exist as dipolar ions called zwitterions.

## 16.3 Amino Acids as Acids and Bases

**Learning Goal:** Draw the ionized structure of an amino acid at pH values above or below pI.

Amino acids exist as dipolar ions called zwitterions, as positive ions at low pH, and as negative ions at high pH levels. At their isoelectric points, zwitterions are neutral.

## 16.4 Formation of Peptides

**Learning Goal:** Draw the structure of a dipeptide.

Peptides form when an amide (peptide) bond links the carboxyl group of one amino acid to the amino group of a second amino acid. Long chains of amino acids are called proteins.

## 16.5 Levels of Protein Structure

**Learning Goal:** Identify the structural levels of a protein.

The primary structure of a protein is its sequence of amino acids. In the secondary structure, hydrogen bonds between peptide groups produce a characteristic shape such as an $\alpha$ helix, $\beta$-pleated sheet, or a triple helix. A tertiary structure is stabilized by attractions between side chains that form hydrogen bonds, disulfide bonds, and salt bridges as well as interactions that move hydrophobic side chains to the inside and hydrophilic side chains to the outside surface. In a quaternary structure, two or more tertiary subunits needed for biological activity are held together by the same interactions found in tertiary structures.

Denaturation of a protein occurs when high temperatures, acids or bases, organic compounds, metal ions, or agitation destroy the secondary, tertiary, or quaternary structures of a protein with a loss of biological activity.

## 16.6 Enzymes

**Learning Goal:** Describe how enzymes function as catalysts and give their names.

Enzymes act as biological catalysts by lowering activation energy and accelerating the rate of cellular reactions. The names of most enzymes ending in *ase* describe the compound or reaction catalyzed by the enzyme. Enzymes are classified by the main type of reaction they catalyze, such as oxidoreductase, transferase, or isomerase.

## 16.7 Enzyme Action

**Learning Goal:** Describe the role of an enzyme in an enzyme-catalyzed reaction.

Within the tertiary structure of an enzyme, a small pocket called the active site binds the substrates. In the lock-and-key model, a substrate precisely fits the shape of the active site. In the induced-fit model, substrates induce the active site to change structure to give an optimal fit by the substrate. In the enzyme–substrate complex, catalysis takes place when amino acid side chains react with a substrate. The products are released and the enzyme is available to bind another substrate molecule.

## 16.8 Factors Affecting Enzyme Activity

**Learning Goal:** Describe the effect of temperature, pH, concentration of substrate, and inhibitors on enzyme activity.

Enzymes are most effective at optimum temperature, usually 37 °C, and optimum pH, usually 7.4. The rate of an enzyme-catalyzed reaction decreases as temperature and pH values go above or below the optimum. An increase in substrate concentration increases the reaction rate of an enzyme-catalyzed reaction. If an enzyme is saturated, adding more substrate will not increase the rate further. An inhibitor reduces the activity of an enzyme or makes it inactive. A competitive inhibitor has a structure similar to the substrate and competes for the active site. When the active site is occupied, the enzyme cannot catalyze the reaction of the substrate. A noncompetitive inhibitor attaches elsewhere on the enzyme, changing the shape of both the enzyme and its active site. As long as the noncompetitive inhibitor is attached, the altered active site cannot bind with substrate.

## 16.9 Enzyme Cofactors

**Learning Goal:** Describe the types of cofactors found in enzymes.

Simple enzymes are biologically active as a protein only, whereas other enzymes require small organic molecules or metals ions called cofactors. A cofactor may be a metal ion such as $Cu^{2+}$ or $Fe^{2+}$ or an organic molecule called a coenzyme. A vitamin is a small organic molecule needed for health and normal growth. Vitamins are obtained in small amounts through foods in the diet. The water-soluble vitamins B and C function as coenzymes.

# KEY TERMS

**acidic amino acid** An amino acid that has a carboxylic acid side chain (—COOH), which ionizes as a weak acid.

**active site** A pocket in a part of the tertiary enzyme structure that binds substrate and catalyzes a reaction.

**activity** The rate at which an enzyme catalyzes the reaction that converts substrate to product.

**$\alpha$ (alpha) helix** A secondary level of protein structure, in which hydrogen bonds connect the NH of one peptide bond with the C═O of a peptide bond later in the chain to form a coiled or corkscrew structure.

**amino acid** The building block of proteins, consisting of an amino group, a carboxylic acid group, and a unique side group attached to the alpha carbon.

**basic amino acid** An amino acid that contains an amino (—NH₂) group that can ionize as a weak base.

**β (beta)-pleated sheet** A secondary level of protein structure that consists of hydrogen bonds between peptide links in parallel polypeptide chains.

**C terminal** The end amino acid in a peptide chain with a free —COO⁻ group.

**coenzyme** An organic molecule, usually a vitamin, required as a cofactor in enzyme action.

**cofactor** A nonprotein metal ion or an organic molecule that is necessary for a biologically functional enzyme.

**collagen** The most abundant form of protein in the body, composed of fibrils of triple helices with hydrogen bonding between —OH groups of hydroxyproline and hydroxylysine.

**competitive inhibitor** A molecule with a structure similar to the substrate that inhibits enzyme action by competing for the active site.

**denaturation** The loss of secondary and tertiary protein structure, caused by heat, acids, bases, organic compounds, heavy metals, and/or agitation.

**disulfide bonds** Covalent —S—S— bonds that form between the —SH groups of cysteines in a protein to stabilize the tertiary structure.

**enzymes** Substances including globular proteins that catalyze biological reactions.

**enzyme–substrate (ES) complex** An intermediate consisting of an enzyme that binds to a substrate in an enzyme-catalyzed reaction.

**fat-soluble vitamins** Vitamins that are not soluble in water and can be stored in the liver and body fat.

**fibrous protein** A protein that is insoluble in water; consists of polypeptide chains with α helices or β-pleated sheets that make up the fibers of hair, wool, skin, nails, and silk.

**globular proteins** Proteins that acquire a compact shape from attractions between side chains of the amino acid in the protein.

**hydrogen bonds** Attractions between polar side chains such as —OH, —NH₂, and —COOH of amino acids in a polypeptide chain.

**hydrophilic interactions** The attractions between water and polar side chains on the outside of the protein.

**hydrophobic interactions** The attractions between nonpolar side chains on the inside of a globular protein.

**induced-fit model** A model of enzyme action in which the shapes of the substrate and its active site are modified to give an optimal fit.

**inhibitors** Substances that make an enzyme inactive by interfering with its ability to react with a substrate.

**isoelectric point (pI)** The pH at which an amino acid exists as a zwitterion (dipolar ion).

**isoenzymes** Enzymes with different combinations of polypeptide subunits that catalyze the same reaction in different tissues of the body.

**lock-and-key model** A model of enzyme action in which the substrate is like a key that fits the specific shape of the active site.

**N terminal** The end amino acid in a peptide with a free —NH₃⁺ group.

**noncompetitive inhibitor** A type of inhibitor that alters the shape of an enzyme as well as the active site so that the substrate cannot bind properly.

**nonpolar amino acids** Amino acids with nonpolar side chains that are not soluble in water.

**optimum pH** The pH at which an enzyme is most active.

**optimum temperature** The temperature at which an enzyme is most active.

**peptide** The combination of two or more amino acids joined by peptide bonds; dipeptide, tripeptide, and so on.

**peptide bond** The amide bond that joins amino acids in polypeptides and proteins.

**polar amino acids** Amino acids with polar side chains that are soluble in water.

**primary structure** The specific sequence of the amino acids in a protein.

**protein** A term used for biologically active polypeptides that have many amino acids linked together by peptide bonds.

**quaternary structure** A protein structure in which two or more protein subunits form an active protein.

**salt bridge** The attraction between ionized side groups of basic and acidic amino acids in the tertiary structure of a protein.

**secondary structure** The formation of an α helix, a β-pleated sheet, or a triple helix.

**simple enzyme** An enzyme that is active as a polypeptide only.

**substrate** The molecule that reacts in the active site in an enzyme-catalyzed reaction.

**tertiary structure** The folding of the secondary structure of a protein into a compact structure that is stabilized by the interactions of side chains.

**triple helix** The protein structure found in collagen, consisting of three polypeptide chains woven together like a braid.

**vitamins** Organic molecules, which are essential for normal health and growth, obtained in small amounts from the diet.

**water-soluble vitamins** Vitamins that are soluble in water; cannot be stored in the body; are easily destroyed by heat, ultraviolet light, and oxygen; and function as coenzymes.

**zwitterion** The dipolar form of an amino acid consisting of two oppositely charged ionic regions, —NH₃⁺ and —COO⁻.

# UNDERSTANDING THE CONCEPTS

**16.61** Ethylene glycol (HO—CH₂—CH₂—OH) is a major component of antifreeze. In the body, it is first converted to HOOC—CHO (oxoethanoic acid) and then to HOOC—COOH (oxalic acid), which is toxic.

    **a.** What class of enzyme catalyzes both of the reactions of ethylene glycol?

    **b.** The treatment for the ingestion of ethylene glycol is an intravenous solution of ethanol. How might this help prevent toxic levels of oxalic acid in the body?

**16.62** Adults who are lactose intolerant cannot break down the disaccharide in milk products. To help digest dairy food, a product

known as Lactaid can be added to milk and the milk then refrigerated for 24 hours.

a. What enzyme is present in Lactaid, and what is the major class?
b. What might happen to the enzyme if the digestion product were stored at 55 °C?

**16.63** Aspartame, which is used in artificial sweeteners, contains the following dipeptide:

a. What are the amino acids in aspartame?
b. How would you name the dipeptide in aspartame?

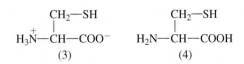

**16.64** If cysteine, an amino acid prevalent in hair, has a pI of 5.0, which structure would it have in solutions with the following pH values?

$$CH_2-SH \qquad CH_2-SH$$
$$H_2N-CH-COO^- \qquad H_3\overset{+}{N}-CH-COOH$$
$$(1) \qquad\qquad (2)$$

$$CH_2-SH \qquad\qquad CH_2-SH$$
$$H_3\overset{+}{N}-CH-COO^- \qquad H_2N-CH-COOH$$
$$(3) \qquad\qquad\qquad (4)$$

a. pH = 2      b. pH = 5      c. pH = 9

**16.65** Identify the amino acids and type of cross-link that occur between the following side groups in tertiary protein structures:

a.
$$-CH_2-\overset{\overset{O}{\|}}{C}-NH_2 \quad and \quad HO-CH_2-$$

b.
$$-CH_2-\overset{\overset{O}{\|}}{C}-O^- \quad and \quad H_3\overset{+}{N}-(CH_2)_4-$$

c.
$$-CH_2-SH \quad and \quad HS-CH_2-$$

d.
$$-CH_2-\overset{\overset{CH_3}{|}}{CH}-CH_3 \quad and \quad CH_3-$$

**16.66** Fresh pineapple contains the enzyme bromelain that degrades proteins.

a. The directions on a Jello® package say not to add fresh pineapple. However, canned pineapple where pineapple is heated to high temperatures can be added. Why?
b. Fresh pineapple is used in a marinade to tenderize tough meat. Why?

# ADDITIONAL QUESTIONS AND PROBLEMS

**16.67** Seeds and vegetables are often deficient in one or more essential amino acids. Using the following table, state whether the following combinations would provide all the essential amino acids:

| Source | Lysine | Tryptophan | Methionine |
|---|---|---|---|
| Oatmeal | No | Yes | Yes |
| Rice | No | Yes | Yes |
| Garbanzo beans | Yes | No | Yes |
| Lima beans | Yes | No | No |
| Cornmeal | No | Yes | Yes |

    **a.** rice and garbanzo beans
    **b.** lima beans and cornmeal
    **c.** a salad of garbanzo beans and lima beans
    **d.** rice and lima beans
    **e.** rice and oatmeal
    **f.** oatmeal and lima beans

**16.68** **a.** Draw the structure of Val-Ala-Leu.
    **b.** Would you expect to find this segment at the center or at the surface of a globular protein? Why?

**16.69** What are some differences between the following pairs?
    **a.** secondary and tertiary protein structures
    **b.** essential and nonessential amino acids
    **c.** polar and nonpolar amino acids
    **d.** dipeptides and tripeptides
    **e.** an ionic bond (salt bridge) and a disulfide bond
    **f.** fibrous and globular proteins
    **g.** $\alpha$ helix and $\beta$-pleated sheet
    **h.** tertiary and quaternary structures of proteins

**16.70** **a.** Where is collagen found?
    **b.** What type of secondary structure is used to form collagen?

**16.71** **a.** Draw the structure of Ser-Lys-Asp.
    **b.** Would you expect to find this segment at the center or at the surface of a globular protein? Why?

**16.72** What type of interaction would you expect between the following in a tertiary structure?
    **a.** threonine and asparagine
    **b.** valine and alanine
    **c.** arginine and aspartic acid

**16.73** If serine were replaced by valine in a protein, how would the tertiary structure be affected?

**16.74** If you eat rice, what other vegetable protein source(s) could you eat to ingest all essential amino acids?

**16.75** How do enzymes differ from catalysts used in chemical laboratories?

**16.76** Why do enzymes function only under mild conditions?

**16.77** Lactase is an enzyme that hydrolyzes lactose to glucose and galactose.
    **a.** What are the reactants and products of the reaction?
    **b.** Draw an energy diagram for the reaction with and without lactase.
    **c.** How does lactase make the reaction go faster?

**16.78** Maltase is an enzyme that hydrolyzes maltose to two glucose molecules.
    **a.** What are the reactants and products of the reaction?
    **b.** Draw an energy diagram for the reaction with and without maltase.
    **c.** How does maltase make the reaction go faster?

**16.79** Indicate whether each of the following would be a substrate (S) or an enzyme (E):
    **a.** lactose            **b.** lactase
    **c.** urease            **d.** trypsin
    **e.** pyruvate         **f.** transaminase

**16.80** Indicate whether each of the following would be a substrate (S) or an enzyme (E):
    **a.** glucose          **b.** hydrolase
    **c.** maleate isomerase   **d.** alanine
    **e.** amylose         **f.** amylase

**16.81** Give the substrate of each of the following enzymes:
    **a.** urease
    **b.** lactase
    **c.** aspartate transaminase
    **d.** tyrosinase

**16.82** Give the substrate of each of the following enzymes:
    **a.** maltase         **b.** fructose oxidase
    **c.** phenolase      **d.** sucrase

**16.83** How would the lock-and-key theory explain that sucrase hydrolyzes sucrose, but not lactose?

**16.84** How does the induced-fit model of enzyme action allow an enzyme to catalyze a reaction of a group of substrates?

**16.85** If a blood test indicates a high level of LDH and CK, what could be the cause?

**16.86** If a blood test indicates a high level of ALT, what could be the cause?

**16.87** Indicate whether an enzyme is saturated or unsaturated in each of the following conditions:
    **a.** Adding more substrate does not increase the rate of reaction.
    **b.** Doubling the substrate concentration doubles the rate of reaction.

**16.88** Indicate whether each of the following enzymes would be functional:
    **a.** pepsin, a digestive enzyme, at pH 2
    **b.** an enzyme at 37 °C, if the enzyme is from a type of thermophilic bacteria that have an optimum temperature of 100 °C

**16.89** Does each of the following statements describe a simple enzyme or an enzyme that requires a cofactor?
    **a.** contains $Mg^{2+}$ in the active site
    **b.** has catalytic activity as a tertiary protein structure
    **c.** requires folic acid for catalytic activity

**16.90** Does each of the following statements describe a simple enzyme or an enzyme that requires a cofactor?
    **a.** contains riboflavin or vitamin $B_2$
    **b.** has four subunits of polypeptide chains
    **c.** requires $Fe^{3+}$ in the active site for catalytic activity

# CHALLENGE QUESTIONS

**16.91** Indicate the charge of each of the following amino acids at the following pH values as:
　**a.** 0　　**b.** +1　　**c.** +2　　**d.** −2　　**e.** −1
　**1.** serine at pH 6.0
　**2.** glutamic acid at pH 10
　**3.** arginine at pH 2
　**4.** histidine at pH 7.6
　**5.** isoleucine at pH 3
　**6.** leucine at pH 9

**16.92** Consider a mixture of the amino acids lysine, valine, and aspartic acid at pH 6.0 that is subjected to an electric current.

 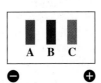

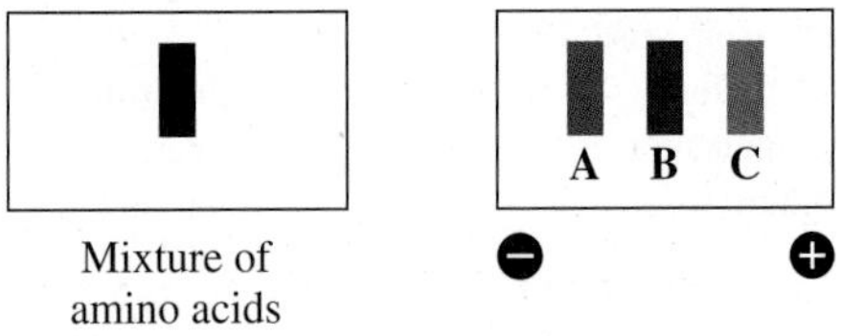

Mixture of amino acids

**a.** Indicate which amino acid would migrate toward the positive electrode (+), the negative electrode (−), or remain stationary.
**b.** If the mixture is the result of hydrolyzing a tripeptide, what are the sequences if one molecule of each amino acid is present?
**c.** If present in an enzyme, which of these amino acids would
　**1.** be found in hydrophobic regions
　**2.** be found in hydrophilic regions
　**3.** form hydrogen bonds
　**4.** form disulfide bonds
　**5.** form salt bridges

# ANSWERS

## Answers to Study Checks

**16.1** hormones

**16.2** **a.** nonpolar
**b.** polar

**16.3**
$$NH_3^+ - CH - C - OH$$
with side chain $CH_2 - OH$ and carbonyl O

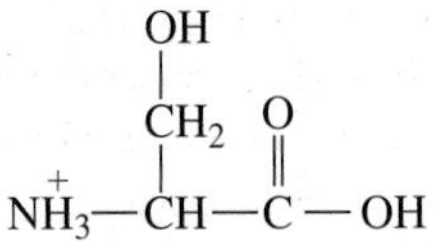

**16.4** threonylleucylphenylalanine

**16.5** Both are nonpolar and would be found on the inside of the tertiary structure.

**16.6** quaternary

**16.7** lipase

**16.8** In the lock-and-key model, the shape of a substrate fits the shape of the active site exactly. In the induced-fit model, the substrate and the active site adjust shape to provide the best fit.

**16.9** At a pH lower than the optimum pH, denaturation will decrease the activity of urease.

**16.10** competitive inhibition

**16.11** vitamin $B_6$

## Answers to Selected Questions and Problems

**16.1** **a.** transport
**b.** structural
**c.** structural
**d.** enzyme

**16.3** All amino acids contain a carboxylic acid group and an amino group on the $\alpha$ carbon.

**16.5** **a.**
$$H_3N^+ - CH - CO^-$$
with $CH_3$ side chain

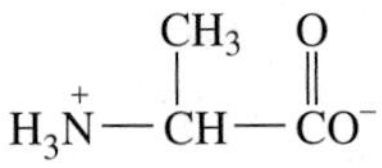

**b.**
$$H_3N^+ - CH - CO^-$$
with $H - C - OH$ and $CH_3$ side chain

**c.**
$$H_3N^+ - CH - CO^-$$
with side chain $CH_2 - CH_2 - C$ (O, $O^-$)

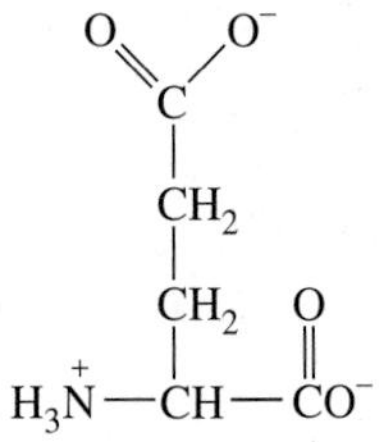

**d.**
$$H_3N^+ - CH - CO^-$$
with $CH_2 -$ benzene ring side chain

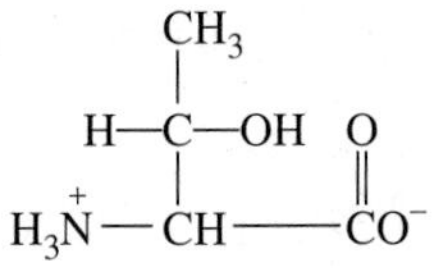

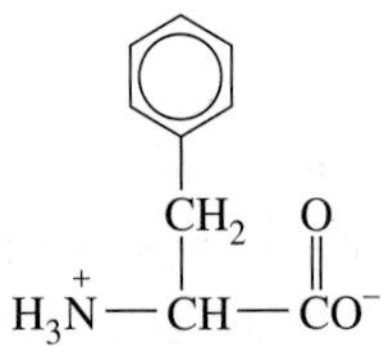

**16.7** **a.** hydrophobic (nonpolar)
**b.** hydrophilic (polar, neutral)
**c.** acidic
**d.** hydrophobic (nonpolar)

**16.9** **a.** alanine
**b.** valine
**c.** lysine
**d.** cysteine

**16.11 a.**

$$H_3\overset{+}{N}-\overset{\displaystyle COO^-}{\underset{\displaystyle CH}{\overset{|}{\underset{|}{C}}}}-H$$
$$\underset{H_3C\quad CH_3}{}$$

**b.**

$$H-\overset{\displaystyle COO^-}{\underset{\displaystyle CH_2SH}{\overset{|}{\underset{|}{C}}}}-\overset{+}{N}H_3$$

**16.13 a.**

$$H_3\overset{+}{N}-\overset{\displaystyle H}{\underset{\phantom{|}}{\overset{|}{CH}}}-\overset{\displaystyle O}{\overset{\|}{C}}O^-$$

**b.**

$$H_3\overset{+}{N}-\overset{\displaystyle SH}{\underset{\displaystyle CH_2}{\overset{|}{\underset{|}{CH}}}}-\overset{\displaystyle O}{\overset{\|}{C}}O^-$$

**c.**

$$H_3\overset{+}{N}-\overset{\displaystyle OH}{\underset{\displaystyle CH_2}{\overset{|}{\underset{|}{CH}}}}-\overset{\displaystyle O}{\overset{\|}{C}}O^-$$

**d.**

$$H_3\overset{+}{N}-\overset{\displaystyle CH_3}{\underset{\phantom{|}}{\overset{|}{CH}}}-\overset{\displaystyle O}{\overset{\|}{C}}O^-$$

**16.15 a.**

$$H_3\overset{+}{N}-\overset{\displaystyle H}{\underset{\phantom{|}}{\overset{|}{CH}}}-\overset{\displaystyle O}{\overset{\|}{C}}OH$$

**b.**

$$H_3\overset{+}{N}-\overset{\displaystyle SH}{\underset{\displaystyle CH_2}{\overset{|}{\underset{|}{CH}}}}-\overset{\displaystyle O}{\overset{\|}{C}}OH$$

**c.**

$$H_3\overset{+}{N}-\overset{\displaystyle OH}{\underset{\displaystyle CH_2}{\overset{|}{\underset{|}{CH}}}}-\overset{\displaystyle O}{\overset{\|}{C}}OH$$

**d.**

$$H_3\overset{+}{N}-\overset{\displaystyle CH_3}{\underset{\phantom{|}}{\overset{|}{CH}}}-\overset{\displaystyle O}{\overset{\|}{C}}OH$$

**16.17 a.** above pI  **b.** below pI  **c.** at pI

**16.19**

**a.**

$$H_3\overset{+}{N}-\overset{\displaystyle O}{\underset{\displaystyle CH_3}{\overset{|}{\underset{|}{CH}}}}-\overset{\|}{C}-NH-\overset{\displaystyle CH_2}{\underset{\displaystyle SH}{\overset{|}{\underset{|}{CH}}}}-\overset{\displaystyle O}{\overset{\|}{C}}O^-$$

Ala-Cys

**b.**

$$H_3\overset{+}{N}-\overset{\displaystyle CH_2}{\underset{\displaystyle OH}{\overset{|}{\underset{|}{CH}}}}-\overset{\displaystyle O}{\overset{\|}{C}}-NH-CH-\overset{\displaystyle O}{\overset{\|}{C}}O^-$$

Ser-Phe

**c.**

$$H_3\overset{+}{N}-\overset{\displaystyle H}{\underset{\phantom{|}}{\overset{|}{CH}}}-\overset{\displaystyle O}{\overset{\|}{C}}-NH-\overset{\displaystyle CH_3}{\overset{|}{CH}}-\overset{\displaystyle O}{\overset{\|}{C}}-NH-CH-\overset{\displaystyle O}{\overset{\|}{C}}O^-$$

Gly-Ala-Val

**d.**

$$H_3\overset{+}{N}-CH-\overset{\displaystyle O}{\overset{\|}{C}}-NH-CH-\overset{\displaystyle O}{\overset{\|}{C}}-NH-CH-\overset{\displaystyle O}{\overset{\|}{C}}O^-$$

Val-Ile-Trp

**16.21** Peptide bonds (amide bonds) connect the amino acids in the primary structure of a protein.

**16.23** Val-Ser-Ser, Ser-Val-Ser, or Ser-Ser-Val

**16.25** The primary structure remains unchanged and intact as hydrogen bonds form between carbonyl oxygen atoms and amino hydrogen atoms.

**16.27** In the $\alpha$ helix, hydrogen bonds form between the carbonyl oxygen atom and the amino hydrogen atom of the fourth amino acid in the sequence. In the $\beta$-pleated sheet, hydrogen bonds occur between parallel peptides or across sections of a long polypeptide chain.

**16.29 a.** a disulfide bond
**b.** salt bridge
**c.** hydrogen bond
**d.** hydrophobic interaction

**16.31 a.** cysteine
**b.** Leucine and valine will be found on the inside of the protein because they are hydrophobic.
**c.** The cysteine and aspartic acid would be on the outside of the protein because they are polar.
**d.** The order of the amino acids (the primary structure) provides side chains that interact to determine the tertiary structure of the protein.

**16.33 a.** Placing an egg in boiling water disrupts hydrogen bonds and hydrophobic interactions, which changes secondary and tertiary structures and causes loss of overall shape.
**b.** Using an alcohol swab disrupts hydrophobic interactions and changes the tertiary structure.
**c.** The heat from an autoclave will disrupt hydrogen bonds and hydrophobic interactions, which changes secondary and tertiary structures of the protein in any bacteria present.
**d.** Heat will cause changes in the secondary and tertiary structure of surrounding protein, which results in the formation of solid protein that helps to close the wound.

**16.35** The chemical reactions can occur without enzymes, but the rates are too slow. Catalyzed reactions, which are many times faster, provide the amounts of products needed by the cell at a particular time.

**16.37 a.** galactose  **b.** lipid  **c.** aspartic acid

**16.39 a.** hydrolase  **b.** oxidoreductase
**c.** isomerase  **d.** transferase

**16.41 a.** enzyme
**b.** enzyme–substrate complex
**c.** substrate

**16.43 a.** $E + S \rightleftharpoons ES \longrightarrow E + P$
**b.** The active site is a region or pocket within the tertiary structure of an enzyme that accepts the substrate, aligns the substrate for reaction, and catalyzes the reaction.

**16.45** Isoenzymes are slightly different forms of an enzyme that catalyze the same reaction in different organs and tissues of the body.

**16.47** A doctor might run tests for the enzymes CK, LDH, and AST to determine if the patient had a heart attack.

**16.49 a.** The rate would decrease.
**b.** The rate would decrease.
**c.** The rate would decrease.

**16.51** pepsin, pH 2; urease, pH 5; trypsin, pH 8

**16.53** **a.** competitive        **b.** noncompetitive
      **c.** competitive        **d.** noncompetitive
      **e.** competitive

**16.55** **a.** methanol, $CH_3OH$; ethanol, $CH_3CH_2OH$
      **b.** Ethanol has a structure similar to methanol and could compete for the active site.
      **c.** Ethanol is a competitive inhibitor of methanol oxidation.

**16.57** **a.** an enzyme that requires a cofactor
      **b.** an enzyme that requires a cofactor
      **c.** a simple enzyme

**16.59** **a.** pantothenic acid (vitamin $B_5$)
      **b.** folic acid
      **c.** niacin (vitamin $B_3$)

**16.61** **a.** oxidoreductase
      **b.** At high concentration, ethanol, which acts as a competitive inhibitor of ethylene glycol, would saturate the alcohol dehydrogenase enzyme to allow ethylene glycol to be removed from the body without producing oxalic acid.

**16.63** **a.** aspartic acid and phenylalanine
      **b.** aspartylphenylalanine

**16.65** **a.** asparagine and serine; hydrogen bond
      **b.** aspartic acid and lysine; salt bridge
      **c.** cysteine and cysteine; disulfide bond
      **d.** leucine and alanine; hydrophobic interaction

**16.67** **a.** yes        **b.** yes        **c.** no
      **d.** yes        **e.** no         **f.** yes

**16.69** **a.** The secondary structure of a protein depends on hydrogen bonds to form a helix or a pleated sheet; the tertiary structure is determined by the interactions of side chains such as disulfide bonds and salt bridges.
      **b.** Nonessential amino acids can be synthesized by the body; essential amino acids must be supplied in the diet.
      **c.** Polar amino acids have hydrophilic side groups, whereas nonpolar amino acids have hydrophobic side groups.
      **d.** Dipeptides contain two amino acids, whereas tripeptides contain three.
      **e.** An ionic bond is an interaction between a basic and an acidic side group; a disulfide bond links the sulfides of two cysteines.
      **f.** Fibrous proteins consist of $\alpha$ helices coiled like a rope. Globular proteins form a compact spherical shape.
      **g.** The $\alpha$ helix is the secondary shape like a staircase or corkscrew. The $\beta$-pleated sheet is a secondary structure that is formed by many protein chains side by side like a pleated sheet.
      **h.** The tertiary structure of a protein is its three-dimensional structure. The quaternary structure involves the grouping of two or more peptide units for the protein to be active.

**16.71** **a.**

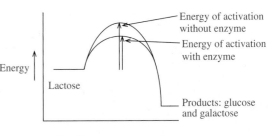

      **b.** This segment contains polar side chains, which would be found on the surface of a globular protein where they hydrogen bond with water.

**16.73** Serine is a polar amino acid, whereas valine is nonpolar. Valine would be in the center of the tertiary structure. However, serine would pull that part of the chain to the outside surface of the protein where it would form hydrophilic bonds with water.

**16.75** In chemical laboratories, reactions are often run at high temperatures using catalysts that are strong acids or bases. Enzymes, which function at physiological temperatures and pH, are denatured rapidly if high temperatures or acids or bases are used.

**16.77** **a.** The reactant is lactose, and the products are glucose and galactose.
      **b.**

```
        ┌── Energy of activation
        │       without enzyme
        │  ┌── Energy of activation
        │  │    with enzyme
Energy
   Lactose
              └── Products: glucose
                   and galactose

        Reaction progress ──→
```

      **c.** By lowering the energy of activation, the enzyme furnishes a lower energy pathway by which the reaction can take place.

**16.79** **a.** S    **b.** E    **c.** E    **d.** E    **e.** S    **f.** E

**16.81** **a.** urea                **b.** lactose
      **c.** aspartate            **d.** tyrosine

**16.83** Sucrose fits the shape of the active site in sucrase, but lactose does not.

**16.85** A heart attack may be the cause.

**16.87** **a.** saturated            **b.** unsaturated

**16.89** **a.** requires a cofactor
      **b.** simple enzyme
      **c.** requires a cofactor (coenzyme)

**16.91** **1.** a    **2.** d    **3.** c    **4.** a    **5.** b    **6.** e

# 17

# Nucleic Acids and Protein Synthesis

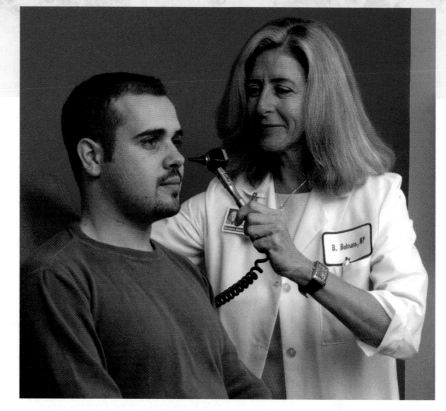

the **Chemistry** place

Visit **www.chemplace.com** for extra quizzes, interactive tutorials, career resources, PowerPoint slides for chapter review, math help, and case studies.

*"I run the Hepatitis C Clinic, where patients are often anxious when diagnosed," says Barbara Behrens, nurse practitioner, Hepatitis C Clinic, Kaiser Hospital. "The treatment for hepatitis C can produce significant reactions such as a radical drop in blood count. When this happens, I get help to them within 24 hours. I monitor our patients very closely, and many call me whenever they need to."*

*Hepatitis C is a RNA virus, or retrovirus, that causes liver inflammation, often resulting in chronic liver disease. Unlike many viruses to which we eventually develop immunity, the hepatitis C virus undergoes mutations so rapidly that scientists have not been able to produce vaccines. People who carry the virus are contagious throughout their lives and are able to pass the virus to other people.*

N

ucleic acids are large molecules in our cells that store information and direct activities for cellular growth and reproduction.
Deoxyribonucleic acid (DNA), the genetic material in the nucleus
of a cell, contains all the information needed for the development
of a complete living system. The way you grow, your hair, your eyes, your physical appearance, the activities of the cells in your body are all determined by a
set of directions contained in the DNA of your cells.

All of the genetic information in the cell is called the *genome*. Every time a
cell divides, the information in the genome is copied and passed on to the new
cells. This replication process must duplicate the genetic instructions exactly.
Some sections of DNA called *genes* contain the information to make a particular protein.

As a cell requires protein, another type of nucleic acid, RNA, translates the
genetic information in DNA and carries that information to the ribosomes, where
the synthesis of protein takes place. However, mistakes sometimes occur that
lead to mutations that affect the synthesis of a certain protein.

## LEARNING GOAL

Describe the bases and ribose sugars that make up the nucleic acids
DNA and RNA.

# 17.1 COMPONENTS OF NUCLEIC ACIDS

There are two closely related types of nucleic acids: *deoxyribonucleic acid* (**DNA**), and
*ribonucleic acid* (**RNA**). Both are unbranched polymers of repeating monomer units
known as *nucleotides*. A DNA molecule may contain several million nucleotides; smaller
RNA molecules may contain up to several thousand. Each nucleotide has three components: a base, a five-carbon sugar, and a phosphate group. (See Figure 17.1.)

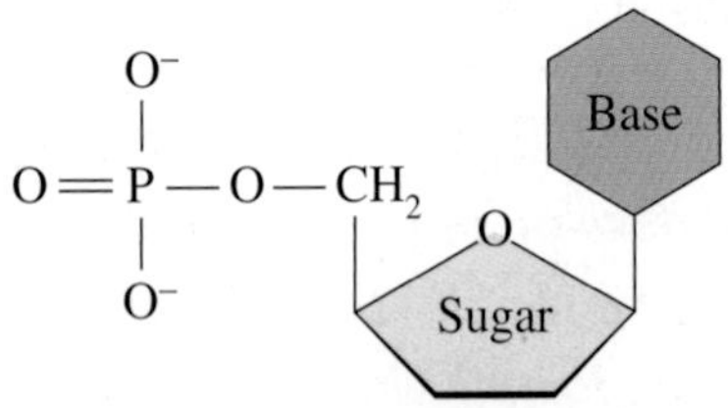

## Bases

The **bases** in nucleic acids are derivatives of *pyrimidine* or *purine*.

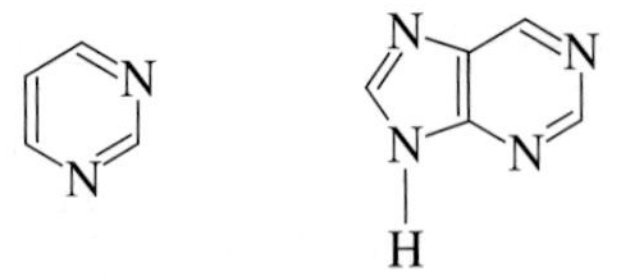

Pyrimidine          Purine

In DNA, the bases are two purines, adenine (A) and guanine (G); and two pyrimidines, cytosine (C) and thymine (T). RNA contains the same bases, except thymine
(5-methyluracil) is replaced by uracil (U). (See Figure 17.2.)

## Ribose and Deoxyribose Sugars

In RNA, the five-carbon sugar is *ribose*, which gives the letter R in the abbreviation
RNA. The atoms in the pentose sugars are numbered with primes ($1'$, $2'$, $3'$, $4'$, and $5'$)

FIGURE 17.1 A diagram of the
general structure of a nucleotide
found in nucleic acids.
**Q** In a nucleotide, what types
of groups are bonded to a
5-carbon sugar?

## Pyrimidines

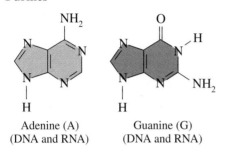

Cytosine (C)
(DNA and RNA)

Thymine (T)
(DNA only)

Uracil (U)
(RNA only)

## Purines

Adenine (A)
(DNA and RNA)

Guanine (G)
(DNA and RNA)

**FIGURE 17.2** DNA contains the bases A, G, C, and T; RNA contains A, G, C, and U.
**Q** Which bases are found in DNA?

to differentiate them from the atoms in the bases. (See Figure 17.3.) In DNA, the five-carbon sugar is *deoxyribose*, which is similar to ribose except that there is no hydroxyl group (—OH) on C2′. The *deoxy* prefix means "without oxygen" and provides the D in DNA.

### SAMPLE PROBLEM 17.1

#### ■ Components of Nucleic Acids

Identify each of the following bases as a purine or a pyrimidine.

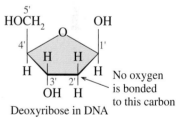

**SOLUTION**

**a.** Guanine is a purine.
**b.** Uracil is a pyrimidine.

**STUDY CHECK**

Indicate if the bases in Sample Problem 17.1 are found in RNA, DNA, or both.

**Pentose sugars in RNA and DNA**

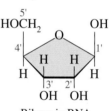

Ribose in RNA

No oxygen is bonded to this carbon

Deoxyribose in DNA

**FIGURE 17.3** The 5-carbon pentose sugar found in RNA is ribose and in DNA, deoxyribose.
**Q** What is the difference between ribose and deoxyribose?

## Nucleosides and Nucleotides

A **nucleoside** is produced when a pyrimidine or a purine forms a glycosidic bond to C1′ of a sugar, either ribose or deoxyribose. For example, adenine, a purine, and ribose form a nucleoside called aden**osine**.

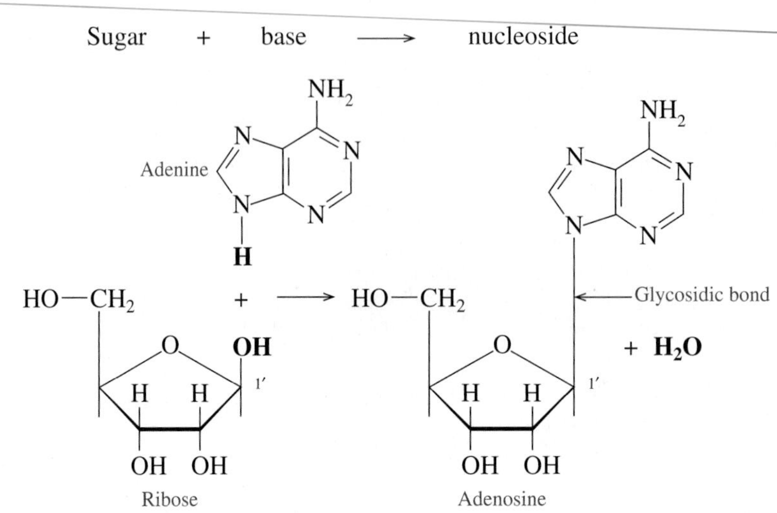

Sugar    +    base    ⟶    nucleoside

**Nucleotides** are formed when the C5′—OH group of ribose or deoxyribose in a nucleoside forms a phosphate ester. Other hydroxyl groups on ribose can form phosphate esters too, but only the 5′-monophosphate nucleotides are found in RNA and DNA. All the nucleotides in RNA and DNA are shown in Figure 17.4.

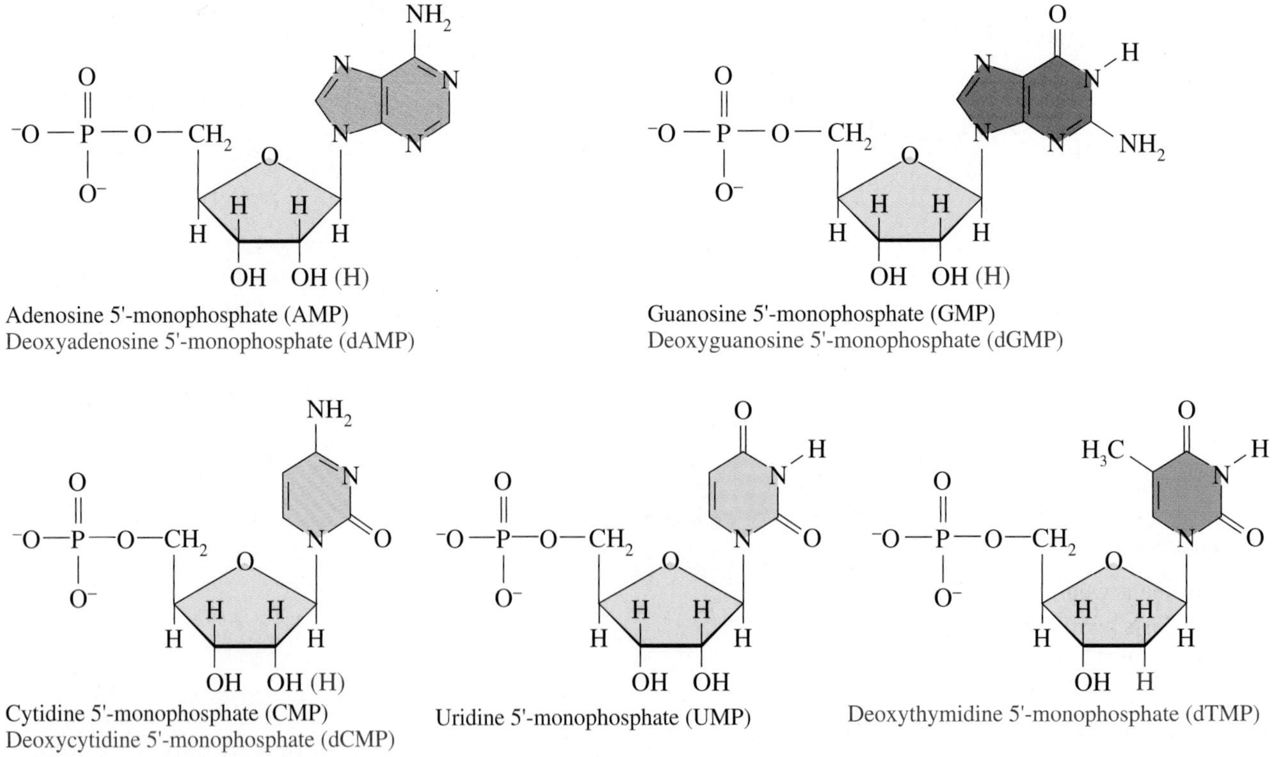

Adenosine 5'-monophosphate (AMP)
Deoxyadenosine 5'-monophosphate (dAMP)

Guanosine 5'-monophosphate (GMP)
Deoxyguanosine 5'-monophosphate (dGMP)

Cytidine 5'-monophosphate (CMP)
Deoxycytidine 5'-monophosphate (dCMP)

Uridine 5'-monophosphate (UMP)

Deoxythymidine 5'-monophosphate (dTMP)

**FIGURE 17.4** The nucleotides of RNA are identical to those of DNA, except in DNA the sugar is deoxyribose and deoxythymidine replaces uridine.
**Q** What are two differences in the nucleotides of RNA and DNA?

## Naming Nucleosides and Nucleotides

The name of a nucleoside that contains a purine ends with *osine* whereas a nucleoside that contains a pyrimidine ends with *idine*. The names of nucleosides DNA add *deoxy* to the beginning of their names. The corresponding nucleotides in RNA and DNA are named by adding 5′-monophosphate. Although the letters A, G, C, U, and T represent the bases, they are often used in the abbreviations of the respective nucleosides and nucleotides. The names and abbreviations of the bases, nucleosides, and nucleotides in DNA and RNA are listed in Table 17.1.

**TABLE 17.1** Nucleosides and Nucleotides in DNA and RNA

| Base | Nucleosides | Nucleotides |
| --- | --- | --- |
| **RNA** | | |
| Adenine (A) | Adenosine (A) | Adenosine 5′-monophosphate (AMP) |
| Guanine (G) | Guanosine (G) | Guanosine 5′-monophosphate (GMP) |
| Cytosine (C) | Cytidine (C) | Cytidine 5′-monophosphate (CMP) |
| Uracil (U) | Uridine (U) | Uridine 5′-monophosphate (UMP) |
| **DNA** | | |
| Adenine (A) | Deoxyadenosine (A) | Deoxyadenosine 5′-monophosphate (dAMP) |
| Guanine (G) | Deoxyguanosine (G) | Deoxyguanosine 5′-monophosphate (dGMP) |
| Cytosine (C) | Deoxycytidine (C) | Deoxycytidine 5′-monophosphate (dCMP) |
| Thymine (T) | Deoxythymidine (T) | Deoxythymidine 5′-monophosphate (dTMP) |

**SAMPLE PROBLEM 17.2**

■ **Nucleotides**

Identify which nucleic acid (DNA or RNA) contains each of the following nucleotides; state the components of each nucleotide:

**a.** deoxyguanosine 5′-monophosphate (dGMP)
**b.** adenosine 5′-monophosphate (AMP)

**SOLUTION**

**a.** This DNA nucleotide consists of deoxyribose, guanine, and phosphate.
**b.** This RNA nucleotide contains ribose, adenine, and phosphate.

**STUDY CHECK**

What is the name and abbreviation of the DNA nucleotide of cytosine?

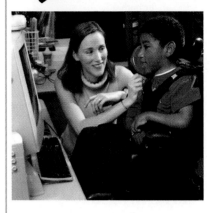

*Career Focus*

### Occupational Therapist

"Occupational therapists teach children and adults the skills they need for the job of living," says occupational therapist Leslie Wakasa. "When working with the pediatric population, we are crucial in educating disabled children, their families, caregivers, and school staff in ways to help them be as independent as they can be in all aspects of their daily lives. It's rewarding when you can show children how to feed themselves, which is a huge self-esteem issue for them. The opportunity to help people become more independent is very rewarding."

A combination of technology and occupational therapy helps children who are nonverbal to communicate and interact with their environment. By leaning on a red switch, Alex is learning to use a computer.

## QUESTIONS AND PROBLEMS

### Components of Nucleic Acids

**17.1** Identify each of the following bases as a purine or pyrimidine:
  **a.** thymine
  **b.**

**17.2** Identify each of the following bases as a purine or pyrimidine:
  **a.** guanine
  **b.**

**17.3** Identify the bases in problem 17.1 as present in RNA, DNA, or both.

**17.4** Identify the bases in problem 17.2 as present in RNA, DNA, or both.

**17.5** What are the names and abbreviations of the four nucleotides in DNA?

**17.6** What are the names and abbreviations of the four nucleotides in RNA?

**17.7** Identify each of the following as a nucleoside or nucleotide:
 **a.** adenosine          **b.** deoxycytidine
 **c.** uridine           **d.** cytidine 5′-monophosphate

**17.8** Identify each of the following as a nucleoside or nucleotide:
 **a.** deoxythymidine
 **b.** guanosine
 **c.** adenosine
 **d.** uridine 5′-monophosphate

**17.9** Draw the structure of deoxyadenosine 5′-monophosphate (dAMP).

**17.10** Draw the structure of uridine 5′-monophosphate (UMP).

## LEARNING GOAL

Describe the primary structures of RNA and DNA.

**WEB TUTORIAL**
**DNA and RNA Structure**

# 17.2 PRIMARY STRUCTURE OF NUCLEIC ACIDS

The **nucleic acids** consist of polymers of many nucleotides in which the 3′-OH group of the sugar in one nucleotide bonds to the phosphate group on the 5′-carbon atom in the sugar of the next nucleotide. This phosphate link between the sugars in adjacent nucleotides is referred to as a **phosphodiester bond**. As more nucleotides are added through phosphodiester bonds, a backbone forms that consists of alternating sugar groups and phosphate groups.

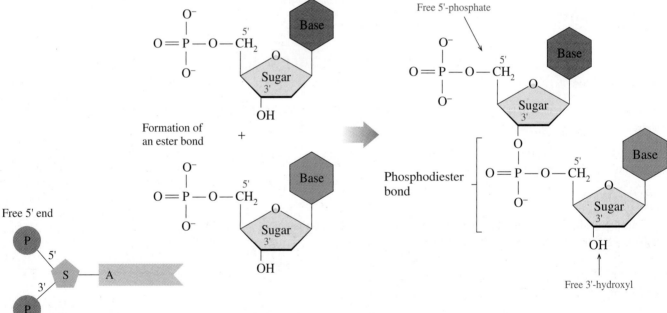

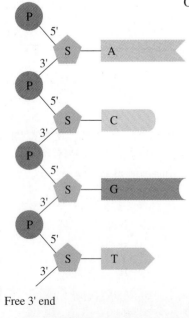

Each nucleic acid has its own unique sequence of bases, which is known as its **primary structure**. It is this sequence of bases that carries the genetic information from one cell to the next. Along a DNA or RNA chain, the bases attached to each of the sugars extend out from the nucleic acid backbone. In any nucleic acid, the sugar at the one end has an unreacted or free 5′-phosphate terminal end, and the sugar at the other end has a free 3′-hydroxyl group.

A nucleic acid sequence is read from the sugar with free 5′-phosphate to the sugar with the free 3′-hydroxyl group. The order of nucleotides is often written using only the letters of the bases. For example, the nucleotide order starting with adenine (free 5′-phosphate end) in the section of RNA shown in Figure 17.5 is 5′—A—C—G—U—3′.

**RNA (ribonucleic acid)**

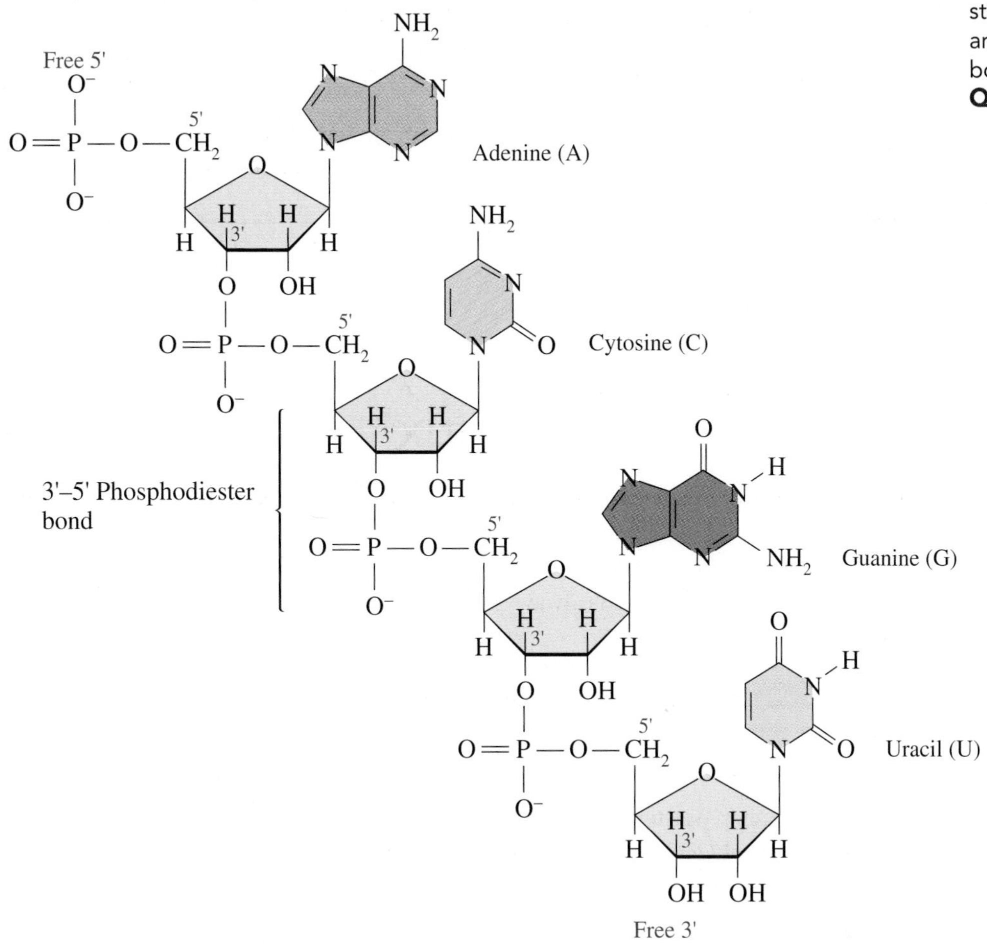

**FIGURE 17.5** In the primary structure of a RNA, A, C, G, and U are linked by 3′−5′-phosphodiester bonds.

**Q** Where are the free 5′-phosphate and 3′-hydroxyl groups of ribose?

## SAMPLE PROBLEM   17.3

### ■ Bonding of Nucleotides

Draw the structure of a RNA dinucleotide formed by two cytidine monophosphates.

SOLUTION

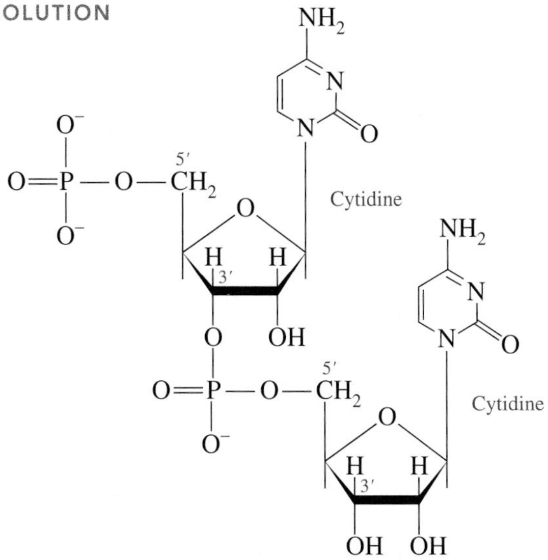

STUDY CHECK

In the dinucleotide of cytidine shown in the solution to Sample Problem 17.3, identify the free 5′-phosphate group and the free 3′-hydroxyl (—OH) group.

## QUESTIONS AND PROBLEMS

### Primary Structure of Nucleic Acids

**17.11** How are the nucleotides held together in a nucleic acid chain?

**17.12** How do the ends of a nucleic acid polymer differ?

**17.13** Write the structure of the dinucleotide GC that would be in RNA.

**17.14** Write the structure of the dinucleotide AT that would be in DNA.

---

**LEARNING GOAL**

Describe the double helix of DNA.

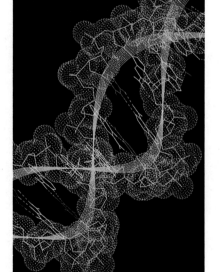

**FIGURE 17.6** A computer-generated model of a DNA molecule.

**Q** What is meant by the term *double helix*?

# $17.3$ DNA DOUBLE HELIX

In the 1940s, biologists determined that DNA in a variety of organisms had a specific relationship between bases: the amount of adenine (A) was equal to the amount of thymine (T), and the amount of guanine (G) was equal to the amount of cytosine (C). Eventually, it was determined that adenine is always paired (1:1) with thymine, and guanine is always paired (1:1) with cytosine.

Number of purine molecules = number of pyrimidine molecules

$$A = T$$
$$G = C$$

In 1953, James Watson and Francis Crick proposed that DNA is a **double helix** that consists of two polynucleotide strands winding about each other like a spiral staircase. (See Figure 17.6.) The sugar–phosphate backbones are the outside railings with the bases arranged like steps along the inside. The two strands run in opposite directions. One strand goes from the 5′ to 3′ direction, and the other strand goes in the 3′ to 5′ direction.

## Complementary Base Pairs

Each of the bases along one polynucleotide strand forms hydrogen bonds to a specific base on the opposite DNA strand. Adenine bonds only to thymine, and guanine bonds only to cytosine. (See Figure 17.7.) The pairs A—T and G—C are called **complementary base pairs**. The specific pairing of the bases is due to the fact that adenine and thymine can form only two hydrogen bonds, while cytosine and guanine can form three hydrogen bonds. This explains why DNA has equal amounts of A and T bases and equal amounts of G and C bases.

---

### SAMPLE PROBLEM 17.4

■ **Complementary Base Pairs**

Write the base sequence of the complementary segment for the following segment of a strand of DNA with a base sequence —A—C—G—A—T—C—T—.

SOLUTION

In the complementary segment of DNA, A pairs with T, and G pairs with C.

Given segment of DNA:      —A—C—G—A—T—C—T—

Complementary segment:   —T—G—C—T—A—G—A—

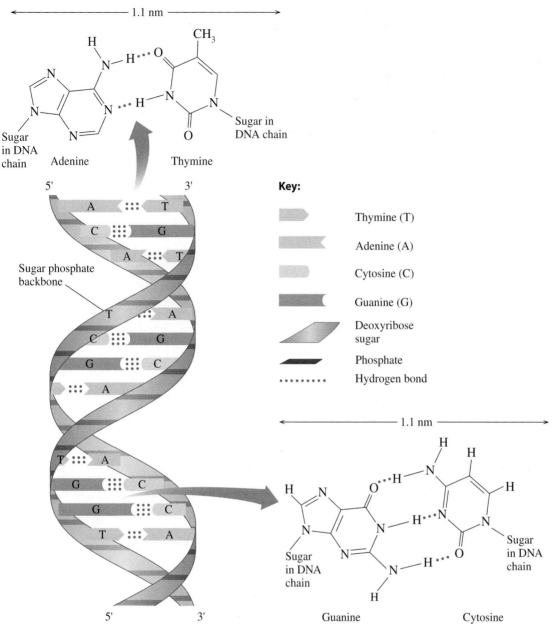

**Key:**

- Thymine (T)
- Adenine (A)
- Cytosine (C)
- Guanine (G)
- Deoxyribose sugar
- Phosphate
- Hydrogen bond

**FIGURE 17.7** Hydrogen bonds between complementary base pairs hold the polynucleotide strands in the double helix of DNA.

**Q** Why are G—C base pairs more stable than A—T base pairs?

---

**STUDY CHECK**

What is the sequence of bases that is complementary to a segment of DNA with a base sequence of —G—G—T—T—A—A—C—C—?

---

## DNA Replication

In DNA **replication**, the strands in the parent DNA separate, allowing the synthesis of complementary strands. The replication process begins when an enzyme called *helicase* catalyzes the unwinding of a portion of the double helix by breaking the hydrogen bonds

**WEB TUTORIAL**
**DNA Replication**

between the complementary bases. These single strands now act as templates for the synthesis of new complementary strands. (See Figure 17.8.)

After the base pairs are formed, *DNA polymerase* catalyzes the formation of phosphodiester bonds between the nucleotides. Eventually the entire double helix of the parent DNA is copied. In each new DNA molecule, one strand of the double helix is from the original DNA and one is a newly synthesized strand. This process produces two new DNAs called *daughter DNA* that are identical to each other and exact copies of the original parent DNA. In the process of DNA replication, complementary base pairing ensures the correct placements of bases in the new DNA strands.

**FIGURE 17.8** In DNA replication, the separate strands of the parent DNA are the templates for the synthesis of complementary strands, which produces two exact copies of DNA.

**Q** How many strands of the parent DNA are in each of the new double-stranded copies of DNA?

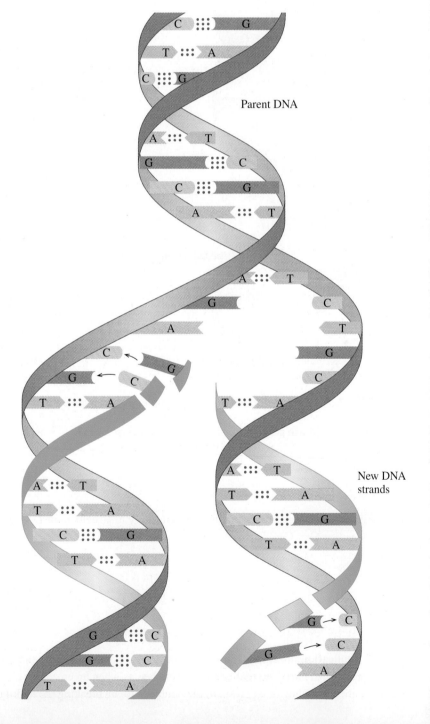

Parent DNA

New DNA strands

## Health Note

### DNA Fingerprinting

In a process called DNA fingerprinting, enzymes are used to cut DNA into smaller sections. The resulting DNA fragments are then separated by size by placing them on a gel. The gel is treated with a radioactive isotope that adheres to specific base sequences in the fragments. A piece of X-ray film that is placed over the gel is exposed by the radiation. The pattern of dark and light bands on the film is known as a DNA fingerprint. It has been estimated that the odds of two persons who are not identical twins producing the same DNA fingerprint is less than one in a billion.

One application of DNA fingerprinting is in forensic science, where DNA from samples such as blood, hair, or semen is used to connect a suspect with a crime. Recently, it has been used to gain the release of persons who were wrongly convicted. Other applications of DNA fingerprinting are to determine the biological parents of a child, establish the identity of a deceased person, and to match recipients with organ donors.

#### HUMAN GENOME PROJECT

In the 1970s, scientists began to map the location of *genes* within the DNA of the *genome,* which contains the hereditary information of an organism. By 1987, the genome of *E. coli* was determined. More recently, these techniques combined with new computer programs have compiled the map of the human genome, which

contains about 30 000 genes. It appears that most of the genome is not functional and perhaps is carried from generation to generation for millions of years. Large blocks of genes are copied from one human chromosome to another even though they no longer code for needed proteins. Thus, the coding portions of the genes seem to make up only about 1% of the total genome. The results of the genome project will help us identify defective genes that lead to genetic disease. Today DNA fingerprinting is used to screen for genes responsible for genetic diseases such as sickle-cell disease, cystic fibrosis, breast cancer, colon cancer, Huntington's disease, and Lou Gehrig's disease.

---

## SAMPLE PROBLEM 17.5

### ■ DNA Replication

In an original DNA strand, there is a base sequence —A—G—T—. What nucleotides are placed in a growing DNA daughter strand complementary to this sequence?

#### SOLUTION

Only one possible nucleotide can pair with each base in the original sequence. Thymine will pair only with adenine, cytosine with guanine, and adenine with thymine to give the complementary base sequence —T—C—A—.

#### STUDY CHECK

What is the complementary section of DNA for the base sequence —G—C—A—A—T—C—?

---

## QUESTIONS AND PROBLEMS

### DNA Double Helix

**17.15** How are the two strands of nucleic acid in DNA held together?

**17.16** What is meant by complementary base pairing?

**17.17** Write the base sequence in a complementary DNA segment if the original has the following base sequence:
  **a.** —A—A—A—A—A—A—
  **b.** —G—G—G—G—G—G—

  **c.** —A—G—T—C—C—A—G—G—T—
  **d.** —C—T—G—T—A—T—A—C—G—T—T—A—

**17.18** Write the base sequence in a complementary DNA segment if the original has the following base sequence:
  **a.** —T—T—T—T—T—T—
  **b.** —C—C—C—C—C—C—C—C—
  **c.** —A—T—G—G—C—A—
  **d.** —A—T—A—T—G—C—G—C—T—A—A—A—

# 17.4 RNA AND THE GENETIC CODE

Ribonucleic acid, RNA, which makes up most of the nucleic acid found in the cell, is involved with transmitting the genetic information needed to operate the cell. Similar to DNA, RNA molecules are unbranched polymers of nucleotides. However, there are several important differences.

1. The sugar in RNA is ribose rather than the deoxyribose found in DNA.
2. The base uracil replaces thymine.
3. RNA molecules are single, not double stranded.
4. RNA molecules are much smaller than DNA molecules.

## Types of RNA

There are three major types of RNA in the cells: *messenger RNA*, *ribosomal RNA*, and *transfer RNA*. Ribosomal RNA (**rRNA**), the most abundant type of RNA, is combined with proteins in the ribosomes. Ribosomes, which are the sites for protein synthesis, consist of two subunits, a large subunit and a small subunit. (See Figure 17.9.) Cells that synthesize large numbers of proteins have thousands of ribosomes.

Messenger RNA (**mRNA**) carries genetic information from the DNA in the nucleus to the ribosomes in the cytoplasm for protein synthesis. Each gene, a segment of DNA, produces a separate mRNA molecule when a certain protein is needed in the cell, but then the mRNA is broken down quickly. The size of a mRNA depends on the number of nucleotides in that particular gene.

Transfer RNA (**tRNA**), the smallest of the RNA molecules, interprets the genetic information in mRNA and brings specific amino acids to the ribosome for protein synthesis. Only the tRNAs can translate the genetic information into amino acids for proteins. There are one or more different tRNAs for each of the 20 amino acids. The structures of the transfer RNAs are similar, consisting of 70–90 nucleotides. Hydrogen bonds between some of the complementary bases in the chain produce loops that give some double-stranded regions.

Although the structure of tRNA is complex, we draw tRNA as a cloverleaf to illustrate its features. All tRNA molecules have a 3′ end with the nucleotide sequence ACC, which is known as the *acceptor stem*. An enzyme attaches an amino acid by forming an ester bond with the free —OH at the end of the acceptor stem. Each tRNA contains an

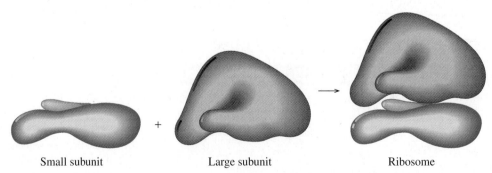

Small subunit　　　　Large subunit　　　　Ribosome

**FIGURE 17.9** A typical ribosome consists of a small subunit and a large subunit.
**Q** Why would there be many thousands of ribosomes in a cell?

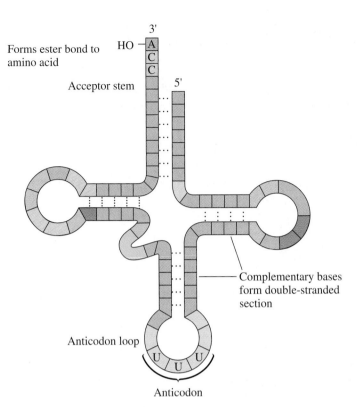

Forms ester bond to amino acid

Acceptor stem

Anticodon loop

Anticodon

**FIGURE 17.10** A typical tRNA molecule has an acceptor stem at the 3′ end of the nucleic acid, where an amino acid attaches, and an anticodon loop that complements three bases on mRNA.

**Q** Why will different tRNAs have different bases in the anticodon loop?

Complementary bases form double-stranded section

**anticodon**, which is a series of three bases that complements three bases on a mRNA. (See Figure 17.10.) Table 17.2 summarizes the three types of RNA.

**TABLE 17.2** Types of RNA Molecules

| Type | Abbreviation | Percentage of Total RNA | Function in the Cell |
|---|---|---|---|
| Ribosomal RNA | rRNA | 75 | Major component of the ribosomes |
| Messenger RNA | mRNA | 5–10 | Carries information for protein synthesis from the DNA in the nucleus to the ribosomes |
| Transfer RNA | tRNA | 10–15 | Brings amino acids to the ribosomes for protein synthesis |

**SAMPLE PROBLEM    17.6**

**■ Types of RNA**

What is the function of mRNA in a cell?

**SOLUTION**

mRNA carries the instructions for the synthesis of a protein from the DNA in the nucleus to the ribosomes.

**STUDY CHECK**

What is the function of tRNA in a cell?

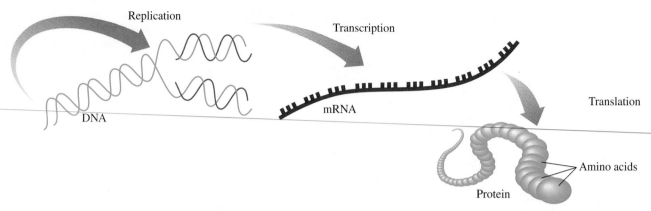

**FIGURE 17.11** The genetic information in DNA is replicated in cell division and is used to produce messenger RNAs that code for the amino acids needed for protein synthesis.
**Q** What is the difference between transcription and translation?

## RNA and Protein Synthesis

We now look at the overall process involved in transferring genetic information encoded in the DNA to the production of proteins. In the nucleus, genetic information for the synthesis of a protein is copied from a gene in DNA to make messenger RNA (mRNA), a process called **transcription**. The mRNA molecules move out of the nucleus into the cytoplasm where they combine with ribosomes. Then in a process called **translation**, tRNA molecules convert the information in the mRNA into amino acids, which are placed in the proper sequence to synthesize a protein. (See Figure 17.11.)

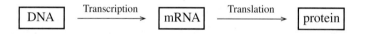

**WEB TUTORIAL**
Transcription
Overview of Protein Synthesis

## Transcription: Synthesis of mRNA

Transcription begins when the section of a DNA that contains the gene to be copied unwinds. Within this unwound DNA, a RNA polymerase enzyme uses one of the strands as a template to synthesize a mRNA. Just as in DNA synthesis, C pairs with G, T pairs with A, but in mRNA, U (not T) pairs with A. The RNA polymerase moves along the DNA template strand, forming bonds between the bases. When the RNA polymerase reaches the termination point, transcription ends and the new mRNA is released. The unwound section of the DNA returns to its double helix structure. (See Figure 17.12.)

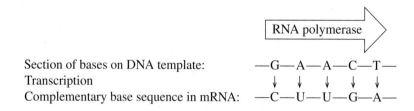

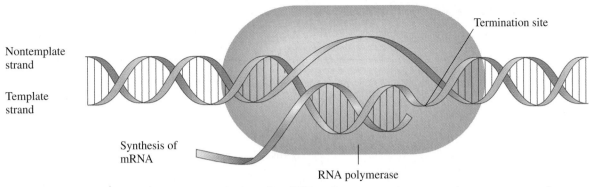

**FIGURE 17.12** DNA undergoes transcription when RNA polymerase makes a complementary copy of a gene using only one of the DNA strands as the template.
**Q** Why is the mRNA synthesized on only one DNA strand?

---

### SAMPLE PROBLEM 17.7

■ **RNA Synthesis**

The sequence of bases in a part of the DNA template for mRNA is —C—G—A—T—C—A—. What is the corresponding mRNA produced?

**SOLUTION**

To form the mRNA, the bases in the DNA template are paired with their complementary bases: G with C, C with G, T with A, and A with U.

Portion of DNA template:  —C—G—A—T—C—A—

Complementary bases in mRNA:  —G—C—U—A—G—U—

**STUDY CHECK**

What is the DNA template that codes for a mRNA segment that has the sequence —G—G—G—U—U—U—A—A—A—?

---

## The Genetic Code

The **genetic code** consists of a series of three nucleotides (triplet) in mRNA called **codons** that specify the amino acids and their sequence in the protein. Early work on protein synthesis showed that repeating triplets of uracil (UUU) produced a polypeptide that contained only phenylalanine. Therefore, a sequence of —UUU—UUU—UUU— codes for three phenylalanines.

Codons in mRNA          —UUU—UUU—UUU—
Translation
Amino acid sequence     — Phe — Phe — Phe—

Codons have been determined for all 20 amino acids. A total of 64 codons are possible from the triplet combinations of A, G, C, and U. Three of these, UGA, UAA, and UAG, are stop signals that code for the termination of protein synthesis. All the other three-base codons shown in Table 17.3 specify amino acids. Thus one amino acid can have several codons. For example, glycine has four codons: GGU, GGC, GGA, and GGG. The triplet AUG has two roles in protein synthesis. At the beginning of a mRNA, the codon AUG signals the start of protein synthesis. In the middle of a series of codons, the AUG codon specifies the amino acid methionine.

**TABLE 17.3** mRNA Codons: The Genetic Code for Amino Acids

| First Letter | Second Letter | | | | Third Letter |
|---|---|---|---|---|---|
| | U | C | A | G | |
| U | UUU ⎱ Phe<br>UUC ⎰<br>UUA ⎱ Leu<br>UUG ⎰ | UCU ⎱<br>UCC ⎰ Ser<br>UCA ⎰<br>UCG ⎰ | UAU ⎱ Tyr<br>UAC ⎰<br>UAA STOP<br>UAG STOP | UGU ⎱ Cys<br>UGC ⎰<br>UGA STOP<br>UGG Trp | U<br>C<br>A<br>G |
| C | CUU ⎱<br>CUC ⎰ Leu<br>CUA ⎰<br>CUG ⎰ | CCU ⎱<br>CCC ⎰ Pro<br>CCA ⎰<br>CCG ⎰ | CAU ⎱ His<br>CAC ⎰<br>CAA ⎱ Gln<br>CAG ⎰ | CGU ⎱<br>CGC ⎰ Arg<br>CGA ⎰<br>CGG ⎰ | U<br>C<br>A<br>G |
| A | AUU ⎱<br>AUC ⎰ Ile<br>AUA ⎰<br>ᵃAUG Met/start | ACU ⎱<br>ACC ⎰ Thr<br>ACA ⎰<br>ACG ⎰ | AAU ⎱ Asn<br>AAC ⎰<br>AAA ⎱ Lys<br>AAG ⎰ | AGU ⎱ Ser<br>AGC ⎰<br>AGA ⎱ Arg<br>AGG ⎰ | U<br>C<br>A<br>G |
| G | GUU ⎱<br>GUC ⎰ Val<br>GUA ⎰<br>GUG ⎰ | GCU ⎱<br>GCC ⎰ Ala<br>GCA ⎰<br>GCG ⎰ | GAU ⎱ Asp<br>GAC ⎰<br>GAA ⎱ Glu<br>GAG ⎰ | GGU ⎱<br>GGC ⎰ Gly<br>GGA ⎰<br>GGG ⎰ | U<br>C<br>A<br>G |

ᵃ Codon that signals the start of a peptide chain. STOP codons signal the end of a peptide chain.

---

**SAMPLE PROBLEM 17.8**

**■ Codons**

What is the sequence of amino acids coded by the following codons in mRNA?

—GUC—AGC—CCA—

**SOLUTION**

According to Table 17.3, GUC codes for valine, AGC for serine, and CCA for proline. The sequence of amino acids is Val-Ser-Pro.

**STUDY CHECK**

The codon UGA does not code for an amino acid. What is its function in mRNA?

---

## QUESTIONS AND PROBLEMS

### RNA and the Genetic Code

**17.19** What are the three different types of RNA?

**17.20** What are the functions of each type of RNA?

**17.21** What is meant by the term "transcription"?

**17.22** What bases in mRNA are used to complement the bases A, T, G, and C in DNA?

**17.23** Write the segment of mRNA produced from the following section of a DNA template:

—C—C—G—A—A—G—G—T—T—C—A—C—

**17.24** Write the segment of mRNA produced from the following section of a DNA template:

—T—A—C—G—G—C—A—A—G—C—T—A—

**17.25** What is a codon?

**17.26** What is the genetic code?

**17.27** What amino acid is coded for by each codon?
  **a.** CUU    **b.** UCA
  **c.** GGU    **d.** AGG

**17.28** What amino acid is coded for by each codon?
  **a.** AAA    **b.** GUC
  **c.** CGG    **d.** GCA

**17.29** When does the codon AUG signal the start of a protein, and when does it code for the amino acid methionine?

**17.30** The codons UGA, UAA, and UAG do not code for amino acids. What is their role as codons in mRNA?

# 17.5 PROTEIN SYNTHESIS

Once the mRNA is synthesized, it migrates out of the nucleus into the cytoplasm to the ribosomes. In the *translation* process, tRNA molecules, amino acids, and enzymes convert the codons on mRNA to make a protein.

## Activation of tRNA

Each tRNA molecule contains a loop called an *anticodon*, which is a triplet of bases that complements a codon in a mRNA. (See Figure 17.13.) A tRNA is activated for protein synthesis by an enzyme called tRNA synthetase. The synthetase uses the anticodon to attach the correct amino acid to the acceptor stem of the tRNA. Each synthetase checks the attachment of an amino acid to a tRNA and hydrolyzes any incorrect combinations.

## Initiation of Protein Synthesis

Protein synthesis begins when a mRNA combines with a ribosome. The first codon in a mRNA is a *start* codon, AUG. Therefore, a tRNA with an anticodon of UAC and the amino acid methionine forms hydrogen bonds with the AUG codon. The tRNA that carries the second amino acid bonds to the second codon on the mRNA. With two amino acids close together, a peptide bond forms. (See Figure 17.14.) Then the first tRNA detaches from the ribosome and the ribosome shifts to the next available codon on the mRNA, a process called *translocation*. As the ribosome moves along the mRNA a peptide bond joins each new amino acid to the growing polypeptide chain. Sometimes

## LEARNING GOAL

Describe the process of protein synthesis from mRNA.

**WEB TUTORIAL**
Translation

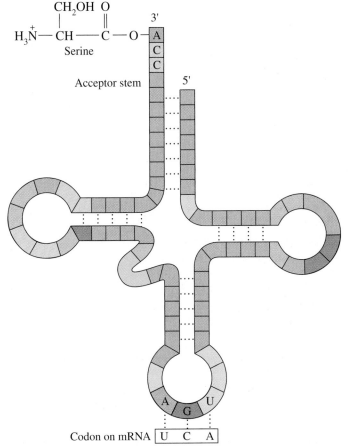

**FIGURE 17.13** An activated tRNA with anticodon AGU bonds to serine at the acceptor stem.
**Q** What is the codon for serine for this tRNA?

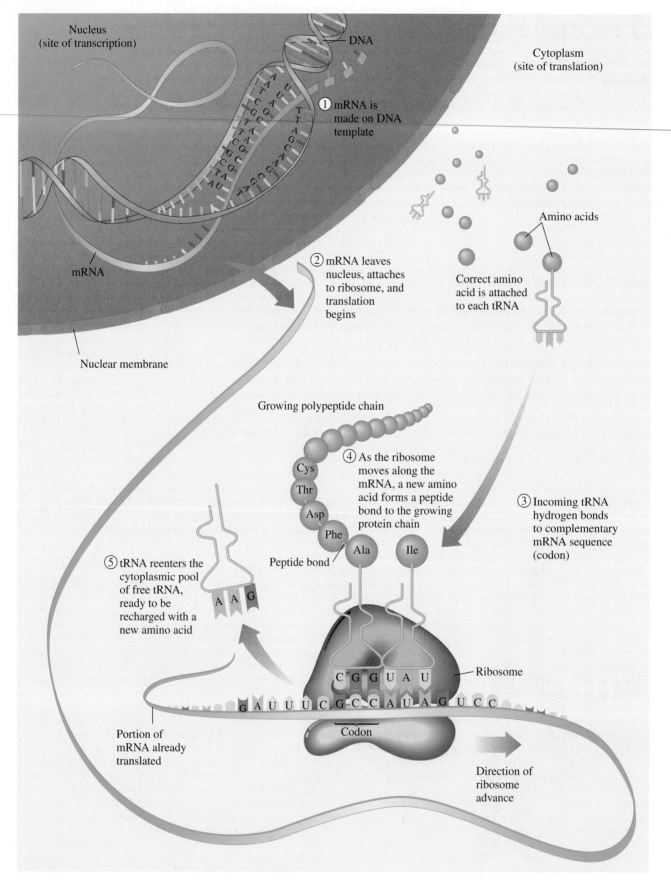

**FIGURE 17.14** In the translation process, the mRNA synthesized by transcription attaches to a ribosome and tRNAs pick up their amino acids and place them in a growing peptide chain.

**Q** How is the correct amino acid placed in the peptide chain?

several ribosomes, called a polysome, translate the same strand of mRNA at the same time to produce several copies of the peptide chain at the same time.

## Termination

Eventually, the ribosome encounters a stop codon, and protein synthesis ends. The polypeptide chain is released from the ribosome. The initial amino acid methionine is often removed from the beginning of the polypeptide chain. Now the side chains of the amino acids in the polypeptide form cross-links to give the tertiary structure of a biologically active protein.

### SAMPLE PROBLEM 17.9

■ **Protein Synthesis: Translation**

What order of amino acids would you expect in a peptide for the mRNA sequence of —UCA—AAA—GCC—CUU—?

**SOLUTION**

Each of the codons specifies a particular amino acid. Using Table 17.3, we write a peptide with the following amino acid sequence:

mRNA codons:  —UCA—AAA—GCC—CUU—
            ↓    ↓    ↓    ↓
Amino acid sequence:  Ser — Lys — Ala — Leu

**STUDY CHECK**

Where would protein synthesis stop in the following series of bases in a mRNA?

—GGG—AGC—AGU—UAG—GUU—

---

### Health Note

#### Many Antibiotics Inhibit Protein Synthesis

Several antibiotics stop bacterial infections by interfering with the synthesis of proteins needed by the bacteria. Some antibiotics act only on bacterial cells by binding to the ribosomes in bacteria but do not act on human cells. A description of some of these antibiotics is given in Table 17.4.

**TABLE 17.4** Antibiotics That Inhibit Protein Synthesis in Bacterial Cells

| Antibiotic | Effect on Ribosomes to Inhibit Protein Synthesis |
|---|---|
| Chloramphenicol | Inhibits peptide bond formation and prevents the binding of tRNA |
| Erythromycin | Inhibits peptide chain growth by preventing the translocation of the ribosome along the mRNA |
| Puromycin | Causes release of an incomplete protein by ending the growth of the polypeptide early |
| Streptomycin | Prevents the proper attachment of tRNAs |
| Tetracycline | Prevents the binding of tRNAs |

---

## QUESTIONS AND PROBLEMS

### Protein Synthesis: Translation

**17.31** What is the difference between a *codon* and an *anticodon*?

**17.32** Why are there at least 20 different tRNAs?

**17.33** What amino acid sequence would you expect from each of the following mRNA segments?
  a. —AAA—AAA—AAA—
  b. —UUU—CCC—UUU—CCC—
  c. —UAC—GGG—AGA—UGU—

**17.34** What amino acid sequence would you expect from each of the following mRNA segments?
  a. —AAA—CCC—UUG—GCC—
  b. —CCU—CGA—AGC—CCA—UGA—
  c. —AUG—CAC—AAA—GAA—GUA—CUU—

**17.35** How is a peptide chain extended?

**17.36** What is meant by "translocation"?

**17.37** The following portion of DNA is in the template DNA strand:
  —GCT—TTT—CAA—AAA—
  a. What is the corresponding mRNA section?
  b. What are the anticodons of the tRNAs?
  c. What amino acids will be placed in the peptide chain?

**17.38** The following portion of DNA is in the template DNA strand:
  —TGT—GGG—GTT—ATT—
  a. What is the corresponding mRNA section?
  b. What are the anticodons of the tRNAs?
  c. What amino acids will be placed in the peptide chain?

# 17.6 GENETIC MUTATIONS

A **mutation** is a change in the DNA nucleotide sequence that alters the sequence of amino acids, which may alter the structure and function of a protein in a cell. Mutations may result from X-rays, overexposure to sun (ultraviolet, or UV, light), chemicals called mutagens, and possibly some viruses. If a change in DNA occurs in a somatic cell (a cell other than a reproductive cell) the altered DNA will be limited to that cell and its daughter cells. If there is uncontrolled growth, the mutation could lead to cancer. If the mutation occurs in germ cell DNA (egg or sperm), then all the DNA produced in a new individual will contain the same genetic change. If a mutation alters the formation of important structural proteins or enzymes, the new cells may not survive or the person may exhibit a genetic disease.

## Types of Mutations

Consider a triplet of bases CCG in the coding strand of DNA, which produces the codon GGC in mRNA. At the ribosome, tRNA would place the amino acid glycine in the peptide chain. (See Figure 17.15a.) Now, suppose that T replaces the first C in the DNA triplet, which gives TCG as the triplet. Then the codon produced in the mRNA is AGC, which brings the amino acid serine to the peptide chain. The replacement of one base in the coding strand of DNA with another is called a **substitution** mutation. The change in the codon can lead to the insertion of a different amino acid at that point in the polypeptide. Substitution is the most common way in which mutations occur. (See Figure 17.15b.)

In a **frameshift mutation**, a base is added to or deleted from the normal order of bases in the coding strand of DNA. Suppose that an A is deleted from the triplet AAA, giving a new triplet of AAC. The next triplet becomes CGA rather than CCG and so on. All the triplets shift over by one base, which changes all the codons that follow and leads to a different sequence of amino acids from that point. Figure 17.15c illustrates a frameshift mutation by deletion.

## Effect of Mutations

When a mutation causes a change in the amino acid sequence, the structure of the resulting protein can be altered severely and it may lose biological activity. If the protein is an enzyme, it may no longer bind to its substrate or react with the substrate at the active site. When an altered enzyme cannot catalyze a reaction, certain substances may accumulate until they act as poisons in the cell or substances vital to survival may not be synthesized. If a defective enzyme occurs in a major metabolic pathway or in the building of a cell membrane, the mutation can be lethal. When a protein deficiency is hereditary the condition is called a **genetic disease**.

X-rays,
UV sunlight,
mutagens,
viruses
DNA $\longrightarrow$ alteration of $\longrightarrow$ defective $\longrightarrow$ genetic disease (germ cells)
DNA protein or cancer (somatic cells)

---

**SAMPLE PROBLEM 17.10**

■ **Mutations**

A mRNA has the sequence of codons —CCC—AGA—GCC—. If a base substitution in the DNA changes the mRNA codon of AGA to GGA, how is the amino acid sequence affected in the resulting protein?

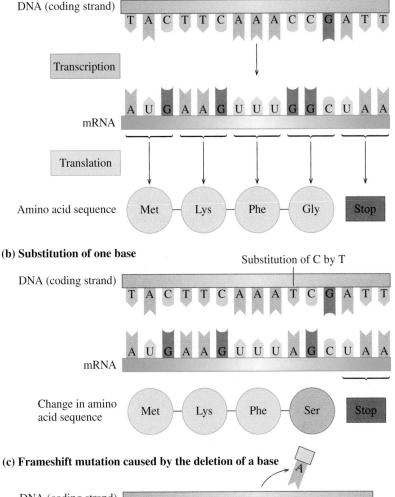

**(a) Normal DNA and protein synthesis**

DNA (coding strand)

T A C T T C A A A C C G A T T

Transcription

mRNA

A U G A A G U U U G G C U A A

Translation

Amino acid sequence

Met — Lys — Phe — Gly — Stop

**(b) Substitution of one base**

Substitution of C by T

DNA (coding strand)

T A C T T C A A A T C G A T T

mRNA

A U G A A G U U U A G C U A A

Change in amino acid sequence

Met — Lys — Phe — Ser — Stop

**(c) Frameshift mutation caused by the deletion of a base**

DNA (coding strand)

A

T A C T T C A A C C G A T T

mRNA

A U G A A G U U G G C U A A .....

Changes in amino acid sequence

Met — Lys — Leu — Ala .....

**FIGURE 17.15** An alteration in the DNA coding strand (template) produces a change in the sequence of amino acids in the protein, which may lead to a mutation. **(a)** A normal DNA leads to the correct amino acid order in a protein. **(b)** The substitution of a base in DNA leads to a change in the mRNA codon and a change in the amino acid. **(c)** The deletion of a base causes a frameshift mutation, which changes the subsequent amino acid order.

**Q** When would a substitution mutation cause protein synthesis to stop?

*Explore Your World*

## A Model for DNA Replication and Mutation

1. Cut out 16 rectangular pieces of paper. Using 8 rectangular pieces for strand 1, write each of the following nucleotide symbols twice: A═, T═, G≡, and C≡.

2. Using the other 8 rectangular pieces for strand 2, write each of the following nucleotide symbols twice: ═A, ═T, ≡G, and ≡C.

3. Place the pieces for strand 1 in random order.

4. Complete the DNA segment in part 3, by selecting the correct complementary bases of strand 2.

5. Using the nucleotides from parts 1 and 2, make a DNA segment with strand 1 of

   —A—T—T—G—C—C—

6. In the original, change the G to an A. How does this change the complementary strand? If the original strand 1 is the template, how would this change the mRNA codons and lead to a mutation?

**SOLUTION**

The initial mRNA sequence of —CCC—AGA—GCC— codes for the amino acids proline, arginine, and alanine. When the mutation occurs, the new sequence of mRNA codons —CCC—GGA—GGG— are for proline, glycine, and alanine. The amino acid arginine is replaced by glycine.

**STUDY CHECK**

How might the protein made from this mRNA be affected by this mutation?

**FIGURE 17.16** This peacock with albinism does not produce the melanin needed to make the bright colors of its feathers.

**Q** Why are traits such as albinism related to the gene?

# Genetic Diseases

A genetic disease is the result of a defective enzyme caused by a mutation in its genetic code. For example, phenylketonuria, PKU, results when DNA cannot direct the synthesis of the enzyme phenylalanine hydroxylase, required for the conversion of phenylalanine to tyrosine. In an attempt to break down the phenylalanine, other enzymes in the cells convert it to phenylpyruvate. The accumulation of phenylalanine and phenylpyruvate in the blood can lead to severe brain damage and mental retardation. If PKU is detected in a newborn baby, a diet is prescribed that eliminates all the foods that contain phenylalanine. Preventing the buildup of the phenylpyruvate ensures normal growth and development.

The amino acid tyrosine is needed in the formation of melanin, the pigment that gives the color to our skin and hair. If the enzyme that converts tyrosine to melanin is defective, no melanin is produced and a genetic disease known as albinism results. Persons and animals with no melanin have no skin or hair pigment. (See Figure 17.16.) Table 17.5 lists some other common genetic diseases and the type of metabolism or area affected.

$$\underset{\text{Phenylalanine}}{\text{C}_6\text{H}_5-\text{CH}_2-\underset{\overset{\displaystyle|}{\text{NH}_3^+}}{\text{CH}}-\text{COO}^-} \longrightarrow \underset{\text{Phenylpyruvate}}{\text{C}_6\text{H}_5-\text{CH}_2-\underset{\overset{\displaystyle\|}{\text{O}}}{\text{C}}-\text{COO}^-} \searrow \text{Phenylketonuria (PKU)}$$

Phenylalanine hydroxylase

$$\underset{\text{Tyrosine}}{\text{HO}-\text{C}_6\text{H}_4-\text{CH}_2-\underset{\overset{\displaystyle|}{\text{NH}_3^+}}{\text{CH}}-\text{COO}^-} \xrightarrow{\quad\times\quad} \text{Melanin (pigments)} \searrow \text{Albinism}$$

**TABLE 17.5** Some Genetic Diseases

| Genetic Disease | Result |
|---|---|
| Galactosemia | The transferase enzyme required for the metabolism of galactose-1-phosphate is absent. Accumulation of Gal-1-P leads to cataracts and mental retardation. |
| Cystic fibrosis | The most common inherited disease. Thick mucus secretions make breathing difficult and block pancreatic function. |
| Down syndrome | The leading cause of mental retardation, occurring in about 1 of every 800 live births. Mental and physical problems including heart and eye defects are the result of the formation of three chromosomes, usually number 21, instead of a pair of chromosomes. |
| Familial hypercholesterolemia | A mutation of a gene on chromosome 19 results in high cholesterol levels that lead to early coronary heart disease in people 30–40 years old. |
| Muscular dystrophy (Duchenne) | One of 10 forms of MD. A mutation in the X chromosome results in the low or abnormal production of *dystrophin*. This muscle-destroying disease appears at about age 5, with death by age 20, and it occurs in about 1 of 10 000 males. |
| Huntington's disease (HD) | Appearing in middle age, HD affects the nervous system, leading to total physical impairment. It is the result of a mutation in a gene on chromosome 4, which can now be mapped to test people in families with HD. |
| Sickle-cell anemia | Defective hemoglobin from a mutation in a gene on chromosome 11 decreases the oxygen-carrying ability of red blood cells, which take on a sickled shape, causing anemia and plugged capillaries from red blood cell aggregation. |
| Hemophilia | One or more defective blood-clotting factors lead to poor coagulation, excessive bleeding, and internal hemorrhages. |
| Tay-Sachs disease | Hexosaminidase A is defective, causing an accumulation of gangliosides, resulting in mental retardation, loss of motor control, and early death. |

## QUESTIONS AND PROBLEMS

### Genetic Mutations

**17.39** What is a substitution mutation?

**17.40** How does a substitution mutation in the genetic code for an enzyme affect the order of amino acids in that protein?

**17.41** What is the effect of a frameshift mutation on the amino acid sequence in the polypeptide?

**17.42** How can a mutation decrease the activity of a protein?

**17.43** How is protein synthesis affected if the normal base sequence TTT in the DNA template is changed to TTC?

**17.44** How is protein synthesis affected if the normal base sequence CCC is changed to ACC?

**17.45** Consider the following portion of mRNA produced by the normal order of DNA nucleotides:

$$-ACA-UCA-CGG-GUA-$$

　**a.** What is the amino acid order produced for normal DNA?
　**b.** What is the amino acid order if a mutation changes UCA to ACA?
　**c.** What is the amino acid order if a mutation changes CGG to GGG?
　**d.** What happens to protein synthesis if a mutation changes UCA to UAA?
　**e.** What happens if a G is added to the beginning of a chain?
　**f.** What happens if the A is removed from the beginning of a chain?

**17.46** Consider the following portion of mRNA produced by the normal order of DNA nucleotides:

$$-CUU-AAA-CGA-GUU-$$

　**a.** What is the amino acid order produced for normal DNA?
　**b.** What is the amino acid order if a mutation changes CUU to CCU?
　**c.** What is the amino acid order if a mutation changes CGA to AGA?
　**d.** What happens to protein synthesis if a mutation changes AAA to UAA?

**17.47 a.** A base substitution changes a codon for an enzyme from GCC to GCA. Why is there no change in the amino acid order in the protein?
　**b.** In sickle-cell anemia, a base substitution in hemoglobin replaces glutamic acid (a polar amino acid) with valine. Why does the replacement of one amino acid cause such a drastic change in biological function?

**17.48 a.** A base substitution for an enzyme replaces leucine (a nonpolar amino acid) with alanine. Why does this change in amino acids have little effect on the biological activity of the enzyme?
　**b.** A base substitution replaces cytosine in the codon UCA with adenine. How would this substitution affect the amino acids in the protein?

# 17.7 VIRUSES

**Viruses** are small particles of 3–200 genes that cannot replicate without a host cell. A typical virus contains a nucleic acid, DNA or RNA, but not both, inside a protein coat. It does not have the necessary material such as nucleotides and enzymes to make proteins and grow. The only way a virus can replicate is to invade a host cell and take over the materials necessary for protein synthesis and growth. Some infections caused by viruses invading human cells are listed in Table 17.6. There are also viruses that attack bacteria, plants, and animals.

A viral infection begins when an enzyme in the protein coat makes a hole in the host cell, allowing the nucleic acids to enter and mix with the materials in the host cell. (See Figure 17.17.) If the virus contains DNA, the host cell begins to replicate the viral DNA in the same way it would replicate normal DNA. Viral DNA produces viral RNA, which proceeds to make the proteins for more coats. So many virus particles are synthesized that the cell bursts and releases new viruses to infect more cells.

Vaccines are inactive forms of viruses that boost the immune response by causing the body to produce antibodies to the virus. Several childhood diseases such as polio, mumps, chicken pox, and measles can be prevented through the use of vaccines.

### Reverse Transcription

A virus that contains RNA as its genetic material is a **retrovirus**. Once inside the host cell, it must first make viral DNA using a process known as reverse transcription. A retrovirus contains a polymerase enzyme called *reverse transcriptase* that uses the viral RNA template to synthesize complementary strands of DNA. Once produced, the DNA

## LEARNING GOAL

Describe the methods by which a virus infects a cell.

**TABLE 17.6** Some Diseases Caused by Viral Infection

| Disease | Virus |
| --- | --- |
| Common cold | Coronavirus (over 100 types) |
| Influenza | Orthomyxovirus |
| Warts | Papovavirus |
| Herpes | Herpesvirus |
| Leukemia, cancers, AIDS | Retroviruses |
| Hepatitis | Hepatitis A virus (HAV), hepatitis B virus (HBV), hepatitis C (HCV) |
| Mumps | Paramyxovirus |
| Epstein–Barr | Epstein–Barr virus (EBV) |

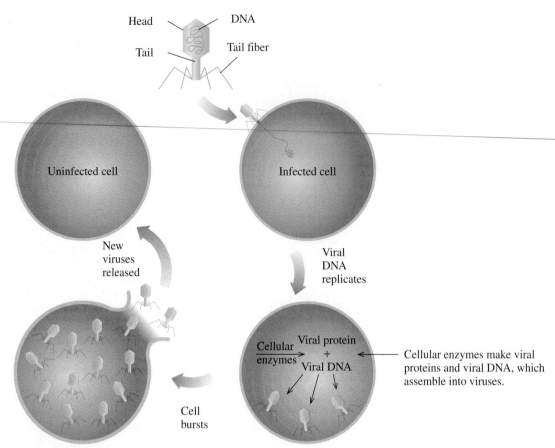

**FIGURE 17.17** After a virus attaches to the host cell, it injects its viral DNA and uses the host cell's amino acids to synthesize viral protein. It uses the host cell's nucleic acids, enzymes, and ribosomes to make viral RNA. When the cell bursts, the new viruses are released to infect other cells.
**Q** Why does a virus need a host cell for replication?

strands form double-stranded DNA using the nucleotides present in the host cell. This newly formed viral DNA, called a *provirus*, joins the DNA of the host cell. A protease produces a protein coat to form a viral particle that leaves the cell to infect other cells. (See Figure 17.18.)

**FIGURE 17.18** After a retrovirus injects its viral RNA into a cell, it forms a DNA strand by reverse transcription. The DNA forms a double-stranded DNA called a provirus, which joins the host cell DNA. When the cell replicates, the provirus produces the viral RNA needed to produce more virus particles.
**Q** What is reverse transcription?

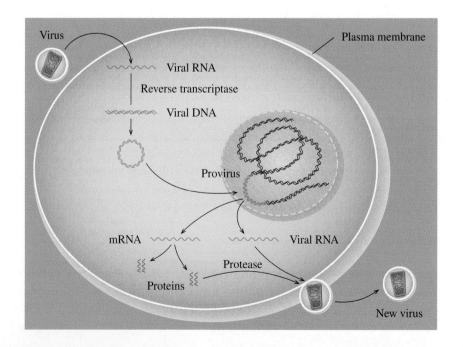

# AIDS

In the early 1980s, a disease called AIDS (acquired immune deficiency syndrome) began to claim an alarming number of lives. An HIV-1 virus (human immunodeficiency virus type 1) is now known to be the AIDS-causing agent. (See Figure 17.19.) HIV is a retrovirus that infects and destroys T4 lymphocyte cells, which are involved in the immune response. After the HIV-1 virus binds to receptors on the surface of a T4 cell, the virus injects viral RNA into the host cell. As a retrovirus, the genes of the viral RNA direct the formation of viral DNA. The gradual depletion of T4 cells reduces the ability of the immune system to destroy harmful organisms. The AIDS syndrome is characterized by opportunistic infections such as *Pneumocystis carinii*, which causes pneumonia, and *Kaposi's sarcoma*, a skin cancer.

Treatment for AIDS is based on attacking the HIV-1 at different points in its life cycle, including reverse transcription and protein synthesis. Nucleoside analogs mimic the structures of the nucleosides used for DNA synthesis. For example, the drug AZT (azidothymine) is similar to thymine, and ddI (dideoxyinosine) is similar to guanosine. Two other drugs include dideoxycytidine (ddC) and didehydro-3'-deoxythymidine (d4T). Such compounds are found in the "cocktails" that are providing extended remission of HIV infections. When a nucleoside analog is incorporated into viral DNA, the lack of a hydroxyl group on the 3'-carbon in the sugar prevents the formation of the sugar–phosphate bonds and stops the replication of the virus.

WEB TUTORIAL
HIV Reproductive Cycle

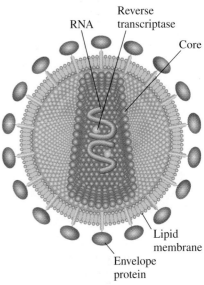

**FIGURE 17.19** The HIV virus that causes AIDS destroys the immune system in the body.
**Q** Is the HIV virus a DNA virus or a RNA retrovirus?

AZT
Azidothymine

Dideoxyinosine (ddI)

Dideoxycytidine (ddC)

Didehydro-3'-deoxythymidine (d4T)

The newest and most powerful anti-HIV drugs are the protease inhibitors such as saquinavir (Invirase), indinavir, and ritonavir. The inhibition of protease prevents the synthesis of proteins needed to make more copies of the virus. Researchers are not yet certain how long protease inhibitors will be beneficial for a person with AIDS, but they are encouraged by the current studies.

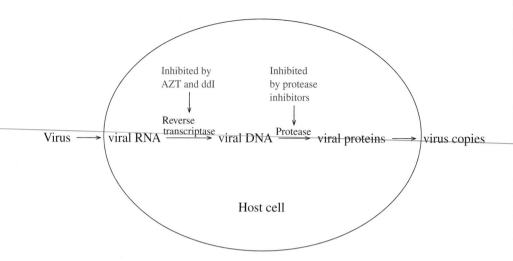

## Cancer

In an adult body, many cells do not continue to reproduce. When cells in the body begin to grow and multiply without control, they invade neighboring cells and appear as a tumor or growth (neoplasm). When tumors interfere with normal functions of the body, they are cancerous. If they are limited, they are benign. Cancer can be caused by chemical and environmental substances, by radiation, or by oncogenic viruses.

Some reports estimate that 70–80% of all human cancers are initiated by chemical and environmental substances. A carcinogen is any substance that increases the chance of inducing a tumor. Known carcinogens include aniline dyes, cigarette smoke, and asbestos. More than 90% of all persons with lung cancer are smokers. A carcinogen causes cancer by reacting with molecules in a cell, probably DNA, and altering the growth of that cell. Some known carcinogens are listed in Table 17.7.

Radiant energy from sunlight or medical radiation is another type of environmental factor. Skin cancer has become one of the most prevalent forms of cancer. It appears that DNA damage in the exposed areas of the skin causes mutations. The cells lose their ability to control protein synthesis, and uncontrolled cell division leads to cancer. The incidence of malignant melanoma, one of the most serious skin cancers, has been rapidly increasing. Some possible factors for this increase may be the popularity of suntanning as well as the reduction of the ozone layer, which absorbs much of the harmful radiation from sunlight.

Oncogenic viruses cause cancer when cells are infected. Several viruses associated with human cancers are listed in Table 17.8. Some cancers such as retinoblastoma and breast cancer appear to occur more frequently within families. There is some indication that a missing or defective gene may be responsible.

**TABLE 17.7** Some Chemical and Environmental Carcinogens

| Carcinogen | Tumor Site |
| --- | --- |
| Asbestos | Lungs, respiratory tract |
| Arsenic | Skin, lungs |
| Cadmium | Prostate, kidneys |
| Chromium | Lungs |
| Nickel | Lungs, sinuses |
| Aflatoxin | Liver |
| Nitrites | Stomach |
| Aniline dyes | Bladder |
| Vinyl chloride | Liver |

**TABLE 17.8** Human Cancers Caused by Oncogenic Viruses

| Virus | Disease |
| --- | --- |
| **RNA viruses** | |
| Human T-cell lymphotropic virus-type 1 (HTLV-1) | Leukemia |
| **DNA viruses** | |
| Epstein–Barr virus (EBV) | Epstein–Barr |
| | Burkitt's lymphoma (cancer of white blood B cells) |
| | Nasopharyngeal carcinoma |
| | Hodgkin's disease |
| Hepatitis B virus (HBV) | Liver cancer |
| Herpes simplex virus (type 2) | Cervical and uterine cancer |
| Papilloma virus | Cervical and colon cancer, genital warts |

## SAMPLE PROBLEM 17.11

■ **Viruses**

Why are viruses unable to replicate on their own?

SOLUTION

Viruses contain only packets of DNA or RNA, but not the necessary replication machinery that includes enzymes and nucleosides.

STUDY CHECK

How do protease inhibitors affect the life cycle of the HIV-1 virus?

## QUESTIONS AND PROBLEMS

### Viruses

**17.49** What type of genetic information is found in a virus?

**17.50** Why do viruses need to invade a host cell?

**17.51** A virus contains viral RNA.
   **a.** Why would reverse transcription be used in the life cycle of this type of virus?
   **b.** What is the name of this type of virus?

**17.52** What is the purpose of a vaccine?

**17.53** How do nucleoside analogs disrupt the life cycle of the HIV-1 virus?

**17.54** How do protease inhibitors disrupt the life cycle of the HIV-1 virus?

## CONCEPT MAP

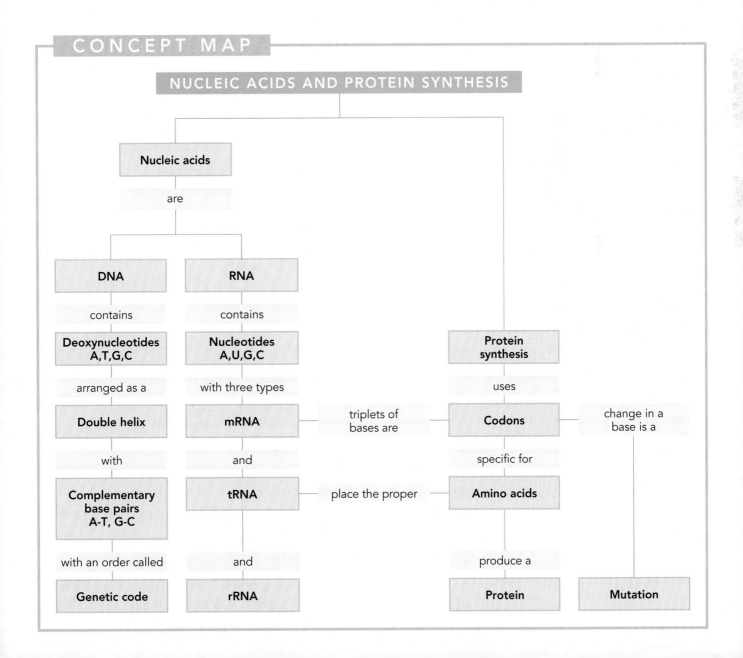

# CHAPTER REVIEW

## 17.1 Components of Nucleic Acids

**Learning Goal:** Describe the bases and ribose sugars that make up the nucleic acids DNA and RNA.

Nucleic acids, deoxyribonucleic acid (DNA) and ribonucleic acid (RNA), are polymers of nucleotides. A nucleoside is a combination of a pentose sugar and a base. A nucleotide is composed of three parts: a base, a sugar, and a phosphate group. In DNA, the sugar is deoxyribose and the base can be adenine, thymine, guanine, or cytosine. In RNA, the sugar is ribose and uracil replaces thymine.

## 17.2 Primary Structure of Nucleic Acids

**Learning Goal:** Describe the primary structures of RNA and DNA.

Each nucleic acid has its own unique sequence of bases known as its primary structure. In a nucleic acid polymer, the 3′-OH of each ribose sugar in RNA or deoxyribose sugar in DNA forms a phosphodiester bond to the phosphate group of the 5′-carbon atom group of the sugar in the next nucleotide to give a backbone of alternating sugar and phosphate groups. There is a free 5′-phosphate at one end of the polymer and a free 3′-OH group at the other end.

## 17.3 DNA Double Helix

**Learning Goal:** Describe the double helix of DNA.

A DNA molecule consists of two strands of nucleotides that are wound around each other like a spiral staircase. The two strands are held together by hydrogen bonds between complementary base pairs, A with T and G with C. During DNA replication, new DNA strands are made along each of the original DNA strands that serve as templates. Complementary base pairing ensures the correct pairing of bases to give identical copies of the original DNA.

## 17.4 RNA and the Genetic Code

**Learning Goal:** Identify the different RNAs; describe the synthesis of mRNA.

The three types of RNA differ by function in the cell: ribosomal RNA makes up most of the structure of the ribosomes, messenger RNA carries genetic information from the DNA to the ribosomes, and transfer RNA places the correct amino acids in the protein. Transcription is the process by which RNA polymerase produces mRNA from one strand of DNA. The bases in the mRNA are complementary to the DNA, except U is paired with A in RNA. The production of mRNA occurs when certain proteins are needed in the cell. The genetic code consists of a sequence of three bases (triplet) that specifies the order for the amino acids in a protein. The codon AUG signals the start of transcription and codons UAG, UGA, and UAA signal it to stop.

## 17.5 Protein Synthesis

**Learning Goal:** Describe the process of protein synthesis from mRNA.

Proteins are synthesized at the ribosomes in a translation process that includes three steps: initiation, translocation, and termination. During translation, tRNAs bring the appropriate amino acids to the ribosome and peptide bonds form. When the polypeptide is released, it takes on its secondary and tertiary structures and becomes a functional protein in the cell.

## 17.6 Genetic Mutations

**Learning Goal:** Describe some ways in which DNA is altered to cause mutations.

A genetic mutation is a change of one or more bases in the DNA sequence that may alter the structure and ability of the resulting protein to function properly. In a substitution, a different base replaces the base in a codon. A frameshift mutation inserts or deletes a base in the DNA sequence, which alters the mRNA codons after the mutation.

## 17.7 Viruses

**Learning Goal:** Describe the methods by which a virus infects a cell.

Viruses containing DNA or RNA invade host cells where they use the machinery within the cell to synthesize more viruses. For a retrovirus containing RNA, a viral DNA is synthesized by reverse transcription using the nucleotides and enzymes in the host cell. In the treatment of AIDS, nucleoside analogs inhibit the reverse transcriptase of the HIV-1 virus, and protease inhibitors disrupt the catalytic activity of protease needed to produce proteins for the synthesis of more viruses.

# KEY TERMS

**anticodon** The triplet of bases in the center loop of tRNA that is complementary to a codon on mRNA.

**base** Purine and pyrimidine compounds found in DNA and RNA: adenine (A), thymine (T), cytosine (C), guanine (G), and uracil (U).

**codon** A sequence of three bases in mRNA that specifies a certain amino acid to be placed in a protein. A few codons signal the start or stop of transcription.

**complementary base pairs** In DNA, adenine is always paired with thymine (A—T or T—A), and guanine is always paired with cytosine (G—C or C—G). In forming RNA, adenine is paired with uracil (A—U).

**DNA** Deoxyribonucleic acid; the genetic material of all cells containing nucleotides with deoxyribose sugar, phosphate, and the four bases adenine, thymine, guanine, and cytosine.

**double helix** The helical shape of the double chain of DNA that is like a spiral staircase with a sugar–phosphate backbone on the outside and base pairs like stair steps on the inside.

**frameshift mutation** A mutation that inserts or deletes a base in a DNA sequence.

**genetic code** The information in DNA that is transferred to mRNA as a sequence of codons for the synthesis of protein.

**genetic disease** A physical malformation or metabolic dysfunction caused by a mutation in the base sequence of DNA.

**mRNA** Messenger RNA; produced in the nucleus by DNA to carry the genetic information to the ribosomes for the construction of a protein.

**mutation** A change in the DNA base sequence that alters the formation of a protein in the cell.

**nucleic acids** Large molecules composed of nucleotides, found as a double helix in DNA and as the single strands of RNA.

**nucleoside** The combination of a pentose sugar and a base.

**nucleotides** Building blocks of a nucleic acid, consisting of a base, a pentose sugar (ribose or deoxyribose), and a phosphate group.

**phosphodiester bond** The phosphate link that joins the 3'-hydroxyl group in one nucleotide to the phosphate group on the 5'-carbon atom in the next nucleotide.

**primary structure** The sequences of nucleotides in nucleic acids.

**replication** The process of duplicating DNA by pairing the bases on each parent strand with their complementary base.

**retrovirus** A virus that contains RNA as its genetic material and synthesizes a complementary DNA strand inside a cell.

**RNA** Ribonucleic acid, a type of nucleic acid that is a single strand of nucleotides containing adenine, cytosine, guanine, and uracil.

**rRNA** Ribosomal RNA; the most prevalent type of RNA; a major component of the ribosomes.

**substitution** A mutation that replaces one base in a DNA with a different base.

**transcription** The transfer of genetic information from DNA by the formation of mRNA.

**translation** The interpretation of the codons in mRNA as amino acids in a peptide.

**tRNA** Transfer RNA; a RNA that places a specific amino acid into a peptide chain at the ribosome. There is one or more tRNA for each of the 20 different amino acids.

**virus** Small particles containing DNA or RNA in a protein coat that require a host cell for replication.

# UNDERSTANDING THE CONCEPTS

**17.55** Answer the following questions for the given section of DNA.
   **a.** Complete the bases for the parent and new strands.

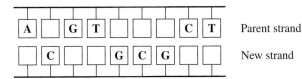

   **b.** Using the new strand as a template, write the mRNA sequence.

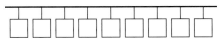

   **c.** Write the 3-letter symbols of the amino acids that would go into the peptide from the mRNA you wrote in part b.

**17.56** Suppose a mutation occurs in the DNA section in problem 17.55 and the first base in the parent chain, adenine, is replaced by guanine.
   **a.** What type of mutation has occurred?
   **b.** Using the new strand that results from this mutation as a template, write the order of bases in the altered mRNA.

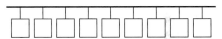

   **c.** Write the 3-letter symbols of the amino acids that would go into the peptide from the mRNA you wrote in part b.

   **d.** What effect, if any, might this mutation have on the structure and/or function of the resulting protein?

# ADDITIONAL QUESTIONS AND PROBLEMS

**17.57** Identify each of the following bases as a pyrimidine or a purine:
   **a.** cytosine          **b.** adenine
   **c.** uracil            **d.** thymine
   **e.** guanine

**17.58** Indicate if each of the bases in problem 17.57 are found in DNA only, RNA only, or both DNA and RNA.

**17.59** Identify the base and sugar in each of the following nucleosides:
   **a.** deoxythymidine    **b.** adenosine
   **c.** cytidine          **d.** deoxyguanosine

**17.60** Identify the base and sugar in each of the following nucleotides:
   **a.** CMP               **b.** dAMP
   **c.** dGMP              **d.** UMP

**17.61** How do the bases of thymine and uracil differ?

**17.62** How do the bases of cytosine and uracil differ?

**17.63** What is similar about the primary structure of RNA and DNA?

**17.64** What is different about the primary structure of RNA and DNA?

**17.65** Write the complementary base sequence for each of the following DNA segments:
   **a.** —G—A—C—T—T—A—G—G—C—
   **b.** —T—G—C—A—A—A—C—T—A—G—C—T—
   **c.** —A—T—C—G—A—T—C—G—A—T—C—G—

**17.66** Write the complementary base sequence for each of the following DNA segments:
   **a.** —T—T—A—C—G—G—A—C—C—G—C—
   **b.** —A—T—A—G—C—C—C—T—T—A—C—
        T—G—G—
   **c.** —G—G—C—C—T—A—C—C—T—T—A—
        A—C—G—A—C—G—

**17.67** Match the following statements with rRNA, mRNA, or tRNA.
   **a.** is the smallest type of RNA
   **b.** makes up the highest percent of RNA in the cell
   **c.** carries genetic information from the nucleus to the ribosomes

**17.68** Match the following statements with rRNA, mRNA, or tRNA.
  **a.** combines with proteins to form ribosomes
  **b.** brings amino acids to the ribosomes for protein synthesis
  **c.** acts as a template for protein synthesis

**17.69** What are the possible codons for each of the following amino acids?
  **a.** threonine
  **b.** serine
  **c.** cysteine

**17.70** What are the possible codons for each of the following amino acids?
  **a.** valine
  **b.** proline
  **c.** histidine

**17.71** What is the amino acid for each of the following codons?
  **a.** AAG    **b.** AUU    **c.** CGG

**17.72** What is the amino acid for each of the following codons?
  **a.** CAA    **b.** GGC    **c.** AAC

**17.73** Endorphins are polypeptides that reduce pain. What is the amino acid order for the following mRNA that codes for a pentapeptide that is an endorphin called leucine enkephalin?
  —AUG—UAC—GGU—GGA—UUU—CUA—UAA—

**17.74** Endorphins are polypeptides that reduce pain. What is the amino acid order for the following mRNA that codes for a pentapeptide that is an endorphin called methionine enkephalin?
  —AUG—UAC—GGU—GGA—UUU—AUG—UAA—

**17.75** What is the anticodon on tRNA for each of the following codons in a mRNA?
  **a.** AGC    **b.** UAU    **c.** CCA

**17.76** What is the anticodon on tRNA for each of the following codons in a mRNA?
  **a.** GUG    **b.** CCC    **c.** GAA

## CHALLENGE QUESTIONS

**17.77** Oxytocin is a nonapeptide with nine amino acids. How many nucleotides would be found in the mRNA for this protein?

**17.78** A protein contains 35 amino acids. How many nucleotides would be found in the mRNA for this protein?

**17.79** **a.** If the DNA double helix in salmon contains 28% adenine, what is the percent of thymine, guanine, and cytosine?

  **b.** If the DNA double helix in humans contains 20% cytosine, what is the percent of guanine, adenine, and thymine?

**17.80** Why are there no base pairs in DNA between adenine and guanine, or thymine and cytosine?

**17.81** What is the difference between a DNA virus and a retrovirus?

## ANSWERS

### Answers to Study Checks

**17.1** **a.** Guanine is found in both RNA and DNA.
  **b.** Uracil is found only in RNA.

**17.2** deoxycytidine 5′-monophosphate (dCMP)

**17.3**

**17.4** —C—C—A—A—T—T—G—G—

**17.5** —C—G—T—T—A—G—

**17.6** Each type of tRNA brings specific amino acids to the ribosome for protein synthesis.

**17.7** —C—C—C—A—A—A—T—T—T—

**17.8** UGA is a stop codon that signals the termination of translation.

**17.9** at UAG

**17.10** If the substitution of an amino acid in the polypeptide affects an interaction essential to functional structure on the binding of a substrate, the resulting protein could be less effective or nonfunctional. In this example, glycine, a nonpolar amino acid, replaced arginine, a basic amino acid. The shape of this protein is likely to change because there will be disruptions of cross-links such as salt bridges and hydrophobic and hydrophilic attractions.

**17.11** A protease inhibitor interferes with the enzymes that synthesize proteins needed to produce more viruses.

### Answers to Selected Questions and Problems

**17.1** **a.** pyrimidine
  **b.** pyrimidine

**17.3 a.** DNA
**b.** both DNA and RNA

**17.5** deoxyadenosine 5′-monophosphate (dAMP), deoxythymidine 5′-monophosphate (dTMP), deoxycytidine 5′-monophosphate (dCMP), and deoxyguanosine 5′-monophosphate (dGMP)

**17.7 a.** nucleoside    **b.** nucleoside
    **c.** nucleoside    **d.** nucleotide

**17.9**

**17.11** The nucleotides in nucleic acids are held together by phosphodiester bonds between the 3′-OH of a sugar (ribose or deoxyribose) and a phosphate group on the 5′-carbon of another sugar.

**17.13**

**17.15** The two DNA strands are held together by hydrogen bonds between the bases in each strand.

**17.17 a.** —T—T—T—T—T—T—
    **b.** —C—C—C—C—C—C—
    **c.** —T—C—A—G—G—T—C—C—A—
    **d.** —G—A—C—A—T—A—T—G—C—A—A—T—

**17.19** Ribosomal RNA, messenger RNA, and transfer RNA

**17.21** In transcription, the sequence of nucleotides on a DNA template (one strand) is used to produce the base sequences of a messenger RNA.

**17.23** —G—G—C—U—U—C—C—A—A—G—U—G—

**17.25** A three-base sequence in mRNA that codes for a specific amino acid in a protein.

**17.27 a.** leucine    **b.** serine
    **c.** glycine    **d.** arginine

**17.29** When AUG is the first codon, it signals the start of protein synthesis. Thereafter, AUG codes for methionine.

**17.31** A codon is a base triplet in the mRNA. An anticodon is the complementary triplet on a tRNA for a specific amino acid.

**17.33 a.** —Lys—Lys—Lys—
    **b.** —Phe—Pro—Phe—Pro—
    **c.** —Tyr—Gly—Arg—Cys—

**17.35** The new amino acid is joined by a peptide bond to the peptide chain. The ribosome moves to the next codon, which attaches to a tRNA carrying the next amino acid.

**17.37 a.** —CGA—AAA—GUU—UUU—
    **b.** GCU, UUU, CAA, AAA
    **c.** Using codons in mRNA: —Arg—Lys—Val—Phe—

**17.39** A base in DNA is replaced by a different base.

**17.41** In a frameshift mutation, a base is lost or gained, which changes the codons and therefore the amino acids in the remaining polypeptide chain.

**17.43** The normal triplet TTT forms a codon AAA, which codes for lysine. The mutation TTC forms a codon AAG, which also codes for lysine. There is no effect on the amino acid sequence.

**17.45 a.** —Thr—Ser—Arg—Val—
    **b.** —Thr—Thr—Arg—Val—
    **c.** —Thr—Ser—Gly—Val
    **d.** —Thr—STOP. Protein synthesis would terminate early. If this occurs early in the formation of the polypeptide, the resulting protein will probably be nonfunctional.
    **e.** The new protein will contain the sequence —Asp—Ile—Thr—Gly—.
    **f.** The new protein will contain the sequence —His—His—Gly—.

**17.47 a.** GCC and GCA both code for alanine.
    **b.** A cross-link in the tertiary structure of hemoglobin cannot be formed when the polar glutamic acid is replaced by nonpolar valine.

**17.49** DNA or RNA, but not both.

**17.51 a.** Viral RNA is used to synthesize viral DNA, which produces the mRNA to make the protein coat that allows the virus to replicate and leave the cell.
    **b.** retrovirus

**17.53** Nucleoside analogs such as AZT and ddI are similar to the nucleosides required to make viral DNA in reverse transcription. When they are incorporated into viral DNA, the lack of a hydroxyl group on the 3′-carbon in the sugar prevents the formation of the sugar–phosphate bonds and stops the replication of the virus.

**17.55 a.**

**b.** | A | G | G | U | C | G | C | C | U |

**c.** Arg — Ser — Pro

**17.57**  **a.** pyrimidine
**b.** purine
**c.** pyrimidine
**d.** pyrimidine
**e.** purine

**17.59**  **a.** thymine and deoxyribose
**b.** adenine and ribose
**c.** cytosine and ribose
**d.** guanine and deoxyribose

**17.61**  They are both pyrimidines, but thymine has a methyl group.

**17.63**  They are both polymers of nucleotides connected through phosphodiester bonds between alternating sugar and phosphate groups with bases extending out from each sugar.

**17.65**  **a.** —C—T—G—A—A—T—C—C—G—
**b.** —A—C—G—T—T—T—G—A—T—C—G—A—
**c.** —T—A—G—C—T—A—G—C—T—A—G—C—

**17.67**  **a.** tRNA          **b.** rRNA
**c.** mRNA

**17.69**  **a.** ACU, ACC, ACA, and ACG
**b.** UCU, UCC, UCA, UCG, AGU, and AGC
**c.** UGU and UGC

**17.71**  **a.** lysine
**b.** isoleucine
**c.** arginine

**17.73**  START—Tyr—Gly—Gly—Phe—Leu—STOP

**17.75**  **a.** UCG
**b.** AUA
**c.** GGU

**17.77**  Three nucleotides are needed for each amino acid, plus a start and stop triplet, making a minimum total of 33 nucleotides.

**17.79**  **a.** Because A bonds with T, T is also 28%. Because A and T = 56%, there is 44% for the other nucleotides or 22% G and 22% C.
**b.** Because C bonds with G, G is also 20%. Because C and G = 40%, there is 60% for the other nucleotides or 30% A and 30% T.

**17.81**  A DNA virus attaches to a cell and injects viral DNA that uses the host cell to produce copies of DNA to make viral RNA. A retrovirus injects viral RNA from which complementary DNA is produced by reverse transcription.

# 18

# Metabolic Pathways and Energy Production

Visit **www.chemplace.com** for extra quizzes, interactive tutorials, career resources, PowerPoint slides for chapter review, math help, and case studies.

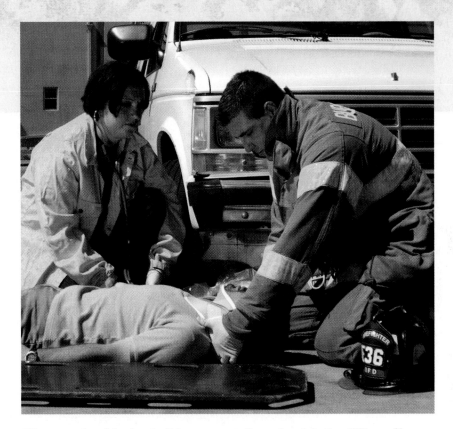

*"I am trained in basic life support. I work with the ER staff to assist in patient care," says Mandy Dornell, emergency medical technician at Seaton Medical Center. "In the ER, I take vital signs, do patient assessment, and do CPR. If someone has a motor vehicle accident, I may suspect a neck or back injury. Then I may use a backboard or a cervical collar, which prevents the patient from moving and causing further damage. When people have difficulty breathing, I insert an airway—nasal or oral—to assist ventilation. I also set up and monitor IVs, and I am trained in childbirth."*

*When someone is critically ill or injured, the quick reactions of emergency medical technicians (EMTs) and paramedics provide immediate medical care and transport to an ER or trauma center.*

All the chemical reactions that take place in living cells to break down or build molecules are known as *metabolism*. In a metabolic pathway, reactions are linked together in a series, each catalyzed by a specific enzyme. In this chapter, we will look at these pathways and the way they produce energy and cellular compounds.

When we eat food such as a tuna fish sandwich, the polysaccharides, lipids, and proteins are digested to smaller molecules that are absorbed into the cells of our body. As these molecules of glucose, fatty acids, and amino acids are broken down further, energy is released. This energy is used in the cells to synthesize high-energy adenosine triphosphate (ATP). Our cells utilize ATP energy when they do work such as contracting muscles, synthesizing large molecules, sending nerve impulses, and moving substances across cell membranes.

In the *citric acid cycle*, a series of metabolic reactions in the mitochondria oxidize acetyl CoA to carbon dioxide, which releases energy to produce NADH and $FADH_2$. These reduced coenzymes enter *electron transport* where they provide hydrogen ions and electrons that combine with oxygen ($O_2$) to form $H_2O$. The energy released during electron transport is used to synthesize ATP from ADP.

Although glucose is the primary fuel for the synthesis of ATP, lipids and proteins also play an important role in metabolism and energy production. Almost all our energy is stored in the form of triacylglycerols in fat cells. When our caloric intake exceeds the nutritional and metabolic needs of the body, excess carbohydrates and fatty acids are converted to triacylglycerols, which are stored in our fat cells.

The degradation of proteins provides amino acids that are used to synthesize nitrogen-containing compounds in our cells. Although amino acids are not considered a primary source of fuel, energy can be extracted from amino acids if carbohydrate and fat reserves have been depleted.

## 18.1 METABOLISM AND ATP ENERGY

The term **metabolism** refers to all the chemical reactions that provide energy and the substances required for continued cell growth. There are two types of metabolic reactions: catabolic and anabolic. In **catabolic reactions**, complex molecules are broken down to simpler ones with an accompanying release of energy. **Anabolic reactions** utilize energy available in the cell to build large molecules from simple ones.

We can think of the catabolic processes in metabolism as consisting of three stages. (See Figure 18.1.) Consider a tuna fish sandwich for our example. In stage 1 of metabolism, the processes of **digestion** break down the large macromolecules into small monomer units. The polysaccharides in bread break down to monosaccharides, the lipids in the mayonnaise break down to glycerol and fatty acids, and the proteins from the tuna yield amino acids. These digestion products diffuse into the bloodstream for transport to cells. In stage 2, they are broken down to two- and three-carbon compounds such as

## Stages of Metabolism

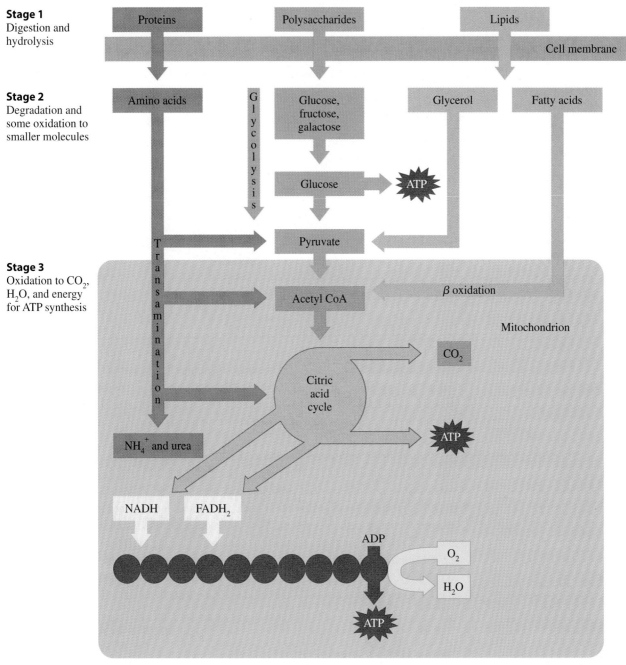

**Stage 1**
Digestion and hydrolysis

**Stage 2**
Degradation and some oxidation to smaller molecules

**Stage 3**
Oxidation to $CO_2$, $H_2O$, and energy for ATP synthesis

**FIGURE 18.1** In the three stages of catabolic metabolism, foods are digested and degraded into smaller molecules, which are oxidized to produce energy.
**Q** Where is most of the ATP energy produced in the cells?

pyruvate and acetyl CoA. Stage 3 begins with the oxidation of the two-carbon acetyl CoA in the citric acid cycle, which produces several reduced coenzymes. As long as the cells have oxygen, the hydrogen ions and electrons are transferred from the coenzymes to electron transport, where most of the energy for the cell is produced.

## Cell Structure for Metabolism

To understand the relationship of metabolic reactions, we need to look at where metabolic reactions take place in the cells. (See Figure 18.2.)

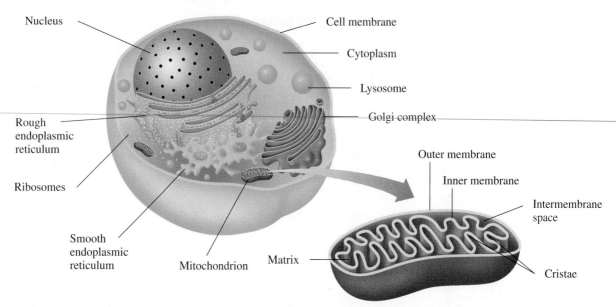

**FIGURE 18.2** The diagram illustrates the major components of a typical animal cell.
**Q** What is the cytoplasm in a cell?

In animals, a *membrane* separates the materials inside the cell from the aqueous environment surrounding the cell. The *nucleus* contains the genes that control DNA replication and protein synthesis of the cell. The **cytoplasm** consists of all the materials between the nucleus and the cell membrane. The **cytosol**, which is the fluid part of the cytoplasm, is an aqueous solution of electrolytes and enzymes that catalyze many of the cell's chemical reactions.

Within the cytoplasm are specialized structures called *organelles* that carry out specific functions in the cell. We have already seen (Chapter 17) that the *ribosomes* are the sites of protein synthesis. The **mitochondria** are the energy-producing factories of the cells. A mitochondrion consists of an outer membrane and an inner membrane, with an intermembrane space between them. The fluid section surrounded by the inner membrane is called the *matrix*. Enzymes located in the matrix and along the inner membrane catalyze the oxidation of carbohydrates, fats, and amino acids. All of these oxidation pathways lead to $CO_2$, $H_2O$, and energy, which is used to form energy-rich compounds. Table 18.1 summarizes some of the functions of the cellular components in animal cells.

SAMPLE PROBLEM 18.1

■ **Metabolism and Cell Structure**

Identify the following as catabolic or anabolic reactions:

**a.** Digestion of polysaccharides
**b.** Synthesis of proteins
**c.** Oxidation of glucose to $CO_2$ and $H_2O$

SOLUTION

**a.** The breakdown of large molecules is a catabolic reaction.
**b.** The synthesis of large molecules requires energy and involves anabolic reactions.
**c.** The breakdown of monomers such as glucose involves catabolic reactions.

STUDY CHECK

What is the difference between the cytoplasm and cytosol?

**TABLE 18.1** Locations and Functions of Components in Animal Cells

| Component | Description and Function |
| --- | --- |
| Cell membrane | Separates the contents of a cell from the external environment and contains structures that communicate with other cells |
| Cytoplasm | Consists of all of the cellular contents between the cell membrane and nucleus |
| Cytosol | Is the fluid part of the cytoplasm that contains enzymes for many of the cell's chemical reactions |
| Endoplasmic reticulum | Rough type processes proteins for secretion and synthesizes phospholipids; smooth type synthesizes fats and steroids |
| Golgi complex | Modifies and secretes proteins from the endoplasmic reticulum and synthesizes cell membranes |
| Lysosome | Contains hydrolytic enzymes that digest and recycle old cell structures |
| Mitochondrion | Contains the structures for the synthesis of ATP from energy-producing reactions |
| Nucleus | Contains genetic information for the replication of DNA and the synthesis of protein |
| Ribosome | Is the site of protein synthesis using mRNA templates |

## ATP and Energy

In our cells, the energy released from the oxidation of the food we eat is used to form a "high-energy" compound called *adenosine triphosphate* (**ATP**). The ATP molecule is composed of the nitrogen base adenine, a ribose sugar, and three phosphate groups. When ATP undergoes hydrolysis, the products are adenosine diphosphate (**ADP**), a phosphate group abbreviated as $P_i$, and energy of 7.3 kcal per mole of ATP.

**WEB TUTORIAL ATP**

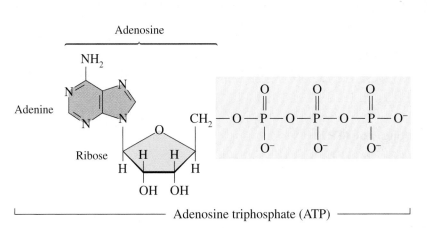

Adenosine triphosphate (ATP)

We can write this reaction as

$$ATP + H_2O \longrightarrow ADP + P_i + 7.3 \text{ kcal/mole}$$

Every time we contract muscles, move substances across cellular membranes, send nerve signals, or synthesize an enzyme, we use energy from ATP hydrolysis. In a cell that is doing work (anabolic processes), 1–2 million ATP molecules may be hydrolyzed in one second. The amount of ATP hydrolyzed in one day can be as much as our body mass, even though only about 1 gram of ATP is present in all our cells at any given time.

When we take in food, the resulting catabolic reactions provide energy to regenerate ATP in our cells. Then 7.3 kcal/mole is used to make ATP from ADP and $P_i$.

$$ADP + P_i + 7.3 \text{ kcal/mole} \longrightarrow ATP$$

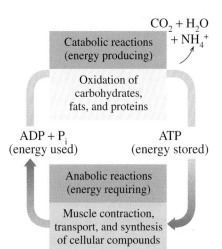

# *Health Note*

## ATP Energy and Ca$^{2+}$ Needed to Contract Muscles

Our muscles consist of thousands of parallel fibers. Within these muscle fibers are fibrils composed of two kinds of proteins, called filaments. Arranged in alternating rows, the thick filaments of myosin overlap the thin filaments containing actin. During a muscle contraction, the thin filaments slide inward over the thick filaments, causing a shortening of the muscle fibers.

Calcium ion, Ca$^{2+}$, and ATP play an important role in muscle contraction. An increase in the Ca$^{2+}$ concentration in the muscle fibers causes the filaments to slide, while a decrease stops the process. In a relaxed muscle, the Ca$^{2+}$ concentration is low. However, when a nerve impulse reaches the muscle, calcium channels in the membrane open, and Ca$^{2+}$ flows into the fluid surrounding the filaments. The muscle contracts as myosin binds to actin and pulls

the filaments inward. The energy for the contraction is provided by splitting ATP to ADP + P$_i$. Muscle contraction continues as long as both ATP and Ca$^{2+}$ levels are high around the filaments. When the nerve impulse ends, the calcium channels close. The Ca$^{2+}$ concentration decreases as energy from ATP pumps Ca$^{2+}$ out of the filaments, which causes the muscle to relax. In rigor mortis, Ca$^{2+}$ concentration remains high within the muscle fibers causing a continued state of rigidity. After approximately 24 hours, Ca$^{2+}$ decreases because of cellular deterioration and muscles relax.

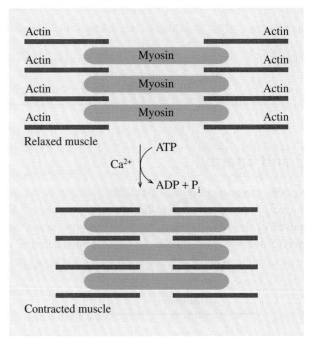

---

## SAMPLE PROBLEM 18.2

### ■ Hydrolysis of ATP

Write an equation for the hydrolysis of ATP.

**SOLUTION**

The hydrolysis of ATP produces ADP, P$_i$, and energy.

$$ATP + H_2O \longrightarrow ADP + P_i + energy$$

**STUDY CHECK**

What are the components of the ATP molecule?

---

## QUESTIONS AND PROBLEMS

### Metabolism and ATP Energy

**18.1** What stage of metabolism involves the digestion of polysaccharides?

**18.2** What stage of metabolism involves the conversion of small molecules to CO$_2$, H$_2$O, and energy?

**18.3** What is meant by a catabolic reaction in metabolism?

**18.4** What is meant by an anabolic reaction in metabolism?

**18.5** Why is ATP considered an energy-rich compound?

**18.6** How much energy is obtained from the hydrolysis of ATP?

# 18.2 DIGESTION OF FOODS

In the first stage of catabolism, foods undergo **digestion**, a process that converts large molecules to smaller ones that can be absorbed by the body.

## Digestion of Carbohydrates

We begin the digestion of carbohydrates as soon as we chew food. **Amylase**, an enzyme produced in the salivary glands, hydrolyzes some of the $\alpha$-glycosidic bonds in amylose and amylopectin, producing maltose, glucose, and smaller polysaccharides called dextrins, which contain three to eight glucose units. After swallowing, the partially digested starches enter the acid environment of the stomach, where the low pH soon stops further carbohydrate digestion. (See Figure 18.3.)

In the small intestine, which has a pH of about 8, an $\alpha$-amylase produced in the pancreas hydrolyzes the remaining polysaccharides to maltose and glucose. Then enzymes produced in the mucosal cells that line the small intestine hydrolyze maltose as well as lactose and sucrose. The hydrolysis reactions for the three common dietary disaccharides are written as follows:

$$\text{Lactose} + H_2O \xrightarrow{\text{Lactase}} \text{galactose} + \text{glucose}$$

$$\text{Sucrose} + H_2O \xrightarrow{\text{Sucrase}} \text{fructose} + \text{glucose}$$

$$\text{Maltose} + H_2O \xrightarrow{\text{Maltase}} \text{glucose} + \text{glucose}$$

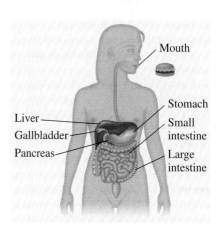

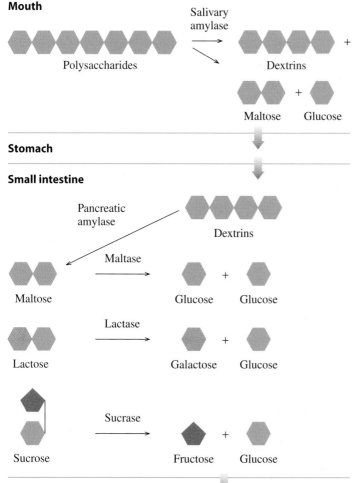

**FIGURE 18.3** In stage 1 of metabolism, the digestion of carbohydrates begins in the mouth and is completed in the small intestine.

**Q** Why is there little or no digestion of carbohydrates in the stomach?

## Carbohydrate Digestion

1. Obtain a cracker or small piece of bread and chew it for 4–5 minutes. During that time, observe any change in the taste.
2. Some milk products contain Lactaid, which is the lactase that digests lactose. Look for the brands of milk and ice cream that contain Lactaid or lactase enzyme.

### QUESTIONS

**1a.** How does the taste of the cracker or bread change after you have chewed it for 4–5 minutes? What could be an explanation?
**1b.** What part of carbohydrate digestion occurs in the mouth?
**2a.** Write an equation for the digestion of lactose.
**2b.** Where does lactose undergo digestion?

---

The monosaccharides are absorbed through the intestinal wall into the bloodstream, which carries them to the liver, where fructose and galactose are converted to glucose.

**SAMPLE PROBLEM 18.3**

### ■ Digestion of Carbohydrates

Indicate the carbohydrate that undergoes digestion in each of the following sites:
**a.** mouth        **b.** stomach        **c.** small intestine

#### SOLUTION

**a.** Starches amylose and amylopectin.
**b.** Essentially no carbohydrate digestion occurs in the stomach.
**c.** Dextrins, maltose, sucrose, and lactose.

#### STUDY CHECK

Describe the digestion of amylose, a polymer of glucose molecules joined by $\alpha$-glycosidic bonds.

## Digestion of Fats

The digestion of dietary fats begins in the small intestine, when the hydrophobic fat globules mix with bile salts released from the gallbladder. In a process called *emulsification*, the bile salts break the fat globules into smaller droplets called micelles. Then *pancreatic lipases* released from the pancreas hydrolyze the triacylglycerols in the micelles to yield monoacylglycerols and free fatty acids. These digestion products are absorbed into the intestinal lining where they recombine to form triacylglycerols, which are coated with proteins to form lipoproteins called *chylomicrons*. The chylomicrons transport the triacylglycerols through the lymph system and into the bloodstream to be carried to cells of the heart, muscle, and adipose tissues. (See Figure 18.4.) In the cells, enzymes hydrolyze the triacylglycerols to yield glycerol and free fatty acids, which can be used for energy production. We can write the overall equation for the digestion of triacylglycerols as follows:

$$\text{Triacylglycerols} + 3\text{H}_2\text{O} \xrightarrow{\text{Lipase}} \text{glycerol} + 3 \text{ fatty acids}$$

The products of fat digestion, glycerol and fatty acids, diffuse into the bloodstream and bind with plasma proteins (albumin) to be transported to the tissues. Most of the glycerol goes into the liver where it is converted to glucose.

**SAMPLE PROBLEM 18.4**

### ■ Fats and Digestion

What are the sites, enzymes, and products for the digestion of triacylglycerols?

#### SOLUTION

The digestion of triacylglycerols takes place in the small intestine where pancreatic lipase catalyzes their hydrolysis to monoacylglycerols and fatty acids.

#### STUDY CHECK

What happens to the products from the digestion of triacylglycerols in the membrane of the small intestine?

**Small intestine**

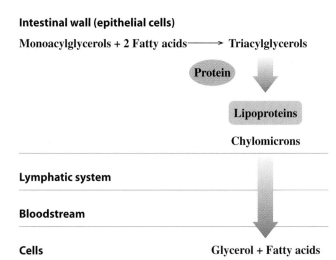

Triacylglycerol $\xrightarrow{\text{Lipase}}$ Monoacylglycerol

$$\begin{array}{ccc}
\text{H}_2\text{C} - \text{Fatty acid} & & \text{H}_2\text{C} - \text{OH} \\
| & & | \\
\text{HC} - \text{Fatty acid} & \xrightarrow[\text{Lipase}]{+ 2\text{H}_2\text{O}} & \text{HC} - \text{Fatty acid} \\
| & & | \quad + 2 \text{ Fatty acids} \\
\text{H}_2\text{C} - \text{Fatty acid} & & \text{H}_2\text{C} - \text{OH}
\end{array}$$

**Triacylglycerol**          **Monoacylglycerol**

**Intestinal wall (epithelial cells)**

**Monoacylglycerols + 2 Fatty acids** ⟶ **Triacylglycerols**

**Protein**

**Lipoproteins**

**Chylomicrons**

**Lymphatic system**

**Bloodstream**

**Cells**          **Glycerol + Fatty acids**

**FIGURE 18.4** The triacylglycerols that re-form in the intestinal wall bind to proteins for transport through the lymphatic system and bloodstream to the cells.
**Q** What kinds of enzymes are secreted from the pancreas into the small intestine to hydrolyze triacylglycerols?

## Digestion of Proteins

In stage 1, the digestion of proteins begins in the stomach, where hydrochloric acid (HCl) at pH 2 denatures proteins and activates enzymes such as pepsin that begin to hydrolyze peptide bonds. Polypeptides from the stomach move into the small intestine, where trypsin and chymotrypsin complete the hydrolysis of the peptides to amino acids. The amino acids are absorbed through the intestinal walls into the bloodstream for transport to the cells. (See Figure 18.5.)

**Stomach**

Pepsinogen $\xrightarrow{\text{HCl}}$ Pepsin

Proteins ⟶ Polypeptides

**Small intestine**

Trypsin
Chymotrypsin

Amino acids

**Intestinal wall**

**Bloodstream**

**FIGURE 18.5** Proteins are hydrolyzed to polypeptides in the stomach and to amino acids in the small intestine.
**Q** What enzyme, secreted into the small intestine, hydrolyzes peptides?

---

**SAMPLE PROBLEM 18.5**

### ▇ Digestion of Proteins

What are the sites and end products for the digestion of proteins?

**SOLUTION**

Proteins begin digestion in the stomach and complete digestion in the small intestine to yield amino acids.

**STUDY CHECK**

What is the function of HCl in the stomach?

## QUESTIONS AND PROBLEMS

### Digestion of Foods

**18.7** What is the general type of reaction that occurs during the digestion of carbohydrates?

**18.8** Why is α-amylase produced in the salivary glands and in the pancreas?

**18.9** Complete the following equation by filling in the missing words:
  **a.** _____ + $H_2O$ ⟶ galactose + glucose
  **b.** Sucrose + $H_2O$ ⟶ _____ + _____
  **c.** Maltose + $H_2O$ ⟶ _____ + _____

**18.10** Give the site and the enzyme for each of the reactions in problem 18.9.

**18.11** What is the role of bile salts in lipid digestion?

**18.12** How are insoluble triacylglycerols transported to the tissues?

**18.13** Where do dietary proteins undergo digestion in the body?

**18.14** What is the purpose of digestion in stage 1?

---

# 18.3 IMPORTANT COENZYMES IN METABOLIC PATHWAYS

**LEARNING GOAL**

Describe the components and functions of the coenzymes FAD, NAD$^+$, and coenzyme A.

Before we look at the metabolic reactions that extract energy from the food we digest, we need to review some ideas about oxidation and reduction reactions (Chapters 5 and 12). An *oxidation* reaction involves the loss of hydrogen or electrons by a substance. When an enzyme catalyzes oxidation, hydrogen atoms are removed from a substrate as hydrogen ions, $2H^+$, and electrons, $2\,e^-$.

$$2 \text{ H atoms (removed in oxidation)} \longrightarrow 2H^+ + 2\,e^-$$

*Reduction* is the gain of hydrogen or electrons. When hydrogen ions and electrons are picked up by a coenzyme, it is reduced. As we saw in Chapter 16, the structures of many coenzymes include the water-soluble B vitamins we obtain from the foods in our diets. Now we can look at the structures of several important coenzymes in their oxidized and reduced forms.

## NAD$^+$

NAD$^+$ (nicotinamide adenine dinucleotide) is an important coenzyme in which the vitamin *niacin* provides the *nicotinamide* group, which is bonded to adenosine diphosphate (ADP). (See Figure 18.6.) The NAD$^+$ coenzyme participates in reactions that produce a carbon–oxygen (C=O) double bond, such as in the oxidation of alcohols to aldehydes and ketones. The NAD$^+$ is reduced when the carbon in nicotinamide accepts a hydrogen ion and two electrons, leaving one $H^+$. Let's look at the reactions that take place when ethanol is oxidized in the liver to acetaldehyde using NAD$^+$.

**Overall oxidation–reduction reaction**

$$CH_3{-}CH_2{-}OH + NAD^+ \xrightarrow[\text{dehydrogenase}]{\text{Alcohol}} CH_3{-}\overset{\displaystyle O}{\overset{\|}{C}}{-}H + NADH + H^+$$

Ethanol                                              Acetaldehyde

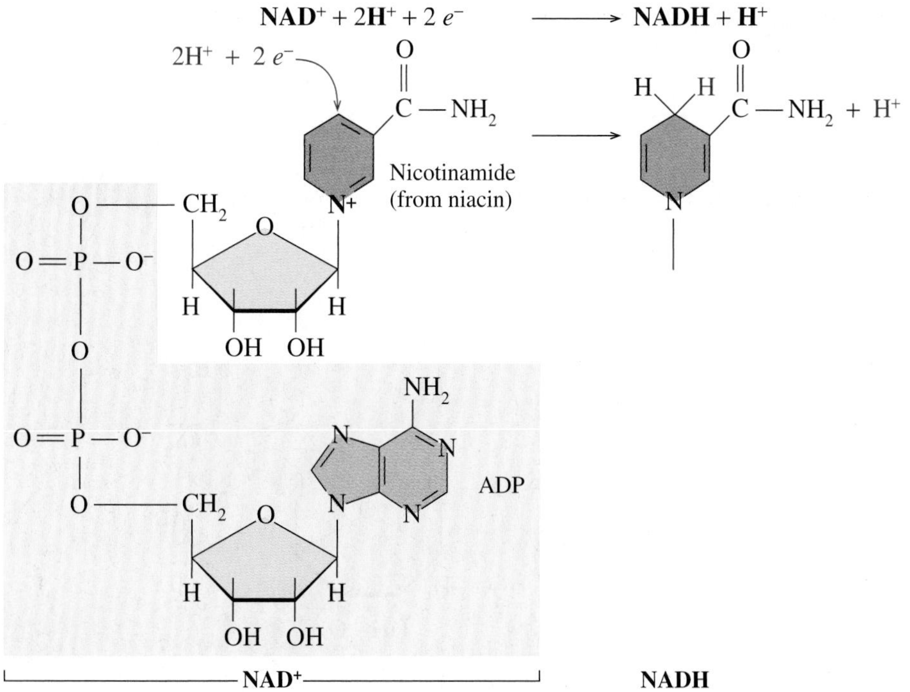

**FIGURE 18.6** The coenzyme $NAD^+$ (nicotinamide adenine dinucleotide), which consists of a nicotinamide portion from the vitamin niacin, ribose, and adenine diphosphate, is reduced to NADH + $H^+$.

**Q** Why is the conversion of $NAD^+$ to NADH and $H^+$ called a reduction?

## FAD

**FAD** (flavin adenine dinucleotide) is a coenzyme that contains adenosine diphosphate (ADP) and riboflavin. Riboflavin, also known as vitamin $B_2$, consists of ribitol (a sugar alcohol) and flavin. As a coenzyme, two nitrogen atoms in the flavin part of the FAD coenzyme accept the hydrogen, which reduces the FAD to $FADH_2$. (See Figure 18.7.) FAD typically participates in oxidation reactions that produce a carbon–carbon (C=C) double bond.

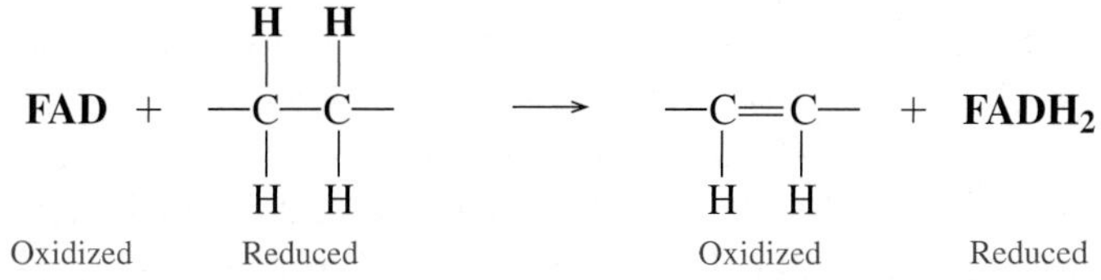

## Coenzyme A

**Coenzyme A (CoA)** is made up of several components: pantothenic acid (vitamin $B_5$), adenosine diphosphate (ADP), and aminoethanethiol. (See Figure 18.8.) One of the main functions of coenzyme A is to prepare small groups such as the acetyl group for enzyme action.

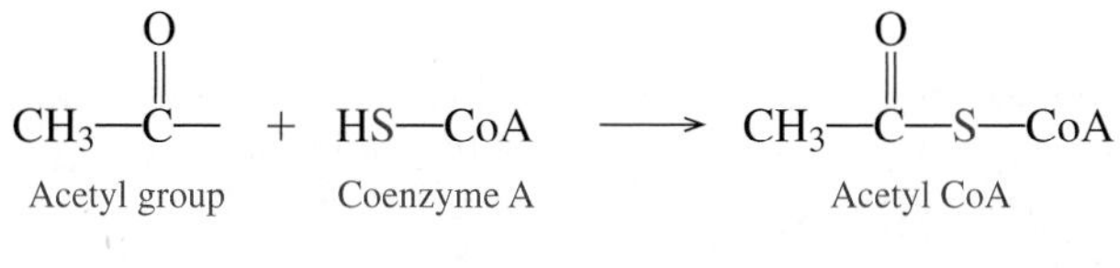

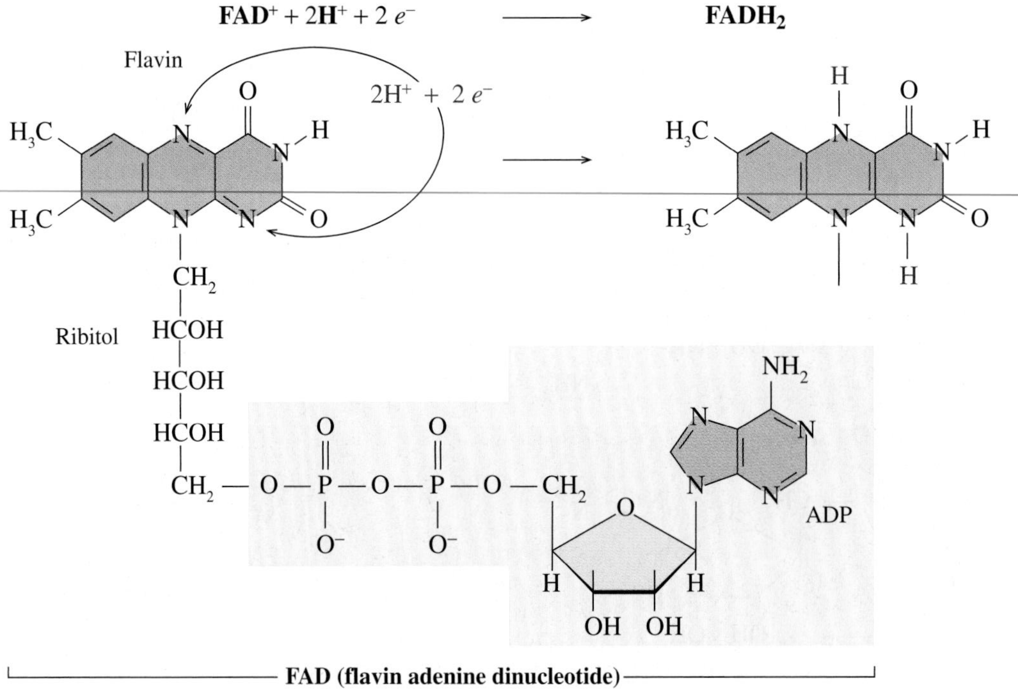

FIGURE 18.7  The coenzyme FAD (flavin adenine dinucleotide) made from riboflavin (vitamin $B_2$) and adenine diphosphate is reduced to $FADH_2$.
Q  What is the type of reaction in which FAD accepts hydrogen?

---

SAMPLE PROBLEM 18.6

### ■ Coenzymes in Metabolic Pathways

What vitamin is part of the coenzyme FAD?

SOLUTION

FAD, flavin adenine dinucleotide, contains the vitamin riboflavin.

STUDY CHECK

What is the abbreviation of the reduced form of FAD?

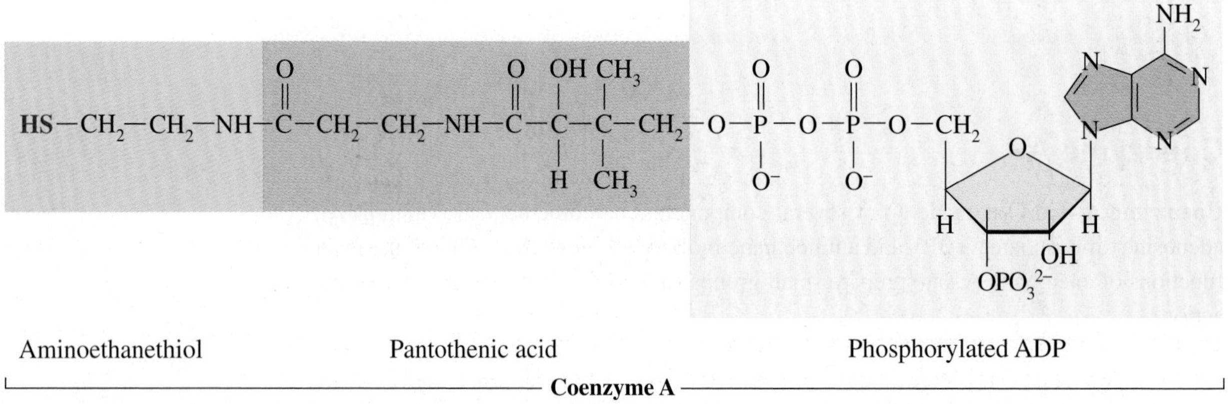

---

FIGURE 18.8  Coenzyme A is derived from a phosphorylated adenine diphosphate (ADP) and pantothenic acid bonded by an amide bond to aminoethanethiol, which contains the —SH reactive part of the molecule.
Q  What part of coenzyme A reacts with a two-carbon acetyl group?

## Health Note

### Lactose Intolerance

The disaccharide in milk is lactose, which is broken down by *lactase* in the intestinal tract to monosaccharides that are a source of energy. Infants and small children produce lactase to break down the lactose in milk. It is rare for an infant to lack the ability to produce lactase. However, the production of lactase decreases as many people age, which causes lactose intolerance. This condition affects approximately 25% of the people in the United States. A deficiency of lactase occurs in adults throughout the world, but in the United States it is prevalent among the African-American, Hispanic, and Asian populations.

When lactose is not broken down into glucose and galactose, it cannot be absorbed through the intestinal wall and remains in the intestinal tract. In the intestines, the lactose undergoes fermentation to products that include lactic acid and gases such as methane ($CH_4$) and $CO_2$. Symptoms of lactose intolerance, which appear approximately $\frac{1}{2}$ to 1 hour after ingesting milk or milk products, include nausea, abdominal cramps, and diarrhea. The severity of the symptoms depends on how much lactose is present in the food and how much lactase a person produces.

#### TREATMENT OF LACTOSE INTOLERANCE

One way to reduce the reaction to lactose is to avoid products that contain lactose. However, it is important to consume foods that provide the body with calcium. Many people with lactose intolerance seem to tolerate yogurt, which is a good source of calcium. Although

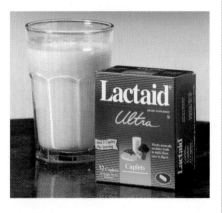

there is lactose in yogurt, the bacteria in yogurt may produce some lactase, which helps to digest the lactose. A person who is lactose intolerant should also know that some foods that may not seem to be dairy products contain lactose. For example, baked goods, cereals, breakfast drinks, salad dressings, and even lunchmeat can contain lactose in their ingredients. Labels must be read carefully to see if the ingredients include milk or lactose.

The enzyme lactase is now available in many forms such as tablets that are taken with meals, drops that are added to milk, or as additives in many dairy products. When lactase is added to milk that is left in the refrigerator for 24 hours, the lactose level is reduced by 70–90%. Lactase pills or chewable tablets are taken when a person begins to eat a meal that contains dairy foods. If taken too far ahead of the meal, too much of the lactase will be degraded by stomach acid. If taken following a meal, the lactose will have entered the lower intestine.

---

## QUESTIONS AND PROBLEMS

### Important Coenzymes in Metabolic Pathways

**18.15** Give the abbreviation for the following:
   **a.** the reduced form of $NAD^+$
   **b.** the oxidized form of $FADH_2$

**18.16** Give the abbreviation for the following:
   **a.** the reduced form of FAD
   **b.** the oxidized form of NADH

**18.17** What coenzyme picks up hydrogen when a carbon–carbon double bond is formed?

**18.18** What coenzyme picks up hydrogen when a carbon–oxygen double bond is formed?

---

# 18.4 GLYCOLYSIS: OXIDATION OF GLUCOSE

### LEARNING GOAL

Describe the conversion of glucose to pyruvate in glycolysis and the subsequent conversion of pyruvate to acetyl CoA or lactate.

Our major source of energy is the glucose produced when we digest the carbohydrates in our food, or from glycogen, a polysaccharide stored in the liver and skeletal muscle. Glucose in the bloodstream enters our cells for further degradation in a pathway called glycolysis. Early organisms used glycolysis to produce energy from simple nutrients long before there was any oxygen in Earth's atmosphere. Glycolysis is an **anaerobic** process; no oxygen is required.

In **glycolysis**, which is one of the metabolic pathways in stage 2, a six-carbon glucose molecule is broken down to yield two three-carbon pyruvate molecules. In the first

WEB TUTORIAL
Glycolysis

five reactions (1–5), the energy of 2 ATP is used to add phosphate groups to form sugar phosphates. In reactions 4 and 5, the six-carbon sugar phosphate is split to yield two three-carbon sugar phosphate molecules. In the last five reactions (6–10), the energy of 4 ATP is produced when the phosphate groups in these energy-rich trioses are hydrolyzed. (See Figure 18.9.)

**FIGURE 18.9** In glycolysis, the six-carbon glucose molecule is degraded to yield two three-carbon pyruvate molecules. A net of two ATPs are produced along with two NADH.

**Q** Where in the glycolysis pathway is glucose cleaved to yield two three-carbon compounds?

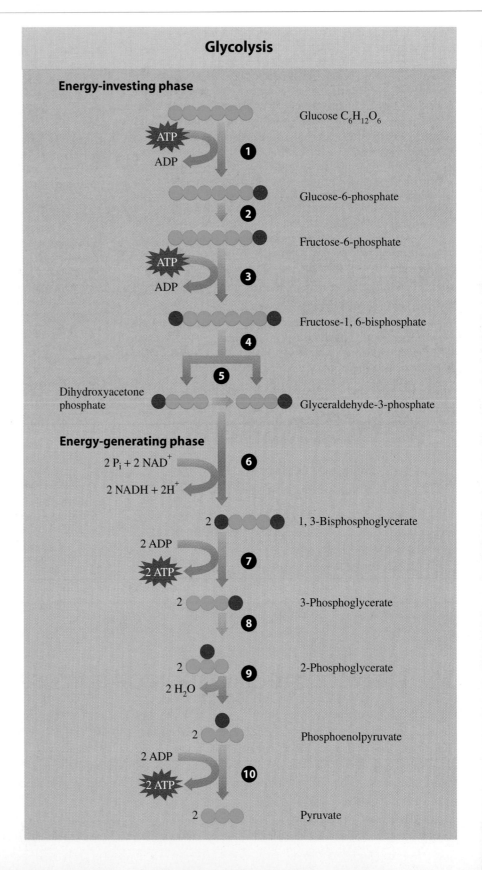

# Energy-Investing Reactions 1–5

### Reaction 1    Phosphorylation: first ATP invested

Glucose is converted to glucose-6-phosphate by a reaction with ATP catalyzed by *hexokinase*.

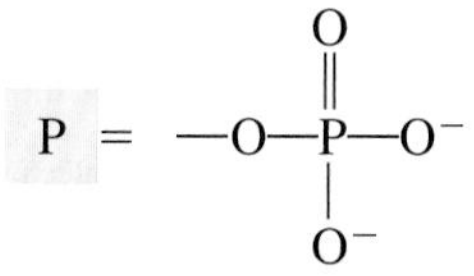

$$P = \quad -O-\overset{\overset{\displaystyle O}{\|}}{\underset{\underset{\displaystyle O^-}{|}}{P}}-O^-$$

### Reaction 2    Isomerization

The enzyme *phosphoglucoisomerase* converts glucose-6-phosphate, an aldose, to fructose-6-phosphate, a ketose.

### Reaction 3    Phosphorylation: second ATP invested

A second ATP reacts with fructose-6-phosphate to give fructose-1,6-bisphosphate. The word *bisphosphate* is used to show that the phosphates are on different carbons in fructose and not connected to each other.

### Reaction 4    Cleavage: two trioses form

Fructose-1,6-bisphosphate is split into two triose phosphates—dihydroxyacetone phosphate and glyceraldehyde-3-phosphate—catalyzed by *aldolase*.

### Reaction 5    Isomerization of a triose

In this reaction, *triosephosphate isomerase* converts one of the triose products, dihydroxyacetone phosphate to a second molecule of glyceraldehyde-3-phosphate. Now all 6 carbon atoms from glucose are in two identical triose phosphates.

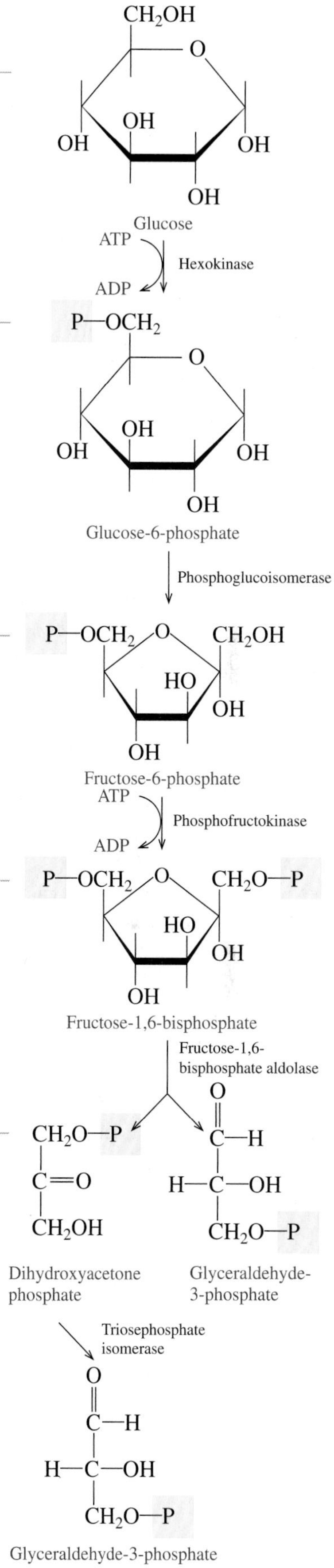

Glucose

ATP

Hexokinase

ADP

Glucose-6-phosphate

Phosphoglucoisomerase

Fructose-6-phosphate

ATP

Phosphofructokinase

ADP

Fructose-1,6-bisphosphate

Fructose-1,6-bisphosphate aldolase

Dihydroxyacetone phosphate

Glyceraldehyde-3-phosphate

Triosephosphate isomerase

Glyceraldehyde-3-phosphate

## Energy-Generating Reactions 6–10

### Reaction 6   First energy-rich compound

Now we look at reactions 6–10 that involve the two three-carbon compounds from the initial glucose. In reaction 6, the aldehyde group of glyceraldehyde-3-phosphate is oxidized and phosphorylated by *glyceraldehyde-3-phosphate dehydrogenase*. The product is a high-energy compound, 1,3-bisphosphoglycerate. The coenzyme $NAD^+$ is reduced to NADH and $H^+$.

### Reaction 7   Formation of first ATP

Now a *phosphoglycerate kinase* transfers a phosphate group from 1,3-bisphosphoglycerate to ADP to form ATP. This process is called a substrate-level phosphorylation. At this point in glycolysis, the reaction of two 1,3-bisphosphoglycerate molecules pays back the debt of two ATP invested in reactions 1 and 3.

### Reaction 8   Formation of 2-phosphoglycerate

A *phosphoglycerate mutase* transfers the phosphate group from carbon 3 of 3-phosphoglycerate to carbon 2 to yield two molecules of 2-phosphoglycerate.

### Reaction 9   Second energy-rich compound

An *enolase* catalyzes the removal of water from two phosphoglycerate molecules to give two high-energy molecules of phosphoenolpyruvate.

### Reaction 10   Formation of second ATP

In a second direct substrate phosphorylation, *pyruvate kinase* transfers phosphate groups from two phosphoenolpyruvate to two ADPs to give two pyruvates and two ATPs.

## Summary of Glycolysis

In the glycolysis pathway, glucose is converted to two pyruvates. Initially, two ATP molecules are required to form sugar phosphate. Later, four ATPs are generated, which gives a net gain of two ATPs. Overall, glycolysis yields two ATPs and two NADHs for each glucose that is converted to two pyruvates.

$$C_6H_{12}O_6 + 2\,NAD^+ \xrightarrow[\;2\,ADP\; +\; 2\,P_i\;\;\;\;\;\;2\,ATP\;]{} 2\,CH_3-\overset{\displaystyle O}{\overset{\|}{C}}-COO^- + 2\,NADH + 4H^+$$

Glucose                                    Pyruvate

---

**SAMPLE PROBLEM  18.7**

### ■ Glycolysis

What are the reactions in glycolysis that generate ATP?

**SOLUTION**

ATP is produced when phosphate groups are transferred directly to ADP from 1,3-bis-phosphoglycerate (7) and from phosphoenolpyruvate (10).

**STUDY CHECK**

If four ATP molecules are produced in glycolysis, why is there a net yield of two ATPs?

---

## Pathways for Pyruvate

The pyruvate produced from glucose can now enter pathways that continue to extract energy. The available pathway depends on whether there is sufficient oxygen in the cell. During **aerobic** conditions, oxygen is available to convert pyruvate to acetyl coenzyme A (CoA). When oxygen levels are low, pyruvate is reduced to lactate.

### Aerobic Conditions

In glycolysis, two ATP molecules were generated when glucose was converted to pyruvate. However, much more energy is still available. The greatest amount of the energy is obtained from glucose when oxygen levels are high in the cells. Under aerobic conditions, pyruvate is oxidized and a carbon atom is removed from pyruvate as $CO_2$. The coenzyme $NAD^+$ is required for the oxidation. The resulting two-carbon acetyl compound is attached to CoA, producing **acetyl CoA**, an important intermediate in many metabolic pathways. (See Figure 18.10.)

$$CH_3-\overset{\displaystyle O}{\overset{\|}{C}}-\overset{\displaystyle O}{\overset{\|}{C}}-O^- + HS-CoA + NAD^+ \xrightarrow[\text{dehydrogenase}]{\text{Pyruvate}} CH_3-\overset{\displaystyle O}{\overset{\|}{C}}-S-CoA + CO_2 + NADH$$

Pyruvate                                                          Acetyl CoA

### Anaerobic Conditions

When we engage in strenuous exercise, the oxygen stored in our muscle cells is quickly depleted. Under anaerobic conditions, pyruvate is reduced to lactate. $NAD^+$ is produced

**FIGURE 18.10** Pyruvate is converted to acetyl CoA under aerobic conditions and to lactate under anaerobic conditions.
**Q** During vigorous exercise, why does lactate accumulate in the muscles?

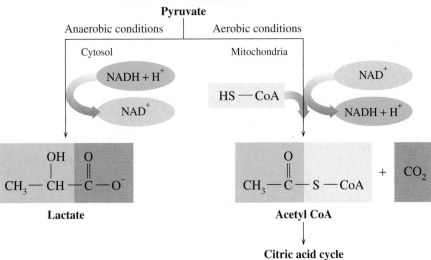

and is used to oxidize more glyceraldehyde-3-phosphate in the glycolysis pathway, which produces a small but needed amount of ATP.

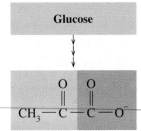

The accumulation of lactate causes the muscles to tire rapidly and become sore. After exercise, a person continues to breathe rapidly to repay the oxygen debt incurred during exercise. Most of the lactate is transported to the liver, where it is converted back into pyruvate. Under anaerobic conditions, the only ATP production in glycolysis occurs during the steps that phosphorylate ADP directly, giving a net gain of only two ATP molecules:

$$C_6H_{12}O_6 + 2\,ADP + 2\,P_i \longrightarrow 2\,CH_3{-}\underset{\underset{\text{Lactate}}{|}}{\overset{\overset{OH}{|}}{CH}}{-}COO^- + 2\,ATP$$

Glucose

Bacteria also convert pyruvate to lactate under anaerobic conditions. In the preparation of kimchee and sauerkraut, cabbage is covered with salt brine. The glucose from the starches is converted to lactate. This acid environment acts as a preservative that prevents the growth of other bacteria. The pickling of olives and cucumbers gives similar products. When cultures of bacteria that produce lactate are added to milk, the acid denatures the milk proteins to give sour cream and yogurt.

SAMPLE PROBLEM 18.8

■ Fates of Pyruvate

Is each of the following products from pyruvate produced under anaerobic or aerobic conditions?

**a.** acetyl CoA          **b.** lactate

SOLUTION

**a.** aerobic conditions     **b.** anaerobic conditions

STUDY CHECK

After strenuous exercise, some lactate is oxidized back to pyruvate by lactate dehydrogenase using $NAD^+$. Write an equation to show this reaction.

---

## QUESTIONS AND PROBLEMS

### Glycolysis: Oxidation of Glucose

**18.19** What is the starting compound of glycolysis?

**18.20** What is the end product of glycolysis?

**18.21** How is ATP used in the initial steps of glycolysis?

**18.22** How many ATP molecules are used in the initial steps of glycolysis?

**18.23** How does direct phosphorylation account for the production of ATP in glycolysis?

**18.24** Why are there two ATP molecules formed for one molecule of glucose?

**18.25** How many ATP or NADH are produced (or required) in each of the following steps in glycolysis?
**a.** glucose to glucose-6-phosphate
**b.** glyceraldehyde-3-phosphate to 1,3-bisphosphoglycerate
**c.** glucose to pyruvate

**18.26** How many ATP or NADH are produced (or required) in each of the following steps in glycolysis?

**a.** 1,3-bisphosphoglycerate to 3-phosphoglycerate
**b.** fructose-6-phosphate to fructose-1,6-bisphosphate
**c.** phosphoenolpyruvate to pyruvate

**18.27** What condition is needed in the cell to convert pyruvate to acetyl CoA?

**18.28** What coenzymes are needed for the oxidation of pyruvate to acetyl CoA?

**18.29** Write the overall equation for the conversion of pyruvate to acetyl CoA.

**18.30** What is the product of pyruvate under anaerobic conditions?

**18.31** How does the formation of lactate permit glycolysis to continue under anaerobic conditions?

**18.32** After running a marathon, a runner has muscle pain and cramping. What might have occurred in the muscle cells to cause this?

---

# 18.5 THE CITRIC ACID CYCLE

As a central pathway in metabolism, the citric acid cycle uses acetyl CoA from the degradation of carbohydrates, lipids, and proteins. The **citric acid cycle** is a series of reactions that degrades acetyl CoA to yield $CO_2$, $NADH + H^+$, and $FADH_2$.

## Overview of the Citric Acid Cycle

There are a total of 8 reactions in the citric acid cycle, which we can separate into two parts. In part 1, a two-carbon acetyl group bonds with a four-carbon oxaloacetate to yield

**LEARNING GOAL**

Describe the oxidation of acetyl CoA in the citric acid cycle.

**FIGURE 18.11** In part 1 of the citric acid cycle, two carbon atoms are removed as $CO_2$ from six-carbon citrate to give four-carbon succinyl CoA, which is converted in part 2 to four-carbon oxaloacetate.

**Q** What is the difference in part 1 and part 2 of the citric acid cycle?

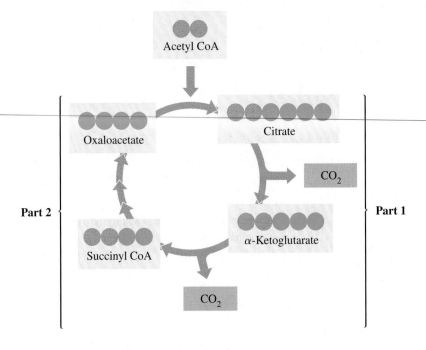

**WEB TUTORIAL**
Krebs Cycle

citrate. (See Figure 18.11.) Then decarboxylation reactions remove two carbon atoms as $CO_2$, producing a four-carbon compound. Then the reactions in part 2 convert the four-carbon compound back to oxaloacetate, which bonds with another acetyl CoA and goes through the cycle again. In one turn of the citric acid cycle, four oxidation reactions provide hydrogen ions and electrons, which are used to reduce FAD and $NAD^+$ coenzymes.

### Reaction 1    Formation of Citrate

In the first reaction, the two-carbon acetyl group in acetyl CoA bonds with the four-carbon oxaloacetate to yield citrate and CoA. (See Figure 18.12.)

### Reaction 2    Isomerization to Isocitrate

In order to continue oxidation, citrate undergoes isomerization to yield isocitrate. This is necessary because the secondary hydroxyl group (Chapter 12) in isocitrate can be oxidized in the next reaction, while the tertiary hydroxyl group in citrate cannot be oxidized.

### Reaction 3    First Oxidative Decarboxylation ($CO_2$)

This is the first time in the citric acid cycle that both an oxidation and a decarboxylation occur together. The oxidation converts the hydroxyl group to a ketone, and **decarboxylation** removes a carbon as a $CO_2$ molecule. The loss of $CO_2$ shortens the carbon chain to yield a five-carbon $\alpha$-ketoglutarate. The energy from the oxidation is used to transfer hydrogen ions and electrons to $NAD^+$. We can summarize the reactions as follows:

1. The hydroxyl group ($-OH$) is oxidized to a ketone ($C=O$).
2. $NAD^+$ is reduced to yield NADH.
3. A carboxylate group ($COO^-$) is removed as $CO_2$.

### Reaction 4    Second Oxidative Decarboxylation ($CO_2$)

In this reaction, a second $CO_2$ is removed as $\alpha$-ketoglutarate undergoes oxidative decarboxylation. The resulting four-carbon group combines with coenzyme A to form succinyl CoA, and hydrogen ions and electrons are transferred to $NAD^+$. The two reactions are as follows:

1. A second carbon is removed as $CO_2$.
2. $NAD^+$ is reduced to yield NADH.

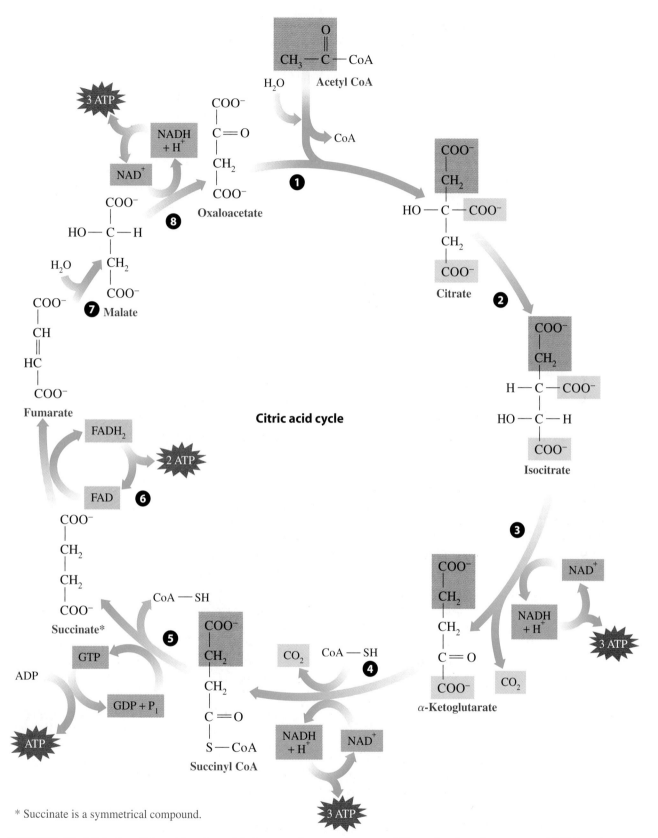

**Citric acid cycle**

* Succinate is a symmetrical compound.

**FIGURE 18.12** In the citric acid cycle, oxidation reactions produce two $CO_2$ and reduced coenzymes NADH and $FADH_2$, and it also regenerates oxaloacetate.

**Q** How many reactions in the citric acid cycle produce a reduced coenzyme?

### Reaction 5    Hydrolysis of Succinyl CoA

The energy released by the hydrolysis of succinyl CoA is used to add a phosphate group ($P_i$) directly to GDP (guanosine diphosphate). The products are succinate and GTP, a high-energy compound similar to ATP.

The hydrolysis of GTP is used to add a phosphate group to ADP, which regenerates GDP for the citric acid cycle. This is the only time in the citric acid cycle that a direct substrate phosphorylation is used to produce ATP.

$$GTP + ADP \longrightarrow GDP + ATP$$

### Reaction 6    Dehydrogenation of Succinate

In this oxidation reaction, hydrogen is removed from two carbon atoms in succinate, which produces fumarate, a compound with a trans double bond. This is the only place in the citric acid cycle where FAD is reduced to $FADH_2$.

### Reaction 7    Hydration

In a hydration reaction, water adds to the double bond of fumarate to yield malate.

### Reaction 8    Dehydrogenation Forms Oxaloacetate

In the last step of the citric acid cycle, the hydroxyl (—OH) group in malate is oxidized to yield oxaloacetate, which has a ketone group. The coenzyme $NAD^+$ is reduced to $NADH + H^+$.

## Summary of Products from the Citric Acid Cycle

The products from one turn of the citric acid cycle are:

2 $CO_2$
3 NADH
1 $FADH_2$
1 GTP used to form ATP
1 CoA
2 $H^+$

We can write the overall chemical equation for one turn of the citric acid cycle as follows:

$$\text{Acetyl CoA} + 3\,NAD^+ + FAD + GDP + P_i + 2H_2O \longrightarrow$$

$$2CO_2 + 3\,NADH + 2H^+ + FADH_2 + CoA + GTP$$

---

### SAMPLE PROBLEM    18.9

■ **Citric Acid Cycle**

When one acetyl CoA completes the citric acid cycle, how many of each of the following is produced?

**a.** NADH
**b.** ketone group
**c.** $CO_2$

SOLUTION

**a.** One turn of the citric acid cycle produces three molecules of NADH.
**b.** Two ketone groups form when the secondary alcohol groups in isocitrate and malate are oxidized by $NAD^+$.

c. Two molecules of $CO_2$ are produced by the decarboxylation of isocitrate and α-ketoglutarate.

**STUDY CHECK**

What is the substrate in the first reaction of the citric acid cycle, as well as a product in the last reaction?

---

## QUESTIONS AND PROBLEMS

### The Citric Acid Cycle

**18.33** What are the products from one turn of the citric acid cycle?

**18.34** What compounds are needed to start the citric acid cycle?

**18.35** Which reactions of the citric acid cycle involve oxidative decarboxylation?

**18.36** Which reactions of the citric acid cycle involve a hydration reaction?

**18.37** Which reactions of the citric acid cycle reduce $NAD^+$?

**18.38** Which reactions of the citric acid cycle reduce FAD?

**18.39** Where does a direct substrate phosphorylation occur?

**18.40** What is the total NADH and total $FADH_2$ produced in one turn of the citric acid cycle?

**18.41** Refer to the diagram of the citric acid cycle to answer each of the following:
  a. What are the six-carbon compounds?
  b. How is the number of carbon atoms decreased?
  c. What are the five-carbon compounds?
  d. Which reactions are oxidation reactions?
  e. In which reactions are secondary alcohols oxidized?

**18.42** Refer to the diagram of the citric acid cycle to answer each of the following:
  a. What is the yield of $CO_2$ molecules?
  b. What are the four-carbon compounds?
  c. What is the yield of GTP molecules?
  d. What are the decarboxylation reactions?
  e. Where does a hydration occur?

---

# 18.6 ELECTRON TRANSPORT

In **electron transport**, which occurs in the mitochondria, hydrogen ions and electrons from NADH and $FADH_2$ pass from one electron carrier to the next until they combine with oxygen to form $H_2O$.

## Electron Carriers

There are four types of electron carriers that make up the electron transport system.

1. **Fe–S clusters** are groups of proteins in electron transport that contain iron ions, inorganic sulfides, and several cysteine —SH side chains. The iron in the Fe-S clusters is reduced to $Fe^{2+}$ and oxidized to $Fe^{3+}$ as electrons are accepted and lost. (See Figure 18.13.)

**LEARNING GOAL**

Describe the electron carriers involved in electron transport.

**WEB TUTORIAL**
Electron Transport

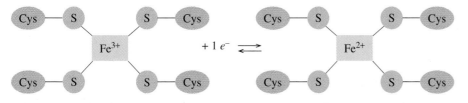

**Typical Fe–S cluster**

**FIGURE 18.13** In a typical iron–sulfur cluster, an iron ion bonds to sulfur atoms in the thiol (—SH) groups of four cysteines in proteins.
**Q** In an iron–sulfur cluster, what are the ionic charges of the oxidized and reduced iron ions?

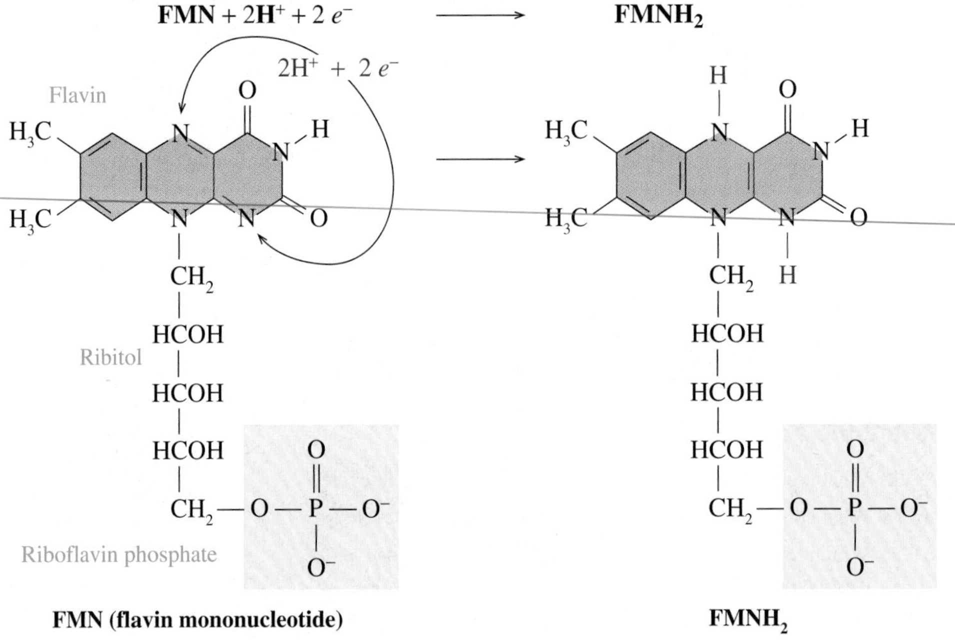

**FIGURE 18.14** The electron carrier FMN consists of a flavin ring system containing the reactive center, ribitol, and a phosphate group.
**Q** What part of the FMN molecule is reduced when hydrogen ions and electrons are accepted?

2. **FMN (flavin mononucleotide)** is a coenzyme derived from riboflavin (vitamin $B_2$). In riboflavin, the ring system is attached to ribitol, the sugar alcohol of ribose. (See Figure 18.14.) The reduced product is $FMNH_2$.
3. **Coenzyme Q (Q or CoQ)** is derived from quinone, which is a six-carbon cyclic compound with two double bonds and two keto groups attached to a long carbon chain. (See Figure 18.15.) Coenzyme Q is reduced when the keto groups of quinone accept hydrogen ions and electrons.
4. **Cytochromes (cyt)** are proteins that contain an iron ion in a heme group. The different cytochromes are indicated by the letters following the abbreviation for cytochrome (cyt): cyt $b$, cyt $c_1$, cyt $c$, cyt $a$, and cyt $a_3$. In each cytochrome, the $Fe^{3+}$ accepts a single electron to form $Fe^{2+}$, which is oxidized back to $Fe^{3+}$ when the electron is passed to the next cytochrome. (See Figure 18.16.)

$$Fe^{3+} + 1\ e^- \rightleftarrows Fe^{2+}$$

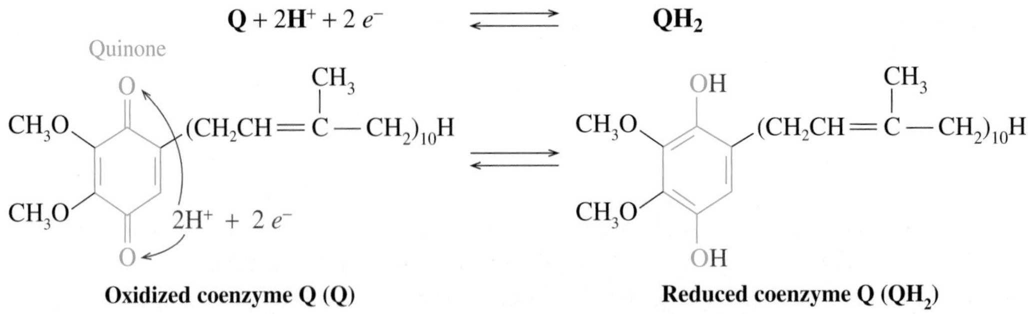

**FIGURE 18.15** The electron carrier coenzyme Q accepts electrons from $FADH_2$ and $FMNH_2$ and passes them to the cytochromes.
**Q** How does reduced coenzyme Q differ from the oxidized form?

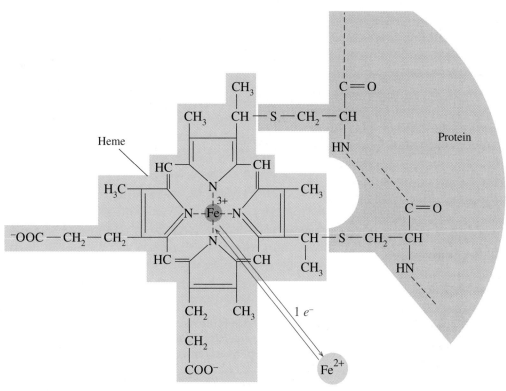

**Simplified structure of cytochromes $c$ and $c_1$**

**FIGURE 18.16** The iron-containing proteins known as cytochromes are identified as $b$, $c$, $c_1$, $a$, and $a_3$.

**Q** What are the reduced and oxidized forms of the cytochromes?

---

SAMPLE PROBLEM 18.10

### ■ Oxidation and Reduction

Identify the following steps in electron transport as oxidation or reduction:

**a.** $FMN + 2H^+ + 2 e^- \longrightarrow FMNH_2$
**b.** Cyt $c$ $(Fe^{2+}) \longrightarrow$ cyt $c$ $(Fe^{3+}) + e^-$

SOLUTION

**a.** The gain of hydrogen is reduction.
**b.** The loss of electrons is oxidation.

STUDY CHECK

Identify each of the following as oxidation or reduction:
**a.** $QH_2 \longrightarrow Q$
**b.** Cyt $b$ $(Fe^{3+}) \longrightarrow$ cyt $b$ $(Fe^{2+})$

---

## Electron Transfer

A mitochondrion consists of an outer membrane, an intermembrane space, and an inner membrane that surrounds the matrix. Along the highly folded inner membrane are the enzymes and electron carriers required for electron transport.

Two electron carriers, coenzyme Q and cytochrome $c$, are not firmly attached to the membrane. They function as mobile carriers shuttling electrons between the protein complexes that are tightly bound to the membrane. (See Figure 18.17.)

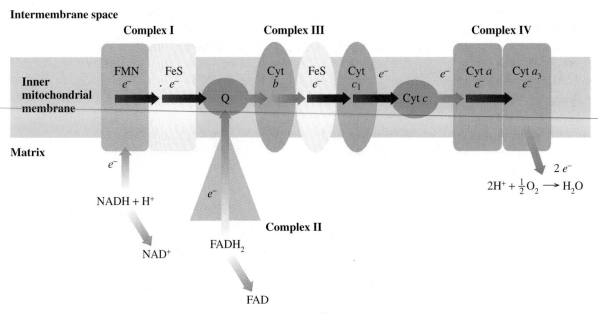

**FIGURE 18.17** Electron carriers in electron transport are found in protein complexes on the inner membrane of the mitochondria. CoQ and Cyt *c* are mobile carriers that carry electrons between the protein complexes. **Q** What is the function of the electron carriers coenzyme Q and cytochrome *c*?

## Complex I: NADH Dehydrogenase

At complex I, NADH transfers hydrogen ions and electrons to FMN. The reduced $FMNH_2$ forms, while NADH is reoxidized to $NAD^+$; this returns to oxidative pathways such as the citric acid cycle to oxidize more substrates.

$$NADH + H^+ + FMN \longrightarrow NAD^+ + FMNH_2$$

Within complex I, the NADH electrons are transferred to Fe–S clusters and then to coenzyme Q.

$$FMNH_2 + Q \longrightarrow QH_2 + FMN$$

The overall reaction sequence in complex I can be written as follows:

$$NADH + H^+ + Q \longrightarrow QH_2 + NAD^+$$

## Complex II: Succinate Dehydrogenase

Complex II is specifically used when $FADH_2$ is generated by the conversion of succinate to fumarate in the citric acid cycle. The electrons from $FADH_2$ are transferred to coenzyme Q to yield $QH_2$. Because complex II is at a lower energy level than complex I, the electrons from $FADH_2$ enter electron transport at a lower energy level than those from NADH.

$$FADH_2 + Q \longrightarrow FAD + QH_2$$

## Complex III: Coenzyme Q-Cytochrome *c* Reductase

The mobile carrier $QH_2$ transfers electrons from NADH and $FADH_2$ to an iron–sulfur (Fe–S) cluster and then to cytochrome *b*, the first cytochrome in complex III.

$$QH_2 + 2 \text{ cyt } b \text{ (Fe}^{3+}) \longrightarrow Q + 2 \text{ cyt } b \text{ (Fe}^{2+}) + 2H^+$$

From cyt $b$, the electron is transferred to an Fe–S cluster, then to cytochrome $c_1$, and then to cytochrome $c$. Each time $Fe^{3+}$ accepts an electron, it is reduced to $Fe^{2+}$, and then oxidized back to $Fe^{3+}$. Cytochrome $c$, another mobile carrier, moves the electron from complex III to complex IV.

## Complex IV: Cytochrome c Oxidase

At complex IV, electrons are transferred from cytochrome $c$ to cytochrome $a$, and then to cytochrome $a_3$, the last cytochrome. In the final step of electron transport, electrons and hydrogen ions combine with oxygen ($O_2$) to form water.

$$2H^+ + 2e^- + \frac{1}{2}O_2 \longrightarrow H_2O$$

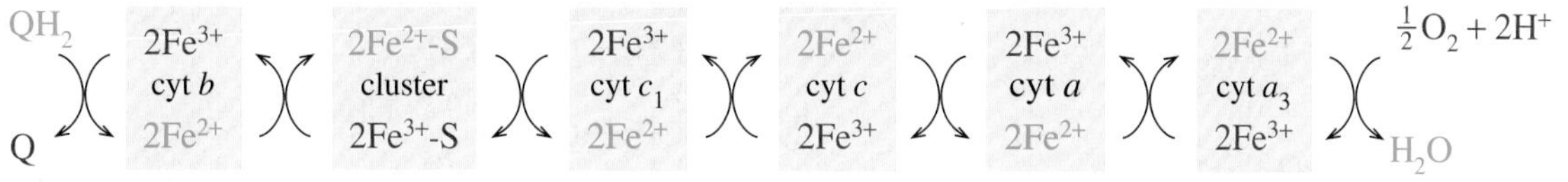

---

### SAMPLE PROBLEM  18.11

■ **Electron Transport**

Identify which of the following electron carriers are mobile carriers:

**a.** Cyt $c$                **b.** FMN
**c.** Fe–S clusters          **d.** Q

**SOLUTION**

**a.** and **d.**

**STUDY CHECK**

What is the final substance that accepts electrons in electron transport?

---

## QUESTIONS AND PROBLEMS

### Electron Transport

**18.43** Is cyt $b$ ($Fe^{3+}$) the abbreviation for the oxidized or reduced form of cytochrome $b$?

**18.44** Is $FMNH_2$ the abbreviation for the oxidized or reduced form of flavin mononucleotide?

**18.45** Identify the following as oxidation or reduction:
  **a.** $FMNH_2 \longrightarrow FMN + 2H^+ + 2e^-$
  **b.** $Q + 2H^+ + 2e^- \longrightarrow QH_2$

**18.46** Identify the following as oxidation or reduction:
  **a.** Cyt $c$ ($Fe^{3+}$) $+ e^- \longrightarrow$ cyt $c$ ($Fe^{2+}$)
  **b.** $Fe^{2+}$–S cluster $\longrightarrow Fe^{3+}$–S cluster $+ e^-$

**18.47** What reduced coenzymes provide the electrons for electron transport?

**18.48** What happens to the energy level as electrons are passed along in electron transport?

**18.49** Arrange the following in the order they appear in electron transport: cytochrome $c$, cytochrome $b$, FAD, and coenzyme Q.

**18.50** Arrange the following in the order they appear in electron transport: $O_2$, $NAD^+$, cytochrome $a_3$, and FMN.

**18.51** How are electrons carried from complex I to complex III?

**18.52** How are electrons carried from complex III to complex IV?

**18.53** How is NADH oxidized in electron transport?

**18.54** How is $FADH_2$ oxidized in electron transport?

**18.55** Complete the following reactions in electron transport:
  **a.** NADH $+ H^+ +$ _____ $\longrightarrow$ _____ $+ FMNH_2$
  **b.** $QH_2 + 2$ cyt $b$ ($Fe^{3+}$) $\longrightarrow$ _____ $+$ _____ $+ 2H^+$

**18.56** Complete the following reactions in electron transport:
  **a.** Q $+$ _____ $\longrightarrow$ _____ $+ FAD$
  **b.** $2$ cyt $a$ ($Fe^{2+}$) $+ 2$ cyt $a_3$ ($Fe^{3+}$) $\longrightarrow$ _____ $+$ _____

## Health Note

### Toxins: Inhibitors of Electron Transport

Several substances can inhibit the electron carriers in the electron transport system. Rotenone, a product from a plant root used in South America to poison fish, blocks electron transport between the NADH dehydrogenase (complex I) and coenzyme Q. The barbiturates amytal and Demerol also inhibit the NADH dehydrogenase complex. Another inhibitor is the antibiotic antimycin A, which blocks the flow of electrons between cytochrome $b$ and cytochrome $c_1$ (complex III). Another group of compounds, including cyanide and carbon monoxide, inhibit cytochrome $c$ oxidase (complex IV). The toxic nature of these compounds makes it clear that an organism relies heavily on the process of electron transport.

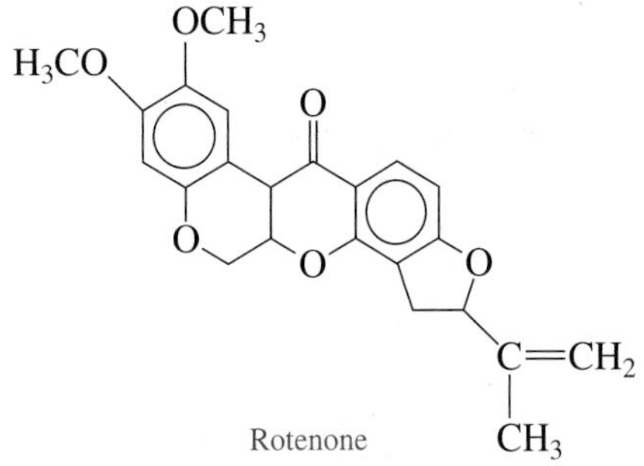

Rotenone

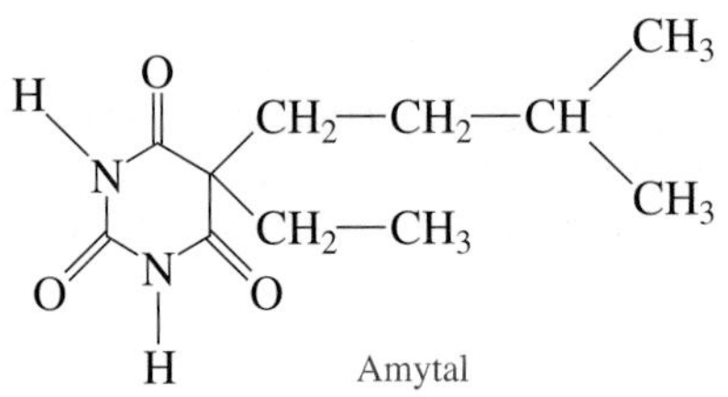

Amytal

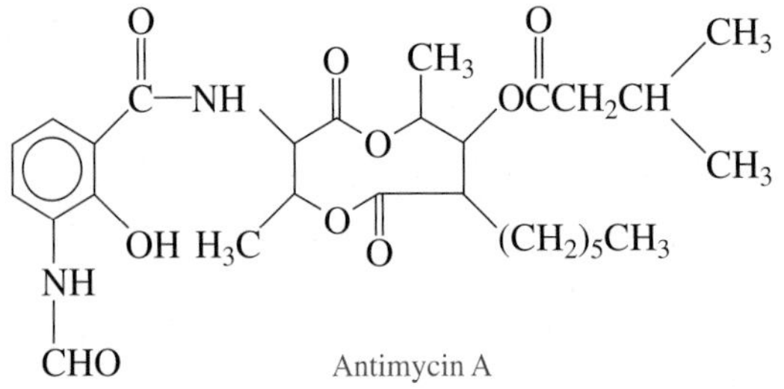

Antimycin A

When an inhibitor blocks a step in electron transport, the carriers preceding that step are unable to transfer electrons and remain in their reduced forms. All the carriers after the blocked step remain oxidized without a source of electrons. Thus, any of these inhibitors can shut down the flow of electrons in the electron transport system.

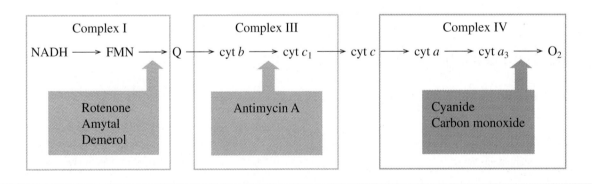

## 18.7 OXIDATIVE PHOSPHORYLATION AND ATP

### LEARNING GOAL

Describe the process of oxidative phosphorylation in ATP synthesis; calculate the ATP from complete oxidation of glucose.

We have seen that energy is generated when electrons from the oxidation of substrates flow through electron transport. Now we will look at how that energy is used in the production of ATP for the cell, which is a process called **oxidative phosphorylation**.

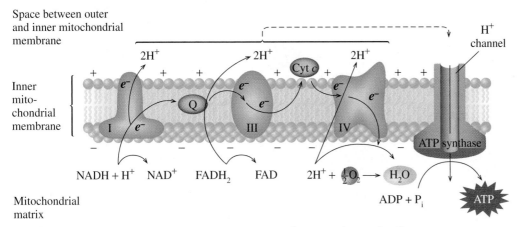

**FIGURE 18.18** In electron transport, protein complexes oxidize and reduce coenzymes to provide electrons and protons that move into the intermembrane space where they create a proton gradient that drives ATP synthesis.
**Q** What is the major source of NADH for electron transport?

## Chemiosmotic Model

In 1978, Peter Mitchell received the Nobel Prize in chemistry for his theory called the **chemiosmotic model**, which links the energy from electron transport to a proton gradient that drives the synthesis of ATP. In this model, three of the complexes (I, III, and IV) extend through the inner membrane with one end of each complex in the matrix and the other end in the intermembrane space. In the chemiosmotic model, each of these complexes act as a **proton pump** by pushing protons (H$^+$) out of the matrix and into the intermembrane space. This increase in protons in the intermembrane space lowers the pH and creates a proton gradient. Because protons are positively charged, both the lower pH and the electrical charge of the proton gradient make it an electrochemical gradient. (See Figure 18.18.)

To equalize the pH between the intermembrane space and the matrix, there is a tendency by the protons to return to the matrix. However, protons cannot diffuse through the inner membrane. The only way protons can return to the matrix is to pass through a protein complex called **ATP synthase**. We might think of the flow of protons as a stream or river that turns a waterwheel. As the protons flow through ATP synthase, energy generated from the proton gradient is used to drive the ATP synthesis. Thus the process of oxidative phosphorylation couples the energy from electron transport to the synthesis of ATP from ADP and P$_i$.

$$\text{ADP} + \text{P}_i + \text{energy} \xrightarrow{\text{ATP synthase}} \text{ATP}$$

## Electron Transport and ATP Synthesis

We have seen that oxidative phosphorylation couples the energy from electron transport with the synthesis of ATP. Because NADH enters electron transport at complex I, energy is released from the oxidation of NADH to synthesize three ATP molecules. However, FADH$_2$, which enters at a lower energy level at complex II, provides energy to produce only two ATPs. (See Figure 18.19.) The overall equation for the oxidation of NADH and FADH$_2$ can be written as follows:

$$\boxed{\textbf{NADH} + \textbf{H}^+} + \tfrac{1}{2}\text{O}_2 + 3\text{ADP} + 3\text{P}_i \longrightarrow \text{NAD}^+ + \text{H}_2\text{O} + \boxed{\textbf{3 ATP}}$$

$$\boxed{\textbf{FADH}_2} + \tfrac{1}{2}\text{O}_2 + 2\text{ADP} + 2\text{P}_i \longrightarrow \text{FAD} + \text{H}_2\text{O} + \boxed{\textbf{2 ATP}}$$

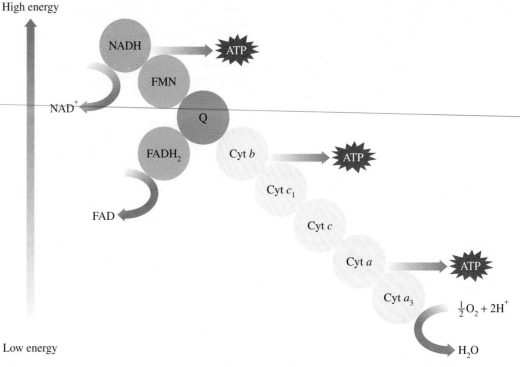

**FIGURE 18.19** As the energy levels decrease in the flow of electrons along the major electron carriers, electron transfers release sufficient energy to drive ATP synthesis.
**Q** Why does electron transfer from FADH₂ provide less energy than from NADH and H⁺?

---

**SAMPLE PROBLEM 18.12**

### ■ ATP Synthesis

Why does the oxidation of NADH provide energy for the formation of three ATP molecules, whereas $FADH_2$ produces two ATPs?

#### SOLUTION

Electrons from the oxidation of NADH enter electron transport at a higher energy level than $FADH_2$. NADH provides energy to pump three pairs of protons into the intermembrane to form three ATPs, whereas energy from $FADH_2$ pumps two pairs of protons to form two ATPs.

#### STUDY CHECK

Which complexes on the inner membrane act as proton pumps?

---

## ATP from Glycolysis

In glycolysis, the oxidation of glucose stores energy in two NADH molecules as well as two ATP molecules from direct substrate phosphorylation. However, glycolysis occurs in the cytoplasm, and the NADH produced cannot pass through the mitochondrial membrane.

Therefore the hydrogen ions and electrons from NADH in the cytoplasm are transferred to compounds that can enter the mitochondria. In this shuttle system, $FADH_2$ is produced. The overall reaction for the shuttle is

$$NADH + H^+ + FAD \longrightarrow NAD^+ + FADH_2$$

Cytoplasm                                    Mitochondria

Therefore the transfer of electrons from NADH in the cytoplasm to $FADH_2$ produces only two ATPs, rather than three. In glycolysis, glucose yields a total of six ATP; four ATPs from two NADHs and two ATPs from direct phosphorylation.

$$Glucose \longrightarrow 2 \text{ pyruvate} + 2 \text{ ATP} + 2 \text{ NADH} (\longrightarrow 2 \text{ } FADH_2)$$
$$Glucose \longrightarrow 2 \text{ pyruvate} + 6 \text{ ATP}$$

## ATP from the Oxidation of Two Pyruvates

Under aerobic conditions, pyruvate enters the mitochondria, where it is oxidized to give acetyl CoA, $CO_2$, and NADH. Because glucose yields two pyruvates, two NADHs enter electron transport. The oxidation of two pyruvates leads to the production of six ATPs.

$$2 \text{ Pyruvate} \longrightarrow 2 \text{ acetyl CoA} + 6 \text{ ATP}$$

## ATP from the Citric Acid Cycle

One turn of the citric acid cycle produces two $CO_2$, three NADH, one $FADH_2$, and one ATP molecule by direct substrate phosphorylation.

$$3 \text{ NADH} \times 3 \text{ ATP} = 9 \text{ ATP}$$
$$1 \text{ } FADH_2 \times 2 \text{ ATP} = 2 \text{ ATP}$$
$$\underline{1 \text{ GTP} \times 1 \text{ ATP} = 1 \text{ ATP}}$$
$$\text{Total (one turn)} \quad = 12 \text{ ATP}$$

*Health Note*

## ATP Synthase and Heating the Body

Some types of compounds, called *uncouplers*, separate the electron transport system from the ATP synthase. They do this by disrupting the proton gradient needed for the synthesis of ATP. The electrons are transported to $O_2$ in electron transport, but ATP is not formed by ATP synthase.

Some uncouplers transport the protons through the inner membrane, which is normally impermeable to protons; others block the channel in the ATP synthase. Compounds such as dicumarol and 2,4-dinitrophenol (DNP) are hydrophobic and bind with protons to carry them across the inner membrane. An antibiotic, oligomycin, blocks the channel, which does not allow any protons to return to the matrix. By removing protons or blocking the channel, there is no proton flow to generate energy for ATP synthesis.

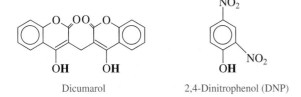

Dicumarol                    2,4-Dinitrophenol (DNP)

When there is no mechanism for ATP synthesis, the energy of electron transport is released as heat. Certain animals that are adapted to cold climates have developed their own uncoupling system, which allows them to use electron transport energy for heat production.

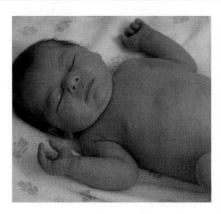

These animals have large amounts of a tissue called brown fat, which contains a high concentration of mitochondria. This tissue is brown because of the color of iron in the cytochromes of the mitochondria. The proton pumps still operate in brown tissue, but a protein embedded in the cell wall allows the protons to bypass ATP synthase. The energy that would be used to synthesize ATP is released as heat. In newborn babies, brown fat is used to generate heat because newborns have not stored much fat. The brown fat deposits are located near major blood vessels, which carry the warmed blood to the body. Infants have a small mass but a large surface area and need to produce more heat than adults. Most adults have little or no brown fat, although someone who works outdoors in a cold climate will develop some brown fat deposits.

Plants also use uncouplers. In early spring, heat is used to warm early shoots of plants under the snow, which helps them melt the snow around the plants. Some plants, such as skunk cabbage, use uncoupling agents to volatilize fragrant compounds that attract insects to pollinate the plants.

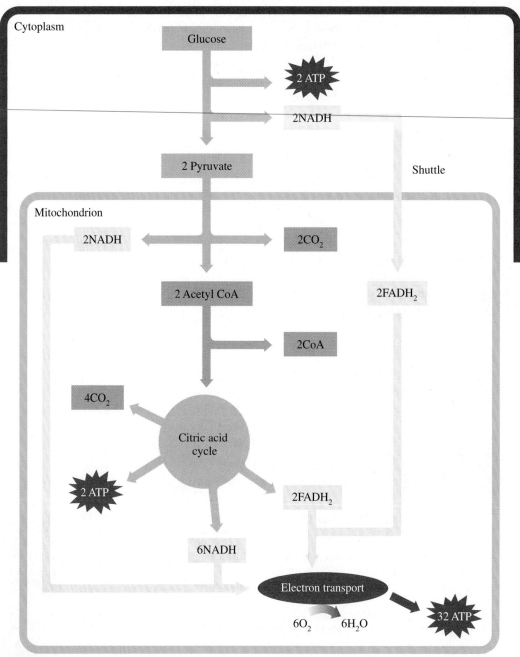

**FIGURE 18.20** The complete oxidation of glucose to $CO_2$ and $H_2O$ yields a total of 36 ATPs.
**Q** What metabolic pathway produces most of the ATP from the oxidation of glucose?

Because one glucose produces two acetyl CoAs, two turns of the citric acid cycle produce 24 ATPs.

Acetyl CoA $\longrightarrow$ $2CO_2$ + 12 ATP (one turn of citric acid cycle)

2 Acetyl CoA $\longrightarrow$ $4CO_2$ + 24 ATP (two turns of citric acid cycle)

## ATP from the Complete Oxidation of Glucose

The total ATP for the complete oxidation of glucose is calculated by combining the ATP produced from glycolysis, the oxidation of pyruvate, and the citric acid cycle. (See Figure 18.20.) The ATP produced for these reactions is given in Table 18.2.

**TABLE 18.2** ATP Produced by the Complete Oxidation of Glucose

| Reaction | ATP for One Glucose |
|---|:---:|
| **ATP from Glycolysis** | |
| Activation of glucose | −2 ATP |
| Oxidation of glyceraldehyde-3-phosphate (2 NADH) | 6 ATP |
| Transport of 2 NADHs across membrane | −2 ATP |
| Direct ADP phosphorylation (two triose phosphate) | 4 ATP |
| Summary: $C_6H_{12}O_6 \longrightarrow$ 2 pyruvate + $2H_2O$<br>　　　　Glucose | 6 ATP |
| **ATP from Pyruvate** | |
| 2 Pyruvate $\longrightarrow$ 2 acetyl CoA (2 NADH) | 6 ATP |
| **ATP from Two Turns of the Citric Acid Cycle** | |
| Oxidation of 2 isocitrate (2 NADH) | 6 ATP |
| Oxidation of 2 $\alpha$-ketoglutarate (2 NADH) | 6 ATP |
| 2 Direct substrate phosphorylations (2 GTP) | 2 ATP |
| Oxidation of 2 succinate (2 $FADH_2$) | 4 ATP |
| Oxidation of 2 malate (2 NADH) | 6 ATP |
| Summary: 2 Acetyl CoA $\longrightarrow$ $4CO_2$ + $2H_2O$ | 24 ATP |
| **Overall ATP Production for One Glucose** | |
| $C_6H_{12}O_6$ + $6O_2$ + 36 ADP + $36P_i \longrightarrow 6CO_2 + 6H_2O$ + 36 ATP<br>　Glucose | |

---

## SAMPLE PROBLEM 18.13

### ■ ATP Production

Indicate the amount of ATP produced by each of the following oxidation reactions:

**a.** pyruvate to acetyl CoA
**b.** glucose to acetyl CoA

#### SOLUTION

**a.** The oxidation of pyruvate to acetyl CoA produces one NADH, which yields three ATPs.
**b.** Six ATPs are produced from the oxidation of glucose to two pyruvates. Another six ATPs result from the oxidation of two pyruvates to two acetyl CoAs. A total of 12 ATPs are produced from the oxidation of glucose to two acetyl CoAs.

#### STUDY CHECK

What are the sources and amounts of ATP for one turn of the citric acid cycle?

When glucose is not immediately used by the cells for energy, it is stored as glycogen in the liver and muscles. When the levels of glucose in the brain or blood become low, the glycogen reserves are hydrolyzed and glucose is released into the blood. If glycogen stores are depleted, some glucose can be synthesized from noncarbohydrate sources. It is the balance of all these reactions that maintains the necessary blood glucose level available to our cells and provides the necessary amount of ATP for our energy needs.

## QUESTIONS AND PROBLEMS

### Oxidative Phosphorylation and ATP

**18.57** What is meant by oxidative phosphorylation?

**18.58** How is the proton gradient established?

**18.59** According to the chemiosmotic theory, how does the proton gradient provide energy to synthesize ATP?

**18.60** How does the phosphorylation of ADP occur?

**18.61** How are glycolysis and the citric acid cycle linked to the production of ATP by electron transport?

**18.62** Why does $FADH_2$ yield two ATPs, using the electron transport system, but NADH yields three ATPs?

**18.63** What is the energy yield in ATP molecules associated with each of the following?
**a.** $NADH \longrightarrow NAD^+$
**b.** glucose $\longrightarrow$ 2 pyruvate
**c.** 2 pyruvate $\longrightarrow$ 2 acetyl CoA + $2CO_2$
**d.** acetyl CoA $\longrightarrow$ $2CO_2$

**18.64** What is the energy yield in ATP molecules associated with each of the following?
**a.** $FADH_2 \longrightarrow FAD$
**b.** glucose + $6O_2 \longrightarrow 6CO_2 + 6H_2O$
**c.** glucose $\longrightarrow$ 2 lactate
**d.** pyruvate $\longrightarrow$ lactate

---

## LEARNING GOAL

Describe the metabolic pathway of $\beta$ oxidation; calculate the ATP from the complete oxidation of a fatty acid.

# 18.8 OXIDATION OF FATTY ACIDS

A large amount of energy is obtained when fatty acids undergo oxidation in the mitochondria to yield acetyl CoA. In stage 2 of fat metabolism, fatty acids undergo **beta oxidation ($\beta$ oxidation)**, which removes two-carbon segments, one at a time, from a fatty acid.

Each cycle in $\beta$ oxidation produces acetyl CoA and a fatty acid that is shorter by two carbons. The cycle repeats until the original fatty acid is completely degraded to two-carbon acetyl CoA units. Each acetyl CoA can then enter the citric acid cycle in the same way as the acetyl CoA units derived from glucose.

## Fatty Acid Activation

Fatty acids are prepared for transport through the inner membrane of the mitochondria by adding an acyl group to the fatty acid in the cytosol. This process, called *activation*, combines a fatty acid with coenzyme A to yield fatty acyl CoA. The energy for the activation is obtained from the hydrolysis of ATP to give AMP (adenosine monophosphate) and two inorganic phosphates (2 $P_i$). This is equivalent to the energy released from the hydrolysis of two ATPs to two ADPs.

## Reactions of $\beta$-Oxidation Cycle

In the matrix, fatty acyl CoA molecules undergo $\beta$ oxidation, which is a cycle of four reactions that convert the $-CH_2-$ of the $\beta$ carbon to a $\beta$-keto group. Once the $\beta$-keto group is formed, a two-carbon acetyl group can be split from the carbon chain, which shortens the fatty acyl group.

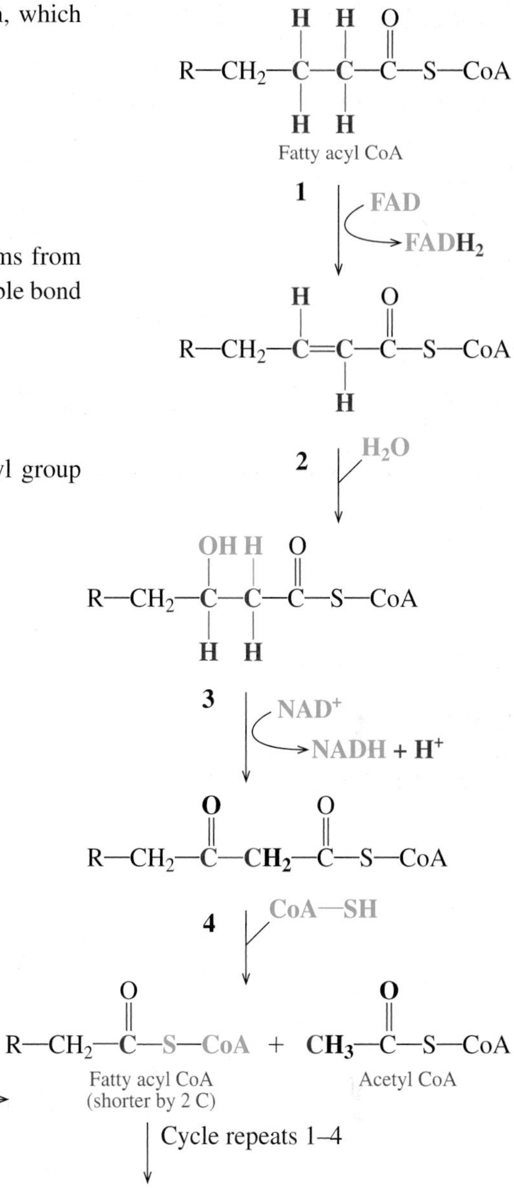

## β-Oxidation Pathway

### Reaction 1  Oxidation (Dehydrogenation)

In the first reaction of β oxidation, the FAD coenzyme removes hydrogen atoms from the $\alpha$ and $\beta$ carbons of the activated fatty acid to form a trans carbon–carbon double bond and $FADH_2$.

### Reaction 2  Hydration

A water molecule now adds to the trans double bond, which places a hydroxyl group (—OH) on the $\beta$ carbon.

### Reaction 3  Oxidation (Dehydrogenation)

The hydroxyl group on the $\beta$ carbon is oxidized to yield a ketone. The hydrogen atoms removed in the dehydrogenation reduce coenzyme $NAD^+$ to $NADH + H^+$. At this point, the $\beta$ carbon has been oxidized to a keto group.

### Reaction 4  Cleavage of Acetyl CoA

In the final step of β oxidation, the $C_\alpha$—$C_\beta$ bond splits to yield free acetyl CoA and a fatty acyl CoA molecule that is shorter by two carbon atoms. This shorter fatty acyl CoA is ready to go through the β-oxidation cycle again.

The reaction for one cycle of β oxidation is written as follows:

$$R-CH_2-CH_2-\overset{\overset{\displaystyle O}{\|}}{C}-S-CoA + NAD^+ + FAD + H_2O + HS-CoA \longrightarrow$$

Fatty acyl CoA

$$R-\overset{\overset{\displaystyle O}{\|}}{C}-S-CoA + CH_3-\overset{\overset{\displaystyle O}{\|}}{C}-S-CoA + NADH + H^+ + FADH_2$$

New fatty acyl (−2 C) CoA     Acetyl CoA

Fatty acyl CoA

1 | FAD → FADH₂

2 | H₂O

3 | NAD⁺ → NADH + H⁺

4 | CoA—SH

Fatty acyl CoA (shorter by 2 C) + Acetyl CoA

Cycle repeats 1–4

---

## SAMPLE PROBLEM  18.14

### ■ β Oxidation

Match each of the following with reactions in the β-oxidation cycle:
(1) first oxidation  (2) hydration  (3) second oxidation  (4) cleavage

**a.** Water is added to a trans double bond.  **b.** An acetyl CoA is removed.
**c.** FAD is reduced to $FADH_2$.  **d.** $NAD^+$ is reduced to NADH.

#### SOLUTION

**a.** (2) hydration  **b.** (4) cleavage
**c.** (1) first oxidation  **d.** (3) second oxidation

#### STUDY CHECK

Which coenzyme is needed in reaction 3 when a β-hydroxyl group is converted to a β-keto group?

*Explore Your World*

## Fat Storage and Blubber

Obtain two medium-sized plastic baggies and a can of Crisco or other type of shortening used for cooking. You will also need a bucket or large container with water and ice cubes. Place several tablespoons of the shortening in one of the baggies. Place the second baggie inside and tape the outside edges of the bag. With your hand inside the inner baggie, move the shortening around to cover your hand. With one hand inside the double baggie, submerge both your hands in the container of ice water. Measure the time it takes for one hand to feel uncomfortably cold. Experiment with different amounts of shortening.

### QUESTIONS

1. How effective is the bag with "blubber" in protecting your hand from the cold?
2. How does blubber help an animal survive starvation?
3. How would twice the amount of shortening affect your results?
4. Why would animals in warm climates, such as camels and migratory birds, need to store fat?
5. If you placed 300 g of shortening in the baggie, how many moles of ATP could it provide if used for energy production (assume it produces the same ATP as myristic acid)?

## Fatty Acid Length Determines Cycle Repeats

The number of carbon atoms in a fatty acid determines the number of times the cycle repeats and the number of acetyl CoA units it produces. For example, the complete $\beta$ oxidation of myristic acid ($C_{14}$) produces seven acetyl CoA groups, which is equal to one-half the number of carbon atoms in the fatty acid. Because the final turn of the cycle produces two acetyl CoA groups, the total number of times the cycle repeats is one less than the total number of acetyl groups it produces. Therefore, the $C_{14}$ fatty acid goes through the cycle six times.

| Fatty Acid | Number of Acetyl CoA | $\beta$-Oxidation Cycles |
|---|---|---|
| Myristic acid $C_{14}$ | 7 | 6 |
| Palmitic acid $C_{16}$ | 8 | 7 |
| Stearic acid $C_{18}$ | 9 | 8 |

We can write an overall equation for the complete oxidation of myristyl CoA as follows:

$$\text{Myristyl CoA} + 6\text{CoA} + 6\text{FAD} + 6\text{NAD}^+ + 6\text{H}_2\text{O} \longrightarrow$$
$$7 \text{ acetyl CoA} + 6\text{FADH}_2 + 6\text{NADH} + 6\text{H}^+$$

## ATP from Fatty Acid Oxidation

We can now determine the total energy yield from the oxidation of a particular fatty acid. In each $\beta$-oxidation cycle, one NADH, one $FADH_2$, and one acetyl CoA are produced. Each NADH generates sufficient energy to synthesize three ATPs, whereas $FADH_2$ leads to the synthesis of two ATPs. However, the greatest amount of energy produced from a fatty acid is generated by the production of the acetyl CoA units that enter the citric acid cycle. Earlier, we saw that one acetyl CoA leads to the synthesis of a total of 12 ATP.

So far we know that the $C_{14}$ acid produces seven acetyl CoA units and goes through six turns of the cycle. We also need to remember that activation of the myristic acid requires the equivalent of two ATP. We can set up the calculation as follows.

**ATP Production for Myristic Acid**

| | |
|---|---|
| **Activation** | $-2$ ATP |
| **7 acetyl CoA** | |
| 7 acetyl CoA $\times$ 12 ATP/acetyl CoA | 84 ATP |
| **6 $\beta$-oxidation cycles** | |
| 6 $FADH_2 \times$ 2 ATP/$FADH_2$ | 12 ATP |
| 6 NADH $\times$ 3 ATP/NADH | 18 ATP |
| Total | 112 ATP |

# Health Note

## Stored Fat and Obesity

The storage of fat is an important survival feature in the lives of many animals. In hibernating animals, large amounts of stored fat provide the energy for the entire hibernation period, which could be several months. In camels, such as dromedary camels, large amounts of food are stored in the camel's hump, which is actually a huge fat deposit. When food resources are low, the camel can survive months without food or water by utilizing the fat reserves in the hump. Migratory birds preparing to fly long distances also store large amounts of fat. Whales are kept warm by a layer of body fat called "blubber" under their skin, which can be as thick as 2 feet. Blubber also provides energy when whales must survive long periods of starvation. Penguins also have blubber, which protects them from the cold and provides energy when they are sitting on a nest of eggs.

Humans also have the capability to store large amounts of fat, although they do not hibernate or usually have to survive for long periods of time without food. When humans survived on sparse diets that were mostly vegetarian, the fat content was about 20%. Today, a typical diet includes more dairy products and foods with high fat levels, which increases the daily fat intake to as much as 60% of the diet. The U.S. Public Health Service now estimates that in the United States, more than one-third of adults are obese. Obesity is defined as a body weight that is more than 20 percent over an ideal weight. Obesity is a major factor in health problems such as diabetes, heart disease, high blood pressure, stroke, and gallstones as well as some cancers and some forms of arthritis.

At one time, we thought that obesity was simply a problem of

eating too much. However, research now indicates that certain pathways in lipid and carbohydrate metabolism may cause excessive weight gain in some people. In 1995, scientists discovered that a hormone called *leptin* is produced in fat cells. When fat cells are full, high levels of leptin signal the brain to limit the intake of food. When fat stores are low, leptin production decreases, which signals the brain to increase food intake. Some obese persons have high levels of leptin, which means that leptin does not cause them to decrease how much they eat.

Research on obesity has become a major research field. Scientists are studying differences in the rate of leptin production, degrees of resistance to leptin, and possible combinations of these factors. After a person has dieted and lost weight, the leptin level drops. This decrease in leptin may cause slow metabolism, increased hunger, and increased food intake, which starts the weight-gain cycle all over again. Currently, studies are being made to assess the safety of leptin therapy following weight loss.

## SAMPLE PROBLEM 18.15

### ■ ATP Production from β Oxidation

How much ATP will be produced from the β oxidation of palmitic acid, a $C_{16}$ saturated fatty acid?

#### SOLUTION

A 16-carbon fatty acid will produce 8 acetyl CoA units and go through 7 β-oxidation cycles. Each acetyl CoA can produce 12 ATPs, by way of the citric acid cycle. In electron transport, each $FADH_2$ produces 2 ATPs, and each NADH produces 3 ATPs.

**ATP Production for Palmitic Acid ($C_{16}H_{32}O_2$)**

| | |
|---|---|
| Activation of palmitic acid to palmityl CoA | −2 ATP |
| 8 acetyl CoA × 12 ATP (citric acid cycle) | 96 ATP |
| 7 FADH$_2$ × 2 ATP (electron transport) | 14 ATP |
| 7 NADH × 3 ATP (electron transport) | 21 ATP |
| Total | 129 ATP |

**STUDY CHECK**

In $\beta$ oxidation, why is more ATP produced from acetyl CoA than from reduced coenzymes?

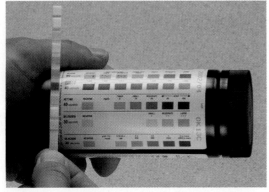

CASE STUDY
Diabetes and Blood Glucose

A test strip determines ketone bodies in a urine sample.

## Ketone Bodies

When carbohydrates are not available to meet energy needs, the body breaks down body fat. However, the oxidation of large amounts of fatty acids can cause acetyl CoA molecules to accumulate in the liver. Then acetyl CoA molecules combine to form keto compounds called **ketone bodies**. (See Figure 18.21.)

Ketone bodies are produced mostly in the liver and transported to cells in the heart, brain, and skeletal muscle, where small amounts of energy can be obtained by converting acetoacetate or $\beta$-hydroxybutyrate back to acetyl CoA.

$$\beta\text{-Hydroxybutyrate} \longrightarrow \text{acetoacetate} + 2\,\text{CoA} \longrightarrow 2\,\text{acetyl CoA}$$

## Ketosis

When ketone bodies accumulate, they may be incompletely metabolized by the body. This may lead to a condition called **ketosis**, which is found in severe diabetes, diets high in fat and low in carbohydrates, and starvation.

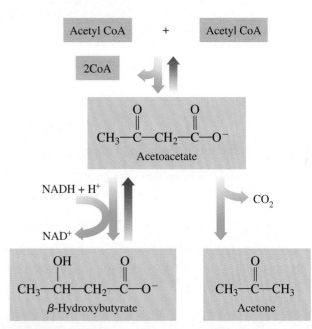

**FIGURE 18.21** In ketogenesis, acetyl CoA molecules combine to produce ketone bodies: acetoacetate, $\beta$-hydroxybutyrate, and acetone.

**Q** What condition in the body leads to the formation of ketone bodies?

Because two of the ketone bodies are acids, they can lower the blood pH below 7.4, which is **acidosis**, a condition that often accompanies ketosis. A drop in blood pH can interfere with the ability of the blood to carry oxygen and cause breathing difficulties.

## Fatty Acid Synthesis

When the body has met all its energy needs and the glycogen stores are full, acetyl CoA from the breakdown of carbohydrates and fatty acids is used to form new fatty acids. Two-carbon acetyl units are linked together to give a 16-carbon fatty acid, palmitic acid. Several of the reactions are the reverse of the reactions we discussed in fatty acid oxidation. However, fatty acid oxidation occurs in the mitochondria and uses FAD and $NAD^+$, whereas fatty acid synthesis occurs in the cytosol and uses the reduced coenzyme NADPH. NADPH is similar to NADH, except it has a phosphate group. The new fatty acids are attached to glycerol to make triacylglycerols, which are stored as body fat.

## Health Note

### Ketone Bodies and Diabetes

Blood glucose is elevated within 30 minutes following a meal containing carbohydrates. The elevated level of glucose stimulates the secretion of the hormone insulin from the pancreas, which increases the flow of glucose into muscle and adipose tissue for the synthesis of glycogen. As blood glucose levels drop, the secretion of insulin decreases. When blood glucose is low, another hormone, glucagon, is secreted by the pancreas, which stimulates the breakdown of glycogen in the liver to yield glucose.

In *diabetes mellitus*, glucose cannot be utilized or stored as glycogen because insulin is not secreted or does not function properly. In type 1, *insulin-dependent diabetes*, which often occurs in childhood, the pancreas produces inadequate levels of insulin. This type of diabetes can result from damage to the pancreas by viral infections or from genetic mutations. In type 2, *insulin-resistant diabetes*, which usually occurs in adults, insulin is produced, but insulin receptors are not responsive. Thus a person with type 2 diabetes does not respond to insulin therapy. *Gestational diabetes* can

occur during pregnancy, but blood glucose levels usually return to normal after the baby is born. Mothers with diabetes tend to gain weight and have large babies.

In all types of diabetes, insufficient amounts of glucose are available in the muscle, liver, and adipose tissue. As a result, liver cells synthesize glucose from noncarbohydrate sources (gluconeogenesis) and break down fat, elevating the acetyl CoA level. Excess acetyl CoA undergoes ketogenesis, and ketone bodies accumulate in the blood. As the level of acetone increases, its odor can be detected on the breath of a person with uncontrolled diabetes who is in ketosis.

In uncontrolled diabetes, the concentration of blood glucose exceeds the ability of the kidney to reabsorb glucose, and glucose appears in the urine. High levels of glucose increase the osmotic pressure in the blood, which leads to an increase in urine output. Symptoms of diabetes include frequent urination and excessive thirst. Treatment for diabetes includes a change to a diet limiting carbohydrate intake and may require medication such as a daily injection of insulin or pills taken by mouth.

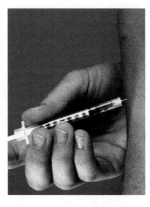
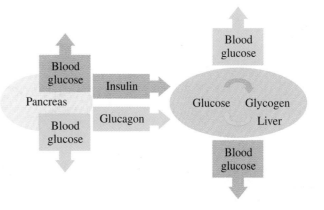

## QUESTIONS AND PROBLEMS

### Oxidation of Fatty Acids

**18.65** Where in the cell is a fatty acid activated?

**18.66** What coenzymes are required for $\beta$ oxidation?

**18.67** Capric acid, $CH_3-(CH_2)_8-COOH$, is a $C_{10}$ fatty acid.
   **a.** Write the formula of the activated form of capric acid.
   **b.** Indicate the $\alpha$ and $\beta$ carbon atoms in the fatty acid.
   **c.** Write the overall equation for the first cycle of $\beta$ oxidation for capric acid.
   **d.** Write the overall equation for the complete $\beta$ oxidation of capric acid.

**18.68** Arachidic acid, $CH_3-(CH_2)_{18}-COOH$, is a $C_{20}$ fatty acid.
   **a.** Write the formula of the activated form of arachidic acid.
   **b.** Indicate the $\alpha$ and $\beta$ carbon atoms in the fatty acid.
   **c.** Write the overall equation for the first cycle of $\beta$ oxidation for arachidic acid.
   **d.** Write the overall equation for the complete $\beta$ oxidation of arachidic acid.

**18.69** Why is the energy of fatty acid activation from ATP to AMP considered the same as the hydrolysis of 2ATP $\longrightarrow$ 2ADP?

**18.70** What is the number of ATP obtained from one acetyl CoA in the citric acid cycle?

**18.71** Consider the complete oxidation of capric acid, $CH_3-(CH_2)_8-COOH$, a $C_{10}$ fatty acid.
   **a.** How many acetyl CoA units are produced?
   **b.** How many cycles of $\beta$ oxidation are needed?
   **c.** How many ATPs are generated from the oxidation of capric acid?

**18.72** Consider the complete oxidation of arachidic acid, $CH_3-(CH_2)_{18}-COOH$, a $C_{20}$ fatty acid.
   **a.** How many acetyl CoA units are produced?
   **b.** How many cycles of $\beta$ oxidation are needed?
   **c.** How many ATPs are generated from the oxidation of arachidic acid?

**18.73** When are ketone bodies produced in the body?

**18.74** If a person is fasting, why would he or she have high levels of acetyl CoA?

**18.75** What are some conditions that characterize ketosis?

**18.76** Why do diabetics produce high levels of ketone bodies?

LEARNING GOAL

Describe the reactions of transamination, oxidative deamination, and the entry of amino acid carbons into the citric acid cycle.

# 18.9 DEGRADATION OF AMINO ACIDS

When dietary protein exceeds the nitrogen needed by the body, the excess amino acids are degraded. The $\alpha$-amino group is removed to yield a keto acid, which can be converted to an intermediate of other metabolic pathways. The carbon atoms from amino acids are used in the citric acid cycle as well as the synthesis of fatty acids, ketone bodies, and glucose. Most of the amino groups are converted to urea.

## Transamination

The degradation of amino acids occurs primarily in the liver. In a **transamination** reaction, an $\alpha$-amino group is transferred from an amino acid to an $\alpha$-keto acid, usually $\alpha$-ketoglutarate. A new amino acid and a new $\alpha$-keto acid are produced. We can write an equation to show the transfer of the amino group from alanine to $\alpha$-ketoglutarate to yield glutamate, the new amino acid, and the $\alpha$-keto acid pyruvate.

$$\underset{\text{Alanine}}{CH_3-\overset{\overset{+}{N}H_3}{\underset{|}{C}}H-COO^-} \; + \; \underset{\alpha\text{-Ketoglutarate}}{{}^-OOC-\overset{O}{\overset{||}{C}}-CH_2-CH_2-COO^-} \; \xrightarrow{\overset{\text{Alanine}}{\text{aminotransferase}}}$$

$$\underset{\text{Pyruvate}}{CH_3-\overset{O}{\overset{||}{C}}-COO^-} \; + \; \underset{\text{Glutamate}}{{}^-OOC-\overset{\overset{+}{N}H_3}{\underset{|}{C}}H-CH_2-CH_2-COO^-}$$

## Oxidative Deamination

In a process called **oxidative deamination**, the amino group in glutamate is removed as an ammonium ion, $NH_4^+$. The reaction catalyzed by *glutamate dehydrogenase* uses either $NAD^+$ or $NADP^+$ as a coenzyme.

$$\overset{\overset{+}{N}H_3}{\underset{\text{Glutamate}}{^-OOC-CH-CH_2-CH_2-COO^-}} + H_2O + NAD^+ \text{ (or } NADP^+\text{)} \xrightarrow{\text{Glutamate dehydrogenase}}$$

$$\underset{\alpha\text{-Ketoglutarate}}{^-OOC-\overset{\overset{O}{\|}}{C}-CH_2-CH_2-COO^-} + NH_4^+ + NADH \text{ (or } NADPH\text{)} + H^+$$

Therefore, the amino group from any amino acid can be used to form glutamate, which undergoes oxidative deamination, converting the amino group to an ammonium ion. Then the ammonium ion is converted to urea.

### SAMPLE PROBLEM 18.16

■ **Transamination and Oxidative Deamination**

Indicate whether each of the following represents a transamination or an oxidative deamination:

**a.** Glutamate is converted to $\alpha$-ketoglutarate and $NH_4^+$.
**b.** Alanine and $\alpha$-ketoglutarate react to form pyruvate and glutamate.
**c.** A reaction is catalyzed by glutamate dehydrogenase, which requires $NAD^+$.

SOLUTION

**a.** oxidative deamination
**b.** transamination
**c.** oxidative deamination

STUDY CHECK

How is $\alpha$-ketoglutarate regenerated to participate in more transamination reactions?

## Urea Cycle

The ammonium ion, which is the end product of amino acid degradation, is toxic if it is allowed to accumulate. Therefore, a series of reactions, called the **urea cycle**, detoxifies ammonium ion ($NH_4^+$) by forming urea, which is excreted in the urine.

$$2NH_4^+ + CO_2 \longrightarrow \underset{\text{Urea}}{H_2N-\overset{\overset{O}{\|}}{C}-NH_2} + 2H^+ + H_2O$$

In one day, a typical adult may excrete about 25–30 g of urea in the urine. This amount increases when a diet is high in protein. If urea is not properly excreted, it builds up quickly to a toxic level. To detect renal disease, the blood urea nitrogen (BUN) level is

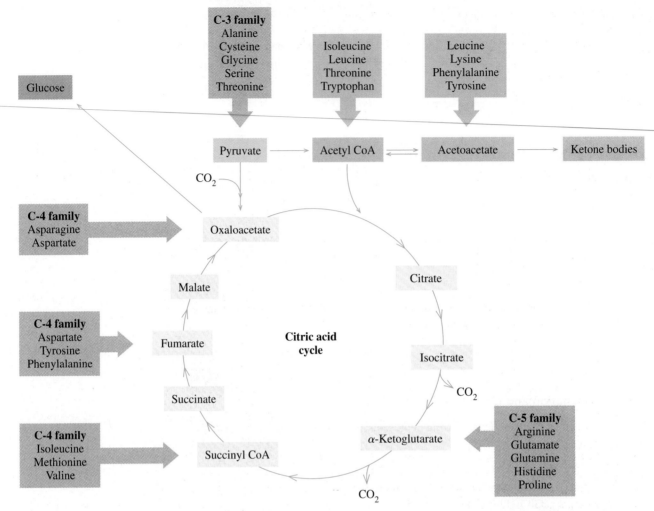

**FIGURE 18.22** Carbon atoms from degraded amino acids are converted to the intermediates of the citric acid cycle or other pathways.

**Q** What compound in the citric acid cycle is obtained from the carbon atoms of alanine and glycine?

measured. If the BUN is high, protein intake must be reduced, and hemodialysis may be needed to remove toxic nitrogen waste from the blood.

## Fates of the Carbon Atoms from Amino Acids

The carbon skeletons from the transamination of amino acids are used as intermediates of the citric acid cycle or other metabolic pathways. We can classify the amino acids according to the number of carbon atoms in those intermediates. (See Figure 18.22.) The amino acids with three carbons are converted to pyruvate. The four-carbon group consists of amino acids that are converted to oxaloacetate, and the five-carbon group provides α-ketoglutarate. Some amino acids are listed twice because they can enter different pathways to form citric acid cycle intermediates.

## Energy from Amino Acids

Normally, only a small amount (about 10%) of our energy needs is supplied by amino acids. However, more energy is extracted from amino acids in conditions such as fasting or starvation, when carbohydrate and fat stores are exhausted. If amino acids remain the only source of energy for a long period of time, the breakdown of body proteins eventually leads to a destruction of essential body tissues.

---

SAMPLE PROBLEM 18.17

■ **Carbon Atoms from Amino Acids**

What degradation products of amino acids are citric acid intermediates?

SOLUTION

Carbon atoms from amino acids enter the citric acid cycle as acetyl CoA, $\alpha$-ketoglutarate, succinyl CoA, fumarate, or oxaloacetate.

STUDY CHECK

Which amino acids provide carbon atoms that enter the citric acid as $\alpha$-ketoglutarate?

---

## Overview of Metabolism

In this chapter, we have seen that catabolic pathways degrade large molecules to small molecules that are used for energy production, using the citric acid cycle and electron transport. We have also indicated that anabolic pathways lead to the synthesis of larger molecules in the cell. In the overall view of metabolism, there are several branch points from which compounds may be degraded for energy or used to synthesize larger molecules. For example, glucose can be degraded to acetyl CoA for the citric acid cycle to produce energy or converted to glycogen for storage. When glycogen stores are depleted, fatty acids are degraded for energy. Amino acids normally used to synthesize nitrogen-containing compounds in the cells can also be used for energy after they are degraded to intermediates of the citric acid cycle. In the synthesis of certain amino acids, $\alpha$-keto acids of the citric acid cycle enter a variety of reactions that convert them to amino acids through transamination by glutamate. (See Figure 18.23.)

---

## QUESTIONS AND PROBLEMS

### Degradation of Amino Acids

**18.77** What are the reactants and products in transamination reactions?

**18.78** What types of enzymes catalyze transamination reactions?

**18.79** Write the structure of the $\alpha$-keto acid produced from each of the following in transamination:

$$\overset{+}{N}H_3$$
**a.** $H-\underset{|}{CH}-COO^-$  Glycine

$$\overset{+}{N}H_3$$
**b.** $CH_3-\underset{|}{CH}-COO^-$  Alanine

**18.80** Write the structure of the $\alpha$-keto acid produced from each of the following in transamination:

$$\overset{+}{N}H_3$$
**a.** $^-OOC-CH_2-\underset{|}{CH}-COO^-$  Aspartate

$$CH_3 \quad \overset{+}{N}H_3$$
**b.** $CH_3-CH_2-\underset{|}{CH}-\underset{|}{CH}-COO^-$  Isoleucine

**18.81** Why does the body convert $NH_4^+$ to urea?

**18.82** What is the structure of urea?

**18.83** What metabolic substrate(s) can be produced from the carbon atoms of each of the following amino acids?
a. alanine
b. aspartate
c. valine
d. glutamine

**18.84** What metabolic substrate(s) can be produced from the carbon atoms of each of the following amino acids?
a. leucine
b. asparagine
c. cysteine
d. arginine

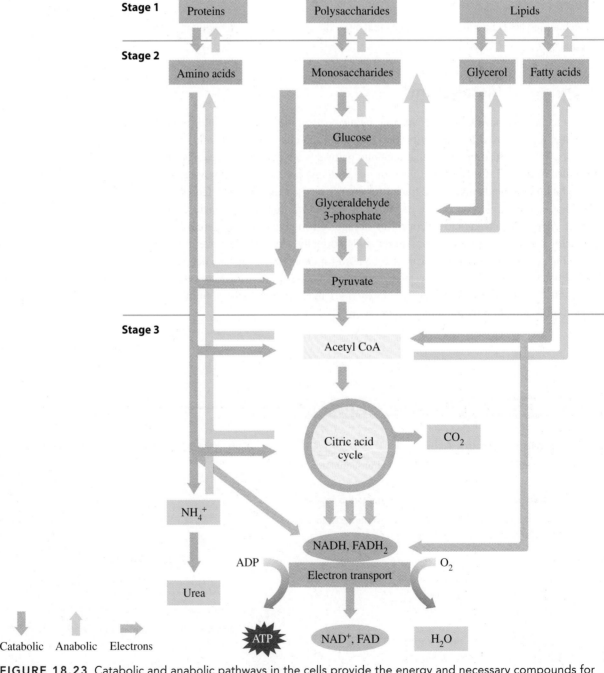

**FIGURE 18.23** Catabolic and anabolic pathways in the cells provide the energy and necessary compounds for the cells.
**Q** Under what conditions in the cell are amino acids degraded for energy?

# CONCEPT MAP

## METABOLIC PATHWAYS AND ENERGY PRODUCTION

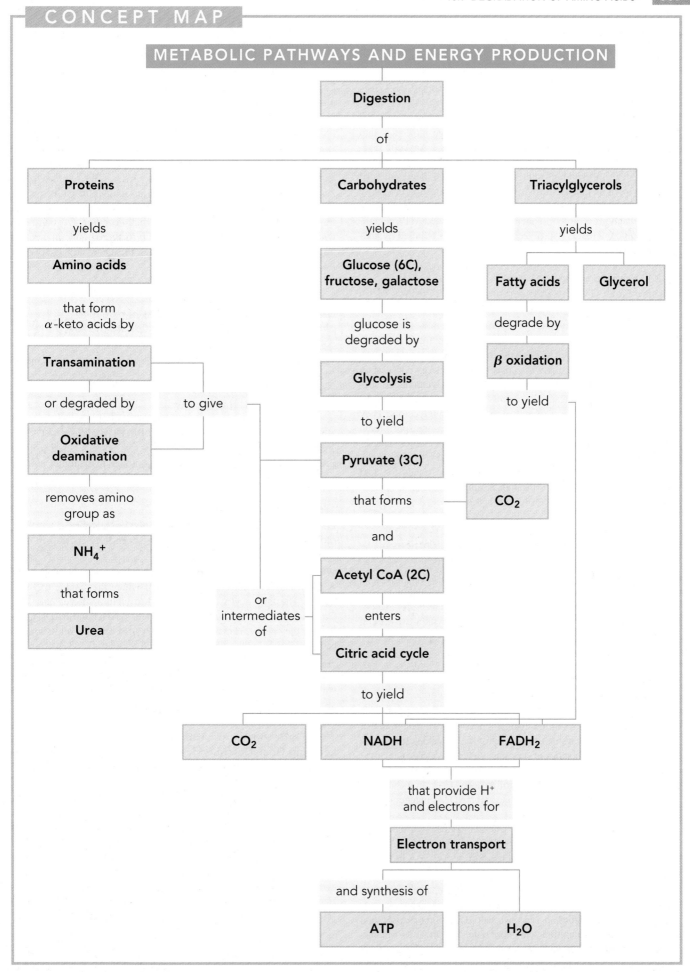

# CHAPTER REVIEW

## 18.1 Metabolism and ATP Energy
**Learning Goal:** Describe three stages of metabolism and the role of ATP.

Metabolism includes all the catabolic and anabolic reactions that occur in the cells. Catabolic reactions degrade large molecules into smaller ones with an accompanying release of energy. Anabolic reactions require energy to synthesize larger molecules from smaller ones. The three stages of metabolism are digestion of food, degradation of monomers such as glucose to pyruvate, and the extraction of energy from the two- and three-carbon compounds. Energy obtained from catabolic reactions is stored in adenosine triphosphate (ATP), a high-energy compound that is hydrolyzed when energy is required by anabolic reactions.

## 18.2 Digestion of Foods
**Learning Goal:** Give the sites and products of digestion of carbohydrates, triacylglycerols, and proteins.

The digestion of carbohydrates is a series of reactions that breaks down polysaccharides into glucose, galactose, and fructose. These monomers can be absorbed through the intestinal wall into the bloodstream to be carried to cells where they provide energy and carbon atoms for synthesis of new molecules. Triacylglycerols are hydrolyzed in the small intestine to monoacylglycerol and fatty acids, which enter the intestinal wall and form new triacylglycerols. They bind with proteins to form chylomicrons, which transport them through the lymphatic system and bloodstream to the tissues. The digestion of proteins, which begins in the stomach and continues in the small intestine, involves the hydrolysis of peptide bonds to yield amino acids that are absorbed through the intestinal wall and transported to the cells.

## 18.3 Important Coenzymes in Metabolic Pathways
**Learning Goal:** Describe the components and functions of the coenzymes FAD, NAD$^+$, and coenzyme A.

FAD and NAD$^+$ are the oxidized forms of coenzymes that participate in oxidation–reduction reactions. When they pick up hydrogen ions and electrons, they are reduced to FADH$_2$ and NADH + H$^+$. Coenzyme A contains a thiol group that usually bonds with a two-carbon acetyl group (acetyl CoA).

## 18.4 Glycolysis: Oxidation of Glucose
**Learning Goal:** Describe the conversion of glucose to pyruvate in glycolysis and the subsequent conversion of pyruvate to acetyl CoA or lactate.

Glycolysis, which occurs in the cytosol, consists of 10 reactions that degrade glucose (six carbons) to two pyruvate molecules (three carbons each). The overall series of reactions yields two molecules of the reduced coenzyme NADH and two ATP. Under aerobic conditions, pyruvate is oxidized in the mitochondria to acetyl CoA. In the absence of oxygen, pyruvate is reduced to lactate and NAD$^+$ is regenerated for the continuation of glycolysis.

## 18.5 The Citric Acid Cycle
**Learning Goal:** Describe the oxidation of acetyl CoA in the citric acid cycle.

In a sequence of reactions called the citric acid cycle, an acetyl group is combined with oxaloacetate to yield citrate. Citrate undergoes oxidation and decarboxylation to yield two CO$_2$, GTP, three NADH, and FADH$_2$, with the regeneration of oxaloacetate. The direct phosphorylation of ADP by GTP yields ATP.

## 18.6 Electron Transport
**Learning Goal:** Describe the electron carriers involved in electron transport.

Electron carriers that transfer hydrogen ions and electrons include FMN, iron–sulfur proteins, coenzyme Q, and several cytochromes. Both iron–sulfur proteins and cytochromes contain iron ions that are reduced to Fe$^{2+}$ and reoxidized to Fe$^{3+}$ as electrons are accepted and then passed to the next electron carrier. The reduced coenzymes NADH and FADH$_2$ from various metabolic pathways are oxidized to NAD$^+$ and FAD when their protons and electrons are transferred to electron transport. The final acceptor, O$_2$, combines with protons and electrons to yield H$_2$O. At three points in electron transport, the energy decrease provides the necessary energy for ATP synthesis.

## 18.7 Oxidative Phosphorylation and ATP
**Learning Goal:** Describe the process of oxidative phosphorylation in ATP synthesis; calculate the ATP from complete oxidation of glucose.

The protein complexes in electron transport act as proton pumps to move protons into the inner membrane space, which produces a proton gradient. As the protons return to the matrix by way of ATP synthase, energy is generated. This energy is used to drive the synthesis of ATP in a process known as oxidative phosphorylation. The oxidation of NADH yields three ATP molecules, and FADH$_2$ yields two ATP. The energy from the NADH produced in the cytoplasm is used to form FADH$_2$. Under aerobic conditions, the complete oxidation of glucose yields a total of 36 ATPs.

## 18.8 Oxidation of Fatty Acids
**Learning Goal:** Describe the metabolic pathway of $\beta$ oxidation; calculate the ATP from the complete oxidation of a fatty acid.

When needed as an energy source, fatty acids are linked to coenzyme A and transported into the mitochondria where they undergo $\beta$ oxidation. The fatty acyl CoA is oxidized to yield a shorter fatty acyl CoA, acetyl CoA, and reduced coenzymes NADH and FADH$_2$. Although the energy from a particular fatty acid depends on its length, each oxidation cycle yields 5 ATPs with another 12 ATPs from the acetyl CoA that enters the citric acid cycle. When high levels of acetyl CoA are present in the cell, they enter the ketogenesis pathway, forming ketone bodies such as acetoacetate, which cause ketosis and acidosis.

## 18.9 Degradation of Amino Acids
**Learning Goal:** Describe the reactions of transamination, oxidative deamination, and the entry of amino acid carbons into the citric acid cycle.

When the amount of amino acids in the cells exceeds that needed for synthesis of nitrogen compounds, the process of transamination converts them to $\alpha$-keto acids and glutamate. Oxidative deamination of glutamate produces ammonium ions and $\alpha$-ketoglutarate. Ammonium ions from oxidative deamination are converted to urea. The carbon atoms from the degradation of amino acids can enter the citric acid cycle or other metabolic pathways. Certain amino acids are synthesized when amino groups from glutamate are transferred to an $\alpha$-keto acid obtained from glycolysis or the citric acid cycle.

# SUMMARY OF KEY REACTIONS

### HYDROLYSIS OF ATP

$$ATP + H_2O \longrightarrow ADP + P_i + 7.3 \text{ kcal/mole}$$

### FORMATION OF ATP

$$ADP + P_i + 7.3 \text{ kcal/mole} \longrightarrow ATP$$

### REDUCTION OF FAD AND NAD$^+$

$$FAD + 2H^+ + 2\,e^- \longrightarrow FADH_2$$

$$NAD^+ + 2H^+ + 2\,e^- \longrightarrow NADH + H^+$$

### DIGESTION OF PROTEINS

$$\text{Protein} + H_2O \xrightarrow{H^+} \text{amino acids}$$

### GLYCOLYSIS

$$C_6H_{12}O_6 + 2ADP + 2P_i + 2NAD^+ \rightarrow 2CH_3\overset{\overset{\displaystyle O}{\|}}{C}-COO^- + 2ATP + 2NADH + 4H^+$$

Glucose    Pyruvate

### OXIDATION OF PYRUVATE TO ACETYL CoA

$$CH_3\overset{\overset{\displaystyle O}{\|}}{C}-COO^- + NAD^+ + HS-CoA \longrightarrow CH_3\overset{\overset{\displaystyle O}{\|}}{C}-S-CoA + NADH + CO_2$$

Pyruvate    Acetyl CoA

### CITRIC ACID CYCLE

$$\text{Acetyl CoA} + 3NAD^+ + FAD + GDP + P_i + 2H_2O \longrightarrow 2CO_2 + 3NADH + 2H^+ + FADH_2 + CoA + GTP$$

### ELECTRON TRANSPORT

$$NADH + H^+ + 3ADP + 3P_i + \tfrac{1}{2}O_2 \longrightarrow NAD^+ + 3ATP + H_2O$$

$$FADH_2 + 2ADP + 2P_i + \tfrac{1}{2}O_2 \longrightarrow FAD + 2ATP + H_2O$$

### COMPLETE OXIDATION OF GLUCOSE

$$C_6H_{12}O_6 + 6O_2 + 36ADP + 36P_i \longrightarrow 6CO_2 + 6H_2O + 36ATP$$

### REDUCTION OF PYRUVATE TO LACTATE

$$CH_3\overset{\overset{\displaystyle O}{\|}}{C}-COO^- + NADH + H^+ \longrightarrow CH_3\overset{\overset{\displaystyle OH}{|}}{CH}-COO^- + NAD^+$$

Pyruvate    Lactate

### $\beta$ OXIDATION OF FATTY ACID

$$\text{Myristyl CoA} + 6CoA + 6FAD + 6NAD^+ + 6H_2O \longrightarrow 7 \text{ acetyl CoA} + 6FADH_2 + 6NADH + 6H^+$$

### HYDROLYSIS OF DISACCHARIDES

$$\text{Lactose} + H_2O \xrightarrow{\text{Lactase}} \text{galactose} + \text{glucose}$$

$$\text{Sucrose} + H_2O \xrightarrow{\text{Sucrase}} \text{fructose} + \text{glucose}$$

$$\text{Maltose} + H_2O \xrightarrow{\text{Maltase}} \text{glucose} + \text{glucose}$$

### DIGESTION OF TRIACYLGLYCEROLS

$$\text{Triacylglycerols} + 3H_2O \xrightarrow{\text{Lipases}} \text{glycerol} + 3 \text{ fatty acids}$$

## TRANSAMINATION

$$\underset{\text{Alanine}}{CH_3-\underset{\underset{\overset{|}{NH_3^+}}{}}{CH}-COO^-} + \underset{\alpha\text{-Ketoglutarate}}{{}^-OOC-\underset{\overset{\|}{O}}{C}-CH_2-CH_2-COO^-} \xrightarrow{\text{Alanine aminotransferase}}$$

$$\underset{\text{Pyruvate}}{CH_3-\underset{\overset{\|}{O}}{C}-COO^-} + \underset{\text{Glutamate}}{{}^-OOC-\underset{\underset{\overset{|}{NH_3^+}}{}}{CH}-CH_2-CH_2-COO^-}$$

## OXIDATIVE DEAMINATION

$$\underset{\text{Glutamate}}{{}^-OOC-\underset{\underset{\overset{|}{NH_3^+}}{}}{CH}-CH_2-CH_2-COO^-} + H_2O + NAD^+ \text{(NADP)} \xrightarrow{\text{Glutamate dehydrogenase}}$$

$$\underset{\alpha\text{-Ketoglutarate}}{{}^-OOC-\underset{\overset{\|}{O}}{C}-CH_2-CH_2-COO^-} + NH_4^+ + NADH \text{ (NADPH)} + H^+$$

## UREA CYCLE

$$2NH_4^+ + CO_2 \longrightarrow H_2N-\underset{\overset{\|}{O}}{C}-NH_2 + 2H^+ + H_2O$$

# KEY TERMS

**acetyl CoA** A two-carbon acetyl unit from oxidation of pyruvate that bonds to coenzyme A.

**acidosis** Low blood pH resulting from the formation of acidic ketone bodies.

**ADP** Adenosine diphosphate, formed by the hydrolysis of ATP, consists of adenine (A), a ribose sugar, and two phosphate groups ($P_i$).

**aerobic** An oxygen-containing environment in the cells.

**amylase** An enzyme that hydrolyzes the glycosidic bonds in polysaccharides during digestion.

**anabolic reaction** A metabolic reaction that requires energy.

**anaerobic** A condition in cells when there is no oxygen.

**ATP** Adenosine triphosphate, a high-energy compound that stores energy in the cells and that consists of adenine, a ribose sugar, and three phosphate groups.

**ATP synthase** An enzyme complex that uses the energy released by protons returning to the matrix to synthesize ATP from ADP and $P_i$.

**beta ($\beta$) oxidation** The degradation of fatty acids that removes two-carbon segments from the fatty acid at the oxidized $\beta$ carbon.

**catabolic reaction** A metabolic reaction that produces energy for the cell by the degradation and oxidation of glucose and other molecules.

**chemiosmotic model** The conservation of energy from electron transport by pumping protons into the intermembrane space to produce a proton gradient that provides the energy to synthesize ATP.

**citric acid cycle** A series of oxidation reactions in the mitochondria that convert acetyl CoA to $CO_2$ and yield NADH and $FADH_2$. It is also called the tricarboxylic acid cycle and the Krebs cycle.

**coenzyme A (CoA)** A coenzyme that transports acyl and acetyl groups.

**coenzyme Q (CoQ, Q)** A mobile carrier that transfers electrons from NADH and $FADH_2$ to cytochrome *b* in complex III.

**cytochromes (cyt)** Iron-containing proteins that transfer electrons from $QH_2$ to oxygen.

**cytoplasm** The material in eukaryotic cells between the nucleus and the plasma membrane.

**cytosol** The fluid of the cytoplasm, which is an aqueous solution of electrolytes and enzymes.

**decarboxylation** The loss of a carbon atom in the form of $CO_2$.

**digestion** The processes in the gastrointestinal tract that break down large food molecules to smaller ones that pass through the intestinal membrane into the bloodstream.

**electron transport** A series of reactions in the mitochondria that transfer electrons from NADH and $FADH_2$ to electron carriers, which are arranged from higher to lower energy levels, and finally to $O_2$, which produces $H_2O$. Energy changes during three of these transfers provide energy for ATP synthesis.

**FAD** A coenzyme (flavin adenine dinucleotide) for dehydrogenase enzymes that form carbon–carbon double bonds.

**Fe–S (iron–sulfur) clusters** Proteins containing iron and sulfur, in which the iron ions accept/release electrons during electron transport.

**FMN (flavin mononucleotide)** An electron carrier derived from riboflavin (vitamin $B_2$) that transfers hydrogen ions and electrons from NADH to electron transport.

**glycolysis** The ten oxidation reactions of glucose that yield two pyruvate molecules.

**ketone bodies** The products of ketogenesis: acetoacetate, $\beta$-hydroxybutyrate, and acetone.

**ketosis** A condition in which high levels of ketone bodies cannot be metabolized, leading to lower blood pH.

**metabolism** All the chemical reactions in living cells that carry out molecular and energy transformations.

**mitochondria** The organelles of the cells where energy-producing reactions take place.

**NAD$^+$** The hydrogen acceptor used in oxidation reactions that form carbon–oxygen double bonds.

**oxidative deamination** The loss of ammonium ion when glutamate is degraded to $\alpha$-ketoglutarate.

**oxidative phosphorylation** The synthesis of ATP from ADP and $P_i$, using energy generated by the oxidation reactions during electron transport.

**proton pumps** The enzyme complexes I, III, and IV that move protons from the matrix into the intermembrane space, creating a proton gradient.

**transamination** The transfer of an amino group from an amino acid to an $\alpha$-keto acid.

**urea cycle** The process in which ammonium ions from the degradation of amino acids are converted to urea.

# UNDERSTANDING THE CONCEPTS

**18.85** Lauric acid, $CH_3$—$(CH_2)_{10}$—$COOH$, which is found in coconut oil, is a saturated fatty acid.

**a.** Write the formula of the activated form of lauric acid.

**b.** Indicate the $\alpha$ and $\beta$ carbon atoms in the fatty acyl molecule.

**c.** Write the overall equation for the complete $\beta$ oxidation for lauric acid.

**d.** How many acetyl CoA units are produced?

**e.** How many cycles of $\beta$ oxidation are needed?

**f.** Account for the total ATP yield from $\beta$ oxidation of lauric acid ($C_{12}$ fatty acid) by completing the following calculation:

| | | |
|---|---|---|
| activation | $\longrightarrow$ | $-2$ ATP |
| _____ acetyl CoA | $\longrightarrow$ | _____ ATP |
| _____ FADH$_2$ | $\longrightarrow$ | _____ ATP |
| _____ NADH | $\longrightarrow$ | _____ ATP |
| | **Total** | _____ ATP |

**18.86** A hiker expends 450 kcal. How many moles of ATP does this require?

**18.87** Identify the type of food as carbohydrate, fat, or protein that gives each of the following digestion products:

**a.** glucose      **b.** fatty acid

**c.** maltose      **d.** glycerol

**e.** amino acids      **f.** dextrins

**18.88** Identify each of the following as a 6-carbon or a 3-carbon compound and arrange them in the order in which they occur in glycolysis.

**a.** 3-phosphoglycerate      **b.** pyruvate

**c.** glucose-6-phosphate      **d.** glucose

**e.** fructose-1,6-bisphosphate

## ADDITIONAL QUESTIONS AND PROBLEMS

**18.89** Write an equation for the hydrolysis of ATP to ADP.

**18.90** At the gym, you expend 300 kcal riding the stationary bicycle for 1 h. How many moles of ATP will this require?

**18.91** How and where does lactose undergo digestion in the body? What are the products?

**18.92** How and where does sucrose undergo digestion in the body? What are the products?

**18.93** What are the reactant and product of glycolysis?

**18.94** What is the general type of reaction that takes place in the digestion of carbohydrates?

**18.95** When is pyruvate converted to lactate in the body?

**18.96** When pyruvate is used to form acetyl CoA, the product has only two carbon atoms. What happened to the third carbon?

**18.97** What is the main function of the citric acid cycle in energy production?

**18.98** Most metabolic pathways are not considered cycles. Why is the citric acid cycle considered to be a metabolic cycle?

**18.99** If there are no reactions in the citric acid cycle that use oxygen, $O_2$, why does the cycle operate only in aerobic conditions?

**18.100** What products of the citric acid cycle are needed for electron transport?

**18.101** In the chemiosmotic model, how is energy provided to synthesize ATP?

**18.102** What is the effect of proton accumulation in the intermembrane space?

**18.103** How many ATPs are produced when glucose is oxidized to pyruvate, compared to when glucose is oxidized to $CO_2$ and $H_2O$?

**18.104** What metabolic substrate(s) can be produced from the carbon atoms of each of the following amino acids?
**a.** leucine
**b.** isoleucine
**c.** cysteine
**d.** phenylalanine

## CHALLENGE QUESTIONS

**18.105** One cell at work may break down 2 million (2 000 000) ATP molecules in one second. Some researchers estimate that the human body has about $10^{13}$ cells.
**a.** How much energy in kcal could be produced by the cells in the body in one day?
**b.** If ATP has a molar mass of 507 g/mole, how many grams of ATP are hydrolyzed?

**18.106** State if each of the following processes produce or consume ATP:
**a.** citric acid cycle
**b.** glucose forms two pyruvates
**c.** pyruvate forms acetyl CoA
**d.** glucose forms glucose-6-phosphate
**e.** oxidation of $\alpha$-ketoglutarate
**f.** transport of NADH across the mitochondrial membrane
**g.** activation of a fatty acid

**18.107** Match the following ATP yields to reactions a–g.

| 2 ATP | 3 ATP | 6 ATP | 12 ATP |
|-------|-------|-------|--------|
| 18 ATP | 36 ATP | 44 ATP | |

**a.** Glucose forms two pyruvates.
**b.** Pyruvate forms acetyl CoA.
**c.** Glucose forms two acetyl CoAs.
**d.** Acetyl CoA goes through one turn of the citric acid cycle.
**e.** Caproic acid ($C_6$) is completely oxidized.
**f.** NADH + $H^+$ is oxidized to $NAD^+$.
**g.** $FADH_2$ is oxidized to FAD.

**18.108** Identify each of the reactions **a.** to **e.** in the $\beta$ oxidation of palmitic acid ($C_{14}$), a fatty acid, as
**(1)** activation
**(2)** first dehydrogenation (oxidation)
**(3)** hydration
**(4)** second dehydrogenation
**(5)** cleavage of acetyl CoA
**a.** Palmityl CoA and FAD form $\alpha$, $\beta$-unsaturated palmityl CoA and $FADH_2$.
**b.** $\beta$-Keto palmityl CoA forms myristyl CoA and acetyl CoA.
**c.** Palmitic acid and ATP form palmityl CoA.
**d.** $\alpha$, $\beta$-Unsaturated palmityl CoA and $H_2O$ form $\beta$-hydroxy palmityl CoA.
**e.** $\beta$-Hydroxy palmityl CoA and $NAD^+$ form $\beta$-keto palmityl CoA and NADH + $H^+$.

**18.109** Which of the following molecules will produce the most ATP per mole?
**a.** glucose or maltose
**b.** myristic acid, $CH_3$—$(CH_2)_{12}$—COOH, or stearic acid, $CH_3$—$(CH_2)_{16}$—COOH
**c.** glucose or two acetyl CoA
**d.** glucose or caprylic acid ($C_8$)
**e.** citrate or succinate in one turn of the citric acid cycle

**18.110** If acetyl CoA has a molar mass of 809 g/mole, how many moles of ATP are produced when 1.0 $\mu$g acetyl CoA completes the citric acid cycle?

# ANSWERS

## Answers to Study Checks

**18.1** Cytoplasm is all the cellular material between the cell membrane and the nucleus. Cytosol, the aqueous part of the cytoplasm, is a solution of electrolytes and enzymes.

**18.2** Adenine, ribose, and three phosphate groups.

**18.3** The digestion of amylose begins in the mouth when salivary amylase hydrolyzes some of the glycosidic bonds. In the small intestine, pancreatic amylase hydrolyzes more glycosidic bonds, and finally maltose is hydrolyzed by maltase to yield glucose.

**18.4** In the membrane of the small intestine, monoacylglycerols and fatty acids recombine to form new triacylglycerols that bind with proteins to form chylomicrons for transport to the lymphatic system and the bloodstream.

**18.5** HCl denatures proteins and activates enzymes such as pepsin.

**18.6** $FADH_2$

**18.7** In the initial reactions of glycolysis, energy in the form of two ATPs is invested to convert glucose to fructose-1,6-bisphosphate.

**18.8**

$$CH_3-\underset{\underset{OH}{|}}{CH}-\underset{\underset{O}{\|}}{C}-O^- + NAD^+ \xrightarrow{\text{Lactate dehydrogenase}}$$

$$CH_3-\underset{\underset{O}{\|}}{C}-\underset{\underset{O}{\|}}{C}-O^- + NADH + H^+$$

**18.9** oxaloacetate

**18.10 a.** oxidation  **b.** reduction

**18.11** Oxygen ($O_2$) is the last substance that accepts electrons.

**18.12** The protein complexes I, III, and IV pump protons from the matrix to the intermembrane space.

**18.13** One turn of the citric acid cycle produces a total of 12 ATPs: 9 ATPs from 3 NADHs, 2 ATPs from $FADH_2$, and 1 ATP from a direct phosphorylation.

**18.14** $NAD^+$

**18.15** In $\beta$ oxidation, each acetyl CoA enters the citric acid cycle to produce 3 NADHs, $FADH_2$s, and GTPs for a total of 12 ATPs. The reduced coenzymes NADH and $FADH_2$ enter electron transport to give 5 ATPs.

**18.16** The process of oxidative deamination regenerates $\alpha$-ketoglutarate from glutamate.

**18.17** Arginine, glutamate, glutamine, histidine, and proline provide carbon atoms for $\alpha$-ketoglutarate.

## Answers to Selected Questions and Problems

**18.1** The digestion of polysaccharides takes place in stage 1.

**18.3** In metabolism, a catabolic reaction breaks apart large molecules, releasing energy.

**18.5** The hydrolysis of the phosphodiester bond (P—O—P) in ATP releases energy that is sufficient for energy-requiring processes in the cell.

**18.7** Hydrolysis is the main reaction involved in the digestion of carbohydrates.

**18.9 a.** lactose  **b.** glucose and fructose
  **c.** glucose and glucose

**18.11** The bile salts emulsify fat to give small fat globules for lipase hydrolysis.

**18.13** The digestion of proteins begins in the stomach and is completed in the small intestine.

**18.15 a.** NADH
  **b.** FAD

**18.17** FAD

**18.19** glucose

**18.21** ATP is required in phosphorylation reactions.

**18.23** ATP is produced in glycolysis by transferring a phosphate from 1,3-bisphosphoglycerate and from phosphoenolpyruvate directly to ADP.

**18.25 a.** 1 ATP required
  **b.** 1 NADH is produced for each triose.
  **c.** 2 ATPs and 2 NADHs are produced.

**18.27** Aerobic (oxygen) conditions are needed.

**18.29** The oxidation of pyruvate converts $NAD^+$ to NADH and produces acetyl CoA and $CO_2$.
Pyruvate + $NAD^+$ + CoA $\longrightarrow$
  acetyl CoA + $CO_2$ + NADH + $H^+$

**18.31** When pyruvate is reduced to lactate, the $NAD^+$ is used to oxidize glyceraldehyde-3-phosphate, which recycles NADH.

**18.33** $2CO_2$, 3NADH, $FADH_2$, GTP (ATP), CoA, and $2H^+$.

**18.35** Two reactions, 3 and 4, involve oxidative decarboxylation.

**18.37** $NAD^+$ is reduced in reactions 3, 4, and 8 of the citric acid cycle.

**18.39** In reaction 5, GDP undergoes a direct substrate phosphorylation.

**18.41 a.** citrate and isocitrate
  **b.** In decarboxylation, a carbon atom is lost as $CO_2$.
  **c.** $\alpha$-ketoglutarate
  **d.** isocitrate $\longrightarrow$ $\alpha$-ketoglutarate; $\alpha$-ketoglutarate $\longrightarrow$ succinyl CoA; succinate $\longrightarrow$ fumarate; malate $\longrightarrow$ oxaloacetate
  **e.** steps 3, 8

**18.43** oxidized

**18.45 a.** oxidation
  **b.** reduction

**18.47** NADH and $FADH_2$

**18.49** FAD, coenzyme Q, cytochrome $b$, cytochrome $c$

**18.51** The mobile carrier Q transfers electrons from complex I to III.

**18.53** NADH transfers electrons to FMN in complex I to give $NAD^+$.

**18.55 a.** NADH + $H^+$ + FMN $\longrightarrow$ $NAD^+$ + $FMNH_2$
  **b.** $QH_2$ + 2cyt $b$ ($Fe^{3+}$) $\longrightarrow$ Q + 2cyt $b$ ($Fe^{2+}$) + $2H^+$

**18.57** In oxidative phosphorylation, the energy from the oxidation reactions in electron transport is used to drive ATP synthesis.

**18.59** Protons return to a lower energy in the matrix by passing through ATP synthase, which releases energy to drive the synthesis of ATP.

**18.61** The reduced coenzymes NADH and FADH$_2$ from glycolysis and the citric acid cycle transfer electrons to electron transport, which generates energy to drive the synthesis of ATP.

**18.63 a.** 3 ATP
**b.** 6 ATP
**c.** 6 ATP
**d.** 12 ATP

**18.65** In the cytosol at the outer mitochondrial membrane

**18.67 a. and b.** $CH_3—(CH_2)_6—CH_2—CH_2—\overset{\overset{O}{\|}}{C}—S—CoA$
$\qquad\qquad\qquad\qquad\quad \beta \quad\ \alpha$

**c.**

$CH_3—(CH_2)_8—\overset{\overset{O}{\|}}{C}—S—CoA + NAD^+ + FAD + H_2O + HS—CoA \longrightarrow$

$CH_3—(CH_2)_6—\overset{\overset{O}{\|}}{C}—S—CoA + CH_3—\overset{\overset{O}{\|}}{C}—S—CoA + NADH + H^+ + FADH_2$

**d.**
$CH_3—(CH_2)_8—COOH + 5CoA + 4FAD + 4NAD^+ + 4H_2O \longrightarrow$
$\qquad\qquad 5\ \text{acetyl CoA} + 4FADH_2 + 4NADH + 4H^+$

**18.69** The hydrolysis of ATP to AMP removes two inorganic phosphates from ATP, which provides the same amount of energy as the hydrolysis of 2ATP to 2ADP.

**18.71 a.** 5 acetyl CoA units
**b.** 4 cycles of $\beta$ oxidation
**c.** 60 ATP from 5 acetyl CoA (citric acid cycle) + 12 ATP from 4 NADH + 8 ATP from 4 FADH$_2$ − 2 ATP (activation) = 80 − 2 = 78 ATP

**18.73** Ketone bodies form in the body when excess acetyl CoA results from the breakdown of large amounts of fat.

**18.75** High levels of ketone bodies lead to ketosis, a condition characterized by acidosis (a drop in blood pH values), excessive urination, and strong thirst.

**18.77** The reactants are an amino acid and an $\alpha$-keto acid, and the products are a new amino acid and a new $\alpha$-keto acid.

**18.79 a.** $\overset{\overset{O}{\|}}{H—C—COO^-}$    **b.** $CH_3—\overset{\overset{O}{\|}}{C}—COO^-$

**18.81** NH$_4^+$ is toxic if allowed to accumulate in the body.

**18.83 a.** pyruvate    **b.** oxaloacetate, fumarate
**c.** succinyl CoA    **d.** $\alpha$-ketoglutarate

**18.85 a. and b.** $CH_3—(CH_2)_8—CH_2—CH_2—\overset{\overset{O}{\|}}{C}—CoA$
$\qquad\qquad\qquad\qquad\qquad \beta \qquad \alpha$

**c.** Lauryl CoA + 5CoA + 5FAD + 5NAD$^+$ + 5H$_2$O $\longrightarrow$
$\qquad$ 6acetyl CoA + 5FADH$_2$ + 5NADH + 5H$^+$

**d.** Six acetyl CoA units are produced.
**e.** Five cycles of $\beta$ oxidation are needed.

**f.** 

| | | |
|---|---|---|
| activation | $\longrightarrow$ | −2 ATP |
| 6 acetyl CoA | × 12 ATP $\longrightarrow$ | 72 ATP |
| 5 FADH$_2$ | × 2 ATP $\longrightarrow$ | 10 ATP |
| 5 NADH | × 3 ATP $\longrightarrow$ | 15 ATP |
| | **Total** | 95 ATP |

**18.87 a.** carbohydrate
**b.** fat
**c.** carbohydrate
**d.** fat
**e.** protein
**f.** carbohydrate

**18.89** ATP + H$_2$O $\longrightarrow$ ADP + P$_i$ + 7.3 kcal/mole

**18.91** Lactose undergoes digestion in the mucosal cells of the small intestine to yield galactose and glucose.

**18.93** Glucose is the reactant, and pyruvate is the product of glycolysis.

**18.95** Pyruvate is converted to lactate when oxygen is not present in the cell (anaerobic) to regenerate NAD$^+$ for glycolysis.

**18.97** The oxidation reactions of the citric acid cycle produce a source of reduced coenzymes for electron transport and ATP synthesis.

**18.99** The oxidized coenzymes NAD$^+$ and FAD needed for the citric acid cycle are regenerated by the electron transport system.

**18.101** Energy released as protons flow through ATP synthase and then back to the matrix is utilized for the synthesis of ATP.

**18.103** The oxidation of glucose to pyruvate produces 6 ATPs whereas the oxidation of glucose to CO$_2$ and H$_2$O produces 36 ATPs.

**18.105 a.** 21 kcal
**b.** 1500 g of ATP

**18.107 a.** 6 ATP/glucose
**b.** 3 ATP/pyruvate
**c.** 12 ATP/glucose
**d.** 12 ATP/acetyl CoA
**e.** 44 ATP/C$_6$ acid
**f.** 3 ATP/NADH
**g.** 2 ATP/FADH$_2$

**18.109 a.** maltose
**b.** stearic acid
**c.** glucose
**d.** caprylic acid
**e.** citrate

# Combining Ideas from Chapters 16 to 18

**CI.29** Beano® contains an enzyme that breaks down polysaccharides into mono- and disaccharides that are more digestible. It is used to diminish gas formation that can occur after eating foods such as vegetables, grains, and high-fiber foods.

**a.** The label on Beano says "contains alpha-galactosidase." What class of enzymes is present in Beano?
**b.** What is the substrate for the enzyme?
**c.** The directions indicate you should not heat or cook with Beano. Why?

**CI.30** Kevlar® is a lightweight polymer used in tires and bulletproof vests. Part of the strength of Kevlar is due to hydrogen bonds between polymer chains.
**a.** Draw the structures of the carboxylic acid and amine that are polymerized to make Kevlar.
**b.** What feature of Kevlar will produce hydrogen bonds between the polymer chains?
**c.** What type of secondary protein structure would have bonds that are similar to those in Kevlar?

Portion of a polymer chain in Kevlar

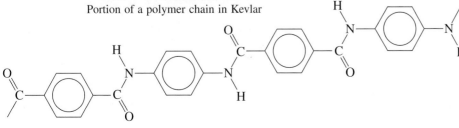

**CI.31** Identify each of the following as a substance that is part of the citric acid cycle, electron transport, or both.
**a.** succinate
**b.** $QH_2$
**c.** FAD
**d.** cyt $c$ ($Fe^{2+}$)
**e.** cytochrome $c$ oxidase
**f.** $H_2O$
**g.** malate
**h.** $NAD^+$

**CI.32** From direct sources and electron transport, use the value of 7.3 kcal per mole of ATP and determine the total kcal stored as ATP from each of the following.
**a.** 1 mole of glucose in glycolysis
**b.** the oxidation of 2 moles of pyruvate to 2 moles of acetyl CoA
**c.** complete oxidation of 1 mole of glucose to $CO_2$ and $H_2O$
**d.** 1 mole of lauric acid, a $C_{12}$ fatty acid, in $\beta$ oxidation
**e.** 1 mole of glutamate (from protein) in the citric acid cycle

**CI.33** Acetyl coenzyme A (CoA) is the fuel for the citric acid cycle. It has the formula $C_{23}H_{38}N_7O_{17}P_3S$.
**a.** What are the components of coenzyme A?
**b.** What is the function of CoA?
**c.** Where does the acetyl group attach in CoA?
**d.** What is the molar mass to 3 significant figures for acetyl CoA?

**CI.34** Behenic acid is a saturated 22-carbon fatty acid found in peanut and canola oils.

**a.** Draw the structure of activated behenic acid.
**b.** Indicate the $\alpha$ and $\beta$ carbon atoms in the fatty acyl formula in **a**.
**c.** Write the overall equation for the complete $\beta$ oxidation of behenic acid.
**d.** How many acetyl CoA units are produced?

e. How many cycles of $\beta$ oxidation are needed?

f. What is the total ATP yield from the $\beta$ oxidation of arachidic acid?

| | | |
|---|---|---|
| activation | $-2$ | ATP |
| _____ acetyl CoA | _____ | ATP |
| _____ FADH$_2$ | _____ | ATP |
| _____ NADH | _____ | ATP |
| **Total** | _____ | **ATP** |

**CI.35** Butter is a fat that contains 80% triacylglycerols, and the rest is water. Assume the triacylglycerol in butter is glyceryl tripalmitate.

a. Write an equation for the hydrolysis of glyceryl tripalmitate.

b. What is the molar mass of glyceryl tripalmitate, $C_{51}H_{98}O_6$?

c. Calculate the ATP yield from the fatty acids in 1 mole of palmitic acid.

d. How many kcal are released from the palmitic acid in a 0.50-oz pat of butter?

e. If running for 1 h uses 750 kcal, how many pats of butter would provide the energy (kcal) for a 45-min run?

**CI.36** Thalassemia is an inherited genetic mutation that limits the production of the beta chain needed for hemoglobin formation. With low levels of beta hemoglobin, there is a shortage of red blood cells (anemia); as a result, the body does not have sufficient amounts of oxygen. In one form of thalassemia, a single nucleotide is deleted in the DNA that codes for the beta chain. The mutation involves the deletion of thymine (T) from section 91 in the following segment of normal DNA.

89    90    91    92    93    94
— AGT — GAG — CTG — CAC — TGT — GAC — A . . . .

a. Write the complementary strand for this normal DNA section.

b. Write the mRNA sequence from normal DNA using the complementary strand in part **a**.

c. What amino acids are placed in the beta chain by this portion of mRNA?

d. What is the order of nucleotides in the mutation?

e. Write the complementary strand for the mutant DNA section.

f. Write the mRNA sequence of mutant DNA, using the complementary strand in part **e**.

g. What amino acids are placed in the beta chain by the mutation?

h. What type of mutation occurs in this form of thalassemia?

i. How might the properties of this section of the beta chain be different from the properties of the normal protein?

j. How might the level of structure in hemoglobin be affected if beta chains are not produced?

# ANSWERS

**CI.29** a. An alpha-galactosidase is a hydrolase.

b. The substrate is the $\alpha$-1,4-glycosidic bond of galactose.

c. High temperatures will denature the hydrolase so it no longer functions.

**CI.31** a. citric acid cycle
b. electron transport
c. both
d. electron transport
e. electron transport
f. both
g. citric acid cycle
h. both

**CI.33** a. aminoethanethiol, pantothenic acid, phosphorylated adenosine diphosphate

b. Coenzyme A carries an acetyl group to the citric acid cycle for oxidation.

c. The acetyl group links to the S atom in the aminoethanethiol part of CoA.

d. 809 g/mole

**CI.35** a.

b. glyceryl tripalmitate, 807 g/mole
c. 129 ATP
d. 40. kcal
e. 14 pats of butter

# CREDITS

Unless otherwise acknowledged, all photographs are the property of Pearson Education, Benjamin Cummings Publishers.

# GLOSSARY/INDEX

**Heat** The energy associated with the motion of particles in a substance. 53

**Heating curve** A diagram that shows the temperature changes and changes of state of a substance as it is heated. 72

**Heat of fusion** The energy required to melt exactly 1 g of a substance at its melting point. For water, 80. cal are needed to melt 1 g of ice; 80. cal are released when 1 g of water freezes. 68

**Heat of vaporization** The energy required to vaporize 1 g of a substance at its boiling point. 70–72

**Hemodialysis** A mechanical cleansing of the blood by an artificial kidney using the principle of dialysis. 273

**Hemolysis** A swelling and bursting of red blood cells in a hypotonic solution because of an increase in fluid volume. 271

**Henry's law** The solubility of a gas in a liquid is directly related to the pressure of that gas above the liquid. 253–254

**Heterogeneous mixture** A mixture of two or more substances that are not mixed uniformly. 84, 85, 112, 267

effect of, 610
frameshift, 610–611, 618, 621
substitution, 610–611, 613
types of, 610
Myelin sheath, 531
Mylanta, 302
Myocardial infarction, 514. *See also* Heart attacks
Myofacial release, 558
Myoglobin, 565, 570
Myosin, 553, 628
Myrcene, 386
Myristic acid, 517, 523, 658

## N

**N terminal** The end amino acid in a peptide with a free —$NH_3^+$ group. 559
Nabumetone (Relafen), 521
**NAD$^+$ (nicotinamide adenine dinucleotide)** The hydrogen acceptor used in oxidation reactions that form carbon–oxygen double bonds. 632–634
NADH dehydrogenase, 648
NADP, 663
Nail-polish remover, 406, 423
Nails, 566
Nandrolone, 539, 540
*nano-*, 27
Naprosyn, 521
Naproxen, 521
Naphthalene, 398
Nasopharyngeal carcinoma, 616
Natural gas, 3, 55, 139, 314, 367, 368
detecting gas leaks, 410
Natural opiate, in the body, 561
Natural radioactivity, 318–322
Negative ions, 122–123
Neon, 40, 89, 90, 92, 105, 122, 125, 139
symbol for, 87
Neon lights, 105
Neoplasm, 616
Neo-Synephrine, 464, 477
Neotame™, 501
Nerve impulse, 542, 624, 628
Neutral solution, 290, 292, 298
**Neutralization** A reaction between an acid and a base to form a salt and water. 300
of acids, 300
of amines, 466
balancing neutralization equations, 300–301
of carboxlic acids, 453–454
of hydroxide, 300
Neutralization end point, 302
Neutralization equations, balancing, 300–301
**Neutron** A neutral subatomic particle having a mass of 1 amu and found in the nucleus of an atom; its symbol is $n$ or $n^0$. 96–101
calculating number of, 100
in isotopes, 102
Newton, Isaac, 7
Niacin, 582
Nickel, 84, 326, 390, 616
symbol for, 87
Nickel-60, 326
Nicotinamide, 390, 582, 632, 633
Nicotinamide adenine dinucleotide (NAD$^+$), 582
Nicotinamide adenine dinucleotide phosphate (NADP$^+$), 582
Nicotine, 373, 433, 448, 467
Night vision, cis–trans isomers for, 390
Nitrate, 134, 281
Nitric acid, 281, 287
Nitride, 130

Nitrite, 134, 281, 616
Nitrogen, 2, 40, 66, 99, 105, 121, 125, 134, 210, 233, 351, 369
in atmosphere, 210, 211
the bends, 234
in blood, 234
in body, 90
diatomic molecule, 139
symbol for, 87
Nitrogen-13, 327, 337
Nitrogen dioxide, 194, 203, 239, 299
Nitrogen gas ($N_2$), 2
Nitrogen oxide, 2, 141, 203
Nitrogen trichloride, 142
Nitrous acid, 281, 287
**Noble gas** An element in Group 8A (18) of the periodic table, generally unreactive and seldom found in combination with other elements. 93
uses for, 125
Noguchi, Mark, 224
*non-*, 351
Nonane, 355
**Noncompetitive inhibitor** A type of inhibitor that alters the shape of an enzyme as well as the active site so that the substrate cannot bind properly. 578
**Nonelectrolyte** A substance that dissolves in water as molecules; its solution will not conduct an electrical current. 246, 247–248
solutions of, 247
**Nonmetal** An element with little or no luster that is a poor conductor of heat and electricity. The nonmetals are located to the right of the zigzag line in the periodic table. 92
bonding in covalent compounds, 140
characteristics of, 93
compounds of, 121
electronegativity values of, 154
and ionic compounds, 121
ions of, 121–123
**Nonpolar amino acid** An amino acid with a nonpolar side chain that is not soluble in water. 555
**Nonpolar covalent bond** A covalent bond in which the electrons are shared equally between atoms. 145–146
**Nonpolar molecule** A molecule that has only nonpolar bonds or in which the bond dipoles cancel. 149–150
Nonpolar solute, 245–246, 275
Nonsteroidal anti-inflammatory drugs (NSAID), 521
Nonstick coatings, 394
Noradrenaline. *See* Norepinephrine
Norepinephrine, 432, 464
Norethindrone, 538
Nuclear equations, guide to completing, 323–324
Nuclear fission, 340
Nuclear fusion, 340–342
Nuclear medicine:
establishment of field, 318
radiation doses in diagnostic and therapeutic procedures, 339
Nuclear medicine technologist, 317
Nuclear power plants, 342
Nuclear radiation, 317–349. *See also* Radiation
**Nuclear reactions**, 323–329
alpha decay, 323
beta decay, 325–326
gamma emission, 327
positron emission, 326
radioactive decay, 323
radioactive isotopes, producing, 327–328
Nucleases, 572
**Nucleic acids** Large molecules composed of nucleotides, found as a double helix in DNA and as the single strands of RNA. 592
bases, 592
components of, 592–596, 596

# METRIC AND SI UNITS AND SOME USEFUL CONVERSION FACTORS

| Length    SI unit meter (m) | Volume    SI unit cubic meter (m³) | Mass    SI unit kilogram (kg) |
|---|---|---|
| 1 meter (m) = 100 centimeters (cm) | 1 liter (L) = 1000 milliliters (mL) | 1 kilogram (kg) = 1000 grams (g) |
| 1 meter (m) = 1000 millimeters (mm) | 1 mL = 1 cm³ | 1 g = 1000 milligrams (mg) |
| 1 cm = 10 mm | 1 L = 1.06 quart (qt) | 1 kg = 2.20 lb |
| 1 kilometer (km) = 0.621 mile (mi) | 1 qt = 946 mL | 1 lb = 454 g |
| 1 inch (in.) = 2.54 cm (exact) | | 1 mole = $6.02 \times 10^{23}$ particles |
| | | **Water** |
| | | density = 1.00 g/mL |

| Temperature    SI unit kelvin (K) | Pressure    SI unit pascal (Pa) | Energy    SI unit joule (J) |
|---|---|---|
| °F = 1.8(°C) + 32 | 1 atm = 760 mmHg | 1 calorie (cal) = 4.184 J |
| $°C = \dfrac{(°F - 32)}{1.8}$ | 1 atm = 760 torr | 1 kcal = 1000 cal |
| | 1 mole (STP) = 22.4 L | **Water** |
| K = °C + 273 | | Heat of fusion = 80. cal/g |
| | | Heat of vaporization = 540 cal/g |
| | | Specific heat = 4.184 J/g°C |

# PREFIXES FOR METRIC (SI) UNITS

| Prefix | Symbol | Power of Ten |
|---|---|---|
| **Values greater than 1** | | |
| tera | T | $10^{12}$ |
| giga | G | $10^{9}$ |
| mega | M | $10^{6}$ |
| kilo | k | $10^{3}$ |
| **Values less than 1** | | |
| deci | d | $10^{-1}$ |
| centi | c | $10^{-2}$ |
| milli | m | $10^{-3}$ |
| micro | $\mu$ | $10^{-6}$ |
| nano | n | $10^{-9}$ |
| pico | p | $10^{-12}$ |

# FORMULAS AND MOLAR MASSES OF SOME TYPICAL COMPOUNDS

| Name | Formula | Molar Mass (g/mole) | Name | Formula | Molar Mass (g/mole) |
|---|---|---|---|---|---|
| Ammonia | $NH_3$ | 17.0 | Hydrogen chloride | HCl | 36.5 |
| Ammonium chloride | $NH_4Cl$ | 53.5 | Iron(III) oxide | $Fe_2O_3$ | 159.8 |
| Ammonium sulfate | $(NH_4)_2SO_4$ | 132.1 | Magnesium oxide | MgO | 40.3 |
| Bromine | $Br_2$ | 159.8 | Methane | $CH_4$ | 16.0 |
| Butane | $C_4H_{10}$ | 58.0 | Nitrogen | $N_2$ | 28.0 |
| Calcium carbonate | $CaCO_3$ | 100.1 | Oxygen | $O_2$ | 32.0 |
| Calcium chloride | $CaCl_2$ | 111.1 | Potassium carbonate | $K_2CO_3$ | 138.2 |
| Calcium oxide | CaO | 56.1 | Propane | $C_3H_8$ | 44.0 |
| Carbon dioxide | $CO_2$ | 44.0 | Sodium chloride | NaCl | 58.5 |
| Chlorine | $Cl_2$ | 71.0 | Sodium hydroxide | NaOH | 40.0 |
| Copper(II) sulfide | CuS | 95.7 | Sulfur dioxide | $SO_2$ | 64.1 |
| Hydrogen | $H_2$ | 2.0 | Water | $H_2O$ | 18.0 |